LINEAR CIRCUITS

TIME DOMAIN, PHASOR, AND LAPLACE TRANSFORM APPROACHES

THIRD EDITION

Raymond A. DeCarlo
Purdue University

Pen-Min Lin
Purdue University

Kendall Hunt
publishing company

Cover image © Shutterstock, Inc.

Kendall Hunt
publishing company

www.kendallhunt.com
Send all inquiries to:
4050 Westmark Drive
Dubuque, IA 52004-1840

Printed in the United States of America
10   9   8   7

DEDICATION

To our wives Chris and Louise.

TABLE OF CONTENTS

PREFACE

For the last several decades, EE/ECE departments of US universities have typically required two semesters of linear circuits during the sophomore year for EE majors and one semester for other engineering majors. Over the same time period discrete time system concepts and computer engineering principles have become required fare for EE undergraduates. Thus we continue to use Laplace transforms as a vehicle for understanding basic concepts such as impedance, admittance, filtering, and magnetic circuits. Further, software programs such as PSpice, MATLAB and its toolboxes, Mathematica, Maple, and a host of other tools have streamlined the computational drudgery of engineering analysis and design. MATLAB remains a working tool in this 3rd edition of Linear Circuits.

In addition to a continuing extensive use of MATLAB, we have removed much of the more complex material from the book and rewritten much of the remaining book in an attempt to make the text and the examples more illustrative and accessible. More importantly, many of the more difficult homework exercises have been replaced with more routine problems often with numerical answers or checks.

Our hope is that we have made the text more readable and understandable by today's engineering undergraduates.

C H A P T E R

1

Charge, Current, Voltage and Ohm's Law

CHAPTER OUTLINE

1. Role and Importance of Circuits in Engineering
2. Charge and Current
3. Voltage
4. Circuit Elements
5. Voltage, Current, Power, Energy, Relationships
6. Ideal Voltage and Current Sources
7. Resistance, Ohm's Law, and Power (a Reprise)
8. V-I Characteristics of Ideal Resistors, Constant Voltage, and Constant Current Sources
   Summary
   Terms and Concepts
   Problems

CHAPTER OBJECTIVES

1. Introduce and investigate three basic electrical quantities: charge, current, and voltage, and the conventions for their reference directions.
2. Define a two-terminal circuit element.
3. Define and investigate power and energy conversion in electric circuits, and demonstrate that these quantities are conserved.
4. Define independent and dependent voltage and current sources that act as energy or signal generators in a circuit.
5. Define Ohm's law, $v(t) = R\, i(t)$, for a resistor with resistance R.
6. Investigate power dissipation in a resistor.
7. Classify memoryless circuit elements by their terminal voltage-current relationships.
8. Explain the difference between a device and its circuit model.

1. ROLE AND IMPORTANCE OF CIRCUITS IN ENGINEERING

Are you curious about how fuses blow? About the meaning of different wattages on light bulbs? About the heating elements in an oven? And how is the presence of your car sensed at a stoplight? Circuit theory, the focus of this text, provides answers to all these questions.

When you learn basic circuit theory, you learn how to harness the power of electricity, as is done, for example, in

- an electric motor that runs the compressor in an air conditioner or the pump in a dishwasher;
- a microwave oven;
- a radio, TV, or stereo;
- an iPod;
- a car heater.

In this text, we define and analyze common circuit elements and describe their interaction. Our aim is to create a modular framework for analyzing circuit behavior, while simultaneously developing a set of tools essential for circuit design. These skills are, of course, crucial to every electrical engineer. But they also have broad applicability in other fields. For instance, disciplines such as bioengineering and mechanical engineering have similar patterns of analysis and often utilize circuit analogies.

WHAT IS A CIRCUIT?

A circuit is an energy or signal/information processor. Each circuit consists of interconnections of "simple" circuit elements, or devices. Each circuit element can, in turn, be thought of as an energy or signal/information processor. For example, a circuit element called a "source" produces a voltage or a current signal. This signal may serve as a power source for the circuit, or it may represent information. Information in the form of voltage or current signals can be processed by the circuit to produce new signals or new/different information. In a radio transmitter, electricity powers the circuits that convert pictures, voices, or music (that is, information) into electromagnetic energy. This energy then radiates into the atmosphere or into space from a transmitting antenna. A satellite in space can pick up this electromagnetic energy and transmit it to locations all over the world. Similarly, a TV reception antenna or a satellite dish can pick up and direct this energy to a TV set. The TV contains circuits (Figure 1.1) that reconvert the information within the received signal back into pictures with sound.

FIGURE 1.1 Cathode ray tube with surrounding circuitry for converting electrical signals into pictures.

2. CHARGE AND CURRENT

CHARGE

Charge is an electrical property of matter. Matter consists of atoms. Roughly speaking, an atom contains a nucleus that is made up of positively charged protons and neutrons (which have no charge). The nucleus is surrounded by a cloud of negatively charged electrons. The accumulated charge on 6.2415×10^{18} electrons equals -1 coulomb (C). Thus, the charge on an electron is $-1.602176 \times 10^{-19}$ C.

Particles with opposite charges attract each other, whereas those with similar charges repel. The force of attraction or repulsion between two charged bodies is inversely proportional to the square of the distance between them, assuming the dimensions of the bodies are very small compared with the distance of separation. Two equally charged particles 1 meter (m) apart in free space have charges of 1 C each if they repel each other with a force of 10^{-7} c^2 Newtons (N), where $c = 3 \times 10^8$ m/s is the speed of light, by definition. The force is attractive if the particles have opposite charges. Notationally, Q will denote a fixed charge, and q or $q(t)$, a time-varying charge.

Exercise. How many electrons have a combined charge of -53.406×10^{-12} C?
ANSWER: 333,391,597

Exercise. Sketch the time-dependent charge profile $q(t) = 3(1-e^{-2t})$ C, $t \geq 0$, present on a metal plate. MATLAB is a good tool for such sketches.

A **conductor** refers to a material in which electrons can move to neighboring atoms with relative ease. Metals, carbon, and acids are common conductors. Copper wire is probably the most common conductor. An **ideal conductor** offers zero resistance to electron movement. Wires are assumed to be ideal conductors, unless otherwise indicated.

Insulators oppose electron movement. Common insulators include dry air, dry wood, ceramic, glass, and plastic. An **ideal insulator** offers infinite opposition to electron movement.

CURRENT

Current refers to the net flow of charge across any cross section of a conductor. The net movement of 1 **coulomb** (1 C) of charge through a cross section of a conductor in 1 second (1 sec) produces an electric current of 1 ampere (1 A). The **ampere** is the basic unit of electric current and equals 1 C/s.

The direction of current flow is taken by convention as opposite to the direction of electron flow, as illustrated in Figure 1.2. This is because early in the history of electricity, scientists erroneously believed that current was the movement of only positive charges, as illustrated in Figure 1.3. In metallic conductors, current consists solely of the movement of electrons. However, as our understanding of device physics advanced, scientists learned that in ionized gases, in electrolytic solutions, and in some semiconductor materials, movement of positive charges constitutes part or all of the total current flow.

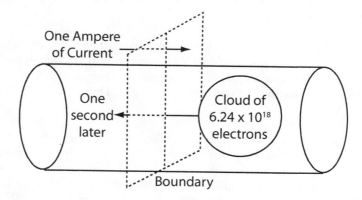

FIGURE 1.2 A cloud of negative charge moves past a cross section of an ideal conductor from right to left. By convention, the positive current direction is taken as left to right.

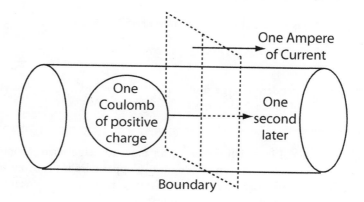

FIGURE 1.3 In the late nineteenth century, current was thought to be the movement of a positive charge past a cross section of a conductor, giving rise to the conventional reference "direction of positive current flow."

Both Figures 1.2 and 1.3 depict a current of 1 A flowing from left to right. In circuit analysis, we do not distinguish between these two cases: each is represented symbolically, as in Figure 1.4(a). The arrowhead serves as a reference for determining the true direction of the current. A positive value of current means the current flows in the same direction as the arrow. A current of negative value implies flow is in the opposite direction of the arrow. For example, in both Figures 1.4a and b, a current of 1 A flows from left to right.

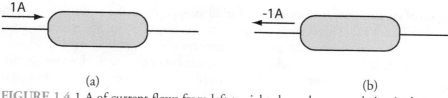

(a)   (b)

FIGURE 1.4 1 A of current flows from left to right through a general circuit element.

In Figure 1.4, the current is constant. The wall socket in a typical home is a source of alternating current, which changes its sign periodically, as we will describe shortly. In addition, a current direction may not be known *a priori*. These situations require the notion of a negative current.

EXAMPLE 1.1.

Figure 1.5 shows a slab of material in which the following is true:
  1. Positive charge carriers move from left to right at the rate of 0.2 C/s.
  2. Negative charge carriers move from right to left at the rate of 0.48 C/s.

Given these conditions,
  a)   Find I_a and I_b;
  b)   Describe the charge movement on the wire at the boundaries A and B.

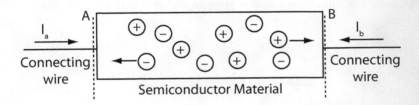

FIGURE 1.5 Material through which positive and negative charges move.

SOLUTION

  a)   The current from left to right, due to the movement of the positive charges, is 0.2 A. The current from left to right, due to the movement of the negative charges, is 0.48 A. Therefore, I_a, the total current from left to right, is 0.2 + 0.48 = 0.68 A. Since I_b is the current from right to left, its value is then –0.68 A.

  b)   The wire is a metallic conductor in which only electrons move. Therefore, at boundaries A and B, negative charges (carried by electrons) move from right to left at the rate of 0.68 C/s.

Exercise. In Example 1.1, suppose positive-charge carriers move from right to left at the rate of 0.5 C/s, and negative carriers move from left to right at the rate of 0.4 C/s. Find I_a and I_b.
ANSWER: I_a = –0.9 A; I_b = 0.9 A

If a net charge Δq crosses a boundary in a short time frame of Δt (in seconds), then the approximate current flow is

$$I = \frac{\Delta q}{\Delta t}$$

(1.1)

where I, in this case, is a constant. The instantaneous (time-dependent) current flow is the limiting case of Equation 1.1, i.e.,

$$i(t) = \frac{dq(t)}{dt}$$

(1.2)

Here $q(t)$ is the amount of charge that has crossed the boundary in the time interval $[t_0, t]$. The equivalent integral counterpart of Equation 1.2 is

$$q(t) = \int_{t_0}^{t} i(\tau)d\tau$$

(1.3)

EXAMPLE 1.2

The charge crossing a boundary in a wire is given in Figure 1.6(a) for $t \geq 0$. Plot the current $i(t)$ through the wire.

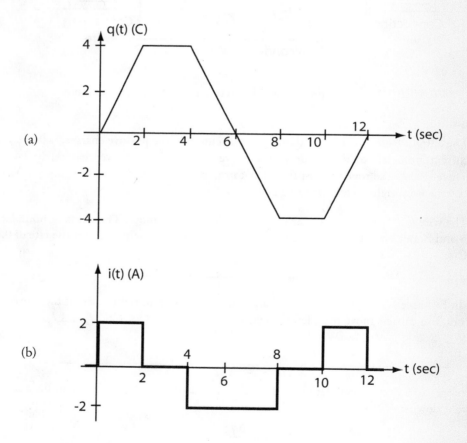

FIGURE 1.6 (a) Charge crossing a hypothetical boundary; (b) current flow associated with the charge plot of (a).

SOLUTION
As per Equation 1.2, the current is the time derivative of $q(t)$. The slopes of the straight-line segments of $q(t)$ in Figure 1.6(a) determine the piecewise constant current plotted in Figure 1.6(b).

Exercise. The charge crossing a boundary in a wire varies as $q(t) = \dfrac{1 - \cos(\omega t)}{\omega}$ C, for $t \geq 0$. Compute the current flow.
ANSWER: $\sin(\omega t)$ A, for $t \geq 0$

Exercise. Repeat the preceding exercise if $q(t) = 5e^{-0.5t}$ C, for $t \geq 0$.
ANSWER: $-2.5e^{-0.5t}$ A, for $t \geq 0$

EXAMPLE 1.3
Find $q(t)$, the charge transported through a cross section of a conductor over $[0, t]$, and also the total charge Q transported, if the current through the conductor is given by the waveform of Figure 1.7(a).

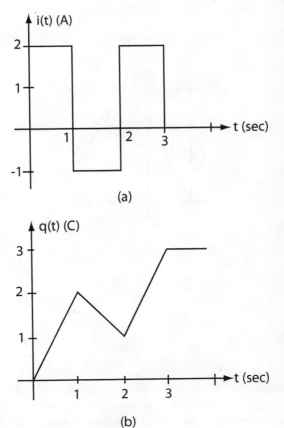

(a)

(b)

FIGURE 1.7 (a) Square-wave current signal; (b) $q(t)$ equal to the integral of $i(t)$ given in (a).

SOLUTION
From Equation 1.3, for $t \geq 0$,

$$q(t) = \int_0^t i(\tau)d\tau$$

Thus, $q(t)$ is the running area under the $i(t)$ versus t curve. Since $i(t)$ is piecewise constant, the integral is piecewise linear because the area either increases or decreases linearly with time, as shown in Figure 1.7(b). Since $q(t)$ is constant for $t \geq 3$, the total charge transported is $Q = q(3) = 3$ C.

Exercise. If the current flow through a cross section of conductor is $i(t) = \cos(120\pi t)$ A for $t \geq 0$ and 0 otherwise, find $q(t)$ for $t \geq 0$.

ANSWER: $q(t) = \dfrac{\sin(120\pi t)}{120\pi}$ C for $t \geq 0$

Exercise. Suppose the current through a cross section of conductor is given in Figure 1.8. Find $q(t)$ for $t \geq 0$.

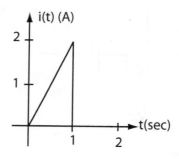

FIGURE 1.8

ANSWER: $q(t) = t^2$ C for $0 \leq t \leq 1$; $q(t) = 1$C for $t \geq 1$

TYPES OF CURRENT

There are two very important current types: **direct current (dc)** and **alternating current (ac)**. Constant current (i.e., $dq/dt = I$ is constant) is called **direct current,** which is illustrated graphically in Figure 1.9(a). Figure 1.9(b) shows an **alternating current,** generally meaning a sinusoidal waveform, i.e., current of the form $A\sin(\omega t + \phi)$, where A is the peak magnitude, ω is the angular frequency, and ϕ is the phase angle of the sine wave. With alternating current, the instantaneous value of the waveform changes periodically through negative and positive values, i.e., the direction of the current flow changes regularly as indicated by the + and − values in Figure 1.9(b). Household current is ac.

Lastly, Figure 1.9(c) shows a current that is neither dc nor ac, but that nevertheless will appear in later circuit analyses. There are many other types of waveforms. Interestingly, currents inside computers, CD players, TVs, and other entertainment devices are typically neither dc nor ac.

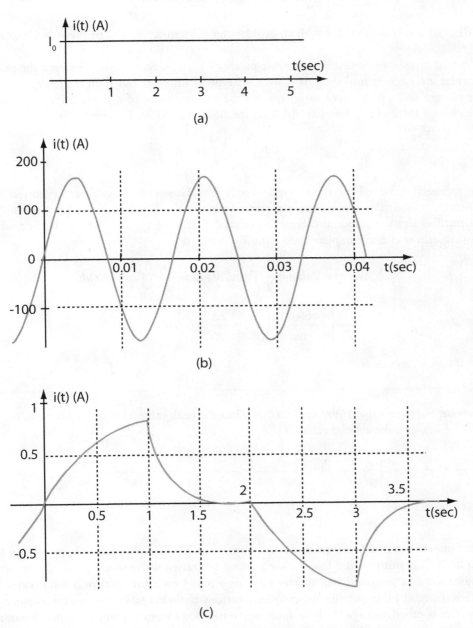

FIGURE 1.9 (a) Direct current, or dc; $i(t) = I_0$; (b) alternating current, or ac; $i(t) = 120\sqrt{2}\sin(120t)$ A; (c) neither ac nor dc.

Because the value of an ac waveform changes with time, ac is measured in different ways. Suppose the instantaneous value of the current at time t is $K\sin(\omega t + \phi)$. The term **peak value** refers to K in $K\sin(\omega t + \phi)$. The **peak-to-peak value** is $2K$. Another measure of the alternating current, indicative of its heating effect, is the **root mean square (rms)**, or **effective value**. The rms or effective value is related to the peak value by the formula

$$\text{rms} = \frac{\sqrt{2}}{2} \times \text{peak–value} = 0.7071\,K \tag{1.4}$$

A derivation of Equation 1.4 with an explanation of its meaning will be given in Chapter 11.

A special instrument called an **ammeter** measures current. Some ammeters read the peak value, whereas some others read the rms value. One type of ammeter, based on the interaction between the current and a permanent magnet, reads the **average value** of a current. From calculus, F_{ave}, the average value of any function $f(t)$, over the time interval $[0, T]$ is given by

$$F_{ave} = \frac{1}{T}\int_0^T f(t)\,dt \tag{1.5}$$

For a general ac waveform, the average value is zero. However, ac signals are often rectified, i.e., converted to their absolute values, in power-supply circuits. For such circuits, the average value of the rectified signal is important. From Equation 1.5, the average value of the absolute value of an ac waveform over one complete cycle with $T = 2\pi/\omega$, is

$$Average\ Value = \frac{K}{T}\int_0^T |\sin(\omega t)|\,dt = \frac{2K}{T}\int_0^{0.5\,T} \sin(\omega t)\,dt$$

$$= \frac{2K}{T}\left[\frac{-\cos(\omega t)}{\omega}\right]_0^{0.5T} = \frac{2K}{\pi} = 0.636K \tag{1.6}$$

i.e., 0.636 × peak value.

Exercise. Suppose $i(t) = 169.7\sin(50\pi t)$ A. Find the peak value, the peak-to-peak value, the rms value of $i(t)$, and the average value of $|i(t)|$.
ANSWER: 169.7, 339.4, 120, and 107.93 A, respectively

---

3. VOLTAGE

What causes current to flow? An analogous question might be, What causes water to flow in a pipe or a hose? Without pressure from either a pump or gravity, water in a pipe is still. Pressure from a water tower, a pressured bug sprayer tank, or a pump on a fire truck will force water flow. In electrical circuits, the "pressure" that forces electrons to flow, i.e., produces a current in a wire or a device, is called **voltage**. Strictly speaking, water flows from a point of higher pressure—say, point A—to a point of lower pressure—say, point B—along a pipe. Between the two points A and B, there is said to be a pressure drop. In electrical circuits, a voltage drop from point A to point B

along a conductor will force current to flow from point *A* to point *B*; there is said to be a voltage drop from point *A* to point *B* in such cases.

Gravity forces the water to flow from a higher elevation to a lower elevation. An analogous phenomenon occurs in an electric field, as illustrated in Figure 1.10(a). Figure 1.10(a) shows two conducting plates separated by a vacuum. On the top plate is a fixed amount of positive static charge. On the bottom plate is an equal amount of negative static charge. Suppose a small positive charge were placed between the plates. This small charge would experience a force directed toward the negatively charged bottom plate. Part of the force is due to repulsion by the positive charges on the top plate, and part is due to the attraction by the negative charges on the bottom plate. This repulsion and attraction marks the presence of an electric field produced by the opposite sets of static charges on the plates.

The electric field indicated in Figure 1.10 sets up an "electric pressure" or voltage drop from the top plate to the bottom plate, which forces positive charges to flow "downhill" in the way that water flows from a water tower to your faucet. Unlike water flow, negative charges are forced "uphill" from the negatively charged bottom plate to the positively charged top plate. As mentioned in the previous section, this constitutes a net current flow caused by the bilateral flow of positive and negative charges. The point is that current flow is induced by an electric pressure called a voltage drop.

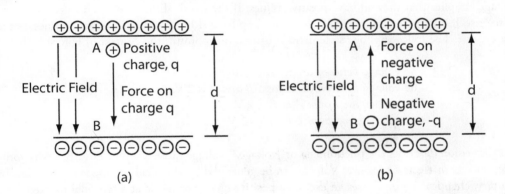

FIGURE 1.10 (a) Positive charge in a (uniform) electric field; (b) negative charge in a uniform electric field.

As mentioned, in Figure 1.10, the positive charge *q* at *A* tends to move toward *B*. We say, *qualitatively*, that point *A* in the electric field is at a *higher potential* than point *B*. Equivalently, point *B* is at a *lower potential* than point *A*. An analogy is now evident: a positive charge in an electric field "falls" from a higher potential point to a lower potential point, just as a ball falls from a higher elevation to a lower elevation in a gravitational field.

Note, however, that if we turn the whole setup of Figure 1.10(a) upside down, the positive charge *q* still moves from point *A* to point *B*, an upward spatial movement. Similarly, if a negative charge *–q* is placed at *B*, as in Figure 1.10(b), then the negative charge experiences an upward-pulling force, moving from the lower potential, point *B*, to the higher potential, point *A*.

Again, consider Figure 1.10(a). As the charge q moves from point A toward B, it picks up velocity and gains kinetic energy. Just before q hits the bottom plate, the kinetic energy gained equals the (constant) force acting on q multiplied by the distance traveled *in the direction of the force*. The kinetic energy is proportional to q and to d, the "distance traveled." Therefore,

$$\text{energy converted} = \text{kinetic energy gained} \propto q$$

The missing proportionality constant in this relationship is defined as the **potential difference** or **voltage** between **A** and **B**. The term "voltage" is synonymous with "potential difference." Mathematically,

$$\text{voltage} = \text{potential difference} = \frac{\text{energy converted}}{\text{magnitude of charge}} \tag{1.8}$$

The standard unit for measuring potential difference or voltage is the volt (V). *According to Equation 1.8, if 1 joule (J) of energy is converted from one form to another when moving 1 C of charge from point A to point B, then the potential difference, or voltage, between A and B is 1 V.* In equation form, with standard units of V, J, and C, we have

$$1 \text{ V} = 1 \text{ }^{\text{J}}\!/_{\text{C}} \tag{1.9}$$

The use of terms such as "elevation difference," "energy converted," "potential difference," or "voltage" implies that they all have positive values. If the word "difference" is changed to "drop" (or to "rise"), then potential drop and elevation drop have either positive or negative values, as the case may be. The following four statements illustrate this point in the context of Figure 1.10:

$$\left\{ \begin{array}{l} \text{The voltage } \textit{between} \text{ (or across) } A \textit{ and } B \text{ is 2 V.} \\ \text{The voltage } \textit{between} \text{ (or across) } B \textit{ and } A \text{ is 2 V.} \end{array} \right.$$
$$\left\{ \begin{array}{l} \text{The voltage } \textit{drop from } A \textit{ to } B \text{ is 2 V.} \\ \text{The voltage } \textit{drop from } B \textit{ to } A \text{ is } -2 \text{ V.} \end{array} \right.$$

This discussion describes the phenomena of "voltage." Voltage causes current flow. But what produces voltage or electric pressure? Voltage can be generated by chemical action, as in batteries. In a battery, chemical action causes an excess of positive charge to reside at a terminal marked with a plus sign and an equal amount of negative charge to reside at a terminal marked with a negative sign. When a device such as a headlight is connected between the terminals, the voltage causes a current to flow through the headlight, heating up the tiny wire and making it "light up." Another source of voltage/current is an electric generator in which mechanical energy used to rotate the shaft of the generator is converted to electrical energy using properties of electro-magnetic fields.

All types of circuit analysis require knowledge of the potential difference between two points, say A and B, and specifically whether point A or point B is at a higher potential. To this end, we speak of the *voltage drop* from point A to point B, conveniently denoted by a double-subscript, as V_{AB}. If the value of V_{AB} is positive, then point A is at a higher potential than point B. On the other hand, if V_{AB} is negative, then point B is at a higher potential than point A. Since V_{BA} stands for the voltage drop from point B to point A, $V_{AB} = -V_{BA}$.

The double-subscript convention is one of three methods commonly used to unambiguously specify a voltage drop. Using this convention requires labeling all points of interest with letters or integers so that V_{AB}, V_{AC}, V_{12}, or V_{13} make sense. A second, more-common convention uses + and – markings on two points, together with a variable or numerical labeling of the voltage drop from the point marked + to the point marked –. Figure 1.11 illustrates this second convention, where V_0 denotes the voltage drop from A (marked +) to B (marked –). If V_0 is positive, then A is at a higher potential than B. On the other hand, if V_0 is negative, then B is at a higher potential than A. The value of V_0, together with the markings + and –, stipulates which terminal is at a higher potential; neither alone can do this. For a general circuit element, the (+, –) markings—that is, the reference directions—can be assigned *arbitrarily*. A third method for specifying a voltage drop, using a single subscript, will be discussed in Chapter 2.

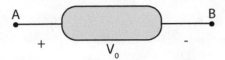

FIGURE 1.11 The + and – markings establish a reference direction for voltage drop. For accuracy, always place the (+, –) markings reasonably close to the circuit element to avoid uncertainty.

The following example illustrates the use of the double subscript and the (+, –) markings for designating voltage drops.

EXAMPLE 1.4

Figure 1.12 shows a circuit consisting of four general circuit elements, with voltage drops as indicated. Suppose we know that $V_{AB} = 4$ V, and $V_{AD} = 9$ V. Find the values of V_x, V_y, V_z, V_{BC}, and V_{CD}.

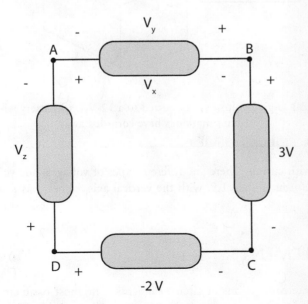

FIGURE 1.12 Arbitrary circuit elements for exploring the use of (+, –) for specifying a voltage drop.

SOLUTION

The meaning of the double subscript notation and the (+, −) markings for a voltage imply that

$$V_x = V_{AB} = 4 \text{ V}, \; V_y = V_{BA} = -V_{AB} = -V_x = -4 \text{ V}$$
$$V_z = V_{DA} = -V_{AD} = -9 \text{ V}$$
$$V_{BC} = 3 \text{ V}$$
$$V_{CD} = -V_{DC} = -(-2) = 2 \text{ V}$$

Exercise. In Figure 1.12, find V_{CB} and V_{DC}.
ANSWER: −3 V; −2 V

Exercise. The convention of the (+, −) markings is commonly used as described. Figure 1.13 shows an old 12-V automobile battery whose (+, −) markings cannot be seen because of the corrosion of the terminals. A digital voltmeter (DVM) is connected across the terminals, as shown. The display reads −12 V. Figure out the (+, −) marking of the battery terminals.
ANSWER: left terminal, −; right terminal, +

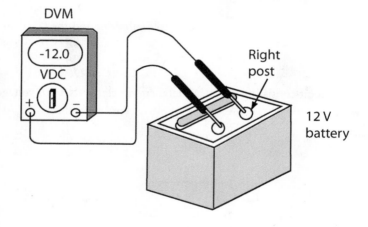

FIGURE 1.13 Digital voltmeter connected to a 12-V (car) battery whose plus and minus markings have corroded away.

One final note: As with current, there are different types of voltages—dc voltage, ac voltage, and general voltage waveforms. Figure 1.9, with the vertical axis relabeled as $v(t)$, illustrates different voltage types.

4. CIRCUIT ELEMENTS

Circuits consist of interconnections of circuit elements. The most basic circuit element has two terminals, and is called a two-terminal circuit element, as illustrated in Figure 1.14. A circuit element called a source provides either voltage, current, or both. The battery is a very common source, providing nearly constant voltage and the usually small current needed to operate small electronic devices. Car batteries, for example, are typically 12 volts and can produce large currents during starting. The wall outlet in a home can be thought of as a 110-volt ac source. Figure 1.14(a) shows a (battery) voltage $v(t)$ across a general undefined circuit element. A current $i(t)$ flows through the element. Recall from our earlier intuitive discussion that voltage is analogous to water pressure: pressure causes water to flow through pipes; voltage causes current to flow through circuit elements. Total water into a pipe equals total water out of the pipe. Analogously, *the current entering a two-terminal device must, by definition, equal the current leaving the two-terminal device.*

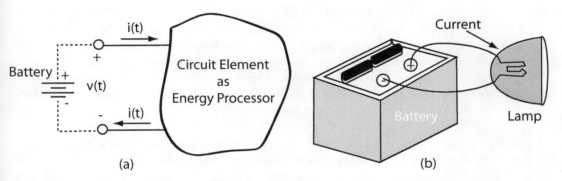

FIGURE 1.14 (a) General circuit element (connected to a battery) as an energy or signal processor: $v(t)$ is the voltage developed across the circuit element, and $i(t)$ is the current flowing through the circuit element; (b) practical example of a general circuit element (car headlight) connected to a car battery.

The circuit element of Figure 1.14(a) has a specific labeling: the current $i(t)$ flows from the plus terminal to the minus terminal through the circuit element. Such a labeling of the voltage-current reference directions is called the **passive sign convention**. In contrast, the current $i(t)$ flows from the minus terminal to the plus terminal through the battery; this labeling is conventional for sources but not for non-source circuit elements.

In addition to sources, there are other common two-terminal circuit elements:
- The resistor
- The capacitor
- The inductor

For a resistor, the amount of current flow depends on a property called **resistance**; the smaller the resistance, the larger the current flow for a fixed voltage across the resistor. A small-diameter pipe offers more resistance to water flow than a large-diameter pipe. Similarly, different types of conductors offer different resistances to current flow. A conductor that is designed to have a specific resistance is called a **resistor**. If the device is an ideal resistor, then $v(t) = Ri(t)$, where R is a constant of resistance. More on this shortly.

The circuit elements called the capacitor and the inductor will be described later in the text. Also, future chapters will describe the operational amplifier and the transformer that are circuit elements having more than two terminals.

5. VOLTAGE, CURRENT, POWER, ENERGY, RELATIONSHIPS

The relationship between voltage across and current through a two-terminal element determines whether power (and, thus, energy) is delivered or absorbed. The heating element in an electric oven can be thought of as a resistor. The heating element absorbs electric energy and converts it into heat energy that cooks, among other things, turkey dinners.

In Figure 1.14(a), a battery is connected to a circuit element. Figure 1.14(b) concretely illustrates this with a 12-V car battery connected to a headlight. With reference to Figures 1.14(a) and 1.14(b), suppose $v(t) = 12$ V, and $i(t) = 5$ A: 5 A of current flows through the headlight. The headlight converts electrical energy into heat and light. Power (in watts) is the rate at which the energy is converted. At each instant of time, the electrical power delivered to (absorbed by) the headlight is $p(t) = v(t)i(t) = 12 \times 5 = 60$ watts. Similarly, at each instant of time, the battery can be viewed as delivering 60 watts of power to the headlight. Inside the battery, the stored potential energy of the chemicals and metals undergoes a chemical reaction that produces the electrical potential difference and the current flow to the headlight: chemical energy is converted into electrical energy that is converted into light and heat.

Figure 1.15 depicts a more general scenario: a circuit element is connected to its surrounding circuit at points A and B. (One, of course, could imagine that the "remainder of circuit" is a battery, and circuit element 1 is a headlight.) Suppose there is a *constant* voltage drop from A to B, denoted by V_{AB}. Also assume that a *constant* current I_{AB} flows from terminal A to terminal B *through circuit element 1*, as shown.

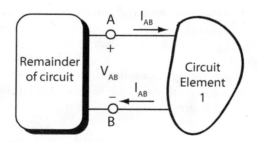

FIGURE 1.15 A general circuit in which a two-element circuit element is extracted and labeled according to the passive sign convention.

For discussion purposes, assume $V_{AB} > 0$ and $I_{AB} > 0$. During a time interval of T s, $(V_{AB} \times T)$ C of charge moves through circuit element 1 from A to B. In "falling" from a higher potential, point A, to a lower potential, point B, the charge loses electric potential energy. The lost potential energy is converted within element 1 into some other form of energy—heat or light being two of several possibilities. According to Equation 1.8, the amount of energy *converted* (*absorbed by the element*) is $V_{AB}(I_{AB} \times T) > 0$. The **power** *absorbed* by element 1 is, by definition, the rate at which it converts or absorbs energy. This rate equals

$$\frac{V_{AB}(I_{AB} \times T)}{T} = V_{AB}I_{AB} > 0.$$

Exercise. In Figure 1.15, the current $I_{AB} = 5$ mA, and $V_{AB} = 400$ V. What is the energy absorbed by circuit element 1 in one minute? What is the power absorbed by circuit element 1?
ANSWER: W = 120 J; P = 2 watts

With respect to Figure 1.15, for constant (direct) voltages and currents, we arrive at a very simple relationship:

$$P_1 = V_{AB}I_{AB} \qquad\qquad (1.10)$$

where P_1 is the power (in W) absorbed by the circuit element. Consequently, the energy, W_1 (in J), absorbed during the time interval T is

$$W_1 = P_1 \times T \qquad\qquad (1.11)$$

Now, let us reconsider Figure 1.15. One can think of $-I_{AB}$ as flowing from A to B through the remainder of the circuit. In this case, $P_0 = V_{AB}(-I_{AB}) = -V_{AB}I_{AB} < 0$. This means that the remainder of the circuit absorbs negative power or equivalently delivers $|V_{AB}(-I_{AB})| = V_{AB}I_{AB}$ to circuit element 1. As such, the remainder of the circuit is said to *generate* electric energy. By definition, the electric power *generated* by the remainder of the circuit is the rate at which it generates electric energy. From Equation 1.8, this rate equals

$$\frac{V_{AB}(I_{AB} \times T)}{T} = V_{AB}I_{AB} > 0.$$

Observe that the rate at which the remainder of the circuit generates power precisely equals the rate at which circuit element 1 absorbs power. This equality is called the **principle of conservation of power**: total power generated equals total power absorbed. Equivalently, the sum of the powers absorbed by all the circuit elements must add to zero, $P_1 + P_0 = V_{AB}I_{AB} + V_{AB}(-I_{AB}) = 0$.

Exercise. In Figure 1.15, $P_0 = -10$ watts, i.e., the remainder of the circuit absorbs -10 watts of power. How much power does circuit element 1 absorb?
ANSWER: 10 watts

In general, whenever a two-terminal general circuit element is labeled according to the passive sign convention, as in Figure 1.15, then $P = V_{AB}I_{AB} > 0$ means the element absorbs (positive) power, whereas if $P = V_{AB}I_{AB} < 0$, it absorbs negative power or delivers (positive) power to whatever it is connected. As a general convention, non-source circuit elements are labeled according to the passive sign convention. Usually, sources are labeled with the current leaving the terminal labeled with "+". For such labeling of sources, if the product of the source voltage and the current leaving the "+" terminal is positive, then the source is delivering power to the network.

RULE FOR CALCULATING ABSORBED POWER

The power absorbed by any circuit element (Figure 1.16) with terminals labeled A and B is equal to the voltage drop from A to B multiplied by the current through the element from A to B, i.e., $P = V_{AB}I_{AB}$.

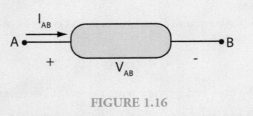

FIGURE 1.16

Exercise. Compute the power absorbed by each of the elements in Figure 1.17.

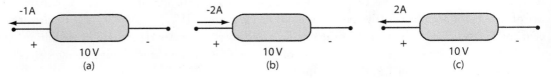

FIGURE 1.17

ANSWER: (a) 10 W; (b) –20 W; (c) 20 W

As mentioned, **power** is the rate of change of work per unit of time. The ability to determine the power absorbed by each circuit element is highly important because using a circuit element or some device beyond its power-handling capability could damage the device, cause a fire, or result in a serious disaster. This is why households use circuit breakers to make sure electrical wiring is not overloaded.

Exercise. In Figure 1.18, a car heater is attached to a 12-volt DC voltage source. How much power can the car heater absorb before the 20-amp fuse blows.

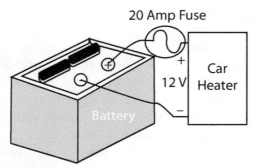

FIGURE 1.18 Car heater connected to a 12-volt car battery through a 20-amp fuse.

ANSWER: 240 watts

As mentioned earlier, the calculated value of absorbed power P may be negative. If the absorbed power P is negative, then the circuit element actually generates power or, equivalently, delivers power to the remainder of the circuit. In any circuit, some elements will have positive absorbed powers, whereas some others will have negative absorbed powers. If one adds up the absorbed powers of *ALL* elements, the sum is zero! This is a universal property called **conservation of power**.

PRINCIPLE OF CONSERVATION OF POWER

The sum of the powers absorbed by all elements in a circuit is zero at any instant of time. Equivalently, the sum of the absorbed powers equals the sum of the generated powers at each instant of time.

The 2nd edition of this text contains a rigorous proof of this principle. For the present, we will simply use it to solve various problems. The following example will help clarify the sign conventions and illustrate the principle of conservation of power.

EXAMPLE 1.5

Light bulbs come in all sorts of shapes, sizes, and wattages. **Wattage** measures the power consumed by a bulb. Typical wattages include 15, 25, 40, 60, 75, and 100 W. Power consumptions differ because the current required to light a higher-wattage (and brighter) bulb is larger for a fixed outlet voltage: a higher-wattage bulb converts more electric energy into light energy. In Figure 1.19, the source delivers 215 watts of power. What is the wattage of the unlabeled bulb?

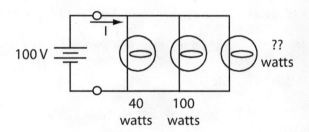

FIGURE 1.19. Three bulbs connected to a 100-V battery.

SOLUTION

From conservation of power, the total power delivered by the battery equals the total power absorbed by all the bulbs. Therefore, the power absorbed by the unknown bulb is

$$215 - 40 - 100 = 75 \text{ watts}$$

Exercise. Determine the current I leaving the battery in Example 1.5.
ANSWER: 2.15 amps

EXAMPLE 1.6

An electroplating apparatus uses electrical current to coat materials with metals such as copper or silver. In Figure 1.20, suppose a 220-V electrical source supplies 10 A dc to the electroplating apparatus.

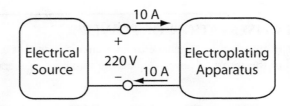

FIGURE 1.20 Electrical source operating an electroplating apparatus.

a) What is the power consumed by the apparatus?
b) If electric energy costs 10 cents per kilowatt-hour (kWh), what will it cost to operate the apparatus for a single 12-h day?

SOLUTION

Step 1. From Equation 1.10, the power consumed is

$$P = 220 \times 10 = 2200 \text{ W, or } 2.2 \text{ kW}$$

Step 2. According to Equation 1.11, the energy consumed per 12-h period is

$$W = 2.2 \times 12 = 26.4 \text{ kWh}$$

Step 3. Therefore, the cost to operate is

$$26.4 \times .01 = \$2.64 \ / \text{ day}$$

Exercise. Suppose the electroplating apparatus of Example 1.6 draws 12 A DC at the same voltage. What is the cost of operation for a single 12-h day? What is the cost of operating for a 20 workday month?
ANSWER: \$3.168; \$63.36

EXAMPLE 1.7

Each box in the circuit of Figure 1.21 is a two-terminal element. Compute the power absorbed by each circuit element. Which elements are delivering power? Verify the conservation of power principle for this circuit.

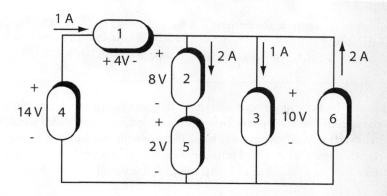

FIGURE 1.21 Circuit containing several general circuit elements.

SOLUTION

Step 1. *Compute power absorbed by each element.* Using either Equation 1.10 or the power consumption rule, the power absorbed by each element is

a) For element 1: $P_1 = 4 \times 1 = 4$ W
b) For element 2: $P_2 = 8 \times 2 = 16$ W
c) For element 3: $P_3 = 10 \times 1 = 10$ W
d) For element 4: $P_4 = 14 \times (-1) = -14$ W
e) For element 5: $P_5 = 2 \times 2 = 4$ W
f) For element 6: $P_6 = 10 \times (-2) = -20$ W

Step 2. *Verify conservation of power.* Since P_4 and P_6 are negative, element 4 delivers 14 W, and element 6 delivers 20 W of power. The remaining four elements absorb power. Observe that the sum of the six absorbed powers, $4 + 16 + 10 - 14 + 4 - 20 = 0$, as expected from the principle of conservation of power. Equivalently, the total positive generated power, $(14 + 20) = 34$ W, equals the total positive absorbed power, $(4 + 16 + 10 + 4) = 34$ W.

Exercise. In Figure 1.22, find the powers absorbed by elements 1, 2, and 3.

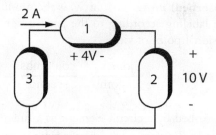

FIGURE 1.22

ANSWER: 8 W, 20 W, −28 W; element 3 equivalently delivers 28 W

Exercise. In Figure 1.22, suppose the current 2 A were changed to −4 A. What is the new power absorbed by element 3?
ANSWER: 56 watts

If the power absorbed by a circuit element is positive, the exact nature of the element determines the type of energy conversion that takes place. For example, a circuit element called a **resistor** (to be discussed shortly) converts electric energy into heat. If the circuit element is a battery that is being charged, then electric energy is converted into chemical energy within the battery. If the circuit element is a dc motor turning a fan, then electrical energy is converted into mechanical energy.

NON-DC POWER AND ENERGY CALCULATIONS

Consider Figure 1.23, where $i(t)$ is an arbitrary time-varying current entering a general two-terminal circuit element, and $v(t)$ is the time-varying voltage across the element. Because voltage and current are functions of time, the power $p(t) = v(t)i(t)$ is also a function of time. For any specific value of $t = t_1$, the value $p(t_1)$ indicates the power absorbed by the element at that particular time—hence, the terminology **instantaneous power** for $p(t)$.

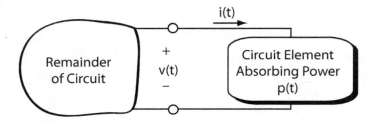

FIGURE 1.23 Calculation of absorbed power for time-varying voltages and currents for circuit elements labeled with the passive sign convention; here, power is $p(t) = v(t)i(t)$.

Equation 1.12 extends Equation 1.10 in the obvious way,

$$p(t) = v(t)i(t) \tag{1.12}$$

i.e., the **instantaneous (absorbed) power** $p(t)$, in W, is the product of the voltage $v(t)$, in V, and the current $i(t)$, in A, with labeling according to the passive sign convention. This product also makes sense from a dimensional point of view:

$$\text{volts} \times \text{amps} = \frac{\text{joules}}{\text{coulomb}} \times \frac{\text{coulombs}}{\text{second}} = \frac{\text{joules}}{\text{second}}$$

Knowing the power $p(t)$ absorbed by a circuit element as a function of t allows one to compute the energy $W(t_0, t)$ absorbed by the element during the time interval $[t_0, t > t_0]$. $W(t_0, t)$ (J) is the integral of $p(t)$ (W) with respect to t over $[t_0, t]$, i.e.,

$$W(t_0,t) = \int_{t_0}^{t} p(\tau)d\tau$$

$$(1.13)$$

where t_0, the lower limit of the integral, could possibly be $-\infty$. For the dc case, $p(t) = P$ (a constant). From Equation 1.13,

$$W(t_0,t) = \int_{t_0}^{t} p(\tau)d\tau = P\int_{t_0}^{t} d\tau = P(t - t_0) = P \times T$$

where $T = t - t_0$, as given in Equation 1.11. If, in Equation 1.13, $t_0 = -\infty$, then W$(-\infty, t)$ becomes a function only of t which, for convenience, is denoted by

$$W(t) = \int_{-\infty}^{t} p(\tau)d\tau$$

$$(1.14)$$

$W(t) = W(-\infty, t)$, in joules, represents the total energy absorbed by the circuit element from the beginning of time to the present time t when $p(t)$ is in watts.

Exercise. a) Suppose the power absorbed by a circuit element over $[0,\infty)$ is $p(t) = 2e^{-0.5t}$ watts. Find $W(0, \infty)$. b) Now suppose the absorbed power of the circuit element is

$$p(t) = \begin{cases} 2e^{0.5t} & -\infty < t \le 0 \\ 1 & t > 0 \end{cases}$$. Find for $W(-\infty,t)$ for $t > 0$.

ANSWER: 4 J; (4+t) J

Since energy is the integral of power, power is the rate of change (derivative) of energy. Differentiating both sides of Equation 1.14 yields the expected equation for instantaneous power,

$$v(t)i(t) = p(t) = \frac{dW(t)}{dt}$$

$$(1.15a)$$

or, equivalently, for $t > t_0$,

$$v(t)i(t) = p(t) = \frac{dW(t_0,t)}{dt}$$

$$(1.15b)$$

Exercise. Suppose that for t > 0, the work done by an electronic device satisfies $W(t) = 10(1 - e^{-t})$ J. If the voltage supplied by the device is 10 V, then for t > 0, find the power and current supplied by the device, assuming standard labeling, i.e., the passive sign convention.

ANSWER: $p(t) = 10e^{-t}$ watts; $i(t) = e^{-t}$ A

EXAMPLE 1.8

In the circuit of Figure 1.23, the current $i(t)$ and voltage $v(t)$ have the waveforms graphed in Figure 1.24. Sketch $p(t)$, the instantaneous power absorbed by the circuit element, and then sketch $W(0, t)$, the energy absorbed over the interval $[0, t]$.

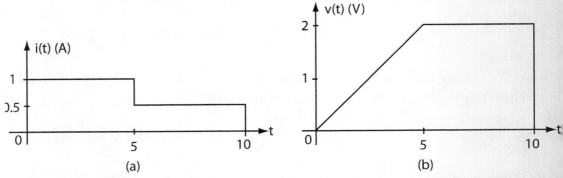

FIGURE 1.24 (a) Current and (b) voltage profiles with respect to t for circuit of Figure 1.23.

SOLUTION

A simple graphical multiplication of Figures 1.24(a) and (b) yields the sketch of the curves in instantaneous power shown in Figure 1.25(a). From Equation 1.13 with $t_0 = 0$, we have, for $0 \leq t \leq 5$,

$$W(0,t) = \int_0^t p(\tau)d\tau = \int_0^t \frac{2\tau}{5}d\tau = \frac{t^2}{5}$$

and for $t \geq 5$,

$$W(0,t) = \int_0^t p(\tau)d\tau = \int_0^5 p(\tau)d\tau + \int_5^t p(\tau)d\tau = 5 + \int_5^t d\tau = 5 + (t-5) = t$$

Figure 1.25(b) presents the resulting graph.

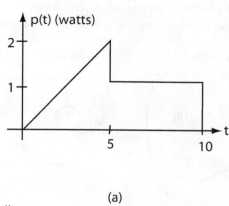

(a)

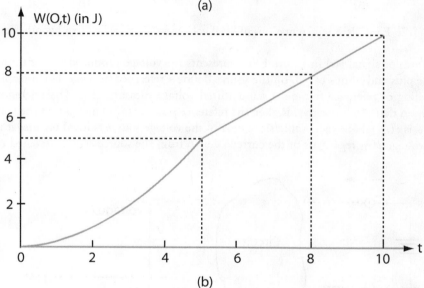

(b)

FIGURE 1.25 (a) Profile of the instantaneous power $p(t) = v(t)i(t)$ for the current and voltage waveforms of Figure 1.24; (b) associated profile of energy versus time.

6. IDEAL VOLTAGE AND CURRENT SOURCES

Two-terminal circuit elements may be classified according to their terminal voltage-current relationships. The goal of this section is to define **ideal voltage** and **current sources** via their terminal voltage-current relationships.

The wall socket of a typical home represents a practical voltage source. After flipping the switch on an appliance plugged into a wall socket, a current flows through the internal circuitry of the appliance, which, for a vacuum cleaner or dishwasher, converts electrical energy into mechanical energy. For modest amounts of current draw (below the fuse setting), the voltage nearly maintains its nominal pattern of $120 \sqrt{2} \sin(120 \pi t) = 169.7 \sin(120 \pi t)$ V. This practical situation is idealized in circuit analysis by the **ideal voltage source** symbol shown in Figure 1.26(a), a circle with a ± reference inside. The symbol is more commonly referred to as **independent voltage source.**

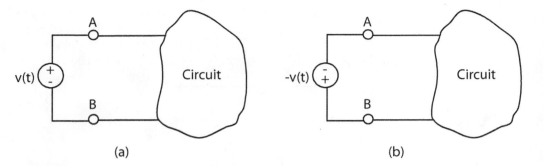

FIGURE 1.26 Equivalent representations of ideal voltage source attached to a hypothetical circuit.

The waveform or signal $v(t)$ in Figure 1.26 represents the voltage produced by the source at each time t. The plus and minus (+, −), on the source define a reference polarity. The reference polarity is a labeling or reference frame for standardized voltage measurement. The reference polarity does not mean that $v(t)$ is positive. Rather, the reference polarity (+, −) means that the voltage drop from + to − is $v(t)$, whatever its value/sign. Finally, the voltage source is **ideal** because it maintains the given voltage $v(t)$, regardless of the current drawn from the source by the attached circuit.

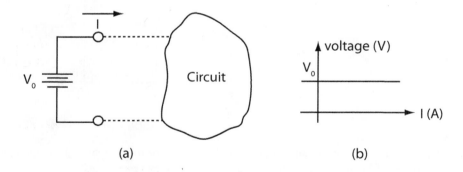

FIGURE 1.27 (a) Ideal battery representation of ideal voltage source; (b) v-i characteristic of ideal battery.

Figure 1.27(a) shows a source symbol for an **ideal battery**. The voltage drop from the long-dash side to the short-dash side is V_0, with $V_0 > 0$. In commercial products, the terminal marked with a + sign corresponds to the long-dash side of Figure 1.27(a). An ideal battery produces a constant voltage under all operating conditions, i.e., regardless of current drawn from an attached circuit or circuit element, as indicated by the v-i characteristic of Figure 1.27(b). Real batteries are not ideal but approximate the ideal case over a manufacturer-specified range of current requirements.

Practical sources (i.e., non-ideal); voltage sources, such as commercial dc and ac generators; and real batteries deviate from the ideal in many respects. One important respect is that the terminal voltage depends on the current delivered by the source. The most common generators convert mechanical energy into electrical energy, while batteries convert chemical energy into electrical energy. There are two general battery categories: nonrechargeable and rechargeable. A discussion of the dramatically advancing battery technology is beyond the scope of this text.

Besides batteries and ideal voltage sources, devices called **ideal** or **independent current sources** maintain fixed current waveforms into a circuit, as illustrated in Figure 1.28. The symbol of an ideal current source is a circle with an arrow inside, indicating a reference current direction. An ideal current source produces and maintains the current $i(t)$ under all operating conditions. Of course, the current $i(t)$ flowing from the source can be a constant (dc), sinusoidal (ac), or any other time-varying function.

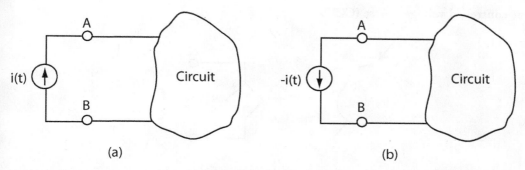

FIGURE 1.28 Equivalent ideal current sources whose current $i(t)$ is maintained under all operating conditions of the circuit.

In nature, lightning is an example of an approximately ideal current source. When lightning strikes a lightning rod, the path to the ground is almost a short circuit, and very little voltage is developed between the top of the rod and the ground. However, if lightning strikes a tree, the path of the current to the ground is impeded by the trunk of the tree. A large voltage then develops from the top of the tree to the ground.

Independent sources have conventional labeling, as shown in Figure 1.29, which is different from that of the passive sign convention. Here the source delivers power if $p(t) = v(t)i(t) > 0$ and would absorb power if $p(t) = v(t)i(t) < 0$. A complicated circuit called a battery charger can deliver energy to a drained car battery. The car battery, although usually a source delivering power, exemplifies a source absorbing power from the charger.

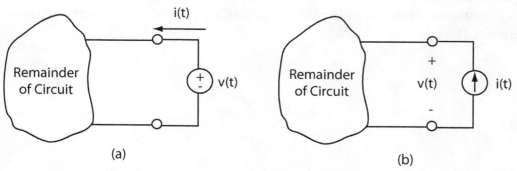

FIGURE 1.29 Common voltage and current source labeling.

Another type of ideal source is a dependent source. A **dependent source or a controlled source** produces a current or voltage that depends on a current through or voltage across some other element in the circuit. Such sources model real-world devices that are used in real circuits. In the text, the symbol for a dependent source is a diamond. If a ± appears inside the diamond, it is a **dependent voltage source**, as illustrated in Figure 1.30. If an arrow appears inside the diamond, it is a **dependent current source**, as illustrated in Figure 1.31. In Figure 1.30, the voltage across the diamond-shaped source, $v(t)$, depends either on a current, labeled i_x, through some other circuit device, or on the voltage v_x across it. If the voltage across the source depends on the voltage v_x, i.e., $v(t) = \mu\, v_x$, then the source is called a **voltage-controlled voltage source (VCVS)**. If the voltage across the source depends on the current i_x, i.e., $v(t) = r_m i_x$, then the source is called a **current-controlled voltage source (CCVS)**.

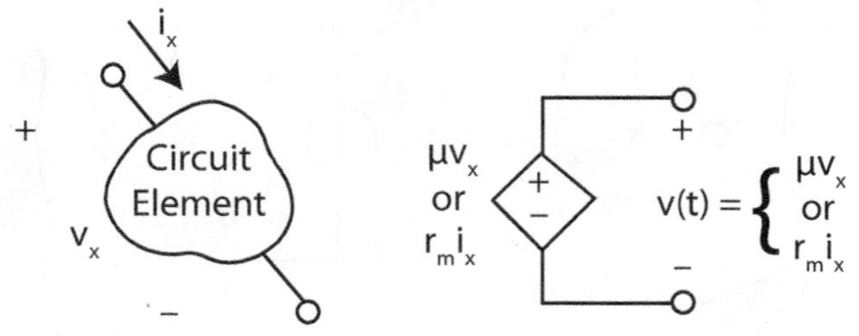

FIGURE 1.30 The right element is a voltage-controlled voltage source (VCVS) if $v(t) = \mu v_x$ (μ is here dimensionless), or a current-controlled voltage source (CCVS) if $v(t) = r_m i_x$ (r_m here has units of ohm).

Exercise. The voltage across a particular circuit element is $v_x = 5$ V, and the current through the element is $i_x = 0.5$ A, using the standard labeling.
   a)   If a VCVS (Figure 1.30) with $\mu = 0.4$ were associated with the controlled-source branch, find $v(t)$.
   b)   If a CCVS (Figure 1.30) with $r_m = 3\ \Omega$ were associated with the controlled branch, find $v(t)$.
ANSWER: a) 2 V; b) 1.5 V

There is dual terminology for dependent current sources. The configuration of Figure 1.31 shows a **voltage-controlled current source (VCCS)**, i.e., $i(t) = g_m v_x$, or a **current-controlled current source (CCCS),** for which $i(t) = \beta i_x$.

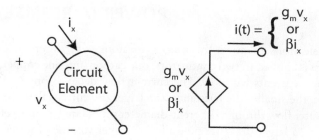

FIGURE 1.31 The right element is a voltage-controlled current source (VCCS) if $i(t) = g_m v_x$ (g_m has units of siemens) or a current-controlled current source (CCCS) if $i(t) = \beta i_x$ (β is dimensionless).

Source voltages or currents are called **excitations**, **inputs**, or **input signals**. A constant voltage will normally be denoted by an uppercase letter, such as V, V_0, V_1, V_A, and so on. A constant current will typically be denoted by I, I_0, I_1, I_A, and so on. The units are volts, amperes, and so on. Smaller and larger quantities are expressed by the use of prefixes, as defined in Standard Engineering Notation Table 1.1.

Exercise. The voltage across a particular circuit element is $v_x = 5$, and the current through the element is $i_x = 0.5$ A using the standard labeling.

   a)   If a VCCS (Figure 1.31) with $g_m = 0.1$ S were associated with the controlled-source branch, find $i(t)$.
   b)   If a CCCS (Figure 1.31) with $\beta = 0.5$ were associated with the controlled-source branch, find $i(t)$.
ANSWER: a) 0.5 A; b) 0.25 A

TABLE 1.1. Engineering Notation for Large and Small Quantities

Name	Prefix	Value
femto	f	10^{-15}
pico	p	10^{-12}
nano	n	10^{-9}
micro	μ	10^{-6}
milli	m	10^{-3}
kilo	k	10^{3}
mega	M	10^{6}
giga	G	10^{9}
tera	T	10^{12}

7. RESISTANCE, OHM'S LAW, AND POWER (A REPRISE)

Different materials allow electrons to move from atom to atom with different levels of ease. Suppose the same dc voltage is applied to two conductors, one carbon and one copper, of the same size and shape. Two different currents will flow. The current flow depends on a property of the conductor called **resistance**: the smaller the resistance, the larger the current flow for a fixed voltage. The idea is similar to water flow through different-diameter pipes (analogous to electrical conductors): for a given pressure, a larger-diameter pipe allows a larger volume of water to flow and, therefore, has a smaller resistance than a pipe with, say, half the diameter.

A conductor designed to have a specific resistance is called a **resistor**. Hence, a resistor is a device that impedes current flow. Just as dams impede water flow and provide flood control for rivers, resistors provide a means to control current flow in a circuit. Further, resistors are a good approximate model to a wide assortment of electric devices such as light bulbs and heating elements in ovens. Figure 1.32(a) shows the standard symbol for a resistor, where the voltage and current reference directions are marked in accordance with the **passive sign convention**. Figure 1.32(b) pictures a resistor connected to an ideal battery.

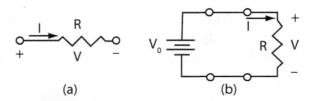

(a)　　　　　　　　　　　　　(b)

FIGURE 1.32 (a) Symbol for a resistor with reference voltage polarity and current direction consistent with the passive sign convention; (b) resistor connected to an ideal battery.

In 1827, Ohm observed that for a connection like that of Figure 1.32(b), the direct current through the conductor/resistor is proportional to the voltage across the conductor/resistor, i.e., $I \approx V$. Inserting a proportionality constant, one can write

$$I = \frac{1}{R}V = GV \qquad (1.16a)$$

or, equivalently,

$$V = RI \qquad (1.16b)$$

The proportionality constant R is the **resistance** of the conductor in ohms. The resistance R measures the degree to which the device impedes current flow. For conductors/resistors, the ohm (Ω) is the basic unit of resistance. *A two-terminal device has a 1-Ω resistance if a 1-V excitation causes 1-A of current to flow.* In Equation 1.16(a), the proportionality constant is the reciprocal of R, i.e., $G = 1/R$, which is called the **conductance** of the device. The unit for conductance according to the International System of Units (SI) system is the **siemen**, S. In the United States, the older term for the unit of conductance is the **mho** ℧, that is, ohm spelled backward, which is still widely used. In this text, we try to adhere to the SI system. If a device or wire has zero resistance ($R = 0$) or infinite conductance ($G = \infty$), it is termed a **short circuit**. On the other hand, if a device or wire has infinite resistance (zero conductance), it is called an **open circuit**. Technically speaking, a resistor means a real physical device, with resistance being the essential property of the device. In most of the literature on electronic circuits, resistor and resistance are used synonymously, and we will continue this practice.

OHM'S LAW

Ohm's law, as observed for constant voltages and currents, is given by Equation 1.16(b), with its equivalent form in Equation 1.16(a). However, it is true for all time-dependent waveforms exciting a linear resistor. Thus, we can generalize Equation 1.16 as

$$v(t) = Ri(t) \tag{1.17a}$$

or

$$i(t) = \frac{1}{R}v(t) = Gv(t) \tag{1.17b}$$

according to Figure 1.33, whose voltage-current labeling is consistent with the passive sign convention.

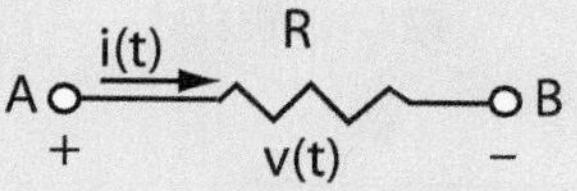

FIGURE 1.33

If either the voltage or the current direction is reversed, but not both, then Ohm's law becomes $v(t) = -Ri(t)$. As an aid in writing the correct v-i relationship for a resistor, Ohm's law is stated here in words:

For a resistor connected between terminals A and B, the voltage drop from A to B is equal to the resistance multiplied by the current flowing from A to B through the resistor.

Exercise. Find the resistance R for each of the resistor configurations in Figure 1.34.
ANSWER: (a) 12 Ω; (b) 3 Ω; (c) 6 Ω

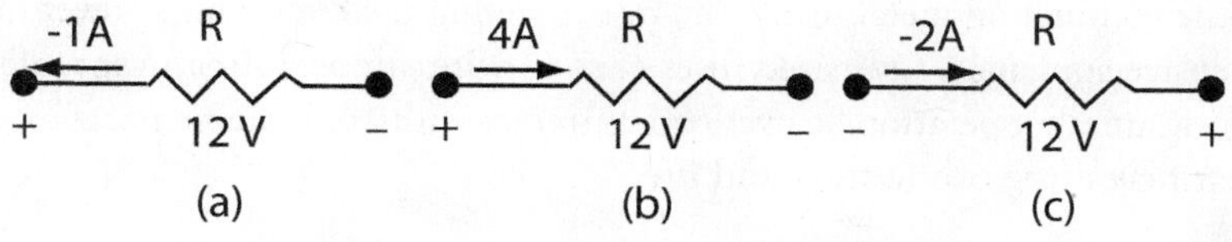

FIGURE 1.34

Once the voltage and the current associated with a resistor are known, the power absorbed by the resistor is easily calculated. Assuming the passive sign convention, then combining Equation 1.12 for power and Ohm's law (Equation 1.17), the instantaneous absorbed power is

$$p(t) = v(t)i(t) = i^2(t)R = \frac{v^2(t)}{R} \tag{1.18a}$$

which for the dc case reduces to

$$P = VI = I^2R = \frac{V^2}{R} \tag{1.18b}$$

Exercise. Find the power absorbed by each of the resistors in Figure 1.35.

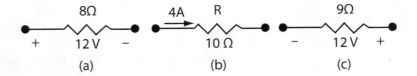

FIGURE 1.35

ANSWER: (a) 18 W; (b) 160 W; (c) 16 W

Equations 1.18(a) and (b) bring out a very important property: a resistor *always* absorbs power, dissipating it as heat. Intuitively speaking, electrons that flow through the resistor collide with other particles along the way. The process resembles the action in a pinball game: the pinball successively collides with various pegs as it rolls from a higher to a lower elevation. With each collision, part of the electron's kinetic energy is converted into heat as the voltage pressure continues to reaccelerate the electron.

Electrical energy that is converted to heat or used to overcome friction is usually called a **loss**. Such losses are termed *I*-squared-*R* (I^2R) losses because of the form of Equation 1.18. On the other hand, a stove's heating element purposely converts to heat as much electric energy as possible, in which case, the I^2R loss is desirable. This heating effect also proves useful as the basis for the operation of fuses. A **fuse** is a short piece of inexpensive conductor with a very low resistance and a predetermined current-carrying capacity. When inserted in a circuit, it carries the current of the equipment or appliances it must protect. When the current rises above the fuse rating, the generated heat melts the conducting metal inside the fuse, opening the circuit and preventing damage to the more-expensive appliance. Oversized fuses or solid-wire jumpers circumvent safe fuse operation by permitting unsafe operation at overload currents, with consequent electrical damage to the appliance that may cause overheating and fire.

Resistance of a conductor depends on the material and its geometrical structure. For a specific temperature, R is proportional to the length L of a conductor and inversely proportional to its cross-sectional area A,

$$R = \rho\frac{L}{A} \tag{1.19}$$

where the proportionality constant ρ is the **resistivity** in ohm-meters ($\Omega \cdot$ m). The resistivity of copper at 20°C is 1.7×10^{-8} $\Omega \cdot$ m. Table 1.2 lists the relative resistivities of various materials with respect to copper.

Table 1.2 Resistivities of Various Materials Relative to Copper.[*]

Silver	0.94	Chromium	1.8	Tin	6.7
Copper	1.00	Zinc	3.4	Carbon	2.4×10^3
Gold	1.4	Nickel	5.1	Aluminum	1.6

*The resistivity of copper at 20 degrees C is 1.7×10^{-8} Ωxm.

EXAMPLE 1.9

Sixteen-gauge (16 AWG) copper wire has a resistance of 4.094 Ω for every 1,000 feet of wire. Find the resistance of 100 feet of 16 AWG aluminum wire and 100 feet of 16 AWG nickel wire. Then find the voltage across each wire and the power absorbed (given off as heat) by each wire if a 10-A direct current flows through 100 feet of each wire.

SOLUTION

The resistivities of aluminum and nickel wire relative to copper are 1.6 and 5.1, respectively. Hence, 100 feet of aluminum/nickel wire has a resistance of

(aluminum)$1.6 \times 0.4094 = 0.655$ Ω

(nickel) $5.1 \times 0.4094 = 2.088$ Ω

Given a 10-A current flowing through 100 feet of copper, aluminum, and nickel wire, Ohm's law implies

(copper) $V = RI = 0.4094 \times 10 = 4.094$ V

(aluminum) $V = RI = 0.655 \times 10 = 6.55$ V

(nickel) $V = RI = 2.088 \times 10 = 20.88$ V

Finally, from Equation 1.18(b), the absorbed power given off as heat is

(copper) $P = VI = RI^2 = 0.4094 \times 100 = 40.94$ W

(aluminum) $P = VI = RI^2 = 0.655 \times 100 = 65.5$ W

(nickel) $P = VI = RI^2 = 2.088 \times 100 = 208.8$ W

Notice that every 100 feet of 16 AWG aluminum wire would absorb $65.5 - 40.9 = 24.6$ W more power than copper. And nickel wire absorbs even more power:

$$5.1 = \frac{208.8}{40.94}$$

times more power than copper per unit length. This absorbed power, given off as heat, is why nickel wire is used for heating elements in toasters and ovens.

Exercise. (a) If a constant current of 10 A flows through 1,000 feet of (16 AWG) copper wire, how many watts of heat are generated by the wire?
(b) If the wire of part (a) were changed to (16 AWG) aluminum, how many watts of heat would be generated?

ANSWER: (a) 409.4 watts; (b) 655.04 watts

Temperature also affects resistance. For example, light bulbs have a "cold" resistance and a "hot" resistance of more importance during lighting. For most metallic conductors, resistance increases with increasing temperature—except carbon, which has a decrease in resistance as temperature rises. Since resistors absorb power dissipated as heat, they should have adequate physical dimensions to better radiate the heat or there must be some external cooling to prevent overheating.

EXAMPLE 1.10

The hot resistance of a light bulb is 120 Ω. Find the current through and the power absorbed by the bulb if it is connected across a constant 90-V source, as illustrated in Figure 1.36.

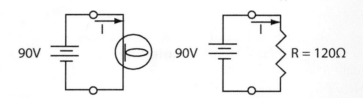

FIGURE 1.36 Light bulb and equivalent resistive circuit model.

SOLUTION

Step 1. From Ohm's law, Equation 1.16(a),

$$I = \frac{V}{R} = \frac{90}{120} = 0.75 \text{ A}$$

Step 2. By Equation 1.18(b), the power absorbed by the lamp is

$$P = 0.75^2 \times 120 = 67.5 \text{ W}$$

Step 3. *Check conservation of power.* The power delivered by the source is 90 × 0.75 = 67.5 W. Therefore, the power delivered by the source equals the power absorbed by the resistor. This verifies conservation of power for the circuit.

Exercise. In Example 1.10, suppose the battery voltage is cut in half to 60 V. What is the power absorbed by the lamp? What is the power delivered by the battery? Repeat with the battery voltage changed to 120 V.

ANSWER: 30 watts; 120 watts

The following example illustrates power consumption for a parallel connection of light bulbs.

EXAMPLE 1.11

Figure 1.37 shows four automobile halogen light bulbs connected in **parallel** across a 12-V battery. Find the following:
   (a)   The effective "hot" resistance of each bulb
   (b)   The total power delivered by the source
   (c)   After 700 hours of operation, the current supplied by the source drops to 11.417 A. Discover which light bulb has burned out.

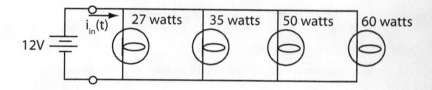

FIGURE 1.37 Parallel connection of light bulbs.

SOLUTION

(a) From Equation 1.18(b), $P = V^2/R$,

$$R_{27W} = \frac{12^2}{27} = 5.33\,\Omega \qquad R_{35W} = \frac{144}{35} = 4.114\,\Omega$$

$$R_{50W} = \frac{144}{50} = 2.88\,\Omega \qquad R_{60W} = \frac{144}{60} = 2.4\,\Omega$$

(b) The power delivered by the source equals the sum of the powers consumed by each bulb, which is 172 W.

(c) Since the current supplied by the source has dropped to 11.417 A, then the power delivered by the source drops to $P_{source,new} = 12 \times 11.417 = 137$ watts, which is 35 watts less than the earlier-delivered power of 172 watts. Hence, the 35-watt bulb has gone dark.

Exercise. Repeat Example 1.11(a) with the battery voltage changed to 48 V and a new set of light bulbs whose operating voltage is 48 V.
ANSWER: $R_{27W} = 85.333\,\Omega$; $R_{35W} = 65.83\,\Omega$; $R_{50W} = 46.08\,\Omega$; $R_{60W} = 38.4\,\Omega$.

EXAMPLE 1.12

When connected to a 120-volt source, halogen light bulb number 1 uses 40 watts of power. When similarly connected, halogen light bulb 2 uses 60 watts of power.
   (a)   Find the hot resistance of each bulb.
   (b)   If the two bulbs are connected in a series, as in Figure 1.38 and placed across the 120-V source, find the power absorbed by each bulb and the power delivered by the source, assuming the hot resistances computed in part (a) do not change.

(c)   Find the voltage V_1 and V_2 across each bulb.

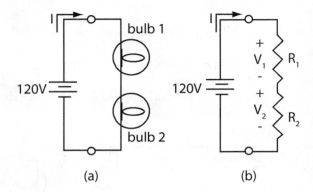

(a)   (b)

FIGURE 1.38 Series connection of two light bulbs and equivalent resistive circuit model.

SOLUTION

Step 1. *Find the hot resistances.* The hot resistances of each bulb are given by

$$R_1 = \frac{V_{source}^2}{P_{bulb1}} = \frac{120^2}{40} = 360 \text{ and } R_2 = \frac{V_{source}^2}{P_{bulb1}} = \frac{120^2}{60} = 240$$

Step 2. *Find the current through each bulb, the power absorbed by each bulb, and the power delivered by the source.* The circuit of Figure 1.38(a) has the equivalent representation in terms of resistances in Figure 1.38(b). By definition, in a two-terminal circuit element, the current entering each resistor equals the current leaving. Therefore, the current through each resistor in the series connection is the same, and is denoted I. So the new power dissipated by each bulb/resistor is

$$P_{1,new} = R_1 I^2 \text{ and } P_{2,new} = R_2 I^2$$

To calculate these values, we need to know I. By conservation of power, the power delivered by the source is the sum of the absorbed powers, i.e.,

$$P_{source} = 120 \times I = P_{1,new} + P_{2,new} = R_1 I^2 + R_2 I^2$$

Hence, dividing through by I,

$$120 = R_1 I + R_2 I = (R_1 + R_2)I = 600 I$$

Therefore,

$$I = \frac{120}{600} = 0.2 \text{ A}$$

Hence,

$$P_{1,new} = 360 \times 0.2^2 = 14.4 \text{ W}, \ P_{2,new} = 240 \times 0.2^2 = 9.6 \text{ W}, \ P_{source} = P_{1,new} + P_{2,new} = 24 \text{ W}$$

Step 3. *Find voltages across each bulb.* From Ohm's law,

$$V_1 = R_1 I = 72 \text{ V and } V_2 = R_2 I = 48$$

Although involved, the solution of this problem uses the definition of a two-terminal circuit element and conservation of power to arrive at the result in a roundabout way. In Chapter 2, we can more directly arrive at the answers by using Kirchhoff's voltage and current laws.

A potential problem with series connections of light bulbs is circuit failure. If one bulb burns out, i.e., the filament in the bulb open-circuits, then all other lights are extinguished. Parallel circuits continue to operate in the presence of open-circuit failures and are easier to fix: only the unlit bulb must be replaced.

8. V-I CHARACTERISTICS OF IDEAL RESISTOR, CONSTANT VOLTAGE, AND CONSTANT CURRENT SOURCES

The **ideal (linear) resistor** is a device that satisfies Ohm's law. **Ohm's law** is a relationship between the current through the linear resistor and the voltage across it. A graph of this relationship is known as the ***v-i* characteristic** of the resistor. The ideal resistor studied in this chapter has the *v-i* characteristic given in Figure 1.39. The slope of the line in the *v-i* plane is the value of the resistance.

Recall that an **ideal voltage source** maintains a given voltage, irrespective of the current demands of the attached circuit. For constant-voltage sources, as shown in Figure 1.40(a), this property is depicted graphically by a constant horizontal line (slope equals 0) in the *v-i* plane

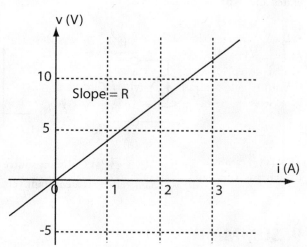

FIGURE 1.39 Linear resistor characteristic in which voltage is the constant R times the current through the resistor.

(Figure 1.40(b)). This means that the "internal" resistance of an ideal voltage source is zero. Further, if $V_s = 0$, the voltage source looks like a short circuit because the current flow, generated by the remaining circuit, will induce no voltage across the source. For now, we must be content with this brief discussion. Chapter 2 will reiterate and expand on these ideas.

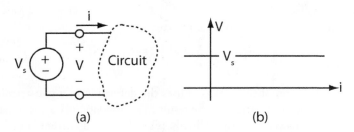

FIGURE 1.40 (a) Constant source V_s attached to circuit; (b) v-i characteristic is a constant horizontal line in the v-i plane.

Analogously, an **ideal current source** maintains the given current, irrespective of the voltage requirements of the attached circuit. For constant-current sources, as in Figure 1.41(a), this property is depicted by a constant vertical line (infinite slope) in the v-i plane (Figure 1.41(b)). This means that an ideal current source has infinite "internal" resistance. Further, if $I_s = 0$, the current source looks like an open circuit because no current will flow, regardless of any voltage generated by the rest of the circuit. Again, we must be content with this brief discussion until Chapter 2 reiterates and expands on the ideas.

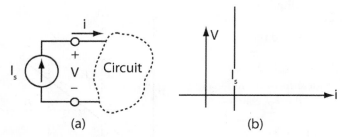

FIGURE 1.41 (a) Constant source I_s attached to circuit;
(b) v-i characteristic is a constant vertical line in the v-i plane.

9. SUMMARY

Building on a simplified physics of charge (coulombs), electric fields, and charge movement, this chapter set forth the notions of current, $i(t)$ or I for dc, and voltage, $v(t)$ or V for constant voltages. A rigorous treatment would require field theory and quantum electronics. More specifically, the notions of current, current direction, voltage, and voltage polarity, a two-terminal circuit element (the current entering equals the current leaving), the passive sign convention, power consumption [$p(t) = v(t)i(t)$ assuming the passive sign convention], and dissipated energy (the integral of power) were all defined. In general, we can say that every circuit element does one of the following:
- Absorbs energy
- Stores energy
- Delivers energy, or
- Converts energy from one form to another

The chapter subsequently introduced ideal independent and dependent voltage and current sources: the voltage-controlled voltage source (VCVS), the current-controlled voltage source (CCVS), the voltage-controlled current source (VCCS), and the current-controlled current source (CCCS). A dependent source produces a voltage or current proportional to a voltage across or a current through some other element of the circuit. The various types of dependent sources are summarized in Table 1.4.

TABLE 1.4 Summary of the Four Possible Dependent Sources.

VCVS (Voltage-Controlled Voltage Source, μ is dimensionless)	
CCVS (Current-Controlled Current Source, r_m is in ohms)	
VCCS (Voltage-Controlled Voltage Source, g_m is in S)	
CCCS (Current-Controlled Current Source, β is dimensionless)	

The chapter keynoted a special two-terminal element, called a resistor, whose terminal voltage and current satisfied Ohm's law, $v(t) = Ri(t)$, where $v(t)$ is the voltage in volts, R is the resistance in ohms, and $i(t)$ is the current in amperes. The resistor, as defined in this chapter, is a passive element, meaning that it always absorbs power, $p(t) = v(t)i(t) = v^2(t)/R = Ri^2(t) > 0$ since $R > 0$. This absorbed power is dissipated as heat. Hence, the (passive) resistor models the heating elements in a stove or toaster oven quite well. In addition, the resistor models the hot resistance of a light bulb. Throughout the text, the resistor will often represent a fixed electrical load. In a later chapter, we will discover that it is possible to construct a device with a negative resistance, $R < 0$, which can generate power. However, such a device is rather complex to build and requires such things as the operational amplifier covered in Chapter 4.

The various quantities defined and used throughout the chapter have various units. The quantities and their units are summarized as follows:

TABLE 1.5 Summary of Units

Charge	Current	Voltage	Resistance	Conductance	Power	Energy
Coulomb (C)	Ampere (A)	Volt (V)	Ohm (Ω)	S (Siemens) mho ℧	watt = volt x amp	Joule (J)

Throughout this chapter, a number of examples illustrated the various concepts that were introduced. Some simple resistive circuits were analyzed. To analyze more complex circuits, one needs Kirchhoff's voltage and current laws, which specify how circuit elements interact in a complex circuit. These basic laws of circuit theory are set forth in the next chapter.

10. TERMS AND CONCEPTS

Alternating current: a sinusoidally time-varying current signal having the form $K\sin(\omega t+\phi)$.

Battery: a device that converts chemical energy into electrical energy, and maintains approximately a constant voltage between its terminals.

Charge: an electric property of matter, measured in Coulombs. Like charges repel, and unlike charges attract each other. Each electron carries the smallest known indivisible amount of charge equal to -1.6×10^{-19} Coulomb.

Conductance: reciprocal of resistance, with siemens (S) (or formerly, mhos) as its unit.

Conductor: a material, usually a metal, in which electrons can move to neighboring atoms with relative ease.

Conservation of power (energy): the sum of powers generated by a group of circuit elements is equal to the sum of powers absorbed by the remaining circuit elements.

Current: the movement of charges constitutes an electric current. Current is measured in Amperes. One Ampere means movement of charges through a surface at the rate of 1 Coulomb per second.

Current source: a device that generates electrical current.

Dependent (controlled) current source: a current source whose output current depends on the voltage or current of some other element in the circuit.

Dependent (controlled) voltage source: a voltage source whose output voltage depends on the voltage or current of some other element in the circuit.

Direct current: a current constant with time.

Ideal conductor: offers zero resistance to electron movement.

Ideal insulator: offers infinite resistance to electron movement.

Independent (ideal) current source: an ideal device that delivers current as a prescribed function of time, e.g., {2 cos(t) + 12}A, no matter what circuit element is connected across its terminals.

Independent (ideal) voltage source: an ideal device whose terminal voltage is a prescribed function of time, e.g., {2 cos(t) + 12}V, no matter what current goes through the device.

Instantaneous power: the value of $p(t) = v(t)i(t)$ at a particular time instant.

Insulator: a material that opposes easy electron movement.

Mho: historical unit of conductance equal to the reciprocal of an ohm.

Ohm: unit of resistance. One ohm equals the ratio of 1V to 1A.

Ohm's law: for a linear conductor, the current through the conductor at any time t is proportional to the voltage across the conductor at the same time.

Open circuit: connection of infinite resistance or zero conductance.

Passive sign convention: voltage and current reference directions, indicated by +, −, and an arrow, which conform to that shown in Figure 1.15.

Peak-to-peak value: equals $2K$ in $K \sin(\omega t + \phi)$ of the ac waveform.

Peak value: refers to K in $K \sin(\omega t + \phi)$ of the ac waveform.

Power: rate of change of work per unit of time.

Resistance: for a resistor, $v(t) \propto i(t)$. The proportionality constant R is called the resistance, i.e., $v(t) = Ri(t)$. Resistance is measured in ohms: 1 ohm means the voltage is 1 V when the current is 1 A.

Resistivity: the resistance of a conductor is proportional to its length and inversely proportional to its cross-sectional area. The proportionality constant ρ is called the resistivity of the material. The resistivity of copper at 20°C is 1.7×10^{-8} ohm-meters.

Resistor: physical device that obeys Ohm's law. There are commercially available nonlinear resistors that do not obey Ohm's law. Resistors convert electric energy into heat.

Root mean square (rms) or **effective value:** measure of ac current, which is related to the peak value by the formula rms = $0.7071K$, where $K \sin(\omega t + \phi)$ is the ac waveform.

Short circuit: connection of zero resistance or infinite conductance.

Siemens: unit of conductance (formerly, mho) or inverse ohms.

***v-i* characteristic:** graphical or functional representation of a memoryless circuit element.

Voltage (potential difference): positive charge, without obstruction, will move from a higher potential point to a lower potential point, accompanied by a conversion of energy. Voltage is measured in volts; 1 volt between two points A and B means that the energy converted when moving 1 Coulomb of charge between A and B is 1 joule.

Voltage source: device that generates an electric voltage or potential difference.

Wattage: measure of power consumption.

PROBLEMS

CHARGE AND CURRENT PROBLEMS

1. Consider the diagram of Figure P1.1a.
   (a) Determine the charge on 7.573×10^{17} electrons.
   (b) If this number of electrons moves uniformly from the left end of a wire to the right in 1 ms (milli second), what current flows through the wire?
   (c) How many electrons must pass a given point in 1 minute to produce a current of 10 Amperes?
   (d) If the charge profile across the cross-section of a conductor from left to right is given by $q(t) = t + 0.2e^{-5t} - 0.2$ C for $t \geq 0$ and zero for $t < 0$, plot the profile of the current that flows across the boundary. In what direction would the current flow?

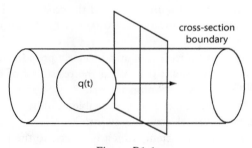

Figure P1.1a

   (e) Repeat part (d) for the charge waveform (in coulombs) sketched in Figure P.1.1b.

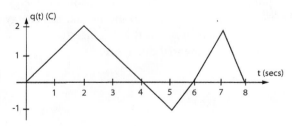

Figure P1.1b

ANSWER: (a) -0.1213 C; (b) 121.3 A; (c) 3.75 $\times 10^{21}$; (d) 1 - exp(-5t) for $t > 0$ and 0 for $t < 0$, from left to right; (e) line segments joining (0,0), (0,1), (2,1), (2,-1), (5,-1), (5,1), (6,1), (6,2), (7,2), (7,-2), (8,-2).

2. For the following questions, draw diagrams whenever necessary.
   (a) Determine the charge on 6.023×10^{15} electrons.
   (b) If this number of electrons moves uniformly from the left end of a wire to the right in 1 ms (milli second), what current flows through the wire?
   (c) How many electrons must pass a given point in 1 minute to produce a current of 5 Amperes?
   (d) The charge profile residing in a volume V = 10 cm³ is given by $q(t) = t + 0.5 \sin(\pi t)$ C for $t \geq 0$ and zero for $t < 0$. Plot the current that flows across the boundary of the volume for $0 < t < 2$ sec. In what direction would the current flow at $t = 1$ second? Explain.

3. Reconsider Figure 1.5 in the text in which I_a is changed to $i_a(t)$ and I_b is changed to $i_b(t)$. Suppose
   (i) Positive charge carriers move from left to right at the rate of 2cos(10t) C/s
   (ii) Negative charge carriers move from right to left at the rate of 6 cos(10t) C/s
   (a) Find $i_a(t)$ and $i_b(t)$ as functions of time.
   (b) Describe the charge movement on the wire at the boundaries A and B.

4. (a) Suppose the charge transported across the cross section of a conductor for $t \geq 0$ is $q(t) = e^{-1} \sin(120\pi t)$ C. Find the current, $i(t)$, $t \geq 0$, flowing in the conductor.
   (b) The charge crossing a boundary in a wire is given in Figure P1.4 for $t \geq 0$. Plot the current $i(t)$ through the wire. See Example 1.2.

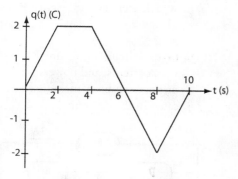

Figure P1.4 Charge crossing a hypothetical boundary.

5. (a) The current in an ideal conductor is given by $i(t) = 5 - 3e^{-2t} - 2e^{-4t}$ A for $t \geq 0$. Determine the charge transferred, $q(t)$, as a function of time for $t \geq 0$.

(b) Repeat part (a) for the current plot sketched in Figure P1.5.

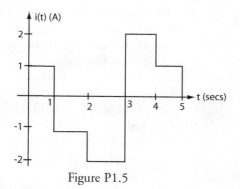

Figure P1.5

6. A plot of the current flowing past point A is shown on the graph of Figure P1.6. Find the net positive charge transferred in the direction of the current arrow during the interval $0 \leq t \leq 6$ sec, in Coulombs.

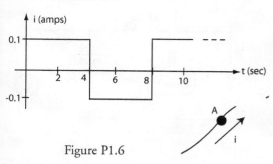

Figure P1.6

ANSWER: Q = integral of current. Hence, Q = 0.4 - 0.2 = 0.2.

7. (a) The current in an ideal conductor is given by $i(t) = 2 - e^{-2t} - \cos(2t)$ A for $t \geq 0$ and 0 for $t < 0$. Determine the charge transferred, $q(t)$, as a function of time for $t \geq 0$.

(b) Now suppose the charge transferred across some surface for $t \geq 0$ from left to right is $q(t) = 2 - e^{-2t} - \cos(2t)$ C. Find the current $i(t)$ through the surface for $t \geq 0$ from left to right.

(c) Repeat part (a) for the current plot sketched in Figure P1.7. Again, the current is zero for $t < 0$.

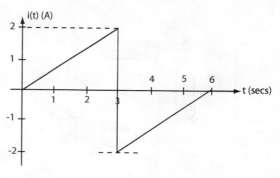

Figure P1.7

CHECK: (c) For, $6 \geq t > 3$, $q(t) = \dfrac{t^2}{3} - 4t + 12$

8. Find $i(t)$ when the charge transported across a surface cutting a conductor is shown in Figure P1.8.

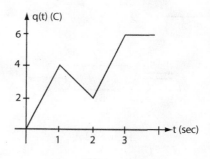

Figure P1.8

9. (a)   Find the average value of the voltage, $v(t) = K \cos(\omega t)$ over one period,

$$T = \frac{2\pi}{\omega}.$$

Hint: See Equation 1.6.

(b)   Find the average value of the absolute value of the voltage, $v(t) = K \cos(\omega t)$ over one period, $T = 2\pi/\omega$.

VOLTAGE, CURRENT, POWER, ENERGY

10. In Figure P1.10, suppose we know that $V_{AB} = 8$ V and $V_{AD} = 18$ V. Find the values of V_x, V_y, V_z, V_{BC}, and V_{CD}.

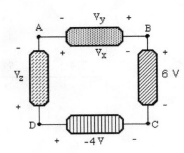

Figure P1.10

11. (a)   Which of the three elements in Figure P1.11 is labeled with the passive sign convention?

(b)   Find the absorbed powers for each circuit element in Figure P1.11.

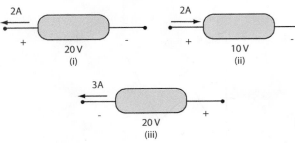

Figure P1.11

ANSWERS (b): (i) -40 W; (ii) 20 W; (iii) 60 W

12.(a)   Which elements in Figure P1.12 are labeled with the passive sign convention?

(b)   In the circuit of Figure P1.12, voltages, currents, and powers of some elements have been measured and indicated in the diagram.

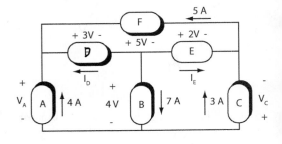

Figure P1.12

(i)   If element A generates 28 W power, find V_A.

(ii)   Find the power absorbed by element B.

(iii) If element C generates 6 W power, find V_C.

(iv)   If element D absorbs 27 W power, find I_D.

(v)   If element E absorbs 4 W power, find I_E.

(vi)   Find the power absorbed by element F.

13. Consider the circuit of Figure P1.13.

(a)   Find the power absorbed by the circuit elements 1 and 2.

(b)   Show that the algebraic sum of the absorbed powers is zero. Be careful of sign.

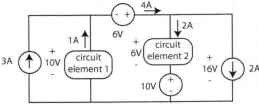

Figure P1.13

14. (a) Determine the power absorbed by each of the circuit elements in Figures P1.14a below.

(b) Show that the algebraic sum of the absorbed powers is zero. Be careful of sign.

(c) Repeat parts (a) and (b) for the circuit of Figure P1.14b, where $v_1(t) = 3 - 3e^{-2t}$ for $t \geq 0$, $v_2(t) = 1 + 3e^{-2t}$ for $t \geq 0$, $i_1(t) = 2 + e^{-2t}$ for $t \geq 0$, and $i_2(t) = e^{-2t}$ for $t \geq 0$.

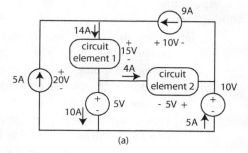

(a)

Figure P1.14a

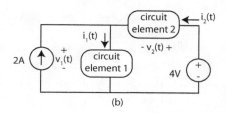

(b)

Figure P1.14b

15. In the circuit of Figure P1.15, there are three independent sources and five ordinary resistors.

(a) Determine which of the circuit elements are sources and which are resistors.

(b) Determine the value of the resistance for each of the resistors.

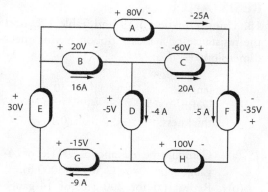

Figure P1.15

CHECK: $R_E = \dfrac{30}{9} = 3.334$, $R_H = 20$ in ohms.

16. In the circuit shown in Figure P1.16, $i(t) = 100(1 - e^t)$ mA for $t \geq 0$.

(a) How much energy does the element A absorb for the interval $[0, t]$?

(b) If element B is a 5 Ω resistor, determine the power absorbed at time t, and the energy absorbed for the interval $[0, t]$.

(c) What is the energy delivered by the source over the interval $[0, t]$?

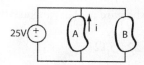

Figure P1.16

CHECK: (b) 125 W, 125t J

17. Suppose energy cost in Indiana is 10 cents per kwh.

(a) How much does it cost to run a 100-watt TV set 8 hours per day for 30 days?

(b) How many 100-watt light bulbs run for 6 hours a day are needed to use $9.00 of energy every 30 days?

ANSWER: (a) 8 cents per day; $2.40 per month; (b) 5 bulbs

RESISTANCE

18. Using Equation 1.19 and Table 1.2, find the resistance of a nickel ribbon having these dimensions:

length:	40 m
width:	1.5 cm
thickness:	0.1 cm

19. (a) Compute the resistance of 800 feet of 14-gauge copper wire ($2.575\ \Omega$/1000 ft).
    (b) Repeat (a) for 200 feet of 14-gauge nickel wire.
    (c) If one end of the copper wire is soldered to one end of the nickel wire, find the total resistance of the 1000 feet of wire. Can you justify your answer?

20. The resistance of a conductor is function of the temperature T (in °C). Over a range of temperature that is not too distant from 20°C, the relationship between $R(T)$ and T is linear and can be expressed as $R(T) = R(20)[1 + \alpha(T-20)]$ where α is called the temperature coefficient of the conducting material. For copper $\alpha = 0.0039$ per °C. If the resistance of a coil of wire is 21 Ohms at -10° C, what is the resistance when the wire is operating at 10° C?

ANSWER: 22.85 Ω

AVERAGE VALUE, POWER, AND ENERGY CALCULATIONS

21. The current through a 500 Ω resistor is given in Figure P1.21 where $I_m = 6$ mA.
    (a) How much total charge is transferred over the time interval $t = 0$ to $t = 2$ seconds?
    (b) How much total energy must a source deliver over the time interval $t = 0$ to $t = 2$ seconds?
    (c) If $i(t)$ in Figure P1.21 is periodic, with period equal to 2 seconds, i.e., the indicated waveform is replicated every two seconds, find the average power absorbed by the resistor. Use intuitive reasoning.

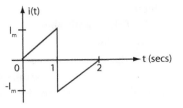

Figure P1.21

22. For the circuit of Figure P1.22, $R = 1\Omega$, and the input current to the circuit of is $i_s(t) = 400\sin(20\pi t)$ mA for $t \geq 0$ and zero for $t < 0$.
    (a) Compute the instantaneous power delivered by the source. Using a graphing program, graph the power delivered as a function of time for $0 \leq t \leq 0.5$ s.
    (b) Now compute and graph an expression for the energy dissipated in the resistor as a function of t for $0 \leq t \leq 0.5$ s.

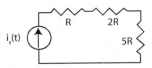

Figure P1.22

23. The switch S in Figure P1.23 is assumed to be ideal, i.e., it behaves as a short circuit when closed, and as an open circuit when open. Suppose the switch is repeatedly closed for 1 ms and opened for 1ms.
    (a) What is the average value of $i(t)$?
    (b) What is the average power delivered by the source?

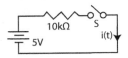

Figure P1.23

ANSWERS: 0.25 mA. 1.25mW

24. Repeat problem 23 when the switch is repeatedly closed for 3 ms and opened for 1 ms.

CHECK: P_{ave} = 1.875 mW

25. In Figure P1.25, V_0 = 10 V, and the switch S alternately stays at position A for 4 ms and at position B for 1 ms. Find the average value of $i(t)$.

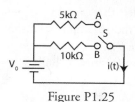

Figure P1.25

CHECK: Average current is 1.8 mA.

APPLICATIONS OF OHM'S LAW

26 (a) What is the safe maximum current of a 0.25 W, 277 kΩ metal film resistor used in a radio receiver?
(b) What is the safe maximum current of a 1 W, 130 Ω resistor?
(c) What is the safe maximum current of a 2 kW, 2 Ω resistor used in an electric power station?

ANSWERS: (a) 0.95 mA, (b) 87.7 mA, (c) 31.6 A

27. In Figure P1.27, V_0 = 120 V.
(a) The power absorbed by the bulb in the circuit shown in Figure P1.27a is 60 watts. Find the value of the hot resistance of the bulb.
(b) The power delivered by the source to the parallel connection of two identical bulbs in Figure 1.27b is 150 watts. Find the hot resistance of each bulb.

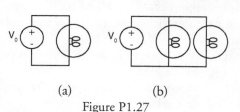

(a)   (b)

Figure P1.27

ANSWER: (b) 192 Ω

28. In Figure P1.28, V_0 = 125 V.
(a) Suppose bulb A and bulb B each use 100 watts of power. Find I_0 and the hot resistance of each bulb.
(b) Suppose bulb A uses 40 watts of power and bulb B 60 watts of power. Find I_0 and the hot resistance of each bulb.

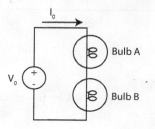

Figure P1.28

ANSWER: (b) I_0 = 0.8 A, R_A = 62.5 Ω, R_B = 93.75 Ω

29. The power delivered by the source in the circuit of Figure P1.29 is 750 watts.
(a) If I_s = 5 A, determine the value of R.
(b) Suppose now that R = 11 Ω. Find I_s. Hint: What is the power consumed in each resistor as a function of I_s?

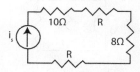

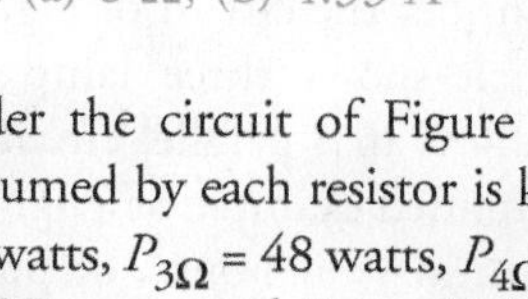

Figure P1.29

ANSWER: (a) 6 Ω, (b) 4.33 A

30. Consider the circuit of Figure P1.30. The power consumed by each resistor is known to be $P_{2\Omega}$ = 392 watts, $P_{3\Omega}$ = 48 watts, $P_{4\Omega}$ = 64 watts, $P_{5\Omega}$ = 3075.2 watts, and $P_{6\Omega}$ = 1944 watts.
(a) For each resistor, determine the indicated voltage or current.
(b) Determine the total power delivered by the two sources.

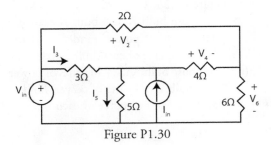

Figure P1.30

CHECK: (a) V_2 = 28 V, V_6 = 108 V, (b) 5523.2 watts

31. The power absorbed by the resistor R in the circuit of Figure P1.31 is 100 watts and V_0 = 20 V.

   (a) Find the value of R.

   (b) Find the value of the current flowing through R and determine its direction as per the passive sign convention.

   (c) Find the power absorbed by the 20 Ω resistor.

   (d) Find the power delivered by the source and the value of I_0.

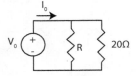

Figure P1.31.

CHECK: I_0 = 6 A.

32. (a) Consider the circuit of Figure P1.32a, which shows three lamps, AA, BB, and CC in a parallel circuit. This is a simplified example of a light circuit on a car, in your house, or possibly on a Christmas tree. Halogen bulb AA uses 35 watts when lit, the Halogen Xenon bulb BB uses 36 watts when lit, and the incandescent bulb CC bulb uses 25 watts when lit.

   (i) Determine the current through each bulb.

   (ii) Determine the total power delivered by the battery.

   (iii) Determine the current, I_{in}, delivered by the battery.

   (iv) Assuming that each bulb behaves as a resistor, determine the hot resistance of each bulb.

  (b) Determine the number of AA bulbs in parallel that would be required to blow the 15-amp fuse in the circuit of Figure P1.32b.

  (c) Repeat (b) for CC bulbs.

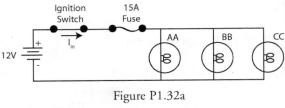

Figure P1.32a

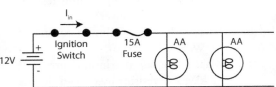

Figure P1.32b

ANSWERS: (a) (i) 2.9167, 3, 2.08; (ii) 6.2 A; (iii) 96; (iv) 4.11, 4, 5.76; (b) n ≥ 6; (c) n ≥ 8

33. An automobile battery has a terminal voltage of approximately 12 V when the engine is not running and the starter motor is not engaged. A car with such a battery is parked at a picnic. For music, the car stereo is playing, using 240 watts, and some of the lights are on using 120 watts. With this load, the battery will supply approximately 3 MJ of energy before it will have insufficient stored energy to start the car.

  (a) What power does the battery supply to the load?

  (b) What current does the battery supply to the load?

(c) Approximately how long can the car remain parked with the stereo and lights on and still start the car?

CHECK: (c) 2.31 hours

34. In Figure P1.34, $V_0 = 24$ V, $R_1 = 4 \Omega$, the unknown resistance, R_2, consumes 20 watts of power. Find I_x and R_2. (How many possible solutions are there?)

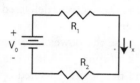

Figure P1.34

CHECK: 1 A, 20 Ω or ?????

35. In Figure P1.35, $V_0 = 48$ V.
    (a) Determine the value(s) of the current I_x in the DC resistive circuit of Figure P1.35a given that the unknown devices absorb the powers indicated.
    (b) What value of V_0 results in a unique solution for I_x?
    (c) If the circuit is modified as shown in Figure P1.35b, determine the two new values of the current I_x.

ANSWER: (a) 10, 2 A; (b) 35.77 V

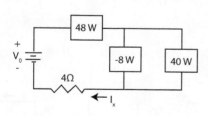

Figure P1.35a

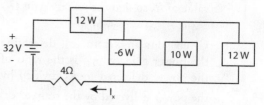

Figure P1.35b

DEPENDENT SOURCE PROBLEMS

36. Consider the circuit in Figure P1.36.
    (a) If $V_s = 6$ V, find V_L and the power in watts absorbed by the load R_L. What is the power delivered by each source?
    (b) If the power absorbed by the load resistor is $P_L = 80$ watts, then find V_L, I_1 and V_s.

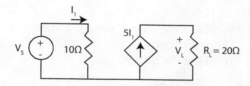

Figure P1.36

CHECK: (a) $V_L = 60$ V, $P_{V\text{-}source} = 3.6$ watts; (b) $V_s = 4$ V

37. For the circuit in Figure P1.37, determine I_{out} and P_{R2} in terms of I_{in}, R_1, R_2 and μ.

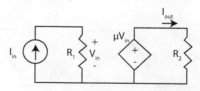

Figure P1.37

38. Consider the circuit of Figure P1.38.
    (a) Determine an expression for V_{out} and the voltage gain

$$G_V = \frac{V_{out}}{V_{in}}$$

in terms of R_1, R_2, α, and V_{in}.
    (b) If $R_1 = 4 \Omega$ and α = 0.8, determine the

value of R_2 so that the voltage gain G_V = 4.

(c) Given your answer to (b), determine the power gain, which is the ratio of the power delivered to R_2 divided by the power delivered by the source.

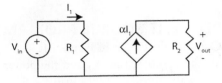

Figure P1.38

39. For the circuit of Figure P1.39, suppose I_{in} = 100 mA, R_1= 50 Ω, R_2= 10 Ω, and R_3= 100 Ω.

(a) Find the output voltage and output current.
(b) Find the current gain,

$$G_I = \left| \frac{I_{out}}{I_{in}} \right|, \text{ the voltage gain}$$

$$G_V = \left| \frac{V_{out}}{V_{in}} \right|, \text{ and the power gain}$$

$$G_P = \frac{P_{out}}{P_{in}}.$$

(c) Find the power absorbed by each resistor.

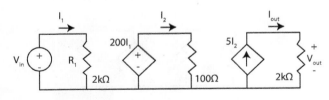

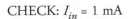

Figure P1.39

40. In problem 39, suppose R_1= 1 k Ω, R_2= 10 Ω, and R_3= 20 Ω and P_{R3}= 80 watts. Find V_{out}, I_{out}, V_2, V_{in} and I_{in}.

CHECK: I_{in} = 1 mA

41. For the circuit of Figure P1.41, suppose V_{in} = 10 V.

(a) Find the output voltage and output current.
(b) Find the voltage gain

$$G_V = \left| \frac{V_{out}}{V_{in}} \right|, \text{ and the power gain}$$

$$G_P = \frac{P_{out}}{P_{in}}.$$

(c) Find the power delivered by each source.
(d) Suppose the power absorbed by the 2 k Ω output resistor were 80 watts. Find the power delivered by the input source, P_{in}, and the voltage V_{in}.

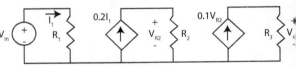

Figure P1.41

CHECK: (b) G_V= 10

42. For the circuit of Figure P1.42, suppose R_3= $2R_2$, and R_2= $10R_1$. Find the resistor values so that $G_V = \dfrac{V_{out}}{V_{in}} = 1000$.

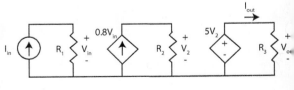

Figure P1.42

CHECK: R_3= 5 k Ω

C H A P T E R 2

Kirchhoff's Current & Voltage Laws and Series-Parallel Resistive Circuits

A CAR HEATER FAN SPEED-CONTROL APPLICATION

One use of resistors in electronic circuits is to control current flow, just as dams control water flow along rivers. Ohm's law, $V = RI$, gauges the ability of resistors to control this current flow: for a fixed voltage, high values of resistance lead to small currents, whereas low values of resistance lead to higher currents. This property underlies the adjustment of the blower (fan) speed for ventilation in a typical car, as represented in the following diagram.

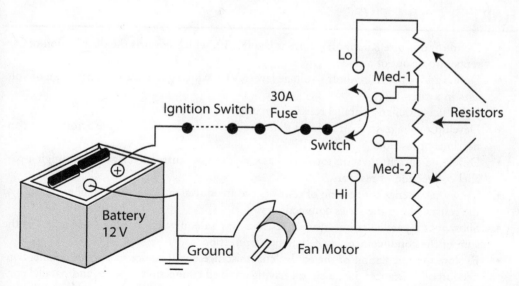

In this diagram, three resistors are connected in series, and their connecting points are attached to a switch. As we will learn in this chapter, the resistance of a series connection is the sum of the resistances. So with the switch in the low position, the 12-V car battery sees three resistors in series with the motor. The series connection of three resistors represents a "large" resistance and heavily restricts the current through the motor. With less current, there is less power, and the fan motor speed is slow. When the switch moves to the Med-1 position, a resistor is bypassed, producing less resistance in the series circuit and allowing more current to flow. More current flow increases the fan motor speed. Each successive switch position removes resistance from the circuit, and the fan motor speed increases accordingly.

Analysis of such practical circuits builds on the principles set forth in this chapter.

CHAPTER OUTLINE

1. Introduction and Terminology: Parallel, Series, Node, Branch, and so on
2. Kirchhoff's Current Law
3. Kirchhoff's Voltage Law
4. Series Resistances and Voltage Division
5. Parallel Resistances and Current Division
6. Series-Parallel Interconnections
7. Dependent Sources Revisited
8. Model for a Non-ideal Battery
9. Non-ideal Sources
   Summary
   Terms and Concepts
   Problems

CHAPTER OBJECTIVES

- Define and utilize Kirchhoff's current law (KCL), which governs the distribution of currents into or out of a node.
- Define and utilize Kirchhoff's voltage law (KVL), which governs the distribution of voltages in a circuit.
- Introduce series and parallel resistive circuits.
- Develop a voltage division formula that specifies how voltages distribute across series connections of resistors.
- Develop a current division formula that specifies how currents distribute through a parallel connection of resistors.
- Show that a series connection of resistors has an equivalent resistance equal to the sum of the resistances in the series connection.
- Show that a parallel connection of resistors has an equivalent conductance equal to the sum of the conductances in the parallel connection.
- Explore the calculation of the equivalent resistance/conductance of a series-parallel connection of resistances, i.e., a circuit having a mixed connection of series and parallel connections of resistors.
- Explore the calculation of voltages, currents, and power in a series-parallel connection of resistances.
- Revisit the notion of a dependent source and use a VCCS to model an amplifier circuit.
- Describe a practical battery source and look at a general practical source model.

1. INTRODUCTION AND TERMINOLOGY: PARALLEL, SERIES, NODE, BRANCH, AND SO ON

The circuits studied in Chapter 1 were interconnections of resistors and sources that were two-terminal circuit elements. This chapter sets forth **Kirchhoff's voltage law (KVL)** and **Kirchhoff's current law (KCL)**. These laws govern the voltage relationships and the current relationships, respectively, of interconnections of two-terminal circuit elements.

FIGURE 2.1

Some new terminology underpins the statements of KVL and KCL. Figure 2.2a shows a series **circuit** consisting of a sequential connection of two-terminal circuit elements (resistors) end to end. The common connection point between any elements is called a **node**. In general, a **node** is the connection point of one or more circuit elements. Figure 2.2b shows a **parallel circuit**, in which the top terminals and the bottom terminals of each resistor are wired together. The common connection point of the top terminals is a node, as is the common connection point of the bottom terminals.

An important property of the series connection of Figure 2.2a is that all the two-terminal elements carry the same current, in this case i_R, because the input current for each two-terminal element must equal the exit current. Similarly, in a parallel connection, such as Figure 2.2b, the same voltage, in this case, v_R, appears across every circuit element.

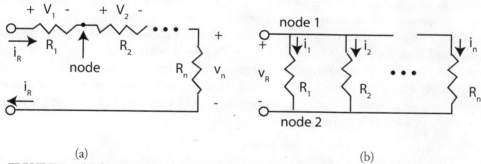

(a)   (b)

FIGURE 2.2 (a) Series connection of resistors with the property that each resistor carries the same current; (b) parallel connection of resistors with the property that the same voltage appears across each resistor.

Sources interconnected with circuit elements produce currents through the elements and voltages across the elements. For example, a voltage source connected across Figure 2.2a would generate a current i_R and the voltages v_1 through v_n. **Kirchhoff's voltage law (KVL)** governs the distribution of voltages around loops of circuit elements, as shown in Figure 2.2a. Similarly, a current source connected across the circuit of Figure 2.2b would produce the voltage v_R and the currents i_1 through i_n. **Kirchhoff's current law (KCL)** governs the flow of currents into and out of a common connection point or **node**, as in the top and bottom connections of Figure 2.2b. This chapter sets forth precise statements of these laws and illustrates their application.

A proper statement of KVL and KCL requires the additional notion of branch. A **branch** of a circuit is a generic name for a two-terminal circuit element and is denoted by a line segment, as in Figure 2.3. The endpoints of a branch (the terminals of the circuit element) are called **nodes**, as in Figure 2.3a. Ordinarily, however, **node** means a common connection point of two or more circuit elements (branches), as shown in Figure 2.3b.

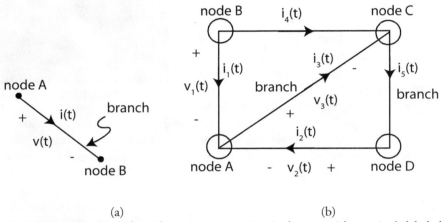

(a)   (b)

FIGURE 2.3 (a) Single branch representing a circuit element with terminals labeled as nodes A and B; (b) interconnection of branches (circuit elements) with common connection points labeled as nodes A through D.

The voltage polarity and current direction for the branches in Figures 2.2 and 2.3 are labeled in accordance with the **passive sign convention**: the arrowhead on a branch denotes the reference current direction, which is from plus to minus. Recall that the + to – does not mean that the voltage is always positive if measured from the plus-sign to the minus-sign. In general, reference directions can be assigned arbitrarily. The conventional assignment of voltage polarity and current direction to voltage and current sources is given in Figure 2.4, which is different from the passive sign convention. Note that with these conventional assignments, the (instantaneous) power delivered by a source is $p_{del}(t) = v_{in}(t)i_{in}(t)$, and the power absorbed by a source is $p_{abs}(t) = -p_{del}(t)$.

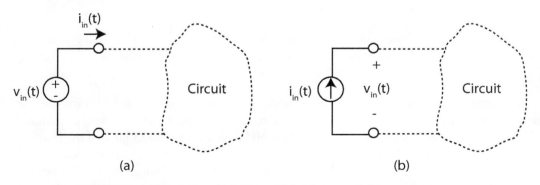

(a)   (b)

FIGURE 2.4 Conventional labeling of (a) voltage, and (b) current sources.

2. KIRCHHOFF'S CURRENT LAW

Imagine a number of branches connected at a common point, as at node A of Figure 2.3b. The current through each branch has a reference direction indicated by an arrow. If the arrow points toward the node, the reference direction of the current is entering the node; if the arrow points away from the node, the reference direction of the current is leaving. If a current is referenced as leaving a node, then the negative of the current enters the node, and conversely.

KIRCHHOFF'S CURRENT LAW (KCL)

STATEMENT 1: The algebraic sum of the currents entering a node is zero for every instant of time.
STATEMENT 2: Equivalently, the algebraic sum of the currents leaving a node is zero for every instant of time.

The two statements of KCL are equivalent because the negative of the sum of the currents entering a node corresponds to the sum of the currents leaving the node. Further, from physics we know that charge is neither created nor destroyed. Thus, the charge transported into the node must equal the charge leaving the node because charge cannot accumulate at a node. KCL expresses the conservation of charge law in terms of branch currents. Moreover, KCL specifies how branch currents interact at a node, regardless of the type of element connected to the node.

Referring to Figure 2.3b, KCL at node A requires that $i_1(t) + i_2(t) - i_3(t) = 0$ for all t. KCL at node B requires that $i_4(t) = -i_1(t)$. Finally, KCL at node D requires that $i_5(t) = i_2(t)$.

EXAMPLE 2.1
For the node shown in Figure 2.5, find $i_R(t)$.

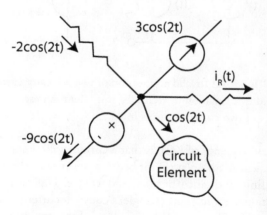

FIGURE 2.5 Connection of five circuit elements at a single node.

SOLUTION

By KCL, the sum of the currents entering the node must be zero. Hence, the current $i_R(t) =$ $9\cos(2t) - 3\cos(2t) - \cos(2t) - 2\cos(2t) = 3\cos(2t)$ A.

Exercise. 1. Suppose the current through the voltage source in Figure 2.5 is changed to $-2\cos(2t)$. Find $i_R(t)$.
ANSWER: $-4\cos(2t)$ A.

2. Three branches connect at a node. All branch currents have reference directions leaving the node. If $I_1 = I_2 = 2$ A, then find I_3.
ANSWER: -4 A

Two implications of KCL are of immediate interest. First, as a general rule, KCL forbids the series connection of current sources. Figure 2.6a shows an invalid connection of two arbitrary current sources $i_1(t)$ and $i_2(t)$, where $i_1(t) \neq i_2(t)$. It is invalid because KCL requires that $i_1(t) = i_2(t)$. On the other hand, a parallel connection of two current sources can be combined to form an equivalent source, as in Figure 2.6b, where $i_{eq}(t) = i_1(t) + i_2(t)$.

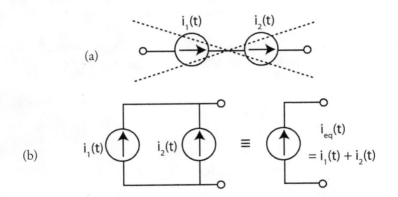

FIGURE 2.6 (a) Invalid connection of two arbitrary current sources when $i_1(t) \neq i_2(t)$. Avoid this violation of KCL; (b) equivalent representation of a parallel connection of two current sources in which $i_{eq}(t) = i_1(t) + i_2(t)$.

A second immediate consequence of KCL is that a current source supplying zero current [$i(t) = 0$ in Figure 2.7] is equivalent to an open circuit because the current through an open circuit is zero. An open circuit has infinite resistance, or zero conductance. This means that a **current source** has **infinite internal resistance**. From another angle, a constant current source is represented by a vertical line in the *iv* plane (see Figure 1.41b). The slope of the vertical line, which is infinite, determines the internal resistance of the source.

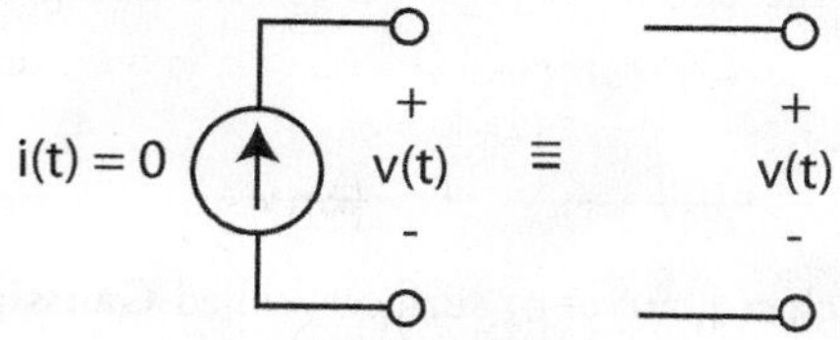

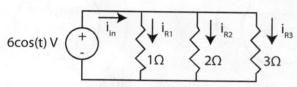

FIGURE 2.7 Ideal current source with $i(t) = 0$ is an open circuit.

A typical application of KCL is given in the following example.

EXAMPLE 2.2

In the parallel resistive circuit of Figure 2.8, the voltage across each resistor is $6\cos(t)$ V. Find the current through each resistor and the current, $i_{in}(t)$, supplied by the voltage source.

FIGURE 2.8 Parallel resistive circuit for Example 2.2.

SOLUTION

By Ohm's law,

$$i_{R1}(t) = 6\cos(t) \text{ A}$$

$$i_{R2}(t) = \frac{6\cos(t)}{2} = 3\cos(t) \text{ A}$$

$$i_{R3}(t) = \frac{6\cos(t)}{3} = 2\cos(t) \text{ A}$$

By KCL,

$$i_{in}(t) = i_{R1}(t) + i_{R2}(t) + i_{R3}(t) = 6\cos(t) + 3\cos(t) + 2\cos(t) = 11\cos(t) \text{ A}$$

Exercise. 1. In Figure 2.8, suppose the source voltage is changed to a constant, labeled V_{in}. Find I_{in} in terms of V_{in}.

ANSWER: $\dfrac{11}{6} V_{in}$

2. Suppose the source voltage in the circuit of Figure 2.8 were changed to $-12 \cos(2t)$ V. Find $i_{in}(t)$.

ANSWER: $-22 \cos(2t)$ A

Kirchhoff's current law holds for closed curves or surfaces, called **Gaussian curves or surfaces**. A Gaussian curve or surface is a closed curve (such as a circle in a plane) or a closed surface (such as a sphere or ellipsoid in three dimensions). *A Gaussian curve or surface has a well-defined inside and outside.* Figure 2.9 illustrates the idea of a Gaussian curve for three (planar) situations.

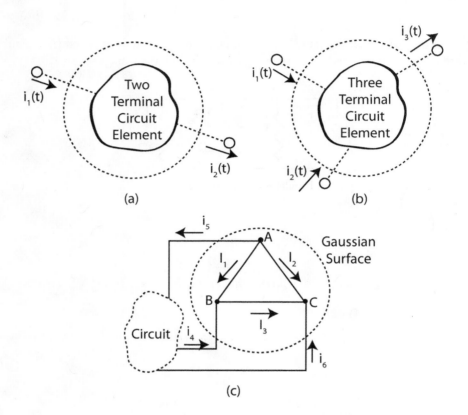

FIGURE 2.9 Illustrations of Gaussian curves:
(a) enclosure of a two-terminal element; (b) enclosure of a three-terminal device, such as a transistor; (c) enclosure of a three-node interconnection with an arbitrary circuit.

For the two-terminal circuit element of Figure 2.9a, KCL for Gaussian curves implies that $i_1(t) = i_2(t)$, which is precisely the definition of a **two-terminal circuit element**. For the three-terminal device of Figure 2.9b, KCL for Gaussian curves implies that $i_3(t) = i_1(t) + i_2(t)$. Finally, for Figure 2.9c, $i_4 - i_5 + i_6 = 0$. From these illustrations, one might imagine that the use of Gaussian surfaces might simplify or provide a short cut to certain branch current computations. The general statement of KCL for Gaussian surfaces is next followed by an example that demonstrates its use for computing branch currents.

KCL FOR GAUSSIAN CURVES OR SURFACES

The algebraic sum of the currents leaving (or entering) a Gaussian curve (or surface) is zero for every instant of time.

EXAMPLE 2.3

This example shows how the use of a Gaussian curve or surface can sometimes simplify a calculation. Figure 2.10 portrays a complicated circuit whose branch currents and voltages are not solvable by methods learned so far. Our objective is to find the current I_R without having to solve a set of complex circuit equations.

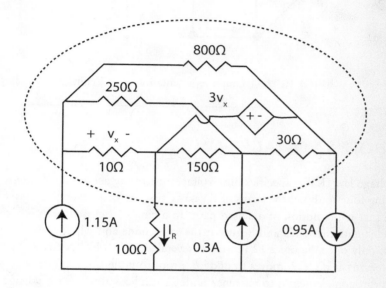

FIGURE 2.10 Circuit for Example 2.3, showing a Gaussian surface to compute I_R directly.

SOLUTION

Using KCL for the indicated Gaussian curve, $-1.15 + I_R - 0.3 + 0.95 = 0$. Equivalently, $I_R = 1.15 + 0.3 - 0.95 = 0.5$ A.

In the next chapter, circuits such as the one in Figure 2.10 are analyzed using a technique called nodal analysis.

Exercise. 1. Draw a Gaussian surface on the circuit in Figure 2.10 that is different from the surface given but still allows one to compute I_R.

ANSWER: One choice is a circle enclosing the bottom node.

2. Draw an appropriate Gaussian curve to find I in the graphical circuit representation in Figure 2.11.

ANSWER: 2 A

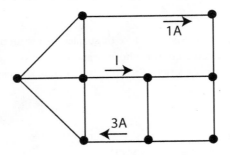

FIGURE 2.11 Graph representation of a circuit.

3. KIRCHHOFF'S VOLTAGE LAW

Kirchhoff's voltage law (KVL) specifies how voltages distribute across the elements of a circuit. Before conveying four equivalent versions of KVL, we first set forth several necessary background concepts. The first is the notion of a closed path. In a circuit, a **closed path** is a connection of two-terminal elements that ends and begins at the same node and which traverses each node in the connection only once. Figure 2.12 illustrates several closed paths. One closed path is *A-B-C-D-E-A*, i.e., it begins at node *A*, moves to node *B*, drops to node *C*, moves through element 4 to node *D*, down through element 6 to reference node *E*, and back through the voltage source to *A*. A second closed path is *A-B-C-E-A*, and a third is *B-D-C-B*.

A second concept pertinent to our KVL statements is that of a node voltage with respect to a reference. A **node voltage** of a circuit is the voltage drop from a given node to a reference node. The reference node is usually indicated on the circuit or is taken as **ground**. The circuit of Figure 2.12 has branches labeled 1 through 6 and nodes labeled *A* through *E*, with node *E* taken as the reference node. The associated node voltages are denoted by v_A, v_B, v_C and v_D. The voltage v_A denotes the voltage drop from node *A* to node *E*; v_D denotes the voltage drop from node *D* to node *E*, and similarly for the remaining node voltages. Node *E*, being the reference node, has zero as its node voltage.

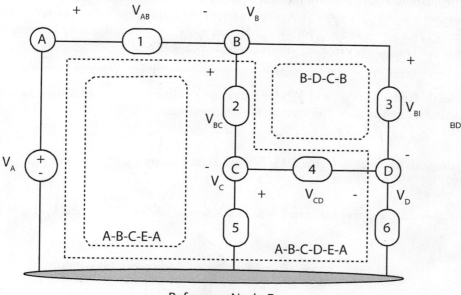

FIGURE 2.12 Circuit diagram illustrating (i) three closed paths (A-B-C-D-E-A); (ii) the concept of node voltages with respect to a given reference node E, (v_A and v_B); (iii) the concept of branch voltages (v_{AB} and v_{BC}).

The concept of a closed path and the concept of a node voltage allow us to state our first two versions of Kirchhoff's voltage law.

KIRCHHOFF'S VOLTAGE LAW (KVL)

Kirchhoff's voltage law can be stated in different ways. Following are two equivalent statements of the law.

STATEMENT 1: The algebraic sum of the voltage drops around any closed path is zero at every instant of time.

STATEMENT 2: For any pair of nodes j and k, the voltage drop v_{jk} from node j to node k is given by

$$v_{jk} = v_j - v_k$$

at every instant of time, where v_j is the voltage at node j with respect to the reference and v_k is the voltage at node k with respect to reference. Here j and k stand for arbitrary node indices. For example, in Figure 2.12, j, k can be any of the nodes A, B, C, D, or E.

Referring back to Figure 2.12, for the closed path A-B-C-D-E-A, statement 1 of KVL requires that $v_{AB} + v_{BC} + v_{CD} + v_{DE} + v_{EA} = 0$. Again, for Figure 2.12, from statement 2 of KVL, the branch voltages $v_{AB} = v_A - v_B$ and $v_{CD} = v_C - v_D$. Hence, $v_{EA} = -v_{AE} = -v_A$. Thus, by knowing the node voltages of a circuit, one can easily compute the branch voltages.

Exercise. 1. Find V_{AB}, V_{BC}, and V_{AC} for the circuit of Figure 2.13 in which we have introduced the ground symbol at node E to identify the reference node.
ANSWERS: $V_{AB} = -3$ V, $V_{BC} = 21$ V, $V_{AC} = 18$ V

2. Again, with reference to Figure 2.13, find the node voltages V_A, V_B, V_C and V_D.
ANSWERS: 2 V, 5 V, –16 V, –6 V

3. In Figure 2.13, suppose the branch labeled 6 V is now labeled –12 V. Find V_{CB}.
ANSWER: –3 V

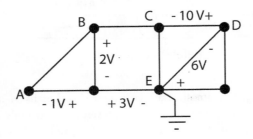

FIGURE 2.13

A third concept needed for two further equivalent statements of KVL is that of a closed node sequence. A **closed node sequence** is a finite sequence of nodes that begins and ends at the same node. A closed node sequence generalizes the notion of a closed path. Finally, we define the notion of a connected circuit. In a **connected circuit,** each node can be reached from any other node by some path through the circuit elements. Figures 2.12 and 2.14 show connected circuits. However, in Figure 2.14, the sequence of nodes A-B-C-D-E-A is a **closed node sequence** but not a **closed path** because there is no circuit element between nodes B and C.

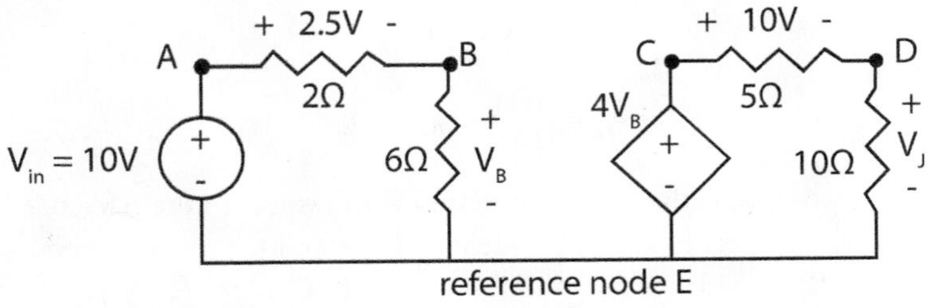

FIGURE 2.14 Simple dependent source circuit for illustrating the concepts of a connected circuit and a closed node sequence.

This brings us to our last two equivalent statements of KVL.

KIRCHHOFF'S VOLTAGE LAW (KVL)

Following are two additional equivalent statements of KVL.

STATEMENT 3: For connected circuits and any node sequence, say A-D-B- ... -G-P, the voltage drop

$$v_{AP} = v_{AD} + v_{DB} + \ldots + v_{GP}$$

at every instant of time.

STATEMENT 4: For connected circuits, the algebraic sum of the node-to-node voltages for any closed node sequence is zero for every instant of time.

Referring back to Figure 2.12, statement 3 of KVL implies that $V_A = V_{AB} + V_{BC} + V_{CD} + V_{DE} = V_{AB} + V_{BC} + V_C$. Referring to Figure 2.14, for the closed node sequence E-A-B-E, $V_{in} = 10 = V_{AB} + V_B = 2.5 + V_B$ and $V_B = 7.5$ V. Now, consider the closed node sequence E-C-D-E. For this sequence, $V_{CE} = 4V_B = V_{CD} + V_D$. Equivalently, $30 = 10 + V_D$, in which case $V_D = 20$ V. Finally, consider the closed node sequence, E-B-C-E, which is not a **closed path** because there is no circuit element between nodes B and C. Nevertheless, by statement 4 of KVL, $-V_B + V_{BC} + V_C = 0$ or equivalently that $V_{BC} = V_B - V_C = 7.5 - 30 = -22.5$ V.

Exercise. 1. In Figure 2.14, suppose $V_{in} = 20$ V, $V_{AB} = 5$ V, and $V_{CD} = 20$ V. Find V_B, V_C, V_{CB}, and V_D.
ANSWER: $V_B = 15$ V, $V_C = 60$ V, $V_{CB} = 45$ V, and $V_D = 40$ V

2. (a) In the circuit of Figure 2.15, suppose $v_C = 10$ V and $v_D = -3$ V. Find v_{DC}.
(b) Suppose $v_B = 120 \cos(120\pi t)$, $v_{BD} = 18 \cos(120\pi t)$ and $v_C = 32 \cos(120\pi t)$. Find v_{CD} at $t = 0.5$ s.
(c) Find v_{AD} when $v_A = 100$ V, $v_{DC} = -10$ V and $v_C = 25$V.
SCRAMBLED ANSWER: 85 V, -13 V, -70 V

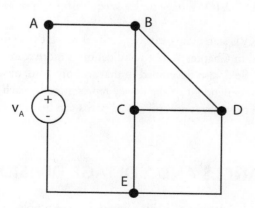

FIGURE 2.15 Circuit with nodes labeled A through E. Node E is taken as the reference node.

Two further implications of the KVL are of immediate interest. First, as a general rule, KVL forbids the parallel connection of two voltage sources—say, $v_1(t)$ and $v_2(t)$—for which $v_1(t) \neq v_2(t)$, as illustrated in Figure 2.16a. On the other hand, two voltage sources in series can be combined to form a single source, as illustrated in Figure 2.16b, where $v_{eq}(t) = v_1(t) + v_2(t)$.

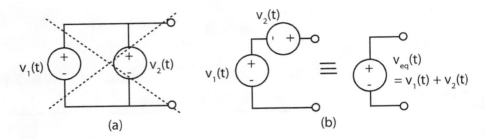

FIGURE 2.16 (a) An improper connection of voltage sources when $v_1(t) \neq v_2(t)$; (b) an equivalent representation of two voltage sources connected in series in which $v_{eq}(t) = v_1(t) + v_2(t)$.

Second, a voltage source supplying 0 V is equivalent to a **short circuit**, as illustrated in Figure 2.17. Also, the **internal resistance** of a **voltage source** is zero. One can see this by referring to the fact that in the iv plane, an ideal dc voltage source is represented by a horizontal line, as was illustrated in Figure 1.40. The slope of the line is zero and represents the resistance of the source. These ideas are dual to those expressed for current sources earlier.

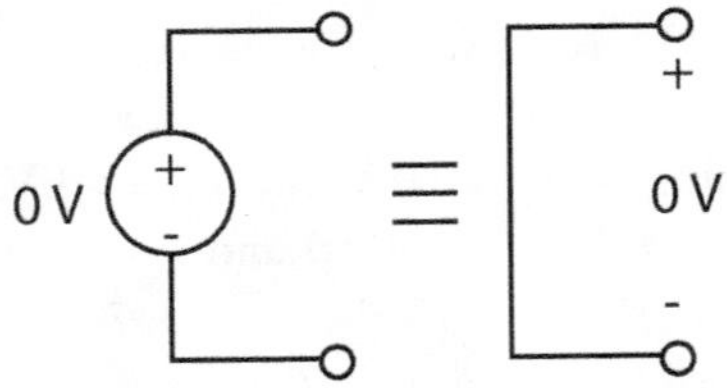

FIGURE 2.17 A 0 V voltage source is equivalent to a short circuit.

Finally, note that all four KVL statements can be justified using the definition and the notation for "voltage" drop presented in Chapter 1. The justification is more readily comprehended via the analogy of the gravitational field, also developed in that section. Also, observe that KVL holds for all closed node sequences, independent of the device represented by each branch of the connected circuit. The distribution of voltages around closed paths can be viewed as a special case of this general statement.

4. SERIES RESISTANCES AND VOLTAGE DIVISION

During holidays, one often sees strings of lights hanging between poles or trees. Sometimes these strings consist of a series connection of light bulbs. Each light bulb contains a filament, a coil of wire, that gives off an intense light when hot. In a circuit's perspective, the filament acts as a resistor and has an equivalent hot resistance. The series connection of bulbs can be modeled by a series connection of resistors, with each resistor paired with a specific bulb. Computing the voltage across each light (a very important type of calculation) would then be equivalent to finding the voltage across each of the resistors in the equivalent circuit model. It is quite common to model electrical loads, such as a light, by resistors.

EXAMPLE 2.4

Figure 2.18a shows a voltage source $v_{in}(t)$ connected to three resistors in series. The objectives of this example are to compute $i_{in}(t)$, the voltages $v_j(t)$, $j = 1, 2, 3$, across each resistor, and R_{eq} the equivalent resistance seen by the voltage source.

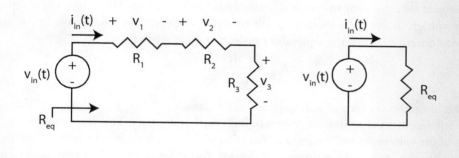

(a)   (b)

FIGURE 2.18 (a) Three series resistors connected across a voltage source. By the definition of a two-terminal resistor or by the KCL, the current through each resistor is $i_{in}(t)$; (b) equivalent resistance $R_{eq} = R_1 + R_2 + R_3$ seen by the source, i.e., $v_{in}(t) = R_{eq}i_{in}(t)$.

SOLUTION

Step 1. *Express the voltage across each resistor in terms of the input current.* For the circuit of Figure 2.18a, the current through each resistor is $i_{in}(t)$ by KCL. From Ohm's law, the voltage across each resistor is

$$v_j(t) = R_j i_{in}(t)$$

for $j = 1, 2, 3$.

Step 2. *Express $v_{in}(t)$ in terms of $i_{in}(t)$, solve for $i_{in}(t)$, and then compute an expression for $v_j(t)$ in terms of the R_j and $v_{in}(t)$.* By KVL, the source voltage equals the sum of the resistor voltages, i.e.,

$$v_{in}(t) = v_1(t) + v_2(t) + v_3(t) = (R_1 + R_2 + R_3)i_{in}(t) \qquad (2.1)$$

where we have substituted $R_j i_{in}(t) = v_j(t)$. Dividing Equation 2.1 by $(R_1 + R_2 + R_3)$ yields

$$i_{in}(t) = \frac{v_{in}(t)}{R_1 + R_2 + R_3}$$

Since $v_j(t) = R_j i_{in}(t)$ for $j = 1, 2, 3$,

$$v_j(t) = R_j i_{in}(t) = \frac{R_j}{R_1 + R_2 + R_3} v_{in}(t)$$

$$(2.2a)$$

Equation 2.2a is a **voltage division formula** for a three-resistor series circuit. This formula implies that if a resistance R_j is small relative to the other resistances in the series circuit, then only a small portion of the source voltage develops across it. On the other hand, if a resistance R_j is large relative to the other resistances, then a larger portion of the source voltage will develop across it. One concludes that the voltage distributes around a loop of resistors in proportion to the value of each resistance. The proportion is simply the ratio of the branch resistance R_j to the total series resistance.

Step 3. *Compute the equivalent resistance R_{eq}, seen by the voltage source.* The equivalent resistance seen by the voltage source for a resistive circuit is implicitly defined by Ohm's law, i.e., $v_{in}(t) = R_{eq} i_{in}(t)$. For nonzero currents, the **equivalent resistance** is defined as

$$R_{eq} = \frac{v_{in}(t)}{i_{in}(t)}$$

Figure 2.18b illustrates the idea of the equivalence. By Equation 2.1, $v_{in}(t) = R_{eq} i_{in}(t) = (R_1 + R_2 + R_3) i_{in}(t)$ implies that the equivalent resistance is $R_{eq} = R_1 + R_2 + R_3$. This means that from the perspective of the voltage source, the series connection of resistors is equivalent to a single resistor of value equal to the sum of the resistances. A formal discussion of equivalent resistance and its generalization (the Thévenin resistance) is taken up in Chapter 6.

Exercise. In Figure 2.18, suppose $R_3 = 3R_1$ and $R_2 = 2R_1$. Find R_{eq}, v_1, and v_3.
ANSWER: $R_{eq} = 6R_1$, $v_1 = v_{in}/6$, and $v_3 = 0.5v_{in}$

Example 2.4 suggests some generalizations. Consider Figure 2.19. The first is that the equivalent resistance R_{eq} seen by the source is the sum of the resistors. This means that *resistances in series add*, i.e., resistors in series can be combined into a single resistor whose resistance is the sum of the individual resistances.

$$R_{eq} = R_1 + R_2 + \cdots + R_n$$

Further, since $v_j(t) = R_j i_{in}(t)$, a **general voltage division formula** can be derived as

$$v_j(t) = \frac{R_j}{R_1 + R_2 + \cdots + R_n} v_{in}(t)$$

$$(2.2b)$$

for $j = 1, \dots , n$.

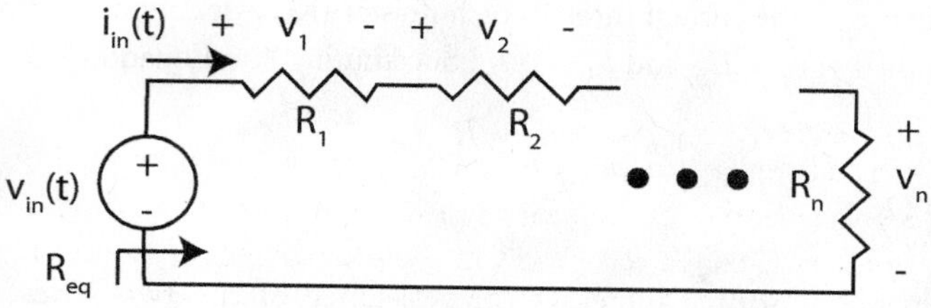

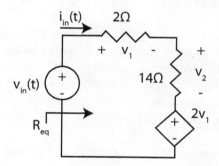

FIGURE 2.19 Series circuit of n resistors driven by a voltage source.

Exercise. In Figure 2.19, suppose each resistor has value R_0. Find the equivalent resistance seen by the source and the voltage across each resistor in terms of the source voltages.
ANSWER: nR_0, $v_{in}(t)/n$

EXAMPLE 2.5

Find the equivalent resistance seen by the source and the voltages v_1 and v_2 for the circuit of Figure 2.20. What is the power dissipated in the 14-Ω resistor if $v_{in}(t) = 2$ V?

FIGURE 2.20 Series circuit containing a dependent voltage source.

SOLUTION

From the preceding discussion, R_{eq} is defined by Ohm's law, i.e., $v_{in}(t) = R_{eq}i_{in}(t)$.

Step 1. *Express* v_{in} *in terms of the remaining branch voltages.* From KVL,

$$v_{in} = v_1 + v_2 + 2v_1 = 3v_1 + v_2$$

$$(2.3)$$

Step 2. *Express the branch voltages in terms of i_{in} and substitute into Equation 2.3.* To express v_1 and v_2 in terms of i_{in}, observe that i_{in} is the current through each resistor (KCL or definition of two-terminal circuit element) and use Ohm's law: $v_1 = 2i_{in}$ and $v_2 = 14i_{in}$. Substituting into Equation 2.3 yields

$$v_{in} = 20i_{in} = R_{eq}i_{in}$$

Therefore, $R_{eq} = 20\ \Omega$.

Notice that the dependent source increases the resistance of the two series resistors by 4 Ω. Dependent sources can increase or decrease the resistance of the circuit. With dependent sources, it is even possible to make the equivalent resistance negative.

Step 3. *Find the power absorbed by the 14-Ω resistor.* To find the power absorbed by the 14-Ω resistor when $v_{in} = 2$ V, first compute i_{in} via Ohm's law: $i_{in} = v_{in}/R_{eq} = 2/20 = 0.1$ A. It follows that $P = I^2R = 0.01 \times 14 = 0.14$ W.

Exercise. Suppose the dependent source in the circuit of Figure 2.20 has its value changed to $2(v_1 + v_2)$. Find R_{eq}.
ANSWER: 48 Ω

---

5. PARALLEL RESISTANCES AND CURRENT DIVISION

Many of the electrical outlets in the average home are connected in parallel. When too many appliances are connected to the same outlet or set of outlets on the same fused circuit, a fuse will blow or a circuit breaker will open. Although each appliance uses only a portion of the maximum allowable current for the (fused) circuit, together, the total current exceeds the allowable limit. Because of this common occurrence, an engineering student ought to know how current distributes through a parallel connection of loads (resistors).

To keep the analysis simple, consider a set of three parallel resistors driven by a current source.

EXAMPLE 2.6
Figure 2.21a shows a circuit of three parallel resistors driven by a current source. Our objectives are to find expressions for $v_{in}(t)$, $i_j(t)$ in terms of the input current $i_{in}(t)$ and the circuit conductances (the reciprocal of the circuit resistances) and the equivalent resistance seen by the current source.

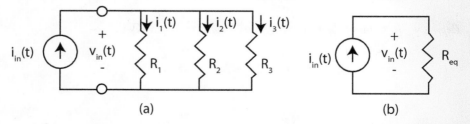

(a)   (b)

FIGURE 2.21 (a) Three parallel resistors driven by a current source;
(b) equivalent resistive circuit as seen from source.

SOLUTION

Step 1. *Find expressions for $i_j(t)$ in terms of $v_{in}(t)$.* The variable that links the branch current $i_j(t)$ to the input current $i_{in}(t)$ is the voltage $v_{in}(t)$, which by KVL appears across each resistor. Since $v_{in}(t)$ appears across each of the resistors, Ohm's law implies that each resistor current is

$$i_j(t) = \frac{v_{in}(t)}{R_j} = G_j v_{in}(t)$$

$$(2.4)$$

where $G_j = 1/R_j$ is the conductance in siemens and $j = 1, 2, 3$.

Step 2. *Compute $v_{in}(t)$ in terms of $i_{in}(t)$.* Applying KCL to the top node of the circuit yields

$$i_{in}(t) = i_1(t) + i_2(t) + i_3(t)$$

Using Equation 2.4 to substitute for each $i_j(t)$ and then solving for v_{in} yields

$$v_{in}(t) = R_{eq} i_{in}(t) = \frac{1}{\dfrac{1}{R_1} + \dfrac{1}{R_2} + \dfrac{1}{R_3}} i_{in} = \frac{1}{G_1 + G_2 + G_3} i_{in}(t) = \frac{1}{G_{eq}} i_{in}(t)$$

$$(2.5)$$

Step 3. *Compute $i_j(t)$ in terms of $i_{in}(t)$.* To obtain a relationship between $i_{in}(t)$ and $i_j(t)$, substitute Equation 2.5 into Equation 2.4 to obtain

$$i_j(t) = \frac{\dfrac{1}{R_j}}{\dfrac{1}{R_1} + \dfrac{1}{R_2} + \dfrac{1}{R_3}} i_{in} = \frac{G_j}{G_1 + G_2 + G_3} i_{in}(t) = \frac{G_j}{G_{eq}} i_{in}(t)$$

$$(2.6)$$

Equation 2.6 is called a **current division formula.** It says that currents distribute through the branches of a parallel resistive circuit in proportion to the conductance of the particular branch G_j relative to the total conductance of the circuit $G_{eq} = G_1 + G_2 + G_3$. The greater the conductance, i.e., the smaller the resistance, the larger the proportion of current flow through the associated branch.

Step 4. *Compute the equivalent resistance R_{eq} seen by the source.* As in Example 2.5, Ohm's law, $i_{in} = G_{eq} v_{in}$, defines G_{eq} or, equivalently, R_{eq}. From Equation 2.5, $G_{eq} = G_1 + G_2 + G_3$ is the **equivalent conductance** of the parallel circuit, and the **equivalent resistance** is

$$R_{eq} = \frac{1}{\dfrac{1}{R_1} + \dfrac{1}{R_2} + \dfrac{1}{R_3}} = \frac{1}{G_1 + G_2 + G_3} = \frac{1}{G_{eq}}$$

The idea is illustrated in Figure 2.21b.

Exercise. In Figure 2.21a, suppose $R_1 = 1\ \Omega$, $R_2 = 0.5\ \Omega$, and $R_3 = 0.5\ \Omega$. Find the current through R_1 if $i_{in}(t) = 10e^{-t}$ A.
ANSWER: $i_1(t) = 2e^{-t}$ A

The Example 2.6 suggests a very important property. Since

$$i_{in} = \left(G_1 + G_2 + G_3\right)v_{in} = G_{eq}v_{in}$$

in addition to implying that $G_{eq} = G_1 + G_2 + G_3$ is the equivalent conductance seen by the source of the parallel circuit, one can further interpret this to mean that *conductances in parallel add to form equivalent conductances.* This parallels the property that resistors in series add to form equivalent resistances. On the other hand, *resistances in parallel do not add*, and *conductances in series do not add.* We can conclude that from the perspective of the source, the parallel circuit of Figure 2.21a has the equivalent representations given in Figure 2.21b.

These ideas generalize to n resistors in parallel, as illustrated in Figure 2.22. In particular, the **equivalent resistance** R_{eq} of the parallel set of resistors in Figure 2.22 is

$$R_{eq} = \frac{1}{\dfrac{1}{R_1} + \dfrac{1}{R_2} + \cdots + \dfrac{1}{R_n}} = \frac{1}{G_1 + G_2 + \cdots + G_n} = \frac{1}{G_{eq}}$$

$$(2.7)$$

i.e., $G_{eq} = G_1 + G_2 + \ldots + G_n$ is the **equivalent conductance**. Further, the current through each branch satisfies the general **current division** formula

$$i_j(t) = \frac{\dfrac{1}{R_j}}{\dfrac{1}{R_1} + \dfrac{1}{R_2} + \cdots + \dfrac{1}{R_n}} i_{in}(t) = \frac{G_j}{G_1 + G_2 + \cdots + G_n} i_{in}(t) = \frac{G_j}{G_{eq}} i_{in}(t)$$

$$(2.8)$$

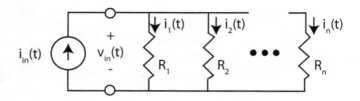

FIGURE 2.22 Parallel connection of n resistors driven by current source.

Exercise. Consider Figure 2.22. Suppose ten 10 Ω resistors are in parallel. Find R_{eq} and the current through each resistor.

ANSWER: $R_{eq} = 1\ \Omega$ and each current is $0.1 i_{in}(t)$

EXAMPLE 2.7

Consider the circuit of Figure 2.23 exhibiting a current source driving two parallel resistors. Show that

$$R_{eq} = \frac{R_1 R_2}{R_1 + R_2},\ i_1(t) = \frac{R_2}{R_1 + R_2} i_{in}(t),\ \text{and}\ i_2(t) = \frac{R_1}{R_1 + R_2} i_{in}(t).$$

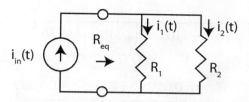

FIGURE 2.23 Two resistors in parallel driven by a current source.

SOLUTION

Step 1. *Find the equivalent resistance seen by the current source.* From Equation 2.7, with $n = 2$, it follows that

$$R_{eq} = \frac{1}{\dfrac{1}{R_1} + \dfrac{1}{R_2}} = \frac{R_1 R_2}{R_1 + R_2}$$

This formula, called the **product over sum rule**, is quite useful in many calculations.

Step 2. *Find $i_1(t)$ and $i_2(t)$.* From Equation 2.8, with $n = 2$, it follows that

$$i_1(t) = \frac{G_1}{G_1 + G_2} = \frac{\dfrac{1}{R_1}}{\dfrac{1}{R_1} + \dfrac{1}{R_2}} = \frac{R_2}{R_1 + R_2} i_{in}(t)$$

and

$$i_2(t) = \frac{G_2}{G_1 + G_2} = \frac{\dfrac{1}{R_2}}{\dfrac{1}{R_1} + \dfrac{1}{R_2}} = \frac{R_1}{R_1 + R_2} i_{in}(t)$$

EXAMPLE 2.8

For the circuit of Figure 2.24, find the input voltage $v_{in}(t)$, the current $i_2(t)$ through R_2, and the instantaneous power absorbed by R_2 when

$$i_{in}(t) = \begin{cases} 5e^{-t} & t \geq 0 \\ 0 & t < 0 \end{cases}$$

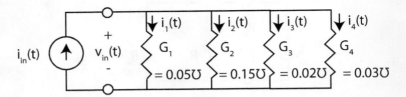

FIGURE 2.24 Parallel connection of four resistors.

SOLUTION

Step 1. *Compute the equivalent conductance and equivalent resistance of the circuit.* Since conductances in parallel add,

$$G_{eq} = G_1 + G_2 + G_3 + G_4 = 0.25 \text{ S}$$

and

$$R_{eq} = \frac{1}{G_{eq}} = 4 \ \Omega$$

Step 2. *Compute $v_{in}(t)$.* From Ohm's law, the voltage across the current source is

$$v_{in}(t) = R_{eq}i_{in}(t) = \begin{cases} 20e^{-t} \text{ V} & t \geq 0 \\ 0 & t < 0 \end{cases}$$

Step 3. *Compute the current $i_2(t)$.* Using the current division formula of Equation 2.8 yields

$$i_2(t) = \frac{G_2}{G_{eq}} i_{in}(t) = \frac{0.15}{0.25} i_{in}(t) = \begin{cases} 3e^{-t} \text{ A} & t \geq 0 \\ 0 & t < 0 \end{cases}$$

Step 4. *Compute the power absorbed by R_2.* To compute the power absorbed by R_2 for $t \geq 0$,

$$p_2(t) = v_{in}(t) \times i_2(t) = \left[i_2(t)\right]^2 R_2 = 60e^{-2t} \text{ W}$$

Exercises. 1. For the circuit of Figure 2.24, find $i_4(t)$ and the power absorbed by R_4.
ANSWER: $0.6e^{-t}$ A, $12\, e^{-2t}$ W

2. In the circuit of Figure 2.24, suppose each conductance is doubled and $i_{in}(t) = 100$ mA. Find R_{eq}, $v_{in}(t)$, and the power absorbed by the new G_3.
ANSWER: 2 Ω, 200 mV, and 1.6 mW

---

6. SERIES-PARALLEL INTERCONNECTIONS

The last two sections covered series and parallel resistive networks. Suppose we take a series circuit and connect it in parallel with another series circuit; this is a parallel connection of two series circuits. Alternately, we could take two parallel circuits and connect them in series. This would result in a series connection of parallel circuits. We could also put a series connection of two parallel subcircuits in parallel with a replica of itself or some other series or parallel circuit. Many other interconnections are possible. Arbitrary series and parallel connections of such subcircuits are called **series-parallel circuits**. This section explores the calculation of the equivalent resistance of series-parallel circuits by repeated use of formulas for series and parallel resistance computation. Related voltage and current computation is also explored. Example 2.11 presents a practical application of series-parallel concepts.

EXAMPLE 2.9

Find the equivalent resistance, R_{eq}, and the voltage across the source, V_{in}, the voltages V_1, V_2, the power absorbed by the 6 kΩ resistor, and the power delivered by the source for the circuit of Figure 2.25, when $I_{in} = 20$ mA.

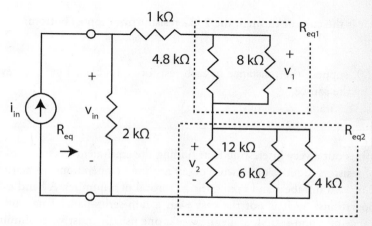

FIGURE 2.25 Series-parallel resistive circuit.

SOLUTION

Step 1. *Compute R_{eq}.* To compute R_{eq}, first compute R_{eq1} and R_{eq2}:

$$R_{eq1} = \frac{8 \times 4.8}{8 + 4.8} = \frac{38.4}{12.8} = 3 \text{ k}\Omega$$

and

$$R_{eq2} = \frac{1}{\dfrac{1}{12} + \dfrac{1}{6} + \dfrac{1}{4}} = \frac{1}{\dfrac{6}{12}} = 2 \text{ k}\Omega$$

The resistance in parallel with the 2kΩ resistor is, say,

$$R_{eq3} = 1000 + R_{eq1} + R_{eq2} = 6 \text{ k}\Omega$$

Finally,

$$R_{eq} = 2k\Omega // R_{eq3} = \frac{2 \times 6}{2 + 6} = 1.5 \text{ k}\Omega$$

Step 2. *Compute V_{in}.* From Ohm's law,

$$V_{in} = R_{eq} I_{in} = 20 \times 1.5 = 30 \text{ V}$$

Step 3. *Compute V_1 and V_2.* By voltage division,

$$V_1 = \frac{R_{eq1}}{R_{eq3}} V_{in} = \frac{3}{6} V_{in} = 0.5 V_{in} = 15 \text{ V} \quad \text{and} \quad V_2 = \frac{R_{eq2}}{R_{eq3}} V_{in} = \frac{2}{6} V_{in} = 10 \text{ V}$$

Step 4. *Compute the power absorbed by the 6 kΩ resistor.*

$$P_{6k\Omega} = \frac{(V_2)^2}{6000} = \frac{100}{6000} = \frac{1}{60} \text{ W}$$

Step 5. *Compute the power delivered by the source.*

$$P_{source} = V_{in} I_{in} = 30 \times 0.02 = 0.6 \text{ W}$$

Exercise. 1. What is the current through the 2 kΩ resistor from top to bottom?
ANSWER: 50 mA

2. In Example 2.9, suppose the resistance of each resistor is doubled. Find the new R_{eq} and the power delivered by the source.
ANSWER: 3 kΩ, 1.2 watts

This example points out a very interesting fact: finding the equivalent resistance of a series-parallel connection of resistors requires only two types of arithmetic operations no matter the network complexity: adding two numbers and taking the reciprocal of a number. A hand calculator easily executes both operations. Such is not the case with a non-series-parallel network. To find the equivalent resistance of a non-series-parallel network, one usually must write simultaneous equations and evaluate determinants, a topic detailed in Chapter 3.

It is then important to recognize when a problem belongs to the series-parallel category in order to take advantage of the simple arithmetic operations. In the previous series-parallel examples, one—and only one—independent source was specified on the circuit diagram. This is part of the definition of a series-parallel network. The independent source must be indicated, or, equivalently, the pair of input terminals to which the source is connected must be specified. The specification of the input terminals determines whether or not a network is series-parallel. The following example illustrates the effect of different input terminal designations on the computation of equivalent resistance.

EXAMPLE 2.10
For the circuit of Figure 2.26a, determine whether or not the network is series-parallel as seen from each of the following terminal pairs:

1. Case 1: (A, B)
2. Case 2: (A, C)
3. Case 3: (C, D)

If the answer is affirmative, give an expression and compute the numerical value for the equivalent resistance, using the notation // (double slash) for combining resistances in parallel, i.e., $R_1//R_2$ means R_1 and R_2 are in parallel, and $R_1//R_2//R_3$ means R_1 is in parallel with R_2, which is in parallel with R_3.

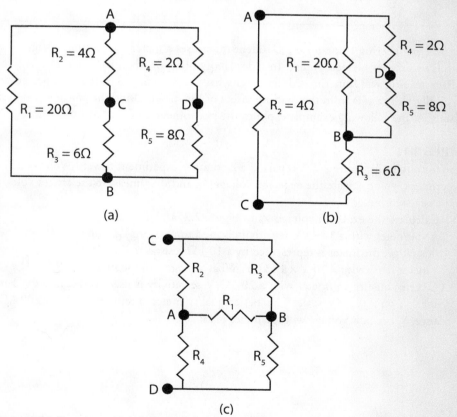

(a)   (b)   (c)

FIGURE 2.26 (a) From terminals (C, D) the network is not series-parallel. However, from terminals (A, B) the network is a series-parallel one. (b) Redrawing of the network of (a) as seen from terminals (A, C); the resulting network is series-parallel.(c) Non-series-parallel network seen from (C, D).

SOLUTION

Case 1. Find equivalent resistance seen at (A, B). With an independent source connected to nodes A and B, the source sees a series-parallel network. By inspection of Figure 2.26a, the equivalent resistance is

$$R_{eq} = R_1 // [(R_2 + R_3) // (R_4 + R_5)] = 20 // [(4 + 6) // (2 + 8)] = 4\ \Omega$$

Case 2. Find equivalent resistance seen at (A, C). With (A, C) as the input terminal pair, the network is again series-parallel. This is made apparent by redrawing the network, as shown in Figure 2.26b, from which

$$R_{eq} = R_2 // \{R_3 + [(R_4 + R_5) // R_1]\} = 4 // \{6 + [(2 + 8) // 20]\} = 3.04\ \Omega$$

Case 3. Find equivalent resistance seen at (C, D). With (C, D) as the input terminal pair, the network is not series-parallel, as can be garnered from Figure 2.26c. The calculation of R_{eq} for this case requires methods to be discussed in Chapter 3 and is omitted.

Exercise. 1. In Figure 2.26b, suppose R_1 is changed to 40 Ω. Find R_{eq}.
ANSWER: 3.11 Ω

In electrical engineering laboratories, a student often uses a meter to measure voltages associated with a piece of electronic equipment. In older laboratories, or when using an inexpensive meter, the voltage reading will sometimes differ from what the student calculated or expected to measure. Typically, this results from the **loading effect** of the meter. Using the concept of series-parallel resistances, the following example explores the phenomenon of loading.

EXAMPLE 2.11

Suppose the circuit in Figure 2.27a is part of a laboratory experiment to verify voltage division. In this experiment, you calculate the expected voltage V_0 and then measure the circuit voltage using an inexpensive voltmeter.

(a) Calculate the expected voltage V_0 in Figure 2.27a.
(b) A voltmeter with a 1-kΩ/V sensitivity is used to measure V_0. You use a 0-10-V range. In this range, the meter is represented by a 10-kΩ resistance, i.e., 10 kΩ = full-scale reading $\times$ meter sensitivity = 10 V $\times$ 1 kΩ/V. What voltage will the meter read?
(c) A better-quality voltmeter with a 20-kΩ/V sensitivity is used to measure the same voltage V_0, on a 0-10-V scale. This better-quality meter is represented by a 200-kΩ resistance. What new voltage will the meter read?

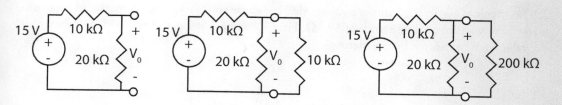

FIGURE 2.27 Three circuits for exploring the effect of loading on a circuit: (a) circuit for validating voltage division; (b) circuit of (a) with an attached voltmeter having an internal resistance of 10 kΩ; (c) circuit of (a) with an attached voltmeter having an internal resistance of 200 kΩ.

SOLUTION

(a) Voltage division on the circuit of Figure 2.27a yields

$$V_0 = \frac{20}{10 + 20} 15 = 10 \text{ V}$$

(b) On the 0-10-V range, the voltmeter internal resistance between the probes is 10 kΩ, as stated. This represents a 10-kΩ load connected in parallel with the 20-kΩ resistance, as shown in Figure 2.27b. The voltage V_0 will now change because the 15-V source no longer sees 10 kΩ in series with 20 kΩ. Rather, the source sees 10 kΩ in series with 6.67 kΩ = 20 kΩ//10 kΩ. By voltage division,

$$V_0 = \frac{6.67}{10 + 6.67} 15 = 6 \text{ V}$$

This is a 40% deviation from the true answer, $V_0 = 10$ V, as calculated in part (a).

(c) On the 0-10-V range with the better voltmeter, the internal resistance between the probes is 200 kΩ. As before, this represents a 200-kΩ load connected in parallel with the 20-kΩ resistance, as shown in Figure 2.27c. 20 kΩ//200 kΩ = 18.18 kΩ. By voltage division, this yields

$$V_0 = \frac{18.18}{10 + 18.18} 15 = 9.677 \text{ V}$$

This 3.23% deviation is within a reasonable tolerance of the precise answer of 10 V.

Example 2.11 demonstrates the effect of loading due to a measuring instrument, emphasizing the importance of choosing a good voltmeter with adequate sensitivity. Although modern-day voltmeters typically have sensitivities better than 20 kΩ/V, a meter with a sensitivity of 1 kΩ/V is used in the example to dramatize the effect of loading.

Exercise. 1. Repeat Example 2.11 if the 20-kΩ resistance is changed to 40 kΩ.
ANSWER: 12 V, 6.667 V, 11.538 V

2. The circuit of Figure 2.28 shows a voltage divider whose voltage V_0 is to be measured by a voltmeter having an internal resistance of 80 kΩ. Find V_0 without the meter attached, and then find the value of V_0 measured by the meter.

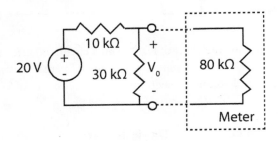

FIGURE 2.28 Voltage divider circuit.

ANSWER: 15 V, 13.71 V

7. DEPENDENT SOURCES REVISITED

Chapter 1 introduced the notion of a **dependent** or **controlled source** whose voltage or current depends on the voltage or current in another branch of the network, i.e., each source has a controlling voltage or current and an output voltage or current. Figure 2.29 depicts the four types of controlled sources designated by a diamond containing either a ± or an arrow:

1.   Voltage-controlled voltage source   (VCVS)
2.   Voltage-controlled current source   (VCCS)
3.   Current-controlled voltage source   (CCVS)
4.   Current-controlled current source   (CCCS)

An arrow inside the diamond indicates a controlled current source having the reference current direction given by the arrow. A ± inside the diamond specifies a controlled voltage source, with the reference voltage polarity given by the ± sign. A parameter value completes the specification of a *linear* controlled source. In Figure 2.29 the (constant) parameters are μ, g_m, r_m, and β. These symbols are common to many electronic circuit texts and have useful physical interpretations to practicing engineers and technicians. For consistency, a g_m-type controlled source is a VCCS and a μ-type source is a VCVS, and so on.

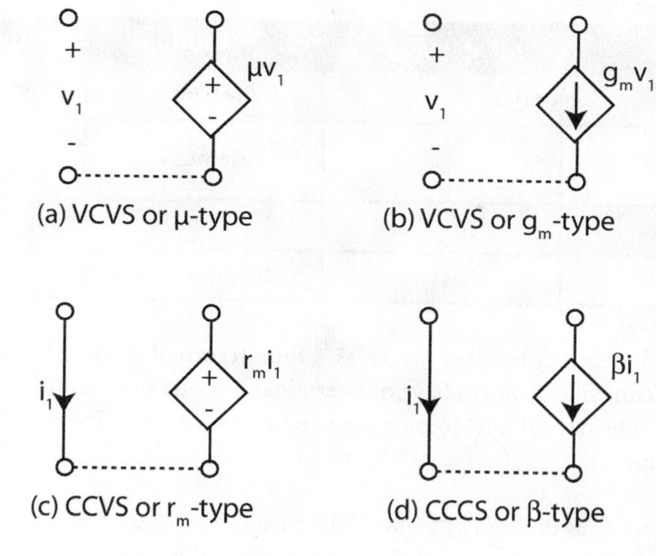

(a) VCVS or μ-type

(b) VCVS or g_m-type

(c) CCVS or r_m-type

(d) CCCS or β-type

FIGURE 2.29 Designations for the various controlled sources.

In practical controlled sources, the controlling voltage (v_1 in Figure 2.29a and b) or current (i_1 in Figure 2.29c and d) is ordinarily associated with a particular circuit element, but not always. For generality, the controlling voltage v_1 in Figure 2.29a and b is shown across a pair of nodes. Also, in Figure 2.29c and d, the controlling current i_1 is shown to flow through a short circuit. (Strictly speaking, neither an open circuit nor a short circuit is a circuit element.) In a real circuit, the current may be flowing through an actual circuit element, such as a resistor or even a source.

In Figure 2.29b, once the controlling voltage v_1 is known, the right-hand source behaves as an independent current source of value $g_m v_1$. Since the unit for $g_m v_1$ is amperes and the unit for v_1 is volt, the unit for g_m is amperes per volt, or siemens. Since g_m has units of conductance, and the controlling and controlled variables belong to two different network branches, g_m is called a **transfer conductance**, or transconductance. The other controlled sources have a similar interpretation. The parameter r_m has the unit of resistance, ohms, and is called a **transfer resistance**. The parameter μ is dimensionless because the controlling voltage v_1 has units of volts and the output variable μv_1 must have units of volts. Similarly, the parameter β is dimensionless. The units and association are set forth in Table 2.1.

Table 2.1 Units and Association.

Type	Parameter	Unit	Appellation
VCVS	μ	dimensionless	Voltage gain
VCCS	g_m	siemens	Transfer conductance
CCVS	r_m	ohm	Transfer resistance
CCCS	β	dimensionless	Current gain

Figure 2.29 portrays each controlled source as a **four-terminal device**. In practical circuits, the great majority of controlled sources have one terminal or node in common, making them **three-terminal devices.** The dashed lines joining the two bottom nodes in Figure 2.29 suggest this quite common configuration.

The controlled sources as defined in Figure 2.29 have linear *v-i* relationships. Controlled sources may also have a **nonlinear** *v-i* relationship. In such cases, the element will be called a **nonlinear controlled source.** This text deals only with linear controlled sources.

The next few examples describe some of the unique features of controlled sources.

Exercise. Find v_1, i_{out}, and the power delivered by each source in Figure 2.30.

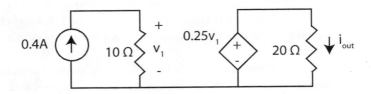

FIGURE 2.30

ANSWER: 4 V, 0.05 A, 1.6 W, 0.05 W

EXAMPLE 2.12

This example analyzes the circuit of Figure 2.31. The independent voltage source in series with the 3-Ω resistor represents a **practical source** discussed at greater length later in this chapter. The circuit within the box of Figure 2.30 approximates a simplified amplifier circuit by a VCCS. The 8-Ω resistor is considered a load and could, for example, model a loudspeaker. Two important quantities of an amplifier circuit are **voltage gain** and **power gain**, which are computed here along with various other quantities.

(a) Find the equivalent resistance R_{eq} seen by the independent voltage source.

(b) Compute I_{in}.

(c)   Compute I_{out}.
(d)   Compute V_{out}.
(e)   Compute the voltage gain V_{out}/V_{in}.
(f)   Compute the power P_{in} delivered to the amplifier.
(g)   Compute the power delivered by the dependent current source.
(h)   Compute P_{out}, the power absorbed by the 8-Ω resistor.
(i)   Compute the power gain, P_{out}/P_{in}.

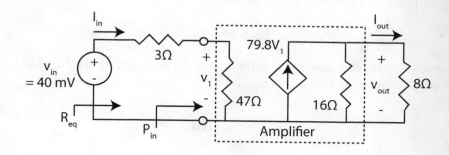

FIGURE 2.31 Practical source (ideal independent voltage source in series with a resistor) driving a simplified VCCS approximation of an amplifier circuit loaded by an 8-Ω resistor.

SOLUTION

(a)   Since resistances in series add, $R_{eq} = 3 + 47 = 50$ Ω.

(b)   By Ohm's law, $I_{in} = V_{in}/R_{eq} = 0.8$ mA.

(c)   To compute I_{out}, one must first compute V_1. Here one can use Ohm's law directly, since we know I_{in}, or one can use voltage division. Doing the calculation by voltage division,

$$V_1 = \frac{47}{50} 40 \times 10^{-3} = 37.6 \times 10^{-3} \text{ V}$$

Using this value of V_1 and current division on the right half of the circuit yields

$$I_{out} = \frac{0.125}{0.125 + 0.0625} \left(79.8 \times 37.6 \times 10^{-3} \right) = 2 \text{ A}$$

(d) V_{out} follows by Ohm's law

$$V_{out} = 2 \times 8 = 16 \text{ V}$$

(e)   The voltage gain with respect to the input signal is

$$\frac{V_{out}}{V_{in}} = \frac{16}{0.04} = 400$$

(f)   By Equation 1.18, the power delivered to the amplifier circuit is

$$P_{in} = V_1 I_{in} = 47 (I_{in})^2 = 47 \times 0.8^2 \times 10^{-6} = 30.08 \ \mu W$$

(g)   The power delivered by the dependent current source is

$$P_{VCCS} = V_{out} \times 79.8 V_1 = 16(79.8 \times 0.0376) = 48.01 \ W$$

(h) P_{out} is simply the product of voltage and current delivered to the load

$$P_{out} = V_{out} I_{out} = 2 \times 16 = 32 \ W$$

(i)   The resulting amplifier power gain is the ratio of the power absorbed by the 8-Ω load to the power delivered to the amplifier, P_{in},

$$G_P = \frac{P_{out}}{P_{in}} = \frac{32}{30.08} 10^6 = 1.064 \times 10^6$$

Exercise. Suppose the 8-Ω load resistor in Figure 2.30 is changed to 16 Ω. Compute I_{out}, V_{out}, and the power gain.
ANSWER: 1.5 A, 24 V, 1.197 $\times$ 10^6

The analysis in Example 2.12 required only KCL, KVL, and simple voltage divider and/or current divider formulas. More complicated linear circuits necessitate a more systematic approach. To see this need, add a resistor between the top of the 47-Ω resistor and the top of the dependent current source in Figure 2.31. The methods of solution used in the example immediately break down because the circuit is no longer series-parallel; hence, one cannot use voltage division to compute V_1. Chapter 3 will explain more systematic methods called nodal and loop analysis.

Unlike a passive element such as a resistor, which always dissipates power as heat, a controlled source may generate power as computed in part (g) of Example 2.12, or may dissipate power in other cases. Since a controlled source has the potential of generating power, it is called an **active element.**

In Example 2.12, the practical voltage source delivers 30.08 μW of power to the circuit, which is easy to accept because the source could have been a small battery. On the other hand, the controlled source generates 48 W. This seems a little puzzling. Where does the power come from? Why not purchase a controlled source at a local electronics store and use it to power, say, a lamp? Here it is important to recognize that a controlled source is not a stand-alone component picked off the shelf like a resistor. A controlled source is usually constructed from one or more semiconductor devices and requires a dc power supply for its operation. The power delivered by the controlled source actually comes from the power supply. Here, we use the controlled source to mathematically model an amplifier and facilitate analysis of the circuit.

With simple series-parallel connections of resistors, the equivalent resistance is always positive. When controlled sources are present, a strange result may happen, as illustrated in the next example.

EXAMPLE 2.13

Find the equivalent resistance $R_{eq} \triangleq V_s/I_s$ for the circuit of Figure 2.32 when (a) $\mu = 0.5$ and (b) $\mu = 2$.

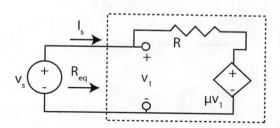

FIGURE 2.32 Calculation of R_{eq} for a circuit with controlled source for two values of μ.

SOLUTION

With μ unspecified, we can apply KVL to the single loop, noting that $V_1 = V_s$. Here,

$$V_s = RI_s + \mu V_1 = RI_s + \mu V_s$$

Consequently, $(1 - \mu)V_s = RI_s$ and

$$R_{eq} = \frac{V_s}{I_s} = \frac{R}{1 - \mu}$$

For $\mu = 0.5$, $R_{eq} = 2R$, which means that the dependent source acts like a resistor of R Ω. In this case, it absorbs power. On the other hand, for $\mu = 2$, $R_{eq} = -R$, a negative equivalent resistance. In this case, the dependent source acts like a -2R-Ω resistor and, in fact, delivers power to the independent source. An important conclusion can be drawn from this example: in the study of linear circuit analysis, controlled sources allow the possibility of negative resistances. Since a negative resistance generates power, it is also an active element.

Exercise. In Figure 2.32, find the values of μ so that $R_{eq} = 0.5R$ and $R_{eq} = 2R$.
ANSWER: −1, 0.5

Exercise. For Figure 2.33, find R_{eq} for the following three values of g_m: 0.5 mS, 1 mS, and 2 mS.

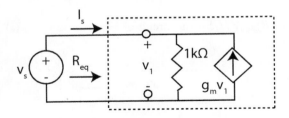

FIGURE 2.33

ANSWER: 2 kΩ, open circuit, -1 kΩ

8. MODEL FOR A NON-IDEAL BATTERY

The **ideal battery** of Figure 1.30, repeated in Figure 2.34a, delivers a constant voltage regardless of the current drawn by a load. The *i-v* plane characteristic is a horizontal line through V_s, as shown in Figure 1.40b and repeated in Figure 2.34b. Ideal batteries do not exist in the real world. The terminal voltage always depends on the supplied current. A more accurate representation of a practical battery, but by no means a fully realistic one, is an ideal battery in series with a resistance, say, R_s, as shown in Figure 2.34c. R_s is termed an internal resistance, which crudely models the effects of chemical action and electrodes inside the battery.

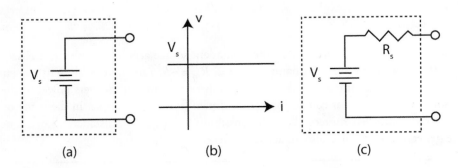

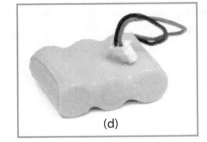

FIGURE 2.34 (a) Ideal battery; (b) i-v battery characteristic; (c) battery model with internal resistance to crudely approximate effects of chemical action and presence of electrodes; (d) nickel-cadmium battery.

EXAMPLE 2.14

This example shows the effect of the internal resistance of a battery on the terminal voltage. Suppose a nickel-cadmium battery has an open circuit terminal voltage of 6 volts. When connected across a 2-Ω resistor, the voltage drops to 5.97 V. Find the internal resistance of the battery.

SOLUTION

Figure 2.35 illustrates the situation. Here, the dashed box represents the battery model with internal resistance R_s. In Figure 2.35a, no load is connected to the battery. Hence, no current flows through the internal resistance, in which case, the terminal battery voltage is 6 V.

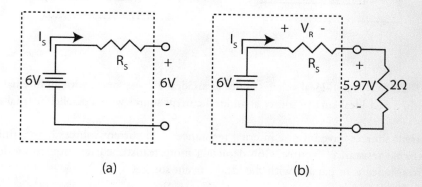

(a)   (b)

FIGURE 2.35 Battery model with internal resistance: (a) open circuited (Is = 0); and (b) connected to a 2-Ω load.

Figure 2.35b shows the battery connected to the 2-Ω resistive load. The measured voltage is 5.97 V. By KVL, the voltage across the internal resistance, R_S is $V_R = 6 - 5.97 = 0.03$ V. From Ohm's law, the current through R_S is $I_s = (5.97/2) = 2.985$ A. Again, by Ohm's law,

$$R_s = \frac{V_R}{I_s} = \frac{0.03}{2.985} = 0.01005 \ \Omega$$

Exercise. In Example 2.14, suppose the internal resistance is known to be $R_s = 0.005 \ \Omega$ and although the load resistance is unknown, the load current is 4 A. What is the voltage across the load resistance, and what is the load resistance?
ANSWER: 5.98 V and 1.495 Ω

---

9. NON-IDEAL SOURCES

Ideal voltage sources have zero internal resistance. Real voltage sources, such as batteries, have an internal resistance. The value of this resistance may change with the current load. There may also be other effects. However, for our purposes, a more realistic model of a voltage source contains a series internal resistance, as illustrated in Figure 2.36a.

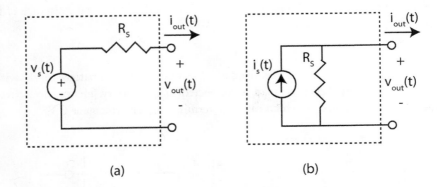

(a)   (b)

FIGURE 2.36 (a) A non-ideal voltage source as an ideal voltage source with an internal series resistance; (b) a non-ideal current source as an ideal current source with a parallel internal resistance.

Ideal current sources have infinite internal resistance. Real current sources have a finite, typically large, internal resistance. Figure 2.36b depicts a more realistic current source model where the internal resistance is in parallel with the ideal current source.

In the case of constant voltage and current sources, ideal and non-ideal source models have a graphical interpretation. The *i-v* (current-voltage) characteristic of an **ideal constant voltage source** $(v_s(t) = V_s)$ is a horizontal straight line. This means that the voltage supplied by the source is fixed for all possible current loads. An **ideal constant current source** $(i_s(t) = I_s)$ has a vertical straight line characteristic, which means that the current is constant for all possible voltages across the source. Figure 2.37 illustrates these relationships graphically.

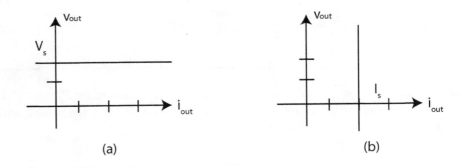

(a)   (b)

FIGURE 2.37 *v-i* characteristics of (a) an ideal constant voltage source, and
(b) an ideal constant current source.

The non-ideal case is quite different. Because of the internal resistance a **non-ideal constant voltage source** i-v characteristic satisfies the linear relationship

$$v_{out} = -R_s i_{out} + V_s$$

(2.10)

and for a **non-ideal constant current source** in which $G_s = 1/R_s$,

$$i_{out} = -G_s v_{out} + I_s$$

(2.11)

Equations 2.10 and 2.11 are illustrated by the graphs in Figure 2.38 when $v_s(t) = V_s$ for the non-ideal voltage source and $i_s(t) = I_s$ for the non-ideal current source. For a voltage source, if the value of R_s is very small in comparison with potential load resistances, as ordinarily expected, then the line in Figure 2.38a approximates a horizontal line, the ideal case. On the other hand, for a current source, the line in Figure 2.38b approximates a vertical line whenever R_s is much much larger than a potential load resistance. This would then approximate the ideal current source case.

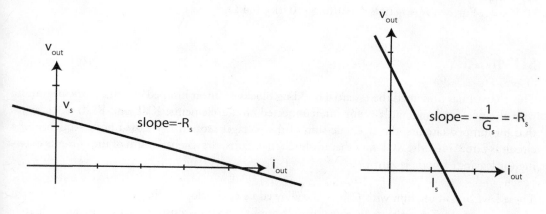

FIGURE 2.38 v-i characteristics of (a) non-ideal constant voltage source, and (b) non-ideal current source.

In a similar way, **non-ideal dependent voltage sources** are a connection of an ideal dependent source with a series resistance. A **non-ideal dependent current source** is a connection of an ideal dependent current source with a parallel resistance.

EXAMPLE 2.16 Figure 2.39 shows the measured voltages of a dc power supply found in an old laboratory. Assuming a non-ideal model of Figure 2.38a, find V_s and the internal resistance R_s of the power supply.

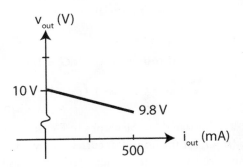

FIGURE 2.39 Graph of measured voltages and currents for a dc power supply.

SOLUTION

From Equation 2.10, we know that $v_{out} = -R_s\, i_{out} + V_s$. From the graph, when $i_{out} = 0$, $v_{out} = 10\text{ V} = V_s$. Further, $R_s = -\, (9.8 - 10)/(0.5 - 0.0) = 0.4\ \Omega$.

SUMMARY

This chapter has presented the essential building blocks of linear lumped circuit theory, beginning with the two fundamental laws for interconnected circuit elements: KVL and KCL. KVL states that for lumped circuits, the algebraic sum of the voltages around any closed node sequence of a circuit is zero. Similarly, KCL says that for lumped circuits, the algebraic sum of the currents entering (or leaving) a node is zero.

These laws in conjunction with Ohm's law allowed us to develop a voltage division and a current division formula. The voltage division formula applies to series-resistive circuits driven by a voltage source. The voltage developed across each resistor was found to be proportional to the resistance of the particular element relative to the equivalent resistance seen by the source. For example, in a two-resistor series circuit, R_1 in series with R_2, we found that

$$v_1 = \frac{R_1}{R_1 + R_2} v_{in}\,.$$

The current division formula applies to parallel-resistive circuits driven by a current source. Here, the current through each resistor with conductance G_i was found to be proportional to G_i divided by the equivalent conductance seen by the source. Since conductance is the reciprocal of resistance, the idea can also be expressed in terms of the resistances of the circuit. For example, in a two-resistor parallel circuit, R_1 is parallel with R_2,

$$i_1 = \frac{G_1}{G_1 + G_2} i_{in} = \frac{R_2}{R_1 + R_2} i_{in}$$

In deriving the voltage division formula, we learned that the resistances of a series connection of resistors may be added together to obtain an equivalent resistance, prompting the phrase "resistors in series add." Analogously, the derivation of the current division formula for parallel circuits led us to conclude that a parallel connection of resistors has an equivalent conductance equal to the sum of conductances. This is sometimes expressed in terms of resistances as the inverse of the sum of reciprocal, i.e., for n resistors in parallel,

$$R_{eq} = \frac{1}{\dfrac{1}{R_1} + ... + \dfrac{1}{R_n}},$$

which leads to the very special formula for two resistors in parallel,

$$R_{eq} = \frac{R_1 R_2}{R_1 + R_2},$$

often referred to as the product over sum rule.

Dependent sources, first introduced in Chapter 1, were re-examined in greater detail. Some practical points were described.

All of the above ideas were applied to the analysis of series-parallel networks that are interconnections of series and parallel groupings of resistors. Our analysis showed us how to compute the equivalent resistance of series-parallel circuits. An example was given that described the application of these ideas to voltage measurement. This was followed by a discussion of battery models and battery usage. Finally, battery modeling ideas were used to describe non-ideal source models.

12. TERMS AND CONCEPTS

Branch: a two-terminal circuit element denoted by a line segment.

Closed node sequence: a finite sequence of nodes that begins and ends with the same node.

Closed path: a connection of devices or branches through a sequence of nodes so that the connection ends on the node where it began and traverses each node in the connection only once.

Connected circuit: one for which any node can be reached from any other node by some path through the circuit elements.

Current division: the current in a branch of a parallel-resistive circuit is equal to the input current times the conductance of the particular resistor, G_j, divided by the total parallel conductance of the circuit, $G_{eq} = G_1 + ... + G_n$.

Dependent (controlled) current source: a current source whose output current depends on the voltage or current of some other element in the circuit.

Dependent (controlled) voltage source: a voltage source whose output voltage depends on the voltage or current of some other element in the circuit.

Gaussian surface: a closed curve in the plane or a closed surface in three dimensions. A Gaussian surface has a well-defined inside and outside.

Kirchhoff's current law (KCL): the algebraic sum of the currents entering a node of a circuit consisting of lumped elements is zero for every instant of time. In general, for lumped circuits, the algebraic sum of the currents entering (leaving) a Gaussian surface is zero at every instant of time.

Kirchhoff's voltage law (KVL): for lumped circuits, the algebraic sum of the voltage drops around any closed path in a network is zero at every instant of time. In general, for lumped connected circuits, the algebraic sum of the node-to-node voltages for any closed node sequence is zero for every instant of time.

Node: the common connection point between each element; in general, a node is a connection point of one or more circuit elements.

Node voltage: the voltage drop from a given node to the reference node.

Parallel circuit: a side-by-side connection of two-terminal circuit elements whose top terminals are wired together and whose bottom terminals are wired together.

Series circuit: a sequential connection of two terminal circuit elements, end-to-end.

Voltage division: each resistor voltage in a series connection is a fraction of the input voltage equal to the ratio of the branch resistance to the total series resistance.

// (double-slash): notation for combining resistances in parallel, i.e., $R_1//R_2$ means R_1 and R_2 are in parallel, and $R_1//R_2//R_3$ means R_1 is in parallel with R_2, which is in parallel with R_3.

PROBLEMS

KIRCHHOFF'S CURRENT LAW

1. (a) Find the value of I_1 for each of the node connections in Figure P2.1a and P2.1b given that $I_2 = 2$ A, $I_3 = 3$ A, and $I_4 = 4$ A.
   (b) Repeat part (a) when $I_2 = I_3 = I_4 = 2$ A.

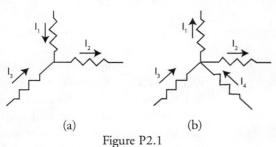

(a)   (b)

Figure P2.1

ANSWERS: (b) 0, 2 A

2. In the circuit of Figure P2.2, each shaded box is a general circuit element.
   (a) Suppose $I_{in1} = 20$ mA, $I_{in2} = 80$ mA, $I_{in3} = 100$ mA, and $I_{in4} = 0.05$ A. Apply KCL to find I_1, I_2, I_3, and I_4.
   (b) Repeat part (a) when $I_{in1} = I_{in2} = I_{in3} = I_{in4} = 100$ mA.

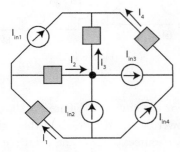

Figure P2.2

ANSWERS: (b) (scrambled) 200, -300, -200, -300 mA

3. (a) For the circuit of Figure P2.3a, find the value of the current I_1 using only a single application of KCL. (Hint: Construct a Gaussian surface.)
   (b) Repeat part (a) for I_2
   (c) True-False: I_3 can be uniquely determined as in part (a) and part (b).

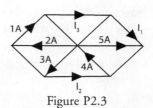

Figure P2.3

ANSWER: (b) 4 A

4. (a) Find the value of I_1 in the circuit of Figure P2.4.
   (b) Find the value of I_2 in the circuit of Figure P2.4 by a single application of KCL.

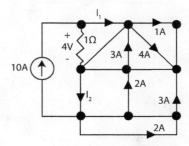

Figure P2.4

ANSWER: (a) 6 A

KIRCHHOFF'S VOLTAGE LAW

5. (a) Consider the circuit of Figure P2.5a where each branch represents a circuit element. Find V_1 and V_2.
   (b) Find V_1 and V_2 for Figure P2.5b. Each unspecified branch represents an unknown circuit element.

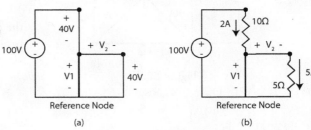

(a)   (b)

Figure P2.5

6. (a) For the circuit of Figure P2.6, determine the voltages v_1, ... , v_4 and the power absorbed by each resistor.

(b)   Now determine the node voltages V_A, V_B, V_C, and V_D with respect to the reference node indicated by the ground symbol.

(c)   Compute V_{CA}.

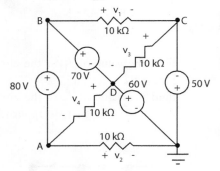

Figure P2.6

ANSWERS: (a) 180 V, 50 V, -110 V, 10 V

7. (a)   Find the values of the voltages V_1, V_2, and V_3 in the circuit graph of Figure P2.7, where each branch represents a circuit element.

(b)   Now determine the node voltages V_A, V_B, V_C, and V_D with respect to the reference node indicated by the ground symbol.

(c)   Compute V_{CA}, V_{BD}, and V_{AD}.

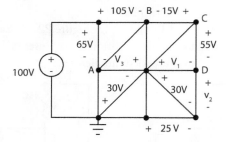

Figure P2.7

SCRAMBLED ANSWERS: (a) -65 V, 15 V, 15 V; (b) 35 V, –45 V, 10 V, –5 V

8. (a)   Use KVL to determine the voltages V_x and V_y in the circuit of Figure P2.8.

(b)   Now compute V_{AB}.

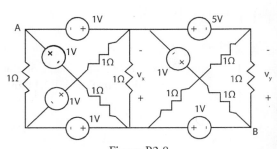

Figure P2.8

SCRAMBLED ANSWERS: (a) 2 V, –2 V

KCL AND KVL

9. (a)   Consider the circuit of Figure P2.9a. Use KCL and KVL to find the voltage across each current source from the arrow head to the arrow tail and the current through each voltage source from minus to plus. Finally, find the power delivered by each source and verify conservation of power.

(b)   For the circuit of Figure P2.9b, find the voltages V_3 and V_4.

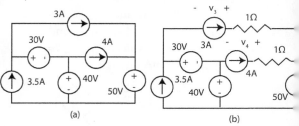

Figure P2.9

ANSWERS: (b) –17 V, 14 V

10.   For the circuit in Figure P2.10, calculate the power delivered by each of the eight independent sources. Verify the principle of conservation of power.

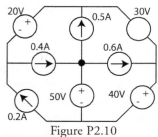

Figure P2.10

ANSWERS: –4, –9, –36, 35, 10, 0, 10, –6 W

11. Four circuit elements and a dependent voltage source are shown in the circuit of Figure P2.11. The current through and the voltage across each element are identified on the diagram. However, one—and only one—voltage (or current) value is labeled incorrectly. Mark the incorrect voltage (or current) on the circuit diagram and give the correct value for this voltage (or current).

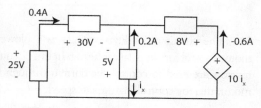

Figure P2.11

12. Find the currents and voltages I_x, V_x, I_y, and V_y in the circuit of Figure P2.12.

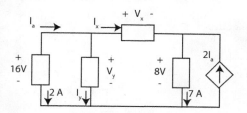

Figure P2.12

KCL, KVL, AND OHM'S LAW

13.(a) For the circuit of figure P2.13, suppose I_{in} = 875 mA and R_L = 80Ω. Find V_{in}, I_1, I_2, and I_3.

(b) Now suppose that I_{in} = 7 A and I_2 = 1 A. Find V_{in}, I_3, R_L, and the power delivered to the load R_L.

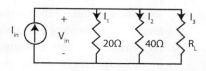

Figure P2.13

CHECK: (b) P_L = 160 watts

14.(a) For the circuit of Figure P2.13, determine I_{in} in terms of I_1, a and R_L.

(b) Supposing that I_1 = 4 A, I_1 = 2 A, and a = 0.25, determine R_L and the power delivered to R_L.

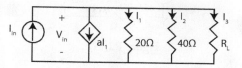

Figure P2.14
ANSWERS: (b) 80 Ω, 20 W

15. Consider the circuit in Figure P2.15 in which I_{in} = 1 A and R_L = 84 Ω. Find the value of R_1 for each of the following cases:

(a) The power delivered by the source is 13.44 watts.

(b) The power absorbed by R_L is 13.44 watts.

(c) The power absorbed by R_1 is 13.44 watts.

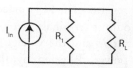

Figure P2.15
SCRAMBLED ANSWERS: 21, 16, 336, 56

16. Consider the circuit of Figure P2.16.
(a) If R = 5 Ω, find V_R.
(b) Find the value of R when V_R = 40 V.

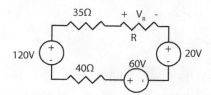

Figure P2.16
ANSWER: (b) 25 Ω

17. For the circuit of Figure 2.17, find
(a) the voltage V_1 and the power absorbed by the 10 Ω resistor;
(b) the voltage V_2;
(c) the power delivered by each source.

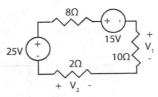

Figure P2.17

ANSWER: (c) 12.5 watts and −7.5 watts

18. Find the power absorbed by the unknown circuit element x and the voltage V_x in the circuit of Figure P2.18.

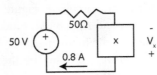

Figure P2.18

19. (a) Find the current I_R and the voltage V_{out} for the circuit of Figure P2.19.
    (b) If a resistor of R Ω is placed across the output terminals, determine the current I_R and the voltage V_{out} and the power delivered by the 10 V source.

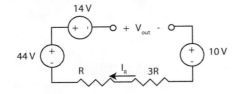

Figure P2.19

CHECK: (b) V_{out} = 4 V

20. (a) In Figure P2.20a, V_2 = 32 V and the power delivered by the source is 80 mW. Compute I_{in}, V_1, V_{in}, and R.
    (b) In Figure P2.20b a dependent voltage source has been added to the circuit of Figure P2.20a. Suppose V_{in} = 40 V. Determine V_1 in terms of a and R. If I_{in} = 0.8 mA and a = 5, find V_1 and R.
    (c) For each circuit of Figure P2.20, determine the resistance seen by the voltage source, V_{in}

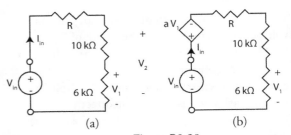

(a)   (b)

Figure P2.20

ANSWERS: (b) $V_1 = \dfrac{6 \times 10^3}{16 \times 10^3 + R + 6a \times 10^3} V_{in}$
4.8 V, 4 kΩ

21. The circuit of in Figure P2.21 is a blower motor control for a typical car heater. In this circuit, resistors are used to control the current through a motor, thereby controlling the fan speed.

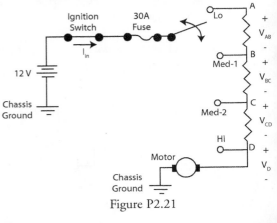

Figure P2.21

(a) With the switch in the Lo position, the current supplied by the battery is 2.5 A. The voltage drops across the resistors and motor are V_{AB} = 6.75 V, V_{BC} = 1.5 V, V_{CD} = 0.625 V, and V_D = 3.125 V. Consider the motor as represented by a load resistance.
    (i) Determine the value of each resistance and the value of the equivalent resistance representing the motor.
    (ii) Determine the power dissipated in each resistor and the power used by the motor.
    (iii) Determine the relative efficiency of the circuit, which is the ratio of the power used by the motor to the power delivered by the battery.

(b) With the switch in Med-1 position, determine:

    (i) The voltage drop across each resistor.

    (ii) The current delivered by the battery.

    (iii) The relative efficiency of the circuit.

(c) Repeat part (b) with the switch in position Med-2.

(d) The switch is in the high position. A winding in the motor shorts out. The fuse blows. What is the largest equivalent resistance of the motor that will cause the fuse to blow?

ANSWERS:

(a) (i) $R_{AB} = 2.7\ \Omega$, $R_{BC} = 0.6\ \Omega$, R_{CD}
    $= 0.25\ \Omega$, $R_{motor} =$
    $1.25\ \Omega$

    (ii) $P_{AB} = 16.875\ W$, $P_{BC} = 3.75\ W$,
    $P_{CD} = 1.5625\ W$, $P_{motor} = 7.8125$
    W

    (iii) 26%

(b) (i) $V_{AB} = 0$, $V_{BC} = 3.43\ V$, $V_{CD} =$
    $1.43\ V$, $V_{motor} = 7.14\ V$

    (ii) 5.71 A

    (iii) 59.5%

(c) (i) $V_{AB} = 0$, $V_{BC} = 0$, $V_{CD} = 2\ V$,
    $V_{motor} = 10\ V$

    (ii) 8 A

    (iii) 83.3%

(d) $0.4\ \Omega$

22. Suppose one has two resistors $R_1 = 20\ \Omega$ and $R_2 = 20\ \Omega$ that can be conected to a source, $V_s = 100\ V$. By connecting the resistors to the source in different ways, what are the different wattages that can be delivered by the source? The different types of connections represent what might occur in an electric space heater having a low, medium, and high setting.

CHECK: There are three possible connections with medium using 500 watts.

23. In Problem 22, find the values of R_1 and R_2 so that the lowest wattage delivered by the 100 V source is 200 W and the highest wattage

delivered is 1250 W. How many possible medium wattages are there and what are they?

CHECK: 10 ohms, 40 ohms

24. Consider the circuit of Figure P2.24.

(a) Suppose $R = 20\ \Omega$, find the power delivered by the current source.

(b) Suppose the power delivered by the current source is 120 watts. Find the value of R.

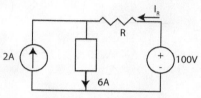

Figure P2.24

CHECK: (b) $8 \le R \le 15$

25. Given that 4 W is absorbed by the 100-Ω resistor, find V_s and the power delivered by the voltage source in the circuit of Figure P2.25.

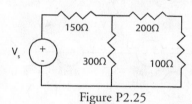

Figure P2.25

26. In the circuit of Figure P2.26, suppose $V_2 = 60\ V$. Find V_s, I_s, and the power delivered by the source.

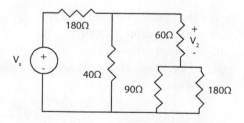

Figure P2.26

SCRAMBLED ANSWERS: 3360, 840, 4

27. Find the power *delivered* by each independent source and the power *absorbed* by each resistor in the circuit of figure P2.27. (Check: Total of delivered power = total of absorbed power.)

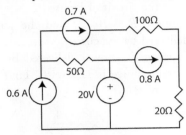

Figure P2.27

SCRAMBLED ANSWERS: 59.5, 9, 8, 49, 0.5, 45, 18

28. For the circuit of Figure P2.28 with the indicated currents and voltages, find
   (a) Currents I_1 through I_4
   (b) Voltages V_1 through V_4
   (c) Power delivered by each independent current source

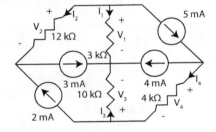

Figure P2.28

CHECK: $V_2 = 12$ V, $I_4 = 1$ mA, $P_{5mA} = 60$mW

29. For the circuit of Figure P2.29, find
   (a) Voltage drop V_1, and
   (b) Voltage drop V_2
   (c) The value of the unspecified resistance, R_x

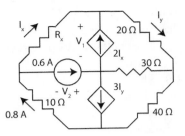

Figure P2.29

CHECK: (a) $45 \le V_1 \le 65$; (b) $-85 \le V_2 \le -65$; (c) $80 \le R_x \le 125$

30. Consider the circuit of Figure P2.30.
   (a) Write an equation for V_s in terms of a and I_x
   (b) If $V_s = 40$ V and $a = 0.5$, find the value of the current I_x
   (c) How much power is delivered by the independent source? How much power is delivered by the dependent source? Verify the principle of conservation of power for this circuit.

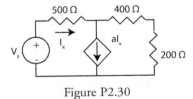

Figure P2.30

CHECK: $I_x = 0.05$ A

VOLTAGE AND CURRENT DIVISION

31. Consider the circuit of Figure P2.31 in which $V_{s1} = 30$ V and $V_{s2} = 20$ V. Find V_1 and V_2 for each of the following cases:
   (a) Switch S_1 is closed and switch S_2 is open.
   (b) Switch S_1 is open and switch S_2 is closed.
   (c) Switch S_1 is closed and switch S_2 is closed.

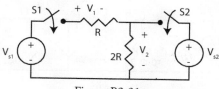

Figure P2.31

32. Construct a series voltage divider circuit whose total resistance is 2400 Ω as illustrated in Figure 2.32.

(a) Suppose $V_1 = 0.75V_s$ and $V_2 = 0.25V_s$. Find the values of R_1, R_2, and R_s.

(b) Suppose $V_1 = 0.8V_s$ and $V_2 = 0.5V_s$. Find the values of R_1, R_2, and R_s.

(c) Suppose $V_1 = 0.8V_s$ and $V_2 = 0.5V_s$. Find the values of R_1, R_2, and R_s.

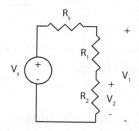

Figure P2.32

ANSWER: (c) $R_1 = R_2 = 960$ Ω, $R_s = 480$ Ω

33. For the circuit of Figure P2.33, suppose $V_s = 48$ V.

(a) Find v_x with the switch in position A, i.e., the switch is open.

(b) Find v_x with the switch in position B, i.e., the switch is closed.

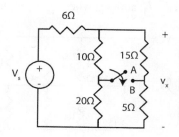

Figure P2.33

CHECK: (b) $25 \le v_x \le 30$ V

34. For the circuit of Figure P2.33, $V_s = 120$ V

(a) Suppose $R_2 = 120$ Ω. Find the value of R_1 that is necessary to achieve $V_1 = 90$ V. Compute V_2.

(b) Find the values of R_1 and R_2 that are necessary to achieve $V_1 = 100$ V and $V_2 = 80$ V.

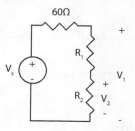

Figure P2.34

CHECK: (b) 60 Ω, 240 Ω

35. Figure P2.35 shows a Wheatstone bridge circuit that is commonly used in a variety of measurement equipment. The bridge circuit is said to be balanced if $R_aR_d = R_bR_c$. In this case, the voltage $V_{out} = 0$ for any voltage V_{in}.

(a) Use voltage division to compute the voltages V_b and V_d. Check:

$$V_d = \frac{R_d}{R_c + R_d}V_{in}.$$

(b) If $V_{out} = 0$, then what must be true about V_b and V_d? Show that $V_{out} = 0$ if and only if $R_aR_d = R_bR_c$.

(c) Suppose that $R_aR_d = R_bR_c$ and a 0.5-Ω resistor is connected across the V_{out} terminals. Then find the current through the 0.5-Ω resistor.

(d) Suppose $V_{in} = 18$ V, $R_a = 3$ Ω, $R_b = 6$ Ω, $R_c = 2$ Ω, and $R_d = 2$ Ω. Find V_{out}.

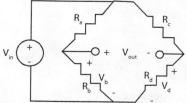

Figure P2.35 Wheatstone bridge circuit.

ANSWER: (d) 3 V

36. Find V_x, V_y, and V_z for the circuit of Figure P2.36 when V_{s1} = 50 V and V_{s2} = 25 V.

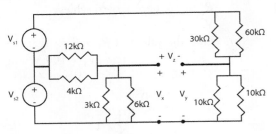

Figure P2.36

SCRAMBLED ANSWERS: −5, 10, 15

37. Consider the circuit of Figure P2.37 in which I_{in} = 0.1 A, G_1 = 2.5 mS,
   (a) With the switch in position A, find G_{eq}, R_{eq}, I_2, V_2, V_{in} and the power delivered by the source.
   (b) Repeat part (a) if the switch is in position B.

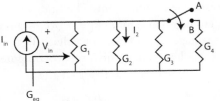

Figure P2.37

SCRAMBLED ANSWERS: (a) 50, 0.5, 5, 50, 20

38. In the circuit of Figure P2.38, it is required that $I_1 = 0.81 I_{in}$. Find R (in Ω), G_{eq}, I_2 in terms of I_{in}, and V_{in} in terms of I_{in}.

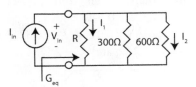

Figure P2.38

CHECK: $V_{in} = 40 I_{in}$ and $I_2 = \dfrac{1}{15} I_{in}$

39. In the circuit of Figure P2.39, I_{in} = 10 mA. Find I_1, I_2, V_d, and the power delivered by the dependent source.

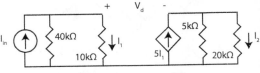

Figure P2.39

CHECK: $P_{dep, \, del}$ = 6.4 watts

40. Find I_1, I_2, I_3, V_{in}, and the power delivered by the source in the circuit of Figure P2.40 when I_{in} = 120 mA.

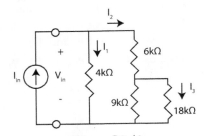

Figure P2.40

CHECK: V_{in} = 360 V

41. Find I_1 and I_2 for the circuit of figure P2.41 when I_{s1} = 10 mA, I_{s2} = 4 mA, and I_{s3} = 14 mA.

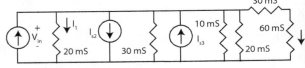

Figure P2.41

42. For the circuit of Figure P2.42, find the currents I_1, I_2, I_3, and V_{in} when I_{in} = 300 mA.

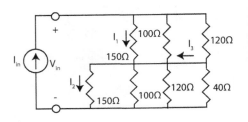

Figure P2.42

SCRAMBLED ANSWERS: −50, 40, 80 (in mA)

R_{EQ} AND RELATED CALCULATIONS OF SERIES-PARALLEL CIRCUITS

43. For each of the circuits of Figure P2.43, find the value of R_{eq} and the power delivered if a 10-V source were connected.

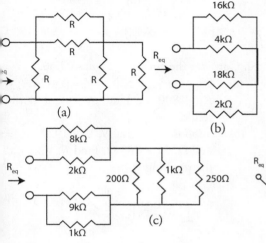

(a)

(b)

(c)

Figure P2.43

ANSWERS: 0.5R, 5 kΩ, 2.6 kΩ

44. Find the value of R_{in} for each of the circuits of Figure P2.44.

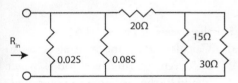

Figure 2.44 (a)

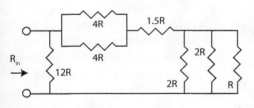

Figure 2.44 (b)

45. For each of the circuits in Figure P2.45, compute the equivalent resistance R_{eq} seen by the source, the input current I_{in}, the power delivered by each source, and V_{out}, when V_{in} = 80 V.

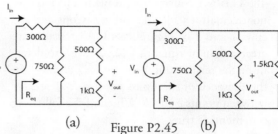

(a)   Figure P2.45   (b)

SCRAMBLED ANSWERS: (b) 29.63, 675, 0.1185

46. Find R_{eq} for each of the circuits in Figure P2.46. Notice that the circuit of (b) is a modification of (a) and that of (c) a modification of (b).

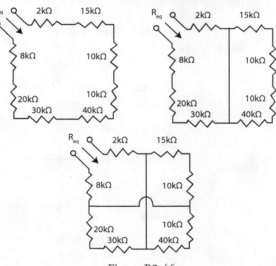

Figure P2.46

SCRAMBLED ANSWERS: 60 kΩ, 22.5 kΩ, 135 kΩ

47. Find R_{eq} in the circuit of Figure P2.47
(a) When the switch is open
(b) When the switch is closed

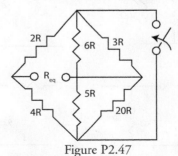

Figure P2.47

CHECK: Answers are the same.

48. This is a conceptual problem and requires no calculations for the answer. Consider circuits 1 and 2 of Figure P2.48. All resistors are greater than or equal to 1 Ω. We wish to determine the relationship between R_{eq1} and R_{eq2} in the presence of the finite positive R-Ω resistor between points a and b. Which of the following statements is true?

(a) $R_{eq1} > R_{eq2}$
(b) $R_{eq1} < R_{eq2}$
(c) $R_{eq1} = R_{eq2}$
(d) There is no general relationship between R_{eq1} and R_{eq2}. Any relationship depends on the value of R.

Explain your reasoning.

Circuit 1

Circuit 2

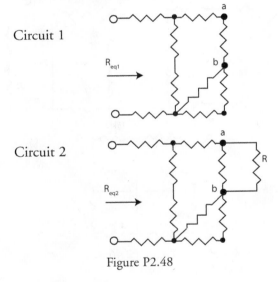

Figure P2.48

♦49. For each of the circuits of Figure P2.49, find the value of R that makes $R_{eq} = 1000\ \Omega$.

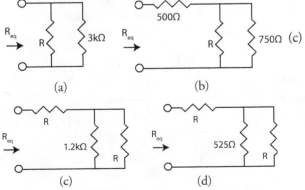

(a)

(b)

(c)

(d)

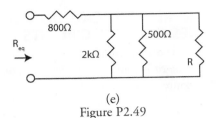

(e)

Figure P2.49

SCRAMBLED ANSWERS: 400, 700, 500, 1500

50. Consider the circuit of Figure P2.50.
(a) Suppose $V_{in} = 320$ V, $V_r = 256$ V, $R_1 = R_2 = 800\ \Omega$. Find R, V_{R2}, and the resulting R_{eq}.
(b) Suppose $V_{in} = 320$ V, $V_r = 192$ V, $R_1 = 400\ \Omega$, $R_2 = 800\ \Omega$. Find R, V_{R2}, and the resulting R_{eq}.

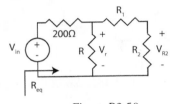

Figure P2.50

SCRAMBLED ANSWERS: 500, 1000, 1600, 400, 170.67, 128

51. For the circuit of Figure P2.51:
(a) Calculate R_{AC}, the equivalent resistance seen at terminals A and C, which would be the reading on an ohmmeter if the two probes were connected to A and C, respectively.
(b) Calculate R_{BC}, the equivalent resistance seen at terminals B and C, which would be the reading on an ohmmeter if the two probes were connected to B and C, respectively.
(c) Can the equivalent resistance R_{AB} be calculated using the series-parallel formulas? State your reasons without performing any calculations.

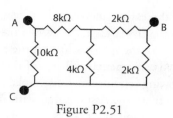

Figure P2.51

52. Some physical problems have models that are infinite ladders of resistors, as illustrated in Figure P2.52.

(a) Find the equivalent resistance R_{eq} at the terminals *a-b* in figure P2.52a. (Hint: Since the resistive network is infinite, the equivalent resistance seen at terminals *a-b* is the same as the equivalent resistance to the right of terminals *c-d*; this means that the network to the right of c-d can be replaced by what???) Evaluate if $R_1 = 1\ \Omega$ and $R_2 = 100\ \Omega$. This type of problem is useful for representing series and shunt conductance in transmission lines.

(b) Find R_{eq} at terminals *a-b* for the ladder network of Figure P2.52b.

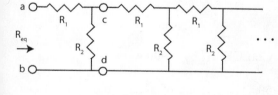

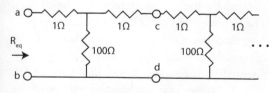

Figure P2.52

NUMERICAL ANSWERS: 10.512, 14.177

53. Consider the circuit of Figure 2.53a in which $R_1 = 0.5\ \Omega$. Suppose each AA-bulb represents a 12-watt fluorescent bulb at approximately 12 volts, having an internal resistance of 12 Ω.

(a) For Figure P2.53a, how many bulbs can be put in parallel before the 15 A fuse blows? Given the maximum number that can be put in parallel without blowing the fuse, find R_{eq} and V_A.

(b) In Figure 2.53b, bulbs BB and CC are 24 watt and 36 watt, respectively, at approximately 12 volts. Find the internal resistances of each bulb. How many CC bulbs can be present before the 15 amp fuse blows? Given this number of CC bulbs, find R_{eq} and V_A.

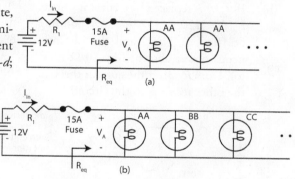

Figure P2.53

CHECK: (a) n = 16; (b) n = 4

54. In the circuit of Figure P2.54, $V_{in} = 330$ V and $R_s = 40\ \Omega$. The switch S_1 closes at $t = 5$ s, S_2 closes at $t = 10$ s, S_3 closes at $t = 15$ sec, and S_4 closes at $t = 20$ sec. Plot $v_{out}(t)$ and calculate $R_{eq}(t)$ for $0 \le t \le 25$ s.

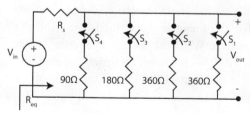

Figure P2.54

55. Consider the circuit of Figure P2.55.

(a) Find $i_1(t)$, max[$i_1(t)$], and the average value of $i_1(t)$.

(b) Find $i_2(t)$, max[$i_2(t)$], and the average value of $i_2(t)$.

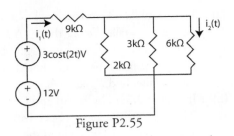

Figure P2.55

56. Consider the circuits of Figure P2.56. In Figure 2.56, $v_{in}(t) = 120 \sin(377t)$ V and $R_s = 5\ \Omega$.
    (a)   Find v_{out}, i_{out}, and the instantaneous power absorbed by 30 Ω resistor.
    (b)   If $v_{in}(t)$ is replaced by a current source, $i_{in}(t) = 120 \sin(377t)$ mA, pointing up, find v_{out}, i_{out}, and the instantaneous power absorbed by 30 Ω resistor. Does R_s affect the current through the other resistors in the circuit?

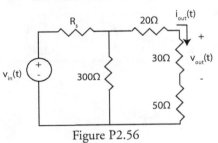

Figure P2.56

CHECK: $P_{30\Omega} = 37.97\sin^2(377t)$ watts

57. The circuit of Figure P2.57 shows a simple scheme to determine R_0, the internal resistance of the battery model. The loading effect due to the digital voltmeter may be neglected (consider that the meter is represented by an infinite resistance). With the switch open, the meter reads 12 V. With the switch (briefly) closed, the reading drops to 11.96 V. Find the value of R_0.

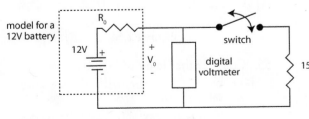

Figure P2.57

ANSWER: 0.0376 Ω

58. With the car engine turned off, you have been listening to the car radio. While the radio is on, you turn the ignition to start the engine. You noticed a momentary silence of the radio. The following circuit analysis explains this effect quantitatively. Assume that with the car engine not running, the 12-V car battery is represented by the model shown in Figure P2.58. The load due to the car radio is represented by an equivalent resistance of 240 Ω. The starter motor draws 150 A of current when the ignition is turned on and before the engine starts. Find V_{out} at the moment when the ignition switch is turned on. Compare this to the voltage V_{out} before the ignition switch is turned on. Why do you think the radio goes silent momentarily?

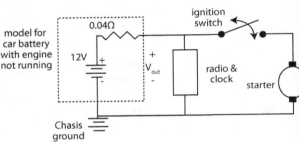

Figure P2.58

CHECK: $P_{starter} = 900$ watts

59. The volume of a car radio is not much affected by the on/off state of the headlights. The following circuit analysis explains this phenomenon quantitatively. Assume that with the car engine running, the 12-V car battery is represented by the model shown in Figure P2.59.
    Notice that the effective voltage of the car battery increases due to the effect of the alternator while the engine is running. The load due to the car radio is represented by an equivalent resistance of 240 Ω. At 12 V dc, each headlight consumes 35 W on low beam and 65 W on high beam.
    (a)   Find the equivalent resistance of each headlight on low beam.

(b)   Find the equivalent resistance of each headlight on high beam.

(c)   Find V_{out} when the headlights are turned off.

(d)   Find V_{out} when the low beams are turned on.

(e)   Find V_{out} when the high beams are turned on.

(f)   How much power does each high beam consume given your answer to part (e)? Why is this value different from 65 watts?

(g)   How much power must the battery deliver to overcome its internal losses and operate the high beams and radio.

CHECK: (a) 4.11 Ω; (f) 195.58 watts

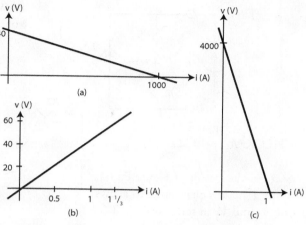

Figure P2.61

DEPENDENT SOURCE PROBLEMS

62. In the circuit of Figure P2.62, determine g_m so that the power delivered to the 5-kΩ load resistor is $100P_{in}$, where P_{in} is the instantaneous power consumed by the 8-kΩ resistor. Equivalently, P_{in} is the power delivered by the non-ideal voltage source.

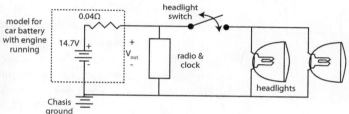

Figure P2.59

60. A 50-cell lead acid storage battery has an open-circuit voltage of 102 V and a total internal resistance of 0.2 Ω.

(a)   If the battery delivers 40 A to a load resistor, what is the terminal voltage?

(b)   What is the terminal voltage when the battery is being charged at a 50 A rate?

(c)   What is the power delivered by the charger in part (b)? How much of the power is lost in the battery as heat?

SCRAMBLED ANSWERS: 500, 112, 5600, 94

61. A non-ideal constant voltage source, an ordinary resistor, and a non-ideal constant current source have the v-i characteristics given in Figure P2.61. Determine the values of the source voltage or current, the value of the source internal resistance, and, finally, the value of the resistance for Figure P2.61.

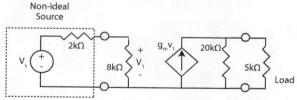

Figure P2.62

ANSWER: 6.25 mS

63. Find the equivalent conductance G_{eq} and then the equivalent resistance R_{eq} "seen" by the current source I_s in the circuit of Figure P2.63 in terms of the literals R_1, R_2, and g_m. Evaluate when $R_1 = 1$ kΩ, $R_2 = 3$ kΩ, $g_m = 0.2$ mS.

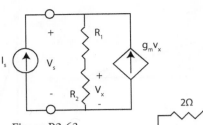

Figure P2.63

CHECK: R_{eq} = 10 kΩ

64. For the circuit of Figure P2.64, write a node equation that allows you to find V_1 in terms of V_{in}. Then find

$$G_{eq} = \frac{I_{in}}{V_{in}} \cdot$$

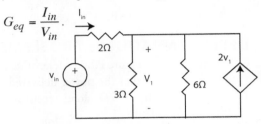

Figure P2.64

CHECK: 0.75 S.

65. In the circuit of Figure P2.65 r_m = 12.5 kΩ and g_m = 12.5 mS:
   (a) Compute the output voltage and output current in terms of V_{in}.
   (b) Compute the voltage gain, $G_V = V_{out}/V_{in}$.

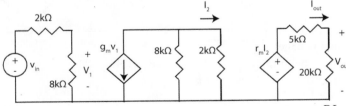

Figure P2.65

MATLAB PROBLEMS

►66. (a) Find the output voltage, V_{load}, the output current (what is its direction), and the power absorbed by the load (8-Ω resistor) for the circuit of Figure P2.66.

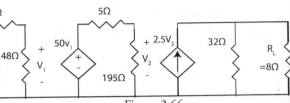

Figure 2.66

(b) Using MATLAB or equivalent, compute and PLOT with appropriate labels the power absorbed by the load, denoted by R_L, as R_L varies from 8 to 64 Ω in increments of 1 Ω. Also plot the current, again using MATLAB, as a function of R_L. At what value of R_L is the absorbed power a maximum? Knowing this is important, for example, when matching loudspeaker resistances to the output resistance of your stereo. For this problem, you should use MATLAB. You will need to turn in an original printout (no copies permitted) of your code and plots.

Hint: Begin your program with the commands listed below. ?? indicates that you should insert the proper number or formula.

RL = 8:1:64;
% *This command generates an array of numbers for RL beginning at 8 and ending at 64 in increments of 1. If you do not end it with a semicolon, it will list every entry of the array.*
V2 = ??
% This value should be precomputed
IL = ??;
PL = RL .* IL .^ 2;

% *Note that because IL and RL are arrays of numbers .^ means to square each number in the array IL and .* means to multiply each number in IL by the corresponding number in the array for RL.*
plot(RL, PL)
grid
% *Plot IL in mA*
plot(RL, IL*1000)
grid
% *typing grid adds a grid to your plot. Always add a grid.*
% *You can put both plots on the same graph as follows:*
plot(RL,IL*1000,RL,PL)
% *The motivated student might investigate using the "hold" command instead.*

Beginning your MATLAB solution:
% *Define element values*
R1= 15; R2= 4; R3= 9; R4= 2; R5=8; R6=18;
% *To find Req start from right side.*
Ra= R4 + R5;
Ga= 1/Ra;
Gb = Ga + 1/R1;
Rb = 1/Gb;
% *Continue these additions and reciprocals until obtaining Req.*

% *To find Vout requires repeated use of voltage and current division formulas.*
Geq = 1/Req;
IRc = 20*Gc/Geq;
VRb = IRc*Rb;
% *Now write down the MATLAB expression for finding Vout.*
ANSWERS: (a) 3 Ω; (b) 24 V

▶67. The analysis of series-parallel circuits with numerical element values can be done with only two types arithmetic operations: adding two numbers and taking the reciprocal of a number. As such, MATLAB is an extremely convenient tool for finding the equivalent resistances and the voltages and currents throughout a series-parallel circuit. This problem illustrates such a use of MATLAB. For the circuit of Figure P2.67

   (a)   Find R_{eq}.
   (b)   Find V_{out}.

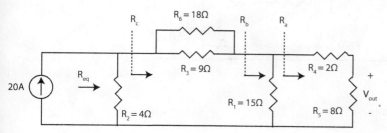

Figure P2.67

68. Use MATLAB to find R_{in} and V_{out} for the circuit of Figure P2.68. Turn in your MATLAB code with your answers. Hint: Label the equivalent seen at each node to facilitate computation of V_{out}.

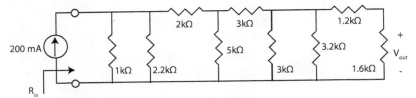

Figure P2.68

ANSWERS: 591.2 Ω, 8.869 V

▶69. Use MATLAB to find R_{in}, V_{out}, and I_1 for the circuit of Figure P2.69.

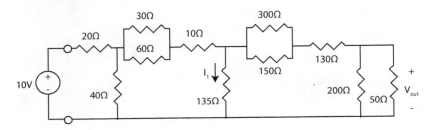

Figure P2.69

▶ 70. Use MATLAB to find R_{in} in the circuit of Figure 2P.70.

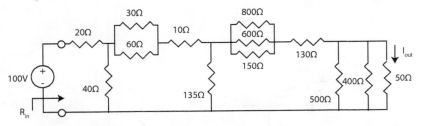

Figure P2.70

ANSWER: 50.53 Ω, 133.8 mA

CHAPTER 3

Nodal and Loop Analyses

HISTORICAL NOTE

For a network consisting of resistors and independent voltage sources, one can apply KCL to the nodes, KVL to the various loops, and Ohm's law to the elements to construct a large set of simultaneous equations whose solution yields all currents and voltages in the circuit. In theory, this approach completely solves the basic analysis problem. In practice, this approach proves impractical because large numbers of equations are required even for a small network. For example, a 6-branch, 4-node network, with each node connected to the other nodes through a single element, leads to a set of 12 equations in 12 unknowns: 3 equations from KCL, 3 equations from KVL, and 6 equations from the element v–i relationships. The 12 unknowns are the 6 branch currents and 6 branch voltages.

Before the advent of digital computers, engineers solved simultaneous equations manually, possibly with the aid of a slide rule, or some primitive mechanical calculating machines. Any technique or trick that reduced the number of equations was highly treasured. In such an environment, Maxwell's mesh analysis technique (1881) received much acclaim and credit. Through the use of a fictitious circulating current, called a mesh current, Maxwell was able to greatly reduce the number of equations. For the above-mentioned network, the number of equations drops from 12 to 3 equations in the unknown mesh currents.

An alternate KCL-based technique (now called nodal analysis) appeared in literature as early as 1901. The method did not gain momentum until the late 1940s, because most problems in the early days of electrical engineering could be solved efficiently using mesh equations in conjunction with some network theorems. With the invention of multi-element vacuum tubes having interelectrode capacitances, some compelling reasons to use the node method appeared: primarily, the node method accounts for the presence of capacitances without introducing more equations, and secondly, those vacuum tubes that behave very much like current sources are more easily accommodated with nodal equations. By the late 1950s, almost all circuit texts presented both the mesh and node methods.

Since the 1960s, many digital computer software programs (SPICE being the most ubiquitous) have been developed for the simulation of electronic circuits that otherwise would defy hand calculation. These software packages use a node equation method over the mesh equation approach. One of several reasons is that a node is easily identifiable, whereas a set of proper meshes is difficult for a computer to recognize.

For resistive networks driven by current sources, writing node equations is straightforward. Certain difficulties arise in writing node equations for circuits containing independent and dependent voltage sources. During the 1970s, a modification of the conventional node method by a research group at IBM resulted in the "modified nodal analysis" (MNA) technique. With the MNA method, the formulation of network equations, even in the presence of voltage sources and all types of dependent sources, becomes very systematic.

This chapter discusses the writing and solution of equations to find pertinent voltages and currents for linear resistive networks.

CHAPTER OUTLINE

1. Introduction, Review, and Terminology
2. The Concepts of Nodal and Loop Analysis
3. Nodal Analysis I: Grounded Voltage Sources
4. Nodal Analysis II: Floating Voltage Sources
5. Loop Analysis
6. Summary
7. Terms and Concepts
8. Problems

CHAPTER OBJECTIVES

1. Describe and illustrate the method of node analysis for the computation of node voltages in a circuit. Knowledge of the node voltages of a circuit allows one to compute all the branch voltages and, thus, with knowledge of the element values, all the branch currents.
2. Define the notion of a mesh or loop current and describe and illustrate the method of mesh or, more generally, loop analysis for the computation of loop currents in a circuit. Knowledge of all the loop currents of a circuit allows one to compute all the branch currents. Thus, in conjunction with the knowledge of the branch element information, one can compute all the branch voltages.
3. Formulate the node analysis and loop analysis equations as matrix equations and use matrix methods in their solution emphasizing the use of existing software for the general solution.
4. Describe and illustrate the modified nodal approach to circuit analysis. This method underlies the general software algorithms available for computer simulation of circuits.

1. INTRODUCTION, REVIEW, AND TERMINOLOGY

Chapter 1 introduced basic circuit elements, Ohm's law, and power calculations. Chapter 2 introduced the important laws of circuit theory, KVL and KCL, and investigated series, parallel, and series-parallel circuits. Recall from Chapter 2 that a **node voltage** is the voltage drop from a given node to a reference node. As a brief review, consider Figure 3.1, which portrays a circuit labeled with nodes A through D having associated node voltages, V_A, V_B, V_C, V_D, and eight branches, one for the current source and one for each of the seven conductances, G_1, ... , G_7. (Since this chapter deals almost exclusively with dc, the uppercase notation for voltages and currents is commonplace.)

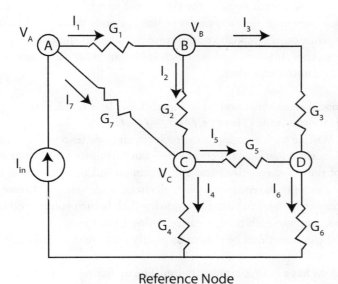

Reference Node

FIGURE 3.1. Diagram of a circuit with labeled node voltages, V_A, V_B, V_C, V_D, with respect to the given reference node.

KVL states that every branch voltage is the difference of the node voltages present at the terminals of the branch: for circuits in this text and all pairs of nodes, j and k, the voltage drop V_{jk}, from node j to node k, is

$$V_{jk} = V_j - V_k$$

at every instant of time, where V_j is the voltage at node j with respect to the reference and V_k is the voltage at node k with respect to reference. Here, j and k stand for arbitrary indices and could be any of the nodes, A, B, C, or D, in Figure 3.1. These statements mean that knowledge of all node voltages in conjunction with device information paints a rather complete picture of the circuit's behavior. This chapter develops techniques for a systematic construction of equations that characterize a circuit's behavior.

One last introductory point: Throughout this chapter and in many subsequent chapters, software programs such as MATLAB facilitate calculations. Constructing sets of equations that characterize the voltages and currents in a circuit is often a challenge. Solving such sets of equations without the use of software tools presents a much greater challenge. Yet facilitated by MATLAB or equivalent, the calculations reduce to a hit of the return key. MATLAB and the circuit simulation program called PSpice or Spice (utilized in Chapter 4) are but two of the many modern and important software tools available to engineers.

2. THE CONCEPTS OF NODAL AND LOOP ANALYSIS

Nodal analysis is an organized means for computing ALL node voltages of a circuit. Nodal analysis builds around KCL, i.e., at each node of the circuit, the sum of the currents leaving (entering) the node is zero. Each current in the sum enters or leaves a node through a branch. Each branch current generally depends on the branch conductance, a subset of the circuit node voltages, and possibly source values. After substituting this branch information for each current in a node's KCL equation, one obtains a **nodal equation.**

As an example, the nodal equation at node A in Figure 3.1 is $I_{in} = I_1 + I_7 = G_1 (V_A - V_B) + G_7 (V_A - V_C)$. The nodal equation at node C is $-I_2 + I_4 + I_5 - I_7 = G_2 (V_C - V_B) + G_4 V_C + G_5 (V_C - V_D) + G_7 (V_C - V_A) = 0$. Writing such an equation at each circuit node (except the reference node) produces a set of independent equations. Of course, one can substitute a KCL equation at the reference node for any of the other equations and still obtain an independent set of nodal equations. The solution of such a set of nodal equations yields all circuit node voltages. Knowing all node voltages permits us to compute all branch voltages. Knowing each branch voltage and each branch conductance allows us to compute each branch current using Ohm's law. The reference node may be chosen arbitrarily and can sometimes be chosen to greatly simplify the analysis.

A set of nodal equations has a matrix representation. The matrix representation permits easy solution for the node voltages using MATLAB or an equivalent software package. A variation of the nodal analysis method, termed **modified nodal analysis**, relies heavily on matrix methods for constructing and solving the circuit equations. The basic principles of this widespread analysis technique are illustrated in Section 4.

Because computer-based circuit analysis packages build on a matrix formulation of the circuit equations and because of the widespread use of matrices in circuits, systems, and control, we will stress a matrix formulation of equations throughout this chapter. The student unfamiliar with matrix methods might look through a calculus text or a linear algebra text for a good explanation of their basic properties and uses.

The counterpart to nodal analysis is **loop analysis**. In loop analysis, the counterpart of a node voltage is a **loop current**, which circulates around a closed path in a circuit. A **loop** or **closed path** in a circuit is a contiguous sequence of branches that begins and ends on the same node and touches no other node more than once. For each loop in the circuit, one defines a **loop current**, as illustrated in Figure 3.2, that depicts three loops or closed paths having corresponding loop currents I_1, I_2, and I_3. Of course, one can draw other closed paths or loops for this circuit and define other loop currents.

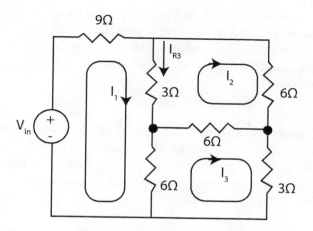

FIGURE 3.2. Simple resistive circuit showing three closed paths (dotted lines) that represent three loop currents, I_1, I_2, and I_3; the branch current $I_{R3} = I_1 - I_2$, which is a difference of the two loop currents through the resistor.

Using a fluid flow analogy, one can think of loop currents as fluid circulating through closed sections of pipe. The fluid in different closed paths may share a segment of pipe. This segment is analogous to a branch of a circuit on which two or more loop currents are incident. The net current in the branch is analogous to the net fluid flow. Note that each branch current can be expressed as a sum of loop currents with due regard to direction. For example, in Figure 3.2, the branch current $I_{R3} = I_1 - I_2$. Using loop currents, element resistance values, and source values, it is possible by KVL and Ohm's law to express the sum of the voltages around each loop in terms of the loop currents. For example, the first loop, labeled I_1 in Figure 3.2, has the loop equation

$$V_{in} = 9I_1 + 3(I_1 - I_2) + 6(I_1 - I_3)$$

We will explore this concept more thoroughly in Section 5. Here we see that loop analysis builds on KVL, whereas node analysis builds on KCL.

3. NODAL ANALYSIS I: GROUNDED VOLTAGE SOURCES

As mentioned earlier, nodal analysis is a technique for finding all node voltages in a circuit. With knowledge of all the node voltages and all the element values, one can compute all branch voltages and currents, and thus the power absorbed or delivered by each branch. This section describes nodal analysis for circuits containing dependent and independent current sources, resistances, and independent voltage sources that are grounded to the reference node (see Figure 3.3). Floating independent or dependent voltage sources (those not directly connected to the reference node) are covered in Section 4.

For the class of circuits discussed in this section, it is possible to write a nodal (KCL) equation at each node not connected to a voltage source. A node connected to a voltage source grounded to the reference node has a node voltage equal to the source voltage. The other node voltages must be computed from the set of nodal equations. Each nodal equation will sum the currents leaving a node. Each current in the sum will be expressed in terms of dependent or independent current sources or branch conductances and node voltages. The set of these equations will have a solution that yields all the pertinent node voltages of the circuit. Examples 3.1 and 3.2 illustrate the basic techniques of nodal analysis.

EXAMPLE 3.1.

The circuit of Figure 3.3a contains an independent voltage source, an independent current source, and five resistances whose conductances in S are G_1 through G_5. The nodes other than the reference are labeled with the node voltages V_a, V_b, and V_c, which respectively denote nodes a, b, and c. The analysis of this circuit illustrates the process of nodal analysis to find the node voltages V_a, V_b, and $V_x = V_a - V_b$.

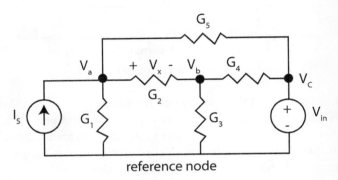

FIGURE 3.3A. Resistive circuit for Example 3.1. Note that node voltage V_c is specified by the voltage source.

SOLUTION.

Step 1. *Consider node c.* A voltage source ties node c to the reference node. Hence, the node voltage V_c is fixed at V_{in}, i.e., $V_c = V_{in}$. Because $V_c = V_{in}$, it is not necessary to apply KCL to this node unless the current through the voltage source is required, for example, when determining the power delivered by the source.

Step 2. *Sum the currents leaving node a.* From KCL, the sum of the currents leaving node a is zero. As per the partial circuit in Figure 3.3b, this requires that

$$G_1 V_a + G_2(V_a - V_b) + G_5(V_a - V_{in}) - I_s = 0$$

Grouping the coefficients of V_a and V_b and moving the source values to the right side of the equation yields our first nodal equation

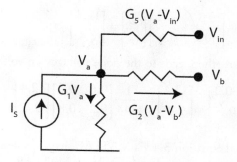

FIGURE 3.3B

$$(G_1 + G_2 + G_5) \ V_a - G_2 V_b = I_s + G_5 V_{in} \tag{3.1}$$

Step 3. *Sum the currents leaving node b.* Applying KCL to node b, reproduced in **Figure 3.3c**, yields the equation

$$G_2(V_b - V_a) + G_3 V_b + G_4(V_b - V_{in}) = 0$$

After regrouping terms, one obtains our second nodal equation:

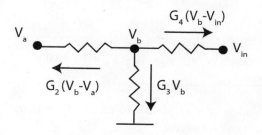

FIGURE 3.3C

$$-G_2 V_a + (G_2 + G_3 + G_4) V_b = G_4 \ V_{in} \tag{3.2}$$

Step 4. *Write set of nodal equations in matrix form.* Equations 3.1 and 3.2 in matrix form are

$$\begin{bmatrix} G_1 + G_2 + G_5 & -G_2 \\ -G_2 & G_2 + G_3 + G_4 \end{bmatrix}\begin{bmatrix} V_a \\ V_b \end{bmatrix} = \begin{bmatrix} I_s + G_5 V_{in} \\ G_4 V_{in} \end{bmatrix} \tag{3.3}$$

Matrix equations organize relevant data into a unified framework. Because many calculators do matrix arithmetic, because of the widespread availability of matrix software packages such as MATLAB, and because equation solution techniques in circuits, systems, and control heavily utilize matrix methods, the matrix equation formulation has widespread and critical importance.

Step 5. *Solve the matrix equation 3.3:* For this part, suppose that the conductance values in S are $G_1 = 0.2$, $G_2 = 0.2$, $G_3 = 0.3$, $G_4 = 0.1$, $G_5 = 0.4$, that $I_s = 2.8$ A, and that $V_{in} = 24$ V. After substitution, equation 3.3 simplifies to

$$\begin{bmatrix} 0.8 & -0.2 \\ -0.2 & 0.6 \end{bmatrix}\begin{bmatrix} V_a \\ V_b \end{bmatrix} = \begin{bmatrix} 12.4 \\ 2.4 \end{bmatrix} \tag{3.4}$$

Solving using the inverse matrix method leads to the node voltages (in volts):

$$\begin{bmatrix} V_a \\ V_b \end{bmatrix} = \begin{bmatrix} 0.8 & -0.2 \\ -0.2 & 0.6 \end{bmatrix}^{-1}\begin{bmatrix} 12.4 \\ 2.4 \end{bmatrix} = \frac{1}{0.44}\begin{bmatrix} 0.6 & 0.2 \\ 0.2 & 0.8 \end{bmatrix}\begin{bmatrix} 12.4 \\ 2.4 \end{bmatrix} = \begin{bmatrix} 18 \\ 10 \end{bmatrix} \text{ V}$$

Alternately, one could have solved equation 3.4 via MATLAB, its equivalent, or the age-old hand method of adding and subtracting equations. For example, in MATLAB

»M=[0.8 -0.2;-0.2 0.6];
»b = [12.4 2.4]';
»NodeV = M\b
NodeV =
 1.8000e+01
 1.0000e+01
»% OR EQUIVALENTLY
»NodeV = inv(M)*b
NodeV =
 18
 10

Step 6. *Compute V_x.* The branch voltage $V_x = V_a - V_b = 18 - 10 = 8$ V.

Exercises. 1. Utilize the solution of Example 3.1 to compute the current leaving and the power delivered by the independent voltage source.
ANSWER: 3.8 A and 91.2 watts

2. Referring to Figure 3.3a and the values set forth in Step 5 of Example 3.1, suppose the value of I_s is cut in half, the value of V_{in} is 24 V, and the value of each of the conductances is also cut in half. What are the new values of the node voltages?
ANSWER: All node voltages are the same.

3. By what single factor must the values of I_s and V_{in} be multiplied so that the node voltages are doubled?
ANSWER: 2

4. Construct a node equation for V_a in Figure 3.4.

ANSWER: $(G_1 + G_2)V_a = I_s - G_2V_{in}$

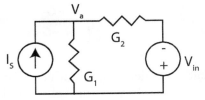

FIGURE 3.4.

EXAMPLE 3.2.

Consider the circuit of Figure 3.5a. Similar to Example 3.1, the objective is to find the node voltages V_a, V_b, and V_c. However, in the circuit of Figure 3.5a, an independent current source has replaced the independent voltage source of Figure 3.3a. This change unfreezes the constraint on the value of V_c present in the circuit Figure 3.3a. There will result three nodal equations in the three unknowns V_a, V_b, and V_c.

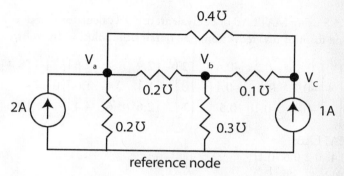

FIGURE 3.5A. Circuit containing two independent current sources and three unknown node voltages V_a, V_b, and V_c.

SOLUTION.

Step 1. Sum currents leaving node a. This step is the same as Step 2 of Example 3.1. By inspection of node a,

$$0.2V_a + 0.2(V_a - V_b) + 0.4(V_a - V_c) - 2 = 0$$

which upon regrouping terms yields

$$0.8V_a - 0.2V_b - 0.4V_c = 2 \tag{3.5}$$

Step 2. Sum currents leaving node b. This step is the same as Step 3 of Example 3.1. Again, by inspection,

$$0.2 (V_b - V_a) + 0.3 V_b + 0.1 (V_b - V_c) = 0$$

Simplification yields

$$-0.2V_a + 0.6V_b - 0.1V_c = 0 \tag{3.6}$$

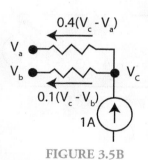

FIGURE 3.5B

Step 3. Sum currents leaving node c. Because a current source drives node c, the similarity to example 3.1 ends, and we must write a third node equation. Summing the currents leaving node c, as shown in Figure 3.5b, yields

$$0.4 (V_c - V_a) + 0.1 (V_c - V_b) - 1 = 0$$

Upon simplification, we have

$$-0.4V_a - 0.1V_b + 0.5V_c = 1 \tag{3.7}$$

Step 4. *Write equations 3.5–3.7 as a matrix equation and solve.* The matrix form of our nodal equations 3.5–3.7 is

$$\begin{bmatrix} 0.8 & -0.2 & -0.4 \\ -0.2 & 0.6 & -0.1 \\ -0.4 & -0.1 & 0.5 \end{bmatrix} \begin{bmatrix} V_a \\ V_b \\ V_c \end{bmatrix} = \begin{bmatrix} 2 \\ 0 \\ 1 \end{bmatrix} \tag{3.8a}$$

Solving equation 3.8 using MATLAB or equivalent, using a calculator that does **matrix operations,** or solving via some form of row reduction, one obtains the solution (in volts)

$$\begin{bmatrix} V_a \\ V_b \\ V_c \end{bmatrix} = \begin{bmatrix} 0.8 & -0.2 & -0.4 \\ -0.2 & 0.6 & -0.1 \\ -0.4 & -0.1 & 0.5 \end{bmatrix}^{-1} \begin{bmatrix} 2 \\ 0 \\ 1 \end{bmatrix} = \begin{bmatrix} 2.9 & 1.4 & 2.6 \\ 1.4 & 2.4 & 1.6 \\ 2.6 & 1.6 & 4.4 \end{bmatrix} \begin{bmatrix} 2 \\ 0 \\ 1 \end{bmatrix} = \begin{bmatrix} 8.4 \\ 4.4 \\ 9.6 \end{bmatrix} \text{ V} \tag{3.8b}$$

Specifically, in MATLAB
»M =[0.8 -0.2 -0.4;-0.2 0.6 -0.1;
-0.4 -0.1 0.5];
»b = [2 0 1]';
»NodeV = M\b
NodeV =
  8.4000e+00
  4.4000e+00
  9.6000e+00

Exercises. 1. Suppose the values of the current sources in Figure 3.5a are doubled. What are the new values of the node voltages? Hint: Consider the effect on equation 3.8.
ANSWER: All node voltages are doubled.

2. Suppose the conductances in the circuit of Figure 3.5a are cut in half, i.e., the resistances are doubled. What are the new node voltages?
ANSWER: Node voltages are doubled.

3. Suppose the conductances in the circuit of Figure 3.5a are cut in half. What happens to the magnitudes of the branch currents? Hint: Express the branch current in terms of the branch conductance and its terminal node voltages.
ANSWER: The magnitudes of the branch currents stay the same.

4. Find two node equations characterizing the circuit of Figure 3.6.

ANSWER: $(G_1 + G_2) - V_a - G_2 V_b = I_{s1}$ and
$- G_2 V_a + (G_2 + G_3) V_b = -I_{s2}$

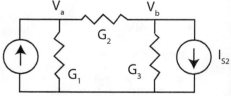

FIGURE 3.6

The matrices in equations 3.3, 3.4, and 3.8a are symmetric. A symmetric matrix, say A, is one whose transpose equals itself, i.e., $A^T = A$; this means that if $A = [a_{ij}]$ is an n × n matrix whose i–j entry is a_{ij}, then A is symmetric if $a_{ij} = a_{ji}$. In words, the off–diagonal entries are mirror images of each other. For example,

$$A = \begin{bmatrix} 0.8 & -0.2 & -0.4 \\ -0.2 & 0.6 & -0.1 \\ -0.4 & -0.1 & 0.5 \end{bmatrix} = A^T$$

When there are no dependent sources present in the circuit, as in Examples 3.1 and 3.2, the coefficient matrix of the node equations (as exemplified in equations 3.3, 3.4, and 3.8) is always symmetric, provided the equations are written in the natural order.

When only resistances, independent current sources, and grounded independent voltage sources are present in the circuit, the value of the entries in the coefficient matrix of the nodal equations can be computed by inspection. The 1–1 entry of the matrix is the sum of the conductances at node a (or 1); the 2–2 entry is the sum of the conductances at node b (or 2). In general, the i–i entry of the coefficient matrix is the sum of the conductances incident at node i. Further, the 1–2 entry of the matrix is the negative of the sum of the conductances between nodes a and b (or between nodes 1 and 2), and the 2–1 entry has the same value. In Example 3.2, the 1–2 entry of –0.2 S is the negative of the sum of the conductances between nodes a and b; the 1–3 entry of –0.4 S is the negative of the sum of the conductances between nodes a and c (or between 1 and 3, if the nodes were so numbered). Thus, whenever the circuit contains no dependent sources, the node equations can be written by inspection. Further, if independent voltage sources are absent, then the right-hand side of the nodal matrix equation can also be written by inspection: the i–th entry is simply the sum of the independent source currents injected into the i–th node at which KCL is applied.

When controlled sources are present in the circuit, the resultant nodal matrix is generally not symmetric, as illustrated in the following two examples.

EXAMPLE 3.3.

The circuit of Figure 3.7 represents a small-signal low-frequency equivalent circuit of an amplifier in which the input signal V_{in} is "amplified" at the output, $V_{out} = V_2$. Small-signal means that the input signal should have a relatively small magnitude so that a LINEAR circuit will adequately represent the amplifier. Similarly, low-frequency means that the frequency of any sinusoidal input must be relatively low for the (resistive) circuit model of the amplifier to remain valid.

The amplifier circuit model contains a current-controlled current source (CCCS) and a voltage-controlled current source (VCCS). These two dependent sources have currents that depend on other circuit parameters and require some special handling when constructing node equations. Our objective is to set forth the methodology for writing the node equations when dependent current sources are part of the circuit and to compute the magnitude of the voltage gain, $|V_{out}/V_{in}|$ $= |V_2/V_{in}|$.

Note that the source voltage, V_{in}, specifies the voltage at the node at the bottom of G_1; hence, a nodal equation at this node is unnecessary. Nodal equations must be written at the remaining nodes, which are labeled with the voltages V_1, V_2 ($= V_{out}$), and V_3. (Numbering and labeling is often a matter of personal preference. In this example, we have chosen 1, 2, and 3 as node labels, in contrast to the previous two examples, where we used a, b, and c.)

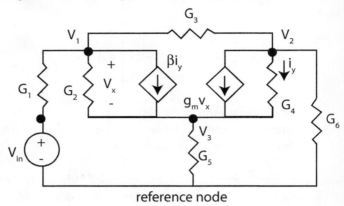

FIGURE 3.7. An equivalent circuit model of an amplifier.

SOLUTION.

Step 1. *Sum the currents leaving node 1.* Summing the current leaving node 1 leads to

$$G_1 (V_1 - V_{in}) + G_2 (V_1 - V_3) + G_3 (V_1 - V_2) + \beta\, i_y = 0$$

or, equivalently, after grouping like terms,

$$(G_1 + G_2 + G_3)V_1 - G_3V_2 - G_2V_3 + \beta i_y = G_1 V_{in} \qquad (3.9)$$

Step 2. *Substitute for i_y in equation 3.9 and simplify.* In equation 3.9, $\beta\, i_y$ accounts for the effect of the CCCS at node 1 and is not given in terms of the circuit node voltages. To specify this term in terms of the circuit node voltages, observe that in Figure 3.7, i_y is the current from node 2 to node 3 through G_4. Hence,

$$\beta i_y = \beta G_4(V_2 - V_3) = \beta G_4 V_2 - \beta G_4 V_3 \qquad (3.10)$$

Substituting equation 3.10 into 3.9, again grouping like terms, one obtains the first nodal equation,

$$(G_1 + G_2 + G_3)V_1 + (\beta G_4 - G_3)V_2 - (G_2 + \beta G_4)V_3 = G_1 V_{in} \qquad (3.11)$$

Step 3. *Sum the currents leaving node 2.* By the usual methods,

$$G_3(V_2 - V_1) + G_6V_2 + G_4(V_2 - V_3) + g_m v_x = 0$$

which, after regrouping terms, reduces to

$$-G_3V_1 + (G_3 + G_6 + G_4)V_2 - G_4V_3 + g_m v_x = 0 \qquad (3.12)$$

Step 4. *Specify $g_m v_x$ in terms of node voltages, substitute into equation 3.12, and simplify.* Inspecting the circuit of Figure 3.7 shows that v_x is the voltage across G_2 from node 1 to node 3. Hence,

$$g_m v_x = g_m(V_1 - V_3) = g_m V_1 - g_m V_3 \tag{3.13}$$

Substituting $g_m v_x$ of equation 3.13 into equation 3.12 leads to our second nodal equation,

$$(g_m - G_3)V_1 + (G_3 + G_4 + G_6)V_2 - (G_4 + g_m)V_3 = 0 \tag{3.14}$$

Step 5. *Sum the currents leaving node 3.* Applying KCL to node 3 yields,

$$G_2(V_3 - V_1) + G_4(V_3 - V_2) + G_5 V_3 - \beta i_y - g_m v_x = 0 \tag{3.15a}$$

Using equations 3.10 and 3.13 for , i_y and $g_m v_x$ respectively, we have

$$0 = G_2(V_3 - V_1) + G_4(V_3 - V_2) + G_5 V_3 - \beta G_4(V_2 - V_3) - g_m(V_1 - V_3)$$

Grouping like terms leads to our third equation in the three unknowns V_1, V_2, and V_3:

$$-(G_2 + g_m)V_1 - (\beta G_4 + G_4)V_2 + (G_2 + G_4 + \beta G_4 + G_5 + g_m)V_3 = 0 \tag{3.15b}$$

Step 6. *Put nodal equations in matrix form.* The three nodal equations 3.11, 3.14, and 3.15b have the matrix form

$$\begin{bmatrix} G_1 + G_2 + G_3 & \beta G_4 - G_3 & -G_2 - \beta G_4 \\ g_m - G_3 & G_3 + G_6 + G_4 & -G_4 - g_m \\ -G_2 - g_m & -G_4 - \beta G_4 & G_2 + G_4 + \beta G_4 + G_5 + g_m \end{bmatrix} \begin{bmatrix} V_1 \\ V_2 \\ V_3 \end{bmatrix} = \begin{bmatrix} G_1 V_{in} \\ 0 \\ 0 \end{bmatrix}$$

Step 7. *Substitute values and solve.* Suppose that the various circuit conductances have the following values in μS: $G_1 = 1,000$, $G_2 = 2.0$, $G_3 = 1.0$, $G_4 = 10$, $G_5 = 20,100$, and $G_6 = 200$. Suppose further that $V_{in} = 2.1$ V, $\beta = 4/1010$ and $g_m = 21,112$ μS. This allows us to generate the following MATLAB code for the solution:

```
»G1 = 1000e-6;G2 = 2e-6; G3 = 1e-6; G4 = 10e-6;
»G5 = 20100e-6; G6 = 200e-6; Vin = 2.1;
»beta = 4/1010; gm = 21112e-6;
»M =[G1+G2+G3   beta*G4-G3   -G2-beta*G4;
gm-G3     G3+G6+G4    -G4-gm;
-G2-gm    -G4-beta*G4    G2+G4+beta*G4+G5+gm];
»b = [G1*Vin    0    0]';
»NodeV = M\b
NodeV =
 2.0000e+00
-1.0000e+02
 1.0000e+00
```

in which case,

$$\begin{bmatrix} V_1 \\ V_2 \\ V_3 \end{bmatrix} = \begin{bmatrix} 2 \\ -100 \\ 1 \end{bmatrix} V$$

Step 8. *Compute the voltage gain.* The voltage gain of the amplifier is given by

$$\left| \frac{V_{out}}{V_{in}} \right| = \left| \frac{V_2}{V_{in}} \right| = \left| \frac{-100}{2.1} \right| = 47.62$$

Exercises. 1. Suppose V_{in} in the circuit of Figure 3.7 is doubled. What are the new node voltages? Hint: Consider the matrix equation of Step 6.
ANSWER: Node voltages are doubled.

2. Suppose all conductances in the circuit of Figure 3.7 are cut in half (resistances are doubled) and β is held constant. How must g_m change for the node voltages to remain at their same values?
ANSWER: g_m must double.

Realistic problems do not permit hand solutions. For hand solutions, the smallest number of equations is generally desired. For matrix solutions using software packages such as MATLAB, more variables with more equations may often be easier to construct and may often result in more reliable numerical calculations. This can be illustrated using the equations of Example 3.3. All the pertinent basic equations of the circuit of Figure 3.7 can be written down as follows: from equations 3.9 and 3.10 we have

$$(G_1 + G_2 + G_3)V_1 - G_3V_2 - G_2V_3 + \beta i_y = G_1 V_{in}$$

and

$$G_4V_2 - G_4V_3 - i_y = 0$$

However, in contrast to the example, we do not substitute 3.10 into 3.9 to obtain 3.11. Rather, we just let them be two independent equations. Further, from equations 3.12, 3.13, and 3.15a, we have

$$-G_3V_1 + (G_3 + G_4 + G_6)V_2 - G_4V_3 + g_m v_x = 0$$

and

$$V_1 - V_3 - v_x = 0$$

By not substituting for i_y and v_x, we avoid unnecessary hand calculation, and if there is an error, it is easier to find. The resulting equations have the matrix form where i_y and v_x now appear as additional unknowns, easily handled by a software program:

$$
\begin{bmatrix}
G_1 + G_2 + G_3 & -G_3 & -G_2 & \beta & 0 \\
0 & G_4 & -G_4 & -1 & 0 \\
-G_3 & G_3 + G_4 + G_6 & -G_4 & 0 & g_m \\
1 & 0 & -1 & 0 & -1 \\
-G_2 & -G_4 & G_2 + G_4 + G_5 & -\beta & -g_m
\end{bmatrix}
\begin{bmatrix}
V_1 \\ V_2 \\ V_3 \\ i_y \\ v_x
\end{bmatrix}
=
\begin{bmatrix}
G_1 V_{in} \\ 0 \\ 0 \\ 0 \\ 0
\end{bmatrix}
$$

As a general rule, we would reorder the equations so that rows 1, 3, and 5 came first, as they correspond to the three nodal equations at V_1, V_2, and V_3. Then we would write the constraint equations for i_y and v_x. Such a reordering leads to certain symmetry properties discussed earlier.

Exercise. Solve the above matrix equation in MATLAB or equivalent, using the numbers of Example 3.3 to verify that $i_y = -1.01 \times 10^{-3}$ A and $v_x = 1$ V.

Matrix methods as used in the above examples and in the ones to follow necessitate the power of a calculator or a software program such as MATLAB for easy solution. Such programs permit a straightforward calculation of the required answers and are not prone to arithmetic errors.

The next example illustrates how to write node equations for circuits containing a voltage-controlled voltage source (VCVS) grounded to the reference node. The analysis of CCVSs grounded to the reference node is similar. The more challenging analysis of circuits containing floating dependent or independent voltage sources is taken up in the next section.

EXAMPLE 3.4.

The circuit of Figure 3.8 models a poor operational amplifier circuit[1] in which the output voltage $V_{out} = V_2$ approximates $-2V_{in}$. For the analysis, let $\mu = 70$. The adjective "poor" arises because μ should have a value much larger than 70.

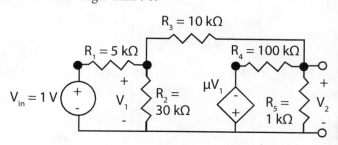

FIGURE 3.8. A two-node (amplifier) circuit containing a grounded VCVS with $\mu = 70$.

SOLUTION.

The circuit contains two nodes labeled V_1 and V_2 (equivalently nodes 1 and 2) not constrained by voltage sources. The goal of our analysis is to find these node voltages by writing two equations in these voltages and solving. As is commonly the case, resistances are in ohms and will be converted to conductances in S for convenience in writing the node equations.

Step 1. *Compute conductance values in S.* Conductances are the reciprocal of resistances, i.e., $G_i = 1/R_i$. Hence,

$$G_1 = 2.0 \ 10^{-3}, \ G_2 = 3.33333 \ 10^{-5}, \ G_3 = 10^{-4}, \ G_4 = 0.01, \text{ and } G_5 = 10^{-3}$$

Step 2. *Write a node equation at node 1.* Summing the currents leaving node 1 yields

$$G_1 \ (V_1 - V_{in}) + G_2 \ V_1 + G_3 \ (V_1 - V_2) = 0$$

Grouping like terms leads to
$$(G_1 + G_2 + G_3)V_1 - G_3V_2 = G_1 V_{in}$$

Inserting numerical quantities yields the first node equation

$$3.3333V_1 - V_2 = 2 \qquad (3.16)$$

Step 3. *Sum currents leaving node 2.* Summing the currents leaving node 2 yields

$$G_3 \ (V_2 - V_1) + G_5 \ V_2 + G_4 \ (V_2 + \mu \ V_1) = 0$$

The dimensionless coefficient μ is placed with the conductance, G_4, while grouping like terms to obtain
$$(\mu G_4 - G_3)V_1 + (G_3 + G_4 + G_5)V_2 = 0$$

Inserting the numerical values produces the second node equation

$$0.6999V_1 + 0.0111V_2 = 0 \qquad (3.17)$$

Step 4. *Write equations 3.16 and 3.17 in matrix form and solve.* In matrix form

$$\begin{bmatrix} 3.3333 & -1 \\ 0.6999 & 0.0111 \end{bmatrix}\begin{bmatrix} V_1 \\ V_2 \end{bmatrix} = \begin{bmatrix} 2 \\ 0 \end{bmatrix}$$

Using the formula for the inverse of a 2 × 2 matrix (interchange the diagonal entries, change the sign on the off diagonals, and divide by the determinant), one obtains

$$\begin{bmatrix} V_1 \\ V_2 \end{bmatrix} = \frac{1}{0.7369}\begin{bmatrix} 0.0111 & 1 \\ -0.6999 & 3.3333 \end{bmatrix}\begin{bmatrix} 2 \\ 0 \end{bmatrix} = \begin{bmatrix} 0.030126 \\ -1.8996 \end{bmatrix}$$

In Example 3.4, observe that $V_{out} = V_2 = -1.8996$, which approximates $-2 \ V_{in}$ since $V_{in} = 1$ V.

Exercises. 1. Write MATLAB code to solve the above example. Check that your code works. Hint: See Example 3.3.
2. If R_2 is changed to 100 kΩ in Example 3.4, show that $V_2 = -1.9063$ V.

---

4. NODAL ANALYSIS II: FLOATING VOLTAGE SOURCES

A **floating** voltage source means that neither node of the source is connected to the reference node. When a floating dependent or independent voltage source is present with respect to a given reference node, a direct application of KCL to either terminal node of the voltage source is unfruitful. There are several ways to handle this situation. One fruitful method is to enclose the source and its terminal nodes by a **Gaussian surface**, i.e., a closed curve, to create what is commonly called a **supernode**, as illustrated in Figure 3.9. One would then write KCL for the supernode as is done in a number of circuit texts. However, there is a conceptually more straightforward approach, which is often called the **modified nodal analysis**, or **MNA**. In MNA, we add an additional current label to each floating voltage source. In Figure 3.9, we have added the current label I_{cb}. This additional current becomes an unknown in a set of nodal equations generated by applying KCL to each node. At this point, further explanation is best done by an example, but the concept is similar to the discussion following Example 3.3.

EXAMPLE 3.5.

Find the node voltages V_a, V_b, V_c, and the unknown current I_{cb} in the circuit of Figure 3.9, when the bottom node is taken as reference.

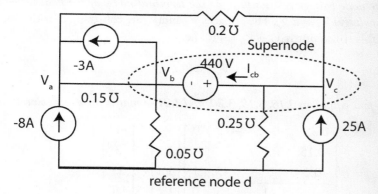

FIGURE 3.9. Resistive circuit containing a floating voltage source for the given reference; generally, the reference node may be chosen arbitrarily.

SOLUTION.

Step 1. *Write a node equation at node a.* Summing the currents leaving node *a* yields

$$8 + 0.15\,(V_a - V_b) + 3 + 0.2\,(V_a - V_c) = 0$$

After grouping terms appropriately, we have

$$(0.15 + 0.2)V_a - 0.15V_b - 0.2V_c + 8 + 3 = 0$$

or, equivalently,

$$0.35V_a - 0.15V_b - 0.2V_c = -11 \tag{3.18}$$

This provides our first equation in four unknowns.

Step 2. *Write a nodal equation at node b.* Here,

$$-3 + 0.15\,(V_b - V_a) + 0.05\,V_b - I_{cb} = 0$$

or equivalently,

$$-0.15\,V_a + (0.15 + 0.05)\,V_b - I_{cb} = 3$$

Simplifying this expression leads to

$$-0.15\,V_a + 0.2\,V_b - I_{cb} = 3 \tag{3.19}$$

Step 3. *Write a nodal equation at node c.* Here,

$$I_{cb} + 0.25\,V_c - 25 + 0.2\,(V_c - V_a) = 0$$

or equivalently,

$$-0.2\,V_a + 0.45\,V_c + I_{cb} = 25 \tag{3.20}$$

Step 4. *Write the node voltage relationship for the terminal nodes of the floating voltage source, i.e., between the voltages V_b and V_c.* The voltages V_b and V_c are constrained by the voltage source. Mathematically, this constraint is $V_c - V_b = 440$, i.e.,

$$-V_b + V_c = 440 \tag{3.21}$$

Step 5. *Write the four equations 3.18, 3.19, 3.20, and 3.21 in matrix form and solve.* In matrix form,

$$\begin{bmatrix} 0.35 & -0.15 & -0.2 & 0 \\ -0.15 & 0.2 & 0 & -1 \\ -0.2 & 0 & 0.45 & 1 \\ \hline 0 & -1 & 1 & 0 \end{bmatrix} \begin{bmatrix} V_a \\ V_b \\ V_c \\ I_{cb} \end{bmatrix} = \begin{bmatrix} -11 \\ 3 \\ 25 \\ \hline 440 \end{bmatrix} \tag{3.22}$$

Because of the extra variable, the equations become too large for hand calculation. Hence, we use MATLAB as follows:

```
»M = [0.35 −0.15 −0.2 0;−0.15 0.2 0 −1;
−0.2 0 0.45 1; 0 −1 1 0];
»b = [−11 3 25 440]';
»x = M\b
x =
−9.0000e+01
−3.1000e+02
1.3000e+02
−5.1500e+01
```

Hence,

$$V_a = -90 \text{ V}, \quad V_b = -310 \text{ V}, \quad V_c = 130 \text{ V}, \quad I_{cb} = -51.5 \text{ A}$$

In a conventional nodal analysis, all unknowns are node voltages. Here we have the additional unknown current, I_{cb}. Because of this additional unknown current, the method is called a modified nodal analysis.

Also, in this example, node d was taken as the reference node. However, one could just as easily take node b as the reference node, in which case, the voltage source would not have been floating. A home problem investigates this choice of reference node.

Exercise. 1. For Example 3.5, compute the voltages V_{db}, V_{ab}, and V_{cb}.
2. For Example 3.5, compute the power absorbed by the 0.15 S resistor.
3. Compute the power delivered by the floating voltage source.
ANSWERS in random order: 220 V, 22.66 kw, 310 V, 440 V, 7260 watts

The next example investigates a circuit having floating independent and dependent voltage sources. By convention, the reference node of this circuit, figure 3.10, and all subsequent circuits, will be the bottom node of the circuit unless stated otherwise.

EXAMPLE 3.6.

The circuit of Figure 3.10 contains a floating independent and a floating dependent voltage source. Find the node voltages V_a, V_b, V_c, and the unknown currents I_{ac} and I_{cb}. Then find the power delivered by the 30 V source and the dependent source.

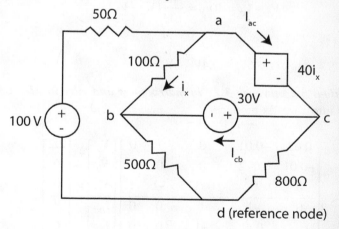

FIGURE 3.10. Resistive circuit containing a floating dependent voltage source and a floating independent voltage when node d is chosen as the reference node.

SOLUTION.

Step 1. *Sum currents leaving node a.* Here,

$$\frac{1}{50}(V_a - 110) + \frac{1}{100}(V_a - V_b) + I_{ac} = 0$$

Equivalently,

$$\frac{3}{100}V_a - \frac{1}{100}V_b + I_{ac} = 2.2 \tag{3.23}$$

Step 2. *Sum currents leaving node b.* Here,

$$\frac{1}{100}(V_b - V_a) + \frac{V_b}{500} - I_{cb} = 0$$

Equivalently,

$$-\frac{1}{100}V_a + \frac{6}{500}V_b - I_{cb} = 0 \tag{3.24}$$

Step 3. *Sum currents leaving node c.* Here,

$$\frac{1}{800}V_c + I_{cb} - I_{ac} = 0 \tag{3.25}$$

Step 4. *Write an equation relating the terminal voltages of the independent voltage source.* Here,

$$V_c - V_b = 30 \tag{3.26}$$

Step 5. *Write an equation relating the terminal voltages of the dependent voltage source.* Here,

$$V_a - V_c = 40i_x = \frac{40}{100}(V_a - V_b)$$

Equivalently,

$$0.6V_a + 0.4V_b - V_c = 0 \tag{3.27}$$

Step 6. *Write equations 3.23 through 3.27 in matrix form and solve in MATLAB.* Combining the above equations into a matrix produces

$$\begin{bmatrix} 0.03 & -0.01 & 0 & | & 1 & 0 \\ -0.01 & 0.012 & 0 & | & 0 & -1 \\ 0 & 0 & 0.00125 & | & -1 & 1 \\ \hline 0 & -1 & 1 & | & 0 & 0 \\ 0.6 & 0.4 & -1 & | & 0 & 0 \end{bmatrix} \begin{bmatrix} V_a \\ V_b \\ V_c \\ I_{ac} \\ I_{cb} \end{bmatrix} = \begin{bmatrix} 2.2 \\ 0 \\ 0 \\ \hline 30 \\ 0 \end{bmatrix}$$

Again, this matrix equation is too large for hand computation. Hence in MATLAB,

```
»M = [0.03 –0.01 0 1 0;
–0.01 0.012 0 0 –1;
0 0 0.00125 –1 1;
0 –1 1 0 0;
0.6 0.4 –1 0 0];
»
```

»b = [2.2 0 0 30 0]';
»x = M\b
x =
  1.0000e+02
  5.0000e+01
  8.0000e+01
 −3.0000e−01
 −4.0000e−01

Hence,

$$\begin{bmatrix} V_a \\ V_b \\ V_c \\ I_{ac} \\ I_{cb} \end{bmatrix} = \begin{bmatrix} 100 \\ 50 \\ 80 \\ -0.3 \\ -0.4 \end{bmatrix}$$

Step 4. *Compute the power delivered by the 30 V source.* The power delivered by the 30-V source is

$$P_{del} = -30I_{cb} = 30 \times 0.4 = 12 \text{ W}$$

Step 5. *Compute the power delivered by the dependent source.* The power delivered by the dependent source is

$$P_{del} = -40i_x I_{ac} = -40 \times \frac{V_{ab}}{100}(-0.3) = 0.12(V_a - V_b) = 6 \text{ W}$$

Exercises. 1. For Example 3.6, compute the voltages V_{dc}, V_{ac}. and the power absorbed by the 800 Ω resistor.
ANSWERS in random order: 8 watts, 20 V, −80 V

2. Suppose the two independent voltage source values in Example 3.6 are doubled. What are the new node voltages? What are the new branch currents?
ANSWERS: Node voltages are doubled and branch currents are doubled.

3. Suppose all resistances in the circuit of Figure 3.10 are doubled and the value of the parameter on the dependent source is also doubled. What are the new branch currents?
ANSWER: All branch currents are cut in half.

The above example increases the number of unknowns beyond the node voltages to include the two currents through the floating voltage sources. However, we could have included additional currents to the set of equations making the dimension even higher. With a tool like MATLAB, this poses no difficulty. However, it does make hand computation a challenge. For example, we could have included $I_x = I_{ab}$ as an additional variable with a corresponding increase in the number of equations. By adding additional unknowns we would simplify the writing of the individual node equations but increase the dimension of the matrix equation. Specifically, the node equation at "a" becomes

$$\frac{V_a}{50} + I_{ac} + I_x = \frac{110}{50}$$

and the resulting larger matrix equation is

$$\begin{bmatrix} 0.02 & 0 & 0 & 1 & 0 & 1 \\ 0 & 0.002 & 0 & 0 & -1 & -1 \\ 0 & 0 & 0.00125 & -1 & 1 & 0 \\ 0 & -1 & 1 & 0 & 0 & 0 \\ 1 & 0 & -1 & 0 & 0 & -40 \\ 1 & -1 & 0 & 0 & 0 & -100 \end{bmatrix} \begin{bmatrix} V_a \\ V_b \\ V_c \\ I_{ac} \\ I_{cb} \\ I_x \end{bmatrix} = \begin{bmatrix} 2.2 \\ 0 \\ 0 \\ 30 \\ 0 \\ 0 \end{bmatrix}$$

This completes our discussion of the standard nodal equation method of circuit analysis. The next section takes up a discussion of an alternative analysis method entitled loop analysis.

5. LOOP ANALYSIS

Loop analysis is a second general analysis technique for computing the voltages and currents in a circuit. **Mesh analysis** is a special type of loop analysis for **planar circuits**, i.e., circuits that can be drawn on a plane without branch crossings. For planar circuits, loops can be chosen as meshes, as illustrated in Figure 3.2, or as in 3.11 below. Associated with each loop is a loop current. **Loop currents** circulate around closed paths (loops) in the circuit. Similarly, for planar circuits, the term **mesh current** is used traditionally for loop current. By KVL, the sum of the voltages across each branch in a loop is zero. By expressing each of these branch voltages in terms of the designated loop currents, one can write an equation in the loop currents for each designated loop in the circuit. For branches that are often common to two or more designated loops, the **branch current** equals the *net flow* of the loop currents incident on the branch. Writing an equation for each loop produces a set of equations called **loop equations**. If sufficient independent loops are defined, one can solve the loop equations for the loop currents. Once the loop currents are known, we can easily compute the branch currents and then the branch voltages in the circuit. Then we can compute any other quantities of interest, such as power absorbed, power delivered, voltage gain, etc.

EXAMPLE 3.7.

Consider the planar circuit of Figure 3.11 with the three specified loops, which are also called meshes. Denote the "loop" currents for each loop by I_1, I_2, and I_3. The objective is to write three equations in the currents I_1, I_2, and I_3 using KVL and solve these equations for their values. Then we will compute the power absorbed by the 2–Ω resistor marked with the voltage v. Suppose the source voltages are $V_{s1} = 40$ V and $V_{s2} = 20$ V.

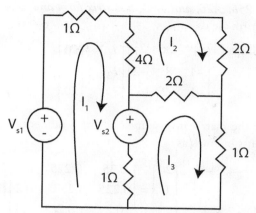

FIGURE 3.11. Resistive circuit containing only independent voltage sources for the loop analysis of Example 3.7.

SOLUTION.

Step 1. *Write a KVL equation based on loop 1 by summing voltages around this loop.* Summing the voltages around loop 1 using Ohm's law and the defined loop currents produces

$$V_{s1} = 40 = I_1 + 4\,(I_1 - I_2) + V_{s2} + (I_1 - I_3) = 6I_1 - 4I_2 - I_3 + 20 \qquad (3.25a)$$

Here, observe that the 4 Ω resistor is incident on two loops; the net current flowing from top to bottom, i.e., with respect to the direction of loop 1, is $I_1 - I_2$. The idea is analogous to a pair of distinct water pipes that share a common length. The common length is analogous to the 4–Ω resistor. The flow rate in each pipe is analogous to the currents I_1 and I_2, which in fact, are rates at which charge flows past a cross sectional area of the conductor. It follows that the net flow through the common length of pipe with respect to the direction of loop 1 is the difference in the net flow rates of pipes 1 and 2, respectively. This is precisely the meaning of $I_1 - I_2$. A similar explanation can be made for the 1–Ω resistor common to loops 1 and 3 for which the net flow rate with respect to the direction of loop 1 is $I_1 - I_3$.

Simplifying equation 3.25a yields

$$6I_1 - 4I_2 - I_3 = 20 \qquad (3.25b)$$

Step 2. *Write a KVL equation based on loop 2 by summing the voltages around this loop.* Applying Ohm's law and KVL to loop 2 produces

$$0 = 4(I_2 - I_1) + 2I_2 + 2(I_2 - I_3) = -4I_1 + 8I_2 - 2I_3 \qquad (3.26)$$

Notice that with respect to the direction of loop 2, the net flow rate through the 4 Ω resistor is $I_2 - I_1$.

Step 3. *Finally, write a KVL equation based on loop 3.* Summing the voltages around loop 3 yields

$$V_{s2} = 20 = 2(I_3 - I_2) + I_3 + (I_3 - I_1) = -I_1 - 2I_2 + 4I_3 \qquad (3.27)$$

Step 4. *Write equations 3.25b, 3.26, and 3.27 in matrix form and solve.* Writing the above three loop equations in matrix form yields

$$\begin{bmatrix} 6 & -4 & -1 \\ -4 & 8 & -2 \\ -1 & -2 & 4 \end{bmatrix} \begin{bmatrix} I_1 \\ I_2 \\ I_3 \end{bmatrix} = \begin{bmatrix} 20 \\ 0 \\ 20 \end{bmatrix} \qquad (3.28)$$

Solving Equation 3.28 by the matrix inverse method (by a numerical algorithm or by Cramer's rule) yields the loop currents in amps as

$$\begin{bmatrix} I_1 \\ I_2 \\ I_3 \end{bmatrix} = \begin{bmatrix} 6 & -4 & -1 \\ -4 & 8 & -2 \\ -1 & -2 & 4 \end{bmatrix}^{-1} \begin{bmatrix} 20 \\ 0 \\ 20 \end{bmatrix} = \begin{bmatrix} 0.35 & 0.225 & 0.2 \\ 0.225 & 0.2875 & 0.2 \\ 0.2 & 0.2 & 0.4 \end{bmatrix} \begin{bmatrix} 20 \\ 0 \\ 20 \end{bmatrix} = \begin{bmatrix} 11 \\ 8.5 \\ 12 \end{bmatrix}$$

Step 5. *Compute the power consumed by the 2 Ω resistor.* Knowledge of the loop currents makes it possible to compute all voltages and currents in the circuit. For our purpose, the voltage

$$v = 2(I_3 - I_2) = 7 \text{ V}$$

and the power absorbed by the 2 Ω resistor is $v^2 / 2 = 24.5$ watts.

Exercises. All exercises are for the circuit of Figure 3.11.
1. Compute the power delivered by the 20 V source.
ANSWER: 20 watts

2. Compute the power absorbed by the 4 Ω resistor.
ANSWER: 25 watts

3. Suppose the source values are doubled. What are the new values of the loop currents?
ANSWER: loop currents are doubled

4. Suppose the resistance values are multiplied by 4. What are the new loop currents? What are the new node voltages?
ANSWERS: Loop currents are 0.25 times their original values, and node voltages are unchanged.

Observe that there are no dependent current or voltage sources in the circuit. Similar to the nodal analysis case, whenever dependent sources are absent and the equations are written in the natural order, the loop (or mesh) equations are symmetric, as illustrated by the coefficient matrix of equation 3.28 where, for example, the 1–2 and 2–1 entries coincide, as do the 1–3 and 3–1 entries, etc. Also, the value of all entries can be computed by inspection. The 1–1 entry of the matrix is the sum of the resistances in loop 1; the 2–2 entry is the sum of the resistances in loop 2, etc. In general, the i–i entry is the sum of the resistances in loop i. The 1–2 entry of the matrix is $\Sigma(\pm R_k)$ (the large sigma means summation), where each R_k is a resistance common to both loops 1 and 2. Use the + sign when both loop currents circulate through R_k in the same direction, and use the –

sign otherwise. Further, if independent current sources are absent, then the right-hand side of the loop equations can also be written by inspection. The i–th entry is simply the net voltage of the sources in the i–th loop that tends to deliver a current in the direction of the loop current.

Exercises. 1. Use the inspection rules described above to write two mesh equations for the circuit of Figure 3.12, when both mesh currents are assigned clockwise direction.

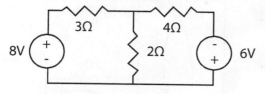

FIGURE 3.12.

2. Use the inspection rules described above to write two mesh equations for the circuit of Figure 3.12, when the left mesh current is clockwise and the right mesh current is counterclockwise.

3. Use the inspection rules described above to determine the right-hand side of the mesh equation for the circuit of Figure 3.13.

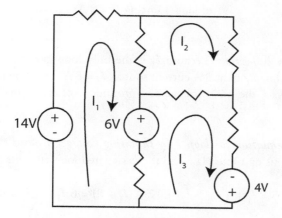

FIGURE 3.13

ANSWERS: 8, 0, 10

A simplifying reduction to the set of loop equations occurs if an independent current source coincides with a single loop current. The analysis becomes simpler because that loop current is no longer an unknown; rather it is equal to the value of the source current if their directions coincide, or to the negative value if their directions are opposing. Because the associated loop current is known, there are fewer loop equations to write and solve. One would apply KVL to such a loop only if it were necessary to compute the voltage across the independent current source, which might be necessary for determining the power delivered by the source. The entire situation is analogous to an independent voltage source tied between a node and the reference in nodal analysis. The following example illustrates the details of this discussion.

EXAMPLE 3.8.

The circuit of Figure 3.14 is a modification of the one of 3.11 in which (i) a 1 ohm resistor on the perimeter of the circuit is replaced by an 8 A independent current source, and (ii) the values of the voltage sources are doubled. The currents for each loop are again denoted by I_1, I_2, and I_3. Our objective is to find all the loop currents, the voltage V_s, and the power delivered by the 8 A source.

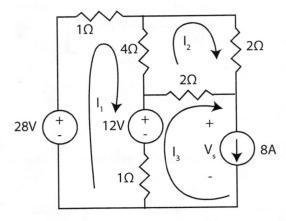

FIGURE 3.14. A resistive circuit containing an independent current source on the perimeter of loop 3 forcing I_3 = 8 A.

SOLUTION.

Step 1. *Solve for I_3 by inspection.* Because I_3 is the only loop current circulating through the branch containing the independent 8 A current source, I_3 = 8 A. This phenomena is similar to the fact that in nodal analysis, the node voltage of a grounded voltage source is fixed at the voltage source value.

Step 2. *Write a KVL equation for loop 1 by summing voltages around this loop.* Summing the voltages around loop 1 using Ohm's law and the designated loop currents produces

$$28 = I_1 + 4\,(I_1 - I_2) + 12 + (I_1 - 8) = 6\,I_1 - 4\,I_2 + 4$$

Hence,

$$6I_1 - 4\,I_2 = 24 \tag{3.29}$$

Step 3. *Write a KVL equation for loop 2 by summing the voltages around this loop.* Applying KVL and Ohm's law to loop 2 produces

$$0 = 4\,(I_2 - I_1) + 2\,I_2 + 2\,(I_2 - 8) = -4\,I_1 + 8\,I_2 - 16$$

Equivalently,

$$-4I_1 - 8\,I_2 = 16 \tag{3.30}$$

Step 4. *Write above loop equations in matrix form and solve.* The matrix form of equations 3.29 and 3.30 is

$$\begin{bmatrix} 6 & -4 \\ -4 & 8 \end{bmatrix}\begin{bmatrix} I_1 \\ I_2 \end{bmatrix} = \begin{bmatrix} 24 \\ 16 \end{bmatrix}$$

Using the inverse matrix technique to compute the solution, we have

$$\begin{bmatrix} I_1 \\ I_2 \end{bmatrix} = \begin{bmatrix} 6 & -4 \\ -4 & 8 \end{bmatrix}^{-1}\begin{bmatrix} 24 \\ 16 \end{bmatrix} = \frac{1}{32}\begin{bmatrix} 8 & 4 \\ 4 & 6 \end{bmatrix}\begin{bmatrix} 24 \\ 16 \end{bmatrix} = \begin{bmatrix} 8 \\ 6 \end{bmatrix}$$

Step 5. *Compute V_s.* By KVL,
$$V_s = 2\,(I_2 - 8) + 12 + (I_1 - 8) = 8 \text{ V}$$

Step 6. *Compute power delivered by 8 A source.* Observe that the 8 A current source is labeled according to the passive sign convention, in which case,

$$P_{del} = -P_{abs} = -(8 \times 8) = -64 \text{ watts}$$

Hence, the source actually absorbs 64 watts.

Exercise. In the circuit of Figure 3.15, two of the three mesh currents coincide with independent source currents. By writing and solving just one mesh equation, find I_1.

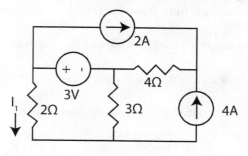

FIGURE 3.15

ANSWER: 3 A

Not only do independent current sources constrain loop currents, but dependent currents sources do also. This situation is illustrated in Example 3.9.

EXAMPLE 3.9.

This example illustrates the writing of loop equations for a simplified small signal equivalent circuit, Figure 3.16, of a two-stage amplifier that contains a current-controlled current source (CCCS) and a current-controlled voltage source (CCVS). This process extends the techniques of Examples 3.7 and 3.8 to find some important characteristics of the amplifier. Specifically, find (a) the input resistance seen by the source, i.e., $R_{in} = v_{in}/i_b$,

(b) the voltage gain, v_o/v_{in}, and

(c) the voltage v across the dependent current source.

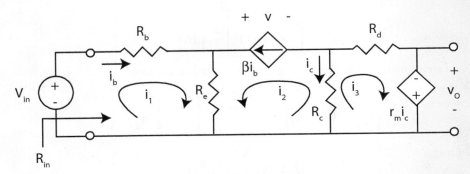

FIGURE 3.16. Small signal equivalent circuit for a two-stage amplifier. Signals in amplifiers are usually time dependent, so we adopt the lowercase notation for voltages and currents.

SOLUTION.

The circuit of Figure 3.16 contains three loop or mesh currents. The direction of the loops is a user-chosen preference. For convenience, we have chosen mesh current i_2 to be consistent with the direction of the arrow in the dependent current source. Because this dependent current source lies on the perimeter of the circuit, it constrains the value of i_2, i.e., $i_2 = \beta\, i_b$. Since the controlling current, $i_b = i_1$, $i_2 =$, $i_b = \beta\, i_1$. This relationship implies that the mesh current of loop 2 depends directly on the mesh current of loop 1. This observation allows us to skip constructing a mesh equation for loop 2. Only equations for loops 1 and 3 are needed, thereby reducing the number of simultaneous equations from three (because there are three loops) to two.

Step 1. *Apply KVL to loop/mesh 1.* Here, by KVL and the observation that $i_2 = \beta\, i_1$,

$$v_{in} = R_b\, i_1 + R_e\, (i_1 + \beta\, i_1) = (R_b + (1 + \beta)\, R_e)\, i_1$$

From this equation, we can immediately compute the input resistance

$$R_{in} = \frac{v_{in}}{i_b} = \frac{v_{in}}{i_1} = R_b + (\beta + 1)\, R_e \qquad (3.31)$$

Step 2. *Apply KVL to loop/mesh 3.* In this case, observe that $i_c = -(i_2 + i_3) = -(\beta\, i_1 + i_3)$. By KVL,

$$R_d\, i_3 - r_m\, i_c + R_c\, (i_3 + i_2) = R_d\, i_3 + r_m\, (\beta\, i_1 + i_3) + R_c\, (i_3 + \beta\, i_1) = 0$$

Combining like terms, it follows that

$$(r_m + R_c)\, \beta\, i_1 + (r_m + R_c + R_d)\, i_3 = 0 \qquad (3.32)$$

Step 3. *Write equations 3.31 and 3.32 in matrix form and solve.* The matrix form of these equations is

$$\begin{bmatrix} R_b + (\beta + 1) R_e & 0 \\ \beta \left(R_c + r_m \right) & R_c + R_d + r_m \end{bmatrix} \begin{bmatrix} i_1 \\ i_3 \end{bmatrix} = \begin{bmatrix} v_{in} \\ 0 \end{bmatrix} \tag{3.33}$$

Because the solution is desired in terms of the literal variables, we solve equation 3.33 using Cramer's rule, which utilizes determinants. In this task, first define

$$\Delta \equiv \det \begin{bmatrix} R_b + (\beta + 1) R_e & 0 \\ \beta \left(R_c + r_m \right) & R_c + R_d + r_m \end{bmatrix} = \left(R_b + (\beta + 1) R_e \right) \left(R_c + R_d + r_m \right)$$

Using the notation Δ for the determinant, Cramer's rule provides the solution for i_1 according to the formula

$$i_1 = \frac{\det \begin{bmatrix} v_{in} & 0 \\ 0 & R_c + R_d + r_m \end{bmatrix}}{\Delta} = \frac{R_c + R_d + r_m}{\Delta} v_{in} \tag{3.34}$$

Applying Cramer's rule for the solution of i_3, yields

$$i_3 = \frac{\det \begin{bmatrix} R_b + (\beta + 1) R_e & v_{in} \\ \beta \left(R_c + r_m \right) & 0 \end{bmatrix}}{\Delta} = \frac{-\beta \left(R_c + r_m \right)}{\Delta} v_{in} \tag{3.35}$$

Step 4. *Compute v_o in terms of v_{in} and then the voltage gain v_o/v_{in}.* As per the circuit of Figure 3.11 and equations 3.34 and 3.35,

$$v_o = -r_m i_c = r_m (\beta i_1 + i_3) = r_m \frac{\beta \left(R_c + R_d + r_m \right) - \beta \left(R_c + r_m \right)}{\Delta} v_{in} = r_m \frac{\beta R_d}{\Delta} v_{in}$$

After substituting for Δ, the voltage gain is

$$\frac{v_o}{v_{in}} = r_m \frac{\beta R_d}{\Delta} = r_m \frac{\beta R_d}{\left(R_b + (\beta + 1) R_e \right) \left(R_c + R_d + r_m \right)}$$

Step 4. *Compute v.* To compute the voltage across the dependent current source, apply KVL to mesh 2 to obtain

$$v = R_e (i_2 + i_1) + R_c (i_2 + i_3) = [\beta (R_e + R_c) + R_e] i_1 + R_c i_3$$

Exercise. Find a simplified loop equation for i_1 in the circuit of Figure 3.17.

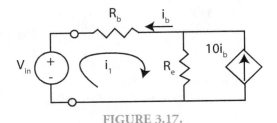

FIGURE 3.17.

ANSWER: $V_{in} = (R_b - 9R_e)\, i_1$

To see the importance of the calculations of the **amplifier** circuit of Example 3.9, suppose two amplifiers are available for use with a non-ideal voltage source. The non-ideal voltage source is modeled by an ideal one-volt source in series with a 100 Ω source resistance. Suppose amplifier 1 has a voltage gain, $v_o/v_{in} = 10$ and $R_{in1} = 100$ kΩ. Suppose amplifier 2 has a voltage gain of 100 and $R_{in2} = 5$ Ω. If amplifier 1 is attached to the non-ideal source, then by voltage division, $v_{in1} = 100,000/(100,000 + 100) = 0.999$ V, whereas in the case of amplifier 2, $v_{in2} = 5/(100 + 5) = 0.0476$ V. In the first case, the gain from v_{in} to v_o is 10, yielding $v_{o1} = 9.99$ V. In the second case, the same gain is 100, yielding $v_{o2} = 4.76$ V. One concludes that amplifier 1 is better suited to this particular application, although it has a lower voltage gain than amplifier 2. Hence, Example 3.9 illustrates the need to know both the voltage gain and the input resistance to determine the output voltage in practical applications. Further, using the literal solution to the example allows us to apply the formulas to different sets of parameter values without repeating the complete analysis.

In the previous two examples, there were current sources on the perimeter of the circuit. Such current sources were incident to only one loop. It often happens that independent and dependent current sources can be common to two or more loops. When this happens, a situation analogous to floating voltage sources in nodal analysis occurs. To handle such cases, many texts define something called a **supermesh** and write a special loop equation for this supermesh. Supermeshes often confuse the beginner. There is an easier way.

Example 3.10 below illustrates how to write "loop" equations when current sources are common to two or more loops. In such cases, we introduce **auxiliary voltage variables** across current sources common to two or more loops. The resulting set of simultaneous equations will contain not only the loop currents as unknowns, but also the auxiliary voltages as unknowns. Because the resulting set of equations contains both loop currents and additional (auxiliary) voltage variables, the equations are called **modified loop equations**. The process of writing modified loop equations is extremely systematic and straightforward. Further, it allows us to avoid explaining the very confusing concept of a supermesh. On the other hand, the presence of auxiliary voltage variables increases the number of "unknowns," i.e., the number of simultaneous equations increases. Because of the availability of software packages such as MATLAB, MATHEMATICA, and MAPLE, this increased dimension is not a hindrance.

EXAMPLE 3.10.

Consider the circuit of Figure 3.18 in which $V_{s1} = 28$ V and $I_{s2} = 0.06$ A. Note that the independent current source is common to loops 1 and 3 and a voltage-controlled current source is common to loops 1 and 2. Find values for the loop currents I_1, I_2, I_3, and the power delivered by each independent source.

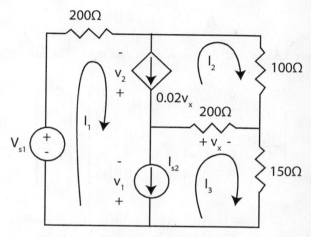

FIGURE 3.18. Circuit containing a current source between loops.

SOLUTION.

To begin the solution, we introduce two auxiliary voltage variables v_1 and v_2 associated with the current sources common to two (or multiple) loops. The purpose of these variables is to facilitate the application of KVL for constructing the loop equations. This will require that we obtain three KVL equations, one for each loop, and two constraint equations, one for each current source.

Step 1. Apply KVL to loop 1. By a clear-cut application of KVL,

$$28 = 200I_1 - v_1 - v_2 \qquad (3.36)$$

Step 2. Apply KVL and Ohm's law to loop 2. Again applying KVL and Ohm's law to loop 2, we obtain $100 I_2 + 200 (I_2 - I_3) + v_2 = 0$. After grouping like terms,

$$300I_2 - 200I_3 + v_2 = 0 \qquad (3.37)$$

Step 3. Apply KVL to loop 3. Applying KVL to loop 3 yields $150 I_3 + v_1 + 200 (I_3 - I_2) = 0$. Equivalently,

$$-200I_2 + 350I_3 + v_1 = 0 \qquad (3.38)$$

Step 4. Write a constraint equation determined by the independent current source. Here, loops 1 and 3 are incident on the independent current source so that

$$0.06 = I_1 - I_3 \qquad (3.39)$$

Step 5. *Write a constraint equation determined by the dependent current source.* In a straightforward manner, we have

$$I_1 - I_2 = 0.02v_x = 0.02\left[200\left(I_3 - I_2\right)\right] = 4I_3 - 4I_2$$

After simplification,

$$I_1 + 3I_2 - 4I_3 = 0 \tag{3.40}$$

Step 6. *Write equations 3.36 to 3.40 in matrix form and solve.* The matrix form of these equations is

$$
\begin{bmatrix}
200 & 0 & 0 & -1 & -1 \\
0 & 300 & -200 & 0 & 1 \\
0 & -200 & 350 & 1 & 0 \\
1 & 0 & -1 & 0 & 0 \\
1 & 3 & -4 & 0 & 0
\end{bmatrix}
\begin{bmatrix}
I_1 \\
I_2 \\
I_3 \\
v_1 \\
v_2
\end{bmatrix}
=
\begin{bmatrix}
28 \\
0 \\
0 \\
0.06 \\
0
\end{bmatrix}
\tag{3.41}
$$

Solving equation 3.41 by the matrix inverse method or by an available software package yields the solution (currents in A and voltages in V) given by equation 3.42 below:

$$
\begin{bmatrix}
I_1 \\
I_2 \\
I_3 \\
v_1 \\
v_2
\end{bmatrix}
=
\begin{bmatrix}
200 & 0 & 0 & -1 & -1 \\
0 & 300 & -200 & 0 & 1 \\
0 & -200 & 350 & 1 & 0 \\
1 & 0 & -1 & 0 & 0 \\
1 & 3 & -4 & 0 & 0
\end{bmatrix}^{-1}
\begin{bmatrix}
28 \\
0 \\
0 \\
0.06 \\
0
\end{bmatrix}
=
\begin{bmatrix}
0.1 \\
0.02 \\
0.04 \\
-10 \\
2
\end{bmatrix}
\tag{3.42}
$$

which can be obtained using the following MATLAB code:

```
M = [ 200 0 0 -1 -1
0 300 -200 0 1
0 -200 350 1 0
1 0 -1 0 0
1 3 -4 0 0];

»b = [28 0 0 0.06 0]';

»LoopIplus= M\b
LoopIplus =
  1.0000e-01
  2.0000e-02
  4.0000e-02
 -1.0000e+01
  2.0000e+00
```

Step 7. *Compute the powers delivered by the independent sources.* First, the power delivered by the independent voltage source is

$$P_{V\text{-source}} = 28 \, I_1 = 2.8 \text{ watts}$$

The power delivered by the independent current source is

$$P_{I\text{-source}} = 0.06 \, v_2 = -0.6 \text{ watts}$$

This last value indicates that the independent current source actually absorbs power from the circuit.

Exercise. For the circuit of Figure 3.19, write the modified loop equations having two unknowns I_1 and v, following the procedure described in Example 3.10. Solve the equations and find the power absorbed by the 2–Ω resistor.

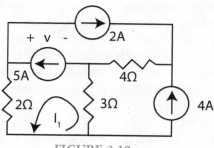

FIGURE 3.19.

ANSWER: 18 watts

One final point before closing our discussion of loop analysis. Loops can be chosen in different ways. Cleverly choosing loops can sometimes simplify the solution of the associated equations. For example, by choosing a loop that passes through a current source so that no other loop is common to the source, the loop current is automatically specified by that current source.

6. SUMMARY

This chapter introduced the technique of nodal analysis. Nodal analysis is a technique for writing a set of equations whose solution yields all node voltages in a circuit. With knowledge of all the node voltages and all the element values, one can compute all branch voltages and currents. As mentioned, whenever there are no dependent sources present, the coefficient matrix of the node equations is always symmetric. Hence, whenever dependent sources are absent, it is possible to write the nodal equation coefficient matrix by inspection. Further, if independent voltage sources are absent, then the right-hand side of the matrix form of the nodal equations can also be written by inspection: the entry is simply the sum of the independent source currents injected into the node at which KCL is applied. When VCCSs are present, the steps for writing nodal equations are the same as illustrated in Example 3.3. Generally, in such cases, the resultant coefficient matrix is not symmetric.

When floating dependent or independent voltage sources are present with respect to a given reference node, we introduce new current variable through these floating sources as unknowns. The node equations then incorporate these additional unknown currents, as was illustrated in Examples 3.5 and 3.6. This method increases the number of equations but simplifies the construction of the individual equations. With a tool like MATLAB to compute solutions, there is no difficulty, although hand computation may become more difficult. This concept is the basis of the modified nodal analysis method used in circuit simulation programs like SPICE.

Loop/mesh analysis, an approach dual to nodal analysis, was introduced in Section 5. Mesh analysis is a special case of loop analysis for planar circuits when the loops are chosen to be the obvious meshes, similar in geometry to a fish net. In loop analysis, one sums the voltages around a loop or mesh to zero. Each of the branch voltages in the loop is expressed as a product of resistances and (fictitious) loop currents that circulate through the branch resistance, as illustrated in Figures 3.10, 3.14, and 3.16. The branch current of the circuit are equal to the net flow of the loop currents incident on a particular branch, meaning that each branch current is expressible as a sum of loop currents. The desired set of loop equations is produced by summing the voltages around each loop, expressing these voltages either as source values or as resistances times loop currents. One solves the loop equations for the loop currents. Once the loop currents are known, we can then compute the individual branch currents and then the branch voltages, and thus any other pertinent current, voltage, or power. Whenever there are no dependent sources present, the coefficient matrix of the loop equations is always symmetric. Whenever dependent sources are absent, it is possible to easily write the loop matrix by inspection.

As the size of an arbitrary circuit grows larger, there are two good reasons for choosing the nodal method over the loop method: (i) the number of nodal equations is usually smaller than the number of loop equations, and (ii) the formulation of nodal equations for computer solution is easier than methods based on loop equations. Writing nodal equations is particularly easy if the circuit contains only resistances, independent current sources, and VCCSs – for short, an R–I_s–g_m network. For an R–I_s–g_m network, one simply applies KVL to every node (except the reference node) and obtains a set of node equations directly. For floating independent or dependent voltage sources, the task is more complex. Examples 3.5 and 3.6 illustrate cases where, besides the node voltages, additional unknown auxiliary currents are added. By adding additional auxiliary variables to the formulation of the nodal equations, we described the concept behind the **modified nodal analysis** (MNA) method. The MNA method retains the simplicity of the nodal method while removing its limitations and is the most commonly used method in present-day computer-aided circuit analysis programs.

7. TERMS AND CONCEPTS

Connected circuit: every pair of nodes in the circuit is joined by some set of branches.
Cramer's rule: a method for solving a linear matrix equation for the unknowns, one by one, through the use of determinants; the method has serious numerical problems when implemented on a computer, but is often convenient for small, 2 x 2 or 3 x 3, hand calculations.
Floating source: neither node of the source is connected to the reference node.

Gaussian surface: a closed curve or a closed surface surrounding two or more nodes.

Linear matrix equation: an equation of the form $Ax = b$, where A is a $n \times n$ matrix, x is an n–vector of unknowns, and b is an n–vector of constants.

Loop (closed path): a contiguous sequence of branches that begins and ends on the same node and touches no node more than once.

Loop analysis: an organized method of circuit analysis for computing loop currents in a circuit. Knowledge of the loop currents allows one to compute the individual element currents and, consequently, the element voltages.

Loop current: a (fictitious) current circulating around a closed path in a circuit.

Matrix inverse: the inverse, if it exists, of an n n matrix A, denoted by A^{-1}, satisfies the equation $A A^{-1} = A^{-1} A = I,$. where I is the $n \times n$ identity matrix; the solution of the linear matrix equation, $Ax = b$, is given by $x = A^{-1}b$.

Mesh: After drawing a planar graph without branch crossing, the boundary of any region with finite area is called a mesh. Intuitively, meshes resemble the openings of a fish net.

Mesh analysis: the special case of loop analysis for planar circuits in which the loops are chosen to be the meshes.

Mesh current: a fictitious current circulating around a mesh in a planar circuit.

Modified nodal analysis: a modification of the basic nodal analysis method in which the unknowns are the usual nodal voltages plus some naturally occurring auxiliary currents.

Nodal analysis: an organized method of circuit analysis built around KCL for computing all node voltages of a circuit.

Node voltage: the voltage drop from a given node to a reference node.

Symmetric matrix: a matrix whose transpose is itself. If $A = [a_{ij}]$ is a $n \times n$ matrix whose i–j entry is a_{ij}; then A is symmetric if $a_{ij} = a_{ji}$.

PROBLEMS

SINGLE NODE PROBLEMS

1. For the circuit of Figure P3.1, write a single node equation in G_1, G_2, G_3, V_{s1}, and V_{s2}. For a fixed R > 0, $R_1 = R$, $R_2 = 2R$, $R_3 = 2R$. Compute V_1 in terms of R and V_{s1} if $V_{s2} = 4V_{s1}$.

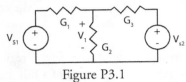

Figure P3.1

ANSWER: $V_1 = 1.5V_{S1}$

2. The battery of your car has been dealt a sudden death by the sub-zero North wind and a faulty alternator. Unable to fight the elements, you wait a few days hoping for a thaw, which comes. You replace the alternator. Then, using your roommate's car, you attempt a jump-start. Nothing happens. You let it sit for a while with your roommate's car running juice into your battery for 20 minutes. Still, nothing happens. Why won't your car start? Consider the circuit of Figure P3.2. Notice that your "dead" batter is labeled as V_0. Your roommate's battery is labeled 12 V. Each battery has an internal resistance of 0.02 Ω and the starter, an internal resistance of 0.2 Ω. The starter motor requires 50 A to crank the engine. Find the minimum value of voltage V_0 needed before the starter can draw 50 A and work.

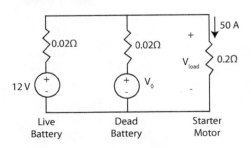

Figure P3.2

3. For the circuit of Figure P3.3, suppose $I_s = 1.2$ A. Write a single node equation in the voltage V_x and solve.

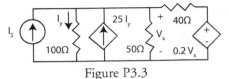

Figure P3.3

ANSWER: −6 V

MULTIPLE NODE PROBLEMS

4. The purpose of this problem is to write the nodal equations directly by inspection of the circuit diagram of Figure P3.4. Recall that when the network has only independent current sources and resistors, the nodal equation matrix is symmetric and the entries can be written down by inspection as per the discussion following Example 3.2. Construct the nodal equations in matrix form for the circuit of Figure 3.4 by inspection.

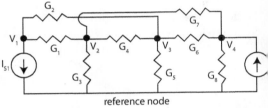

reference node
Figure P3.4

A FEW ANSWERS: The 3-3 entry is $G_2 + G_4 + G_5 + G_6$, and the 2-1 entry is $-G_1$.

5. Consider the circuit of Figure P3.5 in which $I_{s1} = 0.5$ A and $V_{s2} = 40$ V. Further, let $G_1 = 5$ mS, $G_2 = 2.5$ mS, $G_3 = 2.5$ mS, and $G_4 = 12.5$ mS.

    (a)   By inspection, what is the value of V_c?

    (b)   Write a minimum number of node equations and put in matrix form.

    (c)   Solve the node equations for the voltages V_a and V_b using MATLAB or the formula for the inverse of a 2 × 2 matrix:

$$\begin{bmatrix} a & b \\ c & d \end{bmatrix}^{-1} = \frac{1}{ad - bc} \begin{bmatrix} d & -b \\ -c & a \end{bmatrix}$$

(d)   Find V_x, V_{da} and V_{db}.

(e)   Find the power delivered by each source and the power absorbed by each resistor. Verify the principle of conservation of power.

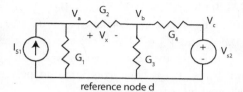

reference node d

Figure P3.5

6. (a)   In the circuit of Figure P3.6, $I_{s1} = 8$ mA, $V_{s2} = 40$ V. Further, $R_1 = 5$ kΩ, $R_2 = R_3 = R_4 = 20$ kΩ, and $R_5 = 10$ kΩ. Find the node voltages, V_A and V_B, and also the voltage, V_x. Compute the power absorbed by R_4 and the power delivered by each of the sources. It is suggested that you write your equations in matrix form and solve using MATLAB or the formula for a 2 × 2 inverse given in problem 5.

(b)   Repeat part (a) when all resistances are cut in half.

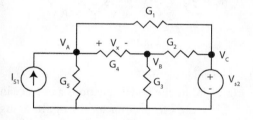

Figure P3.6

7. In the circuit of Figure P3.7, $V_{s1} = 10$ V, $V_{s2} = 4$ V, and $I_{s3} = 1$ mA. Further, in mS, $G_1 = 0.4$, $G_2 = 2$, $G_3 = 3$, and $G_4 = 5$. Use nodal analysis to find V_1 and V_2. Then compute the power delivered by the independent sources and the power absorbed by G_2.

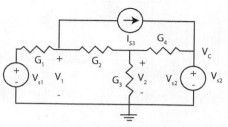

Figure P3.7

8. The circuit of Figure P3.8 is an experimental measurement circuit for determining temperature inside a cavern underneath the Polar ice cap. The cavern is heated by a fissure leading to some volcanic activity deep in the earth. The resistor R_{sensor} changes its value linearly from 15 kΩ to 65 kΩ as a function of temperature over the range −25°C to +25°C. The nominal temperature of the cavern is 0°C. In this type of circuit, the voltage $V_C - V_B$ is a measure of how the temperature changes. Suppose that $V_s = 50$ V, and in kΩ, $R_1 = 20$, $R_2 = 44$, $R_3 = 20$, and $R_4 = 12.5$. Note that the 44 kΩ resistor is a result of manufacturing tolerances that often permit deviations from a nominal of, say, 40 kΩ, by as much as 20%. As usual, it is cost versus precision.

(a)   Write a set of nodal equations in the variables V_B and V_C.

(b)   Assuming $R_{sensor} = 40$ kΩ at 0°C, put the nodal equations in matrix form and solve for the node voltages, V_B and V_C.

(c)   Determine the power delivered by the source.

(d)   Use MATLAB to solve for all the node voltages as R_{sensor} varies from 15 to 65 kΩ in 1 kΩ increments. Do not print out. Then find the linear equation relating R_{sensor} to temperature. Plot $V_B - V_C$ as a function of temperature, i.e., over the range −25°C to +25°C. Over what range of temperatures about 0 degrees would the sensor be reasonably accurate? Why and why not?

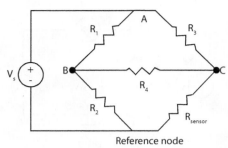

Reference node

Figure P3.8

9. In the circuit of Figure P3.9, all resistances are 1 kΩ, except $R_8 = 500$ Ω. Suppose $V_{s1} = 100$ V and $I_{s2} = 0.3$ A. Compute all the node voltages of the circuit. You may want to use MATLAB or a calculator that inverts matrices to compute the answer. Compute the power delivered by the independent sources.

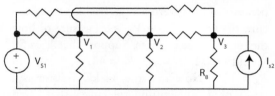

Figure P3.9

ANSWERS IN RANDOM ORDER: 11, 7, 0.6, 7

10. In the circuit of Figure P3.10, $V_{s1} = 30$ V, $I_{s2} = 1.2$ A, and $I_{s3} = 0.6$ A. Use nodal analysis on the circuit below, as indicated:

   (a)   Write a nodal equation at node A.
   (b)   Write a nodal equation at node B.
   (c)   Write a third nodal equation at node C.
   (d)   Solve the 3 equations in 3 unknowns by hand, with your calculator, using MATLAB, or using some other software program to obtain all the node voltages. Show ALL work/procedures.
   (e)   Find the power delivered by the independent voltage source.

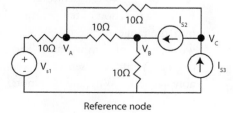

Reference node

Figure P3.10

11. Consider the circuit of Figure P3.11. Choose node D as the reference node. This choice eliminates the floating voltage source and hence the nodal equations can be written without the need of a so-called supernode. Let $G_1 = 0.08$ S, $G_2 = 0.08$ S, $G_3 = 0.01$ S, $G_4 = 0.02$ S, $G_5 = 0.02$ S, $I_{s1} = 0.3$ A, $I_{s2} = 0.2$ A, $I_{s3} = 0.3$ A, and $V_{s4} = 50$ V. Write and solve a set of nodal equations for the voltages $V_A = V_{AD}$ and $V_B = V_{BD}$. Then compute the powers delivered by each of the sources.

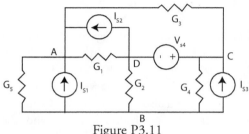

Figure P3.11

12. In the circuit of Figure P3.12, $R_1 = 5$ kΩ, $R_2 = 20$ kΩ, $R_3 = 20$ kΩ, $g_m = 0.55 \times 10^{-3}$, $I_{in} = 5$ mA. Find V_o and the power delivered by the dependent current source.

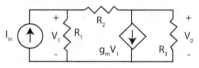

Figure P3.12

13. Consider Figure P3.13.
   (a)   Write the nodal equations and place in matrix form prior to solving. In doing this, let $G_i = 1 / R_i$.
   (b)   With $V_s = 150$ V, $R_s = 1$ kΩ, $R_1 = 5$ kΩ, $R_2 = 10$ kΩ, $R_3 = 10$ kΩ, and $g_m = 7.5$ mS, find V_A, V_B the power delivered by V_s, and the power absorbed by R_3.
   (c)   Compute I_2, the current through R_2 from left to right.

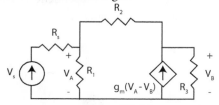

Figure P3.13

14. Consider the circuit of Figure P3.14
   (a) Write two node equations in terms of the literal variables in Figure P3.14 and put in matrix form.
   (c) Solve the node equations for the voltages V_A and V_B when $I_{s1} = 0.1$ A, $I_{s2} = 0.2$ A, $g_{m1} = 7$ mS; $g_{m2} = 2$ mS, $R_1 = 500\ \Omega$; $R_2 = 333.33\ \Omega$; and $R_3 = 1\ \Omega$.
   (d) Determine V_0.
   (e) Determine the power delivered by each source. (Be careful of sign.)

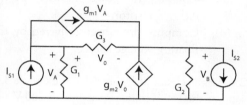

Figure P3.14

15. Consider the circuit of Figure P3.15 in which $V_s = 20$ V and $R_s = 0$; node voltage C is V_s.
   (a) Write the two nodal equations in terms of the literal variables.
   (b) Suppose $\mu = 6$ and the following in S are given: $G_1 = 0.5$, $G_2 = 1$, $G_3 = 0.5$, $G_4 = 4$, $G_5 = 1$. Solve for V_A and V_{out}. Check: 20 and 10 volts.
   (c) Find I_{in} and then find the equivalent resistance seen by the independent voltage source.
   (d) Find the power delivered by the independent source and the dependent source.
   (e) What is the power absorbed by the G_5 output resistor?

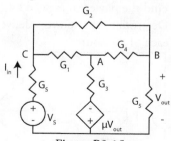

Figure P3.15

16. Redo problem 15 with $G_s = 0.25$ S and $V_s = 60$ V.

17. Use nodal analysis to find the voltages V_A, V_B, V_C, and V_x in the circuit of Figure P3.17. Suppose $R_1 = 20\ \Omega$, $R_2 = 10\ \Omega$, $R_3 = 4\ \Omega$, $g_m = 0.1$ S, $r_m = 10\ \Omega$, and $I_s = 6$ A. Note that in solving this problem, you are to generate three (nodal) equations in which the unknowns are V_A, V_B, V_C; you could eliminate the equation for V_C, but this problem is to illustrate that such elimination is not necessary. Finally, determine the equivalent resistance seen by the independent current source.

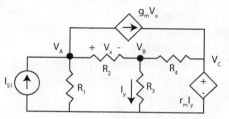

Figure P3.17

18. Consider the circuit of Figure P3.18. By choosing node C as the reference node, we eliminate a floating voltage source. Write an appropriate set of nodal equations, with node C as the reference node. Solve the nodal equations, specify the voltages V_{AC}, V_{BC}, V_{DC} and V_{AB}, and the power delivered by the sources. Finally, find the equivalent resistance seen by the current source. Let $R_1 = 9$ kΩ, $R_2 = 18$ kΩ, $R_3 = 6$ kΩ, $R_4 = 9$ kΩ, $r_m = 3000\ \Omega$, and $I_s = 20$ mA.

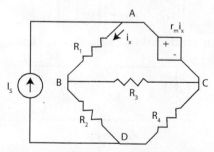

Figure P3.18. By choosing node C as the reference node, it is possible to simplify the construction of the node equations.

19. Consider the circuit of Figure P3.19 in which $R_1 = 20\ \Omega$, $R_2 = 20\ \Omega$, $R_3 = 30\ \Omega$, $R_4 = 60\ \Omega$, $r_{m1} = r_{m2} = 20\ \Omega$, $V_{s1} = 12$ V, and $I_{s2} =$

0.6 A. The point of this problem is to illustrate how a good choice of reference node may simplify the calculation of node voltages, whereas a poor choice may lead to a complicated formulation of the node equations.

(a)   Choose a reference node so that there are no floating voltage sources. Write three equations in the unknown voltages. Solve for the node voltages. CHECK: $i_x = 24$ A and $i_y = -30$ A.

(b)   Determine the power delivered by each source.

(c)   Determine the power absorbed by each resistor.

(d)   Verify conservation of power using the results of parts (b) and (c).

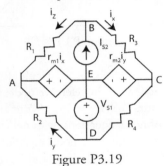

Figure P3.19

20. The nodal equations for the circuit in Figure P3.20 are

$$\begin{bmatrix} 0.03 & -0.01 \\ 0.09 & 0.04 \end{bmatrix} \begin{bmatrix} V_1 \\ V_0 \end{bmatrix} = \begin{bmatrix} I_{in} \\ 0 \end{bmatrix}$$

Compute the values of R_1, R_2, R_3, and g_m.

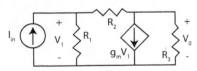

Figure P3.20

21. Consider the circuit of Figure P3.15, which has nodal equations ($R_s \neq 0$) given by

$$\begin{bmatrix} 0.008 & 0.019 & -0.005 \\ -0.001 & 0.0055 & -0.002 \\ -0.005 & -0.002 & 0.107 \end{bmatrix} \begin{bmatrix} V_A \\ V_B \\ V_C \end{bmatrix} = \begin{bmatrix} 0 \\ 0 \\ 0.1 V_s \end{bmatrix} = \begin{bmatrix} 0 \\ 0 \\ G_s V_s \end{bmatrix}$$

Compute the values of R_1, R_2, R_3, R_4, R_5, and μ.

FLOATING VOLTAGE SOURCE PROBLEMS

22. For example 3.5 suppose all resistance values are doubled, the floating voltage source remains the same at 440 V, and all current sources are scaled down to one-half of their original values.

(a)   Compute all node voltages and the current I_{cb}.

(b)   Compute the voltages V_{db}, V_{ab}, and V_{cb}.

(c)   Compute the power absorbed by the 0.075 S resistor.

(d)   Compute the power delivered by the floating voltage source.

23. For the circuit of Figure 3.10 in Example 3.6, suppose the 110 V source is changed to 200 V and the 50 Ω resistor is changed to 500 Ω. Find the node voltages V_a, V_b, V_c, and the unknown currents I_{ac} and I_{cb}. Then find the powers delivered by the 30 V source and the dependent source.

24. Consider the circuit of Figure P3.24 in which $V_{s1} = 200$ V, $V_{s2} = 50$ V, $R_1 = 50$ Ω, $R_2 = 20$ Ω, $R_3 = 50$ Ω, and $R_4 = 40$ Ω.

(a)   Identify the floating voltage source and add a current label through the source.

(b)   Write modified nodal equations, which include both node voltages and unknown currents through any floating voltage sources.

(c)   Solve the equations for the node voltages V_B and V_C and the current through the 50 V source. CHECK: $V_B = 50$ V and $V_C = 100$ V.

(d)   Find the power consumed by R_3.

(e)   Determine the power delivered by each of the sources.

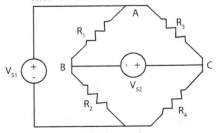

Reference node

Figure P3.24

25. Consider the circuit of Figure P3.25 in which $V_{s1} = 250$ V, $V_{s2} = 50$ V, $R_1 = 50$ Ω , $R_2 = 20$ Ω, $R_3 = 50$ Ω, $R_4 = 40$ Ω , and $R_s = 10$ Ω.
   (a) Identify the floating voltage source and add a current label through the source.
   (b) Write modified nodal equations, which include both node voltages and unknown currents through any floating voltage sources.
   (c) Solve the equations for the node voltages V_B and V_C and the current through the 50 V source. CHECK: $V_B = 50$ V and $V_C = 100$ V.
   (d) Find the power consumed by R_3 .
   (e) Determine the power delivered by each of the sources.

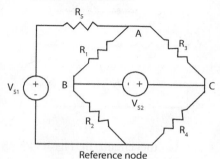

Reference node
Figure P3.25

26. Consider the circuit of Figure P3.26. $R_1 = 10$ Ω , $R_2 = 100$ Ω, $R_3 = 100$ Ω, $R_4 = 50$ Ω , $V_{s1} = 100$ V, $V_{s2} = 60$ V, $V_{s3} = 100$ V, $I_{s4} = 14$ A. Label appropriate currents I_{AB} and I_{BC} through the floating voltage sources.
   (a) Write the modified nodal equations for the three unknown node voltages and two unknown currents.
   (b) Solve for the five unknowns (in MAT-LAB).
   (c) Find the power delivered by each of the sources.

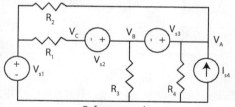

Reference node
Figure P3.26

27. The modified nodal equations for the circuit of Figure P3.27 are

$$\begin{bmatrix} 0.004 & -0.001 & -0.002 & 0 \\ -0.001 & 0.001 & 0 & -1 \\ -0.002 & 0 & 0.004 & 1 \\ 2 & -1 & -1 & 0 \end{bmatrix} \begin{bmatrix} V_A \\ V_B \\ V_C \\ I_{CB} \end{bmatrix} = \begin{bmatrix} 0 \\ I_s \\ 0 \\ 0 \end{bmatrix}$$

Compute all four resistor values and β. Hint: Find all the conductances first and then convert to resistances.

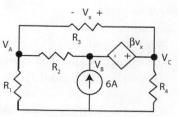

Figure P3.27

28. For the circuit of Figure P3.28, $G_1 = 0.02$ S, $G_2 = 0.025$ S, $G_3 = 0.2$ S, $r_m = 10$ Ω, $I_{s1} = 0.4$ A, and $V_{s2} = 12$ V. Use nodal analysis to find all node voltages, the current I_x , the power absorbed by G_2, and the power delivered by the two sources.
   (a) Determine V_C.
   (b) Label the current I_{AB}.
   (c) Using V_A, V_B, I_{AB}, and I_x as unknowns, write a 4 × 4 matrix set of nodal equations.
   (d) Solve the nodal equations for V_A, V_B, I_{AB}, and I_x.
   (e) Determine the power absorbed by G_2 .
   (f) Determine the power delivered by all sources.

ANSWERS (D) IN RANDOM ORDER: 12, 0.24, 9.6, 0.16

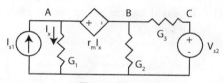

Figure P3.28

29. Repeat Problem 28, except this time write only three nodal equations in the variables V_A, V_B, and I_{AB}. Notice that you must express I_x in

terms of V_A and the appropriate conductance. One can even reduce the number of equations to two using the so-called supernode approach, which is the subject of other texts.

30. For the circuit of Figure P3.30, $R_1 = 100\ \Omega$, $R_2 = 20\ \Omega$, $R_3 = 20\ \Omega$, $G_4 = 0.09\ S$, $r_m = 300\ \Omega\ A$, $I_{s1} = 2\ A$, $V_{s2} = 50\ V$. Use nodal analysis to find all node voltages, the current I_x, the power absorbed by the 20 Ω resistor between nodes B and C, and the power delivered by the independent sources as follows:
   (a) Determine V_C.
   (b) Write a set of modified nodal equations that contain extra current variables including I_{AB} and I_x.
   (c) Solve your nodal equations for the unknowns. CHECK: $V_B = 180\ V$.
   (d) Compute the power delivered by the independent sources.
   (e) Compute the power absorbed by the 20 Ω resistor between nodes B and C.

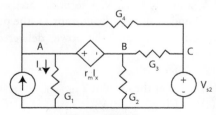

Figure P3.30

31. Use nodal analysis on the circuit of Figure P3.31 as indicated. $V_{s1} = 100\ V$, $I_{s2} = 1\ A$, and all resistors are 10 Ω.
   (a) Write modified nodal equations including the extra variable I_{BC}.
   (b) Solve the modified nodal equations in MATLAB.
   (c) Find the power delivered by each of the sources.

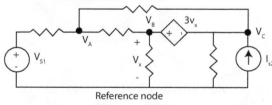

Reference node

Figure P3.31

32 Consider the circuit of Figure P3.32, where $R_1 = 4\ k\Omega$, $R_2 = 1\ k\Omega$, $R_3 = 4/3\ k\Omega$, $g_m = 0.75$ mS, $\mu = 4\ S$, $V_{s1} = 160\ V$, and $I_{s2} = 40\ mA$.
   (a) Specify V_A.
   (b) Write modified nodal equations.
   (c) Solve the modified nodal equations in MATLAB.
   (d) Find the power delivered by each of the sources.
   (e) Compute the power absorbed by each resistor.
   (f) Verify conservation of power.

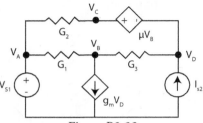

Figure P3.32

33. Consider the circuit in Figure P3.33 in which $V_{in} = 60\ V$, $G_1 = 0.1\ S$, $G_2 = 0.1\ S$, $G_3 = 0.3\ S$, $G_4 = 0.4\ S$, $G_5 = 0.1\ S$, $G_6 = 0.1\ S$, $G_7 = 7/480\ S$, $\mu = 3$, and $\beta = 2$.
   (a) Write the modified nodal equations using V_A, V_B, V_C and I_{AC} as unknowns.
   (b) Solve the modified nodal equations in MATLAB.
   (c) Find $R_{eq} = V_{in}/I_{in}$, the equivalent resistance seen by the independent voltage source.

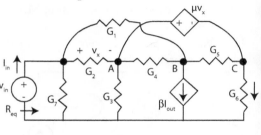

Figure P3.33

SINGLE LOOP-EQUATION PROBLEMS

34. In the circuit of Figure P3.34, $R_1 = 400\ \Omega$, $R_2 = 50\ \Omega$, and $\mu = 0.5$. If $V_{in} = 50$ V, find I_{in}, the power delivered by the independent and dependent voltage sources, and the equivalent resistance, R_{eq}, seen by the independent source.

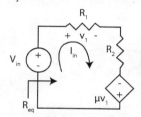

Figure P3.34

CHECKS: $R_{eq} = 250\ \Omega$, and $P_{deps} = 8$ watts.

35. In the circuit of Figure P3.35, $R_1 = 400\ \Omega$, $R_2 = 50\ \Omega$, and $r_m = 50\ \Omega$, and $\mu = 0.5$. If $V_{in} = 50$ V, find I_{in}, the power delivered by the independent and dependent voltage sources, and the equivalent resistance, R_{eq}, seen by the independent source.

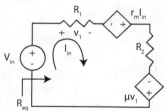

Figure P3.35

CHECKS: $R_{eq} = 200\ \Omega$, and $P_{indeps} = 12.5$ watts.

36. In the circuit of Figure P3.36, $V_{s1} = 200$ V and $I_{s2} = 20$ mA. Find I_{in}. Then find the power delivered by each of the independent sources. Finally, find the power absorbed by each resistor and verify conservation of power for this circuit.

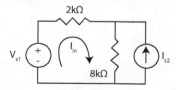

Figure P3.36

CHECK: $I_1 = 4$ mA.

37. Consider the circuit of Figure P3.37
(a) Suppose $R_1 = 200\ \Omega$, $R_3 = 300\ \Omega$, $R_2 = 500\ \Omega$, $I_{s1} = 750$ mA and $I_{s2} = 100$ mA. Find I_1 and the power delivered by each of the independent sources.

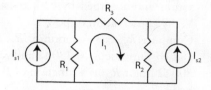

Figure P3.37

CHECK: $I_1 = 100$ mA.
(b)   Now suppose $I_{s1} = 400$ mA and $I_{s2} = 100$ mA and the loop equation for I_1 written in the standard way directly yields $2000I_1 = 60$. Find R_1 and R_2 if $R_3 = 600\ \Omega$. Note: If the equations are not written in the standard way, the solution is not unique. For example, multiplying both sides of the above equation by 0.5 yields a different answer in which $R_1 = 140\ \Omega$, as opposed to the correct answer of $R_1 = 400\ \Omega$.

38. In the circuit of Figure P3.38, $V_{s1} = 56$ V, $I_{s2} = 100$ mA, $I_{s3} = 200$ mA, and $I_{s4} = 100$ mA mA. Find I_1 and the power delivered by each independent source.

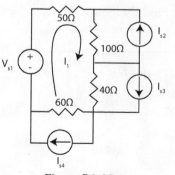

Figure P3.38

CHECK: Sum of powers delivered by the sources is 15.68 watts.

39. Consider the circuit of Figure P3.39.
    (a) Suppose $R_1 = 250\ \Omega$, $R_2 = 500\ \Omega$, $V_s = 100\ V$, and $\beta = 0.5$. Use loop analysis to find I_1 and R_{eq}.
    (b) Compute the power delivered by each source and absorbed by each resistor. Verify conservation of power.
    (c) Compute R_{eq} as a function of R_1, R_2, and β. Suppose $R_1 = 250\ \Omega$ and $R_2 = 500\ \Omega$, plot R_{eq} as a function of β, $0 \le \beta \le 2$.

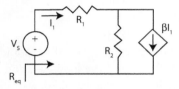

Figure P3.39

CHECK: Power absorbed by resistors is 15 watts and $R_{eq} > 450\ \Omega$.

40. (a) For the circuit of Figure P3.40, $R_1 = 1\ k\Omega$, $R_2 = 5\ k\Omega$, $R_3 = 4\ k\Omega$, $I_s = 100\ mA$ and $g_m = 4 \times 10^{-4}\ S$. Find I_1 and V_x by writing two equations in the two unknowns I_1 and V_x. The first equation is the usual loop equation and the second determines the relationship of I_1 and V_x.
    (b) Given your answer to (a), find the equivalent resistance, R_{eq}, seen by the independent source.
    (c) Find the power delivered by the dependent source.

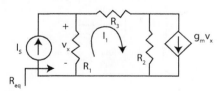

Figure P3.40

MULTIPLE LOOP PROBLEMS

41. The circuit of Figure P3.41 represents two non-ideal batteries $V_{s1} = 21\ V$ and $V_{s2} = 24\ V$

with internal resistances $R_1 = 20\ \Omega$ and $R_2 = 80\ \Omega$ (faulty connection) respectively connected in parallel to supply power to a load of $R_3 = 80\ \Omega$. Compute the power absorbed by the load R_3 and the power delivered by each independent source. Which battery supplies more current to R_3 and hence more power to the load? How much power is wasted by the internal resistances of the battery?

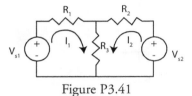

Figure P3.41

CHECK: $P_{s1} = 3.15$ watts and $P_{s2} = 1.8$ watts.

42. Reconsider the circuit of Problem 3.41, redrawn with different loop currents in Figure 3.42a and 3.42b. The point of this problem is to verify that different sets of independent loop equations produce the same element currents and branch voltages.
    (a) Write the new loop equations and find I_1 and I_2 for the circuit of Figure 3.42a, and then find the voltage across and the power consumed by R_3.
    (b) Write the new loop equations and find I_1 and I_2 for the circuit of Figure 3.42b, and then find the voltage across and the power consumed by R_3.

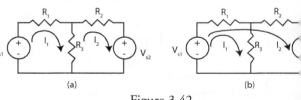

(a)   (b)

Figure 3.42

43. The matrix loop equation of the circuit of Figure P3.43 is

$$\begin{bmatrix} 150 & -40 & -100 \\ -40 & 140 & 0 \\ -100 & 0 & 150 \end{bmatrix} \begin{bmatrix} I_1 \\ I_2 \\ I_3 \end{bmatrix} = \begin{bmatrix} 100 \\ -20 \\ 20 \end{bmatrix}$$

Find the value of each resistance and each source in the circuit.

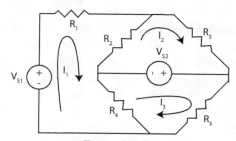

Figure P3.43

CHECKS: $V_{s2} = 20$ V, $R_2 = 40$ Ω.

44. The mesh equations for the circuit of Figure P3.44 are

$$\begin{bmatrix} 40 & -80 & -10 \\ -30 & 130 & -50 \\ -10 & -50 & 70 \end{bmatrix} \begin{bmatrix} i_1 \\ i_2 \\ i_3 \end{bmatrix} = \begin{bmatrix} v_1 \\ -v_2 \\ 0 \end{bmatrix}$$

Find R_1, R_2, R_3, R_4, and r_m.

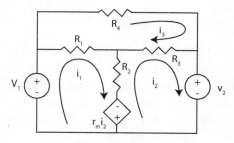

Figure P3.44

45. Figure P3.45a shows an electric locomotive propelled by a dc motor. The locomotive pulls a train of 12 cars. The motor behaves like a 590 V battery in series with a 1.296 Ω resistor. Suppose the train is midway between stations, West side and East side, where 660 V dc sources provide electricity. The resistance of the rails affects the current received by the locomotive. The equivalent circuit diagram is given by Figure P3.45b, where $R = 0.15$ Ω. Using mesh analysis find

(a)   the currents I_1 and I_2

(b)   the current in the locomotive motor and the power absorbed by the locomotive

(c)   repeat parts (a) and (b) when the locomotive is 1/3 distant from either station

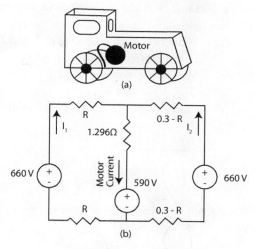

Figure P3.45

46. Reconsider Problem 3.45. Let $V_m = 590$ V and $R_m = 1.296$ Ω. This time, suppose there are two locomotives on the track. One is 1/3 distant from the East side station, and the other is 1/3 distant from the West side station.

(a)   Determine the resistance R in Figure P3.46.

(b)   Using the indicated currents, write a set of three mesh equations and solve for I_1, I_2, and I_3.

(c)   Determine the two motor currents.

(d)   Determine the power delivered by each of the 660 V sources.

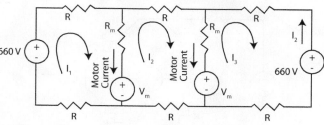

Figure P3.46

47. Reconsider the Problem 3.5 and the circuit of Figure P3.5. Draw two loop currents and

solve for these currents. Then compute the node voltages V_a, V_b, and V_c.

48. In Figure P3.48, let $R_1 = 9$ kΩ, $R_2 = 18$ kΩ, $R_3 = 6$ kΩ, $R_4 = 9$ kΩ, $r_m = 3000$ Ω, and $I_s = 20$ mA.
   (a) Write two mesh equations in I_1 and I_2. Put in matrix form and solve.
   (b) Specify the voltages V_{AB}, V_{BD}, V_{AC}, V_{BC}, V_{CD}, and V_{AD}.
   (c) Find the power delivered by each of the sources.

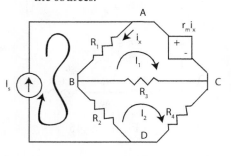

Figure P3.48

49. Consider the circuit of Figure P3.49 in which $V_{s1} = 250$ V, $V_{s2} = 50$, V, $R_1 = 50$ Ω, $R_2 = 20$ Ω, $R_3 = 50$ Ω, $R_4 = 40$ Ω, and $R_s = 10$ Ω.
   (a) Write three standard loop equations and put in matrix form.
   (b) Solve the equations for the loop currents and determine the node voltages V_A, V_B and V_C.
   (c) Find the power consumed by R_3.
   (d) Determine the power delivered by each of the sources.

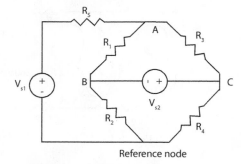

Figure P3.49

50. Consider the circuit of Figure P3.50 in which $V_{s1} = 40$ V and $V_{s2} = 20$ V. Write a set of three loop equations by inspection. Refer to Example 3.7 and the discussion following the example. Solve the loop equations using matrix methods via your calculator or an appropriate software program. Compute the voltage v. Note that I_1 and I_2 should have values identical with those in example 3.7. Finally, find the power delivered by each of the sources.

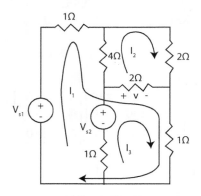

Figure P3.50

51. Consider the circuit of Figure P3.51.
   (a) Write two mesh equations and put in matrix form in terms of the literal parameters.
   (b) Solve the mesh equations for the unknown currents assuming $R_1 = 100$ Ω, $R_2 = 40$ Ω, $R_3 = 20$ Ω, $r_m = 80$ Ω, A, and $V_{s2} = 80$ V.
   (c) Find V_A and V_B.
   (d) Find the power delivered by the independent sources.
   (e) Find the power delivered by the dependent source.

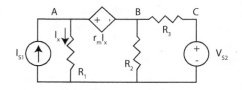

Figure P3.51

52. Repeat Problem 51 when $R_1 = 100\ \Omega$, $R_2 = 40\ \Omega$, $R_3 = 80\ \Omega$, $r_m = 60\ \Omega$, $I_{s1} = 1$ A, and $V_{s2} = 40$ V.

53. Consider the circuit of Figure P3.53.
  (a) Write two mesh equations in terms of the literal parameter values.
  (b) Solve the mesh equations assuming $R_1 = 100\ \Omega$, $R_2 = 40\ \Omega$, $R_3 = 80\ \Omega$, $R_4 = 80\ \Omega$, $r_m = 60\ \Omega$, $I_{s2} = 1$ A, and $V_{s1} = 40$ V.
  (c) Compute V_{R2}, V_B and V_C.
  (d) Find the power delivered by all the sources in the circuit.

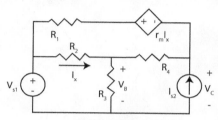

Figure P3.53

54. Repeat Problem 53 when $R_1 = 40\ \Omega$, $R_2 = 30\ \Omega$, $R_3 = 40\ \Omega$, $R_4 = 20\ \Omega$, $r_m = 10\ \Omega$, $I_{s2} = 0.25$ A, and $V_{s1} = 60$ V.

MODIFIED LOOP ANALYSIS PROBLEMS

55. Consider the circuit of Figure P3.55. The objective of this example is to illustrate a numerical approach to loop analysis where the number of variables to be found is quite large, but the equations are quite easy to write and do not require multiple substitutions.
  (a) If the loop equation matrix is of the form below, compute the undetermined entries.

$$\begin{bmatrix} R_2 + ?? & 1 & 1 + ?? \\ ?? & -1 & 0 \\ R_3 & ???? & -1 \end{bmatrix} \begin{bmatrix} I_1 \\ V_x \\ V_y \end{bmatrix} = \begin{bmatrix} V_s - R_2 I_{in} \\ 0 \\ 0 \end{bmatrix}$$

  (b) If $R_1 = 500\ \Omega$, $R_2 = 100\ \Omega$, $R_3 = 400\ \Omega$, $R_4 = 100\ \Omega$, $V_s = 150$ V, $I_{in} = 0.5$ A, find the three unknowns.

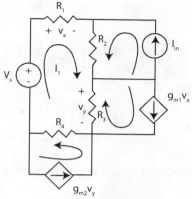

Figure P3.55

CHECK: $I_1 = 0.1$ A, $V_x = 50$ V, and $V_y = 20$ V.

56. Consider the circuit of Figure P3.56.
  (a) Write the modified loop equations (using the indicated loops) in matrix form.
  (b) If $V_{s1} = 200$ V, $I_{s2} = 0.3$ A, $R_1 = R_3 = 100\ \Omega$, $R_2 = 400\ \Omega$, and $\mu = 0.5$, find I_1, I_2, and V_{s2}.
  (c) What is the power delivered by the three sources?

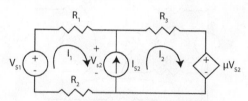

Figure P3.56

57. Consider the circuit of Figure P3.57.
  (a) Write the modified loop equations in matrix form.
  (b) If $V_{s1} = 210$ V, $V_{s2} = 150$ V, $I_{s3} = 0.1$ A, $R_1 = 200\ \Omega$, $R_2 = 400\ \Omega$, $R_3 = 100\ \Omega$, $R_4 = 500\ \Omega$, and $R_5 = 1150\ \Omega$, compute I_1, I_2, I_3, and V_{s3}.
  (c) Compute the power delivered by the independent sources.

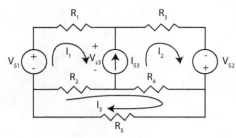

Figure P3.57

CHECK: $I_1 = 0.4$ A and $V_{s3} = 50$ V.

58. Consider the circuit of Figure P3.58.
    (a) Write the modified loop equations in matrix form in terms of the literal values.
    (b) If $V_{s1} = 400$ V, $V_{s2} = 200$ V, $R_1 = 30 \ \Omega$, $R_2 = 20 \ \Omega$, $R_3 = 270 \ \Omega$, $R_4 = 80 \ \Omega$, $R_5 = 140 \ \Omega$, and compute I_1, I_2, I_3, and V_3.
    (c) Compute the power delivered by all sources.
    (d) Compute the power absorbed by each resistor and verify conservation of power.

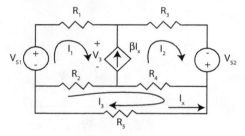

Figure P3.58

CHECKS: $V_3 = 265$ V and $P_{dep} = -397.5$ watts.

CHAPTER 4

The Operational Amplifier

Amplification of voice allows announcers at sports events to convey their comments on the play-by-play action to the crowd. At concerts, high-powered amplifiers project a singer's voice and the instrumental music into a crowded auditorium. Electronic amplifiers make this possible. One of the simplest and most common amplifiers is the operational amplifier, the subject of this chapter. The word "operational," though, suggests a purpose beyond simple amplification. Often one must sum signals to produce a new signal, or take the difference of two signals. Sometimes one must decide whether one dc signal is larger than another. The operational amplifier is operational precisely because it can be configured to do these things and many other tasks, as we will see later in the text.

CHAPTER OBJECTIVES

1. Introduce the notion of an ideal operational amplifier, called an op amp.
2. Describe and analyze basic op amp circuits.
3. Describe and illustrate a simple method for designing a general summing amplifier.
4. Describe and illustrate the phenomenon of saturation in op amp circuits and describe circuits that utilize saturation for their operation.

SECTION HEADINGS

1. **Introduction**
2. **The Idealized Operational Amplifier: Definition and Circuit Analysis**
3. **The Design of General Summing Amplifiers**
4. **Saturation and the Active Region of the Op Amp**
5. **Summary**
   Terms and Concepts
   Problems

1. INTRODUCTION

Chapters 1 and 2 defined and discussed independent and dependent voltage and current sources. Chapter 3 investigated the nodal and loop analysis of resistive circuits containing such sources. Often, dependent sources supply energy and power to a circuit, making them so-called **active** elements. On the other hand, resistors are **passive** elements because they only absorb energy. Circuit models of real amplifiers (see Examples 3.3 and 3.4 with associated Figures 3.7 and 3.8, respectively) contain controlled sources that underlie their analysis and performance evaluation. Indeed, the VCVS is the core component of the operational amplifier (op amp), the main focus of this chapter. Thus, the op amp is an active circuit element whose analysis is done with the techniques of Chapters 1 through 3.

A real op amp is a semiconductor device consisting of nearly two dozen transistors and a dozen resistors sealed in a package from which a small number of terminals protrude, as shown in Figure 4.1(a). Despite its apparent internal complexity, advances in integrated circuit manufacturing technology have made the op amp only slightly more expensive than a single discrete transistor. Its simplicity, utility, reliability, and low cost have made the op amp an essential basic building block in communication, control, and the instrumentation circuits that can be found in all undergraduate EE laboratories.

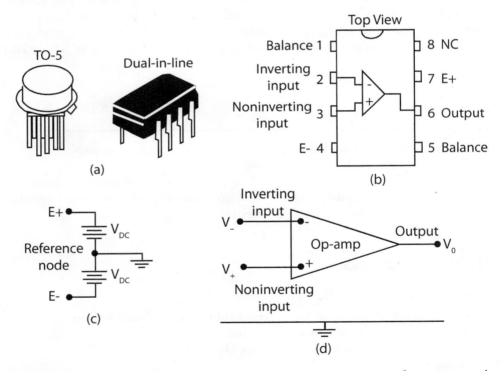

FIGURE 4.1 (a) Typical op amp packages; (b) typical terminal arrangement of an op amp package; (c) dual power supply notation; (d) essential terminals for circuit analysis.

Figure 4.1(b) shows a typical arrangement of terminals for a dual-in-line op amp package. The terminal markings and the symbol shown in Figure 4.1(b) do not appear on the actual device, but are included here for reference. In Figure 4.1(b), the terminal labeled "NC" (no connection) is not used. The E+ and E- terminals (Figure 4.1(b)) are connected to a dual power supply, illustrated in Figure 4.1(c), where V_{DC} typically ranges between 3 V and 15 V, depending on the application; adequate voltage is required for proper operation. The three terminals in Figure 4.1(b) marked "inverting input," "non-inverting input," and "output" interact with a surrounding circuit, and correspond to V_-, V_+, and V_0 in Figure 4.1(d). The two terminals labeled *balance* or *off-set* have importance only when the op am is part of a larger circuit: resistors of appropriate values are connected to these terminals to make sure the output voltage is zero when the input voltage is zero. This "balancing process" is best discussed in a laboratory session.

This chapter sketches the basic properties of the op amp: just enough to understand some of the interesting applications. The ideal op amp model and the saturation model are described. Using these models and the principles of analysis covered in Chapters 1 through 3, we then analyze the behavior of some widely used op amp configurations. These application examples hint at the importance of the op amp and furnish motivation for the study of electronic circuits.

Several of the examples include a SPICE simulation of the circuit being analyzed. SPICE is a sophisticated circuit simulation program. Behind the user-interface, SPICE uses complex models of the real operational amplifier. Our purpose in using SPICE simulation is to verify or test the theoretical analysis set forth in the examples. What we show is that the simplified theoretical analysis provides a very good approximation to the actual circuit behavior represented in the SPICE simulation results. Industrial circuit designers often use SPICE to visualize the expected behavior of very complex circuits. Later chapters cover some of the more complex op amp applications.

2. THE IDEALIZED OPERATIONAL AMPLIFIER

This section analyzes resistive circuits containing an operational amplifier. Figure 4.2 explicitly shows an op amp embedded in a surrounding resistive circuit.

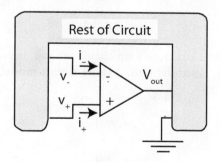

FIGURE 4.2

One possibility for analyzing op amp circuits is to represent the op amp by one of the simplified models shown in Figure 4.3 that do not account for saturation effects. The first model of Figure 4.3(a) consists of an input resistor, R_{in}, an output resistor, R_{out}, and a VCVS with finite gain A. Of practical import is the idealization of this model (Figure 4.3(a)) to the one of Figure 4.3(b) by

(1) letting R_{in} become infinite, setting up an open circuit condition at the input terminals; (2) letting R_{out} become zero, making the output voltage of the op amp equal to that of the VCVS; and (3) letting the gain A approach infinity. These conditions are idealizations because (1) with R_{in} infinite, there is no loading to a circuit attached to the input; (2) with $R_{out} = 0$, the full output voltage appears across any circuit connected to the output; and (3) $A \rightarrow \infty$ leads to a simplification of the associated analysis. These conditions, stated below as equation 4.1, define the so-called **ideal operational amplifier**:

$$A \rightarrow \infty \ \text{(infinite gain)} \tag{4.1a}$$

$$R_{in} \rightarrow \infty \ \text{(infinite input resistance)} \tag{4.1b}$$

$$R_{out} \rightarrow 0 \ \text{(zero output resistance)} \tag{4.1c}$$

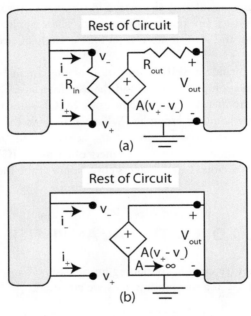

FIGURE 4.3

To see how this idealization simplifies op amp circuit analysis, consider an equivalent set of conditions for the ideal op amp, called the **virtual short circuit model**:

$$i_+ = 0 \tag{4.2a}$$

$$i_- = 0 \tag{4.2b}$$

$$v_+ = v_- \tag{4.2c}$$

From Figure 4.3(b), the conditions that $i_+ = 0$ and $i_- = 0$ follow directly from the open circuit condition at the input terminals. The condition that $v_+ = v_-$ (hence, the term "virtual short circuit") will be discussed later, but occurs because $A \rightarrow \infty$, forcing $(v_+ - v_-) \rightarrow 0$.

The recommended way to analyze circuits containing op amps is to replace any ideal op amp by the model of Figure 4.3(b), the virtual short circuit model of equation 4.2. The following examples illustrate the use of the virtual short circuit model.

EXAMPLE 4.1. This example investigates the **inverting amplifier** of Figure 4.4, which is used in a wide range of commercial circuits. The objective is to compute v_{out} in terms of R_1, R_f and v_{in}.

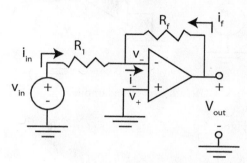

FIGURE 4.4 Inverting amplifier, assuming an ideal op amp in which $v_{out} = -\dfrac{R_f}{R_{in}} v_{in}$.

SOLUTION

Step 1. *Compute v_+ and v_-.* Since the + terminal is grounded, $v_+ = 0$. From the virtual short property of the ideal op amp, $v_- = v_+ = 0$.

Step 2. *Compute i_{in}.* Since $v_- = 0$, the voltage across R_1 is v_{in}. From Ohm's law,

$$i_{in} = \frac{v_{in}}{R_1}$$

Step 3. *Compute i_f.* Again, since $v_- = 0$, the voltage across R_f is v_{out}. From Ohm's law,

$$i_f = \frac{v_{out}}{R_f}$$

Step 4. *Relate the currents i_{in} and i_f, and substitute the results of Steps 2 and 3.* From KCL, $i_{in} - i_-$ $+ i_f = 0$. From the properties of the ideal op amp, $i_- = 0$, in which case, $i_f = -i_{in}$. This implies that

$$\frac{v_{out}}{R_f} = -\frac{v_{in}}{R_1}$$

Hence, the voltage gain relationship of the inverting op am circuit is

$$v_{out} = -\frac{R_f}{R_{in}} v_{in} \qquad (4.3)$$

Equation 4.3 shows that the input and output voltages are always of opposite polarity, hence the name **inverting amplifier**. One also observes that by choosing proper values for R_f and R_1, a voltage gain of any magnitude is possible, in theory. In practice, other factors limit the range of obtainable gains.

Exercises. 1. Find v_{out} for the circuit of Figure 4.5.

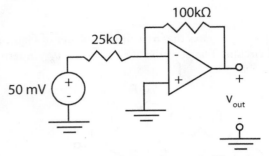

FIGURE 4.5 Inverting amplifier.

2. Find v_{out} for the circuit of Figure 4.6.

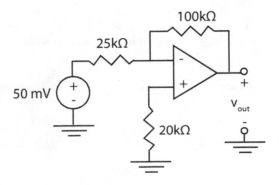

FIGURE 4.6 Inverting amplifier with additional resistor.

ANSWER FOR BOTH: $v_{out} = -200$ mV.

A few remarks are in order. Op amp configurations in which one of the input terminals is grounded, as is the non-inverting terminal in Figure 4.4, are said to operate in the *single-ended mode*. The input terminal can be grounded directly or through a resistor, as in Exercise 2 above. Also, since $v_- = v_+$ or $v_- - v_+ = 0$, the terminals are *virtually* short circuited even though there is no hard-wired direct connection between them. This condition is called a **virtual short circuit**. Further, if one of the terminals is grounded, then the other terminal is said to be **virtually grounded**, as is the case in Figures 4.4, 4.5, and 4.6. Specifically, in Figure 4.4, there is a **virtual ground** at the inverting input terminal.

The next example continues the investigation of the ideal **inverting amplifier** for the two-input, single-output op amp circuit of Figure 4.7. The solution again makes use of the virtual ground and virtual short circuit properties of the ideal op amp.

EXAMPLE 4.2. For the circuit of Figure 4.7, our objective is to compute v_{out} in terms of R_1, R_2, R_f and the two input voltages v_{s1}, and v_{s2}.

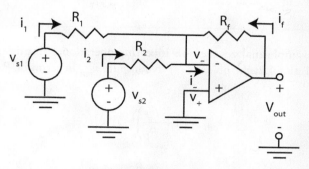

FIGURE 4.7 Inverting (ideal) amplifier with two inputs for which

$$v_{out} = -\frac{R_f}{R_1}v_{s1} - \frac{R_f}{R_2}v_{s2}.$$

SOLUTION. As in Example 4.1, and by the same reasoning described there, $v_- = v_+ = 0$.

Step 1. *Compute i_1 and i_2.* Since $v_- = v_+ = 0$, the voltage across R_1 is v_{s1} and the voltage across R_2 is v_{s2}. From Ohm's law,

$$i_1 = \frac{v_{s1}}{R_1} \text{ and } i_2 = \frac{v_{s2}}{R_2}$$

Step 2. *Compute i_f.* Again, since $v_- = 0$, the voltage across R_f is v_{out} and from Ohm's law,

$$i_f = \frac{v_{out}}{R_f}$$

Step 3. *Relate the currents i_1, i_2 and i_f and then substitute the results of Steps 1 and 2.* From KCL, $i_1 + i_2 - i_- + i_f = 0$. From the properties of the ideal op amp, $i_- = 0$, in which case, $i_f = -(i_1 + i_2)$. This implies that

$$\frac{v_{out}}{R_f} = -\frac{v_{s1}}{R_1} - \frac{v_{s2}}{R_2}$$

Hence,

$$v_{out} = -\frac{R_f}{R_1}v_{s1} - \frac{R_f}{R_2}v_{s2}$$

$$(4.4)$$

Exercise. In Figure 4.7, suppose $R_f = 100$ kΩ. Find R_1 and R_2 so that $v_{out} = -4v_{s1} - 2v_{s2}$.
ANSWER: $R_1 = 25$ kΩ and $R_2 = 50$ kΩ

EXAMPLE 4.3. This example analyzes the **non-inverting** operational **amplifier** circuit of Figure 4.8. As in Examples 4.1 and 4.2, the objective is to compute v_{out} in terms of R_1, R_2, and v_{in}. We show that

$$v_{out} = \left(1 + \frac{R_f}{R_1}\right) v_{in}.$$

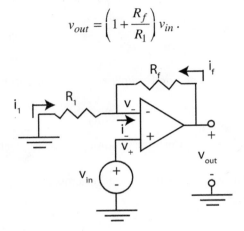

FIGURE 4.8 A non-inverting op amp circuit.

SOLUTION
Step 1. *Compute v_+ and v_-.* Since the + terminal is connected to the input voltage source, $v_+ = v_{in}$. From the virtual short property of the ideal op amp, $v_- = v_+ = v_{in}$.

Step 2. *Compute i_1.* Since $v_- = v_{in}$, the voltage across the resistor R_1 is v_{in}. Observe that the current, i_1, has reference direction different from the passive sign convention. Hence, from Ohm's law,

$$i_1 = -\frac{v_{in}}{R_1}$$

Step 3. *Compute i_f.* Again, since $v_- = v_{in}$, the voltage across R_f is $(v_{out} - v_{in})$. From Ohm's law,

$$i_f = \frac{v_{out} - v_{in}}{R_f}$$

Step 4. *Relate the currents i_1 and i_f, and substitute the results of Steps 1 and 2.* From KCL, $i_1 - i_- + i_f = 0$. From the ideal op amp property of equation 4.2, $i_- = 0$, forcing $i_f = -i_1$. This implies that

$$\frac{v_{out} - v_{in}}{R_f} = \frac{v_{in}}{R_1}$$

Hence, the input-output voltage relationship

$$v_{out} = \left(1 + \frac{R_f}{R_1}\right) v_{in} \tag{4.5}$$

From equation 4.5, the voltage gain is greater than 1, i.e., $\left(1 + \frac{R_f}{R_1}\right) > 1$, and v_{out} and v_{in} always have the same polarity; the circuit is naturally called a **non-inverting amplifier**.

Exercise. For the non-inverting amplifier of Figure 4.8, find R_1 and R_f so that the gain is 2, and when v_{in} = 5 V, the power absorbed by R_f is 5 mW.
ANSWER: R_f = 5 kΩ and R_1 = 5 kΩ

EXAMPLE 4.4. This example analyzes the ideal general **difference amplifier** circuit of Figure 4.9. We show that

$$v_{out} = \left(1 + \frac{R_f}{R_1}\right)\left(\frac{R_g}{R_g + R_2}\right) v_{s2} - \frac{R_f}{R_1} v_{s1}$$

In a basic **difference amplifier**, the output is the difference of two input voltages. For the general difference amplifier of this example, the output is a difference of the scaled input voltages, $v_{out} = a_2 v_{s2} - a_1 v_{s1}$ for appropriate positive a_1 and a_2.

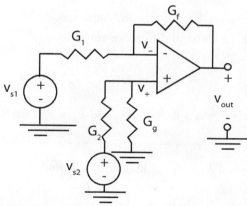

FIGURE 4.9 A general difference amplifier circuit.

SOLUTION. From the ideal op amp property of equations 4.2, $v_- = v_+$ and no current enters the inverting and non-inverting op amp input terminals.

Step 1. *Write a node equation at the non-inverting input terminal of the op amp.* Summing the currents leaving the + node of the op amp yields

$$G_2(v_+ - v_{s2}) + G_g v_+ = 0$$

Solving for v_+ leads to

$$v_+ = \frac{G_2}{G_2 + G_g} v_{s2} \tag{4.6a}$$

Step 2. *Write a node equation at the inverting input terminal of the op amp.* Recall $v_- = v_+$. The sum of the currents leaving the – node satisfies

$$G_1(v_+ - v_{s1}) + G_f(v_+ - v_{out}) = 0$$

Thus,

$$v_{out} = -\frac{G_1}{G_f} v_{s1} + \left(1 + \frac{G_1}{G_f}\right) v_+ \tag{4.6b}$$

Step 3. *Combine Steps 1 and 2.* Substituting equation 4.6a into 4.6b yields

$$v_{out} = \left(1 + \frac{G_1}{G_f}\right)\left(\frac{G_2}{G_2 + G_g}\right) v_{s2} - \frac{G_1}{G_f} v_{s1} \tag{4.7a}$$

or, in terms of resistances,

$$v_{out} = \left(1 + \frac{R_f}{R_1}\right)\left(\frac{R_g}{R_g + R_2}\right) v_{s2} - \frac{R_f}{R_1} v_{s1} \tag{4.7b}$$

Equations 4.7 have the desired form: $v_{out} = a_2 v_{s2} - a_1 v_{s1}$ for appropriate positive constants a_2 and a_1, which can be obtained by proper choices of the resistors.

Two special cases of Example 4.4 are of practical importance. First, if $R_f = R_1$ and $R_g = R_2$, then equation 4.7b reduces to the **classical difference amplifier** equation,

$$v_{out} = K(v_{s2} - v_{s1})$$

with $K = 1$ and for an arbitrary $K > 0$, $R_f = KR_1$ and $R_g = KR_2$ fits the bill.

Exercises. 1. In Figure 4.9, if $R_1 = R_2 = 5\ \text{k}\Omega$ and $K = 2$, find R_f, and R_g.
ANSWER: $R_f = R_g = 10\ \text{k}\Omega$

2. Using the circuit of Figure 4.9, design a difference amplifier so that $v_{out} = 4(v_{s2} - v_{s1})$ and the feedback resistance $R_f = 20\ \text{k}\Omega$.
ANSWER: $R_1 = R_2 = 5\ \text{k}\Omega$ and $R_g = 20\ \text{k}\Omega$

3. In Exercise 1, suppose R_1 and R_f are scaled by a positive constant K_1, i.e., $R_{new} = K_1 R_{old}$ and R_2 and R_g are scaled by a positive constant K_2. Determine the new input-output relationship.
ANSWER: $v_{out} = K(v_{s2} - v_{s1})$, with K the same as in Exercise 1

The point of Exercise 3 is that the group $\{R_1, R_f\}$ can be independently scaled by K_1 and the group $\{R_2, R_g\}$ independently scaled by K_2 without affecting the gain of equation 4.7b.

EXAMPLE 4.5. This example analyzes a special case of the non-inverting amplifier called the **buffer** or **isolation amplifier,** shown in Figure 4.10, where $v_{out} = v_{in}$. When connected between two circuits, the buffer amplifier prevents one circuit from having a loading effect on the other.

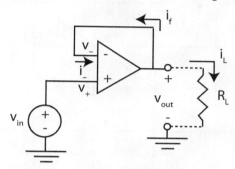

FIGURE 4.10 The buffer or isolation amplifier for which $v_{out} = v_{in}$.

SOLUTION. From the connection shown in Figure 4.10, $v_{in} = v_+$ and $v_{out} = v_-$. From the properties of the ideal op amp, $v_+ = v_-$, in which case $v_{out} = v_{in}$.

Exercise. Compute the power delivered by the source in Figure 4.10 and the power delivered to the load R_L.

ANSWER: 0 and $\dfrac{v_{in}^2}{R_L}$

The circuit of Figure 4.10 is called an isolation or buffer amplifier, because no current is drawn from the source v_{in}. However, the op amp does supply current (and power) directly to the load by maintaining $v_{out} = v_{in}$ under the condition that i_L not exceed the manufacturer's maximum output current rating. Since $v_{out}(t) = v_{in}(t)$, the circuit is also called a **voltage follower**.

Figure 4.11 shows a SPICE simulation that verifies the behavior arrived at in Example 4.5. Here a dc voltage sweep, $0 \le v_{in} \le 12$ V, was input to a highly accurate SPICE model of a Burr Brown 741 connected to ±10 V power supply. Observe in Figure 4.11 that the output follows the input up to the 10-volt value, after which, the output remains at 10 V despite increased input values. This non-ideal phenomenon, called **saturation,** is due to the power supply voltage level and is discussed in Section 4.

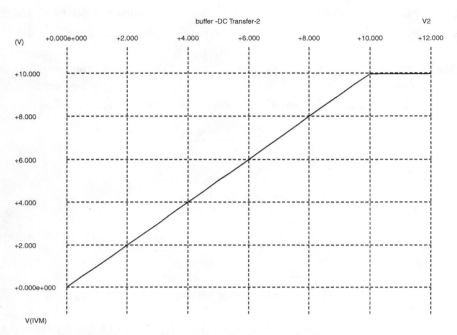

FIGURE 4.11 Spice simulation of voltage-follower circuit.

Exercise. Find v_{out} for the circuit of Figure 4.12, the power supplied by the source v_s, and the power supplied to the 12 kΩ load.

ANSWERS: $v_{out} = v_s, 0, \dfrac{v_s^2}{12 \times 10^3}$

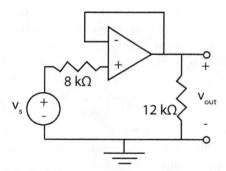

FIGURE 4.12 Isolation of load from source using buffer amplifier.

3. THE DESIGN OF GENERAL SUMMING AMPLIFIERS[1]

Often data acquisition equipment and active filters require multi-input single-output amplifiers having a more general summing characteristic, such as

$$V_{out} = -(\alpha_1 V_{a1} + \alpha_2 V_{a2}) + (\beta_1 V_{b1} + \beta_2 V_{b2}) \tag{4.8}$$

where the constants $\alpha_i > 0$ and $\beta_i > 0$. The inverting and non-inverting amplifier configurations (Examples 4.1, 4.2, and 4.3), as well as the difference amplifier configuration of Example 4.4, are special cases of equation 4.8. With a little cleverness, it is possible to design by inspection an op amp circuit whose input-output characteristic is precisely equation 4.8. The op amp circuit of Figure 4.13 having the four inputs V_{a1}, V_{a2}, V_{b1}, V_{b2} accomplishes this. The circuit looks ordinary except for the presence of one additional conductance, ΔG, incident on the inverting terminal of the op amp. The dashed lines in Figure 4.13 are present because this conductance may or may not be needed. Computation of the values of ΔG and G_g are explained in design Step 2, below.

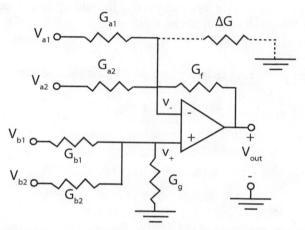

FIGURE 4.13 A general op amp circuit that realizes equation 4.8.

Design Choices for the General Summing Circuit of Figure 4.13

The first two design steps constitute a preliminary or prototype design, meaning that the feedback resistor is normalized to 1 Ω, or equivalently, 1 S. After completing the prototype design, an engineer would scale the resistances to more practical values without changing the gain characteristics. The scaling procedure is explained in Step 3.

Design Step 1. *Prototype design.* Set $G_f = 1$ S, $G_{a1} = \alpha_1$ S, $G_{a2} = \alpha_2$ S, $G_{b1} = \beta_1$ S, and $G_{b2} = \beta_2$ S. For the design to remain simple, the total conductance incident on the inverting terminal must equal the total conductance incident on the non-inverting terminal. This is achieved by proper choice of ΔG and/or G_g. The proper choices are given in Step 2.

Design Step 2. *Prototype design continued: Computation of G_g and/or ΔG so that the total conductance incident at the inverting terminal of the op amp equals the total conductance incident at the non-inverting terminal.*

To achieve this equality, recall that in design Step 1, $G_f = 1$ S, $G_{a1} = \alpha_1$ S, and $G_{b1} = \beta_1$ S. Define a numerical quantity

$$\delta = (1 + \alpha_1 + \alpha_2) - (\beta_1 + \beta_2)$$

The sign of δ leads to two cases:

Case 1: If $\delta > 0$, then set $G_g = \delta$ and $\Delta G = 0$.

Case 2. If $\delta \leq 0$, set G_g to some value, for example, $G_g = 1$ S and $\Delta G = |\delta| + G_g$.

Design Step 3. *Scaling to achieve practical element values.* Multiply all the resistances (divide all conductances) incident at the inverting input terminal of the op amp by a constant K_{ma}. Similarly, multiply all resistances (divide all conductances) incident at the non-inverting terminal of the op amp by K_{mb}. It is permissible to choose $K_{ma} = K_{mb}$, but this is not necessary.

EXAMPLE 4.6. Design an op amp circuit having the input-output relationship

$$v_{out} = -7v_{a1} - 3v_{a2} + 2v_{b1} + 4v_{b2}$$

$$(4.9)$$

SOLUTION

Step 1. *Prototype design.* Using Figure 4.13, choose $G_f = 1$ S, $G_{a1} = 7$ S, $G_{a2} = 3$ S, $G_{b1} = 2$ S, and $G_{b2} = 4$ S.

Step 2. *Equalization of total conductances at inverting and non-inverting terminals.* Since $\delta = (1 + 7 + 3) - (2 + 4) = 5 > 0$, set $\Delta G = 0$ and $G_g = \delta = 5$ S. The circuit in Figure 4.14(a) exemplifies the prototype design.

Step 3. *Scaling.* To have practical element values, let us choose $K_{ma} = K_{mb} = 10^5$. This scaling leads to a design with resistances $R_f = 100$ kΩ, $R_{a1} = 14.28$ kΩ, $R_{a2} = 33.33$ kΩ, $R_{b1} = 50$ kΩ, $R_{b2} = 25$ kΩ and $R_g = 20$ kΩ.

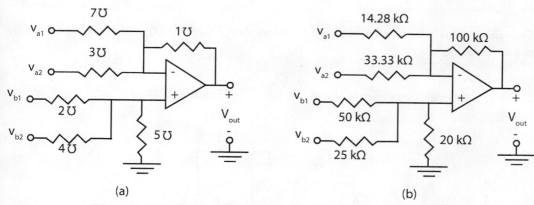

FIGURE 4.14. (a) Prototype design of equation 4.9; (b) final design after scaling with $K_{ma} = K_{mb} = 10^5$.

EXAMPLE 4.7. Design an op amp circuit to have the input-output relationship:

$$v_{out} = -2v_{a1} - 4v_{a2} + 7v_{b1} + 5v_{b2} \tag{4.10}$$

SOLUTION

Step 1. *Prototype design.* Again, using Figure 4.13, choose $G_f = 1$ S, $G_{a1} = 2$ S, $G_{a2} = 4$ S, $G_{b1} = 7$ S, and $G_{b2} = 5$ S.

Step 2. *Equalization of total conductances at inverting and non-inverting terminals.* $\delta = (1 + 2 + 4) - (7 + 5) = -5 < 0$; set $G_g = 1$ S, $\Delta G = |\delta| + G_g = 5 + 1 = 6$ S. This prototype design is given in Figure 4.15(a).

Step 3. *Scaling.* To have practical element values, let us again choose $K_{ma} = K_{mb} = 10^5$. This scaling leads to a design with resistances $R_f = 100$ kΩ, $R_{a1} = 50$ kΩ, $R_{a2} = 25$ kΩ, $R_{b1} = 14.28$ kΩ, $R_{b1} = 20$ kΩ, $R_g = 100$ kΩ, $\Delta R = 16.67$ kΩ. The final design is set forth in Figure 4.15(b).

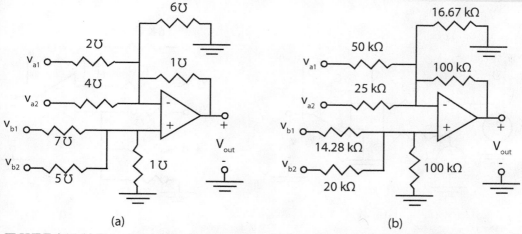

FIGURE 4.15 (a) Prototype design of equation 4.10; (b) final design after scaling with $K_{ma} = K_{mb} = 10^5$.

Exercise. 1. Obtain an alternative design for Example 4.7 such that $G_g = 0$, implying the saving of one resistor.
ANSWER: In prototype design, $\Delta G = 5$ mho.

2. Design a difference amplifier so that $v_{out} = 5(v_{s1} - v_{s2})$, with $R_f = 10$ kΩ.
ANSWER: See Figure 4.16.

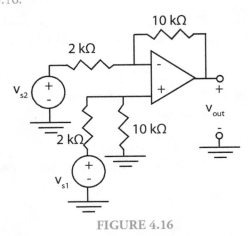

FIGURE 4.16

At this point, the reader may wonder how this simple procedure is derived. The derivation of this procedure is beyond the scope of the light edition[2]. The interested reader is directed to the 2nd edition of this text.

Exercise. 1. Find v_{out} in terms of v_{s1}, v_{s2}, and the G_i for the circuit in Figure 4.17(a).
ANSWER:
$$v_{out} = -\frac{G_1}{G_3}v_{s1} - \frac{G_2}{G_3}v_{s2}$$

2. Find $v_1, ..., v_4$ in terms of for the circuit in Figure 4.17(b).
ANSWER: $v_{out} - 6v_1 - 7v_2 + 4v_3 + 8v_4$

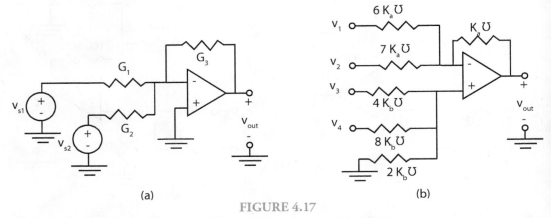

(a)   (b)

FIGURE 4.17

4. SATURATION AND THE ACTIVE REGION OF THE OP AMP

In the previous sections, we assumed the op am functioned ideally: $v_+ = v_-$ and $i_+ = i_- = 0$. For the inverting amplifier of Example 4.1, this led to the very simple gain formula,

$$v_{out} = -\frac{R_f}{R_{in}}v_{in}.$$

Thus, as the input voltage increases, the output voltage increases proportionately. For real circuits, this proportional relationship holds only when $|v_{out}| \leq V_{sat}$ for some value of V_{sat} that is associated with the power supply voltage. Intuitively speaking, an op amp cannot generate an output voltage beyond that of its power supply voltage, typically less than or equal to 15 V. When the V_{sat}-limit is reached, further increases in the magnitude of v_{in} produce no change in the value of v_{out}. This behavior is called **saturation**.

To explain this saturation behavior, we refer to Figure 4.18. In Figure 4.18, $f(v_+ - v_-)$ represents a nonlinear controlled voltage source, as opposed to the linear relationship $A(v_+ - v_-)$, shown in Figure 4.4(a). However, because the op amp functions more or less linearly until reaching its saturation limits, we can approximate $f(v_+ - v_-)$ by the three-segment piecewise linear relationship shown in Figure 4.19(a), wherein the saturation effects are captured by segments II and III. One observes that when $v_d > \dfrac{V_{sat}}{A}$, the voltage $f(v_+ - v_-)$ clamps at V_{sat}, and when $v_d < -\dfrac{V_{sat}}{A}$, the voltage $f(v_+ - v_-)$ clamps at $-V_{sat}$. As observed, the critical threshold voltages of v_d at which saturation occurs are $\dfrac{\pm V_{sat}}{A}$. If $V_{sat} = 15$ V and $A = 10^5$, the critical threshold voltages are ± 0.15 mV; if A is infinite, as in Figure 4.19(b), then saturation occurs when $|v_d| > 0$.

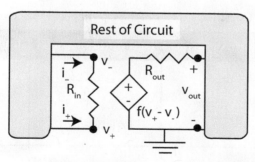

FIGURE 4.18 Practical op amp model with a nonlinear controlled voltage source.

The linear relationship, $f(v_+ - v_-)=A(v_+ - v_-)$, holds for segment 1 in Figure 4.19(a), which is said to be the linear region or **active region** of the op amp, denoted by

$$-\frac{V_{sat}}{A} < v_d = (v_+ - v_-) < \frac{V_{sat}}{A}.$$

Typical values of finite A range from 10^4 to 10^6. The active region is the ordinary region of operation. In the active region, the op amp provides a very high (open loop) voltage gain A, the slope of segment I. The phrase "open loop" gain means that there is no connection through a wire, a resistor, or some other device back to the input terminals.

Models of the three operating regions of the op amp are summarized in Table 4.1.

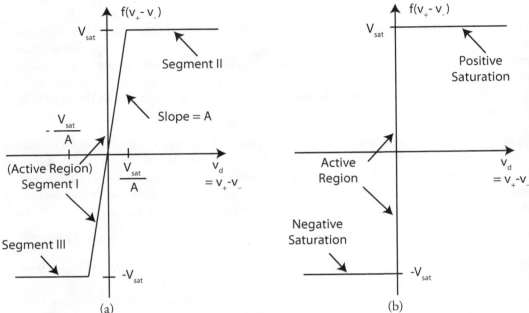

(a)   (b)

FIGURE 4.19 A piecewise linear (three-segment) curve for the op amp that specifies the active and positive/negative saturation regions of operation: (a) finite gain A, and (b) (ideal) infinite gain A.

TABLE 4.1 Operating Regions of the Op Amp with Associated Models

CURVE SEGMENT	NAME OF REGION	DEFINING EQUATIONS	IDEALIZED CIRCUIT MODEL
I	Active	$v_d = \dfrac{f(v_d)}{A}$ and $\|f(v_d)\| < V_{sat}$	
II	Positive saturation	$v_d \geq \dfrac{V_{sat}}{A}$ and $f(v_d) = V_{sat}$	
III	Negative saturation	$v_d \leq -\dfrac{V_{sat}}{A}$ and $f(v_d) = -V_{sat}$	

The use of a three-segment curve in Figure 4.19 is different from the techniques of earlier chapters. The operating point, $(v_d, f(v_d))$, determines the proper segment to be used for analysis. If the input is small, one reasonably assumes the operation is in the active region, segment I. However, when the input magnitude is large, one must "guess and check" to determine the appropriate operating region. For example, should the guess be incorrect, then the model for one of the other regions must be used and the analysis repeated until a valid solution (and operating region) is obtained. The following example illustrates the approach.

EXAMPLE 4.8. The purpose of this example is to illustrate that an op amp may operate in any of three regions and also to illustrate that the determination of the region of operation using the "guess and check" method. Recall the **inverting amplifier** of Figure 4.5. Suppose A is infinite, R_f = 50 kΩ, R_1 = 10 kΩ and V_{sat} = 15 V. Find v_{out} and v_d for the following three cases: (a) v_{in} = 0.5 V; (b) v_{in} = 4 V; and (c) v_{in} = −5 V. Finally, verify the theoretical analysis using SPICE.

SOLUTION
(a) Assume the op amp operates in its active region. From equation 4.3 in Example 4.1, the output voltage is

$$v_{out} = -\frac{R_f}{R_1}v_{in} = -\frac{50}{10} \times 0.5 = -2.5 \text{ V}$$

Since $|-2.5| < V_{sat}$ = 15 V, the op amp operates in its active region; the answers v_{out} = −2.5 V and v_d = 0 are valid.

(b) With v_{in} = 4 V, assuming operation in the active region, $v_{out} = -\frac{R_f}{R_1}v_{in} = -\frac{50}{10} \times 4 = -20$ V.

However, since $|-20| > V_{sat}$ = 15 V, the op amp does not operate in its active region. Therefore, v_{out} = −20 V is invalid, but does suggest operation in the negative saturation region. The negative saturation model of Table 4.1 yields the circuit of Figure 4.20 in which v_{out} = −15 V.

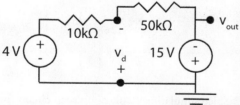

FIGURE 4.20 Op amp operating in negative saturation region.

By writing and solving a single node equation at the inverting input terminal designated by the minus sign in Figure 4.20, we obtain v_d = −0.83 V.

(c) With v_{in} = −5 V, assuming operation in the active region, $v_{out} = -\frac{R_f}{R_1}v_{in} = -\frac{50}{10} \times (-5) = 25$ V

This result suggests that the op amp is really operating in the positive saturation region. Using the positive saturation model of Table 4.1, Figure 4.21 shows the proper circuit configuration with

v_{out} = 15 V. As in the previous case, by writing and solving a single node equation at the inverting terminal designated by the minus sign in Figure 4.21, v_d = 1.667 V. In this case and in case (b) above, $v_d \neq 0$, as we were not in the active region of operation, and it was necessary to change the guessed region of operation to obtain valid results.

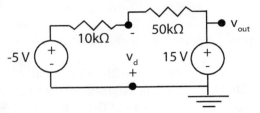

FIGURE 4.21 Op amp operating in positive saturation region.

A SPICE simulation was used to validate the theoretical analysis[3]. A DC sweep, $-4 \leq v_{in} \leq 4$ V, is an adequate input to demonstrate the saturation effects. In the SPICE simulation, an accurate model for a 741 op amp manufactured by Burr Brown was used. The resulting dc transfer curve is shown below in Figure 4.22.

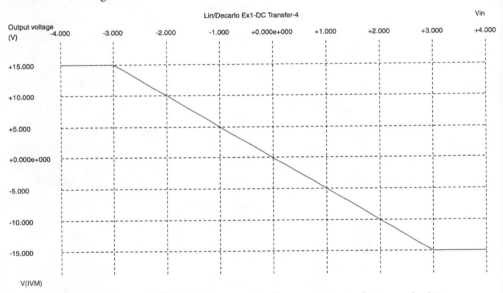

FIGURE 4.22 Spice simulation result of op amp circuit for example 4.8.

From this curve, one can see that the op amp saturates for input voltages v_{in} such that $|v_{in}| > 3$, and the op amp operates in its linear region whenever $|v_{in}| \leq 3$. As hoped, the simplified three-segment model in Table 4.1 yields very good results in all regions of operation relative to the realistic SPICE simulation.

One can conclude from the above example that for the purpose of faithfully amplifying an input signal, the input should not be so large as to drive the op amp into saturation. Driving an op amp into saturation distorts the output signal relative to the input. On the other hand, for some special applications, such as the **comparator**, saturation is precisely the property to be utilized. Figure

4.23 shows two comparator circuits. A **comparator** circuit compares the input voltage v_{in} with a reference voltage v_{ref} (or some multiple of v_{ref}). Only two different output voltages are produced, one for $v_{in} > v_{ref}$, and the other for $v_{in} < v_{ref}$.

EXAMPLE 4.9. For the **comparator** circuits shown in Figure 4.23, each op amp has infinite gain and a saturation voltage $V_{sat} = 15$ V[4]:

(a) Find the v_{out} vs. v_{in} relationship for the comparator of Figure 4.23(a).

(b) Repeat part (a) for Figure 4.23(b).

Note that in both circuits, there is no connection between the output and inverting input terminals, a departure from all the previous circuit configurations. Because of this, for almost all voltages, $v_+ \neq v_-$, and $v_d = v_+ - v_- \neq 0$.

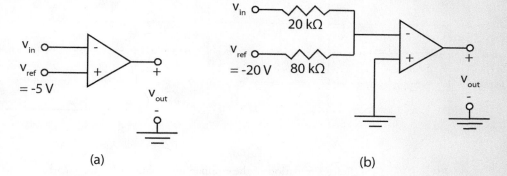

(a)   (b)

FIGURE 4.23 Two comparator circuits that are used to determine when an input voltage is above or below a reference voltage.

SOLUTION

(a) For $v_{in} > -5$ V, the voltage $v_d = v_+ - v_- = -5 - v_{in} < 0$. Referring to Figure 4.19(b), $v_{out} = -v_{sat} = -15$ V. Similarly, for $v_{in} < -5$ V, the voltage $v_d = v_+ - v_- = -5 - v_{in} > 0$, and hence $v_{out} = v_{sat} = 15$ V.

(b) By the fact that no current flows into the input terminals of Figure 4.23(b), using nodal analysis, we have that

$$\frac{v_- - v_{in}}{20 \times 10^3} + \frac{v_- - v_{ref}}{80 \times 10^3} = 0$$

in which case,

$$v_- = -4 + 0.8 v_{in}$$

For $v_{in} > -5$ V, the voltage $v_d = v_+ - v_- = 0 - (-4 + 0.8 v_{in}) < 0$. Here, referring again to the saturation curve of Figure 4.19(b), $v_{out} = -v_{sat} = -15$ V. Similarly, when $v_{in} < 5$ V, the voltage $v_d = v_+ - v_- = 0 - (-4 + 0.8 v_{in}) > 0$; hence, $v_{out} = v_{sat} = 15$ V.

To verify this analysis, the circuit of Figure 4.23(b) was simulated in B2Spice using a Burr Brown 741 op amp model. The results of the simulation are given in Figure 4.24. The theoretical analysis based on the simplified models of Table 4 .1 shows a very good match with the more realistic SPICE simulation results.

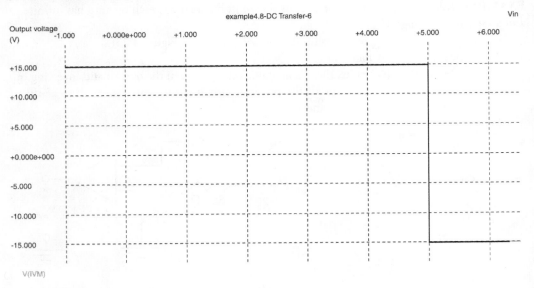

FIGURE 4.24 B2Spice simulation of the comparator circuit of Figure 4.23(b).

Exercise. For the circuit of Figure 4.25, suppose $v_{s1} = 12$ V, find the range of v_{s2} for which the op amp is in positive saturation. Then find the range of v_{s2} for negative saturation.

ANSWER: $v_{out} = v_{sat}$ when $v_{s2} < -4$ V, and $v_{out} = -v_{sat}$ when $v_{s2} > -4$ V

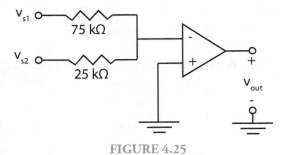

FIGURE 4.25

5. SUMMARY

This chapter has introduced the operational amplifier and a number of practical circuits that utilize this new device. These circuits include the inverting and non-inverting amplifiers, the buffer amplifier, the difference amplifier, and the general summing amplifier. With regard to the general summing amplifier, a simple design algorithm is described and exemplified. The analysis of these circuits builds on the definition of an ideal op amp, meaning that, when properly configured, no current enters the input terminals and the voltage across the input terminals is zero; these properties are referred to as the **virtual short circuit model** of the op amp, i.e., the ideal op amp has infinite input resistance, zero output resistance, and an infinite internal gain, A. (See equations 4.1 and 4.2.) Practically speaking, the gain A, is not infinite, but ranges between 10^4 and 10^6.

After exploring properties of the ideal op amp, we discussed the phenomena of output voltage saturation. By introducing output saturation, the ideal model of the op amp gives way to a more realistic one, characterized by three regions of operation, each having its own "ideal" model, as set forth in Table 4.1. In practical design and applications, output saturation is either to be avoided or utilized to some advantage, as in the case of the comparator circuit studied in Example 4.8. For a faithful amplification of an input signal, saturation is to be avoided.

6. TERMS AND CONCEPTS

Active element: A circuit element that requires an outside power supply for proper operation and has the capability of delivering net power to a circuit such as is the case for an op amp or negative resistance.

Buffer: A circuit designed to prevent the loading effect in a multistage amplifier. It isolates two successive amplifier stages. Characteristics of an ideal buffer are infinite input impedance, zero output impedance, and constant voltage gain.

Comparator: an op amp circuit that compares the input voltage v_{in} with a reference voltage v_{ref} (or some multiple of v_{ref}); only two different output voltages are produced, one for $v_{in} < v_{in}$, and the other for $v_{in} < v_{in}$.

Difference amplifier: given two inputs, v_{s1} and v_{s2}, a difference amplifier produces the output $v_{out} = k(v_{s1} - v_{s2})$ for an appropriate constant k, often taken as 1.

General summing amplifier: an op amp circuit having the input-output relationship $v_{out} = -(\alpha_1 v_{a1} + ... + \alpha_n v_{an}) + (\beta_1 v_{b1} + ... + \beta_m v_{bm})$ for positive constant α_i and β_i.

Ideal op amp: An operational amplifier with infinite input resistance and infinite open-loop gain.

Inverting amplifier: An operational amplifier connected to provide a negative voltage gain at dc.

Linear active region: In the op amp output vs. input transfer characteristic, the region where the curve is essentially a straight line through the origin is called the linear active region.

Non-inverting amplifier: An operational amplifier connected to provide a positive voltage gain at dc.

Open-loop gain: The ratio of the output voltage (loaded, but without any feedback connection) to the voltage across the two input terminals of an op amp. The slope, μ, of the straight line in the active region of an op amp is the open loop gain under no load condition. When a load R_L is present, the open loop gain is reduced to $\mu R_L/(R_L + R_o)$, where R_o is the output resistance of the op amp.

Operational amplifier (abbreviated op amp): A multi-stage amplifier with very high voltage gain (exceeding 10^4) used as a single circuit element.

Passive elements: a circuit element that cannot deliver net power to a circuit such as a resistor.

Saturation regions: In the op amp output vs. input transfer characteristic, the region where the curve is essentially a horizontal line is called the saturation region. There are two such regions: one for positive input voltage, and the other for negative input voltage.

SPICE: Acronym for Simulation Program with Integrated Circuit Emphasis. It is a very sophisticated software tool for simulating electronic circuit behavior.

Virtual ground: When an ideal op amp has one of its input terminals grounded, and is operating in the active region, then the other input terminal is also held at the ground potential because of the virtual short effect (see below). Such a condition is called a virtual ground (in contrast to a *physical* ground).

Virtual short circuit: When an ideal op amp is operating in the active region, the voltage across the two input terminals is zero, even though the two terminals are not hard-wired together. Such a condition is called a virtual short circuit (in contrast to a *physical* short circuit).

Voltage follower: A voltage-controlled voltage source with gain equal to 1, often utilized to separate stages of amplification in a multi-stage amplifier device.

---

[1] The circuit proposed in this section is a modification of one proposed in W. J. Kerwin, L. P. Huesman, and R. W. Newcomb, "State-Variable Synthesis for Insensitive Integrated Circuit Transfer Functions," IEEE Jr. of Solid State Circuits, Vol. SC-2, pp. 87-92, Sept. 1967. The modification consists of an additional resistor, which greatly simplifies the design calculations and was published by P. M. Lin as "Simple Design Procedure for a General Summer," Electron. Eng., vol. 57, no. 708, pp. 37-38, Dec. 1985.

[2] See *Linear Circuit Analysis* by DeCarlo and Lin, 2nd edition, New York: Oxford University Press, 2002.

[3] Any of the SPICE or PSPICE software programs available by a variety of vendors will suffice to obtain the indicated curve.

[4] An op amp and a comparator as seen in a parts catalog are essentially the same, except that the comparator device has a modified output stage that makes it compatible with digital circuits.

Problems

ANALYSIS USING IDEAL OP AMP MODEL

1. Consider the inverting amplifier circuit of Figure P4.1, in which $v_{in} = 4$ V.
   (a) If $R_1 = 2$ kΩ, find R_2 so that the power delivered to $R_L = 100$ Ω is 4 W.
   (b) Now suppose $R_2 = 12$ kΩ. Find R_1 so that the power delivered to $R_L = 2$ kΩ is 450 mW. Then find the power consumed in R_1 and R_2.

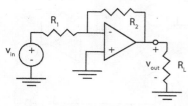

Figure P4.1

Check: (b) $1500 \le R_1 \le 2000$, and $P_{R1} = 10$ mW

2. Consider the non-inverting circuit of Figure P4.2, in which $v_{in} = 4$ V.
   (a) If $R_1 = 2$ kΩ, find R_2 so that the power delivered to $R_L = 100$ Ω is 4 W.
   (b) Now suppose $R_2 = 13$ kΩ. Find R_1 so that the power delivered to $R_L = 2$ kΩ is 450 mW. Then find the power consumed in R_1 and R_2.

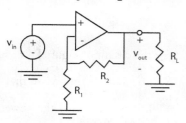

Figure P4.2

Check: (b) $1500 \le R_1 \le 3000$

3. For the circuit of Figure P4.3, find the voltage gains,

$$G_{v_+} = \frac{v_+}{v_{in}} \text{ and } G_v = \frac{v_{out}}{v_{in}}$$

in terms of the literal resistor values.

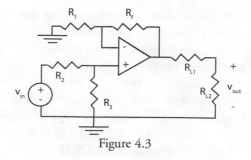

Figure 4.3

4. In the circuits of Figure P4.4, a source is represented by an ideal voltage source, $v_{in}(t) = 4$ V, in series with $R_s = 10$ Ω resistor. The loadin both cases is $R_L = 40$ Ω.
   '(a) With the load connected directly to the source, as shown in Figure P4.4(a), find the load voltage, the load current, the source current, and the power delivered to the load.
   (b) As in Figure 4.4(b), a buffer amplifier separates the source and the load. Again, find the load voltage, the load current, the source current, and the power delivered by the op amp to the load.

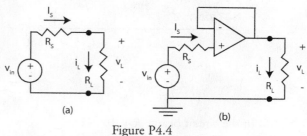

Figure P4.4

SCRAMBLED ANSWERS: (a) 0.256, 3.2, 0.08

5. Figure P4.5 contains three circuits that explore loading and the elimination of loading effects using either a dependent source or an equivalent buffering op amp circuit.
   (a) For the circuit of Figure P4.5(a), compute v_1 and v_{out} in terms of v_s. Observe that the 80-Ω-240 Ω resistor combination loads down the 320-Ω resistor.
   (b) For the circuit of Figure P4.5(b), compute v_1 and v_{out} in terms of v_s. Notice that v_1 is different from the answer

computed in part (a) because the 80-Ω-240 Ω resistor combination is isolated from the 320–Ω resistor.

(c)   For the circuit of Figure P4.5(c), again compute v_1 and v_{out} in terms of v_s. Your answers should be the same as those in part (b). The buffering op amp circuit again isolates the 80-Ω-240 Ω resistor comination from the 320 Ω resistor.

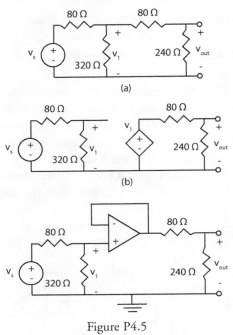

Figure P4.5

ANSWER: (a) $\frac{2}{3}V_s$, $0.5V_s$

6. In the circuit below, R_f= 10 kΩ.
   (a)   Find R_1 and R_2 so that $v_{out} = -2v_{s1} - 5v_{s2}$.
   (b)   Given correct answers to part (a), suppose a 1 kΩ resistor is attached as a load. Find the power delivered to the load if v_{s1} = 200 mV and v_{s2} = –600 mV.

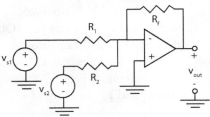

Figure P4.6

ANSWER: (a) R_1 = 5 kΩ; (b) 6.76 mW

7. (a)   For the op amp circuit of Figure P4.7(a), find v_{out} as a function v_{s1}, v_{s2}, and the R_i.
   (b)   Repeat part (a) for the circuit of Figure P4.7(b).
   (c)   If for Figure 4.7(b), R_f= 12 kΩ, R_1 = 3 kΩ , R_3 = 4 kΩ, R_2 = 1 kΩ, v_{s1} = 1.5 V, and v_{s2} = 2 V, find the power delivered to the load R_L = 100 Ω.

ANSWER: (c) 0.04 watts

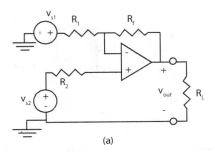

(a)

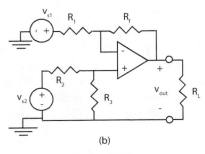

(b)

Figure P4.7

8. (a)   For the circuit of Figure P4.8, suppose R_2 = 6 kΩ, and find R_1 so that $\left|\dfrac{V_{out}}{V_s}\right|$ = 20. If v_s = 0.6, find the power delivered to the 8–Ω load.
   (b)   Now suppose R_2 = 5 kΩ, and find R_2 so that $\left|\dfrac{V_{out}}{V_s}\right| = 20$
   (c)   Finally, suppose R_1 = R_2, and find their common value so that $\left|\dfrac{V_{out}}{V_s}\right| = 20$

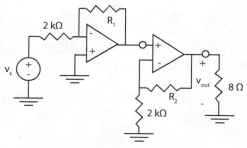

Figure P4.8

ANSWERS (in random order): 14 kΩ, 8 kΩ, 10 kΩ, 18

9. In the circuit below, R_f= 12 kΩ and R_3 > 1 kΩ.
   (a)   Find R_1 and R_2 so that v_{out} = $-10v_{s1}$ − $20v_{s2}$.
   (b)   Given correct answers to part (a), find the power delivered to the load if v_{s1} = −200 mV and v_{s2} = 600 mV.

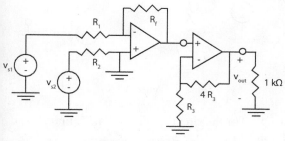

Figure P4.9
CHECK: (b) P_L = 0.1 watt

10. Consider the circuit of Figure P4.10.
   (a)   Find the value of $R = R_1 = R_2$ so that the power delivered to R_L = 1.25 kΩ is 0.5 watt when v_s = 1 V.
   (b)   Suppose $6R_1 = R_2$ and v_s = 2 V. Find R_1 and R_2 so that the power delivered to R_L = 1.25 kΩ is 2 watts.

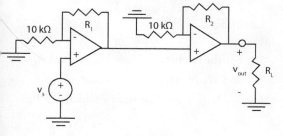

Figure P4.10

ANSWER: (a) 40 kΩ; (b) R_1 = 15 kΩ

11. Consider the circuit shown in Figure P4.11.
   (a)   Find v_{out} in terms of v_{s1} and v_{s2}.
   (b)   If v_{s1} = 250 mV and v_{s2} = 500 mV, find the power delivered to the 1 kΩ load resistance.

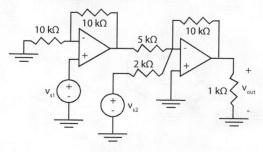

Figure P4.11
CHECK: (b) P_L = 12.25 mW.

12. (a)   For the circuit of Figure P4.12a, the input voltage v_{s1} = 2 V and the input voltage v_{s2} = 3 V. Find V_{out} and the power delivered to the 1 kΩ load resistor.
     (b)   Repeat part (a), when v_{s1} = 4 V and the input voltage v_{s2} = 2 V.
     (c)   Reconsider part (b). Find the minimum value of R so that the maximum amount of power consumed in either R-ohm input resistors is 2 mW.

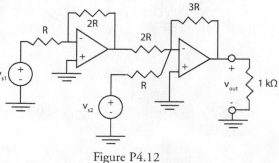

Figure P4.12

13. (a)   For the circuit of Figure P4.13(a), the input voltage v_{s1} = 1 V, and the input voltage v_{s2} = 500 mV. Find R_1 in terms of R so that v_{out} = 10 V.

(b)   Repeat part (a) for the circuit of Figure
      P4.13(b), given that v_{s3} = 2.5 V.

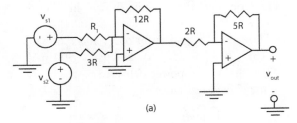

(a)

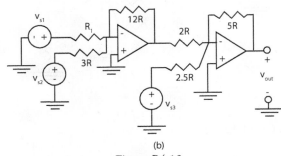

(b)

Figure P4.13

ANSWERS: (a) 6R; (b) 3R

14. For the circuit in Figure P4.14, the input
voltages are v_{s1} = 2 V, v_{s2} = -1.5 V, and v_{s3} = 2
V.
      (a)   Find v_{out}.
      (b)   If R = 10 kΩ, find the power delivered
            by each of the operational amplifiers.

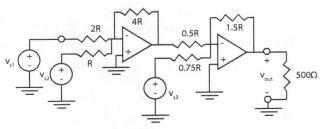

Figure P4.14

CHECK: $P_{del,1}$ = 0.9 mW, $P_{del,2}$ = 0.20667
watts

15. For the circuit of Figure P4.15, the input voltage v_{s1} = 5 V, and the input voltage v_{s2} = -1 V.
      (a)   If R_1 = 8R and R_2 = R_3, find v_{out}.
      (b)   If R_1 = 8R and R_2 = 4R_3, find v_{out}.
      (c)   If R_2 = R_3, find R_1 in terms of R so
            that v_{out} = 10V.

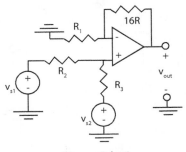

Figure P4.15

CHECK: (a) v_{out} = 6 V

16. For the circuit of Figure P4.16, find v_{out} in
terms of v_{s1} V, v_{s2} V, and v_{s3} V.

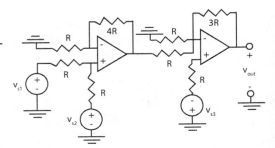

Figure 4.16

17. For the circuit of Figure P4.17, find R_1 and
R_2, in terms of R so that v_{out} = 5v_{s1} + 4v_{s2}.

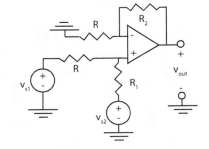

Figure 4.17

CHECK: R_2 = 8R

18. For the circuit of Figure P4.18, find R_1 and
R_2, and R_3 in terms of R so that v_{out} = 8v_{s1} +
10v_{s2} - 2v_{s3}. Hint: Consider Problem 17 first.

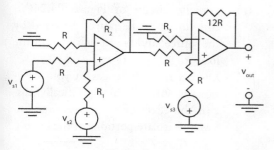

Figure 4.18

ANSWERS: $R_3 = 4R$, $R_2 = 8R$, $R_1 = 0.8R$

NON-IDEAL OP AMP–SATURATION EFFECTS

19. The op amp in Figure 4.19(a) has $V_{sat} = 15$ V, $R_1 = 4$ kΩ, and $R_2 = 20$ kΩ.
    (a) Plot the v_{out} versus v_s for v_s given in Figure 4.19(b).
    (b) Plot $v_{out}(t)$ for $0 \le t \le 6$ s for $v_s(t)$ in Figure 4.19(c).
    (c) Verify your analysis in part (a) using SPICE. Assume that the op amp is a type 741 whose model should be available within your SPICE program.

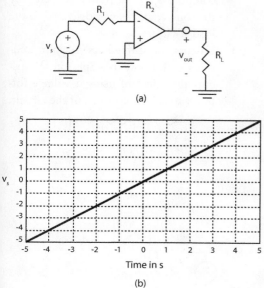

(a)

(b)

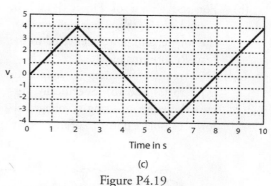

Time in s

(c)

Figure P4.19

20. Repeat Problem 19 for the op amp circuit of Figure P4.20 when $R_1 = 4$ kΩ, and $R_2 = 20$ kΩ.

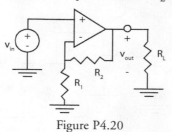

Figure P4.20

21. For the circuit of Figure P4.21, each amplifier saturates at $V_{sat} = 15$ V.
    (a) Suppose the input voltage $v_{s1} = 5$ V and the input voltage $v_{s2} = -2.5$ V. Find v_1 and v_{out}.
    (b) If $v_{s2} = -2.5$ V and $v_{out} = 15$ V, find v_{s1} so that no amplifier saturates.

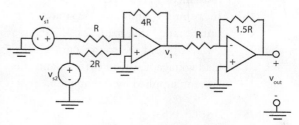

Figure P4.21

ANSWER: (a) 15 V; (b) $V_{s1} = 3.75$ V

22. For the circuits of Figure P4.22, suppose $R_1 = 40$ kΩ, and $R_L = 120$ Ω.
    (a) For the circuit of Figure P4.22(a), compute v_{out}, the power delivered by the source, and power delivered to the load R_L in terms of v_{in}.

(b)   For the circuit of Figure P4.22(b), compute v_{out}, the power delivered by the source, and power delivered to the load R_L in terms of v_{in}.

(c)   Discuss the differences in your solutions to (a) and (b). Specifically, discuss the effect of using a voltage follower to isolate portions of the circuit.

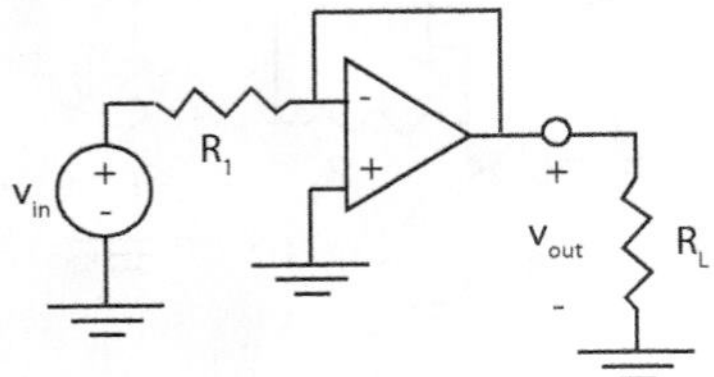

(b)   For the circuit of Figure P4.24(b), compute v_{out}, the power delivered by the source, and power delivered to the load R_L in terms of v_{in}.

(c)   Discuss the differences in your solutions to (a) and (b). Specifically, discuss the effect of using a voltage follower to isolate portions of the circuit.

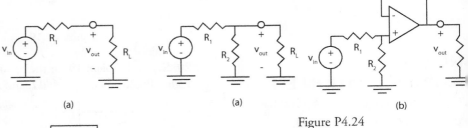

Figure P4.24

25. For the circuits of Figure P4.25, suppose $R_1 = 20\ \Omega$, and $R_2 = 160\ \Omega$, $R_3 = 40\ \Omega$, and $R_L = 120\ \Omega$.

(a)   For the circuit of Figure P4.25(a), compute v_{out}, the power delivered by the source, and power delivered to the load R_L in terms of v_{in}.

(b)   For the circuit of Figure P4.25(b), compute v_{out}, the power delivered by the source, and power delivered to the load R_L in terms of v_{in}.

(c)   Discuss the differences in your solutions to (a) and (b). Specifically, discuss the effect of using a voltage follower to isolate portions of the circuit.

Figure P4.22

23. In the circuits of Figure P4.22, all resistances are 100 Ω, and $v_{in} = 1$ V.

(a)   For the circuit of Figure P4.22(a), find the load voltage, the load current, and the source current again when all resistances are 100 Ω.

(b)   If a buffer amplifier separates the source and the load, as in Figure P4.22(b), find the source current, the load voltage, the load current, and the current supplied by the op amp.

ANSWERS: (a) 0.5V, 5 mA, 5 mA; (b) 0 A, 1 V, 10 mA, 10 mA

24. For the circuits of Figure P4.24, suppose $R_1 = 40\ \Omega$, and $R_2 = R_L = 120\ \Omega$.

(a)   For the circuit of Figure P4.24(a), compute v_{out}, the power delivered by the source, and power delivered to the load R_L in terms of v_{in}.

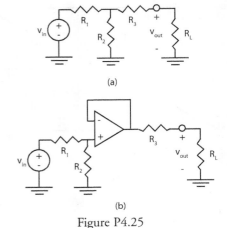

Figure P4.25

ANSWERS: (a) $P_L = 0.003v^2_{in}$; (b) $P_L = 3.7037 \times 10^{-3}v^2_{in}$

26. Figure P4.26 contains three circuits that explore loading and the elimination of loading effects using either a dependent source or an equivalent buffering op amp circuit.

   (a) For the circuit of Figure P4.26(a), compute v_1 and v_{out}. Observe that the 8-Ω-24 Ω resistor combination loads down the 32–Ω resistor.

   (b) For the circuit of Figure P4.26(b), compute v_1 and v_{out}. Notice that v_1 is different from the answer computed in part (a) because the 8-Ω-24 Ω resistor combination is isolated from the 32–Ω resistor.

   (c) For the circuit of Figure P4.26(c), again compute v_1 and v_{out}. Your answers should be the same as those in part (b). The buffering op amp circuit again isolates the 8-Ω-24 Ω resistor combination from the 32 Ω resistor.

27. Two non-ideal voltage sources are each represented by a connection of a (grounded) independent voltage source and a series resistor. Denote the parameters of each connection by (v_{s1}, R_{s1}) and (v_{s2}, R_{s2}). Design an op amp circuit such that the output voltage with respect to ground is

$$v_{out} = 4(v_{s1} - v_{s2})$$

for all values of R_{s1} and R_{s2}. R_2 is to be greater than or equal to 100 kΩ so that only small amounts of current are drawn from the buffer amplifiers. Note that the general difference amplifier circuit of the chapter will not work here because of the presence of the resistances R_{s1} and R_{s2}. To achieve such a design, it is necessary to isolate the (practical) sources from the difference amplifier inputs using buffer amplifiers, as shown in Figure P4.27. Explore your design for various values of R_{s1} and R_{s2} using SPICE. Do the SPICE simulations verify that the output is independent of the values of R_{s1} and R_{s2}?

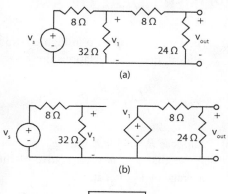

(a)

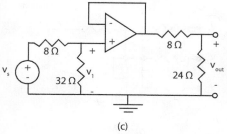

(b)

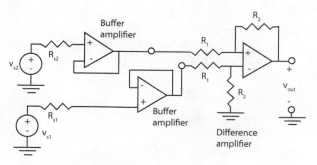

Figure P4.27

28. Following Example 4.9, for the comparator circuit of Figure P4.28, suppose the op amp has infinite gain and a saturation voltage $V_{sat} = 15$ V. Find the v_{out} versus v_{in} relationship and plot as a function of v_{in}. Verify your analysis using SPICE. Assume that the op amp is a type 741 whose model should be available within your SPICE program.

(c)

Figure P4.26

ANSWERS: (a) 0.6665 V_s, 0.5 V_s; (b) and (c) 0.8 V_s, 0.6 V_s

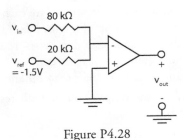

Figure P4.28

29. (a) Find the v_{out} versus v_{in} relationship for the comparator circuit of Figure P4.29. Specifically, show that when

$$v_{in} < -\frac{R_?}{R_?}v_{ref}$$

then $v_{out} = V_{sat}$ and when $v_{in} > ??$, then $v_{out} = -V_{sat}$.

(b) Now suppose $R_1 = R_2 = 100$ kΩ and $v_{ref} = 20$ V. Find the v_{out} versus v_{in} characteristic if $V_{sat} = 15$ V. Verify your analysis using SPICE. Assume that the op amp is a type 741 whose model should be available within your SPICE program.

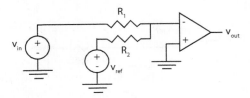

Figure P4.29

ANSWERS: (b) $v_{out} = 15$ V if $v_{in} < 2$ V, and $v_{out} = \Omega$ 15 V if $v_{in} > 2$ V

30. Using a 1.5 V battery, an op amp with $V_{sat} = 10$ V, and some resistors, design a comparator circuit such that $v_{out} = 10$ V when $v_{in} < 1$ V, and $v_{out} = -10$ V when $v_{in} > 1$ V. Check your design using SPICE. Use part (a) of Problem 29 as a guide.

31. Using a 1.5-V battery, an op amp with $v_{sat} = 10$ V, and some resistors, design a comparator circuit such that $v_{out} = V_{sat} = 10$ V when $v_{in} > 1$ V, and $v_{out} = -V_{sat} = -10$ V when $v_{in} < 1$ V.

Check your design using SPICE. Hint: How can the circuit of Figure P4.29 be modified to achieve the correct polarities?

ANSWER: $\dfrac{R_1}{R_2} = \dfrac{1}{1.5}$

32. Find the v_{out} versus v_{in} relationship for the comparator circuit of Figure P4.32. Specifically, show that when

$$v_{in} > \frac{R_1}{R_2}v_{ref}, \text{ then } v_{out} = V_{sat}, \text{ and when}$$

$$v_{in} < \frac{R_1}{R_2}v_{ref}, \text{ then } v_{out} = -V_{sat}.$$

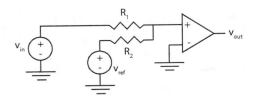

Figure P4.32

GENERAL SUMMING AMPLIFIER (IDEAL OP AMP MODEL)

33.(a) Assuming the op amp in Figure P4.33 is ideal, derive the relationship

$$v_{out} = -\frac{R_f}{R_1}V_1 - \frac{R_f}{R_2}V_2 - \frac{R_f}{R_3}V_3$$

(b) Suppose $R_f = 25$ kΩ. Find R_1, R_2, and R_3 so that v_{out} is the negative of the average of V_1, V_2, and V_3.

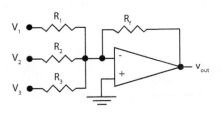

Figure P4.33

ANSWER: (b) $R_i = 75$ kΩ

34. Using the topology of Figure 4.13, design an op amp circuit to have the input-output relationship

$$v_{out} = -4v_{a1} - 3v_{a2} + 2v_{b1} + 4v_{b2}$$

Two different designs are to be produced for comparison and selection:
- (a) Design 1: $R_f = 100$ kΩ.
- (b) Design 2: $R_f = 50$ kΩ. Specify all final values in terms of Ω.

CHECK: $R_{a1fnl} = 25$ kΩ, and $R_{a2fnl} = 33.33$ kΩ

35. Using the topology of Figure 4.13, design an op amp circuit to have the input-output relationship

$$v_{out} = -4v_{a1} - 2v_{a2} + 5v_{b1} + 4v_{b2}$$

Two different designs are to be produced for comparison and selection:

- (a) Design 1: $R_f = 100$ kΩ.
- (b) Design 2: $R_f = 50$ kΩ. Specify all final values in terms of Ω.

36. Using the topology of Figure 4.13, design an op amp circuit to have the input-output relationship

$$v_{out} = -2v_{a1} - 5v_{a2} + 8v_{b1} + 4v_{b2}$$

Your design must have $R_f = 10$ kΩ in the final circuit, and all other resistors should be within the range 2 kΩ to 20 kΩ.

37. Using the topology of Figure 4.13, design an op amp circuit to have the input-output relationship

$$v_{out} = -2v_{a1} - 5v_{a2} + 8v_{b1} + 4v_{b2}$$

Your design must have all resistors, including R_f, in the range 5 kΩ to 25 kΩ.

38. Using the topology of Figure 4.13, design an op amp circuit to have the input-output relationship

$$v_{out} = -4v_{a1} - 2v_{a2} + 5v_{b1} + 4v_{b2}$$

Two different designs are to be produced for comparison and selection:
- (a) Design 1: $R_f = 50$ kΩ.
- (b) Design 2: $R_f = 100$ kΩ.

39. Generalizing the topology of Figure 4.13, design an op amp circuit to have the input-output relationship

$$v_{out} = -v_{a1} - 2v_{a2} - 4v_{a3} + 4v_{b1}$$

In the final circuit $R_f = 40$ kΩ.

40. Generalizing the topology of Figure 4.13, design an op amp circuit to have the input-output relationship

$$v_{out} = -4v_{a1} + 2v_{b1} + v_{b2} + 4v_{b3}$$

In the final circuit, $R_f = 40$ kΩ.

VARIABLE GAIN AMPLIFIERS

41. The circuit of Figure P4.41 is a modification of the basic non-inverting amplifier. In the modification, a potentiometer R_p is connected between the output terminal and R_0, with the sliding contact between points A and B, as shown. Show that as the sliding contact of the potentiometer is moved between positions A and B, the range of voltage gain achievable is

$$1 \le \frac{v_{out}}{v_{in}} \le 1 + \frac{R_p}{R_0}$$

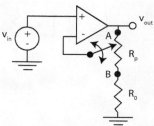

Figure P4.41 Variable gain non-inverting amplifier.

42. The circuit of Figure P4.42 is a simple modification of the basic inverting amplifier circuit in which a potentiometer is connected to the feedback resistor R_f, as shown. Show that the range of gains achievable by this circuit is

$$-\frac{R_f}{R_1} \geq \frac{v_{out}}{v_{in}} \geq -\frac{R_f + R_p}{R_1}.$$

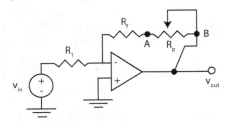

Figure P4.42

43. The circuit of Figure P4.43 is another modification of the basic inverting amplifier to obtain a variable gain amplifier. Show that as the sliding contact of the potentiometer is moved between the two extreme positions, the range of achievable voltage gain is

$$-\frac{R_f}{R_1} \geq \frac{v_{out}}{v_{in}} \geq -\alpha\frac{R_f}{R_1}$$

where $\alpha = 1 + \dfrac{R_p}{R_0} + \dfrac{R_p}{R_f} = 1 + \dfrac{R_p}{R_0 // R_f}$.

Hint: Apply KCL to the non-inverting input terminal, and make use of the virtual ground property of an ideal op amp.

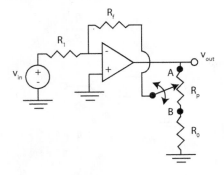

Figure P4.43 Variable-gain inverting amplifier circuit.

SIMULATION OF CONTROLLED SOURCES USING OP AMPS

44. Design an op amp circuit to simulate the grounded VCVS in Figure P4.44 when μ > 1. Hint: Consider the non-inverting amplifier of Example 4.3.

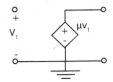

Figure P4.44 Grounded VCVS.

45. Design an op amp circuit to simulate the grounded VCVS in Figure P4.44 for any μ > 0. Hint: Try a voltage follower in cascade with two inverting op amp circuits.

46. Reconsider the design of Problem 45 so that only two op amps are used. In this case, one still needs the voltage follower. Why? Hint: Consider using a voltage divider followed by a non-inverting amplifier circuit.

47. Design an op amp circuit to simulate the grounded VCVS in Figure P4.44 when μ < 0. Hint: Consider an inverting amplifier configuration in conjunction with a buffer amplifier.

48. For the circuit of Figure P4.48, show that the load current I_L equals V_i/R_a, which is independent of the load resistance R_L. Hence, this op amp circuit converts a grounded voltage source into a floating current source. (This is sometimes called a **voltage-to-current converter**.)

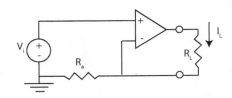

Figure P4.48 Op amp circuit simulating a floating current source.

49. In Problem 4.48, since I_L depends on V_i and R_a only, the load need not be a resistor. For example, R_L may be replaced by an LED (light-emitting diode), as shown in Figure P4.49. Then by turning the knob of the 10-kΩ potentiometer, one can control the brightness of the LED. The current through the load is supplied by the op amp. The potentiometer, which controls the brightness of the LED, uses a low-voltage part of the circuit. Find the magnitude of the LED current if the potentiometer is set at (a) $R_1 = 5$ kΩ and (b) $R_1 = 8$ kΩ.

ANSWERS: 1.32 mA, 2.1 mA

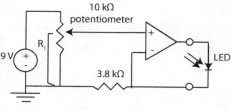

Figure P4.49

50. This problem is a variation of Problem 4.49 in which the load current flows in the opposite direction. For the circuit of Figure P4.50, show that the load current I_L equals v_{in}/R_a, which is independent of the load resistance R_L. Hence, this op amp circuit converts a grounded voltage source into a floating current source in which the current enters the op amp output terminal. (This is sometimes called a **voltage-to-current converter.**)

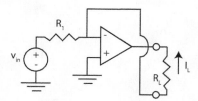

Figure P4.50 Op amp circuit simulating a float-ing current source.

CHAPTER 5

Linearity, Superposition, and Source Transformation

HISTORICAL NOTE

In the mid-nineteenth century, before the introduction of the alternating current (ac), electricity was available mainly as direct current (dc). This time period saw the evolution of basic laws for the analysis of electrical circuits composed of dc voltage sources and resistors: Ohm's law, KVL, and KCL. Application of these laws to the analysis of circuits led to the development of the mesh and nodal techniques requiring the solution of simultaneous equations. Before the computer age, manual solution of a (large) set of equations was very difficult. To circumvent this difficulty, researchers developed a number of network theorems that (i) simplified the aforementioned manual analysis, (ii) reduced the need for repeated solution of the same set of equations, and (iii) provided insight into the behavior of circuits. These network theorems remain useful even in the present day of high-powered computing.

CHAPTER OBJECTIVES

1. Introduce and apply the property of linearity.
2. State and explore the two consequences of linearity called superposition and proportionality to simplify response computation.
3. Use superposition and proportionality to simplify manual analysis and to gain better insight into circuit behavior.
4. Introduce and apply the source transformation theorem to again simplify manual analysis.

SECTION HEADINGS

1. **Introduction**
2. **Linearity**
3. **Linearity Revisited: Superposition and Proportionality**
4. **Source Transformations**
5. **Equivalent Networks**
6. **Summary**

1. INTRODUCTION

Chapter 3 covered nodal and loop/mesh analyses. Node voltage or loop current calculation proceeds by constructing a set of simultaneous node or loop equations and solving them by hand, by MATLAB, or with some equivalent software package. Few of us will attempt a paper-and-pencil solution of four equations in four unknowns. Yet, MATLAB, Mathematica, or some other computational software program, can easily and reliably crunch numbers, relieving us of tedious hand calculations. Nevertheless, manual analysis in some form remains important for a deeper understanding or insight into a circuit's behavior, as well as a way to check the validity of a program output.

Experience teaches us that manual analysis is ordinarily practical only for small circuits. Fortunately, the network theorems studied in this chapter and the next can often reduce seemingly complex circuits to simpler ones amenable to manual analysis. They also provide shortcuts for computing outputs and allow us to obtain deeper insights into a circuit's behavior.

This chapter talks about linearity and superposition, which are motivated by the following questions: What is the effect on the circuit output (voltage or current) of a single independent voltage source, say v_k, *acting alone*. "Acting alone" means that the independent source, v_k, has a nonzero value, while all other independent sources are set to zero. A deactivated voltage source acts as short circuit (see Chapter 2), and a deactivated current source acts as an open circuit (again, see Chapter 2). Is there a shortcut to computing the response if v_k is doubled in value?

To answer the above questions and others, our discussion begins with the important property of linearity. Linearity relates the values of independent sources to a circuit output with a very compact equation. This equation defines the effect of any independent source on a circuit output. After studying linearity, we discuss two special consequences called superposition and proportionality. Each of these concepts helps reduce manual computation of responses, and each provides insight into circuit behavior. Next, we state the source transformation theorem and show how this method can reduce a complex circuit to a more simple form. Finally, we set forth the notion of an equivalent two-terminal network and then outline a proof of the source transformation theorem.

2. LINEARITY

This section investigates the circuit property of linearity, which we introduce with a motivating example.

EXAMPLE 5.1. For the circuit of Figure 5.1, find the outputs I_A and V_B in terms of the source current I_{s1}, and the source voltage V_{s2}. We will derive the relationships $V_B = 40I_{s1} + \dfrac{2}{3}V_{s2}$ and

$$I_A = \frac{2}{3}I_{s1} - \frac{1}{180}V_{s2}.$$

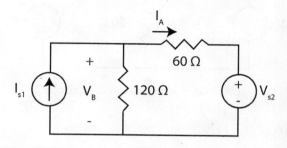

FIGURE 5.1. Resistive circuit driven by current and voltage sources.

SOLUTION

Step 1. *Find the voltage* V_B. A node equation at the top of the current source is

$$\frac{V_B}{120} + \frac{V_B - V_{s2}}{60} = I_{s1}$$

Solving for V_B yields

$$V_B = 40I_{s1} + \frac{2}{3}V_{s2}$$

Here, V_B appears as a constant times I_{s1}, plus another constant times V_{s2}, a so-called linear combination.

Step 2. *Find the current* I_A. From Step 1, we know V_B. The current I_A satisfies

$$I_A = \frac{V_B - V_{s2}}{60} = \frac{1}{60}\left(40I_{s1} + \frac{2}{3}V_{s2} - V_{s2}\right) = \frac{2}{3}I_{s1} - \frac{1}{180}V_{s2}$$

Similar to Step 1, the output current I_A is a constant times I_{s1} plus another constant times V_{s2}, a *linear combination*.

Exercise. 1. In Example 5.1, suppose the 60 Ω resistor is changed to 120 Ω. Find the outputs I_A and V_B in terms of the sources, I_{s1} and V_{s2}.
ANSWER: $V_B = 60I_{s1} + 0.5V_{s2}$ and $I_A = 0.5I_{s1} - \dfrac{1}{240}V_{s2}$

2. For the circuit of Figure 5.2, find V_B in terms of V_{s1} and V_{s2}.

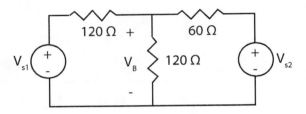

FIGURE 5.2 Resistive circuit for Exercise 2.

ANSWER: $V_B = 0.25V_{s1} + 0.5V_{s2}$

In the above example and exercises, the desired output voltage or current was a so-called linear combination of the independent source values. This is, in fact, a quite general phenomena, as indicated by the linearity theorem below.

LINEARITY THEOREM

For all practical linear resistive circuits, as per Figure 5.3, any output voltage, v_A, or any current, i_B, can be related linearly to the independent source values, as in the following equations:

$$v_A = \alpha_i V_{s1} + \dots + \alpha_n V_{sn} + \beta_1 i_{s1} + \dots + \beta_m i_{sm} \tag{5.1a}$$

or

$$i_B = \alpha_1 V_{s1} + \dots + \alpha_n V_{sn} + \beta_1 i_{s1} + \dots + \beta_m i_{sm} \tag{5.1b}$$

where the α_i and β_i are properly dimensioned constants.

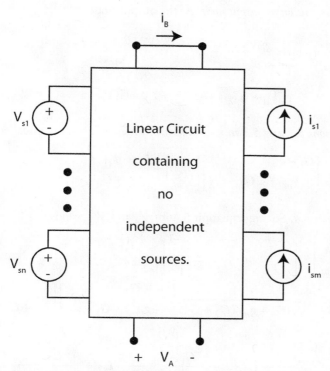

FIGURE 5.3. A linear circuit driven by n independent voltage sources and m independent current sources with outputs of v_A and i_B.

A rigorous proof of the linearity theorem entails solving a set of modified nodal or loop equations using matrix algebra and is beyond the scope of this text.

EXAMPLE 5.2. For the circuit of Figure 5.4, our objective in this example is to express V_{out} as a linear combination of I_{s1}, I_{s2}, and V_{s3}, as per equation 5.1a. In doing this, we review **nodal analysis**.

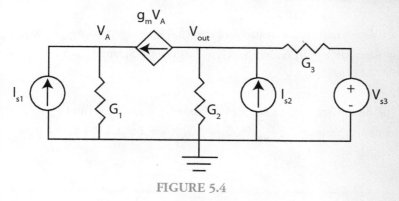

FIGURE 5.4

SOLUTION

Step 1. *Write nodal equation at A.* For node A,

$$(G_1 - g_m)V_A = I_{s1} \tag{5.2}$$

Step 2. *Write nodal equation at output node.* At the output node,

$$g_m V_A + G_2 V_{out} + G_3(V_{out} - V_{s3}) = I_{s2}$$

or equivalently,

$$g_m V_A + (G_2 + G_3) V_{out} = I_{s2} + G_3 V_{s3} \qquad (5.3)$$

Step 3. *Write equations 5.2 and 5.3 in matrix form.* In matrix form, the nodal equations are

$$\begin{bmatrix} G_1 - g_m & 0 \\ g_m & G_2 + G_3 \end{bmatrix} \begin{bmatrix} V_A \\ V_{out} \end{bmatrix} = \begin{bmatrix} I_{s1} \\ I_{s2} + G_3 V_{s3} \end{bmatrix} \qquad (5.4)$$

Step 4. *Solve equation 5.4.* Solving equation 5.4 for V_A and V_{out} yields

$$\begin{bmatrix} V_A \\ V_{out} \end{bmatrix} = \begin{bmatrix} G_1 - g_m & 0 \\ g_m & G_2 + G_3 \end{bmatrix}^{-1} \begin{bmatrix} I_{s1} \\ I_{s2} + G_3 V_{s3} \end{bmatrix}$$

$$= \frac{1}{(G_1 - g_m)(G_2 + G_3)} \begin{bmatrix} G_2 + G_3 & 0 \\ -g_m & G_1 - g_m \end{bmatrix} \begin{bmatrix} I_{s1} \\ I_{s2} + G_3 V_{s3} \end{bmatrix}$$

It follows that

$$V_{out} = \frac{-g_m}{(G_1 - g_m)(G_2 + G_3)} I_{s1} + \frac{G_1 - g_m}{(G_1 - g_m)(G_2 + G_3)} I_{s2} + \frac{(G_1 - g_m)G_3}{(G_1 - g_m)(G_2 + G_3)} V_{s3} \qquad (5.5)$$

as set forth in equation 5.1(a).

Exercise. 1. In equation 5.5, suppose $G_1 = 1$ S, $G_2 = 2$ S, $G_3 = 3$ S, and $g_m = 5$ S. Find the numerical expression for V_{out}.
ANSWER: $V_{out} = 0.25 I_{s1} + 0.2 I_{s2} + 0.6 V_{s3}$

2. Suppose the dependent current source in Figure 5.4 is changed from $g_m V_A$ to $g_m V_{out}$. Compute the new expression for V_{out} if $G_1 = 1$ S, $G_2 = 2$ S, $G_3 = 3$ S, and $g_m = 5$ S.
ANSWER: $V_{out} = 0.1 I_{s2} + 0.3 V_{s3}$

3. Suppose the dependent current source in Figure 5.4 is changed from $g_m V_A$ to $g_m(V_A + V_{out})$. Compute the new expression for V_{out} if $G_1 = 1$ S, $G_2 = 2$ S, $G_3 = 3$ S, and $g_m = 5$ S.

ANSWER: $V_{out} = \frac{1}{3} I_{s1} + \frac{4}{15} I_{s2} + \frac{4}{5} V_{s3}$

EXAMPLE 5.3. A linear resistive circuit has two inputs v_{s1} and i_{s2} with output v_{out}, as shown in Figure 5.5. Rows 1 and 2 of Table 5.1 list the results of two sets of measurements taken in a laboratory. The measurements are taken in a practical way by first setting the value of the current source to zero, i.e., $i_{s2} = 0$ and exciting v_{s1} with a dc power supply set to 5 V; then the voltage source is removed and replaced by a short circuit using a jumper cable, i.e., $v_{s1} = 0$, and the current source is excited by a power supply producing a constant current of $i_{s2} = 0.2$ A.

    (a)   Derive the linear relationship $v_{out} = 0.8v_{s1} + 50i_{s2}$ using the data in Table 5.1.

    (b)   Find v_{out} when $v_{s1} = 10$ V and $i_{s2} = 0.5$ A, i.e., complete the third row of Table 5.1.

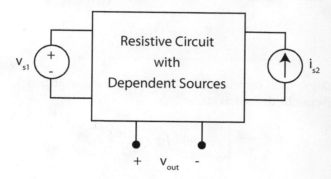

FIGURE 5.5. Linear resistive circuit driven by two sources.

TABLE 5.1. Two Sets of Measurements of a Linear Circuit in which One is Allowed to Set Each Source Value to 0

v_{s1} (volts)	i_{s2} (amperes)	v_{out} (volts)
5	0	4
0	0.2	10
10	0.5	??

SOLUTION

From the linearity equation 5.1(a),

$$v_{out} = \alpha_1 v_{s1} + \beta_2 i_{s2}$$

for appropriate α_1 and β_2. From the data in rows 1 and 2 of Table 5.1,

$$4 = \alpha_1 \times 5 + \beta_2 \times 0 = 5\alpha_1 \Rightarrow \alpha_1 = 0.8$$

and

$$10 = \alpha_1 \times 0 + \beta_2 \times 0.2 = 0.2\beta_2 \Rightarrow \beta_2 = 50$$

in which case,

$$v_{out} = 0.8v_{s1} + 50i_{s2} \tag{5.7}$$

So from row 3 of Table 5.1, if $v_{s1} = 10$ V and $i_{s2} = 0.5$ A, we have that

$$v_{out} = 0.8 \times 10 + 50 \times 0.5 = 33 \text{ V.}$$

Exercise. 1. For Example 5.4, suppose v_{s1} = 50 V and i_{s2} = 0.4 A. Find v_{out}.

ANSWER: 60 V

2. Suppose the data in row 1, column 3, of Table 5.1 is changed to 10 V. Find v_{out} when v_{s1} = 50 V and i_{s2} = 0.4 A.

ANSWER: 120 V

Comparing the development of equation 5.7 in Example 5.3 with equation 5.1 suggests that the coefficients α_1 and β_2 can be defined as ratios:

$$\alpha_1 = \left.\frac{v_{out}}{v_{s1}}\right]_{i_{s2}=0} \quad \text{and} \quad \beta_2 = \left.\frac{v_{out}}{i_{s2}}\right]_{v_{s1}=0}$$

Example 5.3 and these equations suggest the algorithm for finding the coefficients in equation 5.1 by setting all inputs to zero except the input associated with the desired coefficient. This approach is sometimes impractical. It is not always possible to set an independent source voltage or current source to zero: imagine turning off a generator for downtown Manhattan to obtain a coefficient. The following example illustrates an alternate approach.

EXAMPLE 5.4. Consider Figure 5.6, which has two inputs v_{s1} and i_{s2} with output i_{out}. Table 5.2 lists measurement data taken in a laboratory. Row 1 of Table 5.2 lists the nominal operating conditions of the circuit. Rows 2 and 3 illustrate measurements in which one source has its value only slightly changed (although the change may be arbitrary) while keeping the other source value the same. From the linearity theorem, we know $i_{out} = \alpha_1 v_{s1} + \beta_2 i_{s2}$. Compute α_1 and β_2, and then find i_{out} to complete row 4 of Table 5.2.

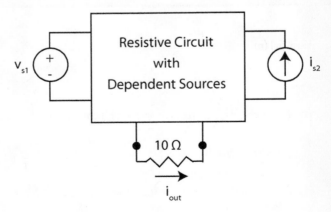

FIGURE 5.6. Linear resistive circuit driven by two sources.

TABLE 5.2. Two sets of measurements of a linear circuit.

v_{s1} (volts)	i_{s2} (amperes)	i_{out} (amps)
5	0.25	−1
5+0.1	0.25	−1.03
5	0.25+0.05	−0.9
15	0.5	????

SOLUTION

From rows 1 and 2 of Table 5.2,

$$-1 = \alpha_1 \times 5 + \beta_2 \times 0.25 \tag{5.8a}$$

and

$$-1.03 = \alpha_1 \times (5 + 0.1) + \beta_2 \times 0.25 \tag{5.8b}$$

Subtracting equation 5.8(a) from equation 5.8(b), we have

$$-0.03 = \alpha_1 \times 0.1 \implies \alpha_1 = -0.3$$

Similarly, from row 3, we have that

$$-0.9 = \alpha_1 \times 5 + \beta_2 \times (0.25 + 0.05) \tag{5.8c}$$

Again, subtracting equation 5.8(a) from equation 5.8(c), we have

$$0.1 = \beta_2 \times 0.05 \implies \beta_2 = 2$$

Equation 5.1 for the given data has the linear form

$$i_{out} = -0.3v_{s1} + 2i_{s2} \tag{5.9}$$

Hence, for row 4 of Table 5.2, we have that

$$i_{out} = -0.3 \times 15 + 2 \times 0.5 = -3.5 \text{ V}$$

Exercise. Find the unknown value in Table 5.3 using linearity.

TABLE 5.3. Two Sets of Measurements of a Linear Circuit

v_{s1} (volts)	i_{s2} (mA)	i_{out} (amps)
20	100	15
22	100	15.9
20	110	15.6
28	80	???

ANSWER: 17.4 A

As a final comment on linearity, we note that by simply using the data of rows 1 and 2 of Table 5.2, one can solve for the coefficients by solving simultaneous equations. Specifically, using the data of rows 1 and 2 of Table 5.2, we have the following matrix equation

$$\begin{bmatrix} 5 & 0.25 \\ 5.1 & 0.25 \end{bmatrix}\begin{bmatrix} \alpha_1 \\ \beta_1 \end{bmatrix} = \begin{bmatrix} -1 \\ -1.03 \end{bmatrix}$$

whose solution yields the proper coefficients of equation 5.9.

$$\begin{bmatrix} \alpha_1 \\ \beta_1 \end{bmatrix} = \begin{bmatrix} 5 & 0.25 \\ 5.1 & 0.25 \end{bmatrix}^{-1}\begin{bmatrix} -1 \\ -1.03 \end{bmatrix} = -40\begin{bmatrix} 0.25 & -0.25 \\ -5.1 & 5 \end{bmatrix}\begin{bmatrix} -1 \\ -1.03 \end{bmatrix} = \begin{bmatrix} -0.3 \\ 2 \end{bmatrix}$$

Exercise. Find the unknown entry in Table 5.4 after finding α_1 and β_1 in the equation $v_{out} = \alpha_1 v_{s1} + \beta_1 i_{s2}$.

TABLE 5.4. Two Sets of Measurements of a Linear Circuit

v_{s1} (volts)	i_{s2} (mA)	v_{out} (volts)
10	100	15
20	100	20
30	150	???

ANSWER: $v_{out} = 0.5v_{s1} + 100i_{s2}$ and 30 V

3. LINEARITY REVISITED: SUPERPOSITION AND PROPORTIONALITY

The linearity principle of equation 5.1 has the more simple form

$$y = a_1 u_1 + \dots + a_n u_n \tag{5.10}$$

Here, y denotes an output, whether it be current or voltage, and each u_i denotes a source input, whether it be voltage or current. A special consequence of the linearity principle is the **superposition property**. Equation 5.10 says that the total response y is the sum of the responses "$a_i u_i$". Each "$a_i u_i$" is the response of the circuit to u_i acting alone, i.e., when all other independent sources are set to zero. Although implied by linearity, this property is so important that we single it out.

THE SUPERPOSITION PROPERTY

For almost all linear resistive circuits containing more than one independent source, any output (voltage or current) in the circuit may be calculated by adding together the contributions due to each independent source acting alone with the remaining independent sources deactivated, i.e., their source values are set to zero.

EXAMPLE 5.5. A linear resistive circuit has two inputs V_{s1} and V_{s2} with output V_{out} as shown in Figure 5.7, where $R_1 = 2\ \Omega$, $R_2 = 2.5\ \Omega$, and $R_3 = 10\ \Omega$. Find V_{out} by the principle of superposition. Then, compute the power absorbed by the 10 Ω resistor. We show that $V_{out} = V_{out}^1 + V_{out}^2 = 0.5 V_{s1} + 0.4 V_{s2}$, where V_{out}^k is the contribution of the source V_{sk} acting alone for k = 1, 2.

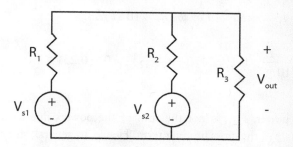

FIGURE 5.7 Linear resistive circuit driven by two voltage sources;
$R_1 = 2\ \Omega$, $R_2 = 2.5\ \Omega$, and $R_3 = 10\ \Omega$.

SOLUTION

Step 1. *Find the contribution to V_{out} due only to V_{s1}.* Denote this contribution by V_{out}^1. With $V_{s2} = 0$, the equivalent circuit is shown in Figure 5.8(a). Here, the 2.5 Ω and 10 Ω resistors are in parallel, yielding an equivalent resistance of 2 = 2.5 × 10/12.5 Ω. By voltage division,

$$V_{out}^1 = \frac{2}{2+2} V_{s1} = 0.5 V_{s1}$$

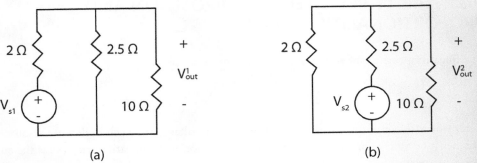

(a)   (b)

FIGURE 5.8 (a) Circuit equivalent to Figure 5.7 when $V_{s2} = 0$;
(b) circuit equivalent to Figure 5.7 when $V_{s1} = 0$.

Step 2. *Find the contribution to V_{out} due to V_{s2}.* Denote this contribution by V_{out}^2. With $V_{s1} = 0$, the equivalent circuit is shown in Figure 5.8(b). Here, the 2 Ω and 10 Ω resistors are in parallel, yielding an equivalent resistance of $5/3 = 2 \times 10/12$ Ω. By voltage division,

$$V_{out}^2 = \frac{\dfrac{5}{3}}{2.5 + \dfrac{5}{3}} V_{s2} = 0.4 V_{s2}$$

Step 3. *Compute V_{out} by superposition.* Using superposition,

$$V_{out} = V_{out}^1 + V_{out}^2 = 0.5 V_{s1} + 0.4 V_{s2}$$

Step 4. *Compute the power absorbed by the $R_3 = 10$ Ω resistor.*

$$P_{R3} = \frac{(V_{out})^2}{10} = 0.1\left(0.5 V_{s1} + 0.4 V_{s2}\right)^2 = 0.1\left(0.25 V_{s1}^2 + 0.2 V_{s1} V_{s2} + 0.16 V_{s2}^2\right)$$

Note that the total power, P_{R3}, is not the sum of the powers due to each source acting alone because of the presence of the cross product term. Hence, in general, superposition does not apply to the calculation of power.

> *For dc circuit analysis, the principle of superposition does NOT apply to power calculations.*

Exercise. Reconsider Figure 5.7 in which $R_1 = 2$ Ω, $R_2 = 4$ Ω, and $R_3 = 4$ Ω. Find V_{out} by the principle of superposition.
ANSWER: $V_{out} = 0.5 V_{s1} + 0.25 V_{s2}$

The next example adds a controlled source to the circuit of Figure 5.7 and repeats the superposition analysis.

EXAMPLE 5.6. For the circuit of Figure 5.9, suppose $R_1 = 2\ \Omega$, $R_2 = 5\ \Omega$, $R_3 = 10\ \Omega$, $g_m = 0.2$ S. Using superposition, find v_{out} in terms of v_{s1} and v_{s2}. We will use the superposition theorem to show that $v_{out} = v_{out}^1 + v_{out}^2 = 0.5v_{s1} + 0.4v_{s2}$, where v_{out}^k is the contribution from the source v_{sk} acting alone for k = 1, 2.

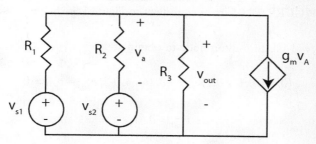

FIGURE 5.9 Circuit containing a dependent source for illustrating the principle of superposition.

SOLUTION

Step 1. *Compute the contribution due only to* v_{s1}. Setting $v_{s2} = 0$ leads to the circuit of Figure 5.10, where we note that $v_a = v_{out}^1$.

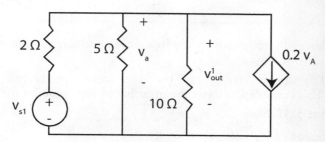

FIGURE 5.10 Circuit equivalent to Figure 5.9 when $v_{s2} = 0$.

Applying KCL to the top node yields

$$0.5(v_{out}^1 - v_{s1}) + (0.2 + 0.1 + 0.2)v_{out}^1 = 0$$

Therefore,

$$v_{out}^1 = 0.5v_{s1}$$

Step 2. *Compute the contributions due only to* v_{s2}. Setting $v_{s1} = 0$ in Figure 5.9 leads to the circuit of Figure 5.11, where this time, $v_a = v_{out}^2 - v_{s2}$.

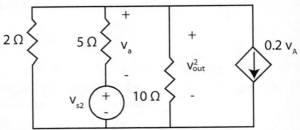

FIGURE 5.11 Circuit equivalent to Figure 5.9 when $v_{s1} = 0$.

As in Step 1, we apply KCL to the top node to obtain

$$(0.5 + 0.1 + 0.2)v_{out}^2 - 0.2v_{s2} + 0.2(v_{out}^2 - v_{s2}) = 0$$

Therefore,

$$v_{out}^2 = 0.4v_{s2}$$

Step 3. Using superposition, add up the contributions due to each independent source acting alone.

$$v_{out} = v_{out}^1 + v_{out}^2 = 0.5v_{s1} + 0.4v_{s2} \tag{5.11}$$

Exercise. Repeat Example 5.6 with $R_1 = R_2 = R_3 = 4\ \Omega$ and $g_m = 0.25$ S.

ANSWER: $v_{out} = 0.25v_{s1} + 0.5v_{s2}$

The above examples used voltage division and superposition to compute an output voltage due to two independent voltage sources.

EXAMPLE 5.7. This example illustrates the principle of superposition for the three-input op amp circuit of Figure 5. 12. Show that $V_{out} = V_{out}^1 + V_{out}^2 + V_{out}^3 = -4V_{s1} + 2.5V_{s2} + 2.5V_{s3}$, where V_{out}^k is the contribution of V_{sk} acting alone for k = 1, 2, 3.

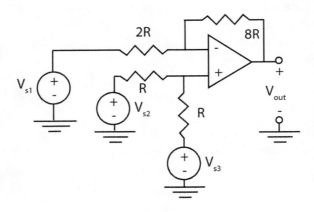

FIGURE 5.12 Three-input op amp circuit.

SOLUTION

Step 1. *Find the contribution to V_{out} due only to V_{s1}.* Denote this output by V_{out}^1. With $V_{s2} = V_{s3} = 0$, the circuit of Figure 5.12 reduces to that of Figure 5.13(a). The properties of an ideal op amp ensure that $i_+ = 0$, making $v_+ = -0.5Ri_+ = 0$. Thus, $v_- = v_+ = 0$ implies

$$V_{out}^1 = -\frac{8R}{2R}V_{s1} = -4V_{s1}$$

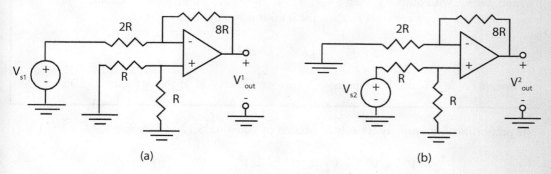

(a)   (b)

FIGURE 5.13

Step 2. *Find the contribution to V_{out} due only to V_{s2}.* With $V_{s1} = V_{s3} = 0$, the equivalent circuit is shown in Figure 5.13(b) where we denote the output as V_{out}^2. From op amp properties and voltage division,

$$v_- = v_+ = \frac{R}{R+R}V_{s2} = 0.5V_{s2}$$

Hence, from Example 4.3,

$$V_{out}^2 = \frac{10R}{2R}v_- = 2.5V_{s2}$$

Step 3. *Find the contribution to V_{out} due only to V_{s3}.* The equivalent circuit in this case is the same as that of Figure 5.13(b) with V_{s2} replaced by V_{s3}. Therefore, the output due to source V_{s3} acting alone is

$$V_{out}^3 = 2.5V_{s3}$$

Step 4. *Sum up contributions due to each source.* By the principle of superposition,

$$V_{out} = V_{out}^1 + V_{out}^2 + V_{out}^3 = -4V_{s1} + 2.5V_{s2} + 2.5V_{s3} \qquad (5.12)$$

Exercise. 1. For Example 5.7, suppose $V_{s1} = V_{s2} = V_{s3} = 2$ V. Find V_{out}.
ANSWER: 2 V

2. Now suppose $V_{s1} = 8$ V, $V_{s2} = V_{s3} = 2$, and the op amp saturates at $|V_{out}| = 12$ V; compute V_{out}.
ANSWER: −12 V

The above examples have generated the linearity formula, equation 5.1, using superposition, i.e., the response of a circuit is the sum of the responses due to each source acting alone. The technique is equivalent to that described in Example 5.3. However, superposition alone is not equivalent to linearity. Linearity is equivalent to the properties of superposition AND proportionality, which is now stated.

THE PROPORTIONALITY PROPERTY

For almost all linear resistive circuits, when any one of the independent sources is acting alone, say u_1, with output y, then $y = a_1 u_1$ for some constant a_1. Proportionality says that if u_1 is multiplied by a constant K, then the output is multiplied by K, i.e.,

$$y^{new} = \left(Ky^{old}\right) = a_1\left(Ku_1\right).$$

However, for dc analysis, the proportionality property does NOT apply for power calculations.

The proportionality property is easily illustrated by equation 5.12 of Example 5.7:

$$V_{out} = V_{out}^1 + V_{out}^2 + V_{out}^3 = -4V_{s1} + 2.5V_{s2} + 2.5V_{s3}$$

If $V_{s1} = K_1 V_0$ and $V_{s2} = V_{s3} = 0$, then $V_{out}^{new} = -4(KV_{s1}) = K(-4V_{s1}) = KV_{out}^{old}$.

Exercises. For certain nonlinear circuits, the principle of superposition may be satisfied, but proportionality not satisfied, or vice versa. This exercise explores these distinctions.

1. If a circuit has input-output relationship $v_{out} = a_1 v_{s1}^3 + a_2 v_{s2}^3$, show that the principle of superposition is satisfied, but proportionality is not satisfied.
2. If a circuit has input-output relationship $v_{out} = a_1 v_{s1} + a_2 v_{s2} + a_3 v_{s1} v_{s2}$, show that the principle of proportionality is satisfied, but superposition is not satisfied.

A very interesting and significant application of the proportionality property occurs in the analysis of a resistive ladder network. A resistive **ladder network** is one having the patterned structure shown in Figure 5.14, where each box represents a resistor.

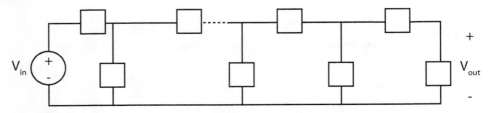

FIGURE 5.14 A ladder network.

A typical analysis problem follows:

Given V_{in} and all resistances in Figure 5.14, find all node voltages. One can, of course, solve the problem by writing and solving a set of mesh equations or node equations. A simple trick using the proportionality property allows us to solve arbitrarily long ladder networks without simultaneous equations, as follows: assume $V_{out}^{old} = 1$ V. We can sequentially compute currents and voltages in a backwards fashion to obtain the required source value to yield $V_{out}^{old} = 1$ V. Suppose we call this voltage V_s^{old}. Define $K = \dfrac{V_s}{V_s^{old}}$ to be the proportionality constant, where is the actual source voltage. Then the correct output voltage is $V_{out} = V_{out}^{new} = KV_{out}^{old}$.

EXAMPLE 5.8. Find all the voltages V_i, $i = 1, \ldots, 6$ in the resistive ladder network of Figure 5.15.

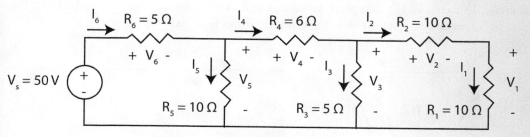

FIGURE 5.15 A simple resistive ladder network.

SOLUTION
Assume $V_1 = 1$ V. Repeatedly apply Ohm's law, KCL, and KVL as follows: (Ω, V and A are used throughout):

$$I_1 = \frac{V_1}{R_1} = 0.1 \qquad \text{(Ohm's law)}$$

$$I_2 = I_1 = 0.1 \qquad \text{(KCL)}$$

$$V_2 = R_2 I_2 = 10 \times 0.1 = 1 \qquad \text{(Ohm's law)}$$

$$V_3 = V_1 + V_2 = 2 \qquad \text{(KVL)}$$

$$I_3 = \frac{V_3}{R_3} = 0.4 \qquad \text{(Ohm's law)}$$

$$I_4 = I_2 + I_3 = 0.1 + 0.4 = 0.5 \qquad \text{(KCL)}$$

$$V_4 = R_4 I_4 = 6 \times 0.5 = 3 \qquad \text{(Ohm's law)}$$

$$V_5 = V_3 + V_4 = 5 \qquad \text{(KVL)}$$

$$I_5 = \frac{V_5}{R_5} = 0.5 \qquad \text{(Ohm's law)}$$

$$I_6 = I_4 + I_5 = 0.5 + 0.5 = 1 \qquad \text{(KCL)}$$

$$V_6 = R_6 I_6 = 5 \qquad \text{(Ohm's law)}$$

$$V_s = V_5 + V_6 = 10 \qquad \text{(KVL)}$$

We conclude that if $V_1 = 1$ V, the source voltage must be $V_s = 10$ V. But the actual source voltage is 50 V. Define $K = \dfrac{V_s^{new}}{V_s^{old}} = \dfrac{50}{10} = 5$. By the proportionality property, if $V_s = 50$ V, then $V_1 = K \times 1 = 5$ V. Similarly, $V_2 = 5$ V, $V_3 = 10$ V, $V_4 = 15$ V, and $V_5 = 25$ V.

In the solution given above, we have separated the expressions into calculation blocks to empha-size the repetitive pattern. For example, the expressions in block #3 are simply obtained from block #2 by increasing all subscripts by 2. When the ladder network has more elements, the sequence of expressions contains more blocks, each of which entails two additions and two mul-tiplications. This method then allows us to straightforwardly solve ladder networks of any size without writing or solving simultaneous equations.

Exercise. In Example 5.7, change all resistances to 2 Ω and find V_1. Would it make any differ-ence in the voltage V_1 if all the resistors were changed to R ohms?

ANSWER: $V_1 = \dfrac{50}{13} = 3.85$ V, and no difference.

---

4. SOURCE TRANSFORMATIONS

The words "source transformation" refer to the conversion of a voltage source in series with an R-ohm resistor to a current source in parallel with an R-ohm resistor, and/or vice versa. This section explains the details of such transformations and how they can simplify analysis. But first we must recall from Chapter 2 that voltage sources in series add together (such as batteries added to a flash-light) and that current sources in parallel combine into an equivalent single current source. This is illustrated for multiple voltage sources in series in Figure 5.16. Similarly, Figure 5.17 shows how multiple current sources combine into a single source.

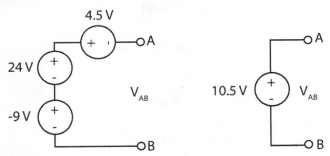

FIGURE 5.16 (a) Three voltage sources in series; (b) equivalent single voltage source.

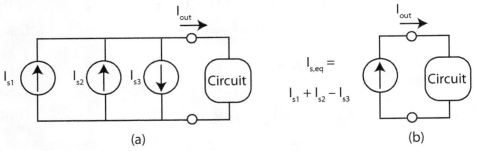

FIGURE 5.17 (a) Three independent current sources in parallel;
(b) equivalent single source circuit.

SOURCE TRANSFORMATION THEOREM FOR INDEPENDENT SOURCES

A 2-terminal network consisting of a series connection of an independent voltage source V_a and a nonzero finite resistance R is equivalent to a 2-terminal network consisting of an independent current source, $I_b = V_a/R$ in parallel with R.

Conversely, a 2-terminal network consisting of a parallel connection of an independent current source I_b and a nonzero finite resistance R, is equivalent to a 2-terminal network consisting of an independent voltage source, $V_a = RI_b$, in series with R. The reference directions for voltages and currents are as shown in Figure 5.18.

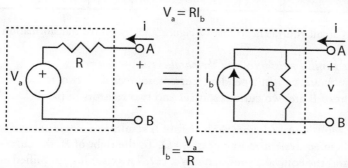

FIGURE 5.18 Illustration of source transformation theorem for independent sources.

A justification for the source transformation theorem will be given in the next section. Practically speaking, it can save significant computational effort. For example, in the circuit of Figure 5.19 in Example 5.9 below, a solution approach using mesh analysis requires writing and solving three simultaneous equations. Nodal analysis at A and B requires writing and solving two simultaneous equations. Applying the *source transformation theorem* is a third avenue that avoids all simultaneous equations.

EXAMPLE 5.9. Find I_{AB} in Figure 5.19 by repeated applications of the source transformation theorem. Then find the power absorbed by the 4 kΩ resistor.

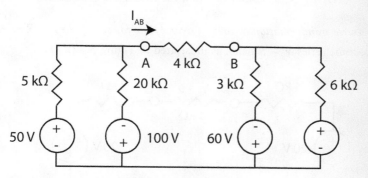

FIGURE 5.19 Circuit for Example 5.9.

SOLUTION
Step 1. Substitute all series V_s - R combinations by their parallel I_s - R equivalents, where in each case, $I_s = \dfrac{V_s}{R}$. Applying the source transformation theorem four times results in Figure 5.20.

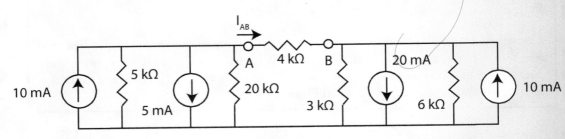

FIGURE 5.20 Circuit equivalent to that of Figure 5.19 by source transformation theorem.

Step 2. *Combine the parallel resistances and the parallel current sources.*
To the left of point A are two independent current sources and two resistors, all in parallel. Similarly, to the right of B are two current sources and two resistors in parallel.

Combining current sources and resistors to the left of A results in a single current source of 5 mA directed upward and an equivalent resistance of 4 kΩ. To the right of B, the current sources cancel each other out, and the equivalent resistance is 2 kΩ. The resulting simplified circuit is shown in Figure 5.21.

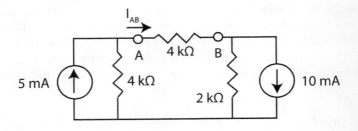

FIGURE 5.21 Simplification of the circuit in Figure 5.20.

Step 3. *Apply the source transformation theorem a second time to each of the I_s – R pairs. These parallel I_s – R pairs become series V_s – R pairs, as illustrated in Figure 5.22.*

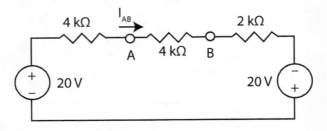

FIGURE 5.22 Further simplification of Figure 5.21.

Step 4. *Find I_{AB} and $P_{4k\Omega}$.* From Ohm's law,

$$I_{AB} = \frac{20 + 20}{4 + 4 + 2} = 4 \text{ mA}$$

Thus, $P_{4k\Omega} = 4000 \times (0.004)^2 = 64$ mW.

Exercises. 1. For the circuit of Figure 5.23(a), I_{s1} = 50 mA and R_a = 500 Ω. Convert the parallel $I_s - R$ combination to a series $V_s - R$ combination, where V_s = ? and R_{series} = ? in Figure 5.23(b).
ANSWER: 25 V, 500 Ω

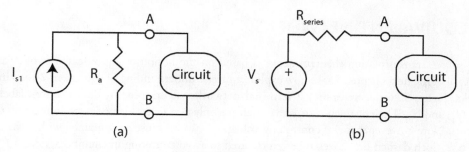

(a)   (b)

FIGURE 5.23

2. For the circuit of Figure 5.24(a), I_{s1} = 50 mA and R_a = 500 Ω, while I_{s2} = 150 mA and R_a = 300 Ω. Convert the two parallel $I_s - R$ combination to a single series $V_s - R$ combination, where V_s = ? and R_{series} = ? in Figure 5.24(b).
ANSWER: –20 V, 800 Ω

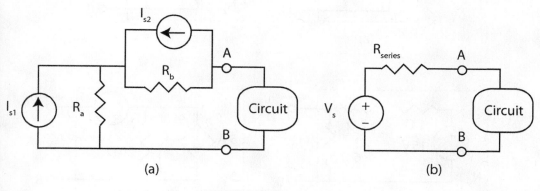

(a)   (b)

FIGURE 5.24

3. Consider the circuit in Figure 5.25(a). Using a source transformation and resistance combinations, determine the values of I_s and R_{para} in Figure 5.25(b).

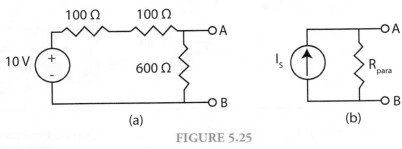

FIGURE 5.25

ANSWER: 150 Ω, 50 mA

---

5. EQUIVALENT NETWORKS

The source transformation theorem above is based on the notion of equivalent networks, as is the material of the next chapter. So we now explore a precise understanding of equivalent **2-terminal networks**. Figure 5.26 illustrates four **2-terminal networks,** all enclosed by dashed boxes, labeled N_1, N_2, N_3, and N_4. Their characteristic is that there are only two accessible nodes for connection to other circuits. Note however that any controlling voltage or current must be contained within the dashed-line box. Such dashed-line boxes are often omitted to avoid cluttering in circuit diagrams.

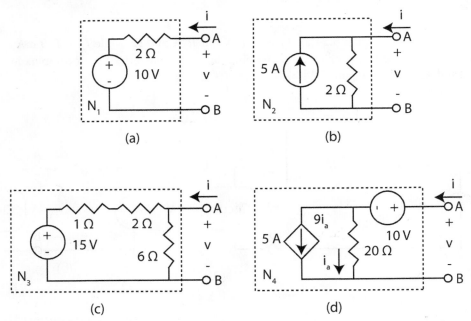

FIGURE. 5.26 Examples of 2-terminal networks, i.e., networks in which only two terminals are available for connection to other networks.

Observe that networks N_1 and N_2 in Figure 5.26(a) and (b) have the same terminal characteristics: at the terminals of N_1, the $v - i$ characteristic is

$$v = 2i + 10$$

At the terminals of N_2, the characteristic is

$$i = \frac{v}{2} - 5 \implies v = 2i + 10$$

The two equations are identical. We then say that a pair of 2-terminal networks are **equivalent** if they have the same terminal characteristics. Therefore, N_1 and N_2 are equivalent.

Now, observe that networks N_3 and N_4 are also equivalent to N_1 and N_2. To see this, note that for N_3,

$$i = \frac{v}{6} + \frac{v - 15}{3} \implies 6i = 3v - 30 \implies v = 2i + 10$$

And for N_4, first observe that $i = 10i_a$ from KCL, in which case $i_a = 0.1i$; further, from KVL,

$$v = 10 + 20i_a \implies v = 2i + 10$$

as was to be shown.

Because equivalent 2-terminal networks have the same terminal $v - i$ characteristic, if one network is interchanged with its equivalent, all currents and voltages *outside the box* remain the same as illustrated in Figure 5.27; i.e., all voltages and currents in the "rest of the circuit" are the same as before.

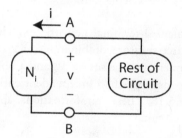

FIGURE 5.27 The networks denoted N_i, $i = 1,2$, are equivalent when the v-i values at the terminals are identical; logically then, all voltages and currents inside the "Rest of Circuit" remain the same.

These examples allow us to justify the source transformation theorem as follows. Both 2-terminal networks in Figure 5.18 have the same $v - i$ relationship: $v = Ri + V_a$ and

$$i = \frac{v}{R} - i_b \implies v = Ri + Ri_b = Ri + V_a.$$

Therefore, the two networks of Figure 5.18 are equivalent, and the source transformation is a valid analysis technique.

6. SUMMARY

This chapter covers the notions of linearity, superposition, proportionality, and source transformations. Linearity states that for any linear resistive circuit, any output voltage or current, denoted as y, is related to the independent sources by the formula $y = a_1 u_1 + ... + a_n u_n$, where u_1 through u_n are the voltage and current values of the independent sources, and a_1 through a_n are appropriate constants. Once values for the a_i are known, one can compute the output for any (new) set of input values without having to resolve the circuit equations, a tremendous savings in time and effort. A special consequence of linearity is the widely used principle of superposition.

Superposition means that in any linear resistive circuit containing more than one independent source, any output (voltage or current) can be calculated by adding together the contributions due to each independent source acting alone with the remaining independent source values set to zero. Practically speaking, this is the customary path to computing the coefficients, a_i, in the linearity formula.

Proportionality, another consequence of linearity, means that if a single input is scaled by a constant, with the other inputs set to zero, then the output is scaled by the same constant. This property led to a clever technique for analyzing ladder networks without writing simultaneous equations. Since power is proportional to the square of a voltage or current, $P = \dfrac{V^2}{R} = RI^2$ for dc resistive circuits, the principle of linearity and its consequences, superposition and proportionality, DO NOT APPLY for power calculations.

Using the notion of an equivalent 2-terminal network, the chapter set forth the theorems on source transformations for source-resistor combinations: a 2-terminal network consisting of a series connection of an independent voltage source V_a and a nonzero finite resistance R is equivalent to a 2-terminal network consisting of an independent current source, $I_b = V_a/R$, in parallel with R, as illustrated in Figure 5.18. These transformations, applied multiple times to a circuit, often simplify the analysis of a circuit.

7. TERMS AND CONCEPTS

2-terminal network: an interconnection of circuit elements inside a box having only 2 accessible terminals for connection to other networks. The concept is extendible to n-terminal networks.

Equivalent 2-terminal networks: two 2-terminal networks having the same terminal voltage-current relationship. If two 2-terminal networks N_1 and N_2 are **equivalent**, then one can be substituted for the other without affecting the voltages and currents in any attached network.

Linearity property: let the responses due to inputs u_1 and u_2, each acting alone, be y_1 and y_2. When the scaled inputs $\alpha_1 u_1$ and $\alpha_2 u_2$ are applied simultaneously, the response is $y = a_1 y_1 + \alpha_2 y_2$. Linearity implies both superposition and proportionality, and vice versa.

Linear resistive element: a 2-terminal circuit element whose terminal voltage and current relationships is described by Ohm's law.

Linear resistive circuit/network: a network consisting of linear resistive elements, independent voltage and current sources, op amps, and controlled sources.

Proportionality property: when an input to a linear resistive network is acting alone, multiplying the input by a constant, K, implies that the response is multiplied by K.

Source transformation: a 2-terminal network consisting of an independent voltage source in series with a resistance is equivalent to another 2-terminal network consisting of an independent current source in parallel with a resistance of the same value.

Superposition property: when a number of inputs are applied to a linear resistive network simultaneously, the response is the sum of the responses due to each input acting alone.

Problems

LINEARITY

1. Consider the circuit of Figure P5.1 in which $R_1 = 5\ \Omega$ and $R_2 = 20\ \Omega$.

   (a) Using linearity, $v_{out}(t)$ may be expressed as $v_{out}(t) = \alpha_1 v_1(t) + \beta_2 I_2$. Compute α_1 and β_2.

   (b) If $v_1(t) = 10\ \cos(10t)$ V and $I_2 = 2$ A, find $v_{out}(t)$.

   (c) Redo part (a), but this time express α_1 and β_2 in terms of the literals

   $$G_1 = \frac{1}{R_1} \text{ and } G_2 = \frac{1}{R_2}.$$

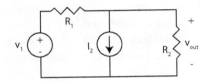

Figure P5.1

ANSWER: (b) $8\ \cos(10t) - 8$ V

2. For the circuit of Figure P5.2,

   (a) find V_B in terms of V_{s1}, V_{s2}, G_1, G_2, and G_3, and

   (b) find I_B in terms of V_{s1}, V_{s2}, G_1, G_2, and G_3.

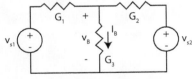

Figure P5.2

3. For the circuit of Figure P5.3,

   (a) find I_B in terms of I_{s1}, I_{s2}, R_1, R_2, and R_3, and

   (b) find V_B in terms of I_{s1}, I_{s2}, R_1, R_2, and R_3.

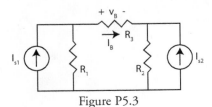

Figure P5.3

4. Consider the circuit shown in Figure P5.4.

   (a) Find the coefficients α_1 and β_1 in the linear relationship $v_{o1} = \alpha_1 V_{s1} + \beta_1 I_{s2}$.

   (b) Find the coefficients α_1 and β_1 in the linear relationship $v_{o2} = \alpha_2 V_{s1} + \beta_2 I_{s2}$.

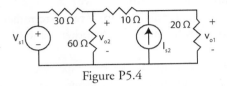

Figure P5.4

5. Consider the circuit of Figure P5.5.

   (a) Find the linear relationship between V_{out} and the input sources V_{s1} and I_{s2}.

   (b) If $V_{s1} = 20$ V and $I_{s2} = 0.5$ A, find V_{out}.

   (c) **(Challenge)** What is the effect of doubling all resistance values on the coefficients of the linear relationship found in part (a)?

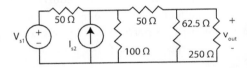

Figure P5.5

CHECK: $v_{out} = 0.25 V_{s1} + ????\ I_{s2}$

6. For the circuit of Figure P5.6, find the linear relationship between I_{out} and the independent sources. Hint: Write a single loop equation.

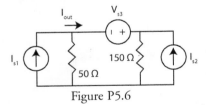

Figure P5.6

7. Consider the circuit shown in Figure P5.7 in which $R_1 = 80\ \Omega$. $R_2 = 20\ \Omega$, $R_3 = 80\ \Omega$. and $r_m = 20$.

   (a) Find the linear relationship between v_{out} and the two independent sources. Hint: Write two loop equations.

   (b) If $V_{s1} = 20$ V and $I_{s2} = 0.125$ A, compute the power delivered by the dependent source.

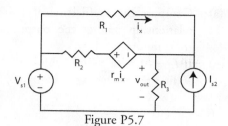

Figure P5.7

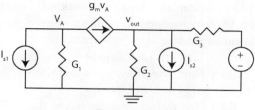

Figure P5.9

CHECK: (a) v_{out}= ????? V_{s1} + 16I_{s2}, 39 mW ≥ P_{dep} ≥ 36 mW

8. Consider the circuit of Figure P5.8. in which R_1 = 18 Ω. R_2 = 9 Ω, R_3 = 18 Ω, R_4 = 36 Ω, and R_5 = 18 Ω.

   (a)   Find the linear relationship between $v_{out}(t)$ and the four independent sources.

   (b)   **(Challenge)** If each of the resistances is doubled, what is the new linear relationship. (Reason your way to the answer without having to resolve the circuit. Hints: Investigate the effect of changing the resistance in Ohm's law for fixed current. Investigate the effect of equal changes in all resistances on a voltage divider formula.)

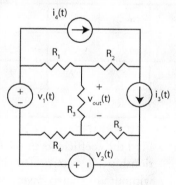

Figure P5.8

CHECK: $v_{out} = \dfrac{3}{8}v_1 + ??\,i_4 - \dfrac{27}{4}i_3 + ??\,v_2$

9. For the circuit of Figure P5.9, express V_{out} as a linear combination of I_{s1}, I_{s2}, and V_{s3} as per equation 5.1(a). Assume G_1 = 0.4 S, G_2 = G_3 = 0.05 S, and g_m = 0.1 S.

10. A linear resistive circuit has two independent sources, as shown in Figure P5.10. If $i_{s1}(t)$ = 0 with $v_{s2}(t)$ = 10cos(2t) V, then $v_{out1}(t)$ = 20 cos(2t) V. On the other hand, if $i_{s1}(t)$ = 10cos(2t) mA with $v_{s2}(t)$ = 0, then $v_{out}(t)$ = 2 cos(2t) V. Find the linear relationship between $v_{out}(t)$ and the inputs, $i_{s1}(t)$ and $v_{s2}(t)$. Now compute $v_{out}(t)$ when $i_{s1}(t)$ = 20cos(2t) mA and $v_{s2}(t)$ = 20 V.

ANSWER: $v_{out}(t)$ = 40 + 4cos(2t) V

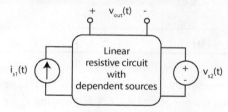

Figure P5.10

11. Again, consider the configuration of Figure P5.10. If $i_{s1}(t)$ = 0 with $v_{s2}(t)$ = 10 V, then $v_{out}(t)$ = 55 V. On the other hand, if $i_{s1}(t)$ = 4cos(2t) A with $v_{s2}(t)$ = 0, then $v_{out}(t)$ = −2cos(2t) V.

   (a)   If $i_{s1}(t)$ = 2cos(2t) A and $v_{s2}(t)$ = −10cos(2t) V, find $v_{out}(t)$.

   (b)   If $i_{s1}(t)$ = −4cos(5t) A and $v_{s2}(t)$ = 20cos(5t) V, find $v_{out}(t)$.

CHECK: (b) $v_{out}(t)$ = 108cos(5t) V

12. Consider again Figure P5.10. Suppose the measured data are as follows: (i) v_{out} = 15 V when i_{s1} = 2 A and v_{s2} = 10 V, and (ii) v_{out} = 10 V when i_{s1} = 3 A and v_{s2} = 5 V.

   (a)   Determine the linear relationship v_{out} = $\alpha_1 i_{s1}$ + $\beta_2 v_{s2}$.

   (b)   Find v_{out} when i_{s1} = 1 A and v_{s2} = 5 V

ANSWER: 7.5 V

13. Consider the linear network of Figure P5.13, which contains, at most, resistors and linear dependent sources. Measurement data is given in Table P5.13.

   (a)   Find the linear relationship $i_{out} = \alpha_1 v_{s1} + \beta_2 i_{s2}$.

   (b)   Find the power consumed by the 10 Ω resistor when $v_{s1} = 20$ V and $i_{s2} = 500$ mA.

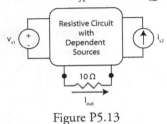

Figure P5.13

Table P5.13

v_{s1} (volts)	i_{s2} (amperes)	i_{out} (amps)
5	0.4	−1
10	1	2

14. Reconsider Figure P5.13 Two separate dc measurements are taken. In the first experiment, $v_{s1} = 7$ V and $i_{s2} = 3$ A, yielding $i_{out} = 1$ A. In the second experiment, $v_{s1} = 9$ V and $i_{s2} = 1$ A, yielding $i_{out} = 3$ A.

   (a)   Find the coefficients of the linear relationship $i_{out} = \alpha_1 v_{s1} + \beta_2 i_{s2}$.

   (b)   Given the equation found in part (a), compute P_{out} when $v_{s1} = 15$ V and $i_{s2} = 5$ A.

ANSWER: (b) 90 watts

15. The box in the circuit of Figure P5.15 contains resistors and dependent sources. $R_L = 100$ Ω. Table P5.15 contains various data measurements.

Table P5.15

	I_{s1} (mA)	V_{s2} (V)	V_{s3} (V)	V_{out} (V)
Case 1	50	−2	5	−13
Case 2	0	3	5	2
Case 3	0	2	4	0

   (a)   Compute the coefficients of a linear relationship among the output and three inputs.

   (b)   If $I_{s1} = -1$ A, $V_{s2} = 40$ V, and $V_{s3} = 10$ V, find the power absorbed by R_L.

ANSWER: 16 W

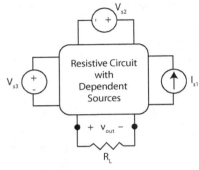

Figure P5.15

16. Again consider Figure P5.15. Suppose the data measurements are given in Table P5.16.

Table P5.16

	I_{s1} (mA)	V_{s2} (V)	V_{s3} (V)	V_{out} (V)
Case 1	30	2	−1	11.5
Case 2	40	2	−1	13
Case 3	30	2.2	−1	11.6
Case 4	30	2	−0.9	11.9
Case 5	40	8	10	??

   (a)   Find the coefficients in the linear relationship $V_{out} = \alpha_1 I_{s1} + \alpha_2 V_{s2} + \alpha_3 V_{s3}$ without any matrix inversions.

   (b)   Find the power consumed by R_L for the data in Case 5.

ANSWER: 25 watts

17. Again consider Figure P5.15. Suppose the data measurements are given in Table P5.17.

Table P5.17

	I_{s1} (mA)	V_{s2} (V)	V_{s3} (V)	V_{out} (V)
Case 1	30	2	−1	11.5
Case 2	−20	4	2	27
Case 3	−10	−3	1	−14
Case 4	40	10	10	???

Table P5.18

i_{s4} (mA)	v_{out} (V)
1	6
2	10
5	?
?	0

(a) Find the coefficients in the linear relationship $V_{out} = \alpha_1 I_{s1} + \alpha_2 V_{s2} + \alpha_3 V_{s3}$ using a matrix inversion.

(b) Find the power consumed by R_L for the data in Case 4.

ANSWER: 102.01 watts

18. The linear resistive circuit of Figure P5.18 has four independent sources. Three of these sources have fixed values. Only one, i_{s4}, is adjustable. In a laboratory, the data set forth in rows 1 and 2 of Table P5.18 were taken. Complete the last two rows of Table P5.18 using linearity and the data from the first two rows. For the data in row 3, find the power delivered by the current source i_{s4}. Hint: To solve this problem, recall from the linearity equation 5.1

$$v_{out} = a_1 v_{s1} + a_2 v_{s2} + a_3 i_{s3} + a_4 i_{s4} + (a_1 v_{s1} + a_2 v_{s2} + a_3 i_{s3}) = a_4 i_{s4} + K$$

We have used the fact here that the term $(a_1 v_{s1} + a_2 v_{s2} + a_3 i_{s3})$ is constant because the associated source values are constant. Thus, $(a_1 v_{s1} + a_2 v_{s2} + a_3 i_{s3}) = K$ for some K. Hence, one can use the data from the first two rows of Table P5.18 to solve for a_4 and K.

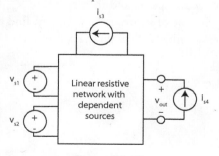

Figure P5.18

SUPERPOSITION AND PROPORTIONALITY

19. For the circuit of Figure P5.19, $R_1 = 200 \ \Omega$, $R_2 = 50 \ \Omega$, $v_{s1} = 12$ V, and $i_{s2} = 60$ mA.

(a) Find v_{out} using superposition. Specifically, first find v^1_{out} due to v_{s1} acting alone, and v^2_{out} due to i_{s2} acting alone.

ANSWER: 2.4 V, 2.4 V, 4.8 V

(b) Find v_{out} in terms of the literals R_1, R_2, v_{s1}, and i_{s2} and then compute the specific numerical relationship.

(c) If $v_{s1} = 10 \times 12$ V and $i_{s2} = 5 \times 60$ mA, determine v_{out} using the proportionality theorem by first computing v^1_{out} due to the modified v_{s1} acting alone, and v^2_{out} due to the modified i_{s2} acting alone.

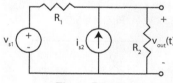

Figure P5.19

20. Consider the circuit of Figure P5.20 in which $R_1 = 20 \ \Omega$, $R_2 = 60 \ \Omega$, $R_3 = 20 \ \Omega$.

(a) Find the coefficients of the linear relationship $v_{out} = a_1 v_{s1} + a_2 i_{s2} + a_3 i_{s3}$ by superposition. Specifically, first find v^1_{out} due to v_{s1} acting alone, due to i_{s2} acting alone, and v^3_{out} due to i_{s3} acting alone.

(b) Repeat part (a), but express your answers in terms of the literals R_i, $i = 1, 2, 3$.

(c) Find v_{out}^1, v_{out}^2, v_{out}^3, and v_{out} and the power delivered to R_3 when $v_{s1} = 100$ V, $i_{s2} = 2$ A, and $i_{s3} = 4$ A.

(d) Repeat part (c) when v_{s1} is doubled, i_{s2} is tripled, and i_{s3} is halved.

ANSWER: (c) 100 V, 500 watts

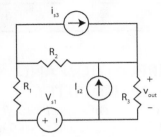

Figure P5.20

21. In the circuit shown in Figure P5.21, $R_1 = 180$ Ω, $R_2 = 360$ Ω, $R_3 = 90$ Ω, an $R_4 = 720$ Ω.

(a) Find the coefficients of the linear relationship $v_{out} = v_{out}^1 + v_{out}^2 = a_1 v_{s1} + a_2 i_{s2}$ by superposition.

(b) Repeat part (a), but express your answers in terms of the literals $G_i = 1/R_i$, $i = 1, 2, 3, 4$.

(c) Find v_{out} and the power absorbed by R_4 when $v_{s1} = 100$ V and $v_{s2} = 50$ V.

ANSWER: $v_{out} = 60$ V; $P_{R4} = 5$ watts

(d) Repeat part (c) when $v_{s1} = 0.5 \times 100$ V and $v_{s2} = -10 \times 2$ V.

Figure P5.21

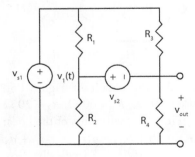

22. For the circuit of Figure P5.22, suppose $R_1 = 20$ Ω, $R_2 = 50$ Ω and $R_3 = 100$ Ω.

(a) Using superposition, find v_{out}^1 due to v_{s1} acting alone, and then find v_{out}^2 due to v_{s2} acting alone. What is v_{out}?

(b) Redo part (a) using the literals $G_i = \dfrac{1}{R_i}$.

ANSWER: (a) $v_{out}^1 = \dfrac{5}{8} v_{s1}$, $v_{out}^2 = \dfrac{7}{8} v_{s2}$

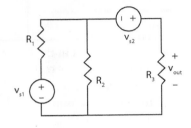

Figure P5.22

23. For the circuit of Figure P5.23, suppose $R_1 = 20$ Ω, $R_2 = 50$ Ω, $R_3 = 100$ Ω and $g_m = 0.02$ S.

(a) Using superposition, find v_{out}^1 due to v_{s1} acting alone, and then find v_{out}^2 due to v_{s2} acting alone. What is v_{out}?

ANSWER: $v_{out} = 0.5 v_{s1} + 0.9 v_{s2}$

(b) Redo part (a) using the literals $G_i = \dfrac{1}{R_i}$ and g_m.

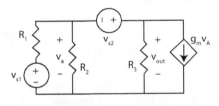

Figure P5.23

24. For the circuit of Figure P5.24, find the contribution to V_{out}^i from each independent source acting alone, and then compute V_{out} by the principle of superposition. Finally, find the power absorbed by the 900 Ω resistor.

ANSWER: 38 V and 1.6 watts

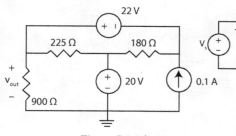

Figure P5.24

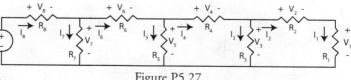

Figure P5.27

25. For the circuit shown in Figure P5.25, V_s = 160 V. Find V_{out}. Then find the intermediate node voltages. Hint: Assume V_{out} = 1 V and use proportionality, as per Example 5.11.

CHECK: Answer is an integer.

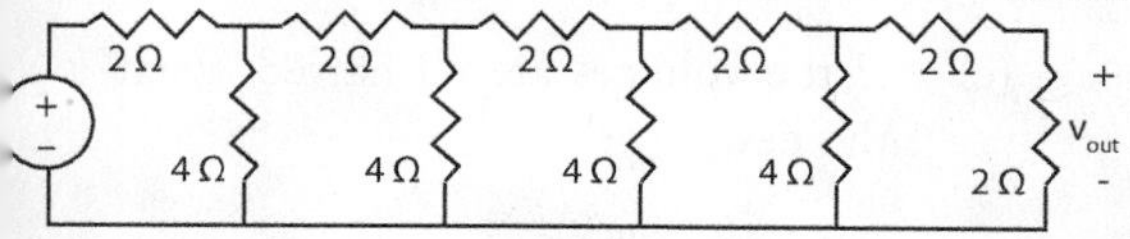

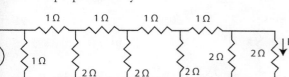

Figure P5.25

26. For the circuit shown in Figure P5.26, i_s = 64 mA. Find I_{out}. Hint: Assume I_{out} = 1 A and then use proportionality.

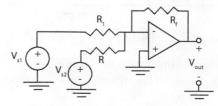

Figure 5.26

27.(a) For the circuit shown in Figure P5.27, If V_1 = 1 V, find V_s by writing a MATLAB program to solve the problem, given that R_1 = 10 Ω, R_2 = 10 Ω, R_3 = 5 Ω, R_4 = 6 Ω, R_5 = 10 Ω, R_6 = 5 Ω, R_7 = 20 Ω, and R_8 = 5 Ω.

(b) If it is known that V_s = 175 volts, find V_1.

(c) Find the equivalent resistance seen by the voltage source.

(d) Suppose R_1 is changed from 1 to 10 Ω in steps of 0.25 Ω. Obtain a plot of V_s vs R_1 by modifying the MATLAB code of (a). Assume V_1 = 1 V.

28. (a) For the circuit shown in Figure P5.28, If I_1 = 1 A, find I_s by writing a MATLAB program.

(b) If it is known that I_s = 200 mA, find I_1.

(c) Find the equivalent resistance seen by the current source.

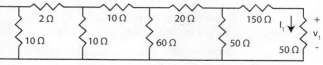

Figure P5.28

LINEARITY AND OP AMP CIRCUITS

29. Consider the circuit in Figure P5.29.
(a) Find the contribution to V_{out} due only to V_{s1}.
(b) Find the contribution to V_{out} due only to V_{s2}.
(c) Find V_{out} by superposition.

Figure P5.29

30. Consider the circuit in Figure P5.30.
(a) Find the contribution to V_{out} due only to V_{s1}.
(b) Find the contribution to V_{out} due only to V_{s2}.
(c) Find V_{out} by superposition.

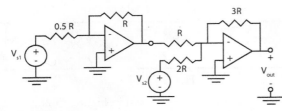

Figure P5.30

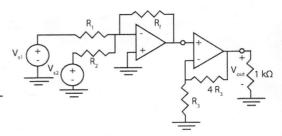

Figure P5.32

31. (a) For the circuit of Figure P5.31, find V_{out}^k, the voltage due to each source V_k acting alone in terms of the literal values.
    (b) Find V_{out} in terms of the literals.
    (c) Now suppose that $R_1 = 2R_0$, $R_2 = 3R_0$, $R_3 = 4R_0$, $R_f = 12R_0$, and $R_L = 100$ Ω. Suppose each voltage source has value 2 V: (i) Find the power absorbed by the load to each source acting alone, and (ii) the actual power delivered to the load when all sources are active.

33. (a) For the circuit in Figure P5.33, find the linear relationship between V_{out} and V_{s1}, V_{s2}, and V_{s3}.
    (b) If the input voltages are $V_{s1} = 5$ V, $V_{s2} = -2.5$ V and $V_{s3} = 2$ V, determine V_{out}.
    (c) If the voltages are all halved, what is the new V_{out}?

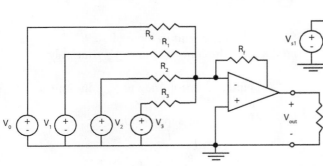

Figure P5.31

32. Consider the circuit in Figure P5.32.
    (a) Find the contribution to V_{out} due only to V_{s1}.
    (b) Find the contribution to V_{out} due only to V_{s2}.
    (c) Find V_{out} by superposition.
    (d) If $R_1 = R_2 = R_3 = 0.5R_f = 5$ kΩ and $V_{s1} = 2V_{s2} = 4$ V, find the power delivered to the 1 kΩ load.

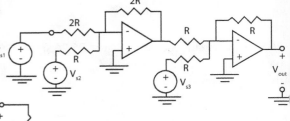

Figure P5.33

34. (a) For the circuit in Figure P5.34, find the linear relationship between V_{out} and V_{s1}, V_{s2}, and V_{s3}.
    (b) If the input voltages are $V_{s1} = 0.25$ V, $V_{s2} = -0.5$ V and $V_{s3} = 2$ V, determine V_{out}.
    (c) If the voltages are all halved, what is the new V_{out}?

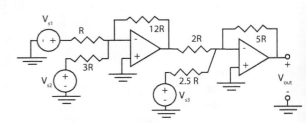

Figure P5.34

35. Consider the circuit in Figure P5.35.
   (a) Find the contribution to V_{out} due only to V_{s1}.
   (b) Find the contribution to V_{out} due only to V_{s2}.
   (c) Find V_{out} by superposition.

   (d) Find the power delivered to R_L when V_{s1} is acting alone, i.e., $V_{s2} = 0$ and then find the power delivered to R_L when V_{s2} is acting alone when $R_2 = 4$ R and $R_1 = R$.
   (e) Find the total power delivered to R_L when $R_2 = 4$ R and $R_1 = R$.

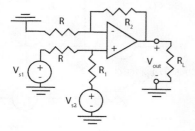

Figure P5.35

36. Consider the circuit in Figure P5.36 in which $R_1 = 0.25 R_f$.
   (a) Find the contribution to V_{out} due only to V_{s1}.
   (b) Find the contribution to V_{out} due only to V_{s2}.
   (c) Find V_{out} by superposition.
   (d) Find the power delivered to R_L when V_{s1} is acting alone, i.e., $V_{s2} = 0$, and then find the power delivered to R_L when V_{s2} is acting alone when $R_f = 4R_1$.
   (e) Find the total power delivered to R_L when $R_f = 4R_1$.

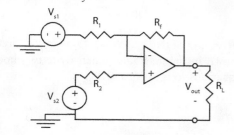

Figure P5.36

37. Consider the circuit in Figure P5.37 in which $R_1 = 0.25 R_f = R_2 = R_3$.
   (a) Find the contribution to V_{out} due only to V_{s1}.
   (b) Find the contribution to V_{out} due only to V_{s2}.
   (c) Find V_{out} by superposition.

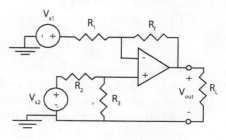

Figure P5.37

38. Consider the circuit in Figure P5.38.
   (a) Find the contribution to V_{out} due only to V_{s1}.
   (b) Find the contribution to V_{out} due only to V_{s2}.
   (c) Find the contribution to V_{out} due only to V_{s3}.
   (d) Find V_{out} by superposition.

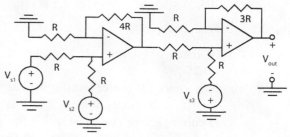

Figure P5.38

SOURCE TRANSFORMATIONS

39. Use a series of source transformations to simplify the circuit of Figure P5.39 into one consisting of a single voltage source in series with a single resistance.

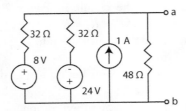

Figure P5.39
ANSWER: 6 V source in series with 12 Ω resistor

40. Consider the circuit of Figure P5.40 in which I_{s1} = 10 mA, V_{s2} = 20 V, and V_{s3} = 80 V.
  (a) Use a series of source transformations to find a single voltage source in series with a resistance that is in series with the 9.6 kΩ resistor.
  (b) Then find the power absorbed by the 9.6 kΩ resistor.

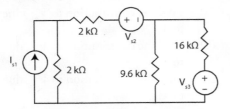

Figure P5.40
ANSWER: (a) 48 V in series with 3.2 kΩ; (b) 135 mW

41. In the circuit of Figure 5.41, V_{s1} = 240 V and I_{s2} = 0.25 A.
  (a) Use a series of source transformations to reduce the circuit of figure P5.41 to a current source in parallel with a single resistor in parallel with the 20 Ω resistor across which V_{out} appears. Find V_{out}.
  (b) Find the power dissipated in the 20 Ω resistor.
  (c) If both sources have their values increased by a factor of two, compute the new value of V_{out}. Can you do this by inspection? Explain.

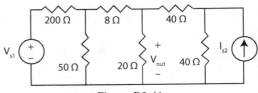

Figure P5.41
ANSWER: 13.5 and 9.1125

42. Use source transformations on the circuit of Figure P5.42, to compute the value of V_s needed to deliver a current of I_s = 0.25 A.

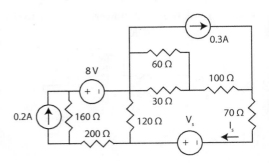

Figure P5.42
ANSWER: 28 V

43. For Figure P5.43, use a series of source transformations to find the value of V_s so that the power delivered to R_L is 16 watts.

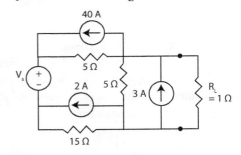

Figure P5.43

44. Apply source transformations to the circuit shown in Figure P5.44. Then write two nodal equations to find V_1 and V_2.

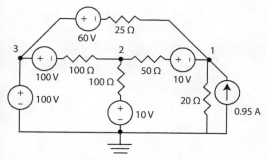

Figure P5.44 Source transformations simplify
writing node equations.
ANSWER: 25 V, 20 V

45. Apply source transformations to the circuit
of Figure P5.45. Then write two nodal equations to find V_1 and V_2.

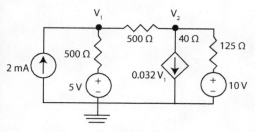

Figure P5.45
ANSWER: 2.8 V, −0.4 V

C H A P T E R

6

Thevenin, Norton, and Maximum Power Transfer Theorems

HISTORICAL NOTE

In the early days of electricity, engineers wanted to know how much voltage or current could be delivered to a load, such as a set of street lamps, through a complex transmission network. Before the days of computer-aided circuit simulation, simplification of complex circuits allowed engineers to analyze these very complex circuits manually. In 1883, a French telegraph engineer, M. L. Thevenin, first stated that a complex (passive) network could be replaced by an equivalent circuit consisting of an independent voltage source in series with a resistor. Although stated only for passive networks, the idea of a Thevenin equivalent evolved to include active networks. Its widespread use has simplified the homework of students for many years now and probably will continue to do so for many years to come.

A more recent but quite similar idea is the Norton equivalent circuit consisting of an independent current source in parallel with a resistance. At the time of E. L. Norton (a scientist with Bell Laboratories), the invention of vacuum tubes made independent current sources a realistic possibility. Many electronic circuits were modeled with independent and dependent *current sources*. The appearance of Norton's equivalent circuit was a natural outcome of advances in technology.

CHAPTER OBJECTIVES

1. Define and construct the Thevenin and Norton equivalent circuits for passive networks.
2. Define and construct the Thevenin and Norton equivalent circuits for active networks containing dependent sources or op amps.
3. Illustrate several different techniques for constructing the Thevenin and Norton equivalent circuits.
4. Investigate maximum power transfer to a load using Thevenin or Norton equivalents.

SECTION HEADINGS

1. INTRODUCTION

Practicing electrical engineers often want to know the power absorbed by one particular load. The load may be a large machine in a factory or a lighting network in the electrical engineering building. Simple resistances often represent such loads. Usually the load varies over time in which different resistances are used at different times to represent the load. What is the effect of this load variation on the absorbed power and on the current drawn by the load? To simplify analysis, the rest of the linear network (exclusive of the load) is replaced by a simple equivalent circuit consisting of just one resistance and one independent source.

For our purposes, a (resistive) load is a two-terminal network defined in Chapter 1, meaning that the current entering one of the terminals equals the current leaving the other. More generally, a two-terminal network is any circuit for which there are only two terminals available for connection to other networks. (See Figure 6.1.) The important question for our work in this chapter is: *How does one characterize a two-terminal network?* As is shown in Figure 6.1(a), there is a voltage $v(t)$ across the terminals and a current $i(t)$ entering one terminal and leaving the other. The relationship between the voltage $v(t)$ and the current $i(t)$ characterizes the two-terminal network. For example, if $v(t) = Ri(t)$, we would recognize the terminal network as an equivalent resistance R. Or, if $v(t) = Ri(t) + V_0$, we might recognized this equation as that of a resistance in series with a voltage source. In fact, this equation could be represented as graph, e.g. Figure 6.1(b).

This leads to our next question: *When are two 2-terminal networks equivalent?* As developed in Chapter 5, two 2-terminal networks are said to be equivalent when their terminal v-i characteristics are the same. Of particular interest for this chapter is an equivalent network consisting of a voltage source in series with a resistance, called the Thevenin equivalent network, and a current source in parallel with a resistance, called a Norton equivalent network. Figure 6.1c shows a Thevenin equivalent for a linear resistive circuit.

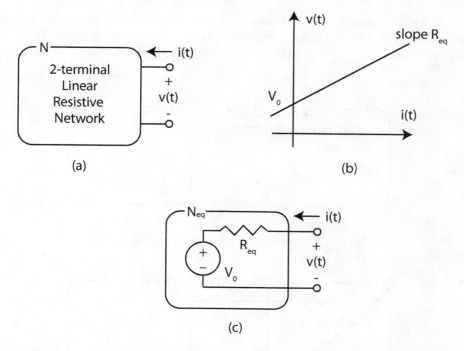

FIGURE 6.1. (a) a 2-terminal linear network with terminal voltage $v(t)$ and current $i(t)$;
(b) graphical representation of the equation $v(t) = R_{eq}i(t) + V_0$;
(c) Thevenin equivalent network N_{eq} having the same terminal $v(t)$ and $i(t)$ relationship as (b).

This chapter investigates the replacement of a network N by its **Thevenin equivalent** or its **Norton equivalent**. The first section describes the Thevenin and Norton equivalent theorems for passive networks, those containing only independent sources and resistors. Following that, we generalize the statements to include active networks. However, because op amps have peculiar properties, Thevenin and Norton equivalents of circuits with op amps are explored exclusively in Section 4. Following this, in Section 5, we describe how to obtain a Thevenin or Norton equivalent from measured data without having to know anything about the internal circuit structure. This is particularly useful when one has equipment such as a power supply but no schematic diagram of the internal circuitry. Unfortunately, not all linear devices have a well-defined Thevenin or Norton equivalent. The homework exercises illustrate a few cases. Section 6 explores the problem of maximum power transfer to a load in the context of the Thevenin equivalent circuit, which ends the chapter.

2. THEVENIN AND NORTON EQUIVALENT CIRCUITS FOR LINEAR PASSIVE NETWORKS

Our first objective is to develop and illustrate the celebrated Thevenin theorem for passive networks. Then we will state and illustrate Norton's theorem, dual to Thevenin's theorem.

To develop Thevenin's theorem, consider Figure 6.2(a) consisting of two 2-terminal networks, N and N_L, joined at A and B. Only resistors and independent sources make up N, while N_L contains arbitrary even nonlinear elements. Suppose N_L undergoes various changes as part of an experiment, while N, complicated in its own right, remains unchanged. To simplify repeated calculations, N is replaced by its Thevenin equivalent, as illustrated in Figure 6.2(b). The more simple Thevenin equivalent consists of a single voltage source, $v_{oc}(t)$, in series with a single resistance, R_{th}.

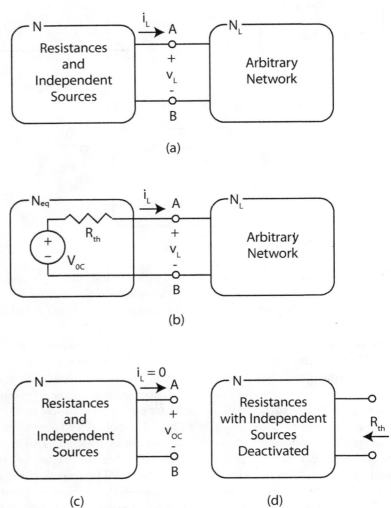

FIGURE 6.2 (a) Network N attached to an arbitrary network load, N_L; (b) N replaced by its so-called Thevenin equivalent, N_{eq}, still attached to N_L; (c) circuit for computing v_{oc}; (d) circuit for computing R_{th} in which all independent sources inside N are deactivated.

This brings us to a formal statement of Thevenin's theorem for passive networks.

THEVENIN'S THEOREM FOR PASSIVE NETWORKS

Given an arbitrary 2-terminal linear network, N, consisting of resistances and independent sources, then, for almost all such N, there exists an equivalent 2-terminal network consisting of a resistance, R_{th}, in series with an independent voltage source, $v_{oc}(t)$. The voltage, $v_{oc}(t)$, called the open-circuit voltage, is what appears across the 2 terminals of N. R_{th}, called the Thevenin equivalent resistance, is the equivalent resistance of N when all independent sources are deactivated. Figure 6.2(c) shows the appropriate polarity for $v_{oc}(t)$.

In the above theorem, "for almost all" means there are exceptions. For example, an independent current source does not have a Thevenin equivalent. More generally, any two-terminal network characterized by $i(t)$ = constant does not have a Thevenin equivalent. This leads us to suggest that there ought to be an equivalent current source formulation of an equivalent network. From Chapter 5, the source transformation theorem tells us that the Thevenin equivalent of Figure 6.2(b) when $R_{th} \neq 0$ is equivalent to a current source in parallel with R_{th}, as in Figure 6.3(b). Figure 6.3 leads us to a formal statement of the so-called Norton theorem.

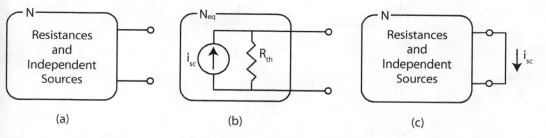

FIGURE 6.3 (a) Arbitrary 2-terminal linear network of resistors and independent sources; (b) Norton equivalent circuit; (c) circuit for computing $i_{sc}(t)$ with R_{th} computed, as per Figure 6.2(d).

NORTON'S THEOREM FOR PASSIVE NETWORKS

Given an arbitrary 2-terminal linear network, N, consisting of resistances and independent sources, then for almost all such N, there exists an equivalent 2-terminal network consisting of a resistance, R_{th}, in parallel with an independent current source, $i_{sc}(t)$. The current, $i_{sc}(t)$, called the short circuit current, is what flows through a short circuit of the 2 terminals of N, as per Figure 6.3(c). R_{th}, as before, is the **Thevenin equivalent resistance** of N computed when all independent sources are deactivated.

A single voltage source does not have a Norton equivalent, and—as mentioned—a single current source does not have a Thevenin equivalent. Both Thevenin and Norton equivalents exist for a 2-terminal linear circuit when $R_{th} \neq 0$ and is finite. When both the Thevenin and Norton equivalents exist for the same network, the source transformation theorem and Ohm's law imply that

$$v_{oc}(t) = R_{th}i_{sc}(t) \tag{6.1a}$$

and when $i_{sc}(t) \neq 0$, then

$$R_{th} = \frac{v_{oc}(t)}{i_{sc}(t)} \tag{6.1b}$$

This formula turns out to be useful in calculating R_{th} for a variety of circuits, especially op amp circuits.

EXAMPLE 6.1. For the circuit of Figure 6.4, using literals, find the open circuit voltage, v_{oc}, the short circuit current, i_{sc}, and the Thevenin equivalent resistance, R_{th}. Then, if $R_1 = 50\ \Omega$, $R_2 = 200\ \Omega$, $v_{s1} = 100$ V, and $i_{s2} = 2$ A, construct the Thevenin and Norton equivalent circuits.

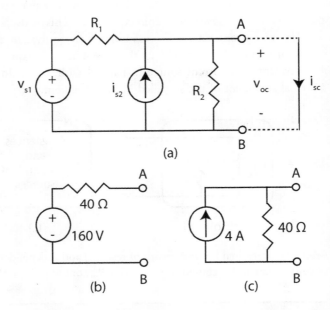

FIGURE 6.4. (a) Resistive 2-terminal network; (b) Thevenin equivalent; (c) Norton equivalent.

SOLUTION
Step 1. *Find* v_{oc}. Using superposition, we have by voltage division and Ohm's law,

$$v_{oc} = \frac{R_2}{R_1 + R_2} v_{s1} + \frac{R_1 R_2}{R_1 + R_2} i_{s2}$$

Substituting the given values into this formula yields

$$v_{oc} = 0.8 \times 100 + 40 \times 2 = 160\ \text{V}$$

Step 2. *Find* i_{sc}. As per Figure 6.4, with terminals A and B shorted together, all the current from flows through the short circuit. From superposition, $i_{sc} = i_{s2} + \frac{1}{R_1} v_{s1}$. Substituting numbers into this formula yields

$$i_{sc} = 2 + 0.02 \times 100 = 4\ \text{A}$$

Step 3. *Find R_{th}.* Replacing v_{s1} by a short circuit and i_{s2} by an open circuit implies that

$$R_{th} = \frac{R_1 R_2}{R_1 + R_2} = 40 \ \Omega$$

Step 4. *Determine the Thevenin and Norton equivalent circuits.* The Thevenin equivalent circuit follows from Steps 1 and 3 and is illustrated in Figure 6.4(b). The Norton equivalent circuit follows from Steps 2 and 3 and is illustrated in Figure 6.4(c). We also note that

$$R_{th} = \frac{v_{oc}}{i_{sc}} = \frac{160}{4} = 40 \ \Omega$$

as expected.

It is important to note that for many circuits, especially when the deactivated circuit is a series-parallel connection of resistances, one can obtain the Thevenin equivalent by a series of source transformations.

Exercises. 1. Redo Example 6.1 using a series of source transformations.
2. In Example 6.1, suppose $R_1 = 100 \ \Omega$, $R_2 = 400 \ \Omega$, $v_{s1} = 100$ V, and $i_{s2} = 2$ A. Find R_{th}, v_{oc}, and i_{sc}.
ANSWER: 80 Ω, 240 V, 3 A

Among the three quantities, R_{th}, v_{oc}, and i_{sc}, if two have been calculated, then the remaining one follows easily from Equation 6.1. In some cases, the choice of which two to find first either increases or decreases the amount of calculation. The following exercises illustrate this point.

Exercises. 1. For the circuit of Figure 6.5, $R_1 = 200 \ \Omega$, $R_2 = 50 \ \Omega$, $R_3 = 10 \ \Omega$, $v_{s1} = 100$ V, and $v_{s2} = 50$ V. Find R_{th}, i_{sc}, and v_{oc} in this order.
ANSWERS: 8 Ω, 1.5 A, 12 V

2. For the circuit of Figure 6.5 with the same values as in Exercise 1, find v_{oc}, i_{sc}, and R_{th} in this order.
ANSWER: Same as in 1, but v_{oc} is harder to find.

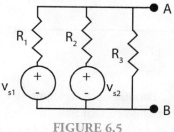

FIGURE 6.5

3. For the circuit of Figure 6.5, find the Thevenin equivalent circuit using a series of source transformations.

The next example illustrates the computation of the Thevenin and Norton equivalent circuits using loop analysis.

EXAMPLE 6.2. Find the Thevenin and Norton equivalent circuits seen at the terminals A-B for the circuit depicted in Figure 6.6, where v_{s1} = 100 V and i_{s2} = 3.2 A. We show that R_{th} = 400 Ω, v_{oc} = 200 V, and i_{sc} = 0.5 A.

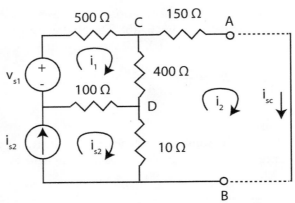

FIGURE 6.6 Two-source circuit for Example 6.2 with loop currents shown; v_{s1} = 100 V and i_{s2} = 3.2 A.

SOLUTION

Step 1. *Compute R_{th}.* To compute R_{th}, we set all source values to zero. Each voltage source becomes a short, and each current source becomes an open. This leads to the circuit of Figure 6.7. Here, we have a 500 Ω in series with 100 Ω, yielding 600 Ω. Since this 600 Ω resistance is in parallel with 400 Ω, the resulting equivalent resistance is 240 Ω. Hence, R_{th} = (150 + 240 + 10) = 400 Ω.

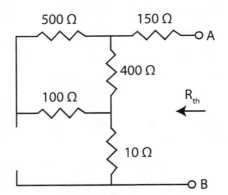

FIGURE 6.7 The circuit of Figure 6.6 with all independent sources deactivated.

Step 2. *Compute an expression for v_{oc}.* Because we are computing v_{oc}, the short across the terminals A-B is NOT present. Hence, i_2 = 0 and no current flows through the 150 Ω resistor. This means its voltage drop is zero. (One often says that the 150 Ω resistor is dangling.) Thus, from KVL we have

$$v_{oc} = 400i_1 + 10i_{s2} \qquad (6.2)$$

Step 3. *Compute* i_1. The only unknown in Equation 6.2 is i_1, since i_{s2} = 3.2 A. Hence, around loop 1,

$$v_{s1} = 900i_1 + 100(i_1 - i_{s2})$$

in which case, $v_{s1} + 100i_{s2} = 1000i_1$ and

$$i_1 = 0.001v_{s1} + 0.1i_{s2}$$

Thus, from Equation 6.2,

$$v_{oc} = 400(0.001v_{s1} + 0.1i_{s2}) + 10i_{s2} = 0.4v_{s1} + 50i_{s2} = 40 + 160 = 200 \text{ V} \tag{6.3}$$

Step 4. *Construct the Thevenin and Norton equivalent circuits.* Equation 6.3 with R_{th} = 400 Ω yields the Thevenin equivalent of Figure 6.8(a). Further, from the source transformation theorem,

$$i_{sc} = \frac{v_{oc}}{R_{th}} = \frac{200}{400} = 0.5 \text{ A} \tag{6.4}$$

Equation 6.4 leads to the Norton equivalent circuit of Figure 6.8(b).

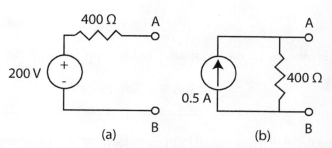

FIGURE 6.8 (a) Thevenin equivalent of circuit of Figure 6.6;
(b) Norton equivalent of Figure 6.6.

Step 5. *Compute* i_{sc} *directly so as to verify the above calculation.* This step is merely given to illustrate the direct calculation of i_{sc} and is unnecessary at this point to the solution of the problem. Referring again to Figure 6.6 and assuming that the short across A-B is present, then $i_2 = i_{sc}$. Hence, around loop 1,

$$v_{s1} = 500i_1 + 400(i_1 - i_{sc}) + 100(i_1 - i_{s2})$$

in which case,

$$v_{s1} + 100i_{s2} = 1000i_1 - 400i_{sc} = 420 \text{ V}$$

Around loop 2 we have

$$560i_{sc} - 400i_1 = 10i_{s2} = 32 \text{ V}$$

In matrix form, the pertinent equations are

$$\begin{bmatrix} 1000 & -400 \\ -400 & 560 \end{bmatrix} \begin{bmatrix} i_1 \\ i_{sc} \end{bmatrix} = \begin{bmatrix} 420 \\ 32 \end{bmatrix}$$

Thus,

$$
\begin{bmatrix} i_1 \\ i_{sc} \end{bmatrix} = \begin{bmatrix} 1000 & -400 \\ -400 & 560 \end{bmatrix}^{-1} \begin{bmatrix} 420 \\ 32 \end{bmatrix} = \begin{bmatrix} 0.62 \\ 0.5 \end{bmatrix} \text{ A}
$$

Consequently, $i_{sc} = 0.5$ A as was found earlier using the easier method of $i_{sc} = \dfrac{v_{oc}}{R_{th}}$.

Exercises. 1. Suppose all source values in the circuit of Figure 6.6 are doubled. What is the new v_{oc}? Does R_{th} change?
ANSWER: $v_{oc} = 400$ V, no

2. Suppose all resistances in the circuit of Figure 6.6 are multiplied by 4 and the independent current source is changed to 0.6 A. Find v_{oc}, R_{th}, and i_{sc}. Hint: For v_{oc}, in equation 6.3, the value "50" is in ohms, so if the resistances are multiplied by four, what is the new value?
ANSWER: $v_{oc} = 160$ V, $R_{th} = 4 \times 400 = 1600$ Ω, and $i_{sc} = 0.1$ A

3. A 400 Ω resistor is connected in series with terminal A of the circuit of Figure 6.6. Find the new v_{oc}, R_{th}, and i_{sc}.
ANSWER: $v_{oc, old} = v_{oc, new} = 200$ V, $R_{th, new} = 800$ Ω, and $i_{sc, new} = 0.25$ A

4. A 400 Ω resistor is connected across terminals A and B of the circuit of Figure 6.6. Find the new i_{sc}, R_{th}, and v_{oc}.
ANSWER: $i_{sc, old} = i_{sc, new} = 0.5$ A, $R_{th, new} = 200$ Ω, and $v_{oc} = 100$ V

In the above two examples, deactivation of all independent sources led to a series-parallel network. Calculation of R_{th} was then straightforward. In fact, we can state a corollary to Thevenin and Norton's theorems.

COROLLARY TO THEVENIN AND NORTON'S THEOREMS FOR PASSIVE NETWORKS

When a network contains no independent sources, $v_{oc}(t) = i_{sc}(t) = 0$, and the Thevenin or Norton equivalent consists of a single resistance R_{th}. For a series-parallel network, R_{th} can be computed by straightforward resistance combinations.

---

3. A GENERAL APPROACH TO FINDING THEVENIN AND NORTON EQUIVALENTS

Consider Figure 6.10(a) where we have a network N connected to the remainder of a larger circuit. Our goal is to replace the network N by its Thevenin equivalent, as shown in Figure 6.10(b).

The terminal v-i characteristics of the network N and its Thevenin equivalent must be the same. Consider that the v-i characteristic at A-B of the Thevenin equivalent of N is

$$v_{AB} = R_{th}i_A + v_{oc} \qquad (6.5)$$

while the Norton equivalent of N as per Figure 6.10(c) has the v-i relationship

$$i_A = \frac{1}{R_{th}}v_{AB} - i_{sc} = G_{th}v_{AB} - i_{sc} \qquad (6.6)$$

These relationships tell us that if we have a linear network and assume there is a voltage v_{AB} across its terminals and a current i_A entering the network, as shown in Figure 6.10(a), then obtaining an equation of the form

$$v_{AB} = [?????]i_A + [?????] \qquad (6.7)$$

or of the form

$$i_A = [?????]v_{AB} - [?????] \qquad (6.8)$$

allows us to match the coefficients of equations 6.7 and 6.5 to determine R_{th} and v_{oc}, or to match the coefficients of equations 6.8 and 6.6 to determine $G_{th} = \dfrac{1}{R_{th}}$ and i_{sc}. This sometimes proves an easier approach for non-simple circuits, as the next two examples illustrate.

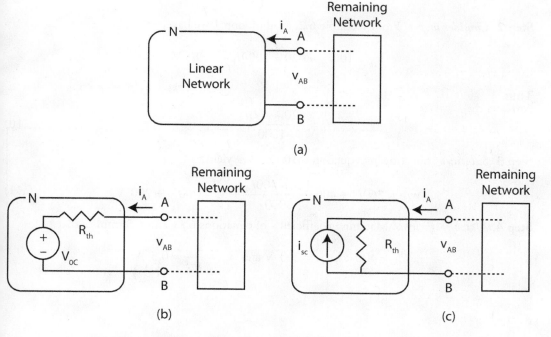

FIGURE 6.10 (a) Network N attached to an unknown network; (b) the Theveinin equivalent of N attached to the unknown network; (c) the Norton equivalent of N attached to the unknown network.

EXAMPLE 6.3. This example revisits Example 6.2 using the new approach. Again, we find the Thevenin and Norton equivalent circuits seen at the terminals A-B for the circuit depicted in Figure 6.11, where $v_{s1} = 100$ V and $i_{s2} = 3.2$ A. Our goal is to find the v-i characteristic at the terminals A-B.

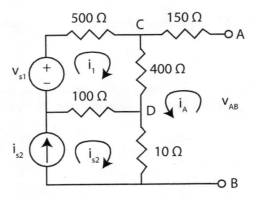

FIGURE 6.11 Two-source circuit for Example 6.2 with loop currents shown; $v_{s1} = 100$ V and $i_{s2} = 3.2$ A.

SOLUTION

Step 1. *Consider i_A loop.* Around the loop for i_A, we have

$$v_{AB} = 560i_A + 400i_1 + 10i_{s2} = 560i_A + 400i_1 + 32 \tag{6.9}$$

Step 2. *Consider loop 1.* From Example 6.2, around loop 1 we have,

$$v_{s1} = 100i_{s2} = 420 = 1000i_1 + 400i_A \text{ V}$$

Thus,

$$i_1 = \frac{420 - 400i_A}{1000} \tag{6.10}$$

Step 3. *Substitute.* Substituting equation 6.10 into 6.9 yields

$$v_{AB} = 560i_A + 400\frac{420 - 400i_A}{1000} + 32 = 400i_A + 200 \text{ V} \tag{6.11}$$

Step 4. *Match coefficients.* Matching coefficients of equations 6.11 and 6.5 implies that

$$R_{th} = 400 \text{ }\Omega, v_{oc} = 200 \text{ V and } i_{sc} = \frac{v_{oc}}{R_{th}} = 0.5 \text{ A.}$$

EXAMPLE 6.4. For the circuit of Figure 6.12, find the Thevenin equivalent of the 2-terminal Network N defined by the dashed line box. We show that $v_{oc} = 9.6$ V, $R_{th} = 4.4$ Ω, and $i_{sc} = 2.1818$ A.

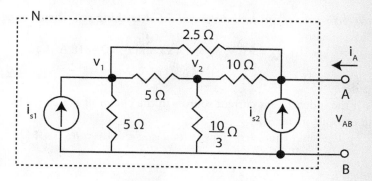

FIGURE 6.12 A current source is is attached to N for computing R_{th} and v_{oc}.

SOLUTION

Our objective is to compute the relationship of the form of equation 6.7 using Nodal analysis and then match coefficients with equation 6.5 to obtain R_{th} and v_{oc}. Assume $i_{s1} = 2$ A and $i_{s2} = 1$ A.

Step 1. *Write nodal equations.* For writing the equations of this circuit, the reader might first review Example 3.2. Alternately, using the inspection method, the matrix nodal equations are

$$\begin{bmatrix} 0.8 & -0.2 & -0.4 \\ -0.2 & 0.6 & -0.1 \\ -0.4 & -0.1 & 0.5 \end{bmatrix} \begin{bmatrix} v_1 \\ v_2 \\ v_{AB} \end{bmatrix} \triangleq M \begin{bmatrix} v_1 \\ v_2 \\ v_{AB} \end{bmatrix} = \begin{bmatrix} i_{s1} \\ 0 \\ i_{s2} + i_A \end{bmatrix} \tag{6.12}$$

Step 2. *Solve equation 6.12 for v_{AB} using Crammer's rule.* First, we note that

$$\det(M) = \det \begin{bmatrix} 0.8 & -0.2 & -0.4 \\ -0.2 & 0.6 & -0.1 \\ -0.4 & -0.1 & 0.5 \end{bmatrix} = 0.1$$

From Crammer's rule,

$$v_{AB} = \frac{\det \begin{bmatrix} 0.8 & -0.2 & i_{s1} \\ -0.2 & 0.6 & 0 \\ -0.4 & -0.1 & i_{s2} + i_A \end{bmatrix}}{\det(M)} \text{ V}$$

which from the properties of determinants becomes

$$v_{AB} = \frac{\det \begin{bmatrix} 0.8 & -0.2 & 0 \\ -0.2 & 0.6 & 0 \\ -0.4 & -0.1 & 1 \end{bmatrix}}{\det(M)} i_A + \frac{\det \begin{bmatrix} 0.8 & -0.2 & 1 \\ -0.2 & 0.6 & 0 \\ -0.4 & -0.1 & 0 \end{bmatrix}}{\det(M)} i_{s1} + \frac{\det \begin{bmatrix} 0.8 & -0.2 & 0 \\ -0.2 & 0.6 & 0 \\ -0.4 & -0.1 & 1 \end{bmatrix}}{\det(M)} i_{s2} \tag{6.13}$$

$$= 4.4 i_A + 2.6 i_{s1} + 4.4 i_{s2} = 4.4 i_A + 9.6$$

Equation 6.13 shows that v_{AB} is calculated finding four determinants numerically using MATLAB or equivalent.

Step 3. *Match coefficients of equations 6.13 and 6.5.* Matching coefficients of equation 6.13 with equation 6.5, we obtain

$$v_{oc} = 9.6 \text{ V}, R_{th} = 4.4 \text{ } \Omega, \text{ and } i_{sc} = 2.1818 \text{ A}$$

Exercises. 1. If the independent current sources in the circuit of Figure 6.11 are set to zero, find the Thevenin equivalent circuit.

ANSWER: The Thevenin equivalent consists of a single resistor, $R_{th} = 4.4 \text{ } \Omega$.

2. Find v_{oc} when $i_{sc} = 10$ A and $i_{s2} = 5$ A.

ANSWER: 48 V

3. A 4.4 Ω resistor is connected in series with terminal A of the circuit of Figure 6.12. Find the new v_{oc}, R_{th}, and i_{sc}.

ANSWER: $v_{oc, old} = v_{oc, new} = 9.6$ V, $R_{th, new} = 8.8 \text{ } \Omega$, and $i_{sc, new} = 1.0909$ A

4. A 4.4 Ω resistor is connected across terminals A and B of the circuit of Figure 6.12. Find the new i_{sc}, R_{th}, and v_{oc}.

ANSWER: $i_{sc, old} = i_{sc, new} = 2.1818$ A, $R_{th, new} = 2.2 \text{ } \Omega$, and $v_{oc} = 4.8$ V

At this point, we end our development in this section with an example that shows how to compute a Thevenin equivalent from measured, e.g., in a laboratory setting where there is a power supply with an adjustable voltage.

EXAMPLE 6.5. Consider Figure 6.13, which shows the Thevenin equivalent of an unknown network N attached to a variable voltage, v_{AB}, power supply, which also shows the current delivered to the unknown network N, i.e., i_A. Two measurements of the unknown network N are taken, and the data is displayed in Table 6.1. Find the Thevenin equivalent of N.

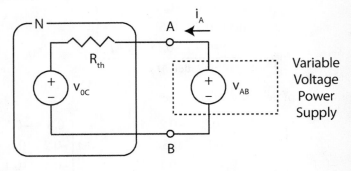

FIGURE 6.13 Thevenin equivalent of an unknown network N connected to a variable voltage power supply.

TABLE 6.1

i_A (mA)	v_{AB} (V)
10	24
20	40

SOLUTION

Substituting the measured data in Table 6.1 into equation 6.5 ($v_{AB} = i_A R_{th} + v_{oc}$) yields

$$24 = 0.01 R_{th} + v_{oc}$$

from row 1 of Table 6.1, and

$$40 = 0.02 R_{th} + v_{oc}$$

from row 2 of Table 6.1. In matrix form,

$$\begin{bmatrix} 0.01 & 1 \\ 0.02 & 1 \end{bmatrix} \begin{bmatrix} R_{th} \\ v_{oc} \end{bmatrix} = \begin{bmatrix} 24 \\ 40 \end{bmatrix}$$

Solving produces

$$\begin{bmatrix} R_{th} \\ v_{oc} \end{bmatrix} = \begin{bmatrix} 0.01 & 1 \\ 0.02 & 1 \end{bmatrix}^{-1} \begin{bmatrix} 24 \\ 40 \end{bmatrix} = -100 \begin{bmatrix} 1 & -1 \\ -0.02 & 0.01 \end{bmatrix} \begin{bmatrix} 24 \\ 40 \end{bmatrix} = \begin{bmatrix} 1600 \\ 8 \end{bmatrix}$$

Hence, $R_{th} = 1600 \, \Omega$ and $v_{oc} = 8$ V.

Thus, one can use the technique of Example 6.5 to determine Thevenin equivalent circuits in the laboratory.

4. THEVENIN AND NORTON EQUIVALENT CIRCUITS FOR ACTIVE NETWORKS

Constructing Thevenin and Norton equivalents for active networks, those containing dependent sources and op amps, presents us with some unique challenges. Except with one extra condition, Thevenin and Norton's theorems and their corollary are valid for active networks. Because active networks contain dependent sources, the extra condition is that all controlling voltages or currents be within the 2-terminal network whose Thevenin/Norton equivalent are being sought.

THEVENIN AND NORTON'S THEOREMS FOR ACTIVE NETWORKS

For almost every 2-terminal linear network, N, as in Figure 6.14(a), consisting of resistances, independent sources, and dependent sources whose controlling voltages and currents are contained within N^1, there is an equivalent 2-terminal network consisting of either (i) a resistance, R_{th}, in series with an independent voltage source, $v_{oc}(t)$, called the **Thevenin equivalent** (Figure 6.14(b)), or (ii) a resistance, R_{th}, in parallel with an independent current source, $i_{sc}(t)$, called the **Norton equivalent** (Figure 6.14(c)). In most cases, both the Thevenin and Norton equivalent circuits exist. Computation of v_{oc} is characterized by Figure 6.14(a), computation of R_{th} by Figure 6.14(d), and computation of i_{sc} by Figure 6.14(e).

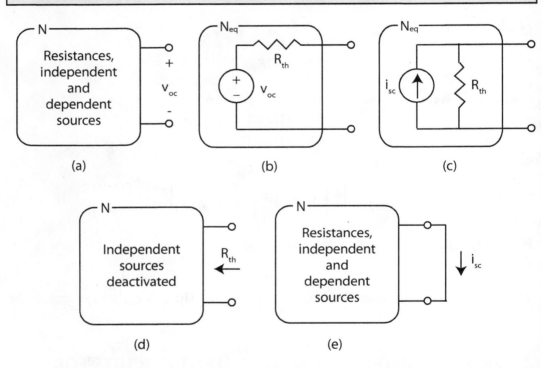

FIGURE 6.14 (a) Arbitrary linear network N; (b) Thevenin equivalent of N; (c) Norton equivalent of N; (d) N with independent sources deactivated for calculating R_{th}; (e) N with short circuited terminals for calculating i_{sc}.

As in the previous section, a corollary to Thevenin and Norton's theorems is that if the network N has no internal independent sources, then the Thevenin and Norton equivalent circuit consists of a single resistance R_{th}. However, in contrast to passive networks, R_{th} can be negative. As a first example illustrating the above theorems, we consider an active network containing no internal independent sources.

EXAMPLE 6.6. Find the Thevenin equivalent circuit for the 2-terminal network (marked by dashed line box) in Figure 6.15(a) using the method of Section 3. (The dependent source acts as a voltage amplifier.)

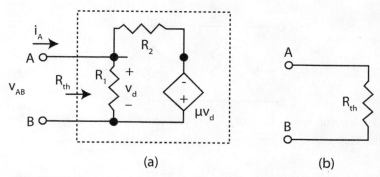

(a)   (b)

FIGURE 6.15. (a) circuit with terminal voltage v_{AB} and input current i_A;

$$(b) \quad R_{th} = R_1 // \frac{R_2}{(\mu+1)}.$$

SOLUTION

Step 1. *Since there are no independent internal sources, the Thevenin equivalent consists of a single resistance, R_{th}, i.e.,* $v_{oc} = i_{sc} = 0$.

Step 2. *Write a nodal equation.* Writing a single node equation we have

$$i_A = G_1 v_{AB} + G_2(v_{AB} + \mu v_d) = (G_1 + (\mu + 1)G_2)v_{AB}$$

Step 3. *Match coefficients with equation 6.6.* Matching coefficients implies that $G_{th} = (G_1 + (\mu + 1)G_2)$ in which case,

$$R_{th} = \frac{1}{\left(\dfrac{1}{R_1} + (\mu+1)\dfrac{1}{R_2}\right)} = \frac{R_1 R_2}{R_2 + (\mu+1)R_1} = \frac{R_1 \dfrac{R_2}{(\mu+1)}}{R_1 + \dfrac{R_2}{(\mu+1)}} \qquad (6.14)$$

We recognize equation 6.14 as the parallel combination of the resistance R_1 and $\dfrac{R_2}{(\mu+1)}$. To illustrate a typical calculation, suppose $\mu = 199$, $R_1 = 100\ \Omega$, and $R_2 = 4\ k\Omega$. Then

$$R_{th} = 500 // 20 = 19.23 \cong 20 = \frac{R_2}{\mu+1}$$

Exercises. 1. For the above example, suppose $\mu = 99$, $R_1 = 500\ \Omega$, and $R_2 = 1\ k\Omega$. Find R_{th}.

ANSWER: 10 Ω

2. For the circuit of Figure 6.16, find the Thevenin equivalent resistance by obtaining i_A in terms of v_{AB}.

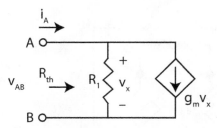

FIGURE 6.16 A circuit having no independent sources, in which case $v_{oc} = i_{sc} = 0$ and the Thevenin equivalent consists only of a single resistance, R_{th}.

ANSWER: $i_A = \left(\dfrac{1}{R_1} + g_m\right) v_{AB} = G_{th} v_{AB}$ and $R_{th} = \dfrac{1}{G_{th}}$

EXAMPLE 6.7. Find the Thevenin and Norton equivalent circuits seen at the terminals A-B in Figure 6.17 when $i_s = 50$ mA. Our computations will proceed using loop analysis to find the terminal v-i characteristic A-B.

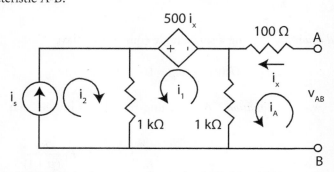

FIGURE 6.17 Arbitrary network for finding Thevenin equivalent.

SOLUTION
We first note that $i_x = i_A$ and $i_2 = 0.05$ A.

Step 1. *Write a set of loop equations for the circuit of Figure 6.17.*

For loop 1, we obtain

$$0 = 1000(i_1 + i_2) + 1000(i_1 - i_A) - 500i_A$$

which simplifies to

$$-5 = 2000i_1 - 1500i_A$$

For loop A, we have

$$v_{AB} = 100\,i_A + 1000(i_A - i_1) = 1100i_A - 1000i_1$$

Step 2. *Write the loop equations in matrix form and solve.* Writing the loop equations in matrix form yields

$$\begin{bmatrix} 2000 & -1500 \\ -1000 & 1100 \end{bmatrix} \begin{bmatrix} i_1 \\ i_A \end{bmatrix} = \begin{bmatrix} -50 \\ v_{AB} \end{bmatrix}$$

Solving for i_A using for example Crammer's rule produces

$$i_A = \frac{\det \begin{bmatrix} 2000 & -50 \\ -1000 & v_{AB} \end{bmatrix}}{\det \begin{bmatrix} 2000 & -1500 \\ -1000 & 1100 \end{bmatrix}} = -50 \frac{\det \begin{bmatrix} 2000 & 1 \\ -1000 & 0 \end{bmatrix}}{700 \times 10^3} + v_{AB} \frac{\det \begin{bmatrix} 2000 & 0 \\ -1000 & 1 \end{bmatrix}}{700 \times 10^3}$$

$$= -\frac{50}{700} + v_{AB} \frac{2}{700} \qquad (6.15)$$

or equivalently,

$$v_{AB} = 350\, i_A + 25 \qquad (6.16)$$

Step 3. *Match coefficients of equation 6.15 with equation 6.6 or equation 6.16 with equation 6.5 to obtain*

$$i_{sc} = \frac{50}{700} = \frac{1}{14} \text{ A}, \; G_{th} = \frac{2}{700} = \frac{1}{350} \text{ S}, \; R_{th} = 350 \; \Omega, \text{ and } v_{oc} = 25 \text{ V}$$

Exercises. 1. In Example 6.7, if $i_s = 5$ mA, find i_{sc}, G_{th}, R_{th}, and v_{oc}. Hint: Use proportionality.

ANSWER: $i_{sc} = \frac{1}{140}$ A, $G_{th} = \frac{1}{350}$ S, $R_{th} = 350 \; \Omega$, and $v_{oc} = 2.5$ V

2. In Example 6.7, if $i_s = 5$ mA and the 100 Ω resistor is replaced by a short circuit, find i_{sc}, G_{th}, R_{th}, and v_{oc}. Hint: We have removed the dangling resistor in this case.

ANSWER: $v_{oc} = 2.5$ V, $R_{th} = 250 \; \Omega$, $G_{th} = \frac{1}{250}$ S, and $i_{sc} = \frac{1}{100} = 0.01$ A

3. Find the Norton equivalent at the terminals A-B of the circuit of Figure 6.18 when mA.
ANSWER: $R_{th} = 200 \; \Omega$ and $i_{sc} = 0.125$ A

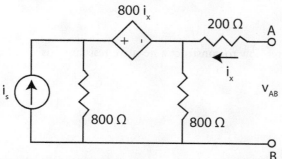

FIGURE 6.18 Modification of the circuit of Figure 6.17.

5. THEVENIN AND NORTON EQUIVALENT CIRCUITS FOR OP AMP CIRCUITS

Op am circuits are active circuits. However, because the op amp is a device with special properties, such as the virtual short circuit in the ideal case and such as output saturation in the non-ideal case, their discussion warrants special consideration. Our discussion begins with a Thevenin equivalent of a non-inverting amplifier with a dangling resistor at the output terminal.

EXAMPLE 6.8. Find the Thevenin equivalent seen at the terminals A-B for the op amp circuit of Figure 6.19.

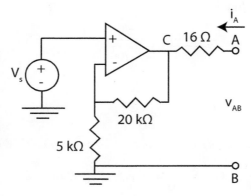

FIGURE 6.19 Non-inverting op amp circuit with series resistance.

SOLUTION
Step 1. *Find* v_{CB}.

$$v_{CB} = \frac{20+5}{5}V_s = 5V_s$$

Step 2. *Find* v_{AB}. By inspection,

$$v_{AB} = 16i_A + 5V_s$$

Step 3. *Match coefficients with equation 6.5.* Matching coefficients we observe that

$$R_{th} = 16 \ \Omega, v_{oc} = 5V_s, \text{ and } i_{sc} = \frac{5V_s}{16}$$

Our next example illustrates how to construct a negative resistance using an ideal op amp.

EXAMPLE 6.9. Find the Thevenin equivalent seen at the terminals A-B for the (ideal) op amp circuit of Figure 6.20.

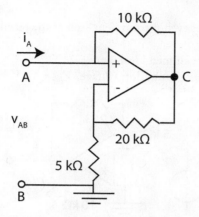

FIGURE 6.20 A resistance converter circuit.

SOLUTION
By V-division and the properties of the op amp,

$$v_{CB} = \left(1 + \frac{20}{5}\right) v_{AB} = 5 v_{AB}$$

Thus, computing i_A we have

$$i_A = \frac{v_{AB} - v_C}{10 \times 10^3} = \frac{-4}{10 \times 10^3} v_{AB}$$

Matching coefficients with equation 6.6 we have

$$i_{sc} = 0 \,, \; G_{th} = \frac{-4}{10 \times 10^3} \; \text{S}, \; R_{th} = -2.5 \; \text{k}\Omega, \text{ and } v_{oc} = 0$$

Exercise. For the circuit of Figure 6.21, find the Thevenin equivalent circuit at A-B.

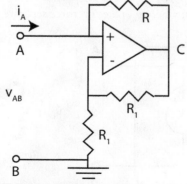

FIGURE 6.21 Resistance converter circuit.

ANSWER: $v_{oc} = 0$ and $R_{th} = -R$

Our third example constructs a Thevenin equivalent of the standard inverting op amp configuration with a terminal resistance. However, we will consider both the ideal and non-ideal cases.

EXAMPLE 6.10. Find the Thevenin equivalent seen at the terminals A-B for the op amp circuit of figure 6.22 when
(a)   when the op amp is assumed ideal, and
(b)   when the op amp has a saturation voltage, $V_{sat} = 15$ V.

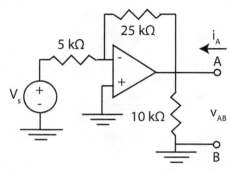

FIGURE 6.22 Inverting amplifier configuration.

SOLUTION

Step 1. *Find the Thevenin equivalent seen at the terminals* A-B *assuming an ideal op amp.* The properties of an ideal op amp imply that $v_{AB} = -5V_s$.

On the other hand, V_s with set to zero, $v_{AB} = 0$. In fact, $v_{AB} = 0$ for all possible currents, i_A, injected into node A. Hence, $v_{AB} = R_{th} \times i_A + v_{oc} = 0 \times i_A - 5V_s$ implies $R_{th} = 0$ and $v_{oc} = -5V_s$. The Thevenin equivalent seen at the terminals A-B consists only of a voltage source of value $v_{oc} = -5V_s$ for the ideal op amp case.

Step 2. *Find the Thevenin equivalent seen at the terminals* A-B *assuming an op amp with output saturation.* When the non-ideal op amp operates in its linear region, the Thevenin equivalent by Step 1 is a voltage source having value $v_{AB} = -5V_s$. When, $|-5V_s| \geq V_{sat} = 15$ V , or equivalently, when $|V_s| \geq 3$ V, then the op amp saturates at $\pm$ 15 V. Specifically, when $V_s \geq 3$ V, then $v_{oc} = -15$ V and when $V_s \leq -3$ V, then $v_{oc} = 15$ V. The Thevenin equivalent for an op amp with output saturation is summarized in Figure 6.23, where v_{oc} takes on three separate values depending on the region of operation of the amplifier.

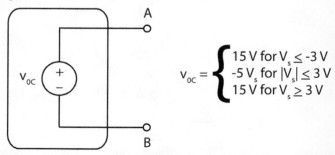

FIGURE 6.23 Thevenin equivalent at output terminals of an inverting amplifier (Figure 6.22) with non-ideal op amp.

This ends our investigation of Thevenin equivalents of op amp circuits. There are many more interesting examples that are beyond the scope of this text.

7. MAXIMUM POWER TRANSFER THEOREM

Figure 6.24 shows the Thevenin equivalent of a network N connected to a variable load designated R_L. The load voltage, v_L, the load current, i_L, and the power, p_L, delivered to the load are all functions of R_L. The main objective of this section is to show that for fixed R_{th}, maximum power is transferred to the load when $R_L = R_{th}$. We illustrate this assertion with an example that shows the power delivered to R_L as a function of R_L. Throughout this section, it is assumed that all resistances are non-negative.

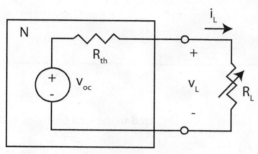

FIGURE 6.24 Thevenin equivalent of network N connected to a variable load, R_L.

EXAMPLE 6.11. For the circuit of Figure 6.24, suppose that $R_{th} = 20\ \Omega$ and $v_{oc} = 20$ V. Plot the power delivered to the load R_L as a function of R_L.

SOLUTION
The power delivered to the load R_L is

$$p_L = v_L i_L = R_L i_L^2 = R_L \left(\frac{v_{oc}}{R_L + R_{th}} \right)^2 = \frac{R_L}{\left(R_L + R_{th} \right)^2} v_{oc}^2 \qquad (6.17)$$

Plugging in the known values yields

$$p_L = \frac{R_L}{\left(R_L + 20 \right)^2} (20)^2$$

To obtain the plot we use the following MATLAB code, resulting in the plot of **Figure 6.25.**

```
»voc = 20; Rth = 20;
»RL = 0:0.25:100;
»PL = RL .* voc^2 ./ ((RL + Rth) .^2);
»plot(RL,PL)
»grid
```

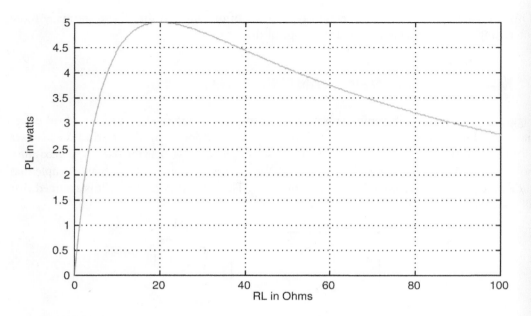

FIGURE 6.25 Plot of power delivered to the load in Figure 6.24 as a function of R_L.

From the curve, maximum power is transferred at $R_L = R_{th} = 20\ \Omega$. In a neighborhood of $R_L = 20\ \Omega$, the curve remains fairly flat. At $R_L = 40\ \Omega$ and $R_L = 10\ \Omega$, the curve shows that about 88% of maximum power is transferred.

This experimentally observed fact, that maximum power transfer occurs when $R_L = R_{th}$, plays an important role when matching speaker "resistances" to the output "resistance" of a stereo amplifier or when trying to get as much power as possible out of an antenna and into a receiver.

MAXIMUM POWER TRANSFER THEOREM

Let a two-terminal linear network, N, represented by its Thevenin equivalent, as in Figure 6.24, be connected to a variable load, R_L. For fixed R_{th}, maximum instantaneous power is transferred to the load when

$$R_L = R_{th}$$

and the maximum instantaneous power is given by

$$p_{L,\text{max}} = \frac{v_{oc}^2}{4\,R_{th}}$$

In the dc case, the instantaneous power is a constant for all t.

A verification of the maximum power transfer theorem proceeds using differential calculus. From equation 6.17, the power absorbed by the load is

$$p_L = \frac{R_L}{(R_L + R_{th})^2} v_{oc}^2$$

Following the standard procedure of calculus for determining a maximum/minimum, we compute the derivative of p_L with respect to R_L, set to zero, and solve for R_L.

$$\frac{dp_L}{dR_L} = \frac{d}{dR_L}\left(\frac{R_L}{(R_L + R_{th})^2} v_{oc}^2\right) = \frac{v_{oc}^2}{(R_L + R_{th})^2} - 2\frac{R_L v_{oc}^2}{(R_L + R_{th})^3}$$

$$= v_{oc}^2 \frac{R_{th} - R_L}{(R_L + R_{th})^3} = 0$$

from which $R_L = R_{th}$ and $R_L = \infty$ are the only possible solutions. But, if $R_L = \infty$, then $p_L = 0$. Hence, because equation 6.17 is positive for $R_L > 0$, $R_L = R_{th}$ produces maximum power, p_L, delivered to the load.

Further, substituting $R_L = R_{th}$ into equation 6.17 yields

$$p_{L,max} = \frac{R_{th}}{(R_{th} + R_{th})^2} v_{oc}^2 = \frac{v_{oc}^2}{4 R_{th}} \qquad (6.18)$$

This completes the verification of the maximum power transfer theorem.

EXAMPLE 6.12. Consider the circuit of Figure 6.26a. Find (i) the value of R_L for maximum power transfer and (ii) the corresponding p_{Lmax}.

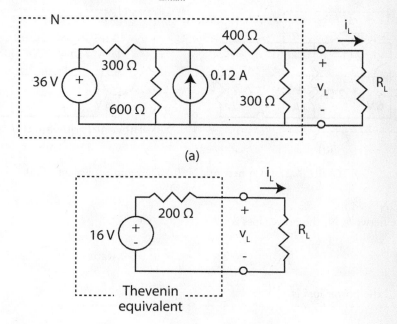

(a)

(b)

FIGURE 6.26 (a) A network N connected to a load R_L; (b) Thevenin equivalent of N connected to R_L.

SOLUTION

Step 1. *To compute R_{th}, the independent voltage source becomes a short and the independent current source becomes an open.* Finding the equivalent resistance seen at the terminals produces R_{th} = 200 Ω. Hence, maximum power is transferred when R_L = 200 Ω.

Step 2. v_{oc} *may be computed by any of the methods discussed throughout this chapter.* For example, by repeated source transformations, the network N reduces to its Thevenin equivalent shown in Figure 6.26(b) with v_{oc} = 16 V. In fact, this approach would have found R_{th} and v_{oc} at the same time.

Plugging v_{oc} = 16 V and R_L = R_{th} = 200 Ω into equation 6.18 yields

$$p_{L,\max} = \frac{v_{oc}^2}{4\,R_{th}} = \frac{(16)^2}{800} = 320 \text{ mW}$$

Exercise. Suppose the 400 Ω resistor in Figure 6.26(a) is changed to 100 Ω. Find v_{oc}, R_{th}, and $p_{L,\max}$.
ANSWERS: 24 V, 150 Ω, 0.96 watts

EXAMPLE 6.13. This example shows that the Thevenin equivalent cannot be used to calculate power consumption within the network N it represents. For this demonstration, consider the network N given in Figure 6.27(a) with its Thevenin equivalent given in Figure 6.27(b). Compute the power loss within the actual network N and within its Thevenin equivalent. We show that these are different.

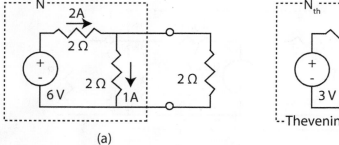
(a)

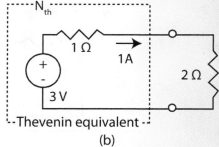
(b)

FIGURE 6.27 (a) A network N; (b) Thevenin equivalent of (a).

SOLUTION

Within the network N, the power loss is

$$p_{N,\ actual} = 2 \times 2^2 + 2 \times 1^2 = 10 \text{ watts}$$

Within N_{th}, the power loss is

$$p_{Nth} = 1 \times 1^2 = 1 \neq p_N$$

This means that the Thevenin equivalent is not, in general, representative of power relationships within the network, i.e., the losses that are dissipated as heat, for example.

When a network N is a voltage source in series with a resistance R_s, and hence is its own Thevenin equivalent, one may ask about maximum power transfer when R_s is variable and the load R_L is fixed, assuming v_s is also fixed. The following example is an experiment for investigating this situation.

EXAMPLE 6.13. For the circuit of Figure 6.28, suppose $R_L = 20\ \Omega$ and $v_s = 20$ V. Plot the power delivered to the load as a function of R_s along with the power loss, p_{loss}, in R_s.

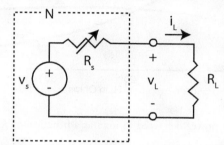

FIGURE 6.28 A network N in which R_s can be adjusted with v_s and R_L fixed.

SOLUTION
The power delivered to the load R_L is

$$p_L = v_L i_L = \frac{R_L}{\left(R_L + R_s\right)^2} v_s^2 = \frac{20 \times 400}{\left(20 + R_s\right)^2}$$

To obtain the plots, we use the following MATLAB code, resulting in the plot of Figure 6.29.

```
»vs = 20; RL = 20;
»Rs = 0:.25:50;
»PL = RL .* vs^2 ./ ((RL + Rs) .^2);
»plot(Rs,PL)
»grid
»hold
»Ploss = Rs .* vs^2 ./ ((RL + Rs) .^2);
»plot(Rs,Ploss,'b')
```

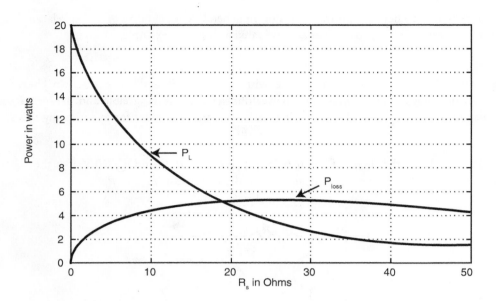

FIGURE 6.29 Plot of power delivered to load as a function of R_L for circuit of Figure 6.28.

According to Figure 6.29, the maximum power of 20 watts is delivered when $R_s = 0$. Observe that if $R_s \leq R_L$ (the usual case), then minimizing p_{loss} maximizes p_L. However, if $R_s > R_L$, minimizing p_{loss} does not maximize p_L.

The proof for the maximum power transfer theorem given earlier considers R_L as the independent variable and sets $\dfrac{dp_L}{dR_L}$ to zero, standard practice in calculus. There is, however, an alternate approach whose derivation is simpler mathematically, but is more meaningful for applications in the sense that the load can be a general 2-terminal linear network, N_L, instead of a single resistor.

For this alternate derivation, refer to Figure 6.30. We ask the question, *What v-i characteristic should the load network N_L have so that maximum power is transferred from N to N_L?*

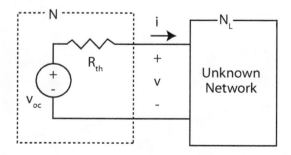

FIGURE 6.30 The Thevenin equivalent of a network N connected to a loading network N_L.

To find the value of v, we note that the power p_L transferred from N to N_L is

$$p_L = vi = v\frac{v_{oc} - v}{R_{th}}$$

To find the value of v that maximizes p_L, we differentiate p_L with respect to v and set the result to zero:

$$\frac{dp_L}{dv} = \frac{v_{oc} - 2v}{R_{th}} = 0$$

Solving for v yields

$$v = 0.5v_{oc} \tag{6.19a}$$

at which value

$$i = \frac{v_{oc} - v}{R_{th}} = \frac{0.5v_{oc}}{R_{th}} \tag{6.19b}$$

which are the conditions on v and i for maximum power transfer to N_L. If N_L consists of a single resistor R_L, it has the v-i characteristic of equations 6.19. Then from Ohm's law,

$$R_L = \frac{v}{i} = R_{th}$$

At $v = 0.5v_{oc}$, the corresponding maximum power is

$$p_{L,max} = v \times i = \frac{v_{oc}^2}{4R_{th}} \tag{6.20}$$

EXAMPLE 6.14. In the circuit of Figure 6.31, $v_{oc} = 2$ V and $R_{th} = 2\ \Omega$. Find the value of that v_L maximizes power transfer to the network N_L.

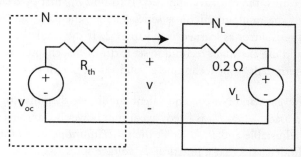

FIGURE 6.31 Circuit for illustrating equations 6.19.

SOLUTION
According to equation 6.19(a), maximum power transfer occurs when $v = 0.5v_{oc} = 1$ V and from

equation 6.19(b), $i = \dfrac{0.5v_{oc}}{R_{th}} = 0.5$ A. Thus,

$$v = 1 = 0.2i + v_L = 0.1 + v_L \implies v_L = 0.9 \text{ V}$$

Exercises. 1 If the 0.2 Ω resistor is changed 4 Ω, find the value of v_L that maximizes power transfer to the network N_L.

ANSWER: −1 V

2. If the 0.2 Ω resistor is changed 2 Ω, find the value of v_L that maximizes power transfer to the network N_L.

ANSWER: 0 V

3. If the 0.2 Ω resistor is variable and v_L = 0.5 V, find the new value of the 0.2 Ω resistor that maximizes power transfer to the N_L.

ANSWER: 1 Ω

8. SUMMARY

This chapter has set forth a powerful strategy for analyzing complex networks by replacing portions of the network by their simpler Thevenin and Norton equivalents. The Thevenin and Norton theorems assure us that almost any 2-terminal linear network, no matter the number of internal elements, is equivalent to a simple network consisting of an independent source either in series with or in parallel with a resistance. Of course, an independent current source does not have a Thevenin equivalent, and an independent voltage source does not have a Norton equivalent. More generally, there are some circuits that have one but not the other. Further, some circuits have neither.

The chapter has illustrated various techniques for constructing the Thevenin and Norton equivalents. For passive networks, the ordinary approach is to find R_{th} first by deactivating all internal independent sources. If the resultant circuit is series-parallel, then R_{th} can be found by combining series and parallel resistances as learned in Chapter 2. If the resultant network is not series-parallel, then one should use the main technique set forth in this chapter, which is to find the v-i characteristic of the terminals. This technique is valid for all circuit types.

With the ideas of a Thevenin and Norton equivalent circuit, we then investigated the problem of transferring power to a load. When R_{th} is fixed, maximum power is transferred when R_L is adjusted to be R_{th}. If R_{th} is adjustable and R_L is fixed, then maximum power is transferred when R_{th} = 0. It is important to understand that a practical dc voltage source (such as a battery in an automobile) is designed to provide nearly constant output voltage for the intended load current. Accordingly, it has a rather small source resistance R_s. Any attempt to transfer the maximum power from such a source continuously will overload the source and may cause damage to its internal structure. For example, in a lead acid battery, the plates may warp or the solution bubble. Hence, maximum power transfer is not of critical importance for power transmission networks, whereas for communication networks, maximum power transfer is important.

9. TERMS AND CONCEPTS

2-terminal network: an interconnection of circuit elements inside a box having only 2 accessible terminals for connection to other networks.

Deactivating an independent current source: replacing the source by an open circuit.

Deactivating an independent voltage source: replacing the source by a short circuit.

Equivalent n-terminal networks: two n-terminal networks having the same terminal voltage-current relationships. Alternately, two n-terminal networks N_1 and N_2 are **equivalent** when substituting one for the other in every possible network N; the voltages and currents in N are unaffected.

i_{sc}**:** the current through a short circuit placed across the output terminals of a 2-terminal network.

Maximum Power Theorem: let an adjustable load resistor R_L be connected to the Thevenin equivalent of a 2-terminal linear network. Maximum power is absorbed by the resistor when $R_L = R_{th}$.

Norton's equivalent circuit: any 2-terminal network consisting of independent sources and linear resistive elements is equivalent to an independent current source in parallel with a resistance.

R_{th} (Thevenin's equivalent resistance): the resistance that appears in the Thevenin equivalent circuit of a 2-terminal linear network. It is also the equivalent resistance of the 2-terminal network when all internal independent sources are deactivated.

Thevenin's equivalent circuit: any 2-terminal network consisting of independent sources and linear resistive elements is equivalent to an independent voltage source in series with a resistance.

v_{oc}**:** the open circuit voltage of a 2-terminal network N when no load is connected.

---

[1] For a generalization of this condition to the case where the controlling voltage or current is outside of N, see the article by Peter Aronheim entitled "Frequency Domain Methods" in *The Circuits and Filters Handbook*, BocaRaton, Fl.: CRC Press, 1995, pp. 682-691.

Problems

THEVENIN/NORTON FOR PASSIVE CIRCUITS

1. For the circuit of Figure P6.1, find R_{th}, i_{sc}, and v_{oc} in terms of the literals. Hint: Consider using $G_i = \dfrac{1}{R_i}$.

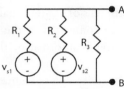

Figure P6.1

2. Find the Thevenin and Norton equivalent circuits seen at the terminals A-B for the circuit depicted in Figure P6.2.

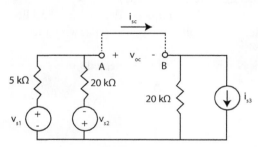

Figure P6.2

3. For the circuit of Figure P6.3, $R_1 = 3$ kΩ, $R_2 = 6$ kΩ, $v_{s1} = 30$ V, and $i_{s2} = 10$ mA.
   (a) Find the Norton and Thevenin equivalents.
   (b) Suppose a variable load resistor R_L is attached across A-B. Plot using MATLAB or equivalent the power absorbed by R_L when $100 \leq R_L \leq 4$ kΩ.

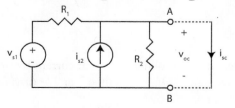

Figure P6.3

ANSWER: $R_{th} = 2$ kΩ, $i_{sc} = 20$ mA, $v_{oc} = 40$ V

4. In the circuit of Figure P6.4, $v_{s1} = 12$ V, $i_{s2} = 0.4$ A, $R_1 = 60$ Ω, $R_2 = 60$ Ω, and $R_3 = 40$ Ω.
   (a) Find the Thevenin and Norton equivalents seen at the terminals A-B.
   (b) If a load resistor of 90 Ω is connected to A-B, find the power absorbed by this resistor.
   (c) Repeat (b) for a 30 Ω resistor. Which resistor, 30 Ω or 90 Ω, absorbs the most power?

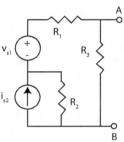

Figure P6.4

5. In the circuit of Figure P6.5, $R_1 = 2$ kΩ, $R_2 = 8$ kΩ, $R_L = 6$ kΩ, $v_{s1} = 60$ V, $i_{s2} = 19$ mA and $i_{s3} = 5$ mA. Find the Thevenin and Norton equivalents of the circuit in the dashed box. Then find i_L and the power absorbed by R_L.

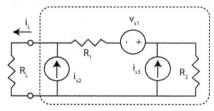

Figure P6.5

CHECK: 200 V, 10 kΩ

6. For the circuit of Figure P6.6, $R_1 = 18$ kΩ, $R_2 = 9$ kΩ, $R_3 = 3$ kΩ, $R_4 = 6$ kΩ, and $v_s = 48$ V. Find the Norton and Thevenin equivalents.

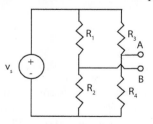

Figure P6.6

CHECK: $i_{sc} = 2$ mA

7. Find the Thevenin equivalent seen at A-B of the circuit of Figure P6.7, where R_1 = 18 kΩ, R_2 = 9 kΩ, R_3 = 3 kΩ, R_4 = 6 kΩ, R_5 = 3.6 kΩ, R_6 = 32 kΩ, v_{s1} = 48 V, and i_{s2} = 8 mA. Hint: Use the result of Problem 6 to find the Thevenin equivalent of the network between C and D.

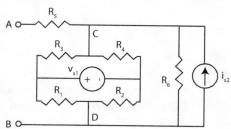

Figure P6.7

CHECK: R_{th} = 10 kΩ

8. (a) Find the Thevenin equivalents for the circuit of Figure P6.8 in terms of the literals v_{s1} and i_{s2}.
   (b) If the A-B is terminated in a 15 kΩ load, and v_{s1} = 30 V, and i_{s2} = 10 mA, find the power delivered to the load.
   (c) What is the proper resistance across the terminals A-B for maximum power transfer and what is the resultant power delivered to the load?

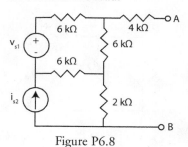

Figure P6.8

ANSWER: R_{th} = 10 kΩ, v_{oc} = 204 V

9. Consider the circuit Figure P6.9, in which v_s = 120 V and R = 300 Ω.
   (a) Find the Thevenin equivalent circuit to the left of the terminals A-B.
   (b) For R_L = 300 Ω, 600 Ω, and 1200 Ω, find the power absorbed by R_L. Does the use of a Thevenin equivalent reduce the effort needed to obtain these answers?

(c) Find the value of R_L for maximum power transfer and the resultant power delivered to the load.
(d) If the value of v_s is doubled, what is the power delivered to the load under the condition of maximum power transfer?

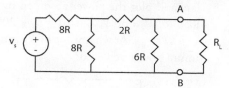

Figure P6.9

CHECKS: (a) 900 Ω, 30 V; (c) 250 mW, (d) 1

10. Find the Thevenin equivalent seen at A-B of the circuit of Figure P6.10. Hint: For this type of problem, the more natural solution technique is source transformations. Why?

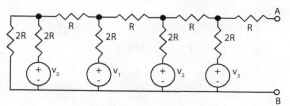

Figure P6.10

ANSWER: R_{th} = 2R, v_{oc} = $\dfrac{V_0}{16} + \dfrac{V_1}{8} + \dfrac{V_2}{4} + \dfrac{V_3}{2}$

11. Find the Thevenin equivalent of the circuit of Figure P6.11 enclosed in the dashed-line box. Then compute I_L and the power absorbed by the 2 kΩ resistor. Hint: What resistances are extraneous to the solution?

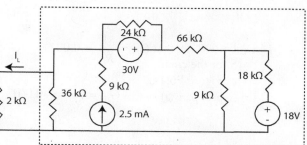

Figure P6.11

ANSWER: 52 V, 24 kΩ, 2 mA

12.(a) Find the Thevenin equivalent circuit for the 2-terminal non-series-parallel network shown in Figure P6.12. Use the general method.

(b) If a load resistance R_L is connected to terminals A-B, use MATLAB to calculate and plot the power absorbed by the load for $V_s = 30$ V, and $10 \le R_L \le 200$ Ω in 5 Ω steps. At what value of R_L is maximum power achieved?

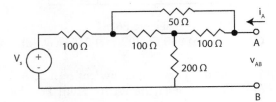

Figure P6.12

ANSWER: For (a), $v_{oc} = 0.6667 v_s$, $R_{th} = 100$ Ω

THEVENIN/NORTON FOR ACTIVE CIRCUITS

13.(a) Find the value of g_m so that the Thevenin equivalent resistance of the circuit shown in Figure P6.13 is 5 kΩ.

(b) Repeat part (a) for the case when $R_{th} = -250$ Ω. CHECK: (b) 6.25 mS

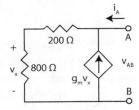

Figure P6.13

ANSWER: $g_m = 1000$ μS

14.(a) Find α so that $G_{th} = \dfrac{1}{R_{th}} = 0.001$ S for the circuit of Figure P6.14.

(b) Repeat part (a) so that $G_{th} = \dfrac{1}{R_{th}} = -0.001$

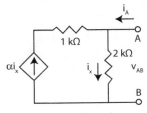

Figure P6.14

15.(a) Find α so that $R_{th} = 3$ kΩ for the circuit of Figure P6.15.

(b) Repeat part (a) so that $R_{th} = -1$ kΩ. Hint: Do problem 14 first, and then modify the Thevenin resistance appropriately.

CHECK: (b) $\alpha = 4000$ Ω

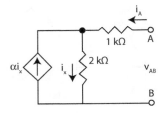

Figure P6.15

16.(a) Find α so that $R_{th} = 5$ kΩ for the circuit of Figure P6.16.

(b) Repeat part (a) so that $R_{th} = -1$ kΩ .

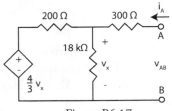

Figure P6.16

17. For the circuit shown in Figure P6.17, find the Norton equivalent circuit.

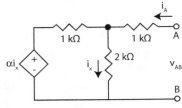

Figure P6.17

ANSWER: $- 600$ Ω, $i_{sc} = 0$

18. Use loop analysis to compute the Thevenin equivalent for the circuit shown in Figure P6.18. What is the Norton equivalent?

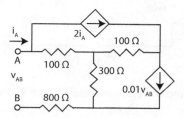

Figure P6.18

ANSWER: $v_{oc} = 0$, $R_{th} = 250\ \Omega$

19.(a) Find the Norton and Thevenin equivalents of the circuit of Figure P6.19.
(b) If a load resistor R_L is attached across the output terminals, plot the power absorbed by the load for $1 \leq R_L \leq 24\ \Omega$ usingMATLAB or equivalent. For what value of R_L does the load absorb maximum power? Determine the power delivered to the load at maximum power transfer.

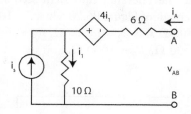

Figure P6.19

ANSWER: $v_{oc} = 6i_s$, $i_{sc} = 0.5i_s$, $P_{Lmax} = \dfrac{3i_s^2}{4}$

20. Find the Thevenin equivalent of the circuit in Figure P6.20 where $i_s = 0.2$ A.

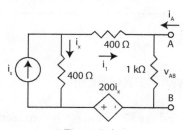

Figure P6.20

ANSWER: $v_{oc} = 60$ V, $R_{th} = 500\ \Omega$

21.(a) Find the Thevenin and Norton equivalent circuits for the network shown in Figure P6.21, assuming that $k = 0.025$ S and $v_s = 20$ V.
(b) For what value of k is the open circuit voltage zero. For this value of k, determine R_{th}.

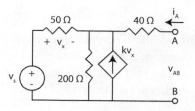

Figure P6.21

ANSWER: $R_{th} = 60\ \Omega$ and $v_{oc} = 18$ V

22. For the circuit shown in Figure P6.22, $b = -0.02$ S and $a = 25\ \Omega$. Find the Norton equivalent circuit.

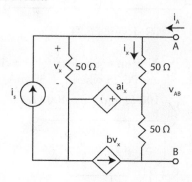

Figure P6.22

ANSWER: $R_{th} = 100\ \Omega$, $v_{oc} = 50i_s$

23. Consider the circuit shown in Figure P6.23.
(a) Find the Thevenin equivalent.
(b) If a load resistor R_L is connected across terminals A-B, determine R_L for maximum power transfer and determine the maximum power delivered to R_L.
(c) If a resistor R_1 were added in series with terminal A of figure P6.23, what is the Thevenin equivalent resistance of the augmented network.

ANSWER: (a) $R_{th} = R$, $v_{oc} = \dfrac{2}{3}V_s$

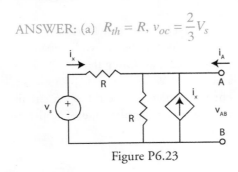

Figure P6.23

24.(a) Find the Thevenin equivalent for the network shown in Figure P6.24.

(b) If the values of each source are cut in half, what is the new v_{oc}?

ANSWER: (a) 1.6 kΩ, – 260 V; (b) $v_{oc} = -130$ V

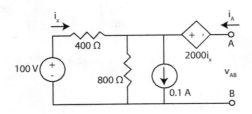

Figure P6.24

25. Find the Norton equivalent for the circuit shown in Figure P6.25 when $i_s = 30$ mA, $g_m = 0.04$ S, $R_1 = 100$ Ω and $R_2 = 400$ Ω.

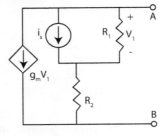

Figure P6.25

CHECK: $i_{sc} = 0.09$ A

26. Consider the circuit of Figure P6.26, where $v_s = 32$ V, $R_1 = 80$ Ω, $R_2 = 240$ Ω, $R_3 = 60$ Ω, and $\beta = 2$.

(a) Replace the circuit to the left of nodes A and B with its Thevenin equivalent.

(b) Given your answer to (a), assume that $R_L = 150$ Ω, and find i_L and the power consumed by R_L.

CHECK: $R_{th} = 30$ Ω

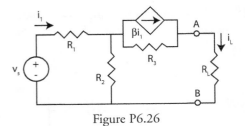

Figure P6.26

27. Find the Thevenin and Norton equivalents of the circuit of Figure P6.27 when $i_s = 20$ mA.

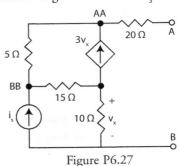

Figure P6.27

ANSWER: $v_{oc} = 12.5$ V, $R_{th} = 650$ Ω

28. Find the Thevenin equivalent seen at A-B for the circuit of Figure P6.28 when $i_s = 10$ mA, $R_1 = 2400$ Ω and $R_2 = 14000$ Ω, $R_3 = 5600$ Ω, $R_4 = 1500$ Ω, $R_5 = 1000$ Ω, and $g_m = 0.25 \times 10^{-3}$ S.

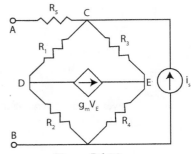

Figure P6.28

CHECK: $i_{sc} = 8$ mA

OP AMP PROBLEMS

29.(a) Find the Thevenin equivalent seen at the terminals A-B for the op amp circuit of Figure P6.29.

(b) What is the value of a load resistor R_L attached across the terminals A-B for maximum power transfer. What is the power absorbed by this R_L?

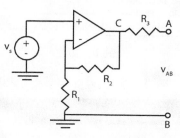

Figure P6.29

CHECK: $R_{th} = R_3$

30. (a) Find the Thevenin and Norton equivalents seen at the terminals A-B for the op amp circuit of Figure P6.30.

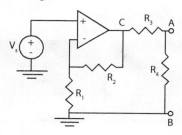

Figure P6.30

CHECK: $i_{sc} = \dfrac{1}{R_3}\left(\dfrac{R_1 + R_2}{R_1}V_s\right)$

31. (a) Find the Thevenin equivalent to the right of the terminals A-B for the (ideal) op amp circuit of Figure P6.31.
   (b) If the practical source indicated in the figure is attached to A-B, find the current I_s in terms of V_s.

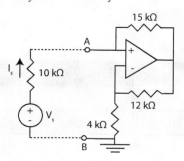

Figure P6.31

CHECK: $R_{th} = 5000\ \Omega$

32. (a) Find the Thevenin and Norton Equivalent circuits of the op amp configuration of Figure P6.32 seen at A-B.
   (b) Repeat (a) for the terminals C-B.

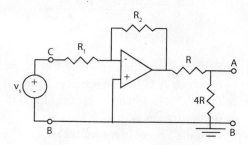

Figure P6.32

ANSWER: (b) $R_{th} = R_1$, $v_{oc} = 0$

33. (a) Find the Thevenin and Norton Equivalent circuits of the op amp configuration of Figure P6.33 seen at A-B.
   (b) Determine the value of a load resistor R_L connected across the terminals A-B for maximum power transfer. If $v_{s1} = 4$ V and $v_{s2} = 5$ V, determine the maximum power transferred to this R_L.

CHECK: (b) $P_{max} = 0.9$ watts

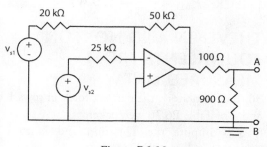

Figure P6.33

34. Find the Thevenin equivalent seen at the terminals A-C for the op amp circuit of Figure P6.34 when the op amp has output saturation, $V_{sat} = 15$ V.

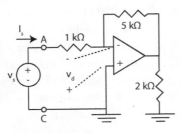

Figure P6.34

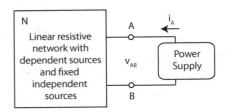

Figure P6.36

35.(a) Find the Thevenin equivalent seen at the terminals A-C for the op amp circuit of Figure P6.35 when the op amp has output saturation, $V_{sat} = 12$ V.
   (b) Find the Thevenin equivalent seen to the left of the terminals B-C and the maximum power that will be absorbed by the 24 kΩ resistor for all variations in V_s.

Table P6.36

i_A (mA)	v_{AB} (V)
1	6
4	12

CHECK: (b) $P_{max} = 2$ mW

37. Repeat Problem 36 with the data given in Table P6.37.

Table P6.37

i_A (mA)	v_{AB} (V)
10	54
40	66

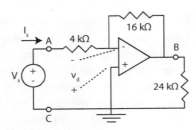

Figure P6.35

CHECK: $P_{max} = \dfrac{144}{2800} = 6$ mW

CHECK: (b) $P_{max} = 1.5625$ W

THEVENIN AND NORTON EQUIVALENTS FROM MEASURED DATA

36. In a laboratory, the data set forth in rows 1 and 2 of Table P6.36 were taken.
   (a) Compute the Thevenin and Norton equivalents of N.
   (b) After the power supply is removed, what resistance, R_L, should be connected across A-B for maximum power transfer? What is P_{max}?

38. The data listed in Table P6.38 was taken for the network N of Figure P6.38.
   (a) Fill in the values for the third column of Table P6.38 and find the Thevenin and Norton equivalents of the linear resistive network N.
   (b) To what resistance should R_L be changed to achieve maximum power transfer? What is P_{max}?

Table P6.38

R_L (kΩ)	v_{AB} (V)	i_A (mA)
2	4	?
10	10	?

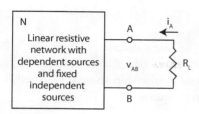

Figure P6.38

CHECK: P_{max} = 10.667 mW

39. Repeat Problem 38 using the data in Table P6.39.

Table P6.39

R_L (Ω)	v_{AB} (V)	i_A (mA)
200	2	?
1200	6	?

CHECK: P_{max} = 31.25 mW

40. The data listed in Table P6.40 was taken for the network N of Figure P6.40 with a voltmeter (VM) whose internal resistance is 10 MΩ. Fill in the values for the third column of Table 6.36 and find the Thevenin equivalent of the linear resistive network N.

Table P6.40

R_L (MΩ)	v_{AB} (V)	i_A (μA)
2	0.4	?
10	1	?

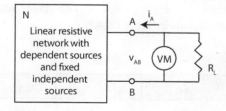

Figure P6.40

CHECK: v_{oc} = 4 V

41. This problem is the first of two problems that outline a laboratory measurement procedure for finding the Thevenin equivalent of a linear resistive 2-terminal network. For this problem, consider Figure P6.41 in which the circuit under test contains no independent sources. The experimental apparatus includes a resistance decade box, denoted R, a dc voltmeter with internal resistance R_m, and a signal generator having known internal resistance, R_s. To begin the procedure, one sets $R = R_1 = 0$ and adjusts the dc level, V_s, of the signal generator to obtain a reasonable meter reading, say $V_{m1} = E_0$, where the subscript "1" indicates our first meter reading. (For an analog meter, the reading should be almost full scale.) Leaving the signal generator set at this value of V_s, increase R until the meter reading drops to $V_{m2} = 0.5\ E_0$. Record this value of R as R_2.

(a) Suppose $R_m = \infty$ and $R_s = 0$. Show that $R_{th} = R_2$.

(b) Now suppose $R_m = \infty$ and $R_s \neq 0$. Show that $R_{th} = R_2 - R_s$.

(c) Finally, suppose R_m and R_s are nonzero and finite. Show that $\dfrac{R_{th}R_m}{R_{th} + R_m} = R_2 - R_s$ and then solve for R_{th}.

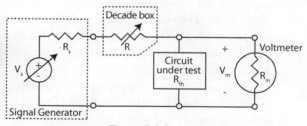

Figure P6.41

42. This problem is the second of two problems that outline a laboratory measurement procedure for finding the Thevenin equivalent of a linear resistive 2-terminal network. For this problem, consider the new configuration of Figure P6.42 in which the circuit under test contains independent sources and has a non-zero v_{oc}. The experimental apparatus includes a resistance decade box, denoted R, and a dc voltmeter with internal resistance

R_m. All devices are connected in parallel. Because $v_{oc} \neq 0$, a signal generator is not needed, as in Problem 41. To begin the procedure, open circuit the decade box, or equivalently set $R = R_1 = \infty$, and set the scale on the voltmeter to obtain a reasonable meter reading, say $V_{m1} = E_0$, where the subscript "1" indicates our first meter reading. (For an analog meter, the reading should be almost full scale.) Next, reconnect the decade and decrease R until the meter reading drops to $V_{m2} = 0.5 E_0$. Record this value of R as R_2.

(a) Suppose $R_m = \infty$. Show that $R_{th} = R_2$.

(b) Suppose R_m is nonzero and finite. Show that $\dfrac{R_{th} R_m}{R_{th} + R_m} = R_2$ and then solve for R_{th}. Then show that $v_{oc} = \left(1 + \dfrac{R_{th}}{R_m}\right) E_0$

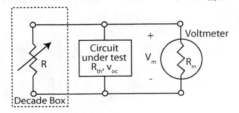

Figure P6.42

43. The Thevenin equivalent of a linear resistive network containing no independent sources is to be found experimentally using the method of Problem 41. The voltmeter has an internal resistance $R_m = 20$ kΩ. The dc signal generator has an internal resistance, $R_s = 2$ kΩ. The following measurements are taken: (i) with $R = 0$, V_s is adjusted until the voltmeter reads 4 V; (ii) keeping V_s fixed, the decade box is adjusted until the voltmeter reads 2 V. For this voltage, the decade box shows $R = 6$ kΩ. Find R_{th}.
ANSWER: 5 kΩ

44. The Thevenin equivalent of a linear resistive network containing independent sources is to be found experimentally by the procedure of Problem 42. The voltmeter has an input resistance $R_m = 1$ MΩ. The following measurements are taken: (i) when R is open-circuited, the voltmeter reads 4 V, and (ii) when R is decreased to 800 kΩ, the voltmeter reads 2 V. Find R_{th} and v_{oc}.

ANSWER: 4 MΩ and 20 V

45. The linear circuit shown in Figure P6.45 is found experimentally to have the voltage and current relationship shown. Find its Thevenin Norton equivalent.

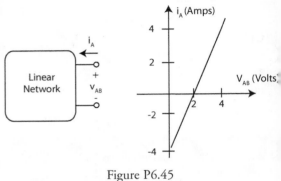

Figure P6.45
ANSWER: 0.5 Ω, 4 A

46. Repeat Problem 45 for the measurement curve shown in Figure P6.46. Then determine the value of a load resistor for maximum power transfer and compute P_{max}.

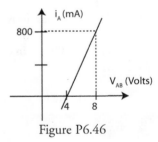

Figure P6.46

47. The i-v curve of the network N in Figure P6.47a is measured in a laboratory, and is approximated by the straight-line segments shown in Figure P6.47b. The meter readings are shown in Table P6.47.

Table P6.47

A	0.2 V	0.1 mA
B	0.7 V	10.1 mA

(a) Find the Thevenin equivalent for the range $0 \leq i < 0.1$ mA.

(b) Find the Thevenin equivalent for the range $0.1 \leq i < 10.1$ mA.

(c) If $R = 500 \; \Omega$, $v_s(t) = 50 \sin(1000 \; t)$ mV, and $v_b = 100$ mV, find $i(t)$. Hint: Use a suitable Thevenin equivalent for N.

(d) If $R = 50 \; \Omega$, $v_s(t) = 200 \sin(1000 \; t)$ mV, and $v_b = 500$ mV, find $i(t)$.

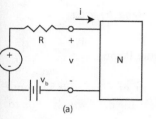
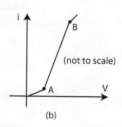

(a)   (b)

Figure P6.47

ANSWER: (a) $v_{oc} = 0$, $R_{th} = 2 \; \text{k}\Omega$; (b) $v_{oc} = 0.195\text{V}$, $R_{th} = 50 \; \Omega$; (c) $0.04 + 0.02 \sin (1000 \; t)$ mA; (d) $3.05 + 2 \sin(1000 \; t)$ mA

MAXIMUM POWER TRANSFER

48. For the circuit of Figure P6.48, $R_1 = 160 \; \Omega$, $R_2 = 480 \; \Omega$, and $v_s = 80$ V. Find the value of R_L for maximum power transfer and P_{max}.

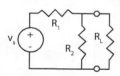

Figure P6.48

CHECK: 7.5 watts

49. For the circuit of Figure P6.49, $R_1 = 900 \; \Omega$, $R_2 = 180 \; \Omega$, $R_3 = 50 \; \Omega$, $i_{s1} = 60$ mA, and $v_{s2} = 21$ V. Find the value of R_L for maximum power transfer and P_{max}.

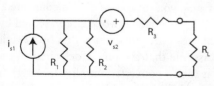

Figure P6.49

50. For the circuits of Figure P6.50, $v_{s1} = 10$ V and $v_{s2} = 15$ V. Find the load resistance R_L needed for maximum power transfer, the asso-

ciated voltage V_L, and the power delivered to the load.

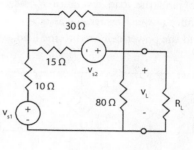

Figure P6.50

51 (a) For the circuits of Figure P6.51, find the load resistance R_L needed for maximum power transfer, the associated voltage V_L, and the power delivered to the load.

(b) If the load resistance is constrained as $5 \; \text{k}\Omega \leq R_L \leq 10 \; \text{k}\Omega$, repeat part (a).

(c) If the load resistance is constrained as $15 \; \text{k}\Omega \leq R_L \leq 20 \; \text{k}\Omega$, repeat part (a).

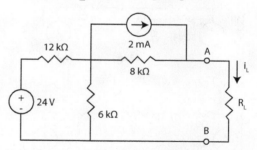

Figure P6.51

52. Consider the circuit of Figure P6.52.
(a) Find the value of R_L for maximum power transfer to the three-resistor load.
(b) Find the power delivered to each load resistor, i.e., to R_L, $R_L/2$, and $R_L/3$.

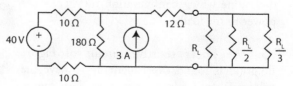

Figure P6.52

53. For the circuits of Figure P6.53, $R_1 = 200\ \Omega$, $R_2 = 1000\ \Omega$, $R_3 = 400\ \Omega$, $g_m = 8$ mS, and $i_{s1} = 0.4$ A. Find the load resistance R_L needed for maximum power transfer, the associated voltage, V_L, and the power delivered to the load.

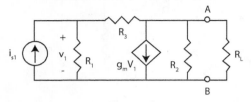

Figure P6.53

54. The circuits of Figure P6.54 have the load resistor R_L connected in different ways. For each circuit, (i) compute the value of R_L which leads to maximum power transfer, (ii) the voltage across the load, and (iii) the power absorbed by the load. Which configuration absorbs more power?

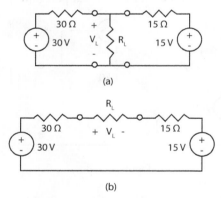

(a)

(b)

Figure P6.54
CHECK: (a) $P_{max} = 10$ watts

55. Suppose the polarity of the 15-V-source in Problem 54 is reversed. Repeat Problem 54 and determine which configuration transfers more power to the load.

56. The linear resistive circuit of Figure P6.56(a) is found experimentally to have the voltage-current relationship plotted in Figure P6.56(b). Find the maximum power that can be absorbed by placing a load resistor across terminals a-b?

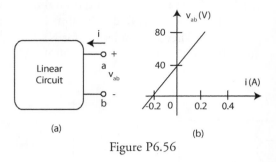

(a)

(b)

Figure P6.56

57. (a) For the circuit of Figure P6.57 compute, (i) the value of R which leads to maximum power transfer to the load, (ii) the voltage across the load, and (iii) the power absorbed by the load. Hint: In MATLAB, the roots of a quadratic, $a_0 x^2 + a_1 x + a_2$, are given by "roots([a0 a1 a2])".

(b) To verify the results of (a) write a MATLAB program to calculate and plot the power absorbed by the load as R varies from $0\ \Omega$ to $400\ \Omega$ in $2\ \Omega$ increments.

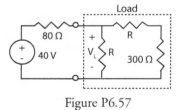

Figure P6.57

58. The i-v relationship of certain type of LED (light emitting diode) in its operating range of 1.7V–3V is represented by a 2 V voltage-source in series with a 50 Ω resistance. The load consists of a network of n such diodes connected in parallel. The source network is represented by a 5 V voltage-source in series with a 50 Ω resistance. Assume that the power delivered to each diode is totally converted into light. Determine how many LEDs should be connected in parallel for maximum brightness. What is power dissipated by each diode?

ANSWER: n = 5, $P_{diode} = 25$ mW

C H A P T E R 7

Inductors and Capacitors

CAPACITIVE SMOOTHING IN POWER SUPPLIES

Every non-portable personal computer contains a power supply that converts the sinusoidal voltage of the ordinary household outlet to a regulated dc voltage. "Regulated" means that the output voltage stays within very tight limits of its nominal value (e.g., 12 ± 0.1 V) over a wide range of power requirements. Engineers design power supply circuits with regulators that produce voltages with a small oscillation because to generate a truly dc voltage is practically impossible.

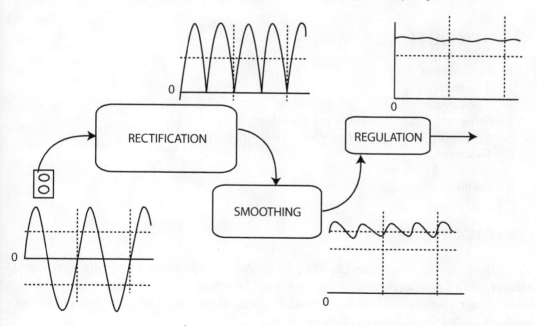

This process of converting ac to dc has three stages: First, the ac waveform is rectified into its absolute value. Then a smoothing operation takes place that reduces the variation in the voltage to a reasonable but still unacceptable level. This first level of smoothing is necessary because the voltage regulator is a precision subcircuit that requires a fairly constant voltage for its proper operation. The partially smoothed waveform is fed into a voltage regulator, which limits the voltage oscillation between critical levels even when the load drawn by any connected device (e.g., your computer) varies in the course of its operation.

As mentioned, the rectified sine wave is smoothed before entering the voltage regulator. A crude smoothing can be accomplished with a capacitor, a device studied in this chapter. Intuitively, capacitors resist voltage changes and are designed to steady the voltage at a constant level. In this chapter, we will study the capacitor and investigate a simplified smoothing operation for a power supply.

CHAPTER OBJECTIVES

1. Define the notion of inductance and introduce the inductor, whose terminal voltage is proportional to the time derivative of the current through it.
2. Investigate the ability of an inductor to store energy and the computation of the equivalent inductance of series-parallel connections.
3. Define the notion of capacitance and introduce the capacitor, whose current is proportional to the time derivative of its terminal voltage.
4. Investigate the ability of a capacitor to store energy and the computation of the equivalent capacitance of series-parallel connections.
5. Define and illustrate the principle of conservation of charge.

CHAPTER OUTLINE

1. Introduction
2. The Inductor
3. The Capacitor
4. Series and Parallel Inductors and Capacitors
5. Smoothing Property of a Capacitor in a Power Supply
6. Summary
7. Terms and Concepts
8. Problems

1. INTRODUCTION

This chapter introduces two new circuit elements, the linear inductor and the linear capacitor, hereafter referred to as an inductor and a capacitor. The inductor, shown in Figure 7.3, is a device whose voltage is proportional to the time rate of change of its current with a constant of proportionality L, called the inductance of the device, i.e.

$$v_L(t) = L\frac{di_L(t)}{dt},$$

as set forth in equation 7.1. The unit of the inductance L, is the henry, denoted by H. Macroscopically, inductance measures the magnitude of the voltage induced by a change in the current through the inductor.

The capacitor, shown in Figure 7.15, is a device whose current is proportional to the time rate of change of its voltage, i.e.,

$$i_C(t) = C \frac{dv_C(t)}{dt}$$

as set forth in equation 7.5. Here, the constant of proportionality, C, is the capacitance of the device with unit farad, denoted by F. Capacitance measures the device's ability to produce a current from changes in the voltage across it.

By adding the inductor and the capacitor to the previously studied devices (the resistor, independent and dependent sources, etc.), one discovers an entire panorama of possible circuit responses, to be explored in the next four chapters. Together, these devices allow one to design radios, transmitters, televisions, stereos, tape decks, and other electronic equipment. In this chapter, our goal is to understand the basic operation of inductors and capacitors.

2. THE INDUCTOR

Some Physics

In Figure 7.1, a changing current flowing from point A to point B through an ideal conductor induces a voltage v_{AB} between points A and B according to Faraday's law. Joseph Henry independently observed the same phenomenon at about 1831. The induced voltage, v_{AB}, was found to be proportional to the rate of change of current, i.e., $v_{AB} \approx \dfrac{di}{dt}$.

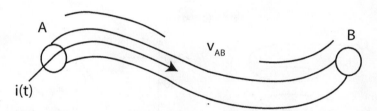

FIGURE 7.1 A time-varying current flowing through an ideal conductor.

The following experiment illustrates the idea. Suppose the conductor in Figure 7.1 is 6 feet of #22 copper with resistance 16.5 Ω/1,000 ft. The 6-foot length has a resistance of about 0.1 Ω. Using a current generator, we apply a pair of ramp currents (shown in Figure 7.2a) to the conductor, as per Figure 7.2b. The measured responses are shown in Figure 7.2c and, as expected, satisfy Ohm's law.

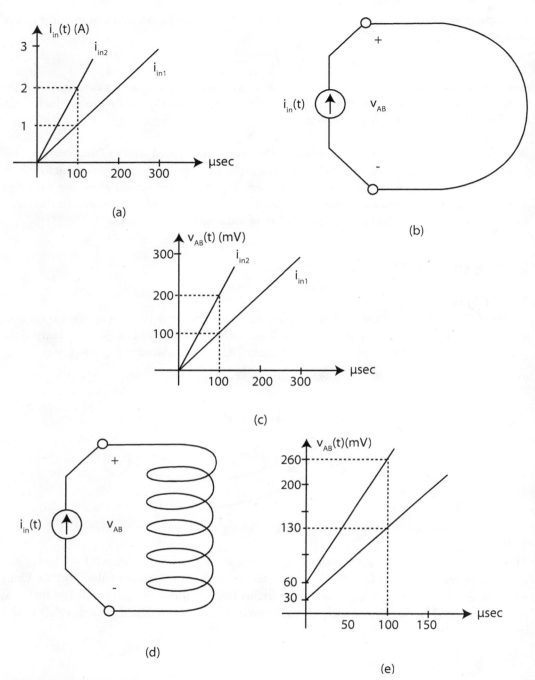

FIGURE 7.2 (a) Ramp current inputs to uncoiled and coiled wire. (b) Six feet of #22 wire attached to a current generator. (c) Voltage responses to ramp current inputs of uncoiled wire. (d) Six feet of #22 wire coiled into 45 turns 1" long and 1" in diameter. (e) Voltage responses to ramp current inputs of coiled wire.

Now suppose the wire is coiled into a cylinder 1" in diameter and 1" long, as in Figure 7.2d. Apply the same ramp currents of Figure 7.2a to the coiled wire. This time, the measured responses are as shown in Figure 7.2e. These responses have the same shape as those of Figure 7.2c, except for the offsets of 30 mV and 60 mV, respectively. These offset voltages are proportional to the derivatives of the input currents, i.e.,

$$\text{Offset} = L \frac{di_{in_k}}{dt}$$

for $k = 1, 2$, where L is the proportionality constant, called the inductance of the coil. Since the derivative of $i_{in_1}(t)$ is 10^4 A/sec, and the derivative of $i_{in_2}(t)$ is 2×10^4 A/sec, the inductance L of the coil can be computed as

$$3 \times 10^{-6} = \frac{\text{Offset}}{\dfrac{di_{in_k}}{dt}} = \frac{0.03}{10^4} = \frac{0.06}{2 \times 10^4} \text{ henries}$$

As mentioned earlier, the **henry**, equal to 1 volt-sec/amp and abbreviated H, is the unit of inductance. Also, from the above experiment, one concludes that the inductance of a cylindrical coil of wire is much greater than the inductance of a straight piece of wire, which in the above experiment was not measurable by our apparatus.

The physics of the preceding interaction is governed by **Maxwell's equations**, which describe the interaction between electric and magnetic fields. A time-varying current flow through a wire creates a time-varying magnetic field around the wire. The magnetic field in turn sets up a time-varying electric field, i.e., an electric potential or voltage. One can verify the presence of this magnetic field by bringing a compass close to a wire carrying a current. The magnetic field surrounding the wire will cause the compass needle to deflect. Physically speaking, a changing current causes a change in the storage of energy in the magnetic field surrounding the conductor. The energy transferred to the magnetic field requires work and, hence, power. Because power is the product of voltage and current, it follows that there is an induced voltage between the ends of the conductor. What is even more interesting is that if a second wire is immersed in the changing magnetic field of the first wire, a voltage will be induced between the ends of the second wire. A proper (mathematical) explanation of this phenomenon is left to a fields course. For our purposes, three facts are important: (1) energy storage occurs, (2) the induced voltage is proportional to the derivative of the current, and (3) the constant of proportionality is called the inductance of the coil and is denoted by L.

As mentioned, a straight wire has a very small inductance, whereas a cylindrical coil of the same length of wire has a much greater inductance. This inductance can be increased many times over, possibly several thousand times, simply by putting an iron bar in the center of a cylindrical coil. Alas, the calculation of inductance is the proper subject of more advanced texts, e.g., on field theory or transmission line theory. Nevertheless, there are empirical formulas for estimating the inductance of a single-layer air-core coil as described in the homework exercises.

BASIC DEFINITION AND EXAMPLES

DEFINITION OF THE LINEAR INDUCTOR

The linear **inductor**, symbolized by a coiled wire as shown in Figure 7.3, is a two-terminal energy storage device whose voltage is proportional to the derivative of the current passing through it. The constant of proportionality, denoted by L, has the unit of **Henry** (H), equal to 1 volt-sec/amp. L is said to be the **inductance** of the coil. The specific voltage-current relationship of the linear inductor is given by

$$v_L(t) = L\frac{di_L(t)}{dt} \tag{7.1}$$

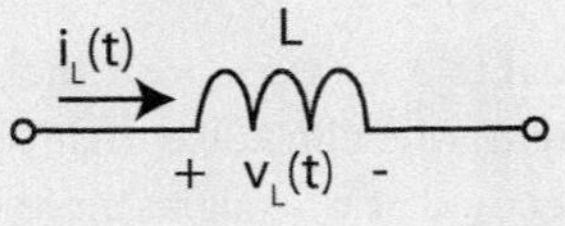

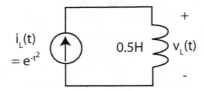

FIGURE 7.3 The inductor and its differential voltage-current relationship as per the passive sign convention.

EXAMPLE 7.1

Compute $v_L(t)$ for the inductor circuit of Figure 7.4 when $i_L(t) = e^{-t^2}$.

$$i_L(t) = e^{-t^2} \qquad 0.5H \qquad v_L(t)$$

FIGURE 7.4 A 0.5 H inductor driven by a current source.

SOLUTION

From equation 7.1, direct differentiation of the inductor current $i_L(t)$ leads to

$$v_L(t) = 0.5\frac{d(e^{-t^2})}{dt} = 0.5(-2t)e^{-t^2} = -te^{-t^2}\ \text{V}$$

Exercises. 1. In Example 7.1, suppose $i_L(t) = 0.5\sin(20t + \pi/3)$ A. Compute $v_L(t)$.
ANSWER: $5\cos(20t + \pi/3)$ V.

2. In Example 7.1, suppose $i_L(t) = (1 - e^{-200t})$ V for $t \geq 0$ and 0 otherwise. Find $v_L(t)$, $t \geq 0$.
ANSWER: $v_L(t) = 100e^{-200t}$ V.

The differential equation 7.1 has a dual integral relationship. Safely supposing that at $t = -\infty$ the inductor had not yet been manufactured, one can take $i_L(-\infty) = 0$, in which case

$$i_L(t) = \frac{1}{L}\int_{-\infty}^{t} v_L(\tau)d\tau = \left[\frac{1}{L}\int_{-\infty}^{t_0} v_L(\tau)d\tau\right] + \frac{1}{L}\int_{t_0}^{t} v_L(\tau)d\tau$$

$$= i_L(t_0) + \frac{1}{L}\int_{t_0}^{t} v_L(\tau)d\tau \tag{7.2}$$

The time t_0 represents an initial time that is of interest or significance, e.g., the time when a switch is thrown or a source excitation is activated. The quantity

$$i_L(t_0) = \frac{1}{L}\int_{-\infty}^{t_0} v_L(\tau)d\tau$$

specifies the initial current flowing through the inductor at t_0. This quantity, $i_L(t_0)$, sums up the entire past history of the voltage excitation across the inductor. Because of this, the inductor is said to have **memory**.

EXAMPLE 7.2

For the circuit of Figure 7.5a, determine $i_L(0)$ and $i_L(t)$ for $t \geq 0$ when $v_L(t) = e^{-|t|}$ V as plotted in Figure 7.5b.

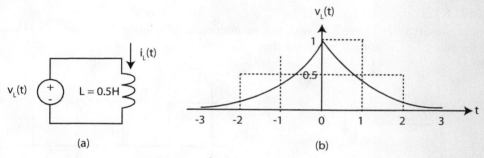

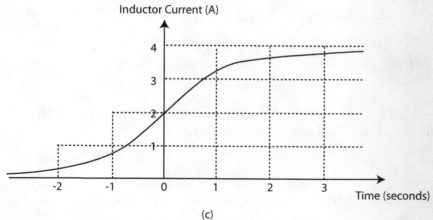

FIGURE 7.5 (a) Simple inductor driven by a voltage source. (b) Source waveform $v_L(t)$. (c) Resulting inductor current $i_L(t)$.

SOLUTION

A direct application of equation 7.2 leads to

$$i_L(t) = \frac{1}{L}\int_{-\infty}^{0} e^{\tau}d\tau + \frac{1}{L}\int_{0}^{t} e^{-\tau}d\tau = \frac{1}{L} + \frac{1}{L}[1 - e^{-t}] \text{ A}$$

It follows that

$$i_L(0) = \frac{1}{L} = 2 \text{ A and } i_L(t) = \frac{1}{L}\left[2 - e^{-t}\right] = 4 - 2e^{-t} \text{ A, } t \geq 0$$

The graph of $i_L(t)$ for all t is given in Figure 7.5c.

Exercises. 1. In Example 7.2, compute an expression for $i_L(t)$ for $t < 0$.
ANSWER: $i_L(t) = 2e^t$ A for $t < 0$.

2. Repeat Example 7.2 with $L = \frac{1}{4\pi}$ H and with $v_L(t) = \cos(2\pi t)$ V for $t \geq -0.25$ sec and zero otherwise.
ANSWER: $i_L(0) = 2$ A, $i_L(t) = 2 + \sin(2\pi t)$ A for $t \geq 0$.

EXAMPLE 7.3

Consider the circuit of Figure 7.6a with voltage excitation $v_L(t)$ shown in Figure 7.6b. Find the inductor current $i_L(t)$ for $t \geq 0$, assuming that $i_L(0) = 0$.

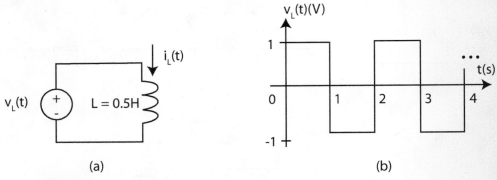

(a)   (b)

FIGURE 7.6 (a) Voltage source driving inductor. (b) Square wave excitation $v_L(t)$.

SOLUTION

It is necessary to apply equation 7.2 to each interval, [0, 1], [1, 2], ... , [n, n + 1], ... For this we need to first specify the initial conditions for each interval.

Step 1. *Compute* $i_L(1)$. From equation 7.2,

$$i_L(1) = i_L(0) + \frac{1}{L}\int_{0}^{1} v_L(\tau)d\tau = \frac{1}{L} \times \text{Net Area} = 2$$

Step 2. *Compute* $i_L(2)$.

$$i_L(2) = i_L(0) + \frac{1}{L}\int_0^2 v_L(\tau)d\tau = \frac{1}{L} \times \text{Net Area} = 0$$

Step 3. *Compute the initial condition for the interval* $[n, n+1]$ *for n even.* Again from equation 7.2, with $t = n$ and n even, we have

$$i_L(n) = i_L(0) + \frac{1}{L}\int_0^n v_L(\tau)d\tau = \frac{1}{L} \times \text{Net Area} = 0$$

Hence $i_L(n) = 0$ for all even values of n.

Step 4. *Compute the initial condition for the interval* $[n, n+1]$ *for n odd.* From equation 7.2, with $t = n$ and n odd, we have, utilizing steps 1 and 3,

$$i_L(n) = i_L(0) + \frac{1}{L}\int_0^{n-1} v_L(\tau)d\tau + \frac{1}{L}\int_{n-1}^n v_L(\tau)d\tau = \frac{1}{L}\int_{n-1}^n v_L(\tau)d\tau = 2$$

since $n - 1$ is even.

Step 5. *Compute* $i_L(t)$ *over* $[n, n+1]$ *with n even.* If n is even, then the value of the inductor current over the interval $[n, n+1]$ is

$$i_L(t) = i_L(n) + \frac{1}{L}\int_n^t d\tau = i_L(n) + 2(t - n) = 2(t - n) \text{ A}$$

Observe that $i_L(t) = 2t - 2n$ A is the equation of a straight line having slope +2 and y-intercept $-2n$.

Step 6. *Compute* $i_L(t)$ *over* $[n, n+1]$ *with n odd.* If n is odd, then for the interval $[n, n+1]$, the inductor current is

$$i_L(t) = i_L(n) + \frac{1}{L}\int_n^t d\tau = i_L(n) - 2(t - n) = 2 - 2(t - n) \text{ A}$$

Here, $i_L(t) = 2 + 2n - 2t$ is the equation of a straight line, with slope -2 and y-intercept $2+2n$.

Step 7. *Piece segments from steps 5 and 6 together.* Thus the segments computed in steps 5 and 6 intercept the t-axis at the same points. Figure 7.7 sketches the resulting triangular response for $t \geq 0$.

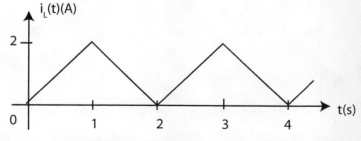

FIGURE 7.7 Triangular shape of inductor current for the square wave voltage excitation of Figure 7.6b applied to the circuit of Figure 7.6a.

Exercises. (All time is in seconds.) 1. Again consider the circuit of Figure 7.6a. Compute $i_L(t)$ for (i) $0 \leq t < 1$, (ii) $1 \leq t < 3$, and (iii) $3 \leq t$ for the waveform of Figure 7.8a, assuming $i_L(0) = 0$. 2. Again consider the circuit of Figure 7.6a. Compute $i_L(t)$ for (i) $0 \leq t < 1$, (ii) $1 \leq t < 3$, (iii) $3 \leq t < 4$, and (iv) $4 \leq t$, for the waveform of Figure 7.8b, assuming $i_L(0) = 0$.

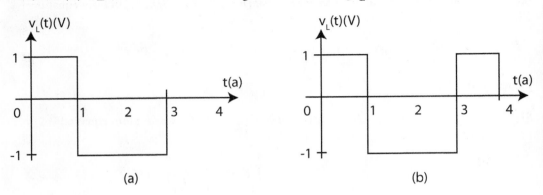

FIGURE 7.8 Voltage excitations for Exercises 1 and 2.

It is important to recognize that the square wave voltage input of Figure 7.6b is discontinuous but the current waveform of Figure 7.7 is continuous. Integration (computation of "area") is a smoothing operation: it smoothes simple discontinuities. This means that the inductor current is a continuous function of t, even for discontinuous inductor voltages, provided that the voltages are bounded. A voltage or current is **bounded** if the absolute value of the excitation remains smaller than some fixed finite constant for all time. Thus, equation 7.2 leads to the **continuity property of the inductor**: if the voltage $v_L(t)$ across an inductor is bounded over the time interval $t_1 \leq t \leq t_2$, then the current through the inductor is continuous for $t_1 < t < t_2$. In particular, if $t_1 < t_0 < t_2$, then $i_L(t_0^-) = i_L(t_0^+)$, even when $v_L(t_0^-) \neq v_L(t_0^+)$. The notation "−" and "+" on t_0 is used to distinguish the moments immediately before and after t_0. For example, in Figure 7.9, $t = 2$ shows a discontinuity of $v_L(t)$. The value of $v_L(2^-)$ is 1 and the value of $v_L(2^+)$ is −1. The value $v_L(2^+)$ can be seen as the limiting value of $v_L(t)$ when approaching $t \to 2$ from the right, whereas $v_L(2^-)$ can be seen as the limiting value of $v_L(t)$ when approaching $t \to 2$ from the left.

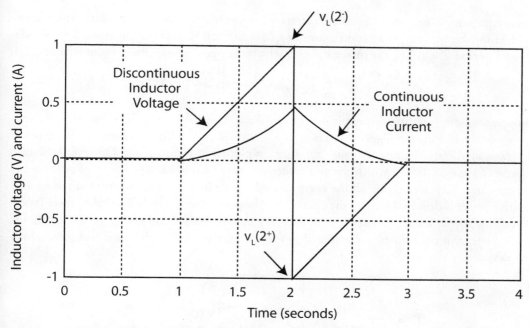

FIGURE 7.9 A possible discontinuous voltage $v_L(t)$ appearing across an inductor of 1 H, and the resulting continuous inductor current.

Power and Energy

Recall that the **instantaneous power** absorbed by a device is the product of the voltage across and the current through the device assuming the passive sign convention. For an inductor,

$$p_L(t) = v_L(t)i_L(t) = Li_L(t)\frac{di_L(t)}{dt} \text{ watts,}$$

where $v_L(t)$ is in volts, $i_L(t)$ in amps, and L in henries.

Since energy (absorbed or delivered) is the integral of the instantaneous power over a given time interval, it follows that the net **energy** $W(t_0, t_1)$ stored[1] over the interval $[t_0, t_1]$ in the magnetic field around the inductor is

$$W_L(t_0, t_1) = L\int_{t_0}^{t_1} \left(i_L(\tau)\frac{di_L(\tau)}{d\tau}\right) d\tau = L\int_{i_L(t_0)}^{i_L(t_1)} i_L \, di_L$$

$$= \frac{1}{2}L\left[i_L^2(t_1) - i_L^2(t_0)\right] \text{ joules,}$$

(7.3)

for L in henries and i_L in amps. From equation 7.3, whenever the current waveform is bounded, the net energy stored in the inductor over the interval $[t_0, t_1]$ depends only on the value of the inductor current at times t_1 and t_0, i.e., on $i_L(t_1)$ and $i_L(t_0)$, respectively. This means that the stored energy is independent of the particular current waveform between t_0 and t_1.

If the current waveform is **periodic**, i.e., if $i_L(t) = i_L(t + T)$ for some constant $T > 0$, then over any time interval of length T, the net stored energy in the inductor is zero because $i_L(t_0) = i_L(t_0 + T)$

forces equation 7.3 to zero. To further illustrate this property, consider Figure 7.10a, which shows a 0.1 H inductor driven by a periodic current $i_L(t) = \sin(2\pi t)$ V. This current signal has a **fundamental period** $T = 1$, i.e., the smallest T over which the signal repeats itself. From equation 7.3,

$$W_L(0,1) = \int_0^1 p_L(t)dt = \frac{1}{2}Li_L^2(1) - \frac{1}{2}Li_L^2(0) = 0$$

However, we can interpret this result in terms of the waveform of $p_L(t)$. First note that the voltage across the inductor in Figure 7.10a is $v_L(t) = 0.2\pi\cos(2\pi t)$ V. Hence, the instantaneous power is $p_L(t) = v_L(t)i_L(t) = 0.2\pi\cos(2\pi t)\sin(2\pi t)$ watts, as plotted in Figure 7.10b. Observe the shaded regions of Figure 7.10b in which the area under the power curve has equal parts of positive and negative area. This means that all the energy stored by the inductor over the part of the cycle of positive power is delivered back to the circuit over the portion of the cycle when the power is negative. This is true for all periodic signals over any period. Because no energy is dissipated, and because energy is only stored and returned to the circuit, the (ideal) inductor is said to be a **lossless** device.

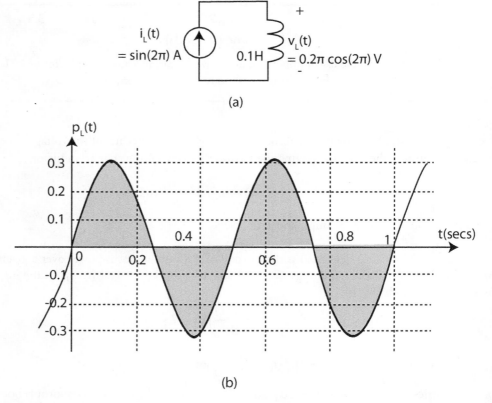

(a)

(b)

FIGURE 7.10 (a) Inductor excited by periodic current. (b) Plot of the power absorbed by the inductor.

It is convenient to define the **instantaneous stored energy** in an inductor as

$$W_L(t) = \frac{1}{2}Li_L^2(t) \tag{7.4}$$

for all t. Equation 7.4 can be viewed as a special case of equation 7.3 in which $t_0 = -\infty$ and $i_L(-\infty) = 0$. Thus, equation 7.4 can be interpreted as the change in stored energy in the inductor over the interval $(-\infty, t]$.

EXAMPLE 7.4

Find the instantaneous energy stored in each inductor of the circuit of Figure 7.11a for the source waveform given in Figure 7.11b. In Figure 7.11b, note that $i_s(t) = 0$ for $t \leq 0$.

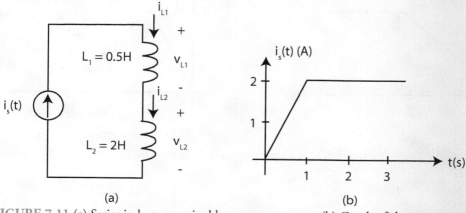

(a)   (b)

FIGURE 7.11 (a) Series inductors excited by a source current. (b) Graph of the source current.

SOLUTION

From KCL, $i_s(t) = i_{L1}(t) = i_{L2}(t)$ for all t. Since $i_s(t) = 2t$ A for $0 \leq t \leq 1$ and $i_s(t) = 2$ A for $t \geq 1$, equation 7.3 or 7.4 immediately yields the instantaneous stored energies (in J) as plotted in Figure 7.12:

$$W_1(t) = \frac{1}{2}L_1 i_{L1}^2(t) = \begin{cases} t^2 & 0 \leq t < 1 \\ 1 & 1 \leq t \end{cases} \text{ and } W_2(t) = \frac{1}{2}L_2 i_{L2}^2(t) = \begin{cases} 4t^2 & 0 \leq t < 1 \\ 4 & 1 \leq t \end{cases}$$

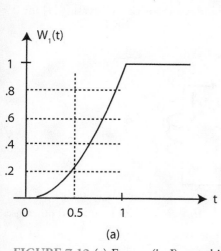

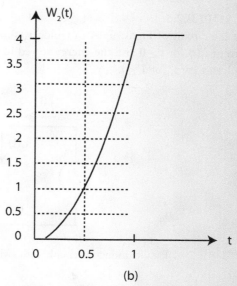

(a)   (b)

FIGURE 7.12 (a) Energy (in J) stored in inductor L_1. (b) Energy (in J) stored in inductor L_2.

Exercises. 1. For the circuit of Figure 7.11a, find analytic expressions for the instantaneous stored energy for the current excitation in Figure 7.13a for $t \geq 0$.

2. Repeat Exercise 1 for Figure 7.13b.

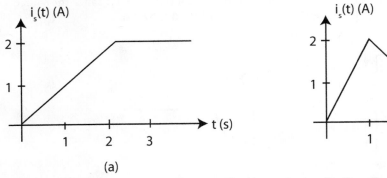
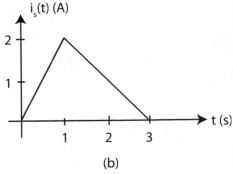

(a)   (b)

FIGURE 7.13 Current excitation for Exercise 2.

ANSWERS:

1.
$$W_1(t) = \begin{cases} 0.25t^2 & 0 \leq t < 2 \\ 1 & 2 \leq t \end{cases} \text{ and } W_2(t) = \begin{cases} t^2 & 0 \leq t < 2 \\ 4 & 2 \leq t \end{cases}$$

2.
$$W_1(t) = \begin{cases} t^2 & 0 \leq t < 1 \\ 0.25(9 - 6t + t^2) & 1 \leq t < 3 \\ 0 & 3 \leq t \end{cases} \text{ and } W_2(t) = \begin{cases} 4t^2 & 0 \leq t < 1 \\ (9 - 6t + t^2) & 1 \leq t < 3 \\ 0 & 3 \leq t \end{cases}$$

EXAMPLE 7.5

For the circuit of Figure 7.14 in which $v_s(t) = \cos(t)$ V for $t \geq 0$ and 0 otherwise, find the input current $i_s(t)$ for $t \geq 0$ and the energy stored in each of the inductors for the intervals [0, t] for $0 \leq t \leq 1$ and [0, t] for $1 \leq t$.

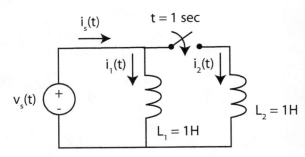

FIGURE 7.14 Parallel inductive circuit with switch in which $v_s(t) = \cos(t)$ V for $t \geq 0$ and 0 otherwise.

SOLUTION

Step 1. Since no voltage is applied to either inductor for $t < 0$, $i_1(0) = 0$. Further, no voltage appears across the second inductor until $t \geq 1$. Hence, $i_2(1) = 0$.

Step 2. Equation 7.2 implies that, for $0 \leq t < 1$,

$$i_s(t) = i_1(t) = \frac{1}{L_1} \int_0^t v_s(\tau) d\tau = \frac{1}{L_1} \int_0^t \cos(\tau) d\tau = \sin(t) \text{ A}$$

Step 3. At $t = 1$, the switch closes. The two inductors are then in parallel, and the source voltage appears across each. Hence, by equation 7.2,

$$i_1(t) = i_1(1) + \int_1^t \cos(\tau) d\tau = \sin(1) + \sin(t) - \sin(1) = \sin(t) \text{ A}$$

Also, equation 7.2 applied to L_2 implies

$$i_2(t) = i_2(1) + \int_1^t \cos(\tau) d\tau = \sin(t) - \sin(1) \text{ A}$$

From the KCL, the input current $i_s(t) = i_1(t) + i_2(t) = 2\sin(t) - \sin(1)$ A for $t \geq 1$.

Step 4. *Compute the energy stored in the inductors over the interval* $[0, t]$. From equation 7.3, it follows that for $0 \leq t \leq 1$, $W_{L1}(0, t) = 0.5 \sin^2(t)$ joules, whereas $W_{L2}(0, t) = 0$.

Step 5. *Compute the energy stored in the inductors over the interval* $[0, t]$ *for* $1 \leq t$. Again from equation 7.3, for $1 \leq t$, $W_{L1}(0, t) = 0.5 \sin^2(t)$ joules and $W_{L2}(0, t) = 0.5[\sin^2(t) - 2 \sin(1) \sin(t) + \sin^2(1)]$ joules.

Exercise. Repeat the calculations of Example 7.5 for $v_s(t) = 2 \sin(t)$ V for $t \geq 0$ and 0 otherwise. ANSWERS: For $0 \leq t < 1$, $W_{L1}(0, t) = [2 - 2 \cos(t)]^2$ J, whereas $W_{L2}(0, t) = 0$; for $1 \leq t$, $W_{L1}(0, t) = [2 - 2 \cos(t)]^2$ J and $W_{L2}(0, t) = [1.0806 - 2 \cos(t)]^2$ J.

---

3. THE CAPACITOR

Definitions and Properties

DEFINITION OF THE CAPACITOR

Like the inductor, the capacitor, denoted by Figure 7.15a, is an energy storage device. Physically, one can think of a capacitor as two metal plates separated by some insulating material (called a **dielectric**) such as air, as illustrated in Figure 7.15b. Placing a voltage across the plates of the capacitor will cause positive charge to accumulate on the top plate and an equal amount of negative charge on the bottom plate. This generates an electric field between the plates that stores energy. Hence, for a capacitor,

$$i_C(t) = \frac{dq(t)}{dt} = C\frac{dv_C(t)}{dt} \qquad (7.5)$$

where $q(t)$ is the accumulated charge on the top plate, which is proportional to the voltage $v_C(t)$ across the plates; thus $q(t) = Cv_C(t)$, with proportionality constant C denoting capacitance and having the unit of Farad (F). One Farad equals 1 amp-sec/volt. The **capacitance** C is a measure of the capacitor's potential to store energy in an electric field.

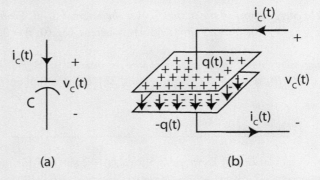

(a)   (b)

FIGURE 7.15 (a) The symbol for the capacitor with conventional voltage and current directions. (b) Illustration of electric field between plates of a parallel-plate capacitor.

Modern-day capacitors take on all sorts of shapes and sizes and materials. In keeping with tradition, the parallel-plate concept remains the customary perspective. Calculating the capacitance of two arbitrarily shaped conducting surfaces separated by a dielectric is, in general, very difficult. Fortunately, the ordinary capacitor of a practical circuit is of the parallel-plate variety, with the plates separated by a thin dielectric. The two plates are often rolled into a tubular form, and the complete structure is sealed.

EXAMPLE 7.6

For the capacitor circuit of Figure 7.16a, compute $i_C(t)$ when $v_{in}(t) = e^{-500t}\sin(1000t)$ V for $t > 0$.

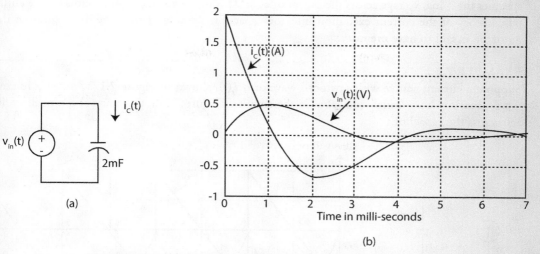

FIGURE 7.16 (a) A 2 mF capacitor connected to a voltage source.
(b) Plots of capacitor voltage and current waveforms.

SOLUTION

A direct application of equation 7.5 yields

$$i_C(t) = C\frac{dv_C(t)}{dt} = -e^{-500t}\sin(1000t) + 2e^{-500t}\cos(1000t) \;\; \text{A}$$

Exercises. 1. In Figure 7.16, suppose $v_{in}(t) = e^{-25t}$ V for $t \geq 0$. Compute $i_C(t)$ for $t > 0$. Sketch (preferably in MATLAB) $i_C(t)$ for $0 \leq t \leq 0.5$ sec.
ANSWER: $-0.05e^{-25t}$ A.

2. Repeat Exercise 1 with $v_{in}(t) = e^{-25t}\cos(100t)$ V for $t \geq 0$ but plot over the time interval $[0, 0.15 \text{ sec}]$.
ANSWER: $-e^{-25t}[0.05\cos(100t) + 0.2\sin(100t)]$ A.

The differential relationship of equation 7.5 has the equivalent integral form

$$v_C(t) = \frac{1}{C}\int_{-\infty}^{t} i_C(\tau)\,d\tau = \frac{1}{C}\int_{-\infty}^{t_0} i_C(\tau)\,d\tau + \frac{1}{C}\int_{t_0}^{t} i_C(\tau)\,d\tau$$

$$(7.6)$$

$$= v_C(t_0) + \frac{1}{C}\int_{t_0}^{t} i_C(\tau)\,d\tau$$

where $v_C(t)$ is in volts, $i_C(t)$ is in amps, and C is in farads, and where we have taken $v_C(-\infty) = 0$ because the capacitor was not manufactured at $t = -\infty$. The time t_0 represents an initial time of interest or significance, e.g., the time when the capacitor is first used in a circuit. The quantity

$$v_C(t_0) = \frac{1}{C}\int_{-\infty}^{t_0} i_C(\tau)\,d\tau$$

specifies the initial voltage across the capacitor at t_0. This initial voltage, $v_C(t_0)$, sums up the entire past history of the current excitation into the capacitor. Because of this, the capacitor, like the inductor, is said to have **memory**.

EXAMPLE 7.7

Suppose a current source with sawtooth waveform $i_s(t)$, shown in Figure 7.17b, drives a relaxed 0.5 F capacitor (zero initial voltage) as in the circuit of Figure 7.17a. Compute and plot the voltage across the capacitor.

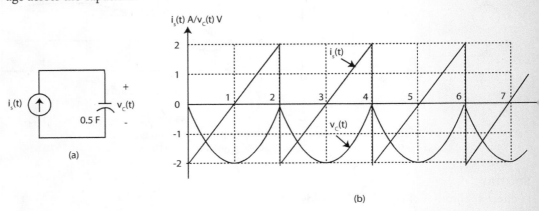

FIGURE 7.17 (a) Current source driving a capacitor.
(b) Sawtooth current waveform and voltage response of a 0.5 F capacitor.

SOLUTION

The input waveform is periodic in that it repeats itself every 2 sec. Therefore, the solution will proceed on a segment-by-segment basis.

Step 1. *Consider the interval* $0 \leq t < 2$. For this interval $i_s(t) = (2t - 2)$ A. With $v_C(0) = 0$, it follows from equation 7.6 that

$$v_C(t) = \frac{1}{C}\int_0^t (2\tau - 2)d\tau = 2(t^2 - 2t) \text{ V for } 0 \leq t < 2$$

Step 2. *Consider the interval* $2 \leq t < 4$. Observe that at $t = 2$, $v_C(2) = 0$; hence, the capacitor voltage over the interval $2 \leq t < 4$ is simply a right-shifted version of the voltage over the first interval. Right-shifting is achieved by replacing t with $t - 2$. In other words,

$$v_C(t) = 2[(t - 2)^2 - 2(t - 2)] \text{ V for } 2 \leq t < 4$$

Step 3. *Consider the general interval* $2k \leq t < 2(k + 1)$. For interval $2k \leq t \leq 2(k + 1)$,

$$v_C(t) = 2[(t - 2k)^2 - 2(t - 2k)], \, k = 0, 1, 2, \dots$$

Lastly, observe that the voltage across the capacitor, as illustrated in Figure 7.17b, is continuous despite the discontinuity of the capacitor current. Again, this follows because the capacitor voltage is the integral (a smoothing operation) of the capacitor current supplied by the source.

Exercise. Consider the capacitor circuit of Figure 7.18. Suppose the current source is $i_s(t) = e^{-t}$ A for $t \geq 0$ and $v_C(0) = 1$ V. Compute the capacitor voltage $v_C(t)$, the resistor voltage $v_R(t)$, and the voltage $v_s(t)$ across the current source for $t \geq 0$.

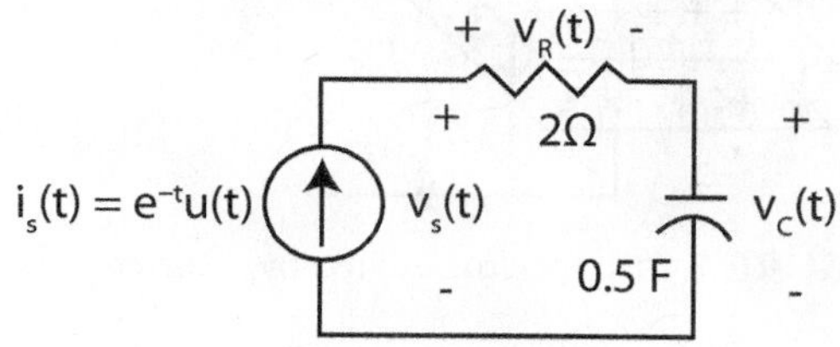

FIGURE 7.18 Series RC circuit driven by a current source for accompanying exercise.

ANSWERS: $v_C(t) = 3 - 2e^{-t}$ for $t \geq 0$, $v_R(t) = 2e^{-t}$ for $t \geq 0$, and, by KVL, $v_s(t) = 3$ V for $t \geq 0$.

It is important to emphasize that the sawtooth current input depicted in Figure 7.17b is a discontinuous function, but the associated voltage waveform is continuous because integration (equation 7.6) is a smoothing operation. This means that the capacitor voltage is a continuous function of t even for discontinuous capacitor currents, provided they are bounded. This observation leads to the **continuity property of the capacitor**: if the current $i_C(t)$ through a capacitor is bounded over the time interval $t_1 \leq t \leq t_2$, then the voltage across the capacitor is continuous for $t_1 < t < t_2$. In particular, for bounded currents, if $t_1 < t_0 < t_2$, then $v_C(t_0^-) = v_C(t_0^+)$, even when $i_C(t_0^-) \neq i_C(t_0^+)$.

At the macroscopic level, there appear to be some exceptions to the continuity property of the capacitor voltage, e.g., when two charged capacitors or one charged and one uncharged capacitor are instantaneously connected in parallel. In such cases, KVL takes precedence and will force an "instantaneous" equality in the capacitor voltages, subject to the principle of conservation of charge, to be discussed shortly. Another example is when capacitors and some independent voltage sources form a loop. When any of the voltage sources has an instantaneous jump, so will the other capacitor voltages. Upon closer examination, however, we see that there is really no exception to the stated continuity rule: it can be shown that in all of the cases where the capacitor voltage jumps instantaneously, an "impulse" current flows in the circuit. Physically, an impulse current is one that is very large (infinite from an ideal viewpoint) and of very short duration. The current is not bounded, and consequently, the capacitor voltage may jump instantaneously. This jump does not violate the rule, which presumes that the current is bounded.

Relationship of Charge to Capacitor Voltage and Current

We have defined the capacitance of a two-terminal device strictly from its terminal voltage-current relationship—the differential equation 7.5 and the integral equation 7.6, which is now repeated:

$$v_C(t) = v_C(t_0) + \frac{1}{C}\int_{t_0}^{t} i_C(\tau)\,d\tau$$

Physically speaking, the integral of $i_C(t)$ over $[t_0, t]$ represents the amount of charge passing through the top wire in Figure 7.19 over $[t_0, t]$.

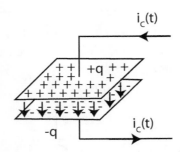

FIGURE 7.19 Capacitor excited by a current.

Because of the insulating material (the dielectric), this charge cannot pass through to the other plate. Instead, a charge of $+q(t)$ is stored on the top plate, as shown in Figure 7.19. By KCL, if $i_C(t)$ flows into the top plate, then $-i_C(t)$ must flow into the bottom plate. This causes a charge of $-q(t)$ to be deposited on the bottom plate. The positive and negative charges on these two plates, separated by the dielectric, produce a voltage drop $v_C(t)$ from the top plate to the bottom plate. For a linear capacitor, the only type studied in this text, the value of $v_C(t)$ is proportional to the charge $q(t)$. The proportionality constant is the capacitance of the device. Specifically,

$$q(t) = Cv_C(t) \tag{7.7}$$

where $q(t)$ is in coulombs, C is in farads, and $v_C(t)$ is in volts. Thus, equation 7.6 has the following physical interpretation: the first term, $v_C(t_0)$, is the capacitor voltage at t_0; the integral in the second term,

$$\int_{t_0}^{t} i_C(\tau)\,d\tau,$$

represents the additional charge transferred to the capacitor during the interval $[t_0, t]$. Dividing this integral by C gives the additional voltage attained by the capacitor during $[t_0, t]$. Therefore, the sum of these two terms, i.e., equation 7.6, is the voltage of the capacitor at t. Since $q(t) = Cv_C(t)$, it follows directly that

$$i_C(t) = \frac{dq(t)}{dt} = C\frac{dv_C(t)}{dt} \tag{7.8}$$

The Principle of Conservation of Charge

It is important in terms of modern trends in circuit applications to further investigate the relationship of charge to capacitor voltages and currents. The principle of **conservation of charge** requires that *the total charge transferred into a junction (or out of a junction) be zero.*[2] This is a direct consequence of KCL. To exemplify, consider the junction of four capacitors shown in Figure 7.20.

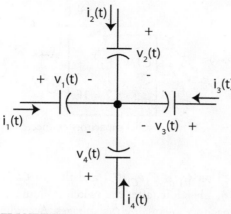

FIGURE 7.20 Junction of four capacitors.

By KCL

$$i_1(t) + i_2(t) + i_3(t) + i_4(t) = 0$$

Since the integral of current with respect to time is charge, the integral of this equation over $(-\infty, t]$ is

$$\int_{-\infty}^{t} \left(i_1(\tau) + i_2(\tau) + i_3(\tau) + i_4(\tau) \right) d\tau = q_1(t) + q_2(t) + q_3(t) + q_4(t) = 0 \qquad (7.9)$$

where $q_k(t)$ is the charge transferred to capacitor k. By equation 7.6, at every instant of time,

$$q_i(t) = C_i v_i(t) \qquad (7.10)$$

which defines the relationship between transported charge, capacitance, and the voltage across the capacitor. Hence, from equations 7.9 and 7.10, at every instant of time,

$$C_1 v_1(t) + C_2 v_2(t) + C_3 v_3(t) + C_4 v_4(t) = 0$$

This simple equation relates voltages, capacitances, and charge transport. The following example provides an application of these ideas.

EXAMPLE 7.8

This example shows that under idealized conditions, capacitor voltages can change instantaneously. Consider the circuit of Figure 7.21, in which $v_{C1}(0^-) = 1$ V and $v_{C2}(0^-) = 0$ V. Find $v_{C1}(t)$ and $v_{C2}(t)$ for $t \geq 0$.

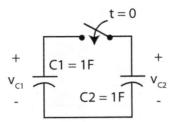

FIGURE 7.21 Two parallel capacitors connected by a switch.

SOLUTION
At $t = 0^-$, the charge stored on C_1 is $C_1 v_{C1}(0^-)$ and that of C_2 is $C_2 v_{C2}(0^-)$. For $t > 0$, KVL requires that $v_{C1}(t) = v_{C2}(t)$. Therefore, after the switch is closed at $t = 0$, some charge must be transferred between the capacitors to equalize the voltages. According to the principle of conservation of charge, the total charge before and after the transfer is the same. Thus, conservation of charge requires that

$$q_1(0^-) + q_2(0^-) = q_1(0^+) + q_2(0^+)$$

Equivalently, $q_1(0^+) - q_1(0^-) + q_2(0^+) - q_2(0^-) = 0$. From equation 7.10,

$$C_1 \left[v_{C1}(0^+) - v_{C1}(0^-) \right] + C_2 \left[v_{C2}(0^+) - v_{C2}(0^-) \right] = 0$$

Since $v_{C1}(0^-) = 1$ V, $v_{C2}(0^-) = 0$, and from KVL $v_{C1}(0^+) = v_{C2}(0^+)$, it follows that

$$C_1 \left[v_{C1}(0^+) - 1 \right] + C_2 \left[v_{C1}(0^+) - 0 \right] = 0$$

Hence, $(C_1 + C_2) v_{C1}(0^+) = 1$ implies that $v_{C1}(0^+) = v_{C2}(0^+) = 0.5$ V.

Exercises. 1. In Example 7.8, make $C_1 = 0.75$ F and $C_2 = 0.25$ F, and compute $v_{C1}(0^+)$.
ANSWER: $v_{C1}(0^+) = 0.75$ V.

2. In Example 7.8, suppose $v_{C1}(0^-) = 10$ V and $v_{C2}(0^-) = -8$ V. Also let $C_1 = 0.75$ F and $C_2 = 0.25$ F. Compute $v_{C1}(0^+)$.
ANSWER: $v_{C1}(0^+) = 5.5$ V.

Example 7.8 is illustrative of a charge transport that is germane to switched capacitor circuits, which are of fundamental importance in the industrial world.

Energy Storage in a Capacitor
As with all devices, the energy stored or utilized in a capacitor is the integral of the power absorbed by the capacitor. The net energy entering the capacitor over the interval $[t_0, t_1]$ is

$$W_C(t_0, t_1) = \int_{t_0}^{t_1} p_C(\tau) d\tau = \int_{t_0}^{t_1} v_C(\tau) i_C(\tau) d\tau$$

$$= C \int_{t_0}^{t_1} \left(v_C(\tau) \frac{dv_C(\tau)}{d\tau} \right) d\tau = C \int_{v_C(t_0)}^{v_C(t_1)} v_C dv_C \qquad (7.11)$$

$$= \frac{1}{2} C \left[v_C^2(t_1) - v_C^2(t_0) \right]$$

for C in farads, v_C in volts, and energy in joules (J). From equation 7.11, the change in energy stored in the capacitor over the interval $[t_0, t_1]$ depends only on the values of the capacitor voltages at times t_0 and t_1, i.e., on $v_C(t_0)$ and $v_C(t_1)$. This means that the change in stored energy is independent of the particular voltage waveform between t_0 and t_1. If the voltage waveform is periodic, i.e., if $v_C(t) = v_C(t + T)$ for some $T > 0$, then over any time interval $[t, t + T]$, the change in the stored energy in the capacitor is zero because $v_C(t_0 + T) = v_C(t_1) = v_C(t_0)$ forces equation 7.11 to zero. Analogous to the inductor, for all periodic voltages, the capacitor stores energy and then returns it to the circuit and is thus called a lossless device.

As with the inductor, it is convenient to define the **instantaneous stored energy** in a capacitor as

$$W_C(t) = \frac{1}{2} C v_C^2(t) \qquad (7.12)$$

which is really the integral of power over the interval $(-\infty, t]$, assuming that all voltages and currents are zero at $t = -\infty$.

EXAMPLE 7.9

Consider the circuit of Figure 7.22, in which $v_C(0) = 0$. It is known that for $t \geq 0$, the source current is $i_s(t)$ and the voltage across the capacitor is $v_C(t) = 4R \left(1 - e^{-\frac{t}{RC}} \right)$ V for $t \geq 0$. Compute (i) the energy, in joules, stored in the capacitor for $t > 0$, (ii) $i_C(t)$, and (iii) $i_s(t)$.

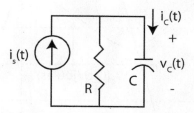

FIGURE 7.22 Parallel RC circuit.

SOLUTION

(i) Since $v_C(0) = 0$, from equation 7.11 (or 7.12),

$$W_C(0,t) = \frac{1}{2} C v_C^2(t) = 8CR^2 \left(1 - e^{-\frac{t}{RC}} \right)^2$$

(ii) To find the capacitor current, recall

$$i_C(t) = C \frac{dv_C(t)}{dt} = 4 \frac{RC}{RC} e^{-\frac{t}{RC}} = 4 e^{-\frac{t}{RC}} \text{ A}$$

(iii) To find $i_s(t)$, we first compute $i_R(t)$, flowing from top to bottom:

$$i_R(t) = \frac{v_C(t)}{R} = 4\left(1 - e^{-\frac{t}{RC}}\right) \text{ A}$$

Thus

$$i_s(t) = i_R(t) + i_C(t) = 4\left(1 - e^{-\frac{t}{RC}}\right) + 4e^{-\frac{t}{RC}} = 4 \text{ A}$$

EXAMPLE 7.10

For the circuit of Figure 7.23a, it is known that the voltage across the capacitor is $v_C(t) = 20\sin(2t + \pi/6)$ V for $t \geq 0$. Compute and plot the instantaneous power absorbed by the capacitor and the energy stored by the capacitor during the time interval $[0, t]$.

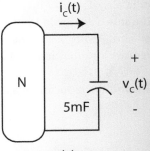

(a)

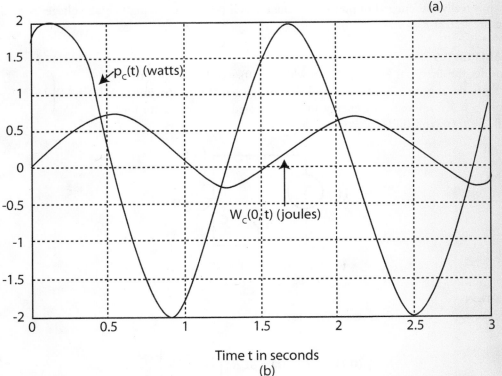

Time t in seconds
(b)

FIGURE 7.23 (a) Capacitor with known voltage $v_C(t)$ connected to a network N. (b) Plot of power, $p_C(t)$, and the net energy, $W_C(0, t)$, stored over the interval $[0, t]$.

SOLUTION

Step 1. *Compute $i_C(t)$.* From equation 7.5, for $t \geq 0$

$$i_C(t) = C\frac{dv_C(t)}{dt} = 0.2\cos\left(2t + \frac{\pi}{6}\right) \text{ A}$$

Step 2. *Compute $p_C(t)$.* By direct multiplication and a standard trig identity,

$$p_C(t) = v_C(t)i_C(t) = 20\sin\left(2t + \frac{\pi}{6}\right) \times 0.2\cos\left(2t + \frac{\pi}{6}\right) = 2\sin\left(4t + \frac{\pi}{3}\right) \text{ watts}$$

Step 3. *Compute $W_C(0, t)$.* From equation 7.11 with $v_C(0) = 20\sin(\pi/6) = 10$ V, we obtain

$$W_C(0,t) = 0.5\,C\,v_C^2(t) - 0.5\,C\,v_C^2(0) = \sin^2\left(2t + \frac{\pi}{6}\right) - 0.25 \text{ J}$$

Plots of $p_C(t)$ and $W_C(0, t)$ are given in Figure 7.23b. Notice that $W_C(0, t)$ can be negative, because $W_C(-\infty, 0) = 0.25$ joules, meaning that at $t = 0$, there is an initial stored energy that can be returned to the circuit at a later time . Figure 7.23b substantiates this.

4. SERIES AND PARALLEL INDUCTORS AND CAPACITORS

Series Inductors

Just as resistors in series combine to form an equivalent resistance, inductors in series combine to form an equivalent inductance. As it turns out, series inductances combine in the same way as series resistances.

EXAMPLE 7.11.

Compute the equivalent inductance of the series connection of three inductors illustrated in Figure 7.24. Then find the voltages v_{Li} as a fraction of the applied voltage v_{Leq}.

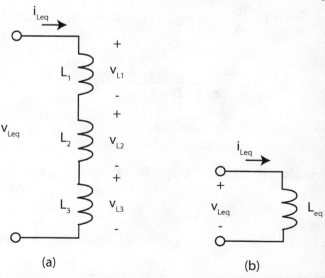

(a)   (b)

FIGURE 7.24 (a) Series connection of three inductors. (b) Equivalent inductance.

SOLUTION

First we must answer the question of what it means to be an equivalent inductance. Earlier, we defined the inductor in terms of its terminal voltage-current relationship. Two 2-terminal inductor circuits have the same inductance if each circuit has the same terminal voltage-current relationship as defined in equation 7.1.

Step 1. The voltage labeled v_{Leq} appears across the series connection, and, by KCL, the current i_{Leq} flows through each of the inductors, i.e., $i_{Leq} = i_{L1} = i_{L2} = i_{L3}$. The equivalent inductance, L_{eq}, is defined by the relationship

$$v_{Leq} = L_{eq}\frac{di_{Leq}}{dt} \qquad (7.13)$$

Our goal is to express L_{eq} in terms of L_1, L_2, and L_3.

Step 2. *Find v_{Leq} in terms of i_{Leq}.* To obtain such an expression, observe that, by KVL,

$$v_{Leq} = v_{L1} + v_{L2} + v_{L3}$$

Since each inductor satisfies the *v-i* relationship

$$v_{Li} = L_i\frac{di_{Li}}{dt},$$

it follows that

$$v_{Leq} = (L_1 + L_2 + L_3)\frac{di_{Leq}}{dt}.$$

Hence, the series inductors of figure 7.24a can be replaced by a single inductor with inductance $L_{eq} = L_1 + L_2 + L_3$.

Finally, since $v_{Li} = L_i\dfrac{di_{Li}}{dt} = L_i\dfrac{di_{Leq}}{dt}$ and $v_{Leq} = (L_1 + L_2 + L_3)\dfrac{di_{Leq}}{dt}$, it follows that

$$v_{Li} = L_i\frac{di_{Leq}}{dt} = \frac{L_i}{(L_1 + L_2 + L_3)}v_{Leq}$$

which is analogous to the voltage divider formula for resistances.

Exercises. 1. If, in Example 7.11, $L_1 = 2$ mH, $L_2 = 5$ mH, and $L_3 = 1$ mH, find L_{eq}.
ANSWER: $L_{eq} = 8$ mH.

2. Find v_{L2} in terms of v_{Leq}.

ANSWER: $v_{L2} = \dfrac{5}{8}v_{Leq}$.

Extension of the formulas in the above example to n inductors is fairly clear, and we state the results without rigorous proof: the formula for series inductances is

$$L_{eq} = L_1 + L_2 + ... + L_n \qquad (7.14a)$$

and the formula for voltage division of series inductances is

$$v_{Li} = \frac{L_i}{L_1 + L_2 + ... + L_n} v_{Leq} \qquad (7.14b)$$

Inductors in Parallel

The same basic question as with inductors in series arises with a parallel connection of inductors: what is the equivalent inductance? Rather than derive the general formula, let us consider the case of three inductors in parallel, as illustrated in Figure 7.25a.

EXAMPLE 7.12

For this example our goal is to show that the equivalent inductance of the circuit of Figure 7.25a is given by the reciprocal of the sum-of-reciprocals formula,

$$L_{eq} = \frac{1}{\dfrac{1}{L_1} + \dfrac{1}{L_2} + \dfrac{1}{L_3}} \qquad (7.15)$$

We then show a formula for current division.

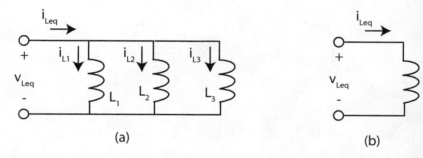

(a)   **(b)**

FIGURE 7.25 (a) Parallel connection of three inductors. (b) Equivalent inductance.

SOLUTION

Once again, equation 7.13 defines the relationship for the equivalent inductance:

$$v_{Leq} = L_{eq} \frac{di_{Leq}}{dt}$$

The goal is to construct L_{eq} in terms of L_1, L_2, and L_3 in a way that satisfies equation 7.13. This will produce equation 7.15.

Step 1. Write KCL for the parallel connection shown in Figure 7.25a. Here, by KCL,

$$i_{Leq} = i_{L1} + i_{L2} + i_{L3}$$

Differentiating both sides with respect to time yields

$$\frac{di_{Leq}}{dt} = \frac{di_{L1}}{dt} + \frac{di_{L2}}{dt} + \frac{di_{L3}}{dt}$$

Step 2. Find $\frac{di_{Lk}}{dt}$ in terms of v_{Lk} and L_k. From equation 7.1, for each inductor

$$\frac{di_{Li}}{dt} = \frac{v_{Li}}{L_i}$$

Substituting into the result of step 1 and noting that $v_{Leq} = v_{L1} = v_{L2} = v_{L3}$ yields

$$\frac{di_{Leq}}{dt} = \frac{v_{L1}}{L_1} + \frac{v_{L2}}{L_2} + \frac{v_{L3}}{L_3} = \left(\frac{1}{L_1} + \frac{1}{L_2} + \frac{1}{L_3}\right) v_{Leq}$$

This has the form of equation 7.13, which implies equation 7.15, i.e.,

$$L_{eq} = \frac{1}{\frac{1}{L_1} + \frac{1}{L_2} + \frac{1}{L_3}}$$

To generate a current division formula we first note that $v_{Leq} = v_{L1} = v_{L2} = v_{L3}$,

$$i_{Lk}(t) = \frac{1}{L_k} \int_{-\infty}^{t} v_{Lk}(\tau)d\tau = \frac{1}{L_k} \int_{-\infty}^{t} v_{eq}(\tau)d\tau \text{ and } L_{eq} \, i_{Leq}(t) = \int_{-\infty}^{t} v_{Leq}(\tau)d\tau$$

Thus

$$i_{Lk}(t) = \frac{1}{L_k} \int_{-\infty}^{t} v_{eq}(\tau)d\tau = \frac{L_{eq}}{L_k} i_{Leq}(t) = \frac{\frac{1}{L_k}}{\frac{1}{L_1} + \frac{1}{L_2} + \frac{1}{L_3}} i_{Leq}(t)$$

Exercises. 1. If, in Example 7.12, L_1 = 2.5 mH, L_2 = 5 mH, and L_3 = 1 mH, find L_{eq}.
ANSWER: L_{eq} = 0.625 mH.

2. Find i_{L2} in terms of i_{Leq}.
ANSWER: i_{L2} = 0.125 i_{Leq}

The above arguments easily generalize. Suppose there are n inductors, $L_1, L_2, \ldots, L_n$, connected in parallel. Then the equivalent inductance is given by the reciprocal of the sum-of-reciprocals formula,

$$L_{eq} = \frac{1}{\frac{1}{L_1} + \frac{1}{L_2} + \ldots + \frac{1}{L_n}} \tag{7.16a}$$

and the current division formula,

$$i_{Lk}(t) = \frac{\dfrac{1}{L_k}}{\dfrac{1}{L_1}+\dfrac{1}{L_2}+L+\dfrac{1}{L_n}}\, i_{Leq}(t) \tag{7.16b}$$

Exercise. For two inductors L_1 and L_2 in parallel, show that the equivalent inductance satisfies the formula

$$L_{eq} = \frac{L_1 L_2}{L_1 + L_2} \tag{7.17}$$

Series-Parallel Combinations

This subsection examines series-parallel connections of inductors. This allows us to use the formulas developed above in an iterative way.

EXAMPLE 7.13

Find the equivalent inductance, L_{eq}, of the circuit of Figure 7.26.

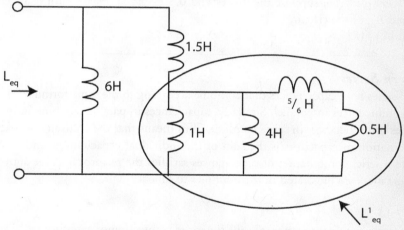

FIGURE 7.26 Series-parallel combinations of inductors.

SOLUTION

Step 1. In the circuit of Figure 7.26, several inductors are enclosed by an ellipse. Let L^1_{eq} denote the equivalent inductance of this combination. Observe that the series inductance of the 5/6 H and 0.5 H inductors equals 4/3 H. This inductance is in parallel with a 1 H and a 4 H inductance. Hence,

$$L^1_{eq} = \frac{1}{\dfrac{1}{1}+\dfrac{1}{4}+\dfrac{3}{4}} = \frac{1}{2}\ \text{H}$$

Step 2. The equivalent circuit at this point is given by Figure 7.27. This figure consists of a series combination of a 1.5 H and a 0.5 H inductor connected in parallel with a 6 H inductor. It follows that

$$L_{eq} = \cfrac{1}{\cfrac{1}{0.5+1.5} + \cfrac{1}{6}} = \frac{6}{4} = 1.5 \text{ H}$$

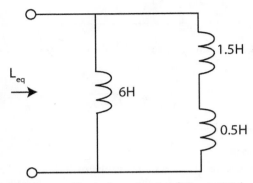

FIGURE 7.27 Equivalent circuit of Figure 7.26.

Exercise. In Example 7.13, suppose the 5/6 H and 0.5 H inductors are both changed to 0.4 H inductors. Find L_{eq} of the circuit.
ANSWER: 1.443 H.

Capacitors in Series

Capacitors in series have capacitances that combine according to the same formula for combining resistances or inductances in parallel. Similarly, capacitances in parallel combine in the same way that resistances or inductances in series combine. This means that the equivalent capacitance of a parallel combination of capacitors is the sum of the individual capacitances, and the equivalent capacitance of a series combination of capacitances satisfies the reciprocal of the sum-of-reciprocals rule. These ideas are illustrated in the examples to follow.

EXAMPLE 7.14

Compute the equivalent capacitance, C_{eq}, of the series connection of capacitors in Figure 7.28a.

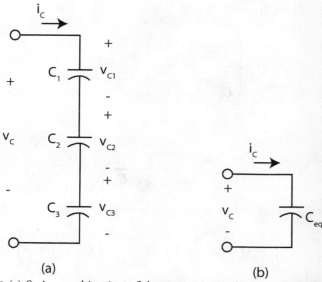

(a)

(b)

FIGURE 7.28 (a) Series combination of three capacitors. (b) Equivalent capacitance, C_{eq}.

SOLUTION

The equivalent capacitance denoted in Figure 7.28b is defined implicitly by the current-voltage terminal conditions according to equation 7.5, i.e.,

$$i_C = C_{eq} \frac{dv_C}{dt}$$

Our goal is to express this same terminal $v\text{-}i$ relationship in terms of the capacitances, C_1, C_2, and C_3. After this we set forth a formula for voltage division.

Step 1. *Set forth the i-v relationship for each capacitor.* For each capacitor, $k = 1, 2, 3$,

$$i_{Ck} = C_k \frac{dv_{Ck}}{dt}$$

But, by KCL, $i_C = i_{Ck}$. Hence,

$$\frac{dv_{Ck}}{dt} = \frac{i_C}{C_k}$$

Step 2. *Apply KVL.* From KVL,

$$v_C = v_{C1} + v_{C2} + v_{C3}$$

Differentiating this expression with respect to time and using the result of step 1 yields

$$\frac{dv_C}{dt} = \frac{dv_{C1}}{dt} + \frac{dv_{C2}}{dt} + \frac{dv_{C3}}{dt} = \left(\frac{1}{C_1} + \frac{1}{C_2} + \frac{1}{C_3} \right) i_C$$

Step 3. *Compute C_{eq}.* From the result of step 2, solve for i_C to obtain

$$i_C = \left(\frac{1}{\dfrac{1}{C_1} + \dfrac{1}{C_2} + \dfrac{1}{C_3}} \right) \frac{dv_C}{dt} = C_{eq} \frac{dv_C}{dt}$$

It follows that

$$C_{eq} = \frac{1}{\dfrac{1}{C_1} + \dfrac{1}{C_2} + \dfrac{1}{C_3}}$$

To set forth a formula for voltage division, we first note that $i_C = i_{C1} = i_{C2} = i_{C3}$,

$$v_{Ck}(t) = \frac{1}{C_k} \int_{-\infty}^{t} i_{Ck}(\tau)d\tau = \frac{1}{C_k} \int_{-\infty}^{t} i_C(\tau)d\tau \text{ and } C_{eq} v_C(t) = \int_{-\infty}^{t} i_C(\tau)d\tau$$

Thus

$$v_{Ck}(t) = \frac{1}{C_k} \int_{-\infty}^{t} i_C(\tau)d\tau = \frac{C_{eq}}{C_k} v_C(t) = \frac{\dfrac{1}{C_k}}{\dfrac{1}{C_1} + \dfrac{1}{C_2} + \dfrac{1}{C_3}} v_C(t)$$

Exercises. 1. In Example 7.14, suppose C_1 = 5 µF, C_2 = 20 µF, and C_3 = 16 µF. Compute C_{eq}.
ANSWER: 3.2 µF.

2. Find v_{C2} in terms of v_C
ANSWER: v_{C2} = $0.16v_C$

Generalizing the result of Example 7.14, we may say that capacitors in series satisfy the reciprocal of the sum-of-reciprocals rule. Thus, for n capacitors C_1, C_2, ... , C_n, connected in series, the equivalent capacitance is

$$C_{eq} = \frac{1}{\dfrac{1}{C_1} + \dfrac{1}{C_2} + ... + \dfrac{1}{C_n}} \qquad (7.18a)$$

and the general voltage division formula is

$$v_{Ck}(t) = \frac{\dfrac{1}{C_k}}{\dfrac{1}{C_1} + \dfrac{1}{C_2} + ... + \dfrac{1}{C_n}} v_C(t) \qquad (7.18b)$$

Exercise. Show that if two capacitors C_1 and C_2 are connected in series, then

$$C_{eq} = \frac{C_1 C_2}{C_1 + C_2} \qquad (7.19)$$

Capacitors in Parallel

If two capacitors are connected in parallel as in Figure 7.29a, there results an equivalent capacitance $C_{eq} = C_1 + C_2$ and a simple current division formula to be derived.

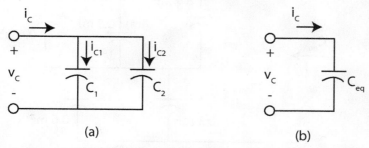

(a)   **(b)**

FIGURE 7.29 (a) Parallel combination of two capacitors. (b) Equivalent capacitance, C_{eq}.

Since the voltage v_C appears across each capacitor, and since $i_C = i_{C1} + i_{C2}$, by KCL it follows that

$$i_C = i_{C1} + i_{C2} = C_1 \frac{dv_C}{dt} + C_2 \frac{dv_C}{dt} = (C_1 + C_2) \frac{dv_C}{dt}$$

Hence,

$$C_{eq} = C_1 + C_2$$

One surmises from the above example that, in general, capacitors in parallel have capacitances that add. And, indeed, this is the case: if there are n capacitors $C_1, C_2, \ldots, C_n$ in parallel, the equivalent capacitance is

$$C_{eq} = C_1 + C_2 + \ldots + C_n \qquad (7.20a)$$

Exercise. Show that in the above derivation

$$i_{Ck} = \frac{C_k}{C_{eq}} i_C = \frac{C_k}{C_1 + C_2} i_C$$

and that for n capacitors in parallel,

$$i_{Ck} = \frac{C_k}{C_1 + C_2 + \ldots + C_n} i_C \qquad (7.20b)$$

Series-Parallel Combinations

We round out our discussion of capacitance by considering a simple series-parallel interconnection.

EXAMPLE 7.15

Consider the circuit of Figure 7.30. Compute the equivalent capacitance, C_{eq}.

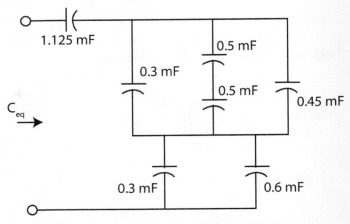

FIGURE 7.30 Series-parallel combination of capacitors.

SOLUTION

Step 1. *Combine series capacitances.* Observe that the two series capacitances of 0.5 mF and 0.5 mF combine to make a 0.25 mF capacitance.

Step 2. *Combine parallel capacitances.* First, as a result of step 1, the three capacitances, 0.3 mF, 0.25 mF, and 0.45 mF, add to an equivalent capacitance of 1 mF. Further, the two parallel capacitances, 0.3 mF and 0.6 mF, at the bottom of the circuit, add to make a 0.9 mF capacitance. The new equivalent circuit is shown in Figure 7.31.

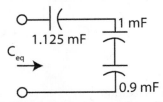

FIGURE 7.31 Circuit equivalent to that of Figure 7.30.

Step 3. *Combine series capacitances.* From equation 7.18,

$$C_{eq} = \frac{1}{\dfrac{1}{1.125} + \dfrac{1}{1} + \dfrac{1}{0.9}} = \frac{1}{3}\,\text{mF}$$

Exercise. Suppose the two 0.5 mF capacitors in Figure 7.30 are changed to 2.5 mF capacitors. Find the new C_{eq}.
ANSWER: 0.4 mF.

5. SMOOTHING PROPERTY OF A CAPACITOR IN A POWER SUPPLY

As mentioned in the chapter opener, a power supply converts a sinusoidal input voltage to an almost constant dc output voltage. Such devices are present in televisions, transistor radios, stereos, computers, and a host of other household electronic gadgets. Producing a truly constant dc voltage from a sinusoidal source is virtually impossible, so engineers design special circuits called **voltage regulators** that generate a voltage with only a small variation between set limits for a given range of variation in load. The voltage regulator is a precision device whose input must be fairly smooth for proper operation. A capacitor can provide a crude, inexpensive **smoothing** function that is often sufficient for the task. This section explores the design of a capacitive smoothing circuit. In practice, such a circuit is used only for low-power applications.

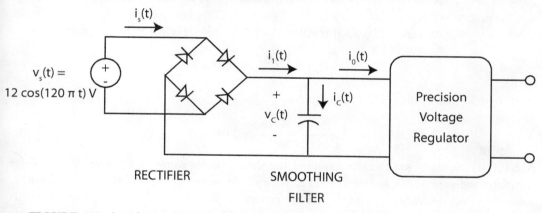

FIGURE 7.32 Simple power supply with capacitive smoothing for low- power applications.

Consider, for example, the circuit shown in Figure 7.32. The four (ideal) diodes are arranged in a configuration called a *full-wave bridge rectifier circuit*. An ideal **diode** allows current to pass only in the direction of the arrow. The diode configuration ensures that $i_1(t)$ remains positive, regardless of the sign of the source current. Specifically, the diodes ensure that $i_1(t) = |i_s(t)|$. Using the integral relationship (equation 7.6) of the capacitor voltage and current, it follows that

$$v_C(t) = v_C(t_0) + \frac{1}{C}\int_{t_0}^{t} i_C(\tau)d\tau = v_C(t_0) + \frac{1}{C}\int_{t_0}^{t}\left[|i_s(\tau)| - i_0(\tau)\right]d\tau \qquad (7.21)$$

Because of the difference $|i_s(t)| - i_0(t)$ inside the integrand of the integral, $i_s(t)$ tends to increase the capacitor voltage, whereas $i_0(t)$ tends to decrease the capacitor voltage. Further, because the diodes are assumed ideal, it follows that

$$v_C(t) \geq |v_s(t)| \qquad (7.22)$$

To see this, suppose the opposite were true; i.e., suppose $|v_s(t)| > v_C(t)$. One of the diodes would then have a positive voltage across it in the direction of the arrow. The diode is said to be *forward biased*. But this is impossible, because an ideal diode behaves like a short circuit when forward biased. The consequence is that $v_C(t)$ will be 12 V whenever $|v_s(t)|$ is 12 V. This occurs every 1/120 of a second. Thus, the rectifier output will recharge the capacitor every 1/120 of a second. Between charging times, the current, $i_0(t)$, will tend to discharge the capacitor and diminish its voltage.

The design problem for the capacitive smoothing circuit is to select a value for C that guarantees that $v_C(t)$ is sufficiently smooth to ensure proper operation of the voltage regulator. Here, "sufficiently smooth" means that the maximum and minimum voltages differ by less than a prescribed amount. To be specific, suppose that $v_C(t)$ must remain between 8 V and 12 V. Recall that $i_s(t)$ tends to increase the capacitor voltage, while $i_0(t)$ tends to decrease it. The design requires selecting a value for C to ensure that $i_s(t)$ can keep up with $i_0(t)$, so that the capacitor voltage remains fairly constant. The value for $i_0(t)$ is obtained from the specification sheet of the voltage regulator. Suppose this value is a constant 1 A. It remains to select C so as to ensure that $v_C(t)$ remains above 8 V between charging times. From equation 7.21, it is necessary that

$$v_C(t) = v_C(t_0) + \frac{1}{C}\int_{t_0}^{t} \left[|i_s(\tau)| - i_0(\tau)\right]d\tau \geq 8$$

Now we need consider only values for t between 0 and $1/120$, because the capacitor will recharge and the process will repeat itself every $1/120$ of a second. Thus, because $i_s(t)$ will only increase the capacitor voltage, to ensure that $v_C(t)$ remains above 8 V, it is sufficient to require that

$$v_C(t_0) - \frac{1}{C}\int_0^{\frac{1}{120}} i_0(\tau)d\tau \geq 8$$

With $i_0(t) = 1$ and $v_C(0) = 12$, it follows that

$$C \geq \frac{1 \text{ A} \times \frac{1}{120}\text{ sec}}{4 \text{ V}} = 2.083 \text{ mF}$$

A 2,100 µF capacitor satisfies this requirement. A method for computing the capacitor voltage waveform is described in Chapter 22 of of 2nd edition. However, using SPICE or one of the other available circuit simulation programs, one can generate a plot of the time-varying capacitor voltage produced by this circuit, as shown in Figure 7.33. In the figure, it is seen that the capacitor voltage varies between 12 and 9.02 V, which is smaller than the allowed variation of (12 − 8) V. Two factors contribute to this conservative design: (1) we used C = 2,100 µF instead of the calculated value, C = 2,083 µF, and (2) the increase in the capacitor voltage due the charging current i_s is not included in the calculation.

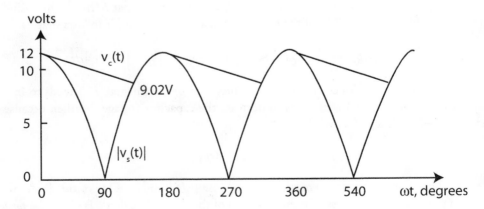

FIGURE 7.33 Time-varying capacitor voltage generated by the circuit in Figure 7.32 when C = 2,100 µF.

The preceding brief introduction made several simplifying assumptions to clarify the basic use of a capacitor as a smoothing or filtering device. Practical power supply design is a challenging field. A complete design would need to consider many other issues, some of which are the nonzero resistance of the source, the non-ideal nature of the diodes, the current-handling ability of the components, protection of the components from high-voltage transients, and heat-sinking of the components.

6. SUMMARY

This chapter has introduced the notions of a capacitor and an inductor, each of which is a lossless energy storage device whose voltage and current satisfy a differential equation. The inductor has a voltage proportional to the derivative of the current through it; the constant of proportionality is the inductance L. The capacitor has a current proportional to the derivative of the voltage across it; the constant of proportionality is the capacitance C. It is interesting to observe that the roles of voltage and current in the capacitor are the reverse of their roles in the inductor. Because of this reversal, the capacitor and the inductor are said to be dual devices.

That the (ideal) inductor and the (ideal) capacitor are lossless energy storage devices means that they can store energy and deliver it back to the circuit, but they can never dissipate energy as does a resistor. The inductor stores energy in a surrounding magnetic field, whereas the capacitor stores energy in an electric field between its conducting surfaces. Unlike energy in a resistor, the energy stored in an inductor over an interval $[t_0, t_1]$ is dependent only on the inductance L and the values of the inductor current $i_L(t_0)$ and $i_L(t_1)$. Likewise, the energy stored in a capacitor over an interval $[t_0, t_1]$ is dependent only on the capacitance C and the values of the capacitor voltage $v_C(t_0)$ and $v_C(t_1)$.

Both the inductor and the capacitor have memory. The inductor has memory because at a particular time t_0, the inductor current depends on the past history of the voltage across the inductor. The capacitor has a voltage at, say, time t_0 that depends on the past current excitation to the capacitor. The concept of memory stems from the fact that the inductor current is proportional to the integral of the voltage across the inductor and the capacitor voltage is proportional to the integral of the current through the capacitor. This integral relationship gives rise to the important properties of the continuity of the inductor current and the continuity of the capacitor voltage under bounded excitations.

The dual capacitor and inductor relationships are set forth in Table 7.1.

Finally, we investigated the smoothing action of a capacitor in a power supply.

TABLE 7.1. Summary of the Dual Relationships of the Capacitor and Inductor

$i_C(t) \rightarrow$ + $v_C(t)$ –	$i_L(t) \rightarrow$ + $v_L(t)$ –
$i_C(t) = C\dfrac{dv_C(t)}{dt}$	$v_L(t) = L\dfrac{di_L(t)}{dt}$
$v_C(t) = v_C(t_0) + \dfrac{1}{C}\displaystyle\int_{t_0}^{t} i_C(\tau)\,dt$	$i_L(t) = i_L(t_0) + \dfrac{1}{L}\displaystyle\int_{t_0}^{t} v_L(\tau)\,d\tau$
$W_C(t_0,t_1) = \dfrac{1}{2}Cv_C^2(t_1) - \dfrac{1}{2}Cv_C^2(t_0)$	$W_L(t_0,t_1) = \dfrac{1}{2}Li_L^2(t_1) - \dfrac{1}{2}Li_L^2(t_0)$

7. TERMS AND CONCEPTS

Bounded voltage or current: voltage or current signal whose absolute value remains below some fixed finite constant for all time.

Capacitance of a pair of conductors: a property of conductors separated by a dielectric that permits the storage of electrically separated charge when a potential difference exists between the conductors. Capacitance is measured in stored charge per unit of potential difference between the conductors.

Capacitor (linear): a two-terminal device whose current is proportional to the time derivative of the voltage across it.

Coil: another name for an inductor.

Conservation-of-charge principle: principle that the total charge transferred into a junction (or out of a junction) is zero.

Continuity property of the capacitor: property such that if the current $i_C(t)$ through a capacitor is bounded over the time interval $t_1 \le t \le t_2$, then the voltage across the capacitor is continuous for $t_1 < t < t_2$. In particular, if $t_1 < t_0 < t_2$, then $v_C(t_0^-) = v_C(t_0^+)$, even when $i_C(t_0^-) \ne i_C(t_0^+)$.

Continuity property of the inductor: property such that if the voltage $v_L(t)$ across an inductor is bounded over the time interval $t_1 \le t \le t_2$, then the current through the inductor is continuous for $t_1 < t < t_2$. In particular, if $t_1 < t_0 < t_2$, then $i_L(t_0^-) = i_L(t_0^+)$, even when $v_L(t_0^-) \ne v_L(t_0^+)$.

Coulomb: quantity of charge that, in 1 second, passes through any cross section of a conductor maintaining a constant 1 A current flow.

Dielectric: an insulating material often used between two conducting surfaces to form a capacitor.

Farad: a measure of capacitance in which a charge of 1 coulomb produces a 1 V potential difference.

Faraday's law of induction: law asserting that, for a coil of wire sufficiently distant from any magnetic material, such as iron, the voltage induced across the coil by a time-varying current is proportional to the time derivative of the current; the constant of proportionality, denoted L, is the inductance of the coil. Faraday's law is usually stated in terms of flux and flux linkages, which are discussed in physics texts.

Henry: the unit of inductance; equal to 1 V-sec/amp.

Inductance: property of a conductor and its local environment (a coil with an air core or iron core) that relates the time derivative of a current through the conductor to an induced voltage across the ends of the conductor.

Inductor (linear): a two-terminal device whose voltage is proportional to the time derivative of the current through it.

Instantaneous power: $p(t) = v(t)i(t)$, in watts when $v(t)$ is in volts and $i(t)$ in amps.

Lossless device: device in which energy can only be stored and retrieved and never dissipated.

Lossy device: a device, such as a resistor (with positive R), that dissipates energy as some form of heat or as work.

Maxwell's equations: a set of mathematical equations governing the properties of electric and magnetic fields and their interaction.

Memory: property of a device whose voltage or current at a particular time depends on the past operational history of the device; e.g., the current through an inductor at time t_0 depends on the history of the voltage excitation across the inductor for $t \leq t_0$.

Unbounded voltage or current: a voltage or current whose value approaches infinity as it nears some instant of time, possibly $t = \infty$.

Voltage regulator: circuit that produces a voltage having only a small variation between set limits for a given range of load variation from a fairly smooth input signal.

---

[1] The word "stored" emphasizes that the energy in the inductor is not dissipated as heat and can be recovered by the circuit, whereas the word "absorbed" is used to mean that the energy cannot be returned to the circuit. In a resistor, energy absorbed is dissipated as heat.

[2] More generally, conservation of charge says that the total charge transferred into a Gaussian surface (or out of a Gaussian surface) is zero.

Problems

THE INDUCTOR AND ITS PROPERTIES

1. If the length of a single-layer air coil is greater than or equal to 0.4 times its diameter, then its inductance is approximately given by the formula

$$L = \frac{4 \times 10^{-5} (diameter)^2 (\# \; of \; turns)^2}{18 (diameter) + 40 (lengths)}$$

where L is in henries, and the diameter and length of the coil are in meters. A 2 cm diameter coil has 48 turns wound at 12 turns/cm. Compute the approximate value of the inductance.

CHECK: $18 \; \mu H \leq L \leq 20 \; \mu H$.

2. (a) Find and plot for $0 < t < 5$ sec the inductor voltage $v_L(t)$ for the circuit of Figure P7.2a driven by the current waveform of Figure P7.2b.
   (b) Find and plot the instantaneous stored energy.
   (c) Find and plot the stored energy $W(1,t)$ as a function of time for $5 \geq t \geq 1$.

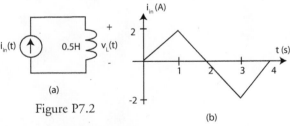

(a)

Figure P7.2

(b)

3. Repeat Problem 2 for:
   (a) $i_{in}(t) = 4te^{-2t} u(t)$, and
   (b) the waveform sketched in Figure P7.3.

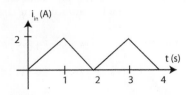

Figure P7.3

4. (a) For $i_s(t) = 10\sin(2000t)$ mA in Figure P7.4, calculate and sketch $i_{out}(t)$ for $0 \leq t \leq 15$ ms assuming both inductor currents are zero at $t = 0$.
   (b) What is the instantaneous power delivered by the dependent source?
   (c) Compute and sketch the energy stored in the 2 mH inductor for $0 \leq t \leq 15$ msec.

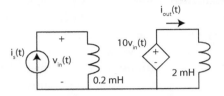

Figure P7.4

5. For $v_S(t)$ sketched in Figure P7.5a, compute and sketch $v_{out}(t)$ for the circuit of Figure P7.5b. What is the instantaneous power delivered by the dependent source?

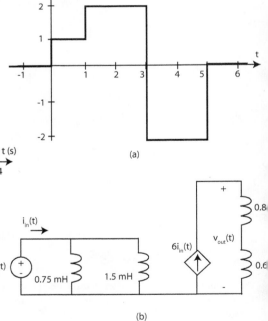

(a)

(b)

Figure P7.5

6. (a) Find and plot for $0 \leq t \leq 6$ sec the inductor current $i_L(t)$ for the circuit of Figure

P7.6a driven by the voltage waveform of Figure P7.6b.

(b) Find and plot the instantaneous stored energy.

(c) Find and plot the stored energy $W(1,t)$ as a function of time for $5 \geq t \geq 1$.

(d) Find and plot $v_1(t)$

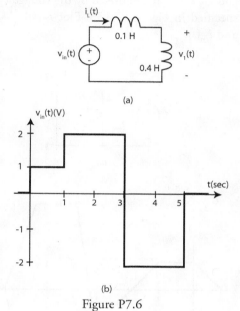

(a)

(b)

Figure P7.6

7. Repeat Problem 6 for
   (a) $v_{in}(t) = 4e^{-2t}u(t)$, and
   (b) the voltage waveform in Figure P7.7.

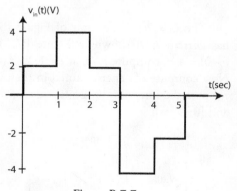

Figure P 7.7

8. For the circuit in Figure P7.8a, suppose $L_1 = 0.8$ H and $L_2 = 0.2$ H. Compute and plot the waveforms $i_1(t)$ and $i_{in}(t)$ for

(a) the voltage waveform sketched in Figure P7.8b, and

(b) the voltage waveform sketched in Figure P7.8c.

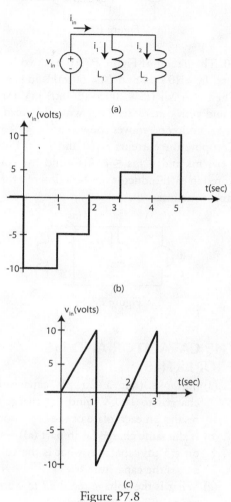

(a)

(b)

(c)

Figure P7.8

9. Consider the circuit in Figure P7.9 in which $L_1 = 0.2$ H, $L_2 = 0.5$ H, and $v_{in}(t) = 100\sin(0.25\pi t)$ mV for $t \geq 0$ and zero otherwise.

   (a) Find the current $i_{in}(t)$ for $t \geq 0$.

   (b) Compute the energy stored in each inductor over the intervals $0 \leq t < 2$ sec and $2 \leq t$.

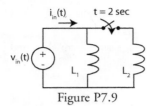

Figure P7.9

10. The circuit of Figure P7.10 has two inductors, $L_1 = 20$ mH and $L_2 = 50$ mH, in parallel. The input is $v_s(t) = 200\cos(500\pi t)$ mV for $t \geq 0$ and zero otherwise. The switch between the two inductors moves down at $t = 4$ ms. Compute the currents $i_{L1}(t)$ and $i_{L2}(t)$ for $0 \leq t < 4$ ms and 4 ms $\leq t$. Also find the energy stored in each inductor as a function of t for the same time intervals.

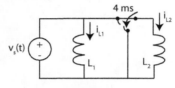

Figure P7.10

THE CAPACITOR AND ITS PROPERTIES

11.(a) Suppose that a 20 µF capacitor is charged to 100 V. Find the charge that resides on each plate of the capacitor.

(b) If the same charge (as in part (a)) resides on a 5 µF capacitor, what is the voltage across the capacitor?

(c) What is the voltage required to store 50 µC on a 2 µF capacitor?

(d) Find the energy required to charge a 20 µF capacitor to 100 V.

12.(a) The $C = 2$ µF capacitor of Figure P7.12 has a terminal voltage of $v_C(t) = 100[1 + \cos(1000\pi t)]$ V. Find the current $i_C(t)$ through the capacitor.

(b) Now suppose the voltage is $v_C(t) = 10\sin(2000t)$ V and $i_C = 10\cos(2000t)$ mA. Find the capacitance C.

Figure P7.12

13. In Figure P7.13a, the capacitors $C_1 = 4$ mF and $C_2 = 12$ mF are driven by the voltage $v_{in}(t)$ specified in Figure P7.13b. Plot $i_{C1}(t)$, $i_{C2}(t)$, and $i_{in}(t)$.

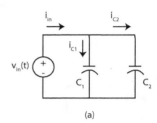

(a)

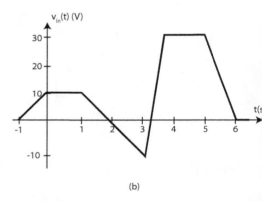

(b)

Figure P7.13

14. The $C = 2$ µF capacitor of Figure P7.14a has current $i_C(t)$ shown in Figure P7.14b. If $v_C(0) = 4$ V, compute $v_C(t)$ at $t = 1, 2, 3, 4$ ms. Now compute the energy stored in the capacitor over the intervals, [0, 2 ms], [2 ms, 3 ms], and [0, 4 ms].

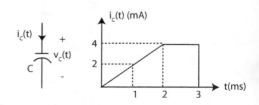

Figure P7.14

15. Suppose $i_{in}(t)$, as specified for all time in Figure P7.15a, excites the circuit of Figure P7.15b, in which $C_1 = 0.2$ µF and $C_2 = 0.1$ µF.
   (a) Plot $v_{C1}(t)$ and $v_{C2}(t)$ for $0 \le t \le 8$ msec.
   (b) Compute and plot the energy stored in the 0.2 µF and 0.1 µF capacitors for $0 \le t \le 8$. Hint: use MATLAB to plot the answers.
   (c) Find $v_{C1}(t)$ and $v_{C2}(t)$ as $t \to \infty$.

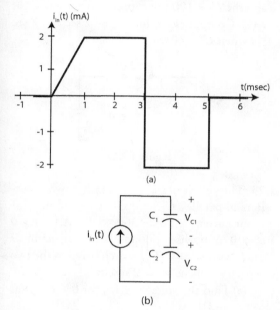

(a)

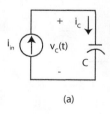

(b)

Figure P7.15

16. For the circuit in Figure P7.16a, $C = 0.25$ mF. Compute and plot the waveforms of the voltage, $v_C(t)$, given $i_{in}(t)$ as sketched in Figures P7.16b and c.

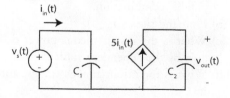

(a)

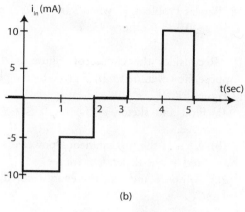

(b)

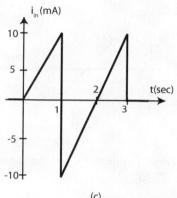

(c)

Figure P7.16

17.(a) Consider the circuit sketched in Figure P7.17 in which $C_1 = 20$ µF and $C_2 = 0.1$ mF. Suppose $v_s(t) = 5\sin(2000t)$ V for $t \ge 0$ and suppose $v_{out}(0) = 10$ V. Find $v_{out}(t)$ for $t > 0$. Is the output voltage independent of the initial voltage on C_1? Why?

(b) What is the instantaneous power delivered by the dependent source?
(c) Find the energy stored in C_2 over the interval $[0, t]$.

Figure P7.17

18. Repeat Problem 17 when $v_s(t) = 10e^{-1000t}$ V for $t \geq 0$ and 0 otherwise.

19. Reconsider the circuit of Figure P7.17. Suppose, however, that $v_s(t)$ is given by the plot in Figure P7.19.
    (a) Find and sketch $v_{out}(t)$ for $0 \leq t \leq 6$ msec.
    (b) What is the instantaneous power delivered by the dependent source?
    (c) Compute and plot the energy stored in C_2.

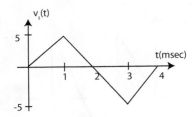

Figure P7.19

20. Repeat Problem 17 for the waveform of Figure P7.20.

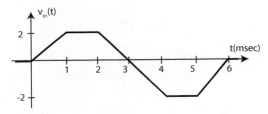

Figure P7.20

21. For the circuit of Figure P7.21, suppose $C_1 = 0.6$ mF, $C_2 = 1.2$ mF, $C_3 = 0.4$ mF, $C_4 = 1.6$ mF, $i_s(t) = 120\sin(100t)$ mA for $t \geq 0$ and 0 for $t < 0$.
    (a) Find the equivalent capacitances C_{eq1} for the series combination and C_{eq2} for the parallel combination.
    (b) Find and sketch $i_{out}(t)$.
    (c) What is the instantaneous power delivered by the dependent source?
    (d) What is the instantaneous energy stored in C_4?

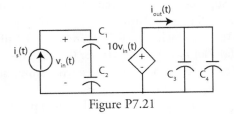

Figure P7.21

22. In the circuit of Figure P7.22 $v_s(t) = 25$ V and the $C_2 = 100$ mF capacitor is uncharged at $t = 0$. Compute $v_C(t)$ for $0 \leq t < 2$ sec and 2 sec $\leq t$ when $C_1 = 400$ mF.

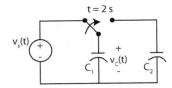

Figure P7.22

23. The circuit of Figure P7.23 has two capacitors in parallel, $C_1 = 30$ mF, $C_2 = 50$ mF. The input current is $i_s(t) = 360e^{-100t}$ mA for $t \geq 0$ and 0 otherwise. Suppose each capacitor is uncharged at $t = 0$. The switch between the two capacitors opens at $t = 2$ msec.
    (a) Find the voltage, $v_C(t)$, for $0 < t < 2$ ms and 2 ms $\leq t$.
    (b) Compute the energy stored in each capacitor as a function of t for the same time intervals.
    (c) Compute the current through each capacitor over each time interval.

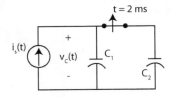

Figure P7.23

MIXED CAPACITOR AND INDUCTOR PROBLEMS

24. Consider the circuit of Figure P7.24, $L_1 = 2.5$ H, $C_1 = 1$ mF, which is excited by the cur-

rent waveform $i_{in}(t) = 200te^{-10t}$ mA for $t \geq 0$ and 0 otherwise.

(a) Compute and sketch $v_L(t)$, $v_C(t)$, and $v_{in}(t)$.

(b) Compute and sketch the energy stored in the inductor for $t \geq 0$.

(c) Compute and sketch the energy stored in the capacitor for $t \geq 0$.

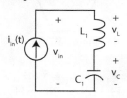

Figure P7.24

25. In the circuit of Figure P7.25, suppose $L_1 = 0.25$ H, $C_1 = 2.5$ mF, $i_s(t) = 20\sin(400t)$ mA for $t \geq 0$ and 0 otherwise. All initial conditions are zero at $t = 0$.

(a) Find $v_L(t)$.

(b) Find $v_C(t)$.

(c) Find the instantaneous stored energy in the capacitor.

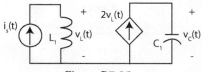

Figure P7.25

26.(a) In the circuit of Figure P7.26, $\beta = 10$, $C_1 = 20$ μF, $C_2 = 80$ mF, $L_1 = L_2 = 20$ mH are initially uncharged. If $v_s(t) = 10t\sin(20t)$ V for $t \geq 0$ and 0 otherwise,

(a) Find $i_s(t)$ for $t > 0$.

(b) Now find $v_{out}(t)$ for $t > 0$.

(c) Compute the energy stored in the 20 mH inductor for $t > 0$.

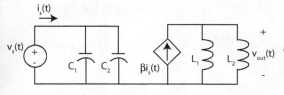

Figure P7.26

27. For the circuit of Figure P7.27,

(a) Compute $i_{C2}(t)$ as a function of $i_s(t)$ and the capacitances C_1 and C_2.

(b) Now find $v_{out}(t)$ in terms of $i_s(t)$ and the circuit parameter values.

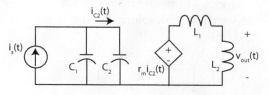

Figure P7.27

28. For the circuit of Figure P7.28, compute $v_{C2}(t)$, $i_{C2}(t)$, and $v_{out}(t)$ as a function of $v_{in}(t)$ and the circuit parameters.

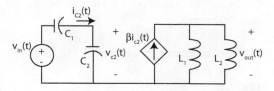

Figure P7.28

SERIES-PARALLEL INDUCTORS

29. In the circuit of Figure P7.29, all inductors are initially relaxed at $t = 0$ and $L_1 = 6$ mH, $L_2 = 38.5$ mH, $L_3 = 22$ mH. A voltage $v_{in}(t) = 200te^{-t}$ mV is applied for $t \geq 0$. Find, L_{eq}, $v_{L1}(t)$, and $v_{L2}(t)$. **Challenge:** Find $i_{in}(t)$.

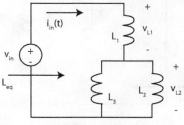

Figure P7.29

CHECKS: 20 mH and $140te^{-t}$ mV.

30. Consider the circuit of Figure P7.30. Suppose $L_1 = 3$ mH, $L_2 = 12$ mH, $L_3 = 36$ mH, and $i_{in}(t) = 120\cos(1000t)$ mA.
   (a) Find L_{eq} and $v_{in}(t)$.
   (b) Find $i_{L2}(t)$.
   (c) Plot the instantaneous power delivered by the source for $0 \le t \le 14$ msec.

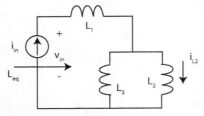

Figure P7.30

CHECK: 12 mH, $-1.44 \sin(1000t)$ V, 90 $\cos(1000t)$ mA.

31. For the circuit of Figure P7.31, $L_1 = 260$ mH, $L_2 = 26$ mH, $L_3 = 39$ mH, and $i_{in}(t) = 10e^{-t^2}$ mA.
   (a) Find L_{eq}, i_{L1}, and i_{L2}.
   (b) Compute $v_{in}(t)$ and $v_{L2}(t)$.
   (c) Compute the instantaneous energy stored in L_1 as a function of t.

CHECKS: 52 mH, 0.2 $i_{in}(t)$, 0.8 $i_{in}(t)$, $v_{in}(t) = -0.104te^{-t^2}$ V, $v_{L2}(t) = 0.4v_{in}(t)$

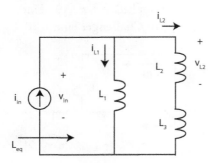

Figure P7.31

32. Repeat Problem 31 for the waveform of Figure P7.32.

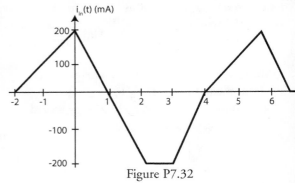

Figure P7.32

33. In Figure P7.33, $L_1 = 5$ mH, $L_2 = 20$ mH, $L_3 = 20$ mH, $L_4 = 80$ mH, and $i_s(t) = 10\sin(1000t)$ mA for $t \ge 0$ and 0 otherwise.
   (a) With the switch in position C, find the equivalent inductance, L_{eq}, v_{L2}, and v_{L4}.
   (b) Repeat part (a) with the switch in position D.

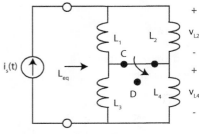

Figure P7.33

34. Consider the circuit of Figure P7.34a with voltage source excitation given in Figure P7.34b. Let the inductor values be those given in Problem 33. Suppose the switch is in position C. Note that each inductor is relaxed at $t = 0$.
   (a) Find L_{eq}.
   (b) Compute and sketch $i_{in}(t)$ for $t \ge 0$.
   (c) Find the instantaneous (total) energy stored in the set of four inductors as a function of time.
   (d) Compute and sketch $i_{L2}(t)$.

CHECK: $L_{eq} = 20$ mH; $i_{in}(1) = i_{in}(3) = 0.8$ A while $i_{in}(2) = i_{in}(4) = 0$

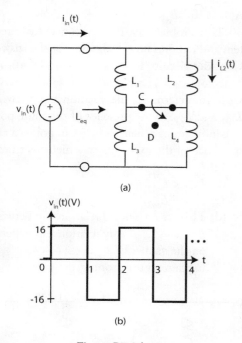

(a)

(b)

Figure P7.34

35. Find L_{eq} for the circuit of Figure P7.35, (a) when the switch is open, and (b) when the switch is closed. The unit of L is henries.

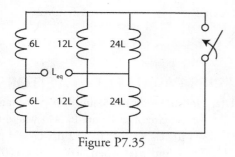

Figure P7.35

36. Find L_{eq} for each of the circuits in Figure P7.36, where $L_1 = 5$ mH, $L_2 = 20$ mH, $L_3 = 40$ mH, $L_4 = 150$ mH, $L_5 = 50$ mH, $L_6 = 180$ mH, $L_7 = 120$ mH, $L_8 = 35$ mH. Notice that the circuit of (b) is a modification of (a) and that of (c) is a modification of (b). Connections can create interesting behaviors.

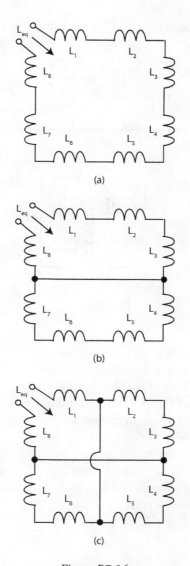

(a)

(b)

(c)

Figure P7.36

SCRAMBLED ANSWERS: 0.1, 0.08, 0.6 (in H)

37. Three 60 mH inductors are available for interconnection. List all equivalent inductances obtainable over all possible interconnections of these elements.

CHECK: There should be seven different values.

38. Find L_{eq} for each of the circuits in Figure P7.38.

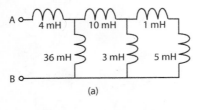

(a)

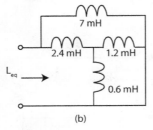

(b)

Figure P7.38

ANSWER: (a) 13 mH; (b) 2 mH

39. This is a conceptual problem and requires no calculations for the answer. Consider circuits 1 and 2 of Figure P7.39. All inductors are 1 H except the one labeled L. We wish to determine the relationship between L_{eq1} and L_{eq2} in the presence of the finite positive inductor of L henries between points a and b. Which of the following statements is true?

(a) $L_{eq1} > L_{eq2}$.
(b) $L_{eq1} < L_{eq2}$.
(c) $L_{eq1} = L_{eq2}$.
(d) There is no general relationship between L_{eq1} and L_{eq2}. Any relationship depends on the value of L.

Explain your reasoning.

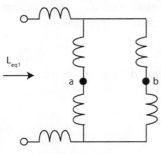

Figure P7.39

ANSWER: $L_{eq1} = L_{eq2}$ for all L values.

40. Like Problem 39, this is a conceptual problem and requires no calculations for the answer. Consider circuits 1 and 2 of Figure P7.40. All inductors are 1 H except the one labeled L. We wish to determine the relationship between L_{eq1} and L_{eq2} in the presence of the finite positive inductor of L henries between points a and b. Which of the following statements is true?

(a) $L_{eq1} > L_{eq2}$.
(b) $L_{eq1} < L_{eq2}$.
(c) $L_{eq1} = L_{eq2}$.
(d) There is no general relationship between L_{eq1} and L_{eq2}. Any relationship depends on the value of L.

Explain your reasoning.

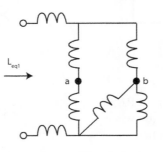
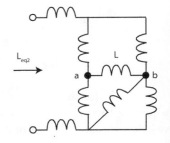

Circuit 1   Circuit 2

Figure P7.40

ANSWER: $L_{eq2} < L_{eq1}$

SERIES-PARALLEL CAPACITORS

41. (a) Find the indicated equivalent capacitance for the circuit of Figure P7.41a where C_1 = 4 µF, C_2 = 3 µF, C_3 = 2 µF, C_4 = 4 µF. Then find $v_s(t)$ when $i_s(t)$ = $10\cos(10^4 t)$ mA for $t \geq 0$ and 0 otherwise.

(b) Repeat for Figure P7.41b in which C_1 = 60 µF, C_2 = 18 µF, C_3 = 18 µF, C_4 = 36 µF, and C_5 = 10.8 µF. Then find $v_s(t)$ when $i_s(t)$ = $10\sin(10^5 t)$ mA for $t \geq 0$ and 0 otherwise.

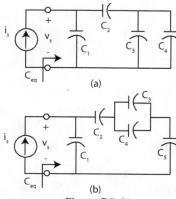

(a)

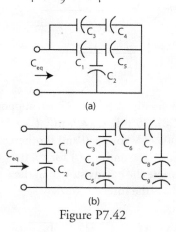

(b)

Figure P7.41

CHECKS: 6 μF, 66 μF

42.(a) Find the indicated equivalent capacitance for the circuit of Figure P7.42a assuming $C_1 = 48$ μF, $C_2 = 16$ μF, $C_3 = 20$ μF, $C_4 = 80$ μF, and $C_5 = 8$ μF.

(b) Repeat for Figure P7.42b assuming $C_1 = 3$ μF, $C_2 = 6$ μF, $C_3 = 3.6$ μF, $C_4 = 6$ μF, $C_5 = 4.5$ μF, $C_6 = 48$ μF, $C_7 = 48$ μF, $C_8 = 24$ μF, $C_9 = 24$ μF.

(a)

(b)

Figure P7.42

43. Find C_{eq} for the circuit of Figure P7.43, (a) when the switch is open, and (b) when the switch is closed, assuming that $C_1 = C_4 = 12$ μF, $C_2 = C_5 = 40$ μF, $C_3 = C_6 = 20$ μF. (c) Repeat parts (a) and (b) for $C_1 = 12$ μF, $C_2 = 40$ μF, $C_3 = 20$ μF, $C_4 = 40$ μF, $C_5 = 20$ μF, $C_6 = 100$ μF.

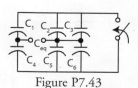

Figure P7.43

44. For the circuit of Figure P7.44, $C_1 = 8$ mF, $C_2 = 6$ mF, $C_3 = 12$ mF, and $i_{in}(t) = 240\sin(200t)$ mA for $t \geq 0$ and 0 otherwise.

(a) Find C_{eq}.

(b) Find $v_{in}(t)$. Note: All capacitors are initially uncharged. Why?

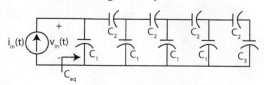

Figure P7.44

45. Three 12 μF capacitors are available for interconnection. List all equivalent capacitances obtainable over all possible interconnections of these capacitors.

46. This is a conceptual problem and requires no calculations for the answer. Consider circuits 1 and 2 of Figure P7.46. All capacitors are 1 F except the one that is labeled C. We wish to determine the relationship between C_{eq1} and C_{eq2} in the presence of the finite positive C F capacitor between points a and b. Which of the following statements is true?

(a) $C_{eq1} > C_{eq2}$.

(b) $C_{eq1} < C_{eq2}$.

(c) $C_{eq1} = C_{eq2}$.

(d) There is no general relationship between C_{eq1} and C_{eq2}. Any relationship depends on the value of C.

Explain your reasoning.

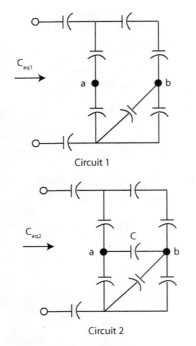

Circuit 1

Circuit 2

Figure P7.46

ANSWER: $C_{eq1} < C_{eq2}$

47. In Figures P7.47a and b, the charge on C_1, C_2 and C_3 is $Q = 72 \times 10^{-3}$ C. In Figure P7.47a, the voltages on C_1, C_2 and C_3 are 2 V, 3 V, and 4 V, respectively; and in Figure 7.47b $v_C = 4$ V while the charges on C_1, C_2 and C_3 are $Q_1 = 48 \times 10^{-3}$ C, $Q_2 = 60 \times 10^{-3}$ C, and $Q_3 = 72 \times 10^{-3}$ C, respectively.
    (a) Find C_{eq} for the circuit of Figure P7.47a.
    (b) Find C_{eq} for the circuit of Figure P7.47b.

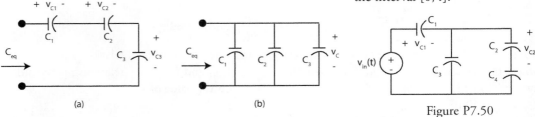

(a)   (b)

Figure P7.47

ANSWER: (a) $C_{eq} = 8$ mF

48. In the circuit of Figure P7.48, $C_1 = 6$ mF, $C_2 = 12$ mF, $C_3 = 36$ mF, $v_{in}(t) = 20(1 - e^{-20t})$ V for $t \geq 0$ and 0 otherwise. If all capacitor volt-

ages at $t = 0^-$ are zero, find all three voltages, $v_{Ci}(t)$, $i = 1,2,3$, for $t > 0$. Then find the instantaneous stored energy at $t = 0.05$ sec.

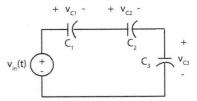

Figure P7.48

CHECK: $v_{C2}(t) = 6(1 - e^{-20t})$ V

49. In the circuit of Figure P7.49, suppose $C_1 = 3$ μF, $C_3 = 0.5$ μF, $C_2 = 1.5$ μF, and $v_{in}(t) = 10\sin(400t)$ mV for $t \geq 0$ and 0 otherwise.
    (a) Find $v_{C1}(t)$ and $v_{C2}(t)$ for $t \geq 0$.
    (b) Find an expression for the energy stored in C_1 and C_2 over the interval $[0, t]$.

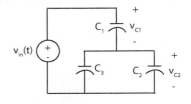

Figure P7.49

50. In the circuit of Figure P7.50, suppose $C_1 = 5$ mF, $C_2 = 20$ mF, $C_3 = 4$ mF, $C_4 = 80$ mF, $v_{in}(t) = 100e^{-2t}$ V for $t \geq 0$ and 0 otherwise.
    (a) Find $v_{C1}(t)$ and $v_{C2}(t)$ for $t > 0$.
    (b) Compute the energy stored in the C_2 over the interval $[0, t]$.

Figure P7.50

51. In the circuit of Figure P7.51, suppose $C_1 = 4$ mF, $C_2 = 80$ mF, $C_3 = 20$ mF, and $i_{in}(t) = 100e^{-2t}$ mA for $t \geq 0$ and 0 otherwise.
    (a) Find $i_{C1}(t)$ and $i_{C2}(t)$ for $t > 0$.
    (c) Compute the energy stored in C_2 over the interval $[0, t]$.

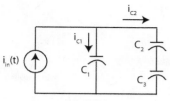

Figure P7.51

MISCELLANEOUS

52. Find and sketch $v_{out}(t)$ for $0 \leq t \leq 4$ sec for the circuit of Figure P7.52a, assuming all capacitors are initially at rest for the excitation of Figure P7.52b. Are any of the capacitors redundant as far as v_{out} is concerned?

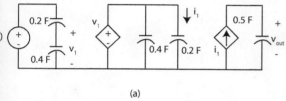

(a)

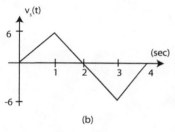

(b)

Figure P7.52

CHECK: $v_{out}(t) = \dfrac{2}{15} v_s(t)$

53. Using the circuit given in Figure 7.32, select a capacitor value to filter the voltage for a regulator requiring $14\,V < v_C(t) < 20\,V$. Use $v_s(t) = 20\cos(200\pi t)$ and $i_0(t) = 2\,A$. CHECK: $C \geq 1.667$ mF.

54. When driving a car into a left-hand turn lane, one often sees a large circular or hexagonal cut in the concrete. Embedded in these cuts is a coil of wire. When your car (containing a large percentage of iron) passes over this coil, its inductance changes. This change of inductance can be used as a sensor to activate a circuit that stops oncoming traffic and lights

the green left turn signal. The interesting variation is that the v-i inductor relationship is different for time-varying inductances:

$$v_L(t) = \frac{d}{dt}\left[L(t)i_L(t)\right]$$

The following highly simplified circuit illustrates the principle of operation, although the configuration and values may not be what are actually used. Consider the circuit of Figure P7.54a consisting of an inductor driven by a current source. When the car with its steel frame moves over the coil of wire, the inductance of the coil changes from L_1 to some larger value $L_2 = 3L_1$ as illustrated in Figure P7.54b, where the time T_1 depends on the speed at which the car is slowing down and

$$L(t) = L_1\left[1 + 4\left(\frac{t}{T_1} - \frac{t^2}{2T_1^2}\right)\right]$$

Plot $v_L(t)$ for $t > 0$ assuming $i_L(t) = I_0$, a constant value, and that the front edge of the car begins to cross the first edge of the coil at $t = 0$. Explain how this voltage signal might be used to control the traffic light.

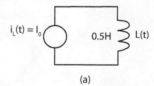

(a)

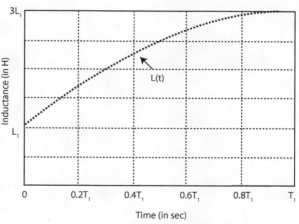

(b)

Figure P7.54

C H A P T E R 8

First Order RL and RC Circuits

When watching a manufacturing process, a visitor might see a pair of robotic arms assemble an engine or machine a block of metal with perfectly timed maneuvers. Timing is a critical aspect of a manufacturing process. In TV transmitters there is a signal called the raster, which is critical to the generation of the screen image. In an oscilloscope a timing signal called a horizontal sweep acts as a time base, which allows one to view measured input signals as a function of time. All these applications utilize a signal having sawtooth shape and called a linear voltage sweep. The linear voltage sweep is nicknamed the sawtooth. This sawtooth is pictured here together with an approximating exponential curve for comparison.

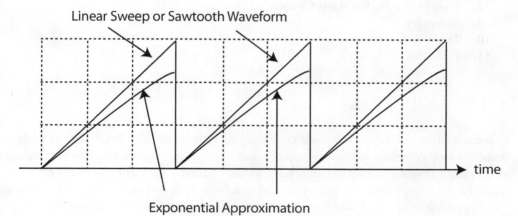

Linear Sweep or Sawtooth Waveform

Exponential Approximation

time

Ideally, the sawtooth voltage increases linearly with time until reaching a threshold where it immediately drops to zero, which reinitiates the process. The threshold voltage corresponds to a fixed unit of time. The linear voltage increase then acts as an electronic second hand, ticking off the smaller units of time. In practice, the linear increase in voltage is approximated by the "linear" part of an exponential response of an RC circuit. When the voltage across the capacitor reaches a certain threshold, an electronic switch changes the equivalent circuit seen by the capacitor, allowing the capacitor to discharge very quickly, i.e., the capacitor voltage drops to zero almost instantaneously. Once the voltage nears zero, the electronic switch reinstates the earlier circuit structure, causing the capacitor to charge up again. The process repeats itself indefinitely.

CHAPTER OBJECTIVES

1. Explore the use of a constant-coefficient first-order linear differential equation as a model for first-order RL and RC circuits.
2. Derive from the differential equation model, the exponential response form (voltage or current) of first-order RL and RC circuits without sources and with constant excitations.
3. Interpret the solution form of the differential equation model in terms of the circuit time constant and the initial and final values of the capacitor voltage or inductor current.
4. Develop techniques to handle switching and piecewise constant excitations within first-order RL and RC circuits.
5. Investigate waveform generation and RC op amp circuits.

SECTION HEADINGS

1. **Introduction**
2. **Some Mathematical Preliminaries**
3. **Source-Free or Zero-Input Response**
4. **DC or Step Response of First-Order Circuits**
5. **Superposition and Linearity**
6 **Response Classifications**
7. **Further Points of Analysis and Theory**
8. **First-Order RC Op Amp Circuits**
9. **Summary**
10. **Terms and Concepts**
11. **Problems**

1. INTRODUCTION

Our study prior to Chapter 7 focused exclusively on resistive circuits. Recall that all nodal equations and loop equations for resistive circuits lead to (algebraic) matrix equations whose solution yields node voltages and loop currents, respectively. Chapter 7 then introduced the capacitor and the inductor. Interconnections of sources, resistors, capacitors, and inductors lead to new and fascinating circuit behaviors. How? Inductors and capacitors have differential or integral voltage-current relationships. Interconnecting resistors and capacitors or resistors and inductors leads to circuits that must satisfy both algebraic (KVL, KCL, and Ohm's law) and differential or integral relationships for L and C values. When only one inductor or one capacitor is present along with resistors and sources, these relationships lead to first-order RL and RC circuits. When the sources are dc, such circuits have voltages and currents of the form $A + Be^{-\lambda t}$ for constants A, B, and λ. The main purpose of this chapter is to develop techniques for computing the exponential responses of first-order RC and RL circuits driven by dc sources. A simple example serves to explain some of these points.

In the series RC circuit of Figure 8.1, suppose an initial voltage $v_C(0^-)$ is present on the capacitor, where 0^- designates the instant immediately before zero. Often we distinguish among 0^-, 0, and 0^+ when switching occurs or when discontinuities of excitation functions occur at $t = 0$.

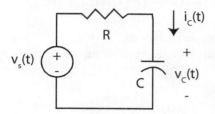

FIGURE 8.1 Series RC circuit.

A loop equation for the series RC circuit leads to

$$v_s(t) = Ri_C(t) + v_C(t) \tag{8.1}$$

Since $i_C(t) = C\dfrac{dv_C(t)}{dt}$, equation 8.1 becomes

$$v_s(t) = RC\frac{dv_C(t)}{dt} + v_C(t)$$

Dividing through by RC yields the constant-coefficient **first-order linear differential equation**

$$\frac{dv_C(t)}{dt} + \frac{1}{RC}v_C(t) = \frac{1}{RC}v_s(t) \tag{8.2}$$

subject to the initial condition $v_C(0^-)$. This equation says that the derivative of the capacitor voltage plus $1/RC$ times the capacitor voltage equals $1/RC$ times the source voltage. The equation enforces constraints on the capacitor voltage, its derivative, and the source voltage, and is different from the algebraic node or loop equations studied earlier. The terminology *first-order* differential equation applies because only the first derivative appears. Equation 8.2 is linear because it comes from a linear circuit. Our goal is to find capacitor voltage waveforms that satisfy the constraints imposed by the differential equation 8.2.

Exercise. For the circuit of Figure 8.1, show that the capacitor current $i_C(t)$ satisfies a differential equation of the form

$$\frac{di_C(t)}{dt} + \frac{1}{RC}i_C(t) = \frac{1}{R}\frac{dv_s(t)}{dt}$$

Our scope in this chapter is limited to circuits containing one inductor or one capacitor—equivalently, first-order *RL* or *RC* circuits. Within this category we further constrain our investigation to circuits with no sources but nonzero initial conditions, circuits driven by constant (dc) sources, circuits driven by piecewise constant sources, and circuits containing switches. First-order circuits driven by arbitrary source excitations are covered in later chapters using the Laplace transform method.

2. SOME MATHEMATICAL PRELIMINARIES

Very often our interest is in source excitations such as $v_s(t) = 2e^{-2t}$ V for $t \geq 0$ and 0 otherwise. To conveniently represent such time-restricted waveforms, we define a signal called the **unit step function**, denoted by $u(t)$, as

$$u(t) = \begin{cases} 1 & t \geq 0 \\ 0 & t < 0 \end{cases}$$

The unit step function is a universally used function and will appear many times in the remainder of this text. MATLAB code for specifying the step function is

```
function f = ustep(t)
t = t + 1e-12;
f = (sign(t)+1)*0.5;
```

With the unit step so defined, $v_s(t) = 2e^{-2t}u(t)$ V, and both relations are plotted in Figure 8.2.

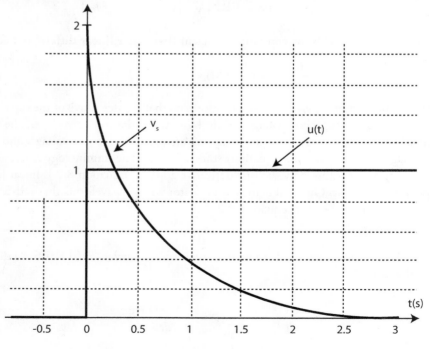

FIGURE 8.2 Unit step function and $v_s(t) = 2e^{-2t}u(t)$ V.

Further, if $v_s(t) = 2e^{-2t}$ V for $t \geq t_0$ and 0 for $t < t_0$, then $v_s(t) = 10e^{-2t}u(t - t_0)$ would be the proper representation because the shifted unit step function, $u(t - t_0)$, means

$$u(t - t_0) = \begin{cases} 1 & t \geq t_0 \\ 0 & t < t_0 \end{cases}$$

Plots of $v_s(t) = 2e^{-2t}u(t - t_0)$ and $u(t - t_0)$ are given in Figure 8.3 for $t_0 = 0.5$.

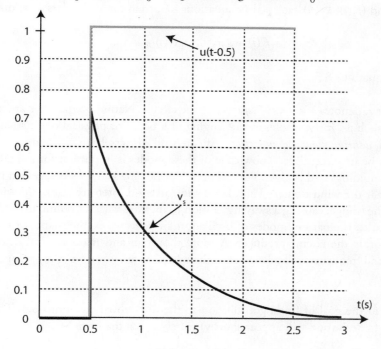

FIGURE 8.3. Plots of $u(t - 0.5)$ and $v_s(t) = 2e^{-2t}u(t - 0.5)$.

Exercise. Plot $u(-t)$ and $u(t_0 - t)$. Hint: For what values of t are the functions zero and for what values are they 1?

A working model of a physical system underlies an engineer's ability to methodically analyze, design, or modify its behavior. Linear circuits are physical systems that have differential equation models. The RL and RC circuits investigated in this chapter have differential equation models of the form

$$\frac{dx(t)}{dt} = \lambda x(t) + f(t), \quad x(t_0) = x_0 \tag{8.3a}$$

or, equivalently,

$$\frac{dx(t)}{dt} - \lambda x(t) = f(t), \quad x(t_0) = x_0 \tag{8.3b}$$

valid for $t \geq t_0$, where $x(t_0) = x_0$ is the initial condition on the differential equations 8.3. The term $f(t)$ denotes a forcing function. Usually, $f(t)$ is a linear function of the input excitations to the circuit.

Before proceeding, it is appropriate to explore the intuitive nature of a differential equation. Equations 8.3 are first-order constant-coefficient *linear differential equations*. They are first order because of the presence of only the first derivative of some unknown function $x(t)$. For example, in equation 8.3a the derivative of $x(t)$ equals a constant λ times $x(t)$ plus a known forcing function $f(t)$, where $f(t)$ incorporates the effect of all the circuit excitations. Rigorously speaking, "linear" means that under the assumption of zero initial conditions, if the pairs of voltage waveforms $(f_1(t), x_1(t))$ and $(f_2(t), x_2(t))$ each satisfy equations 8.3, then for any scalars a_1 and a_2, the pair

$$(a_1 f_1(t) + a_2 f_2(t),\ a_1 x_1(t) + a_2 x_2(t))$$

also satisfies equations 8.3.

The parameter λ denotes a *natural frequency* of the circuit. Natural frequencies are natural modes of oscillation such as, for example, in the ringing of a bell. For physical objects natural frequencies are called *natural modes of vibration*. All physical objects have a vibrational motion even though it may be imperceptible. Knowledge of these modes is important for the safety and reliability of large buildings and bridges. For example, the Tacoma Narrows Bridge had natural modes of vibration that the wind excited. Undulations in the wind intensity resonated with the natural vibrations of the bridge, causing a swaying motion to increase without bound until the bridge collapsed. In circuits, the natural modes of oscillation are reflected in the shapes of the voltage and current waveforms the circuit produces. A more thorough and mathematical discussion of the notion of natural frequency will take place in the next chapter, when we study second-order (*RLC*) circuits.

Let us return to the goal of finding a solution to the differential equations 8.3. The solution to equations 8.3 (a derivation will appear shortly) for $t \geq t_0$ has the form

$$x(t) = e^{\lambda(t-t_0)} x(t_0) + \int_{t_0}^{t} e^{\lambda(t-\tau)} f(\tau)\, d\tau \tag{8.4}$$

This means that the expression on the right-hand side of the equal sign (1) satisfies the differential equations 8.3 [its derivative equals λ times itself plus $f(t)$], and (2) it satisfies the correct initial condition, $x(t_0) = x_0$. A simple example illustrates this point.

EXAMPLE 8.1.

Compute and verify the solution of equation 8.3a using equation 8.4.

SOLUTION

Suppose in equation 8.3a, $f(t) = u(t - 1)$, a shifted unit step function, $\lambda = -1$, $t_0 = 1$, and $x(1) = 10$, in which case

$$\frac{dx(t)}{dt} = -x(t) + u(t - 1) \ , \ x(1) = 10$$

From equation 8.4, for $t \geq 1$,

$$x(t) = 10 e^{-(t-1)} + \int_{1}^{t} e^{-(t-\tau)} u(\tau - 1)\, d\tau = 10 e^{-(t-1)} + \left[1 - e^{-(t-1)}\right] = 9 e^{-(t-1)} + 1$$

To verify that $[9e^{-(t-1)} + 1]$ does indeed satisfy the differential equation, observe that for $t > 1$,

$$\frac{dx(t)}{dt} = \frac{d}{dt}\left[9e^{-(t-1)} + 1\right] = -9e^{-(t-1)} = -\left[9e^{-(t-1)} + 1\right] + 1 = -x(t) + 1$$

Further, at $t = 1$, $[9e^{-(t-1)} + 1] = 10$, which is the mandatory initial condition. Thus, $x(t) = 9e^{-(t-1)} + 1$ is a valid solution for $t \geq 1$.

Example 8.1 spells out the application of the solution (equation 8.4) to the differential equation 8.3a. It also verifies that the computed solution satisfies the differential equation and the proper initial condition. Although not shown, equation 8.4 also satisfies equation 8.3b. A formal derivation of the solution of equation 8.4 requires the use of the **integrating factor method**, the subject of a differential equations course. Briefly, the first step of this method entails multiplying both sides of equation 8.3a or 8.3b by a so-called integrating factor $e^{-\lambda t}$. For equation 8.3b, this results in

$$e^{-\lambda t}\frac{dx(t)}{dt} - \lambda e^{-\lambda t}x(t) = e^{-\lambda t}f(t) \tag{8.5}$$

By the product rule for differentiation, the sum on the left equals

$$\frac{d}{dt}\left[e^{-\lambda t}x(t)\right]$$

in which case equation 8.5 becomes

$$\frac{d}{dt}\left[e^{-\lambda t}x(t)\right] = e^{-\lambda t}f(t) \tag{8.6}$$

One can integrate both sides of equation 8.6 from t_0 to t as follows:

$$\int_{t_0}^{t}\frac{d}{d\tau}\left[e^{-\lambda\tau}x(\tau)\right]d\tau = e^{-\lambda\tau}x(\tau)\Big|_{t_0}^{t}$$

$$= e^{-\lambda t}x(t) - e^{-\lambda t_0}x(t_0) = \int_{t_0}^{t}e^{-\lambda\tau}f(\tau)d\tau \tag{8.7}$$

Bringing the term $e^{-\lambda t_0}x(t_0)$ to the right-hand side of equation 8.7 and multiplying through by $e^{\lambda t}$ results in the solution to the differential equation 8.3a or 8.3b, given by equation 8.4. This completes the derivation of the very powerful formula of equation 8.4.

There are four points to remember about the preceding discussion: (1) circuits have behaviors modeled by differential equations such as equations 8.3; (2) the solution to a first-order differential equation is a waveform (also called a signal or response) satisfying equation 8.4; (3) the formula of equation 8.4 works for all continuous and piecewise continuous time functions $f(t)$; and (4) a solution to a differential equation means that the waveform satisfies the given differential equation with the proper initial condition.

Exercise. Show that the function $x(t) = (1 - e^{-t})$, for $t \geq 0$, is a solution to the differential equation $\dfrac{dx(t)}{dt} = -x(t) + u(t)$ with initial condition $x(0) = 0$ by showing that $x(t)$ satisfies the differential equation and has the proper initial condition at $t = 0$.

3. SOURCE-FREE OR ZERO-INPUT RESPONSE

Figure 8.4 depicts the most basic (undriven or source-free) *RL* or *RC* circuit: a parallel connection of a resistor with an inductor or a capacitor without a source. In these circuits, one assumes the presence of an initial inductor current or initial capacitor voltage. The complication introduced by a voltage or current source is taken up later. Once the source-free or **zero-input** behavior is understood, one can understand more easily the responses resulting from constant source excitations.

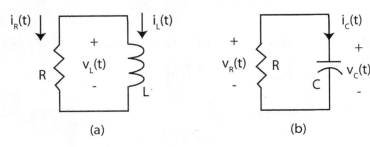

FIGURE 8.4

Our first goal is to derive differential equation models for the *RL* and *RC* circuits of Figures 8.4a and 8.4b, respectively. We do this in parallel.

(1) At the top node of Figure 8.4a, KCL implies

$$i_R(t) = -i_L(t)$$

(2) However,

$$i_R(t) = \frac{v_L(t)}{R} = \frac{L}{R}\frac{di_L(t)}{dt}$$

(3) Making the obvious substitution and multiplying by R/L yields the differential equation model

$$\frac{di_L(t)}{dt} = -\frac{R}{L}i_L(t) \qquad (8.8a)$$

with $i_L(t_0)$ a given initial condition.

(1) Similarly, for Figure 8.4b, KVL implies

$$v_R(t) = v_C(t)$$

(2) However,

$$v_R(t) = -Ri_C(t) = -RC\frac{dv_C(t)}{dt}$$

(3) Making the obvious substitution and dividing by RC yields the differential equation model

$$\frac{dv_C(t)}{dt} = -\frac{1}{RC}v_C(t) \qquad (8.8b)$$

with $v_C(t_0)$ a given initial condition.

Both differential equation models have the same general form,

$$\frac{dx(t)}{dt} = \lambda x(t) = -\frac{1}{\tau}x(t) \tag{8.9}$$

i.e., the derivative of $x(t)$ is a constant, $\lambda = -1/\tau$, times itself. Applying equation 8.4 to equation 8.9 implies that both equations 8.8a and 8.8b have solutions given by

$$x(t) = e^{\lambda(t-t_0)}\, x(t_0) = e^{-\left(\frac{t-t_0}{\tau}\right)}x(t_0) \tag{8.10}$$

where τ is a special constant called the **time constant** of the circuit. Equation 8.10 means that the responses for $t \geq t_0$ of the undriven RL and RC circuits are, respectively, given by

$$i_L(t) = e^{-\frac{R}{L}(t-t_0)}\, i_L(t_0) \qquad v_C(t) = e^{-\frac{1}{RC}(t-t_0)}\, v_C(t_0) \tag{8.11}$$

where the time constant of the RL circuit is $\tau = \dfrac{L}{R}$ and the time constant of the RC circuit is $\tau = RC$.

The time constant of the circuit is the time it takes for the source-free circuit response to drop to $e^{-1} = 0.368$ of its initial value. Roughly speaking, the response value must drop to a little over one-third of its initial value. This is a good rule of thumb for approximate calculations involving decaying exponentials.

The mathematics that underlie the solution to the differential equation 8.9 given in equation 8.10 is nothing more than elementary calculus. To see this, consider the exponential solution form

$$x(t) = Ke^{-t/\tau} \tag{8.12}$$

where K is an arbitrary constant. The function $e^{-t/\tau}$ has the property that its derivative is $-\dfrac{K}{\tau}e^{-t/\tau}$. This is precisely what equation 8.9 requires. Therefore equation 8.12 satisfies the differential equation 8.9 and is said to be a solution. To completely specify $x(t)$ it only remains to identify the proper value of K from the initial condition. Evaluating $x(t)$ at $t = t_0$ yields

$$x(t_0) = Ke^{-t_0/\tau}$$

in which case

$$K = e^{t_0/\tau}x(t_0)$$

Substituting this value of K into equation 8.12 produces the solution given in equation 8.10, which is adapted to specific RL and RC circuits in equations 8.11. Figure 8.5 plots equation 8.12 for arbitrary K and $\tau > 0$. This plot proves instructive for understanding how the response decays as a function of the time constant.

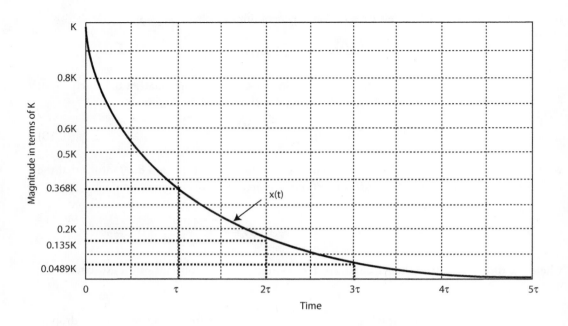

FIGURE 8.5 Plot of equation 8.12. For $t = \tau$, one time constant, $x(t)$ decays to 0.368 of its maximum value.

In summary, the circuits of Figure 8.4 motivate the development of the rudimentary machinery for constructing solutions to undriven *RL* and *RC* circuits. For more general circuits, those containing multiple resistors and dependent sources, it is necessary to use the Thevenin equivalent resistance seen by the inductor or capacitor in place of the *R* in equation 8.11. Figure 8.6 illustrates this idea.

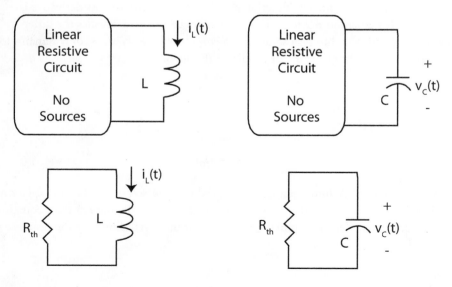

FIGURE 8.6 Replacement of "resistive" part of circuit by its Thevenin equivalent.

These facts imply that the general formulas for computing the responses of undriven RL and RC circuits have the structures

$$i_L(t) = e^{-\frac{R_{th}}{L}(t-t_0)} i_L(t_0) \qquad\qquad v_C(t) = e^{-\frac{1}{R_{th}C}(t-t_0)} v_C(t_0) \qquad (8.13)$$

The difference between equations 8.11 and 8.13 is that in equations 8.13 R_{th} is the Thevenin equivalent resistance seen by the inductor or capacitor.

EXAMPLE 8.2

For the circuit of Figure 8.7, find $i_L(t)$ and $v_L(t)$ for $t \geq 0$ given that $i_L(0^-) = 10$ A and the switch S closes at $t = 0.4$ sec. Then compute the energy dissipated in the 5 Ω resistor over the time interval $[0.4, \infty)$.

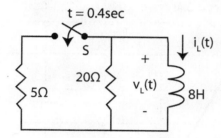

FIGURE 8.7 Parallel RL circuit containing a switch.

SOLUTION

Step 1. *With switch S open, compute the response for $0 \leq t \leq 0.4$ sec.* From the continuity property of the inductor current, $i_L(0^+) = i_L(0^-) = 10$ A. Using equation 8.13,

$$i_L(t) = e^{-\frac{R_{th}}{L}t} i_L(0^+) = 10e^{-2.5t} \text{ A}$$

We note that $i_L(0.4) = 3.679$ A.

Step 2. *With switch S closed, compute the response for $t \geq 0.4$ sec.* For this time interval the Thevenin equivalent resistance seen by the inductor is $R_{th} = 20\|5 = 4\ \Omega$, i.e., the equivalent resistance of a parallel 20 Ω and 5 Ω combination. According to equation 8.13, the response for $t \geq t_0 = 0.4$ sec is

$$i_L(t) = e^{-\frac{R_{th}}{L}(t-t_0)} i_L(t_0) = i_L(0.4)e^{-0.5(t-0.4)} = 3.679e^{-0.5(t-0.4)} \text{ A}$$

Step 3. *Write the complete response as a single expression using step functions:*

$$i_L(t) = 10e^{-2.5t}[u(t) - u(t-0.4)] + 3.69e^{-0.5(t-0.4)}u(t-0.4) \text{A} \qquad (8.14)$$

Step 4. *Plot the complete response.* To plot this using MATLAB, we use the following m-file along with the code given earlier for the unit step function:

```
»t = 0:0.005:1.4;
»iL = 10*exp(-2.5*t) .* (ustep(t).* ustep(t – 0.4)) + 3.679*exp(-0.5*(t–0.4)) .* ustep(t – 0.4);
»plot(t,iL)
»grid
```

Using this code, Figure 8.8 illustrates the complete response, showing the two different time constants. The 0.4 sec time constant has a much faster rate of decay than the lengthy 2 sec time constant.

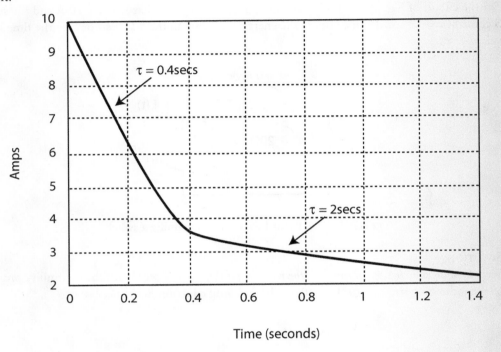

FIGURE 8.8 Sketch of response $i_L(t)$ for Example 8.2.

Step 5. *Compute $v_L(t)$.* It is a simple matter now to compute $v_L(t)$ since

$$v_L(t) = -R_{th}i_L(t)$$

In particular, $v_L(0^+)$ = -200 V. Hence for $0 \le t < 0.4$

$$v_L(t) = e^{-\frac{R_{th}}{L}t}v_L(0^+) = -200e^{-2.5t} \text{ V}$$

For $t \ge 0.4$, however, the circuit structure changes and R_{th}= 4 Ω, in which case $v_L(0.4^+)$ = 4 × 3.679 = 14.716 V. Thus,

$$v_L(t) = e^{-\frac{R_{th}}{L}(t-0.4)}v_L(0.4^+) = -14.716e^{-0.5(t-0.4)} \text{ V}$$

Step 6. *Compute the energy dissipated in the 5 Ω resistor over the interval* [0.4, ∞). The power absorbed by the 5 Ω resistor for $0.4 \leq t$ is

$$p_{5\Omega}(t) = \frac{v_L^2(t)}{5} = \frac{\left[-14.716e^{-0.5(t-0.4)}\right]^2}{5} = 43.312e^{-(t-0.4)} \text{ watts}$$

The energy dissipated over [0.4, ∞) is given by

$$W_{5\Omega}(0.4, \infty) = \int_{0.4}^{\infty} p_{5\Omega}(q)dq = 43.312 \int_{0.4}^{\infty} e^{-(q-0.4)}dq = 43.312 \text{ J}$$

Exercises. 1. Plot $v_L(t)$ using the above m-file, ustep, and the appropriate code.

2. Repeat the calculations of Example 8.2 with the 8 H inductor changed to 8 mH and a switch closing time of 0.4 ms.

ANSWER: $i_L(t) = 10e^{-2500t}u(t)u(0.4 \times 10^{-3} - t) + 3.679e^{-500(t-0.4 \times 10^{-3})}u(t - 0.4 \times 10^{-3})$ A

EXAMPLE 8.3

Find $v_C(t)$ for $t \geq 0$ for the circuit of Figure 8.9 given that $v_C(0) = 9$ V.

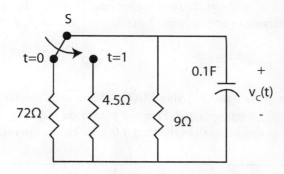

FIGURE 8.9 Parallel *RC* circuit with switch.

SOLUTION

Because there is a switch that changes position at $t = 1$ sec, there are two time intervals to consider.

Step 1. *Compute the response for* $0 \leq t < 1$. Over this time interval, the equivalent circuit is a parallel *RC* circuit, as shown in Figure 8.10a.

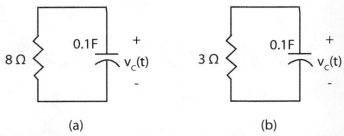

(a)   (b)

FIGURE 8.10 Equivalent circuits for Figure 8.9: (a) $0 \leq t < 1$ and (b) $1 \leq t$.

By the continuity of the capacitor voltage, $v_C(0^+) = v_C(0^-) = 9$ V. Therefore from equation 8.11,

$$v_C(t) = e^{-\frac{1}{R_{th}C}t} v_C(0^+) = 9e^{-1.25t} \text{ V}$$

Step 2. *Compute the response for $t \geq 1$.* Figure 8.10b depicts the pertinent equivalent circuit. Observe that $v_C(1^-) = v_C(1^+) = 2.58$ V and $R_{th} = 3$ Ω. Again by equation 8.11, for $t > t_0 = 1$,

$$v_C(t) = e^{-\frac{1}{R_{th}C}(t-t_0)} v_C(t_0^+) = 2.58e^{-\frac{t-1}{0.3}} \text{ V}$$

Step 3. *Use step functions to specify the complete response.* By using the shifted unit step function, the two expressions obtained previously can be combined into a single expression:

$$v_C(t) = 9e^{-1.25t}[u(t) - u(t-1)] + 2.58e^{-\frac{t-1}{0.3}}u(t-1) \text{ V}$$

Step 4. *Obtain a plot of the response.* Using MATLAB and code similar to that used in Example 8.2, the plot in Figure 8.11 was obtained. Here the part of the response with the 0.3 sec time constant shows a greater rate of decay than the longer 0.8 sec time constant.

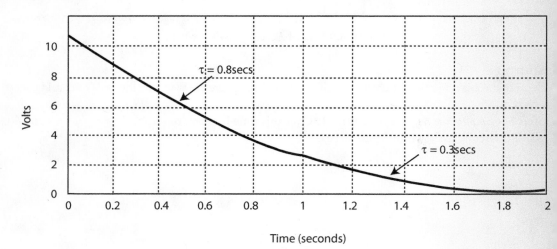

FIGURE 8.11 Response, $v_C(t)$, for the circuit of Figure 8.9.

Exercises. 1. Show that $u(t) - u(t-1) = u(t)u(1-t)$.

2. Suppose that in Example 8.3 the switch moves to the 4.5 Ω resistor at $t = 0.5$ sec instead of 1 sec. Compute the value $v_C(t)$ at $t = 1.2$ sec.

ANSWER: 0.4671 V

For all of these examples $\tau > 0$ and the response is a decaying exponential. Intuitively, the response decays because the resistor dissipates as heat the energy initially stored in the inductor or capacitor. One of the homework exercises will ask the student to show that the total energy dissipated in the resistor from t_0 to ∞ equals the decrease in energy initially stored in the inductor or capacitor at t_0. When controlled sources are present, R_{th} may be negative, in which case $\tau < 0$. Here the negative resistance supplies energy to the circuit and the source-free response will grow exponentially. This is illustrated in the next example.

EXAMPLE 8.4

Find $v_C(t)$ for the circuit of Figure 8.12, assuming that $g_m = 0.75$ S and $v_C(0^-) = 10$ V.

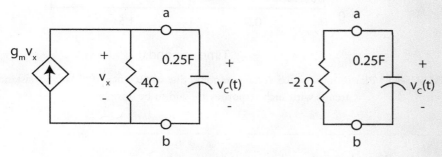

FIGURE 8.12 Parallel *RC* circuit with dependent current source.

SOLUTION

It is straightforward to show that the Thevenin equivalent seen by the capacitor is a **negative resistance**, $R_{th} = -2$ Ω, as shown in Figure 8.12b. Again, by equation 8.11,

$$v_C(t) = e^{-\frac{1}{R_{th}C}t} v_C(0^+) = 10e^{2t}u(t) \text{ V}$$

Because of the negative resistance, this response grows exponentially, as shown in Figure 8.13. A circuit having a response that increases without bound is said to be **unstable**. Practically speaking, an unstable circuit will destroy itself or exhibit a nonlinear phenomenon that clamps the voltage at a finite value, as in the case of saturation in an op amp.

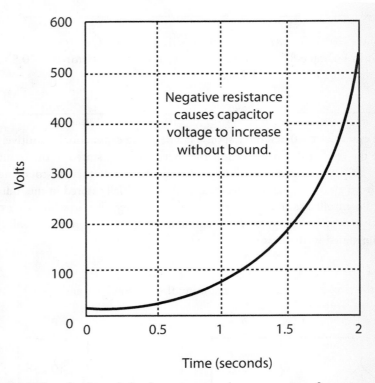

FIGURE 8.13 Plot of unbounded voltage response due to presence of **negative resistance**. Circuits with such responses are said to be unstable.

Exercises. 1. For Example 8.3 show that

$$i_C(t) = e^{-\frac{1}{R_{th}C}t} \quad i_C(0^+) = 5e^{2t}u(t)\,\text{A}$$

2. In Example 8.3, let $g_m = 0.125$ S. Find the equivalent resistance seen by the capacitor and $v_C(t)$, $t \geq 0$.
ANSWERS: 8 Ω, $10e^{-0.5t}u(t)$ V

3. Show that in general, for $t > t_0$, the form of the capacitor current is similar to the voltage form. Hint: Apply the capacitor *v-i* relationship to equation 8.11.

4. DC OR STEP RESPONSE OF FIRST-ORDER CIRCUITS

The circuits of the previous section had no source excitations. This section takes up the calculation of voltage and current responses when constant-voltage or constant-current sources are present. It is instructive to start with the basic series *RL* and *RC* circuits as shown in Figure 8.14.

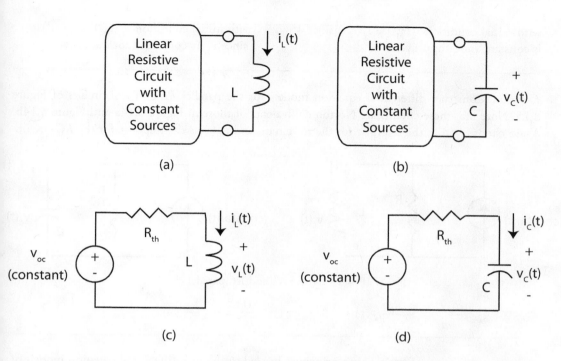

FIGURE 8.14 (a) Driven first-order *RL* circuit. (b) Driven first-order *RC* circuit. (c) Thevenin equivalent representation of (a). (d) Thevenin equivalent representation of (b).

Given these basic circuit representations and initial conditions at t_0, what is the structure of a differential equation model that governs their voltage and current behavior for $t \geq t_0$? The first objective is to derive the "differential equation" models characterizing each circuit's voltage and current responses. It is convenient to use $i_L(t)$ as the desired response for constructing the differential equation for the series *RL* circuit (Figure 8.14c), whereas for the series *RC* circuit (Figure 8.14d), $v_C(t)$ is the more convenient variable.

(i) The circuit model for the inductor is

$$v_L(t) = L \frac{di_L(t)}{dt}$$

(ii) By KVL and Ohm's law,

$$v_L(t) = v_{oc} - R_{th} i_L(t)$$

(iii) Substituting for $v_L(t)$ leads to the differential equation model

$$\frac{di_L(t)}{dt} = -\frac{R_{th}}{L} i_L(t) + \frac{1}{L} v_{oc} \quad (8.15a)$$

(i) The circuit model for the capacitor is

$$i_C(t) = C \frac{dv_C(t)}{dt}$$

(ii) By KCL and Ohm's law,

$$i_C(t) = \frac{v_{oc} - v_C(t)}{R_{th}}$$

(iii) Substituting for $i_C(t)$ leads to the differential equation model

$$\frac{dv_C(t)}{dt} = -\frac{1}{R_{th}C} v_C(t) + \frac{1}{R_{th}C} v_{oc} \quad (8.15b)$$

with initial condition $i_L(t_0^-) = i_L(t_0^+)$ since V_{oc} is constant (not impulsive).

with initial condition $v_C(t_0^-) = v_C(t_0^+)$ since v_{oc} is constant (not impulsive).

Exercise. Construct differential equation models for the parallel *RL* and *RC* circuits of Figure 8.15. Note that these circuits are Norton equivalents of those in Figure 8.14a and Figure 8.14b. Again choose $i_L(t)$ as the response for the *RL* circuit and $v_C(t)$ as the response for the *RC* circuit.

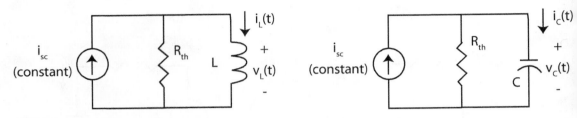

FIGURE 8.15 Driven *RL* and *RC* parallel circuits.

ANSWERS: $\dfrac{di_L(t)}{dt} = -\dfrac{R_{th}}{L} i_L(t) + \dfrac{R_{th}}{L} i_{sc}$ and $\dfrac{dv_C(t)}{dt} = -\dfrac{1}{R_{th}C} v_C(t) + \dfrac{1}{C} i_{sc}$.

A simple application of basic circuit principles has led to the two differential equation models of equations 8.15. The next important question is: What do these two differential equation models tell us about the behavior of each circuit? Equivalently, how do we find a solution to the equations? Observe that both differential equations 8.15 have the same structure:[1]

$$\frac{dx(t)}{dt} = -\frac{1}{\tau} x(t) + F \tag{8.16a}$$

where the time constant $\tau = \dfrac{L}{R_{th}}$ for *RL* circuits and $\tau = R_{th}C$ for *RC* circuits, and $F = v_{oc}/L$ for *RL* circuits and $F = v_{oc}/(R_{th}C)$ for the *RC* case. This equation is valid for $t \geq t_0$. Equation 8.4, rewritten here with $f(q) = F$, presents the general formula for solving the differential equation 8.16a:

$$x(t) = e^{\lambda(t-t_0)}x(t_0^+) + \int_{t_0}^{t} e^{\lambda(t-q)} F \, dq \tag{8.16b}$$

where $\lambda = -\dfrac{1}{\tau}$ is a natural frequency of the circuit, and where we have emphasized the use of the initial condition at t_0^+. Note that as long as $x(t)$ is a capacitor voltage or inductor current, the initial condition is continuous, i.e., $x(t_0^-) = x(t_0^+)$, because F is a constant (non-impulsive) **forcing function**. A straightforward evaluation of the integral of equation 8.16b yields

$$x(t) = e^{-\left(\frac{t-t_0}{\tau}\right)} x(t_0^+) + F e^{-\left(\frac{t}{\tau}\right)} \int_{t_0}^{t} e^{-\left(\frac{q}{\tau}\right)} dq = e^{-\left(\frac{t-t_0}{\tau}\right)} x(t_0^+) + F\tau \left[1 - e^{-\left(\frac{t-t_0}{\tau}\right)} \right] \tag{8.16c}$$

Some rearranging of terms in equation 8.16c produces the desired formula

$$x(t) = F\tau + \left[x(t_0^+) - F\tau\right]e^{-\left(\frac{t-t_0}{\tau}\right)} = -\frac{F}{\lambda} + \left[x(t_0^+) + \frac{F}{\lambda}\right]e^{\lambda(t-t_0)} \tag{8.17}$$

which is valid for $t \geq t_0$. After some interpretation, this formula will serve as a basis for computing the response to RL and RC circuits driven by constant sources. A homework exercise will ask for a different and direct derivation of this formula.

At this point it is helpful to interpret the quantity $F\tau$ in equation 8.17. For RL circuits, when $x(t) = i_L(t)$, equation 8.15a implies that $F = v_{oc}/L$, $\tau = L/R_{th}$, and hence $F\tau = v_{oc}/R_{th} = i_{sc}$. For RC circuits when $x(t) = v_C(t)$, equation 8.15b implies that $F = v_{oc}/R_{th}C$, $\tau = R_{th}C$, and hence $F\tau = v_{oc}$. This interpretation is valid for both positive and negative values of τ. If $\tau > 0$, then

$$x(\infty) = \lim_{t \to \infty} x(t) = \lim_{t \to \infty} \left(F\tau + \left[x(t_0^+) - F\tau\right]e^{-\frac{t-t_0}{\tau}}\right) = F\tau = \begin{cases} i_{sc} = \dfrac{v_{oc}}{R_{th}} & \text{for } RL \text{ case} \\ v_{oc} & \text{for } RC \text{ case} \end{cases} \tag{8.18}$$

This means that for the RL case, $i_L(\infty) = i_{sc} = v_{oc}/R_{th}$, and for the RC case, $v_C(\infty) = v_{oc}$. Of course, i_{sc} is computed by replacing the inductor with a short circuit, and v_{oc} is computed by replacing the capacitor with an open circuit. See Chapter 6 for details. Mathematically, any constant, such as $x(t) = $ constant, that satisfies a differential equation is called an **equilibrium state** of that differential equation. Since the constant $x(t) = F\tau$ satisfies the differential equation 8.16, $F\tau$ is an **equilibrium state** of the differential equation 8.16.

Whenever $\tau > 0$, equation 8.18 implies that the formula (equation 8.17) for the solution of equation 8.16 given constant or dc excitation becomes

$$x(t) = x(\infty) + \left[x(t_0^+) - x(\infty)\right]e^{-\frac{t-t_0}{\tau}} \tag{8.19a}$$

and when $x(t) = i_L(t)$,

$$i_L(t) = i_L(\infty) + \left[i_L(t_0^+) - i_L(\infty)\right]e^{-\frac{R_{th}}{L}(t-t_0)} \tag{8.19b}$$

and when $x(t) = v_C(t)$,

$$v_C(t) = v_C(\infty) + \left[v_C(t_0^+) - v_C(\infty)\right]e^{-\frac{t-t_0}{R_{th}C}} \tag{8.19c}$$

Note that $\tau > 0$ is true whenever $R_{th} > 0$, $C > 0$, and $L > 0$, i.e., the circuit is said to be **passive**. This allows us to state a nice physical interpretation of equation 8.19a:

$$x(t) = [\textit{Final value}] + ([\textit{Initial value}] - [\textit{Final value}])e^{-\frac{\textit{elapsed time}}{\textit{time constant}}}$$

Graphically, equation 8.19a is depicted in Figure 8.16 for $x(\infty) > x(t_0)$.

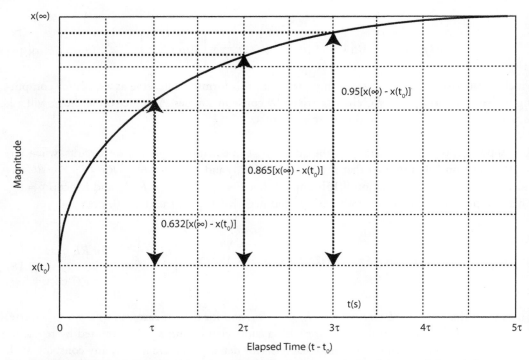

FIGURE 8.16 Graphical interpretation of equation 8.19a for the case $x(\infty) > x(t_0)$.

Exercise. Redo the curve of Figure 8.16 for the case $x(\infty) < x(t_0)$.

The initial value $x(t_0^+)$ is computed from initial conditions and possibly the value of the source excitation, or it can be computed from past excitations up to t_0^+. Several examples will now illustrate the use of equation 8.19.

EXAMPLE 8.5

For the circuit of Figure 8.17, suppose a 10 V unit step excitation is applied at $t = 1$ when it is found that the inductor current is $i_L(1^-) = 1$ A. The 10 V excitation is represented mathematically as $v_{in}(t) = 10u(t - 1)$ V for $t \geq 1$. Find $i_L(t)$ for $t \geq 1$.

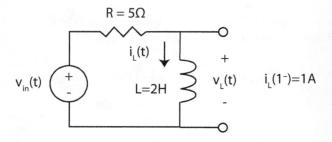

FIGURE 8.17 Driven series *RL* circuit for Example 8.5 with $i_L(1^-) = 1$ A.

SOLUTION

Step 1. *Determine the circuit's differential equation model.* Since the circuit of Figure 8.17 is a driven series *RL* circuit, equation 8.15a implies that the differential equation model of the circuit valid for $t \geq 1$ is

$$\frac{di_L(t)}{dt} = -\frac{R}{L}i_L(t) + \frac{1}{L}V_s = -\frac{1}{0.4}i_L(t) + \frac{10}{2}u(t-1)$$

where the time constant $\tau = 0.4$ sec.

Step 2. *Determine the form of the response.* Since $i_L(1^-) = i_L(1^+)$, equation 8.19b implies that

$$i_L(t) = i_L(\infty) + \left[i_L(1^+) - i_L(\infty)\right]e^{-\left(\frac{t-1}{\tau}\right)}u(t-1) \text{ A}$$

Here the presence of $u(t - 1)$ emphasizes that the response is valid only for $t \geq 1$.

Step 3. *Compute $i_L(\infty)$ and set forth the final expression for $i_L(t)$.* Since $\tau = 0.4 > 0$, we replace the inductor in the circuit of Figure 8.17 with a short circuit to compute $i_{sc} = i_L(\infty) = 2$ A. It follows that

$$i_L(t) = \left[2 + (1-2)e^{-2.5(t-1)}\right]u(t-1) = \left(2 - e^{-2.5(t-1)}\right)u(t-1) \text{ A}$$

Step 4. *Plot $i_L(t)$.* One cannot presume that the response is zero for $t < 1$. Hence, using MATLAB or the equivalent, one can construct the graph of $i_L(t)$ for $t \geq 1$ as given in Figure 8.18.

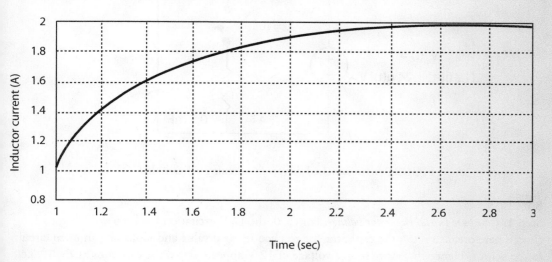

FIGURE 8.18 Plot of $i_L(t)$ for Example 8.5.

Step 5. *Compute $v_L(t)$.* Given the expression for the inductor current in step 3, it follows that for $t > 1$,

$$v_L(t) = L\frac{di_L(t)}{dt} = -\frac{L}{\tau}\left[i_L(1^+) - i_L(\infty)\right]e^{-\frac{t-1}{\tau}}u(t-1^+) = 5e^{-2.5(t-1)}u(t-1^+) \text{ V}$$

Exercises. 1. Verify that in Example 8.5 $v_L(t)$ can be obtained without differentiation by $v_L(t) = v_{oc} - R_{th}i_L(t)$.

2. In Example 8.5, suppose R is changed to 4 Ω. Find $i_L(t)$ at t = 2 sec.
ANSWER: 1.8647 A

Note that we have used the differential equation 8.16 (or equations 8.15) to obtain the solution form of equation 8.19. However, when using equations 8.19, it is not necessary to reconstruct the differential equation of the *RL* or *RC* circuit. Specifically, we need only compute $x(t_0^+)$, $x(\infty)$, and the time constant $\tau = L/R_{th}$ or $R_{th}C$.

The method described for computing final values can also be used to find the *initial values* of v_C and i_L at $t = t_0$ if dc excitations have been applied to the circuit for a long time before $t = t_0$. The next example illustrates this technique and extends the preceding discussion.

EXAMPLE 8.6

The source in the circuit of Figure 8.19 furnishes a 12 V excitation for $t < 0$ and a 24 V excitation for $t < 0$, denoted by $v_{in}(t) = [12u(-t) + 24u(t)]$ V. The switch in the circuit closes at t = 10 sec. First determine the value of the capacitor voltage at $t = 0^-$, which by continuity equals $v_C(0^+)$. Next determine $v_C(t)$ for all $t \geq 0$.

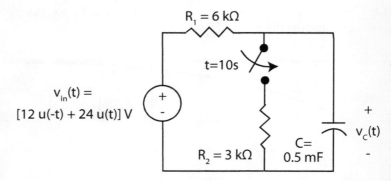

FIGURE 8.19 Switched driven *RC* circuit for Example 8.6.

SOLUTION

Step 1. *Compute initial capacitor voltage.* For $t < 0$, the 12 V excitation has been applied for a long time. Therefore, at $t = 0^-$, the capacitor has reached its final value and looks like an open circuit to the source. Hence the entire source voltage of 12 V appears across the capacitor at $t = 0^-$, i.e., $v_C(0^-) = v_C(0^+) = 12$ V by the continuity property of the capacitor voltage.

Step 2. *Use equation 8.19c to obtain* $v_C(t)$ *for* $0 \leq t \leq 10$ sec. Equation 8.19c requires only that we know $v_C(0^+)$ (step 1), τ, and $v_C(\infty)$. For $0 \leq t \leq 10$ sec, $\tau = R_1C = 3$ sec. It is important to realize here that for $0 \leq t \leq 10$ sec the circuit behaves as if the switch were not present. Hence, the computation of $v_C(\infty)$ proceeds as if no switching would take place at $t = 10$ sec. Here $v_{oc} = v_C(\infty) = 24$ V. Hence, for $0 \leq t \leq 10$, equation 8.19c implies

$$v_C(t) = v_C(\infty) + \left[v_C(0^+) - v_C(\infty)\right]e^{-\frac{t}{R_{th}C}} = 24 + (12 - 24)e^{-\frac{t}{3}} = 24 - 12\,e^{-\frac{t}{3}} \text{ V} \qquad (8.20)$$

Step 3. *Compute the initial condition for the interval* $10 < t$, *i.e.,* $v_C(10^+)$. Plugging into equation 8.20 and using the continuity property of the capacitor voltage yields

$$v_C(10^-) = v_C(10^+) = 24 - 12e^{-10/3} = 23.57 \text{ V}$$

Step 4. *Find* $v_C(t)$ *for* $t > 10$. For $t > 10$, the resistive part of the circuit can be replaced by its Thevenin equivalent, which yields Figure 8.20.

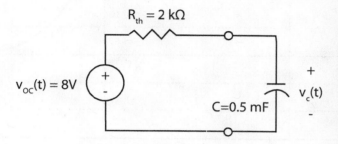

FIGURE 8.20 Circuit equivalent to that of Figure 8.19 for $t \geq 10$.

Here, equation 8.19c applies again. The value for $v_C(\infty)$, however, is now 8 V and the new time constant is $R_{th}C = 1$ sec. Hence, for $t > 10$,

$$v_C(t) = v_C(\infty) + \left[v_C(10^+) - v_C(\infty)\right]e^{-\frac{t-10}{R_{th}C}} = 8 + (23.57 - 8)e^{-(t-10)} = 8 + 15.57e^{-(t-10)} \text{ V}$$

Step 5. *Set forth the complete response using step functions.* Using step functions, the response $v_C(t)$ for $t \geq 0$ is

$$v_C(t) = 24 - 12e^{-\frac{t}{3}}[u(t) - u(t-10)] + \left[8 + 15.57e^{-(t-10)}\right]u(t-10) \text{ V}$$

Step 6. *Plot* $v_C(t)$. Plotting $v_C(t)$ yields the graph of Figure 8.21.

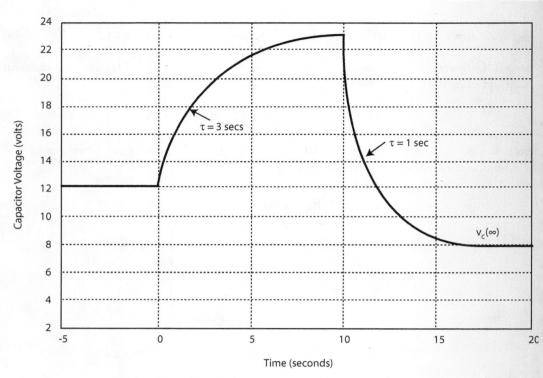

FIGURE 8.21 Capacitor voltage $v_C(t)$ for $t \geq 0$.

Exercise. Suppose the switch in Example 8.6 opens again at $t = 20$ sec. Find $v_C(t)$ at $t = 25$ sec.
ANSWER: 20.98 V

EXAMPLE 8.7 The circuit of Figure 8.22a has a capacitor voltage given by the curve in Figure 8.22b. We note that $v_C(0.1) = 7.057$ V. Find, V_0, $v_C(0)$, the time constant $\tau = RC$, the exact value of $v_C(0.25)$, and the value of C if $R = 100\Omega$.

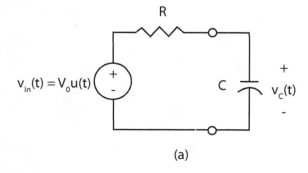

(a)

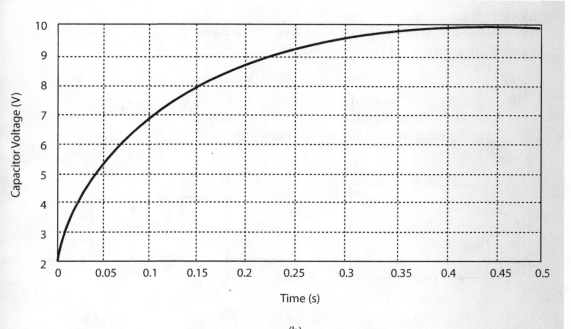

(b)

FIGURE 8.22 (a) Series *RC* circuit. (b) Capacitor voltage, $v_C(t)$.

SOLUTION. A simple inspection of the graph indicates that $v_C(0) = 2$ V. One recalls that

$$v_C(t) = v_C(\infty) + \left[v_C(0^+) - v_C(\infty) \right] e^{-\frac{t}{\tau}}$$

Hence as $t \to \infty$, $v_C(t) \to v_C(\infty) = 10$ V. Since the capacitor looks like an open circuit at $t = \infty$, $V_0 = v_C(\infty) = 10$ V. From the given problem data,

$$v_C(0.1) = 7.057 = 10 - 8e^{-0.1/\tau}$$

Simplifying yields

$$\tau = \frac{-0.1}{\ln\left(\dfrac{10 - 7.057}{8} \right)} = 0.1 \text{ s}$$

Therefore $C = 0.01$ F and $v_C(0.25) = 9.3433$ V.

When switching occurs frequently, or the excitation changes its constant level frequently, then hand analysis, as in Example 8.6, becomes very tedious. For such problems a SPICE simulation (or the equivalent) proves useful and saves time. The next example uses SPICE to compute the waveform of a simple *RC* circuit whose input excitation is a square wave. Like the previous example, the solution is broken down into time intervals such that during each time interval inputs are constant. Because no switching occurs, the time constants for all time intervals are the same. In applying equation 8.19c, the quantities that vary from one time interval to the next are the initial values and final values.

EXAMPLE 8.8

The first-order *RC* circuit of Figure 8.23a is excited by the 50 Hz square wave input voltage of Figure 8.23b given that the capacitor is initially relaxed.

(a)   Plot $v_C(t)$ for $0 < t < 60$ ms, using SPICE or equivalent software.

(b)   Find the initial value and the final value in equation 8.19 when t is very large, for example, at the beginning and end of the interval $1 < t < 1.01$ sec. Plot the $v_C(t)$ wave for this interval using MATLAB or the equivalent.

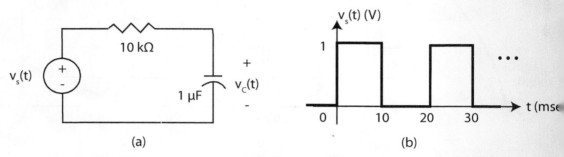

(a)   (b)

FIGURE 8.23 (a) Series *RC* circuit excited by the 50 Hz square wave of (b).

SOLUTION

Part (a)

Doing a SPICE or equivalent simulation gives rise to the response curve shown in Figure 8.24, over which the square wave input is superimposed. Observe that the response $v_C(t)$ has an approximate tri-angular shape. What is happening is that from zero to 10 msec, the circuit sees a step and hence the capacitor voltage rises toward one volt. At 10 msec, the square goes to zero for 10 msec. The capacitor then discharges its stored energy through the resistor, causing a decrease in its voltage value. The decrease does not go to zero, however. So when the square wave again is at 1 volt the capacitor voltage begins to rise again and achieves a slightly higher value at $t = 30$ msec compared to $t = 10$ msec. In fact, one notices in Figure 8.24 that the peak and minimum values are increasing slightly as time increases. Eventually the peak and minimum values will reach their respective fixed values, called steady-state values. To find these values, a simulation program could require a very lengthy simulation interval, which often proves impractical. The steady-state values can be computed analytically as in part (b).

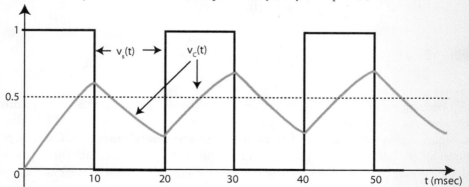

FIGURE 8.24 Response of circuit of Figure 8.23a calculated using SPICE. For reference, the input square wave excitation is superimposed on the plot.

Part (b)

Let $t_0 = mT$, where $T = 20$ msec is the period of the square wave and m is some large integer. Then,

$$v_C(t_0 + 0.5\,T) = 1 + (v_C(t_0) - 1)\,e^{-\frac{0.5T}{RC}} = 1 + [v_C(t_0) - 1]e^{-1} \qquad (8.21)$$

Further, in steady state, $v_C(t_0 + T) = v_C(t_0)$, which implies that

$$v_C(t_0 + T) = v_C(t_0 + 0.5\,T)\,e^{-\frac{0.5T}{RC}} = v_C(t_0 + 0.5\,T)e^{-1} = v_C(t_0) \qquad (8.22a)$$

Equivalently, equation 8.22a implies that

$$v_C(t_0 + 0.5\,T) = v_C(t_0)e^{1} \qquad (8.22b)$$

Substituting equation 8.22b into equation 8.21 yields

$$v_C(t_0)e^{1} = 1 + [v_C(t_0) - 1]e^{-1}$$

the solution of which is

$$v_C(t_0) = \frac{1}{1 + e} = 0.2689 \text{ V}$$

It follows that

$$v_C(t_0 + 0.5\,T) = v_C(t_0)e^{1} = 0.7311 \text{ V}$$

An examination of the response in Figure 8.24 shows that the minimum and peak values are approaching the steady-state values of 0.2689 V and 0.7311 V, respectively.

Exercise. Based on the response in Figure 8.24, roughly sketch the capacitor current, $i_C(t)$. At what time instants is the capacitor current discontinuous?

---

5. SUPERPOSITION AND LINEARITY

Superposition, a special case of linearity, helps simplify the analysis of resistive circuits, as discussed in Chapter 5. Recall that linear resistive circuits are interconnections of resistors and sources, both dependent and independent. Does superposition still apply when capacitors and inductors are added to the circuit? The answer is yes, provided one properly accounts for initial conditions.

In order to justify the use of **superposition** for RC and RL circuits, consider that resistors satisfy Ohm's law, a linear algebraic equation. Capacitors satisfy the differential relationship

$$i_C = C\frac{dv_C}{dt}$$

which is also a linear equation. To see linearity in this i-v relationship, suppose voltages v_{C1} and v_{C2} individually excite a relaxed capacitor producing the respective currents

$$i_{C1} = C\frac{dv_{C1}}{dt}, \quad i_{C2} = C\frac{dv_{C2}}{dt}$$

Let i_{C3} be the current induced by a voltage equal to the sum of v_{C1} and v_{C2}, i.e.,

$$i_{C3} = C\frac{d}{dt}(v_{C1} + v_{C2})$$

However, the linearity of the derivative implies the property of superposition:

$$i_C = C\frac{dv_{C1}}{dt} + C\frac{dv_{C2}}{dt} = i_{C1} + i_{C2}$$

By the same arguments, the current due to the input excitation $v_{C3} = a_1 v_{C1} + a_2 v_{C2}$ is $i_{C3} = a_1 i_{C1} + a_2 i_{C2}$.

On the other hand, suppose two separate currents i_{C1} and i_{C2} individually excite a relaxed capacitor C. Each produces a voltage given by the integral relationship

$$v_{C1}(t) = \frac{1}{C}\int_{-\infty}^{t} i_{C1}(\tau)\, d\tau, \quad v_{C2}(t) = \frac{1}{C}\int_{-\infty}^{t} i_{C2}(\tau)\, d\tau$$

By the distributive property of integrals, the combined effect of the input, $a_1 i_{C1} + a_2 i_{C2}$, would be a voltage,

$$v_{C3}(t) = \frac{1}{C}\int_{-\infty}^{t}\left[a_1 i_{C1}(\tau) + a_2 i_{C2}(\tau)\right] d\tau$$

$$= a_1\left[\frac{1}{C}\int_{-\infty}^{t} i_{C1}(\tau)\, d\tau\right] + a_2\left[\frac{1}{C}\int_{-\infty}^{t} i_{C2}(\tau)\, d\tau\right]$$

$$= a_1 v_{C1}(t) + a_2 v_{C2}(t)$$

Thus linearity and, hence, superposition hold.

Arguments analogous to the preceding imply that a relaxed inductor satisfies a linear relationship, and thus superposition is valid, whether the inductor is excited by currents or by voltages.

The interconnection of linear capacitors and linear inductors with linear resistors and sources satisfying KVL and KCL produces linear circuits because KVL and KCL are linear algebraic constraints on the linear element equations. Hence, the property of linearity is maintained, and as a consequence superposition holds for the interconnected circuit.

To cap off this discussion we must account for the presence of initial conditions on the capacitors and inductors of the circuit. For first-order RC and RL circuits, this need is clearly indicated by the first term of equation 8.4. For a general linear circuit, one can view each initial condition as being set up by an input that shuts off the moment the initial condition is established. Hence the effect of the initial condition can be viewed as the effect of some input that turns off at the time the initial condition is specified. This means that when using superposition on a circuit, one first looks at the effect of each independent source on a circuit having no initial conditions. Then one sets all independent sources to zero and computes the response due to each initial condition with all other initial conditions set to zero. The sum of all the responses to each of the independent sources plus the individual initial condition responses yields the complete circuit response, by the principle of superposition. A rigorous justification of this principle is given in a later chapter using the Laplace transform method.

The following example illustrates the application of these ideas.

EXAMPLE 8.9
The linear circuit of Figure 8.25 has two source excitations applied at $t = 0$, as indicated by the presence of the step functions. The initial condition on the inductor current is $i_L(0^-) = -1$ A. Compute the response $i_L(t)$ for $t \geq 0$ using superposition.

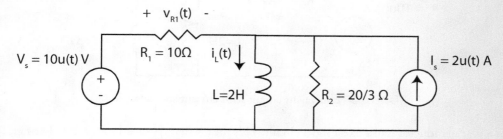

FIGURE 8.25 Driven RL circuit excited by two sources.

SOLUTION
Because the circuit is linear, having a linear differential equation, superposition is but one of several methods for obtaining the solution. An alternative approach is to find the Thevenin equivalent circuit seen by the inductor. As we will see, the superposition approach sometimes has an advantage over the Thevenin approach.

Superposition must be carefully applied, however. First one computes the response due only to the initial condition with the sources set to zero. Second, one computes the response due to V_S with all initial conditions and all other sources set to zero. Third, one computes the response due to I_S with all initial conditions and all other sources set to zero. Finally, one adds these three responses together to obtain the complete circuit response.

Step 1. *Compute the part of the circuit response due only to the initial condition, with all independent sources set to zero.* With both sources set to zero, there results the equivalent circuit given by Figure 8.26. The Thevenin equivalent resistance is $R_{th} = 4$ Ω, resulting from the parallel combination of R_1 and R_2. Figure 8.26 depicts the equivalent undriven RL circuit having response

$$i_L^1(t) = e^{-\frac{R}{L}t} i_L(0^+) = -e^{-2t} u(t)$$

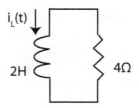

FIGURE 8.26 Equivalent parallel RL circuit when sources are set to zero.

Step 2. *Determine the response due only to $V_S = 10u(t)$ V.* In this case Figure 8.27 portrays the equivalent circuit. The initial condition is set to zero as its effect was accounted for by the calculations of step 1.

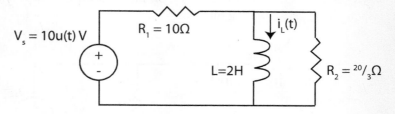

FIGURE 8.27 Equivalent circuit with current source set to zero.

At $t = \infty$, the inductor looks like a short circuit. This means that $i_L(\infty) = V_S/R_1 = 1$ A. Further, $R_{th} = R_1 // R_2 = 4\ \Omega$, as before. Thus the contribution to the complete response due only to the source V_S is

$$i_L^2(t) = i_L^2(\infty) + \left[i_L^2(0^+) - i_L^2(\infty)\right]e^{-\frac{R_{th}}{L}t}u(t) = 1 - e^{-2t}\,u(t)\ \text{A}$$

Step 3. *Compute the response due only to the current source $I_S = 2u(t)$ A.* Here the equivalent circuit is given by Figure 8.28, where again $i_L(0^-) = 0$ as its effect was accounted for in step 1. For this equivalent circuit $i_L(\infty) = 2$ A and $R_{th} = 4\ \Omega$, as before. Hence

$$i_L^3(t) = i_L^3(\infty) + \left[i_L^3(0^+) - i_L^3(\infty)\right]e^{-\frac{R_{th}}{L}t}u(t) = 2\left(1 - e^{-2t}\right)u(t)\ \text{A}$$

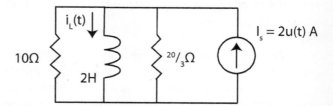

FIGURE 8.28 Equivalent circuit when voltage source is set to zero.

Step 4. *Apply the principle of superposition.* By superposition, the complete response $i_L(t)$ is given by the sum of the individual responses. Hence,

$$i_L(t) = i_L{}^1(t) + i_L{}^2(t) + i_L{}^3(t)$$

$$= \underset{\substack{\text{due to} \\ \text{initial} \\ \text{condition}}}{-e^{-2t}u(t)} \quad + \quad \underset{\substack{\text{due to} \\ \text{source } V_S}}{\left(1-e^{-2t}\right)u(t)} \quad + \quad \underset{\substack{\text{due to} \\ \text{source } I_S}}{2\left(1-e^{-2t}\right)u(t)} \; \mathbf{A}$$

This of course simplifies to $i_L(t) = (3 - 4e^{-2t})u(t)$ A.

The question remains as to why the superposition principle holds an advantage over the Thevenin equivalent method. The reasoning lies in the answers to the following questions:

Question 1. What is the new response if the initial condition is changed to $i_L(0^-) = 5$ A?
With the complete response decomposed as $i_L(t) = i_L{}^1(t) + i_L{}^2(t) + i_L{}^3(t)$, only the part due to the initial condition changes, i.e., $i_L{}^1(t)$. Since the circuit is linear, this change is linear, i.e., $i_L(0^-)_{\text{new}}$ $= -5 \times i_L(0^-)_{\text{old}}$. Therefore the part of the response due to the initial condition must be scaled by a factor of -5. Hence the new response is given by

$$i_L{}^{new}(t) = i_L{}^{1new}(t) + i_L{}^2(t) + i_L{}^3(t) = -5i_L{}^{1old}(t) + i_L{}^2(t) + i_L{}^3(t)$$

$$= 5e^{-2t}u(t) + \left[1 - e^{-2t}\right]u(t) + 2\left[1 - e^{-2t}\right]u(t) \; \text{A}$$

Question 2. What is the new response if the voltage source V_S is changed to $5u(t)$ V, with all other parameters held constant at their original values?
In this case the value of the voltage source is cut in half, i.e., scaled by a factor of 0.5. Therefore the contribution to the response due only to V_S must be scaled by 0.5, i.e.,

$$i_L{}^{new}(t) = i_L{}^1(t) + i_L{}^{2new}(t) + i_L{}^3(t) = i_L{}^1(t) + 0.5i_L{}^{2old}(t) + i_L{}^3(t)$$

$$= -e^{-2t}u(t) + 0.5\left[1 - e^{-2t}\right]u(t) + 2\left[1 - e^{-2t}\right]u(t) \; \text{A}$$

Question 3. What is the new response if the initial condition is changed to 5 A, the voltage source V_S is changed to $5u(t)$ V, and the current source I_S is changed to $8u(t)$ A?
   Again, by linearity it is necessary to scale each of the individual responses by the appropriate scale factor. Specifically,

$$i_L{}^{new}(t) = i_L{}^{1new}(t) + i_L{}^{2new}(t) + i_L{}^{3new}(t) = -5i_L{}^{1old}(t) + 0.5i_L{}^{2old}(t) + 4i_L{}^{3old}(t)$$

$$= 5e^{-2t}u(t) + 0.5\left[1 - e^{-2t}\right]u(t) + 8\left[1 - e^{-2t}\right]u(t) \; \text{A}$$

This discussion argues that linearity permits one to decompose a complete circuit response into the superposition of the response due only to the initial condition and the responses due to each of the source excitations taken alone. Changes in any source or in the initial condition are easily reflected in the new response without the need to resolve the circuit equations. This allows one to explore easily a circuit's behavior over a wide range of excitations and initial conditions.

An approach based on the Thevenin equivalent circuit seen by the inductor would allow one to quickly compute the complete response, but not in a way that identifies the contributions due to each of the individual sources. Answers to the preceding three questions would have required repeated solutions to the circuit equations. However, if one keeps the source values in literal form, then the Thevenin equivalent approach would be as efficient.

6. RESPONSE CLASSIFICATIONS

Having gained some understanding of the form of the behavior of *RL* and *RC* circuits, it is instructive to classify the responses into categories. The **zero-input response** of a circuit is the response to the initial conditions when all the inputs are set to zero. The **zero-state response** of a circuit is the response to a specified input signal or set of input signals given that the initial conditions are all set to zero. By linearity, the sum of the zero-input and zero-state responses is the **complete response** of the circuit. This categorization is the convention in advanced linear systems and linear control texts.

Frequently circuits texts include two other notions of response, the **natural response** and the **forced response.** However, decomposition of the complete response into the sum of a natural and a forced response applies only when the input excitation is (1) dc, (2) real exponential, (3) sinusoidal, or (4) exponentially modulated sinusoidal. Further, the exponent of the input excitation, for example, a in $f(t) = e^{at}u(t)$, must be different from that appearing in the zero-input response. Under these conditions it is possible to define the natural and forced responses as follows: (1) the natural response is that portion of the complete response that has the same exponents as the zero-input response, and (2) the forced response is that portion of the complete response that has the same "exponent" as the input excitation provided the input excitation has exponents different from that of the zero-input response.

This decomposition is important for two reasons. First, it agrees with the classical method of solving linear ordinary differential equations with constant coefficients where the natural response corresponds to the **complementary function** and the forced response corresponds to the **particular integral**. Students fresh from a course in linear differential equations will feel quite at home with these concepts. The second reason is that the forced response is easily calculated for dc inputs. For general systems this type of decomposition is not used.

7. FURTHER POINTS OF ANALYSIS AND THEORY

In deriving equations 8.17 and 8.19a, the quantity $x(t)$ was thought of as a capacitor voltage or an inductor current. It turns out that any voltage or current in an *RC* or *RL* first-order linear circuit with constant input has the form

$$x(t) = X_e + \left[x(t_0^+) - X_e \right] e^{-\frac{(t-t_0)}{\tau}} \tag{8.23}$$

For τ negative or positive equation 8.23 is identical to equation 8.17 with $F\tau = X_e$. Further, τ is the circuit time constant and X_e is that voltage or current of interest computed under the condition that the inductor is replaced by a short circuit for the *RL* case or the capacitor is replaced by an open circuit for the *RC* case.

How do we justify the form of equation 8.23 for all variables? We invoke the linearity theorem of Chapter 5 and the source substitution theorem of Chapter 6 of 2nd edition. Suppose in a first-order *RC* circuit we have found $v_C(t)$. By the source substitution theorem, the capacitor can then be replaced by a voltage source whose voltage is the computed $v_C(t)$. This new circuit consists of constant independent sources, one independent source of value $v_C(t)$, and resistors and dependent sources. By linearity, any voltage or current in the circuit has the form

$$x(t) = K_1 + K_2 v_C(t)$$

for appropriate K_1 and K_2. By equation 8.17,

$$v_C(t) = K_3 + K_4 e^{-\frac{(t-t_0)}{\tau}}$$

which implies that

$$x(t) = K_5 + K_6 e^{-\frac{(t-t_0)}{\tau}}$$

for appropriate K_i. This is the same form as equation 8.23 for proper choices of K_5 and K_6.

Exercise. Show that, for $\tau > 0$, $K_5 = X_e = x(\infty)$ and $K_6 = x(t_0{}^+) - X_e$.

Note that equation 8.23 requires that the initial value be evaluated at $t = t_0{}^+$ instead of $t = t_0{}^-$. This is because only the inductor currents and the capacitor voltages are guaranteed to be continuous from one instant to the next for constant input excitations. The capacitor current and the inductor voltage as well as other circuit voltages and currents may not behave continuously.

EXAMPLE 8.10

This example illustrates the application of equation 8.23. For the circuit of Figure 8.29, $v_{in}(t) = -18u(-t) + 9u(t)$ V. Find $i_{in}(t)$ for $t > 0$.

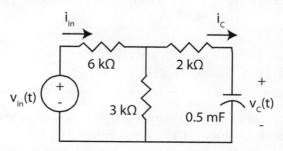

FIGURE 8.29 *RC* circuit with $i_{in}(t)$ as the desired response.

SOLUTION

Step 1. *Compute* $i_{in}(0^+)$. To obtain $i_{in}(0^+)$, we first compute $v_C(0^-) = v_C(0^+)$. Since for $t < 0$, -18 V has excited the circuit for a long time, the capacitor looks like an open circuit. By voltage division,

$$v_C(0^+) = v_C(0^-) = \frac{3}{9}(-18) = -6 \text{ V}$$

Thus at $t = 0^+$, the equivalent circuit is as shown in Figure 8.30.

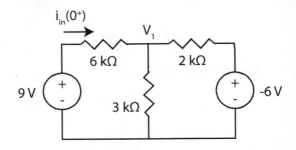

Figure 8.30 Circuit equivalent to that of Figure 8.29 at $t = 0^+$.

Application of superposition to Figure 8.30 shows that $V_1 = -1.5$ V. Hence,

$$i_{in}(0^+) = \frac{9 - (-1.5)}{6 \times 10^3} = 1.75 \text{ mA}$$

Step 2. *Find the circuit time constant and the equilibrium value of* $i_{in}(t)$. From Figure 8.29, the equivalent resistance seen by the capacitor is $R_{eq} = 4$ kΩ. Hence, the time constant is $\tau = 2$ sec. Further, since $t > 0$, $X_e = i_{in}(\infty)$, which is computed when the capacitor is replaced by an open circuit. In this case,

$$X_e = i_{in}(\infty) = \frac{9}{9 \times 10^3} = 1 \text{ mA}$$

Step 3. *Apply equation 8.23.* Using equation 8.23, we have, for $t > 0$,

$$i_{in}(t) = i_{in}(\infty) + \left[i_{in}(0^+) - i_{in}(\infty)\right]e^{-\frac{t}{\tau}} = 1 + 0.75e^{-0.5t} \text{ mA}$$

Exercise. In Example 8.10, find $i_C(0^-)$, $i_C(0^+)$, and $i_C(t)$ for $t > 0$ using equation 8.23 directly.
ANSWERS: 0, 2.25 mA, $2.25e^{-0.5t}$ mA

Note that in Example 8.10 we used $i_{in}(0^+)$ instead of $i_{in}(0^-)$ to obtain the correct answer. Some straightforward arithmetic shows that $i_{in}(0^-) = -2$ mA. Since $i_{in}(0^+) = 1.75$ mA, the input current is discontinuous at $t = 0$, unlike the capacitor voltage, which is continuous at $t = 0$. This emphasizes the need to use $x(t_0^+)$ in equation 8.23.

In several of the examples of sections 3 and 4, the circuits contain switches that operate at *pre-scribed* time instants. In some electronic circuits, the switch is a semiconductor device whose on/off state is determined by the value of a controlling voltage somewhere else in the circuit. If the controlling voltage is below a certain level, the electronic switch is off; if the voltage moves above a fixed level, the electronic switch is on. The time it takes for a controlling voltage to rise (or fall) from one level to another is very important because timing is as critical in electronic circuits as is scheduling for large organizations. For first-order linear networks with constant excitations, cal-culation of the time for a voltage or current to rise (or fall) from one level to another is straight-forward because all waveforms are exponential functions, as per equation 8.19. The situation is illustrated in Figure 8.31.

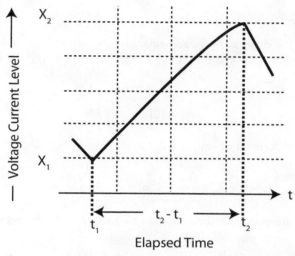

FIGURE 8.31 First-order response showing a rise from the voltage or current level X_1 to the voltage or current level X_2, for which the elapsed time is $t_2 - t_1$.

In equation 8.19a, let $x(t_1) = X_1$ and $x(t_2) = X_2$ be the two levels of interest. A straightforward manipulation of equation 8.19a leads to the **elapsed time formula** for first-order circuits,

$$t_2 - t_1 = \tau \ln\left[\frac{X_1 - x(\infty)}{X_2 - x(\infty)}\right]$$

(8.24)

EXAMPLE 8.11

This example uses the elapsed time formula of equation 8.24 for the circuit of Figure 8.32. The switch in this circuit is used to produce two different "final" capacitor voltages. When the switch is open, the final capacitor voltage is 12 V. When the switch closes, the final capacitor voltage, by V-division, changes to 4 V. Thus the switch causes the capacitor to charge and discharge repeat-edly. For our purposes we show that the choice of resistances produces an approximate **triangular** waveform.

For the purposes of this example, suppose the switch in Figure 8.32 is controlled electronically so that it closes when v_C rises to 9 V and opens when v_C falls to 5 V. Find and plot $v_C(t)$ for several switchings.

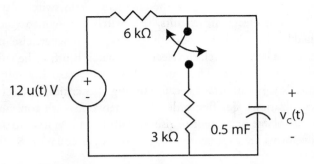

FIGURE 8.32 Switched driven RC circuit used to generate an approximate triangular waveform.

SOLUTION

Suppose the switch first closes at $t = t_a$, subsequently opening at $t = t_b$ and closing again at $t = t_c$, and so on. For $0 < t < t_a$, the time constant $\tau = 3$ sec, $v_C(0) = 0$, and $v_C(\infty) = 12$ V. From equation 8.19c,

$$v_C(t) = 12\left(1 - e^{-\frac{t}{3}}\right) \text{ V}$$

From the elapsed time formula of equation 8.24,

$$t_a - 0 = 3\ln\left(\frac{0-12}{9-12}\right) = 3 \times 1.386 = 4.159 \text{ s}$$

Now, for $t_a < t < t_b$, there is a new time constant $\tau = 1$ sec, $v_C(t_a) = 9$, and $v_C(\infty) = 4$. Again using equation 8.19c,

$$v_C(t) = 4 + 5e^{-(t-t_a)} \text{ V}$$

From the elapsed time formula, equation 8.24,

$$t_b - t_a = \ln\left(\frac{9-4}{5-4}\right) = \ln(5) = 1.61 \text{ s}$$

Finally, for the time interval $t_b < t < t_c$, $\tau = 3$ sec, $v_C(t_b) = 5$, and $v_C(\infty) = 12$. Using equation 8.19c,

$$v_C(t) = 12 - 7e^{-\frac{(t-t_b)}{3}}$$

and from the elapsed time formula,

$$t_c - t_b = 3\ln\left(\frac{5-12}{9-12}\right) = 3\ln\left(\frac{7}{3}\right) = 2.54 \text{ s}$$

The waveform of $v_C(t)$ for $0 < t < t_c$ is plotted in Figure 8.33.

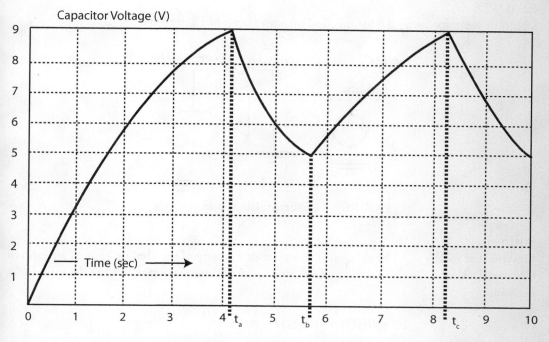

FIGURE 8.33 Generation of a nearly triangular waveform.

From the preceding solution $t_a = 4.16$ sec, $t_b = 4.16 + 1.61 = 5.77$ sec, and $t_c = 5.77 + 2.54 = 8.31$ sec. If we proceed to calculate the waveform for $t > t_c$, the waveform begins to repeat itself, as is evident from Figure 8.33. Practically speaking, the first cycle of a periodic, approximately triangular waveform occurs in the time interval $[t_a, t_c]$, and the period is $t_c - t_a = 8.31 - 4.16 = 4.15$ sec. Note that the triangular waveform has a frequency

$$f = \frac{1}{period} = \frac{1}{(t_c - t_b) + (t_b - t_a)} = \frac{1}{2.54 + 1.61} = \frac{1}{4.15} = 0.241 \text{ Hz}$$

The waveform in figure 8.33 is approximately *triangular*. This is due the fact that two time constants, 1 and 3 s, have the same order of magnitude. If we select the resistances so that the charging time constant is much larger than the discharging time constant, then the capacitance voltage waveform will look more like a sawtooth waveform. Sawtooth waveforms are used to drive the horizontal sweep of the electronic beam in an oscilloscope or a TV picture tube.

8. FIRST-ORDER RC OP AMP CIRCUITS

RC op amp circuits have some singular characteristics that set them apart from standard passive *RC* and *RL* types of circuits. Specifically, because of the nature of the operational amplifier, the time constant of the circuit will often depend only on some of the resistances. We present four important examples to illustrate the behavior of *RC* op amp circuits.

EXAMPLE 8.12

Compute the response $v_{out}(t)$ for the ideal op amp circuit of Figure 8.34.

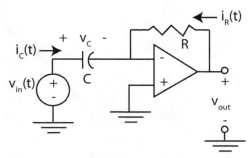

FIGURE 8.34 Differentiating op amp circuit.

SOLUTION

Observe that $v_C(t) = v_{in}(t)$ by the virtual short-circuit property of the ideal op amp, as set forth in Chapter 4. Also, $i_C(t) = -i_R(t) = v_{out}(t)/R$. Hence, from these equalities and the definition of a capacitor,

$$v_{out}(t) = Ri_R(t) = -Ri_C(t) = -RC\frac{dv_C(t)}{dt} = -RC\frac{dv_{in}(t)}{dt} \qquad (8.25)$$

Since the output is a negative constant (user chosen) times the derivative of the input, the circuit is called a **differentiator**.

Exercise. Suppose $v_{in}(t) = \cos(250t)$. Find R for the circuit of Figure 8.34 so that $v_{out}(t) = \sin(250t)$ V and $C = 1\ \mu F$.
ANSWER: 4 kΩ

EXAMPLE 8.13

Compute the response $v_{out}(t)$ for the ideal op amp circuit of Figure 8.35 assuming $v_C(0^-) = v_C(0^+) = 0$.

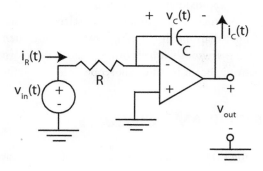

FIGURE 8.35 Integrating op amp circuit.

SOLUTION

Observe that $i_R(t) = v_{in}(t)/R$ by the virtual short circuit property of the ideal op amp. Also, $i_C(t) = -i_R(t)$. Hence, from these equalities and the integral v-i relationship of a capacitor,

$$v_{out}(t) = v_C(t) = v_C(0^+) + \frac{1}{C}\int_0^t i_C(\tau)\,d\tau$$

$$= -\frac{1}{C}\int_0^t i_R(\tau)\,d\tau = -\frac{1}{RC}\int_0^t v_{in}(\tau)\,d\tau$$

(8.26)

Since the output is a negative constant (user chosen) times the integral of the input, the circuit is called an **integrator**.

Exercise. Suppose $v_{in}(t) = \cos(250t)$ V. Find R for the circuit of Figure 8.35 so that $v_{out}(t) = \sin(250t)$ V and $C = 1$ μF.
ANSWER: 4 kΩ

EXAMPLE 8.14

This example considers the so-called **leaky integrator circuit** of Figure 8.36, which contains an ideal op amp. The input for all time is $v_s(t) = -5u(t)$ V. R_2 represents the leakage resistance of the capacitor. Given C and R_2, the resistance R_1 is chosen to achieve a dc gain of 10. The objective is to compute the response $v_{out}(t)$ assuming $v_C(0^-) = 0$ and compare it to a pure integrator having a gain of 1. This problem is reconsidered in Chapter 13.

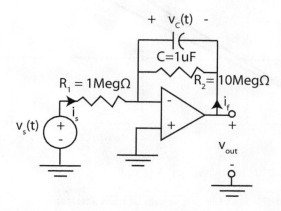

FIGURE 8.36 Leaky integrator op amp circuit in which $v_s(t) = -5u(t)$ V.

SOLUTION

Because there is only one capacitor, the circuit of Figure 8.36 is a first-order linear circuit. Because the inverting terminal of the op amp is at virtual ground, $v_{out}(t) = -v_C(t)$, and the capacitor sees an equivalent resistance $R_{th} = R_2$. Hence $\tau = R_2C = 10$ sec > 0. Equation 8.19 implies that

$$v_{out}(t) = v_{out}(\infty) + \left[v_{out}(0^+) - v_{out}(\infty)\right]e^{-\frac{t}{R_2C}} u(t) \text{ V} \qquad (8.27)$$

Because the voltage source $v_s(t) = 0$ for $t < 0$, $v_{out}(0^-) = -v_C(0^-) = 0$. For $t > 0$, $v_s(t) = -5$ V. Since the source voltage is constant, the capacitor looks like an open circuit at $t = \infty$ having final value

$$v_{out}(\infty) = -\frac{R_2}{R_1}(-5) = 50 \text{ V}$$

Entering numbers into equation 8.27 yields

$$v_{out}(t) = 50 + (0 - 50)e^{-0.1t} = 50\left(1 - e^{-0.1t}\right)u(t) \text{ V}$$

A plot of the op amp output voltage appears in Figure 8.37 along with that of an ideal integrator. One observes that the more realistic leaky integrator circuit approximates an ideal integrator only for $0 \le t \le 0.15\tau$ before the error induced by the feedback resistor R_2 becomes noticeable. Such integrators need to be reinitialized periodically by resetting the capacitor voltage to zero.

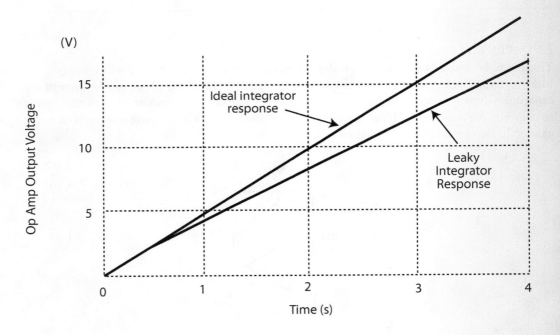

FIGURE 8.37 Output voltage of leaky integrator that approximates an ideal integrator.

So far we have assumed an ideal op amp. In practice, the output voltage will saturate at a level determined by the power supply voltage and the specs of the particular amplifier used. Further, practical op amps have complex models. To evaluate the preceding analysis, Figure 8.38 shows a SPICE simulation using the standard 741 op amp.

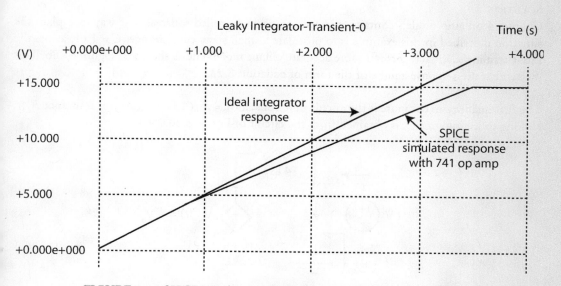

FIGURE 8.38 SPICE simulation of leaky integrator circuit of Figure 8.35.

Observe that the response approximates the ideal up to about $0.15\tau = 1.5$ sec, which corroborates our analysis using the ideal op amp. Note, however, that the simulation accounts for saturation present in practical op amps but absent from the ideal.

EXAMPLE 8.15

In trying to build an inverting amplifier in a laboratory, a student inadvertently reverses the connection of the two input terminals, which results in the circuit of Figure 8.39a. Assume a practical op amp model with $V_{sat} = 15$ V and a finite gain of $A = 10^4$. Instead of seeing the expected output of $v_{out} = -4$ V, the student observes a 15 V output. Explain how this 15 V output could possibly exist.

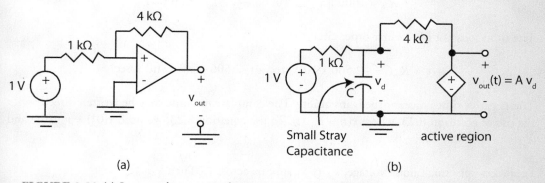

FIGURE 8.39 (a) Incorrectly connected inverting amplifier. (b) Circuit model, including a stray capacitance $C = 1$ pF and a finite gain $A = 10^4$.

SOLUTION

Chapter 4 op amp models contain only resistors and controlled sources. One way to explain the situation described in this example is to postulate a small **stray capacitance**, $C = 1$ pF, across the input terminals. In fact, this is a more accurate circuit model and is shown in Figure 8.38b. This means that the response will be of the form of equation 8.23.

The first quantity to compute is the circuit time constant $\tau = R_{th}C$. The equivalent resistance R_{eq1}, looking to the right of C, is obtained from the circuit of Figure 8.40.

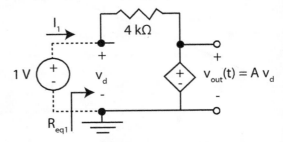

FIGURE 8.40 Circuit for computing R_{eq1}; note the artificial 1 V excitation.

From Figure 8.40 and our knowledge of constructing Thevenin equivalents,

$$I_1 = \frac{1 - 10^4}{4000} \text{ A}$$

Hence,

$$R_{eq1} = \frac{1}{I_1} = \frac{4000}{1 - 10^4} = -2.5 \ \Omega$$

Observe that in the actual circuit, R_{eq1} is in parallel with 1 kΩ. Hence, the Thevenin equivalent resistance seen by C is

$$R_{th} = 1000 \ || R_{eq1} = 1000|| \ (-2.5) = -2.506 \ \Omega$$

The time constant of the first-order circuit is

$$\tau = R_{th}C = -2.506 \times 10^{-12} \text{ sec, or -2.506 picosecond (psec)}$$

The negative time constant spells instability. The complete response may be written directly with the use of equation 8.23, where $x(t) = v_{out}(t)$. To use equation 8.23, we need $x(0^+) = v_{out}(0^+)$ and X_e.

Suppose a very small noise voltage, $v_d = \varepsilon$ V, appears across C. Then

$$v_{out}(0^+) = 10^4 \varepsilon \text{ V}$$

To compute the equilibrium output voltage X_e, we open-circuit the capacitor and compute v_{out}:

$$X_e = v_{out} \text{ with } C \text{ open-circuited}$$

$$= \left(1 \times \frac{-2.5}{1000 - 2.5}\right) 10^4 = -25.063 \text{ V}$$

From equation 8.23, the complete response is

$$v_{out}(t) = -25.063 + \left(10^4 \varepsilon + 25.063\right) e^{0.339 \times 10^{12} t} \tag{8.28}$$

For any small positive initial capacitor voltage ε, equation 8.28 implies that the output would increase exponentially, had the op amp been ideal. Because this particular real op amp saturates at 15 V, the output more or less instantaneously saturates at 15 V, the phenomenon observed by the student. Had the initial capacitance voltage been sufficiently negative,

$$\varepsilon < \frac{-25.063}{10^4} \text{ V},$$

equation 8.28 implies that v_{out} would saturate at -15 V.

9. SUMMARY

This chapter has explored the behavior of first-order RL and RC circuits (1) without sources for given ICs, (2) for constant excitations (dc), (3) for piecewise constant excitations, and (4) with switching under constant excitations. In general, first-order RL and RC circuits have only one capacitor or one inductor present, although there are special conditions when more than one inductor or capacitor can be present. Our discussion has presumed only one capacitor or one inductor is present in the circuit.

Using a first-order constant-coefficient linear differential equation model of the circuit, the chapter sets forth two types of exponential responses, the source-free response and the response when constant independent sources are present. The source-free responses for the RL and RC circuits have the exponential forms

$$i_L(t) = e^{-\frac{R_{th}}{L}(t-t_0)} i_L(t_0) \qquad\qquad v_C(t) = e^{-\frac{1}{R_{th}C}(t-t_0)} v_C(t_0)$$

where $i_L(t_0)$ is the initial condition for the inductor and $v_C(t_0)$ the initial condition on the capacitor. For an RC circuit, the time constant $\tau = R_{th}C$ where R_{th} is the Thevenin equivalent resistance seen by the capacitor. For an RL circuit, the time constant $\tau = L/R_{th}$, where R_{th} now is the Thevenin equivalent resistance seen by the inductor.

When independent sources are present in the circuit, the response of a first-order RC or RL circuit has the general form

$$x(t) = x(\infty) + \left[x(t_0^+) - x(\infty)\right] e^{-\frac{t-t_0}{\tau}}$$

for $\tau > 0$. Stated in words, this formula is

$$x(t) = [Final\ value] + ([Initial\ value] - [Final\ value])e^{-\frac{elapsed\ time}{time\ constant}}$$

provided the time constant $\tau > 0$. When the time constant $\tau < 0$, then it is necessary to modify the interpretation as discussed in section 7, with equation 8.23 identifying the form of any voltage or current in the circuit:

$$x(t) = X_e + \left[x(t_0^+) - X_e\right]e^{-\left(\frac{t-t_0}{\tau}\right)}$$

The time constants of a circuit can be changed by switching within the circuit. By changing time constants in a circuit, one can generate different types of waveforms such as the triangular waveform of Figure 8.32. As mentioned at the beginning of the chapter, wave shaping is an important application of circuit design. When inductors, resistors, and capacitors are present in the same circuit, many other wave shapes can be generated. *RLC* circuits are the topic of the next chapter and allow even greater freedom in waveform construction.

As a final application of the concepts of this chapter, we looked at the leaky integrator op amp circuit. Integrators are present in a host of signal processing and control applications. Unfortunately, ideal integrators do not exist in practice. The leaky integrator circuit of Figure 8.35 provides a reasonable model of an ideal integrator.

10. TERMS AND CONCEPTS

Complete response: sum of zero-input and zero-state responses.

Differential equation of a circuit: equation in which a weighted sum of derivatives of an important circuit variable (e.g., a voltage or current) is equated to a weighted sum of derivatives of the source excitations to the circuit.

Differentiator circuit: op amp circuit whose output is a constant times the derivative of the input.

Equilibrium state of a differential equation: constant, say $x(t) = X_e$, that satisfies the differential equation in the variable $x(t)$.

First-order differential equation of a circuit: differential equation whose highest derivative is first order.

Forced response: that portion of a complete response that has the same "exponent" as the input excitation, provided the input excitation has exponents different from that of the zero-input response, under the condition that the input excitation is either (1) dc, (2) real exponential, (3) sinusoidal, or (4) exponentially modulated sinusoidal.

Integrating factor method: mathematical technique for finding the solution of a differential equation in which multiplication by the integrating factor $e^{-\lambda t}$ on both sides of the differential equation leads to a new equation that can be explicitly integrated for a solution.

Integrator circuit: op amp circuit whose output is a constant times the integral of the input.

Leaky integrator circuit: op amp circuit having a response approximating an ideal integrator, as described in Example 8.13.

Natural frequency of a circuit: natural mode of "oscillation" of the circuit. For a first-order circuit having a response proportional to $e^{\lambda t}$, it is the coefficient λ in the exponent.

Natural response: that portion of the complete response that has the same exponents as the zero-input response.

Passive *RLC* circuit: circuit consisting of resistors, inductors, and capacitors that can only store and/or dissipate energy.

Sawtooth waveform: triangular waveform resembling the teeth on a saw blade and typically used to drive the horizontal sweep of the electronic beam in an oscilloscope or a TV picture tube.

Source-free response: response of a circuit in which sources are either absent or set to zero.

Step response: response, for $t \geq 0$, of a relaxed single-input circuit to a unit step, i.e., a constant excitation of unit amplitude.

Stray capacitance: small capacitance always present between a conductor and ground. It usually can be ignored, but as Example 8.14 shows, it can critically affect the response of a circuit.

Superposition: in linear *RC* and *RL* circuits, the complete response is the superposition of the relaxed circuit responses due to each source with all other sources set to zero, plus the responses to each initial condition when all other initial conditions are set to zero and all independent sources are set to zero.

Time constant: in a source-free first-order circuit, the time it takes for the circuit response to drop to $e^{-1} = 0.368$ of its initial value. Roughly speaking, the response value must drop to a little over one-third of its initial value or rise to within one-third of its final value. For *RL* circuits $\tau = L/R_{th}$ and for *RC* circuits $\tau = R_{th}C$.

Unit step function: function denoted $u(t)$ whose value is 1 for $t \geq 0$ and 0 for $t < 0$.

Unstable response: response whose magnitude increases without bound as t increases. The time constant for first-order circuits is negative for an unstable response.

Zero-input response: response in which all sources are set to zero.

Zero-state response: response to a specified input signal or set of input signals given that the initial conditions are all set to zero.

---

[1] It happens that all variables in a first-order *RL* or *RC* circuit satisfy a differential equation of the same form. The interpretation of the solution is somewhat different from what follows. A detailed explanation of the general solution is presented in section 7.

Problems

UNDRIVEN RESPONSE WITH GIVEN INITIAL CONDITIONS

1. For the *RC* circuit of Figure P8.1, $R = 2.5$ kΩ and $C = 50$ μF.
   (a) If $v_C(0) = 10$ V, find $v_C(t)$. Plot your answer for $0 \le t \le 5\tau$, where τ is the circuit time constant.
   (b) If $v_C(0) = 10$ V, find $i_R(t) = -i_C(t)$ without differentiating your answer to part (a). Plot your answer for $0 \le t \le 5\tau$, where τ is the circuit time constant. At what time is the energy stored in the capacitor about 1% of its initial value?
   (c) Compute $i_R(t)$ for $v_C(0) = 5$ V and $v_C(0) = 20$ V without doing any further calculations, i.e., by using the principle of linearity.

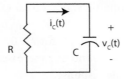

Figure P8.1

2. Consider the *RL* circuit of Figure P8.2 in which $R = 50$ Ω and $L = 0.1$ mH.
   (a) If the energy stored in the inductor at $t = 0$ is 2 μJ, find $i_L(0)$ and $i_L(t)$. Plot your answer for $0 \le t \le 5\tau$, where τ is the circuit time constant.
   (b) If the energy stored in the inductor at $t = 0$ is 2 μJ, find $v_L(t)$ without differentiating your answer to part (a). Plot your answer for $0 \le t \le 5\tau$, where τ is the circuit time constant.
   (c) Repeat part (b) for $i_L(0) = 50$ mA and $i_L(0) = 250$ mA. Hint: What principle makes this a straightforward calculation given your answer to part (a)?

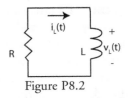

Figure P8.2

◆3. In Figure P8.1, suppose $R = 25$ kΩ and $v_C(0) = 20$ V.
   (a) Find C so that $v_C(0.25) = 2.7067$ V.
   (b) Given your answer to part (a), find $v_C(t)$.
   CHECK: $C = 5$ μF

◆4. In Figure P8.2, suppose $R = 2.5$ kΩ and $i_L(0) = 20$ mA.
   (a) Find L so that at $t_1 = 1$ msec, $i_L(t_1) = 2.7067$ mA.
   (b) Given your answer to part (a), find $i_L(t)$.
   CHECK: $L = 1.25$ H

5. The response of an undriven parallel *RC* circuit is plotted in Figure P8.5. Find the time constant of the circuit, at least approximately. If $C = 0.25$ mF, find R.

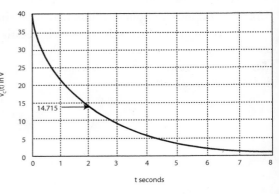

Figure P8.5

6. In the circuit of Figure P8.6, suppose $R_1 = 50$ Ω, $R_2 = 200$ Ω, $L = 2$ H, $i_L(0) = 100$ mA, and the switch opens at $t = 50$ msec.
   (a) Find $i_L(t)$ for $t \ge 0$. Plot your answer in MATLAB for $0 \le t \le 0.1$ sec.
   (b) Find $v_R(0)$ and $v_R(t)$, $t \ge 0$. Plot your answer in MATLAB for $0 \le t \le 0.1$ sec.

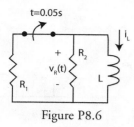

t=0.05s

Figure P8.6

7. In the circuit of Figure P8.7, suppose $R_1 = 5$ kΩ, $R_2 = 20$ kΩ, $C = 50$ μF H, $v_C(0) = 20$ V, and the switch opens at $t = 0.4$ sec.
   (a) Find $v_C(t)$ for $t \geq 0$. Plot your answer in MATLAB for $0 \leq t \leq 4$ sec.
   (b) Find $i_R(0)$ and $i_R(t)$, $t \geq 0$. Plot your answer in MATLAB for $0 \leq t \leq 4$ sec.

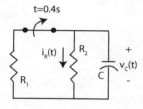

t=0.4s

Figure P8.7

8. Consider the circuit of Figure P8.8.
   (a) Find the value of R and the initial condition $i_L(0)$ so that $i_L(0.05$ msec$) = 9.197$ mA and $i_L(0.15$ msec$) = 1.2447$ mA.
   (b) Given your answer to part (a), compute and plot $i_L(t)$ for $0 \leq t \leq 5\tau$, where τ is the circuit time constant.

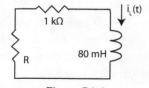

Figure P8.8

ANSWER: (a) 600 Ω, 25 mA

9. Consider the circuit of Figure P8.9, in which $R_1 = 100$ Ω and $R_2 = 50$ Ω. Let $v_C(0) = 500$ mV.
   (a) Let $\alpha = 7$. Compute the equivalent resistance seen by the capacitor. Find $v_C(t)$ for $t \geq 0$. Plot $v_C(t)$ for $0 \leq t \leq 5\tau$

where τ is the circuit time constant.
   (b) Let $\alpha = -11$. Compute the equivalent resistance seen by the capacitor. Find $v_C(t)$ for $t \geq 0$. Plot $v_C(t)$ for $0 \leq t \leq 2\tau$ where τ is the circuit time constant.
   (c) Find the range of α for which the time constant is positive.

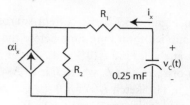

Figure P8.9

CHECK: (c) $\alpha > -3$

10. In the circuit of Figure P8.10, $R_1 = 100$ Ω, $R_2 = 20$ Ω, $\beta = 200$, $L = 0.5$ H, and $i_L(0) = 250$ mA. The switch opens at $t = 0.03$ sec. Find the Thevenin equivalent resistance seen by the inductor before the switch opens, and then compute $i_L(t)$ and $v_{R1}(t)$ for $t \geq 0$.

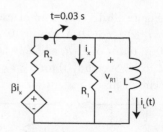

t=0.03 s

Figure P8.10

CHECK: $R_{th} = -25$ Ω

11. Consider the circuit of Figure P8.11 in which $R_1 = 25$ Ω, $R_2 = 50$ Ω, and $L = 2.5$ H. Suppose $i_L(0) = 2$ A.
   (a) With $\alpha = 0.1$, compute the Thevenin equivalent resistance seen by the inductor; then compute $i_L(t)$ for $t \geq 0$. Plot in MATLAB for $0 \leq t \leq 5\tau$, where τ is the time constant of the circuit.
   (b) With $\alpha = 0.1$, compute $v_{R2}(t)$, $t \geq 0$.
   (c) Repeat part (a) for $\alpha = 0.02$. Determine the time, say t_1, when the inductor has lost 99% of its initial stored energy.

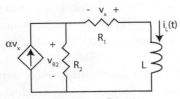

Figure P8.11

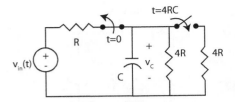

Figure P8.14

12. In Figure P8.12, the current source I_s has been applied for a long time before the switch opens at $t = 0$. Find $i_L(0^+)$ and $i_L(t)$ for $t > 0$ in terms of I_s, R, and L, where I_s is in A, R in Ω, and L in H. Sketch $i_L(t)$ for $0 \le t \le 4\tau$ where τ is the circuit time constant for $t > 0$.

15. In Figure P8.15, the current excitation is given by $v_{in}(t) = V_s u(-t)$ V. Find $v_C(0^+)$ and $v_C(t)$ for $t > 0$ in terms of V_s, R, and C, where R is in Ω and C in F. Sketch $v_C(t)$ for $0 \le t \le 3\tau$, where τ is the circuit time constant for $t > 0$.

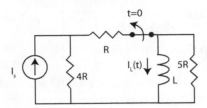

Figure P8.12

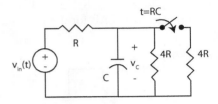

Figure P8.15

13. In Figure P8.13, the current excitation is given by $i_{in}(t) = I_s u(-t)$ A. Find $i_L(0^+)$ and $i_L(t)$ for $t > 0$ in terms of I_s, R, and L, where R is in Ω and L in H. Sketch $i_L(t)$ for $0 \le t \le 4\tau$, where τ is the circuit time constant for $t > 0$.

16. Repeat Problem 14, except find $i_C(0^+)$ and $i_C(t)$ for $t \ge 0$.

17. Repeat Problem 15, except find $i_C(0^+)$ and $i_C(t)$ for $t \ge 0$.

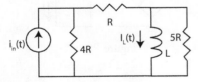

Figure P8.13

RESPONSE OF DRIVEN CIRCUITS

18. Consider the RC circuit of Figure P8.18 in which $R = 10 \text{ k}\Omega$ and $C = 0.4$ mF.

14. In Figure P8.14 the voltage source $v_{in}(t) = V_S$ has been applied for a long time before the switch opens at $t = 0$. Find $v_C(0^+)$ and $v_C(t)$ for $t > 0$ in terms of V_s, R, and C, where V_s is in V, R in Ω, and C in F. Sketch $v_C(t)$ for $0 \le t \le 3\tau$, where τ is the circuit time constant for $t > 0$.

(a) If $v_C(0) = 0$ and $v_{in}(t) = 20u(t)$ V, find $v_C(t)$. Plot your answer for $0 \le t \le 4\tau$, where τ is the circuit time constant.

(b) If $v_C(0) = 10$ V and $v_{in}(t) = 0$, find $v_C(t)$. Plot your answer for $0 \le t \le 4\tau$, where τ is the circuit time constant.

Now making use of linearity and its associated properties, compute the indicated responses without any further circuit analysis.

(c) If $v_C(0) = 10$ V and $v_{in}(t) = 20u(t)$ V, find $v_C(t)$.

(d) If $v_C(0) = -20$ V and $v_{in}(t) = -10u(t)$ V, find $v_C(t)$.

(e) If $v_C(0) = 10$ V and $v_{in}(t) = 20u(t)$ V, find $i_C(t)$ without differentiating your answer to part (c). Plot your answer for $0 \le t \le 4\tau$, where τ is the circuit time constant.

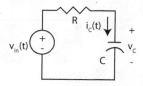

Figure P8.18

19. Consider the *RL* circuit of Figure P8.19. Suppose $R = 100$ Ω, $L = 0.2$ H.

(a) If $i_L(0) = 0$ and $v_{in}(t) = 20u(t)$ V, find $i_L(t)$. Plot your answer for $0 \le t \le 5\tau$, where τ is the circuit time constant.

(b) If $i_L(0) = -50$ mA and $v_{in}(t) = 0$, find $i_L(t)$. Plot your answer for $0 \le t \le 5\tau$, where τ is the circuit time constant. Now making use of linearity and its associated properties, compute the indicated responses without any further circuit analysis.

(c) If $i_L(0) = -50$ mA and $v_{in}(t) = 20u(t)$ V, find $i_L(t)$. Plot your answer for $0 \le t \le 5\tau$, where τ is the circuit time constant.

(d) If $i_L(0) = 25$ mA and $v_{in}(t) = -10u(t)$ V, find $i_L(t)$. Plot your answer for $0 \le t \le 5\tau$, where τ is the circuit time constant.

(e) If $i_L(0) = -50$ mA and $v_{in}(t) = 20u(t)$ V, find $v_L(t)$ without differentiating your answer to part (c). Plot your answer for $0 \le t \le 5\tau$, where τ is the circuit time constant.

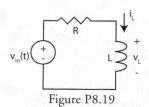

Figure P8.19

20. In Figure P8.20, $R_1 = 50$ Ω, $R_2 = 200$ Ω, $C = 2.5$ mF, and the voltage excitation is given by $v_{in}(t) = V_{s1}u(-t) + V_{s2}u(t)$, where $V_{s1} = -10$ V and $V_{s2} = 20$ V.

(a) Find $v_C(0^+)$ and $v_C(t)$ for $t > 0$.

(b) Sketch $v_C(t)$ for $0 \le t \le 5\tau$, where τ is the circuit time constant for $t > 0$.

(c) Identify the zero-input response ($t \ge 0$) and the zero-state response ($t \ge 0$) for the answer computed in part (a).

(d) Now compute $i_C(t)$ for $t > 0$ assuming the switch opens at $t = 0.25$ sec. Plot your result for $0 \le t \le 0.5$ sec.

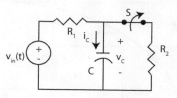

Figure P8.20

21. In Figure P8.21, $R_1 = 50$ Ω, $R_2 = 200$ Ω, $L = 2$ H, and the voltage excitation is given by $v_{in}(t) = V_{s1}u(-t) + V_{s2}u(t)$, where $V_{s1} = -10$ V and $V_{s2} = 20$ V.

(a) Find $i_L(0^+)$ and $i_L(t)$ for $t > 0$.

(b) Sketch $i_L(t)$ for $0 \le t \le 4\tau$, where τ is the circuit time constant for $t > 0$.

(c) Identify the zero-input response ($t \ge 0$) and the zero-state response ($t \ge 0$) for the answer computed in part (a).

(d) Now compute $v_L(t)$ for $t > 0$ assuming the switch opens at $t = 0.04$ sec. Plot your result for $0 \le t \le 0.2$ sec.

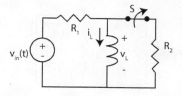

Figure P8.21

22. Consider the *RC* circuit of Figure P8.22a in which $v_{in}(t) = V_0u(t)$, where $V_0 = 100$ V.

(a) Find $v_C(0^+)$, the time constant of the cir-

cuit, R_1, and R_2 so that the circuit response $v_C(t)$ is given by Figure P8.22b. Assume $C = 0.25$ mF.

(b) Find R_1 and C so that the circuit response $v_C(t)$ is given by Figure P8.22b. Assume $R_2 = 10$ kΩ.

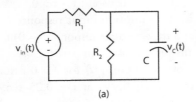

(a)

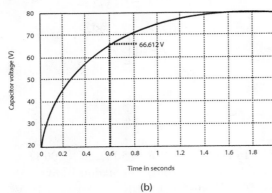

(b)

Figure P8.22

23. Consider the RL circuit of Figure P8.23a in which $v_{in}(t) = V_0 u(t)$, where $V_0 = 100$ V.

(a) If $L = 2$ H, find $i_L(0^+)$, the circuit time constant, R_1, and R_2 so that the circuit response $i_L(t)$ is given by Figure P8.23b.

(b) If $R_2 = 2$ kΩ, find $i_L(0^+)$, R_1, and L so that the circuit response $i_L(t)$ is given by Figure P8.23b.

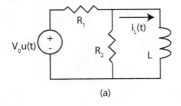

(a)

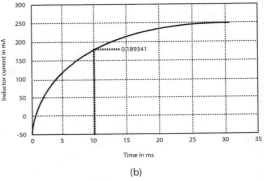

(b)

Figure P8.23

24. In the circuit of Figure P8.24, $v_C(0^-) = 25$ V, $i_{s1}(t) = 50u(t)$ mA, and $i_{s1}(t) = 25u(t)$ mA

(a) Find the zero-input response, i.e., the response due only to the initial condition.

(b) Find $v_C(t)$ for $t > 0$ due only to $i_{s1}(t)$.

(c) Find $v_C(t)$ for $t > 0$ due only to $i_{s2}(t)$.

(d) Find the zero-state response.

(e) Find the complete response $v_C(t)$, for $t > 0$.

(f) Suppose the initial condition is doubled and each independent source is cut in half. Find the new complete response using linearity.

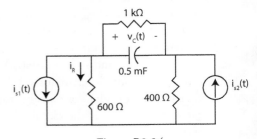

Figure P8.24

25. In Figure P8.25 $R_1 = 200$ Ω, $R_2 = 600$ Ω, $R_3 = 650\Omega$, $L = 20$ H, $v_{s1}(t) = -100u(-t) + 50u(t)$ V, and $v_{s2}(t) = 50u(t - 0.5)$ V. Compute $i_L(t)$ for $t \geq 0$. Plot your answer using MATLAB or its equivalent for $0 \leq t \leq 8\tau$.

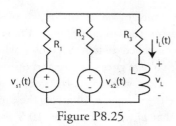

Figure P8.25

26. Repeat Problem 25, except compute $v_L(t)$ for $t \geq 0$.

27. In Figure P8.27 $R_1 = 200\ \Omega$, $R_2 = 600\ \Omega$, $R_3 = 850\Omega$, $C = 2.5$ mF, $v_{s1}(t) = -50u(-t) + 100u(t)$ V, and $v_{s2}(t) = -50u(t-5)$ V. Compute $v_C(t)$ for $t \geq 0$. Plot your answer using MATLAB or its equivalent for $0 \leq t \leq 6\tau$.

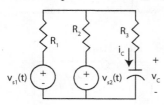

Figure P8.27 ·

28. Repeat Problem 27, except compute the capacitor current $i_C(t)$ for $t \geq 0$.

29. For the circuit of Figure P8.29, $v_{in}(t) = -20u(-t) + 20u(t)$ V. Find $v_C(0^-)$ and $v_C(t)$ for $t \geq 0$. Plot $v_C(t)$ for $0 \leq t \leq 40$ msec.

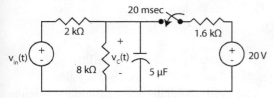

Figure P8.29

30. Consider the circuit of Figure P8.30 in which $R_1 = 300\Omega$, $R_2 = 800\ \Omega$, $R_3 = 600\ \Omega$, $L = 4$ H, $v_{s1}(t) = -24u(-t) + 24u(t)$ V, and $v_{s2}(t) = 24u(t)$ V.

(a) Compute the response $i_L(t)$ for $t \geq 0$. Plot for $0 \leq t \leq 4\tau$ where τ is the circuit time constant.

(b) Find the inductor voltage $v_L(t)$ for $t > 0$

directly without differentiating your answer to part (a).

(c) What are the new responses if the value of each source is doubled?

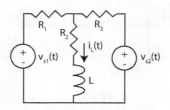

Figure P8.30

31. The switch in the circuit of Figure P8.31 has been open for a long time before it is closed at $t = 0$. Suppose $R_1 = 6R$, $R_2 = 30R$, $R_3 = 20R$, and $v_{in}(t) = V_0u(T - t)$. In terms of V_0, R, C, and $T = 6RC$,

(a) Find $v_C(0^-)$ and $v_C(0^+)$.

(b) Find the Thevenin equivalent seen by the capacitance for $0 < t < T$.

(c) Using the Thevenin equivalent found in part (b), find an expression for $v_C(t)$ valid over $0 < t < T$.

(d) Find the expressions for $v_C(T^-)$ and $v_C(T^+)$.

(e) Find the time constant valid for $t > T$.

(f) Find $v_C(t)$ for $t > T$.

(g) Plot $v_C(t)$ for $0 \leq t \leq 4T$ using MATLAB.

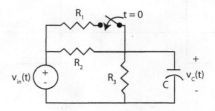

Figure P8.31

32. For the circuit of Figure P8.32, $-V_0 = -10$ V, $V_1 = 20$ V, $R_1 = 600\ \Omega$, $R_2 = 200\ \Omega$, and $C = 12.5\ \mu$F.

(a) Find $v_C(0^+)$.

(b) Using the initial condition computed in part (a), find $v_{out}(t)$. Plot the result for $0 \leq t \leq 160$ msec.

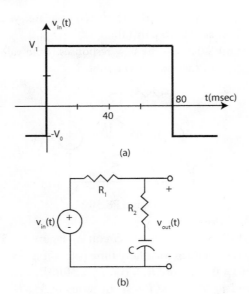

(a)

(b)

Figure P8.32 (a) Pulse driving
RC circuit of part (b).

33. For the circuit of Figure P8.33, , $-V_0 = -10$ V,
$V_1 = 20$ V, $R_1 = 80$ Ω, $R_2 = 20$ Ω, and $L = 4$ H.
(a) Find $i_L(0^+)$.
(b) Using the initial condition computed in
part (a), find $v_{out}(t)$. Plot the result for 0
$\le t \le 160$ msec.

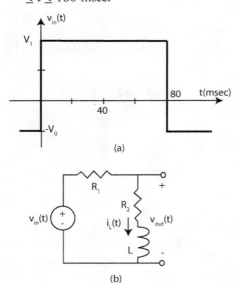

(a)

(b)

Figure P8.33

34. The voltage waveform $v_{in}(t)$ of Figure
P8.34a drives the circuit of Figure P8.34b. The
voltage-controlled switch S1 closes when the
capacitor voltage goes positive and opens when
the capacitor voltage $v_C(t)$ goes negative.
Compute the voltage $v_C(t)$ across the capacitor.
Assume that $v_{in}(t)$ has been at -10 V for $t < 0$
for a very long time. Hint: Use the elapsed time
formula as needed.

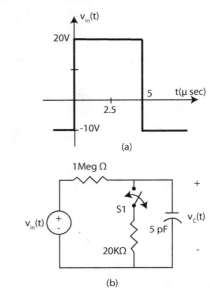

(a)

(b)

Figure P8.34 (a) Pulse waveform exciting
RC circuit in part (b).

35. Repeat Problem 34, except find $i_C(t)$ for $t \ge$
0.

36. Consider the circuit of Figure P8.36. Suppose
$v_C(0) = 0$ and find $v_C(t)$ for $t \ge 0$ as follows:
(a) Find the Thevenin equivalent circuit
seen by the capacitor.
(b) Find the complete response $v_C(t)$ for $t \ge$
0. What is $v_C(\infty)$?

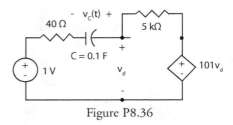

Figure P8.36

OP AMP CIRCUITS

37. In the circuit of Figure P8.37, $v_{in}(t) = K \sin(\omega t)u(t)$ V and all capacitor voltages are zero at $t = 0$. Find $v_1(t)$ and $v_{out}(t)$ for $t \geq 0$ in terms of R, C, K, and ω.

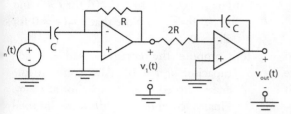

Figure P8.37

38. In the circuit of Figure P8.38, $R = 10$ kΩ, $C = 10$ μF, $v_{in}(t) = 10 \sin(50t)u(t)$ mV, and all capacitor voltages are zero at $t = 0$. Find $v_{out}(t)$ and plot it for $0 \leq t \leq 6$ sec.

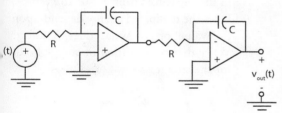

Figure P8.38

39. Repeat Problem 38 for $v_{in}(t) = 100e^{-2t}u(t)$ mV. But plot from 0 to 6 seconds.

40. In the circuit of Figure P8.40, $R_1 = 10$ kΩ, $R_2 = 40$ kΩ, $C = 12.5$ μF, and $v_s(t) = -100u(t)$ mV.
   (a) If $v_C(0^-) = 0$, compute $v_{out}(t)$ for $t \geq 0$.
   (b) Repeat part (a) for $v_C(0^-) = 50$ mV.

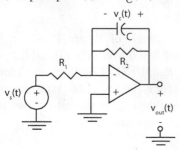

Figure P8.40

41. Figure P8.41a shows an op amp integrator with positive gain, and Figure P8.41b shows a differentiator with positive gain for the constant $K > 0$.
   (a) For each of the circuits find a literal expression for $v_{out}(t)$ in terms of $v_{in}(t)$.
   (b) For $R_1 = 10$ kΩ and $C = 0.1$ mF, find $v_{out}(t)$ when $v_{in}(t) = 100\sin(20t)u(t)$ mV assuming that $v_C(0^-) = 0$ in each case.

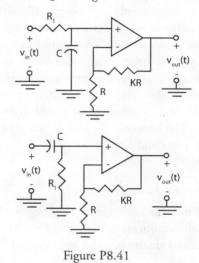

Figure P8.41

MISCELLANEOUS

42. Although most of the first-order circuits considered in the text have only one capacitor or one inductor, it is possible to have a first-order circuit containing more than one energy storage element. Consider the situation depicted in Figures P8.42a and b. Here $v_{C1}(0^+)$ and $v_{C2}(0^+)$ are given. The networks N_1 and N_2 are equivalent under two conditions:

$$C_{eq} = \frac{C_1 C_2}{C_1 + C_2} \quad \text{and} \quad v_{Ceq}(0^+) = v_{C1}(0^+) + v_{C2}(0^+)$$

Prove this equivalence using the integral relationship of a capacitor to show that the i-v terminal conditions are the same for both N_1 and N_2.

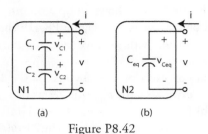

(a)   (b)

Figure P8.42

43. As mentioned in Problem 42, although most of the first-order circuits considered in the text have only one inductor or one capacitor, it is possible to have a first-order circuit containing more than one energy storage element. Consider the situation depicted in Figures P8.43a and b. Here $i_{L1}(0^+)$ and $i_{L2}(0^+)$ are given. The networks N_1 and N_2 are equivalent under two conditions:

$$L_{eq} = \frac{L_1 L_2}{L_1 + L_2} \quad \text{and} \quad i_{Leq}(0^+) = i_{L1}(0^+) + i_{L2}(0^+)$$

Prove this equivalence using the integral relationship of an inductor to show that the i-v terminal conditions are the same for both N_1 and N_2.

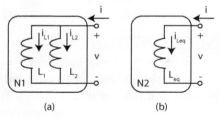

(a)   (b)

Figure P8.43

44. In the circuit of Figure P8.44, suppose $R_1 = 50\ \Omega$, $R_2 = 200\ \Omega$, $C_1 = 0.06$ F, $C_2 = 0.3$ F, $v_{C1}(0^-) = 15$ V, and $v_{C2}(0^-) = 5$ V. Let $v_{in}(t) = 40u(t)$ V. Use the equivalence set forth in Problem 42 to compute $v_{out}(t)$ for $t \geq 0$.

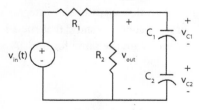

Figure P8.44

45. In the circuit of Figure P8.45, $C_1 = C_2 = 2$ F and the switch closes at time $t = 0$. The initial conditions on the two capacitors are $v_{C1}(0^-) = 4$ V and $v_{C2}(0^-) = 0$ V.
   (a) For $R = 0.5\ \Omega$, find an expression for the current $i_R(t)$ for $t \geq 0$.
   (b) For $R = 0.5\ \Omega$, find $v_{C1}(t)$ for $t \geq 0$ and $v_{C2}(t)$ for $t \geq 0$. Note that $v_{C2}(t) \neq 0$ for all $t > 0$.
   (c) Compute the energy stored in each capacitor at $t = 0^+$. Also compute the energy stored in each capacitor at $t = \infty$. Finally, compute the decrease in total energy stored in the capacitors from $t = 0^+$ to $t = \infty$.
   (d) Compute the energy dissipated in the $0.5\ \Omega$ resistor from $t = 0^+$ to $t = \infty$. Verify that the energy dissipated in the resistor equals the decrease in total energy stored in the capacitors from $t = 0^+$ to $t = \infty$.
   (e) Does the dissipated energy depend upon the value of R? What does R affect? Verify that conservation of energy holds for the circuit.

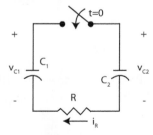

Figure P8.45

46. Repeat Problem 45 for the circuit of Figure P8.45 when $C_1 = 1$ F, $C_2 = 0.25$ F, $v_{C1}(0^-) = 3$ V, and $v_{C2}(0^-) = 8$ V.

47. In the circuit of Figure P8.47, $v_{in}(t) = 110u(-t) + 220u(t)$ mV, $L_1 = 110$ mH, $L_2 = 11$ mH, and $R = 10\ \Omega$. Compute and plot the waveforms for $i_{L1}(t)$ and $i_{L2}(t)$. Hint: Adapt the results of Problem 43 to the case of two parallel inductors with initial currents.

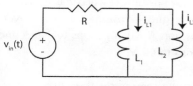

Figure P8.47

48. Repeat Problem 47 for the case where $v_{in}(t)$ = $220u(t)$ mV, $i_{L1}(0^+)$ = 44 mA, and $i_{L2}(0^+)$ = 11 mA.

49. Consider the circuit of Figure P8.49 in which C_1 = 1 F, C_2 = 4 F, v_{in} = 10 V, and R = 2 Ω.

   (a) Compute $v_R(0^-)$ and $v_R(0^+)$. Hint: How does the charge distribute over the two capacitors at $t = 0^+$?
   (b) Compute $v_R(t)$ for $t > 0$.

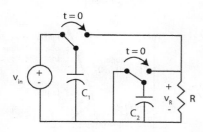

Figure P8.49

50. The circuit of Figure P8.50 contains two capacitors.

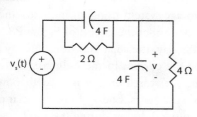

Figure P8.50

   (a) Suppose $v_s(t) = K_0u(t)$ V. Show that for $t > 0$, the voltage $v(t)$ satisfies the first-order differential equation

   $$\frac{dv}{dt} = -K_1v + K_2$$

for appropriate K_1 and K_2. Find K_1 and K_2 in terms of K_0.

   (b) Suppose the input $v_s(t)$ = $-12u(-t)$ + $24u(t)$. Find $v(0^-)$ by inspection of the circuit, and $v(0^+)$ by the principle of conservation of charge.

   (c) Use equation 8.17 to write down directly the answer for $v(t)$, $t > 0$. Had $v(0^-)$ been used, would the answer still be correct?

51. The solution to the basic RL or RC differential equation in this chapter, equation 8.3, builds on the integral solution of equation 8.4, which is valid for arbitrary $f(t)$. This powerful formula will be studied in a course on differential equation theory. When $f(t) = F$, a constant, it is possible to develop an alternative derivation of equation 8.17 using no more than some basic knowledge of calculus. Since the solution to the source-free case is the exponential, it is reasonable to expect (or to try) a solution for the constant input case of the form

$$x(t) = K_1e^{-\frac{t}{\tau}} + K_2 \qquad (1)$$

where K_1 and K_2 are two constants to be determined. The constant K_2 arises from the constant input, suggesting that the response would intuitively contain a constant term.

   (a) Substitute equation 1 into equation 8.16. You should obtain the result K_2 = $F\tau$.

   (b) With K_2 determined, evaluate $x(t)$ at $t = t_0^+$ to obtain an expression for K_1. Your result should be

$$K_1 = \left[x(t_0^+) - F\tau\right]e^{\frac{t_0}{\tau}} \qquad (2)$$

   (c) Finally, substitute K_2 = $F\tau$ and equation 2 into equation 1.

APPLICATIONS

52. An approximate sawtooth waveform can be produced by charging and discharging a capacitor with widely different time constants. The circuit of Figure P8.52 illustrates the idea. $v_{in}(t)$ = 10 V, R_1 = 20 kΩ, R_2 = 1 kΩ, and C = 10 μF. The switch S is operated as follows: S has been at position B for a long time, and S is moved to position A at t = 0 to charge the capacitor. When v_C increases to 9 V, switch S is moved to position B to discharge the capacitor. When v_C decreases to 1 V, switch S is moved to position A to charge the capacitor again. The process repeats indefinitely.

    (a) Compute the waveform of $v_C(t)$ for four switchings.

    (b) Plot $v_C(t)$. Is the name "sawtooth waveform" appropriate? What is the frequency in hertz of the sawtooth waveform?

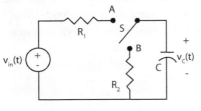

Figure P8.52

53. The sawtooth waveform is used in TV sets and oscilloscopes to control the horizontal motion of the electron beam that sweeps across the screen. One method of generating such a waveform is to repeatedly charge a capacitor with a large time constant and then repeatedly discharge it with a very small time constant. The circuit in the shaded box of Figure P8.53a is a crude functional model for the neon bulb in Figure P8.53b (type 5AB, costing about 75 cents), whose i-v characteristic is shown in Figure P8.53c. The switch S in the model operates as follows:

    (i) S is at position A when v is less than 90 V and *increasing*, and it moves to B when v reaches 90 V (the **breakdown voltage**).

    (ii) S is at position B when v is greater than 60 V and *decreasing*, and it moves to position A when i drops to 1 mA and v drops to 60 V.

Assume that at t = 0, switch S is at A, and $v_{out}(0)$ = 60 V. Find $v_{out}(t)$ for one cycle of operation (i.e., charging and discharging the capacitor), and roughly sketch the waveform. What is the frequency of the sawtooth waveform?

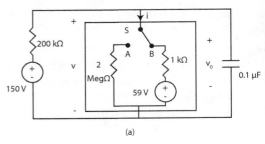

(a)

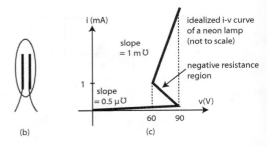

(b)   (c)

Figure P8.53

54. This problem is based on a hypothetical energy storage system using an inductor and a solar cell. Consider the circuit of Figure P8.54. During the day, the solar cell stores energy by increasing the current in the inductor. During the night, the stored energy is used to power lights and appliances. Energy from the solar cell is stored in the inductor during $0 \leq t < T_1$. At t = 0, the beginning of the storage interval, $i_L(0^-)$ = 0. At $t = T_1$, the device L_{store} is switched from storing energy in the solar cell via the source V_{solar} to powering a light denoted by R_1. Note that there is some overlap in the switching movement; this is to ensure continuity of the inductor current. At $t = T_2$, the TV, represented by R_2, is also turned on.

Remark: All answers to parts (a) to (f) should be in terms of V_{solar}, R_{solar}, L_{store}, R_{store}, R_1, and R_2.

(a) Draw a simplified equivalent circuit with three circuit elements for $0 \le t < T_1$, indicating all device values.

(b) Construct an expression for $i_L(t)$, $0 \le t < T_1$.

(c) Draw a simplified equivalent circuit with two circuit elements for $T_1 \le t < T_2$ indicating all device values.

(d) Construct an expression for $i_L(t)$, $T_1 \le t < T_2$. You will need a value or expression for $i_L(T_1^-)$. After obtaining your expression, for simplicity, let i_{T1-} denote $i_L(T_1^-)$.

(e) Draw a simplified equivalent circuit with two circuit elements for $T_2 \le t$, indicating all device values.

(f) Construct an expression for $i_L(t)$, $T_2 \le t$. You will need an expression for $i_L(T_2^-)$. After obtaining your expression, for simplicity, let I_{T2-} denote $i_L(T_2^-)$.

Remark: For the remaining parts, all answers are to be given in terms of V_{solar}, R_{solar}, L_{store}, R_{store}, R_1, and R_2, and $i_L(t)$. This is to prevent the substitution of possibly incorrect answers from prior parts for $i_L(t)$.

(g) For each of the four devices V_{solar}, R_{solar}, L_{store}, and R_{store}, write down an expression for the power absorbed at time t. Call the results P_{Vsolar}, P_{Rsolar}, P_{Lstore}, and P_{Rstore}.

(h) For time t give an expression for the energy $W_L(t)$ stored in the inductor if $W_L(t = 0) = 0$.

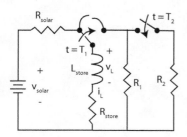

Figure P8.54

55. The circuit of Figure P8.55 is a **transistor photo timer** used for timing the light in photographic enlarger and printing boxes. Briefly, the circuit operates as follows. When the relay contact closes, the lamp is lit. When the contact opens, the lamp is turned off. The relay has a 4000 Ω dc resistance and a negligible inductance. The pickup current is 2 mA, and the dropout current is 0.5 mA; i.e., the contact closes when the relay current increases from zero to 2 mA, and it opens when the current drops below 0.5 mA.

To use the timer, switch S_2 is closed first. Switch S_1 is normally in the B position. When it is thrown momentarily to position A, the battery charges the 1000 µF electrolytic capacitor C to 1.5 V. When S_1 is then thrown back to position B at $t = 0$, the capacitor discharges and produces a current i_b, which, after amplification by the transistor, actuates the relay and turns on the lamp. At some later instant t_1, the amplified current drops below a point for the relay to open and the lamp is turned off.

Compute t_1 if the 10 kΩ potentiometer is set at the middle of its full range (i.e., only 5 kΩ is used in the circuit).

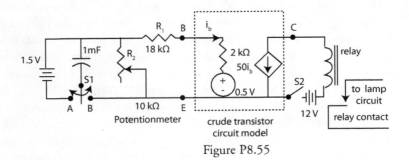

Figure P8.55

56. The circuit of Figure P8.56 suggests a way of generating a sustained sinusoidal oscillation. All op amps are assumed to be ideal. Capacitors, $C = 0.1\ \mu F$ are uncharged at $t = 0$. The first two op amps are differentiators and the last is an inverting amplifier.

(a) With switch S at position A and $v_s(t) = \sin(1000t)$ V, find $v_a(t)$, $v_b(t)$, and $v_{out}(t)$ for $t \geq 0$.

(b) If at a later instant switch S is quickly moved to position B, what would you expect $v_{out}(t)$ to be?

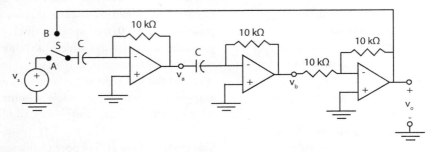

Figure P8.56

C H A P T E R 9

Second Order Linear Circuits

Warming up snacks in a microwave oven is a common activity in student dormitories. It works much faster than a conventional oven: heating a sandwich takes about 30 seconds. How does the microwave oven do this? While a precise explanation is beyond the scope of this text, the basic principle can be understood through the properties of a simple *LC* circuit.

Recall that two conducting plates separated by a dielectric (insulating material) form a capacitor. Suppose some food were placed between the plates in place of an ordinary dielectric. The food itself would act as a dielectric. Ordinary food contains a great number of water molecules. Each water molecule has a positively charged end and a negatively charge end, with their orientations totally random for uncharged plates, as illustrated in part (a) of the figure below. Applying a sufficiently high dc voltage to the plates sets up an electric field produced by the charge deposited on the plates. This causes the water molecules to align themselves with the field as illustrated in part (b). If the polarity of the dc voltage is reversed, the molecules will realign in the opposite direction as illustrated in part (c). If the polarity of the applied voltage is reversed repeatedly, then the water molecules will repeatedly flip their orientations. In doing so, the water molecules encounter considerable friction, resulting in a buildup of heat, which cooks the food. Microwave cooking is therefore very different from conventional cooking. Instead of heat coming from the outside, the heat is generated inside the food itself.

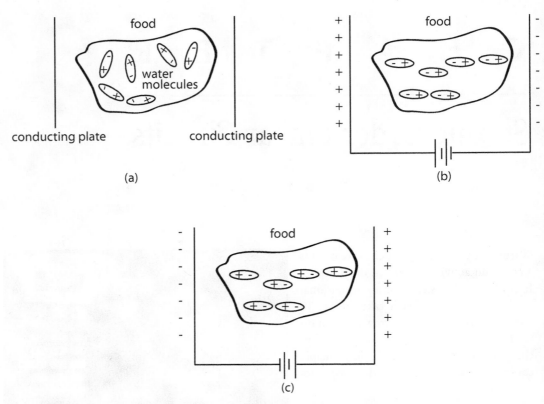

(a)

(b)

(c)

Reversal of the polarity of the applied voltage at a low frequency can be easily achieved with the circuit elements studied in earlier chapters: the resistor, the capacitor, and the inductor. However, the friction-induced heat production is inefficient at low frequency. To produce a useful amount of heat for cooking purposes, very high frequencies must be used. The typical frequency used in a microwave oven is 2.45 gigahertz, i.e., the water molecules reverse their orientations $2 \times 2.45 \times 10^9$ times per second. At such a high frequency, capacitors and inductors are quite different in their behavior from their conventional forms. For example, the LC circuit becomes a "resonant cavity" and the connecting wire becomes a "waveguide." These microwave components will be studied in a future field theory course. The theory studied in this chapter will enable us to understand the low-frequency version of the phenomenon, i.e., how a connected inductor and capacitor can produce oscillatory voltage and current waveforms.

CHAPTER OBJECTIVES

1. Investigate the voltage-current interactions that occur when an ideal inductor is connected to an ideal capacitor with initial stored charge.
2. Use a second-order differential equation for modeling the series RLC and parallel RLC circuits.
3. Learn to solve a second-order differential equation circuit model by first finding the natural frequencies of the circuit, then looking up the general solution form, and finally determining the associated arbitrary constants.

4. Define and understand the concepts of underdamped, overdamped, and critically damped responses.

5. Investigate and understand the underlying principles of various oscillator circuits.

SECTION HEADINGS

1. **Introduction**
2. **Discharging a Capacitor through an Inductor**
3. **Source-Free Second-Order Linear Networks**
4. **Second-Order Linear Networks with Constant Inputs**
5. **Oscillator Application**
6. **Summary**
7. **Terms and Concepts**
8. **Problems**

1. INTRODUCTION

The previous chapter developed techniques for computing the responses of first-order linear networks, either without sources or with dc (constant) sources, having first-order linear differential equation models. Recall that the source-free response contains only real exponential terms.

This chapter focuses on second-order linear networks having second-order linear differential equation models. Usually, but not always, a second-order network contains two energy storage elements, either (L, C), (C, C), or (L, L). Second-order circuits have a wide variety of response waveforms: exponentials $(A_1 e^{s_1 t} + A_2 e^{s_2 t})$, sinusoids $(A_1 \cos(\omega_d t) + A_2 \sin(\omega_d t))$, exponentially damped sinusoids, and exponentially growing sinusoids, among others. Tables 9.1 and 9.2 catalogue the various response types. With no sources or with constant-value sources, some straightforward extensions of the solution methods of Chapter 8 are sufficient to compute the various responses. The behavior of second-order circuits is a microcosm of the behavior of higher-order circuits and systems. Many higher-order systems can be broken down into cascades of second-order systems or sums of second-order systems. This suggests that our exploration of second-order circuits can build a core knowledge base for understanding the behavior of higher-order, more complex physical systems.

Many introductory texts discuss only parallel and series RLC circuits, stating separate formulas for the responses of each. Our approach seeks a unified treatment. To this end, we formulate a basic second-order differential equation circuit model. The associated solution techniques become applicable to any second-order linear network and, for that matter, to second-order mechanical systems.

An oscillator circuit (section 5) motivates our study of second-order linear networks. The chapter contains several other practical examples illustrative of the wide variety of second-order circuit applications. Some advanced applications pertinent to higher-level courses include low-pass, high-pass, and bandpass filtering (covered later in the text); dc motor analysis; position control; and many others. Most important, the concepts presented in this chapter are common to a host of engineering problems and disciplines. Hence, the techniques and concepts described here will prove useful time and time again.

2. DISCHARGING A CAPACITOR THROUGH AN INDUCTOR

Chapter 8 showed that the voltage of an initially charged capacitor in parallel with a resistor decreases exponentially to zero: the capacitor discharges its stored energy through the resistor. When an inductor replaces the resistor, as in Figure 9.1a, very different voltage waveforms emerge for $v_C(t)$ and $i_L(t)$. In order to construct these new waveforms, we first develop a differential equation model of the LC circuit.

EXAMPLE 9.1. The goal of this example is to develop a differential equation model of the circuit in Figure 9.1b. In Figure 9.1a, with the switch S in position A, the voltage source charges the capacitor to V_0 volts. At $t = 0$, the switch moves to position B, resulting in the new circuit of Figure 9.1b, valid for $t \geq 0$.

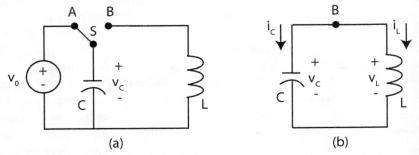

(a)   (b)

FIGURE 9.1 (a) A voltage source charges a capacitor. (b) An LC second-order linear network in which the energy stored in the capacitor in part (a) is passed back and forth to the inductor.

SOLUTION.
Step 1. *Write down the terminal i-v relationship for the capacitor and inductor; then apply KCL and KVL, respectively.* Using the *i-v* relationships for L and C (see Chapter 7) in conjunction with KCL and KVL, it follows that

$$\underbrace{\frac{dv_C}{dt} = \frac{i_C}{C}}_{C\ definition} \underbrace{= -\frac{i_L}{C}}_{KCL}$$

(9.1a)

and

$$\underbrace{\frac{di_L}{dt} = \frac{v_L}{L}}_{L\ definition} \underbrace{= \frac{v_C}{L}}_{KVL}$$

(9.1b)

Step 2. *Obtain a differential equation in the capacitor voltage, $v_C(t)$. For this, first differentiate equation 9.1a to obtain*

$$\frac{d^2 v_C}{dt} = -\frac{1}{C}\frac{di_L}{dt}$$

Substituting equation 9.1b into this equation yields

$$\frac{d^2 v_C}{dt} = -\frac{1}{LC}v_C \qquad (9.2)$$

Equation 9.2 is a second-order linear differential equation circuit model of Figure 9.1b in terms of the unknown capacitor voltage, $v_C(t)$. Equation 9.2 stipulates that the second-order derivative of the unknown function, $v_C(t)$, must equal the function itself multiplied by a *negative constant*, $\frac{-1}{LC}$.

Step 3. *Obtain a differential equation in the current, $i_L(t)$. An alternative circuit model in i_L is obtained by first differentiating equation 9.1b and substituting equation 9.1a into the result to produce*

$$\frac{d^2 i_L}{dt} = -\frac{1}{LC}i_L \qquad (9.3)$$

Equation 9.3 has precisely the same form as 9.2: the second-order derivative of the unknown function, $i_L(t)$, equals the function itself multiplied by a *negative constant*, $\frac{-1}{LC}$. This similarity suggests a similarity of solutions, which we shall pursue further.

Exercise. Fill in the details of the derivation of equation 9.3 from 9.1.

Our next goal is to construct the waveforms $v_C(t)$ and $i_L(t)$, which are the solutions of the differential equations 9.2 and 9.3. Although differential equations are not usually part of the common background of students in a beginning course on circuits, the solutions of equations 9.2 and 9.3 do not demand this background. Some elementary knowledge of differential calculus is sufficient. Specifically, recall the differential properties of the sine and cosine functions:

$$\frac{d}{dt}\sin(\omega t + \theta) = \omega\cos(\omega t + \theta) \text{ and } \frac{d}{dt}\cos(\omega t + \theta) = -\omega\sin(\omega t + \theta)$$

Differentiating a second time yields

$$\frac{d^2}{dt^2}\sin(\omega t + \theta) = -\omega^2\sin(\omega t + \theta) \text{ and } \frac{d^2}{dt^2}\cos(\omega t + \theta) = -\omega^2\cos(\omega t + \theta)$$

In both cases, the second derivative equals the function itself multiplied by a *negative constant*. This is precisely the property required by equations 9.2 and 9.3. Thus one reasonably assumes that the solutions of equations 9.2 and 9.3 have the general forms

$$v_C(t) = K\cos(\omega t + \theta) \qquad (9.4a)$$

and

$$i_L(t) = K \cos(\omega t + \theta) \tag{9.4b}$$

These forms are general because the cosine function can be replaced by the sine function with a proper change in the phase angle. Specifically, we note that $K \sin(\omega t + \phi) = K \cos(\omega t + \phi - 0.5\pi)$ $= K \cos(\omega t + \theta)$ with $\theta = \phi - 0.5\pi$. Computing values for ω, K, and θ specifies the solutions to equations 9.2 and 9.3.

EXAMPLE 9.2. Find K and θ for the capacitor voltage in equation 9.4a.
SOLUTION.
Step 1. Differentiate equation 9.4a to obtain

$$\frac{dv_C(t)}{dt} = -K\omega \sin(\omega t + \theta) \tag{9.5}$$

Step 2. *Differentiate a second time.* Differentiating equation 9.5 (the second derivative of 9.4) yields

$$\frac{d^2 v_C}{dt^2} = -K\omega^2 \cos(\omega t + \theta) = -\omega^2 v_C \tag{9.6}$$

Step 3. *Match the coefficients of equation 9.6 with those of 9.2 to specify ω.* Under this matching,

$$\omega^2 = \frac{1}{LC} \text{ or } \omega = \frac{1}{\sqrt{LC}} \tag{9.7}$$

Equation 9.7 specifies ω, the angular frequency of oscillation, in rad/sec, of the capacitor voltage.

Step 4. *Compute K and θ in equation 9.4a.* These two constants depend on the initial conditions as follows: when the switch is at position A, the capacitor is charged up to V_0 volts and the inductor current is zero; immediately after the switch moves to position B, i.e., at $t = 0^+$, the continuity property of the capacitor voltage ensures that $v_C(0^+) = V_0$ and the continuity property of the inductor current ensures that $i_L(0^+) = 0$. The initial value, $\frac{dv_C(0^+)}{dt}$,[1] is now calculated from equation 9.1a as

$$\frac{dv_C(0^+)}{dt} = \frac{i_C(0^+)}{C} = \frac{-i_L(0^+)}{C} = 0$$

Evaluating equations 9.4a and 9.5 at $t = 0^+$, we have

$$v_C(0^+) = K \cos(\theta) = V_0 \tag{9.8a}$$

and

$$\frac{dv_C(0^+)}{dt} = -K\omega \sin(\theta) = 0 \tag{9.8b}$$

From equation 9.8b, $\theta = 0$. Consequently from 9.8a, $K = V_0$. Hence the capacitor voltage, i.e., the solution of the second-order differential equation 9.2, is

$$v_C(t) = V_0 \cos\left(\frac{t}{\sqrt{LC}}\right) \tag{9.9}$$

As per equation 9.1a, one can obtain $i_L(t)$ directly by differentiating equation 9.9 and multiplying by $-C$. However, one could also solve equation 9.3 by repeating the above steps to arrive at the same answer.

Exercise. Assuming that $i_L(t) = K\cos(\omega t + \theta)$, solve for ω, K, and θ in terms of the initial conditions, and show that

$$i_L(t) = V_0\sqrt{\frac{C}{L}}\sin\left(\frac{t}{\sqrt{LC}}\right)$$

Several very interesting and significant facts about this parallel LC circuit and the solution method are apparent:

(1) For the source-free LC circuit of Figure 9.1, the voltage and current responses are sinusoidal waveforms with an angular frequency equal to $\dfrac{1}{\sqrt{LC}}$. Since the amplitude of sinusoidal oscillations remains constant (i.e., does not damp out), the circuit is said to be **undamped.**

(2) The frequency, ω, depends on the values of L and C only, while the amplitude K and the phase angle θ depend on L, C, and the initial values of the capacitor voltage and inductor current.

(3) Although the instantaneous energy stored in the capacitor, $W_C(t)$, and the instantaneous energy stored in the inductor, $W_L(t)$, both vary with time, their sum is constant. (This is investigated in a homework exercise.) Physically there is a continuous exchange of the energy stored in the magnetic field of the inductor and that stored in the electric field of the capacitor, with no *net* energy loss. This is analogous to a frictionless hanging mass-spring system: because of the absence of friction, the up-and-down motion of the mass never stops; in such a mechanical system there is a continuous interchange between potential and kinetic energy.

Figure 9.1 shows what is, in theory, the simplest circuit that generates sinusoidal waveforms. Such an electronic circuit is an (idealized) **oscillator circuit.** Oscillator circuits play an important role in many communication and instrumentation systems.

---

3. SOURCE-FREE SECOND-ORDER LINEAR NETWORKS

Unlike their ideal counterparts, real capacitors and inductors have resistances. A better understanding of a realistic oscillator entails the analysis of an RLC circuit. This section investigates source-free RLC circuits having two energy storage elements. Our investigation begins with the development of the differential equation models of the series and parallel RLC circuits. Both models are special cases of an undriven general second-order linear differential equation. Hence we will discuss the solution of a general second-order linear differential equation and adapt the solution to the series and parallel RLC circuits. We will also illustrate the theory with a second-order circuit that is not a parallel or series RLC.

Development of Differential Equation Models for Series/Parallel RLC Circuit

The first goal of this section is to develop differential equation models for series and parallel *RLC* circuits as detailed in the following example.

EXAMPLE 9.3. For the series and parallel *RLC* circuits shown in Figure 9.2, develop two second-order differential equation models (one in i_L and one in v_C) for each circuit.

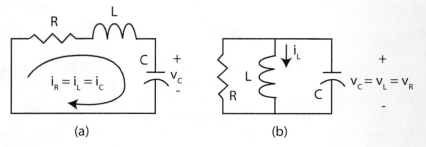

(a)   (b)

FIGURE 9.2 (a) Series *RLC* circuit. (b) Parallel *RLC* circuit. Passive sign convention is assumed as usual.

SOLUTION
We do this in "parallel" rather than in "series."

Step 1. *Apply KVL to the series RLC.*

$$v_R + v_L + v_C = 0$$

Step 2. *Choose i_L as circuit variable and express v_R, v_L, and v_C in terms of i_L:*

$$R i_L + L\frac{di_L}{dt} + \frac{1}{C}\int_{-\infty}^{t} i_L(\tau)d\tau = 0$$

Differentiate, rearrange terms, and divide by L to obtain

$$\frac{d^2 i_L}{dt^2} + \frac{R}{L}\frac{di_L}{dt} + \frac{1}{LC}i_L = 0$$

Step 3. *Choose v_C as the circuit variable and express v_R and v_L in terms of v_C*
Again using the KVL of step 1,

$$R i_C + L\frac{di_C}{dt} + v_C = 0$$

Step 1. *Apply KCL to the parallel RLC.*

$$i_R + i_L + i_C = 0$$

Step 2. *Choose v_C as circuit variable and express i_R, i_L, and i_C in terms of v_C:*

$$\frac{v_C}{R} + \frac{1}{L}\int_{-\infty}^{t} v_C(\tau)d\tau + C\frac{dv_C}{dt} = 0$$

Differentiate, rearrange terms, and divide by C to obtain

$$\frac{d^2 v_C}{dt^2} + \frac{1}{RC}\frac{dv_C}{dt} + \frac{1}{LC}v_C = 0$$

Step 3. *Choose i_L as the circuit variable and express i_R and i_C in terms of i_L.*
Again using the KCL of step 1,

$$\frac{v_L}{R} + i_L + C\frac{dv_L}{dt} = 0$$

Hence, substituting for $i_C = C\, dv_C/dt$, rearranging, and dividing through by LC yields

$$\frac{d^2 v_C}{dt^2} + \frac{R}{L}\frac{dv_C}{dt} + \frac{1}{LC}v_C = 0$$

Hence, substituting for $v_L = L\, di_L/dt$, rearranging, and dividing through by LC yields

$$\frac{d^2 i_L}{dt^2} + \frac{1}{RC}\frac{di_L}{dt} + \frac{1}{LC}i_L = 0$$

Each circuit has two second-order differential equation models, one each for i_L and v_C as the unknown quantity.

Exercise. Show that i_C and v_L satisfy second-order differential equations similar to those derived in Example 9.3.

Solution of the General Second-Order Differential Equation Model

The final differential equations of Example 9.3 force the current i_L or the voltage v_C to satisfy certain differential constraints. All four (differential) equations have the general form

$$\frac{d^2 x}{dt^2} + b\frac{dx}{dt} + cx = 0 \qquad (9.10)$$

for appropriate constants b and c, where x is either i_L or v_C. Equation 9.10 stipulates that the second derivative of the function $x(t)$ plus b times the first derivative of $x(t)$ plus c times $x(t)$ itself adds to zero at all times, t. Unlike the example of section 2, where a sinusoidal solution was easily predicted, the present differential equation requires a more careful mathematical analysis. Recall from elementary calculus that the derivative of an exponential is an exponential. Thus the first and second derivatives of an exponential are proportional to the original exponential. This suggests postulating a solution of the form $x(t) = Ke^{st}$ where we make no *a priori* assumptions about s. If it is truly a solution, it must satisfy equation 9.10. Under what conditions will $x(t) = Ke^{st}$ satisfy equation 9.10?

EXAMPLE 9.4. Determine conditions under which the postulated solution $x(t) = Ke^{st}$ satisfies equation 9.10.

SOLUTION.

Step 1. Substituting Ke^{st} for $x(t)$ in equation 9.10 produces

$$K\frac{d^2 e^{st}}{dt^2} + bK\frac{de^{st}}{dt} + cKe^{st} = Ke^{st}\left(s^2 + bs + c\right) = 0 \qquad (9.11)$$

Step 2. *Interpret equation 9.11.* For nontrivial solutions, K is nonzero. The function e^{st} is always different from zero. Hence the quadratic in s on the right side of equation 9.11 must be zero. This necessarily constrains s to be a root of

$$s^2 + bs + c = 0 \qquad (9.12)$$

Step 3. *Solve equation 9.12.* From the quadratic formula, the roots of equation 9.12 are

$$s_1, s_2 = \frac{-b \pm \sqrt{b^2 - 4c}}{2} \tag{9.13}$$

CONCLUSION: $x(t) = Ke^{st}$ satisfies equation 9.10 provided s takes on values given by equation 9.13. Equation 9.10 does not constrain K; however, the initial conditions will.

Equations such as 9.12 whose solution is given by equation 9.13 are a common characteristic of second-order networks. Hence, equation 9.12 is called the **characteristic equation** of the second-order linear circuit. The associated roots, equation 9.13, are called the **natural frequencies** of the circuit. These are the "natural" or intrinsic frequencies of the circuit response and are akin to the natural frequencies of oscillations of a pendulum (for small swings) or of a bouncing ball.

From elementary algebra, a quadratic equation (the above characteristic equation) can have distinct roots or equal roots. Distinct roots can be real or complex. Thus s_1 and s_2 can be two distinct real roots, two distinct conjugate complex roots, or two repeated (equal) roots, depending on whether the discriminant, $b^2 - 4c$, is greater than, less than, or equal to zero. This trifold grouping separates the solution of equation 9.10 into three categories, listed below as cases 1, 2, and 3:

Case 1. Real and distinct roots, i.e., $b^2 - 4c > 0$. If the roots are real and distinct, then for arbitrary constants K_1 and K_2, both

$$x(t) = x_1(t) = K_1 e^{s_1 t}$$

and

$$x(t) = x_2(t) = K_2 e^{s_2 t}$$

satisfy the second-order linear differential equation 9.10, i.e., are solutions to the differential equation. Since equation 9.10 is a *linear* differential equation, by superposition the sum $x(t) = x_1(t) + x_2(t)$ is also a solution, a fact easily verified by direct substitution. Therefore, whenever $s_1 \neq s_2$, the most general form of the solution to equation 9.10 is

$$x(t) = K_1 e^{s_1 t} + K_2 e^{s_2 t} \tag{9.14}$$

The constants K_1 and K_2 depend on the initial conditions of the differential equation, which depend on the initial capacitor voltages and inductor currents. For example, if $x(0^+)$ and $x'(0^+)$ are known, then from equation 9.14,

$$x(0^+) = \left[K_1 e^{s_1 t} + K_2 e^{s_2 t} \right]_{t=0^+} = K_1 + K_2$$

and

$$x'(0^+) = \left. \frac{dx(t)}{dt} \right]_{t=0^+} = s_1 K_1 + s_2 K_2$$

These are simultaneous equations solvable for K_1 and K_2.

If s_1 and s_2 are negative, the response given by equation 9.14 decays to zero for large t and the circuit is said to be **overdamped**.

Case 2. *The roots, s_1 and s_2, of the characteristic equation are distinct but complex, i.e., $b^2 - 4c < 0$.* Since $s_1 \neq s_2$, the most general form of the solution to equation 9.10 is again given by equation 9.14, i.e.,

$$x(t) = K_1 e^{s_1 t} + K_2 e^{s_2 t}$$

with complex s_1 and s_2 given by

$$s_1, s_2 = \frac{-b}{2} \pm j \frac{\sqrt{4c - b^2}}{2} = -\sigma \pm j\omega_d \tag{9.15}$$

where $\sigma = b/2$ and $\omega_d = \dfrac{\sqrt{4c - b^2}}{2}$. Since s_1 and s_2 are conjugates, so are $e^{s_1 t}$ and $e^{s_2 t}$ in equation 9.14.

For $x(t)$ to be real, the constants K_1 and K_2 in equation 9.14 must also be complex conjugates, i.e., $K_2 = K_1^*$. Using **Euler's formula,**

$$e^{jy} = \cos y + j \sin y$$

the two terms in equation 9.14 combine to yield a real time function:

$$x(t) = K_1 e^{s_1 t} + K_2 e^{s_2 t} = K_1 e^{(-\sigma + j\omega_d)t} + K_1^* e^{(-\sigma - j\omega_d)t}$$

$$= e^{-\sigma t}\left[K_1 \cos(\omega_d t) + jK_1 \sin(\omega_d t)\right] + e^{-\sigma t}\left[K_1^* \cos(\omega_d t) - jK_1^* \sin(\omega_d t)\right]$$

$$= e^{-\sigma t}\left[(K_1 + K_1^*)\cos(\omega_d t) + (jK_1 + jK_1^*)\sin(\omega_d t)\right]$$

Thus the solution to equation 9.14 with s_1 and s_2 complex is given by the (damped) sinusoidal response

$$x(t) = e^{-\sigma t}\left[A \cos(\omega_d t) + B \sin(\omega_d t)\right] \tag{9.16}$$

where $A = K_1 + K_1^* = 2\,\mathrm{Re}[K_1] = K_1 + K_2$ and $B = jK_1 - j K_1^* = -2\,\mathrm{Im}K_1] = jK_1 - jK_2$ are *real* constants and where Re[·] denotes the real part and Im[·] denotes the imaginary part. The solution expressed in equation 9.16 is completed by specifying A and B. As before, A and B depend on the initial conditions, $x(0^+)$ and $x'(0^+)$ as follows:

$$x(0^+) = \left(e^{-\sigma t}\left[A\cos(\omega_d t) + B\sin(\omega_d t)\right]\right)\Big|_{t=0^+} = A$$

$$x'(0^+) = \left[-\sigma\, e^{-\sigma t}\Big(A\cos(\omega_d t) + B\sin(\omega_d t) \Big) + e^{-\sigma t}\Big(-\omega_d A\sin(\omega_d t) + \omega_d B\cos(\omega_d t) \Big) \right]_{t=0^+}$$

$$= -\sigma A + \omega_d B$$

and

which are easily solved for A and B.

Making use of a standard trigonometric identity, the general solution of equation 9.16 has the equivalent form

$$x(t) = e^{-\sigma t}\,[A\cos(\omega_d t) + B\sin(\omega_d t)] = K e^{-\sigma t}\cos(\omega_d t + \theta) \qquad (9.17a)$$

where

$$K = \sqrt{A^2 + B^2}\,, \quad \theta = \tan^{-1}\!\left(\frac{-B}{A}\right) \qquad (9.17b)$$

and the quadrant of θ is determined by the signs of $-B$ and A. In MATLAB, one uses the command "atan2(–B,A)" to obtain the angle in the proper quadrant. Note that the response waveforms have oscillations with angular frequency ω_d. These oscillations are bounded by the envelope $\pm K e^{-\sigma t}$. If $\mathrm{Re}[s_1] = -\sigma < 0$, the amplitude of the oscillations decays to zero and the response is said to be **underdamped**. If $\mathrm{Re}[s_1] = -\sigma > 0$, the amplitude of the oscillations grows to infinity.

Case 3. *The roots are real and equal, i.e.,* $b^2 - 4c = 0$. When the two roots of the characteristic equation are equal, equation 9.14 does not represent the general solution form because if $s_1 = s_2$, the two terms collapse into a single term. However, the general solution for $s_1 = s_2$ is

$$x(t) = \left(K_1 + K_2 t \right) e^{s_1 t} \qquad (9.18)$$

(This is investigated in a homework exercise.) Calculation of K_1 and K_2 in equation 9.18 is straightforward:

$$x(0^+) = K_1$$

and

$$\dot{x}(0^+) = s_1 K_1 + K_2$$

Substituting the value of K_1 into $x'(0^+)$ yields a simple calculation for K_2.

If $s_1 = s_2$ is negative, the response decays to zero and is said to be **critically damped**. "Critically damped" defines the boundary between overdamped and underdamped. This means that with a slight change in circuit parameters, the response would almost always change to either overdamped or underdamped.

The discussion of the three cases is summarized in Table 9.1.

TABLE 9.1. General Solutions for Source-Free Second-Order Networks

General solution of the homogeneous differential equation

$$\frac{d^2x}{dt^2} + b\frac{dx}{dt} + cx = 0$$

having characteristic equation $s^2 + bs + c = (s - s_1)(s - s_2) = 0$, where

$$s_1, s_2 = \frac{-b \pm \sqrt{b^2 - 4c}}{2}$$

Case 1. *Real and distinct roots, i.e., $b^2 - 4c > 0$:*

$$x(t) = K_1 e^{s_1 t} + K_2 e^{s_2 t}$$

where

$$x(0^+) = K_1 + K_2 \text{ and } x'(0^+) = s_1 K_1 + s_2 K_2$$

Case 2. *The roots, $s_1 = -\sigma + j\omega_d$ and $s_2 = -\sigma - j\omega_d$, of the characteristic equation are distinct but complex, i.e., $b^2 - 4c < 0$:*

$$x(t) = e^{-\sigma t}[A\cos(\omega_d t) + B\sin(\omega_d t)] = Ke^{-\sigma t}\cos(\omega_d t + \theta)$$

where

$$x(0^+) = A, \, x'(0^+) = -\sigma A + \omega_d B$$

and

$$K = \sqrt{A^2 + B^2}, \, \theta = \tan^{-1}\left(\frac{-B}{A}\right)$$

Case 3. *The roots are real and equal, i.e., $s_1 = s_2$ and $b^2 - 4c = 0$:*

$$x(t) = (K_1 + K_2 t)e^{s_1 t}$$

where

$$x(0^+) = K_1 \text{ and } x'(0^+) = s_1 K_1 + K_2$$

Figure 9.3 displays the various response forms described above for the case where $\text{Re}[s_1]$ and $\text{Re}[s_2]$ are negative or zero. Because of their similarity, it is not possible to distinguish between the overdamped and the critically damped responses by merely looking at the waveforms. Both types of response may have at most one zero-crossing.

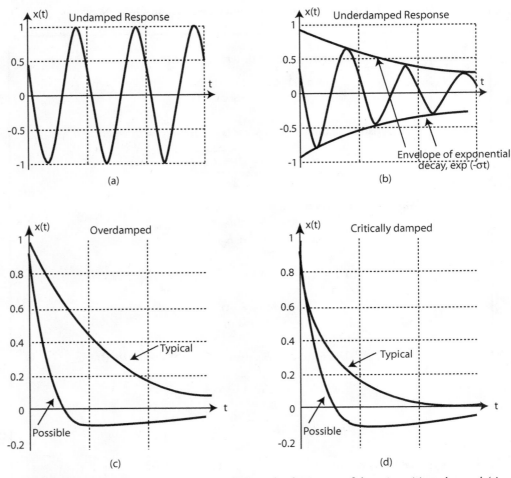

FIGURE 9.3 Generic waveforms corresponding to the four cases of damping: (a) undamped (sinusoidal) response, (b) underdamped (exponentially decaying oscillatory) response, (c) overdamped (exponentially decaying) response, and (d) critically damped (exponentially decaying) response.

The terms "undamped," "underdamped," "overdamped," and "critically damped" stem from an intuitive notion of "damping." The source-free response of an *undamped* second-order linear system, whether electrical or mechanical, has an oscillatory response (waveform) of constant amplitude. *Damping*, due to system elements that consume energy, means a monotonic decrease in the amplitude of oscillation. In electrical circuits, resistances produce the damping effect. In mechanical systems, friction causes damping. When the amount of damping is just enough to prevent oscillation, the system is *critically damped*. Less damping corresponds to the *underdamped* case, where oscillation is present but eventually dies out. A greater amount of damping corresponds to the *overdamped* case, where the waveform is non-oscillatory, and a very small perturbation of any circuit parameter will *not* cause oscillations to occur.

In summary, once the roots of the characteristic equation are found and the expression for the general solution selected from the above cases or Table 9.1, it remains to find the constants K_1 and

K_2 (or A and B) from the initial conditions on the circuit. In the above development, K_1 and K_2 (or A and B) are given in terms of $x(0^+)$ and $x'(0^+)$. Since $x(t)$ represents either a capacitor voltage or an inductor current, its value at $t = 0^+$ is usually given, or can be determined from the past history of the circuit. (See Example 9.5.) The value of $x'(0^+)$, on the other hand, is often unknown and must be calculated. If $x(t) = v_C(t)$, then the capacitor v-i relationship implies that

$$x'(0^+) = v_C'(0^+) = \frac{i_C(0^+)}{C} \ .$$

If $x(t) = i_L(t)$, then the v-i relationship of an inductor implies that

$$x'(0^+) = i_L'(0^+) = \frac{v_L(0^+)}{L} \ .$$

The problem then reduces to finding an unknown capacitor current, $i_C(0^+)$, or an unknown inductor voltage, $v_L(0^+)$.

To find $i_C(0^+)$ or $v_L(0^+)$, we construct an auxiliary resistive circuit valid at $t = 0^+$. Since the initial values, $v_C(0^+)$ and $i_L(0^+)$, are known, we replace (each) capacitor in the original circuit by an independent voltage source of value $v_C(0^+)$, and (each) inductor in the original circuit by an independent current source of value $i_L(0^+)$. Here the current $i_C(0^+)$ retains its original direction and the voltage $v_L(0^+)$ retains its original polarity. After the replacements, the (new) circuit is resistive. Values for $i_C(0^+)$ and $v_L(0^+)$ follow by applying any of the standard methods of resistive circuit analysis learned earlier. This allows us to specify $x(0^+)$ and $x'(0^+)$ in terms of the initial conditions on the circuit. Two equations in the two unknowns K_1 and K_2 (or A and B) result. Example 9.5 and, in particular, Figure 9.4c illustrate this procedure.

Response Calculation of Source-Free Parallel and Series RLC Circuits

Before any additional circuit examples, let us summarize the solution procedure.

Procedure for Solving Second-Order RLC Circuits
Step 1. Determine the differential equation model of the circuit.
Step 2. From the differential equation model, construct the characteristic equation and find its roots using the quadratic root formula.
Step 3. From the nature of the roots (real distinct, real equal, or complex), determine the general form of the solution from Table 9.1; the solution form will contain two unknown parameters.
Step 4. Find the two unknown parameters using the initial conditions on the circuit.

The following example illustrates these calculations for the three cases described in Table 9.1.

EXAMPLE 9.5. In the circuit of Figure 9.4a, the 1 µF capacitor is assumed to be ideal, and the inductor is modeled by a 10 mH ideal inductor in series with a 20 Ω resistor to account for the resistance of the coiled wire. Suppose the switch S in Figure 9.4a has been in position A for a long time. The capacitor becomes charged to 10 V. Then the switch moves to position B at $t = 0$. Find and plot $v_C(t)$ for $t \geq 0$ for the following three cases: (1) $R_2 = 405$ Ω, (2) $R_2 = 0$, (3) $R_2 = 180$ Ω. Each of these cases produces a different response type.

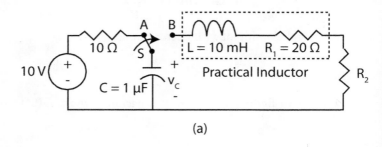

(a)

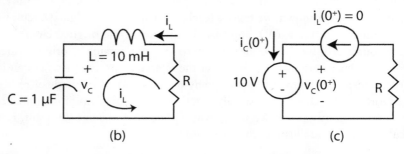

(b)   (c)

FIGURE 9.4 (a) Discharge of a capacitor through a practical inductor in series with a resistance R_2. (b) Equivalent circuit for $t \geq 0$. (c) Equivalent circuit at $t = 0^+$ for calculating $i_C(0^+)$ in which the inductor has been replaced by an independent current source of value $i_L(0^+)$ and the capacitor by an independent voltage source of value $v_C(0^+)$.

SOLUTION

From the problem statements, $v_C(0^-) = 10$ V. When the switch moves to position B, $v_C(0^+) = v_C(0^-) = 10$ V by continuity of the capacitor voltage; the circuit now becomes a series RLC, for $t > 0$, with $R = R_1 + R_2$ as shown in Figure 9.4b. The first step in the calculation of the circuit response is to find a second-order differential equation in the unknown v_C. From Example 9.3, for the series RLC,

$$\frac{d^2 v_C}{dt^2} + \frac{R}{L}\frac{dv_C}{dt} + \frac{1}{LC}v_C = 0 \tag{9.19}$$

Since L and C are known, the *series RLC characteristic equation* is

$$s^2 + \frac{R}{L}s + \frac{1}{LC} = s^2 + (20 + R_2)10^2 s + 10^8 = 0 \tag{9.20}$$

With this framework, we can separately investigate each of the three cases.

Case 1: $R_2 = 405\ \Omega$ or $R = 425\ \Omega$

Step 1. *Find the characteristic equation and the general form of the response using Table 9.1.* If $R_2 = 405\ \Omega$, the characteristic equation of 9.20 is $s^2 + 42{,}500s + 10^8 = 0$. Solving for the roots by the quadratic formula yields

$$s_{1,2} = -21{,}250 \pm 18{,}750 = -2500, -40{,}000\ \text{sec}^{-1}$$

Real distinct roots imply an *overdamped* response of the form

$$v_C(t) = K_1 e^{s_1 t} + K_2 e^{s_2 t} = K_1 e^{-2{,}500t} + K_2 e^{-40{,}000t} \qquad (9.21)$$

valid for $t > 0$.

Step 2. *Find K_1 and K_2.* Evaluating at $t = 0^+$ implies

$$v_C(0^+) = 10 = K_1 + K_2$$

$$\qquad (9.22a)$$

Differentiating equation 9.21 implies

$$v'_C(0^+) = \frac{i_C(0^+)}{C} = -2.5 \times 10^3 K_1 - 40 \times 10^3 K_2 \qquad (9.22b)$$

From the circuit of Figure 9.4c, $v'_C(0^+) = \dfrac{i_C(0^+)}{C} = \dfrac{i_L(0^+)}{C} = \dfrac{i_L(0^-)}{C} = 0$, where $i_L(0^+) = i_L(0^-)$ by the continuity of the inductor current. Solving equations 9.22a and b after substituting the above values yields

$$K_1 = 10.667 \text{ and } K_2 = -0.667$$

Step 3. *Set forth the solution for $v_C(t)$.* For $t > 0$,

$$v_C(t) = 10.667 e^{-2{,}500t} - 0.667 e^{-40{,}000t}\ \text{V}$$

This function is plotted in Figure 9.5.

Exercise. You may verify the above answer with the Student Edition of MATLAB (version 4.0 or later) by typing the command: y = dsolve('D2y+42500*Dy + 1e8*y = 0,y(0) = 10, Dy(0) = 0').

Case 2: $R_2 = 0$ or $R = 20\ \Omega$

Step 1. If $R_2 = 0$, then from equation 9.20, the characteristic equation is $s^2 + 2{,}000s + 10^8 = 0$. Since $b^2 - 4c = -396{,}000{,}000 < 0$, the roots are complex. From the quadratic root formula,

$$s_1 = -1000 + j9950 = -\sigma + j\omega_d \text{ and } s_2 = -1000 - j9950 = -\sigma - j\omega_d$$

From Table 9.1, the underdamped response form is

$$v_C(t) = e^{-\sigma t}[A\cos(\omega_d t) + B\sin(\omega_d t)] = e^{-1000t}[A\cos(9950t) + B\sin(9950t)] \qquad (9.23)$$

Step 2. *Find A and B.* It remains to determine A and B in equation 9.23. From equation 9.23 and its derivative,

$$v_C(0^+) = 10 = A \qquad (9.24a)$$

and

$$v'_C(0^+) = -\sigma A + \omega_d B = -1000A + 9950B \qquad (9.24b)$$

As in case 1, $v_C'(0^+) = \dfrac{i_C(0^+)}{C} = \dfrac{i_L(0^+)}{C} = \dfrac{i_L(0^-)}{C} = 0$. Substituting into equations 9.24 and solving yields

$$A = 10 \text{ and } B = \frac{\sigma A}{\omega_d} = 1.005$$

Step 3. *Set forth the solution for $v_C(t)$.* For $t > 0$,

$$v_C(t) = e^{-1000t}\,[10\,\cos(9950t) + 1.005\,\sin(9950t)] = 10.05 e^{-1000t}\,\cos(9950t + 5.7°)\ \text{V}$$

This waveform is also plotted in Figure 9.5.

Exercise. You may verify the above answer with the Student Edition of MATLAB (version 4.0 or later) by typing the command: y = dsolve('D2y+2000*Dy + 1e8*y = 0,y(0) = 10, Dy(0) = 0').

Case 3: $R_2 = 180\ \Omega$ or $R = 200\ \Omega$

Step 1. If $R_2 = 180\ \Omega$, then the characteristic equation from 9.20 is $s^2 + 20{,}000s + 10^8 = 0$, whose roots are $s_1 = s_2 = -10^4$, implying a critically damped response. From Table 9.1, the general critically damped response form is, for $t \geq 0$,

$$v_C(t) = \left(K_1 + K_2 t\right)e^{s_1 t} = \left(K_1 + K_2 t\right)e^{-10^4 t} \qquad (9.25)$$

Step 2. *Find K_1 and K_2.* From equation 9.25, its derivative, and the known initial conditions from cases 1 and 2,

$$v_C(0^+) = 10 = K_1 \qquad (9.26a)$$

and

$$v'_C(0^+) = s_1 K_1 + K_2 = -10^4 K_1 + K_2 = 0 \qquad (9.26b)$$

Solving equations 9.26 yields

$$K_1 = 10 \text{ and } K_2 = -s_1 K_1 = 10^5$$

Step 3. *Set forth the solution for $v_C(t)$.* For $t > 0$,

$$v_C(t) = \left(10 + 10^5 t\right)e^{-10^4 t}$$

The waveforms of $v_C(t)$ for the three cases (underdamped, critically damped, and overdamped) are plotted in Figure 9.5.

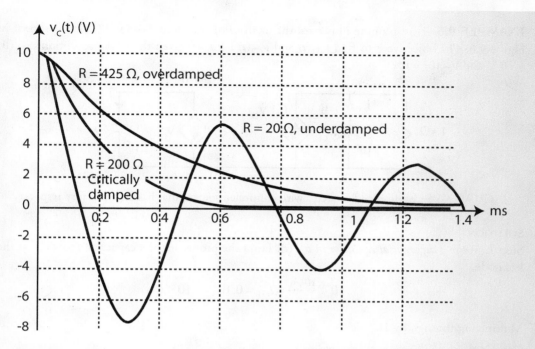

FIGURE 9.5 Waveforms of $v_c(t)$ in Example 9.5 for three different degrees of damping. Critical damping represents the boundary between the overdamped condition and the oscillatory behavior of underdamping.

Exercise. Verify the answer calculated in Example 9.5 using the Student Edition of MATLAB and the "desolve" command.

On a practical note, commercially available resistors come in standard values each with an associated tolerance. Tolerances vary from ±1% (precision resistor) to as much as ±20%. Further, because of heating action over a long period of time, resistance values change. Given the above example, in which the type of response depends on the resistance, one can imagine the care needed in the design of such circuits: without consideration of precision and long-term heating effects, a desired critically damped response could easily become oscillatory.

Not all second-order circuits are *RLC*. Some are only *RC* but with two capacitors and some are *RL* with two inductors. Passive *RC* or *RL* circuits cannot have an oscillatory response. The proof of this assertion can be found in texts on passive network synthesis. However, with controlled sources a second-order *RC* or *RL* circuit can have an oscillatory response that is not characteristic of a first-order circuit, but of a second- or higher-order circuit. The example below illustrates the analysis of a second-order *RC* circuit containing controlled sources that has an oscillatory response.

EXAMPLE 9.6. This example illustrates the analysis of the second-order RC circuit shown in Figure 9.6. The objective is to find $v_{C1}(t)$ and $v_{C2}(t)$ for $t > 0$ given the initial conditions $v_{C1}(0)$ = 10 V and $v_{C2}(0) = 0$.

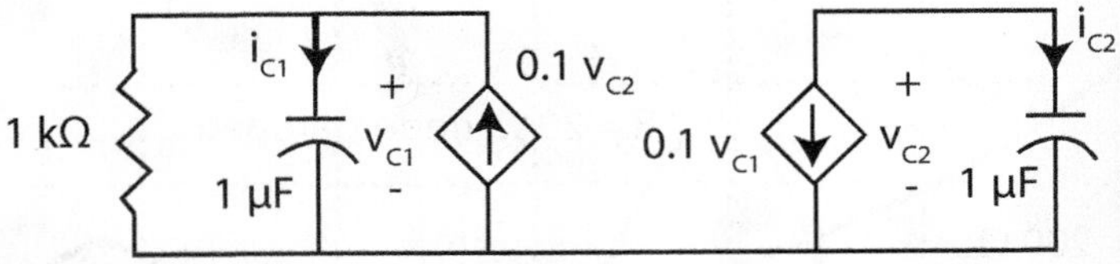

FIGURE 9.6 Second-order RC circuit with controlled sources that has an oscillatory response.

SOLUTION

Step 1. *Write a differential equation in* $v_{C1}(t)$. From the properties of a capacitor and KCL at the left node,

$$10^{-6} \frac{dv_{C1}}{dt} = i_{C1} = 0.1 v_{C2} - 10^{-3} v_{C1}$$

Multiplying through by 10^6 yields

$$\frac{dv_{C1}}{dt} = 10^6 i_{C1} = 10^5 v_{C2} - 10^3 v_{C1} \tag{9.27a}$$

We expect a second-order differential equation, so differentiating a second time yields

$$\frac{d^2 v_{C1}}{dt^2} = 10^5 \frac{dv_{C2}}{dt} - 10^3 \frac{dv_{C1}}{dt} \tag{9.27b}$$

To obtain a differential equation in $v_{C1}, \frac{dv_{C2}}{dt}$ in equation 9.27b must be eliminated. This requires another relationship between v_{C1} and v_{C2}. At the right node,

$$10^{-6} \frac{dv_{C2}}{dt} = i_{C2} = -0.1 v_{C1}$$

or equivalently,

$$\frac{dv_{C2}}{dt} = -10^5 v_{C1} \tag{9.28}$$

Substituting this expression into equation 9.27b produces

$$\frac{d^2 v_{C1}}{dt^2} = -10^{10} v_{C1} - 10^3 \frac{dv_{C1}}{dt}$$

After rearranging terms, we obtain a second-order differential equation in v_{C1}:

$$\frac{d^2 v_{C1}}{dt^2} + 10^3 \frac{dv_{C1}}{dt} + 10^{10} v_{C1} = 0 \tag{9.29}$$

Step 2. *Determine the characteristic equation, its roots, and the form of the response.* The differential equation 9.29 has characteristic equation

$$s^2 + 10^3 s + 10^{10} = 0$$

From the quadratic formula, the complex roots are

$$s_{1,2} = -500 \pm j99,998.75 = -\sigma \pm j\omega d$$

From Table 9.1, complex roots imply an underdamped response of the form

$$v_{C1}(t) = e^{-\sigma t} [A \cos(\omega_d t) + B \sin(\omega_d t)] = e^{-500t} [A \cos(99,998.75t) + B \sin(99,998.75t)] \quad (9.30)$$

Step 3. *Find A and B.* At $t = 0$, $v_{C1}(0) = 10 = A$. Also, from equation 9.27a and the initial conditions,

$$\frac{dv_{C1}(0)}{dt} = 10^5 v_{C2}(0) - 10^3 v_{C1}(0) = -10^4 \quad (9.31)$$

Differentiating equation 9.30, evaluating at $t = 0$, and equating the result with equation 9.31 produces

$$-10^4 = \frac{dv_{C1}(0)}{dt} = B\omega_d$$

in which case $B = 5.0001 \times 10^{-2}$.

Step 4. *Set forth the final form of $v_{C1}(t)$.* The final form of the response is

$$v_{C1}(t) = e^{-500t} [10 \cos(99,998.75t) + 5.0001 \times 10^{-2} \sin(99,998.75t)] \text{ V}$$
$$= 10e^{-500t} \cos(99,998.75t - 0.2865°) \text{ V}$$

Step 5. *Plot the response.* A plot of the (underdamped) response is given in Figure 9.7.

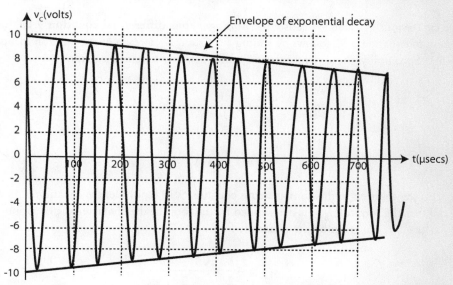

FIGURE 9.7 Underdamped (oscillatory) response showing envelope of exponential decay.

Exercise. Construct a parallel *RLC* circuit to have the same second-order differential equation model as 9.29. Note that there is no unique solution.
CHECK: $RC = 10^{-3}$ and $LC = 10^{-10}$

It is important to observe here that the design of Example 9.6 achieves a second-order *RLC* response without the use of an inductor, which is important for integrated circuit technology.

4. SECOND-ORDER LINEAR NETWORKS WITH CONSTANT INPUTS

The preceding section studied source-free second-order linear networks. When independent sources are present, such as in the circuit of Example 9.7 below, the network (differential) equations are similar to the source-free case except for an additional term that accounts for the effect of the input:

$$\frac{d^2x}{dt^2} + b\frac{dx}{dt} + cx = f(t) \tag{9.32}$$

where $f(t)$ is a scaled sum of the inputs and/or their first-order derivatives. Ordinarily one might expect $f(t)$ to be the value of the input. A homework problem illustrates that $f(t)$ can depend not only on the source input, but also on the derivatives of the input. For general circuits, those not reducible to parallel or series *RLC* circuits, constructing equation 9.32 can be a challenge. Further, the solution of 9.32 for *arbitrary* inputs and initial conditions is no less challenging but is best obtained via the Laplace transform method, which is a topic studied in a second circuits course. However, when the input excitations are constant, $f(t) = F$, the solution to 9.32 is a straightforward modification of the source-free solution, as explained in the remainder of this section.

Since the expressions of Table 9.1 satisfy the **homogeneous differential equation** 9.10, the general solution to equation 9.32 follows by adding a constant X_F to each of the solution forms given in Table 9.1. Specifically, the general solution of the driven differential equation

$$\frac{d^2x}{dt^2} + b\frac{dx}{dt} + cx = F \tag{9.33}$$

is

$$x(t) = x_n(t) + X_F \tag{9.34}$$

where $x_n(t)$ is the solution to the homogeneous equation 9.10 (equivalently, equation 9.33 with $F = 0$). Recall that the form of $x_n(t)$ is determined by the roots of the characteristic equation $s^2 + bs + c = (s - s_1)(s - s_2) = 0$, given by the quadratic formula

$$s_{1,2} = -\frac{b}{2} \pm \frac{\sqrt{b^2 - 4c}}{2}$$

To verify that the structure of equation 9.34 is a solution to 9.33 and to compute the value of X_F, substitute the structure given by equation 9.34 into 9.33. Since $x_n(t)$ satisfies the homogeneous equation 9.10, it contributes zero to the left-hand side. What remains is $cX_F = F$. Therefore, $X_F = \dfrac{F}{c}$ which is independent of the roots of the characteristic equation. However, if $\text{Re}[s_1]$ and $\text{Re}[s_2] < 0$, then $x_n(t)$ tends to zero for large t. Hence $x(t)$ tends to X_F for large t. Consequently X_F is termed the final value of the response.

Because of the trifold structure of $x_n(t)$ as summarized in Table 9.1, the solution form of equation 9.34 once again breaks down into three distinct cases. We summarize this trifold structure for the constant-input case in Table 9.2.

TABLE 9.2 General Solutions for Constant-Source Second-Order Networks

---

General solution of the driven differential equation
$$\frac{d^2x}{dt^2} + b\frac{dx}{dt} + cx = F$$

having characteristic equation $s^2 + bs + c = (s - s_1)(s - s_2) = 0$ with roots
$$s_{1,2} = -\frac{b}{2} \pm \frac{\sqrt{b^2 - 4c}}{2}$$

Case 1. *Real and distinct roots, i.e.,* $b^2 - 4c > 0$:
$$x(t) = K_1 e^{s_1 t} + K_2 e^{s_2 t} + X_F$$
with $X_F = \dfrac{F}{c}$. Further,
$$x(0^+) = K_1 + K_2 + X_F \text{ and } x'(0^+) = s_1 K_1 + s_2 K_2$$

Case 2. *The roots,* $s_1 = -\sigma + j\omega_d$ *and* $s_2 = -\sigma - j\omega_d$ *of the characteristic equation are distinct but complex, i.e.,* $b^2 - 4c < 0$. *The general solution form is*
$$x(t) = e^{-\sigma t}[A\cos(\omega_d t) + B\sin(\omega_d t)] + X_F = Ke^{-\sigma t}\cos(\omega_d t + \theta) + X_F$$
where again $X_F = \dfrac{F}{c}$, with
$$x(0^+) = A + X_F,\ x'(0^+) = -\sigma A + \omega_d B$$
and
$$K = \sqrt{A^2 + B^2},\ \theta = \tan^{-1}\left(\frac{-B}{A}\right)$$

Case 3. *The roots are real and equal, i.e.,* $s_1 = s_2$ *and* $b^2 - 4c = 0$. *The solution form is*
$$x(t) = (K_1 + K_2 t)e^{s_1 t} + X_F$$
where $X_F = \dfrac{F}{c}$, and
$$x(0^+) = K_1 + X_F \text{ and } x'(0^+) = s_1 K_1 + K_2$$

The interpretation $X_F = F/c$. is a mathematical one. When the differential equation describes a linear circuit with constant inputs, there is a physical interpretation of X_F and a circuit theoretic method for computing its value, even without writing the differential equation. Since "$x(t) = X_F$ = a constant" satisfies the differential equation 9.33, it is also a constant solution to the circuit. Hence X_F is either a constant capacitor voltage or a constant inductor current. If a capacitor voltage is constant, its current is zero; this is interpreted as an open circuit. Similarly, if an inductor current is constant, its voltage is zero; this is interpreted as a short circuit. Therefore, X_F is an appropriate (capacitor) voltage or (inductor) current obtained when the capacitor (or capacitors) are open-circuited and the inductor (or inductors) are short-circuited. The value of X_F can be obtained by analyzing the resistive network resulting when all capacitors are open-circuited and all inductors are short-circuited. Recall that if $\mathrm{Re}[s_1]$ and $\mathrm{Re}[s_2] < 0$, then $x(t)$ tends to the constant value X_F. Physically speaking, then, X_F equals either $v_C(\infty)$ or $i_L(\infty)$ when $\mathrm{Re}[s_1]$ and $\mathrm{Re}[s_2] < 0$.

Once the proper general solution structure is ascertained from Table 9.2 and the constant X_F is found, the parameters K_1 and K_2 (or A and B) are computed by the same methods used in the source-free case. The following example illustrates the procedure for a parallel RLC circuit.

EXAMPLE 9.7. A step current input, $i_{in}(t) = u(t)$ A, excites the parallel RLC circuit of Figure 9.8, whose initial conditions satisfy $i_L(0) = 0$ and $v_C(0) = 0$. This simply means that the current source turns on with a value of 1 amp at $t = 0$ and maintains this constant current excitation for all time. The objective is to find the inductor current, $i_L(t)$, for $t \geq 0$, for three values of R: (i) $R = 500\ \Omega$, (ii) $R = 25\ \Omega$, and (iii) $R = 20\ \Omega$.

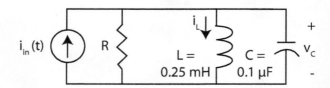

FIGURE 9.8 Parallel RLC circuit excited by a step current input.

SOLUTION
Because the circuit is a parallel RLC, the characteristic equation is

$$s^2 + \frac{1}{RC}s + \frac{1}{LC} = s^2 + \frac{10^7}{R}s + 4 \times 10^{10} = 0 \qquad (9.35)$$

For all positive values of R, the roots of the circuit's characteristic equation have negative real parts. Thus for large t or ideally at "$t = \infty$," the inductor looks like a short circuit and the capacitor like an open circuit. Hence, for all cases of this example, $X_F = i_L(\infty) = 1$ A; note that $v_C(\infty) = 0$ because the inductor looks like a short at $t = \infty$.

Case 1. For $R = 500\ \Omega$, the characteristic equation 9.35 reduces to $s^2 + 20{,}000s + 4 \times 10^{10} = 0$. From the quadratic formula, the roots are

$$s_{1,2} = -1.0 \times 10^4 \pm j1.9975 \times 10^5 = -\sigma \pm j\omega_d$$

which indicates an *underdamped* response of the form (Table 9.2)

$$i_L(t) = e^{-\sigma t}[A\cos(\omega_d t) + B\sin(\omega_d t)] + X_F = Ke^{-\sigma t}\cos(\omega_d t + \theta) + X_F$$

Since $0 = i_L(0^+) = A + X_F = A + 1$, then $A = -1$. Further,

$$0 = \underbrace{\frac{v_c(0^+)}{L} = \frac{v_L(0^+)}{L} = \frac{di_L(t)}{dt}\bigg]_{t=0^+}}_{\text{From physical circuit}} = \underbrace{-\sigma A + \omega_d B}_{\substack{\text{From derivative of}\\ \text{expression for } i_L(t)}}$$

From physical circuit   From derivative of
expression for $i_L(t)$

This implies that

$$B = \frac{\sigma A}{\omega_d} = -5.0063 \times 10^{-2}$$

Hence for $t \geq 0$,

$$i_L(t) = e^{-10,000t}[\cos(1.9975 \times 10^5 t) + 5.0063 \times 10^{-2}\sin(1.9975 \times 10^5 t)] + 1$$
$$= 1.0013e^{-10,000t}\cos(1.9975 \times 10^5 t + 2.866°) + 1 \text{ A}$$

Case 2. For $R = 25\ \Omega$, the characteristic equation 9.35 reduces to $s^2 + 4 \times 10^5 s + 4 \times 10^{10} = 0$. From the quadratic formula, the roots are

$$s_{1,2} = -2.0 \times 10^5$$

indicating a *critically damped* response of the form (Table 9.2)

$$i_L(t) = (K_1 + K_2 t)e^{s_1 t} + X_F$$

Since $0 = i_L(0^+) = K_1 + X_F = K_1 + 1$, then $K_1 = -1$. Further,

$$0 = \frac{v_C(0^+)}{L} = \frac{v_L(0^+)}{L} = \frac{di_L(t)}{dt}\bigg]_{t=0^+} = s_1 K_1 + K_2 = 2 \times 10^5 + K_2$$

This implies that $K_2 = -2 \times 10^5$ and for $t \geq 0$,

$$i_L(t) = -(1 + 2 \times 10^5 t)e^{-2 \times 10^5 t} + 1 \text{ A}$$

Case 3. For $R = 20\ \Omega$, the characteristic equation 9.35 is $s^2 + 5 \times 10^5 s + 4 \times 10^{10} = 0$. From the quadratic formula, the roots are

$$s_1 = -1.0 \times 10^5 \text{ and } s_2 = -4.0 \times 10^5$$

specifying an *overdamped* response of the form (Table 9.2)

$$i_L(t) = K_1 e^{s_1 t} + K_2 e^{s_2 t} + X_F$$

Evaluating this response at $t = 0^+$ yields $i_L(0^+) = K_1 + K_2 + X_F$. Further,

$$0 = \frac{v_C(0^+)}{L} = \frac{v_L(0^+)}{L} = \frac{di_L(t)}{dt}\bigg]_{t=0^+} = s_1K_1 + s_2K_2$$

Equivalently, $-K_1 - 4K_2 = 0$. Solving these two equations yields $K_1 = -\frac{4}{3}$ and $K_2 = \frac{1}{3}$. Therefore the actual response for $t \geq 0$ is

$$i_L(t) = -\frac{4}{3}e^{-10^5 t} + \frac{1}{3}e^{-4\times10^5 t} + 1 \text{ A}$$

Figure 9.9 displays a graph of the response for each of the three cases.

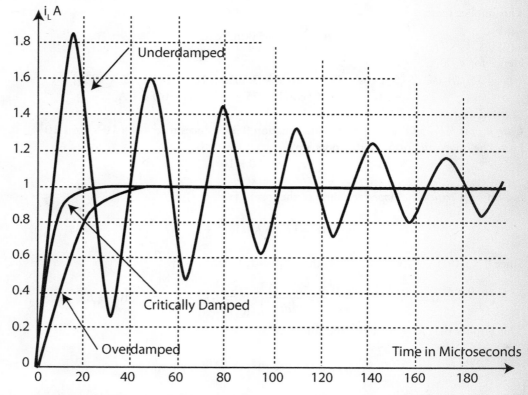

FIGURE 9.9 Underdamped, critically damped, and overdamped response curves for the parallel RLC circuit of Example 9.7.

Exercises. 1. Show that for $t \geq 0$, the differential equation for the circuit of Example 9.7 with $R = 500 \ \Omega$ is

$$\frac{d^2 i_L(t)}{dt^2} + 2\times10^4 \frac{di_L(t)}{dt} + 4\times10^{10} i_L(t) = 4\times10^{10}$$

2. Use MATLAB's "dsolve" command to verify the solution obtained for case 1 in Example 9.7.

In a linear circuit or system, the response to a step input often indicates the quality of the system performance. The problem of measuring a battery voltage using a voltmeter is illustrative of this indicator. Here the battery dc voltage is the input and the output is the meter pointer position. Connecting the meter probes to the battery terminals amounts to applying a step input to the voltmeter circuit that drives a second-order mechanical system consisting of a spring and mass with friction. Naturally, one would like the pointer to settle on the proper voltage reading quickly. If the mechanical system is underdamped, then the pointer oscillates (undesirably) for a short time before resting at its final position. On the other hand, if the mechanical system is overdamped, the pointer will not oscillate but may take a long time to reach its final resting point. This also is undesirable. A near critically damped response is the most desirable one: the pointer will come to rest at the proper voltage as quickly as possible without being oscillatory, and small changes in the mechanical system will not make it oscillatory.

In the next example, we reverse the process of analysis and ask what the original circuit parameters are given a plot of the response that might have been taken in a laboratory.

EXAMPLE 9.8. Consider the circuit of Figure 9.10, which shows the response, $v_C(t)$, of a (relaxed) series RLC circuit to the voltage input $v_{in}(t) = 10u(t)$ V. In laboratory, you have measured the capacitor voltage values (approximately). If the response has the form $v_C(t) = Ke^{-\sigma t} \cos(\omega_d t + \theta) + X_F$, find X_F, ω_d, σ, θ, K, and the values of R and C if it is known that $L = 0.5$ H. Your lab instructor has told you that ω_d and σ are integers.

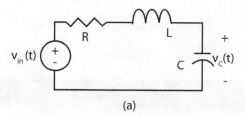

(a)

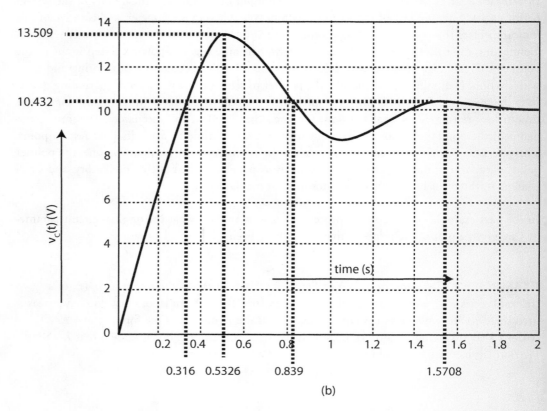

FIGURE 9.10 (a) Series *RLC*; (b) Response to $v_{in}(t) = 10\ u(t)$ V.

TABLE 9.3

Time (sec)	0.316	0.5236	0.839	1.5708
$v_C(t)$ (V)	10	13.509	10	10.432
	First crossing of 10 V	First peak	Second crossing of 10 V	Second peak

SOLUTION

Step 1. *Find X_F.* By inspection, the curve is settling out at $X_F = 10$ V.

Step 2. *Find* ω_d. Now observe that the first two crossings of $v_C(t) = 10$ occur at $t = 0.316$ sec, 0.839 sec (Table 9.3). This means that a full π radians is traversed by the cosine over [0.316, 0.839], which is a half cycle or half period. So the period of the cosine is $T = 2(0.839 - 0.316)$ $= 1.046$ sec, making

$$\omega_d = \frac{2\pi}{T} = 6.007 \cong 6 \text{ rad/sec.}$$

Step 3. *Find* σ. From Table 9.3, we know that two successive "peaks" occur at $t_1 = 0.523$ sec and $t_2 = 1.5708$ sec. This means that for $k = 1, 2$,

$$v_C(t_k) = Ke^{-\sigma t_k} \cos(\omega_d t_k + \theta) + X_F \tag{9.36}$$

After some manipulation, equation 9.36 implies

$$\frac{v_C(t_1) - X_F}{\cos(\omega_d t_1 + \theta)} = Ke^{-\sigma t_1} \text{ and } \frac{v_C(t_2) - X_F}{\cos(\omega_d t_2 + \theta)} = Ke^{-\sigma t_2}$$

Thus

$$\frac{\cos(\omega_d t_1 + \theta)}{\cos(\omega_d t_2 + \theta)} \frac{v_C(t_2) - X_F}{v_C(t_1) - X_F} = e^{-\sigma(t_2 - t_1)} \tag{9.37}$$

Equation 9.37 simplifies because two adjacent positive peaks must be 2π radians apart, i.e., $(\omega_d t_2 + \theta) - (\omega_d t_1 + \theta) = 2\pi$, which means $\cos(\omega_d t_2 + \theta) = \cos(\omega_d t_1 + \theta)$. It follows that

$$\frac{v_C(t_2) - X_F}{v_C(t_1) - X_F} = e^{-\sigma(t_2 - t_1)} \implies \frac{0.432}{3.509} = 0.1231 = e^{-\sigma 1.0472} \tag{9.38}$$

Solving leads to $\sigma = 2$.

Step 4. *Find* θ *and* K. At the first crossing of 10 V, we have

$$0 = Ke^{-2 \times 0.316} \cos(6 \times 0.316 + \theta)$$

Thus $6 \times 0.316 + \theta$ must equal 0.5π or 1.5π radians. We also know that since $v_C(0) = 0 = K \cos(\theta) + 10$, we must have $K \cos(\theta) = -10$. Since $K > 0$ (always by convention), the value of $\cos(\theta)$ must be negative. This means $0.5\pi \leq \theta \leq 1.5\pi$. So therefore, it must be that at the first crossing of 10 V

$$6 \times 0.316 + \theta = \frac{3\pi}{2} \text{ or } \theta = 2.1864 \text{ rad.}$$

Therefore $K = \dfrac{-10}{\cos(\theta)} = 10.553$.

Step 5. *Find* R *and* C. We know that the characteristic equation of the series RLC circuit must be

$$s^2 + \frac{R}{L}s + \frac{1}{LC} = s^2 + 2Rs + \frac{2}{C} = (s+2)^2 + 6^2 = s^2 + 4s + 40$$

Therefore

$$R = 2 \ \Omega \text{ and } C = 0.05 \text{ F}$$

In the previous examples one observe that the characteristic equations are independent of the source values. This is a general property of linear circuits with constant parameters. Hence when constructing the characteristic equation we may without loss of generality set independent source values to zero; i.e., independent voltage sources become short circuits and independent current sources become open circuits. With this operation, some circuits that appear to be non-series/parallel, become series/parallel. This allows us to easily compute the characteristic equation and then use Table 9.2 and physical reasoning to obtain the solution without having to construct the differential equation explicitly. The following example illustrates this procedure for a pseudo-parallel/series RLC. The example will also illustrate the computation of initial conditions due to past excitations and the computation of the complete response when the input changes its dc level.

EXAMPLE 9.9. The circuit of Figure 9.11b is driven by the input of Figure 9.11a, i.e., $v_s(t) = -60u(-t) + 60u(t) + 60u(t-1)$ V. Our goal is to find the response $v_C(t)$, for $t \geq 0$.

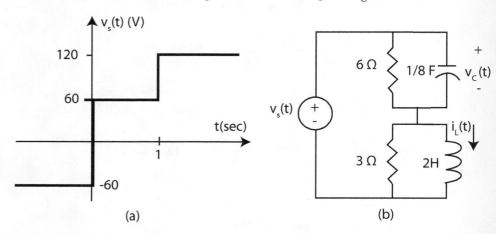

(a)   (b)

FIGURE 9.11 (a) Input excitation whose dc level changes at $t = 0$ and $t = 1$ second. (b) A pseudo-parallel RLC circuit; i.e.,when the voltage source is replaced by a short, the circuit reduces to a parallel RLC whose characteristic equation is $s^2 + \dfrac{1}{RC}s + \dfrac{1}{LC} = 0$.

SOLUTION
Step 1. *Analysis at 0^-.* Here the circuit has been excited by a constant -60 V level for a long time. Therefore at $t = 0^-$, the capacitor looks like an open circuit and the inductor a short circuit. Because the inductor looks like a short, the entire -60 V appears across the 6 Ω resistor, making $v_C(0^-) = -60$ V and $i_L(0^-) = -60/6 = -10$ A.

Step 2. *Analysis at 0^+.* By the continuity of the capacitor voltage and the inductor current, the equivalent circuit at 0^+ is given in Figure 9.12.

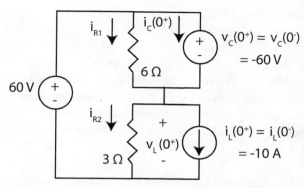

FIGURE 9.12 Equivalent circuit for analysis at 0^+; the capacitor is replaced by a voltage source of value $v_C(0^+) = v_C(0^-)$ and the inductor by a current source of value $i_L(0^+) = i_L(0^-)$.

From the circuit diagram of Figure 9.12, $v_L(0^+) = 60 - (-60) = 120$ V, $i_{R1} = -60/6 = -10$ A, and $i_{R2} = v_L(0^+)/3 = 40$ A. It follows that $i_C(0^+) = -i_{R1} + i_{R2} + i_L(0^+) = 40$ A.

Step 3. *Find the characteristic equation and the form of the response using Table 9.2.* To find the characteristic equation, we set the independent voltage source to zero. The resulting circuit is a parallel *RLC* with characteristic equation

$$s^2 + \frac{1}{RC}s + \frac{1}{LC} = s^2 + 4s + 4 = (s + 2)^2 = 0$$

where $R = 2\ \Omega$ is the parallel combination of 6 Ω and 3 Ω. The characteristic roots are $s_{1,2} = -2$, which correspond to a critically damped response of the form (Table 9.2)

$$v_C(t) = (K_1 + K_2 t)\exp(s_1 t) + X_F$$

Step 4. *Find constants in the response form for $0 \le t < 1$.* The input is constant for $0 \le t < 1$, but changes its value to 120 V at $t = 1$ sec. However, the circuit does not know the input is going to change, and so its response behaves as if the input were to remain at 60 V for all time: the circuit cannot anticipate the future, and thus its response over $0 \le t < 1$ behaves as if no further switching were going to occur. If no further switching were to occur and if the input remained at 60 V, then in Figure 9.10b for large t the capacitor is an open circuit and the inductor is a short circuit; hence $X_F = 60$ V. Under these same conditions we find K_1 and K_2. To find K_1, observe that from step 2, $v_C(0^+) = -60$ V. Evaluating the response form of step 3 yields $v_C(0^+) = K_1 + X_F$. Equating these two expressions produces $-60 = v_C(0^+) = K_1 + X_F$, which implies that $K_1 = -120$. To calculate K_2, observe that from step 2, $i_C(0^+) = 40$ A. Since $C = 0.125$, it follows that

$$\left.\frac{dv_C(t)}{dt}\right|_{t=0^+} = \frac{i_C(0^+)}{C} = 320 = s_1 K_1 + K_2 = 240 + K_2$$

Hence, $K_2 = 80$. Thus, the response of the circuit for $0 \le t < 1$ is

$$v_C(t) = (-120 + 80t)e^{-2t} + 60 \text{ V}$$

Similarly, one can compute, for $0 \leq t < 1$,

$$i_L(t) = (-20 + 20t)e^{-2t} + 10 \text{ A}$$

Step 5. *Analysis at t = 1$^-$.* Although the circuit does not know the input will change at $t = 1$ sec, we do and we must prepare for the analysis for $t \geq 1$. To do this we must evaluate the initial conditions at $t = 1^-$ and then use the continuity of the capacitor voltage and inductor current to obtain the initial conditions at $t = 1^+$. At $t = 1^-$, using step 4 we have $v_C(1^-) = v_C(1^+) = 54.59$ V and $i_L(1^-) = i_L(1^+) = 10$ A.

Step 6. *Analysis at t = 1$^+$.* This step mimics step 2 for $t = 1^+$. The capacitor is replaced again by an independent voltage source and the inductor by an independent current source as shown in Figure 9.13. Here $v_L(1^+) = 120 - 54.59 = 65.41$ V and $i_C(1^+) = -i_{R1} + i_{R2} + i_L(0^+) = \left(\dfrac{-54.59}{6} + \dfrac{120 - 54.59}{3} + 10\right)$ $= -9.098 + 21.8 + 10 = 22.71$ A.

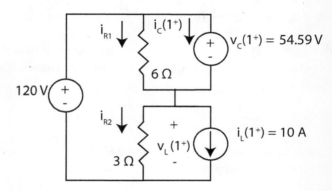

FIGURE 9.13 Equivalent circuit to that of Figure 9.11b at $t = 1^+$ sec.

Step 7. *Computation of the response for t $\geq$ 1.* Because the characteristic equation is independent of the input excitation, the form of the response is almost the same as in step 3, except for the replacement of t by $(t - 1)$; this substitution follows by the time invariance (constant parameter values) of the circuit. Thus, for $t \geq 1$,

$$v_C(t) = [K_1 + K_2(t - 1)] \exp[s_1(t - 1)] + X_F$$

Since the source excitation for $t \geq 1$ is 120 volts, by inspection of Figure 9.11b $X_F = v_C(\infty) = 120$ V. To find K_1, $54.59 = v_C(1^+) = X_F + K_1$. This implies $K_1 = -65.41$. Finally, to find K_2 consider that

$$\left.\frac{dv_C(t)}{dt}\right|_{t=1^+} = \frac{i_C(1^+)}{C} = 181.7 = s_1K_1 + K_2 = 130.8 + K_2$$

which makes $K_2 = 50.9$. Thus, for $t \geq 1$,

$$v_C(t) = [-65.41 + 50.9(t - 1)]e^{-2(t - 1)} + 120 \text{ V} \tag{9.39}$$

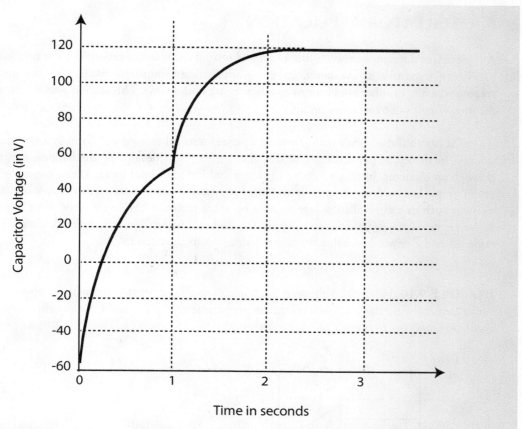

FIGURE 9.14 Complete analytical response of the capacitor voltage for $0 \le t \le 3$ sec.

Exercise. Fill in the details for the computation of $i_L(t) = (-20 + 20t)e^{2t} + 10$ A for $0 \le t < 1$ and then compute $i_L(t)$ for $2.5 > t \ge 1$. Also, compute $i_L(t)$ for $t \ge 2.5$.

Despite the idea illustrated in Example 9.9, many second-order *RLC* circuits are not reducible to series or parallel *RLC* circuits when the independent sources are set to zero. Furthermore, when a dependent source is present, the circuit is generally not reducible to a series or parallel *RLC*. In such cases one ordinarily uses a systematic methodology to compute the circuit's differential equation and, subsequently, the characteristic equation. This systematic procedure is described in more advanced texts and in the second edition of this text. Nevertheless, for some situations one can use the earlier method's integro-differential equations, which must be differentiated again to eliminate the integral. This is illustrated in the example of the next section.

5. OSCILLATOR APPLICATION

An important difference between first-order and second-order linear networks is the possibility of oscillatory responses in the latter. In some applications sinusoidal oscillations are intended responses, while in other applications oscillations are undesirable. This section presents an example of a Wien bridge oscillator circuit.

The goal is to build a circuit that generates a pure sinusoidal voltage waveform at a specified frequency. In theory, as per section 2 of this chapter, this is achievable by discharging a capacitor through an inductor. In practice, both capacitor and inductor have losses. Losses cause the oscillation amplitude to decay eventually to zero. For sustained sinusoidal oscillations, some "active" element such as a controlled source or op amp must replenish the lost energy. Note that these active elements require a dc power supply for their operation. Ultimately the dc power supply replenishes the power losses due to various resistances in the circuit.

EXAMPLE 9.10. Figure 9.15 shows a Wien bridge oscillator constructed with an op amp as the active element. Find the condition on the circuit parameters R_1, R_2, and C for sustained sinusoidal oscillation, and the frequency of oscillation.

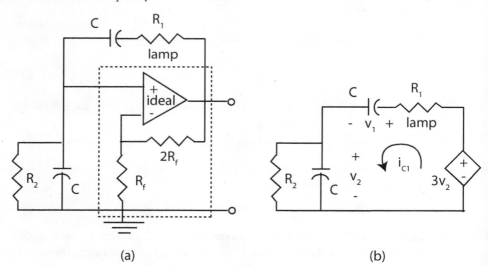

(a)   (b)

FIGURE 9.15 (a) Wien bridge oscillator. (b) Equivalent circuit.

SOLUTION

From the principles described in Chapter 4, the non-inverting amplifier enclosed in the dashed box of Figure 9.15a is equivalent to a voltage-controlled voltage source with a gain equal to $(2R_f + R_f)/R_f = 3$. (See Chapter 4.) Replacing the dashed box with this equivalent yields the simplified circuit of Figure 9.15b. Using the simplified circuit, the first task is to derive the differential equation model of the circuit.

Step 1. *Write a single-loop equation.*

$$R_1 i_{C1} + \frac{1}{C} \int i_{C1} dt + v_2 - 3v_2 = R_1 i_{C1} + \frac{1}{C} \int i_{C1} dt - 2v_2 = 0 \qquad (9.40a)$$

To eliminate the integral, we differentiate again to obtain

$$\frac{di_{C1}}{dt} + \frac{1}{R_1 C} i_{C1} - \frac{2}{R_1}\frac{dv_2}{dt} = 0 \qquad (9.40b)$$

Step 2. *Express i_{C1} in terms of v_2.* By inspection of Figure 9.15b we observe that

$$i_{C1} = C\frac{dv_2}{dt} + \frac{v_2}{R_2} \qquad (9.41a)$$

and thus, differentiating again,

$$\frac{di_{C1}}{dt} = C\frac{d^2 v_2}{dt^2} + \frac{1}{R_2}\frac{dv_2}{dt} \qquad (9.41b)$$

Step 3. *Substitute equations 9.41 into equation 9.40b.* Substituting as indicated yields

$$C\frac{d^2 v_2}{dt^2} + \frac{1}{R_2}\frac{dv_2}{dt} + \frac{1}{R_1 C}\left(C\frac{dv_2}{dt} + \frac{v_2}{R_2}\right) - \frac{2}{R_1}\frac{dv_2}{dt} = 0 \qquad (9.42a)$$

Grouping terms and dividing by C produces

$$\frac{d^2 v_2}{dt^2} + \frac{1}{R_2 C}\frac{dv_2}{dt} + \frac{1}{R_1 C}\frac{dv_2}{dt} + \frac{v_2}{R_1 R_2 C^2} - \frac{2}{R_1 C}\frac{dv_2}{dt} = 0$$

which simplifies to

$$\frac{d^2 v_2}{dt^2} + \left(\frac{1}{R_2 C} - \frac{1}{R_1 C}\right)\frac{dv_2}{dt} + \frac{v_2}{R_1 R_2 C^2} = 0 \qquad (9.42b)$$

Step 4. *Compute the characteristic equation and determine the conditions for sustained oscillations.* The resulting characteristic equation is

$$s^2 + bs + c = s^2 + \left(\frac{1}{R_2 C} - \frac{1}{R_1 C}\right)s + \frac{1}{R_1 R_2 C^2} = 0 \qquad (9.43)$$

For sustained sinusoidal oscillations to occur, the roots must be purely imaginary. Thus the coefficient of s must be zero, i.e.,

$$b = \frac{1}{R_2 C} - \frac{1}{R_1 C} = \frac{R_1 - R_2}{R_1 R_2 C} = 0 \qquad (9.44)$$

Thus the condition for sustained sinusoidal oscillations reduces to $R_1 = R_2$.

Step 6. *Find the frequency of oscillation.* Under the condition $R_1 = R_2$, the roots of the characteristic equation are

$$\pm j\omega_0 = \pm j\sqrt{\frac{1}{R_1 R_2 C^2}} = \pm j\frac{1}{C\sqrt{R_1 R_2}} = \pm j\frac{1}{R_1 C}$$

We conclude that the frequency of oscillation (in rad/sec) is

$$\omega_0 = \frac{1}{R_1 C} \qquad (9.45)$$

An examination of equation 9.44 shows that if $R_1 > R_2$, then $b > 0$ and the unforced response is an exponentially decreasing sinusoid. On the other hand, if $R_1 < R_2$, then $b < 0$, and the unforced response is an exponentially growing sinusoid. For the oscillations to start, the value of R_1 should be designed to be slightly smaller than R_2. Then the value for b in equation 9.58 will be *negative*, producing an exponentially growing sinusoidal response. If all circuit parameters are truly constant, the amplitude of oscillation would theoretically grow to infinity. In real oscillator circuits, such growth is limited to a finite amplitude by saturation effects or nonlinearities that clamp the response when the voltage swing grows large. The resulting waveform then only approximates a pure sine wave. The analysis of this nonlinear effect is beyond the scope of this book. However, the next example illustrates the growing oscillation when $R_1 < R_2$ and also shows the effect of saturation to produce an approximate sinusoidal oscillation.

EXAMPLE 9.11. The circuit of Figure 9.16a is a B2 Spice schematic for the Wein bridge oscillator of Figure 9.15. The op amp is a 741 with $V_{sat} = 15$ V. Suppose that $v_{C1}(0) = 10$ mV and $v_{C2}(0) = 0$. Observe that $R_2 = 10$ k$\Omega > R_1 = 9.5$ kΩ. According to the analysis of Example 9.10, the output voltage labeled IVout should be a growing sinusoid. The output response of Figure 9.16b shows this growth and the saturation effects induced by the op amp. The waveform is not a pure sinusoid due to these saturation effects. Also note that the frequency of oscillation is approximately 16 Hz, which is consistent with equation 9.45, i.e.,

$$f_0 = \frac{\omega_0}{2\pi} \approx \frac{1}{2\pi\sqrt{R_1 R_2}\, C} = 16.3 \text{ Hz}$$

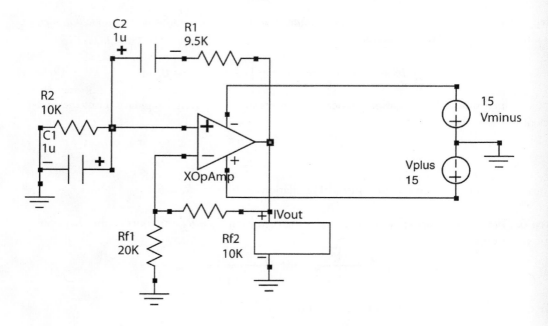

(a)

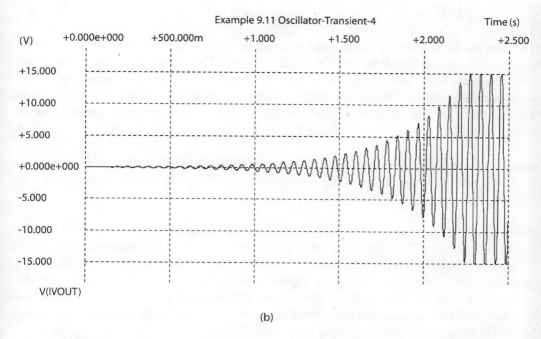

FIGURE 9.16 (a) Schematic diagram of Wein bridge oscillator. (b) Voltage response showing growing oscillation clamped at ±15 V due to saturation effects of op amp.

An alternative approach to initiating oscillations and simultaneously limiting amplitude is to use a temperature-sensitive resistor, R_1, with a positive temperature coefficient. Any incandescent lamp is an example of a temperature-sensitive resistor. For small voltages the temperature of an incandescent lamp is lower than for larger voltages because the dissipated power is lower. Hence the lamp temperature (and thus its resistance) increases with increasing voltage. In the case of our oscillator, we have a desired output voltage swing. The nominal value of R_1 is designed to be slightly less than R_2 when the output voltage swing is below a pre-specified voltage less than V_{sat}. This causes a growing oscillation. As the voltage swing increases, the temperature of R_1 and thus its resistance increase. When the resistance of R_1 reaches R_2, the amplitude will settle (stabilize) at the pre-specified voltage swing, at least theoretically. If R_1 happens to increase beyond R_2, a decaying sinusoid will result, decreasing the temperature and hence the resistance of R_1. Should the amplitude of oscillation decrease for any reason, R_1 will decrease, causing a growing sinusoid. Although the resistance of R_1 may dither about R_2, the amplitude of oscillation will nevertheless restore itself to the equilibrium level. In practice this equilibrium level only approximates the specified value due to imperfections in the circuit parameter values. The resulting waveform is almost a pure sinusoid.

6. SUMMARY

This chapter has explored the differential equation modeling and response computation of second-order linear circuits having either no input or constant input excitation. Such second-order circuits contain at least two dynamic elements, either an *LC*, *CC*, or *LL* combination. Second-order circuits may also contain active elements such as op amps. In contrast to first-order circuits, second-order linear circuits allow for the possibility of damped and undamped sinusoidal oscillations.

Analysis of second-order linear circuits has two phases. Phase 1 entails the formulation of the second-order differential equation circuit model. For simple *LC*, parallel *RLC*, or series *RLC*, the circuit model can be found by inspection.

Phase 2 of the development centers on the solution of the second-order differential equation model of the circuit. The first step here is to compute the (quadratic) characteristic equation and then solve for the two roots. The roots of the characteristic equation determine the type of response. The three types of roots for a quadratic—real distinct, real identical, and complex—specify the three response types of overdamped, critically damped, and underdamped, respectively. These three types of responses characterize all second-order linear differential equation models, be they of electrical circuits, mechanical systems, or electro-mechanical systems.

Since sinusoidal waveforms are germane to many electrical systems, this chapter presented an oscillator circuit that generates a sinusoidal waveform. Of the many types of oscillator circuits, we chose one containing an *RC* circuit built around an op amp, avoiding the use of an inductor.

7. TERMS AND CONCEPTS

Characteristic equation: for a linear circuit described by a second-order differential equation $x''(t) + bx'(t) + cx(t) = f(t)$, the algebraic equation $s^2 + bs + c = 0$ is called its characteristic equation.

Characteristic roots: roots of the characteristic equation, also called the natural frequencies of the linear circuit.

Critically damped circuit: a second-order linear circuit having characteristic roots that are real and identical. The source-free response of such a circuit has a non-oscillatory waveform, but is on the verge of becoming oscillatory.

Damped oscillation frequency: in an underdamped second-order linear circuit, the source-free response has the form $\{Ke^{-\sigma t} \cos(\omega_d t + \theta)\}$. The angular frequency ω_d is the damped oscillation frequency, which is the magnitude of the imaginary part of the characteristic roots.

Homogeneous differential equation: a differential equation in which there are no forcing terms. For example, $x''(t) + bx'(t) + cx(t) = 0$.

Natural frequencies: the characteristic roots.

Oscillator circuit: an electronic circuit designed to produce sinusoidal voltage or current waveforms.

Overdamped circuit: a second-order linear circuit having a characteristic equation whose characteristic roots are real and distinct.

Scaled sum of waveforms: let $f_1(t), \ldots, f_n(t)$ be a set of waveforms. A scaled sum of these waveforms is an expression of the form $f(t) = a_1 f_1(t) + \ldots + a_n f_n(t)$ for real (possibly complex) scalars $a_1, \ldots, a_n$.

Second-order linear circuit: a circuit whose input-output relationship may be expressed by a second-order differential equation of the form $x''(t) + bx'(t) + cx(t) = f(t)$.

Source free: there are no independent sources, or all independent sources have zero values.

Step function: a function equal to zero for $t < 0$ and equal to 1 for $t \geq 0$.

Step response: the response of a circuit to a step function input when all capacitor voltages and inductor currents are initially zero.

Undamped circuit: a second-order linear circuit where the characteristic roots are purely imaginary and the unforced response is purely sinusoidal.

Underdamped circuit: a second-order linear circuit whose characteristic roots are complex with nonzero real part.

---

[1] The notations $v'(t)$ and $\dot{v}(t)$ are used interchangeably in the literature to denote the first derivative of $v(t)$.

PROBLEMS

THEORY RELATED

1. In section 2, the solution to the undriven LC circuit is given by $v_C(t) = K \cos(\omega t + \theta)$ V. Observe that

$$\frac{d^2}{dt^2} \cos(\omega t) = -\omega^2 \cos(\omega t)$$

and

$$\frac{d^2}{dt^2} \sin(\omega t) = -\omega^2 \sin(\omega t)$$

imply that $A \cos(\omega t)$ and $B \sin(\omega t)$ are both solutions to the differential equation

$$\frac{d^2 v_C}{dt^2} = -\frac{1}{LC} v_C$$

By superposition, then, $v_C(t) = A \cos(\omega t) + B \sin(\omega t)$ V. Show that for a given A and B, there exist K and θ such that $v_C(t) = A \cos(\omega t) + B \sin(\omega t) = K \cos(\omega t + \theta)$. One concludes that the two solution forms are equivalent.

2. Find the expressions of the instantaneous energy stored in C and L for the circuit of Figure 9.1b. Show that the sum of $W_C(t)$ and $W_L(t)$ is constant.

3. By direct substitution, show that

$$x(t) = (K_1 + K_2 t)e^{-\lambda t}$$

satisfies the differential equation

$$x''(t) + 2\lambda x'(t) + \lambda^2 x(t) = 0$$

where K_1 and K_2 are arbitrary constants,

4. The voltage or current in a second-order source-free *overdamped* circuit has the general form

$$x(t) = K_1 e^{-s_1 t} + K_2 e^{-s_2 t}$$

where all coefficients are real (but not necessarily positive).

(a) Prove that $x(t) = 0$ at some $t = T$, $0 < T < \infty$, only if K_1 and K_2 have opposite signs.

(b) Prove that the $x(t)$ vs. t curve has at most one zero crossing for $t > 0$.

(c) State the *necessary and sufficient* conditions on the coefficients K_1, K_2, s_1, and s_2 for the presence of one zero crossing.

5. The voltage or current in a second-order source-free *critically damped* circuit has the general form

$$x(t) = (K_1 + K_2 t)e^{-\lambda t}$$

where all parameters are real, but not necessarily positive. Prove that $x(t) = 0$ at some $t = T < \infty$ if and only if K_1 and K_2 have opposite signs.

6. (a) Consider case 2 of Example 9.5. The circuit is underdamped and the response is given by
$v_C(t) = e^{-1000t} [10 \cos(9950t) + 1.005 \sin(9950t)] = 10.05e^{-1000t} \cos(9950t + 5.7°)$
How many cycles of "ringing" occur in the voltage waveform before the peak value drops from its largest value of 10.05 to $10.05/e = 0.368 \times 10.05$?

(b) Suppose the characteristic polynomial is written as $s^2 + 2\sigma s + \sigma^2 + \omega_d^2$ with response form $x(t) = X_{max}e^{-\sigma t} \cos(\omega_d t + \theta)$. Prove that for the underdamped case, the circuit will ring for $N = \omega_d/(2\pi\sigma)$ cycles before the amplitude decreases to $1/e$ of its initial value.

7. When a dc voltage of V_s volts is applied to a series LC circuit with no initial stored energy, the voltage across the capacitor reaches a peak value of twice the source voltage. To investigate

this phenomenon, consider the circuit of Figure P9.7 where switch S is closed at $t = 0$. Assume the inductor current and the capacitor voltage are zero at $t = 0$.

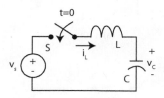

Figure P9.7

Show that for $t \geq 0$

$$i_L(t) = \frac{V_s}{\sqrt{\dfrac{L}{C}}} \sin\left(\frac{1}{\sqrt{LC}} t\right)$$

and

$$v_C(t) = V_s \left[1 - \cos\left(\frac{1}{\sqrt{LC}} t\right)\right]$$

8. The circuit in Figure P9.8 is a dual of the previous problem.

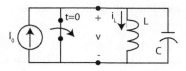

Figure P9.8

The switch S is opened at $t = 0$. Both the inductor current and the capacitor voltage are zero at $t = 0$. Show that for $t \geq 0$

$$v(t) = I_0 \sqrt{\frac{L}{C}} \sin\left(\frac{1}{\sqrt{LC}} t\right)$$

and

$$i_L(t) = I_0 \left[1 - \cos\left(\frac{1}{\sqrt{LC}} t\right)\right]$$

UNDRIVEN RLC PROBLEMS

9. The switch S in the circuit of Figure P9.9 has been closed for a long time and is opened at $t = 0$. Express $v_C(t)$ and $i_L(t)$ for $t > 0$ in terms of the literals R, L, C, and I_0. Also compute the initial stored energy in the inductor and capacitor.

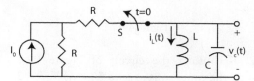

Figure P9.9

10. In the circuit of Figure P9.10, suppose $v_{in}(t) = 10$ V, $R = 10\ \Omega$, $C = 0.4$ mF, $L = 0.25$ H, and the switch opens at $t = 0$.

(a) Compute $v_C(0^-)$, $v_C(0^+)$, $i_L(0^-)$, and $i_L(0^+)$.

(b) Compute the energy stored in the inductor and the capacitor at $t = 0$.

(c) Using only energy considerations, compute the maximum value of $v_C(t)$, $t > 0$.

(d) Find the analytical expression for $v_C(t)$, and verify the maximum value of $v_C(t)$ computed in part (c).

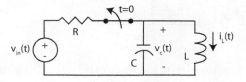

Figure P9.10

11. Consider the circuit of Figure P9.11 in which $v_{in}(t) = -20u(-t)$ V, $R = 10\ \Omega$, $C = 0.4$ mF, and $L = 0.25$ H.

(a) Compute $v_C(0^-)$, $v_C(0^+)$, $i_L(0^-)$, and $i_L(0^-)$.

(b) Compute the energy stored in the inductor and the capacitor at $t = 0$.

(c) Find the analytical expression for $v_C(t)$. Plot using MATLAB for $0 \leq t \leq 200$ msec.

(d) Find the analytical expression for $i_L(t)$. Plot using MATLAB for $0 \le t \le 200$ msec.

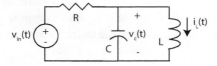

Figure P9.11

12. Reconsider the circuit of Figure P9.11 under the conditions $v_{in}(t) = 50u(-t)$ V, $R = 25$ Ω, $C = 0.8$ mF, and $L = 1$ H. Repeat Problem 11.

13. Reconsider the circuit of Figure P9.11 under the conditions $v_{in}(t) = -50u(-t)$ V, $R = 25$ Ω, $C = 0.8$ mF, and $L = 2$ H. Repeat Problem 11.

14. For the circuit of Figure P9.14, suppose $C = 0.8$ mF and determine L so that the frequency of the sinusoidal response, for $t > 0$, is 2500 rad/sec. Now find $v_C(0^-)$, $v_C(0^+)$, $i_C(0^-)$, $i_C(0^+)$, and $v_C(t)$ for $t > 0$.

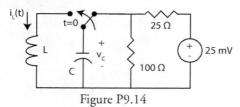

Figure P9.14

15. Consider the circuit of Figure P9.15, in which $v_{in} = 20$ V, $R_s = 2$ Ω, $R_O = 8$ Ω, $R = 2$ Ω, $L = 0.5$ H, and $C = 62.5$ mF.
(a) Figure P9.15 a shows a source-free parallel RLC circuit whose past history is depicted by Figure P9.15b where the switch S has been at position A for a long time before moving to position B at $t = 0$. Find $v_C(0^-)$, $v_C(0^+)$, $i_C(0^-)$, $i_C(0^+)$, and $v_C(t)$ for $t > 0$. Plot $v_C(t)$ using MATLAB for $0 \le t \le 1.25$ sec.
(b) Find $i_L(t)$ for $t > 0$. Plot $v_C(t)$ using MATLAB for $0 \le t \le 1.25$ sec.

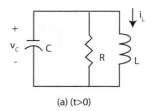

(a) (t>0)

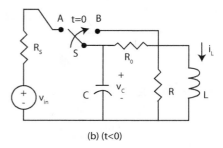

(b) (t<0)

Figure P9.15

16. Figure P9.16 shows an overdamped source-free circuit in which $R = 0.4$ Ω, $L = 0.5$ H, and $C = 0.5$ F.
(a) If $v_C(0) = -2$ V and $i_L(0) = 2.5$ A, find $v_C(t)$ for $t \ge 0$. Use MATLAB or the equivalent to plot the $v_C(t)$ waveform and verify that there is no zero crossing.
(b) If $v_C(0) = 2$ V and $i_L(0) = 2.5$ A, find $v_C(t)$ for $t \ge 0$. Use MATLAB or equivalent to plot the $v_C(t)$ waveform and verify that there is no zero crossing.

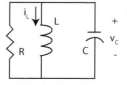

Figure P9.16

17. In Figure P9.17, the switch S has been at position A for a long time and is moved to position B at $t = 0$. Suppose $v_{in}(t) = 100$ mV, $R = 0.5$ Ω, $L = 1$ H, and $C = 0.01$ F.
(a) Find $i_L(0^+)$, $v_C(0^+)$, and $v_C(t)$ for $t > 0$. Plot for $0 \le t \le 50$ sec.
(b) Find $i_L(t)$ for $t > 0$. Plot for $0 \le t \le 50$ sec.

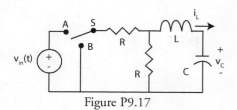

Figure P9.17

18. For the circuit of Figure P9.18, $i_{in}(t) = 0.5u(-t)$ A.
   (a) If $R = 20\ \Omega$, $L = 1$ H, and $C = 8$ mF, find and plot $v_C(t)$ and $i_L(t)$ for $t > 0$.
   (b) Repeat part (a) for $R = 22.5\ \Omega$.

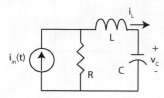

Figure P9.18

19. In Figure P9.19, $v_C(0^-) = 25$ V, $i_L(0^-) = 50$ mA, $R = 2\ k\Omega$, $L = 0.1$ H, and $C = 0.1\ \mu$F. The switch closes at $t = 0$.
   (a) Compute $v_C(0^+)$, $v_L(0^+)$, and $i_L(0^+)$.
   (b) Compute $i_L(t)$, $v_L(t)$, and $v_C(t)$.
   (c) Plot $i_L(t)$ and $v_L(t)$ for $0 < t < 1$ msec.
   (d) Find the energy stored in the circuit over the interval [0, 0.2 msec], i.e., in the capacitor and the inductor over this interval. Is this energy positive or negative? Also compute the energy dissipated in the resistor over this same interval. The sum of the energy dissipated in the resistors, the energy stored in the capacitor, and that stored in the inductor should equal zero. Why?

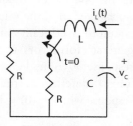

Figure P9.19

20. The voltage across the capacitor for the circuit of Figure P9.20a is given by Figure P9.20b. Suppose $R = 25\ k\Omega$.
   (a) Using the plot, estimate the values of L and C.
   (b) Now estimate the value of the initial capacitor voltage and initial inductor current.
Clearly show and explain all steps in your calculations. Hint: You might assume a general response form $v_C(t) = Ae^{-\sigma t}\cos(\omega_d t + \theta)$ V and then use the plot to estimate σ and ω_d; what is the relation of σ and ω_d in the characteristic polynomial of a parallel RLC?

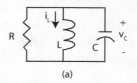

(a)

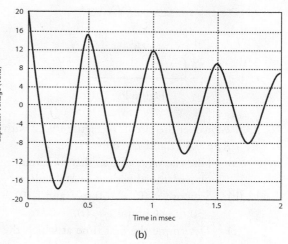

(b)

Figure P9.20

21. In the circuit of Figure P9.21, $R = 4\ \Omega$, $C = 6.25$ mF, and $v_C(t) = (20 - 1200t)e^{-20t}$ mV for $t \geq 0$.
   (a) Compute the value of L.
   (b) Find the value of the initial conditions, $v_C(0^+)$ and $i_L(0^+)$.
   (c) Find $i_L(t)$ for $t > 0$.

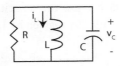

Figure P9.21

22. For the circuit of Figure P9.22, $L = 0.04$ H, $C = 2.5$ mF, and $R_1 = 10$ Ω.
   (a) Find the value of R (in ohms) that makes the circuit of Figure P9.22 critically damped.
   (b) Given this value of R, suppose $v_C(0) = 160$ mV and $i_L(0) = -60$ mA. Compute $v_C(t)$.
   (c) Determine the first time at which the capacitor voltage is zero. Plot your result using MATLAB or the equivalent to verify your calculation.

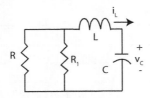

Figure P9.22

23. Reconsider the circuit of Figure P9.22 with $R // R_1 = 0.8$ Ω and $L = 0.04$ H.
   (a) Find the value of C so that the circuit is underdamped with $\omega_d = 100$ rad/sec.
   (b) Suppose $v_C(0) = 160$ mV and $i_L(0) = -30$ mA. Compute $v_C(t)$.
   (c) Determine the first time at which the capacitor voltage is zero. Plot your result using MATLAB or its equivalent to verify your calculation.

24. In the circuit of Figure P9.24, $R = 20$ Ω and
$$i_L(t) = 500e^{-10t} \sin\left(10\sqrt{3}t\right)$$
mA for $t \geq 0$. Find the proper values of L and C to produce this response. Now find $v_C(0^+)$, $v'_C(0^+)$, and $v_C(t)$ for $t > 0$.

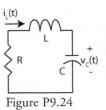

Figure P9.24

25. Figure P9.25 shows a critically damped source-free circuit.

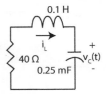

Figure P9.25

   (a) If $v_C(0) = -5$ V and $i_L(0) = 1$ A, find $i_L(t)$ for $t \geq 0$. Determine the differential equation for the circuit. Let $y = i_L$ and use "y = dsolve('D2y + 400*Dy + 40e3*y = 0,y(0) = 1,Dy(0) = -350')" in MATLAB to verify your answer.
   (b) If $v_C(0) = 5$ V and $i_L(0) = 1$ A, find $v_C(t)$ for $t \geq 0$. Plot the $v_C(t)$ waveform and verify that there is one zero crossing. Again use the dsolve command in MATLAB to verify your calculations.

26. The capacitance voltage of a source-free parallel RLC circuit, with $R = 2.4$ Ω, has the form
$$v_C(t) = Ae^{-6t} \cos(8t + \theta)$$
   (a) Find the values of L and C.
   (b) If $v_C(0) = 10$ and $i_L(0) = 0$, determine A and θ.
   (c) If the values of L and C remain unchanged, find the value of R for the circuit to be critically damped, and the general form of the source-free response $v_C(t)$ under this condition. Then determine the source-free response when $v_C(0) = 10$ and $i_L(0) = 0$.

27. Almost 75% of failures in circuits, i.e., situations where a circuit dramatically fails to perform as designed, are due to opens and shorts of individual circuit elements. Heating, cycling a circuit on and off, etc., cause degradation in the circuit parameters, resistances, capacitances, inductances, etc. that often precipitates the short or open situation. For example, the material inside a resistor might become brittle over a period of time and finally crumble, leaving a break in the circuit. On the other hand, the material might congeal or become dense, decreasing the resistance. In the problems below you are to determine the length of time it takes for a circuit to move from an over-damped behavior to an underdamped behavior due to changes in the resistor characteristic as a function of time.

(a) For the parallel *RLC* circuit in Figure P9.27a, suppose $R = R_0 + \exp(t - 5)$ Ω where $t \geq 0$ constitutes time in years. Determine the time t' for which the circuit changes its behavior from over-damped to underdamped.

(b) For the series *RLC* circuit of Figure P9.27b, the resistor satisfies $R = R_0/[1 + \exp(t - 5)]$ Ω, where again t is time in years. Here it is presumed that the circuit is part of a larger piece of electronic apparatus, such as a TV, which is used extensively over a period of years. The time t' then is not connected with the response time of the circuit. Determine the time t' for which the circuit changes its behavior from overdamped to underdamped.

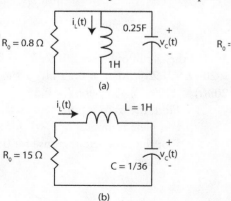

(a)

(b)

Figure P9.27 Parallel and series *RLC* circuits subject to resistor degradation over time.

DRIVEN SERIES AND PARALLEL *RLC* CIRCUITS

28. (Initial condition calculation) For the circuit shown in Figure P9.28, find $v_C(0^+)$, $i_L(0^+)$, $i_C(0^+)$, and $v_L(0^+)$ in two steps:

Step 1. Find $v_C(0^-)$ and $i_L(0^-)$ by open-circuiting C and short-circuiting L.

Step 2. Construct a resistive circuit valid at $t = 0^+$ and from this find $i_C(0^+)$, and $v_L(0^+)$.

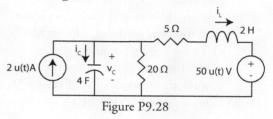

Figure P9.28

29. Consider the circuit of Figure P9.29 in which $v_{in}(t) = -10u(-t) + 20u(t)$ V, $R = 20$ Ω, $C = 0.1$ mF, and $L = 0.25$ H.

(a) Compute $v_C(0^-)$, $v_C(0^+)$, $i_L(0^-)$, and $i_L(0^+)$.

(b) Compute the energy stored in the inductor and the capacitor at $t = 0$.

(c) Find the analytical expressions for the zero-input, zero-state, and complete responses for $v_C(t)$. Identify the transient and steady-state responses. Plot $v_C(t)$ using MATLAB over [0, 40 msec].

(d) Find the analytical expressions for the zero-input, zero-state, and complete responses for $i_L(t)$. Plot $i_L(t)$ using MATLAB over [0, 40 msec].

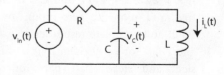

Figure P9.29

30. Reconsider the circuit of Figure P9.29 under the conditions $v_{in}(t) = 50u(-t) + 25u(t)$ V, $R = 25$ Ω, $C = 0.8$ mF, and $L = 2$ H. Repeat Problem 29 but construct plots over [0, 400 ms].

31. Reconsider the circuit of Figure P9.29 under the conditions $v_{in}(t) = -50u(-t) + 25u(t)$ V, $R = 25\ \Omega$, $C = 0.8$ mF, and $L = 0.2$ mH. Repeat Problem 29 but construct plots over [0, 40 ms].

32. In Figure P9.32 $v_{in}(t) = -250u(-t) + 750u(t)$ mV, $R = 0.5\ \Omega$, $L = 1$ H, and $C = 0.01$ F.
    (a) Find $i_L(0^+)$, $v_C(0^+)$, and the zero-input, zero-state, and complete responses of $v_C(t)$ for $t > 0$. Identify the steady-state and transient parts of the complete response. Plot in MAT-LAB for $0 \le t \le 50$ msec.
    (b) Find the zero-input, zero-state, and complete responses of $i_L(t)$ for $t > 0$. Plot in MATLAB for $0 \le t \le 50$ msec.

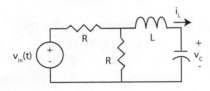

Figure P9.32

33. Repeat Problem 32 for $R = 40\ \Omega$ and $v_{in}(t) = -0.5u(-t) + 2u(t)$ V. Plots in MATLAB should be for $0 \le t \le 800$ msec.

34. Repeat Problem 32 for $R = 50\ \Omega$ and and $v_{in}(t) = -0.5u(-t) - 2u(t)$ V. Plots in MATLAB should be for $0 \le t \le 1$ sec.

35. For the circuit of Figure P9.35, $i_{in}(t) = -0.5u(-t) + 2u(t)$ A.
    (a) If $R = 2\ \Omega$, $L = 1$ H, and $C = 8$ mF, find and plot the zero-input, zero-state, and complete responses of $v_C(t)$ and $i_L(t)$ for $t > 0$. Identify the transient and steady-state parts of the complete response.
    (b) Repeat part (a) for $R = 22.5\ \Omega$.

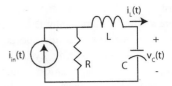

Figure P9.35

36. In Figure P9.36, $i_{in}(t) = -10u(-t) + 40u(t)$ mA, $R = 4$ kΩ, $L = 0.1$ H, and $C = 0.1\ \mu F$. The switch closes at $t = 0$.
    (a) Compute $v_C(0^+)$, $v_L(0^+)$, and $i_L(0^+)$.
    (b) Compute the zero-input, zero-state, and complete responses of $i_L(t)$, $v_L(t)$, and $v_C(t)$. Identify the transient and steady-state parts of the complete response.
    (c) Plot $i_L(t)$ and $v_L(t)$ for $0 < t < 1$ msec.
    (d) Find the energy stored in the circuit over the interval [0, 0.2 msec], i.e., in the capacitor and the inductor over this interval. Is this energy positive or negative? Also compute the energy dissipated in the resistor over this same interval.

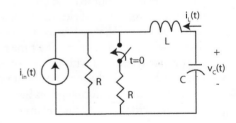

Figure P9.36

37. For the circuit of Figure P9.37, $R_s = 5\ \Omega$, $R_1 = 20\ \Omega$, $C = 2.5$ mF, $L = 0.25$ H, and $v_{in}(t) = 20u(t) - 20u(t-T)$ V, where $T = 0.25$ sec.
    (a) Find the zero-input, zero-state, and complete responses of $v_C(t)$ for $t > 0$. Plot the complete response for $0 < t \le 0.25$ sec.

(b) Find the zero-input, zero-state, and complete responses of $i_L(t)$ for t > 0. Plot the complete response for $0 < t \leq$ 0.25 sec.

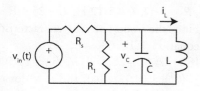

Figure P9.37

38. Repeat Problem 37 for $R_s = 50\ \Omega$, $R_1 = 200\ \Omega$, $C = 0.05$ mF, $L = 0.5$ H, and $v_{in}(t) = -50u(-t) + 50u(t) - 50u(t - T)$ V, where $T = 0.08$ sec. However, only plot the complete responses for $0 < t \leq 200$ msec.

39. Repeat Problem 37 for $R_s = 100\ \Omega$, $R_1 = 100\ \Omega$, $C = 0.25$ mF, $L = 2.5$ H, and $v_{in}(t) = -50u(-t) + 50u(t) - 50u(t - T)$ V, where $T = 100$ msec. However, only plot the complete responses for $0 < t \leq 300$ msec.

40. Consider the *RLC* circuit of Figure P9.40 in which $R_s = 100\ \Omega$, $R_1 = 400\ \Omega$, $C = 0.125$ mF, $L = 0.2$ H, and $v_{in}(t) = 50u(t) - 50u(t - T)$ V, where $T = 0.025$ sec, $v_C(0^-) = -25$ V, and $i_L(0^-) = 10$ mA.
  (a) Find the zero-state, zero-input, and complete responses of $v_C(t)$ for $t \geq 0$. Plot for $0 < t \leq 60$ msec.
  (b) Find the zero-state, zero-input, and complete responses of $v_L(t)$ for $t \geq 0$. Plot for $0 < t \leq 60$ msec.

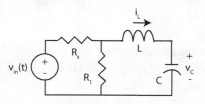

Figure P9.40

41. Repeat Problem 40 for $R_s = 140\ \Omega$ and $R_1 = 360\ \Omega$.

42. Repeat Problem 40 for $R_s = 50\ \Omega$ and $R_1 = 200\ \Omega$.

43. Consider the *RLC* circuit in Figure P9.43 where $R_{s1} = 60\ \Omega$, $R_{s2} = 40\ \Omega$, $L = 4$ H, and $C = 5$ mF.
  (a) $i_{s1}(t) = 100u(-t)$ mA and $v_{s2}(t) = 20u(-t)$ V. Find the response, $v_C(t)$, for $t > 0$. Plot for $0 < t \leq 1$ sec.
  (b) $i_{s1}(t) = 100u(-t) + 500u(t)$ mA and $v_{s2}(t) = 20u(-t)$ V. Find the response, $v_C(t)$, for $t > 0$. Plot for $0 < t \leq 1$ sec.

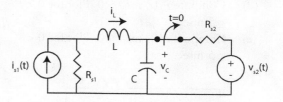

Figure P9.43

44. Repeat Problem 43, except find $v_L(t)$, $t > 0$.

45. Consider the *RLC* circuit in Figure P9.43 where $R_{s1} = 20\ \Omega$, $R_{s2} = 20\ \Omega$, $L = 0.4$ H, $C = 4$ mF.
  (a) $i_{s1}(t) = -u(-t)$ A and $v_{s2}(t) = 40u(-t)$ V. Find the response, $v_C(t)$, for $t > 0$. Plot for $0 < t \leq 0.8$ sec.
  (b) $i_{s1}(t) = -u(-t) + 2u(t) - 2u(t - 0.4)$ A and $v_{s2}(t) = 40u(-t)$ V. Find the response, $v_C(t)$, for $t > 0$. Plot for $0 < t \leq 0.8$ sec.

46. Repeat Problem 45, except find $v_L(t)$, $t > 0$.

47. Consider the *RLC* circuit in Figure P9.43 where $R_{s1} = 16\ \Omega$, $R_{s2} = 32\ \Omega$, $L = 0.4$ H, and $C = 4$ mF.
  (a) $i_{s1}(t) = -0.5u(-t)$ A and $v_{s2}(t) = 24u(-t)$ V. Find the response, $v_C(t)$, for $t > 0$. Plot for $0 < t \leq 0.3$ sec.
  (b) $i_{s1}(t) = -0.5u(-t) + 0.5u(t)$ A and $v_{s2}(t) = 24u(-t)$ V. Find the response, $v_C(t)$, for $t > 0$. Plot for $0 < t \leq 0.3$ sec.

48. Repeat Problem 47, except find $v_L(t)$, $t > 0$.

49. The current source with $i_{s1}(t) = 5u(-t)$ mA and the voltage source $v_{s2}(t) = 10$ V in Figure P9.49 drive the circuit in which $R = 1$ kΩ, $C = 0.5$ μF, and $L = 0.184$ H.
   (a) Find $v_C(0^+)$, $i_C(0^+)$, $i_L(0^+)$, and $v_L(0^+)$.
   (b) Compute $i_L(t)$ for $t > 0$. Plot for $0 < t \le 5$ msec.
   (c) Compute $v_C(t)$ for $t > 0$. Plot for $0 < t \le 5$ msec.
   (d) Compute $v_L(t)$ for $t > 0$. Plot for $0 < t \le 5$ msec.
   (e) Compute $i_C(t)$ for $t > 0$. Plot for $0 < t \le 5$ msec.

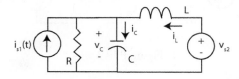

Figure P9.49

50. Repeat Problem 49 for the new source current $i_{s1}(t) = 5u(-t) - 5u(t - 0.005)$ mA. Plot for $0 \le t \le 10$ ms.

51. The switch in the circuit of Figure P9.51 is in position A for a long time and moves to position B at $t = 0$. Find the voltage $v_C(t)$ for $t > 0$ when L equals (a) 0.625 H, (b) 0.4 H, and (c) 0.2 H.

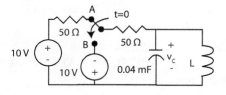

Figure P9.51

PSEUDO SERIES AND PARALLEL *RLC* CIRCUITS

52. Consider the circuit of Figure P9.52 in which $R_1 = 4$ Ω, $R_2 = 4$ Ω, $L = 5/12$ H, $C = 25$ mF.
   (a) Find the roots of the characteristic equation. CHECK: -8, -12
   (b) If $v_{in}(t) = -20u(-t)$ V, find $v_C(0^-)$, $v_C(0^+)$, $i_L(0^-)$, $i_L(0^+)$, and $v_C(t)$ for $t > 0$. Hint: After finding the initial voltages and currents, draw the equivalent circuit valid for $t > 0$. Check your answer for $v_C(t)$ using the "dsolve" command in MATLAB. Plot $v_C(t)$ for $0 \le t \le 2$ sec using MATLAB or the equivalent. Be sure you properly label your plot.
   (c) If $v_{in}(t) = -20u(-t) + 20u(t)$ V, find $v_C(t)$ for $t > 0$.
   (d) If $v_{in}(t) = -20u(-t) + 20u(t) - 20u(t - 1)$ V, find and plot $v_C(t)$ for $0 \le t \le 2$ sec.

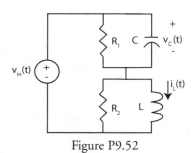

Figure P9.52

53. Repeat Problem 52, but find $i_L(t)$ and $v_L(t)$ without differentiating $i_L(t)$.

54. Repeat Problem 52 for $R_1 = 4$ Ω, $R_2 = 4$ Ω, $L = 0.2$ H, and $C = 0.2$ F.

55. Reconsider the circuit of Figure P9.52 in which $R_1 = 600$ Ω, $R_2 = 120$ Ω, $L = 2$ H, $C = 1$ mF and $v_{in}(t) = -72u(-t) + 72u(t) - 72u(t - 1)$. Find the response $i_L(t)$, for $t \ge 0$ as follows:
   (a) Find $v_C(0^-)$ and $i_L(0^-)$.
   (b) Find $v_C(0^+)$ and $i_L(0^+)$.
   (c) Draw the equivalent circuit valid at 0^+ and find $v_L(0^+)$ and $i_C(0^+)$.

(d) Find the characteristic equation and natural frequencies of the circuit.

(e) Determine the general form of the response, $i_L(t)$, valid for $0 \le t < 1$.

(f) Determine all coefficients in the general form of the response.

(g) Determine the form of the response, $i_L(t)$, for $t \ge 1$. Find the response.

(h) Plot the response $i_L(t)$ for $t > 0$ using MATLAB or the equivalent.

56. Repeat Problem 55, except find $v_C(t)$.

57. Consider the circuit of Figure P9.57 in which $R_1 = 2\,\Omega$, $R_2 = 2\,\Omega$, $L = 0.4$ H, and $C = 0.1$ F.

(a) Find the roots of the characteristic equation.

(b) If $i_{in}(t) = -2u(-t)$ A, find $v_C(0^-)$, $v_C(0^+)$, $i_L(0^-)$, $i_L(0^+)$, and $i_L(t)$ for $t > 0$. Hint: After finding the initial voltages and currents, draw the equivalent circuit valid for $t > 0$. Check your answer for $i_L(t)$ using the "dsolve" command in MATLAB. Plot $i_L(t)$ for $0 \le t \le 1$ sec using MATLAB or equivalent. Be sure you properly label your plot.

(c) If $i_{in}(t) = -2u(-t) + 2u(t)$ A, find $i_L(t)$ for $t > 0$.

(d) If $i_{in}(t) = -2u(-t) + 2u(t) - 2u(t-1)$ A, find and plot $i_L(t)$ for $t > 0$.

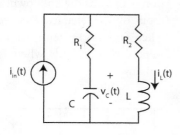

Figure P9.57

58. Repeat Problem 57, except find $v_C(t)$.

59. Repeat Problem 57 with $R_1 = 80\,\Omega$, $R_2 = 20\,\Omega$, $L = 10$ H, and $C = 1/240$ F.

60. Reconsider the circuit of Figure P9.57. Suppose $R_1 = 80\,\Omega$, $R_2 = 40\,\Omega$, $L = 2$ H, and $C = 0.625$ mF with $i_{in}(t) = -150u(-t) + 150u(t)$ mA.

(a) Find $v_C(0^-)$ and $v_C(0^+)$.

(b) Find $i_L(0^-)$ and $i_L(0^+)$.

(c) Find $v_L(0^+)$ and $i_C(0^+)$.

(d) Find the characteristic equation and natural frequencies of the circuit.

(e) Find the response, $v_C(t)$, $t > 0$.

(f) Find the response, $i_L(t)$, $t > 0$.

61. Consider the circuit of Figure P9.61 Suppose $R_1 = 80\,\Omega$, $R_2 = 40\,\Omega$, $L = 2$ H, and $C = 0.625$ mF with $i_{in}(t) = 300u(t)$ mA, $V_O = 50$ V, and $R_3 = 20\,\Omega$.

(a) Find $v_C(0^-)$ and $v_C(0^+)$.

(b) Find $i_L(0^-)$ and $i_L(0^+)$.

(c) Find $v_L(0^+)$ and $i_C(0^+)$.

(d) Find the characteristic equation and natural frequencies of the circuit for $t > 0$.

(e) Find the response, $v_C(t)$, $t > 0$.

(f) Find the response, $i_L(t)$, $t > 0$.

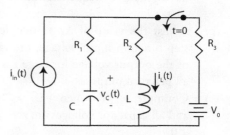

Figure P9.61

62. The switch in the *RLC* circuit of Figure P9.62 opens at $t = 0$ after having been closed for a long time. The purpose of this problem is to find the complete response of the capacitor voltage, $t \ge 0$. Suppose $i_{s1}(t) = 1$ A, $v_{s2}(t) = 20$ V, $C = 4$ mF, and $L = 0.625$ H.

(a) Using a dc analysis, find the initial conditions $i_L(0^-)$ and $v_C(0^-)$.

(b) Find $v_C(0^+)$ and $i_C(0^+)$.

(c) Using a dc analysis, find the final value of the capacitor voltage, $v_C(\infty)$.

(d) Find the characteristic equation and

compute its roots. Given the roots, write down the general form of the response $v_C(t)$.

(e) Solve for the unknown coefficients in the response form of part (d) and write down the exact expression for $v_C(t)$ valid for $t \geq 0$.

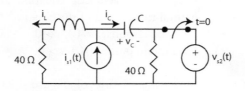

Figure P9.62

GENERAL SECOND-ORDER CIRCUITS

63. In the operational amplifier circuit of Figure P9.63 is a second-order circuit. Suppose $R_2 = R_3 = R_4 = 50$ kΩ.

(a) Determine the values of C_1 and C_2 that produce a characteristic equation having natural frequencies at -5 and -10.

(b) Adjust the value of R_1 so that for a step function input voltage, the value of the output voltage for large t is for all practical purposes is 5.

(c) Compute $v_{out}(t)$ when $v_s(t) = 2u(t)$ and all capacitor voltages are zero at $t = 0$.

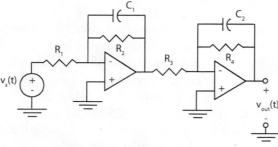

Figure P9.63 Cascade of leaky integrator circuits having a second-order response.

64. Consider the circuit shown in Figure P9.64. Write a second-order differential equa-

tion with $i_L(t)$ as the unknown. Observe that the derivative of $v_{in}(t)$ is present on the right-hand side.

ANSWER: $i''(t) + i'(t) + i(t) = 0.5v'_{in}(t) + 0.5v_{in}(t)$.

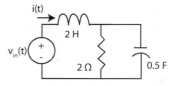

Figure P9.64

65. Consider the circuit shown in Figure P9.65.

(a) Write a second-order differential equation with v_C as the unknown. Find the roots of the characteristic equation.

(b) If $v_{in}(t) = u(t)$ V, find $v_C(t)$ for $t > 0$.

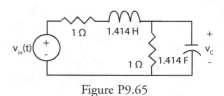

Figure P9.65

66. Consider the circuit shown in Figure P9.66.

(a) Write a second-order differential equation with v_C as the unknown. Find the roots of the characteristic equation. Then find $v_C(t)$ when $i_{in}(t) = u(t)$ A.

(b) Repeat part (a), for when i_L is the unknown.

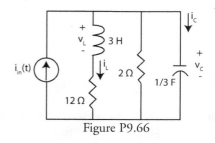

Figure P9.66

67. The second-order circuit shown in Figure P9.67 contains two capacitors.
   (a) Find the second-order differential equation with v_{C1} as the unknown. Give the values of s.
   (b) If $v_{C1}(0) = 2$ V, and $v_{C2}(0) = 4$ V, find $v_{C1}(t)$ for $t > 0$.

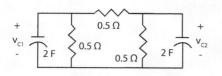

Figure P9.67

68. In the circuit of Figure P9.68, the voltage-controlled voltage source has a gain $A > 0$. Find the ranges of A for the circuit to be (a) over-damped, (b) underdamped, (c) critically damped, and (d) undamped.

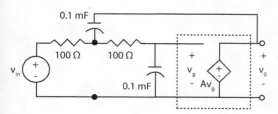

Figure P9.68

69. The second-order circuit shown in Figure P9.69 is of the overdamped type. Find the step response, i.e., the expression for $v_o(t)$ for $t > 0$, when the input is $v_i(t) = u(t)$ V, and the capacitors are initially uncharged. Roughly sketch the waveform of $v_o(t)$. Verify your sketch by doing a SPICE simulation of the circuit.

Remark: The waveform $v_o(t)$ consists of a very fast rise toward 1 V, and then a relatively slow exponential decrease toward 0 V. This can be explained using the first-order RC circuit properties studied in Chapter 8. During the first few microseconds, the 0.1 µF capacitor behaves almost as a short circuit, and the 1 nF capacitor is charged with a time constant of about 10 nsec. After the smaller capacitor is charged to

nearly 1 V in about 5 nsec, it behaves approximately as an open circuit. The 0.1 µF capacitor is then charged up with a time constant of about 0.1 msec. As the larger capacitor is charged up, the output across the smaller one decreases toward zero.

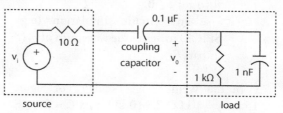

Figure P9.69

70. Find the value of the negative resistance $-R_n$ for the circuit shown in Figure P9.70 required to generate sinusoidal oscillations with constant amplitude.

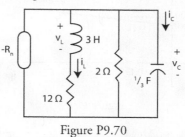

Figure P9.70

71. Refer to the Wien bridge oscillator of Example 9.11. Suppose the op amp has a saturation voltage 15 V. $C = 1$ µF and $R_2 = 500$ Ω. R_1 is a temperature-sensitive resistor whose resistance is a function of the amplitude of the sinusoidal current passing through R_1. The following relationships are given:
$$i_1(t) = I_m \sin(\omega t + \theta)$$
$$R_1 = 500 + 100(I_m - 0.01)$$
   (a) Find the frequency of oscillation ω_n.
   (b) Find the amplitude of voltage at the op amp output terminal (with respect to ground).
   (c) Suppose R_1 is a fixed resistance of 490 Ω, $v_{C1}(0) = 100$ mV, and $v_{C2}(0) = 0$ in the Wein bridge oscillator circuit. Perform a SPICE simulation. Does the circuit behave as expected?

72. In the Wein bridge oscillator example of this chapter, let $R_1 = R_2 = 1$ kΩ, $R_f = 10$ kΩ, and $C = 0.1$ μF.

    (a)   Determine the frequency of oscillation in Hz.

    (b)   If $v_{C1}(0) = 5$ V and $v_{C2}(0) = 0$ V, find $v_1(t)$ for $t \geq 0$.

    (c)   Use any circuit simulation (e.g., SPICE) to verify the waveform of $v_1(t)$ in part (b).

73. In the Wein bridge oscillator of Figure 9.15a, $R_2 = 1$ kΩ, $R_f = 10$ kΩ, and $C = 0.1$ μF. The lamp resistance R_1 is a function of the peak value of the sinusoidal current as shown Figure P9.73.

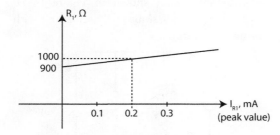

Figure P9.73

    (a)   Determine the frequency of oscillation.

    (b)   Determine the amplitude of the sinusoidal waveform $v_1(t)$.

C H A P T E R 10

Sinusoidal Steady State Analysis by Phasor Methods

A HIGH-ACCURACY PRESSURE SENSOR APPLICATION

The control of high-performance jet engines requires highly accurate pressure measurements, with errors less than one-tenth of 1% of a full-range measurement, over a wide range of temperatures, −65° to 200° F. The pressure range may be as low as 20 psia or as high as 650 psia. In jet (turbine) engine applications, knowing pressure and temperature allows one to compute the mass (volume) air flow, a critical aspect of an engine's performance. A pressure sensor is also a critical component in the regulation of aircraft cabin pressure. Such a sensor is depicted here along with a functional block diagram of its operation. A diaphragm consisting of two fused quartz plates separated by a vacuum has a capacitance that changes as a function of pressure and temperature. This quartz capacitive diaphragm is an element in a bridge circuit. It is this bridge circuit, in conjunction with detailed knowledge of the characteristics of a pair of quartz capacitors over the required operating range of pressure and temperature, that enables accurate pressure measurements.

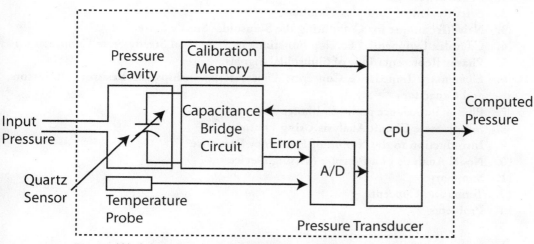

Functional block diagram courtesy of AlliedSignal Aerospace Company.

Because of the small capacitances, on the order of picofarads, associated with the quartz (diaphragm) capacitor, the bridge circuit is driven by an ac source and is called an ac bridge. Driving the bridge by an ac source moves its analysis outside the realm of the dc and step response techniques studied in earlier chapters. New methods of analysis, such as phasor analysis, are necessary. Phasor methods, the primary focus of this chapter, allow us to analyze capacitive and inductive circuits excited by sinusoidal (ac) inputs. In particular, phasor techniques permit us to analyze an ac bridge circuit. Although the analysis of the pressure sensor shown here is beyond the scope of this text, the chapter will end with a simplified pressure sensor circuit based on the one shown.

CHAPTER OBJECTIVES

1. Review and elaborate on the basic arithmetic and essential properties of complex numbers pertinent to sinusoidal steady-state analysis of circuits.
2. Develop two complementary techniques for computing the response of simple *RL*, *RC*, and *RLC* circuits excited by sinusoidal inputs and modeled by differential equations.
3. Define the notion of a (complex) phasor for representing sinusoidal currents and voltages in a circuit.
4. Using the notion of phasor, introduce the notions of impedance, admittance, and a generalized Ohm's law for two-terminal circuit elements having phasor currents and voltages.
5. Utilizing the methods of nodal and loop analysis and the network theorems of Chapters 5 and 6, analyze passive and op amp circuits by the phasor method.
6. Introduce the notion of frequency response for linear circuits, i.e., investigate the behavior of a circuit driven by a sinusoid as its frequency ranges over a given band.

SECTION HEADINGS

1. **Introduction**
2. **Brief Review of Complex Numbers**
3. **Naive Technique for Computing the Sinusoidal Steady State**
4. **Complex Exponential Forcing Functions in Sinusoidal Steady-State Computation**
5. **Phasor Representations of Sinusoidal Signals**
6. **Elementary Impedance Concepts: Phasor Relationships for Resistors, Inductors, and Capacitors**
7. **Phasor Impedance and Admittance**
8. **Steady-State Circuit Analysis Using Phasors**
9. **Introduction to the Notion of Frequency Response**
10. **Nodal Analysis of a Pressure-Sensing Device**
11. **Summary**
12. **Terms and Concepts**
13. **Problems**

1. INTRODUCTION

Perhaps you have experienced the bouncing motion of a car with broken shock absorbers or watched the (mechanical) oscillations of a swinging pendulum. These motions reflect the sinusoidal and damped sinusoidal oscillations in *RLC* circuits with conjugate poles of the characteristic equation, as detailed in Chapter 9. In this chapter we allow sources with sinusoidal forcing functions (such as such as $V_s \cos(\omega t + \theta)$ or perhaps $V_s \sin(\omega t + \theta)$), which almost always result in sinusoidal responses regardless of the root locations of the characteristic equation. A sinusoidal voltage source models the voltage from the ubiquitous wall outlet.

If one hooks up an oscilloscope to measure a voltage in a linear circuit driven by sources with sinusoidal values, the voltage may not look sinusoidal at first. However, if the circuit is **stable**, after a sufficiently long period of time the screen of the scope will trace out a sinusoidal waveform. (Here "stable" means that any zero-input response consists of decaying exponentials or exponentially decreasing sinusoids.) The eventual sinusoidal behavior is not immediately apparent because at startup, stable circuits exhibit a **transient** response. "Transient" means that the circuit response is transitioning—for example, from an initial voltage or current value to another constant value. Flickering lights during a thunderstorm illustrate the phenomenon of transient behavior: lightning may have struck a transmission line or pole, causing the power system to waver briefly from its nominal behavior.

Because sinusoidal excitations and sinusoidal responses are so common, their study falls under the heading of **sinusoidal steady-state (SSS) analysis**. Here "sinusoidal" means that source excitations have the form $V_s \cos(\omega t + \theta)$ or $V_s \sin(\omega t + \theta)$. For consistency with traditional approaches, we take $V_s \cos(\omega t + \theta)$ as the general input excitation, as shown in Figure 10.1, because $V_s \sin(\omega t + \theta) = V_s \cos(\omega t + \theta - \pi/2)$. **Steady state** mean that all transient behavior of the stable circuit has died out, i.e., decayed to zero. Observe that every sinusoidal waveform is periodic with angular argument $(\omega t + \theta)$. In terms of angle, each cycle of the waveform traverses 2π radians. In terms of time, each cycle covers a time interval of $T = 2\pi/\omega$ seconds, called the **period** of the waveform. The number of cycles contained in 1 second is called the **frequency** of the sinusoidal waveform and is denoted by f. The unit for f is the hertz (Hz), meaning "cycles per second." The quantity ω, which specifies the variation of the angular argument $(\omega t + \theta)$ in 1 second, is called the **angular frequency** of the waveform. The unit of ω is radians per second (rad/sec). From these definitions, $f = 1/T = \omega/2\pi$ and $\omega = 2\pi f$.

Stable circuits driven by sinusoidal excitations produce sinusoidal voltages and currents, as illustrated in Figure 10.1. The output excitation in Figure 10.1 has the general form $V_m \cos(\omega t + \phi)$ to distinguish it from the input excitation, $V_s \cos(\omega t + \theta)$. Because of linearity, the circuit can change only the magnitude of the input sinusoid (V_s is changed to V_m) and the phase angle of the input sinusoid (θ is changed to ϕ) while ensuring that the angular frequency ω remains the same. For nonlinear circuits, ω can and usually does change.

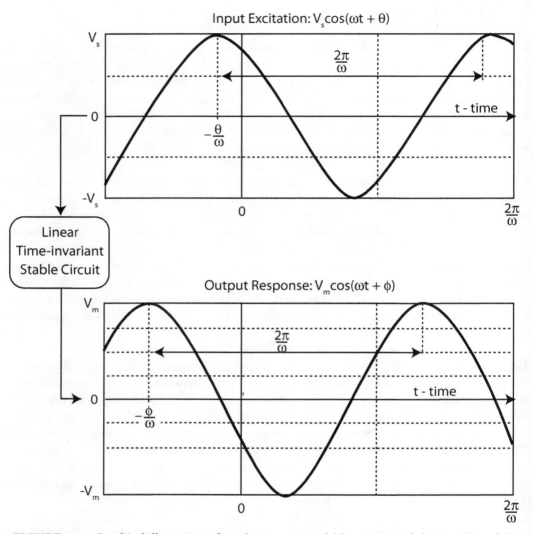

FIGURE 10.1 Graphical illustration of steady-state sinusoidal linear circuit behavior. (V_s and V_m could just as well be I_s and I_m or any combination thereof.) Note that θ and ϕ are often different and that ω is the same for both input and output excitations.

In Figure 10.1 the steady-state (voltage) response is $V_m \cos(\omega t + \phi)$. Alternatively this could have been a current response, $I_m \cos(\omega t + \phi)$. Such waveforms have the equivalent structure $A \cos(\omega t) + B \sin(\omega t)$, deducible from trigonometric identities,

$$V_m \cos(\omega t + \phi) = V_m \cos(\phi) \cos(\omega t) - V_m \sin(\phi) \sin(\omega t)$$

$$= A \cos(\omega t) + B \sin(\omega t) \tag{10.1}$$

where $A = V_m \cos(\phi)$ and $B = -V_m \sin(\phi)$. Conversely, by summing the squares of A and B, one obtains

$$V_m = \sqrt{A^2 + B^2} \qquad (10.2a)$$

By taking the inverse tangent of the ratio of $-B$ and A, one obtains

$$\phi = \tan^{-1}\left(\frac{-B}{A}\right) \qquad (10.2b)$$

In using equation 10.2b it is important to adjust the resulting angle for the proper quadrant of the complex plane. Equations 10.1 and 10.2 turn out to be useful in developing a conceptually simple, although naive, technique for computing the steady-state response using a differential equation model of the circuit, as explained in section 3.

Sinusoidal steady-state (SSS) analysis of circuits draws its importance from several areas. The analysis of power systems normally occurs in the steady state where voltages and currents are sinusoidal. Music is a rhythmic blend of different notes. Mathematically, a musical (voltage) signal can be decomposed into a sum of sinusoidal voltages of different frequencies. The analysis of a sound system typically builds around the steady-state behavior of the microphone, the amplifier, and the loudspeakers driven by sinusoidal excitations whose frequency varies from around 40 Hz to 20 kHz. Indeed, almost any form of speech or music transmission requires an understanding of steady-state circuit behavior. There are many other areas of applicability.

This chapter will introduce three techniques for computing the SSS response. The first two, somewhat naive, approaches map out a natural motivation and path to the third, very powerful technique of **phasor analysis**. Phasor analysis builds on the arithmetic of complex numbers and the basic circuit principles studied thus far. To set the stage for phasor analysis, section 2 reviews the necessary basics of complex number arithmetic. Of course, the student is assumed to have studied complex numbers in high school and in prerequisite calculus courses.

2. BRIEF REVIEW OF COMPLEX NUMBERS

Let $z_1 = a + jb$ be an arbitrary complex number, where $j = \sqrt{-1}$. The real number a is the **real part** of z_1, denoted by $a = \text{Re}[z_1]$. The real number b is the **imaginary part** of z_1, denoted by $b = \text{Im}[z_1]$. It is simple to verify that

$$a = \text{Re}[z_1] = \frac{z_1 + \bar{z}_1}{2}$$

and

$$b = \text{Im}[z_1] = \frac{z_1 - \bar{z}_2}{2j}$$

where $\bar{z}_1 = a - jb$ is the complex conjugate[1] of z_1. The **magnitude** or **modulus** of z_1, denoted by $|z_1|$, satisfies

$$|z_1|^2 = a^2 + b^2$$

The number $z_1 = a + jb$ is said to be represented in **rectangular coordinates.** Another representation of z_1, called polar form or **polar coordinates**, follows from the simple geometry illustrated in Figure 10.2.

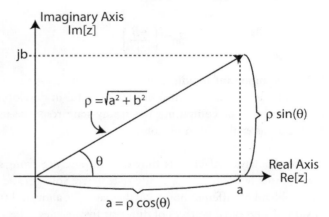

FIGURE 10.2 Diagram showing relationship between polar and rectangular coordinates of a complex number.

In Figure 10.2, the number z_1 can be thought of as a vector of length $\rho = \sqrt{a^2 + b^2} = |z_1|$, which makes an angle $\theta = \tan^{-1}(b/a)$ with the horizontal in the counterclockwise direction. (In computing $\tan^{-1}(b/a)$ it is important to adjust the angle (principal part) to be in the proper quadrant of the complex plane.) Hence $z_1 = a + jb$ is completely specified by its magnitude ρ and angle θ i.e.,

$$z_1 = a + jb = \rho \cos(\theta) + j\rho \sin(\theta) = \rho[\cos(\theta) + j \sin(\theta)] = \rho e^{j\theta} = \rho \angle \theta$$

where $\rho \angle \theta$ is a shorthand notation for $\rho e^{j\theta}$ and

$$e^{j\theta} = \cos(\theta) + j \sin(\theta) \tag{10.3}$$

is the famous **Euler identity**. The Euler identity can be demonstrated by writing the Taylor series for $e^{j\theta}$ and recognizing it as the sum of the Taylor series for $\cos(\theta)$ added to j times the Taylor series for $\sin(\theta)$. Note that the symbol $\angle$ has two meanings, depending on the context of its use: (1) $\angle z$ means angle of the complex number z, and (2) $\rho \angle \theta$ means the complex number whose magnitude is ρ and whose angle is θ. The properties of the exponential immediately imply that

$$e^{j(\theta_1 + \theta_2)} = e^{j\theta_1} e^{j\theta_2} \tag{10.4}$$

Exercises. 1. Compute the polar coordinates of $z_1 = -1 - j$ and $z_2 = 1 + j$.

ANSWERS: $\sqrt{2}e^{-j135^o}$, $\sqrt{2}e^{j45^o}$

2. Let $z = 6\ e^{j\pi/6}$, where $\pi/6$ has units of radians; for example, π rad equals 180°. Find the real and imaginary parts of z.
ANSWER: $z = 5.1962 + j3$

3. Show by direct computation that since $e^{j\theta_1} = \cos(\theta_1) + j\sin(\theta_1)$ and $e^{j\theta_2} = \cos(\theta_2) + j\sin(\theta_2)$, then $e^{j(\theta_1 + \theta_2)} = \cos(\theta_1 + \theta_2) + j\sin(\theta_1 + \theta_2) = e^{j\theta_1}e^{j\theta_2}$.

With these simple definitions, the product of two complex numbers $z_1 = a + jb = \rho_1 e^{j\theta_1}$ and $z_2 = c + jd = \rho_2 e^{j\theta_2}$ can be found using rectangular coordinates as

$$z_1 z_2 = (a + jb)(c + jd) = ac - bd + j(bc + ad)$$

or using polar coordinates as

$$z_1 z_2 = \rho_1 e^{j\theta_1}\rho_2 e^{j\theta_2} = \rho_1\rho_2 e^{j(\theta_1 + \theta_2)}$$

$$= \rho_1\rho_2\cos(\theta_1 + \theta_2) + j\rho_1\rho_2\sin(\theta_1 + \theta_2)$$

which in shorthand notation is

$$z_1 z_2 = \rho_1\rho_2 \angle (\theta_1 + \theta_2)$$

EXAMPLE 10.1. Suppose $z_1 = 3 - j4 = 5\angle-53.13^\circ$ and $z_2 = 8 + j6 = 10\angle36.87^\circ$. Then
$z_1 z_2 = (24 + 24) + j(18 - 32) = 48 - j14$

Equivalently,

$z_1 z_2 = 5\angle-53.13^\circ \times 10\angle36.87^\circ = 50 \angle(-53.13^\circ + 36.87^\circ)$

$= 50 e^{j(-53.13^\circ + 36.87^\circ)}$

$= 50\cos(16.26^\circ) - j50\sin(16.26^\circ) = 48 - j14$

Exercise. Let $z_1 = 2 + j2$ and $z_2 = -2 + j6$.
     (a)   Compute the polar form of z_1 and z_2.
     (b)   Compute $z_1 z_2$ in rectangular coordinates.
     (c)   Compute $z_1 z_2$ in polar coordinates.
ANSWERS: $2.8284 e^{j45^\circ}$, $6.3246 e^{j108.43^\circ}$, $-16 + j8$, $17.8885 e^{j153.43^\circ}$

Similarly, in rectangular coordinates the arithmetic for the division of two complex numbers is

$$\frac{z_1}{z_2} = \frac{a + jb}{c + jd} = \frac{(a + jb)(c - jd)}{(c + jd)(c - jd)} = \frac{(a + jb)(c - jd)}{c^2 + d^2}$$

$$= \frac{(ac + bd) + j(bc - ad)}{c^2 + d^2}$$

In polar coordinates the calculation is more straightforward:

$$\frac{z_1}{z_2} = \frac{\rho_1 e^{j\theta_1}}{\rho_2 e^{j\theta_2}} = \frac{\rho_1}{\rho_2} e^{j(\theta_1 - \theta_2)} = \frac{\rho_1}{\rho_2} \cos(\theta_1 - \theta_2) + j\frac{\rho_1}{\rho_2} \sin(\theta_1 - \theta_2)$$

In our shorthand notation,

$$\frac{z_1}{z_2} = \frac{\rho_1}{\rho_2} \angle(\theta_1 - \theta_2)$$

Exercise. Let $z_1 = 2 + j2$ and $z_2 = -2 + j6$.
  (a) Compute z_1/z_2 in rectangular coordinates.
  (b) Compute z_1/z_2 in polar coordinates.
ANSWERS: $0.2 - j0.4$, $0.4472 e^{j63.435°}$

Of particular concern in this chapter are equations involving mixed representations of complex numbers. For example, suppose an unknown complex number $z = Ve^{j\theta}$ satisfies the equation

$$Ve^{j\theta} (a + jb) = c + jd$$

Then dividing through by $a + jb$ yields

$$Ve^{j\theta} = \frac{c + jd}{a + jb}$$

Since V is the magnitude of the complex number on the right-hand side of the equal sign, it follows that

$$V = \frac{|c + jd|}{|a + jb|} = \frac{\sqrt{c^2 + d^2}}{\sqrt{a^2 + b^2}}$$

Here we have used the fact that a complex number that is the ratio of two other complex numbers has a magnitude equal to the ratio of the magnitudes. To determine the angle θ, one uses the property that θ equals the angle of the complex number in the numerator minus the angle of the complex number in the denominator,

$$\theta = \angle(c + jd) - \angle(a + jb) = \tan^{-1}\left(\frac{d}{c}\right) - \tan^{-1}\left(\frac{b}{a}\right)$$

Exercise. Let $z_1 = 2 - 2j$ and $z_2 = 5.5 + j2.4$. Find V and θ when $Ve^{j\theta} = z_1/z_2$.
ANSWER: $0.471 \angle -68.47°$

EXAMPLE 10.2.

Suppose $v(t) = \text{Re}[Ve^{j(\omega t + \phi)}]$ and

$$-5Ve^{j\phi} + j6Ve^{j\phi} - 3Ve^{j\phi} = 20e^{j45°}$$

Find V, ϕ, and $v(t)$.

SOLUTION

Factoring $Ve^{j\phi}$ out to the left and dividing by $(-8 + j6)$ yields

$$Ve^{j\phi} = \frac{20e^{j45°}}{-8 + j6} = 2e^{-j98.13°}$$

Hence, $V = 2$, $\phi = -98.13°$, and

$$v(t) = \text{Re}[Ve^{j(\omega t + \phi)}] = 2\cos(\omega t - 98.13°) \text{ V}$$

Sometimes a function $v(t)$ is a complex number for each t, such as $v(t) = Ve^{j(\omega t + \phi)}$, and $v(t)$ will satisfy some specific algebraic or differential equation. When this is the case, it is possible to use the differential equation to find values for V and ϕ. The next two examples illustrate this strategy.

EXAMPLE 10.3. Suppose the function

$$v(t) = Ae^{j(\omega t + \phi)}$$

satisfies the differential equation

$$\frac{d^2v}{dt^2} + 2\frac{dv}{dt} + 2v = 10e^{j(\omega t + 60°)}$$

Find the values of A and ϕ if ω is known to be 2 rad/sec.

SOLUTION

Since the function $v(t)$ must satisfy the differential equation, the first step is to substitute into the differential equation. Substituting $Ae^{j(\omega t + \phi)}$ into the differential equation and taking appropriate derivatives yields

$$-\omega^2 Ae^{j(\omega t + \phi)} + j2\omega Ae^{j(\omega t + \phi)} + 2Ae^{j(\omega t + \phi)} = 10e^{j(\omega t + 60°)}$$

The $e^{j\omega t}$ term, which is always nonzero, cancels on both sides, leaving

$$Ae^{j\phi}[2 - \omega^2 + j2\omega] = 10e^{j60°}$$

Since $\omega = 2$, one can equate magnitudes and angles to obtain

$$Ae^{j\phi} = \frac{10e^{j60°}}{2 - \omega^2 + j2\omega} = \frac{10e^{j60°}}{-2 + j4} = 2.236e^{-j56.57°}$$

Exercise. Repeat the above example if $\omega = 3$ and $10e^{j(\omega t + 60°)}$ is changed to $18.44e^{j(\omega t - 81.2°)}$.
ANSWERS: $A = 2$, $\phi = 139.4°$

The techniques of circuit analysis in this chapter will often require complex number arithmetic. The voltages and currents of practical interest are always real. The complex arithmetic is a short-cut to computing "real" voltages and currents. The real quantities are obtained by taking the real part of the complex number or complex function. The various manipulations depend on some general properties related to the real part of complex numbers.

Property 10.1. $\text{Re}[z_1 + z_2] = \text{Re}[z_1] + \text{Re}[z_2]$.

This property has a particularly nice application to summing trigonometric waveforms. Let $v_1(t) = \cos(\omega t + 55°)$ and $v_2(t) = 10 \sin(\omega t - 30°) = 10 \cos(\omega t - 120°)$. Note that a $-90°$ shift converts the sine to a cosine. Hence,

$$v_1(t) + v_2(t) = \cos(\omega t + 55°) + 10 \cos(\omega t - 120°)$$

$$= \text{Re}[e^{j(\omega t + 55°)}] + \text{Re}[10e^{j(\omega t - 120°)}] \text{ by the Euler identity, equation 10.3}$$

$$= \text{Re}[e^{j(\omega t + 55°)} + 10e^{j(\omega t - 120°)}] \text{ by Property 10.1}$$

$$= \text{Re}[e^{j\omega t}(e^{j55°} + 10e^{-j120°})] \text{ by equation 10.4 and then factoring } e^{j\omega t} \text{ to the left}$$

$$= \text{Re}\{e^{j\omega t}[(0.5736 + j0.8192) + (-5 - j8.66)]\} \text{ after conversion to rectangular form}$$

$$= \text{Re}[e^{j\omega t}(-4.426 - j7.841)] = \text{Re}[9e^{j(\omega t - 119.4°)}] \text{ after simplification and con-version back to polar form}$$

$$= 9 \cos(\omega t - 119.4°) \text{ after taking the real part.}$$

This sequence of manipulations shows that the magnitude and phase of two cosines at the same frequency ω can be represented by distinct complex numbers. One can then add the complex numbers and determine the magnitude and angle of a third cosine equal to the sum of the original two cosines. This presents a shortcut for adding two cosines together.

Property 10.2 (*proportionality property*). $\text{Re}[\alpha z_1] = \alpha \text{Re}[z_1]$ for all real scalars α.

Properties 1 and 2 taken together imply

$$\text{Re}[\alpha_1 z_1 + \alpha_2 z_2] = \alpha_1 \text{Re}[z_1] + \alpha_2 \text{Re}[z_2]$$

which is a **linearity property** for complex numbers with multiplication by real scalars. The next property, which underpins the techniques of this chapter, defines how differentiation can be interchanged with the operation $\text{Re}[\cdot]$.

Property 10.3 (*differentiation property*). Let $A = \rho e^{j\theta}$. Then

$$\frac{d}{dt}\text{Re}\left[Ae^{j\omega t}\right] = \text{Re}\left[\frac{d}{dt}\left(Ae^{j\omega t}\right)\right] = \text{Re}\left[j\omega Ae^{j\omega t}\right]$$

Exercise. Find $\text{Re}\left\{\dfrac{d}{dt}\left[(10 + j5)e^{j100t}\right]\right\}$.

ANSWER: $1118\cos(100t + 116.57°)$

Our fourth property tells us the conditions for the equality of two complex-valued time functions.

Property 10.4. For all possibly complex numbers A and B, $\text{Re}[Ae^{j\omega t}] = \text{Re}[Be^{j\omega t}]$ for all t if and only if $A = B$.

Taken together, the preceding properties imply a fifth, very important property. Here note that a complex exponential is sometimes referred to as a complex sinusoid.

Property 10.5. The sum of any number of (1) complex exponentials, say $A_i e^{j\omega t}$, or (2) derivatives of any order of complex exponentials of the same frequency ω, or (3) indefinite integrals of any order of a complex exponential of the same frequency ω, is a complex exponential of the same frequency ω.

This property is another foundation stone on which the phasor analysis of this chapter builds. Table 10.1 summarizes the properties of complex numbers.

TABLE 10.1 Summary of Properties of Complex Numbers

$V_m\cos(\omega t + \theta) = A\cos(\omega t) + B\sin(\omega t)$	$V_m = \sqrt{A^2 + B^2}$, $\phi = \tan^{-1}\left(\dfrac{-B}{A}\right)$
Euler identity	$e^{j\theta} = \cos(\theta) + j\sin(\theta)$
Real part of sum	$\text{Re}[z_1 + z_2] = \text{Re}[z_1] + \text{Re}[z_2]$
Proportionality property	$\text{Re}[\alpha z_1] = \alpha\,\text{Re}[z_1]$ for all real scalars α
Linearity property	$\text{Re}[\alpha_1 z_1 + \alpha_2 z_2] = \alpha_1\,\text{Re}[z_1] + \alpha_2\,\text{Re}[z_2]$
Differentiation property	$\dfrac{d}{dt}\text{Re}\left[Ae^{j\omega t}\right] = \text{Re}\left[\dfrac{d}{dt}\left(Ae^{j\omega t}\right)\right] = \text{Re}\left[j\omega Ae^{j\omega t}\right]$
Equality property	$\text{Re}[Ae^{j\omega t}] = \text{Re}[Be^{j\omega t}]$ for all t if and only if $A = B$
Single-frequency property	Sum of complex exponentials $A_i e^{j\omega t}$, or derivatives, or their indefinite integrals of any order is a complex exponential of the same frequency ω

3. NAIVE TECHNIQUE FOR COMPUTING THE SINUSOIDAL STEADY STATE

Property 10.5 of the previous section suggests a technique for computing the SSS response of a circuit. The technique uses a differential equation model of an *RL*, *RC*, or *RLC* circuit as developed in Chapters 8 and 9. In contrast to the dc sources in those chapters, suppose the source excitations have the form $V_s \cos(\omega t + \theta)$. In addition we assume that the zero-input response consists of (eventually) decaying exponentials or (eventually) exponentially decaying sinusoids to ensure that there is a valid sinusoidal steady state. Thus the form of a first-order circuit differential equation model with a sinusoidal excitation is

$$\frac{dx(t)}{dt} + ax(t) = V_s \cos(\omega t + \theta) \tag{10.5a}$$

or in the second-order case,

$$\frac{d^2 x(t)}{dt^2} + b\frac{dx(t)}{dt} + cx(t) = V_s \cos(\omega t + \theta) \tag{10.5b}$$

where $x(t)$ is a desired voltage or current, such as $v_L(t)$, $i_L(t)$, $v_C(t)$.

Property 10.5 guarantees that the sum of any number of cosines or derivatives of any order of cosines of the same frequency ω is a cosine of the same frequency ω. Hence, the circuit response $x(t)$ in equations 10.5 has a steady-state cosine form of frequency ω. Further, the scaled *sum* of $x(t)$ and its derivatives on the left-hand side of each differential equation 10.5 must equal $V_s \cos(\omega t + \theta)$, the input excitation. This also implies that the steady-state circuit response, $x(t)$, is a cosine of the same frequency as the input, but not necessarily the same magnitude or phase. We conclude that $x(t) = V_m \cos(\omega t + \phi) = A \cos(\omega t) + B \sin(\omega t)$. The SSS response is then specified upon finding A and B. The following example illustrates this calculation.

EXAMPLE 10.4. Let the source excitation to the circuit of Figure 10.3 be $i_s(t) = I_s \cos(\omega t)$. Compute the SSS response $i_L(t)$.

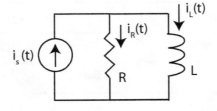

FIGURE 10.3 Parallel *RL* circuit for Example 10.4.

SOLUTION

Step 1. *Determine the differential equation model of the circuit.* From KCL applied to the top node of the circuit, $i_s(t) = i_R(t) + i_L(t)$. Since the resistor and inductor voltages coincide, the *v-i* relationship of the inductor implies that the inductor current satisfies the differential equation

$$\frac{di_L(t)}{dt} + \frac{R}{L}i_L(t) = \frac{R}{L}i_s(t)$$

(10.6)

which has the form of equation 10.5a.

Step 2. *Determine the form of the response.* Since the input is a cosine wave, the SSS response will have the sinusoidal form

$$i_L(t) = A\cos(\omega t) + B\sin(\omega t)$$

(10.7)

Step 3. *Substitute the form of the response (equation 10.7) into the differential equation 10.6.* Inserting equation 10.7 into the differential equation 10.6 and evaluating the derivatives yields

$$\frac{R}{L}I_s\cos(\omega t) = \frac{d}{dt}\left[A\cos(\omega t) + B\sin(\omega t)\right] + \frac{R}{L}\left[A\cos(\omega t) + B\sin(\omega t)\right]$$

$$= -\omega A\sin(\omega t) + \omega B\cos(\omega t) + \frac{RA}{L}\cos(\omega t) + \frac{RB}{L}\sin(\omega t)$$

Step 4. *Group like terms and solve for A and B.* Grouping like terms leads to

$$\left(B\omega + \frac{R}{L}A - \frac{R}{L}I_s\right)\cos(\omega t) + \left(\frac{R}{L}B - A\omega\right)\sin(\omega t) = 0$$

(10.8)

To determine the coefficients A and B, we evaluate equation 10.8 at two distinct time instants. Since equation 10.8 must hold at every instant of time, it must hold at $t = 0$; i.e., at $t = 0$,

$$\left(B\omega + \frac{R}{L}A - \frac{R}{L}I_s\right) = 0$$

or, equivalently,

$$\frac{R}{L}A + \omega B = \frac{R}{L}I_s$$

(10.9a)

In addition, equation 10.8 must hold for $t = \pi/(2\omega)$, in which case

$$-\omega A + \frac{R}{L}B = 0$$

(10.9b)

Solving equations 10.9 simultaneously for A and B yields

$$A = \frac{R^2 I_s}{R^2 + L^2 \omega^2}, \qquad B = \frac{\omega R L I_s}{R^2 + L^2 \omega^2}$$

Step 5. *Determine the steady-state response.* Since A and B are known,

$$i_L(t) = \frac{R^2 I_s}{R^2 + L^2 \omega^2} \cos(\omega t) + \frac{\omega R L I_s}{R^2 + L^2 \omega^2} \sin(\omega t)$$

In the more common alternative form of $i_L(t) = I_m \cos(\omega t + \phi)$ as per equation 10.2a,

$$i_L(t) = \frac{R I_s}{\sqrt{R^2 + L^2 \omega^2}} \cos(\omega t + \phi) \tag{10.10a}$$

where

$$\phi = -\tan^{-1}\left(\frac{\omega L}{R}\right) \tag{10.10b}$$

is adjusted to reflect the proper quadrant of the complex plane.

This example has illustrated a procedure for finding the SSS response of a circuit. Step 1 is to substitute an assumed sinusoidal response form, such as $A\cos(\omega t) + B\sin(\omega t)$, having unspecified constants A and B, into the differential equation and evaluate all derivatives. Step 2 is to group like terms, and step 3 is to compute the constants A and B. After finding A and B, one computes the magnitude, V_m, and phase ϕ of the cosine $V_m \cos(\omega t + \phi)$ via equations 10.2.

The next section offers an alternative approach. Using complex excitation signals of the form $V_s e^{j(\omega t + \theta)}$, one computes V_m and ϕ by a more direct route.

4. COMPLEX EXPONENTIAL FORCING FUNCTIONS IN SINUSOIDAL STEADY-STATE COMPUTATION

Complex exponential forcing functions are simply complex exponential input excitations of the form $V_s e^{j(\omega t + \theta)}$ or $I_s e^{j(\omega t + \theta)}$. From the properties of complex numbers in section 2, we can replace the input excitation $V_s \cos(\omega t + \theta)$ and the assumed circuit response $V_m \cos(\omega t + \phi) = A \cos(\omega t) + B \sin(\omega t)$ with their complex counterparts $V_s e^{j(\omega t + \theta)}$ and $V_m e^{j(\omega t + \phi)}$, respectively, without any penalty. To recover the actual real-valued responses, we simply take the real parts of the complex quantities. Again this is justified by properties 10.1 through 10.5. This process of substitution and subsequent taking of real parts actually simplifies the calculations developed in section 3, because of the simple differential and multiplicative properties of the exponential function. The following example illustrates a more efficient calculation of the steady-state response using complex exponentials.

EXAMPLE 10.5. For the series RC circuit of Figure 10.4 let $v_s(t) = V_s\cos(\omega t)$. Compute the steady-state response $v_C(t)$.

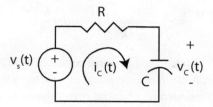

FIGURE 10.4. Series RC circuit for Example 10.5.

Step 1. *Construct the differential equation of the circuit.* Writing a loop equation and substituting for $i_C(t)$ yields

$$RC\frac{dv_C(t)}{dt} + v_C(t) = v_s(t) \tag{10.11}$$

Step 2. *Substitute complex forms of the input and response into the differential equation.* If $v_s(t)$ were to be equal to the complex exponential $V_s e^{j\omega t}$, then the response would be $V_m e^{j(\omega t + \phi)}$. However, if $v_s(t) = \mathrm{Re}[V_s e^{j\omega t}] = V_s\cos(\omega t)$ (as is the case), then $v_C(t) = \mathrm{Re}[V_m e^{j(\omega t + \phi)}]$ from the properties of complex numbers. Hence, for the moment, let us set $v_s(t) = V_s e^{j\omega t}$ and agree that $v_C(t) = V_m e^{j(\omega t + \phi)}$. Later we will take the appropriate real parts.

Substituting the complex expressions into the circuit differential equation 10.11 yields

$$j\omega RC V_m e^{j(\omega t + \phi)} + V_m e^{j(\omega t + \phi)} = V_s e^{j\omega t}$$

After canceling the $e^{j\omega t}$ terms, factoring $V_m e^{j\phi}$ out to the left, and dividing through by $(j\omega RC + 1)$, we obtain

$$V_m e^{j\phi} = \frac{V_s}{1 + j\omega RC} \tag{10.12}$$

Step 3. *Determine the magnitude V_m and the angle ϕ.* Equating magnitudes on both sides of equation 10.12 yields

$$V_m = \frac{V_s}{\sqrt{1 + \omega^2 R^2 C^2}} \tag{10.13a}$$

and equating angles yields

$$\phi = -\tan^{-1}(\omega RC) \tag{10.13b}$$

Step 4. *Determine the steady-state response.* Using equations 10.13 the desired response is computed by taking real parts:

$$v_C(t) = \text{Re}\left[\frac{V_s}{\sqrt{1+\omega^2 R^2 C^2}}\, e^{j[\omega t - \tan^{-1}(\omega RC)]}\right] \tag{10.14}$$

$$= \frac{V_s}{\sqrt{1+\omega^2 R^2 C^2}}\cos[\omega t - \tan^{-1}(\omega RC)]$$

In deriving the relationship 10.12 from the differential equation 10.11, we utilized a complex exponential function as the circuit input. A complex exponential input is not a signal that can be generated in the laboratory. Nevertheless, it is often used in advanced circuit theory to simplify the derivation of many important results, as was done in the preceding example. If one does not mind a more lengthy derivation, then the same result (equations 10.12 through 10.14) can be obtained without the fictitious complex exponential excitation. For example, let the voltage source in Figure 10.4 represent a real signal source

$$v_s(t) = V_s\cos(\omega t) = \text{Re}[V_s e^{j\omega t}]$$

Then the steady-state response has the form

$$v_C(t) = V_m\cos(\omega t + \phi) = \text{Re}[V_m e^{j(\omega t + \phi)}]$$

Substituting these expressions into the differential equation 10.11 yields

$$RC\frac{d}{dt}\left\{\text{Re}\left[V_m e^{j(\omega t+\phi)}\right]\right\} + \text{Re}\left[V_m e^{j(\omega t+\phi)}\right] = \text{Re}\left[V_s e^{j\omega t}\right]$$

Making use of properties 10.2 and 10.3, move the position of the operator Re[] outside the first term to obtain

$$\text{Re}\left[RCV_m\frac{d}{dt}\left(e^{j(\omega t+\phi)}\right)\right] + \text{Re}\left[V_m e^{j(\omega t+\phi)}\right] = \text{Re}\left[V_s e^{j\omega t}\right]$$

Evaluating the derivative and using property 10.2 (linearity) produces

$$\text{Re}\left[RCV_m j\omega e^{j(\omega t+\phi)} + V_m e^{j(\omega t+\phi)}\right] = \text{Re}\left[V_s e^{j\omega t}\right] \tag{10.15}$$

By property 10.4, equation 10.15 holds if and only if

$$RCV_m j\omega e^{j(\omega t+\phi)} + V_m e^{j(\omega t+\phi)} = V_s e^{j\omega t}$$

This is precisely the equation following equation 10.11 that leads to equations 10.12, 10.13, and finally 10.14.

As we can see, the use of complex exponentials does indeed lead to a more direct calculation of the SSS response. However, this method and the method of section 3 require a differential equation model of the circuit. For circuits with multiple sources, dependent sources, and many inter-connections of circuit elements, finding the differential equation model is often a nontrivial task. In the next section we eliminate the need to find a differential equation model of the circuit by introducing the phasor concept.

5. PHASOR REPRESENTATIONS OF SINUSOIDAL SIGNALS

Recall that $A\angle\phi$ is shorthand for $Ae^{j\phi} = A\cos(\phi) + jA\sin(\phi)$. If the frequency ω is known, then the complex number $A\angle\phi$ completely determines the complex exponential $Ae^{j(\omega t+\phi)}$. In turn, if ω is known, then $A\angle\phi$ completely specifies $A\cos(\omega t + \phi) = \text{Re}[Ae^{j(\omega t+\phi)}]$. This means that the complex number $A\angle\phi$ can represent a sinusoidal function $A\cos(\omega t + \phi)$, whenever ω is known. Complex number representations that denote sinusoidal signals at a fixed frequency are called **phasors**. A *phasor* voltage or current will be denoted by a boldface capital letter. A typical voltage phasor is $\mathbf{V} = V_m\angle\phi$ and a typical current phasor is $\mathbf{I} = I_m\angle\phi$. For example, the current $i(t) = 25\cos(\omega t + 45^\circ)$ has the phasor representation $\mathbf{I} = 25\angle45^\circ$. The voltage $v(t) = -15\sin(\omega t + 30^\circ) = 15\cos(\omega t + 120^\circ)$ has the phasor representation $\mathbf{V} = 15\angle120^\circ$.

As all voltages and currents satisfy KVL and KCL, respectively, one might expect phasor voltages and currents to do likewise. This is not patently clear. The following simple example demonstrates why this is true for KCL.

Consider the circuit node drawn in Figure 10.5.

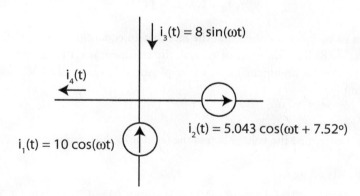

FIGURE 10.5 Single node having four incident branches.

From KCL it follows that

$$i_4(t) = i_1(t) - i_2(t) + i_3(t)$$

$$= 10\cos(\omega t) - 5.043\cos(\omega t + 7.52^\circ) + 8\cos(\omega t - 90^\circ)$$

Using trigonometric identities or property 10.1 to combine terms on the right-hand side leads to

$$i_4(t) = 10\cos(\omega t - 60^\circ)$$

For the corresponding phasors to satisfy KCL, it must follow that

$$10\angle{-60^\circ} = \mathbf{I}_4 = \mathbf{I}_1 - \mathbf{I}_2 + \mathbf{I}_3$$

The right-hand side of this equation requires that

$$\mathbf{I}_1 - \mathbf{I}_2 + \mathbf{I}_3 = 10\angle 0^\circ - 5.043\angle 7.52^\circ + 8\angle{-90^\circ}$$

$$= 10 - (5 + j0.66) + (-j8) = 5 - j8.66 = 10\angle{-60^\circ} = \mathbf{I}_4$$

Thus the phasors (which have both a real and an imaginary part) satisfy KCL. KCL is satisfied because $i_4(t) = i_1(t) - i_2(t) + i_3(t)$ implies

$$\text{Re}[10e^{j(\omega t - 60^\circ)}] = \text{Re}[10e^{j\omega t}] - \text{Re}[5.043e^{j(\omega t + 7.52^\circ)}] + \text{Re}[8e^{j(\omega t - 90^\circ)}]$$

$$= \text{Re}[(10 - 5.043e^{j7.52^\circ} + 8e^{-j90^\circ})e^{j\omega t}] \qquad (10.16)$$

for all t. By property 10.4, equation 10.16 holds if and only if

$$10\angle{-60^\circ} = 10 - 5.043e^{j7.52^\circ} + 8e^{-j90^\circ}$$

In phasor notation this stipulates that

$$\mathbf{I}_4 = \mathbf{I}_1 - \mathbf{I}_2 + \mathbf{I}_3$$

It is the properties of complex numbers and the fact that an equation is true for all t that guarantee that phasors satisfy KCL. Although not general, the argument is sufficient for our present pedagogical purpose. A similar argument implies that phasor voltages satisfy KVL, as illustrated by the following example.

EXAMPLE 10.6. Determine the voltage across the resistor in the circuit of Figure 10.6 using the phasor concept.

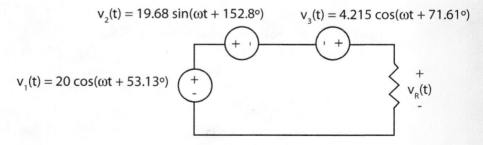

FIGURE 10.6 Resistive circuit with three sources.

SOLUTION

First note that $19.68 \sin(\omega t+152.8°) = 19.68 \cos(\omega t+152.8°-90°) = 19.68 \cos(\omega t+62.8°)$. From KVL,

$$v_R(t) = v_1(t) - v_2(t) + v_3(t)$$

Since voltage phasors must satisfy KVL,

$$\mathbf{V}_R = \mathbf{V}_1 - \mathbf{V}_2 + \mathbf{V}_3 = 20 \angle 53.13° - 19.68 \angle 62.8° + 4.215 \angle 71.6°$$

Changing to rectangular coordinates and adding yields

$$\mathbf{V}_R = 12 + j16 - (9 + j17.5) + 1.33 + j4 = 4.33 + j2.5$$

Equivalently, $\mathbf{V}_R = 5\angle 30°$ V, and

$$v_R(t) = \mathrm{Re}[5e^{j(\omega t + 30°)}] = 5 \cos(\omega t + 30°) \text{ V}$$

Exercise. In Figure 10.6, suppose $v_1(t) = 10 \cos(\omega t)$ V, $v_2(t) = 10 \cos(\omega t - 0.5\pi)$ V, and $v_3(t) = 10\sqrt{2} \cos(\omega t - 0.25\pi)$. Find the phasor $\mathbf{V}_R$ and then $v_R(t)$.

ANSWER: $\mathbf{V}_R = 20 - j20$, $v_R(t) = 20\sqrt{2} \cos(\omega t - 45°)$ V

Given that phasor voltages and currents satisfy KVL and KCL, respectively, it is possible to develop phasor Ohm's law–like relationships for resistors, capacitors, and inductors operating in the SSS. This would allow us to do SSS circuit analysis with techniques similar to resistive dc analysis. The next section takes up this thread by introducing the notion of (phasor) impedance.

6. ELEMENTARY IMPEDANCE CONCEPTS: PHASOR RELATIONSHIPS FOR RESISTORS, INDUCTORS, AND CAPACITORS

Ohm's law–like relationships do exist for resistors, capacitors, and inductors operating in the SSS. The constraint, operating in the SSS, suggests that any Ohm's law–like relationship should be dependent on the sinusoidal frequency.

The first objective of this section is to derive three Ohm's law–like relationships, one each for the resistor, the capacitor, and the inductor. The relationships each take the form $\mathbf{V} = Z(j\omega)\mathbf{I}$, where $\mathbf{V}$ is a phasor voltage, $\mathbf{I}$ is a phasor current, and $Z(j\omega)$ is called the impedance of the device: $Z_R(j\omega)$ for a resistor, $Z_C(j\omega)$ for a capacitor, and $Z_L(j\omega)$ for an inductor. The fact that the phasor voltage $\mathbf{V}$ is a function $Z(j\omega)$ times a phasor current $\mathbf{I}$ indicates a clear kinship with Ohm's law for resistors. Indeed the unit of impedance is the ohm because it is the ratio of phasor voltage to phasor current. The impedance $Z(j\omega)$ explicitly shows that the relationship is potentially frequency dependent.

The derivation of these elementary impedance concepts will build on the assumption that all voltages and currents are complex sinusoids of the same frequency represented by complex phasors. This is permissible because real sinusoids can be recovered from complex sinusoids simply by taking the real part. To this end consider the resistive circuit of Figure 10.7a.

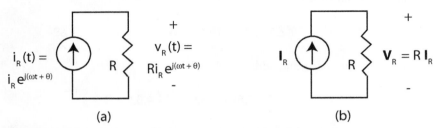

(a)   (b)

FIGURE 10.7 (a) Resistive circuit driven by complex current.
(b) Equivalent phasor representation of the circuit in (a).

From Ohm's law,

$$v_R(t) = Ri_R(t) = RI_R \, e^{j(\omega t + \theta)} = V_R e^{j(\omega t + \theta)}$$

In terms of the phasors $\mathbf{I}_R = I_R e^{j\theta}$ and $\mathbf{V}_R = V_R e^{j\theta}$, this relationship reduces to

$$\mathbf{V}_R = R \, \mathbf{I}_R = Z_R(j\omega) \, \mathbf{I}_R \tag{10.17}$$

where $\mathbf{I}_R = I_R \angle \theta$, $\mathbf{V}_R = V_R \angle \theta$, and $Z_R(j\omega) = R$ is the impedance of the resistor defined by equation 10.17. Ideally the resistor impedance is independent of frequency. Thus $\mathbf{V}_R = RI_R \angle \theta$. If $i_R(t) = I_R \cos(\omega t + \theta) = \text{Re}[I_R e^{j(\omega t + \theta)}]$, then $v_R(t) = RI_R \cos(\omega t + \theta) = \text{Re}[RI_R e^{j(\omega t + \theta)}]$. This phasor relationship restates Ohm's law for complex excitations. The distinctiveness of phasors comes with their application to inductors and capacitors.

Now consider the inductor circuits of Figure 10.8. Assume the circuit of Figure 10.8a is in the steady state.

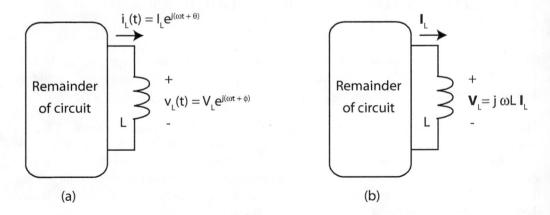

(a)   (b)

FIGURE 10.8 (a) Inductor having complex exponential voltage and current. (b) Phasor relationship of (a).

The complex current and voltage associated with the inductor are, respectively, $i_L(t) = I_L e^{j(\omega t + \theta)}$ and $v_L(t) = V_L e^{j(\omega t + \phi)}$. Substituting these expressions into the defining equation for an inductor yields

$$V_L e^{j(\omega t + \phi)} = L \frac{d}{dt}\left[I_L e^{j(\omega t + \theta)}\right] = j\omega L I_L e^{j(\omega t + \theta)}$$

Canceling out $e^{j\omega t}$ on both sides yields $V_L e^{j\phi} = j\omega L I_L e^{j\theta}$. In terms of the phasors $\mathbf{I}_L = I_L e^{j\theta}$ and $\mathbf{V}_L = V_L e^{j\phi}$, the relationship is

$$\mathbf{V}_L = j\omega L \mathbf{I}_L = Z_L(j\omega)\mathbf{I}_L \qquad (10.18)$$

in which case the **inductor impedance** is derived as $Z_L(j\omega) = j\omega L$. The inductor impedance clearly depends on the value of the radian frequency ω. Specifically, if $\omega = 0$, then the impedance of the inductor is 0, i.e., in SSS the inductor looks like a short circuit to dc excitations. If $\omega = \infty$, the impedance is infinite, i.e., in the steady state the inductor looks like an open circuit to signals of very high frequency.

Equation 10.18 exhibits a frequency-dependent Ohm's law relationship for the inductor. From the properties of the product of two complex numbers, the polar form of the voltage phasor is

$$\mathbf{V}_L = (j\omega L)\mathbf{I}_L = (\omega L I_L) \angle (\theta + 90^\circ)$$

Hence if

$$i_L(t) = \mathrm{Re}[I_L e^{j(\omega t + \theta)}] = I_L \cos(\omega t + \theta) \text{ A}$$

then

$$v_L(t) = \mathrm{Re}[j\omega L I_L e^{j(\omega t + \theta)}] = \mathrm{Re}[\omega L I_L e^{j(\omega t + \theta + 90^\circ)}] = \omega L I_L \cos(\omega t + \theta + 90^\circ) \text{ V}$$

From this relationship one sees that the voltage phase leads the current phase by 90°. Equivalently, one can say that the current lags the voltage by 90°. This leading and lagging takes on a more concrete meaning when one views phasors as vectors in the complex plane, as per Figure 10.9, which shows that the voltage phasor of the inductor always leads the current phasor by 90°.

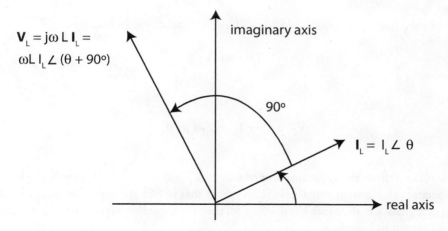

FIGURE 10.9 Diagram of phasor voltage and current for an inductor.

The capacitor has a similar impedance relationship, derived as follows. Assume the circuit of Figure 10.10a is in the steady state.

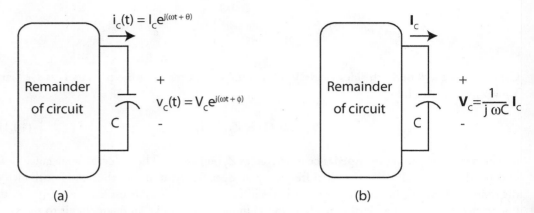

FIGURE 10.10 (a) Capacitor having complex exponential voltage and current. (b) Phasor relationship of (a).

The complex current and voltage associated with the capacitor are, respectively, $i_C(t) = I_C e^{j(\omega t + \theta)}$ and $v_C(t) = V_C e^{j(\omega t + \phi)}$. Substituting these expressions into the defining equation for a capacitor yields

$$I_C e^{j(\omega t + \theta)} = C \frac{d}{dt}\left[V_C e^{j(\omega t + \phi)} \right]$$

$$= j\omega C V_C e^{j(\omega t + \phi)}$$

Canceling out $e^{j\omega t}$ on both sides yields

$$I_C e^{j\theta} = j\omega C V_C e^{j\phi}$$

In terms of the phasors $\mathbf{I}_C = I_C e^{j\theta}$ and $\mathbf{V}_C = V_C e^{j\phi}$, this relationship becomes

$$\mathbf{I}_C = j\omega C \mathbf{V}_C$$

or, equivalently,

$$\mathbf{V}_C = \frac{1}{j\omega C} \mathbf{I}_C = Z_C(j\omega)\mathbf{I}_C \tag{10.19}$$

Equation 10.19 defines the **capacitor impedance** as $Z_C(j\omega) = 1/(j\omega C)$. If $\omega = 0$, the impedance of the capacitor is infinite in magnitude. This means that in SSS the capacitor looks like an open circuit to dc signals. On the other hand, if $\omega = \infty$, then the capacitor has zero impedance and looks like a short circuit to large frequencies.

Looking again at equation 10.19, observe that

$$\mathbf{V}_C = \frac{1}{j\omega C}\mathbf{I}_C = \frac{I_C}{\omega C}\angle(\theta - 90°) \tag{10.20}$$

Equation 10.20 has a vector interpretation in the complex plane, as shown in Figure 10.11.

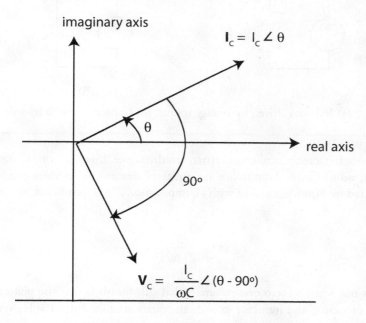

FIGURE 10.11 Diagram of capacitor voltage and current phasors where the voltage phasor lags the current phasor by 90°.

The diagram of Figure 10.11 indicates that the capacitor voltage lags the capacitor current phasor by 90° or that the capacitor current leads the capacitor voltage by 90°, which is the opposite of the case for the inductor.

Exercises. 1. For the circuit of Figure 10.12a, show that

$$\mathbf{I}_L = \frac{1}{j\omega L}\mathbf{V}_L$$

and that

$$i_L(t) = \frac{V_L}{\omega L}\cos(\omega t + \theta - 90°)$$

2. For the circuit of Figure 10.12b, show that

$$\mathbf{I}_C = j\omega C \mathbf{V}_C$$

and that

$$i_C(t) = \omega C V_C \cos(\omega t + \theta + 90^\circ)$$

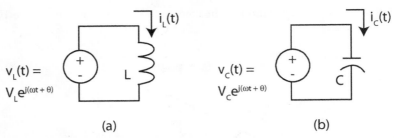

(a)   (b)

FIGURE 10.12 (a) Inductor driven by voltage source. (b) Capacitor driven by voltage source.

Recall that resistance has a reciprocal counterpart, **conductance**. Likewise, impedance has a reciprocal counterpart, **admittance**. Admittance has units of siemens, S, as does conductance. The admittance, denoted by $Y(j\omega)$, associated with an impedance, $Z(j\omega)$, is defined by the inverse relationship

$$Y(j\omega) = \frac{1}{Z(j\omega)} \tag{10.21}$$

provided $Z(j\omega)$ is not equal to zero everywhere. What this means is that the phasor i-v relationship of a resistor, capacitor, and inductor satisfies an equation of the form $\mathbf{I} = Y(j\omega)\mathbf{V}$. Hence, the admittances of the resistor, inductor, and capacitor are respectively given by

$$Y_R(j\omega) = \frac{1}{R}, \quad Y_L(j\omega) = \frac{1}{j\omega L}, \quad Y_C(j\omega) = j\omega C \tag{10.22}$$

The impedance and admittance relationships of the resistor, capacitor, and inductor are summarized in table 10.2.

TABLE 10.2 Summary of Impedance and Admittance Relationships for Resistor, Capacitor, and Inductor

	Impedance	Admittance
resistor	$Z_R(j\omega) = R$	$Y_R(j\omega) = \frac{1}{R}$
capacitor	$Z_C(j\omega) = \frac{1}{j\omega C}$	$Y_C(j\omega) = j\omega C$
inductor	$Z_L(j\omega) = j\omega L$	$Y_L(j\omega) = \frac{1}{j\omega L}$

In the next section the notion of impedance is applied to an arbitrary two-terminal network. This generalization will allow us to consider the impedance and admittance of interconnections of capacitors, inductors, resistors, and dependent sources.

7. PHASOR IMPEDANCE AND ADMITTANCE

For the resistor, the inductor, and the capacitor, the **impedance** equals the ratio of the respective phasor voltage to the phasor current. Analogously, the *impedance* of any two-terminal circuit, as illustrated in Figure 10.13, is the ratio of the phasor voltage to the phasor current, i.e.,

$$Z_{in}(j\omega) = \frac{\mathbf{V}_{in}}{\mathbf{I}_{in}} = R + jX \qquad (10.23a)$$

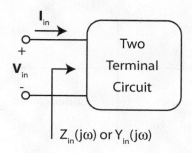

FIGURE 10.13 Two-terminal device with phasor voltage $\mathbf{V}_{in}$, phasor current $\mathbf{I}_{in}$, and input impedance $Z_{in}(j\omega)$.

Because impedance is the ratio of phasor voltage to phasor current, its unit is the ohm. Inverting the relationship of equation 10.23a defines the **admittance** of a two-terminal device as the ratio of phasor current to phasor voltage, i.e.,

$$Y_{in}(j\omega) = \frac{\mathbf{I}_{in}}{\mathbf{V}_{in}} \qquad (10.23b)$$

Provided $Z(j\omega) \neq 0$ for all ω, in contrast to a short circuit, then

$$Y_{in}(j\omega) = \frac{1}{Z_{in}(j\omega)} = G + jB$$

As an example, the impedance of an inductor is $j\omega L$ and its admittance is $1/(j\omega L)$. Historically, impedance and admittance were first defined as per equation 10.23. However, with the wide-spread use and utility of the Laplace transform (Chapter 12) in the past several decades, impedance and admittance have become understood as much broader and more useful concepts than the steady-state presumptions of equation 10.23, as set forth in Chapter 13.

In general, admittances and impedances are rational functions with real coefficients of the complex variable $j\omega$. At each ω the impedance and the admittance are generally complex numbers. Since a complex number has a real part and an imaginary part, we can further classify the real and imaginary parts of an impedance or an admittance. For an impedance $Z(j\omega)$ the expression $\text{Im}[Z(j\omega)] = X$ is called the **reactance** of the two-terminal element, while $\text{Re}[Z(j\omega)] = R$ refers to its **resistance**. Further, for an admittance $Y(j\omega)$, $\text{Im}[Y(j\omega)] = B$ is called the **susceptance** of the two-terminal device whereas $\text{Re}[Y(j\omega)] = G$ is referred to as the **conductance**. These definitions are summarized in Table 10.3.

TABLE 10.3 Summary Definitions of Various Terms

Impedance		Admittance	
$Z(j\omega) = \dfrac{V_{in}}{I_{in}} = R + jX$		$Y(j\omega) = \dfrac{I_{in}}{V_{in}} = G + jB$	
Resistance	Reactance	Conductance	Susceptance
$R = \text{Re}[Z(j\omega)]$	$X = \text{Im}[Z(j\omega)]$	$G = \text{Re}[Y(j\omega)]$	$B = \text{Im}[Y(j\omega)]$

Using equations 10.23, one can compute the equivalent impedance $Z_{in}(j\omega)$ of two devices in series, as in Figure 10.14a. Here $\mathbf{V}_{in} = \mathbf{V}_1 + \mathbf{V}_2$. By Ohm's law for impedances, $\mathbf{V}_1 = Z_1(j\omega)\mathbf{I}_1$ and $\mathbf{V}_2 = Z_2(j\omega)\,\mathbf{I}_2$. But $\mathbf{I}_1 = \mathbf{I}_2 = \mathbf{I}_{in}$. Hence,

$$\mathbf{V}_{in} = [Z_1(j\omega) + Z_2(j\omega)]\mathbf{I}_{in} = Z_{in}(j\omega)\mathbf{I}_{in}$$

i.e.,

$$Z_{in}(j\omega) = \frac{\mathbf{V}_{in}}{\mathbf{I}_{in}} = Z_1(j\omega) + Z_2(j\omega) \tag{10.24}$$

This simple derivation has another consequence: given $Z_{in}(j\omega) = Z_1(j\omega) + Z_2(j\omega)$ and the fact that $\mathbf{V}_i = Z_i(j\omega)\mathbf{I}_{in}$, $i = 1, 2$, a simple substitution yields the **voltage division formula**,

$$\mathbf{V}_i = \frac{Z_i(j\omega)}{Z_1(j\omega) + Z_2(j\omega)}\,\mathbf{V}_{in} \tag{10.25}$$

Equations 10.24 and 10.25 are consistent with our early development of series and parallel resistance.

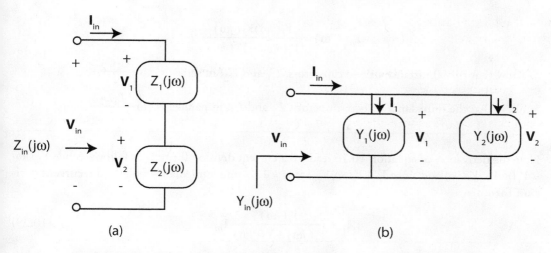

FIGURE 10.14 (a) Two impedances in series. (b) Two admittances in parallel.

Exercises. 1. Duplicate the derivation of equation 10.23 for three impedances in series.
2. Derive a formula for voltage division when there are three impedances in series.

The admittance of two devices in parallel, as sketched in Figure 10.14b, satisfies

$$Y_{in}(j\omega) = \frac{\mathbf{I}_{in}}{\mathbf{V}_{in}} = \frac{\mathbf{I}_1 + \mathbf{I}_2}{\mathbf{V}_{in}} = \frac{\mathbf{I}_1}{\mathbf{V}_1} + \frac{\mathbf{I}_2}{\mathbf{V}_2}$$

since $\mathbf{V}_{in} = \mathbf{V}_1 = \mathbf{V}_2$. Since

$$Y_1(j\omega) = \frac{\mathbf{I}_1}{\mathbf{V}_1} \quad \text{and} \quad Y_2(j\omega) = \frac{\mathbf{I}_2}{\mathbf{V}_2}$$

we conclude that

$$Y_{in}(j\omega) = Y_1(j\omega) + Y_2(j\omega) \tag{10.26}$$

Exercises. 1. Duplicate the derivation of equation 10.26 for three admittances in parallel, i.e., show that $Y_{in}(j\omega) = Y_1(j\omega) + Y_2(j\omega) + Y_3(j\omega)$.
2. Show that the equivalent impedance of two devices, $Z_1(j\omega)$ and $Z_2(j\omega)$, in parallel is given by

$$Z_{eq}(j\omega) = \frac{Z_1(j\omega)Z_2(j\omega)}{Z_1(j\omega) + Z_2(j\omega)} \tag{10.27}$$

3. Show that the equivalent admittance of two devices, $Y_1(j\omega)$ and $Y_2(j\omega)$, in series is given by

$$Y_{eq}(j\omega) = \frac{Y_1(j\omega)Y_2(j\omega)}{Y_1(j\omega) + Y_2(j\omega)}$$

(10.28)

4. Show that the admittance of two capacitors, C_1 and C_2, in series is $j\omega \dfrac{C_1 C_2}{C_1 + C_2}$.

5. Show that the impedance of two inductors, L_1 and L_2, in parallel is $j\omega \dfrac{L_1 L_2}{L_1 + L_2}$.

Now the derivation of equation 10.26 leads to a current division formula as follows. Since $Y_{in}(j\omega)$ = $Y_1(j\omega) + Y_2(j\omega)$ and since $\mathbf{I}_i = Y_i(j\omega)\mathbf{V}_{in}$, for i = 1, 2, one immediately obtains the **current division formula**,

$$\mathbf{I}_i = \frac{Y_i(j\omega)}{Y_1(j\omega) + Y_2(j\omega)} \mathbf{I}_{in}$$

(10.29)

Since devices represented by impedances or admittances must satisfy KVL and KCL in terms of their phasor voltages and currents, and since each device so represented satisfies a generalized Ohm's law, i.e.,

$$\mathbf{V} = Z(j\omega)\mathbf{I} \quad \text{or} \quad \mathbf{I} = Y(j\omega)\mathbf{V}$$

it follows that *impedances can be manipulated in the same manner as resistances, and admittances in the same manner as conductances.* The voltage division formula of equation 10.25 and the current division formula of equation 10.27 illustrate this fact. Example 10.7 further clarifies these statements.

Exercises. 1. Derive a current division formula for three admittances in parallel.

2. Find the admittance and then the impedance of each parallel connection in Figure 10.15.

ANSWERS: Admittances are $\dfrac{1}{j\omega}\left(\dfrac{1}{L_1} + \dfrac{1}{L_2} + \dfrac{1}{L_3}\right)$, $j\omega(C_1 + C_2 + C_3)$.

3. Compute the equivalent inductance for Figure 10.15a and the equivalent capacitance for the circuit of Figure 10.15b.

ANSWERS: $\left(\dfrac{1}{L_1} + \dfrac{1}{L_2} + \dfrac{1}{L_3}\right)^{-1}$, $C_1 + C_2 + C_3$

4. Find $\mathbf{I}_3$ in terms of $\mathbf{I}_{in}$ for each circuit in Figure 10.15.

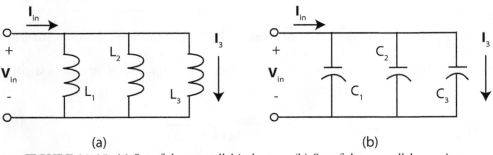

(a)   (b)

FIGURE 10.15. (a) Set of three parallel inductors. (b) Set of three parallel capacitors.

EXAMPLE 10.7. For the circuit of Figure 10.16, compute the input impedance $Z_{in}(j\omega)$ when ω = 500 rad/sec.

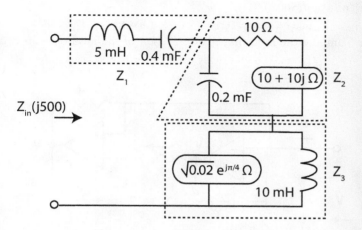

FIGURE 10.16 Series-parallel interconnection of different impedances.

SOLUTION

As shown in Figure 10.16, $Z_{in}(j500)$ can be seen as the sum of three impedances, $Z_1 + Z_2 + Z_3$. Our approach is to first calculate Z_i for each i.

Step 1. *Compute Z_1.* Since this is an LC series combination,

$$Z_1 = j\left(500 \times 0.005 - \frac{1}{500 \times 0.0004}\right) = -j2.5 \ \Omega$$

Step 2. *Compute $Z_2 = 1/Y_2$.* From the property that parallel admittances add and series impedances add,

$$Y_2 = j500 \times 0.0002 + \frac{1}{10 + (10 + j10)} = j0.1 + 0.04 - j0.02 = 0.04 + j0.08$$

Hence, $Z_2 = 1/Y_2 = 5 - j10 \ \Omega$.

Step 3. *Compute $Z_3 = 1/Y_3$.* Here

$$Y_3 = \sqrt{0.02} \ e^{j(\pi/4)} = j\frac{1}{500 \times 0.01} = 0.1 + j0.1 - j0.2 = 0.1 - j0.1$$

Hence, $Z_3 = 5 + j5 \ \Omega$.

Step 4. *Compute Z_{in}.* Adding the three impedances together yields

$$Z_{in} = Z_1 + Z_2 + Z_3 = -j2.5 + 5 - j10 + 5 + j5 = 10 - j7.5 \ \Omega$$
$$= 12.5 \ \angle{-36.87°} \ \Omega$$

Calculations performed in this example are most easily done with an advanced calculator or in MATLAB. For example, in MATLAB the command for computing Z_3 is "Z3 = 1/(sqrt(0.02)*exp(j*pi/4) - j/(500*0.01))."

EXAMPLE 10.8. Compute the input impedance $Z_{in}(j\omega)$ of the ideal op amp circuit of Figure 10.17.

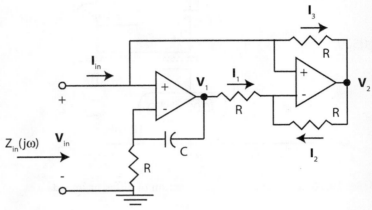

FIGURE 10.17 Op amp circuit called an impedance converter.

SOLUTION

The trick to solving this problem entails full use of the ideal op amp properties discussed in Chapter 4.

Step 1. From the properties of an ideal op amp, from KVL, and from Ohm's law,

$$\mathbf{V}_2 = \mathbf{V}_{in} - R\mathbf{I}_3 = \mathbf{V}_{in} - R\mathbf{I}_{in} \tag{10.30}$$

This follows because the voltage across the input terminals of each ideal op amp is zero and no current enters the + or − terminal of each ideal op amp. This implies that $\mathbf{I}_{in} = \mathbf{I}_3$.

Step 2. Using the phasor voltage division formula of equation 10.25, it follows that

$$\mathbf{V}_{in} = \frac{R}{R + \dfrac{1}{j\omega C}}\mathbf{V}_1$$

or, equivalently,

$$\mathbf{V}_1 = \left(1 + \frac{1}{j\omega RC}\right)\mathbf{V}_{in} \tag{10.31}$$

Here, of course, because of the idealized properties of the op amp, the voltage $\mathbf{V}_{in}$ appears across the resistor R in the leftmost op amp.

Step 3. Writing a node equation at the inverting terminal of the rightmost op amp yields

$$\frac{1}{R}\left(\mathbf{V}_1 - \mathbf{V}_{in}\right) = -\frac{1}{R}\left(\mathbf{V}_2 - \mathbf{V}_{in}\right)$$

again by the properties of an ideal op amp. Simplifying this equation yields

$$2\mathbf{V}_{in} = \mathbf{V}_1 + \mathbf{V}_2 \tag{10.32}$$

Step 4. Substituting equations 10.30 and 10.31 into equation 10.32 yields

$$2\mathbf{V}_{in} = \mathbf{V}_{in} + \frac{\mathbf{V}_{in}}{j\omega RC} + \mathbf{V}_{in} - R\mathbf{I}_{in}$$

Equivalently,

$$Z_{in}(j\omega) = \frac{\mathbf{V}_{in}}{\mathbf{I}_{in}} = j\omega R^2 C \tag{10.33}$$

Equation 10.33 suggests that the op amp circuit of Figure 10.17 can replace a grounded inductor whose impedance is $j\omega L$ with proper choice of R and C, i.e., $L = R^2 C$. In integrated circuit technology it is not possible to build a wire-wound inductor. Instead, inductors are "simulated" by circuits such as that of Figure 10.17.

The next section continues to develop our skill with and deepen our understanding of the phasor technique by computing the steady-state responses of various circuits.

8. STEADY-STATE CIRCUIT ANALYSIS USING PHASORS

This section presents a series of examples that illustrate various aspects of the phasor technique. Our purpose is not only to demonstrate how to compute the SSS, but also to illustrate the phasor counterparts of Thevenin equivalents, nodal analysis, and mesh analysis. Our first example reconsiders the parallel *RL* circuit of Example 10.4, together with the series *RC* circuit of Example 10.6. We will demonstrate the superiority of the phasor technique over the methods presented in sections 3 and 4.

EXAMPLE 10.9. Compute the steady-state voltage $v_C(t)$ for the circuit of Figure 10.18 when $i_s(t)$ = $\cos(100t)$ A.

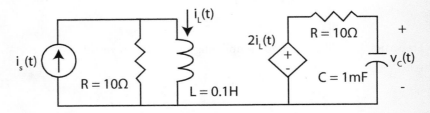

FIGURE 10.18 *RLC* circuit for Example 10.9.

SOLUTION

Step 1. *Determine* $\mathbf{I}_L$. Since the phasor $\mathbf{I}_s = 1\angle 0^\circ$ A, by current division,

$$\mathbf{I}_L = \frac{R}{R + j\omega L}\mathbf{I}_s = \frac{1}{1 + j\dfrac{\omega L}{R}}\mathbf{I}_s = \frac{1}{1 + j} = \frac{1}{\sqrt{2}}\angle -45^\circ \text{ A} \tag{10.34}$$

Step 2. *Use equation 10.34 and voltage division on the RC part of the circuit to compute* $\mathbf{V}_C$. *Using voltage division and equation 10.34, the capacitor voltage phasor is*

$$\mathbf{V}_C = \frac{\dfrac{1}{j\omega C}}{R + \dfrac{1}{j\omega C}}(2\mathbf{I}_L) = \frac{1}{1 + j\omega RC}(2\mathbf{I}_L) = \frac{1}{1 + j}\times\frac{2}{\sqrt{2}}\angle -45^\circ = 1\angle -90^\circ \text{ V} \tag{10.35}$$

Step 3. *Determine* $v_C(t)$. Converting the phasor $\mathbf{V}_C$ of equation 10.35 to its corresponding time function yields

$$v_C(t) = \cos(100t - 90^\circ) = \sin(100t) \text{ V}$$

The next example illustrates voltage division with phasors as well as the basic impedance relationships.

EXAMPLE 10.10. Consider the circuit in Figure 10.19 where R_1 = 5 Ω, C = 0.1 F, R_2 = 4 Ω, L = 2 H, and $v_s(t)$ = 10 $\cos(2t)$ V. Find $v_C(t)$ and $i_L(t)$ in steady state.

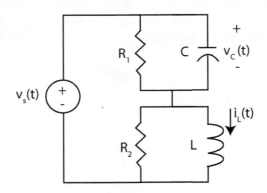

FIGURE 10.19 Pseudo parallel *RLC* circuit.

SOLUTION

Step 1. *Find* $Z_{RC}(j2)$. $Y_{RC}(j2) = \dfrac{1}{R} + j\omega C = 0.2 + j0.2 = 0.2\sqrt{2}\angle 45^o$. Hence

$$Z_{RC}(j2) = \frac{1}{Y_{RC}(j2)} = 2.5\sqrt{2}\angle -45^o = 2.5 - j2.5$$

Step 2. *Find* $Z_{Rl}(j2)$. $Y_{RL}(j2) = \dfrac{1}{R} + \dfrac{1}{j\omega L} = 0.25 - j0.25 = 0.25\sqrt{2}\angle -45^o$. Hence

$$Z_{RL}(j2) = \frac{1}{Y_{RL}(j2)} = 2\sqrt{2}\angle 45^o = 2 + j2$$

Step 3. *Find* $\mathbf{V}_C$ *and* $v_C(t)$. From voltage division

$$\mathbf{V}_C = \frac{Z_{RC}(j2)}{Z_{RC}(j2) + Z_{RL}(j2)}\mathbf{V}_s = \frac{2.5 - j2.5}{2.5 - j2.5 + 2 + j2}10 = 10\frac{2.5 - j2.5}{4.5 - j0.5} = 7.809\angle -38.66^o$$

It follows that $v_C(t) = 7.809 \cos(2t - 38.66^\circ)$ V.

Step 4. *Find* $\mathbf{V}_L$, $\mathbf{I}_L$, *and* $i_L(t)$. From step 3,

$$\mathbf{V}_L = \mathbf{V}_s - \mathbf{V}_C = 10 - 7.809 \angle -38.66^\circ = 10 - (6.098 - j4.878) = 3.9024 + j4.878$$

Hence

$$\mathbf{I}_L = \frac{\mathbf{V}_L}{j\omega L} = \frac{3.9024 + j4.878}{j4} = 1.5617\angle -38.66^o$$

Thus $i_L(t) = 1.5617\cos(2t - 38.66^\circ)$ A.

The next example illustrates the computation of a Thevenin equivalent circuit with the aid of nodal analysis. Because impedances may be manipulated in the same manner as resistances and admittances in the same manner as conductances, the Thevenin theorem, the source transformation theorem (Chapter 5), and node and mesh analysis (Chapter 3) carry over directly.

EXAMPLE 10.11

(a) Find the Thevenin equivalent of the circuit of Figure 10.20 if ω = 4 rad/sec.

(b) Determine the voltage $v_{RL}(t)$ when a 1.2 Ω load resistor is connected across terminals a and b.

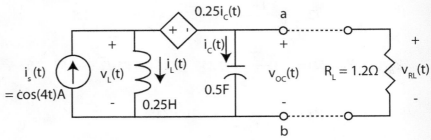

FIGURE 10.20 *LC* circuit for Example 10.11.

SOLUTION

Find the Thevenin equivalent circuit, and then using the Thevenin equivalent, find $v_L(t)$.

Step 1. *Establish nodal equation.* A nodal equation at the left node of Figure 10.20 in terms of phasors is given by

$$\mathbf{I}_s = \frac{1}{j\omega L}\mathbf{V}_L + j\omega C\mathbf{V}_{OC} = -j\mathbf{V}_L + 2j\mathbf{V}_{OC}$$

$$(10.36)$$

Step 2. *Determine the relationship between $\mathbf{V}_L$ and $\mathbf{V}_{oc}$.* The relationship between $\mathbf{V}_L$ and $\mathbf{V}_{oc}$ as determined by the dependent source is

$$\mathbf{V}_L - \mathbf{V}_{oc} = 0.25[j2\mathbf{V}_{oc}]$$

Equivalently,

$$\mathbf{V}_L = (1 + 0.5j)\mathbf{V}_{oc} \qquad\qquad 10.37)$$

Step 3. *Substitute equation 10.37 into equation 10.36.* Substituting yields

$$\mathbf{I}_s = (0.5 + j)\mathbf{V}_{oc}$$

Solving for $\mathbf{V}_{oc}$ with $\mathbf{I}_s = 1\angle 0^\circ$ yields

$$\mathbf{V}_{oc} = \frac{1}{0.5 + j}\mathbf{I}_s = (0.4 - j0.8)\mathbf{I}_s = 0.894\angle -63.43^\circ \text{ V} \qquad (10.38)$$

Step 4. *Compute the Thevenin equivalent impedance $Z_{th}(j4)$.* Consider the circuit of Figure 10.21, which is the phasor version of Figure 10.20 with the output terminals short-circuited. Hence, the short-circuit current phasor is

$$\mathbf{I}_{sc} = 1\angle 0^\circ \text{ A}$$

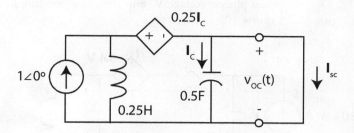

FIGURE 10.21 Phasor version of Figure 10.20 with short-circuited terminals.

Therefore, from equation 10.38,

$$Z_{th}(j4) = \frac{\mathbf{V}_{oc}}{\mathbf{I}_{sc}} = (0.4 - j0.8) \ \Omega$$

Step 5. *Interpret Z_{th} to generate the Thevenin equivalent circuit.* To physically interpret the Thevenin equivalent impedance, consider that

$$Z_{th}(j4) = (0.4 - j0.8) = (R_{th} + 1/j4C) \ \Omega$$

Thus, $R_{th} = 0.4 \ \Omega$ and $C = 0.3125$ F. Hence, the desired Thevenin equivalent circuit (valid at $\omega = 4$ rad/sec) has the form sketched in Figure 10.22.

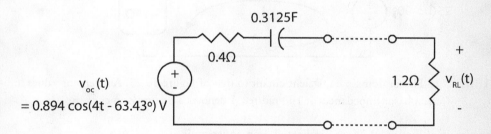

FIGURE 10.22 Thevenin equivalent of Figure 10.20.

Step 6. *Compute $v_{RL}(t)$ by voltage division.* Using voltage division on the circuit of Figure 10.22,

$$\mathbf{V}_{R_L} = \frac{1.2}{1.2 + (0.4 - j0.8)} \mathbf{V}_{oc} = (0.6 + j0.3)(0.894 \angle 63.43°)$$

$$= 0.6 \angle -36.87° \ \text{V}$$

Converting the load voltage phasor to its corresponding time-domain sinusoid yields

$$v_{R_L}(t) = 0.6 \cos(4t - 36.87°) \ \text{V}$$

EXAMPLE 10.12. Determine the phasor voltage V_x and the corresponding time function $v_x(t)$ for the circuit of Figure 10.23 if $\omega = 100$ rad/sec.

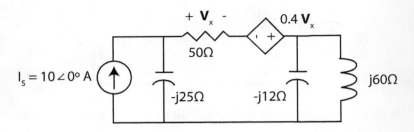

FIGURE 10.23 Phasor domain circuit for Example 10.12. All element values indicate phasor impedances at 100 rad/sec.

SOLUTION
To solve this problem, it is convenient to execute a source transformation on the independent current source and to combine the impedances of the parallel combination of the capacitor and inductor on the right-hand side of the circuit. After executing these two manipulations, one obtains the new circuit of Figure 10.24.

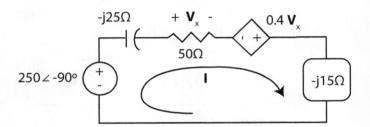

FIGURE 10.24 Phasor domain equivalent circuit to that of Figure 10.23. All element values indicate phasor impedances at 100 rad/sec. **I** denotes a phasor loop current.

For the circuit of Figure 10.24, the indicated loop equation is

$$250\angle{-90^\circ} = (50 - j25)\mathbf{I} - 0.4(50\mathbf{I}) - j15\mathbf{I} = (30 - j40)\mathbf{I}$$

Solving for **I** yields
$$\mathbf{I} = 4 - j3 = 5\angle{-36.87^\circ} \text{ A}$$

Consequently, $\mathbf{V}_x = 50\mathbf{I} = 250\angle{-36.87^\circ}$ V and $v_x(t) = 250\cos(100t - 36.87^\circ)$ V.

9. INTRODUCTION TO THE NOTION OF FREQUENCY RESPONSE

The **frequency response** of a circuit is the graph of the ratio of the phasor output to the phasor input as a function of frequency, i.e., as the frequency varies over some specified range. Since the phasor input and the phasor output are complex numbers, the frequency response consists of two plots: (1) a graph of the *magnitude* of the phasor ratio and (2) a graph of the *angle* of the phasor ratio. Such graphs indicate the magnitude change and the angle change imposed on a sinusoidal input to produce a steady-state output sinusoid. In steady state, the magnitude of the output sinusoid is the product of the magnitude of the input sinusoid and the magnitude of the frequency response at the frequency of the input. Similarly, the phase of the output sinusoid in steady state is the sum of the input phase and the frequency response phase at the input frequency. This property takes on greater importance once one learns that arbitrary input signals can be decomposed into infinite sums of sinusoids of different frequencies, i.e., each signal has a frequency content. This notion is made precise in a signals and systems course, where one studies Fourier series and Fourier transforms. The frequency response of a circuit describes the circuit behavior at each frequency component of the input signal. This permits one to isolate, enhance, or reject certain frequency components of an input signal and thereby isolate, enhance, or reject certain kinds of information.

EXAMPLE 10.13. Plot the frequency response of the *RC* circuit of Figure 10.25.

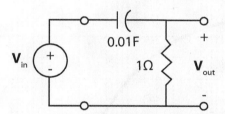

FIGURE 10.25 *RC* circuit passing high-frequency content of an input signal.

SOLUTION

Using voltage division, the ratio of the output phasor voltage $\mathbf{V}_{out}$ to the input phasor voltage $\mathbf{V}_{in}$ is given by

$$\frac{V_{out}}{V_{in}} = \frac{1}{1 + \dfrac{1}{j0.01\omega}} = \frac{j0.01\omega}{1 + j0.01\omega} = H(j\omega)$$

where we have designated this ratio as $H(j\omega)$.

The two universally important frequencies are $\omega = 0$ and $\omega = \infty$. At these frequencies, $H(j0) = 0\angle 90°$ and $H(j\infty) = 1\angle 0°$. Asymptotically then, the magnitude $|H(j\omega)| \rightarrow 1$ as $\omega \rightarrow \infty$ and

$|H(j\omega)| \to 0$ as $\omega \to 0$. With regard to angle, $\angle H(j\omega) \to 0$ as $\omega \to \infty$ and $\angle H(j\omega) \to 90^o$ as $\omega \to$ 0. Also, a close scrutiny of $H(j\omega)$ indicates that $\omega = 100$ rad/sec is also an important frequency. Here $H(j100) = 0.707\angle 45^o$. These values give us a pretty good idea what the magnitude and phase plots look like. Using a computer program, Figure 10.26a and Figure 10.26b show the exact magnitude and phase plots. These plots are consistent with our earlier asymptotic analysis.

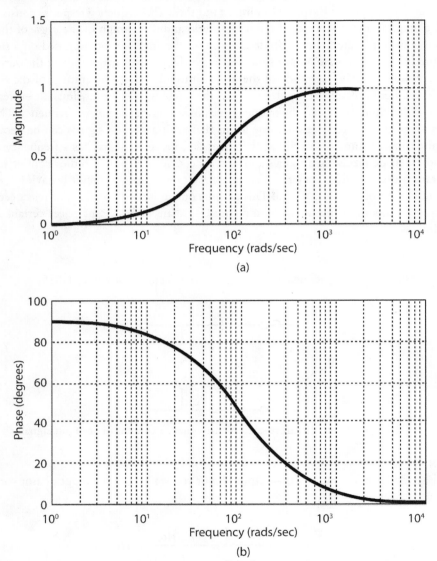

FIGURE 10.26 (a) Magnitude plot of frequency response for Example 10.13.
(b) Phase plot of frequency response.

Do these frequency responses make sense? They should. Going back to the circuit, observe that at $\omega = 0$, the capacitor impedance is infinite. Physically, then, in steady state, the capacitor looks like an open circuit for dc, i.e., at zero frequency. The magnitude plot bears this out. For frequencies close to zero, the capacitor approximates an open circuit and, hence, the magnitude remains small. On the other hand, for large frequencies, the capacitor has a very small impedance. This means that most of the source voltage appears across the output resistor. The gain then approximates 1, as indicated by the magnitude plot. The frequency response of the circuit is such that the high-frequency content of the input signal is passed while the low-frequency content of the input signal is attenuated. Such circuits are commonly called **high-pass circuits**.

EXAMPLE 10.14. Investigate the frequency response of the parallel *RLC* circuit of Figure 10.27.

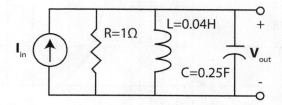

FIGURE 10.27 A parallel *RLC* circuit having a band-pass frequency response.

SOLUTION
The input admittance of the circuit of Figure 10.27 is given by

$$Y_{in}(j\omega) = \frac{1}{R} + \frac{1}{j\omega L} + j\omega C = \frac{\dfrac{1}{LC} - \omega^2 + j\dfrac{\omega}{RC}}{j\dfrac{\omega}{C}}$$

Inverting to obtain the input impedance yields

$$Z_{in}(j\omega) = \frac{j\dfrac{\omega}{C}}{\dfrac{1}{LC} - \omega^2 + j\dfrac{\omega}{RC}} = \frac{j4\omega}{100 - \omega^2 + j4\omega}$$

Clearly, $\mathbf{V}_{out} = Z_{in}(j\omega)\mathbf{I}_{in}$. Hence the ratio of the output phasor to the input phasor is simply $Z_{in}(j\omega)$. Once again, $\omega = 0$ and $\omega = \infty$ are the first two frequencies to look at. Here $Z_{in}(0) = 0\angle 90^{\circ}$ and $Z_{in}(\infty) = 0\angle -90^{\circ}$. Also at $\omega = 10$, the impedance is real, i.e., $Z_{in}(j10) = 1$. These three points provide a rough idea of the magnitude and phase response. Two more points are necessary for a real sense of the frequency response. At what frequency or frequencies does the magnitude drop to 0.707 of its maximum value or when does the phase angle equal $\pm 45^{\circ}$? This will occur when $|100 - \omega^2| = |4\omega|$. This is a quadratic equation. However, because of the absolute values, there are two implicit quadratics, $\omega^2 - 4\omega - 100 = 0$ and $\omega^2 + 4\omega - 100 = 0$. Solving using the quadratic formula yields $\omega = \pm 8.2, \pm 12.2$. Since the magnitude plot is symmetric with respect to the vertical axis ($\omega = 0$ axis), we consider only the positive values of ω. This information provides a good idea of the magnitude and phase plots. A computer program was used to generate the frequency response plots in Figure 10.28a (magnitude) and Figure 10.28b (phase). The magnitude plot shows that frequencies satisfying $8.2 \leq \omega \leq 12.2$ are passed with little attenuation. Frequencies outside this region are attenuated significantly. Such a characteristic is said to be of the band-pass type, and the corresponding circuit is a **band-pass circuit.**

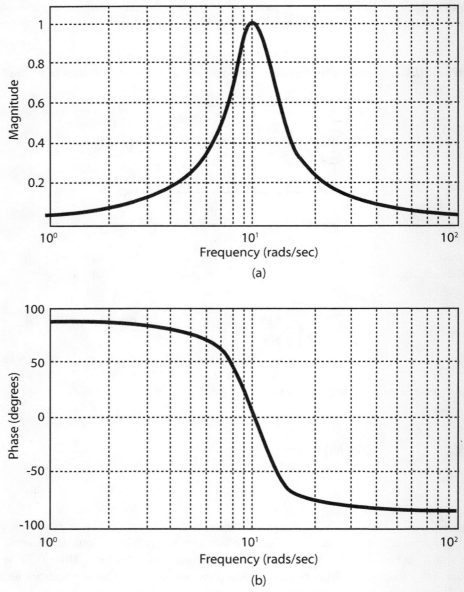

FIGURE 10.28 (a) Magnitude plot of frequency response for band-pass circuit of **Figure** 10.27. (b) Phase plot of frequency response.

EXAMPLE 10.15. As a final example we consider the so-called band-reject circuit of Figure 10.29. A **band-reject** circuit is the opposite of a band-pass circuit. A band-reject circuit has a band of frequencies that are significantly attenuated while it passes with little to no attenuation those frequencies outside the band. In this example our goal is to compute the magnitude and phase of the frequency response of the band-reject circuit of Figure 10.29.

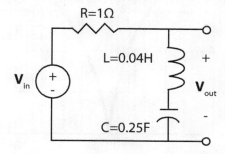

FIGURE 10.29 Band-reject circuit for Example 10.15.

SOLUTION

Once again using voltage division, we obtain the phasor ratio

$$\frac{\mathbf{V}_{out}}{\mathbf{V}_{in}} = \frac{\dfrac{1}{LC} - \omega^2}{\dfrac{1}{LC} - \omega^2 + j\omega\dfrac{R}{L}} = \frac{100 - \omega^2}{100 - \omega^2 + j25\omega} = H(j\omega)$$

At $\omega = 0$ and $\omega = \infty$, $H(j\omega) = 1\angle 0°$. Hence, asymptotically, $|H(j\omega)|$ approaches 1 as ω approaches 0 and ∞. Also at $\omega = 10^-$, $H(j\omega) = 0\angle-90°$ while at $\omega = 10^+$, $H(j\omega) = 0\angle-270° = 0\angle 90°$. For this example, to find the frequencies where $|H(j\omega)|$ drops to $1/\sqrt{2}$ of its maximum value of 1, it is necessary to equate the magnitudes of the real and imaginary parts of the denominator. This produces two quadratics whose positive roots are $\omega = 3.5078$ and $\omega = 28.5078$. At these frequencies the angles of $H(j\omega)$ are $-45°$ and $45°$, respectively. The computer-generated plots of Figures 10.30a and 10.30b are, of course, consistent with these quickly computed values.

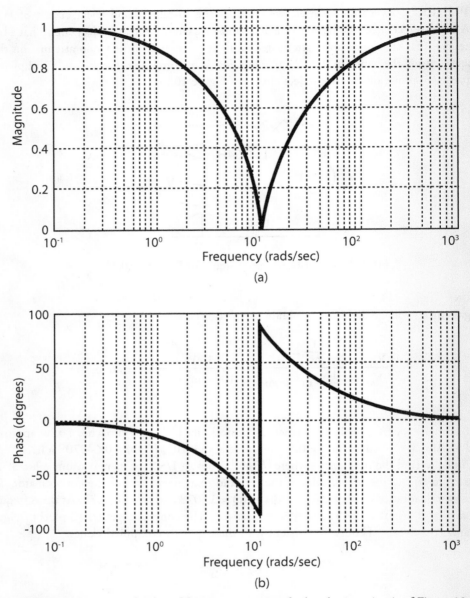

FIGURE 10.30 (a) Magnitude plot of frequency response for band-reject circuit of Figure 10.29. (b) Phase plot of frequency response.

As we can see, a wealth of different kinds of frequency response are obtainable by different interconnections of resistors, inductors, and capacitors. Historically, phasor techniques were the essential tool for the analysis and design of such circuits. Nowadays, engineers ordinarily use either MATLAB or SPICE to obtain frequency response plots. Two examples follow where we use MATLAB, SPICE, or both to obtain the frequency response.

EXAMPLE 10.16. Compute the frequency response of the circuit of Figure 10.31 using MAT-LAB and SPICE.

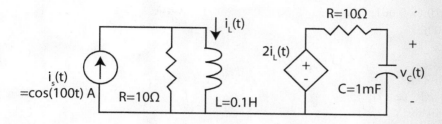

FIGURE 10.31 *RLC* circuit for Example 10.16.

SOLUTION

This circuit was originally analyzed in Example 10.9. You might want to refer to that example before proceeding.

SPICE Part. A SPICE simulation produces the result shown in Figure 10.32.

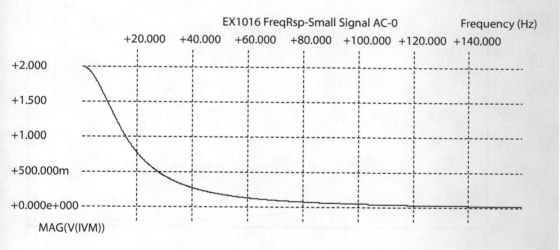

FIGURE 10.32 SPICE plot of capacitor voltage for the circuit of Figure 10.31.

MATLAB Part. Although the analysis appears in Example 10.9, we can use MATLAB to more easily obtain the frequency response. First define $Z_1(j\omega) = j\omega L$ and $Z_2(j\omega) = 1/j\omega C$. Then from current division,

$$\mathbf{I}_L(j\omega) = \frac{R}{R + Z_1(j\omega)} \mathbf{I}_s(j\omega)$$

and from voltage division,

$$\mathbf{V}_C(j\omega) = \frac{Z_2(j\omega)}{R + Z_2(j\omega)} 2\mathbf{I}_L(j\omega) = \frac{1}{R Y_2(j\omega) + 1} 2\mathbf{I}_L(j\omega)$$

Assuming a frequency range of $0 \leq \omega \leq 1000$ rad/sec, the following MATLAB code will result in a suitable magnitude frequency response plot, as shown in Figure 10.33.

```
»L = 0.1;R=10;C=0.001;
»w = 0:1:1000;
»Z1 = j*w*L;
»Y2 = j*w*C;
»IL = R ./(R+Z1);
»VC = 2*IL ./(R*Y2 +1);
»plot(w/(2*pi),abs(VC),'b')
»grid
»xlabel('Frequency in Hz')
»ylabel('Capacitor voltage (V)')
```

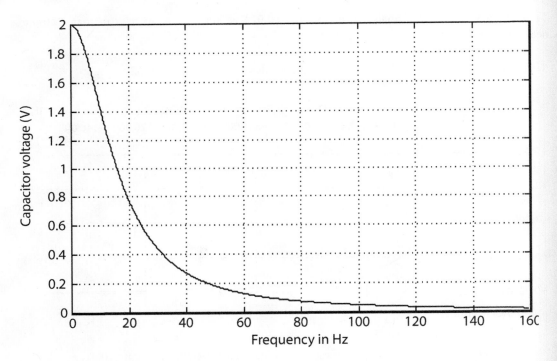

FIGURE 10.33 Magnitude plot of frequency response of capacitor voltage in the circuit of Figure 10.31. The response is of the low-pass type.

Now suppose the inductor in the circuit of Figure 10.31 is replaced by a capacitor $C_1 = 1$ mF with the controlling current changed to $i_{C1}(t)$. The frequency response is easily computed with a single change to the MATLAB code, namely, "Z_1 = 1. ./(j*w*0.001)." The resulting plot shows a band-pass characteristic, as illustrated in Figure 10.34.

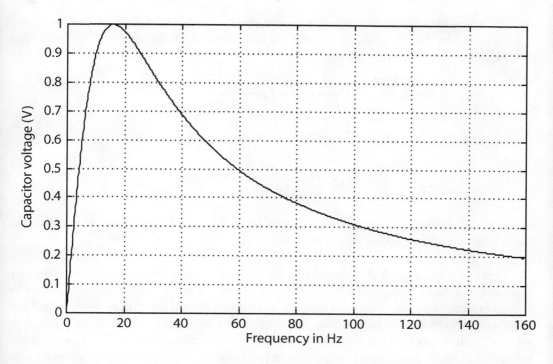

FIGURE 10.34 Magnitude plot of frequency response of capacitor voltage in circuit of Figure 10.31 when inductor is replaced by a 1 mF capacitor. The response is of the band-pass type.

EXAMPLE 10.17. In Chapter 9 we investigated the Wien bridge op amp oscillator circuit, redrawn in B2 Spice in Figure 10.35.

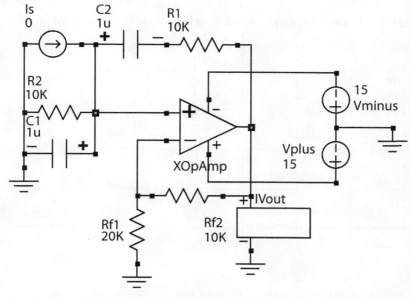

FIGURE 10.35 Wien bridge oscillator circuit drawn in B2 Spice.

Two differences are notable: (1) there is a current source present across the R_2-C_1 combination, and (2) R_1 is now 10 kΩ, as opposed to 9.5 kΩ in Example 9.14, forcing $R_1 = R_2$. This means that the characteristic equation for the circuit is

$$s^2 + bs + c = s^2 + \left(\frac{1}{R_2 C} - \frac{1}{R_1 C}\right)s + \frac{1}{R_1 R_2 C^2} = s^2 + \frac{1}{R_1 R_2 C^2} = 0$$

which indicates a purely sinusoidal oscillation at the frequency

$$f_0 = \frac{\omega_0}{2\pi} = \frac{1}{2\pi\, R_1 C} = 15.92 \text{ Hz}$$

for any initial condition on C_1. In fact one might recall that $R_1 < R_2$ causes a growing oscillation that is limited by the saturation effects of the op amp.

The current source, set at 1 A, is present in Figure 10.35 so that we can obtain the frequency response curve shown in Figure 10.36. In Figure 10.36 observe that the magnitude response peaks at f_0, as expected from the theoretical analysis. In an actual circuit, the current source would not be present. Nevertheless, a sustained sinusoidal oscillation will occur because of the presence of noise. Without going into the analysis, noise contains an infinite number of frequency components, each of which has a minute magnitude. In particular, noise contains frequency components around f_0 that drive the circuit into oscillation. This is precisely what the peak in the frequency response means: a very small (noise) voltage on C_1 will cause a very large-magnitude sinusoid output voltage at f_0. However, the presence of nonlinearities such as saturation keep the magnitude at an acceptable level.

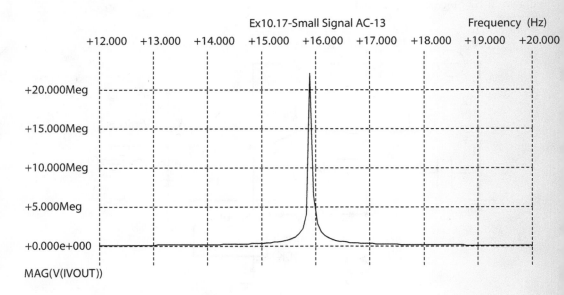

FIGURE 10.36 Frequency response plot of Wien bridge oscillator.

10. NODAL ANALYSIS OF A PRESSURE-SENSING DEVICE

The bridge circuit presented in Figure 10.37, or some variation of this bridge circuit, has been and continues to be a widely used approach to accurate measurement technology. In this section we will analyze the ac bridge circuit of Figure 10.37 as a pressure measurement device. The capacitance C_2 is a diaphragm capacitor consisting of a hollow cylinder capped on either side by fused quartz wafers. Between the wafers is a vacuum. The capacitance of the diaphragm changes with temperature and pressure. For our analysis we will assume that the temperature is constant and that the pressure is constant for a time period greater than five times the longest time constant of the circuit. This will allow the voltages and currents in the circuit to reach steady state and thus allow us to use phasor analysis to compute their values.

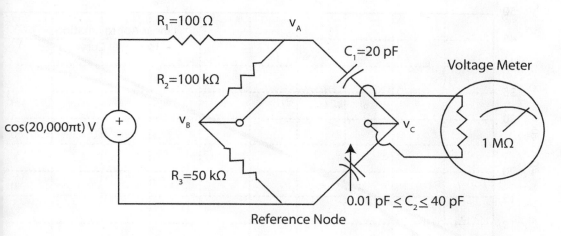

FIGURE 10.37 Bridge circuit diagram of pressure-sensing device. The capacitance C_2 changes as a function of pressure, which causes the voltage $\mathbf{V}_B - \mathbf{V}_C$ to change as a function of pressure. This is registered on the attached voltage meter, which has a 1 MΩ internal impedance.

As a rule of thumb, the capacitance $C_2 \approx 0.224 KA/d$. This means that the capacitance is inversely proportional to the distance d between the plates and proportional to the area A of the plates and to the dielectric constant K of the material between the plates. Increasing the pressure on the diaphragm decreases the distance d between the wafers, increasing the capacitance. Conversely, a decrease in pressure will increase the distance between the wafers, thereby decreasing the capacitance. As the capacitance changes, the magnitude of the ac voltage appearing across the voltage meter will vary accordingly. Hence, two relationships are necessary: (1) the relationship between the capacitance C_2 and the magnitude of the voltage $\mathbf{V}_B - \mathbf{V}_C$, and (2) the relationship between the pressure applied to the diaphragm and the associated capacitance. Our first task will be to specify the relationship between the pressure applied to the diaphragm and the resulting capacitance. Following this, we will use nodal phasor analysis to determine the magnitude of $\mathbf{V}_B - \mathbf{V}_C$ and finally, the relationship between pressure and the magnitude $\mathbf{V}_B - \mathbf{V}_C$.

Pressure is measured in various units. Millimeters of mercury (mm Hg) is a common standard; 1 mm Hg = 1 torr, and 760 torr = 1 atmosphere (atm), where 1 atm is the pressure of the earth's atmosphere at sea level, which supports 760 mm of mercury in a special measuring tube. Suppose it has been found experimentally that the capacitance C_2 (in pF) varies as a function of pressure according to the formula

$$C_2(\Delta P) = C_0 + K \log_{10}\left(\frac{P_0 + \Delta P}{P_0}\right)$$

(10.39)

$$= 26.5 + 68 \log_{10}\left(\frac{760 + \Delta P}{760}\right)$$

A plot of C_2 as a function of ΔP is given in Figure 10.38.

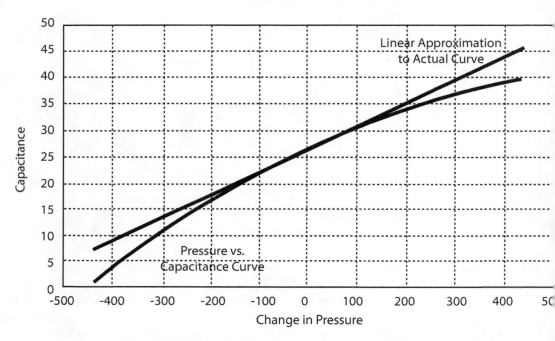

FIGURE 10.38 Plot of capacitance versus pressure.

Our next task is to develop the relationship between the capacitance of the bridge circuit and the magnitude of the phasor voltage $\mathbf{V}_B - \mathbf{V}_C$. In our analysis, $G_1 = (R_1)^{-1}$, $G_2 = (R_2)^{-1}$, $G_3 = (R_3)^{-1}$, and $G_m = 10^{-6}$ S is the conductance of the meter . According to Figure 10.37, $C_1 = 20$ pF. We will let C_2 range as $0 \leq C_2 \leq 40$ pF. Finally, $\omega = 2\pi \times 10^4$ rad/sec. The following phasor analysis will be done symbolically so as not to obscure the methodology.

Summing the phasor currents leaving node A leads to the phasor voltage relationship

$$(G_1 + G_2 + j\omega C_1)\mathbf{V}_A - G_2\mathbf{V}_B - j\omega C_1\mathbf{V}_C = G_1 15$$

Similarly, summing the currents leaving node B leads to the relationship

$$-G_2\mathbf{V}_A + (G_2 + G_3 + G_m)\mathbf{V}_B - G_m\mathbf{V}_C = 0$$

Finally, summing the currents leaving node C produces

$$-j\omega C_1\mathbf{V}_A + j\omega(C_1 + C_2 + G_m)\mathbf{V}_C - G_m\mathbf{V}_B = 0$$

Writing these three equations in matrix form yields

$$\begin{bmatrix} G_1 + G_2 + j\omega C_1 & -G_2 & -j\omega C_1 \\ -G_2 & G_2 + G_3 + G_m & -G_m \\ -j\omega C_1 & -G_m & G_m + j\omega(C_1 + C_2) \end{bmatrix}\begin{bmatrix} \mathbf{V}_A \\ \mathbf{V}_B \\ \mathbf{V}_C \end{bmatrix} = \begin{bmatrix} 15\,G_1 \\ 0 \\ 0 \end{bmatrix} \quad (10.40)$$

The matrix on the left is said to be a nodal admittance matrix. Its entries can be real or complex, as indicated. It is not advisable to solve such a set of equations by hand over the range of possible C_2 values. However, using MATLAB one can solve this matrix equation over the range 0 pF $\leq C_2$ $\leq$ 40 pF to produce the plot of Figure 10.39.

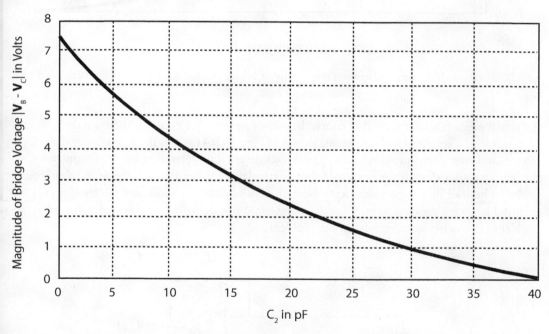

FIGURE 10.39 Plot of the magnitude of the phasor voltage $\mathbf{V}_B - \mathbf{V}_C$ as a function of capacitance.

Of course, one could measure the voltage appearing across the meter, from Figure 10.39 determine the associated value of C_2, refer to Figure 10.38 for ΔP, and then determine $P = 760 + \Delta P$. This is a long route. To complete our analysis, then, we need to develop the relationship between pressure and bridge voltage. As we have the relationship between C_2 and ΔP and the relationship between C_2 and $|\mathbf{V}_B - \mathbf{V}_C|$, it is a matter of using equation 10.39 to derive the value of C_2 in equation 10.40. This is best done with a simple MATLAB routine, which yields the plot given in Figure 10.40.

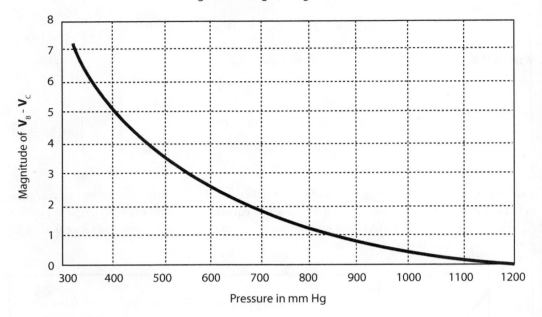

FIGURE 10.40 Relationship between magnitude of bridge output voltage and pressure applied to diaphragm capacitor C_2.

An actual pressure sensor would, of course, be more complex. For example, there would probably be a differential amplifier such as the one shown in Figure 10.41 across the terminals of the bridge circuit, and this would probably drive a peak (ac) detector to determine the maximum value of the ac signal appearing at the output of the differential amplifier. Further, the peak value would probably be read by a digital voltmeter. Nevertheless, our analysis illustrates the basic principles involved in such a measurement. Of course, one could just as easily use loop analysis to solve the problem. This is left as an exercise in the problems.

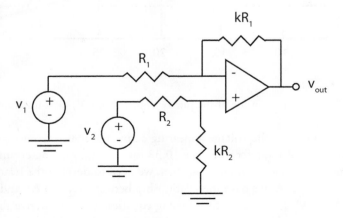

FIGURE 10.41 Differential amplifier having output voltage $v_{out} = k(v_2 - v_1)$.

Exercise. Prove that $v_{out} = k(v_2 - v_1)$ for the differential amplifier of Figure 10.41.

11. SUMMARY

The two primary goals of this chapter were (1) to develop the phasor technique for the analysis of circuits having a sinusoidal steady state and (2) to illustrate how this technique leads to the idea of a circuit frequency response, which characterizes the circuit's behavior in response to the frequency content of an input excitation. In the development, sinusoids were first represented as the real part of a complex sinusoid. As a motivation for the delineation of the phasor method, we showed how the complex sinusoids could be utilized to compute the sinusoidal steady-state response using differential equation circuit models. We then pointed out that a complex (voltage or current) sinusoid is specified by a complex number or phasor representing its magnitude and phase. After introducing the notions of impedance and admittance for the capacitor, the inductor, the resistor, and a general two-terminal circuit element, we showed how the phasor voltage and phasor current for each such element satisfy a frequency-dependent Ohm's law. This allowed us to adapt the analysis techniques and network theorems of Chapters 1 through 6 to the steady-state analysis of circuits excited by sinusoidal inputs. For example, there are voltage division formulas, current division formulas, source transformations, and Thevenin and Norton theorems all valid for phasor representations. This permits us to effectively analyze circuits that have a steady-state response.

The phasor technique opens a door to seeing how circuits behave in response to sinusoids. Given that input excitations are composed of different frequency sinusoids, such as a music signal, phasor analysis shows why a circuit will behave differently toward the different frequencies present in the input signal. This fact prompts the notion of a circuit's frequency response, which is defined as the ratio of the phasor output to the phasor input excitation as a function of ω in the single-input, single-output case. The frequency response consists of two plots. The magnitude plot shows the gain magnitude of the circuit's response to sinusoids of different frequencies, and the phase plot shows the phase shift the circuit introduces to sinusoids of different frequencies. The notion of frequency response will be generalized in Chapter 14 after we introduce the notion of the Laplace transform.

12. TERMS AND CONCEPTS

Admittance: of a two-terminal device, the ratio of the phasor current into the device to the phasor voltage across the device, $Y_{in}(j\omega) = \dfrac{1}{Z_{in}(j\omega)} = \dfrac{\mathbf{I}_{in}}{\mathbf{V}_{in}}$.

Band-pass circuit: circuit in which frequencies within a specified band are passed while frequencies outside the band are attenuated.

Band-reject circuit: circuit in which one band of frequencies is significantly attenuated while those frequencies outside the band are passedes with little to no attenuation.

Complex exponential forcing function: function of the form $v(t) = \mathbf{V}e^{st}$, where $\mathbf{V} = V_m e^{j\phi}$ and $s = \sigma + j\omega$ are complex numbers. A special case ($\sigma = 0$), $v(t) = V_m e^{j(\omega t + \phi)}$, is used throughout the chapter as a shortcut for sinusoidal steady-state analysis.

Conductance: real part of a possibly complex admittance.

Current division: in a parallel connection of admittances driven by a current source, the current through a particular branch is proportional to the ratio of the admittance of the branch to the total parallel admittance.

Euler identity: $e^{j\theta} = \cos(\theta) + j\sin(\theta)$.

Frequency: in a sinusoidal function $A\cos(\omega t + \theta)$ or $B\sin(\omega t + \theta)$, the quantity ω is the angular frequency in radians per second (rad/sec). Equivalently, $A\cos(\omega t + \theta) = A\cos(2\pi f t + \theta)$, where f is the frequency in hertz (Hz or cycles per second). Note that $\omega = 2\pi f$.

Frequency response: (of a circuit) graph of the ratio of the phasor output to the phasor input as a function of frequency. It consists of two parts: (1) a graph of the magnitude of the phasor ratio and (2) a graph of the angle of the phasor ratio.

High-pass circuit: circuit with a frequency response such that the high-frequency content of the input signal is passed while the low-frequency content of the input signal is attenuated.

Imaginary part: the imaginary part of a complex number $z = a + jb$ for real numbers a and b, denoted by Im[z], is b.

Impedance: ordinarily complex frequency-dependent Ohm's law–like relationship of a two-terminal device, defined as $Z(j\omega) = \mathbf{V}/\mathbf{I}$, where $\mathbf{V}$ is the phasor voltage across the device and $\mathbf{I}$ is the phasor current through the device. For the resistor, $Z_R(j\omega) = R$; for the capacitor, $Z_C(j\omega) = 1/(j\omega C)$; and for the inductor, $Z_L(j\omega) = j\omega L$.

Magnitude (modulus): the magnitude of a complex number $z = a + jb$, denoted by $|z|$, is

$$\sqrt{a^2 + b^2}.$$

Phasor: complex number representation denoting sinusoidal signals at a fixed frequency. Boldface capital letters denote phasor voltages or currents; a typical voltage phasor is $\mathbf{V} = V_m\angle\phi$ and a typical current phasor is $\mathbf{I} = I_m\angle\phi$.

Polar coordinates: representation of a complex number z as $\rho e^{j\theta}$, where $\rho > 0$ is the magnitude of z and θ is the angle z makes with respect to the positive horizontal (real) axis of the complex plane.

Reactance: imaginary part of an impedance.

Real part: real part of a complex number $z = a + jb$ for real numbers a and b, denoted by Re[z], is a.

Rectangular coordinates: representation of a complex number z as coordinates in the complex plane, i.e., as $a + jb$ for real numbers a and b.

Resistance: real part of a possibly complex impedance.

Sinusoidal steady-state response: response of a circuit to a sinusoidal excitation after all transient behavior has died out. This definition presumes that the zero-input response of the circuit contains only terms that have an exponential decay.

Stable circuit: circuit such that any zero-input response consists of decaying exponentials or exponentially decaying sinusoids.

Susceptance: imaginary part of an admittance.

Voltage division: in a series connection of impedances driven by a voltage source, the voltage appearing across any one of the impedances is proportional to the ratio of the particular impedance to the total impedance of the connection.

Zero-input response: response of the circuit when all source excitations are set to zero.

---

[1] In the literature, both $\bar{z}$ and z^* are used to denote the conjugate of a complex number z. However, in matrix arithmetic, Z^* usually means the conjugate transpose of the matrix Z. We will sometimes interchange the usage. In MATLAB, * means multiplication and conj(Z) means conjugated. So there is some ambiguity in the usage.

Problems

SOLUTION OF DIFFERENTIAL EQUATIONS WITH COMPLEX EXPONENTIALS

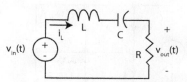

Figure P10.3

ANSWERS: 0.8 sin$(2500t - 36.86°)$, 80 sin$(2500t - 36.86°)$

1. Construct the differential equation model of the series RL circuit of Figure P10.1 in which $L = 0.25$ H and $R = 100\ \Omega$. Then use the method of section 4 to compute the steady-state response when $v_{in}(t) = 20\sqrt{2}\cos(400t)$ V.

4. Construct the differential equation model of the parallel RLC circuit of Figure P10.4 for $R = 100\ \Omega$, $C = 1\ \mu$F, and $L = 40$ mH. Then use the method of section 4 to find the steady-state response when $i_{in}(t) = 20\cos(2500t)$ mA.

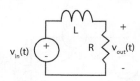

Figure P10.1

ANSWER: $20\cos(400t - \pi/4)$ V

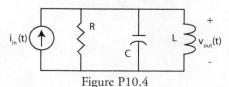

Figure P10.4

ANSWER: $1.6\cos(2500t + 36.87°)$ V

2. Find the differential equation model of the series RC circuit of Figure P10.2 in terms of $v_{in}(t)$ and $v_C(t)$ assuming that $C = 5\ \mu$F and $R = 800\ \Omega$. Write $v_{out}(t)$ as a function of $v_{in}(t)$ and $v_C(t)$. Then use the method of section 4 to determine the steady-state response when $v_{in}(t) = 20\sqrt{2}\sin(250t)$ V.

KCL AND KVL WITH PHASORS

5. Find the phasor current $\mathbf{I}$ and $i(t)$ for each circuit of Figure P10.5 when $\omega = 100\pi$ rad/sec.

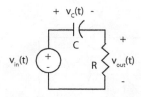

Figure P10.2

ANSWER: $20\sin(250t + 0.25\pi)$ V

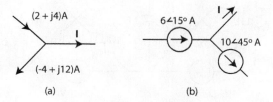

Figure P10.5

ANSWERS: (a) $10\cos(100\pi t - 0.927)$ A, (b) $5.6626\cos(100\pi t - 1.798)$ A

3. Construct the differential equation model of the series RLC circuit of Figure P10.3 in terms of $i_L(t)$ and $v_{in}(t)$, assuming $L = 10$ mH, $C = 4$ μF, and $R = 100\ \Omega$. Then use the method of section 4 to find the steady-state response $i_L(t)$ when $v_{in}(t) = 100\sin(2500t)$ V. Next compute $v_{out}(t)$.

6. The circuit of Figure P10.6 operates in the sinusoidal steady state with the indicated phasor currents when $i_s(t) = 10\cos(1000t)$ A. Find the value of the phasor currents and and the associated $i_1(t)$ and $i_2(t)$.

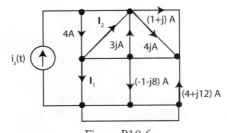

Figure P10.6

CHECK: $i_1(t) = 25 \cos(100t + 0.9273)$ A

7. Suppose that in Figure P10.7, $v_1(t) = 4 \cos(\omega t)$ V and $v_2(t) = 4\sqrt{2} \cos(\omega t - 0.25\pi)$ V. Find $v_L(t) = K \cos(\omega t + \phi)$.

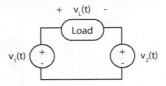

Figure P10.7

ANSWER: $v_L(t) = 4 \cos(\omega t + 90°)$ V

8. Use KVL to determine the phasor voltage $\mathbf{V}_x$ in the circuit of Figure P10.8.

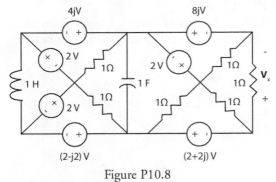

Figure P10.8

ANSWER: −4 V

9. For the circuits of Figures P10.9a, b, and c, compute the indicated phasor currents assuming $R = 2\ \Omega$, $L = 4$ mH, and $C = 1$ mF. If $\omega = 500$ rad/sec, determine the associated time functions.

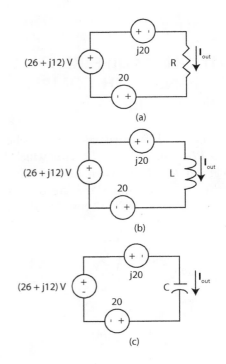

(a)

(b)

(c)

Figure P10.9

10. For the circuit of Figure P10.10, use KCL and KVL to find the phasor voltage $\mathbf{V}_x$ and the phasor current $\mathbf{I}_y$. If the frequency $\omega = 2000\pi$ rad/sec, find the associated voltage and current time functions.

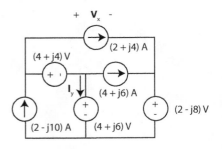

Figure P10.10

ANSWERS: $\mathbf{V}_x = 18.98\angle7156°$ V, $\mathbf{I}_y = 20.4\angle -101.3°$ A

BASIC IMPEDANCE AND ADMITTANCE CONCEPTS

11. (a) A capacitor has an admittance $Y_C = j8$ mS at $\omega = 4000$ rad/sec. Find the

value of C in μF. Compute the value of the capacitor's impedance at $\omega = 500$ rad/sec.

(b) An inductor has an impedance $Z_L = j20$ Ω at $\omega = 4000$ rad/sec. Find the value of L in mH. Compute the value of the inductor's admittance at 10,000 rad/sec.

12. In the circuit of Figure P10.12, $v_{in}(t) = 10 \cos(10t)$ V, $C = 0.2$ F, $L = 0.1$ H, $R = 2.5$ Ω. Determine the phasors $\mathbf{I}_1$, $\mathbf{I}_2$, and $\mathbf{V}_{out}$. Find $v_{out}(t)$.

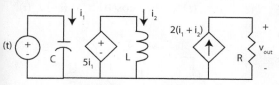

Figure P10.12

13. The circuit of Figure P10.13 is operating in the sinusoidal steady state with $R_1 = R_2 = 20$ Ω, $C = 1$ mF.

(a) Suppose $i_{s1}(t) = 10 \cos(100t + 30°)$ mA and $v_{s2}(t) = 200 \cos(100t)$ mV. Find the phasor $\mathbf{I}_x$ and then the current $i_x(t)$.

(b) Now let $i_{s1}(t) = 10 \cos(50t + 30°)$ mA and $v_{s2}(t) = 200 \cos(100t)$ mV. Find the current $i_x(t)$. How does this part differ from part (a)?

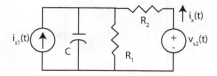

Figure P10.13

14. Find the phasor currents $\mathbf{I}_R$, $\mathbf{I}_C$, $\mathbf{I}_L$, and $\mathbf{I}_{in}$ and then determine $i_{in}(t)$ for the circuit of Figure P10.14 in which $R = 1$ kΩ, $L = 0.5$ H, $C = 1$ μF, and $v_{in}(t) = 20 \cos(1000t + 60°)$ V.

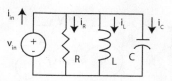

Figure P10.14

ANSWERS: $\mathbf{I}_R = 0.02\angle60°$ A, $\mathbf{I}_L = 0.04\angle -30°$ A, $\mathbf{I}_C = 0.02\angle150°$ A, $i_{in}(t) = 0.0283 \cos(1000t + 15°)$ A

15. Consider the circuit of Figure P10.15 where $R = 200$ Ω, $L = 80$ mH, $C = 1$ μF and $i_{in}(t) = 100 \sin(2500t)$ mA. Find the voltage phasors $\mathbf{V}_R$, $\mathbf{V}_L$, $\mathbf{V}_C$ and $\mathbf{V}_{in}$ and then compute $v_{in}(t)$.

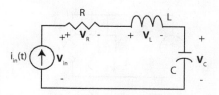

Figure P10.15

CHECK: $v_{in}(t) = 28.28 \cos(2500t - 135°)$ V or $v_{in}(t) = 28.28 \sin(2500t - 45°)$ V

16. In the circuit of Figure P10.16, $G = 0.03$ S, $L = 0.1$ H, $C = 0.4$ mF, $i_1(t) = 1.2 \cos(200t)$ A and $v_{out}(t) = 40 \sin(200t)$ V.

(a) Find the phasors $\mathbf{I}_1$ and $\mathbf{V}_{out}$.

(b) Find the phasor $\mathbf{I}_2$ and the associated time function $i_2(t)$.

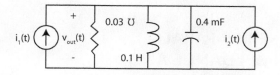

Figure P10.16

ANSWER: (b) $1.2\angle-90°$, $1.2 \sin(200t)$ A

17. In the circuit of Figure P10.17, $R = 6$ Ω, $L = 80$ mH, $C = 0.5$ mF, $v_{s2}(t) = 8 \cos(200t)$ V, and $\mathbf{I}_1 = 0.5\angle90°$ A. Find the source voltage, $v_{s1}(t)$, which operates at the same frequency of 200 rad/sec.

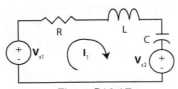

Figure P10.17

ANSWER: 5.8311 cos(200t +30.96°) V

18. The circuit of Figure P10.18 operates in the sinusoidal steady state at a frequency of ω_0 = 2000 rad/sec, R_1 = R_2 = 10 Ω, V_{in} = 50 V, and I_z = 2∠ − 53.13° A. Compute the phasor voltage across R_2 and then find the impedance $Z(j\omega_0)$. Now construct a simple series circuit that represents this impedance at ω_0.

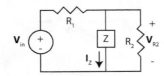

Figure P10.18

ANSWER: Z = 2.5 + j10 Ω

19. (a)   Find the steady-state response of the circuit of Problem 3 using the phasor method. Discuss the relative advantages of the phasor method.

   (b)   Find the steady-state response of the circuit of Problem 14 using the phasor method. Discuss the relative advantages of the phasor method.

SERIES-PARALLEL IMPEDANCE AND ADMITTANCE CALCULATIONS

20. Consider the circuit of Figure P10.20.
   (a)   If C = 0.01 F, find $Z_{in}(j100)$.
   (b)   If $Z_{in}(j100)$ = 25j Ω, find the appropriate value of C.
   (c)   Using the impedance of part (b), if $i_{in}(t)$ = 100 cos(100t + 45°) mA, find $v_C(t)$.

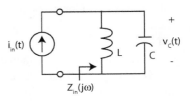

$Z_{in}(j\omega)$

Figure P10.20 Parallel LC circuit.

ANSWERS: (a) −j1.25 Ω, (b) 1.6 mF

21. Consider the circuit of Figure P10.21.
   (a)   Find the impedance at ω = 100 rad/sec.
   (b)   What happens to the impedance as ω gets large?
   (c)   If $v_{in}(t) = 10\sqrt{2}$ cos(100t)V, find $i_{in}(t)$.

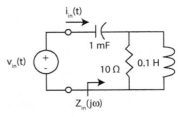

$Z_{in}(j\omega)$

Figure P10.21

ANSWER: (a) 5 − j5 Ω

22. For the circuit of Figure P10.22, suppose R = 100 Ω, L = 0.5 H, C = 5 μF.
   (a)   If $i_{in}(t)$ = 0.1 cos(500t) A, find $v_C(t)$.
   (b)   Find $\omega \neq 0$ in rad/sec so that the input admittance is real.

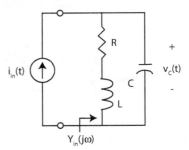

$Y_{in}(j\omega)$

Figure P10.22

23. Consider the circuit of Figure P10.23 in which R_1 = 20 Ω, R_2 = 10 Ω, L = 20 mH.
   (a)   If C = 0.3 mF, find $Y_{in}(j500)$.
   (b)   Find the value of C that makes the input admittance real at ω = 500 rad/sec.

(c) If $C = 0.3$ mF and $i_{in}(t) = 100$ $\cos(500t)$ mA, find $i_{R1}(t)$ and $i_C(t)$ using current division.

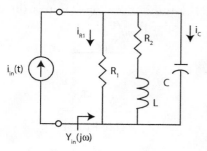

Figure P10.23

ANSWERS: (a) $0.1 + j0.1$ S, (b) 0.1 mF

24. Consider the circuit of Figure P10.24 in which $C = 1$ F. At $\omega = 2$ rad/sec, $Z_{in}(j2) = 4 + j2$ Ω.

(a) Find the appropriate values of R and L.

(b) For $0 \le L < 0.2$ H, specify the range of possible reactance values for $Z_{in}(j2)$.

(c) If $v_{in}(t) = 10 \cos(2t)$ V, find $v_R(t)$.

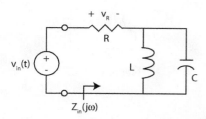

Figure P10.24

ANSWERS: 4 Ω, 0.2 H, 0 to ∞

25. Consider the circuits of Figure P10.25 in which $R = 5$ Ω, $L = 32$ mH, and $C = 5$ μF.

(a) Consider Figure P10.25a. Find $Z_{in}(j\omega)$ as a function of ω. Then compute the frequency at which $Z_{in}(j\omega)$ is purely real, i.e., the reactance is zero. Determine the minimum value of $|Z_{in}(j\omega)|$.

(b) For the circuit of Figure P10.25b, find $Y_{in}(j\omega)$ as a function of ω. Then compute the frequency at which $Y_{in}(j\omega)$ is purely real, i.e., the susceptance is

zero. Determine the minimum value of $|Y_{in}(j\omega)|$.

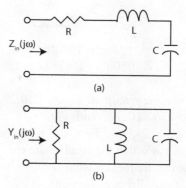

Figure P10.25

26. For a particular two-terminal device, $Y_{in}(j\omega) = 0.002 + j0.002$ at $\omega = 500$ rad/sec. Construct a parallel RC circuit having this admittance at $\omega = 500$ rad/sec. If the circuit is excited by a current source with $i_s(t) = 10$ $\cos(500t)$ mA, find the voltage appearing across the current source.

ANSWER: $C = 4$ μF, $R = 500$ Ω.,
$$v_s(t) = \frac{5}{\sqrt{2}}\cos\left(500t - 45^o\right) \text{ V}$$

27. For a particular two-terminal device, $Z_{in}(j1000) = 2000 + j2000$ Ω. Construct a series RL circuit having this impedance at $\omega = 1000$ rad/sec. If the circuit is excited by a voltage source with $v_s(t) = 10 \cos(1000t)$ V, find the current through the resistor.

ANSWER: $L = 2$ H, $R = 2$ kΩ.,
$$i_s(t) = \frac{5}{\sqrt{2}}\cos\left(1000t - 45^o\right) \text{ mA}$$

28. The circuit of Figure P10.28 operates in the sinusoidal steady state at the frequency $\omega = 5000$ rad/sec with $R = 4$ Ω, $L = 0.4$ mH, and $C = 0.1$ mF. Find $Z_{in}(j5000)$ and $Y_{in}(j5000)$. Construct a simple series circuit that is equivalent to this circuit at $\omega = 5000$ rad/sec. Finally, construct a simple parallel circuit that is equivalent to this circuit at $\omega = 5000$ rad/sec. In both cases specify the element values.

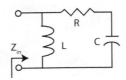

Figure P10.28

CHECK: $Z_{in}(j5000) = 1 + j2\ \Omega$

29. For the circuit of Figure P10.29, let $L = 4$ mH, $C = 10\ \mu F$, and $v_s(t) = V_s \cos(\omega t)$ V. Compute $Z_{in}(j\omega)$. Find the frequency ω (in rad/sec) at which the steady-state current $i_s(t) = 0$. At this frequency, what is the voltage across the LC parallel combination?

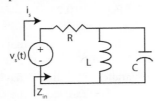

Figure P10.29

ANSWER: $\omega = 5000$ rad/sec

SERIES/PARALLEL IMPEDANCES WITH V/I DIVISION

30. Consider the circuit of Figure P10.30 in which $R = 20\ \Omega$, $L = 4$ H, and $i_{in}(t) = 10\sqrt{2}\cos(5t)$ mA.
   (a) Find the input impedance $Z_{in}(j\omega)$ and the input admittance $Y_{in}(j\omega)$.
   (b) At $\omega = 5$ rad/sec determine the steady-state current $i_L(t)$.

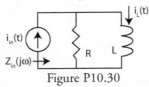

Figure P10.30

ANSWERS: (a) $\dfrac{j20\omega}{5 + j\omega}$, $0.05 - j\dfrac{0.25}{\omega}$;

(b) $10\cos(5t - \pi/4)$ mA

31. (a) For the circuit of Figure P10.31, find the ratio $\mathbf{V}_{out}/\mathbf{V}_{in}$ in terms of R, C, and ω. Express the answer in polar form.

(b) Find the value of ω in terms of R and C at which the phase angle difference between $\mathbf{V}_{in}$ and $\mathbf{V}_{out}$ is $45°$.

(c) At the ω computed in part (b), determine $|\mathbf{V}_{out}/\mathbf{V}_{in}|$.

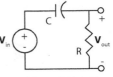

Figure P10.31

32. Consider the circuit of Figure P10.32.
   (a) If $v_s(t) = V_m \cos(t/RC)$, find the sinusoidal steady-state response $v_C(t)$ in terms of V_m, R, and C.
   (b) If $R = 10\ \Omega$ and $v_s(t) = 10\sqrt{2}\cos(10t)$ V, find the value of C so that $v_C(t) = 10\cos(10t + \theta)$ V.
   (c) For the value of C found in part (b), compute the corresponding value of θ.

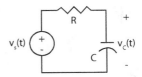

Figure P10.32

ANSWER: (a) $v_C(t) = \dfrac{V_m}{\sqrt{2}}\cos\left(\dfrac{t}{RC} - 45^o\right)$

33. Consider the circuit of Figure P10.33 in which $R = 8\ \Omega$, $L = 8$ mH, and $C = 0.125$ mF.
   (a) Determine the values of the phasors $\mathbf{I}_R$, $\mathbf{I}_L$, and $\mathbf{V}_C$ when $\mathbf{I}_{in} = 2$ A and $\omega = 1000$ rad/sec. Specify the corresponding time functions.
   (b) Repeat part (a) for $\omega = 500$ rad/sec.

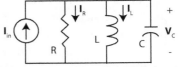

Figure P10.33 Parallel RLC circuit.

34. In Figure P10.33, suppose $R = 500\ \Omega$, $L = 0.5$ mH, $C = 0.125$ mF, and $i_{in}(t) = 10\sqrt{2}\cos(\omega t + 60^o)$ A.

(a) Compute the values of the phasors $\mathbf{I}_R$, $\mathbf{I}_L$, and $\mathbf{V}_C$ when $\mathbf{I}_{in}$ = 2 A and ω = 4000 rad/sec. Specify the corresponding time functions.

(b) Repeat part (a) for ω = 8000 rad/sec.

35. In the circuit of Figure P10.35, suppose R = 20 Ω, L = 0.5 H, and C = 0.625 mF.

(a) If $v_{in}(t)$ = 16 cos(40t) V, find $v_R(t)$, $v_C(t)$ and $v_L(t)$ using phasor voltage division.

(b) If $v_{in}(t)$ = -32sin(40t) V, find $v_R(t)$, $v_C(t)$, and $v_L(t)$ using phasor voltage division. Hint: avoid repeating the calculations of part (a); this can be done by inspection.

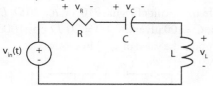

Figure P10.35 Series RLC circuit.

36. Reconsider Figure 10.35 for R = 10 Ω, L = 0.08 H, and C = 0.02 F.

(a) If $v_{in}(t)$ = 10 cos(25t) V, find $v_R(t)$, $v_C(t)$, and $v_L(t)$ using phasor voltage division.

(b) If $v_{in}(t)$ = 16 sin(50t) V, find $v_R(t)$, $v_C(t)$, and $v_L(t)$ using phasor voltage division.

37. In the circuit of Figure P10.37, R_1 = 20 Ω, L = 2 H, and R_2 = 10 Ω. If $|\mathbf{V}_{out}/\mathbf{V}_{in}|$ = 0.2 at ω= 40 rad/sec, find the necessary value(s) of C (in mF).

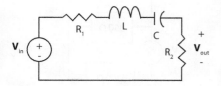

Figure P10.37

ANSWER: 0.625 mF or 0.2083 mF

38. In the circuit of Figure P10.38, R_1 = 50 Ω, C_1 = 1 μF, R_2 = 300 Ω, and C_2 = 0.625 μF.

Using the phasor method, find $v_{out}(t)$ when $v_{in}(t)$ = 50 cos(4000t) mV.

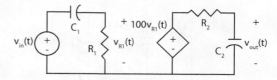

Figure P10.38 Two coupled RC circuits.

39. Consider the circuit of Figure P10.39 in which R_1 = 200 Ω, L = 0.2 H, R_2 = 200 Ω, and C = 0.05 μF. Use the phasor method to find $i_C(t)$ when $i_{in}(t)$ = 10 cos(10^4t) mA.

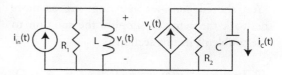

Figure P10.39 Two coupled circuits.

40. Consider the circuit of Figure P10.40 in which R_1 = 500 Ω, L = 0.125 H, R_2 = 100 Ω, and C = 5 μF. Suppose $v_{in}(t)$ = 120 cos(400t) V. Find $v_C(t)$ and $i_L(t)$.

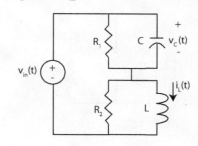

Figure P10.40

41. Consider the circuit of Figure P10.41 in which R_1 = 500 Ω, L = 0.125 H, R_2 = 100 Ω, and C = 5 μF. Suppose $i_{in}(t)$ = 120 cos(400t) mA. Find $v_C(t)$ and $i_L(t)$.

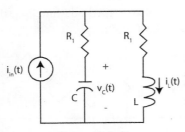

Figure P10.41

NETWORK THEOREMS IN CONJUNCTION WITH V/I DIVISION.

(You should consider applying one or more network theorems to simplify the solution to the problems in this section.)

42. In the circuit of Figure P10.42, R_1 = 60 Ω, R_2 = 40 Ω, and C = 0.1 mF. Find the phasor $\mathbf{I}_1$ and the corresponding steady-state current $i_1(t)$ when $i_s(t) = 5\sqrt{2}\cos(100t)$ mA. This problem can be solved by direct current division or by source transformation and impedance concepts. Which method is easier?

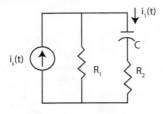

Figure P10.42

43. In the circuit of Figure P10.43, ω = 400 rad/sec, R_1 = R_2 = 2 Ω, L = 5 mH, and C = 625 μF. Find $\mathbf{I}_C$ and the corresponding $i_C(t)$ in steady state. $v_s(t)$ = 12 cos(400t) V

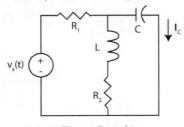

Figure P10.43

ANSWERS: 2∠90°, −2 sin(400t) A

44. For the circuit of Figure P10.44, find the phasor current $\mathbf{I}_L$ and the phasor voltage $\mathbf{V}_C$. If the circuit is known to operate at a frequency of ω = 1000 rad/sec, find $i_L(t)$ and $v_C(t)$, and the values of L and C.

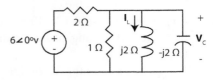

Figure P10.44

ANSWERS: $v_C(t)$ = 2 cos(1000t) V, $i_L(t)$ = cos (1000t − 0.5π) A

45. In the circuit of Figure P10.45, R = 20 Ω, L = 20 mH, C = 100 μF, and $v_s(t) = 20\sqrt{2}\cos(1000t)$V. Compute the value of $v_C(t)$ in steady state.

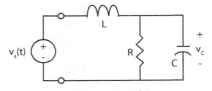

Figure P10.45

ANSWER: 20 cos(1000t − 135°) V

46. Consider the linear circuit of Figure P10.46, which operates at 50 Hz and for which $\mathbf{V}_L = a\mathbf{V}_{s1} + b\mathbf{I}_{s2}$.
   (a) Find the values of a and b.
   (b) If $v_{s1}(t)$ = 10 cos(100πt) V and $i_{s2}(t)$ = 200 sin(100)t mA, find $v_L(t)$.

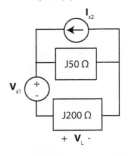

Figure P10.46

ANSWERS (in random order):
−0.8, j40 Ω, -16 cos(100πt) V

47. The *linear* circuit of Figure P10.47 is such that in the *steady state*, if $i_{s1}(t) = 10 \cos(200\pi t)$ A with $v_{s2}(t) = 0$, then $v_1(t) = 20 \cos(200\pi t + 45°)$ V. On the other hand if $i_{s1}(t) = 0$ with $i_{s2}(t) = 10 \cos(200\pi t + 45°)$ V, then $v_1(t) = 5 \cos(200)t + 90°)$ V.

    (a)   Find a linear relationship between i_{s1}, v_{s2}, and v_1.

    (b)   If $i_{s1}(t) = 5 \cos(200\pi t - 45°)$ A and $v_{s2}(t) = 20 \cos(200\pi t)$ V, then in the steady state find $v_1(t)$.

    (c)   Find Z_1 and Z_2. Develop simple circuit realizations of these impedances valid at $\omega = 200\pi$ rad/sec.

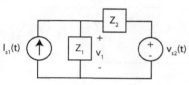

Figure P10.47

ANSWERS: (b) $v_1(t) = 18.46 \cos(200\pi t - 22.5°)$ V; (c) $Z_1 = 0.763 + j2.605 \ \Omega$, $Z_2 = 4 \ \Omega$

48. The circuit of Figure P10.48 is a general Wheatstone bridge circuit (the dc version of which is described in Problem 35 of Chapter 2). Here the circuit is used to measure the value of the unknown inductance L.

    (a)   Suppose $R_s = 0$. Show that the steady-state voltage $v_{out}(t) = 0$ when $R_C R_L = L/C$. *Note*: In general the condition for a null voltage, $v(t) = 0$, in the steady state is that the products of the cross impedances be equal.

    (b)   Again suppose $R_s = 0$. You are given that $R_C C = 2$ sec and that the voltage source $v_{in}(t)$ is a sinusoid with a frequency of 5 rad/sec. With the unknown inductance L inserted in the circuit as shown, you adjust R_L until you reach a sinusoidal steady-state voltage null, $v_{out}(t) = 0$ V. The resulting value for R_L is 3 Ω. Find the value of L.

    (c)   Now suppose $R_s \neq 0$. Show that the condition of part (a) is still valid. You

can do this with some straightforward reasoning without writing any equations.

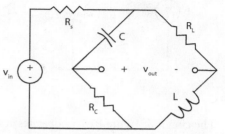

Figure P10.48 AC Wheatstone bridge circuit.

THEVENIN AND NORTON EQUIVALENTS

49. Find the Thevenin equivalent for the circuit of Figure P10.49 when $R = 4 \ \Omega \ L = 20$ mH, $C = 1.25$ mF, $\mathbf{I}_{in} = 2\angle45°$ A and $\omega = 200$ rad/sec. Be sure to express the open-circuit voltage as a time function.

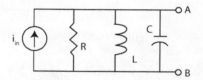

Figure P10.49 Parallel *RLC* circuit.

ANSWERS: $Z_{th} = 4 \ \Omega$, $v_{oc}(t) = 8 \cos(200t + 45°)$ V

50. For the circuit of Figure P10.50, let $L = 10$ mH, $R = 20 \ \Omega$, $C = 20$ μF, and $\mathbf{I}_{in} = 100\angle0°$ mA.

    (a)   Find the Thevenin equivalent at the terminals a and b if $\omega = 2000$ rad/sec.

    (b)   If the circuit is terminated with a load consisting of a series connection of a 20 Ω resistor and a 20 mH inductor, find the sinusoidal steady-state voltage across the load.

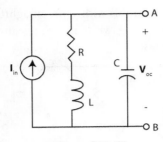

Figure P10.50

51. The circuit of Figure P10.51 operates at ω = 20 krad/sec and $\mathbf{V}_s = 2\angle0°$. Find the Thevenin equivalent circuit (in the phasor domain) at terminals A and B. Use this Thevenin equivalent to find the magnitude of the phasor $\mathbf{V}_{AB}$ when the 10 mH and 1 kΩ series combination load is connected to A and B.

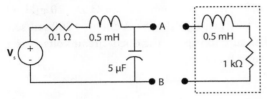

Figure P10.51

CHECK: $190 < |\mathbf{V}_{oc}| < 205$ V, $|\mathbf{V}_{AB}| \cong 0.5|\mathbf{V}_{oc}|$

52. For the circuit of Figure P10.52, R_1 = 1000 Ω, C_1 = 0.2 μF, R_2 = 500 Ω, and C_2 = 1 μF. Find the Norton equivalent circuit when $v_{in}(t)$ = 50 cos(4000t) V.

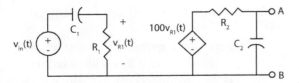

Figure P10.52 Two coupled RC circuits.

53. For the circuit of Figure P10.53 R = 2.5 kΩ. Find the Thevenin equivalent when $i_{in}(t)$ = 10 cos(4000t) mA.

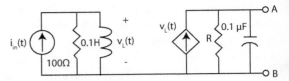

Figure P10.53 Two coupled circuits.

54. The circuit of Figure P10.54 operates in the SSS at ω_0.

(a) Find the Thevenin equivalent impedance Z_{th}.

(b) Find $v_{AB}(t)$ if $i_{in}(t) = I_m \cos(\omega_0 t)$, R = 10 Ω, $a = 5$ Ω, and X = 20 Ω.

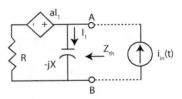

Figure P10.54

55. In the circuit of Figure P10.55, assume ω = 100 rad/sec, L = 40 mH, C = 5 mF, R = 8 Ω, and $a = 4$ Ω. Find the Thevenin impedance seen at terminals A and B. If $v_{in}(t)$ = 20 cos(100t) V, find $i_1(t)$.

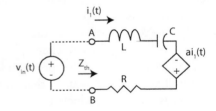

Figure P10.55

GENERAL SSS ANALYSIS (NODE OR LOOP ANALYSIS)

56. (a) Find the phasors $\mathbf{I}_L$ and $\mathbf{V}_{out}$ in the circuit of Figure P10.56 when $\mathbf{V}_{in} = 20\sqrt{2}\angle45°$ V, $R = 4$ Ω, $L = 4$ mH, and ω = 1000 rad/sec. Specify the corresponding time functions.

(b) Determine the value of ω for which the magnitude of the output voltage phasor is 20% of the magnitude of the input voltage.

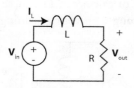

Figure P10.56

CHECKS: 5 A, 20 V

57. In the circuit of Figure P10.57, $Z_L = j30\ \Omega$, $Z_C = -j40\ \Omega$, $V_{s1} = 28$ V, $V_{s2} = 16 \angle 90°$, the sinusoidal sources have been operating for a long time, and $Z = 50 - j40\ \Omega$. Find $\mathbf{V}_Z$.

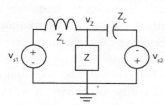

Figure P10.57

58. For the network of Figure P10.58, $a = 20$ Ω, $Z_L = j10\ \Omega$, $Z_C = -j10\ \Omega$. $R = 10\ \Omega$, and $V_S = 20$ V. Find the phasor current $\mathbf{I}_x$.

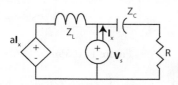

Figure P10.58

ANSWER: $0.6 + 0.2j$

59. Consider the circuit of Figure P10.59 for which $C = 0.8$ mF, $Y_{in}(j100) = 0.01 + j0.04$ S, and $v_{in}(t) = 80 \cos(100t)$ V.
   (a) Find R.
   (b) Find L.
   (c) Find $v_C(t)$.
   (d) Find $i_L(t)$.
   (e) At $\omega = 100$ rad/sec, determine the Thevenin equivalent circuit phasors $\mathbf{V}_{oc}$ and $Z_{th}(j100)$ at terminals A and B.

(f) At $\omega = 100$ rad/sec, determine the Thevenin equivalent circuit in which $Z_{th}(j100)$ is a series combination of two circuit elements seen at terminals A and B.

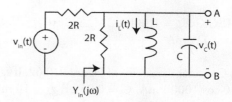

Figure P10.59

60. Consider the circuit of Figure P10.60 in which $v_s(t) = 20 \cos(1000t)$ V, $R_1 = 40\ \Omega$, $R_2 = 20\ \Omega$, $L = 20$ mH, and $C = 75$ μF.
   (a) Write and solve a nodal equation at the top node for $\mathbf{V}_C$. Then write the corresponding time-domain expression for $v_C(t)$.
   (b) Calculate I_L and then write the corresponding time-domain expression for the inductor current $i_L(t)$.
   (c) Find the Thevenin equivalent circuit at the source frequency relative to terminals A and B. Draw the Thevenin equivalent circuit showing the Thevenin impedance as a series circuit of two elements.

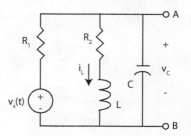

Figure P10.60

ANSWERS: $\mathbf{V}_C = 5 - j5$ V, $\mathbf{I}_L = -j0.25$, $Z_{th} = 10 - j\,10\ \Omega$

61. The circuit of Figure P10.61 operates at $\omega = 2$ krad/sec and $g_m = 1.5$ mS with $\mathbf{V}_s = 100\angle 0°$ V. Find the Thevenin and Norton equivalent circuits seen at terminals A and B.

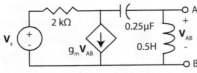

Figure P10.61

CHECK: $Z_{th} = j1000$ Ω, $\mathbf{V}_{oc} = 25 + j25$ V, $\mathbf{I}_{sc} = 25 + j25$ mA

62. Consider the circuit of Figure P10.62. If $i_s(t) = 40 \cos(1000t)$ mA, $C_1 = 0.25$ µF, $g_m = 10$ mS, $R_1 = 1$ kΩ, and $R_2 = 3$ kΩ, find the Thevenin equivalent circuit parameters $\mathbf{V}_{oc}$ and Z_{th}.

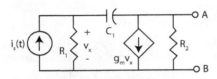

Figure P10.62

CHECK: $Z_{th} = 128 - j337.88$ Ω, $\mathbf{V}_{oc} = 140.25\angle 95.28°$ V

63. This problem tests whether you can synthesize ideas from two different parts of the text. In the circuit of Figure P10.63, $R = 20$ Ω, $L = 1$ H, $v_s(t) = 50 \cos(100t)u(t)$ V (notice the step function), and $i_L(0^+) = 1$ A. If the response for $t \geq 0$ has the form

$$i_L(t) = A \cos(100t + \phi) + Be^{-\lambda t}$$

then determine the constants A, ϕ, λ, and B.

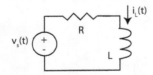

Figure P10.63

FREQUENCY RESPONSE

64. Compute the magnitude and phase functions of the frequency response of the circuit of Figure P10.64 in which $L = 4$ mH and $C = 0.25$ mF. Plot your response in MATLAB ($0 \leq \omega \leq 5000$ rad/sec). Before sketching the responses,

determine the asymptotic behavior for large ω and for at least one other frequency without a computer or calculator. List these properties in writing along with your reasoning.

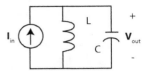

Figure P10.64

65. Compute the magnitude and phase of the frequency response of the circuit of Figure P10.65 where $L = 25$ mH and $R = 50$ Ω. Plot your response in MATLAB ($0 \leq \omega \leq 8000$ rad/sec) and determine the frequency at which the magnitude is $1/\sqrt{2}$ of its maximum value. Before sketching the responses, determine the asymptotic behavior for large ω and for at least one other frequency without a computer or calculator. List these properties in writing along with your reasoning.

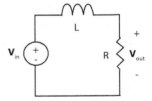

Figure P10.65

66. Inside the black box of Figure P10.66a there is a two-element circuit composed of a resistor of 10 Ω, capacitors, inductors, or some combination of these elements. A variable-frequency voltage $v_s(t) = 10\cos(\omega t)$ V is applied to the box and the voltage $v(t) = V_m \cos(\omega t + \theta)$ is observed. A plot of the magnitude of $v(t)$ with respect to ω is given in Figure P10.66b.

   (a) Draw the circuit contained inside the box. (There are two solutions.)

   (b) Specify the element values.

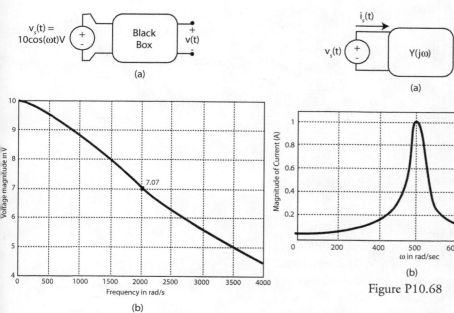

(a)

(b)

Figure P10.66

(b)

Figure P10.68

67. Compute the frequency response of the circuit of Figure P10.67, where $R = 100 \ \Omega$, $L = 10$ mH, $C = 0.1$ mF, and $v_s(t)$ is the output. Use MATLAB or its equivalent to generate the magnitude and phase (in degrees) plots. Consider $0 \leq \omega \leq 3000$ rad/sec.

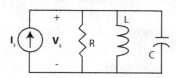

Figure P10.67

68. The box labeled $Y(j\omega)$ in Figure P10.68a contains a single resistor, a single capacitor, and a single inductor. Let $v_s(t)$ be the input excitation and $i_s(t)$ the circuit response. The magnitude frequency response is given by Figure P10.68b. Draw the circuit inside the box and assign component values if it is known that $L = 40$ mH.

69. Reconsider the pressure-sensing example of section 10. Specify a set of mesh currents and write a set of mesh equations that describe the circuit. Solve the equations for 1 pF $\leq C_2 \leq$ 40 pF using MATLAB or some other, equivalent software program. Plot the magnitude of $\mathbf{V}_B - \mathbf{V}_C$ as a function of C_2. Now construct a plot of the magnitude of $\mathbf{V}_B - \mathbf{V}_C$ as a function of pressure in mm Hg.

OP AMP CIRCUITS

70. (a) Compute $v_{out}(t)$ when $v_{in}(t) = \sin(200t)$ mV for the circuit of Figure P10.70a.

(b) For the circuit of Figure P10.70b, find C so that when $v_{in}(t) = \cos(400t)$ mV, $v_{out}(t) = \sin(400t)$ mV.

(c) Find the phasor transfer function, $H(j\omega)$, and plot the magnitude of the frequency response (using MATLAB or the equivalent) as a function of $\omega = 2\pi f$, where f is in Hz and ω in rad/sec.

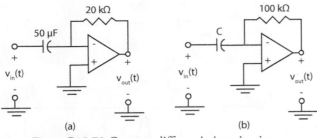

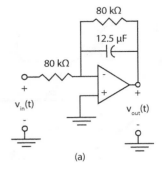

(a)

Figure P10.70 Op amp differentiation circuits.

71. (a) Compute $v_{out}(t)$ when $v_{in}(t)$ = $\sin(400t)$ V for the circuit of Figure P10.71a.
    (b) For the circuit of Figure P10.71b, find C such that when $v_{in}(t) = \sin(500t)$ V, $v_{out}(t) = 5 \cos(500t)$ V. This represents an integration of the input with gain.
    (c) Find the phasor transfer function, $H(j\omega)$, and plot the magnitude of the frequency response (using MATLAB or its equivalent) as a function of $\omega = 2\pi f$, where f is in Hz and ω in rad/sec.

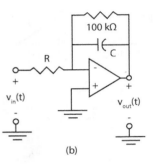

(b)

Figure P10.72 Leaky integrator circuits.

73. (a) At $\omega = 2 \times 10^4$ rad/sec, find the phasor voltage gain $\mathbf{V}_{out}/\mathbf{V}_{in}$ of the op amp circuit of Figure P10.73.
    (b) Find the phasor transfer function, $H(j\omega)$, and plot the magnitude of the frequency response as a function of $\omega = 2\pi f$, where f is in Hz and ω in rad/sec using MATLAB or equivalent software.

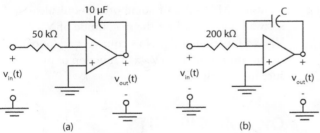

Figure P10.71 Op amp integrators.

72. (a) If an 800 Hz sine wave of unit amplitude excites the leaky integrator circuit of Figure P10.72a, determine the steady-state output voltage.
    (b) For the circuit of Figure P10.72a, find the phasor transfer function, $H(j\omega)$, and plot the magnitude of the frequency response (using MATLAB or its equivalent) as a function of $\omega = 2\pi f$, where f is in Hz and ω in rad/sec.
    (c) If the input to the circuit of Figure P10.72b is $v_{in}(t) = \cos(2000)t)$ V, determine the values of R and C so that $v_{out}(t) = 5\cos(2000\pi t + 135°)$ V.

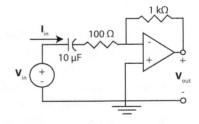

Figure P10.73

74. For the circuit of Figure P10.74, find the expression for the phasor transfer function $H(j\omega) = \mathbf{V}_o/\mathbf{V}_s$. Assume an ideal operational amplifier. Plot the magnitude of the transfer

function as a function of ω using MATLAB or the equivalent, assuming $R = 10$ kΩ and $C = 0.01$ mF.

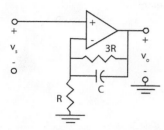

Figure P10.74

75. In the circuit of Figure P10.75 assume the operational amplifier is ideal and that $R_1 = 25$ kΩ, $C_1 = 1$ μF, $R_2 = 5$ kΩ, and $C_2 = 0.2$ μF. Compute the gain of the circuit as a function of ω. Then use MATLAB or the equivalent to plot the magnitude and phase of the frequency response as the logarithm of the frequency for $1 \leq \omega \leq 10^4$ rad/sec.

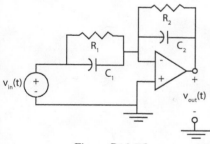

Figure P10.75

76. For the operational amplifier circuit of Figure P10.76, $R_1 = 5$ kΩ, $C_1 = 0.02$ μF, $R_2 = 5$ kΩ, and $C_2 = 0.08$ μF.

    (a) Write two node equations and solve to find a relationship between the output phasor $\mathbf{V}_{out}$ and the input phasor $\mathbf{V}_{in}$ at the frequency $f = 1000$ Hz. Note that the voltage from the minus terminal of the op amp to ground is V_{out}, which equals the voltage from the plus terminal to ground, assuming the op amp is ideal.

    (b) Repeat the calculation at $f = 100$ Hz and $f = 3000$ Hz. What happens as the

frequency increases to infinity? What happens as the frequency decreases to zero?

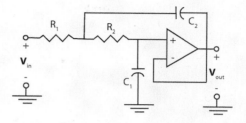

Figure P10.76 Ideal op amp circuit.

77. Consider the circuit of Figure P10.77 in which $R_1 = 200$ Ω, $C_1 = 0.05$ μF, $R_2 = 28$ kΩ, and $C_2 = 0.05$ μF.

    Use nodal analysis to compute the ratio $\mathbf{V}_{out}/\mathbf{V}_{in}$ at $f = 1.34$ kHz. Now use physical reasoning to obtain the approximate the ratio at $f = 1$ Hz and $f = 100$ kHz.

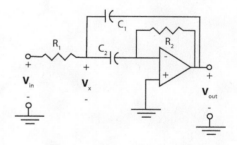

Figure P10.77 Op amp circuit having a band-pass type of response.

Note that $\mathbf{V}_x$ is an intermediary variable useful in the nodal analysis of the circuit.

CHAPTER 11

Sinusoidal Steady State Power Calculations

The AM or FM receiver that is often part of a home stereo system receives signals from radio stations through an attached antenna. The intensity of these signals or radio waves depends on the power radiated into the atmosphere by the broadcasting station, the distance between the receiving and transmitting antennas, and the design of the receiving antenna. The intensity or magnitude of the signals picked up by the receiving antenna is very small. The power available from the antenna and deliverable to the receiver is typically in the microwatt range. Again, this is very small. Hence, it is important to have maximum power transfer from the antenna to the receiver input so that the music signals received can be properly amplified and enjoyed. Since the signals in the antenna are sinusoidal at very high frequencies, the antenna is represented by a phasor Thevenin equivalent circuit as is the input circuit of the receiver. Hence we must understand maximum power transfer in the context of sinusoidal steady-state analysis to describe and analyze this problem. An example at the end of the chapter illustrates some impedance matching techniques to achieve maximum power transfer from an antenna to a receiver.

CHAPTER OBJECTIVES

1. Define and investigate the notion of average power.
2. Define the notion of the effective (rms) value of a periodic voltage or current and its relationship to the average power absorbed by a resistor.
3. Define the notion of complex power and its components—average, reactive, and apparent power—and investigate the significance of each and their relationships.
4. Introduce the notion of power factor associated with a load and describe reasons and a method for improving the power factor.
5. Prove the maximum power transfer theorem for the sinusoidal steady-state case, and illustrate its significance for the input stage of a radio receiver.

SECTION HEADINGS

1. **Introduction**
2. **Instantaneous and Average Powers**
3. **Effective Value of a Signal and Average Power**
4. **Complex Power and Its Components: Average, Reactive, and Apparent Powers**
5. **Conservation of Complex Power in the Sinusoidal Steady State**
6. **Power Factor and Power Factor Correction**
7. **Maximum Power Transfer in the Sinusoidal Steady State**
8. **Summary**
9. **Terms and Concepts**
10. **Problems**

1. INTRODUCTION

Chapter 1 defined the concept of power. The following chapters were primarily devoted to the calculation of voltages and currents. This does not mean that the consideration of power is of secondary importance. The very opposite is true. A homeowner pays for the energy used, not for voltage and current. The integral of power over, say, a 30-day period determines the household energy consumed in a month. Hidden in the homeowner's cost is an adjustment to cover the power losses incurred in transmitting energy from the generating station to the home. Thus power considerations have a significant impact on everyday life.

A second reason for understanding ac power usage is safety. Each appliance, and its cord that plugs into the wall outlet, has a maximum safe power-handling capacity. Misunderstanding such information and/or misusing an appliance can lead to equipment breakdown, fire, or some other life-threatening accident.

Even for electronic equipment in which power consumption is low, such as laptops and handheld PDAs, power consumption and, thus, battery life are important design factors. Power drainage directly determines the PDA's operating time before the battery needs recharging. In fact, optimizing power management in laptops and hybrid electric vehicles is an important research area in today's world.

In this chapter we will investigate different notions of power in ac circuits and discuss their significance and application. The term "ac circuits" has a narrow meaning here. It refers to linear circuits having all sinusoidal sources at the same frequency and consideration of responses only in steady state. The basic analysis tool is the phasor method of Chapter 10.

2. INSTANTANEOUS AND AVERAGE POWERS

Figure 11.1 shows an arbitrary two-terminal circuit element isolated from a larger circuit. With the voltage (in V) and current (in A) having indicated reference directions, the **instantaneous power** (in watts) absorbed by the element is given by equation 11.1:

$$p(t) = v(t)i(t)$$

$$(11.1)$$

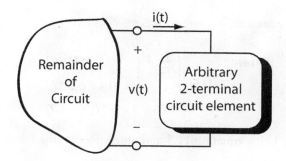

FIGURE 11.1 Instantaneous power delivered to an arbitrary two-terminal element.

Evaluating battery life or the length of operation of your cell phone involves consideration of a quantity called **average power**, P (or P_{ave} for emphasis), defined as the average value of the instantaneous power over an interval $[T_1, T_2]$. The idea is based on the average value of a function, say $f(t)$, which is defined as

$$f_{ave} = \frac{1}{T_2 - T_1} \int_{T_1}^{T_2} f(t)dt$$

Using this idea we define the average power consumed by a two-terminal element as shown in Figure 11.1 over the interval $[T_1, T_2]$ as

$$P_{ave}(T_1, T_2) = \frac{1}{T_2 - T_1} \int_{T_1}^{T_2} p(t)dt \tag{11.2}$$

When the signal is periodic with period T, we speak of the average power consumed by an element over the period T as

$$P_{ave} = \frac{1}{T} \int_0^T p(t)dt = \frac{1}{T} \int_0^T v(t)i(t)dt \tag{11.3}$$

It is not necessary that T be the fundamental period; the evaluation of the integral is the same for any integer multiple of the fundamental period.

EXAMPLE 11.1. Compute the average power absorbed by the resistor R connected to an independent voltage source as shown in Figure 11.2b with the excitation shown in Figure 11.2a.

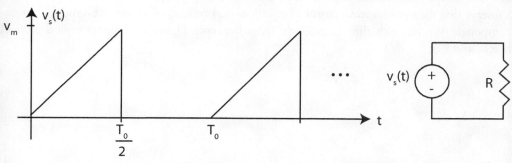

FIGURE 11.2 Triangular voltage waveform $v_s(t)$ driving resistor R.

SOLUTION

Step 1. *Compute the instantaneous power for* $0 \leq t \leq T_0$. Here

$$p(t) = \begin{cases} \dfrac{\left(\dfrac{2V_m t}{T_0}\right)^2}{R} & 0 \leq t < 0.5T_0 \\ 0 & 0.5T_0 \leq t < T_0 \end{cases}$$

Step 2. *Compute* P_{ave}. Using equation 11.3 and observing that the fundamental period is T_0, we have

$$P_{ave} = \frac{1}{T_0} \int_0^{0.5T_0} p(t)dt = \frac{1}{T_0} \int_0^{0.5T_0} \frac{4V_m^2 t^2}{T_0^2 R} dt = \frac{V_m^2}{6R}$$

Exercises. 1. Suppose the sawtooth in Figure 11.2a does not drop to zero at $t = 0.5T_0$, but rather continues to increase until reaching $t = T_0$ when it drops to zero and repeats. Find the average power consumed by R.

ANSWER: $\dfrac{4V_m^2}{3R}$

2. Show that the average power absorbed by an $R\,\Omega$ resistor in parallel with a V_0 V dc source is $\dfrac{V_0^2}{R}$ over any time interval $[T_1, T_2]$.

Of particular importance is the average power consumed by devices in the SSS assuming all excitations are at the same frequency, ω. Consequently, all voltages and currents are sinusoids *at the same frequency*. To compute the average power absorbed by a circuit element as depicted in Figure 11.1 (assuming a linear circuit), suppose $v(t) = V_m \cos(\omega t + \theta_v)$ and $i(t) = I_m \cos(\omega t + \theta_i)$. The associated **instantaneous power** is

$$p(t) = v(t)i(t) = V_m \cos(\omega t + \theta_v) \times I_m \cos(\omega t + \theta_i)$$

$$= \frac{V_m I_m}{2} \cos(\theta_v - \theta_i) + \frac{V_m I_m}{2} \cos(2\omega t + \theta_v + \theta_i)$$

(11.4)

Equation 11.4 follows from the trigonometric identity $\cos(x) \cos(y) = 0.5 \cos(x - y) + 0.5 \cos(x + y)$. Observe that the instantaneous power of equation 11.4 consists of a constant term plus another component varying with time at *twice the input frequency*. Figure 11.3 shows typical plots of $p(t)$, $v(t)$, and $i(t)$.

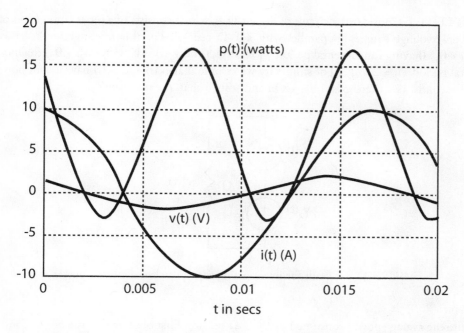

FIGURE 11.3 Plots of $i(t) = 10 \cos(377t)$ A, $v(t) = 2 \cos(377t + 45^\circ)$ V, and $p(t)$.

Using equation 11.3 with $T = 2\pi/\omega$, and observing that the integral of a sinusoid over any period is zero, we obtain the following formula for **average power** in SSS:

$$P_{ave} = \frac{1}{T}\int_0^T \frac{V_m I_m}{2}\cos(\theta_v - \theta_i)dt + \frac{1}{T}\int_0^T \frac{V_m I_m}{2}\cos(2\omega t + \theta_v + \theta_i)dt = \frac{V_m I_m}{2}\cos(\theta_v - \theta_i) \quad (11.5)$$

If the two-terminal element is a resistance R, then $v(t) = Ri(t)$ and $\theta_v - \theta_i = 0$. It follows from equation 11.5 that for a resistor

$$P_{ave,R} = \frac{V_m I_m}{2} = \frac{RI_m^2}{2} = \frac{V_m^2}{2R} \quad (11.6)$$

If the two-terminal element is an inductance L, then $\mathbf{V}_L = (j\omega L)\mathbf{I}_L$ and $\theta_v - \theta_i = 90^\circ$. Hence, $P_{ave,L} = 0$ since $\cos(\pm 90^\circ) = 0$. Similarly, if the two-terminal element is a capacitance C, then $\mathbf{I}_C = (j\omega C)\mathbf{V}_C$ and $\theta_v - \theta_i = -90^\circ$. Hence, $P_{ave,C} = 0$. *This means that the average power consumed by or delivered by a capacitor or an inductor is zero.* Even though an ideal capacitor and an ideal inductor neither consume nor generate *average* power, each may absorb or deliver a large amount of instantaneous power during some particular time interval.

Before closing this section, we need to investigate the question of superposition of average power. Is there a principle of superposition of average power? If so, when is it valid? When is it not valid? The following example provides the answers.

EXAMPLE 11.2. Consider the circuit of Figure 11.4, which consists of a series connection of two (sinusoidal) voltage sources in parallel with a 1 Ω resistor. For this investigation $v_1(t) = V_{m1} \cos(\omega_1 t + \theta_1)$ (having fundamental period $T_1 = 2\pi/\omega_1$) and $v_2(t) = V_{m2} \cos(\omega_2 t + \theta_2)$ (having fundamental period $T_2 = 2\pi/\omega_2$). For simplicity we assume that $v_1(t)$ and $v_2(t)$ have a common period of T seconds, i.e., there exist integers m and n such that $T = nT_1 = mT_2$.

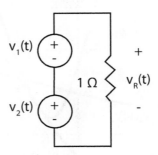

FIGURE 11.4 Circuit for investigating superposition of average power.

SOLUTION

Compute the average power consumed by the 1 Ω resistor. First observe that the power consumed by the 1 Ω resistor with source 1 acting alone, i.e., $v_2(t) = 0$, is

$$P_{ave,1} = \frac{1}{T}\int_0^T v_R(t)i_R(t)dt = \frac{1}{T}\int_0^T v_1^2(t)dt$$

Also note that the power consumed by the 1 Ω resistor with source 2 acting alone, i.e., $v_1(t) = 0$, is

$$P_{ave,2} = \frac{1}{T}\int_0^T v_R(t)i_R(t)dt = \frac{1}{T}\int_0^T v_2^2(t)dt$$

With both sources active, linearity (or KVL) implies that $v_R(t) = v_1(t) + v_2(t)$. By equation 11.3,

$$P_{ave} = \frac{1}{T}\int_0^T v_R(t)i_R(t)dt = \frac{1}{T}\int_0^T \left(v_1(t) + v_2(t)\right)^2 dt$$

$$= \frac{1}{T}\int_0^T v_1^2(t)dt + \frac{1}{T}\int_0^T v_2^2(t)dt + \frac{1}{T}\int_0^T v_1(t)v_2(t)dt$$

$$= P_{ave,1} + P_{ave,2} + \frac{1}{T}\int_0^T v_1(t)v_2(t)dt$$

$$= P_{ave,1} + P_{ave,2} + \frac{V_{m1}V_{m2}}{T}\int_0^T \cos(\omega_1 t + \theta_1)\cos(\omega_2 t + \theta_2)dt$$

$$= P_{ave,1} + P_{ave,2} + \frac{V_{m1}V_{m2}}{2T}\int_0^T \left[\cos\left((\omega_1 + \omega_2)t + (\theta_1 + \theta_2)\right) + \cos\left((\omega_1 - \omega_2)t + (\theta_1 - \theta_2)\right)\right]dt$$

When the integral term in this last equation is zero, then $P_{ave} = P_{ave,1} + P_{ave,2}$, indicating that superposition of average power holds. When this integral term is nonzero, superposition of average power does *not* hold. The next question is, under what circumstances is the integral zero and nonzero? There are three cases to consider. Case 1 is when $\omega_1 \neq \omega_2$, which will result in a zero value of the integral. In this case, the integral consists of two sinusoids integrated over a common period T. The integral of a sinusoid over any period is zero. Thus, the integral is zero and superposition of power holds when $\omega_1 \neq \omega_2$.

Case 2 is when $\omega_1 = \omega_2$ but with $(\theta_1 - \theta_2) = \pm k\pi/2$, k an odd integer. In this case, the integral is again 0. This follows because the first term of the integrand is a sinusoid whose integral is zero over the period T. The second term of the integrand is a constant, $\cos(\theta_1 - \theta_2) = \cos(\pm k\pi/2) = 0$, also resulting in a zero integral. Hence for case 2, superposition of power holds.

Finally, we have case 3, for which $\omega_1 = \omega_2$ but with $(\theta_1 - \theta_2) \neq \pm k\pi/2$, k an odd integer; here superposition of power does *not* hold. The second term of the integrand is a constant, $\cos(\theta_1 - \theta_2) \neq 0$, resulting in a nonzero integral over the period T. So $P_{ave} \neq P_{ave,1} + P_{ave,2}$. For case 3, it is desirable to use the phasor method of Chapter 10 to compute the desired voltage and then use equation 11.5 to compute average power.

Exercises. **1.** In Example 11.2, suppose $v_1(t) = 3\cos(10\pi t)$ V and $v_2(t) = 4\cos(15\pi t + 0.25\pi)$ V. Compute T, a common period for the two sinusoids, and then compute the average power consumed by the 1 Ω resistor.
CHECK: $T = 0.4$ sec will work, and $P_{ave} = 12.5$ watts

2. In Example 11.2, suppose $v_1(t) = 3\cos(10\pi t)$ V and $v_2(t) = 4\sin(10\pi t)$ V. Compute the average power consumed by the 1 Ω resistor.
CHECK: $P_{ave} = 12.5$ watts

3. Now suppose $v_1(t) = 3\cos(10\pi t)$ V and $v_2(t) = 4\cos(15\pi t + 0.25\pi)$ V. Compute the average power consumed by the 1 Ω resistor.
CHECK: $P_{ave} = 20.99$ watts

Equation 11.6 resembles equation 1.18b for the dc power absorbed by a resistor connected to a dc source. However, in equation 11.6 the factor 1/2 is present. With the introduction of a new concept called the **effective value** of a periodic waveform, the formulas for the average power absorbed by a resistor can be made the same for dc, sinusoidal, or any other periodic input waveforms.

3. EFFECTIVE VALUE OF A SIGNAL AND AVERAGE POWER

GENERAL CONSIDERATIONS

From section 2, a resistor of R ohms excited by a periodic voltage or current absorbs an average power, P_{ave}. The **effective value** of any periodic current, $i(t)$, denoted by I_{eff} is a positive constant such that a dc current of value I_{eff} exciting the resistor causes the same amount of average power to be absorbed, i.e., $P_{ave} = RI_{eff}^2$. The same holds for a resistor excited by a periodic voltage $v(t)$. Mathematically,

$$P_{ave,R} = \frac{R}{T}\int_0^T i^2(t)dt = RI_{eff}^2 \tag{11.7a}$$

or

$$P_{ave,R} = \frac{1}{T}\int_0^T \frac{v^2(t)}{R}dt = \frac{V_{eff}^2}{R} \tag{11.7b}$$

Equation 11.7a suggests that

$$I_{eff,R}^2 = \frac{1}{T}\int_{t_0}^{t_0+T} i^2(t)dt$$

Hence, the mathematical definition of the **effective value** of a periodic current $i(t)$ is

$$I_{eff,R} = \sqrt{\frac{1}{T}\int_{t_0}^{t_0+T} i^2(t)dt} \tag{11.8a}$$

and, similarly, the **effective value** of a periodic voltage $v(t)$ is

$$V_{eff,R} = \sqrt{\frac{1}{T}\int_{t_0}^{t_0+T} v^2(t)dt} \tag{11.8b}$$

In general, the **effective value** of any periodic signal $f(t)$ is

$$F_{eff} = \sqrt{\frac{1}{T}\int_{t_0}^{t_0+T} f^2(t)dt} \tag{11.8c}$$

Observe that the expressions under the radical sign in equations 11.8 constitute the average value of the square of the signal. Hence, the expressions give rise to the alternative name for the effective value, the **root-mean-square** (abbreviated **rms**) value of $f(t)$, since F_{eff} is the square *root* of the *mean* value of the *square* of $f(t)$ over one period.

Exercises. 1. Show that the average power absorbed by an $R\,\Omega$ resistor carrying a periodic current $i(t)$ is $P_{ave} = V_{eff}I_{eff}$ with $v(t) = Ri(t)$.

2. Suppose $i(t) = 3\cos(2\pi t) + 4\cos(4\pi t)$ A flows through a 1 Ω resistor. Find P_{ave} and I_{eff}.

ANSWER: $P_{ave} = 12.5$ watts, and $I_{eff} = \dfrac{5}{\sqrt{2}} = 2.5\sqrt{2}$ A

EXAMPLE 11.3. Compute the effective value of the periodic voltage waveform sketched in Figure 11.5.

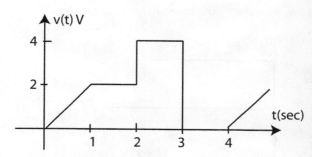

FIGURE 11.5 Periodic waveform having effective value $V_{eff} = 2.3094$ V.

SOLUTION
From equation 18.b,

$$V_{eff,R}^2 = \frac{1}{4}\int_0^4 v^2(t)dt = \frac{1}{4}\int_0^1 (2t)^2\,dt + \frac{1}{4}\int_1^2 (2)^2\,dt + \frac{1}{4}\int_2^3 (4)^2\,dt = \frac{16}{3}$$

Therefore, $V_{eff} = \dfrac{4}{\sqrt{3}} = 2.3094$ V.

Exercise. Repeat the calculation of Example 11.3 for the case where the values on the vertical axis of Figure 11.5 are doubled.

ANSWER: $V_{eff} = \dfrac{8}{\sqrt{3}}$ V

For a sinusoidal signal $f(t) = F_m \cos(\omega t + \theta)$, the effective value can be calculated using the identity

$$\cos^2(x) = 0.5 + 0.5\cos(2x)$$

as follows:

$$f^2(t) = F_m^2 \cos^2(\omega t + \theta) = \frac{F_m^2}{2} + \frac{F_m^2}{2}\cos(2\omega t + 2\theta)$$

Since by assumption $\omega \neq 0$, the average value of the cosine term is zero. The average value of the first (constant) term is itself. Hence, by equation 11.8c,

$$F_{eff} = \sqrt{\frac{F_m^2}{2}} = \frac{F_m}{\sqrt{2}}$$ (11.9)

Thus, for a sinusoidal waveform, the **effective** or **rms value** is always 0.707 times the maximum value or, equivalently, the maximum value divided by $\sqrt{2}$—a basic fact well worth remembering. The ac voltage and current ratings of all electrical equipment, as given on the identification plate, are rms values unless explicitly stated otherwise. For example, the household ac voltage is 110 V, with a maximum voltage of $110 \times \sqrt{2} = 156$ V. A typical appliance such as a coffee maker will have a 110 V rating, ac, at say, 900 watts. The effective values of a few other periodic waveforms are listed in Figure 11.6, with their derivations assigned as exercises.

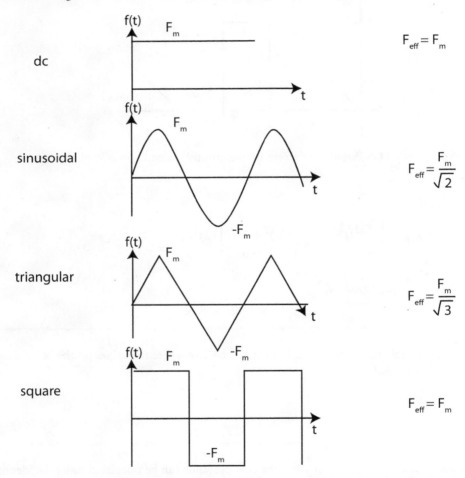

FIGURE 11.6 Effective values of some common periodic waveforms.

Exercises. 1. Derive the formula for the effective value of a triangular waveform shown in Figure 11.6.

2. Compute the effective value of the waveform shown in Figure 11.2a.

ANSWER: $\dfrac{V_m}{\sqrt{6}}$

SINGLE-FREQUENCY ANALYSIS WITH EFFECTIVE VALUES

We return now to the case of single-frequency SSS analysis. The average power as per equation 11.5 absorbed by an arbitrary two-terminal element may now be rewritten in terms of effective values:

$$P_{ave} = \frac{V_m I_m}{2}\cos(\theta_v - \theta_i) = \frac{V_m}{\sqrt{2}}\frac{I_m}{\sqrt{2}}\cos(\theta_v - \theta_i) = V_{eff}I_{eff}\cos(\theta_v - \theta_i) \qquad (11.10)$$

For the remainder of the chapter, all voltage and current phasors will be taken as being effective values unless the subscript *m* or *max* appears, indicating the maximum value. The subscript *eff* will be added sometimes for emphasis, however. This practice is widely accepted in the power engineering literature. Omitting the subscript *eff* in equation 11.10 yields

$$P_{ave} = VI\cos(\theta_v - \theta_i) \triangleq VI\cos(\theta_z) \qquad (11.11)$$

where $\theta_z = \theta_v - \theta_i$, $V = 0.707V_m$, and $I = 0.707I_m$. The angle θ_z is the angle of the impedance $Z(j\omega)$ of the two-terminal element and is also interpreted as the angle by which the voltage phasor *leads* the current phasor.

EXAMPLE 11.4. Figure 11.7 shows two types of household loads connected in parallel to a 110 V, 60 Hz source, $v_{in}(t) = 110\sqrt{2}\cos(120\pi t)$ V. Lamp 1 and lamp 2 have effective hot resistances of 202 Ω and 121 Ω, respectively. The impedance of the fluorescent light is $Z_{fl}(j\omega) = 60 + j70$ Ω.
  (a)   Find the average power consumed by each light.
  (b)   Find the average power delivered by the source.

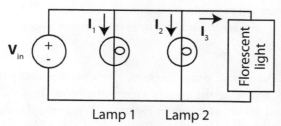

FIGURE 11.7 An example of load current calculation.

SOLUTION

(a) For lamp 1, $Z_1(j\omega) = 202\angle 0°$ Ω. Hence, $\mathbf{I}_1 = \mathbf{V}_{in}/Z_1 = 0.5446\angle 0°$ A. From equation 11.11,

$$P_{1ave} = V_{1eff}I_{1eff}\cos(\theta_{z1}) = 110 \times 0.5446\cos(0°) = 59.9 \text{ watts}$$

This means that lamp 1 is a 60 watt bulb.

Similarly, for lamp 2, $Z_2(j\omega) = 121\angle 0^\circ$ Ω. Hence, $\mathbf{I}_2 = \mathbf{V}_{in}/Z_2 = 0.9091\angle 0^\circ$ A. From equation 11.11,

$$P_{2ave} = V_{2eff}I_{2eff}\cos(\theta_{z2}) = 110 \times 0.9091\cos(0^\circ) = 100 \text{ watts}$$

Finally, for the fluorescent light, $Z_{fl}(j\omega) = 56 + j66 = 86.56\angle 49.7^\circ$ Ω. Hence, $\mathbf{I}_3 = \mathbf{V}_{in}/Z_{fl} = 1.27\angle -49.7^\circ$ A. From equation 11.11,

$$P_{fl,ave} = V_{eff}I_{3eff}\cos(\theta_{z3}) = 110 \times 1.27\cos(49.7^\circ) = 90.4 \text{ watts}$$

(b) For this part we first compute $\mathbf{I}_{in}$ and then apply equation 11.11 to compute the average power delivered by the source. Here by KCL,

$$\mathbf{I}_{in} = \mathbf{I}_1 + \mathbf{I}_2 + \mathbf{I}_3 = 0.5446\angle 0^\circ + 0.9091\angle 0^\circ + 1.27\angle -49.7^\circ$$
$$= 2.2759 - j0.9690 = 2.4736\angle -23.06^\circ \text{ A}$$

By equation 11.11, the average power delivered by the source is

$$P_{ave} = |\mathbf{V}_{in}||\mathbf{I}_{in}|\cos(\theta_v - \theta_i) = V_{in}I_{in}\cos(\theta_v - \theta_i) = 110 \times 2.4736\cos(23.06^o) = 250.35 \text{ watts}$$

Observe that the sum of the individual average powers is 250.3 watts, which equals the power delivered by the source within the accuracy of our calculations, where we have rounded our answers.

4. COMPLEX POWER AND ITS COMPONENTS: AVERAGE, REACTIVE, AND APPARENT POWERS

Recall the notion of a phasor. When all source excitations are sinusoidal at the same frequency, voltages and currents in the SSS can be represented by phasors. Our question here is, can the phasor method aid the computation of power consumption in a circuit? The answer is yes. However, the formulation will bring out several other concepts of power associated with the sinusoidal steady state.

In dc power calculations, the average power consumed by a two-terminal device is the product of the voltage and current, assuming the passive sign convention. In SSS, the complex power absorbed by a two-terminal device, as shown in Figure 11.8, is a complex number defined by the formula

$$\mathbf{S} = \mathbf{V}_{eff}\mathbf{I}_{eff}^* \qquad\qquad (11.12)$$

where $\mathbf{I}_{eff}^*$ is the complex conjugate of $\mathbf{I}_{eff}$.

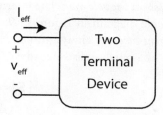

FIGURE 11.8 Two-terminal device with phasor voltage and current consistent with passive sign convention.

The first useful result of this definition is that $P_{ave} = \text{Re}[\mathbf{S}] = \text{Re}\left[\mathbf{V}_{eff}\mathbf{I}_{eff}^*\right]$. To see this result, suppose $v(t) = V_{eff}\sqrt{2}\cos(\omega t + \theta_v)$, which is represented by the phasor $\mathbf{V}_{eff}$. Also suppose $i(t) = I_{eff}\sqrt{2}\cos(\omega t + \theta_i)$, which is represented by the phasor $\mathbf{I}_{eff}$. The average power consumed by a two-terminal device excited by this voltage-current pair (Figure 11.8) is given by equation 11.10 as

$$P_{ave} = V_{eff}I_{eff}\cos(\theta_v - \theta_i)$$

Now observe that

$$\mathbf{S} = \mathbf{V}_{eff}\mathbf{I}_{eff}^* = V_{eff}e^{j\theta_v}\left(I_{eff}e^{j\theta_i}\right)^* = V_{eff}e^{j\theta_v}I_{eff}e^{-j\theta_i} = V_{eff}I_{eff}e^{j(\theta_v - \theta_i)}$$

in which case

$$\text{Re}[\mathbf{S}] = \text{Re}\left[\mathbf{V}_{eff}\mathbf{I}_{eff}^*\right] = \text{Re}\left[V_{eff}I_{eff}e^{j(\theta_v - \theta_i)}\right] = V_{eff}I_{eff}\cos(\theta_v - \theta_i) = P_{ave} \quad (11.13)$$

The curious reader may ask why a conjugate of the current is used in the definition of complex power. Suppose one did not have the conjugate of the current. Then $\text{Re}\left[\mathbf{V}_{eff}\mathbf{I}_{eff}\right] = V_{eff}I_{eff}\cos(\theta_v + \theta_i) \neq P_{ave}$, i.e., the resulting product would have no physical meaning. Now because $\mathbf{S}$ is a complex number, it has an imaginary part, a magnitude, and an angle. The imaginary part of $\mathbf{S}$ defines a quantity called the **reactive power** absorbed by the two-terminal device in Figure 1.18; i.e., *reactive power* is defined as

$$Q = \text{Im}[\mathbf{S}] = \text{Im}\left[\mathbf{V}_{eff}\mathbf{I}_{eff}^*\right] = V_{eff}I_{eff}\sin(\theta_v - \theta_i) = \text{reactive power} \quad (11.14)$$

The unit of reactive power, Q, is VAR, which stands for *volt-amp-reactive*. It follows immediately that

$$\mathbf{S} = \mathbf{V}_{eff}\mathbf{I}_{eff}^* = P + jQ \quad (11.15)$$

where $P = P_{ave}$. Also, the magnitude of $\mathbf{S}$ is defined as the **apparent power** absorbed by the two-terminal device of Figure 1.18, i.e.,

$$|\mathbf{S}| = V_{eff}I_{eff} = |P + jQ| = \sqrt{P^2 + Q^2} = \text{apparent power} \quad (11.16)$$

The unit of apparent power is VA, short for *volt-amp*. The interrelationship of these different powers is illustrated by the right triangle diagram in Figure 11.9, which is often helpful in solving problems. Observe that the *apparent power* is always greater than or equal to the *average power*, with equality applying to the case of a purely resistive load.

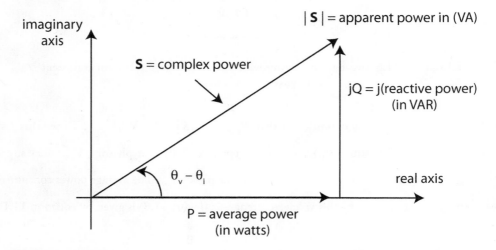

FIGURE 11.9 Relationships among complex, average, reactive, and apparent powers.

The distinction among these various powers is best understood by computing the powers for some basic circuit elements. For simplicity, except when needed or for emphasis, from this point on we will often drop the subscripts *eff* and *ave* as given in equations 11.13–11.16.

EXAMPLE 11.5. This example explores the computation of the various powers for a simple inductor. Given that $i_L(t) = \sqrt{2}I_L \sin(\omega t)$ in the circuit of Figure 11.10, compute $\mathbf{I}_L$, $\mathbf{V}_L$, $\mathbf{S}_L$, P_L, Q_L, the instantaneous absorbed power $p_L(t)$, and the instantaneous stored energy $W_L(t)$ in terms of L, ω, I_L, and V_L. After this show that

(i) $\left. \left| p_L(t) \right| \right|_{max} = \left| Q_L \right|$ and

(ii) $\left. \left| W_L(t) \right| \right|_{max} = \dfrac{\left| Q_L \right|}{\omega}$.

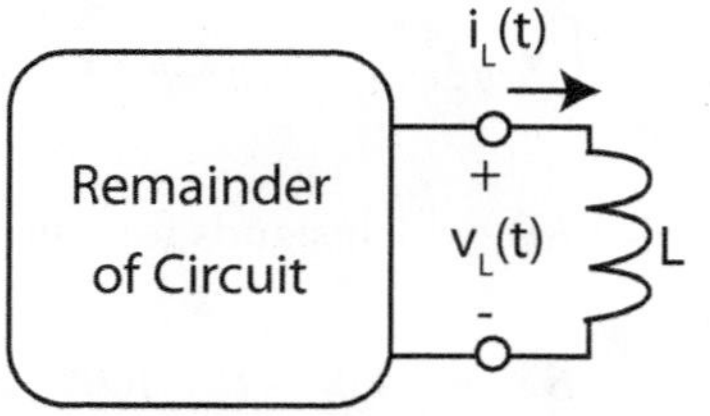

FIGURE 11.10 Isolation of an inductor for investigating the concept of complex power.

Solution

By inspection, and noting that we again presume effective values, $\mathbf{I}_L = -jI_L$, $\mathbf{V}_L = j\omega LI_L = \omega LI_L = V_L$. $\mathbf{S}_L = \mathbf{V}_L\mathbf{I}^*_L = jV_LI_L = P_L + jQ_L$. This implies that $P_L = 0$ and $Q_L = V_LI_L$. Further, the instantaneous absorbed power is $p_L(t) = v_L(t)i_L(t) = \omega L\sqrt{2}I_L\cos(\omega t) \times \sqrt{2}I_L\sin(\omega t) = V_LI_L\sin(2\omega t)$, which

is consistent with equation 11.4. It follows immediately that $\left|p_L(t)\right|_{max} = V_LI_L = \left|Q_L\right|$. Further,

$$W_L(t) = 0.5Li_L^2(t) = Li_L^2\sin^2(\omega t) = 0.5Li_L^2\left[1 - \cos(2\omega t)\right]$$

$$= \omega Li_L^2\frac{\left[1 - \cos(2\omega t)\right]}{2\omega} = V_LI_L\frac{\left[1 - \cos(2\omega t)\right]}{2\omega} = Q_L\frac{\left[1 - \cos(2\omega t)\right]}{2\omega}$$

Since the bracketed quantity varies between 0 and 2, $\left|W_L(t)\right|_{max} = \dfrac{\left|Q_L\right|}{\omega}$, as was to be shown.

EXAMPLE 11.6. This example, like the previous one, investigates the concept of reactive power, but in the case of a capacitor. The calculations will all be dual to those of Example 11.5. Hence, given that $v_C(t) = \sqrt{2}V_C\sin(\omega t)$ in the circuit of Figure 11.11, compute $\mathbf{V}_C$, $\mathbf{I}_C$, $\mathbf{S}_C$, P_C, Q_C, the instantaneous absorbed power $P_C(t)$, and the instantaneous stored energy $W_C(t)$ in terms of C, ω, V_C and I_C. After this show that

(i) $\left|p_C(t)\right|_{max} = \left|Q_C\right|$ and

(ii) $\left|W_C(t)\right|_{max} = \dfrac{\left|Q_C\right|}{\omega}$.

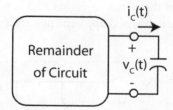

FIGURE 11.11 Isolation of a capacitor for investigating the concept of complex power.

Solution

By inspection, and noting that we again presume effective values, $\mathbf{V}_C = -jV_C$, $\mathbf{I}_C = j\omega CV_C = \omega CV_C = I_C$. $\mathbf{S}_C = \mathbf{V}_C\mathbf{I}^*_C = -jV_CI_C = P_C + jQ_C$. This implies that $P_C = 0$ and $Q_C = -V_CI_C$. Further, the instantaneous absorbed power is $p_C(t) = v_C(t)i_C(t) = \omega C\sqrt{2}V_C\cos(\omega t) \times \sqrt{2}V_C\sin(\omega t) = V_CI_C\sin(2\omega t)$, which is consistent with equation 11.4. It follows immediately that $\left|p_C(t)\right|_{max} = \left|V_CI_C\sin(2\omega t)\right|_{max} = V_CI_C = \left|Q_C\right|$. Further,

$$W_C(t) = 0.5Cv_C^2(t) = Cv_C^2\sin^2(\omega t) = 0.5Cv_C^2\left[1 - \cos(2\omega t)\right]$$

$$= \omega Cv_C^2\frac{\left[1 - \cos(2\omega t)\right]}{2\omega} = V_CI_C\frac{\left[1 - \cos(2\omega t)\right]}{2\omega} = \left|Q_C\right|\frac{\left[1 - \cos(2\omega t)\right]}{2\omega}$$

Since the bracketed quantity varies between 0 and 2, $\left|W_C(t)\right|_{max} = \dfrac{|Q_C|}{\omega}$.

These quantities, $p_L(t)\big|_{max}$, $\left|W_L(t)\right|_{max}$, $\left|p_C(t)\right|_{max}$, and $\left|W_C(t)\right|_{max}$, prove useful for identifying energy storage values in inductors and capacitors in systems where energy is to be recovered and stored, and for modifying the power factor (to be discussed shortly) in networks with motors. Energy storage in systems and power management are important research topics in today's world.

In Examples 11.5 and 11.6, one observes that the inductor absorbs reactive power while the capacitor absorbs negative reactive power or, equivalently, delivers reactive power. This follows from the definition of complex power (equation 11.13, i.e., $\mathbf{S} = \mathbf{V}_{eff}\mathbf{I}^*{}_{eff}$). The structure of equation 11.13 derives from the convention that whenever the phasor current lags the phasor voltage (as with the inductor), the device is considered to absorb reactive power, whereas if the current phasor leads the voltage phasor (as with the capacitor), the device is considered to deliver reactive power. Indeed, the overwhelming majority of loads (toasters, ovens, hair dryers, motors, transformers, TVs, etc.) have lagging currents.

When a two-terminal element absorbs an average power P_{ave}, there is a transformation of electrical energy into other forms of energy—for example, heat or kinetic energy. In contrast, when a two-terminal element absorbs reactive power Q, no energy is expended. The energy transferred into the two-terminal element is merely stored and later returned to the surrounding network. To distinguish it from real (expended) power, we use VAR (volt-ampere-reactive) instead of watt as the unit for the reactive power Q.

EXAMPLE 11.7. This example investigates the computation of the various powers defined above for an RC circuit. Here, consider the circuit of Figure 11.12, where $v_{in}(t) = 100\sqrt{2}\cos(2000\pi t)$ V. Find the complex, average, reactive, and apparent powers absorbed by the load.

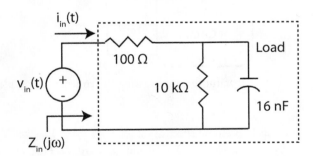

FIGURE 11.12 Simple RC circuit for investigating aspects of complex power.

SOLUTION
Step 1. *Compute $Z_{in}(j\omega)$.*

$$Z_{in}(j2000\pi) = 100 + \cfrac{1}{\dfrac{1}{10^4} + j2000\pi \times 16 \times 10^{-9}} = 5074 - j5000 \ \Omega$$

Step 2. *Compute* I_{in}. Converting $v_{in}(t)$ to a phasor, we have $V_{in} = 100$ V. By Ohm's law,

$$I_{in} = \frac{V_{in}}{Z_{in}} = 10 + j9.85 \ \text{mA}$$

Step 3. *Compute the complex power absorbed by the load.* By equation 11.12,

$$S = V_{eff}I_{eff}^* = 100(10 - j9.85)10^{-3} = 1 - j0.985 \ \text{VA}$$

Step 4. Given the complex power, the average power is

$$P_{ave} = \text{Re}[S] = 1 \ \text{watt}$$

The reactive power is

$$Q = \text{Im}[S] = -0.985 \ \text{VAR}$$

and the apparent power is

$$|S| = 1.404 \ \text{VA}$$

Before doing a more complex example, we will discuss the particulars of the principle of conservation of power in the sinusoidal steady state.

5. CONSERVATION OF COMPLEX POWER IN THE SINUSOIDAL STEADY STATE

Basics and Examples
The basic principle of power conservation is that instantaneous power is conserved.

GENERAL PRINCIPLE OF CONSERVATION OF POWER

In all circuits, linear or not, instantaneous power is conserved; i.e., the sum of the absorbed powers of all the elements in a circuit is zero. If one thinks of sources as generating power and other elements as absorbing power, then we can rephrase this statement as "the sum of generated powers equals the sum of absorbed powers."

The validity of this principle follows from KVL and KCL. This principle leads to the particular fact that complex power is conserved in ac circuits operating in the SSS.

PRINCIPLE OF CONSERVATION OF COMPLEX POWER IN AC CIRCUITS

In ac circuits operating in the SSS, complex power is conserved; i.e., the sum of the absorbed complex powers of all the elements (operating in the steady state) in a circuit is zero. Consequently, average power is conserved and reactive power is conserved.

Note however, that the conservation principle does *not* hold for apparent power, i.e., for the magnitude of the complex power. The following example illustrates a basic use of the conservation law.

EXAMPLE 11.8. This example illustrates the application of the principle of conservation of complex power in determining power delivered by a source and the input current to a circuit. We also show that conservation of apparent power does not hold. Consider the circuit of Figure 11.13. Find the power delivered by the source and the phasor input current, $\mathbf{I}_{in}$, given that $\mathbf{S}_1 = 360 + j160$ VA, $\mathbf{S}_2 = 360 - j120$ VA, $\mathbf{S}_3 = 420 + j540$ VA, $\mathbf{S}_4 = 130 + j80$ VA, $\mathbf{S}_5 = 40 - j100$ VA.

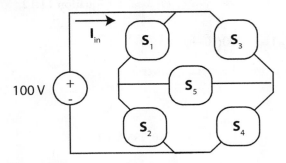

FIGURE 11.13 Bridge circuit where $\mathbf{S}_i$ represents the complex power absorbed by the element.

SOLUTION
By the principle of conservation of power in ac circuits,

$$\mathbf{S}_{source} = \mathbf{S}_1 + \mathbf{S}_2 + \mathbf{S}_3 + \mathbf{S}_4 + \mathbf{S}_5 = 1310 + j560 \text{ VA}$$

This means that the circuit absorbs 1310 watts of average power; the reactive power is 560 VAR, and the apparent power is 1425 VA. Notice that the large component of reactive power makes the apparent (consumed) power larger than the actual consumed power, P_{ave}.

To compute $\mathbf{I}_{in}$, recall that

$$\mathbf{S}_{source} = 100\mathbf{I}_{in}^* = 1310 + j560 \text{ VA}$$

Hence, $\mathbf{I}_{in} = 13.1 - j5.6$ A.

Exercise. Repeat the above example calculations for $\mathbf{S}_1 = 300 + j400$ VA, $\mathbf{S}_2 = 300 - j400$ VA, $\mathbf{S}_3 = 600 + j1000$ VA, $\mathbf{S}_4 = 60 + j80$ VA, $\mathbf{S}_5 = 120 - j160$ VA. What are the average and reactive powers delivered by the source?
ANSWER: $\mathbf{S}_{source} = 1380 + j920$ VA and $\mathbf{I}_{in} = 13.8 + j9.2$ A, 1380 watts and 920 VAR

The next example illustrates the computation of various powers through basic definitions and application of the principle of conservation of power.

EXAMPLE 11.9. Consider the circuit of Figure 11. 14, which depicts a motor connected to a commercial power source. The motor absorbs 50 kW of average power and 37.5 kVAR of reactive power, and has a terminal voltage $\mathbf{V}_m = 230$ V. Find $|\mathbf{I}_s|$, the complex power delivered by the source, $\mathbf{S}_s$, and $|\mathbf{V}_s|$.

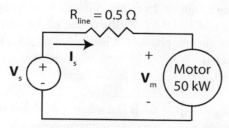

FIGURE 11.14 Motor absorbing 50 kW and 37.5 kVAR at a terminal voltage of 230 V; the value of R_{line} is exaggerated for pedagogical purposes; electrical code requires that the size of the connecting wire be large enough that the voltage drop is only a small percentage of the source voltage.

SOLUTION

Step 1. *Find the apparent power, $|S_m|$, absorbed by the motor.* Since

$$S_m = P_m + jQ_m = 50 + j37.5 \text{ kVA}$$

it follows that $|S_m| = 62.5$ kVA.

Step 2. *Find $|I_s|$.* Here, $|S_m| = |V_m I_s^*| = 230|I_s|$. Hence, $|I_s| = 271.74$ A.

Step 3. *Compute the line loss.*

$$P_{line} = R_{line}|I_s|^2 = 0.5 \times 271.7^2 = 36.92 \text{ kW}$$

Step 4. *Compute the complex power delivered by the source.* From conservation of power,

$$S_s = S_m + S_{line} = S_m + P_{line} = 50\, j37.5 + 36.92 = 86.92 + j37.5 \text{ kVA}$$

Step 5. *Compute $|V_s|$.*

$$|V_s| = \frac{|S_s|}{|I_s^*|} = \frac{|S_s|}{|I_s|} = 348.4 \text{ V}$$

In the above example we choose R_{line} large to illustrate the calculations. In practice a line loss of 36.92 kW for a 50 kW motor operation would not be permitted.

6. POWER FACTOR AND POWER FACTOR CORRECTION

In a resistor, average power is dissipated as heat. In a motor, most of the average consumed power is converted to mechanical power, say, to run a fan or a pump, with a much smaller portion dissipated as heat due to winding resistance and friction. The ratio of the average power to the apparent power is called the **power factor**, denoted by **pf**, i.e.,

$$\text{pf} = \frac{\text{Average Power}}{\text{Apparent Power}} = \frac{P_{ave}}{|S|} = \frac{P_{ave}}{V_{eff} I_{eff}} = \cos(\theta_v - \theta_i) \qquad (11.25)$$

The right-hand portion of equation 11.25 follows directly from equation 11.13. Equation 11.25 specifies the power factor as $\cos(\theta_v - \theta_i)$, i.e., the cosine of the difference between the angles of the voltage phasor $\mathbf{V}$ and the current phasor $\mathbf{I}$. Clearly, $0 \leq \text{pf} \leq 1$. The angle

$$(\theta_v - \theta_i) = \text{power factor angle (pfa)} \tag{11.26}$$

Since $\cos(x) = \cos(-x)$, the sign of $(\theta_v - \theta_i)$ is lost when only the pf is given. In order to carry the relative phase angle information along, the common terminology is *pf lagging* or *pf leading*. A *lagging power factor* occurs when the *current phasor lags the voltage phasor*, i.e., $0 < (\theta_v - \theta_i) < 180^\circ$. A *leading power factor* occurs when the *current phasor leads the voltage phasor*, i.e., $0 < (\theta_i - \theta_v) < 180^\circ$. Practically all types of electrical apparatus have lagging power factors. Some typical power factor values are listed in Table 11.1.

TABLE 11.1. Power Factors for Common Electrical Apparatus

TYPE OF LOAD	POWER FACTOR (LAGGING)
Incandescent lighting	1.0
Fluorescent lighting	0.5–0.95
Single-phase induction motor, up to 1 hp	0.55–0.75, at rated load
Large three-phase induction motor	0.9–0.96, at rated load

To illustrate the idea of leading and lagging pf, consider the circuits of Figure 11.15. Suppose the circuits operate at a frequency of 400 Hz or $\omega = 2513.3$ rad/sec. For the circuit of Figure 11.5a, $\mathbf{I} = (1 - j0.995) \, 10^{-3} \, \mathbf{V} = 1.41 \, 10^{-3}\angle-44.85^\circ \, \mathbf{V}$. Hence, the current phasor lags the voltage phasor, i.e., $(\theta_v - \theta_i) = 44.85^\circ$ and the pf is $\cos(44.85^\circ) = 0.709$ lagging. On the other hand, for the circuit of Figure 11.15b, $\mathbf{I} = (1 + j2.5) \, 10^{-3} \, \mathbf{V} = 2.7 \, 10^{-3}\angle68.3^\circ \, \mathbf{V}$. Hence, the current phasor leads the voltage phasor by 68.3°, i.e., $(\theta_i - \theta_v) = 68.3^\circ$ and the pf is $\cos(68.3^\circ) = 0.688$ leading.

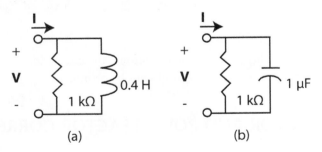

(a)   (b)

FIGURE 11.15 (a) A parallel *RL* circuit illustrating a lagging pf.
(b) A parallel *RC* circuit illustrating a leading pf.

A load with a required average power demand, operating at a fixed voltage with a low pf, say 0.6, has a relatively high reactive power component. This results in a relatively high apparent power.

Since the operating voltage is fixed, the line current needed to drive the load is higher than if the load operated at a higher pf, say 0.95. Relatively speaking, a higher pf has a lower reactive power component with correspondingly lower apparent power. Figure 11.9 helps to visualize the relationships. For fixed line voltage, lower apparent power (higher pf) means lower line current and hence lower power loss in the connecting transmission line. In today's world of energy conservation, it is important to be energy efficient. The following example illustrates how improved pf on a load can reduce line losses and thus decrease cost of operation.

EXAMPLE 11.10. This example reconsiders Example 11.9, involving a motor connected to a commercial power source as illustrated in Figure 11.16. The solution process will emphasize the basic definition of pf and the use of voltage and current phasors. Suppose the motor absorbs 50 kW (about 67 hp) of average power at a pf of 0.8 lagging. The terminal voltage, $\mathbf{V}_m$, is 230 V. The frequency of operation is 60 Hz or $\omega = 120\pi$. For the first part of the example the capacitor in Figure 11.16 is not connected to the motor. In part (c), the capacitor is connected to the motor to improve the pf. This will reduce the magnitude of the current supplied by the source and hence reduce the line losses.

(a)   Find the complex power delivered to the motor.
(b)   Find $\mathbf{I}_s$, $\mathbf{V}_s$, and the power delivered by the source, which might represent the power delivered by the local electric company.
(c)   Correct the power factor of the combined motor-capacitor load to 0.95 lagging by choosing a proper value for C.
(d)   Compute the new power delivered by the source to the combined motor-capacitor load.

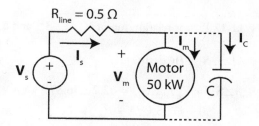

FIGURE 11.16 Motor absorbing 50 kW and 37.5 VAR at a terminal voltage of 230 V. Again, the value of R_{line} is exaggerated for pedagogical purposes; electrical code requires that the size of the connecting wire be large enough that the voltage drop is only a small percentage of the source voltage.

SOLUTION

(a) *Find the complex power delivered to the motor.*

Step 1. *Use the pf of 0.8 lagging and the given average power to find the apparent power.* From the definition of pf,

$$P_{ave} = 50\ \text{kW} = \text{Re}[\mathbf{S}] = |\mathbf{S}_m|\cos(\theta - \phi) = |\mathbf{S}_m| \times pf = |\mathbf{S}_m| \times 0.8$$

As such, the apparent power is

$$|\mathbf{S}_m| = \frac{50}{0.8} = 62.5\ \text{kVA} = V_m I_m \tag{11.27}$$

where $V_m = |\mathbf{V}_m| = 230$ V and $I_m = |\mathbf{I}_m|$.

Step 2. *Compute* $\angle \mathbf{S}_m$. Lagging means that current phase lags behind voltage phase, i.e., $\theta_v - \theta_i$ $= 0 - \theta_i > 0$. Consider the diagram in Figure 11.17, which shows that the current phasor $\mathbf{I}_{eff}$ lags the voltage phasor, i.e., the current phasor makes an angle of $-36.87° = \cos^{-1}(0.8)$ from the voltage phasor. Hence,

$$\angle \mathbf{S}_m = 36.87° > 0$$

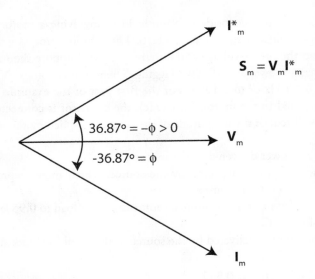

FIGURE 11.17 Phasor relationship of $\mathbf{V}_m$, $\mathbf{I}_m$, $\mathbf{I}_m^*$, and $\mathbf{S}_m$.

Step 3. *Compute the complex power,* $\mathbf{S}_m$. By definition,

$$\mathbf{S}_m = |\mathbf{S}_m| \angle \mathbf{S}_m = V_m I_m \angle \mathbf{S}_m = 62.2 \angle 36.87° \text{ kVA}$$
$$= 50 + j37.5 \text{ kVA} = P_{ave} + jQ$$

(b) *Find* $\mathbf{I}_s$, $\mathbf{V}_s$, *and the power delivered by the source.*

Step 1. *Find* $\mathbf{I}_s$. From equation 11.27 and the fact that $\mathbf{I}_s = \mathbf{I}_m$ for this part,

$$I_s = \frac{|\mathbf{S}_m|}{V_m} = \frac{62.5 \times 10^3}{230} = 271.74 \text{ A}$$

And from Figure 11.17, again since $\mathbf{I}_s = \mathbf{I}_m$, $\angle \mathbf{I}_s = -36.87°$. Hence

$$\mathbf{I}_s = 271.74 \angle -36.87° = 271.74 \angle -0.6435 \text{ rad} = 217.4 - j163 \text{ A}$$

Step 2. *Find* $\mathbf{V}_s$. From KVL and Ohm's law,

$$\mathbf{V}_s = R_{line} \mathbf{I}_s + \mathbf{V}_m = 0.5[217.4 - j163] + 230 = 338.7 - j81.5$$
$$= 384.4 \angle -0.2362 \text{ rad} = 384.4 \angle -13.533° \text{ V}$$

Step 3. *Compute the complex power, , delivered by the source.*

$$\mathbf{S}_s = \mathbf{V}_s\mathbf{I}_s^* = 348.4\angle -13.533^o \times 271.74\angle 36.87^o \text{ VA}$$
$$= 94.664 \angle 23.337^o \text{ kVA} = 94.664 \angle 0.407 \text{ rad kVA}$$
$$= (86.918 + j37.5) \text{ kVA}$$

Note that it takes 86.918 kW to run a 50 kW motor. The difference is the loss in the power line. If we have a way of reducing the magnitude of $\mathbf{I}_{eff}$ this line loss will be reduced. In fact, we do, and this strategy is the goal of the next part of the example.

(c) *Correct the power factor of the combined motor-capacitor load to 0.95 lagging.* Since motors are inductive, a properly chosen capacitor can improve the pf to 0.95 lagging. The new motor configuration is that of Figure 11.16, with the capacitor connected across the motor. The proper value of C must be found.

Step 1. *What does a pf of 0.95 lagging require in terms of complex power absorbed by the motor-capacitor combination?*

$$\mathbf{S}_m^{new} = \frac{50}{0.95}\angle \cos^{-1}(0.95) = 52.63\angle 18.195^o = (50 + j16.4342) \text{ kVA} \qquad (11.28)$$

Recall that

$$\mathbf{S}_m^{old} = (50 + j37.5) \text{ kVA}$$

Step 2. *Find a capacitor value to reduce the reactive power.* For this step consult Figure 11.18.

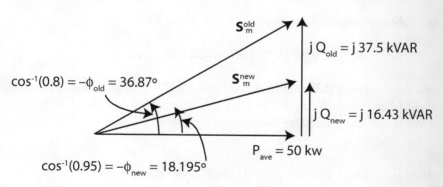

FIGURE 11.18 Relationships between new and old complex powers.

In Figure 11.18, one observes that P_{ave} is the same for both the new and old complex powers since that is what the motor requires for its operation. The reactive powers are different. The new complex power with the 0.95 lagging pf has a smaller reactive power component. The capacitor must be chosen to reduce the old reactive power to this new level. Hence,

$$jQ_{old} - jQ_{new} = j21.07 \text{ kVAR} = -j(\text{reactive power of capacitor}) = -jQ_C$$

Therefore,

$$jQ_C = -j21.07 \text{ kVAR} = \mathbf{V}_m\mathbf{I}_C^* = \mathbf{V}_m\left[Y_C(j\omega)\mathbf{V}_m\right]^* = -j\omega C|\mathbf{V}_m|^2$$

It follows that

$$C = \frac{-Q_C}{\omega|\mathbf{V}_m|} = \frac{21.07 \times 10^3}{120\pi(230)^2} = 1.057 \times 10^{-3} \text{ F}$$

(d) *Compute the new power delivered by the source.*

Step 1. *Compute the complex power, denoted* $\mathbf{S}_m^{new}$, *absorbed by the motor-capacitor load.* The complex power absorbed by the load is the sum of the complex power consumed by the motor and the reactive power of the capacitor, as illustrated in Figure 11.9, i.e.,

$$\mathbf{S}_m^{new} = (50 + j16.43) \text{ kVA}$$

Step 2. *Compute the new* I_s, *denoted* $\mathbf{I}_m^{new}$. Since $\mathbf{S}_m^{new}$ is the complex power of the combined motor-capacitor load,

$$\mathbf{I}_s^{new} = \frac{\left(\mathbf{S}_m^{new}\right)^*}{\mathbf{V}_m^*} = \frac{50 - j16.43}{230}10^3 = (217 + j71.43) \text{ A}$$

Step 3. *Compute .* From KVL and Ohm's law,

$$\mathbf{V}_s^{new} = 0.5\mathbf{I}_s^{new} + \mathbf{V}_m = (338.5 - j35.72) \text{ V}$$

Step 4. *Compute the new complex power delivered by the source.* By definition,

$$\mathbf{S}_s^{new} = \mathbf{V}_s^{new}\left(\mathbf{I}_s^{new}\right)^* = (76 - j16.48) \text{ kVA}$$

Hence, the new average power delivered by the source is 76 kW with pf correction as opposed to 86.9 kW without pf correction. With this pf correction, there is a reduction of 86.9 − 76 = 10.9 kW of power loss in the line connecting the source to the load.

Example 11.10 illustrates how adding a parallel capacitor can improve the pf of a load. The main motivation for improving the pf was to reduce the power loss in R_{line}. However, even if R_{line} is negligible, another strong reason exists for improving the load pf. Example 11.11 illustrates how an improved power factor allows a single generator to run more motors. Example 11.11 will fully utilize the principle of conservation of complex power and the two consequences of equation 11.25.

From equation 11.25 and the fact that $\mathbf{S} = P + jQ$, we can express pf directly in terms of P and Q as follows:

$$\text{pf} = \frac{P}{|\mathbf{S}|} = \frac{P}{|P + jQ|} = \frac{P}{\sqrt{P^2 + Q^2}} \tag{11.29}$$

with a lagging pf for $Q > 0$ and a leading pf for $Q < 0$. Solving for Q from equation 11.29, we obtain

$$Q = \pm P \sqrt{\frac{1}{pf^2} - 1}$$

(11.30)

with $Q > 0$ if pf is lagging and $Q < 0$ if pf is leading.

With these formulas we can simplify the process of power factor correction.

EXAMPLE 11.11. An industrial plant has a 100 kVA, 230 V generator that supplies power to one large motor and several identical smaller motors. The resistance of the connecting line is assumed negligible in the approximate analysis below. The large motor, labeled type A, draws 50 kW at a pf of 0.8 lagging. Each smaller motor, of type B, draws 5 kW at a pf of 0.7 lagging. The configuration is illustrated in Figure 11.19.

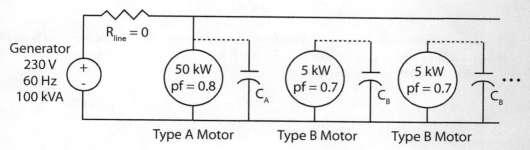

FIGURE 11.19 A generator supplying power to one large motor and several smaller motors.

(a)   Can the generator safely supply power to one large motor and three small motors? What are the generator current (magnitude) and the power factor of the combined loads?

(b)   Compute the number of small motors (besides the one large motor) that can be run simultaneously without exceeding the generator's rating.

(c)   If the power factor for all motors, large or small, is corrected to 0.9 lagging by connecting appropriate parallel capacitors (as done in Example 11.10), how many small motors (besides the one large motor) can be run simultaneously without exceeding the generator's rating?

(d)   Compute the capacitances required in part (c) for the large and the small motors.

SOLUTION

(a) *Compute the reactive power for each motor type.* Using equation 11.30, the reactive power for each type of motor is given as

$$Q_A = P_A \sqrt{\frac{1}{pf_A^2} - 1} = 50 \sqrt{\frac{1}{0.8^2} - 1} = 37.5 \text{ kVA}$$

and

$$Q_B = P_B \sqrt{\frac{1}{pf_B^2} - 1} = 5 \sqrt{\frac{1}{0.7^2} - 1} = 5.101 \text{ kVA}$$

By the principle of conservation of power, the complex power (in kVA) supplied by the generator is

$$S_{gen} = P_A + jQ_A + 3(P_B + jQ_B) = (50 + 15) + j(37.5 + 15.303)$$

$$= 65 + j52.8 = 83.74\angle 39.1° \text{ kVA}$$

By inspection, the apparent power is 83.74 kVA, which is below the generator capacity of 100 kVA, meaning that the generator can safely operate the large motor and three smaller motors.

The magnitude of the generator current is 83,740/230 = 364 A. From equation 11.29, the pf of the combined loads is

$$pf = \frac{P}{\sqrt{P^2 + Q^2}} = \frac{65}{\sqrt{65^2 + 52.8^2}} = 0.7762$$

(b) *Compute the number of small motors (besides the one large motor) that can be run simultaneously.* When one large type A motor and n smaller type B motors are connected in parallel, the complex power delivered by the generator is

$$S_{gen} = P_A + jQ_A + n(P_B + jQ_B) = (50 + n \times 5) + j(37.5 + n \times 5.101) \text{ kVA}$$

The apparent power is

$$\left| S_{gen} \right| = \sqrt{(50 + n \times 5)^2 + (37.5 + n \times 5.101)^2} \text{ kVA}$$

Since the generator has a capacity of 100 kVA, then

$$\left| S_{gen} \right|^2 = (50 + n \times 5)^2 + (37.5 + n \times 5.101)^2 \leq 100^2 = 10^4 \qquad (11.31)$$

Replacing the inequality sign in equation 11.31 by an equality results in the quadratic equation

$$51.020n^2 + 882.5750n - 6,093.8 = 0$$

The resulting zeros are $n_1 = 5.288$ and $n_2 = -22.58$. The largest positive integer satisfying the inequality 11.31 is $n = 5$. Thus, at most, five small motors can be run simultaneously with the large motor without exceeding the generator's capacity.

(c) *If all power factors are corrected to 0.9 lagging, find the number of small motors (besides the one large motor) that can be run simultaneously.* We essentially repeat the calculations of part (b) with the new given power factor of 0.9 lagging:

and

$$Q_A^{new} = P_A \sqrt{\frac{1}{pf_A^2} - 1} = 50\sqrt{\frac{1}{0.9^2} - 1} = 24.216 \text{ kVA}$$

$$Q_B^{new} = P_B \sqrt{\frac{1}{pf_B^2} - 1} = 5\sqrt{\frac{1}{0.9^2} - 1} = 2.4216 \text{ kVA}$$

The complex power (in kVA) supplied by the generator is

$$\mathbf{S}_{gen} = P_A + jQ_A^{new} + n(P_B + jQ_B^{new}) = (50 + n \times 5) + j(24.216 + n \times 2.416)$$

The apparent power (in kVA) is

$$\left| \mathbf{S}_{gen}^{new} \right| = \sqrt{(50 + n \times 5)^2 + (24.216 + n \times 2.4216)^2}$$

As before, n satisfies the inequality

$$(50 + n \times 5)^2 + (24.216 + n \times 2.4216)^2 \leq 10^4$$

To find n, we compute the largest positive root of the quadratic equation

$$30.8641n^2 + 617.2829n - 6{,}913.6 = 0$$

The roots of this quadratic are $n_1 = 8$ and $n_2 = -28$. The largest positive integer that satisfies the above inequality is $n = 8$. Thus, eight small motors, as opposed to five in the earlier case, can be run simultaneously with the large motor without exceeding the generator's capacity.

(d) For the large motor, the capacitor must absorb a negative reactive power equal to $Q_A^{new} - Q_A$. Equivalently, the capacitor must supply a reactive power equal to

$$Q_A - Q_A^{new} = (37.5 - 24.216) \times 1000 = 13284 \text{ VAR} \tag{10.32}$$

From Example 11.6, the reactive power supplied by a capacitor is

$$|Q_{CA}| = I_{CA}V_{CA} = \omega C_A V_{CA}^2 = 2\pi \times 60 C_A \times 230^2 \text{ VAR} \tag{10.33}$$

Equating equations 10.32 and 10.33, we have $2\pi \times 60 C_A 230^2 = 13284$. Solving produces $C_A = 666.16 \times 10^{-6}$ F.

Similarly, for the smaller motors,

$$Q_B - Q_B^{new} = (5.101 - 2.4216) \times 1000 = 2679.4 \text{ VAR}$$

Also, we have

$$|Q_{CB}| = |I_{CB}V_{CB}| = \omega C_B V_{CB}^2 = 2\pi \times 60 C_B \times 230^2 \text{ VAR}$$

Equating these two quantities and solving for C_B, we obtain $C_B = 134.35 \times 10^{-6}$ F. We note that in the power industry, such capacitors are usually specified only by their kVAR rating, with no mention of their actual capacitive value in F.

In the above example the generator capacity was given in terms of VA, the unit of apparent power. The example points out the importance of reducing reactive power to more fully utilize the power capacity of the generator. Use of VA for generator, motor, and transformer capacity arises out of safety considerations. Most ac machinery operates at a specified voltage depending on the insulation strength. The size of the wire and other heat transfer factors determine the maximum allowable current of a machine or transformer. Also, the cost and physical size of most ac equipment are more closely aligned to the VA rating than to other measures. Hence, the VA rating better reflects the safe operating capacity of ac equipment.

Another motivation for improving the power factor is economical. A power company charges a consumer only for the actual electrical energy used. A meter measures this energy usage in units of kWh (kilowatt-hour). As mentioned earlier, most electrical loads have lagging currents. As shown in Examples 11.10 and 11.11, for a given required average power, a higher pf means lower transmission line losses. Also, loads that operate at low pf force power companies to pursue higher kVA ratings of the generator equipment. Thus utilities companies encourage consumers to operate their equipment and appliances at high pfs. Since power companies can supply more power with the same equipment if the pf is high, they adjust their rates so that energy costs are less with a high pf and are greater with a low pf.

7. MAXIMUM POWER TRANSFER IN THE SINUSOIDAL STEADY STATE

Chapter 6 outlined the basics of maximum power transfer for linear resistive networks. Having introduced energy storage elements L and C, and having studied methods for sinusoidal steady analysis, it is time to extend the results on maximum power transfer to general linear networks in the sinusoidal steady state.

MAXIMUM POWER TRANSFER THEOREM FOR AC CIRCUITS

Let a practical ac source be represented by an independent voltage source $\mathbf{V}_S$ *(voltage phasor in rms value) in series with an impedance* $Z_s = R_s + jX_s$. *An adjustable load impedance* $Z_L = R_L + jX_L$, *with* $R_L > 0$, *is connected to the source (Figure 11.20). In steady state, for fixed* Z_S, $\mathbf{V}_S$, *and* ω, *the average power delivered to the load is maximum when* Z_L *is the complex conjugate of* Z_S, *i.e.,*

$$R_L = R_s \qquad (11.33a)$$

and

$$X_L = -X_s \qquad (11.33b)$$

and the maximum average power is given by

$$P_{\max} = \frac{V_{s,eff}^2}{4R_s} \qquad (11.33c)$$

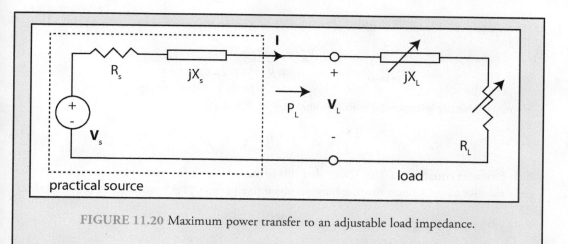

FIGURE 11.20 Maximum power transfer to an adjustable load impedance.

To derive the conditions of the maximum power transfer theorem, observe that the current phasor, $\mathbf{I}$, is

$$\mathbf{I} = \frac{\mathbf{V}_s}{(R_s + R_L) + j(X_s + X_L)} \tag{11.34}$$

Thus the average power delivered to the load is

$$P = P_{ave} = |\mathbf{I}|^2 R_L = \frac{V_s^2 R_L}{(R_s + R_L)^2 + (X_s + X_L)^2} \tag{11.35}$$

Here P_{ave} is a function of two real variables R_L and X_L. To find the conditions for maximum P_{ave}, set the partial derivatives $\dfrac{\partial P}{\partial R_L}$ and $\dfrac{\partial P}{\partial X_L}$ to zero and solve for R_L and X_L. Differentiating equation 11.35 with respect to R_L yields

$$\frac{\partial P}{\partial R_L} = \frac{V_s^2 \left[(R_s + R_L)^2 + (X_s + X_L)^2 - 2R_L(R_s + R_L) \right]}{\left[(R_s + R_L)^2 + (X_s + X_L)^2 \right]^2} = 0 \tag{11.36a}$$

and differentiating with respect to X_L produces

$$\frac{\partial P}{\partial X_L} = \frac{V_s^2 \left[-2R_L(X_s + X_L) \right]}{\left[(R_s + R_L)^2 + (X_s + X_L)^2 \right]^2} = 0 \tag{11.36b}$$

From equation 11.36b, the only physically meaningful solution is

$$X_L = -X_s$$

which is equation 11.33b. Substituting this result into the numerator of equation 11.36a yields

$$\frac{V_s^2(R_s - R_L)}{(R_s + R_L)^3} = 0$$

The only physically meaningful solution here is $V_s \neq 0$ and

$$R_L = R_s$$

which produces equation 11.33a. (Note that this is the condition for maximum power transfer in purely resistive circuits.) Substituting these results into equation 11.35 produces equation 11.33c, i.e.,

$$P_{max} = \frac{V_{s,eff}^2}{4R_s}$$

which verifies the theorem.

The theorem can be established less formally as follows. With any existing Z_L connected to the source, if the total reactance $(X_L + X_s)$ is not zero, we can always increase the magnitude of the current, and hence the power delivered to the load, by "tuning out" the reactance, i.e., by adjusting X_L to be $-X_s$. This implies condition 11.33b. Under such a condition, the circuit becomes resistive, and the maximum power transfer theorem of Chapter 6 may be applied to obtain equations 11.33a and c. The maximum power obtainable with a passive load, given by equation 11.33c, is called the **available power** of the fixed source.

The conditions for maximum power transfer, as given by equation 11.33, are valid when both R_L and X_L are adjustable. If X_L is fixed and only R_L is adjustable, then the condition for maximum power transfer is

$$R_L = \sqrt{R_s^2 + (X_s + X_L)^2} \qquad (11.37)$$

which is obtained by solving equation 11.36a for fixed X_s and X_L.

If the source is a general two-terminal linear network, then its Thevenin equivalent must be found before application of the maximum power transfer theorem. If the source is represented by a Norton equivalent circuit, we can use a source transformation to obtain the Thevenin form and then apply equations 11.33.

As pointed out in Chapter 6, maximum power transfer is not the objective in electric power systems, as the sources usually have very low impedances. On the other hand, it is a very important factor to be considered in the design of many communication circuits, as illustrated in the following example.

EXAMPLE 11.12. The radio receiver shown in Figure 11.21a is connected to an antenna. The antenna intercepts the electromagnetic waves from a broadcast station operating at 1 MHz. For circuit analysis purposes, the antenna is represented by the Thevenin equivalent circuit shown in Figure 11.21b.

(a)   Find the input impedance $R_{in} + jX_{in}$ of the receiver if maximum power is to be transferred from the antenna to the receiver.

(b)   Under the condition of part (a), find the magnitude of the voltage across the receiver terminals, and the average power delivered to the receiver.

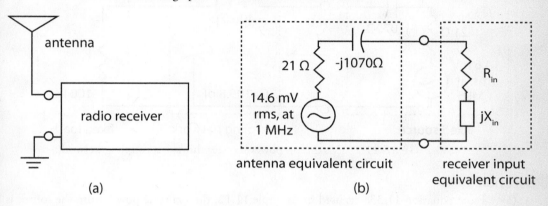

(a)   (b)

FIGURE 11.21 Example of maximum power transfer.

SOLUTION

(a)   From the maximum power transfer theorem, the answers are $R_{in} = 21\ \Omega$ and $X_{in} = 1070$.

(b)   Since the reactances in the circuit have been "tuned out," the input current to the receiver is simply 14.6/(21 + 21) = 0.348 mA. The input impedance has a magnitude

$$|Z_{in}| = \sqrt{21^2 + 1070^2} = 1070.2\,\Omega$$

Therefore the magnitude of the voltage across the receiver terminals is 0.348 × 1070.2 = 372.4 mV (rms). The power transferred from the antenna to the receiver is 0.348^2 × 21 = 2.54 µW.

In the preceding discussions of maximum power transfer, we have assumed that the load is adjustable. In practice the load is often fixed, as for example, in the case of a loudspeaker having a 4 Ω voice coil. In such cases, one designs coupling networks consisting of lossless passive components. These coupling networks transform the fixed load impedance into one whose conjugate matches the fixed source impedance. This permits maximum power transfer to the load. The following example illustrates the principle. A design procedure for some simple coupling networks will be discussed in the second volume of this text.

EXAMPLE 11.13.

A fixed load resistance R_L = 100 kΩ, representing the input resistance of an amplifier, is connected to the source of Example 11.12 through a passive coupling network, i.e., a network that does not generate average power, as shown in Figure 11.22.

(a)   Show that the maximum voltage that can be developed across R_L is 0.504 V.
(b)   Show that the coupling network shown in Figure 11.22 achieves this maximum voltage across R_L .

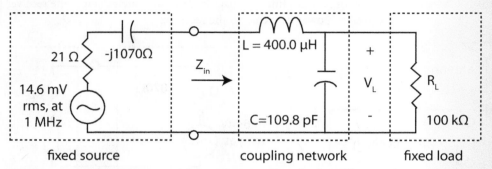

fixed source   coupling network   fixed load

FIGURE 11.22. Maximum power transfer through a coupling network.

SOLUTION

(a)   From equation 11.33c, as used in Example 11.12, the available power from the source is 2.54 µW. If all of the power is delivered to R_L , then the voltage, V_L, must be

$$V_L = \sqrt{P_{max} R_L} = \sqrt{2.54 \times 10^{-6} \times 100000} = 0.504 \ V$$

(b)   The input impedance Z_{in} of the coupling network with load must be the conjugate of the source impedance. Specifically

$$Z_{in} = j\omega L + \cfrac{1}{j\omega C + \cfrac{1}{R_L}}$$

Substituting the values $\omega = 10^6$, $L = 400.9 \times 10^{-6}$, $C = 109.8 \times 10^{-12}$, and $R = 100 \times 10^3$ into the above expression yields

$$Z_{in} = 21 + j1070 \ \Omega$$

which is indeed the conjugate of the source impedance. Since Z_{in} is conjugate-matched to the source impedance, the maximum power of 2.54 µW is transferred to the coupling network. Since the coupling network consists of L and C, neither of which consumes average power, the 2.54 µW power must be transferred out of the coupling network and into the load resistance. The voltage across the load resistor, V_L, is given by

$$V_L = \sqrt{P R_L} = \sqrt{2.54 \times 10^{-6} \times 100000} = 0.504 \ V$$

This verifies that the coupling network of Figure 11.22 enables the largest voltage to appear across the load resistor.

8. SUMMARY

Fundamental to the material in this chapter is the definition of the effective value (rms value) of a periodic voltage or current waveform. For a sine wave, the effective value is the maximum value divided by $\sqrt{2}$. For a general periodic voltage or current, the effective value is the value of a dc waveform that will produce the same amount of heat as the periodic waveform when applied to the same resistance. Using the definition of the effective value of a waveform, formulas for the average power absorbed by a linear two-terminal network in ac steady state were set forth and derived. Recall that for a two-terminal element with sinusoidal voltage $v(t) = V_{eff}\cos(\omega t - \theta_v)$ an current $i(t) = \sqrt{2}\,I_{eff}\cos(\omega t - \theta_i)$, the absorbed average power is $P = V_{eff}I_{eff}\cos(\theta_v - \theta_i)$, assuming the passive sign convention. Next we presented the definition of complex power and its component parts, which include its real part or average power, its imaginary part or reactive power, and its magnitude or apparent power. Various examples illustrating the calculation of these powers were given. Again, for a two-terminal element with sinusoidal voltage $v(t)$ and current $i(t)$ as above, the reactive power absorbed is defined to be $Q = V_{eff}I_{eff}\sin(\theta_v - \theta_i)$ VAR (volt-ampere-reactive). After introducing these different types of power, we proved the principle of conservation of complex power, which implies the conservation of real power and the conservation of reactive power. This was followed by the definition of power factor, pf, the ratio of average power to apparent power, which takes on values between 0 and 1. The need for improving a low power factor and a method for achieving an improved power factor were illustrated with two examples.

The maximum power transfer theorem, first studied in Chapter 6 for the resistive network case, was taken up again in this chapter for the sinusoidal steady-state case. Here, maximum power transfer to the load requires that the load impedance be the conjugate of the Thevenin impedance seen by the load. As pointed out earlier, the theorem has no application in electrical power systems. However, for communication circuits the maximum power transfer theorem is of extreme importance. The power that can be extracted from the antenna of a radio receiver is usually in the microwatt range, a very small value. It is therefore necessary to get as much power as possible from the antenna system. Example 11.13 illustrates this principle.

9. TERMS AND CONCEPTS

Apparent power: the apparent power absorbed by a two-terminal element is $V_{eff}I_{eff}$ assuming the use of a passive sign convention . The unit is VA (volt-ampere).

Average power: the average value of the instantaneous power. For a two-terminal element with sinusoidal voltage $v(t) = \sqrt{2}\,V_{eff}\cos(\omega t + \theta_v)$ and current $i(t) = \sqrt{2}\,V_{eff}\cos(\omega t + \theta_i)$, the absorbed average power is $P = V_{eff}I_{eff}\cos(\theta_v - \theta_i)$, assuming a passive sign convention.

Complex power: for a two-terminal element absorbing average power P and reactive power Q, the complex power is defined to be $\mathbf{S} = P + jQ$. The unit of measurement is VA (volt-ampere). The magnitude of $\mathbf{S}$ is the apparent power.

Conservation of powers: for any network, the sum of the instantaneous powers absorbed by all elements is zero. For any linear network in sinusoidal steady state, the sum of the average powers, reactive powers, or complex powers absorbed by all elements is zero. This property is a consequence of KCL and KVL.

Effective value (rms value): for a sine wave, the effective value is the maximum value divided by $\sqrt{2}$. For a general periodic voltage or current, the effective value is the value of a dc waveform that will produce the same amount of heat as the periodic waveform when applied to the same resistance.

Instantaneous power: the power associated with a circuit element as a function of time. The instantaneous power absorbed by a two-terminal element is $p(t) = v(t)i(t)$, assuming that a passive sign convention is used.

Maximum power transfer theorem: if a variable load $Z_L = R_L + jX_L$ is connected to a fixed source $\mathbf{V}_s$ having a source impedance $Z_s = R_s + jX_s$, then the largest average power is transferred to the load when Z_L is the complex conjugate of Z_s, i.e., $R_L = R_s$ and $X_L = -X_s$.

Power factor: the ratio of average power to apparent power. The pf value lies between 0 and 1. For a passive load, the power factor is said to be lagging when $90^{\circ} > \theta_v - \theta_i > 0$, and leading when $90^{\circ} > \theta_i - \theta_v > 0$.

Real power: in ac circuits, real power means average power. It is the real part of the complex power.

Reactive power: for a two-terminal element with sinusoidal voltage $v(t) = \sqrt{2}\,V_{eff}\cos(\omega t + \theta_v)$ and current $i(t) = \sqrt{2}\,I_{eff}\cos(\omega t + \theta_i)$, the reactive power absorbed, denoted by Q, is defined to be $Q = V_{eff}I_{eff}\sin(\theta_v - \theta_i)$, assuming that a passive sign convention is used. The unit of measurement is VAR (volt-ampere-reactive).

Problems

INSTANTANEOUS AND AVERAGE POWERS

1. For the source current waveform of Figure P11.1a, which drives the circuit of Figure P11.1b, find the average power consumed by the 2 Ω resistor.

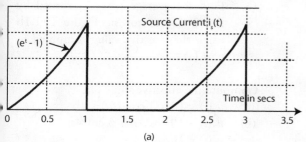

$(e^t - 1)$

(a)

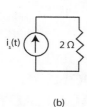

(b)

Figure P11.1

ANSWER. 0.758 watts

2. Compute the average power delivered to a 1 kΩ resistor by a current of the form
   (a) $10 \cos(10t)$ mA
   (b) $10 \, |\cos(10t)|$ mA
   (c) $10 \cos^2(10t)$ mA
   (d)   Plot each of the instantaneous powers for $0 \leq t \leq 1$ sec using MATLAB or its equivalent.

3. (a)   Compute the average power absorbed by a 10 Ω resistor whose voltage is given by each of the waveforms in Figure P11.3.
   (b)   Using MATLAB, plot the instantaneous power associated with each waveform for $0 \leq t \leq 3$ sec.

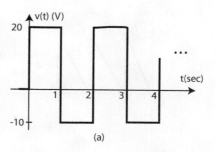

(a)

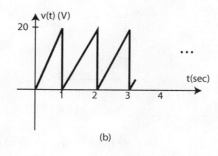

(b)

Figure P11.3 (a) Rectangular waveform. (b) Triangular waveform.

EFFECTIVE VALUE OF NONSINUSOIDAL SIGNALS

4. (a)   Compute the effective value of each of the periodic signals in Figures P11.4a and b.
   (b)   For the circuit of Figure P11.4c, find the power absorbed by R_L if the voltage source $v(t)$ is given by the waveform of Figure P11.4a.
   (c)   Repeat part (b) for the waveform of Figure P11.4b.

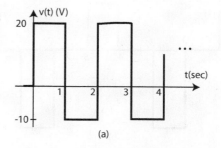

(a)

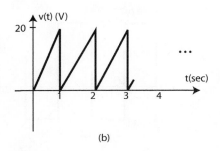

(b)

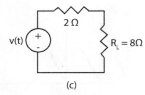

(c)

Figure P11.4

5. (a)   Compute the effective value of each of the periodic signals in Figures P11.5a and b.

   (b)   For the circuit of Figure P11.5c, find the power absorbed by R_L if the current source $i(t)$ is given by the waveform of Figure P11.5a.

   (c)   Repeat part (b) for the waveform of Figure P11.5b.

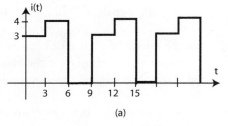

(a)

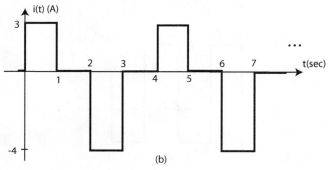

(b)

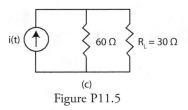

(c)

Figure P11.5

6. (a)   Find the effective value of the source current plotted in Figure P11.1a.

   (b)   Find the average power absorbed by the 2 Ω resistor in Figure P11.1b using the effective value computed in part (a).

7. Compute the effective value of

   (a) $v_1(t) = 10 + 2 \cos(20t)$
   (b) $v_2(t) = 10 \cos(2t) + 5 \cos(4t)$
   (c) $v_3(t) = 10\cos(2t) + 5 \cos(4t) + 5 \cos(4t - 45°)$ V.

AVERAGE POWER CALCULATIONS IN SINUSOIDAL STEADY STATE

8. Using equation 11.10, find the average power absorbed by the resistor in the circuit shown in Figure P11.8, where $v_{in}(t) = 50 \times \sin(5t)$ V, $R = 25$ Ω, $C = 8$ mF.

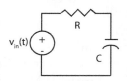

Figure P11.8

ANSWER: 25 watts

9. In the circuit shown in Figure P11.9, $i_{in}(t) =$ 5 cos(30t) A, $R = 5\ \Omega$, and $C = 5$ mF.
   (a) Find $v_L(t)$.
   (b) Using equation 11.10, find the instantaneous and average power absorbed by the load.

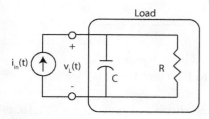

Figure P11.9

ANSWER: (b) 40 watts

10. In the circuit of Figure P11.10, $\mathbf{V}_s = 50\angle{-90°}$ V$_{rms}$, $R = 6\ \Omega$, $Z_L = j12\ \Omega$, and $Z_C = -j4\ \Omega$.
   (a) Find the phasor current $\mathbf{I}_s$ and determine its magnitude.
   (b) Using equation 11.10, find the average power (in watts) delivered by the source.
   (c) Only R absorbs average power. Therefore, once $|\mathbf{I}_s|$ is known, the average power consumed by R is $P_{ave} = R|\mathbf{I}_s|^2$. Check your answer to part (b) using this formula.
   (d) Repeat parts (a) and (b) when $R = 30\ \Omega$, $Z_L = j50\ \Omega$, and $Z_C = -j10\ \Omega$.

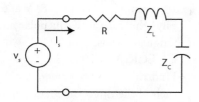

Figure P11.10

ANSWER: (b) 150 watts

11. For the circuit in Figure P11.11, $\mathbf{V}_s = 100\angle{0°}$ V$_{rms}$, $R = 5\ \Omega$, $Z_L = j50\ \Omega$, and $a = 9$.
   (a) Compute the current phasor $\mathbf{I}_L$.

   (b) Compute the average power delivered by each source.
   (c) Repeat parts (a) and (b) for $R = 50\ \Omega$, $Z_L = j50\ \Omega$, and $a = 49$.

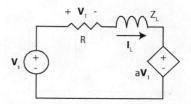

Figure P11.11

ANSWERS: (b) in random order: −90, 100 watts; (c) 4, −3.9 watts

12. Consider the circuit of Figure P11.12, where $\mathbf{V}_s = 120$ V$_{rms}$, $R = 32$ W, $Z_L = j200\ \Omega$, $Z_C = j80\ \Omega$, and $b = 4$.
   (a) Find the average power (in watts) absorbed by the resistor.
   (b) Find the average power delivered by each source.

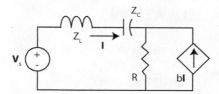

Figure P11.12

ANSWERS: (b) in random order: 57.6, 230.4 watts

13. A coil is modeled by a series connection of L and R. When connected to a 110 V 60 Hz source, the coil absorbs 300 watts of average power. If a 10 Ω resistor is connected in series with the coil and the combination is connected to a 220 V 60 Hz source, the coil also absorbs 300 watts of average power. Find L and R.

ANSWER: R = 0.9901 Ω and L = 16.6 mH.

14. Consider Figure P11.14, where $v_{in}(t) = 220\sqrt{2}\cos(120\pi t)$ V and $R_{coil} = 3\ \Omega$. The inductance L is adjusted so that $|\mathbf{V}_{coil}| = 150$ V$_{rms}$ and $P_{coil} = 250$ watts. Find the magnitude of $\mathbf{I}_{coil}$, the

reactance of the coil, and the values of L and R.

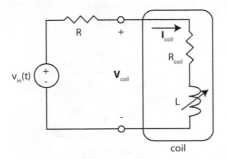

Figure P11.14

ANSWER: $L = 42.9$ mH, $R = 14.883$ Ω

15. Repeat Problem 14 for $R_{coil} = 4$ Ω, $P_{coil} =$ 400 watts, and V_{rms} with all other values the same.

CHECK: $L = 7.958$ mH, $R = 17.794$ Ω

COMPLEX POWER CALCULATIONS

16. For the circuit of Figure P11.16, $R = 5$ Ω $Z_{load} = 3 + j4$ Ω, $Z_C = -j10$ Ω, and $v_{in}(t) = 100\sqrt{2}\cos(120\pi t)$ V.
   (a) Find the complex power absorbed by the load.
   (b) Compute the apparent power in VA, the average power (in watts), and the reactive power in VAR delivered to the load.

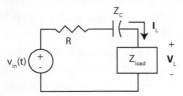

Figure P11.16

CHECK: (b) 300 watts, 400 VAR

17. In the circuit of Figure P11.17, $v_s(t) = 100\sqrt{2}\cos(500t + 30°)$ V, $R_1 = 100$ Ω, $R_2 = 700$ Ω, and $L = 1.2$ H. Find $i_L(t)$, the complex power, average power, and apparent power absorbed by the load.

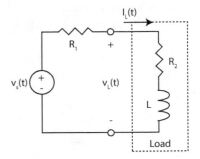

Figure P11.17

CHECK: Complex power is $7+j6$ VA

18. In the circuit shown in Figure P11.18, $\omega = 64$ rad/sec, $V_s = 120\angle60°$ V_{rms}, $R_1 = 20$ Ω, $R_2 = 4$ Ω, $L_1 = 0.375$ H, and $L_2 = 0.125$ H. Find the complex and average powers absorbed by the load.

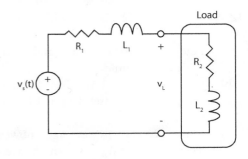

Figure P11.18

CHECK: Complex power is $36 + j72$ VA

19. In the circuit shown in Figure P11.19, $Z_1 = 1 + j$ Ω, $Z_2 = 4 + j22$ Ω, $Z_3 = 2 + j2$ Ω, $V_a = (104 + j50)$ V, and $V_b = (106 + j48)$ V at 60 Hz.
   (a) Find the voltage V_2 in polar and rectangular forms.
      CHECK: $V_2 = 100 + ?$
   (b) Find the complex power absorbed by each of the three impedances and then the power delivered by the two sources.
      CHECK: complex power absorbed by Z_2 is $100 + j550$ VA
   (c) Using the results of part (b), verify conservation of complex power.

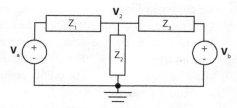

Figure P11.19

CHECK: complex power delivered by source b is $10 + 260j$ VA.

CONSERVATION OF POWER

20. This problem should be done without any phasor voltage or current computations. In the circuit of Figure P11.20, $\mathbf{V}_s = 2300$ V$_{rms}$ at 60 Hz and the following powers (in kVA) are consumed by various impedances and resistances: $\mathbf{S}_1 = 20 + j8$, $\mathbf{S}_2 = 20 + j18$, $\mathbf{S}_3 = 5 + j6$, and $\mathbf{S}_4 = 3 + j4$.

   (a) Find $\mathbf{I}_s$ (rectangular form) and the complex power delivered by the source.
   (b) Determine $\mathbf{I}_s$ in polar form.
   (c) Find the complex power delivered to the group of impedances Z_1, Z_2, and Z_4.
   (d) Find $\mathbf{V}_1$ in rectangular and polar form.
   (e) Find $\mathbf{V}_2$ in rectangular and polar form.

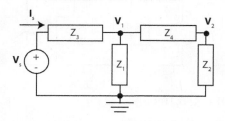

Figure P11.20

CHECKS: (a) Complex power delivered by source: $48 + j36$ kVA; (b) $|\mathbf{I}_s| = 26.087$ A

21. In the circuit of Figure P11.21, $\mathbf{V}_s = 230$ V$_{rms}$ at 60 Hz. The complex powers (in VA) absorbed by the five impedances are $\mathbf{S}_1 = 100 + j100$, $\mathbf{S}_2 = 200 + j100$, $\mathbf{S}_3 = 100 + j50$, $\mathbf{S}_4 = 400 + j250$, and $\mathbf{S}_5 = 800 + j700$. Find

   (a) the complex power, the apparent power, and the average power delivered by the source.
   (b) Find the source current $\mathbf{I}_s$ in rectangular and polar form.

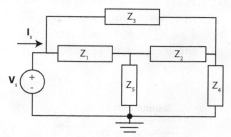

Figure P11.21

CHECKS: $P_{s,ave} = 1600$ watts, $|\mathbf{I}_s| = 8.6957$ A

POWER FACTOR AND POWER FACTOR CORRECTION

22. (a) Find the complex power delivered to a load that absorbs 2 kW of average power with pf = 0.90 lagging.
   (b) Find the complex power delivered to a load that absorbs 4 kW of average power with pf = 0.90 leading.

23. For the circuit of Figure P11.23, Z_1 absorbs 1600 watts at pf = 1, while the apparent power absorbed by Z_2 is 1000 VA at pf = 0.8 lagging, and $v_{in}(t) = 120\sqrt{2}\cos(120\pi t)$ V. Find

   (a) the phasor current $\mathbf{I}_s$ in A$_{rms}$,
   (b) the phasor voltage $\mathbf{V}_1$
   (c) the phasor voltage $\mathbf{V}_2$

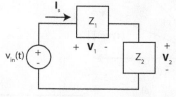

Figure P11.23

CHECK: $\mathbf{I}_s = 20 - 5j$ A$_{rms}$ and $\mathbf{V}_1 = 75.2941 - 18.8235j$

24. The circuit shown in Figure P11.24 is in the sinusoidal steady state. Suppose that Z_L absorbs 3000 W of average power at pf = 0.7905 lagging, R_{line} = 0.1 Ω, and $\mathbf{V}_L$ = 120 V_{rms}.

   (a) Find the average power absorbed in the transmission line resistance (in W).

   (b) Find $v_{in}(t) = A \cos(120\pi t + \theta)$ V and the complex power delivered by the source.

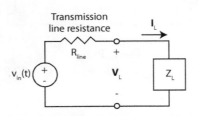

Figure P11.24

CHECK: (a) 100 watts

25. As shown in Figure P11.25b, a capacitor is put in parallel with a motor using average power P_{ave} = 40 kW operating at a power factor of 0.7 lagging to boost it to a power factor of 0.9 lagging. The voltage across the parallel motor-capacitor combination is $230\angle 0°$ V_{rms}. The power relationships are shown in Figure P11.25a. If the frequency of operation is ω = 120π, compute the proper value of the capacitance, C (in mF).

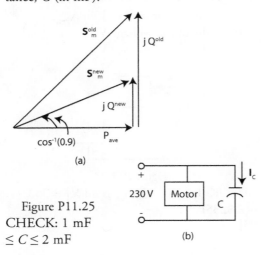

(a)

Figure P11.25

CHECK: 1 mF $\leq C \leq 2$ mF

(b)

26. The circuit in Figure P11.26 operates at ω = 500 rad/sec and $\mathbf{V}_s$ = $100\angle 0°$ V_{rms}. The complex power drawn by the load *without* the capacitor attached is $\mathbf{S}_L$ = $100\angle 30°$ = (86.6 + $j50$) VA. This constitutes a pf of 0.866 lagging.

   (a) Find the values of R and L.

   (b) Find the value of C in μF that produces a pf of 0.95 lagging.

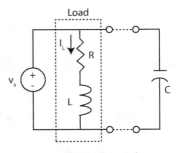

Figure P11.26

CHECK: 3 μF $\leq C \leq 5$ μF

27. The circuit shown in Figure P11.27 is operating in the SSS with $v_s(t) = 120\sqrt{2}\cos(120\pi t)$ V. Device 1 absorbs 360 W with pf = 0.9. Device 2 absorbs 1440 W with a pf of 0.866 lagging. Find the value of the capacitor C such that the magnitude of the source current equals 16 A_{rms}. What is the pf of the two-device-plus-capacitor combination?

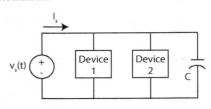

Figure P11.27

CHECK: 0.108 mF

28. A group of induction motors is drawing 7 kW from a 240 V power line at a power factor of 0.65 lagging. Assume ω = 120π rad/sec.

   (a) What is the equivalent capacitance of a capacitor bank needed to raise the power factor to 0.85 lagging?

   (b) What is the kVA rating of the capacitor bank of part (a); i.e., what is the reactive power supplied by the capacitor bank?

(c) Determine the annual savings from installing the capacitor bank if a demand charge (in addition to the charge for the kilowatt-hours used by the induction motors) is applied at $20.00 per kVA per month.

ANSWERS: (a) 0.1771 mF; (b) Minimum kVA rating: 3.8457 kVA; (c) $608.14

29. Consider a source that drives an electric motor that consumes an average power of 94 kW (about 125 hp) at a pf of 0.65 lagging, as shown in Figure P11.29, where $R_{line} = 0.07 \ \Omega$.

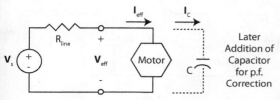

Figure P11.29

The phasor voltage across the motor is $\mathbf{V}_{eff} = 230\angle\ 0$ V. The sinusoidal frequency is 60 Hz.

(a) Find the apparent power delivered to the motor in kVA.

(b) Find the complex power, $\mathbf{S}_m$, delivered to the motor.

(c) Determine the reactive power in VAR delivered to the motor.

(d) Compute $\mathbf{I}_{eff}$.

(e) Compute $\mathbf{V}_s$.

(f) Compute the complex power delivered by the source.

(g) Determine the efficiency of the configuration, i.e., the ratio of average power delivered to the motor to the average power delivered by the source as a percentage.

(h) Add a capacitor across the motor to improve the power factor to 0.98 lagging. Then

    (i) Compute $\mathbf{S}_m^{new}$. Compare with $\mathbf{S}_m^{old}$.

    (ii) Recall that the role of the capacitor is to reduce the reactive power.

By what amount in kVAR must the reactive power be reduced to produce a pf of 0.94 lagging?

    (iii) Compute the needed capacitor current $\mathbf{I}_C$.

    (iv) Compute $Z_C(j\omega)$ as the ratio of the capacitor phasor voltage to the capacitor phasor current, at the indicated frequency.

    (v) Compute the proper value of C in mF.

(i) Compute the new $\mathbf{I}_{eff}^{new}$.

(j) Compute $\mathbf{V}_s^{new}$.

(k) Compute the complex power delivered by the source and the new efficiency.

MAXIMUM POWER TRANSFER

30. Consider the circuit shown in Figure P11.30 in which $R_1 = 100 \ \Omega$, $R_2 = 25 \ \Omega$, $C = 1$ mF.

(a) Find the value of the load impedance Z_L that will absorb maximum power at $\omega = 100$ rad/sec.

(b) Given the conditions of part (a) and $\mathbf{V}_s = 100$ V$_{rms}$, find the average power absorbed by the load.

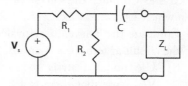

Figure P11.30

CHECK: $20 + j10 \ \Omega$

31. In the circuit of Figure P11.31, $v_s(t) = 100\sqrt{2}\cos(1000t)$ V, $R_1 = 80 \ \Omega$, $R_2 = 20 \ \Omega$, $L = 5$ mH. Find the value of the resistance R_L (in Ω) and the capacitance C (in mF) such that maximum average power is absorbed by the load.

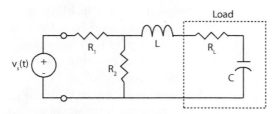

Figure P11.31

ANSWER: $R_L = 16 \, \Omega$ and $C = 0.2$ mF

32. The circuit of Figure P11.32 operates in the sinusoidal steady state with $\omega = 1000$ rad/sec, $R = 1$ kΩ, $C = 1 \, \mu$F, $L = 0.5$ H, $a = 3$ and $\mathbf{I}_s = 2\angle 0° \text{A}_{rms}$.

   (a) Find the value of the load impedance Z_L for maximum average power transfer.

   (b) Find the average power absorbed by the load under the conditions of part (a).

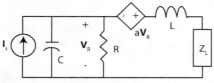

Figure P11.32

ANSWER: $Z_L = 2 + 1.5j$ kΩ, 4000 watts

33. The purpose of this problem is to compute Z_L for maximum power transfer by following a specific procedure. Consider the circuit of Figure P11.33 in which $i_s(t) = 50\sqrt{2}\cos(150t)$ A.

   (a) Compute the Thevenin equivalent of the circuit at terminals A and B:

      (i) Use nodal analysis to compute $\mathbf{V}_{oc}$. Note that there is a floating dependent voltage source.

          CHECK: $\mathbf{V}_{oc} = 28.47 - j91.1$ V

      (ii) Find the Thevenin equivalent impedance, $Z_{th}(j150)$, seen to the left of terminals A and B.

          CHECK: Re$[Z_{th}(j150)]$ = 8.2846Ω

      (iii) Show the phasor form of the Thevenin equivalent circuit. Then show the circuit form of the

Thevenin equivalent, i.e., a voltage source in series with either a series RC or a series RL as appropriate.

   (b) Compute the load impedance, $Z_L(j150)$, necessary for maximum power transfer. Show this load as either a series RL circuit or a series RC circuit. Should it be the opposite of the case in (a)(iii)? Why?

   (c) Compute the average power consumed by the load at maximum power transfer.

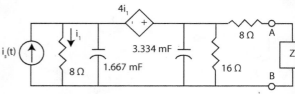

Figure P11.33

34. Consider the circuit of Figure P11.34, which operates at $\omega = 10$ rad/sec. Suppose $R = 10 \, \Omega$, $L = 2$ H, and $i_s(t) = 10\sqrt{2}\cos(10t)$ A.

   (a) Choose the proper values of R_L and C_L to deliver maximum average power to the load. What is this maximum average power?

   (b) If $R_L = 30 \, \Omega$, determine C_L for maximum power transfer to the load. What is this maximum average power?

   (c) If $C_L = 8$ mF, then determine R_L for maximum power transfer to the load. What is this maximum average power?

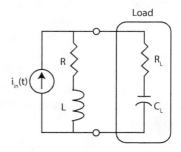

Figure P11.34

ANSWERS: (a) 10 Ω, 5 mF, 1250 watts; (b) 5 mF, 937.5 watts

35. In the circuit of Figure P11.35, $V_s = 100$ V_{rms} and R_L is adjusted to achieve different goals. Assume $R = 60\ \Omega$, $Z_C = -j80\ \Omega$.
   (a) Find the value of R_L that maximizes P_L. (CHECK: 24 Ω.) What is the value of $P_{L,max}$?
   (b) Find the value of R_L that maximizes V_L. What is the value of $|V_L|_{max}$?

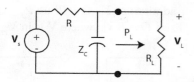

Figure P11.35

ANSWER: (a) 48 Ω, 37.037 watts

36. The circuit shown in Figure P11.36 has $v_{in}(t) = 10\sqrt{2}\cos(60t)$ V, $R_L = 4\ \Omega$, $L = 1/120$ H, and $C_L = 1/30$ F. What value of R should be chosen so that maximum power is delivered to the load? *Note:* It is the source resistor here, not the load resistance, that is being varied. What is this maximum average power consumed by the load?

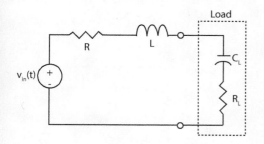

Figure P11.36

ANSWERS: 0, 25 watts

37. The circuit of Figure P11.37 operates in the sinusoidal steady state with $v_s(t) = 50\sqrt{2}\cos(2000t)$ V.
   (a) Choose R and C such that the maximum average power is absorbed in the load resistor $R_L = 5\ \Omega$ when $L = 0.1$ mH. What is this maximum average power?

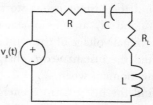

Figure P11.37

   (b) Now suppose $R = 4\ \Omega$, $C = 1$ mF, and $L = 0.1$ mH. Choose R_L for maximum power transfer and find P_{max}.

38. Consider the circuit of Figure P11.38 where $V_s = 0.1\ V_{rms}$ at $\omega = 10^7$ rad/sec. Suppose $R_1 = 100\ \Omega$, $R_2 = 10.1\ k\Omega$, $Z_{C1} = -j1000$.
   (a) Find the impedance Z_L so that P_1 is maximized.
   (b) Find the values of L and C_2 to achieve the impedance Z_L computed in part (a).
   (c) Find the impedance Z_L such that P_2 and $|V_2|$ are maximized.

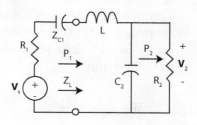

Figure P11.38

ANSWERS: (a) $Z_L = 100 + j1000\ \Omega$; (b) 99 pF and 0.2 mH

39. The series RLC circuit of Figure P11.39 has reached steady state and $v_s(t) = 110\sin(120\pi t)$ V.

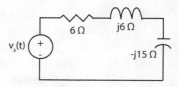

Figure P11.39

(a) Find the instantaneous stored energy W_L and W_C at the moment when the terminal voltage of the source is zero.

(b) Find the instantaneous energy W_L at the moment when $W_C = 0$.

(c) Find the instantaneous energy W_C at the moment when $W_L = 0$.

ANSWERS: in random order (J), 1.14, 1.65, 4.115, 1.27

THEORETICAL PROBLEMS

40. Let the voltage across a capacitance C be $v(t) = V_m \sin(\omega t)$ V.

(a) Find $p(t)$, the instantaneous power delivered to the capacitance and show that $p(t)$ has a peak value of $0.5\omega C(V_m)^2$ watts and an average value of 0.

(b) Find $W_C(t)$, the instantaneous energy store in C (or, rather, in the electric field) and show that $W_C(t)$ has a peak value of $0.5C(V_m)^2$ joules and an average value of $0.25C(V_m)^2$ joules.

(c) Let Q_C be the reactive power absorbed by C. Show that

$$W_{C,ave} = -\frac{Q_C}{2\omega}$$

41. Let the current flowing through an inductance L be $i(t) = I_m \sin(\omega t)$ A.

(a) Find $p(t)$, the instantaneous power delivered to the inductance and show that $p(t)$ has a peak value of $0.5\omega L(I_m)^2$ watts and an average value 0.

(b) Find $W_L(t)$, the instantaneous energy store in L (or, rather, in the magnetic field) and show that $W_L(t)$ has a peak value of $0.5L(I_m)^2$ joules and an average value of $0.25L(I_m)^2$ joules.

(c) Let Q_L be the reactive power absorbed by L. Show that

$$W_{L,ave} = \frac{Q_L}{2\omega}$$

42. (a) For the circuit of Figure P11.42a, show that the powers absorbed by the impedance $Z = R + jX$ are $P_{ave} = R|\mathbf{I}|^2$ and $Q = X|\mathbf{I}|^2$ where $\mathbf{I}$ is in A_{rms}.

(b) For the circuit of Figure P11.42b, show that the powers absorbed by the admittance $Y = G + jB$ are $P_{ave} = G|\mathbf{V}|^2$ and $Q = B|\mathbf{V}|^2$ where $\mathbf{V}$ is in V_{rms}.

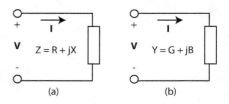

(a)   (b)

Figure P11.42. Two-terminal elements modeled via impedance (a) and admittance (b).

CHAPTER 12

Laplace Transform Analysis I: Basics

HISTORICAL NOTE

The Laplace transform converts a time function into a new function of a complex variable via an integration process. The name Laplace transform comes from the name of a French mathematician, Pierre Simon Laplace (1749–1827). Pierre Laplace adapted the idea from Joseph Louis Lagrange (1736–1813), who in turn had borrowed the notion from Leonhard Euler (1707–1783). These early mathematicians set the stage for converting complicated differential equation models of physical processes into simpler algebraic equations. The Laplace transform technique allows engineers to analyze circuits and to calculate responses quickly and efficiently. In turn engineers became better able to design circuits for radio communication and the telephone, not to mention other, earlier electronic conveniences. This chapter introduces the notion of the Laplace transform, a mathematical tool that is ubiquitous in its application to an army of engineering problems.

CHAPTER OBJECTIVES

1. Explain and illustrate the benefits of using the Laplace transform tool for solving circuits.
2. Develop a basic understanding of the Laplace transform tool and its mathematical properties.
3. Develop some skill in applying the Laplace transform to differential equations and circuits modeled by differential equations.

SECTION HEADINGS

1. Introduction
2. Review and Summary of Deficiencies of "Second-Order" Time Domain Methods
3. Overview of Laplace Transform Analysis
4. Basic Signals
5. The One-Sided Laplace Transform
6. The Inverse Laplace Transform
7. More Transform Properties and Examples

1. INTRODUCTION

This chapter introduces a powerful mathematical tool for circuit analysis and design named the Laplace transform. Later, more advanced courses will describe the design aspects. Use of the Laplace transform is commonplace in engineering, especially electrical engineering. A student might ask why such a potent tool is necessary for the analysis of basic circuits, especially since many texts use an alternative technique called complex frequency analysis. Complex frequency analysis does not permit general transient analysis; rather, it restricts source excitations to sinusoids, exponentials, damped sinusoids, and dc signals. This class of signals is small and does not begin to encompass the broad range of excitations necessary for general circuit analysis and the related area of signal processing. The Laplace transform framework, on the other hand, permits both steady-state and transient analysis of circuits in a single setting. Additionally, it affords general, rigorous definitions of *impedance, transfer function*, and various response classifications pertinent to more advanced courses on system analysis and signal processing. Introducing the Laplace transform early allows students an entire semester to practice using the tool and learn about its many advantages.

Section 2 describes some of the difficulties associated with the methods of circuit analysis introduced in earlier chapters when applied to circuits of order 3 or higher. Following this, we present an overview of Laplace transform analysis in section 3, define important basic signals in section 4, and introduce the formal definition of the one-sided Laplace transform in section 5. The inverse Laplace transform and important properties of the transform process are introduced in sections 6 and 7, with numerous illustrative examples. Section 8 applies the technique to circuits modeled by differential equations. Such models were developed in Chapters 8 and 9.

2. REVIEW AND SUMMARY OF DEFICIENCIES OF "SECOND-ORDER" TIME DOMAIN METHODS

Recall that the output or response of a circuit depends on the independent source excitations, on the initial capacitor voltages, and on the initial inductor currents. Calculation of the output often begins with the writing of an algebraic or a differential equation model of the circuit for the output variable in terms of the source excitations or inputs and element values. For first- and second-order circuits with simple source excitations, such as dc or purely sinusoidal, the solution of the differential equation circuit model has a known general form containing arbitrary constants. See, for example, Tables 9.1 and 9.2. The arbitrary constants depend on the initial conditions and the magnitude of the dc excitation or on the magnitude and phase of the sinusoidal excitation. Specifically, the steps in finding the response of a second-order circuit to a constant input are as follows:

Step 1. *Generate a differential equation model of the circuit.*
Step 2. *Compute the characteristic equation of the circuit/differential equation and then compute its roots (say λ_1 and λ_2) using, for example, the quadratic formula in the second-order case.*
Step 3. *From the location of the roots of the characteristic equation, determine the form of the solution:*

$$Ae^{\lambda_1 t} + Be^{\lambda_2 t} + D$$

or if $\lambda_1 = \lambda_2$,

$$(A + Bt)e^{\lambda_1 t} + D$$

Step 4. *Compute the constant D by shorting inductors, open-circuiting capacitors, and analyzing the resulting resistive circuit.*
Step 5. *Compute the constants A and B using the initial conditions on the circuit.*

For circuits beyond second order, the approach in the above algorithm tends to break down. Example 12.1 demonstrates how the approach breaks down with a simple third-order circuit.

As mentioned earlier, the foregoing technique, although quite useful for simple circuits, has serious drawbacks for circuits with more than two capacitors or inductors. This is because higher-order derivatives of circuit output variables generally have little or no physical meaning. Such derivatives are complicated linear combinations of initial capacitor voltages and initial inductor currents. The following example illuminates the difficulties.

EXAMPLE 12.1. Figure 12.1 shows three RC circuits coupled through the use of dependent voltage sources. The goal of this example is to construct a differential equation model, determine the solution form in terms of arbitrary constants, and demonstrate the difficulties with the simple recipe of the above algorithm by attempting to relate the arbitrary constants to the initial conditions.

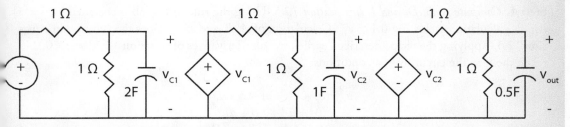

FIGURE 12.1 A cascade of three RC circuits coupled by means of dependent voltage sources. The differential equation model of the circuit is third order.

SOLUTION
Step 1. *Construct the differential equation of the circuit.* For this task, first write a differential equation relating v_{C1} to v_{in}. Then write one relating v_{C2} to v_{C1}, and finally, write one relating v_{out} to v_{C2}. Some straightforward algebra leads to the following three differential equations:

$$\frac{dv_{C1}}{dt} + v_{C1}(t) = 0.5v_{in} \tag{12.1}$$

$$0.5\frac{dv_{C2}}{dt} + v_{C2}(t) = 0.5v_{C1} \tag{12.2}$$

$$0.25\frac{dv_{out}}{dt} + v_{out}(t) = 0.5v_{C2} \tag{12.3}$$

Successively substituting equation 12.1 into equation 12.2 and equation 12.3 into the result produces the input-output differential equation model,

$$\frac{d^3v_{out}}{dt^3} + 7\frac{d^2v_{out}}{dt^2} + 14\frac{dv_{out}}{dt} + 8\,v_{out} = v_{in} \tag{12.4}$$

Step 2. *Compute the characteristic equation and its roots.* The characteristic equation for differential equation 12.4 is

$$s^3 + 7s^2 + 14s + 8 = (s - a)\,(s - b)\,(s - d) = 0$$

which has roots $a = -1$, $b = -2$, and $d = -4$.

Step 3. *Determine the form of the solution.* If $v_{in}(t) = V_{in}u(t)$, then the complete solution has the form

$$v_{out}(t) = Ae^{at} + Be^{bt} + De^{dt} + E = Ae^{-t} + Be^{-2t} + De^{-4t} + E \tag{12.5}$$

for $t \geq 0$.

Step 4. *Compute A, B, D, and E in equation 12.5.* Using the rule of thumb mentioned earlier, a simple calculation yields $E = 0.125V_{in}$. Calculation of A, B, and D specifies the solution in equation 12.5. Applying the recipe described earlier, we take derivatives of equation 12.5, set $t = 0$, and relate them to the circuit initial conditions:

$$v_{out}(0) - V_0 = A + B + D$$
$$\dot{v}_{out}(0) = aA + bB + dD$$
$$\ddot{v}_{out}(0) = a^2A + b^2B + d^2D$$

Again, *one dot over a variable means a first-order time derivative, and two dots denotes a second-order time derivative.* A, B, and D are computed by solving this set of equations. The difficulty is in specifying $\dot{v}_{out}(0)$ and $\ddot{v}_{out}(0)$. First, $v_{out}(0)$ is simply the initial capacitor voltage on the third capacitor. However, $\dot{v}_{out}(t)$ is proportional to the current through it, which depends on all the initial capacitor voltages. Further, what is the physical interpretation of $\ddot{v}_{out}(t)$? And how do $\dot{v}_{out}(0)$ and $\ddot{v}_{out}(0)$ relate to the initial capacitor voltages? The relationship is complex and lacks any meaningful physical interpretation. Finally, even for this simple example, computation and solution of the differential equation 12.4 proves tedious.

Exercise. For Example 12.1, compute expressions for $\dot{v}_{out}(0)$ and $\ddot{v}_{out}(0)$.
ANSWERS: $\dot{v}_{out}(0) = [2v_{C2}(0) - 4\,v_{out}(0)]$; $\ddot{v}_{out}(0) = [16v_{out}(0) - 12v_{C2}(0) + 2v_{C2}(0)]$

One of the advantages of Laplace transform analysis is that it does not destroy the physical meaning of the circuit variables in the analysis process. Chapter 13 addresses how the Laplace transform approach explicitly accounts for initial capacitor voltages and initial inductor currents.

3. OVERVIEW OF LAPLACE TRANSFORM ANALYSIS

Laplace transform analysis is a technique that transforms the time domain analysis of a circuit, system, or differential equation to the so-called **frequency domain**. In the frequency domain, solution of the equations is generally much easier. Hence, obtaining the output responses of a circuit to known inputs proceeds more smoothly.

To apply the technique, one takes the Laplace transform of the time-dependent input signal or signals to produce new signals dependent only on a new frequency variable s. In an intuitive sense, and as precisely derived later, one also takes the Laplace transform of the circuit. Assuming zero initial conditions, one multiplies these two transforms together to produce the Laplace transform of the output signal. Taking the inverse Laplace transform of the output signal by means of known algebraic and table look-up formulas yields the desired response of the circuit. The effect of initial conditions is easily incorporated.

Figure 12.2 is a pictorial rendition of the method. As just mentioned, one transforms the input signal, transforms the circuit to obtain an equivalent circuit in the Laplace transform world, and computes the Laplace transform of the output by "multiplying" the two transforms together. Inverting this (output) transform with the aid of a lookup table or MATLAB produces the desired output signal.

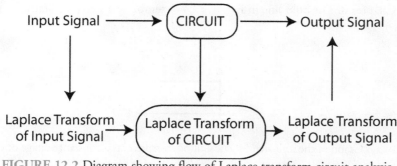

FIGURE 12.2 Diagram showing flow of Laplace transform circuit analysis.

In a mathematical context, one executes the same type of procedure on a differential equation model of a circuit and, indeed, differential equations in general. Figure 12.3 illustrates the idea.

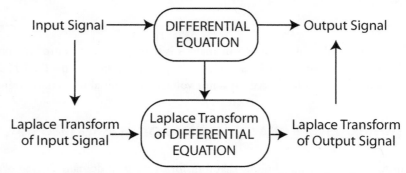

FIGURE 12.3 Diagram showing flow of Laplace transform analysis for solution of differential equations.

The benefit of this type of analysis lies in its numerous uses. Some of these uses include steady-state and transient analysis of circuits driven by complicated as well as the usual basic signals, a straightforward lookup table approach for computing solutions, and explicit incorporation of capacitor and inductor initial conditions in the analysis. The forthcoming sections will flesh out these applications.

4. BASIC SIGNALS

Several basic signals are fundamental to circuit analysis, as well as to future courses in systems analysis. Perhaps the most common signal is the **unit step function**,

$$u(t) = \begin{cases} 1, & t \geq 0 \\ 0, & t < 0 \end{cases} \quad . \tag{12.6}$$

defined in Chapter 8. The bold line in Figure 12.4, resembling a step on a staircase, represents the graph of $u(t)$.

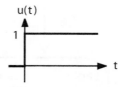

FIGURE 12.4 Graph of the unit step function. It often represents a constant voltage or current level.

The unit step function has many practical uses, including the mathematical representation of dc voltage levels. Any type of sustained, constant physical phenomenon, such as constant pressure, constant heat, or the constant thrust of a jet engine, has a step-like behavior. In the case of jet engine thrust, a pilot sends a command signal through the control panel to the engine requesting a given amount of thrust. The step function models this command signal.

The **shifted step**, shown in Figure 12.5, models a time-delayed unit step signal.

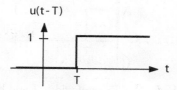

FIGURE 12.5 Graph of a unit step shifted T units to right.
This function is often used to represent a delayed startup.

Shifted steps, $u(t - T)$, often represent voltages that turn on after a prescribed time period T. The *flipped step function*, $u(T - t)$, of Figure 12.6 depicts yet another variation on the unit step. Here the step takes on the value of unity for time $t \le T$. Often it provides an idealized model of signals that have excited the circuit for a long time and turn off at time T. The key to knowing the values of these various step functions is to test whether the argument is non-negative or negative. Whenever the argument is non-negative, the value is 1; when the argument is negative, the value is zero.

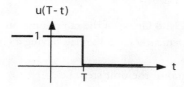

FIGURE 12.6 Graph of flipped and shifted unit step. This function is often used to model signals
that have been on for a long time and turn off at time T.

Exercise. Represent each of the following functions as sums of step functions:

(i) $f(t) = \begin{cases} 1, & 0 \le t \le 2 \\ 0, & \text{otherwise} \end{cases}$

(ii) $f(t) = \begin{cases} 1, & -3 < t < 6 \\ 0, & \text{otherwise} \end{cases}$

(iii) $f(t) = \begin{cases} 1, & t \le -1 \\ 0, & -1 < t < 1 \\ 1, & t \ge 1 \end{cases}$

ANSWERS: in random order: $u(-1 - t) + u(t - 1)$, $u(t) - u(t - 2)$, $u(t + 3) - u(t - 6)$

The *pulse function*, $p_T(t)$, of Figure 12.7 is the product of a step and a flipped step or, equivalently, the difference of a step and a shifted step. Specifically, a pulse of height A and width T is

$$p_T(t) = Au(t)u(T - t) = Au(t) - Au(t - T) \qquad (12.7)$$

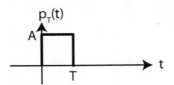

FIGURE 12.7 Pulse of width T and height A.
This function is often used to model signals of fixed magnitude and short duration.

A signal sharing a close kinship with the unit step is the **ramp function** $r(t)$ depicted in Figure 12.8.

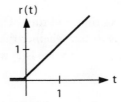

FIGURE 12.8 Graph of the ramp function, $r(t) = tu(t)$.
Ramp functions conveniently model signals having a constant rate of increase.

The ramp function is the integral of the unit step, i.e.,

$$r(t) = \int_{-\infty}^{t} u(\tau) \, d\tau = t \, u(t) \tag{12.8}$$

where τ is simply a dummy variable of integration.

Exercise. Plot $r(-t)$ and $r(t-2)$.
ANSWER: $r(-t)$ is the reflection of $r(t)$ about the vertical axis while $r(t-2)$ is simply the shift of $r(t)$ by two units to the right.

EXAMPLE 12.2
Express Figures 12.9a and b in terms of steps and ramps.

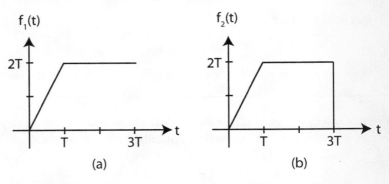

FIGURE 12.9 Two signals to be represented by steps and ramps.

SOLUTION

For the signal $f_1(t)$, observe that the signal begins with a ramp with a slope of 2. Thus we have $f_1(t) = 2r(t) + ?$. At $t = T$, the signal $f_1(t)$ levels off. Since the $2r(t)$ part of the signal continues to increase, the increase must be canceled by another ramp of slope 2, but shifted to the right by T units. Thus, $f_1(t) = 2r(t) - 2r(t - T)$.

The signal $f_2(t)$ replicates $f_1(t)$ up to $3T$. After $3T$, the signal drops to zero. Hence we must subtract a shifted step of height $2T$ from $f_1(t)$. Thus $f_2(t) = f_1(t) - 2Tu(t - 3T) = 2r(t) - 2r(t - T) - 2Tu(t - 3T)$.

Exercises. 1. Figure 12.10 depicts a sawtooth waveform denoted by $f(t)$. Sequences of sawtooth waveforms are used as timing signals in televisions and other electronic devices.

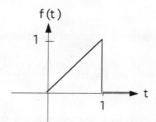

FIGURE 12.10 A sawtooth waveform.

ANSWER: $f(t) = r(t) - r(t - 1) - u(t - 1)$

2. For $f(t)$ of Figure 12.10, plot $f(1 - t)$ and represent the function in terms of steps and ramps.
ANSWER: $f(1 - t) = r(1 - t) - r(-t) - u(-t)$

EXAMPLE 12.3. Express Figure 12.11 in terms of steps and ramps.

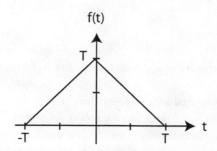

FIGURE 12.11 Triangular waveform to be represented by steps and ramps.

SOLUTION

Observe that the signal $f(t)$ begins with a ramp with a slope of 1 at $t = -T$. Thus $f(t) = r(t + T) + ?$. The signal falls off with a linearly decreasing ramp for $0 \le t \le T$. If we subtract $r(t)$ the signal would become flat for $t \ge 0$. Thus we subtract $2r(t)$ to obtain the linear decrease. Hence, $f(t) =$

$r(t + T) - 2r(t) + ?$. For $t \geq T$, this signal, $r(t + T) - 2r(t)$, continues to linearly decrease. Hence for the signal to be zero for $t \geq T$, we cancel this decrease with an additional ramp. Thus $f(t) = r(t + T) - 2r(t) + r(t - T)$.

Newtonian physics provides a good motivation for defining the ramp signal. Applying a constant force to an object causes a constant acceleration having the functional form $Ku(t)$. The integral of acceleration is velocity, which has the form $Kr(t)$, a ramp function.

A very common and conceptually useful signal is the (Dirac) **delta function**, or **unit impulse function**, implicitly defined by its relationship to the unit step as

$$u(t) = \int_{-\infty}^{t} \delta(q)dq$$

(12.9)

The relationship of equation 12.9 prompts a natural inclination to define

$$\delta(t) = \frac{d}{dt}u(t) = \lim_{h \to 0} \frac{u(t) - u(t - h)}{h}$$

(12.10)

Strictly speaking, the derivative of $u(t)$ does not exist at $t = 0$, due to the discontinuity at that point. Without delving into the mathematics, one typically interprets equations 12.9 and 12.10 as follows: define a set of continuous differentiable functions $u_\Delta(t)$, as illustrated in Figure 12.12a. The derivatives, $\delta_\Delta(t) = \frac{d}{dt}u_\Delta(t)$, of these functions are depicted in Figure 12.12b.

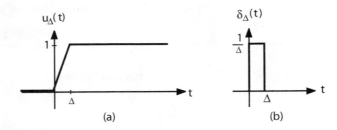

FIGURE 12.12 (a) Continuous differentiable approximation to the unit step.
(b) Derivative of $u_\Delta(t)$: the integral of $\delta_\Delta(t)$ produces $u_\Delta(t)$.

Clearly, $\delta_\Delta(t)$ has a well-defined area of 1, has height $1/\Delta$, and is zero outside the interval $0 \leq t \leq \Delta$. In addition, $u(t) = \lim u_\Delta(t)$, and $\lim \delta_\Delta = \delta(t)$ as $\Delta \to 0$. Hence, although the definition of equation 12.10 is not mathematically rigorous, one can interpret the delta function as the limit of a set of well-behaved functions. In fact, the delta function can be viewed as the limit of a variety of different sets of functions. A problem at the end of the chapter explores this phenomenon.

Despite the preceding mathematics, the delta function is not a function at all, but a distribution,[1] and its rigorous definition (in terms of so-called testing functions) is left to more advanced mathematics courses. Nevertheless, we shall still refer to it as the delta or impulse function. The standard graphical illustration of the delta function appears in Figure 12.13, which shows a pulse of infinite height, zero width, and a well-defined area of unity, as identified by the "1" next to the spike. Visualization of the delta function by means of the spike in the figure will aid our understanding, explanations, and calculations that follow.

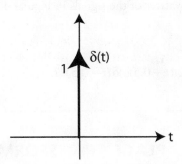

FIGURE 12.13 Standard graphical illustration of a unit impulse function having a well-defined area of 1. The function typically represents an energy transfer, large force, or large impact over a very short time duration, as might occur when a bat hits a baseball.

The unit area property follows from equation 12.9, i.e.,

$$\int_{-\infty}^{\infty} \delta(t)dt = \int_{0^-}^{0^+} \delta(t)dt = u(0^+) - u(0^-) = 1$$

where 0^- is infinitesimally to the left and 0^+ infinitesimally to the right of zero. *If the area is different from unity, a number K alongside the spike will designate the area; i.e., the spike will be a signal $K\delta(t)$.*

One motivation for defining the delta function is its ability to "ideally" represent phenomena in nature involving relative immediate energy transfer (i.e., the elapsed time over which energy transfer takes place is very small compared to the macroscopic behavior of the physical process). An exploding shell inside a gun chamber causing a bullet to change its given initial velocity from zero to some nonzero value "instantaneously" is an example. Another is a batter who hits a pitched ball, "instantaneously" transferring the energy of the swung bat to the ball. Also, the delta function provides a mathematical setting for representing the sampling of a continuous signal. Suppose, for example, that a continuous signal $v(t)$ is to be sampled at discrete time instants $t_1, t_2, t_3, \dots$; i.e., $v(t_i)$ is to be physically measured at these time instants. The mathematical representation of this measuring process is given by the **sifting property** of the delta function,

$$v(t_i) = \int_{-\infty}^{\infty} v(\tau)\delta(\tau - t_i)d\tau \tag{12.11}$$

In other words, the value of the integral is the non-impulsive part of the (continuous) integrand, $v(\tau)$, replaced by that value of τ which makes the argument of the impulse zero, in this case $\tau = t_i$. Verifying equation 12.11 depends on an application of the definition given in equation 12.9. Specifically, if $v(t)$ is continuous at $t = t_i$, then

$$\int_{-\infty}^{\infty} v(\tau)\delta(\tau - t_i)d\tau = \int_{t_i^-}^{t_i^+} v(\tau)\delta(\tau - t_i)d\tau$$

$$= v(t_i)\int_{t_i^-}^{t_i^+} \delta(\tau - t_i)d\tau = v(t_i)$$

by equation 12.9, where $t_i^{\pm}$ are infinitesimally to the right and left of t_i.

Exercises. 1. Compute the derivatives of the signals in Figures 12.9a, 12.9b, and 12.11.
ANSWERS: $\dot{f}_1(t) = 2\big[u(t) - u(t-T)\big]$, $\dot{f}_2(t) = 2\big[u(t) - u(t-T)\big] - 2T\delta(t-3T)$,
$$\dot{f}(t) = u(t+T) - 2u(t) + u(t-T)$$
2. Compute the integral $\displaystyle\int_{-\infty}^{\infty} \sin(2\pi t + 0.5\pi)\delta(t-0.5)dt$.
ANSWER: $\sin(1.5\pi)$

---

5. THE ONE-SIDED LAPLACE TRANSFORM

Intuitively, a transform is like a prism that breaks white light apart into its colored spectral components. The *one-sided* or *unilateral Laplace transform* is an integral mapping, somewhat like a prism, between time-dependent signals $f(t)$ and functions $F(s)$ that are dependent on a complex variable s, called complex frequency.

LAPLACE TRANSFORM
Mathematically, the one-sided Laplace transform of $f(t)$ is

$$\mathcal{L}[f(t)] = F(s) = \int_{0^-}^{\infty} f(t)e^{-st}dt \tag{12.12}$$

where $s = \sigma + j\omega\,(j = \sqrt{-1})$ is a complex variable ordinarily called a complex frequency in the signals and systems literature.

As the equation makes plain, the Laplace transform integrates out time to obtain a new function, $F(s)$, displaying the frequency content of the original time function $f(t)$. In the vernacular, $F(s)$ is the frequency domain counterpart of $f(t)$. Analysis using Laplace transforms is often called frequency domain analysis.

Exercises. 1. Find the Laplace transform of a scaled Dirac delta function, $K\delta(t)$.
ANSWER: K

2. Find $\mathcal{L}[\sin(2\pi t + 0.5\pi)\delta(t-0.5)]$.
ANSWER: $\sin(1.5\pi)e^{-0.5s}$.

A number of questions about the Laplace transform promptly arise:

Question 1: *Why is it called one-sided or unilateral?*

Answer: It is called unilateral because the lower limit of integration is 0^-, as opposed to $-\infty$. If the lower limit of integration were $-\infty$, equation 12.12 would be called the *two-sided Laplace transform,* which is not covered in this text.

Question 2: *Why use 0^- instead of 0^+ or 0 as the lower limit of integration?*

Answer: Our future circuit analysis must account for the effect of "instantaneous energy transfer" and, hence, impulses at $t = 0$. The use of 0^+ would exclude such direct analysis, since the Laplace transform of the impulse function would be zero. Using $t = 0$ is simply ambiguous.

Question 3: *What about functions that are nonzero for $t < 0$?*

Answer: Because the lower limit of integration in equation 12.12 is 0^-, the Laplace transform does not distinguish between functions that are different for $t < 0$ but equal for $t \geq 0$ (e.g., $u(t)$ and $u(t + 1)$ would have the same unilateral Laplace transform). However, since $t = 0$ designates the universal starting time of a circuit or system, the class of signals dealt with will usually be zero for $t < 0$ and thus will have a unique (one-sided) Laplace transform. Conversely, each Laplace transform $F(s)$ will determine a unique time function $f(t)$ with the property that $f(t) = 0$ for $t \leq 0$. Because of this dual uniqueness, the one-sided Laplace transform is said to be bi-unique for signals $f(t)$ with $f(t) = 0$ for $t < 0$.

Question 4: *Does every signal $f(t)$ such that $f(t) = 0$ for $t < 0$ have a Laplace transform?*

Answer: No. For example, the function $f(t) = e^{t^2}u(t)$ does not have a Laplace transform because the integral of equation 12.12 does not exist for this function. To see why, one must study the Laplace transform integral closely, i.e.,

$$\int_{0^-}^{\infty} e^{t^2} e^{-st}\,dt = \int_{0^-}^{\infty} e^{t^2 - st}\,dt = \int_{0^-}^{\infty} e^{t^2 - \sigma t} e^{-j\omega t}\,dt \tag{12.13}$$

Observe that $e^{-j\omega t} = \cos(\omega t) - j\sin(\omega t)$ is a complex sinusoid. As t approaches infinity, the real and imaginary parts of the integrand in equation 12.13 must blow up, due to the $e^{t^2 - \sigma t}$ term. Hence, the area underneath the curve $e^{t^2 - \sigma t}$ grows to infinity, and the integral does not exist for any value of σ.

Whenever $f(t)$ is piecewise continuous, a sufficient condition for the existence of the Laplace transform is that

$$|f(t)| \leq k_1 e^{k_2 t} \tag{12.14}$$

for some constants k_1 and k_2. This bound restricts the growth of a function; i.e., the function cannot rise more rapidly than an exponential. Such a function is said to be *exponentially bounded.* The condition, however, is not necessary for existence. Specifically, the transform exists whenever the integral exists, even if the function $f(t)$ is unbounded. Without belaboring the mathematical rigor underlying the Laplace transform, we will presume throughout the book that the functions we are dealing with are Laplace transformable.

Question 5: *Why does the existence of the Laplace transform integral depend on the value of* σ, *mentioned in the answer to question 4?*

Answer: If the condition in equation 12.14 is satisfied, then there is a range of σ's (recall that $s = \sigma + j\omega$) over which the Laplace transform integral is convergent. This is explained in the following example.

EXAMPLE 12.4. Find the Laplace transform of the unit step. By equation 12.12,

$$\mathcal{L}[u(t)] = U(s) = \int_{0^-}^{\infty} u(t)e^{-st}\,dt = \int_{0^-}^{\infty} e^{-st}\,dt = \int_{0^-}^{\infty} e^{-(\sigma+j\omega)t}\,dt \tag{12.15}$$

$$= -\left.\frac{e^{-\sigma t}e^{-j\omega t}}{\sigma + j\omega}\right]_{0^-}^{\infty} = \frac{1}{\sigma + j\omega} = \frac{1}{s},$$

provided that $\sigma > 0$. Notice that if $\sigma > 0$, then $e^{-\sigma t}e^{-j\omega t} \to 0$ as $t \to \infty$. This keeps the area underneath the curve finite. For $\sigma \leq 0$, the Laplace transform integral will not exist for the unit step. The smallest number σ_0 such that for all $\sigma > \sigma_0$ the Laplace transform integral exists is called the **abscissa of (absolute) convergence**. In the case of the unit step, the integral exists for all $\sigma > 0$; hence, $\sigma_0 = 0$ is the abscissa of convergence. The region $\sigma > 0$ is said to be the **region of convergence (ROC)** of the Laplace transform of the unit step. Figure 12.14 illustrates the ROC for the unit step.

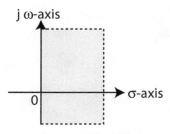

FIGURE 12.14 Region of convergence, $\sigma > 0$, of the Laplace transform of the unit step function (i.e., the Laplace transform integral will exist for all $\sigma > 0$).

Question 6: *Is the unilateral Laplace transform valid only in its region of convergence?*

Answer: Again, the answer is no. There is a method in the theory of complex variables called "analytic continuation" which, although beyond the scope of this text, permits us to uniquely and analytically extend the transform to the entire complex plane.[2] *Analytically* means smoothly and also that the extension is valid everywhere except at the poles (to be discussed later) of the transform. Thus, the region of convergence goes unmentioned in the standard mathematical tables of one-sided Laplace transform pairs.

EXAMPLE 12.5. Find $F(s)$ for $f(t) = Ke^{-at} u(t)$, where K and a are scalars.

SOLUTION
Applying equation 12.12 yields

$$F(s) = \int_{0^-}^{\infty} \left(Ke^{-at} \right) e^{-st} dt = K \int_{0^-}^{\infty} e^{-(s+a)t} dt = \frac{K}{s+a} \qquad (12.16)$$

The integral exists if $\text{Re}[s + a] > 0$. If a is real, then the ROC is $\sigma > -a$. As mentioned in the answer to question 6, by **analytic continuation**, $F(s) = 1/(s + a)$ is valid and analytic in the entire complex plane, except at the point $s = -a$. The point $s = -a$ is a **pole** of the rational function $1/(s + a)$ because as s approaches $-a$, the value of the function becomes infinitely large.

The preceding discussion and examples set up the mathematical framework of the Laplace transform method. Our eventual focus rests on its application to circuit theory, which builds on two fundamental laws: Kirchhoff's voltage law (KVL) and Kirchhoff's current law (KCL). KVL requires that the voltage drops around any closed loop sum to zero, and KCL requires that the sum of all the currents entering a node be zero. For the Laplace transform technique to be useful, it must distribute over such sums of voltages and currents. Fortunately, it does.

Linearity property: The Laplace transform operation is linear. Suppose $f(t) = a_1 f_1(t) + a_2 f_2(t)$. Then

$$\begin{aligned} \mathcal{L}[f(t)] = \mathcal{L}[a_1 f1(t) + a_2 f_2(t)] &= a_1 \mathcal{L}[f_1(t)] + a_2 \mathcal{L}[f_2(t)] \\ &= a_1 F_1(s) + a_2 F_2(t) \end{aligned} \qquad (12.17)$$

This property is easy to verify since integration is linear:

$$\mathcal{L}[a_1 f_1(t) + a_2 f_2(t)] = \int_{0^-}^{\infty} [a_1 f_1(t) + a_2 f_2(t)] e^{-st} dt$$

$$= a_1 \int_{0^-}^{\infty} f_1(t) e^{-st} dt + a_2 \int_{0^-}^{\infty} f_2(t) e^{-st} dt$$

This is precisely what equation 12.17 states. Hence, our curiosity satisfied, we may rest peacefully in the knowledge that the Laplace transform technique conforms to the basic laws of KVL and KCL.

EXAMPLE 12.6. Find $F(s)$ when $f(t) = K_1 u(t) + K_2 e^{-at} u(t)$, for real scalars K_1, K_2, and a.

SOLUTION
The Laplace transform of $u(t)$ is $1/s$ by equation 12.15 and that of $e^{-at} u(t)$ is $1/(s + a)$ by equation 12.16. By the linearity property (equation 12.17),

$$F(s) = \frac{K_1}{s} + \frac{K_2}{s+a},$$

with region of convergence $\{\sigma > 0\} \cap \{\sigma > -a\}$, where $\cap$ denotes intersection. By analytic continuation, the transform is valid in the entire complex plane except at the poles, $s = 0$ and $s = -a$. (Henceforth we will not mention the ROC in our calculations.)

Exercise. Find the Laplace transforms of
- (*i*) $f(t) = e^{at}u(t) + 2e^{-at}u(t) + 2u(t)$,
- (*ii*) $f(t) = -2u(t) + e^{at}u(t) - 2e^{-at}u(t)$, and
- (*iii*) $f(t) = 3u(t) - 4e^{-3t}u(t) + 2e^{-4t}u(t)$

ANSWERS: (*i*) $\dfrac{3s-a}{s^2-a^2} + \dfrac{2}{s}$, (*ii*) $\dfrac{3a-s}{s^2-a^2} - \dfrac{2}{s}$, (*iii*) $\dfrac{3}{s} - \dfrac{4}{s+3} + \dfrac{2}{s+4}$

The transform integral of equation 12.12 has various properties. These properties provide short-cuts in the transform computation of complicated as well as simple signals. For example, the Laplace transform of a right shift of the signal $f(t)$ always has the form $e^{-sT}F(s)$, $T > 0$. Shifts are important for two reasons:

1. Many signals can be expressed as the sum of simple signals and shifts of simple signals.
2. Excitations of circuits are often delayed from $t = 0$.

Hence, provisions for shifts must be built into analysis techniques.

Time shift property: If $\mathcal{L}[f(t)u(t)] = F(s)$, then, for $T > 0$,

$$\mathcal{L}[f(t-T)u(t-T)] = e^{-sT}F(s) \tag{12.18}$$

Verification of this property comes from a direct calculation of the Laplace transform for the shifted function, i.e.,

$$\mathcal{L}[f(t-T)u(t-T)] = \int_{0^-}^{\infty} f(t-T)u(t-T)e^{-st}\,dt = \int_{T^-}^{\infty} f(t-T)e^{-st}\,dt$$

Let $q = t - T$, with $dq = dt$. Noting that the lower limit of integration becomes 0^- with respect to q,

$$\mathcal{L}[f(t-T)u(t-T)] = \int_{0^-}^{\infty} f(q)e^{-sq}e^{-sT}\,dq = e^{-sT}\int_{0^-}^{\infty} f(q)e^{-sq}\,dq = e^{-sT}F(s)$$

Observe that if $T < 0$, the property fails to make sense, since $f(t-T)u(t-T)$ would then shift left. Since the transform ignores information to the left of 0^-, one cannot, strictly speaking, recover $f(t)$ from the resulting transform.

Exercises. 1. Find $\mathcal{L}[f(t-T)]$ when $f(t)$ is (i) $A\delta(t)$, (ii) $Au(t)$, and (iii) $Ae^{-at}u(t)$.
2. $p_T(t) = Au(t) - Au(t-T)$.
ANSWERS: In random order: $\dfrac{e^{-sT}}{s+a}$, $A\dfrac{1-e^{-sT}}{s}$, , e^{-sT}, Ke^{-sT}/s

EXAMPLE 12.7. Using the **time shift** property, find $F(s)$ for the signal $f(t)$ sketched in Figure 12.15.

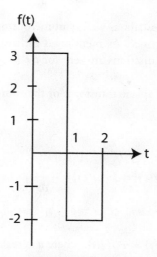

FIGURE 12.15 Signal for Example 12.7.

SOLUTION

Using step functions and shifted step functions, we obtain

$$f(t) = 3u(t) - 5u(t - 1) + 2u(t - 2)$$

Direct application of linearity and the time shift property yields $F(s) = \dfrac{3}{s} - \dfrac{5e^{-s}}{s} + \dfrac{2e^{-2s}}{s}$.

Exercises. 1. Find the Laplace transform of the pulse signal of equation 12.7.

2. Find $F(s)$ when $f(t) = A_1 u(t - T_1) + A_2 u(t - T_2) + A_3 u(t - T_3)$.

ANSWERS: $P_T(s) = A\dfrac{1 - e^{-sT}}{s}$, $F(s) = \dfrac{A_1 e^{-sT_1}}{s} + \dfrac{A_2 e^{-sT_2}}{s} + \dfrac{A_3 e^{-sT_3}}{s}$

One more property allows us to revisit the signals discussed in section 3 and take their Laplace transforms. The new property is multiplication of $f(t)$ by t. This always results in a Laplace transform that is the negative of the derivative of $F(s)$.

Multiplication-by-t property: Let $F(s) = \mathcal{L}[f(t)]$. Then

$$\mathcal{L}[tf(t)] = -\frac{d}{ds}F(s) \qquad (12.19)$$

Verification of this property follows by a direct application of the Laplace transform integral to $tf(t)u(t)$ with the observation that $te^{-st} = -\dfrac{d}{ds}e^{-st}$. In particular,

$$\int_{0^-}^{\infty} tf(t)e^{-st}\,dt = -\int_{0^-}^{\infty} f(t)\left[\frac{d}{ds}e^{-st}\right]dt = -\frac{d}{ds}\int_{0^-}^{\infty} f(t)e^{-st}\,dt = -\frac{d}{ds}F(s)$$

Table 12.1 lists this transform pair, as well as numerous other such pairs, without mention of the underlying region of convergence. As mentioned earlier, we shall dispense with any mention of the ROC, assuming that all functions are zero for $t < 0$.

EXAMPLE 12.8. Find the Laplace transform of the **ramp function**, $r(t) = tu(t)$.

SOLUTION
Using equation 12.19,

$$R(s) = \mathcal{L}[r(t)] = \mathcal{L}[tu(t)] = -\frac{d}{ds}\left(\frac{1}{s}\right) = -\frac{d}{ds}\left(s^{-1}\right) = \frac{1}{s^2} \tag{12.20}$$

EXAMPLE 12.9. Suppose $f(t) = te^{-at}u(t)$, where a is real. Find $F(s)$.

SOLUTION
The quickest way to obtain the answer is to apply equation 12.19. Specifically, since $\mathcal{L}\left[e^{-at}u(t)\right] = \dfrac{1}{s+a}$,

$$\mathcal{L}\left[te^{-at}u(t)\right] = -\frac{d}{ds}\left[\frac{1}{s+a}\right] = -\frac{d}{ds}\left[(s+a)^{-1}\right] = \frac{1}{(s+a)^2} \tag{12.21}$$

An alternative, more tedious approach is to use integration by parts as follows:

$$F(s) = \mathcal{L}\left[e^{-at}u(t)\right] = \int_{0^-}^{\infty} te^{-(s+a)}\,dt = \int_{0^-}^{\infty} v\,du = uv\bigg]_{0^-}^{\infty} - \int_{0^-}^{\infty} u\,dv$$

where $v = t$ and $du = e^{-(a+s)t}\,dt$. Thus,

$$F(s) = -\frac{te^{-(s+a)t}}{s+a}\bigg]_{0^-}^{\infty} + \int_{0^-}^{\infty} \frac{e^{-(s+a)t}}{s+a}\,dt$$

The ROC is $\sigma > -a$, in which case the first term on the right-hand side is zero. Thus, in this ROC, evaluation of the integral term implies that

$$F(s) = \int_{0^-}^{\infty} \frac{e^{-(s+a)t}}{s+a}\,dt = \frac{1}{(s+a)^2}$$

Equation 12.21 is a special case of the more general formula

$$\mathcal{L}[t^n e^{-at}u(t)] = \frac{n!}{(s+a)^{n+1}} \tag{12.22}$$

Exercise. Find the Laplace transform of $f(t) = t^2 e^{-at} u(t)$.

ANSWER: $F(s) = \dfrac{2}{(s+a)^3}$

EXAMPLE 12.10. Find $F(s)$ for the signal depicted in Figure 12.16.

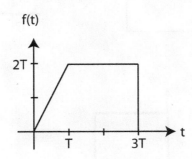

FIGURE 12.16 A signal to be represented by steps and ramps.

SOLUTION
From Example 12.2, $f(t) = 2r(t) - 2r(t - T) - 2Tu(t - 3T)$. Hence by linearity, the time shift property, and equations 12.15 and 12.20,

$$F(s) = \mathcal{L}[2r(t) - 2r(t-T) - 2Tu(t-3T)] = \frac{2 - 2e^{-sT}}{s^2} - \frac{2Te^{-3sT}}{s}$$

Exercises. 1. Note that the sawtooth of Figure 12.17 is $f(t) = t[u(t) - u(t-1)]$. Suppose $g(t) = u(t) - u(t-1)$. Compute $F(s) = -\dfrac{d}{ds} G(s)$.

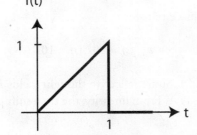

FIGURE 12.17 Sawtooth waveform.

ANSWER: $F(s) = \dfrac{1 - e^{-s}}{s^2} - \dfrac{e^{-s}}{s}$

2. Use equation 12.22 to compute the Laplace transforms of $f(t) = tr(t)$ for the ramp function $r(t)$ and for $f(t) = t^2 r(t)$.

ANSWERS: $\dfrac{2}{s^3}$, $\dfrac{6}{s^4}$

EXAMPLE 12.11. The circuit of Figure 12.18a has two source excitations, $i_1(t)$ and $i_2(t)$, shown in Figures 12.18b and c. Compute $V_{out}(s)$.

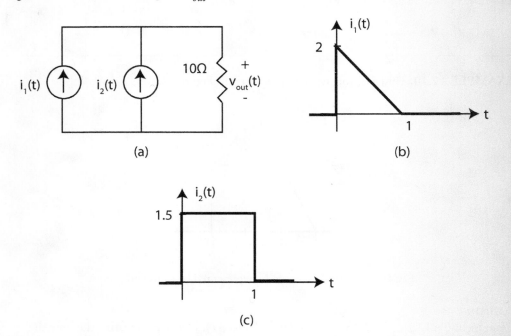

(a)   (b)

(c)

FIGURE 12.18 (a) Resistive circuit driven by two current sources.
(b) Triangular signal, $i_1(t)$, in A. (c) Pulse signal, $i_2(t)$, in A.

SOLUTION
Step 1. *Find the form of* $V_{out}(s)$. By superposition and Ohm's law,

$$v_{out}(t) = 10i_1(t) + 10i_2(t)$$

From the linearity of the Laplace transform,

$$v_{out}(s) = 10I_1(s) + 10I_2(s)$$

Step 2. *Compute* $I_1(s)$ *and* $I_2(s)$. Some reflective thought yields $i_1(t) = 2u(t) - 2r(t) + 2r(t-1)$ A and $i_2(t) = 1.5u(t) - 1.5u(t-1)$ A. From linearity, the time shift property, and the previously computed transforms,

$$I_1(s) = \frac{2}{s} - \frac{2 + 2e^{-s}}{s^2} \quad \text{and} \quad I_2(s) = \frac{1.5 - 1.5e^{-s}}{s}$$

Step 3. *Find* $V_{out}(s)$. Since $V_{out}(s) = 10I_1(s) + 10I_2(s)$, it follows that

$$V_{out}(s) = \frac{35}{s} - \frac{20}{s^2} + e^{-s}\left(\frac{20}{s^2} - \frac{15}{s}\right)$$

Step 4. As an introduction to the next section, by inspection we can compute the time function of the output voltage:

$$v_{out}(t) = 35u(t) - 20r(t) - 15u(t-1) + 20r(t-1)$$

Exercises. 1. Find the Laplace transform of (i) $f_1(t) = e^{at} + e^{-at}$, (ii) $f_2(t) = e^{-at} + e^{-bt} + te^{-bt} + te^{-at}$, and (iii) $f_3(t) = e^{-at}u(t) - e^{-a(t-2)}u(t-2) + u(t-3)$.

ANSWERS: In random order: $\dfrac{1}{s+a} + \dfrac{1}{s+b} + \dfrac{1}{(s+a)^2} + \dfrac{1}{(s+b)^2}$, $\dfrac{1}{(s+a)} + \dfrac{1}{(s-a)}$, $\dfrac{1-e^{-2s}}{s+a} + \dfrac{e^{-3s}}{s}$

2. Recall that $\cos(\omega t) = \dfrac{e^{j\omega t} + e^{-j\omega t}}{2}$. Show that the Laplace transform of $f(t) = \cos(\omega t)u(t)$ is $F(s) = \dfrac{s}{s^2+\omega^2}$.

3. Recall that $\sin(\omega t) = \dfrac{e^{j\omega t} - e^{-j\omega t}}{2j}$. Show that the Laplace transform of $f(t) = \sin(\omega t)u(t)$ is $F(s) = \dfrac{\omega}{s^2+\omega^2}$.

4. Find the time functions associated with $F_1(s) = \dfrac{4}{s}$, $F_2(s) = \dfrac{4}{(s+2)^2}$, $F_3(s) = \dfrac{4e^{-4s}}{s+4}$.

ANSWERS: $f_1(t) = 4u(t)$, $f_2(t) = 4te^{-2t}u(t)$, $f_3(t) = 4e^{-4(t-4)}u(t-4)$

We end this section with Table 12.1, which lists a number of Laplace transform pairs. Some of these will be developed later in the chapter and some in the homework exercises. We will refer to this table in the next section when computing inverse transforms.

TABLE 12.1 Laplace Transform Pairs

Item Number	$f(t)$	$\mathcal{L}[f(t)] = \mathbf{F(s)}$
1	$K\delta(t)$	K
2	$Ku(t)$ or K	K/s
3	$Kr(t)$	K/s^2
4	$t^n u(t)$	$n!/s^{n+1}$
5	$e^{-at}u(t)$	$1/(s + a)$
6	$te^{-at}u(t)$	$1/(s + a)^2$
7	$t^n e^{-at}u(t)$	$n!/(s + a)^{n+1}$
8	$\sin(\omega_0 t)u(t)$	$\omega_0/(s^2 + \omega_0^2)$
9	$\cos(\omega_0 t)u(t)$	$s/(s^2 + \omega_0^2)$
10	$e^{-at}\sin(\omega_0 t)u(t)$	$\dfrac{\omega_0}{(s + a)^2 + \omega_0^2}$
11	$e^{-at}\cos(\omega_0 t)u(t)$	$\dfrac{(s + a)}{(s + a)^2 + \omega_0^2}$
12	$t\sin(\omega_0 t)u(t)$	$\dfrac{2\omega_0 s}{(s^2 + \omega_0^2)^2}$
13	$t\cos(\omega_0 t)u(t)$	$\dfrac{s^2 - \omega_0^2}{(s^2 + \omega_0^2)^2}$
14	$\sin(\omega_0 t + \phi)u(t)$	$\dfrac{s\sin(\phi) + \omega_0\cos(\phi)}{s^2 + \omega_0^2}$
15	$\cos(t + \phi)u(t)$	$\dfrac{s\cos(\phi) - \omega_0\sin(\phi)}{s^2 + \omega_0^2}$
16	$te^{-at}\sin(\omega_0 t)u(t)$	$2\omega_0\dfrac{s + a}{[(s + a)^2 + \omega_0^2]^2}$
17	$te^{-at}\cos(\omega_0 t)u(t)$	$\dfrac{(s + a)^2 - \omega_0^2}{((s + a)^2 + \omega_0^2)^2}$
18	$e^{-at}[\sin(\omega_0 t) - \omega_0 t\cos(\omega_0 t)]u(t)$	$\dfrac{2\omega_0^3}{[(s + a)^2 + \omega_0^2]^2}$
19	$\left[C_1\cos(\omega_0 t) + \left(\dfrac{C_2 - C_1 a}{\omega_0}\right)\sin(\omega_0 t)\right]u(t)$	$\dfrac{C_1 s + C_2}{(s + a)^2 + \omega_0^2}$
20	$2\sqrt{A^2 + B^2}\, e^{-at}\cos\left(\omega_0 t - \tan^{-1}\left(\dfrac{B}{A}\right)\right)$	$\dfrac{A + jB}{s + a + j\omega_0} + \dfrac{A - jB}{s + a - j\omega_0}$
21	$2\sqrt{A^2 + B^2}\, te^{-at}\cos\left(\omega_0 t - \tan^{-1}\left(\dfrac{B}{A}\right)\right)$	$\dfrac{A + jB}{(s + a + j\omega_0)^2} + \dfrac{A - jB}{(s + a - j\omega_0)^2}$

6. THE INVERSE LAPLACE TRANSFORM

For the Laplace transform tool to effectively analyze circuits, one must be able to uniquely reconstruct time functions $f(t)$ from their frequency domain partners $F(s)$. Theoretically, this is attained through the inverse Laplace transform integral.

INVERSE LAPLACE TRANSFORM

Intuitively, if $\mathcal{L}[f(t)] = F(s)$, then $f(t) = \mathcal{L}^{-1}[F(s)]$. Rigorously speaking, the **inverse Laplace transform** integral is a complex line integral defined as

$$f(t) = \mathcal{L}^{-1}[F(s)] = \frac{1}{2\pi j}\int_\Gamma F(s)e^{st}ds \tag{12.23}$$

over a particular path Γ in the complex plane. The path Γ is typically taken to be the vertical line $\sigma_1 + j\omega$ where ω ranges from $-\infty$ to $+\infty$ and σ_1 is any real number greater than σ_0, the abscissa of absolute convergence.

This integral uniquely reconstructs the time structure of $F(s)$ to obtain $f(t)$ in which $f(t)$ is zero for $t < 0$. Conceptually, the process resembles the reverse action of a prism, to produce white light from its spectral components. An appreciation for the power of this integral requires a solid background in complex variables and would not aid our purpose, the analysis of circuits. In fact, the evaluation of the integral is carried out using the famous residue theorem of complex variables. Further discussion is beyond the scope of this text.

Just as the Laplace transform is **linear**, so, too, is the inverse Laplace transform, as its integral structure suggests, i.e., $\mathcal{L}^{-1}[a_1F_1(s) + a_2F_2(s)] = a_1f_1(t) + a_2f_2(t)$. Also, the unilateral transform pair $\{f(t), F(s)\}$ is *unique*, where by unique we mean the following: let $F_1(s) = \mathcal{L}[f_1(t)]$ and $F_2(s) = \mathcal{L}[f_2(t)]$ coincide in any small open region of the complex plane. Then $F_1(s) = F_2(s)$ over their common regions of convergence, and $f_1(t) = f_2(t)$ for almost all $t > 0$. "Almost all" means except for a small or thin set of isolated points that are of no engineering significance. Hence, there is a one-to-one correspondence between time functions $f(t)$ for which $f(t) = 0$ for $t < 0$ and their one-sided Laplace transforms. Linearity and this uniqueness make the Laplace transform technique a productive tool for circuit analysis.

Virtually all the transforms of interest to us have a **rational function** structure; i.e., $F(s)$ is the ratio of two polynomials. Rational functions may be decomposed into sums of simple rational functions. These simple rational functions are called *partial fractions* and their sums are known as **partial fraction expansions**. Two of the more common "simple" terms in partial fraction expansions have the form K/s and $K/(s + a)$. Such simple rational functions correspond to the transforms of steps, exponentials, and the like. Table 12.1 lists these known inverse transforms. With the table, direct evaluation of the line integral in equation 12.23 becomes unnecessary. Our goal is to describe techniques to compute the simple rational functions in a partial fraction of $F(s)$. Once these are found, the transform dictionary in Table 12.1, in conjunction with some well-known properties of the Laplace transform, will allow us quickly to compute the time function $f(t)$.

PARTIAL FRACTION EXPANSIONS: DISTINCT POLES

Our focus will center on proper[3] rational functions, i.e.,

$$F(s) = \frac{a_m s^m + a_{m-1} s^{m-1} + \cdots + a_1 s + a_0}{s^n + b_{n-1} s^{n-1} + \cdots + b_1 s + b_0} = \frac{a_m s^m + a_{m-1} s^{m-1} + \cdots + a_1 s + a_0}{(s - p_1)(s - p_2) \cdots (s - p_n)}$$

where $m \le n$ and $p_1, \ldots, p_n$ are the **zeros** of the denominator polynomial, $s^n + b_{n-1} s^{n-1} + \ldots + b_1 s$ $+ b_0$, and are called the *finite poles* of $F(s)$. For the most part, rational functions are sufficient for the study of basic circuits. There are three cases of partial fraction expansions to consider:

   (1)   the case of distinct poles, i.e., $p_i \ne p_j$ for all $i \ne j$;
   (2)   the case of repeated poles, i.e., $p_i = p_j$ for at least one $i \ne j$; and
   (3)   the case of complex poles. Although case (3) is a subcategory of case (1) or (2) or both, its attributes warrant special recognition.

If $F(s)$ is a proper rational function with *distinct* (equivalently, *simple*) poles $p_1, \ldots, p_n$, then

$$F(s) = K + \frac{A_1}{(s - p_1)} + \frac{A_2}{(s - p_2)} + \cdots + \frac{A_n}{(s - p_n)} \tag{12.24}$$

where

$$K = \lim_{s \to \infty} F(s) \tag{12.25a}$$

The numbers A_i in equation 12.24 are called the *residues* of the pole p_i and can be computed according to the formula

$$A_i = \lim_{s \to p_i} [(s - p_i) F(s)] = [(s - p_i) F(s)]_{s = p_i} \tag{12.25b}$$

The rightmost equality of equation 12.25b is valid only when the numerator factor $(s - p_i)$ has been canceled with the factor $(s - p_i)$ in the denominator of $F(s)$; otherwise, one will obtain zero divided by zero which, in general, is undefined. As intimated earlier, this partial fraction expansion should enable a straightforward reconstruction of $f(t)$. Indeed, from Table 12.1, we immediately conclude

$$f(t) = K\delta(t) + A_1 e^{p_1 t} u(t) + A_2 e^{p_2 t} u(t) + \cdots + A_n e^{p_n t} u(t) \tag{12.26}$$

EXAMPLE 12.12. Find $f(t)$ when

$$F(s) = \frac{1}{s(s + a)}$$

SOLUTION

The solution proceeds by executing a partial fraction expansion (equation 12.24) on $F(s)$ to produce the Laplace transform of two elementary signals, a step and an exponential. Specifically,

$$F(s) = \frac{1}{s(s + a)} = \frac{A}{s} + \frac{B}{s + a}$$

where A/s is the Laplace transform of a weighted step, $Au(t)$, and $B/(s + a)$ is that of a weighted exponential, $Be^{-at}u(t)$. To find A, multiply both sides by s, cancel common numerator and denominator factors, and evaluate the result at $s = 0$, to produce $A = 1/a$. Similarly, to find B, multiply both sides by $s + a$, cancel common numerator and denominator factors, and evaluate the result at $s = -a$, to obtain $B = -1/a$. Recall that, by linearity, $\mathcal{L}^{-1}[aF(s)] = a\mathcal{L}^{-1}[F(s)]$. Hence,

$$f(t) = \frac{1}{a}u(t) - \frac{1}{a}e^{-at}u(t) = \frac{1}{a}(1 - e^{-at})u(t)$$

Exercises. 1. Find $f(t)$ when $F(s) = \dfrac{-2s + 2a}{s(s + a)}$.

ANSWER: $f(t) = 2u(t) - 4e^{-at}u(t)$

2. Find $f(t)$ when $F(s) = \dfrac{s^2 + 3s + 6}{(s + 1)(s + 2)(s + 3)}$.

ANSWER:

$f(t) = [2e^{-t} - 4e^{-2t} + 3e^{-3t}]u(t)$

3. Find a partial fraction expansion of $F(s) = \dfrac{cs^2 + b}{s(s + a)} = K + \dfrac{A}{s} + \dfrac{B}{s + a}$.

ANSWER: $K = c$, $A = \dfrac{b}{a}$, $B = -\dfrac{b + a^2 c}{a}$

4. Find $f(t)$ for $F(s)$ from Exercise 3.

ANSWER: $f(t) = c\delta(t) + \dfrac{b}{a}u(t) - \dfrac{b + a^2 c}{a}e^{-at}u(t)$

EXAMPLE 12.13. Suppose $V_{in}(s) = 10\dfrac{2s^2 + 3s + 2}{s(s + 2)}$ in the circuit of Figure 12.19. Find $V_{out}(s)$ and $v_{out}(t)$. Assume standard units.

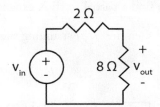

FIGURE 12.19 Series resistive circuit.

SOLUTION

Step 1. *Determine $V_{out}(s)$.* By voltage division, $v_{out}(t) = 0.8v_{in}(t)$, in which case

$$V_{out}(s) = 0.8V_{in}(s) = 8\frac{2s^2 + 3s + 2}{s(s + 2)}$$

Step 2. *Construct a partial fraction expansion of $V_{out}(s)$.* Since the numerator and denominator are both of degree 2,

$$V_{out}(s) = \frac{16s^2 + 24s + 16}{s(s+2)} = K + \frac{A}{s} + \frac{B}{s+2} \qquad (12.27)$$

The value of K in equation 12.27 is determined by the behavior of $F(s)$ at infinity (equation 12.25a), i.e.,

$$K = \lim_{s\to\infty} \left(\frac{16s^2 + 24s + 16}{s(s+2)} \right) = 16$$

To find A in equation 12.27, we use equation 12.25b:

$$\left[\frac{16s^2 + 24s + 16}{(s+2)} \right]_{s=0} = 8 = \left[Ks + A + \frac{Bs}{s+2} \right]_{s=0} = A$$

Similarly,

$$B = \left[\frac{16s^2 + 24s + 16}{s} \right]_{s=-2} = -16$$

Step 3. *Find $v_{out}(t)$.* Using Table 12.1,

$$v_{out}(t) = \mathcal{L}^{-1}\left[16 + \frac{8}{s} - \frac{16}{s+2} \right] = \mathcal{L}^{-1}[16] + \mathcal{L}^{-1}\left[\frac{8}{s} \right] - \mathcal{L}^{-1}\left[\frac{16}{s+2} \right]$$

$$= 16\delta(t) + 8u(t) - 16e^{-2t}u(t) \text{ V}$$

Exercises. 1. Repeat Example 12.13 for $V_{in}(s) = \dfrac{-20s + 20a}{s(s+a)}$.

ANSWER: $v_{out}(t) = 20u(t) - 40e^{-at}u(t)$ V

2. Given the circuit of Figure 12.20, find a partial fraction expansion of

$I_{in}(s) = 5\dfrac{as^2 + (a+b+c)s + b}{s(s+1)}$ and then find $i_2(t)$ assuming standard units.

ANSWER: $i_2(t) = 4a\delta(t) + 4bu(t) + 4ce^{-t}u(t)$ A

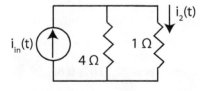

FIGURE 12.20 Parallel resistive circuit.

EXAMPLE 12.14. Compute the inverse transform of the function

$$F(s) = \frac{1-e^{-s}}{s(s+1)} = \frac{1}{s(s+1)} - \frac{e^{-s}}{s(s+1)}$$

SOLUTION
From Example 12.12,

$$\mathcal{L}^{-1}\left[\frac{1}{s(s+1)}\right] = (1-e^{-t})u(t)$$

This result and the shift theorem yield

$$\mathcal{L}^{-1}\left[\frac{e^{-s}}{s(s+1)}\right] = (1-e^{-(t-1)})u(t-1)$$

By the linearity of the inverse Laplace transform,

$$f(t) = (1 - e^{-t})u(t) - (1 - e^{-(t-1)})u(t-1)$$

A sketch of $f(t)$ appears in Figure 12.21..

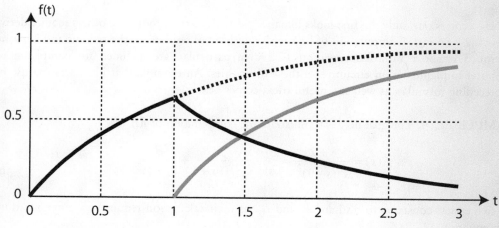

FIGURE 12.21 Sketch of $f(t) = [1 - e^{-t}]u(t) - [1 - e^{-(t-1)}]u(t-1)$.

PARTIAL FRACTION EXPANSIONS: REPEATED POLES

Proper rational functions with repeated roots have a more intricate partial fraction expansion, and calculation of the residues often proves cumbersome. For example, suppose

$$F(s) = \frac{n(s)}{(s-a)^k d(s)}$$

where the denominator factor $(s-a)^k$ specifies a repeated root of order k, $d(s)$ is the remaining factor in the denominator of the rational function $F(s)$, and $n(s)$ is the numerator of $F(s)$. The structure of a partial fraction expansion with repeated roots is

$$F(s) = \frac{A_1}{s-a} + \frac{A_2}{(s-a)^2} + \cdots + \frac{A_k}{(s-a)^k} + \frac{n_1(s)}{d(s)} \tag{12.28}$$

where $A_1, \ldots, A_k$ are unknown constants associated with $s-a, \ldots, (s-a)^k$, respectively, and $n_1(s)$ and $d(s)$ are whatever remains in the partial fraction expansion of $F(s)$. The formulas for computing the A_i of equation 12.28 are

$$A_k = (s-a)^k F(s) \Big]_{s=a} = \frac{n(s)}{d(s)} \Big]_{s=a} \tag{12.29a}$$

$$A_{k-1} = \frac{d}{ds} \left((s-a)^k F(s) \right) \Big]_{s=a} = \frac{d}{ds} \left(\frac{n(s)}{d(s)} \right) \Big]_{s=a} \tag{12.29b}$$

and, in general,

$$A_{k-i} = \frac{1}{i!} \frac{d^i}{ds^i} \left((s-a)^k F(s) \right) \Big]_{s=a} = \frac{1}{i!} \frac{d^i}{ds^i} \left(\frac{n(s)}{d(s)} \right) \Big]_{s=a} \tag{12.29c}$$

Of these expressions, only the first looks like the case with distinct roots; the others require derivatives of $(s - a)^k F(s)$. Computation of high-order derivatives borders on the tedious and is prone to error. The above formulas, equation 12.29c in particular, are included for completeness. Computer implementation circumvents these difficulties. An example that illustrates the use of the preceding formulas, as well as a useful trick, comes next.

EXAMPLE 12.15. The goal here is to illustrate the computation of $f(t)$ when

$$F(s) = \frac{s+2}{s^2(s+1)^2} = \frac{A_1}{s} + \frac{A_2}{s^2} + \frac{B_1}{s+1} + \frac{B_2}{(s+1)^2} \tag{12.30}$$

The two easiest constants to find are A_2 and B_2, as their calculation requires no differentiation. From equation 12.29a,

$$A_2 = s^2 F(s) \Big]_{s=0} = \left[\frac{s+2}{(s+1)^2} \right]_{s=0} = 2$$

and

$$B_2 = (s+1)^2 F(s) \Big]_{s=-1} = \left[\frac{s+2}{s^2} \right]_{s=-1} = 1$$

Finding A_1 and B_1 is more difficult, since formula 12.29b requires some differentiation. According to equation 12.29b,

$$A_1 = \frac{d}{ds} [s^2 F(s)]_{s=0}$$

To implement this formula multiply both sides of equation 12.30 by s^2, take the derivative of the resulting expression with respect to s, and evaluate at $s = 0$:

$$\frac{d}{ds}\left[\frac{s+2}{(s+1)^2}\right]_{s=0} = \frac{d}{ds}\left[A_1 s + A_2 + \frac{B_1 s^2}{s+1} + \frac{B_2 s^2}{(s+1)^2}\right]_{s=0} = A_1$$

Observe that, on the right-hand side, it is *not* necessary to differentiate the terms that contain A_2, B_1, and B_2, since these terms disappear at $s = 0$, as the formula for A_1 requires. Consequently,

$$A_1 = \frac{d}{ds}\left[\frac{s+2}{(s+1)^2}\right]_{s=0} = \left[\frac{1}{(s+1)^2} - 2\frac{s+2}{(s+1)^3}\right]_{s=0} = -3$$

Similarly,

$$B_1 = \frac{d}{ds}\left[(s+1)^2 F(s)\right]_{s=-1} = \frac{d}{ds}\left[\frac{s+2}{s^2}\right]_{s=-1} = \left[\frac{1}{s^2} - 2\frac{s+2}{s^3}\right]_{s=-1} = 3$$

Note that since A_1, A_2, and B_2 were known, a simple trick allows a more direct computation of B_1: merely evaluate equation 12.29 at $s = 1$ (in fact, any value of s, excluding the poles, will do), to obtain

$$0.75 = -3 + 2 + 0.25 + 0.5 B_1$$

As expected, solving yields $B_1 = 3$. Hence

$$F(s) = \frac{s+2}{s^2(s+1)^2} = \frac{-3}{s} + \frac{2}{s^2} + \frac{3}{s+1} + \frac{1}{(s+1)^2}$$

The above result can also be found with the MATLAB command "residue" as follows. Let $F(s) = n(s)/d(s)$. It follows that $n(s) = s + 2$ and $d(s) = s^4 + 2s^3 + s^2$. In MATLAB,

»num = [1 2];
»den = [1 2 1 0 0];
»[r, p, k] = residue(num, den)
The answers from MATLAB are:
r = -3 2 3 1 (the residues associated with the poles)
p = 0 0 -1 -1
and constant
k = 0.

Exercises. 1. Find the partial fraction expansion of $F(s) = \dfrac{2s^4 + 2s^3 + 3s^2 + 3s + 2}{s^2(s+1)^2}$.

ANSWERS (residues in random order): 2, 2, 2, −1, −1

2. The partial fraction expansion of a rational function is given by

$$F(s) = \frac{3s^2 + 10s + 9}{s^3 + 4s^2 + 5s + 2} = \frac{A}{s+2} + \frac{B}{s+1} + \frac{C}{(s+1)^2}$$

Compute A, B, and C.
ANSWERS: In random order: 2, 1, 2

3. Use MATLAB to find a partial fraction expansion of $F(s) = \dfrac{n(s)}{d(s)}$, where $n(s) = 6(s + 2)^3(s - 2)^2$

and $d(s) = s(s + 1)^2(s + 4)^2$. Hint: Use num = 6*poly([-2 -2 -2 2 2]) and den = poly([0 -1 -1 -4 -4]).
ANSWER: [r,p,k] = residue(num,den) yields
r = −44 48 −16 −6 12
p = −4 -4 −1 −1 0
k = 6.
This results in the PFE: $F(s) = \dfrac{-44}{s+4} + \dfrac{48}{(s+4)^2} + \dfrac{-16}{s+1} + \dfrac{-6}{(s+1)^2} + \dfrac{12}{s} + 6$

The derivative formulas of equations 12.29 are often difficult to evaluate for complicated rational functions, such as

$$F(s) = \frac{5s^5 + 95s^4 + 692s^3 + 2369s^2 + 3715s + 2076}{(s+1)(s+2)(s+3)(s+5)^3}$$

$$= \frac{A}{s+1} + \frac{B}{s+2} + \frac{C}{s+3} + \frac{D_1}{s+5} + \frac{D_2}{(s+5)^2} + \frac{D_3}{(s+5)^3}.$$

For these functions, it is very efficient to find A, B, C, and D_3 directly. Then one evaluates $F(s)$ at two values of s, e.g., $s = 0$ and $s = 1$, to obtain two equations in the unknowns D_1 and D_2. Typically, solving the resulting two equations simultaneously is much easier than evaluating D_1 and D_2 directly by equations 12.29. Alternatively, one can use a software program such as MAT-LAB to compute the answers. In particular, in MATLAB:

n =[5 95 692 2369 3715 2076]
d =[1 21 176 746 1665 1825 750]
»[r,p,k]=residue(n,d)
r =
-1.0000e+00
-1.0000e+00
-1.0000e+00
3.0000e+00
2.0000e+00
1.0000e+00
p =
-5.0000e+00
-5.0000e+00
-5.0000e+00
-3.0000e+00
-2.0000e+00
-1.0000e+00
k =
[ ]

PARTIAL FRACTION EXPANSIONS: DISTINCT COMPLEX POLES

Distinct complex roots present challenges different from those for the repeated root case. Since the roots are distinct but not real, the methods of equations 12.25 and 12.29 apply. Unfortunately, the resulting partial fraction expansion has complex residues, and the resulting inverse transform has complex exponentials multiplied by complex constants. Such imaginary time functions lack meaning in the real world unless their imaginary parts cancel to yield real-time functions. When they do, our goal is to find a direct route for computing the associated real-time signals. To do this, consider a rational function having a pair of distinct complex poles as in the following equation:

$$F(s) = \frac{n(s)}{[(s+a)^2 + \omega^2]d(s)} = \frac{n(s)}{(s+a+j\omega)(s+a-j\omega)\,d(s)} \qquad (12.31)$$

Since the poles $-a - j\omega$ and $-a + j\omega$ are distinct, the partial fraction expansion of equation 12.24 is valid. Since the poles are complex conjugates of each other, the residues of each pole are complex conjugates. Therefore, it is possible to write the partial fraction expansion of $F(s)$ as

$$F(s) = \frac{A+jB}{s+a+j\omega} + \frac{A-jB}{s+a-j\omega} + \frac{n_1(s)}{d(s)} \qquad (12.32)$$

for appropriate polynomials $n_1(s)$ and $d(s)$. As per equation 12.25b, the first residue in equation 12.32 is

$$A + jB = (s + a + j\omega)F(s) \,]_{s = -a - j\omega} \qquad (12.33)$$

With A and B known, executing a little algebra on equation 12.32 to eliminate complex numbers results in an expression more amenable to inversion, i.e.,

$$F(s) = \frac{C_1 s + C_2}{(s+a)^2 + \omega^2} + \frac{n_1(s)}{d(s)} = F_0(s) + \frac{n_1(s)}{d(s)} \qquad (12.34)$$

where

$$C_1 = 2A \qquad (12.35a)$$

and

$$C_2 = 2aA + 2\omega B \qquad (12.35b)$$

with A and B specified in equation 12.33. With C_1 and C_2 given by equations 12.35, it is straightforward to show that

$$F_0(s) = \frac{C_1 s + C_2}{(s+a)^2 + \omega^2} = C_1 \frac{s+a}{(s+a)^2 + \omega^2} + \left(\frac{C_2 - C_1 a}{\omega}\right)\frac{\omega}{(s+a)^2 + \omega^2} \qquad (12.36)$$

From Table 12.1, item 19, or a combination of items 10 and 11,

$$f_0(t) = e^{-at}\left[C_1 \cos(\omega t) + \left(\frac{C_2 - C_1 a}{\omega}\right)\sin(\omega t)\right]u(t)$$

Exercise. Suppose $F(s) = \dfrac{C_1 s + C_2}{s^2 + 4}$. Compute $f(t)$.

ANSWER: $C_1 \cos(2t)u(t) + 0.5C_2 \sin(2t)u(t)$

The following example illustrates the algebra for computing C_1 and C_2 without using complex arithmetic.

EXAMPLE 12.16. Find $f(t)$ when

$$F(s) = \frac{3s^2 + s + 3}{(s+1)(s^2 + 4)} = \frac{D}{s+1} + \frac{A + jB}{s + j2} + \frac{A - jB}{s - j2} = \frac{D}{s+1} + \frac{C_1 s + C_2}{s^2 + 4} \qquad (12.37)$$

Step 1. *Compute the coefficients D, C_1, and C_2 in the partial fraction expansion of equation 12.37.* First we find D by the usual techniques:

$$D = \left. \frac{3s^2 + s + 3}{s^2 + 4} \right]_{s=-1} = 1$$

Given that $D = 1$, to find C_2 we evaluate $F(s)$ at $s = 0$, in which case $0.75 = 1 + 0.25C_2$, or $C_2 = -1$. With $D = 1$ and $C_2 = -1$, we evaluate $F(s)$ at $s = 1$ to obtain $0.7 = 0.5 + 0.2(C_1 - 1)$ or, equivalently, $C_1 = 2$. Thus,

$$F(s) = \frac{1}{s+1} + 2\frac{s}{s^2 + 4} - 0.5\frac{2}{s^2 + 4} \qquad (12.38)$$

Step 2. *Compute $f(t)$.* Using Table 12.1, items 8 and 9, to compute the inverse transform yields

$$f(t) = [\, e^{-t} + 2\cos(2t) - 0.5\sin(2t)\,]u(t)$$

Alternative Step 1. *Compute A and B in equation 12.37 by hand or with MATLAB.* In MATLAB,

```
»num = [3,1 3];
»den = conv([1 1],[1 0 4])
den = [1 1 4 4]
»[r, p, k] = residue(num,den)
r =
  1.0000 + 0.2500i
  1.0000 - 0.2500i
  1.0000 + 0.0000i
p =
 -0.0000 + 2.0000i
 -0.0000 - 2.0000i
 -1.0000
k = 0
```

This implies that

$$F(s) = \frac{D}{s+1} + \frac{A+jB}{s+j2} + \frac{A-jB}{s-j2} = \frac{1}{s+1} + \frac{1-j0.25}{s+j2} + \frac{1+j0.25}{s-j2} \qquad (12.39)$$

Alternative Step 2. One must exercise caution here and note the difference between the MAT-LAB output and the form of the partial fraction expansion. From equation 12.39, $\omega = +2$, $A = 1$, and $B = -0.25$. Again using MATLAB to obtain the form needed in item 20 of Table 12.1,

```
»K = 2*sqrt(A^2 + B^2)
K = 2.0616
»theta = atan2(B,A)*180/pi
theta = −14.0362
```

Thus

$$f(t) = [\, e^{-t} + 2.0616 \cos(2t) + 14.04^o \,)]u(t)$$

Example 12.16 illustrates not only the computation of an inverse transform having complex poles, but also the computation of C_1 and C_2 without resorting to complex arithmetic, as was needed in equation 12.32. The trick again was to evaluate $F(s)$ at two distinct s-values different from the poles of $F(s)$. This yields two equations that can be solved for the unknowns C_1 and C_2.

Exercises. 1. Find $f(t)$ when $F(s) = \dfrac{5s^2 - 8s + 4}{s(s^2 + 4)}$.

ANSWER: $f(t) = [1 + 4\cos(2t) - 4\sin(2t)]u(t)$

2. Find $f(t)$ when $F(s) = \dfrac{5s^2 - 2s + 5}{s(s^2 + 2s + 5)}$.

ANSWER: $f(t) = u(t) + 4e^{-t}[\cos(2t) - \sin(2t)]u(t)$

---

7. MORE TRANSFORM PROPERTIES AND EXAMPLES

Another handy property of the Laplace transform is the frequency shift property, which permits one to readily compute the transform of functions multiplied by an exponential. With knowledge of the transforms of $u(t)$, $\sin(\omega t)$, and other functions, computation of $e^{-at}u(t)$ and $e^{-at}\sin(\omega t)u(t)$ becomes quite easy.

Frequency shift property: Let $F(s) = \mathcal{L}[f(t)]$. Then

$$\mathcal{L}[\, e^{-at} f(t)] = F(s + a) \qquad (12.40)$$

This property can be verified by a direct calculation,

$$\mathcal{L}[e^{-at}f(t)] = \int_{0^-}^{\infty} f(t)e^{-(s+a)t}dt = F(s+a)$$

where we have viewed the sum $s + a$ in the integral as a new variable p, which leads to $F(p)$ with p replaced by $s + a$.

EXAMPLE 12.17. Let $f(t) = \sin(\omega t)u(t)$. Define $g(t) = e^{-at}f(t) = e^{-at}\sin(\omega t)u(t)$. Suppose it is known that

$$F(s) = \mathcal{L}[\sin(\omega t)u(t)] = \frac{\omega}{s^2 + \omega^2}$$

Compute $G(s)$.

SOLUTION
By the frequency shift property, $G(s) = F(s + a)$, or

$$G(s) = \mathcal{L}[g(t)] = \mathcal{L}[e^{-at}f(t)] = F(s+a) = \left[\frac{\omega}{s^2 + \omega^2}\right]_{s=s+a} = \frac{\omega}{(s+a)^2 + \omega^2}$$

Exercise. Let $f(t) = \cos(\omega t)u(t)$ for which $F(s) = \dfrac{s}{s^2 + \omega^2}$. Define $g(t) = e^{-at}f(t) = e^{-at}\cos(\omega t)u(t)$. Compute $G(s)$.

ANSWER: $\dfrac{s+a}{(s+a)^2 + \omega^2}$

Another property of particularly widespread applicability is the time differentiation formula. Its utility resides not only in obtaining shortcuts to transforms of signals, but also in the solution of differential equations. Differential equations provide a ubiquitous setting for modeling a large variety of physical systems—mechanical, electrical, chemical, etc. In terms of signal computation, recall that the velocity of a particle is the derivative of its position as a function of time. The acceleration is the derivative of the velocity. After computing the Laplace transform of the position as a function of time, one finds that a differentiation formula allows direct computation of the transforms of the velocity and acceleration. Also, as discussed at the very beginning of this chapter, circuits have differential equation models. For example, weighted sums of derivatives of the response of the circuit are equated to weighted sums of derivatives of the input signal. Therefore, a differentiation formula is an essential ingredient in the analysis of circuits.

First-order time differentiation formula: Let $\mathcal{L}[f(t)] = F(s)$. Then

$$\mathcal{L}\left[\frac{d}{dt}f(t)\right] = sF(s) - f(0^-) \tag{12.41}$$

The differentiation property is validated using integration by parts as follows:

$$\mathcal{L}\left[\frac{d}{dt}f(t)\right] = \int_{0^-}^{\infty}\left(\frac{d}{dt}f(t)\right)e^{-st}dt = f(t)e^{-st}\Big|_{0^-}^{\infty} - \int_{0^-}^{\infty}(-s)f(t)e^{-st}dt = -f(0^-) + sF(s)$$

The following examples explore some clever uses of the first-order time differentiation formula.

EXAMPLE 12.18.

Recall that $\delta(t) = \dfrac{d}{dt}u(t)$. Using the sifting property, a direct calculation yields $\mathcal{L}[\delta(t)] = 1$. Is this consistent with the differentiation property? Interpreting the delta function as the derivative of the step function and applying the differentiation formula yields

$$\mathcal{L}[\delta(t)] = \mathcal{L}\left[\frac{d}{dt}u(t)\right] = s\mathcal{L}[u(t)] - u(0^-) = s\left(\frac{1}{s}\right) = 1 \qquad (12.42)$$

which demonstrates the expected consistency.

Exercises. 1. The Laplace transform of a signal $f(t)$ is $F(s) = \dfrac{2}{s^2 + 4}$. What is the Laplace transform of $e^{-2t}\dfrac{d}{dt}f(t)$?

ANSWER: $\dfrac{2s + 4}{(s+2)^2 + 4}$

2. $\mathcal{L}[\sin(\omega t)\delta(t)]$ = ?
ANSWER: 0

EXAMPLE 12.19.
Suppose $f(t) = \sin(\omega t)u(t)$ and we know (for example, from Table 12.1) that $F(s) = \mathcal{L}[\sin(\omega t)u(t)] = \dfrac{\omega}{s^2 + \omega^2}$. Compute $\mathcal{L}[\cos(\omega t)u(t)]$ using the time differentiation formula.

SOLUTION
Since $\cos(\omega t)u(t) = \dfrac{1}{\omega}\dfrac{d}{dt}\sin(\omega t)u(t)$ and $\sin(\omega t)u(t)|_{t=0} = 0$, the differentiation property immediately implies that

$$\mathcal{L}[\cos(\omega t)u(t)] = \frac{1}{\omega}s\left(\frac{\omega}{s^2 + \omega^2}\right) = \frac{s}{s^2 + \omega^2}$$

Exercises. 1. Express $f(t) = \sin(\omega t)u(t)$ in terms of the derivative of $g(t) = \cos(\omega t)u(t)$. Note the presence of the delta function.

ANSWER: $-\dfrac{1}{\omega}\dfrac{dg(t)}{dt} + \dfrac{1}{\omega}\delta(t) = \sin(\omega t)u(t)$

2. Now suppose it is known that $\mathcal{L}[\cos(\omega t)u(t)] = \dfrac{s}{s^2 + \omega^2}$. Use the result of Exercise 1 and the differentiation property to compute the Laplace transform of $f(t) = \sin(\omega t)u(t)$, noting that $g(0^-) = 0$.

ANSWER: $\mathcal{L}[\sin(\omega t)u(t)] = -\dfrac{1}{\omega}s\left(\dfrac{s}{s^2 + \omega^2}\right) + \dfrac{1}{\omega}g(0^-) + \dfrac{1}{\omega} = \dfrac{1}{\omega}\left[1 - \dfrac{s^2}{s^2 + \omega^2}\right] = \dfrac{\omega}{s^2 + \omega^2}$

EXAMPLE 12.20. Let $f(t)$ and its derivative have the shapes shown in **Figure 12.22**. The goal of this example is to explore the relationship between the Laplace transforms of $f(t)$ and $f'(t)$ in light of the differentiation property.

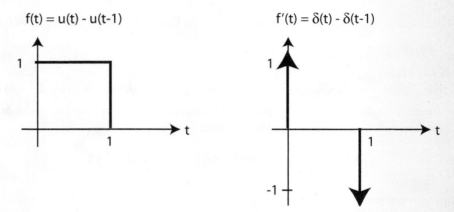

FIGURE 12.22 A pulse and its derivative for Example 12.20. Observe how the derivative of the pulse leads to a pair of delta functions.

Using linearity and the shift theorem on $f(t)$ yields

$$\mathcal{L}[f(t)] = \mathcal{L}[u(t) - u(t-1)] = \mathcal{L}[u(t)] - \mathcal{L}[u(t-1)] = \frac{1}{s}(1 - e^{-s})$$

Applying the linearity of the Laplace transform to $f'(t)$ yields

$$\mathcal{L}[f'(t)] = \mathcal{L}[\delta(t) - \delta(t-1)] = 1 - e^{-s}$$

From the differentiation formula, it must follow that $\mathcal{L}[f'(t)] = s\mathcal{L}[f(t)]$. Thus,

$$\mathcal{L}[f'(t)] = s\mathcal{L}[f(t)] = s\frac{1}{s}(1 - e^{-s}) = 1 - e^{-s}$$

demonstrating consistency.

As might be expected, the formula for the first derivative is a special case of the more general differentiation rule:

$$\mathcal{L}\left[\frac{d^n}{dt^n}f(t)\right] = s^n F(s) - s^{n-1}f(0^-) - s^{n-2}f^{(1)}(0^-) - \ldots - f^{(n-1)}(0^-) \tag{12.43}$$

This rule proves useful in the solution of general nth-order differential equations. Of particular use is the second-order formula:

$$\mathcal{L}[f''(t)] = s^2 F(s) - sf(0^-) - f'(0^-) \tag{12.44}$$

The inverse of differentiation is integration. The following property proves useful for quantities related by integrals.

Integration property: Let $F(s) = \mathcal{L}[f(t)]$. Then for $t \geq 0$,

$$\mathcal{L}\left[\int_{0^-}^{t} f(q)dq\right] = \frac{F(s)}{s} \tag{12.45a}$$

and

$$\mathcal{L}\left[\int_{-\infty}^{t} f(q)dq\right] = \frac{F(s)}{s} + \frac{\int_{-\infty}^{0^-} f(q)dq}{s} \tag{12.45b}$$

As with many of the justifications of the properties, integration by parts plays a key role. By direct computation (using equation 12.16),

$$\mathcal{L}\left[\int_{-\infty}^{t} f(q)dq\right] = \int_{0^-}^{\infty} \left[\int_{-\infty}^{t} f(q)dq\right] e^{-st} dt$$

To use integration by parts, let

$$u = \int_{-\infty}^{t} f(q)dq \quad \text{and} \quad dv = e^{-st} dt$$

Then

$$\mathcal{L}\left[\int_{-\infty}^{t} f(q)dq\right] = uv\Big]_{0^-}^{\infty} - \int_{0^-}^{\infty} v\, du$$

$$= \left[-\frac{e^{-st}}{s} \int_{-\infty}^{t} f(q)dq\right]_{0^-}^{\infty} + \frac{1}{s}\int_{0^-}^{\infty} f(t) e^{-st} dt.$$

For the appropriate region of convergence, the first term to the right of the equal sign reduces to

$$\left[-\frac{e^{-st}}{s} \int_{-\infty}^{t} f(q)dq\right]_{0^-}^{\infty} = \frac{\int_{-\infty}^{0^-} f(q)dq}{s}$$

Since the second term to the right of the equal sign is $F(s)/s$, as per equation 12.45a, the property is verified.

EXAMPLE 12.21. Find the Laplace transform of the signal $f(t)$ sketched in Figure 12.23a using the integration property.

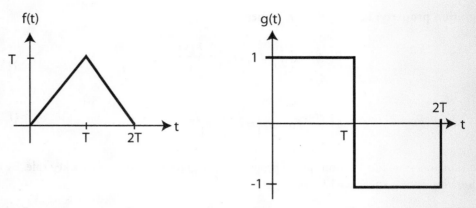

FIGURE 12.23 (a) A triangular signal $f(t)$ for Example 12.21. (b) The derivative of $f(t)$.

SOLUTION

Observe that the triangular waveform $f(t)$ of Figure 12.23a is the integral of the square wave $g(t)$. Since $g(t)$ is easily represented in terms of steps and shifted steps as

$$g(t) = u(t) - 2u(t - T) + u(t - 2T)$$

its Laplace transform follows from an application of linearity and the time shift property:

$$\mathcal{L}[g(t)] = \frac{1}{s}\left[1 - 2e^{-sT} + e^{-2sT}\right]$$

The integration property implies that

$$\mathcal{L}[f(t)] = \mathcal{L}\left[\int_{-\infty}^{t} g(q)dq\right] = \frac{1}{s}\mathcal{L}[g(t)] = \frac{1}{s^2}[1 - 2e^{-sT} + e^{-2sT}]$$

EXAMPLE 12.22. This example explores the voltage-current $(v\text{-}i)$ relationship of a capacitor in the frequency domain by way of the integration property. Recall the integral form of the voltage-current dynamics of a capacitor:

$$v_C(t) = \frac{1}{C}\int_{-\infty}^{t} i_C(\tau)d\tau$$

Taking the Laplace transform of both sides and applying the integration property produces

$$\mathcal{L}[v_C(t)] = \mathcal{L}\left[\frac{1}{C}\int_{-\infty}^{t} i_C(\tau)d\tau\right] = \frac{1}{Cs}I_C(s) + \frac{1}{Cs}\int_{-\infty}^{0^-} i_C(\tau)d\tau$$

But this expression depends on the initial condition $v_C(0^-)$, because

$$v_C(0^-) = \frac{1}{C}\int_{-\infty}^{0^-} i_C(\tau)d\tau$$

Therefore,

$$V_C(s) = \frac{1}{Cs}I_C(s) + \frac{1}{s}v_C(0^-)$$

(12.46)

Equation 12.46 says that the voltage $V_C(s)$ is the sum of two terms: a term dependent on the frequency domain current $I_C(s)$ and a term that looks like a step voltage source and depends on the constant initial condition $v_C(0^-)$. The quantity $Z_C(s) = 1/Cs$ looks like a generalized resistance—"generalized" because it depends on the frequency variable s and a "resistance" because it satisfies an Ohm's law–like relationship, $V_C(s) = Z_C(s)\, I_C(s)$. These analogies prompt a series-circuit interpretation of equation 12.46 as depicted in Figure 12.24. An application of this equivalent frequency domain circuit to general network analysis appears in the next chapter.

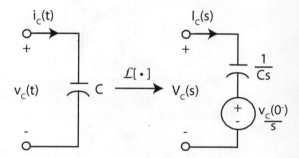

FIGURE 12.24 Equivalent circuit interpretation of a capacitor in the frequency domain. This equivalent is arrived at by applying the integration property of the Laplace transform to the capacitor voltage, seen as the integral of the capacitor current.

A second example interpreting the *v-i* characteristics of the capacitor in the frequency domain ensues from the differentiation rule. Instead of winding up with a series circuit, one obtains a parallel circuit. The interpretation is thus said to be dual to the one just described.

EXAMPLE 12.23. This example has two goals: (i) Verify that equation 12.46 is consistent with the differentiation formula interpretation of the capacitor; (ii) Build a dual frequency domain interpretation of the *v-i* characteristic of a capacitor analogous to that of Example 12.22.

As a first step, recall equation 12.46:

$$V_C(s) = \frac{1}{Cs}I_C(s) + \frac{1}{s}v_C(0^-)$$

which, after some algebra, becomes

$$I_C(s) = CsV_C(s) - Cv_C(0^-)$$

(12.47)

Notice that equation 12.47 is consistent with the application of the derivative formula to $i_C(t) = C[dv_C/dt]$. This consistency offers some reassurance in the accuracy of our development. The interpretation of equation 12.47, however, is quite different from that of equation 12.46. In the latter equation, the current $I_C(s)$ equals the sum of two currents, $CsV_C(s)$ and $-Cv_C(0^-)$. This suggests a nodal interpretation, resulting in an equivalent circuit having two parallel branches. One branch contains a capacitor with voltage $V_C(s)$. The other, parallel branch contains a current source with amperage $Cv_C(0^-)$. The current through the capacitive branch is $(Cs)V_C(s)$, where "Cs" now acts like a generalized conductance because it multiplies a voltage, similar to Ohm's law. "Cs" is generalized because it depends on s. Figure 12.25 presents the equivalent circuit of the capacitor in the frequency domain and is dual to the circuit of Figure 12.24. Chapter 13 covers in detail the role of these equivalent circuits in analysis.

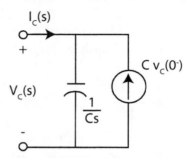

FIGURE 12.25 Equivalent circuit to a capacitor in the frequency domain using the differentiation formula.

The last elementary property of the Laplace transform that we consider in this chapter is the time-scaling property, also called the frequency-scaling property. Its importance is fundamental to network synthesis. Here, numerical problems, such as roundoff, prevent engineers from directly designing a circuit to meet a given set of specifications. Instead, the design engineer will normalize the specifications through a frequency-scaling technique. Once the normalized circuit is designed, frequency-scaling techniques are reapplied in an inverse fashion to obtain a circuit meeting the original specifications.

Time-/Frequency-scaling property: Let $a > 0$ and $\mathcal{L}[f(t)] = F(s)$. Then

$$\mathcal{L}[f(at)] = \frac{1}{a}F\left(\frac{s}{a}\right)$$

$$(12.48)$$

or, equivalently, $F(s/a) = a\mathcal{L}[f(at)]$.

Since the proof of this property is straightforward, it is left as an exercise at the end of the chapter.

EXAMPLE 12.24. Figures 12.26a and b show impulse trains that model sampling in signal-processing applications. The impulse train of Figure 12.26b is the time-scaled counterpart to that of Figure 12.26a.

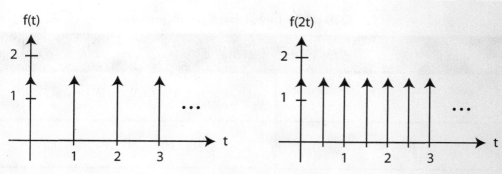

FIGURE 12.26 (a) Unit impulse train. (b) Time-scaled unit impulse train. Unit impulse trains such as these model sampling in signal-processing applications.

The time-scaled impulse train in Figure 12.26b increases the frequency at which the impulses occur (twice as often as in the original signal). This is reflected in the Laplace transforms of the two signals:

$$\mathcal{L}[f(t)] = \mathcal{L}\left[\sum_{k=0}^{\infty} \delta(t-k)\right] = \sum_{k=0}^{\infty} e^{-sk} = \frac{1}{1-e^{-s}} \tag{12.49}$$

By the time-scaling property,

$$\mathcal{L}[f(2t)] = 0.5 \frac{1}{1-e^{-0.5s}} \tag{12.50}$$

Notice that what occurs at, say, s_0 in equation 12.49 now occurs at $2s_0$ in equation 12.50. Hence, time scaling by numbers greater than 1 concentrates more of the frequency content of the signal in the higher frequency bands.

Exercise. Verify, by direct calculation, that $\mathcal{L}[f(2t)]$ is given by the right side of equation 12.50.

Several more properties of the Laplace transform are germane to our purpose. However, these properties have a systems flavor and are postponed until Chapter 13. We close this section by presenting Table 12.2, which lists the Laplace transform properties and the associated transform pairs.

TABLE 12.2 Laplace Transform Properties

Property	Transform Pair
Linearity	$\mathcal{L}[a_1 f_1(t) + a_2 f_2(t)] = a_1 F_1(s) + a_2 F_2(s)$
Time shift	$\mathcal{L}[f(t - T)u(t - T)] = e^{-sT} F(s),\ T > 0$
Multiplication by t	$\mathcal{L}[tf(t)u(t)] = -\dfrac{d}{ds}F(s)$
Multiplication by t^n	$\mathcal{L}[t^n f(t)] = (-1)^n \dfrac{d^n F(s)}{ds^n}$
Frequency shift	$\mathcal{L}[e^{-at}f(t)] = F(s + a)$
Time differentiation	$\mathcal{L}\left[\dfrac{d}{dt}f(t)\right] = sF(s) - f(0^-)$
Second-order differentiation	$\mathcal{L}\left[\dfrac{d^2 f(t)}{dt^2}\right] = s^2 F(s) - sf(0^-) - f^{(1)}(0^-)$
nth-order differentiation	$\mathcal{L}\left[\dfrac{d^n f(t)}{dt^n}\right] = s^n F(s) - s^{n-1} f(0^-) - s^{n-2} f^{(1)}(0^-)$ $- \ldots - f^{(n-1)}(0^-)$
Time integration	(i) $\mathcal{L}\left[\displaystyle\int_{-\infty}^{t} f(q)dq\right] = \dfrac{F(s)}{s} + \dfrac{\displaystyle\int_{-\infty}^{0^-} f(q)dq}{s}$ (ii) $\mathcal{L}\left[\displaystyle\int_{0^-}^{t} f(q)dq\right] = \dfrac{F(s)}{s}$
Time/Frequency scaling	$\mathcal{L}[f(at)] = 1/a\ F(s/a)$

8. SOLUTION OF INTEGRO-DIFFERENTIAL EQUATIONS USING THE LAPLACE TRANSFORM

Differential equations provide a cross-disciplinary mathematical modeling framework. Although differential equation models may represent only the dominant behavioral facets of a circuit or physical process, their widespread utility and importance to circuits and control systems warrant special discussion. To begin, recall the time differentiation formulas of equations 12.41 and 12.43 and the integration formulas of equations 12.45a and 12.45b. Also, recall that a differential equation relates a sum of derivatives of an output signal to a sum of derivatives of an input signal. For example, if the input and output signals are voltages, then the relation

$$\frac{d^n v_{out}}{dt^n} + a_1 \frac{d^{n-1} v_{out}}{dt^{n-1}} + \cdots + a_n v_{out} = \frac{d^m v_{in}}{dt^m} + b_1 \frac{d^{m-1} v_{in}}{dt^{m-1}} + \cdots + b_m v_{in}$$

for constants a_i and b_i might model the behavior of a linear circuit. We may use the following steps to solve this differential equation for $v_{out}(t)$ using the Laplace transform procedure:

1. Take the Laplace transform of both sides of the equation, using the appropriate derivative formulas, equations 12.41 and 12.43.
2. Algebraically solve the resulting expression for $V_{out}(s)$.
3. Compute a partial fraction expansion of the expression for $V_{out}(s)$.
4. Inverse-transform the partial fraction expansion to obtain the time function $v_{out}(t)$.

If the equation is an integro-differential equation, i.e., a mixture of both derivatives and integrals of the input and output signals, then we simply apply the same algorithm, except we use the integral formula where appropriate. Some examples serve to illuminate the procedure.

EXAMPLE 12.25. Consider the pulse current excitation of Figure 12.27a) to the RC circuit of Figure 12.27b. The goals of this example are (i) to use and illustrate Laplace transform techniques to solve a differential equation derived from a simple RC circuit and (ii) to find the response voltage $v_C(t)$, $t \ge 0$, when $v_C(0^-) = 1$ V.

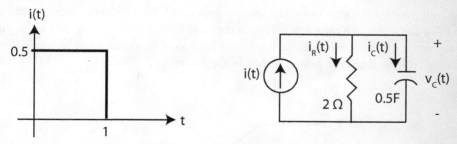

FIGURE 12.27 Excitation current (a) for a simple RC circuit (b) for Example 12.25.

SOLUTION

Step 1. *Find* $\mathcal{L}[i(t)]$. Since $i(t) = 0.5u(t) - 0.5u(t - 1)$,

$$\mathcal{L}[i(t)] = 0.5 \frac{1 - e^{-s}}{s}$$

Step 2. *Find the circuit's differential equation model that links the excitation current $i(t)$ to the response voltage, $v_C(t)$. Since $i_R(t) = 0.5v_C(t)$ and $i_C(t) = 0.5 dv_C/dt$, summing the currents into the top node of the circuit yields*

$$i(t) = i_R(t) + i_C(t) = 0.5v_C(t) + 0.5 \frac{dv_C(t)}{dt}$$

After multiplying through by 2, the desired differential equation circuit model is

$$\frac{dv_C}{dt}(t) + v_C(t) = 2i(t)$$

Step 3. *Take the Laplace transform of both sides, apply the differentiation rule to the left side, and solve for $V_C(s)$. Applying the Laplace transform to both sides yields*

$$sV_C(s) - v_C(0^-) + V_C(s) = 2I(s)$$

Solving for $V_C(s)$ produces

$$V_C(s) = \frac{2}{s+1} I(s) + \frac{v_C(0^-)}{s+1} = \frac{1-e^{-s}}{s(s+1)} + \frac{1}{s+1}$$

Some straightforward calculations show that

$$\frac{1}{s(s+1)} = \frac{1}{s} - \frac{1}{(s+1)}$$

Thus, with the aid of the shift property and the transform pairs of Table 12.1, we obtain

$$v_C(t) = \mathcal{L}^{-1}[V_C(s)] = \mathcal{L}^{-1}\left[\frac{(1-e^{-s})}{s(s+1)} + \frac{1}{s+1} \right]$$

$$= (1-e^{-t})u(t) - \left(1 - e^{-(t-1)}\right)u(t-1) + e^{-t}u(t)$$

$$= u(t) - \left(1 - e^{-(t-1)}\right)u(t-1).$$

Figure 12.28 presents the graph of this response. Because of the initial condition and the magnitude of the pulse input, the capacitor voltage is constant for $0 \le t \le 1$ second. At $t = 1$ second, the pulse magnitude drops to zero, making the circuit equivalent to a source-free RC circuit in which the capacitor voltage decays to zero as shown in the figure.

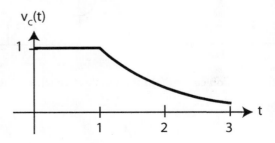

FIGURE 12.28 The response voltage $v_C(t)$ for Example 12.25.

EXAMPLE 12.26. The goal of this example is to compute the response, denoted here by the input current $i_{in}(t)$, to the input voltage excitation $v_{in}(t) = \delta(t)$, given the series *RLC* circuit of Figure 12.29. Suppose the initial conditions are $i_L(0^-) = 1$ A and $v_C(0^-) = -2$ V.

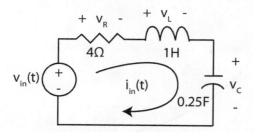

FIGURE 12.29 Series *RLC* circuit for Example 12.26. Here the current $i_{in}(t) = i_L(t)$.

SOLUTION

Step 1. *Compute the Laplace transform of the input.* From the tables or by inspection, $\mathcal{L}[\delta(t)] = 1$.

Step 2. *Compute the integro-differential equation of the circuit of Figure 12.29.* The first task is to sum the voltages around the loop to obtain

$$v_R + v_L + v_C = v_{in}$$

Substituting for each of the element voltages using the mesh current, $i_{in}(t)$, yields the desired integro-differential equation,

$$R i_{in}(t) + L\frac{di_{in}}{dt}(t) + \frac{1}{C}\int_{-\infty}^{t} i_{in}(q)\,dq = v_{in}(t) \tag{12.51}$$

Step 3. *Take the Laplace transform of both sides, substitute for R, L, C, $i_L(0^-)$, $v_C(0^-)$, and $V_{in}(s)$, and solve for $I_{in}(s)$.* With the aid of the differentiation and integration formulas, taking the Laplace transform of both sides of equation 12.51 produces

$$R I_{in}(s) + Ls I_{in}(s) - Li_L(0^-) + \frac{1}{Cs} I_{in}(s) + \frac{v_C(0^-)}{s} = V_{in}(s)$$

This has the form

$$L\frac{s^2 + \frac{R}{L}s + \frac{1}{LC}}{s} I_{in}(s) = V_{in}(s) + Li_L(0^-) - \frac{V_C(0^-)}{s}$$

Plugging in the required quantities and solving for $I_{in}(s)$ produces

$$I_{in}(s) = \frac{s}{s^2 + 4s + 4} + \frac{1}{s+2} = \frac{2}{s+2} - \frac{2}{(s+2)^2}$$

Step 4. *Find $i_{in}(t)$.* Taking the inverse Laplace transform yields the desired result:

$$i_{in}(t) = (2 - 2t)e^{-2t}u(t) \text{ A}$$

A plot of this response appears in Figure 12.30.

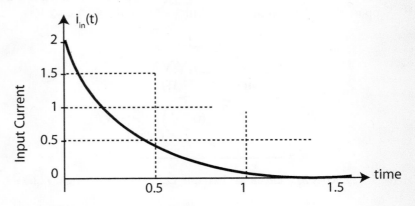

FIGURE 12.30 Plot of $i_{in}(t)$ for Example 12.26.

Exercises. 1. An integro-differential equation for an *LC* circuit is given by

$$\frac{di_C}{dt} + \frac{1}{C} \int_{-\infty}^{t} i_C(q)dq = 0$$

with $C = 1$ F, $i_C(0^-) = 0$, and $v_C(0^-) = -10$ V. Compute $i_C(t)$.

ANSWER: $sI_C(s) + \dfrac{I_C(s)}{s} + \dfrac{v_C(0^-)}{s} = 0 \Rightarrow i_C(t) = 10\sin(t)u(t)$ A

2. If two signals $x(t)$ and $y(t)$ are related by the equations

$$\frac{dx(t)}{dt} + 2y(t) = 4\delta(t) \text{ and } 2x(t) - \int_{0^-}^{t} y(z)dz = 2u(t)$$

where $x(0^-) = 2$, $u(t)$ is the unit step function, and $\delta(t)$ is the Dirac delta function, then find $X(s)$.

ANSWER: $X(s) = \dfrac{2}{s}$

EXAMPLE 12.27. The final example of this chapter looks at the **leaky integrator circuit** of Figure 12.31, which contains an ideal operational amplifier (op amp). R_2 represents the leakage resistance of the capacitor. Given C and R_2, R_1 is chosen to achieve an overall gain constant, in this case, 1. The objective is to compute the response $v_{out}(t)$, assuming that $v_C(0^-) = 0$, and compare it with that of a pure integrator having a gain constant of -1.

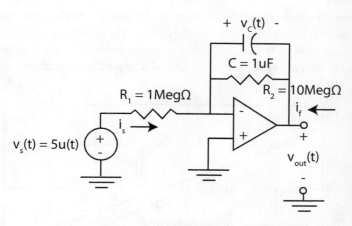

FIGURE 12.31 Leaky integrator op amp circuit.

Solution

First, note that since the op amp is ideal, $-v_C(t) = v_{out}(t)$. The goal, then, is to write a differential equation that relates v_s to v_C and solve for v_C using the Laplace transform method.

Step 1. *Determine the differential equation.* Since the op amp is ideal, it follows that $i_f = -i_s$. From Ohm's law, $i_s = v_s/R_1$. On the other hand,

$$i_f(t) = -\frac{v_C(t)}{R_2} - C\frac{dv_C(t)}{dt}$$

This leads to the differential equation model of the op amp circuit,

$$\frac{dv_{out}(t)}{dt} + \frac{v_{out}(t)}{R_2 C} = -\frac{v_s(t)}{R_1 C}$$

where, as indicated before, $v_{out}(t) = -v_C(t)$. Note that if $R_1 C = 1$ and R_2 is infinite, then the circuit works as a simple integrator. The circuit is called a leaky integrator because $R_2 C$ is large but finite. Since $R_1 C = 1$, one expects the gain constant to be 1 as well.

Step 2. *Substitute values, take the Laplace transform of both sides, and solve for $V_{out}(s)$.* Taking the Laplace transform of both sides, one obtains

$$sV_{out}(s) - v_{out}(0^-) + 0.1V_{out}(s) = -\frac{5}{s}$$

Since $v_{out}(0^-) = v_C(0^-) = 0$, it follows that

$$V_{out}(s) = -\frac{5}{s(s+0.1)} = \frac{-50}{s} + \frac{50}{s+0.1}$$

Step 3. *Invert $V_{out}(s)$ to obtain $v_{out}(t)$.* Solving for $v_{out}(t)$ produces

$$v_{out}(t) = 50(e^{-0.1t} - 1)u(t)$$

A plot of the op amp output voltage appears in Figure 12.32 along with the ideal integrator curve. Observe that the somewhat realistic leaky integrator circuit approximates an ideal integrator only for the approximate time interval $0 \le t \le 1.5$ before the error induced by the feedback resistor R_2 becomes too large. Such integrators need to be reinitialized by setting the capacitor voltage to zero.

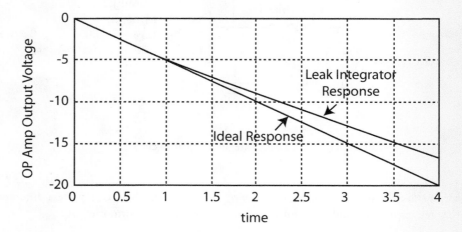

FIGURE 12.32 Output voltage of a leaky integrator that approximates an ideal integrator with a gain of −1.

This concludes our discussion of the basic Laplace transform toolbox pertinent to circuit analysis. The next chapter takes up the application of the tools to the problem of basic circuit analysis.

9. SUMMARY

Chapter 12 has introduced the definitions of the Laplace transform of a signal and the inverse transform using partial fraction expansions. For signals that are zero for $t < 0$, the reconstruction of the original signal from the inverse transform is unique. For the types of signals normally dealt with by electrical engineers, the Laplace transform is typically a rational function or an exponential of the form e^{-sT} times a rational function. If the transform is a rational function, then one performs a partial fraction expansion of the rational function. The partial fraction expansion is a sum of terms that have simple catalogued inverse transforms, as listed in Table 12.1. If the transform is multiplied by e^{-sT}, $T > 0$, then the resulting signal is simply a time-shifted version of the signal.

Various properties aiding the computation of the Laplace transform and its inverse are given in Table 12.2. Of special interest are the derivative and integral properties of the Laplace transform, because these allow one to transform differential and integro-differential equation models of a circuit to the s-domain. One can then solve for the Laplace transform of the output variable and take the inverse transform to obtain the time response of the circuit. This method allows one to easily incorporate the initial conditions of the differential equation into the solution. In the next chapter, we develop a more direct method for computing circuit responses without first developing a differential equation model.

10. TERMS AND CONCEPTS

Abscissa of convergence: the smallest value σ_0 such that when $\text{Re}[s] = \text{Re}[\sigma + j\omega] = \sigma > \sigma_0$, the Laplace transform integral converges.

Analytic continuation: the process of smoothly extending a complex-valued function, analytic in an open region of the complex plane, to the entire complex plane.

Characteristic equation of an ordinary differential equation: a polynomial, derived from an ordinary differential equation, whose zeros determine the characteristic exponents (e.g., p in e^{-pt}) in the response or output.

(Ordinary) Differential equation model: a model of the behavior of a circuit in which the input excitation and the desired response are related by an equation in which a scaled sum of time derivatives of the output signal is set equal to a scaled sum of derivatives of the input signal.

(Dirac) Delta function: see unit impulse function.

Dot notation: one dot over a function means a first-order derivative, two dots a second-order derivative, etc.

Frequency domain: the s-domain, which refers to the analysis of circuits using Laplace transform methods.

Frequency shift: frequency translation of $F(s)$ to $F(s + a)$.

Inverse Laplace transform: a transformation whereby one takes a special complex-valued function $F(s)$ and uses it to compute a time function $f(t)$.

KCL (Kirchhoff's current law): law stating that, whether we are dealing with time signals or the transform of time signals, the sum of the currents entering (or leaving) a node is zero.

KVL (Kirchhoff's voltage law): law stating that, whether we are dealing with time signals or the transform of time signals, the sum of the voltages around a loop is zero.

Laplace transform: a special integral transformation of a time function $f(t)$ to a function $F(s)$, where s is a complex variable.

Partial fraction expansion: the expansion of a rational function into a sum of terms. Each term is a constant (possibly complex) divided by a term of the form $(s + p)^k$.

Pole (simple) of rational function: zero of order 1 in a denominator polynomial.

Poles (finite) of rational function: zeros of a denominator polynomial.

Ramp function, $r(t)$: integral of unit step function where $r(t) = tu(t)$.

Rational function: ratio of two polynomials; also, a polynomial itself is a rational function.

Region of convergence (ROC): the region of the complex plane in which the Laplace transform integral is valid.

Shifted step function, $u(t - T)$: translation of a step function by T units. If T is positive, the translation is to the right; if T is negative, the translation is to the left.

Sifting property: the property of a delta function to simplify an integral, i.e., , provided that $f(t)$ is continuous at $t = T$.

Time shift: time translation of a function $f(t)$ to $f(t - T)$.

Unit impulse function: a so-called distribution whose integral from $-\infty$ to t is the step function; it is non-rigorously interpreted as the derivative of the step function.

Unit step function, $u(t)$: $u(t) = 0$ for $t < 0$, and $u(t) = 1$ for $t \geq 0$.

Zeros (finite) of rational function: values that make a numerator polynomial zero.

---

[1] See A. H. Zemanian, *Distribution Theory and Transform Analysis* (New York: McGraw Hill, 1965).

[2] See John B. Conway, *Functions of One Complex Variable* (New York: Springer-Verlag, 1973).

[3] In the engineering literature, *proper* refers to $m \leq n$ and *strictly proper* to $m < n$; in the mathematics literature, *proper* means $m < n$ and *improper* $m \geq n$.

Problems

BASIC SIGNALS, SIGNAL REPRESENTATION, AND LAPLACE TRANSFORMS

1. Find the Laplace transform of each of the following signals assuming $T_i > 0$.
   - (a) $f_1(t) = f(t + T_0)\delta(t - T_1)$
   - (b) $f_2(t) = f(t - T_0)\delta(t - T_1)$
   - (c) $f_3(t) = e^{-5t}\cos\left(0.5\pi t + \frac{\pi}{4}\right)\delta(2t - T_0)$
   - (d) $f_4(t) = K_1\delta(t) + K_2\delta(t + T_0)$
   - (e) $f_5(t) = K_1\delta(t) + K_2\delta(t - T_0)$
   - (f) $f_6(t) = \cosh(2t)\delta(t - 2) + \sinh(t - 2T_1)\delta(t - T_1)$
   - (g) $f_7(t) = \sin(2\pi t - \pi)\delta(2t - 4)$

2. Find the Laplace transform of each of the following signals. Use Tables 12.1 and 12.2 as needed.
   - (a) $f_1(t) = 2u(t) + u(t - 1) + u(t - 2) - 4u(t - 4)$
   - (b) $f_2(t) = 2r(t) - 2r(t - 1) - r(t - 3) + r(t - 5)$
   - (c) $f_3(t) = 2u(t) - r(t) + r(t - 2)$
   - (d) $f_4(t) = \cos(0.5\pi t)u(t) + \cos(0.5\pi(t - 2))u(t - 2)$
   - (e) $f_5(t) = 4(t - 1)e^{-5(t - 1)}u(t - 1)$

3. Sketch the indicated waveforms and find the Laplace transform. Use Tables 12.1 and 12.2 as needed.
   - (a) $f_1(t) = u(t) - r(t - 1)$.
   - (b) $f_2(t) = u(t) - r(t - 1) + r(t - 2)$
   - (c) $f_3(t) = 2u(t)\,u(2 - t)$.
   - (d) $f_4(t) = 4r(t)\,u(1 - t)$
   - (e) $f_5(t) = 2r(t)u(1 - t) + u(t - 1)r(2 - t)$

4. Find the Laplace transform of each of the following time functions.
   - (a) $f_1(t) = Ktu(t - 1)$
   - (b) $f_2(t) = K(t - 1)u(t)$
   - (c) $f_3(t) = Ktr(t - 1)$
   - (d) $f_4(t) = K(t - 1)r(t)$

5. Find the Laplace transform of each of the following time functions.
   - (a) $f_1(t) = Ke^{-at}u(t - T)$, $T > 0$
   - (b) $f_2(t) = Ke^{-a(t - T)}u(t)$, $T > 0$
   - (c) $f_3(t) = Kte^{-at}u(t - T)$, $T > 0$
   - (d) $f_4(t) = Kte^{-a(t - T)}u(t)$, $T > 0$
   - (e) $f_5(t) = Kt^2e^{-a(t - T)}u(t)$, $T > 0$

6. Compute the Laplace transform for each of the following signals.
   - (a) $f_1(t) = Kt^2[u(t) - u(t - T)]$, $T > 0$
   - (b) $f_2(t) = \sin(2\pi t)u(t) - \sin(2\pi t - 2\pi)u(t - 1)$
   - (c) $f_3(t) = t[\sin(2\pi t)u(t) - \sin(2\pi t - 2\pi)u(t - 1)]$
   - (d) $f_4(t) = 2\sin(4\pi t)u(t)u(2 - t)$

7. Represent each of the following signals using (sums of) steps, ramps, shifts of basic signals, etc. Then find the Laplace transform.

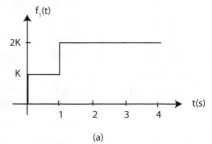

(a)

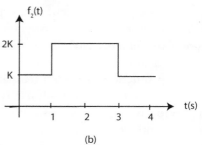

(b)

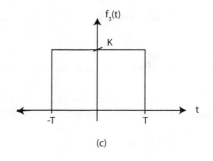

(c)

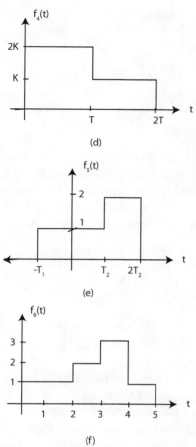

(d)

(e)

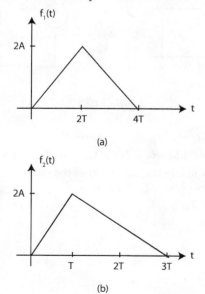

(f)

8. Represent each of the following signals using (sums of) steps, ramps, shifts of basic signals, etc. Then find the Laplace transform.

(a)

(b)

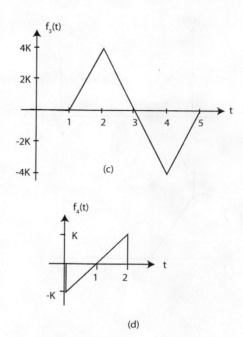

(c)

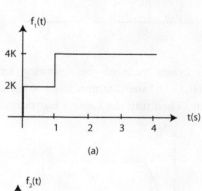

(d)

9. Represent each of the following signals using (sums of) steps, ramps, shifts of basic signals, etc. Then find the Laplace transform.

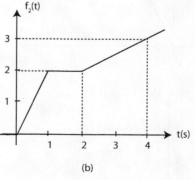

(a)

(b)

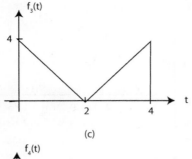

(c)

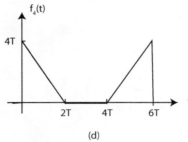

(d)

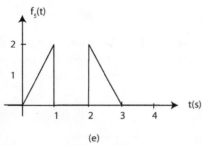

(e)

10. Represent each of the following signals using (sums of) steps, ramps, shifts of basic signals, etc. Then find the Laplace transform.

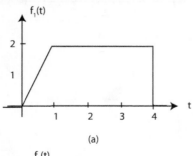

(a)

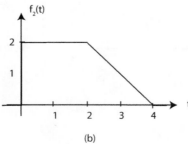

(b)

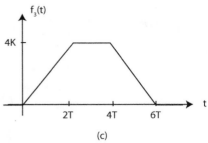

(c)

11. For the circuit of Figure P12.11, suppose $R_1 = 600$ Ω, $R_2 = 1000$ Ω, and $R_3 = 1500$ Ω. Use the Laplace transform tables to compute $I_{out}(s) = \mathcal{L}\left[i_{out}(t)\right]$ for the input $i_{in}(t) = 6u(t) - 12tu(t) + 3e^{-2t}u(t) + 18e^{-t}\sin(2\pi t)u(t)$ A.

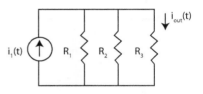

Figure P12.11

12. For the circuit of Figure P12.12b, $R_1 = 600$ Ω, $R_2 = 1000$ Ω, and $R_3 = 2400$ Ω. Use the Laplace transform tables to compute $V_{out}(s) = \mathcal{L}\left[v_{out}(t)\right]$ for the input given in Figure P12.12a.

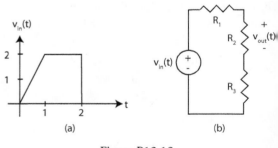

Figure P12.12

13. For the circuit of Figure P12.13, suppose $R_1 = 600$ Ω, $R_2 = 2000$ Ω, and $R_3 = 3000$ Ω. Find the Laplace transform of the voltage $v_{AB}(t)$ when $v_{s1}(t) = 24e^{-2t}u(t)$ V and $v_{s2}(t) = 30e^{-4t}r(t)$ V.

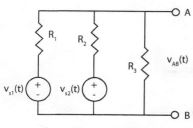

Figure P12.13

LAPLACE TRANSFORMS VIA TABLE 12.1 AND PROPERTIES VIA TABLE 12.2

14. Find the Laplace transform of $f(t) = Ke^{-at}u(t)u(T-t)$, $T > 0$, as follows.
   (a) Express $Ku(t)u(T-t)$ as a difference of step functions.
   (b) Find $\mathcal{L}[Ku(t)u(T-t)]$.
   (c) Apply the frequency shift property on your answer to part (b) to compute $\mathcal{L}[f(t)]$.

15. Prove the time-/frequency-scaling property by direct calculation of the Laplace transform integral.

16.(a) Using the famous *Euler formula,* $e^{j\omega t} = \cos(\omega t) + j\sin(\omega t)$, find an expression for $\cos(\omega t)$ and $\sin(\omega t)$ in terms of the complex exponentials $e^{\pm j\omega t}$.
   (b) Determine the Laplace transform of $e^{\pm j\omega t}$.
   (c) Using the formulas developed in parts (a) and (b), show that

   (i) $\mathcal{L}[\cos(\omega t)] = \dfrac{s}{s^2 + \omega^2}$

   (ii) $\mathcal{L}[\sin(\omega t)] = \dfrac{\omega}{s^2 + \omega^2}$

   (d) Using the formulas of part (c) and the multiplication-by-t property, compute
   (i) $\mathcal{L}[Kt\cos(\omega t)]$
   (ii) $\mathcal{L}[Kt\sin(\omega t)]$

   (e) Using the formulas of part (c) and the frequency shift property, compute
   (i) $\mathcal{L}[Ke^{-at}\cos(\omega t)]$
   (ii) $\mathcal{L}[Ke^{-at}\sin(\omega t)]$

17. Find a simple expression for each of the waveforms shown in Figure P12.17 in terms of sines, cosines, shifted sines, and shifted cosines. Then find the associated Laplace transforms.

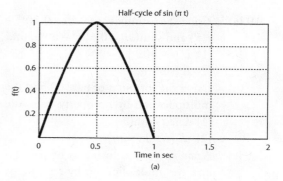

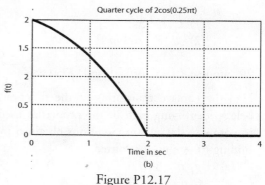

Figure P12.17

18.(a) Represent

   $$g(t) = \begin{cases} \sin(\pi t) & 0 \le t \le 2 \\ 0 & \text{otherwise} \end{cases}$$

   as the difference of a sine and a shifted sine. Find $G(s)$.
   (b) Relate $f(t)$ in Figure P12.18 to $g(t)$ of part (a). Then find $F(s)$ from $G(s)$.

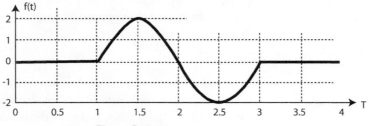

Figure P12.18

19.(a) Using the formulas $\cosh(at) = 0.5(e^{at} + e^{-at})$ and $\sinh(at) = 0.5(e^{at} - e^{-at})$, find
(1) $\mathcal{L}[\cosh(at)]$
(2) $\mathcal{L}[\sinh(bt)]$

(b) Using the formulas of part (a) and the multiplication-by-t property, compute
(1) $\mathcal{L}[Kt\cosh(at)]$
(2) $\mathcal{L}[Kt\sinh(at)]$

(c) Again using the formulas of part (a) and the multiplication-by-t property, compute
(1) $\mathcal{L}[Kt^2\cosh(at)]$
(2) $\mathcal{L}[Kt^2\sinh(at)]$

20. Suppose $f(t) = 0$ for $t < 0$ and $F(s) = \dfrac{2s+4}{s+1}$
Find the Laplace transforms of the functions below, identifying each of the properties used to compute the answer. Solutions obtained by finding $f(t)$ are not permitted.
(a) $g_1(t) = 5\,f(t - T),\ T > 0$
(b) $g_2(t) = 2e^{-at}f(t)$
(c) $g_3(t) = 2e^{-at}f(t - T),\ T > 0$
(d) $g_4(t) = 5tf(t - T),\ T > 0$

21. Use Laplace transform properties to find $G_k(s)$ and $g_k(t)$ as given below when $F(s) = \dfrac{4s+20}{(s+1)^2}$.
State each property that you used. Assume that $f(t) = 0$ for $t < 0$. Solutions obtained by finding $f(t)$ are not permitted.
(a) $g_1(t) = 0.5tf(t)$
(b) $g_2(t) = e^{4t}f(t)$
(c) $g_3(t) = 2e^{4t}g_1(t)$
(d) $g_4(t) = 2g_1(2t - 4)$
(e) $g_5(t) = 2(t - 2)f(t - 2)$
(f) $g_6(t) = 2tf(t - 2)$

22. (a) Consider $f(t)$ in Figure P12.22.
(i) Express $f(t)$ as a sum of appropriate step functions. Compute $F(s)$.
(ii) Compute $\mathcal{L}[g(t)] = \mathcal{L}\left[\dfrac{d}{dt}f(t)\right]$ using the derivative property.
(iii) Compute $\mathcal{L}[h(t)] = \mathcal{L}\left[\int_{0^-}^{t} f(\tau)\,d\tau\right]$ using the integral property.

(b) Repeat part (a) for $f_{new}(t) = f(t + 4)$.

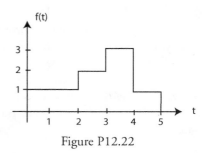

Figure P12.22

23. The Laplace transform of $f(t)$ is given as
$$F(s) = \frac{1 - e^{-(s-a)}}{s - a}.$$
(a) Find the Laplace transform of $e^{-at}f(t)$.
(b) Find the Laplace transform of $tf(2t)$.

ANSWER: (a) $\dfrac{1 - e^{-s}}{s^2} - \dfrac{e^{-s}}{s}$

24. The Laplace transform of $f(t)$ is given as
$$F(s) = \frac{1 - e^{-s}}{s}.$$
(a) Find the Laplace transform of $\dfrac{df(t)}{dt}$ with $f(0^-) = 3$.
(b) Now find the Laplace transform of $tf(t)$.
(c) Finally, find the Laplace transform of $t\,\dfrac{df(t)}{dt}$.

25. Suppose $f(t) = \delta(t) - \delta(t - T)$, $T > 0$.
   (a) Find $\mathcal{L}[f(2t)]$ by direct calculation of the Laplace transform integral.
   (b) Find $\mathcal{L}[f(2t)]$ by computing $F(s)$ and then using the scaling property.

26. Use only Laplace transform properties to answer the following question. Suppose that for $t \leq 0$, $f(t) = e^t u(-t)$, and the one-sided Laplace transform of $f(t)$ is $\mathcal{L}[f(t)] = F(s) = \dfrac{s+1}{s^3}$.

Let $g(t) = f(t)u(t)$. Find the Laplace transform of $v(t)$ when

   (a) $v(t) = 2g''(t) - g'(t)$
   (b) $v(t) = 2f''(t) - f'(t)$
   (c) $v(t) = g'(t) + \int_{-\infty}^{t} g(q)\, dq$
   (d) $v(t) = f'(t) + \int_{-\infty}^{t} f(q)\, dq$

It is not necessary to have the answer be a rational function.

27.(a) Find the Laplace transform of the function $f(t)$ sketched in Figure P12.27a.
   (b) Identify a relationship between $f(t)$ and the function $g(t)$ sketched in Figure P12.27b. Use your answer to part (a) and the appropriate property from Table 12.2 to compute the Laplace transform of $g(t)$.

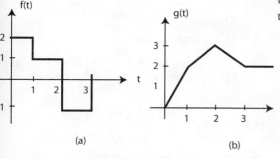

(a)   (b)

Figure P12.27

28. In Figure P12.28, what is the relationship between $f(t)$ and $g(t)$? First find the Laplace

transform of $f(t)$ and then, using the relationship, find the Laplace transform of $g(t)$.

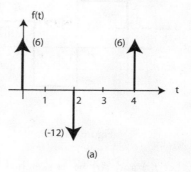

(a)

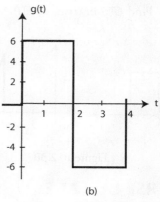

(b)

Figure P12.28

29. Develop a relationship between $f(t)$ and $g(t)$ in Figure P12.29. Find the Laplace transform of $f(t)$ and then $g(t)$ by making use of the relationship between the two functions. Assume that $0 < A < B < C$. Also, determine D and E in terms of A, B, and C.

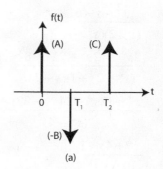

(a)

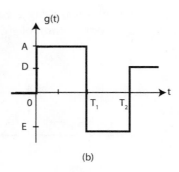

Figure P12.29

30. Let $f(t)$ and $g(t)$ be as sketched in Figure P12.30. Find $G(s)$ in terms of $F(s)$.

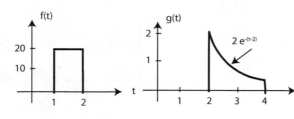

Figure P12.30

INVERSE LAPLACE TRANSFORMS BY PARTIAL FRACTION EXPANSION

31. Using partial fraction expansions and your knowledge of the Laplace transform of simple signals, find $f(t)$ when $F(s)$ equals

(a) $\dfrac{2s^3 + 13s^2 + 30s + 32}{s(s^2 + 6s + 8)}$

(b) $\dfrac{s^3 - s - 6}{(s+2)(s^2-1)}$

(c) $\dfrac{2s^3 + 12s^2 + 22s + 8}{(s^2 + 2s + 1)(s+2)}$

(d) $\dfrac{s^4 + 12s^3 - 24s^2 - 32s + 16}{(s^2-4)^2}$

(e) $\dfrac{s^2 + 4s + 1}{s^3 + s}$

32. Inadvertently left out by the authors.

33. Find (i) the partial fraction expansion and (ii) the inverse Laplace transform for each of the following functions by hand. Show *all* work. (No details, no credit.)

(a) $F_1(s) = \dfrac{b(-as + a^2 - s^2 + bs)}{s(s+a)(s+b)}$

CHECK: One residue is a.

(b) $F_2(s) = \dfrac{2s^3 - 7s^2 + 4s + 2}{(s-1)(s-2)^2}$

CHECK: One residue is −2.

(c) $F_3(s) = \dfrac{2s^4 + 18s^3 + 46s^2 + 44s + 12}{(s+1)^2(s+2)^2}$

CHECK: Two residues are at 2 and −2.

Remark: Check answers using MATLAB. Use the help command to make sure you understand the terms used. For example for part (b),

```
n = [2 -7 4 2];
d = conv([1 -1],[1 -4 4]);
[r,p,k] =residue(n,d)
```

34. Find (i) the *simplified* partial fraction expansion and (ii) the inverse Laplace transform for each of the following functions by hand. Show *all* work.

(a) $F_1(s) = \dfrac{(a+b)s + 2ab}{(s+a)(s+b)}$

Check: One residue is a.

(b) $F_2(s) = \dfrac{(a+b+c)s^2 + (bc+2ab+ac)s + c}{s(s+a)(s+b)}$

CHECK: One residue is a.

(c) $F_3(s) = \dfrac{cs^2 + (a+2ac)s + \left[(1+c)a^2 - a\right]}{(s+a)^2}$

(d) $F_4(s) = \dfrac{s^4 + 18s^3 + 98s^2 + 208s + 144}{(s+2)^2(s+4)^2}$

CHECK: Two residues are at 2 and −2.

(e) $F_5(s) = \dfrac{5s^2 + 144s + 204}{(s+1)\left[(s+2)^2 + 64\right]}$

35. Find $f(t)$ when $F(s)$ equals

(a) $\dfrac{2s+16}{s^2+16}$

(b) $\dfrac{24s-72}{s^2+4s+40}$

(c) $\dfrac{2s^2+88s}{(s+4)(s^2+64)}$

(d) $\dfrac{2s^3+2s^2-2s-6}{\left(s^2+1\right)^2}$

36. Find (i) the partial fraction expansion and (ii) the inverse Laplace transform for each of the following functions by hand. Show *all* work.

(a) $F_1(s)=\dfrac{2s^3+12s^2+23s+17}{(s+1)(s+2)(s+4)}$

CHECK: Residue at $s=-2$ is -1.5.

(b) $F_2(s)=\dfrac{2s^3+9s^2+16s+11}{(s+1)(s+2)^2}$

CHECK: Two of the residues are 2 and 1.

(c) $F_3(s)=\dfrac{s^4+4s^3-2s^2-9s-3}{(s-1)^2(s+2)^2}$

CHECK: Two residues are at 1 and two are at -1.

(d) $F_4(s)=\dfrac{4s^3-12s^2+32s+16}{\left[(s+1)^2+1\right]\left[(s+2)^2+16\right]}$

37. Find the partial fraction expansion and the inverse Laplace transform for each of the indicated output voltages or currents. All answers must be in terms of real functions with real coefficients or symbols. Show *all* work.

(a) For the circuit of Figure P12.37a, find the partial fraction expansion of $V_{out}(s)$ and then find $v_{out}(t)$ for the input voltage

$$V_{in}(s)=20\,\frac{5s^2+8s+16}{s(s^2+4)}\ .$$

(b) For the circuit of Figure P12.37b, find the partial fraction expansion of

$V_{out}(s)$ and then find $v_{out}(t)$ for the input current

$$I_{in}(s)=20\,\frac{s^3-s^2-4s+9}{(s-2)^2(s^2+1)}\ .$$

Check: One residue is at 20.

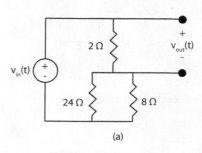

(a)

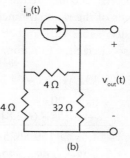

(b)

Figure P12.37

38. Suppose $F(s)=\dfrac{4s-16}{(s+2)^2+9}$; it follows that $f(t)=K_1e^{-at}\cos(\omega t)u(t)+K_2e^{-at}\sin(\omega t)u(t)$. Find a, K_1, K_2, and ω. Now express $f(t)=K_3e^{-at}\cos(\omega t+\theta)u(t)$ by finding K_3, a, ω and θ.

39. The Laplace transform of $f(t)=[K_1e^{-at}\cos(\omega t)+K_2e^{-at}\sin(\omega t)+K_3e^{-bt}]u(t)$ is

$$F(s)=\frac{-52s+228}{(s+4)\left[(s+1)^2+100\right]}$$

Find a, b, K_1, K_2, K_3, and ω.

40. Consider the resistive circuit in Figure P12.40. Use Table 12.1 and the shift property to find $v_{out}(t)$ for each $v_{in}(s)$ below. Sketch $v_{out}(t)$ by hand or with the help of MATLAB.

(a)
$$V_{in}(s) = \frac{10e^{-s} + 5e^{-2s} - 10e^{-3s} - 5e^{-4s}}{s}$$

(b)
$$V_{in}(s) = 10\frac{1 - e^{-2s}}{s^2} - 20\frac{e^{-4s}}{s}$$

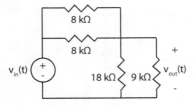

Figure P12.40

41. Use MATLAB to compute the partial fraction expansions of the rational functions listed below. Then use Table 12.1 and MATLAB to obtain the associated time function.

(a) $F_1(s) = \dfrac{3s^3 + 30s^2 + 86s + 64}{s^3 + 8s^2 + 20s + 16}$

(b) $F_2(s) = \dfrac{-46.25s - 692.8125}{s^3 + 14.5s^2 + 169.5625s + 510.25}$

(c) $F_3(s) = \dfrac{-2s^3 + 23s^2 - 68s - 3265}{s^3 + 3.5s^2 + 134s + 797.5}$

(d) $F_4(s) = \dfrac{10.5s^3 + 47.875s^2 + 151.875 - 108.5938}{s^4 + 6.5s^3 + 36.5625s^2 + 101.5625s + 207.0312}$

(e) $F_5(s) = \dfrac{-1.5s^5 - 25.75s^4 - 127.5s^3 - 291.5s^2 - 330s - 143.75}{s^6 + 10.5s^5 + 50s^4 + 141s^3 + 250s^2 + 262.5s + 125}$

CIRCUIT RESPONSES VIA LAPLACE TRANSFORM APPLIED TO DIFFERENTIAL EQUATIONS

42. The input to the circuit of Figure P12.42 is $v_{in}(t) = 10u(t)$ V, valid for $t \geq 0$, and has an initial inductor current of $i_L(0^-) = 0$. Suppose $R = 50$ Ω and $L = 0.2$ H.

    (a)   Show that the differential equation for the circuit is

$$\frac{di_L(t)}{dt} + \frac{R}{L}i_L(t) = \frac{1}{L}v_{in}(t)$$

(b)   Use the Laplace transform method to compute the inductor current, $i_L(t)$, for $t \geq 0$.

(c)   If the input is changed to $v_{in}(t) = 10u(t - T)$ V, where $T = 10$ msec, find $i_L(t)$, for $t \geq 0$. Hint: Your answer should be a shift of the answer computed in part (b).

(d)   If $i_L(0^-) = 100$ mA and $v_{in}(t) = 0$, find $i_L(t)$, for $t \geq 0$.

(e)   If $i_L(0^-) = 100$ mA and $v_{in}(t) = 10u(t)$ V, find $i_L(t)$, for $t \geq 0$. Hint: Can you use superposition?

(f)   If $i_L(0^-) = 100$ mA and $v_{in}(t) = 10u(t - T)$ V, where $T = 10$ msec, find $i_L(t)$, for $t \geq 0$.

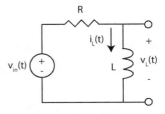

Figure P12.42

43. For the circuit of Figure P12.43, suppose $R = 10$ Ω and $C = 0.01$ F.

    (a)   Show that the differential equation for the circuit is

$$\frac{dv_C(t)}{dt} + \frac{1}{RC}v_C(t) = \frac{1}{RC}v_{in}(t)$$

(b)   Use the Laplace transform method to compute the capacitor voltage, $v_C(t)$, for $t \geq 0$ when $v_{in}(t) = 10u(t)$ V, valid for $t \geq 0$.

(c)   Now suppose the initial capacitor voltage $v_C(0^-) = -10$ V and $v_{in}(t) = 0$.

Find $v_C(t)$, for $t \geq 0$.

(d)   Find $v_C(t)$, for $t \geq 0$, when $v_C(0^-) =$
       -10 and $v_{in}(t) = 10u(t)$ V.

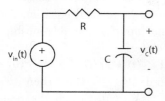

Figure P12.43

44. The circuit of Figure P12.44 has two source
excitations, $v_{s1}(t) = 10u(t)$ V and $i_{s2}(t) = 2u(t)$
A, both applied at $t = 0$. Suppose $R_1 = 5$ Ω, R_2
$= 20$ Ω, and $L = 2$ H. The initial condition on
the inductor current is $i_L(0^-) = -1$ A.
    (a)   Construct a differential equation for
           the circuit, assuming the response is
           $i_L(t)$. Leave the inputs in terms of
           $v_{s1}(t)$ and $i_{s2}(t)$.
    (b)   Find the response due only to $v_{s1}(t)$,
           assuming $i_L(0^-) = 0$.
    (c)   Find the response due only to $i_{s2}(t)$,
           assuming $i_L(0^-) = 0$.
    (d)   Find the response due only to the ini-
           tial condition $i_L(0^-) = -1$ A, assuming
           both inputs are zero.
    (e)   Find the complete response, $i_L(t)$, for t
           ≥ 0 by superposition.

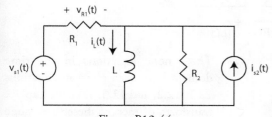

Figure P12.44

45. Consider the LC circuit of Figure P12.45,
for which $i_L(0^-) = I_0$ and $v_C(0^-) = V_0$. Since
there is no resistance present in the circuit,
there is no damping; hence, one expects a purely sinusoidal response. Such circuits are called
lossless.

(a)   Construct a differential equation in
       the capacitor voltage, $v_C(t)$.
(b)   Solve the differential equation by the
       Laplace transform method; i.e., show
       that $v_C(t) = K_1 \sin(\omega t) + K_2 \cos(\omega t)$ for
       $t \geq 0$ and for appropriate constants K_1,
       K_2, and ω, which are to be found in
       terms of I_0, V_0, L and C.

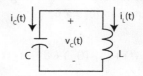

Figure P12.45

ANSWER: (a) $\dfrac{d^2 v_C}{dt^2} + \dfrac{1}{LC} v_C = 0$, (b) $\omega = \dfrac{1}{\sqrt{LC}}$

46. The circuit of Figure P12.46 is a series (lossless) LC circuit driven by a voltage source.
Suppose $v_C(0^-) = 0$ and $i_L(0^-) = 0$.
    (a)   Construct the differential equation of
           the circuit in terms of the capacitor
           voltage, $v_C(t)$.
    (b)   Solve the differential equation using
           the Laplace transform method, and
           show that $v_C(t) = 10 - 10 \cos(0.5\pi t)$
           for $t \geq 0$.

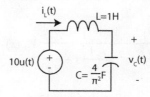

Figure P12.46

47. A pair of (coupled) differential equations
that represent a circuit are given as

$$\frac{dx(t)}{dt} + a_1 x(t) = a_2 y(t)$$

and

$$\frac{dy(t)}{dt} + a_3 y(t) = b_3 f(t)$$

with initial conditions $x(0^-) = 1$ and $y(0^-) = 2$. Suppose $a_1 = 1$, $a_2 = 1$, $a_3 = 1$, $b_3 = 1$, and $f(t) = 2u(t)$. Find $y(t)$ and $x(t)$.

48. Reconsider Problem 47 with $a_1 = 2$, $a_2 = 2$, $a_3 = 4$, $b_3 = 3$, and $f(t) = 2u(t)$.

49. The inductor current $i(t)$ in a second-order RLC circuit satisfies the following integro-differential equation for $t > 0$.

$$4\frac{di_L(t)}{dt} + 12i_L(t) + \left[v_C(0^-) + 8\int_{0^-}^{t} i_L(\tau)d\tau \right] = 8u(t)$$

(a) If $i(0^-) = 8$ A and $v_C(0^-) = -4$ V, find $I_L(s)$.
(b) Use your answer in part (a) to find $i_L(t)$.

50. Consider the circuit of Figure P12.50.
(a) Use KVL and KCL to show that the differential equation relating the input voltage to the capacitor voltage is

$$\frac{d^2v_C}{dt^2} + \frac{1}{RC}\frac{dv_C}{dt} + \frac{1}{LC}v_C = \frac{1}{LC}v_{in}$$

(b) Take the Laplace transform of both sides of this equation to show that

$$V_C(s) = \frac{1}{LC}\frac{1}{s^2 + \frac{1}{RC}s + \frac{1}{LC}}V_{in}(s) + \frac{(s+5)v_C(0^-) + \dot{v}_C(0^-)}{s^2 + \frac{1}{RC}s + \frac{1}{LC}}$$

(c) Assuming that $v_C(0^-) = v'_C(0^-) = 0$, $R = 0.8\ \Omega$, $L = 1$ H, $C = 0.25$ F, and $v_{in}(t) = \delta(t)$, show that

$$v_C(t) = \frac{4}{3}[e^{-t} - e^{-4t}]u(t)\text{ V}$$

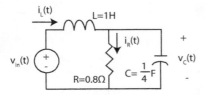

Figure P12.50

51. The op amp in the circuit of Figure P12.51 is assumed to be ideal. $R_1 = 20$ kΩ, $R_2 = 40$ kΩ, and $C = 10$ μF.
(a) Use nodal analysis to construct a first-order differential equation describing the input-output relationship of the voltages.
(b) If $v_{in}(t) = 2u(t)$ V, and $v_C(0) = -1$ V, find $V_{out}(s)$ and then $v_{out}(t) = \mathcal{L}^{-1}[V_{out}(s)]$. Sketch the response in MATLAB.
(c) If $v_{in}(t) = 2e^{-2.5t}u(t)$ V and $v_C(0) = 0$, find $v_{out}(t)$.

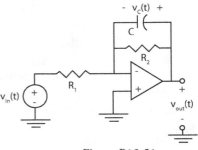

Figure P12.51

52. Reconsider the RC active circuit shown Figure 12.1 of Example 12.1, where we encountered difficulty using the single third-order differential approach. Now we will solve the problem with Laplace transforms applied to three first-order differential equations

(a) Three node equations in the time domain have been given in equations 12.1, 12.2, and 12.3. Take the Laplace transform of each of these three node equations, accounting for initial conditions.
(b) If $v_{in}(t) = 10u(t)$ V and the initial capacitor voltages are $v_{C1}(0) = 12$ V, $v_{C2}(0) = 6$ V, and $v_{C3}(0) = 3$ V, find $V_{out}(s)$.
(c) Now do a partial fraction expansion of $V_{out}(s)$ and determine $v_{out}(t)$ for $t \geq 0$.

C H A P T E R *13*

Laplace Transform Analysis II: Circuit Applications

A FLUORESCENT LIGHT APPLICATION

Fluorescence is a process for converting one type of light into another. In a fluorescent light, an electric current heats up electrodes at each end of a tube. The hot electrodes emit free electrons, which, for a sufficiently high voltage between the electrodes, initiate an arc, causing mercury contained in the tube to vaporize. The energized mercury vapor emits invisible ultraviolet light that strikes a phosphorus coating on the inside of the tube. The phosphorus absorbs this invisible short-wavelength energy and emits light in the visible spectrum.

A starter circuit must quickly generate a sufficient quantity of free electrons and create a sufficiently high voltage to initiate the arc that vaporizes the mercury inside the tube. One type of starter circuit contains a special heat-sensitive switch in series with an inductor. We will model this special switch by an ideal heat-sensitive (bimetal) switch in parallel with a capacitor. The concepts developed in this chapter will allow us to analyze the operation of such a starter circuit as set forth in Example 13.11.

CHAPTER OBJECTIVES

1. In terms of the Laplace transform variable s, define the notion of impedance, denoted $Z(s)$, and the notion of admittance, denoted $Y(s)$. Impedances and admittances will satisfy a type of Ohm's law. These ideas are generalizations of the phasor-based notions of impedance and admittance introduced in Chapter 10.
2. Learn the arithmetic of impedances and admittances in the Laplace transform domain, which is analogous to the arithmetic of resistances and conductances in the time domain.

3. Apply the new concepts of impedance and admittance to redevelop the notions of voltage/current division, source transformations, linearity, and Thevenin and Norton equivalent circuits in the *s*-domain.
4. Define *s*-domain-equivalent circuits of initialized capacitors and inductors for the purpose of transient circuit analysis.
5. Introduce the notion of a transfer function.
6. Define two special types of responses: the impulse and step responses.
7. Redevelop nodal and loop analyses in terms of impedances and admittances.
8. Utilize the Laplace transform technique, especially the *s*-domain-equivalent circuits of initialized capacitors and inductors, for the solution of switched *RLC* circuits.
9. Introduce the notion of a switched capacitor circuit, which has an important place in real-world filtering applications.
10. Set forth a technique for designing general summing integrator circuits.

SECTION HEADINGS

1. **Introduction**
2. **Notions of Impedance and Admittance**
3. **Manipulation of Impedance and Admittance**
4. **Equivalent Circuits for Initialized Inductors and Capacitors**
5. **Notion of Transfer Function**
6. **Impulse and Step Responses**
7. **Nodal and Loop Analysis in the *s*-Domain**
8. **Switching in RLC Circuits**
9. **Switched Capacitor Circuits and Conservation of Charge**
10. **The Design of General Summing Integrators**
11. **Summary**
12. **Terms and Concepts**
13. **Problems**

1. INTRODUCTION

Chapter 12 cultivated the Laplace transform as a mathematical tool particularly useful for circuits modeled by differential equations. This chapter adapts the Laplace transform tool to the peculiar needs and attributes of circuit analysis. With the Laplace transform methods described in this chapter, the intermediate step of constructing a circuit's differential equation, as was done in Chapter 12, can be eliminated.

Available for the analysis of resistive circuits is a wide assortment of techniques: Ohm's law, voltage and **current division**, nodal and loop analysis, linearity, etc. For the sinusoidal **steady-state analysis** of *RLC* circuits, phasors serve as a natural generalization of the techniques of resistive circuit analysis. The Laplace transform tool permits us to extend the sinusoidal steady-state phasor analysis methods to a much wider setting where transient and steady-state analysis are both possible for a broad range of input excitations not amenable to phasor analysis. Recall that transient analysis is not possible with phasors.

The keys to this generalization are the *s*-domain notions of impedance and its inverse, admittance. Instead of defining impedance in terms of *jω*, as in phasor analysis, we will define it in terms of the Laplace transform variable *s*. This definition allows the evolution of a frequency- or *s*-dependent Ohm's law, *s*-dependent voltage and current division formulas, and *s*-dependent nodal and loop analysis; in short, all of the basic circuit analysis techniques have analogous *s*-dependent formulations. What is most important, however, is that with the *s*-dependent formulation, it will be possible to define *s*-dependent equivalents for circuits containing initialized capacitors, inductors, and other linear circuit elements. These equivalent circuits make transient analysis natural in the *s*-domain.

In the final section of the chapter, we introduce the notion of a switched capacitor circuit. Switched capacitor circuits contain switches and capacitors, and possibly some op amps, but no resistors or inductors. Present-day integrated circuit technology allows us to build switches, capacitors, and op amps on chips easily and inexpensively. This has fostered an important trend in circuit design toward switched capacitor circuits. A thorough investigation of switched capacitor circuits is beyond the scope of this text. Nevertheless, it is important to introduce the basic ideas and thereby lay the foundation for more advanced courses on the topic.

2. NOTIONS OF IMPEDANCE AND ADMITTANCE

Chapter 10 introduced an intermediate definition of (phasor) impedance as the ratio of phasor voltage to phasor current, and admittance as the ratio of phasor current to phasor voltage. In the Laplace transform context, impedances and admittances are *s*-dependent generalizations of these phasor notions. Such generalizations do not exist in the time domain. To crystallize this idea, we Laplace-transform the standard differential *v-i* relationship of an inductor,

$$v_L(t) = L\frac{di_L}{dt}(t)$$

to obtain

$$V_L(s) = LsI_L(s), \tag{13.1}$$

assuming $i_L(0^-) = 0$. Here, the quantity $Z_L(s) \triangleq Ls$ multiplies an *s*-domain current, $I_L(s)$, to yield an *s*-domain voltage, $V_L(s)$, in a manner similar to Ohm's law for resistor voltages and currents. The units of $Z_L(s) \triangleq Ls$ are ohms. The quantity Ls depends on the frequency variable *s* and generalizes the concept of a fixed resistance, and it is universally called an **impedance**. This complex-frequency or *s*-domain concept has no time-domain counterpart.

Although the inductor served to motivate *s*-domain impedance, in general an impedance can be defined for any two-terminal device whose input-output behavior is linear and whose parameters do not change with time. A device whose characteristics or parameters do not change with time is called *time invariant*.

IMPEDANCE

The *impedance*, denoted $Z(s)$, of a linear time-invariant two-terminal device, as illustrated in Figure 13.1, relates the Laplace transform of the current, $I(s)$, to the Laplace transform of the voltage, $V(s)$, assuming that all independent sources inside the device are set to zero and that there is no internal stored energy at $t = 0^-$. Under these conditions,

$$V(s) = Z(s)I(s) \tag{13.2a}$$

and, where defined,

$$Z(s) = \frac{V(s)}{I(s)} \tag{13.2b}$$

in units of ohms.

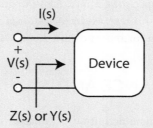

FIGURE 13.1 A two-terminal device having impedance $Z(s)$ or admittance $Y(s)$.

Exercise. For an unknown linear circuit, $V_{in}(s) = \dfrac{8}{s^2 + 4}$ and $I_{in}(s) = 2$. Compute $Z_{in}(s)$.

ANSWER: $\dfrac{4}{s^2 + 4}$

The inverse of resistance is conductance, and the inverse of impedance is admittance. For example, if we divide both sides of equation 13.1 by Ls, we obtain

$$I_L(s) = \frac{1}{Ls}V_L(s) \tag{13.3}$$

This suggests that $1/Ls$ acts as a *generalized conductance* universally called an **admittance**, which is defined as follows.

ADMITTANCE

The *admittance*, denoted $Y(s)$, of a two-terminal linear time-invariant device, as illustrated in Figure 13.1, relates the Laplace transform of the voltage, $V(s)$, across the device to the Laplace transform of the current, $I(s)$, through the device, assuming that all internal independent sources are set to zero and there is no internal stored energy at $t = 0^-$. Under these conditions,

$$I(s) = Y(s)V(s) \qquad\qquad (13.4a)$$

and, where defined,

$$Y(s) = \frac{I(s)}{V(s)} \qquad\qquad (13.4b)$$

in units of S.

From equations 13.2 and 13.4, impedance and admittance satisfy the inverse relationship

$$Y(s) = \frac{1}{Z(s)} \qquad\qquad (13.5)$$

Exercise. For an unknown linear circuit, $V_{in}(s) = \dfrac{8}{(s+2)}$ and $I_{in}(s) = \dfrac{16}{(s+2)(s+4)}$. Compute $Y_{in}(s)$.

ANSWER: $\dfrac{2}{s+4}$

As a first step in deepening our understanding of these notions, we compute the impedances and admittances of the basic circuit elements shown in Figure 13.2.

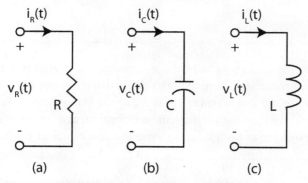

FIGURE 13.2 (a) Resistor. (b) Capacitor. (c) Inductor.

From Ohm's law, the resistor of Figure 13.2a satisfies $v_R(t) = Ri_R(t)$. Laplace-transforming both sides yields the obvious, $V_R(s) = RI_R(s)$. From equations 13.1 and 13.2, the **impedance of the resistor** is

$$Z_R(s) = R$$

and, from equation 13.5, the **admittance of the resistor** is

$$Y_R(s) = \frac{1}{R}$$

Here the kinship of impedance/admittance with resistance/conductance is clear.

The capacitor of Figure 13.2b has the usual current-voltage relationship,

$$i_C(t) = C\frac{dv_C(t)}{dt}$$

Assuming no initial conditions, the Laplace transform relationship is

$$I_C(s) = CsV_C(s)$$

From equation 13.4, the **admittance of the capacitor** is

$$Y_C(s) = Cs$$

and from equation 13.5, the **impedance of the capacitor** is

$$Z_C(s) = \frac{1}{Cs}$$

Repeating this process for the inductor of Figure 13.2c , $\left(v_L(t) = L\frac{di_L(t)}{dt}\right)$ the **impedance** and **admittance of the inductor** are

$$Z_L(s) = Ls, \quad Y_L(s) = \frac{1}{Ls}$$

Exercises. 1. Given the integral form of the *v-i* capacitor relationship, assume no initial stored energy and take the Laplace transform of both sides to derive the impedance of the capacitor. This provides an alternative, more basic means of deriving the impedance characterization.
2. Given the integral form of the *v-i* inductor relationship, assume no initial stored energy and take the Laplace transform of both sides to derive the admittance of the inductor.

Throughout the rest of the text, whenever we refer to an impedance the unit of Ohm is assumed, and similarly, admittance is assumed to have the unit of siemens (S). The units for $V(s)$ and $I(s)$ are usually not shown, although strictly speaking they are volt-second and ampere-second, respectively.

3. MANIPULATION OF IMPEDANCE AND ADMITTANCE

Recall that the Laplace transform is a linear operation with respect to sums of signals, possibly multiplied by constants. KVL and KCL are conservation laws stating, respectively, that sums of voltages around a loop must add to zero and sums of all currents entering (or leaving) a node must add to zero. Since the Laplace transform is linear, it distributes over these sums, so the sum of the Laplace transforms of the voltages around a loop must be zero and the sum of the Laplace transforms of all the currents entering a node must be zero. In other words, complex-frequency domain voltages satisfy KVL and complex-frequency domain currents satisfy KCL. Because of this, and because impedances and admittances generalize the notions of resistance and conductance, one intuitively expects their manipulation properties to be similar. In fact, this is the case.

MANIPULATION RULE. Because impedances map s-domain currents, $I(s)$, to s-domain voltages, $V(s)$, and because all s-domain currents must satisfy KCL and all s-domain voltages must satisfy KVL:

1. Impedances, $Z(s)$, can be manipulated just like resistances and, like resistances, have units of ohms.
2. Admittances, $Y(s)$, can be manipulated just like conductances and, like conductances, have units of S.

This manipulation rule suggests, for example, that admittances in parallel add. The following example verifies this property for the case of two admittances in parallel.

EXAMPLE 13.1. Compute the equivalent admittance, $Y_{in}(s)$, and impedance, $Z_{in}(s)$, of three general admittances, $Y_1(s)$, $Y_2(s)$, and $Y_3(s)$ in parallel, as shown in Figure 13.3. Then develop the current division formula,

$$I_j(s) = \frac{Y_j(s)}{Y_1(s) + Y_2(s) + Y_3(s)} I_{in}(s) \tag{13.6}$$

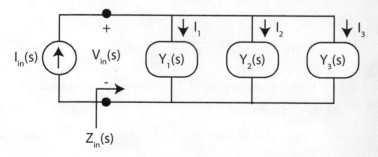

FIGURE 13.3 Three general admittances, $Y_j(s)$, in parallel, having an equivalent admittance $Y_{in}(s)$ or impedance $Z_{in}(s)$.

SOLUTION

We seek the relationship $I_{in}(s) = Y_{in}(s)V_{in}(s)$, which implicitly defines $Y_{in}(s)$. From the definition of the admittance of a two-terminal device, $I_k(s) = Y_k(s)V_{in}(s)$ for $k = 1, 2, 3$. From KCL,

$$I_{in}(s) = I_1(s) + I_2(s) + I_3(s) = (Y_1(s) + Y_2(s)\ Y_3(s))V_{in}(s)$$

This relationship implicitly defines the equivalent admittance as

$$Y_{in}(s) = \frac{I_{in}(s)}{V_{in}(s)} = Y_1(s) + Y_2(s) + Y_3(s)$$

affirming that admittances in parallel add. From the inverse relationship

$$Z_{in}(s) = \frac{1}{Y_1(s) + Y_2(s) + Y_3(s)}$$

Returning to the relationship $I_k(s) = Y_k(s)V_{in}(s)$, we now note that

$$I_k(s) = Y_k(s)V_{in}(s) = Y_k(s)Z_{in}(s)I_{in}(s) = \frac{Y_k(s)}{Y_1(s) + Y_2(s) + Y_3(s)}I_{in}(s)$$

Equation 13.6 has the obvious generalization to any number of parallel elements.

Exercises. 1. Show that for two impedances, $Z_1(s)$ and $Z_2(s)$, in parallel, $Z_{in}(s) = \dfrac{Z_1(s)Z_2(s)}{Z_1(s) + Z_2(s)}$
2. Show that the equivalent impedance of two capacitors in parallel is

$$Z(s) = \frac{1}{C_1 s + C_2 s} = \frac{1}{(C_1 + C_2)s}$$

and that the equivalent capacitance is $C_{eq} = C_1 + C_2$.
3. Derive the following formula for the impedance of two inductors in parallel:

$$Z(s) = \frac{L_1 L_2}{L_1 + L_2} s$$

4. A 2 µF and a 0.5 µF capacitor are in parallel. Find the equivalent capacitance.
ANSWER: 2.5 μF

5. A 2 mH inductor is connected in parallel with a 0.5 mF capacitor. Find the equivalent imped-ance.

ANSWER: $\dfrac{2 \times 10^3 s}{s^2 + 10^{-6}}$

6. In the circuit of Figure 13.3, suppose $Y_1(s) = 1/R$, $Y_2(s) = 1/(Ls)$, and $Y_3(s) = Cs$, a resistance, an inductance, and a capacitance. Derive the relationship

$$I_2(s) = \frac{1}{LCs^2 + \frac{L}{R}s + 1} I_{in}(s)$$

7. In the circuit of Figure 13.3, suppose $Y_1(s) = 0.5$, $Y_2(s) = \frac{1}{4s+2}$, and $Y_3(s) = \frac{s}{2s+1}$. Find the equivalent admittance, $Y_{in}(s)$ and find $I_3(s)$ in terms of $I_{in}(s)$.

ANSWER: $Y_{in}(s) = 1$ and $I_3(s) = \frac{s}{2s+1} I_{in}(s)$

EXAMPLE 13.2. Compute the input impedance of the parallel *RLC* circuit sketched in Figure 13.4.

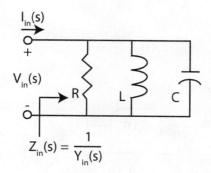

FIGURE 13.4 Parallel *RLC* circuit for Example 13.2.

SOLUTION

For parallel circuits, it is convenient to work with admittances, since *parallel admittances add*. Thus, for the circuit of Figure 13.4,

$$Y_{in}(s) = \frac{1}{R} + \frac{1}{Ls} + Cs = C\frac{s^2 + \frac{1}{RC}s + \frac{1}{LC}}{s}$$

Since impedance is the inverse of admittance,

$$Z_{in}(s) = \frac{1}{Y_{in}(s)} = \frac{1}{C}\frac{s}{s^2 + \frac{1}{RC}s + \frac{1}{LC}} \tag{13.7}$$

which is the equivalent input impedance of a parallel *RLC* circuit.

Exercises. 1. Compute the equivalent impedance of a parallel connection of three inductors having values 4 mH, 5 mH, and 20 mH.
ANSWER: $2 \times 10^{-3}s$

2. Compute the equivalent impedance of a parallel connection of six elements: two resistors, of 6 kΩ and 3 kΩ; two inductors, of 3 mH and 6 mH; and two capacitors, of 0.2 μF and 0.05 μF.
ANSWER: $4 \times 10^6 s/(s^2 + 2 \times 10^3 s + 2 \times 10^9)$

The dual of the parallel circuit of Figure 13.3 is a series connection of three impedances as shown in Figure 13.5. The following example verifies that impedances in series add, and simultaneously develops a voltage division formula.

EXAMPLE 13.3. Compute the equivalent impedance, $Z_{in}(s)$, and admittance, $Y_{in}(s)$, of three general impedances, $Z_1(s)$, $Z_2(s)$, and $Z_3(s)$ in series, as shown in Figure 13.5. Then develop the voltage division formula,

$$V_j(s) = \frac{Z_j(s)}{Z_1(s) + Z_2(s) + Z_3(s)} V_{in}(s)$$

(13.8)

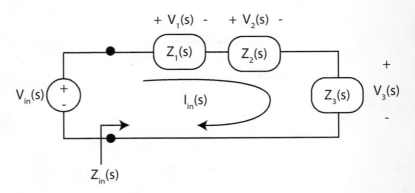

FIGURE 13.5 Series impedance circuit illustrating voltage division.

SOLUTION
Ohm's law tell us that

$$V_i(s) = Z_i(s)I_{in}(s)$$

(13.9)

for $i = 1, 2, 3,$. From KVL

$$V_{in}(s) = V_1(s) + V_2(s) + V_3(s) = [Z_1(s) + Z_2(s) + Z_3(s)]I_{in}(s)$$

(13.10)

Using equation 13.10 and the definition of input impedance, it follows that

$$Z_{in}(s) = \frac{V_{in}(s)}{I_{in}(s)} = Z_1(s) + Z_2(s) + Z_3(s)$$

(13.11)

The voltage division formula of equation 13.8 follows from a modified form of equation 13.10, and equation 13.9, to yield

$$V_i(s) = Z_i(s)I_{in}(s) = \frac{Z_i(s)}{Z_1(s) + Z_2(s) + Z_3(s)} V_{in}(s)$$

The voltage division formula is easily extended to the case of n devices in series:

$$V_i(s) = Z_i(s)I_{in}(s) = \frac{Z_i(s)}{Z_1(s) + Z_2(s) \dots + Z_n(s)} V_{in}(s)$$

Exercises. 1. Compute the equivalent impedance of two capacitors, C_1 and C_2, in series.

ANSWER: $\dfrac{1}{\dfrac{C_1 C_2}{C_1 + C_2} s} = \dfrac{1}{C_{eq} s}$

2. Show that the equivalent admittance of two capacitors, C_1 and C_2, in series is $Y(s) = \dfrac{C_1 C_2}{C_1 + C_2} s$.

3. Suppose $Z_1(s) = 10\ \Omega$, $Z_2(s) = 2s$, and $Z_3(s) = 6s$ in Figure 13.5. Find $Z_{in}(s)$, $V_2(s)$, and $v_2(t)$ if

$V_{in}(s) = \dfrac{10}{s}$.

ANSWERS: $Z_{in}(s) = 10 + 8s$, $V_2(s) = \dfrac{10}{4s + 5}$, $v_2(t) = 2.5e^{-1.25t}u(t)$

4. Suppose $Z_1(s) = 10\ \Omega$, $Z_2(s) = 2s$, and $Z_3(s) = \dfrac{2}{s}$ in Figure 13.5. Find $Z_{in}(s)$ and $V_3(s)$ in terms of $V_{in}(s)$.

ANSWERS: $Z_{in}(s) = \dfrac{2s^2 + 10s + 2}{s}$, $V_3(s) = \dfrac{1}{s^2 + 5s + 1} V_{in}(s)$

5. Verify that the equivalent inductance of two inductors in series is $L_{eq} = L_1 + L_2$.

Of course, there are series-parallel connections of circuit elements that combine the concepts illustrated in Examples 13.1 through 13.3, as set forth next.

EXAMPLE 13.4. Compute the input impedance $Z_{in}(s)$ of a series connection of two pairs of parallel elements, as shown in Figure 13.6, in which $R_1 = 10\ \Omega$, $C = 0.1$ F, $R_2 = 5\ \Omega$, and $L = 1$ H. Then compute $V_2(s)$ in terms of $V_{in}(s)$. If $v_{in}(t) = u(t)$, find $v_2(t)$.

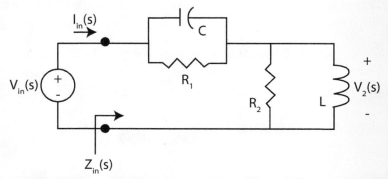

FIGURE 13.6 Series-parallel connection of *RC* elements for Example 13.4.

SOLUTION

Conceptually, view the circuit as shown in Figure 13.7.

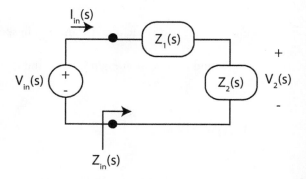

FIGURE 13.7 Conceptual series structure of the circuit in Figure 13.6.

Here

$$Z_1(s) = \frac{1}{0.1 + 0.1s} = \frac{10}{s+1}$$

and

$$Z_2(s) = \frac{2s}{s+2}$$

in which case

$$Z_{in}(s) = \frac{10(s+2) + 2s(s+1)}{(s+1)(s+2)} = \frac{2s^2 + 12s + 20}{(s+1)(s+2)}$$

It follows that

$$V_2(s) = \frac{Z_2(s)}{Z_{in}(s)} V_{in}(s) = \frac{(s+1)(s+2)}{2s^2 + 12s + 20} \times \frac{2s}{s+2} \times V_{in}(s) = \frac{s(s+1)}{s^2 + 6s + 10} V_{in}(s)$$

Finally, if $V_{in}(s) = \dfrac{1}{s}$,

$$V_2(s) = \frac{(s+1)}{s^2 + 6s + 10} = \frac{(s+1)}{(s+3)^2 + 1}$$

From Table 12.1, item 19,

$$v_2(s) = e^{-3t}[\cos(t) - 2\sin(t)]u(t)$$

Exercise. Repeat Example 13.4 with the following changes: $C = 0.01$ F and $R_2 = 10\ \Omega$.

ANSWERS: $Z_{in}(s) = 10\ \Omega$, $V_2(s) = \dfrac{s}{s+10} V_{in}(s)$ $v_2(t) = e^{-10t}u$,

---

Another basic and useful circuit analysis technique is the **source transformation property**, exhibited now in terms of impedances and admittances. The first case we will examine is the voltage-to-current source transformation, illustrated in Figure 13.8.

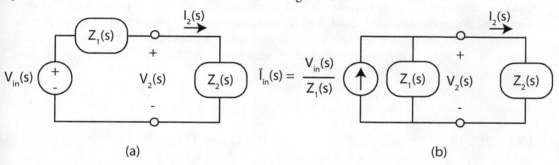

(a)   (b)

FIGURE 13.8 Illustration of the transformation of a voltage source in series with $Z_1(s)$, as shown in part (a), to an equivalent current with a current source in parallel with $Z_1(s)$, as shown in part (b).

Often, voltage-to-current source transformations provide an altered circuit topology that is more convenient for hand or calculator analysis. Mathematically, the goal is to change the structure of a voltage source in series with an impedance to a current source in parallel with an admittance while keeping both $V_2(s)$ and $I_2(s)$ fixed. To justify this, one starts with Figure 13. 8a, in which voltage division implies

$$V_2(s) = \frac{Z_2(s)}{Z_1(s) + Z_2(s)} V_{in}(s) = Z_2(s)I_2(s)$$

Hence, if $Z_1(s) \neq 0$,

$$V_2(s) = \frac{Z_2(s)Z_1(s)}{Z_1(s) + Z_2(s)}\left(\frac{V_{in}(s)}{Z_1(s)}\right) = \frac{1}{Y_1(s) + Y_2(s)}\left(\frac{V_{in}(s)}{Z_{in}(s)}\right) \qquad (13.12)$$

where $Y_i(s) = [Z_i(s)]^{-1}$. This equation identifies the parallel structure of Figure 13.7b; i.e., Figure 13. 8b is a circuit equivalent of equation 13.8.

Reversing these arguments leads to the current-to-voltage source transformation, illustrated in Figure 13.9.

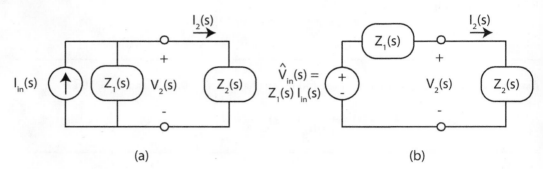

FIGURE 13.9 Illustration of (a) current source to (b) equivalent voltage source transformation.

Clearly, the manipulation of impedances and admittances parallels that of resistances and conductances, as suggested earlier. Indeed, for a rigorous statement of the source transformation technique developed above, refer to the source transformation theorem in Chapter 5 and replace R by $Z(s)$, V_a by $V_a(s)$, and I_b by $I_b(s)$. Indeed, all such values in Chapters 5 and 6 have s-domain counterparts.

This section ends with a demonstration of finding a Thevenin equivalent in the s-domain.

EXAMPLE 13.5. Compute the Thevenin equivalent circuit of Figure 13.10.

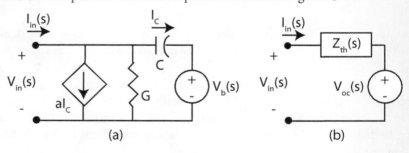

FIGURE 13.10

SOLUTION
From the material in Chapter 6, our new concepts of admittance and impedance, and Figure 13.10b,

$$V_{in}(s) = Z_{th}(s)I_{in}(s) + V_{OC}(s) \qquad (13.13a)$$

or equivalently,

$$I_{in}(s) = \frac{1}{Z_{th}(s)}V_{in}(s) - \frac{1}{Z_{th}(s)}V_{oc}(s) = \frac{1}{Z_{th}(s)}V_{in}(s) - I_{sc}(s) \qquad (13.13b)$$

Now from Figure 13.10a,

$$
\begin{aligned}
I_{in}(s) &= aI_C(s) + I_C(s) + GV_{in}(s) = (a+1)I_C(s) + GV_{in}(s) \\
&= (a+1)sC\left[V_{in}(s) - V_b(s)\right] + GV_{in}(s) \\
&= \left[(a+1)sC + G\right]V_{in}(s) - (a+1)sCV_b(s)
\end{aligned}
\tag{13.14}
$$

Rewriting equation 13.14 in the form of equation 13.13a, we have

$$
V_{in}(s) = \frac{1}{(a+1)sC + G}I_{in}(s) + \frac{(a+1)sC}{(a+1)sC + G}V_b(s)
\tag{13.15}
$$

Comparing equations 13.15 and 13.13a, we identify

$$
Z_{th}(s) = \frac{1}{(a+1)sC + G} \quad \text{and} \quad V_{oc}(s) = \frac{(a+1)sC}{(a+1)sC + G}V_b(s)
$$

Exercises. 1. In Example 13.5, what is the Norton short circuit current, $I_{sc}(s)$?
ANSWER: $I_{sc}(s) = (a+1)sCV_b(s)$

2. Find $Z_{th}(s)$ and $V_{oc}(s)$ for the circuit in Figure 13.11.

ANSWER: $Z_{th}(s) = \left(2 + 2s + \dfrac{2-2a}{s}\right)$, $V_{oc}(s) = \dfrac{2I_b(s)}{s}$

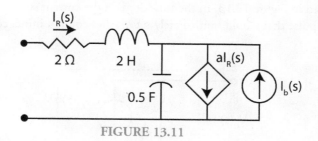

FIGURE 13.11

3. For the circuit of Figure 13.12, use source transformations to find $I_{sc}(s)$ and $Y_{th}(s)$ for the indicated terminals.
ANSWERS: $I_{sc}(s) = 0.2sV_{in}(s)$ and $Y_{th}(s) = 0.2s + \dfrac{1}{s} + 0.4$

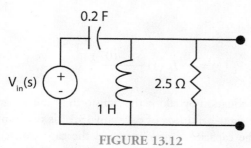

FIGURE 13.12

4. EQUIVALENT CIRCUITS FOR INITIALIZED INDUCTORS AND CAPACITORS

The notions of impedance, admittance, and transfer function do not account for the presence of initial capacitor voltages and initial inductor currents. *How can one incorporate initial conditions into various analysis schemes?* For an answer we look at the transform of an initialized capacitor and inductor and interpret the resulting equation as an equivalent circuit in the complex-frequency domain. For the capacitor and the inductor, two equivalent circuits result for each: a series circuit containing a relaxed (no initial condition) capacitor/inductor in series with a source, and a parallel circuit with a relaxed capacitor/inductor in parallel with a source. Example 12.23 previewed this notion.

The capacitor has the standard voltage-current relationship

$$C\frac{dv_C(t)}{dt} = i_C(t)$$

Taking the Laplace transform and allowing for a nonzero initial condition $v_C(0^-)$ yields

$$CsV_C(s) - Cv_C(0^-) = I_C(s) \tag{13.16}$$

The left side of equation 13.16 is the difference of two currents, one given by the product of the capacitor admittance and the capacitor voltage ($CsV_C(s)$) and the other by $Cv_C(0^-)$. Thus the circuit interpretation of equation 13.16 consists of a relaxed capacitor in parallel with a current source, as illustrated in Figure 13.13. In the time domain the current source of Figure 13.13 corresponds to an impulse that would immediately set up the required initial condition.

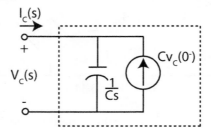

FIGURE 13.13 Parallel form of an equivalent circuit for an initialized capacitor. Here, the capacitor within the dotted box is relaxed while the current source $Cv_C(0^-)$ accounts for the initial condition.

Rearranging equation 13.16 yields

$$V_C(s) = \frac{1}{Cs}I_C(s) + \frac{v_C(0^-)}{s} \tag{13.17}$$

Example 12.22 previewed this equation by taking the transform of the integral relationship of the capacitor. We observe that the right-hand side of equation 13.17 is the sum of two voltages, one of which is the product of the capacitor impedance and the capacitor, current and the other $v_c(0^-)/s$. Thus, the interpretation is a series circuit, as sketched in Figure 13.14.

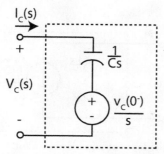

FIGURE 13.14 The series form of an equivalent circuit for an initialized capacitor. Here the capacitor in the dotted box is relaxed, and the voltage source accounts for the effect of the initial condition.

Initialized inductors have similar s-domain equivalent circuits analogous to those of the capacitor. With the voltage and current directions satisfying the passive sign convention, the differential inductor current-voltage relationship is

$$v_L(t) = L\frac{di_L(t)}{dt}$$

Transforming both sides yields

$$V_L(s) = LsI_L(s) - Li_L(0^-) \tag{13.18}$$

Again, this equation consists of a sum of voltages, $LsI_L(s)$ and $-Li_L(0^-)$. Thus equation 13.18 can be interpreted as a series circuit, as depicted in Figure 13.15.

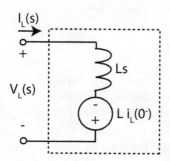

FIGURE 13.15 Series form of equivalent circuit for an initialized inductor. Here the inductor within the dotted box is relaxed; notice the polarity orientation of the voltage source.

To construct a parallel equivalent circuit, divide equation 13.18 by Ls and rearrange to obtain

$$I_L(s) = \frac{1}{Ls}V_L(s) + \frac{i_L(0^-)}{s} \tag{13.19}$$

The right side of Equation 13.19 is a sum of currents that determines a parallel equivalent circuit, as sketched in Figure 13.16.

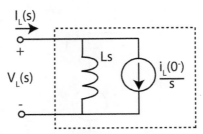

FIGURE 13.16 Parallel form of equivalent circuit for an initialized inductor.
Again, the inductor inside the dotted box is relaxed.

Two examples illustrate the use of these four equivalent circuits for initialized capacitors and inductors.

EXAMPLE 13.6. This example illustrates an s-domain application of superposition. In the RLC circuit of Figure 13.17, suppose $v_C(0^-) = 1$ V, $i_L(0^-) = 2$ A, and $v_{in}(t) = 4u(t)$ V. Find $v_L(t)$ for $t \geq 0$.

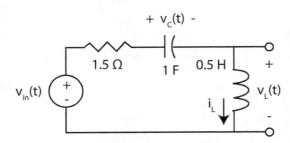

FIGURE 13.17 Circuit for Example 13.6.

SOLUTION
In this example, it is convenient to replace the capacitor by its (series) s-domain voltage source equivalent circuit, because the capacitor is in series with the input voltage source. On the other hand, it is convenient to replace the inductor by its (parallel) s-domain current source equivalent circuit, because the desired output is the inductor voltage. This results in a three-source or multi-input circuit. Once the equivalent circuits are in place, one can apply superposition to obtain the answer, although there are many other ways to solve the problem.

Step 1. *Using the voltage source model for the capacitor and the current source model for the inductor, draw the equivalent s-domain circuit.* Using the equivalent circuits of Figures 13.14 and 13.16, we obtain the circuit of Figure 13.18. Here we note that

$$V_{in}(s) = \frac{4}{s}, \quad \frac{v_C(0^-)}{s} = \frac{1}{s}, \text{ and } \frac{i_L(0^-)}{s} = \frac{2}{s}.$$

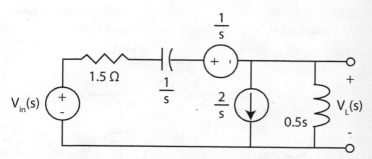

FIGURE 13.18 *s*-domain equivalent accounting for initial conditions of the circuit of Figure 13.17.

Step 2. *Find the contribution to $V_L(s)$ from $V_{in}(s)$. From voltage division,*

$$V_L^1(s) = \frac{0.5s}{1.5 + \dfrac{1}{s} + 0.5s} \times \frac{4}{s} = \frac{4s}{s^2 + 3s + 2}$$

Step 3. *Find the contribution to $V_L(s)$ from $\dfrac{v_C(0^-)}{s} = \dfrac{1}{s}$. Again, from voltage division,*

$$V_L^2(s) = -\frac{0.5s}{1.5 + \dfrac{1}{s} + 0.5s} \times \frac{1}{s} = \frac{-s}{s^2 + 3s + 2}$$

Step 4. *Find the contribution to $V_L(s)$ from $Li_L(0^-) = 1$. Using Ohm's law in the s-domain,*

$$V_L^3(s) = -\frac{0.5s\left(1.5 + \dfrac{1}{s}\right)}{1.5 + \dfrac{1}{s} + 0.5s} \times \frac{2}{s} = \frac{-3s - 2}{s^2 + 3s + 2}$$

Step 5. *Sum the three contributions and take the inverse transform.*

$$V_L(s) = V_L^1(s) + V_L^2(s) + V_L^3(s) = \frac{-2}{s^2 + 3s + 2} = \frac{2}{s+2} - \frac{2}{s+1}$$

in which case

$$v_L(t) = 2e^{-2t}u(t) - 2e^{-t}u(t) \text{ V}$$

Exercise. Find $I_L(s)$ and $i_L(t)$ for the circuit of Figure 13.17 using the equivalent circuits of Figures 13.14 and 13.15. Hint: Write one loop equation.

ANSWER: $I_L(s) = \dfrac{2(s+3)}{(s+1)(s+2)}$, $i_L(t) = 4e^{-t}u(t) - 2e^{-2t}u(t)$

EXAMPLE 13.7. This example illustrates a single-node application of nodal analysis. In the *RLC* circuit of Figure 13.19, suppose $v_C(0^-) = 1$ V, $i_L(0^-) = 2$ A, and $v_{in}(t) = u(t)$ V. Find $v_C(t)$ for $t \geq 0$.

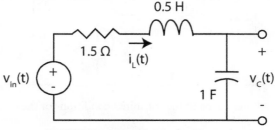

FIGURE 13.19 Circuit for Example 13.7.

SOLUTION

In this example, it is convenient to replace the inductor by its (series) complex-frequency domain voltage source equivalent circuit, because the inductor is in series with the input voltage source. On the other hand, it is convenient to replace the capacitor by its (parallel) complex-frequency domain current source equivalent circuit, because the desired output is the capacitor voltage. This results in a three-source, or multi-input, circuit. Once the equivalent circuits are in place, one can combine the voltage sources and write a single node equation to find $V_C(s)$.

Step 1. *Using the voltage source model for the inductor and the current source model for the capacitor, draw the equivalent complex-frequency domain circuit.* Using the voltage source equivalent for the initialized inductor and the current source equivalent for the capacitor produces the circuit of Figure 13.20a. Combining the voltage sources and the series impedance into single terms results in the circuit shown in Figure 13.20b.

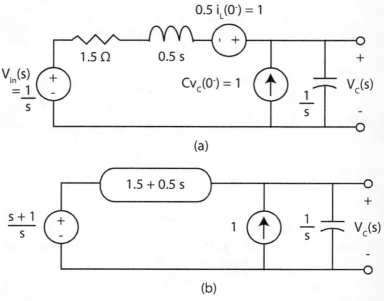

FIGURE 13.20 (a) Complex-frequency domain equivalent accounting for initial conditions of the circuit of Figure 13.19. (b) Circuit equivalent to part (a) with voltage sources combined.

Step 2. *Write a single node equation for $V_C(s)$.* Summing the currents leaving the top node of $V_C(s)$ yields

$$\frac{1}{1.5 + 0.5s}\left(V_C(s) - \frac{s+1}{s}\right) - 1 + sV_C(s) = 0$$

Grouping terms produces

$$\left(\frac{1}{1.5 + 0.5s} + s\right)V_C(s) = \frac{s+1}{s(0.5s + 1.5)} + 1$$

Solving for $V_C(s)$ leads to

$$V_C(s) = \frac{s^2 + 5s + 2}{s(s+1)(s+2)}$$

Step 3. *Execute a partial fraction expansion on $V_C(s)$, and take the inverse transform to obtain $v_C(t)$.* Using the result of step 2,

$$V_C(s) = \frac{s^2 + 5s + 2}{s(s+1)(s+2)} = \frac{1}{s} + \frac{2}{s+1} + \frac{-2}{s+2}$$

Inverting this transform yields the desired time response,

$$v_C(t) = [1 + 2e^{-t} - 2e^{-2t}]u(t) \text{ V}$$

Exercises. 1. In Example 13.7, change the resistance from 1.5 Ω to 2.25 Ω. Find $v_C(t)$ for $t > 0$.
ANSWER: $v_C(t) = [1 + 0.5714e^{-0.5t} - 0.5714e^{-4t}]u(t)$ V

2. Find $I_L(s)$ and $i_L(t)$ for the circuit of Example 13.7, using the equivalent circuits of Figures 13.14 and 13.15. Hint: Write one loop equation.

ANSWER: $I_L(s) = \frac{2s}{(s+1)(s+2)}$, $i_L(t) = -2e^{-t}u(t) + 4e^{-2t}u(t)$

EXAMPLE 13.8. The chapter opened with a discussion of the operation of a fluorescent light with classical starter, common in residential usage. For a fluorescent light to begin operating, there must be a sufficient supply of free electrons in the tube and a sufficiently high voltage between the electrodes to allow arcing to occur. During arcing, mercury particles in the tube vaporize and give off ultraviolet light. The ultraviolet light excites a coating of phosphorus on the inside of the tube that emits light in the visible range.

For a simplified analysis, assume that all resistances are negligible and refer to Figure 13.21. The source $v_{in}(t)$ is 120 V, 60 Hz, i.e., ordinary house voltage, which is too low to cause arcing inside the fluorescent tube. Prior to arcing the gas inside the fluorescent tube acts like a very large resistance between the two electrodes. When the switch is turned on, the starter, a neon bulb with a bimetallic switch inside, lights up and heats the bimetallic strip. This causes the metal to curl and close the contact. The bulb then looks like a short circuit, and a large current, limited by the inductive ballast, flows through the heating electrodes of the fluorescent tube, making them better able to emit electrons. During this time the neon bulb is shorted out and the bimetallic strip cools and opens the circuit after a few seconds. At this point in time, which we will call $t = 0$, the inductor has an initial current $i_L(0^-)$. Because of the LC combination, a very high voltage will then appear across the electrodes of the lamp, resulting in ignition or arcing. After the lamp ignites, the voltage between the electrodes becomes "small" and is insufficient to relight the neon starter lamp. Hence, the ac current flows between the two electrodes inside the fluorescent tube. The ballast again serves to limit the current.

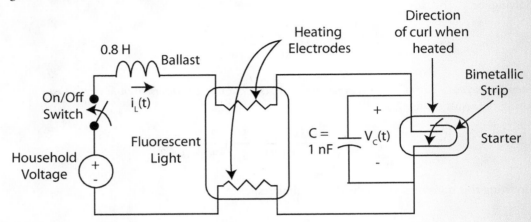

FIGURE 13.21 Wiring diagram of simple fluorescent light circuit, including an inductive ballast, a capacitor, and a starter within which is a neon bulb containing a bimetallic switch.

Suppose $L = 0.8$ H, $C = 1$ nF, and $i_L(0^-) = 0.1$ A. For $t > 0$, we find the component of $v_C(t)$ due to the initial inductor current, i.e., the zero-input response. The other component, the zero-state response, is not as important for ignition purposes. Our strategy will be to use the s-domain equivalent circuit for L, as illustrated in Figure 13.22.

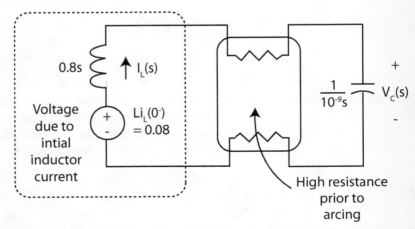

FIGURE 13.22 Equivalent complex-frequency domain circuit immediately prior to arcing and normal lamp operation in fluorescent lighting.

Since we are assuming that all resistances are negligible and that the internal resistance (between electrodes) of the fluorescent lamp prior to arcing approximates infinity, voltage division in terms of impedances yields

$$V_C(s) = -\frac{\frac{1}{Cs}}{\frac{1}{Cs}+Ls}Li_L(0^-) = -\frac{1}{C}\frac{i_L(0^-)}{s^2+\frac{1}{LC}} = -\frac{10^8}{\sqrt{1.25\times10^9}}\frac{\sqrt{1.25\times10^9}}{s^2+1.25\times10^9}$$

$$= -2{,}828\frac{\sqrt{1.25\times10^9}}{s^2+1.25\times10^9}$$

Hence, immediately prior to arcing, the capacitor voltage approximates

$$v_C(t) = -2{,}828\,\sin(35{,}355t)\text{ V}$$

which is sufficiently high to induce arcing and cause the fluorescent lamp to operate.

See the homework exercises for an extension of this analysis to the case where the ballast model includes a resistance of 100 Ω.

5. NOTION OF TRANSFER FUNCTION

Besides impedances and admittances, other quantities such as voltage gains and current gains are critically important in amplifiers and other circuits. The term **transfer function** is a catchall phrase for the different ratios that might be of interest in circuit analysis. Impedances and admittances are special cases of the transfer function concept.

TRANSFER FUNCTION

Suppose a circuit has only one active independent source and only one designated response signal. Suppose further that there is no internal stored energy at $t = 0^-$. The transfer function of such a circuit or system is

$$H(s) = \frac{\mathcal{L}[\text{designated response signal}]}{\mathcal{L}[\text{designated input signal}]} \tag{13.20}$$

Thus if the input is $f(t)$ and the response is $y(t)$, then $Y(s) = H(s)F(s)$, which is a handy formula for computing responses. Notice that if the input is the delta function, then $F(s) = 1$ and $Y(s) = H(s)$. This means that the transfer function is the Laplace transform of the so-called **impulse response** of the circuit, i.e., the response due to an impulse applied at the circuit input source when there are no initial conditions present. The idea is easily extended to multiple inputs and multiple outputs to form a transfer function matrix. This extension, however, is beyond the scope of this text.

Exercise. A transfer function of a particular circuit is $H(s) = \dfrac{s+a+5b}{(s+a)^2+b^2}$. Find the impulse response. Hint: Review Table 12.1.

ANSWER: $e^{-at}[\cos(bt) + 5\sin(bt)]u(t)$

A transfer function, as defined by equation 13.20, has broad applicability to electrical and electro-mechanical systems. For example, the designated output may be a torque while the input might be voltage. However, in the context of circuits, a transfer function is often called a *network function.* The literature distinguishes four special cases: (i) *driving point impedance,* where the input is a current source and the output is the voltage across the current source; (ii) *driving point admittance,* where the input is a voltage source and the output is the current leaving the voltage source; (iii) *transfer impedance,* where the input is a current source and the voltage is across a designated pair of terminals; and (iv) *transfer admittance,* where the input is a voltage source and the output is the current through another branch in the circuit. In cases (i) and (iii), the voltage polarity must be consistent with the conventional labeling of sources as set forth in Chapter 2. In general, however, we will adopt the ordinary language of transfer function.

EXAMPLE 13.9. The circuit of Figure 13.23 has elements with zero initial conditions at $t = 0^-$. Find

$$H(s) = \frac{\mathcal{L}[i_{out}(t)]}{\mathcal{L}[v_{in}(t)]} = \frac{I_{out}(s)}{V_{in}(s)}$$

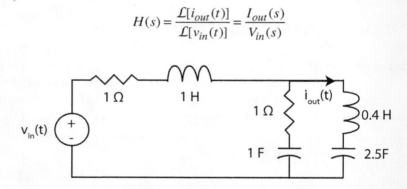

FIGURE 13.23 Circuit for Example 13.9.

SOLUTION

There are many ways to solve this problem. Our approach is to execute a source transformation on the *R-L* impedance in series with the voltage source. After the source transformation, we use current division to obtain the necessary transfer function.

Step 1. *Execute a source transformation to obtain three parallel branches as per Figure 13.24.*

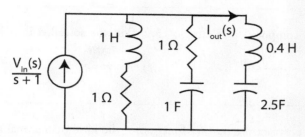

FIGURE 13.24 Circuit equivalent to Figure 13.23 after a source transformation.

This circuit has the parallel structure of Figure 13.25.

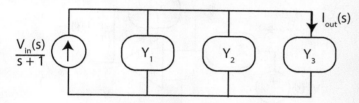

FIGURE 13.25 Parallel admittance form of Figure 13.24.

Step 2. *Use current division.* Since the output current, $I_{out}(s)$, is a current through one of three parallel branches, the current division formula (equation 13.9) applies, producing

$$I_{out}(s) = \left(\frac{Y_3(s)}{Y_1(s) + Y_2(s) + Y_3(s)} \right) \frac{V_{in}(s)}{s+1}.$$

Hence,

$$H(s) = \frac{I_{out}(s)}{V_{in}(s)} = \left(\frac{Y_3(s)}{Y_1(s) + Y_2(s) + Y_3(s)} \right) \frac{1}{s+1}. \qquad (13.21)$$

Step 3. *Compute $Y_1(s)$, $Y_2(s)$, and $Y_3(s)$.* Because impedances in series add, and admittance is the inverse of impedance (equation 13.7), some straightforward algebra yields

$$Y_1(s) = \frac{1}{s+1}; \quad Y_2(s) = \frac{1}{1 + \frac{1}{s}} = \frac{s}{s+1}; \quad Y_3(s) = \frac{1}{0.4s + \frac{0.4}{s}} = \frac{2.5s}{s^2 + 1}$$

Step 4. *Substitute into equation 13.21 to obtain $H(s)$:*

$$H(s) = \frac{\dfrac{2.5s}{s^2+1}}{\dfrac{1}{s+1} + \dfrac{s}{s+1} + \dfrac{2.5s}{s^2+1}} \left(\frac{1}{s+1} \right) = \frac{2.5s}{s^2 + 1 + s(s^2+1) + 2.5s(s+1)}$$

$$= \frac{2.5s}{(s+1)(s^2 + 2.5s + 1)} = \frac{2.5s}{(s+1)(s+0.5)(s+2)}$$

Exercise. For $H(s)$ as computed in Example 13.9, find the so-called impulse response $h(t) = \mathcal{L}^{-1}[H(s)]$.

ANSWER: $h(t) = \left[5e^{-t} - \frac{5}{3}e^{-0.5t} - \frac{10}{3}e^{-2t} \right] u(t)$

EXAMPLE 13.10. Construct the transfer function of the ideal operational amplifier circuit of Figure 13.26, where $Z_f(s)$ and $I_f(s)$ denote a feedback impedance and feedback current, respectively.

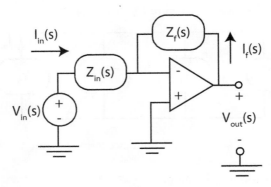

FIGURE 13.26 Simple ideal operational amplifier circuit for Example 13.10.

SOLUTION

Since no current enters the inputs of an ideal op amp, $I_{in}(s) = -I_f(s)$. Further, the voltage at the negative op amp terminal is driven to virtual ground; hence, $V_{in}(s) = Z_{in}(s)I_{in}(s)$, and $V_{out}(s) = Z_f(s)I_f(s)$. Combining these relationships with $I_{in}(s) = -I_f(s)$ yields

$$H(s) = \frac{V_{out}(s)}{V_{in}(s)} = -\frac{Z_f(s)}{Z_{in}(s)} = -\frac{Y_{in}(s)}{Y_f(s)} \tag{13.22}$$

Equation 13.22 is a very handy formula for computing the transfer functions and responses of many op amp circuits.

Exercises. 1. In the circuit of Figure 13.26, suppose $Z_f(s)$ is the impedance of a 0.1 mF capacitor. Find R so that the transfer function is $H(s) = -1/s$, i.e., an inverting integrator.
ANSWER: $R = 10 \text{ k}\Omega$

2. In the circuit of Figure 13.26, now suppose $Z_{in}(s)$ consists of a 10 kΩ resistor in parallel with a 0.2 mF capacitor, and $Z_f(s)$ consists of a 40 kΩ resistor in parallel with a 0.4 mF capacitor. Find the transfer function, $H(s)$, the dc gain, and the gain as $s \to \infty$.
ANSWER: $-(s + 0.5)/(2s + 0.125)$, -4, -0.5

3. Find the value of C for which the transfer function of the op amp circuit in Figure 13.27 is

$$H(s) = -\frac{s}{(s+2)(s+4)}.$$

ANSWER: $C = 0.5$ F

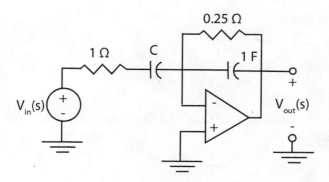

FIGURE 13.27 Op amp circuit.

EXAMPLE 13.11. The ideal op amp circuit of Figure 13.28 is called a *leaky integrator*. If the input to the leaky **integrator circuit** is $v_{in}(t) = e^{-t}u(t)$, find the values of R_1, R_2, and C leading to an output response $v_{out}(t) = -2te^{-t}u(t)$, assuming that $v_C(0^-) = 0$.

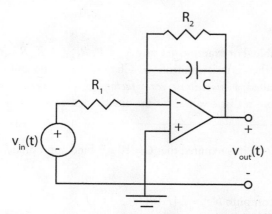

FIGURE 13.28 Ideal operational amplifier circuit known as the leaky integrator.

SOLUTION

Step 1. *From the given data, compute the actual transfer function of the circuit.* By definition of the transfer function,

$$H(s) = \frac{\mathcal{L}[response]}{\mathcal{L}[input]} = \frac{V_{out}(s)}{V_{in}(s)} = -\frac{\dfrac{2}{(s+1)^2}}{\dfrac{1}{s+1}} = -\frac{2}{s+1} \qquad (13.23)$$

Step 2. *Using Figure 13.28, find the transfer function of the circuit in terms of R_1, R_2, and C. Here,* observe that Figure 13.28 has the same topology as Figure 13.26, where

$$Y_f(s) = \frac{1}{Z_f(s)} = Cs + \frac{1}{R_2} \quad and \quad Y_{in}(s) = \frac{1}{Z_{in}(s)} = \frac{1}{R_1}$$

From equation 13.22 of Example 13.10,

$$H(s) = -\frac{Y_{in}(s)}{Y_f(s)} = -\frac{\dfrac{1}{R_1}}{Cs + \dfrac{1}{R_2}} \tag{13.24a}$$

Step 3. *Match coefficients in equations 13.23 and 13.24a to obtain the desired values of R_1, R_2, and C.* Equating the coefficients yields

$$\frac{\dfrac{1}{R_1}}{Cs + \dfrac{1}{R_2}} = \frac{2}{s+1}$$

One possible solution is $R_1 = 0.5\ \Omega$, $R_2 = 1\ \Omega$, and $C = 1$ F. If we rewrite equation 13.24a as

$$H(s) = -\frac{Y_{in}(s)}{Y_f(s)} = -\frac{\dfrac{1}{CR_1}}{s + \dfrac{1}{CR_2}} \tag{13.24b}$$

other solutions are also possible. For example, for any $K_m > 0$, $R_{1new} = K_m R_1$, $R_{2new} = K_m R_2$, $C_{new} = C/K_m$, represents a valid (theoretical) solution. In Chapter 14 we encounter a concept called *magnitude scaling*. K_m is called a *magnitude scale factor*, which leaves this transfer function unchanged but produces more realistic values for the circuit elements.

Exercises. 1. In equation 13.24b, it is required that $C = 10$ µF. Find appropriate values of R_1 and R_2.
ANSWER: $R_1 = 50$ kΩ, $R_2 = 100$ kΩ

2. Given equation 13.24b, compute $h(t) = L[H(s)]$.

ANSWER: $-\dfrac{1}{R_1 C} e^{\left(-\frac{t}{R_2 C}\right)} u(t)$

---

6. IMPULSE AND STEP RESPONSES

Suppose a circuit or system has a transfer function representation $H(s)$, with s-domain input denoted by $F(s)$ and s-domain output given by $Y(s)$ in which case $Y(s) = H(s) F(s)$. Assuming that all initial conditions are zero, if $f(t) = \delta(t)$, then the resulting $y(t)$ is the system **impulse response**. Some simple calculations verify that the transform of the impulse response is the transfer function, i.e.,

$$\mathcal{L}[y(t)] = H(s)\mathcal{L}[\delta(t)] = H(s)$$

Hence, the impulse response of the circuit/system, denoted $h(t)$, is the inverse transform of the transfer function

$$h(t) = \mathcal{L}^{-1}[H(s)] \qquad\qquad (13.25a)$$

and conversely

$$H(s) = \mathcal{L}[h(t)] \qquad\qquad (13.25b)$$

These equivalences represent another use of the transfer function concept.

Exercises. 1. The transfer function of a certain linear network is $H(s) = (s + 3)/[(s + 1)(s + 2)]$. Find the impulse response of the network.
ANSWER: $[2e^{-t} - e^{-2t}]u(t)$

2. If the impulse response of a circuit is a pulse $y(t) = u(t) - u(t - T)$, $T > 0$, compute the transfer function.
ANSWER: $(1 - e^{-sT})/s$

3. Suppose $v(t) = 2\delta(t - 1) - 3\delta(t - 3)$ is the input to a relaxed (zero initial conditions) circuit having an impulse response $h(t) = 2u(t) - 2u(t - 5)$. Find the output $y(t)$.
ANSWER: $y(t) = 2h(t - 1) - 3h(t - 3)$

Why is the impulse response important? As we will see, it is because every linear circuit having constant parameter values for its elements can be represented in the time domain by its impulse response. This is shown in Chapter 15, where we define a mathematical operation called *convolution* and show that the convolution of the input function with the impulse response function yields the zero-state circuit response. In addition to this significant theoretical result, the impulse response is important for identification of linear circuits or systems having unknown constant parameters. Sometimes a transfer function is unavailable or a circuit diagram is lost. In such a predicament, measuring the impulse response on an oscilloscope as the derivative of the step response is quite practical.

What is the step response of a circuit? The *step response* is merely the zero-state response of the circuit to a step function. Observe that if the input $f(t)$ to the circuit is $u(t)$, then $F(s) = 1/s$ and $Y(s) = H(s) (1/s)$. By the integration property of the Laplace transform, it follows that the step response is the integral of the impulse response. Conversely, the derivative of the step response is the impulse response. In lab, many scopes can display the derivative of a trace and hence can display the derivative of the step response, which is the impulse response. Alternatively, a homework problem will suggest a means of directly generating an approximate impulse response.

Exercises. 1. If the transfer function of a circuit is $H(s) = 1/s$, what are the impulse and step responses?

2. If the Laplace transform of the step response of a circuit is given by $Y(s) = 1/[s(s + 1)]$, what is the impulse response?

3. If the step response of a circuit is $y(t) = [1 - 0.5e^{-2t} - 0.5e^{-4t} \cos(2t)]u(t)$, what is the impulse response?

ANSWERS: in random order: $r(t)$, $e^{-t}u(t)$, $u(t)$, $[e^{-2t} + 2.24e^{-4t} \cos(2t + 26.57°)]u(t)$.

EXAMPLE 13.12. Figure 13.29a shows the impulse response of a hypothetical circuit. If an input is $f(t) = \delta(t) + \delta(t - 1)$, compute the response, $y(t)$.

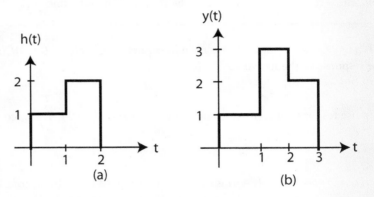

FIGURE 13.29 (a) Impulse response of hypothetical circuit. (b) Response to $\delta(t) + \delta(t - 1)$.

SOLUTION
Since $\mathcal{L}[\delta(t) + \delta(t - 1)] = 1 + e^{-s}$, the response, $y(t)$, is simply the sum of $h(t)$ and $h(t - 1)u(t - 1)$. Doing the addition graphically yields the waveform of Figure 13.25b.

EXAMPLE 13.13. The response of a relaxed circuit to a scaled ramp, $f(t) = 8tu(t)$, is given by $y(t) = (-6 + 4t + 8e^{-t} - 2e^{-2t})u(t)$. Compute the impulse response, $h(t)$.

SOLUTION
The relationship between $f(t)$ and $\delta(t)$ identifies the strategy of the solution. If the step function is the integral of the delta function and the ramp the integral of the step, then the delta function equals the second derivative of the ramp. Hence, $\delta(t) = 0.125f''(t)$. By the linearity of the circuit, the impulse response $h(t) = 0.125y''(t)$, and some straightforward calculations produce

$$y'(t) = [4 - 8e^{-t} + 4e^{-2t}]u(t) + [-6 + 4t + 8e^{-t} - 2e^{-2t}]\delta(t)$$

But the right-hand term is zero. (Why?) Hence, $y''(t) = [8e^{-t} - 8e^{-2t}]u(t)$, and

$$h(t) = [e^{-t} - e^{-2t}]u(t)$$

To see the utility of this approach, try the alternative method of computing $F(s)$, $Y(s)$, and $H(s) = Y(s)/F(s)$. The algebra is straightforward, but tedious and prone to error.

As a final example, we compute a circuit's step response and verify that its derivative is the impulse response.

EXAMPLE 13.14. Compute the step response of the *RLC* circuit of Figure 13.30.

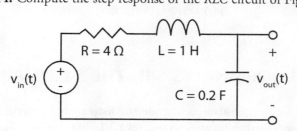

FIGURE 13.30 *RLC* circuit for Example 13.14.

SOLUTION
From voltage division,

$$H(s) = \frac{\dfrac{1}{Cs}}{R + Ls + \dfrac{1}{Cs}} = \frac{1}{LC}\frac{1}{s^2 + \dfrac{R}{L}s + \dfrac{1}{LC}} = \frac{5}{(s+2)^2 + 1} \tag{13.26}$$

From equation 13.26, the Laplace transform of the step response is

$$V_{out}(s) = \frac{H(s)}{s} = \frac{5}{s\left[(s+2)^2 + 1\right]} = \frac{1}{s} + \frac{-s-4}{(s+2)^2 + 1}$$

Rearranging terms yields

$$V_{out}(s) = \frac{1}{s} - \frac{s+2}{(s+2)^2 + 1} - 2\frac{1}{(s+2)^2 + 1}$$

Taking the inverse transform produces the desired step response:

$$v_{out}(t) = [1 - e^{-2t}\cos(t) - 2e^{-2t}\sin(t)]u(t) \tag{13.27}$$

As a check, observe that the derivative of equation 13.27 is

$$\frac{d}{dt} v_{out}(t) = 2e^{-2t}[\cos(t) + 2\sin(t)]u(t) - e^{-2t}[-\sin(t) + 2\cos(t)]u(t) + (1-1)\delta(t)$$

$$= 5e^{-2t}\sin(t)u(t)$$

Thus

$$\mathcal{L}\left[5e^{-2t}\sin(t)u(t)\right] = \frac{5}{(s+2)^2 + 1} = H(s)$$

in which case

$$h(t) = 5e^{-2t}\sin(t)u(t)$$

as expected.

6. NODAL AND LOOP ANALYSIS IN THE S-DOMAIN

This section develops s-domain formulations of node and loop analysis. Nodal analysis of circuits builds around KCL, whereas mesh/loop analysis utilizes KVL. In Chapter 3 and, indeed, in most beginning courses on circuits, loop and nodal analysis are taught first in the context of resistances and conductances and then (in Chapter 10 here) in the phasor context. Recall that KCL requires that the sum of the currents leaving any circuit node be zero. Further, KVL requires that the voltages around any loop of a circuit sum to zero. By linearity, the Laplace transform of a sum is the sum of the individual Laplace transforms. Hence, a KVL equation and a KCL equation have an s-domain formulation where elements are characterized by impedances and/or admittances.

For loop analysis, one writes a KVL equation for each loop in terms of the transformed loop currents and element impedances. The set of all such equations, then, characterizes the circuit's loop currents, which determine the currents through each of the elements. Knowledge of the loop currents and the element impedances permits the computation of any of the element voltages.

In nodal analysis, one writes a KCL equation at each node in terms of the Laplace transform of the node voltages with respect to a reference, the transform of the independent excitations, and the element admittances. The set of all such equations characterizes the node voltages of the circuit in the s-domain. Solving the set of circuit node equations yields the set of transformed node voltages. Knowledge of these permits the computation of any of the element voltages. With knowledge of the element admittances, one may compute all of the element currents. Since nodal analysis has a more extensive application than loop analysis, our focus will be on nodal analysis.

EXAMPLE 13.15. Figure 13.31 shows an ideal operational amplifier circuit called the *Sallen and Key* normalized *low-pass Butterworth filter*. (See Chapter 19 for a full discussion of filters.) A normalized low-pass filter passes frequencies below 1 rad/sec and attenuates higher frequencies. As we will see later in the text, the 1-rad/sec frequency "cutoff" can be changed to any desired value by frequency-scaling the parameter values of the circuit. (See Chapter 14 for a discussion of frequency scaling.) The goal here is to utilize the techniques of nodal analysis to compute the (normalized) transfer function of this circuit.

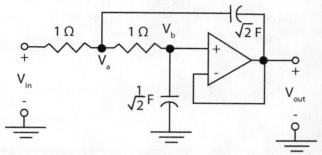

FIGURE 13.31 Sallen and Key normalized Butterworth low-pass filter circuit containing an ideal operational amplifier.

SOLUTION
The solution proceeds in several steps that utilize nodal analysis techniques in conjunction with the properties of an ideal op amp. Recall that for an ideal op amp, the voltage across the input terminals is zero and the current into any of the input terminals is also zero. Finally, note that one does not write a node equation at the output, which appears across a dependent voltage source whose value depends on other voltages in the circuit.

Step 1. *Find V_b.* Because the voltage across the input terminals of an ideal operational amplifier is zero, $V_b = V_{out}$.

Step 2. *Write a node equation at the node identified by the node voltage V_a.* Summing the currents leaving the node yields

$$(V_a - V_{in}) + (V_a - V_b) + \sqrt{2}s(V_a - V_{out}) = 0$$

Substituting V_{out} for V_b and grouping like terms produces

$$\left(\sqrt{2}\, s + 2\right)V_a - \left(\sqrt{2}\, s + 1\right)V_{out} = V_{in} \tag{13.28}$$

Step 3. *Write a node equation at the node identified by the node voltage V_b.* By inspection, the desired node equation is

$$-V_a + \left(\tfrac{1}{\sqrt{2}}s + 1\right)V_{out} = 0 \tag{13.29}$$

Step 4. *Write the foregoing two node equations in matrix form.* In matrix form, equations 13.28 and 13.29 combine to give

$$\begin{bmatrix} \left(\sqrt{2}s + 2\right) & -\left(\sqrt{2}s + 1\right) \\ -1 & \left(\tfrac{1}{\sqrt{2}}s + 1\right) \end{bmatrix} \begin{bmatrix} V_a \\ V_{out} \end{bmatrix} = \begin{bmatrix} V_{in} \\ 0 \end{bmatrix} \tag{13.30}$$

Step 5. *Solve equation 13.30 for* V_{out} *in terms of* V_{in} *using Cramer's rule.* From Cramer's rule,

$$V_{out}(s) = \frac{\det \begin{bmatrix} (\sqrt{2}s+2) & V_{in} \\ -1 & 0 \end{bmatrix}}{\det \begin{bmatrix} (\sqrt{2}s+2) & -(\sqrt{2}s+1) \\ -1 & (\frac{1}{\sqrt{2}}s+1) \end{bmatrix}}$$

The resulting transfer function is

$$H(s) = \frac{V_{out}(s)}{V_{in}} = \frac{1}{(\sqrt{2}s+2)(\frac{1}{\sqrt{2}}s+1)-(\sqrt{2}s+1)} = \frac{1}{s^2+\sqrt{2}s+1}$$

Notice that for small values of $s = j\omega$ (i.e., low frequencies), the magnitude of $H(s)$ approximates 1, and for large values of $s = j\omega$ (i.e., high frequencies, where $|j\omega| \gg 1$), the magnitude of $H(s)$ is small. Since $V_{out}(j\omega) = H(j\omega)V_{in}(j\omega)$, such a circuit blocks high-frequency input excitations and passes low-frequency input excitations. As mentioned at the beginning of the example, the circuit passes low frequencies and attenuates high frequencies.

The preceding example used matrix notation, common to much of advanced circuit analysis. In one sense, *matrix notation* is a shorthand way of writing n simultaneous equations: the n variables are written only once. More generally, matrix notation and the associated matrix arithmetic allow engineers to handle and solve large numbers of equations in numerically efficient ways. Further, the theory of matrices allows one to develop insights into large circuits that would otherwise remain hidden. Hence, many of the examples that follow will utilize the elementary properties of matrix arithmetic.

The next example uses nodal analysis to compute the response to an initialized circuit. The example combines the equivalent circuits for initialized capacitors and inductors with the technique of nodal analysis.

EXAMPLE 13.16. In the circuit of Figure 13.32, suppose $i_{in}(t) = \delta(t)$, $i_L(0^-) = 1$ A, and $v_C(0^-) = 1$ V. Find the voltages $v_C(t)$ and $v_L(t)$.

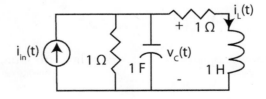

FIGURE 13.32 Two-node *RLC* circuit for Example 13.16.
Given the indicated current direction of $i_L(t)$, what is the implied voltage polarity for $v_L(t)$?

Step 1. *Draw the s-domain equivalent circuit with an eye toward nodal analysis.* Inserting the equivalent current source models for the initialized capacitor and inductor in Figure 13.30, one obtains the s-domain equivalent circuit shown in Figure 13.33.

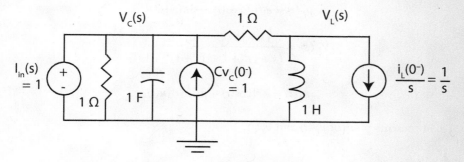

FIGURE 13.33 s-Domain equivalent of the circuit of Figure 13.31.

Step 2. *Write two node equations and put in matrix form.* At the node labeled $V_C(s)$ KCL implies that

$$(1 + s)V_C(s) + [V_C(s) - V_L(s)] = 2$$

Simplifying produces the first node equation:

$$(s + 2)V_C(s) - V_L(s) = 2$$

At the node labeled $V_L(s)$, $[V_L(s) - V_C(s)] + (1/s)V_L(s) = -(1/s)$, or, equivalently,

$$-V_C(s) + \frac{s+1}{s}V_L(s) = -\frac{1}{s}$$

The matrix form of these two node equations is

$$\begin{bmatrix} s+2 & -1 \\ -1 & \dfrac{s+1}{s} \end{bmatrix}\begin{bmatrix} V_C(s) \\ V_L(s) \end{bmatrix} = \begin{bmatrix} 2 \\ -\dfrac{1}{s} \end{bmatrix}$$

Step 3. *Solve the matrix equation of step 2 for the desired voltages.* Using Cramer's rule, computing the inverse, or simultaneously solving the equations gives

$$\begin{bmatrix} V_C(s) \\ V_L(s) \end{bmatrix} = \frac{2}{s^2 + 2s + 2}\begin{bmatrix} \dfrac{s+1}{s} & 1 \\ 1 & s+2 \end{bmatrix}\begin{bmatrix} 2 \\ -\dfrac{1}{s} \end{bmatrix} = \begin{bmatrix} \dfrac{2(s+1)-1}{(s+1)^2 +1} \\ \dfrac{(s+1)-3}{(s+1)^2 +1} \end{bmatrix}$$

(13.31)

Step 4. *Take the inverse Laplace transform to obtain time domain voltages.* Breaking up equation 13.31 into its components yields

$$V_C(s) = \frac{2(s+1)}{(s+1)^2+1} - \frac{1}{(s+1)^2+1}$$

in which case

$$v_C(t) = e^{-t}[2\cos(t) - \sin(t)]u(t)$$

Also,

$$V_L(s) = \frac{(s+1)}{(s+1)^2+1} - \frac{3}{(s+1)^2+1}$$

leading to

$$v_L(t) = e^{-t}[\cos(t) - 3\sin(t)]u(t)$$

Figure 13.34 presents plots of $v_C(t)$ and $v_L(t)$.

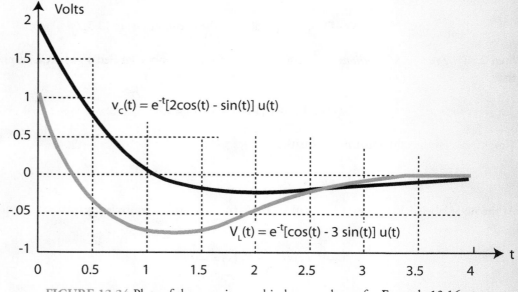

FIGURE 13.34 Plots of the capacitor and inductor voltages for Example 13.16.

Dual to nodal analysis is loop analysis. In loop analysis, one defines loop currents and writes KVL equations in terms of these loop currents. The following example illustrates the method of loop analysis for computing the input impedance of a bridged-T network.

EXAMPLE 13.17. Use loop analysis to compute the input impedance of the bridged-T network illustrated in Figure 13.35.

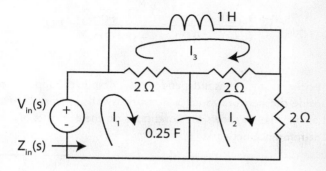

FIGURE 13.35 Bridged-T network for Example 13.17.

SOLUTION

Define $I_{in}(s) = I_1(s)$ in Figure 13.34. Since $Z_{in}(s) = V_{in}(s)/I_{in}(s)$, the goal is to use loop analysis to find $I_{in}(s) = I_1(s)$ in terms of $V_{in}(s)$.

Step 1. *Sum the voltages around loop 1.* Dropping the specific s-dependence for convenience, one obtains

$$2(I_1 - I_3) + \frac{4}{s}(I_1 - I_2) = 2\frac{s+2}{s}I_1 - \frac{4}{s}I_2 - 2I_3 = V_{in}$$

Step 2. *Sum the voltages around loop 2.* By inspection,

$$\frac{4}{s}(I_2 - I_1) + 2(I_2 - I_3) + 2I_2 = -\frac{4}{s}I_1 + 4\frac{s+1}{s}I_2 - 2I_3 = 0$$

Step 3. *Sum the voltages around loop 3.* Again by inspection,

$$2(I_3 - I_1) + sI_3 + 2(I_3 - I_2) = -2I_1 - 2I_2 + (s+4)I_3 = 0$$

Step 4. *Put the three loop equations in matrix form, and solve.* In matrix form, the three loop equations are

$$\begin{bmatrix} 2\dfrac{s+2}{s} & -\dfrac{4}{s} & -2 \\[2mm] -\dfrac{4}{s} & 4\dfrac{s+1}{s} & -2 \\[2mm] -2 & -2 & s+4 \end{bmatrix}\begin{bmatrix} I_1 \\ I_2 \\ I_3 \end{bmatrix} = \begin{bmatrix} V_{in} \\ 0 \\ 0 \end{bmatrix}$$

Using Cramer's rule to solve this equation for $V_{in}(s)$ in terms of $I_{in}(s) = I_1(s)$ yields

$$Z_{in}(s) = \frac{V_{in}(s)}{I_{in}(s)} = \frac{V_{in}(s)}{I_1(s)} = \frac{8\dfrac{s^2 + 4s + 4}{s}}{4\dfrac{(s+2)^2}{s}} = 2\ \Omega$$

It may be a little surprising that $Z_{in}(s)$ is independent of s, despite the appearance of an inductance and a capacitance inside the network. Such a network is called a *constant-resistance network*. A problem at the end of the chapter shows the conditions on the elements of a bridged-T network for it to be a constant-resistance network.

Exercise. The loop equations of the circuit in Figure 13.36 are

$$\begin{bmatrix} 1+2s & 0 & -2s & 1 \\ 0 & 2s+\dfrac{4}{s}+2 & -2 & -1 \\ -2s & -2 & 2s+4 & 0 \\ -1 & 1 & -4 & 0 \end{bmatrix} \begin{bmatrix} I_1 \\ I_2 \\ I_3 \\ \widehat{V} \end{bmatrix} = \begin{bmatrix} V_{s1} \\ 0 \\ 0 \\ 0 \end{bmatrix}$$

Find the values of C and K in Figure 13.36.
ANSWERS: $C = 0.25$ F, $K = -4$

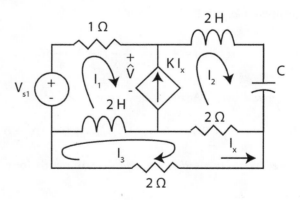

FIGURE 13.36

---

7. SWITCHING IN *RLC* CIRCUITS

Switches control lighting systems, furnaces, car ignitions, traffic lights, and numerous other devices. Switching also takes place inside electronic circuits, as in switched power supplies and switched capacitor filters. All have a functional element called a *switch* that affects and, indeed, shapes the behavior of the circuit. The switch inside an electronic circuit is a special device that we will model simply as an ideal on/off switch. This section investigates the behavior of switching in simple *RLC* circuits, as preparation for the understanding of switching in more elaborate and sophisticated electronic circuits. Our immediate task is to apply the Laplace transform method to compute the responses of switched *RLC* circuits. The following example motivates a general procedure.

EXAMPLE 13.18. In the circuit of Figure 13.37, suppose $R_s = 20\ \Omega$, $R = 4\ \Omega$, $C = 0.25$ F, and $v_{in}(t) = 10e^{-0.1t}\ u(t)$, $v_C(0^-) = 0$. The switch is initially in position A. The switch S moves from position A to position B at $t = 1$ sec and from position B to position A at $t = 2$ sec, and moves back to B at $t = 4$ sec where it remains for all subsequent time. Find $v_C(t)$ for $t \geq 0$.

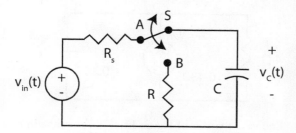

FIGURE 13.37 Switched *RC* circuit ($R_s = 20\ \Omega$, $R = 4\ \Omega$, $C = 0.25$ F) for Example 13.18 in which $v_C(0^-) = 0$; the switch S moves to B at $t = 1$ sec, returns to A at $t = 2$ sec, and moves back to B at $t = 4$ sec, where it remains.

SOLUTION
Because of the switching at $t = 1$ sec, the first step is to determine $v_C(t)$ over the time interval $0 \leq t < 1$. This allows us, in turn, to find $v_C(1^-)$, which will serve as the initial condition over the time interval $1 \leq t < 2$. This then produces $v_C(2^-)$, the initial condition for the interval $2 \leq t \leq 4$, etc.

Step 1. *Compute the response for $0 \leq t < 1$.* Over the interval $0 \leq t < 1$, the circuit of Figure 13.37 is the simple *RC* circuit of Figure 13.38.

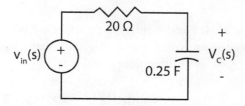

FIGURE 13.38 Equivalent circuit of Figure 13.37 over $0 \leq t < 1$.

For the circuit of Figure 13.37, voltage division implies

$$V_C(s) = \frac{\frac{4}{s}}{20 + \frac{4}{s}} V_{in}(s) = \frac{2}{(s+0.2)(s+0.1)} = \frac{20}{s+0.1} - \frac{20}{s+0.2}$$

Hence, for $0 \leq t < 1$,

$$v_C(t) = 20e^{-0.1t} - 20e^{-0.2t}$$

We note that $v_C(1^-) = 20e^{-0.2} - 20e^{0.1} = 1.722$ V.

Step 2. *Compute the response over $1 \le t < 2$.* After the switch moves from position A to position B, the source is decoupled from the right half of the circuit; the response then depends only on the initial condition at $t = 1^-$, i.e., $v_C(1^-) = 1.722$ V. The goal is to compute $v_C(t)$ over the interval $1 \le t < 2$, or equivalently, over the interval $0 \le t' < 1$ where $t' = t - 1$. The *s*-domain equivalent circuit that models the behavior of the time domain circuit of Figure 13.37 over $1 \le t < 2$ has the form illustrated in Figure 13.39. We note that the value on the current source is $Cv_C(1^-) = 0.4305$.

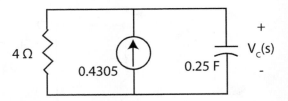

FIGURE 13.39 Equivalent circuit of Figure 13.37 for the time interval $1 \le t < 2$.

The equivalent admittance seen by the current source is $0.25s + 0.25$. Hence by Ohm's law,

$$V_C(s) = \frac{4}{s+1} \times 0.43053 = \frac{1.722}{s+1}$$

For $0 \le t' < 1$, taking the inverse transform yields

$$v_C(t') = 1.722e^{-t'}u(t)$$

or equivalently, for $1 \le t < 2$,

$$v_C(t) = 1.722e^{-(t-1)}u(t-1)$$

We emphasize that this last equation is valid only for $1 \le t < 2$. We note that $v_C(2^-) = 1.722e^{-1} = 0.6335$.

Step 3. *Compute $v_C(t)$ for $2 \le t < 4$.* For this interval, the capacitor is initialized at $t' = t - 2 = 0$. Using the parallel current-source equivalent circuit, we note that $Cv_C(t')\mid_{t'=0} = Cv_C(2^-) = 0.1584$, as shown in Figure 13.40. We further note that

$$v_{in}(t) = v_{in}(t'+2) = \hat{v}_{in}(t') = 10e^{-0.1(t'+2)} = 8.1873e^{-0.1t'}u(t')$$

Thus $\hat{V}_{in}(s)$ in Figure 13.40 is

$$\hat{V}_{in}(s) = \frac{8.1873}{s+0.1}$$

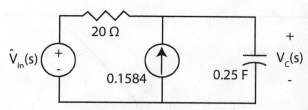

FIGURE 13.40 Equivalent s-domain circuit valid for $2 \leq t < 4$.

Using superposition on the circuit of Figure 13.40,

$$V_C(s) = \frac{\frac{4}{s}}{20 + \frac{4}{s}} \hat{V}_{in}(s) + \frac{0.1584}{0.25s + 0.05} = \frac{0.2 \times 8.1873}{(s + 0.2)(s + 0.1)} + \frac{0.6335}{s + 0.2}$$

Hence, for $0 \leq t' < 2$,

$$v_C(t') = 16.3746e^{-0.1t'} - 15.7411e^{-0.2t'}$$

or equivalently for $2 \leq t < 4$,

$$v_C(t) = 16.3746e^{-0.1(t-2)} - 15.7411e^{-0.2(t-2)}$$

We note that $v_C(4^-) = 16.3746e^{-0.2} - 15.7411e^{-0.4} = 2.8548$ V.

Step 4. *Compute the response over $4 \leq t$.* After the switch moves from position A to position B, again, the source is again decoupled from the right half of the circuit. According to Chapter 8, we can write the answer by inspection: the solution is simply $v_C(t') = 2.8548e^{-t'}u(t')$. Equivalently, for $t \geq 4$, $v_C(t) = 2.8548e^{-(t-4)}$.

Step 5. *Combine results into a single expression and plot.* The combined expressions for $v_C(t)$ are

$$v_C(t) = \begin{cases} 20e^{-0.1t} - 20e^{-0.2t} & 0 \leq t < 1 \\ 1.722e^{-(t-1)}u(t-1) & 1 \leq t < 2 \\ 16.3746e^{-0.1(t-2)} - 15.7411e^{-0.2(t-2)} & 2 \leq t < 4 \\ 2.8548e^{-(t-4)} & 4 \leq t \end{cases}$$

A plot of the response appears in Figure 13.41.

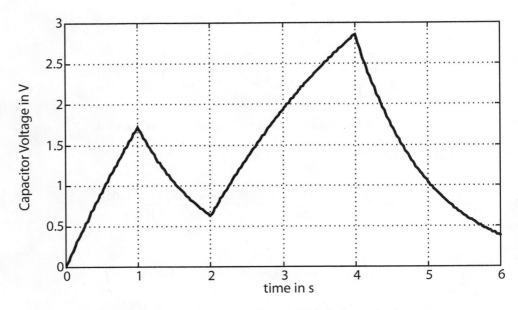

FIGURE 13.41 Plot of capacitor voltage for switched circuit of Figure 13.37.

The extension of this method to more than three switching times is straightforward. Although the preceding example uses an *RLC* circuit, the same strategy or algorithm is applicable to the calculation of switching transients in any linear dynamic circuit. The following is a summary of the general procedure.

Procedure for Applying Laplace Transform Method to Switched Circuits

At $t = 0$, a switching may or may not occur. For $t > 0$, denote the switching time instants *successively* as t_1, t_2, ... , t_n. The goal is to compute the response for $0 \leq t < \infty$.
Divide the time axis into intervals $(0, t_1)$, (t_1, t_2), ... , $(t_n, t_{n+1} = \infty)$, and compute the responses *successively* for each time interval.

Part I. Consider the first time interval, $(0, t_1)$.

   Step 1. Construct the *s*-domain equivalent circuit, making use of the initial conditions at $t = 0^-$, which are either given or calculated from the past history of the circuit for $t < 0$ (see note at the end of the procedure). This equivalent circuit is valid for the time interval $0 \leq t < t_1$.

   Step 2. Find the response by the Laplace transform method for the circuit of step 1.

   Step 3. Evaluate the capacitor voltages and inductor currents at $t = t_1^-$.

Part II. For all remaining time intervals:

   Step 4. Set the value of subscript $i = 1$.

   Step 5. (1) For $t_i \leq t < t_{i+1}$, let $t' = t - t_i$ (i.e., consider the time interval $0 \leq t' < \infty$), and construct the *s*-domain equivalent circuit, making use of the *calculated* initial conditions at $t' = 0^-$ (i.e., at $t = t_i^-$).

(2) Determine the proper form of the input excitation(s) (if there are any), in terms of t'.

(3) Find the response by the Laplace transform method. Note that the time variable is t'. Then obtain the solution in t by substituting $(t - t_i)$ for t'.

(4) If $i = n$, stop. Otherwise, evaluate the capacitor voltages and inductor currents at $t = t^-_{i+1}$.

(5) Increase the subscript value i by 1 and go to the beginning of step 5.

Note: In some situations, the first switching occurs at $t = 0$, but the capacitor voltages and inductor currents at $t = 0^-$ are not given. Instead, the problem specifies that dc and sinusoidal sources have excited the circuit for a long time. If the network is passive—i.e., if it consists of inductors, capacitors, or *lossy* elements–then the circuit will have reached a steady state at $t = 0^-$. The procedure then is first to find the steady-state solution and then to evaluate the capacitor voltages and inductor currents at $t = 0^-$. It is instructive to review the dc and sinusoidal steady-state (phasor) analysis methods studied in a first course. Recall that under certain stability conditions (to be studied in Chapter 15) on the network:

1. For dc steady-state analysis, open-circuit all capacitances and short-circuit all inductors to find the steady-state voltages and currents.

2. For sinusoidal steady-state analysis, use the phasor method to find the steady-state responses.

8. SWITCHED CAPACITOR CIRCUITS AND CONSERVATION OF CHARGE

In addition to its many uses already described, the Laplace transform method is applicable to a special class of circuits called *switched capacitor* (abbreviated SC) networks. These circuits contain only capacitors, switches, independent voltage sources, and possibly some operational amplifiers. No resistors or inductors are present. One can dispense with resistors because it is possible to approximate the effect of a resistor with two switches and a capacitor. Similarly, inductors can be approximated by circuits containing only switches, capacitors, and operational amplifiers. These facts, coupled with the easy and relatively inexpensive fabrication of switches, capacitors, and operational amplifiers in MOS (metal-oxide semiconductor) technology, have made switched capacitor filters an attractive alternative to classical filters. Given this scenario, the purpose of this section is to lay a foundation (i.e., introduce the principles) upon which switched capacitor circuit design builds. More advanced courses delve into the actual analysis and design of real-world switched capacitor circuits.

Besides the Laplace transform approach, an alternative method for analyzing switched capacitor networks builds on the principle of conservation of charge.

> **Principle of conservation of charge:** The total charge transferred into a junction (or out of a junction) of a circuit at any time is zero.

This principle is a direct consequence of Kirchhoff's current law. For example, in Figure 13.42, KCL implies that

$$i_1(t) + i_2(t) + i_3(t) + i_4(t) = 0 \tag{13.32}$$

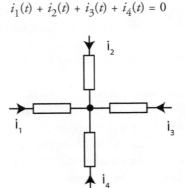

FIGURE 13.42 Node to which KCL applies.

Since charge is the integral of current over a time interval, integrating both sides of equation 13.32 from $-\infty$ to t yields

$$\int_{-\infty}^{t} [i_1(t) + i_2(t) + i_3(t) + i_4(t)]dt = 0$$

or equivalently,

$$q_1(t) + q_2(t) + q_3(t) + q_4(t) = 0$$

which is just another expression of the principle of conservation of charge. A simple switched capacitor circuit will now serve as a test bed for comparing the merits of the foregoing analysis techniques.

EXAMPLE 13.19. Consider the circuit shown in Figure 13.43a. The switch S is closed at $t = 0$. Just before the closing of S, the initial conditions are known to be $v_{C1}(0^-) = 1$ V and $v_{C2}(0^-) = 0$. Compute the voltages $v_{C1}(t)$ and $v_{C2}(t)$ for $t > 0$.

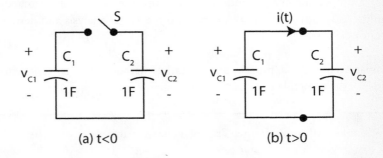

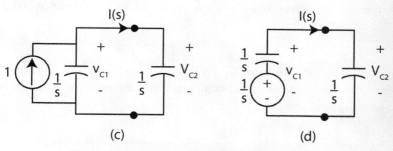

FIGURE 13.43 Equivalent circuits for Example 13.19.

SOLUTION

Method 1. Using the parallel current-source model of a capacitor, we obtain the s-domain equivalent circuit shown in Figure 13.43c. By inspection,

$$V_{C1}(s) = V_{C2}(s) = \frac{1}{s}I(s) = \frac{1}{s}\frac{s}{s+s} = \frac{0.5}{s}$$

Therefore, $v_{C1}(t) = v_{C2}(t) = 0.5u(t)$ V, and $i(t) = 0.5\delta(t)$ A.

Method 2. Using the series voltage-source model of a capacitor, we have the s-domain equivalent circuit shown in Figure 13.43d. Again, by inspection,

$$V_{C1}(s) = V_{C2}(s) = \frac{\frac{1}{s}}{\frac{1}{s}+\frac{1}{s}}\frac{1}{s} = \frac{0.5}{s}; \quad I(s) = \frac{\frac{1}{s}}{\frac{1}{s}+\frac{1}{s}} = 0.5$$

This is the same answer as obtained with method 1.

Method 3. *Conservation-of-charge approach.* For $t > 0$, the network is shown in Figure 13.43b. Clearly, $v_{C1}(t) = v_{C2}(t)$. After S is closed, some charge is transferred from C_1 to C_2. However, according to the principle of conservation of charge, the *total* charge transferred out of the junction must be zero. Hence,

$$C_1[v_{C1}(0^+) - v_{C1}(0^-)] + C_2[v_{C2}(0^+) - v_{C2}(0^-)] = 0$$

and

$$v_{C1}(0^+) = v_{C2}(0^+)$$

Solving these two equations for the two unknowns, $v_{C1}(0^+)$ and $v_{C2}(0^+)$, results in

$$v_{C1}(0^+) = v_{C2}(0^+) = \frac{C_1 v_{C1}(0^-) + C_2 v_{C2}(0^-)}{C_1 + C_2}$$

Since there is no external input applied, the voltages remain constant once the equilibrium condition has been reached. Therefore,

$$v_{C1}(t) = v_{C2}(t) = v_{C1}(0^+) = v_{C1}(0^+)$$

for $t > 0$. For the specific capacitance values given in Figure 13.43a, we obtain $v_{C1}(t) = v_{C2}(t) = 0.5$ V for $t > 0$.

Exercise. In the circuit of Figure 13.43a, let $C_1 = 4$ F and $C_2 = 6$ F, $v_{C1}(0^-) = 8$ V, and $v_{C2}(0^-) = 3$ V. Find, by at least two methods, the capacitor voltages after the switch is closed. (Rework the problem if the answers do not agree.)
ANSWER: 5 V

Computationally, the Laplace transform method is more straightforward. On the other hand, the conservation-of-charge method is more basic and often provides better insight into what happens to the charges stored in various capacitors. It is particularly useful for the purpose of checking answers obtained by other methods: the answers are correct when the conservation-of-charge condition is met at every node.

EXAMPLE 13.20. The initial conditions at $t = 0^-$ of an SC network are shown in Figure 13.44. Switches S_1 and S_2 are closed at $t = 0$, connecting the two dc voltage sources to the network. Find the node voltages for $t > 0$.

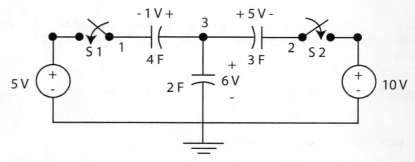

FIGURE 13.44 Switched capacitor network for Example 13.20 at $t = 0^-$.

SOLUTION
We first construct the s-domain equivalent circuit using admittances. The result is shown in Figure 13.45.

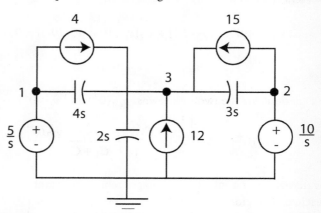

FIGURE 13.45 Equivalent circuit to Figure 13.44, accounting for initial conditions; the admittance of each capacitor is indicated.

Applying KCL to node 3 generates the node equation

$$4s\left(V_3 - \frac{5}{3}\right) + 3s\left(V_3 - \frac{10}{s}\right) + 2sV_3 = 4 + 15 + 12$$

from which we derive $V_3 = 9/s$ and $v_3(t) = 9$ V for $t > 0$. Obviously, $V_1 = 5$ V and $V_2 = 10$ V. To verify that $V_3 = 9$ V is indeed the correct solution, we check for conservation of charge. From $t = 0^-$ to $t = 0^+$, the voltage of the 4 F capacitor jumps from 1 V to 4 V (note that $4 = 9 - 5$), indicating that $4 \times (4 - 1) = 12$ coulombs of charge have been transferred to this capacitor. The voltage of the 2 F capacitor jumps from 6 V to 9 V, indicating that $2 \times (9 - 6) = 6$ coulombs of charge have been transferred here. Finally, the voltage of the 3 F capacitor changes from 5 V to –1 V (note that $-1 = 9 - 10$), indicating that $3 \times (-1 - 5) = -18$ coulombs of charge have been transferred to this capacitor. As a check for conservation of charge, we have $12 + 6 + (-18) = 0$, and the solution is assured to be correct.

Exercise. Solve Example 13.20 again, with all capacitors initially uncharged.
ANSWER: $v_3(t) = 50/9$ V, $t > 0$

The preceding examples considered *idealized* circuits, i.e., no resistances were present. In any *practical* circuit, the connecting wires have some resistance. What is our interest in the analysis of an idealized circuit? The analysis of an idealized circuit is much more straightforward than that of a realistic circuit yet provides relatively accurate answers. As a case in point, reconsider Example 13.20. Suppose we insert a 0.1 Ω resistance in series with every capacitor. The resulting transform analysis would produce a rational function with a cubic denominator polynomial (a third-order network) whose factorization would require the use of a root-finding program. In sharp contrast, the idealized circuit of Example 13.20 was analyzed by writing a single first-order node equation, making a partial fraction expansion unnecessary.

Idealizations of circuit models sometimes lead to phenomena that defy intuitive explanations. An interesting case is given by Example 13.20. Before S is closed, the energy stored in the electric field is $0.5 \times 1^2 + 0.5 \times 0^2 = 0.5$ joule. After S is closed, the stored energy becomes $0.5 \times (0.5)^2 + 0.5 \times (0.5)^2 = 0.25$ joule. Apparently, 0.25 joule of energy has been lost. Since there is no resistance in the circuit to dissipate the energy, what accounts for the lost energy? Is energy not conserved?

An explanation of this paradox is as follows. Instead of considering a zero-resistance circuit, place a resistance R in series with all capacitances. Then analyze the circuit, and let R *approach* zero. The result shows that no matter what value R takes on, the total energy dissipated in the resistance for $0 < t < \infty$ exactly equals the difference of the total stored energies before and after the closing of the switch. This accounts for the apparent lost energy. In actuality, part of the 0.25 joule of energy would be lost in the form of radiated energy. However, the principles of field theory would be necessary to explain the radiation phenomenon.

Note that for idealized SC circuits, as long as the independent voltage sources are *piecewise* constant, all capacitor voltages are *piecewise* constant and all currents in the circuit are impulses. These properties remain valid for more general idealized SC circuits that allow the inclusion of VCVSs, CCCSs, and ideal op amps. The reason is that the parameters characterizing these components are dimensionless and hence do not result in a time constant. All voltage changes are *instantaneous*. On the other hand, if the circuit contains resistances, we have a *lossy switched capacitor circuit*, whose voltages are no longer piecewise constant. The transient analysis of a lossy SC circuit requires the usual Laplace transform analysis.

One reason for our interest in SC circuits is that a SC combination can be used to *approximate* a resistor. As a result, any *RC*–op amp circuit used for signal processing can be *approximated* by an SC–op amp circuit. A study of the general theory of such SC–op amp circuits is beyond the level of this book. We shall merely illustrate the approximation property with a simple integrator circuit.

EXAMPLE 13.21. Consider the *RC*-op amp integrator circuit shown in Figure 13.46.

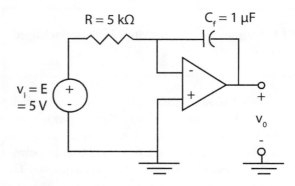

FIGURE 13.46 Op amp integrator of Example 13.21.

Using the result of Example 13.5 and the integration property of the Laplace transform, it is straightforward to show that

$$v_o(t) = -\frac{1}{RC_f} \int_0^t v_i(\tau)d\tau + v_o(0)$$

If the input is a constant voltage $v_i = E = 5$ V, and if $v_o(0) = 0$, then the output waveform is a ramp function $v_o(t) = -1{,}000tu(t)$ (as long as the output has not reached the saturation level), as shown in Figure 13.47.

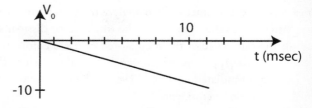

FIGURE 13.47 Ramp output of the integrator in Figure 13.46.

What we would like to do is construct an SC approximation to this circuit. The idea is to replace the resistor with a switch and a capacitor as in Figure 13.48.

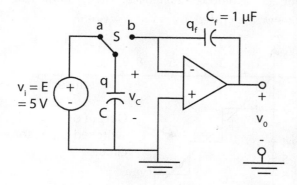

FIGURE 13.48 Switched capacitor equivalent of Figure 13.46.

The switch S is operated in the following manner:

1. At $t = 0$, S is at position a.
2. At $t = T$, S is moved to position b.
3. At $t = 2T$, S is moved to position a.
4. At $t = 3T$, S is moved to position b, etc.

The output waveform may be determined very easily by the principle of conservation of charge as follows: for $0 \leq t < T$, $v_c = E$, $q = CE$, $q_f = 0$, and $v_o = 0$. At $t = T$, switch S is moved to position b. Because the op amp is assumed to be ideal, the voltage across the input terminals is zero, and so is v_c. Thus, C cannot store any charge. The charge CE previously stored on C must be transferred out of C. Since the op amp is ideal, the input impedance is infinity and the input current is zero. Therefore, none of the charge can flow into the op amp. Instead, the charge must be transferred to the capacitor C_f. This leads to $q_f = CE$ and $v_o(T^+) = -CE/C_f$.

At $t = 2T$, switch S is moved back to position a, causing C to be charged to E volts again. Since the charge q_f is "trapped" on C_f the output voltage v_o remains unchanged until S is moved to position b again. At that time, another CE coulombs of charge are transferred to C_f, and v_o is *incremented* by $-CE/C_f$. Subsequent switching is similar.

To make the output waveforms of the circuits of Figures 13.46 and 13.48 *approximately* the same, the *average* of the charges transferred to C_f must be the same in both cases. For Figure 13.46, the current flowing into C_f is at a constant value of E/R. Therefore, every $2T$ sec, the charge transferred to C_f is equal to $2TE/R$. On the other hand, for Figure 13.48, the charge transferred to C_f every $2T$ sec is CE. Equating these two quantities, we have $CE = 2TE/R$, or $RC = 2T$. Thus, there is no unique combination of C and T that will produce the approximate effect of a resistance. A smaller T (i.e., a higher operating frequency of the switch) in Figure 13.48 produces a staircase output waveform that closely "hugs" the ramp output of Figure 13.47. For the purpose of comparison, the output waveform corresponding to $T = 1$ msec and $C = 0.4$ µF is shown in Figure 13.49, together with the ramp output from the RC–op amp integrator. It is worthwhile to note that each of the circuits of Figures 13.46 and 13.48 drains the same average amount of charge from the voltage source and puts the same average amount of charge on the capacitor C_f. The only difference is that in the former the process is *continuous*, whereas in the latter the process occurs in quantized steps.

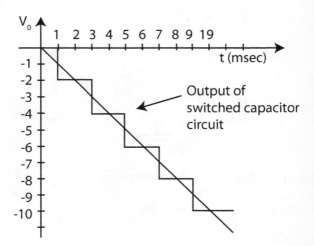

FIGURE 13.49 Responses of circuits in Figures 13.46 and 13.48.

Exercise. Plot the output waveform of the SC circuit of Figure 13.48 for $0 \le t \le 2$ msec if $T = 0.1$ msec and $C = 0.04$ μF. Also, plot the ramp output on the same graph for comparison.

As recently as two decades ago, switched capacitor circuits were considered impractical. No longer! Advances in semiconductor technology have made the fabrication of high-speed electronic switches and op amps as cheap as resistors. Furthermore, large numbers of switches, capacitors, and op amps can be fabricated on a single chip. Consequently, switched capacitor circuits hold an important place in future signal-processing applications. Although we cannot delve into the practical aspects of the design of such circuits, we have at least outlined the basic principles needed for their approximate or exact analysis.

9. THE DESIGN OF GENERAL SUMMING INTEGRATORS

In the design of active filters and in the design of data acquisition equipment, general summing integrators play an important role. This section takes up the design of such op amp circuits from the transfer function perspective. However, to simplify the presentation, we consider only multi-input transfer functions of the form

$$V_{out}(s) = -\frac{a_1}{s}V_{a1} - \frac{a_2}{s}V_{a2} + \frac{b_1}{s}V_{b1} + \frac{b_2}{s}V_{b2} \tag{13.33}$$

where the a_i and b_i are positive gain constants and $V_{a1}(s), \ldots, V_{b2}(s)$ are s-domain inputs. This expression represents the s-domain equivalent of the following time domain equation assuming zero initial conditions at $t = 0^-$ on the variables:

$$v_{out}(t) = -a_1 \int_0^t v_{a1}(\tau)\,d\tau - a_2 \int_0^t v_{a2}(\tau)\,d\tau + b_1 \int_0^t v_{b1}(\tau)\,d\tau + b_2 \int_0^t v_{b2}(\tau)\,d\tau$$

With a little cleverness, it is possible to design by inspection a general integrating operational amplifier circuit whose input-output characteristic is precisely equation 13.33. The four-input op amp circuit of Figure 13.50 accomplishes this. The dashed lines are present because the admittance may or may not be needed. Computation of the values of ΔG_a and ΔG_b are explained in design step 2 below.

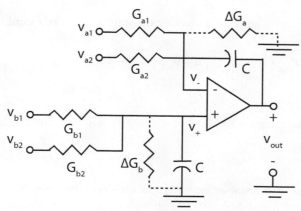

FIGURE 13.50 Structure of general summing integrator.

Design Choices for the General Summing Circuit of Figure 13.50

The first two design steps constitute a preliminary or prototype design, meaning that the two capacitors are normalized to 1 F. After completing the prototype design, an engineer would scale the capacitances and resistances to more practical values without changing the gain characteristics. The scaling procedure is explained in step 3.

Design Step 1. *Prototype design.* Set $C = 1$ F and also set $G_{a1} = a_1$ S, $G_{a2} = a_2$ S, $G_{b1} = b_1$ S, $G_{b2} = b_2$ S.
In the derivation described later on, we conclude that a simplified design requires that the total admittances incident on the inverting terminal equal the total admittance incident on the non-inverting terminal by proper choice of ΔG_a or ΔG_b. The proper choices are given in design step 2.

Design Step 2. *Prototype design (continued): Computation of ΔG_a or ΔG_b such that the total admittance incident at the inverting terminal of the op amp equals the total admittance incident at the non-inverting terminal.* To achieve this equality, recall that in design step 1, $G_{a1} = a_1$ S, $G_{a2} = a_2$ S, $G_{b1} = b_1$ S, $G_{b2} = b_2$ S. Define a numerical quantity $\delta = (a_1 + a_2) - (b_1 + b_2)$

We distinguish two cases:

>   *Case 1:* If $\delta > 0$, set $\Delta G_b = \delta$ and $\Delta G_a = 0$.
>
>   *Case 2:* If $\delta \leq 0$, set $\Delta G_a = -\delta$ and $\Delta G_b = 0$.

Design Step 3. *Scaling to achieve practical element values.* Multiply all the admittances incident at the inverting input terminal of the op amp by a constant K_a. Similarly, multiply all admittances incident at the non-inverting terminal of the op amp by K_b. It is possible to choose $K_a = K_b$, but this is not necessary.

EXAMPLE 13.22. Design an op amp circuit having the input-output relationship

$$V_{out}(s) = -\frac{7}{s}V_{a1} - \frac{3}{s}V_{a2} + \frac{2}{s}V_{b1} + \frac{4}{s}V_{b2} \qquad (13.34)$$

SOLUTION
Step 1. *Prototype design.* Using Figure 13.50, choose $C = 1$ F. With all admittances in S, set

$$G_{a1} = 7 \qquad\qquad G_{a2} = 3, \qquad\qquad G_{b1} = 2, \qquad\qquad G_{b2} = 4$$

Step 2. *Equalization of total admittances at inverting and non-inverting terminals.* Since $\delta = (7 + 3) - (2 + 4) = 4 > 0$, set $\Delta G_b = 4$ S and $\Delta G_a = 0$. The circuit in Figure 13.51a exemplifies the prototype design.

Step 3. *Scaling.* To have practical element values, let us choose $K_a = K_b = 10^{-5}$. This scaling leads to a design with $C = 10$ μF and resistances $R_{a1} = 14.28$ kΩ, $R_{a2} = 33.33$ kΩ, $R_{b1} = 50$ kΩ, $R_{b2} = 25$ kΩ, and $\Delta R_b = 25$ kΩ.

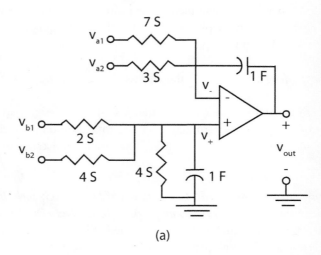

(a)

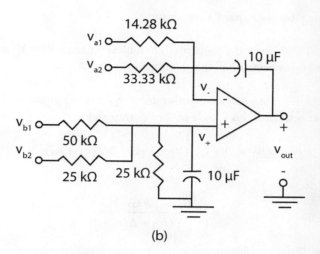

(b)

FIGURE 13.51 (a) Prototype design of equation 13.34. (b) Final design after scaling with $K_a = K_b = 10^{-5}$.

Exercises. 1. Show that the normalized realization of

$$V_{out}(s) = -\frac{2}{s}V_{a1} - \frac{4}{s}V_{a2} + \frac{7}{s}V_{b1} + \frac{5}{s}V_{b2}$$

is given by the circuit of Figure 13.52.

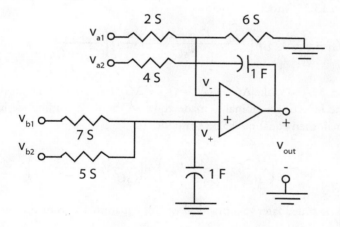

FIGURE 13.52

2. Scale the circuit of Figure 13.51 so that the capacitors become 1 μF.

Derivation of Op Amp Input-Output Characteristic

Referring to Figure 13.50, from the properties of an ideal op amp, $V_+ = V_-$ and no current enters the inverting and non-inverting op amp input terminals.

Derivation Step 1. *Write a node equation at the non-inverting input terminal of the op amp.* Summing the currents leaving the + node of the op amp yields

$$G_{b1}(V_+ - V_{b1}) + G_{b2}(V_+ - V_{b2}) + \Delta G_b V_+ + sC V_+ = 0$$

Solving for V_+ yields

$$V_+ = \frac{G_{b1}V_{b1} + G_{b2}V_{b2}}{G_{b1} + G_{b2} + \Delta G_b + sC} \tag{13.35}$$

Derivation Step 2. *Write a node equation at the inverting input terminal of the op amp.* Recall that $V_- = V_+$. Summing the currents leaving the inverting node yields

$$G_{a1}(V_+ - V_{a1}) + G_{a2}(V_+ - V_{a2}) + sC(V_+ - V_{out}) + \Delta G_a V_+ = 0$$

Thus,

$$V_{out} = -\frac{G_{a1}}{sC}V_{a1} - \frac{G_{a2}}{sC}V_{a2} + \left(\frac{G_{a1} + G_{a2} + \Delta G_a + sC}{sC}\right)V_+ \tag{13.36}$$

Derivation Step 3. *Combine steps 1 and 2 to compute the general input-output relationship.* Substituting equation 13.35 into 13.36 yields

$$V_{out} = -\frac{G_{a1}}{sC}V_{a1} - \frac{G_{a2}}{sC}V_{a2} + \left(\frac{G_{a1} + G_{a2} + \Delta G_a + sC}{sC}\right)\left(\frac{G_{b1}V_{b1} + G_{b2}V_{b2}}{G_{b1} + G_{b2} + \Delta G_b + sC}\right) \tag{13.37}$$

If we choose ΔG_a and ΔG_b to make $\Delta G_a + G_{a1} + G_{a2} = \Delta G_b + G_{b1} + G_{b2}$, i.e., if the total admittance incident on the inverting terminal is made equal to the total admittance incident on the non-inverting terminal, then equation 13.37 simplifies to

$$V_{out} = -\left(\frac{G_{a1}}{sC}V_{a1} + \frac{G_{a2}}{sC}V_{a2}\right) + \left(\frac{G_{b1}}{sC}V_{b1} + \frac{G_{b2}}{sC}V_{b2}\right) \tag{13.38}$$

If we let $C = 1$ F (to be scaled later to a practical value), equation 13.38 further simplifies to

$$V_{out} = -\left(\frac{G_{a1}}{s}V_{a1} + \frac{G_{a2}}{s}V_{a2}\right) + \left(\frac{G_{b1}}{s}V_{b1} + \frac{G_{b2}}{s}V_{b2}\right) \tag{13.39}$$

Equation 13.39 shows that the circuit of Figure 13.50 is a general summing integrating circuit whose gains are proportional to the admittances G_{ai} and/or G_{bi}. The sign of each gain depends on whether the corresponding input is connected to the inverting or non-inverting terminal of the op amp. This completes the derivation of the input-output characteristic of the op amp circuit and is the basis for the prototype design. The extension to more than four inputs is straightforward.

It remains to justify the scaling of step 3 in the design procedure. In step 3, it is stated that all admittances incident at the inverting input terminal of the op amp can be multiplied by a (scaling) constant K_a. Similarly, all admittances incident at the non-inverting terminal may be multiplied by a (scaling) constant K_b. It is possible to choose $K_a = K_b$, but, again, this is not necessary. The verification that these multiplications will not change the input-output characteristic follows directly from equation 13.37. After inserting the scale factors, one immediately sees that they cancel and have no effect on the overall gain.

11. SUMMARY

This chapter has presented the basic principles and techniques of circuit analysis in the s-domain. Impedance, admittance, Thevenin equivalents, superposition, linearity, voltage division, current division, source transformations, and nodal and loop analysis have s-domain forms that allow the analysis of complex circuits excited by a variety of waveform types. Indeed, the simple starter circuit of a fluorescent light points to the usefulness of Laplace transform methods in the analysis and design of simple, everyday electrical conveniences. Various op amp applications were also presented. As subsequent chapters will illustrate, complex circuits and advanced analysis methods build on these basic principles and techniques.

The chapter also introduced the notion of switched capacitor circuits. Integrated circuit technology has made such circuits easy and inexpensive to produce. Applications include speech processing and other types of signal processing. Although a full-scale analysis of such circuits is beyond the scope of this text, the basic principles of their operation are presented as a foundation for more advanced analysis tools.

Finally the chapter introduced a general method for the design of multi-input integrators having both positive and negative gains. Such circuits can be used for implementing active filters and for implementing controllers in practical situations.

12. TERMS AND CONCEPTS

Admittance: the ratio of the Laplace transform of the input current to the Laplace transform of the input voltage with the two-terminal network initially relaxed.

Admittance of capacitor: the quantity Cs.

Admittance of inductor: the quantity $1/Ls$.

Admittance of resistor: the quantity $1/R$.

Current division: a formula for determining how currents distribute through a set of parallel admittances.

Impedance: the ratio of the Laplace transform of the input voltage to the Laplace transform of the input current with the two-terminal network initially relaxed.

Impedance of capacitor: the quantity $1/Cs$.

Impedance of inductor: the quantity Ls.

Impedance of resistor: the quantity R.

Impulse response: the response of a circuit having a single input excitation of a unit impulse; equal to the inverse Laplace transform of the transfer function.

Integrator circuit: usually an op amp circuit with transfer function K/s.

Source transformation property: in the s-domain, voltage sources in series with an impedance are equivalent to the same impedance in parallel with a current source whose value equals the transform voltage divided by the series impedance.

Steady-state analysis: analysis of circuit behavior resulting after excitations have been on for a long time; often refers to finding the sinusoidal or constant parts of the response when the circuit is excited by sinusoids or dc.

Switched capacitor circuit: a circuit containing switches, capacitors, independent voltage sources, and possibly op amps, but no resistors or inductors.

Time-invariant device: a device whose characteristics do not change with time.

Transfer function: the ratio of the Laplace transform of the output quantity to the Laplace transform of the input quantity with the network initially relaxed.

Transient analysis: analysis of circuit behavior for a period of time immediately after independent sources have been turned on.

Voltage division: a formula for determining how voltages distribute around a series connection of impedances.

Problems

IMPEDANCE, ADMITTANCE, VOLTAGE DIVISION, CURRENT DIVISION, SOURCE TRANSFORMATIONS, THEVENIN AND NORTON EQUIVALENTS

1. Consider the circuits in Figure P13.1.

(a)   Find $Z_{in}(s)$ and $Y_{in}(s)$ as rational functions for each of the given circuits.

(b)   Find the s-domain and time domain output responses to a step input for each circuit.

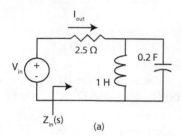

(a)

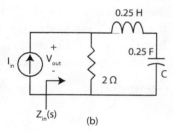

(b)

Figure P13.1

CHECKS: Figure P13.1a:

$$Y_{in}(s) = 0.4 \frac{s^2 + 5}{(s+1)^2 + 2^2}; \text{ Figure P13.1b:}$$

$$Y_{in}(s) = 0.5 \frac{(s+4)^2}{s^2 + 16}$$

2. Find the input impedance and admittance in terms of R, L, and C of each of the networks shown in Figure P13.2. Then find s-domain expressions for the output in terms of the input for each circuit.

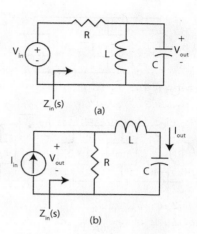

(a)

(b)

Figure P13.2

3. In the (relaxed) circuit of Figure P13.3, if $v_{in}(t) = \delta(t)$, then $I_{in}(s) = \dfrac{s + 20}{s + 40}$

(a)   Compute $Y_{in}(s)$ and $Z_{in}(s)$.

(b)   If $v_{in}(t) = 20u(t)$ V, find $i_{in}(t)$ for $t \geq 0$.

(c)   If $v_{in}(t) = 20e^{-40t}u(t)$ V, find $i_{in}(t)$ for $t \geq 0$.

(d)   If $v_{in}(t) = 20e^{-20t}u(t)$ V, find $i_{in}(t)$ for $t \geq 0$.

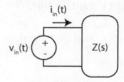

Figure P13.3

4. For each of the circuits in Figure P13.4, find expressions for

(a) $Z_{in}(s)$ and $Y_{in}(s)$

(b) $V_{out}(s)$ and $I_{out}(s)$ (as indicated)

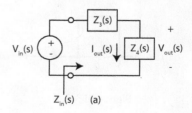

(a)

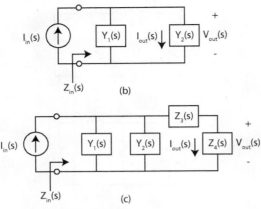

(b)

(c)

Figure P13.4

5. In the circuit of Figure P13.5, suppose $v_C(0^-) = 0$. Find $Y_{in}(s)$, $Z_{in}(s)$, $I_{out}(s)$, and $i_{out}(t)$ when (a) $i_{in}(t) = \delta(t)$ and (b) $i_{in}(t) = 50(1 - e^{-5t})u(t)$ mA.

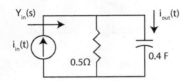

Figure P13.5

6. In the circuit of Figure P13.6a, $R = 40$ Ω and $C = 0.01$ F. Find $Z_{in}(s)$, $Y_{in}(s)$, $V_C(s)$, and $v_C(t)$ for (a) $v_{in}(t) = \delta(t)$ and (b) $v_{in}(t)$ as given in Figure P13.6b.

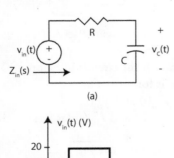

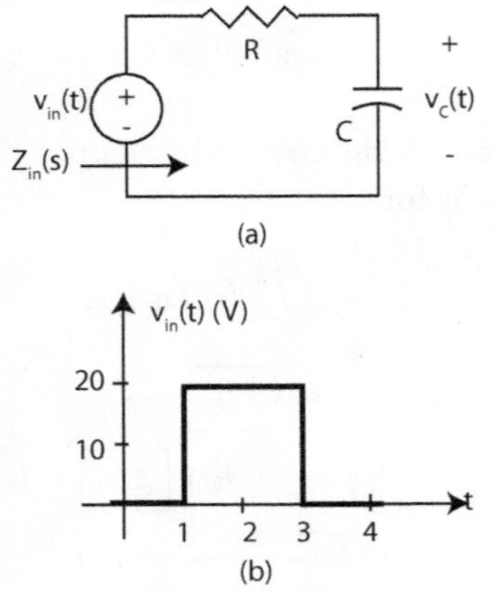

(a)

(b)

Figure P13.6

7. Repeat Problem 6 for $R = 6.25$ Ω, $C = 0.01$ F, and $v_{in}(t) = 32e^{-4t}\cos(4t)u(t)$ V.

8. For the circuit of Figure P13.8, suppose $R = 2$ Ω, $L = 4$ H, and $Z(s) = \dfrac{2s+16}{s+0.5}$. Find $Z_{in}(s)$, $Y_{in}(s)$, $I_{in}(s)$, $V_{out}(s)$, $i_{in}(t)$ and $v_{out}(t)$ when
   (a) $v_{in}(t) = 20e^{-4t}u(t)$ V, assuming zero initial conditions
   (b) $v_{in}(t) = 24e^{-0.5t}u(t)$ V, assuming zero initial conditions

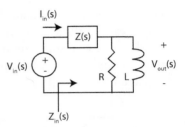

Figure P13.8

9. For the circuit of Figure P13.9, $R_1 = 100$ Ω, $r_{m1} = 100$ Ω, $r_{m2} = 200$ Ω, $R_2 = 100$ Ω, and $C = 1$ mF.
   (a) Compute $I_2(s)$ in terms of $I_1(s)$ and then compute $Z_{in}(s)$ and $Y_{in}(s)$.
   (b) If $v_{in}(t) = e^{10t}u(t)$ V, compute $i_1(t)$.
   (c) If $v_{in}(t) = e^{-10t}u(t)$ V, compute $i_1(t)$.

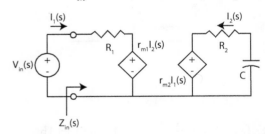

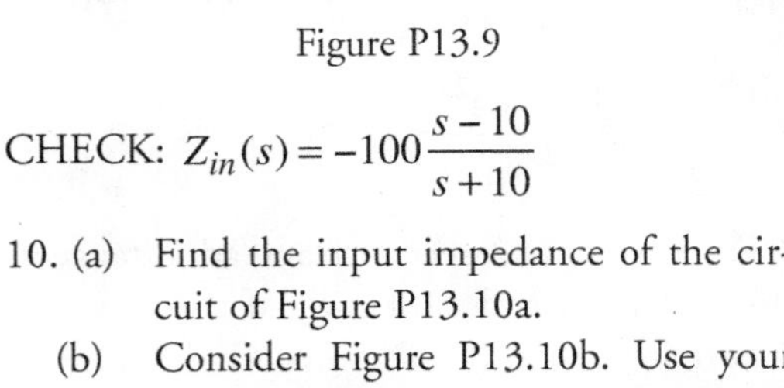

Figure P13.9

CHECK: $Z_{in}(s) = -100\dfrac{s-10}{s+10}$

10. (a) Find the input impedance of the circuit of Figure P13.10a.
    (b) Consider Figure P13.10b. Use your answer from part (a) to compute the responses $i_{in}(t)$ and $v_2(t)$ to a step

input when $R = 30\ \Omega$, $L_1 = L_2 = 1$ H, and $C = 10$ mF.

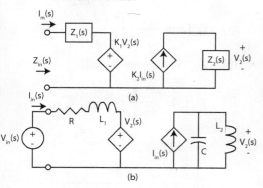

(a)

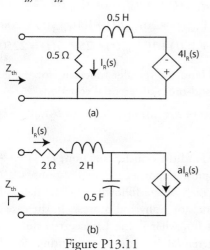

(b)

Figure P13.10

11. Find the Thevenin equivalent impedance of each circuit in Figure P13.11. Hint: Label the terminals $V_{in}(s)$ and an input current to the top terminal as $I_{in}(s)$; determine $V_{in}(s) = Z_{th}(s)I_{in}$ or $I_{in}(s) = Y_{th}(s)V_{in}$.

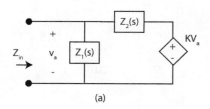

(a)

(b)

Figure P13.11

ANSWER: (b) $\left(2 + 2s + \dfrac{2 - 2a}{s}\right)$

12. Find $Y_{in}(s)$ and $Z_{in}(s)$ for each of the circuits shown in Figure P13.12.

(a)

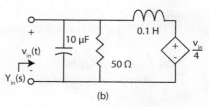

(b)

Figure P13.12

13. For the bridged-T network of Figure P13.13, show that if $Z_1(s)Z_2(s) = R^2$, then $Z_{in}(s) = R$.

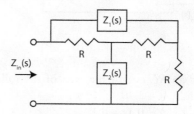

Figure P13.13

14. Consider the circuit of Figure P13.14, in which $R_1 = 10\ \Omega$, $R_2 = 10\ \Omega$, $L = 0.1$ H, and $C = 1$ mF.

(a) Find $Z_{in}(s)$ and $Y_{in}(s)$.

(b) If a voltage source is attached as indicated and $v_s(t) = e^{-100t}u(t)$ V, find $i_s(t)$ assuming the inductor and capacitor are initially relaxed.

(c) If a voltage source is attached as indicated and $v_s(t) = e^{-100t}u(t)$ V, find $v_{out}(t)$ assuming the inductor and capacitor are initially relaxed.

(d) Find the zero-input response if $i_L(0) = 0$ and $v_C(0) = 10$ V.

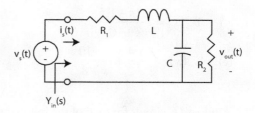

Figure P13.14

CHECK: $Y_{in} = 10\dfrac{s + 100}{s^2 + 200s + ????}$

15. Consider the circuit of Figure P13.15, in which $R_1 = 16\,\Omega$, $R_2 = 4\,\Omega$, $L = 1$ H, and $C = 2$ mF.
  (a) Find $Y_{in}(s)$ and $Z_{in}(s)$.
  (b) Find an expression for $I_{out}(s)$ in terms of $Z_{in}(s)$ and $I_{in}(s)$.

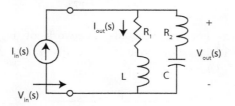

Figure P13.15

16. Find an RC circuit for each of the following:
  (a) $Y_{in}(s) = 0.1s + 0.1$

  (b) $Z_{in}(s) = \dfrac{1}{0.2s + 0.025}$

  (c) $Z_{in}(s) = 10 + \dfrac{10}{0.25s + 0.2}$

  (d) $Z_{in}(s) = \dfrac{2s + 8}{s + 2}$

  (e) $Z_{in}(s) = \dfrac{4s^2 + 36s + 64}{(s + 2)(s + 4)}$

17. Find an RL circuit for each of the following:
  (a) $Z_{in}(s) = 0.1s + 10$

  (b) $Y_{in}(s) = \dfrac{20}{s + 240}$

  (c) $Z_{in}(s) = 0.2s + 20 + \dfrac{s}{20 + 0.04s}$

  (d) $Y_{in}(s) = \dfrac{0.1s + 1.5}{s + 10}$

  (e) $Y_{in}(s) = \dfrac{s + 4}{s + 2} + \dfrac{s + 10}{s + 8}$

18. Find an LC circuit for each of the following:
  (a) $Z_{in}(s) = 0.05s + \dfrac{100}{s}$

  (b) $Y_{in}(s) = 0.02s + \dfrac{20}{s}$

(c) $Z_{in}(s) = \dfrac{0.125s^2 + 4}{0.5s}$

(d) $Y_{in}(s) = \dfrac{0.125s^2 + 4}{0.5s}$

(e) $Z_{in}(s) = \dfrac{50s}{s^2 + 250}$

(f) $Y_{in}(s) = \dfrac{5s}{s^2 + 250}$

(g) $Z_{in}(s) = \dfrac{s^2 + 4}{2s} + \dfrac{0.25s^2 + 4}{0.5s}$

(h) $Y_{in}(s) = \dfrac{s^2 + 16}{4s} + \dfrac{s}{0.25s^2 + 16}$

TRANSFER FUNCTIONS

19. Find the transfer function, $H(s) = \dfrac{V_{out}(s)}{V_{in}(s)}$, impulse response, and step response for each of the following differential equation models of a physical process.
  (a) $\dot{v}_{out}(t) + p_1 v_{out}(t) = K v_{in}(t)$
  (b) $\ddot{v}_{out}(t) + 20\dot{v}_{out}(t) + 100 v_{out}(t) = 200 v_{in}(t)$
  (c) $\ddot{v}_{out}(t) + 10\dot{v}_{out}(t) + 25 v_{out}(t) = 2250 v_{in}(t)$

20. Find the transfer function for the general second-order differential equation

$$\ddot{v}_{out}(t) + a_1 \dot{v}_{out}(t) + a_2 v_{out}(t) = b_2 v_{in}(t) + b_1 \dot{v}_{out}(t)$$

21. Find the transfer function, $H(s) = \dfrac{V_{out}(s)}{V_{in}(s)}$, impulse response, and step response for each of the following differential equation models of a physical process. In the case of (b) and (c), what is the Laplace transform of the differential equations accounting for initial conditions?
  (a) $\dot{v}_{out}(t) + 25 v_{out}(t) + 100\displaystyle\int_{0^-}^{t} v_{out}(\tau)d\tau = 5 v_{in}(t) - 10\displaystyle\int_{0^-}^{t} v_{in}(\tau)d\tau$

  (b) $0.1\dfrac{di_{out}(t)}{dt} + 40\displaystyle\int_{-\infty}^{t} i_{out}(\tau)d\tau = 10 v_{in}(t)$

  (c) $i_{out}(t) + 0.1\dfrac{di_{out}(t)}{dt} + 42.5\displaystyle\int_{-\infty}^{t} i_{out}(\tau)d\tau$
  $= 200 v_{in}(t) + 20\dot{v}_{in}(t)$

ANSWER: (b) $I_{out} = \dfrac{100s}{\left(s^2 + 400\right)}V_{in}$, $0.1sI_{out}$

$$-0.1i_{out}(0^-) + 40\frac{I_{out}}{s} + 40\frac{\displaystyle\int_{-\infty}^{0} i_{out}(\tau)d\tau}{s} = 10V_{in}$$

22. An integro-differential equation for an active circuit driven by a current source with $C = 0.5$ F, $L = 1$ H, $R = 2\ \Omega$ is

$$Ri_L + \frac{1}{C}\int_{-\infty}^{t} i_L(q)dq + L\frac{di_L}{dt} = v_{in} - \frac{dv_{in}}{dt}$$

(a) Find the transfer function $H(s) = \dfrac{I_L(s)}{V_{in}(s)}$.
(b) Find the impulse response.
(c) Find the step response.

23. Suppose two signals $v_{out}(t)$ and $y(t)$ are related by the equations

$$\frac{d^2 v_{out}(t)}{dt^2} + 4v_{out}(t) + 2y(t) = 4\frac{dv_{in}(t)}{dt}$$

and

$$2v_{out}(t) - \int_{0^-}^{t} y(\tau)d\tau = 2\int_{0^-}^{t} v_{in}(\tau)d\tau$$

(a) Assuming $v_{in}(0^-) = 0$, find the transfer function $H(s) = \dfrac{V_{out}(s)}{V_{in}(s)}$.
(b) Find the impulse and step responses.

24. A certain circuit has input $v_{in}(t) = \cos(t)u(t)$ and output $v_{out}(t) = te^{-t}u(t)$. Construct the transfer function of the circuit, assuming that the circuit had no internal stored energy at $t = 0^-$. Then compute the step response.

25. The input to a relaxed (no initial conditions) linear active circuit is $v_{in}(t) = e^{-t}u(t)$ V. The response is measured in volts as $v_{out}(t) = 10e^{-t}\sin(2t)u(t)$.

(a) Find the transfer function
$$H(s) = \frac{V_{out}(s)}{V_{in}(s)} = \frac{n(s)}{d(s)}$$
in which the denominator is in factored form.
(b) Compute the response of the circuit to the new input $v_{in}(t) = 2te^{-t}u(t)$ V.

26. (a) Compute the transfer functions of the op-amp circuit in Figures P13.26a and b.
(b) Given your answers to part (a), compute the step responses.
(c) Assuming $R = 1\ k\Omega$ and $C = 0.5$ mF, compute the zero-state response to the input $v_{in}(t) = 2u(t) - 4u(t - 2)$ V.
(d) For the circuit of Figure P13.26b, assuming $R = 1\ k\Omega$ and $C = 0.5$ mF, compute the output when the input is as given in Figure P13.26c.

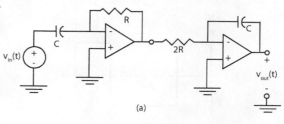

(a)

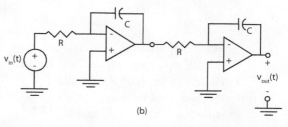

(b)

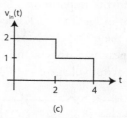

(c)

Figure P13.26

27.(a) Find the transfer function of the (ideal) op amp circuit of Figure P13.27a in terms of R_1, R_2, and C. If it is desired to obtain $H(s) = \dfrac{-2}{s+5}$ with $C = 1\ \mu$F, then find R_1 and R_2.
(b) Find the response to the input function in Figure P13.27b assuming zero for the initial capacitor voltage.

(c) Repeat part (b) assuming the initial capacitor voltage is $v_C(0^-) = -4$ V.

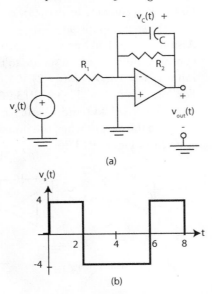

(a)

(b)

Figure P13.27

28. (a) Find the transfer function of the (ideal) op amp circuit of Figure P13.28 in terms of R_1, R_2, and C.

(b) If it is desired to obtain $H(s) = \dfrac{s+8}{s+2}$ with $C = 1$ μF, find R_1 and R_2.

(c) Find the response to the input function in Figure P13.27b assuming zero for the initial capacitor voltage.

(d) Repeat part (b) assuming the initial capacitor voltage is $v_C(0^-) = -4$ V.

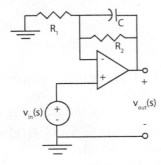

Figure P13.28

CHECK: $H(s) = \dfrac{s + \dfrac{G_1 + G_2}{C}}{s + \dfrac{G_2}{C}}$

29. (a) Find the transfer function of the (ideal) op amp circuit of Figure P13.29 in terms of R_1, R_2, C_1, and C_2. Make the leading coefficient of s in the denominator 1.

(b) If it is desired to obtain $H(s) = -4\dfrac{s+2}{s+4}$ with $C_2 = 100$ μF, find R_1, R_2, and C_1.

(c) Given $H(s) = -4\dfrac{s+2}{s+4}$ find the zero-state response to $v_{in}(t) = \sin(2t)u(t)$ V.

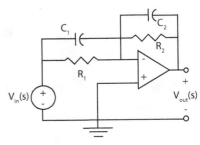

Figure P13.29

30. (a) Find the transfer function of the (ideal) op amp circuit of Figure P13.30 in terms of R_1, G_2, C_1, and C_2.

(b) If it is desired to obtain

$$H(s) = -4\dfrac{s}{(s+2)(s+10)}$$

with $C_2 = 100$ μF, find R_1, R_2, and C_1.

(c) Given the answer to part (b), find the zero-state response to $v_{in}(t) = -2.5e^{-2t}\sin(2t)u(t)$ V.

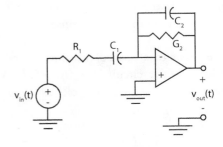

Figure P13.30

CHECK: $H(s) = \dfrac{-1}{C_2 R_1}\dfrac{s}{\left(s + \dfrac{1}{R_1 C_1}\right)\left(s + \dfrac{1}{R_2 C_2}\right)}$

31. (a) Compute the transfer functions of each circuit in Figure P13.31 in terms of C, R_1, R and K.
   (b) Suppose $C = 1\ \mu F$, $K = 3$, and $R = R_1 = 1\ k\Omega$. Find the zero-state response to $v_{in}(t) = 2\sin(500t)u(t)$ for each circuit in Figure P13.31.
   (c) Repeat part (b) for $v_{in}(t) = 2\sin(1500t)u(t)$.

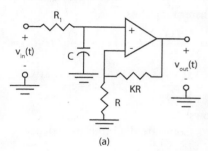

(a)

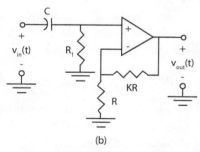

(b)

Figure P13.31

32. (a) Find the transfer function of the circuit of Figure P13.32 in terms of R_1, R_2, R_3, R_4, C_1, and C_2.
   (b) If $C_1 = C_2 = 1\ \mu F$, find values for R_1, R_2, R_3, and R_4 so that
$$H(s) = \frac{-5}{(s+100)} \times \frac{-2}{(s+200)}.$$
   (c) Find the impulse and step response.

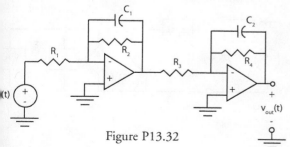

Figure P13.32

33. Construct the transfer function of the circuit of Figure P13.33, assuming that all op amps are ideal, as follows:

   (a) Let $G_1 = \dfrac{1}{R_1}$, $G_2 = \dfrac{1}{R_2}$, $G_3 = \dfrac{1}{R_3}$, and $R_4 = R_5 = R_6 = 1\ \Omega$.

   (b) Compute the ratio $\dfrac{V_{out}(s)}{V_1(s)}$.

   (c) Compute the ratio $\dfrac{V_1(s)}{V_2(s)}$.

   (d) Compute $V_2(s)$ in terms of $V_3(s)$ and $V_{out}(s)$.

   (e) Compute $V_3(s)$ in terms of $V_1(s)$ and $V_{in}(s)$.

   (f) Back-substitute to eliminate intermediate variables and find $H(s) = \dfrac{V_{out}(s)}{V_{in}(s)}$

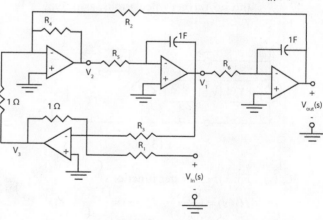

Figure P13.33

ANSWER: $H(s) = \dfrac{G_1}{s^2 + G_3 s + G_2}$

34. (a) For the circuit in Figure P13.34, find the transfer functions
$$H_1(s) = \frac{I_L(s)}{V_{in}(s)} \text{ and } H_2(s) = \frac{V_L(s)}{V_{in}(s)}.$$
   (b) If $Z_s = 10\ \Omega$, $Y_L = 0.1\ S$, $Z_1 = s$, $Y_2 = s$, and $g_m = 20\ S$, compute the impulse and step responses.
   (c) Given the values of part (b), compute the response to the input $v_{in}(t) = 2u(t) - 2u(t - 4)$ V using linearity and no further calculations.

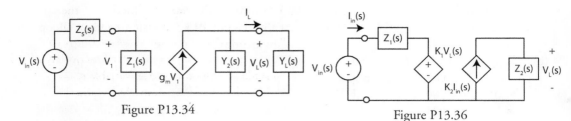

Figure P13.34   Figure P13.36

35. (a) For the circuit in Figure P13.35, find the transfer functions

$$H_1(s) = \frac{I_L(s)}{I_{in}(s)} \text{ and } H_2(s) = \frac{V_L(s)}{I_{in}(s)}.$$

(b) If $Z_2 = \frac{1}{s+1}, Z_L = s+3, Y_1 = s, Y_s = 3,$ and $r_m = 20$ Ω, compute the impulse and step responses.

(c) Given the values of part (b), compute the response to the input $v_{in}(t) = K_1 u(t) - K_2 u(t-T)$ V using linearity and no further calculations, assuming $T > 0$.

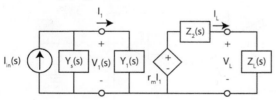

Figure P13.35

36. (a) Find the transfer functions

$$H_1(s) = \frac{V_L(s)}{I_{in}(s)}, H_2(s) = \frac{I_{in}(s)}{V_{in}(s)}, \text{ and}$$

$$H_3(s) = \frac{V_L(s)}{V_{in}(s)} \text{ for the circuit of}$$

Figure P13.36.

(b) Find the impulse and step responses associated with each transfer function if $K_1 = 8, K_2 = 2,$

$$Z_1 = \frac{s}{s+1} \text{ and } Z_2 = \frac{1}{s+5}.$$

37. Consider the circuit of Figure P13.37. Find the transfer functions

$$H_1(s) = \frac{V_1(s)}{V_{in}(s)}, H_2(s) = \frac{V_L(s)}{V_1(s)}, H_3(s) = \frac{I_L(s)}{V_1(s)}$$

$$H_4(s) = \frac{V_L(s)}{V_{in}(s)}, \text{ and } H_5(s) = \frac{I_L(s)}{V_{in}(s)}$$

as follows:

(a) Compute the input impedance

$$Z_{in}(s) = \frac{V_1(s)}{I_{in}(s)}.$$

(b) Given the results from part (a), use voltage division to compute

$$H_1(s) = \frac{V_1(s)}{V_{in}(s)}$$

in terms of $Z_{in}(s)$. Do not substitute your answer to part (a) into the obtained expression.

(c) Compute

$$H_2(s) = \frac{V_L(s)}{V_1(s)}$$

in terms of $Z_{in}(s), Y_2(s), Z_L(s),$ etc.

(d) Compute $$H_3(s) = \frac{I_L(s)}{V_1(s)}.$$

(e) Compute

$$H_4(s) = \frac{V_L(s)}{V_{in}(s)} \text{ and } H_5(s) = \frac{I_L(s)}{V_{in}(s)}$$

and as products of your prior answers.

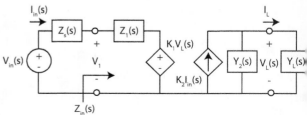

Figure P13.37

CHECK: $Z_{in} = Z_1 + \dfrac{K_1 K_2}{Y_2 + Y_L}$

38. Consider the circuit of Figure P13.38.

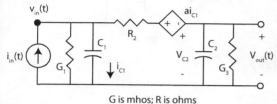

G is mhos; R is ohms

Figure P13.38

(a) Draw the equivalent frequency domain circuit assuming $v_{C1}(0^-) \neq 0$ V and $v_{C2}(0^-) \neq 0$ V.

(b) Define a current $I_D(s)$ from left to right through R_2. Then write three (modified) nodal equations for the circuit of part (a) similar to the text example, but accounting for the initial conditions. Put equations in matrix form (unknowns on the left side and knowns, input, and initial conditions on the right). You should have a 3 by 3 matrix of literals. The unknowns are two (node) voltages and one current, $I_D(s)$.

(c) Determine the transfer function $H(s) = V_{out}(s)/I_{in}(s)$ of the circuit in MATLAB using the following code:

```
syms M MM Iin Vout s G1 C1 C2 G3 G2 R2
a t, H
M = [C1*s+G1 0 1; fill in the rest of your coeffi-
cient matrix]
% Transfer function computation assumes all
% ICs are zero
% Using Cramer's rule, define MM by
% replacing the middle column of M
% by the right-hand side of your matrix nodal
% equation with all IC's set to zero.
MM = [C1*s+G1 Iin 1; fill in the rest of your
matrix]
dt = det(M)
dt = collect(dt)
```

```
Vout = det(MM)/det(M)
% You have now computed the symbolic
% expression for Vout in terms of Iin so you
% can now identify the transfer function as a
% symbolic function of the variable s.
% Now let's do some numerical work.
C1= 1; C2 = 2; G1 = 1; R2 = 1; G3 = 2; a = 0.5;
% Re-enter M above
M = [C1*s+G1 0 1; fill in the rest of your coeffi-
cient matrix]
% check: the roots of the determinant should
% be −1 and −5.
dt = det(M)
factor(dt)
% Re-enter MM above
% Compute the actual transfer function
Vout = det(MM)/det(M)
% Identify the transfer function
```

(d) Find the impulse response h(t) of the circuit as follows:

```
% Rewrite the above expression for Vout with
% out Iin.
syms s t H h
H = -(1/2*s-1)/???????
h = ilaplace(H)
```

(e) Now use the MATLAB command "ilaplace" to compute the step response, again with zero initial conditions.

Remark: Use your newfound MATLAB knowledge of symbolic computation to carry out the following tasks. Hand computation is *not* acceptable.

(f) Find the response due only to the initial condition on C_1 assuming $v_{C1}(0^-) = 16$ V.

(g) Find the response due only to the initial condition on C_2 assuming $v_{C2}(0^-) = -8$ V.

(h) Write down the complete response assuming the input is a step.

39. In the so-called feedback configuration of Figure P13.39, $E(s)$ is the Laplace transform of the error between the reference signal $x_{ref}(t)$ and the response $y(t)$, i.e., $e(t) = x_{ref}(t) - y(t)$. $G_1(s) = n(s)/d(s)$ is called the plant of the system and represents a physical process or special circuit.

(a)   Find the transfer function $H(s) = \dfrac{E(s)}{X_{ref}(s)}$

(b)   Under what conditions does $e(t) \to 0$ as $t \to \infty$ when $x_{ref}(t) = K_0 u(t)$?

(c)   Under what conditions does $e(t) \to K_e \neq 0$ as $t \to \infty$ when $x_{ref}(t) = K_0 u(t)$?

(d)   Under what conditions and how could you determine $e(0^+)$ from $E(s) = H(s)X_{ref}(s)$?

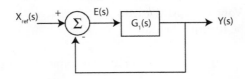

Figure P13.39

CHECK: $E(s) = \dfrac{X_{ref}(s)}{1 + G_1(s)} = \dfrac{d(s)X_{ref}(s)}{d(s) + n(s)}$

40. In the so-called feedback configuration of Figure P13.40, $E(s)$ is the Laplace transform of the error between the reference signal $x_{ref}(t)$ and the response $y(t)$, i.e., $e(t) = x_{ref}(t) - y(t)$. $G_1(s) = n_g(s)/d_g(s)$ is the plant of the system (described in the previous problem) and $F(s) = n_f(s)/d_f(s)$ is a feedback controller to be designed.

(a)   Find the transfer function $H(s) = E(s)/X_{ref}(s)$.

(b)   Under what conditions does $e(t) \to 0$ as $t \to \infty$ when $x_{ref}(t) = K_0 u(t)$?

(c)   Under what conditions does $e(t) \to K_e \neq 0$ as $t \to \infty$ when $x_{ref}(t) = K_0 u(t)$?

(d)   Under what conditions and how could you determine $e(0^+)$ from $E(s) = H(s)X_{ref}(s)$?

(e)   Suppose $d_g(s)$ had a pair of poles on the imaginary axis. How could you design an $F(s)$ to cancel these poles?

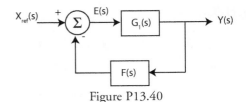

Figure P13.40

CHECK: $E(s) = \dfrac{X_{ref}(s)}{1 + F(s)G_1(s)} = \dfrac{d(s)X_{ref}(s)}{d(s) + n(s)}$

where $F(s)G_1(s) = \dfrac{n_f(s)n_g(s)}{d_f(s)d_g(s)} = \dfrac{n(s)}{d(s)}$

RESPONSE CALCULATION WITH INITIAL CONDITIONS

41. For the circuit of Figure P13.41, $L = 0.5$ H, $C = 0.1$ F, $i_L(0^-) = 2$ A, and $v_C(0^-) = 2$ V. Find $V_C(s)$ and $v_C(t)$ using the equivalent models for initialized L and C in the s-domain.

Figure P13.41

42. For the circuit of Figure P13.42, $i_L(0^-) = 2$ A. Draw an equivalent circuit for the inductor that accounts for the initial condition, write a single node equation in $V_{out}(s)$, and then find $v_{out}(t)$.

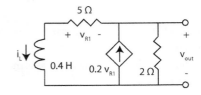

Figure P13.42

ANSWER: $-8e^{-22.5t}u(t)$ V

43. For the circuit of Figure P13.43, assume that $v_{C1}(0^-) = -400$ mV and $v_{C2}(0^-) = 0$ V. Find $v_{out}(t)$, $t \geq 0$.

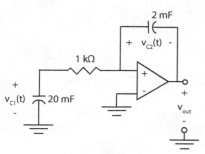

Figure P13.43

44. Consider the circuit of Figure P13.44, in which $C = 20$ mF and $R = 100$ Ω.
   (a) If $v_C(0^-) = 10$ V and $i_{in}(t) = 200e^{-0.5t}u(t)$ mA, find $V_C(s)$ and $v_C(t)$ for $t \geq 0$.
   (b) If $v_C(0^-) = 10$ V and $i_{in}(t) = 200\cos(2t)u(t)$ mA, find $V_C(s)$ and $v_C(t)$ for $t \geq 0$.

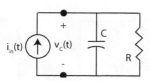

Figure P13.44

45. Consider the circuit of Figure P13.45, in which $R = 200$ Ω.
   (a) Compute the transfer function in terms of L.
   (b) If the response to the input $i_{in}(t) = I_0 tu(t)$ A is $i_L(t) = (2.5t - 0.025 + 0.025e^{-100t})u(t)$ A, assuming $i_L(0^-) = 0$, find the values of L and I_0.
   (c) Assuming $i_L(0^-) = 1$ A and $i_{in}(t) = 2.5\sin(100t)u(t)$ A, find $i_L(t)$.

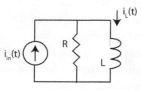

Figure P13.45

CHECK: $I_0 = 2.5$

46. In the circuit of Figure P13.46, $R = 20$ Ω, $C = 0.2$ F, and the capacitor is initially charged at $v_C(0^-) = 10$ V.
   (a) If $v_{in}(t) = 20[u(t) - u(t-20)]$ V, find $I_C(s)$, $V_C(s)$, $i_C(t)$, and $v_C(t)$ for $t > 0$. Plot $v_C(t)$ and $v_{in}(t)$ on the same graph for $0 \leq t \leq 40$ sec using MATLAB or its equivalent.
   (b) If $v_{in}(t) = 20e^{-0.25t}u(t)$ V, find $i_C(t)$ and $v_C(t)$ for $t > 0$. Plot $v_C(t)$ and $v_{in}(t)$ on the same graph for $0 \leq t \leq 20$ sec using MATLAB or its equivalent.

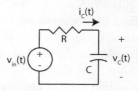

Figure P13.46

47. Consider the circuit of Figure P13.47, in which $R_s = 2$ kΩ, $R_L = 8$ kΩ, and $C = 50$ μF.
   (a) Find the transfer function $H(s)$.
   (b) Compute the step and impulse responses.
   (c) Compute the response to $v_{in}(t) = 10e^{-12.5t}\cos(25t)u(t)$ V.
   (d) Find the response to $v_C(0^-) = 20$ V.
   (e) Using the principle of superposition, find the response to the input of part (c) with the initial condition of part (d).
   (f) From the principle of linearity, what would the response be to the input $v_{in}(t) = 20e^{-12.5t}\cos(12.5t)u(t)$ V and $v_C(0^-) = 10$ V?

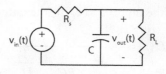

Figure P13.47

CHECK: $H(s) = \dfrac{20}{2s + 25}$

48. Consider the circuit of Figure P13.48, in which $C = 2.5$ mF, $R_1 = 200$ Ω, $R_s = 50$ Ω, $R_L = 5$ Ω, $L = 0.2$ H, and $g_m = 0.5$ S.

(a) Find the transfer functions
$$H_1(s) = \frac{V_C(s)}{V_{in}(s)}, \quad H_2(s) = \frac{I_R(s)}{V_C(s)}, \text{ and}$$
$$H(s) = \frac{I_R(s)}{V_{in}(s)}.$$

(b) Given $H(s)$, find the zero-state response; i.e., assume $i_L(0^-) = 0$ and $v_C(0) = 0$ when $v_{in}(t) = 25(1 - e^{-10t})u(t)$ V.

(c) Find the zero-input response $i_R(t)$ if $i_L(0) = 6$ A and $v_C(0) = -10$ V.

(d) Find the complete response (combine parts (b) and (c)) and identify the transient and steady-state parts of it. (Note that the transient part is the part that is *not* constant or *not* periodic. Usually the transient part converges to zero.)

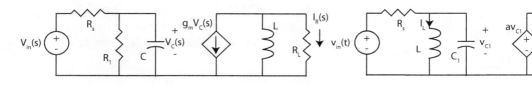

Figure P13.48

49. Consider the circuit of Figure P13.49. Note that the computation of transfer functions presumes no initial internal stored energy.

(a) Use a source transformation and the current divider formula to show that the transfer function between $V_{in}(s)$ (the input) and $I_C(s)$ (the output) is
$$H(s) = \frac{1}{R} \frac{s^2}{s^2 + \frac{1}{RC}s + \frac{1}{LC}}.$$

(b) If $L = 0.2$ H, $C = 25$ mF, $R = 4/3$ Ω, and $v_{in}(t) = 10e^{-10t}u(t)$ V, find $I_C(s)$ and $i_C(t)$.

(c) Find the response $i_C(t)$ when $v_{in}(t) = 0$, $v_C(0) = 10$ V, and $i_L(0) = 0$.

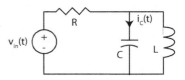

Figure P13.49

50. In the circuit of Figure P13.50, suppose $L = 0.04$ H,
$$C = \frac{1}{9} \text{ F},$$
$R_s = 0.5$ Ω, $R_2 = 5$ Ω, $C_2 = 0.05$ F, and $a = 4$. Compute $V_{C1}(s)$, $V_{out}(s)$, $v_{out}(t)$ assuming that $v_{in}(t) = 36u(t)$ V, $i_L(0^-) = 0$, $v_{C1}(0^-) = -18$ V, and $v_{out}(0^-) = 0$ V. (Hint: You must construct the equivalent circuit in the s-domain, accounting for initial conditions. Consider a source transformation on $v_{in}(t)$, and then draw the equivalent circuit in the s-domain so that you can combine sources in the front half of the circuit.)

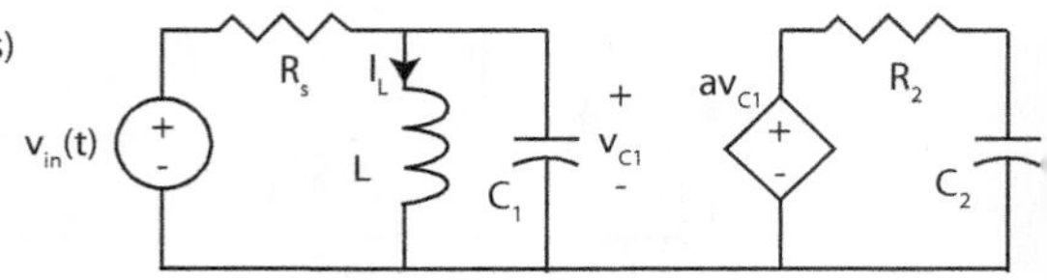

Figure P13.50

51. Consider the circuit of Figure P13.51. Suppose $R_1 = 2$ Ω, $R_2 = 6$ Ω, $R_3 = 3$ Ω, $C = 0.125$ F $L = 1.6$ H.

Suppose $v_{in}(t) = v_1(t) + v_2(t) = -10u(-t) + 10u(t)$ V (plot this input function so that you know what it looks like).

(a) Compute $v_C(0^-)$ and $i_L(0^-)$.

(b) Compute the "zero-input response" for $t \geq 0$; i.e., assume you have the initial conditions computed in step (a) and that the voltage $v_{in}(t) = 0$ for $t \geq 0$. First draw the equivalent s-domain circuit accounting for the initial conditions.

(c) Compute the transfer function $H(s)$ of the circuit valid for $t \geq 0$.

(d) Compute the "zero-state response," i.e., the response to the circuit of the input $v_{in}(t) = v_2(t) = 10u(t)$ *assuming* all initial conditions are zero.

(e) Compute the complete response.

(f) Identify the transient and steady-state parts of the complete response.

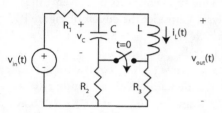

Figure P13.51

52. Consider the circuit of Figure P13.52. Suppose $L = 1$ H, $R_1 = 15$ Ω, $C = 0.01$ F, and $R_2 = 5$ Ω. Suppose $v_C(t)$ is the desired output.

(a) Find the input admittance

$$Y_{in}(s) = \frac{1}{Ls + R_1} + \frac{??}{??}$$

in terms of literals and then substitute numbers. Combine terms to form a rational function.

(b) Find the input impedance

$$Z_{in}(s) = \frac{??}{s^2 + \frac{R_{eq}}{L}s + \frac{1}{LC}}$$

in which case $V_s(s)$ = ????(write down the formula).

(c) Use voltage division to express $V_C(s)$ in terms of $V_s(s)$ and then in terms of $Z_{in}(s)$ and $I_s(s)$. Hence determine the transfer function $H_1(s) = V_C(s)/V_s(s)$. The leading coefficient of the denominator polynomial is to be 1.

(d) Compute the impulse and step responses.

(e) Suppose the input is $i_s(t) = 2e^{-5t}u(t)$ A. Find the partial fraction expansion of $V_C(s)$ and then compute $v_C(t)$ assuming zero initial conditions on L and C.

Plot the resulting time function in MATLAB for $0 \leq t \leq 1$ sec. The partial fraction expansion is most easily computed using MATLAB's "residue" command. (*Note:* Practicing hand calculation is important for the exams.) Now forget about the partial fraction expansion and instead use MATLAB's command "ilaplace" to compute the time function $v_C(t)$. You should define H, Vc, s, t, Iin, and vc as symbols using "syms." The program should be something like the following:

R = ?; L = 1; C = ?;
syms Yin Zin H Vc s t Is vc
Iin = ?
% H will be the transfer function defined in terms of s, a symbol, and R, L, C, and K.
% MATLAB will fill in the numbers.
H = ????
Vc = H*Iin
vc = ilaplace(Vc)

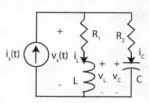

Figure P13.52

53. Repeat Problem 52 using $i_L(t)$ as the desired output.

54. Consider the circuit in Figure P13.54, in which $R_1 = R_2 = R$. Suppose $v_C(t)$ is the desired output.

(a) Find the input impedance in the form

$$Z_{in}(s) = \frac{A}{s - p_1} + \frac{Bs}{s - p_2}$$

in terms of R, L, and C. Identify the poles p_1 and p_2. They should have negative values. Recall that impedances in series add and two impedances in parallel satisfy the product over sum rule.

(b) Compute the transfer function

$$H(s) = \frac{V_C(s)}{V_{in}(s)} = \frac{??}{s^2 + \dfrac{2}{RC}s + \dfrac{1}{LC}}$$

in terms of R, L, and C.

(c) The roots of the characteristic equation (denominator of $H(s)$) are to be at $-2 \pm j4$. If $L = 0.2$ H, determine R and C. Then specify the transfer function with the proper numerical coefficients.

(d) Given the values of part (c),
   (i) Compute the impulse response of the circuit.
   (ii) Compute the step response of the circuit.

(e) Suppose the input is $v_s(t) = 10te^{-10t}u(t)$ A. Find the partial fraction expansion of $V_C(s)$ using MATLAB's "residue" command. Now forget about the partial fraction expansion and instead use MATLAB's command "ilaplace" to compute the time function $v_C(t)$. Note you should define H, Vc, s, t, Vin, and vc as symbols using "syms". For example,

```
R = ?; L = 0.2; C = ?;
K = 1/(R*C)
syms H Vc s t Vin vc
Vin = 10/(s + 10)^2
% H will be the transfer function defined in
terms of s, a symbol, and R, L, C, and K.
% MATLAB will fill in the numbers.
H = ????
Vc = H*Vin
vc = ilaplace(Vc)
```

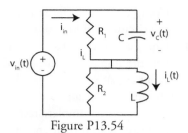

Figure P13.54

55. This problem is to be solved using nodal analysis. Consider the circuit of Figure P13.55, in which $G_1 = G_2 = 1$ S, $G_3 = 4$ S, $L = 1$ H, $C = 2$ F, and $g_m = 2$ S.

(a) Draw the equivalent frequency domain circuit assuming $i_L(0^-) = 2$ A and $v_{out}(0^-) = v_C(0^-) = 2$ V. Since you are using nodal analysis, what type of equivalent circuit might be best to use to account for initial conditions?

(b) Write two nodal equations for the circuit of part (a) in terms of the node voltages $V_{in}(s)$ and $V_{out}(s)$ and, of course, the input $I_{in}(s)$ and initial conditions. Simplify. Put equations in matrix form.

(c) Determine the transfer function of the circuit. Use Cramer's rule.

(d) Find the step response of the circuit.

(e) Find the response due only to the initial condition on the inductor.

(f) Find the response due only to the initial condition on the capacitor.

(g) Find the complete response if the input is a step.

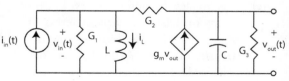

Figure P13.55

56.(a) For the circuit of Figure P13.56, calculate the transfer function.

(b) Suppose $i_{in}(t) = \delta(t)$. Find and plot $i_{out}(t)$.

(c) Now suppose $i_{in}(t) = 16[u(t) - u(t - T)]$ mA, where $T = 10$ msec. Find and plot $i_{out}(t)$.

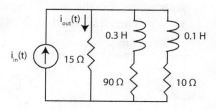

Figure P13.56

CHECK: (a) $(s + 100)(s + 300)/[(s + 200)(s + 400)]$

57. Consider the bridged-T network in Figure P13.57. Let $C = 0.25$ F. Assuming no initial conditions, find $I_1(s)$, $I_2(s)$, and $V_C(s)$ when $v_{in}(t) = 10(1 - e^{-2t})u(t)$ V. You might use Cramer's rule and the symbolic toolbox in MATLAB to solve the problem. Now find $v_C(t)$.

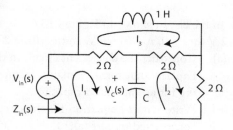

Figure P13.57

58. Consider the circuit of Figure P13.58, in which $R_s = 2$ kΩ, $C = 20$ μF, $R_1 = 20$ Ω, $R_L = 80$ Ω, and $L = 5$ H.

(a) Compute the transfer function
$$H_1(s) = \frac{V_C(s)}{V_{in}(s)}.$$

(b) Compute the transfer function
$$H_2(s) = \frac{V_{out}(s)}{V_C(s)}.$$

(c) Compute the transfer function
$$H(s) = \frac{V_{out}(s)}{V_{in}(s)}.$$

(d) If $v_{in}(t) = 2[u(t) - u(t - 0.5)]$ V, $v_C(0^-) = 5$ V, and $i_L(0^-) = 0$, find $v_{out}(t)$.

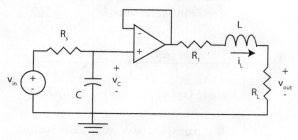

Figure P13.58

59. In the circuit of Figure P13.59, $R_1 = 2$ Ω, $C_1 = 1$ F, $R_2 = 1.75$ Ω, and $C_2 = 2/7$ F.

(a) Show that the transfer function of the circuit in literal form is $H(s) = \dfrac{V_{C2}}{V_{in}} = $

$$\dfrac{\dfrac{1}{R_1 C_2} s}{s^2 + \left(\dfrac{1}{R_2 C_2} + \dfrac{1}{R_1 C_1} + \dfrac{1}{R_1 C_2} \right) s + \dfrac{1}{R_1 C_1 R_2 C_2}}$$

(b) If $v_{in}(t) = 15u(t)$ V and the capacitors are initially relaxed, find $v_{C2}(t)$ for $t \geq 0$. Plot the input and the response on the same graph using MATLAB or the equivalent.

(c) If $v_{in}(t) = 0$, $v_{C2}(0^-) = 0$, and $v_{C1}(0^-) = 15$ V, find $v_{C2}(t)$ for $t \geq 0$.

(d) If $v_{in}(t) = 0$, $v_{C1}(0^-) = 0$, and $v_{C2}(0^-) = 15$ V, find $v_{C2}(t)$ for $t \geq 0$.

(e) If $v_{in}(t) = 15u(t)$ V, $v_{C1}(0^-) = 15$ V, and $v_{C2}(0^-) = 15$ V, find $v_{C2}(t)$ for $t \geq 0$.

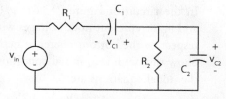

Figure P13.59

ANSWERS: (b) $[7e^{-0.25t} - 7e^{-4t}]u(t)$ V; (c) $[7e^{-0.25t} - 7e^{-4t}]u(t)$ V; (d) $[e^{-0.25t} + 14e^{-4t}]u(t)$ V; (e) sum of parts (b), (c), and (d)

60. In the circuit of Figure P13.60, $R_1 = 0.5\ \Omega$, $L_1 = 1$ H, $R_2 = 1.75\ \Omega$, and $L_2 = 7/8$ H.

   (a) Show that the transfer function of the circuit in literal form is $H(s) = \dfrac{I_{L2}}{V_{in}} =$

$$\dfrac{\dfrac{1}{L_1 G_2 L_2}}{s^2 + \left(\dfrac{1}{G_1 L_1} + \dfrac{1}{G_2 L_2} + \dfrac{1}{G_2 L_1}\right)s + \dfrac{1}{G_1 L_1 G_2 L_2}}$$

   (b) If $v_{in}(t) = 15u(t)$ V and the inductors are initially relaxed, find $i_{L2}(t)$ for $t \geq 0$. Plot the input and the response on the same graph using MATLAB or the equivalent.

   (c) If $v_{in}(t) = 0$, $i_{L2}(0^-) = 0$, and $i_{L1}(0^-) = 15$ A, find $i_{L2}(t)$ for $t \geq 0$.

   (d) If $v_{in}(t) = 0$, $i_{L1}(0^-) = 0$, and $i_{L2}(0^-) = 15$ A, find $i_{L2}(t)$ for $t \geq 0$.

   (e) If $v_{in}(t) = 15u(t)$ V, $i_{L1}(0^-) = 15$ A, and $i_{L2}(0^-) = 15$A, find $i_{L2}(t)$ for $t \geq 0$.

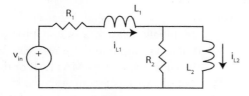

Figure P13.60

ANSWERS: (b) $(30 - 32e^{-0.25t} + 2e^{-4t})u(t)$ A; (c) $[8e^{-0.25t} - 8e^{-4t}]u(t)$ A; (d) $[8e^{-0.25t} + 7e^{-4t}]u(t)$ A; (e) sum of parts(b), (c), and (d).

61.(a) In the circuit of Figure P13.61, $v_C(t)$ and $v_R(t)$ represent node voltages with respect to the bottom/reference node. Show that with zero initial conditions the nodal equations are

$$\frac{1}{s}\begin{bmatrix} 0.8s^2 + 2s + 10 & -10 \\ -10 & s + 10 \end{bmatrix}\begin{bmatrix} V_C \\ V_R \end{bmatrix} = \begin{bmatrix} 2V_{s1} \\ -I_{s2} \end{bmatrix}$$

   (b) Find the response $v_R(t)$ with zero initial conditions and with $v_{s1}(t) = 3u(t)$ V and $i_{s2}(t) = 3u(t)$ A.

   (c) Now suppose $v_C(0^-) = 3$ V, $i_L(0^-) = 0$ A, $v_{s1}(t) = 3u(t)$ V, and $i_{s2}(t) = 3u(t)$ A. Recompute the nodal equations accounting for the initial capacitor voltage. Then find $v_R(t)$.

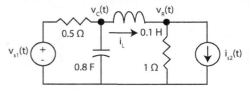

Figure P13.61

62. Reconsider the circuit of Figure P13.61.

   (a) Apply a source transformation to the independent current source and draw an equivalent circuit accounting for the initial conditions $v_C(0^-) = 0$, $i_L(0^-) = 3$ A. Compute the associated mesh equations with $I_L(s)$ as one of the mesh currents.

   (b) If $v_C(0^-) = 0$, $i_L(0^-) = 3$ A, $v_{s1}(t) = 3u(t)$ V, and $i_{s2}(t) = 3u(t)$ A, find $i_L(t)$.

63. In the circuit shown in Figure P13.63, $v_{in}(t) = 12$ V for $t \geq 0$, $v_{C1}(0) = 6$ V, and $v_{C1}(0) = 2$ V.

   (a) Construct the equivalent circuit in the Laplace transform domain, accounting for an initialized capacitor.

   (b) Write a nodal equation for the circuit constructed in part (a).

   (c) Find $V_{C2}(s)$.

   (d) Find $v_{C2}(t)$ for $t \geq 0$.

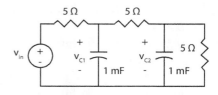

Figure P13.63

CHECK: $v_{C1}(t)$ should contain terms of the form $K_1 e^{-200t}$ and $K_2 u(t)$.

64. Consider the third-order RC circuit of Figure P13.64. Suppose that the initial capaci-

tance voltages are $v_{C1}(0) = 0$, $v_{C2}(0) = 6$ V, $v_{C3}(0) = 2$ V.

(a) Construct the Laplace transform domain equivalent circuit.

(b) Find $V_{C3}(s)$ in terms of $V_{in}(s)$ and the initial conditions.

(c) For the given initial conditions, if $v_{in}(t) = 12u(t)$ V, find $v_{C3}(t)$ for $t \geq 0$.

Hint: After obtaining $V_{C3}(s)$, use the "residue" command in MATLAB to obtain the partial fraction expansion. You might also investigate the use of the command "ilaplace."

equations for the circuit of part (a) only in terms of the variables $V_C(s)$, $V_2(s)$, $I_d(s)$, $V_{s1}(s)$, $I_{s2}(s)$, and the initial conditions. *Simplify each equation.*

(c) Put equations in *matrix* form.

(d) Assuming $i_L(0^-) = 0$, $v_{C1}(0^-) = 5$ V, $v_{s2}(t) = 5\delta(t)$ A, and $v_{s1}(t) = 10\delta(t)$ V, use Cramer's rule to find the current $I_d(s)$ and then $i_d(t)$.

(e) Now suppose that $i_L(0^-) = 0$, $v_{C1}(0^-) = 0$V, $i_{s1}(t) = 0$A, and $v_{s1}(t) = 10u(t)$ V. Find $v_2(t)$.

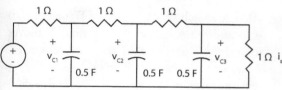

Figure P13.64

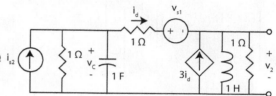

Figure P13.66

65. In the circuit of Figure P13.65, $g_m = 0.25$ S, $C_2 = 0.25$ F, and $R = 1$ Ω. Write a set of modified loop equations and solve for the three loop currents and $\hat{V}$, assuming $v_C(0^-) = 4$ V and $v_{in}(t) = 12u(t)$ V.

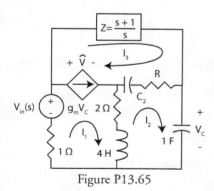

Figure P13.65

66. Consider the circuit of Figure P13.66. This problem is to be solved using (modified) nodal analysis.

(a) Draw the best equivalent frequency domain circuit for nodal analysis accounting for the as-yet-unknown initial conditions $i_L(0^-)$ and $v_{C1}(0^-)$.

(b) Following the procedure explained in the text, write three (modified) nodal

67. Consider the circuit of Figure P13.67. This problem is to be solved using (modified) nodal analysis.

(a) Draw the equivalent frequency domain circuit assuming $i_L(0^-) = 0$ and $v_{C1}(0^-) = 2$ V.

(b) Write three nodal equations for the circuit of part (a) only in terms of the voltages $V_{in}(s)$, $V_{in}(s)$, $I_L(s)$, and the initial conditions. Simplify your equations.

(c) Put equations in atrix form.

(d) Using Cramer's rule, find the transfer function
$$H(s) = \frac{V_{out}(s)}{I_{in}(s)}$$
of the circuit.

(e) Find the impulse response $h(t)$ of the circuit.

(f) Find the response of the circuit to $i_{in}(t) = -8u(t)$ A, assuming the initial conditions are zero.

(g) Find the response due only to the initial condition on the capacitor. A simple observation leads to the answer directly.

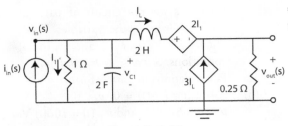

Figure P13.67

ANSWERS: $H(s) = \dfrac{-1}{4s(s+1)}$

and $h(t) =$

$0.25\,(e^{-t} - 1)u(t)$ derived from

$$\begin{bmatrix} (2s+1) & 0 & 1 \\ 0 & 4 & -4 \\ 1 & 1 & 2s \end{bmatrix} \begin{bmatrix} V_{in} \\ V_{out} \\ I_L \end{bmatrix} = \begin{bmatrix} I_{in} + Cv_{C1}(0^-) \\ 0 \\ 0 \end{bmatrix}$$

$$= \begin{bmatrix} I_{in} + 4 \\ 0 \\ 0 \end{bmatrix}$$

Finally $v_{out}(t) = 2(t - 1 + e^{-t})u(t)$ and $Cv_C(0^-)h(t) = (e^{-t} - 1)u(t)$.

SWITCHING PROBLEMS

68. The switch in the circuit of Figure P13.68 is in position A for a very long time and then moves to position B at time $t = 1$ sec, back to A at $t = 2$ sec, and then back to B at $t = 3$ sec, where it remains forever. Suppose $R_S = 500\ \Omega$, $R_L = 1\ k\Omega$, and $C = 2$ mF. Compute $v_{out}(t)$ when $v_{in}(t) = 10u(t) + 5u(t - 1) + 5u(t - 2) - 20u(t - 3)$V. Plot for $0 \le t \le 5$ sec.

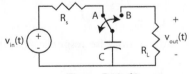

Figure P13.68

69. The switch in the circuit of Figure P13.69 has been in position A for a very long time. At $t = 0$, the switch moves to position B. Compute $v_{out}(t)$, ≥ 0, $p_{R1}(t)$, and the energy dissipated in R_1 over $[0, \infty)$. Assume $R_s = 2\ \Omega$, $R_L = 8\ \Omega$, R_1

$= 190\ \Omega$, $L = 20$ H, and $v_{in}(t) = 100$ V for all time.

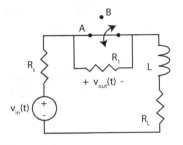

Figure P13.69

70. The switch in the circuit of Figure P13.70 is in position A for a very long time. It moves to position B at time $t = 1$ sec, back to A at $t = 2$ sec, and then back to B at $t = 4$ sec, where it remains forever. Suppose $R_1 = 100\Omega$, $R_2 = 500\ \Omega$, $C_1 = 2$ mF, and $C_2 = 2$ mF. Compute $v_{out}(t)$ when $v_{in}(t) = 20u(t)$ V. Plot for $0 \le t \le 5$ sec.

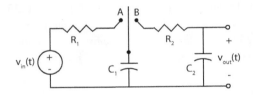

Figure P13.70

71. Repeat Problem 70 for $v_{in}(t) = 20u(t) + 20u(t - 1)$ V.

72. Repeat Problem 70 for $R_1 = 500\ \Omega$ and $v_{in}(t) = 20u(t) + 20u(t - 1)$ V.

73. In the circuit of Figure P13.73,

$$R_S = R_L = \frac{1}{\log_e(2)} = 1.4427\ \Omega,$$

$C_1 = 1$F, and $C_2 = 1$ F. Suppose switches S_1 and S_2 have been closed for a long time. At $t = 0$, S_2 is opened. At $t = 1$ sec, S_2 is closed and S_1 is opened. Will any capacitance voltages change abruptly in this circuit?

   (a)   Find $v_1(0^-)$ and $v_2(0^-)$.

   (b)   Find $v_1(t)$ and $v_2(t)$ for $0 \le t < 1$ sec.

(c)   Find $v_1(1^-)$ and $v_2(1^-)$.

(d)   Find $v_1(1^+)$ and $v_2(1^+)$.

(e)   Find $v_1(t)$ and $v_2(t)$ for $1 \leq t$ sec.

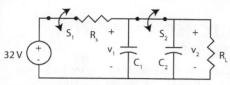

Figure P13.73

74. In the circuit of Figure P13.74, $V_0 = 10$ V, $R_S = 4$ Ω, $R_1 = 3$Ω, $R_2 = 24$Ω, $L_1 = 3$ H, and $L_2 = 6$ H. Suppose switch S_1 has been closed for a long time while S_2 has been open. At $t = 0$ S_2 is closed, after which S_1 is opened. At $t = 1$ sec, S_1 is closed and S_2 is then opened.

(a)   Find $i_1(0^-)$ and $i_2(0^-)$.

(b)   Find $i_1(t)$ and $i_2(t)$ for $0 \leq t < 1$ sec.

(c)   Find $i_1(1^-)$ and $i_2(1^-)$.

(d)   Find $i_1(1^+)$ and $i_2(1^+)$.

(e)   Find $i_1(t)$ and $i_2(t)$ for $1 \leq t$ sec.

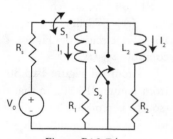

Figure P13.74

CHECK: (b) $I_1(s)$ has a pole at $s = -1$, and I_2 (s) has a pole at $s = -4$; (e) $I_1(s)$ and $I_2(s)$ both have poles at $s = -2$ and $s = -5$.

75. Consider the switched RLC network of Figure P13.75, in which $R_s = 2.5$ Ω, $L = 1$ H, and $C = 0.04$ F.

(a)   Suppose the switch S has been closed for a long time and opens at $t = 0$. If $v_{in}(t) = -40u(-t)$ V, find $v_C(t)$.

(b)   Suppose the switch S has been closed for a long time, and it opens at $t = 0$ and recloses at $t = 1$ sec. If $v_{in}(t) = -40u(t) + 40u(t)$ V, find $v_C(t)$.

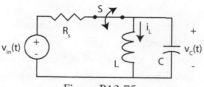

Figure P13.75

76. In the circuit in Figure P13.76, the switch has been in position A for a long time. The switch moves to position B at $t = 0$, back to A at $t = 2$ sec, and finally back to B at $t = 5$ sec, where it remains. Suppose $R_1 = 2.5$ Ω, $L_1 = 1.5625$ H, $R_2 = 2.5$ Ω, $L_2 = 1$ H, $C = 0.04$ F, and $v_{in}(t) = 10$ V for all t.

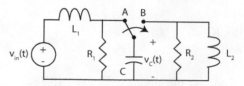

Figure P13.76

(a)   Find $V_C(s)$ and $v_C(t)$ for $0 \leq t < 2$ sec.

(b)   Draw the equivalent frequency domain circuit that is valid for $t \geq 2$ sec.

(c)   Find an expression for $V_C(s)$ with the switch in position B and then determine $v_C(t)$ for $2 < t < 5$ sec.

(d)   Compute $v_C(t)$ for $t > 5$ sec. Hint: This can be done without any further computation.

77. Consider the circuit of Figure P13.77. Suppose $R_1 = 2$Ω, $R_2 = 4$ Ω, $R_3 = 4$ Ω, $C_1 = 0.3$ F, $C_2 = 0.25$ F, and $v_{in}(t) = -30u(-t) + 30u(t) + 30u(t-10)$ V. The switch has been in position A for a long time and moves to position B at $t = 0$.

(a)   Find $v_{C1}(0^-)$ and $v_{C2}(0^-)$.

(b)   Draw the s-domain equivalent circuit valid for $t \geq 0$.

(c)   Write a set of nodal equations in terms of $V_{C1}(s)$ and $V_{out}(s)$.

(d)   Solve the set of nodal equations constructed in part (c) and determine

$v_{out}(t)$ for $0 \leq t < 10$ sec.

(e)   Compute the initial conditions at $t = 10$ sec, i.e., find $v_{C1}(10^-)$ and $v_{C2}(10^-)$. Also assume that at $t = 10$ sec, the switch moves back to position A. Draw the equivalent circuit in the s-domain valid for $t \geq 10$ or $t' \geq 0$. Compute $v_{out}(t)$ for $t \geq 10$ sec.

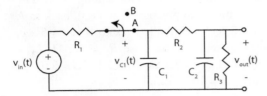

Figure P13.77

CHECK: (d) $v_{out}(t) = [1.846e^{(-2.5t)} - 13.846e^{(-t/3)}]u(t)$.

78. Consider the circuit in Figure P13.78.

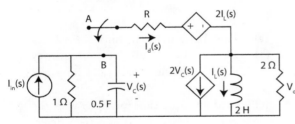

Figure P13.78

Part 1. The switch is in position A at $t = 0$ and $v_C(0^-) = 2$ V, $i_L(0^-) = 0$ and for $0 \leq t < 10$, $i_{in}(t) = 2e^{-2t} u(t)$ A.

(a)   Draw the frequency domain equivalent circuit of the capacitor and compute $V_C(s)$.

(b)   Compute $v_C(t)$. Approximate $v_C(10)$.

(c)   Compute $V_{out}(s)$ and $v_{out}(t)$.

Part 2. At $t = 10$ sec, the switch moves to position B. Assume that the *new* input is given by $i_{in}(t) = 2u(t - 10)$ A.

(d)   Write a set of *three* (modified) nodal equations in the s-domain in terms of $V_C(s)$, $V_{out}(s)$, and $I_d(s)$. Put the equations in matrix form. Your answer should contain R as an undetermined parameter.

(e)   Set up the Cramer's rule equations for solving the equations of part (e) for $V_{out}(s)$.

SWITCHED CAPACITOR NETWORKS

79. Consider the circuit of Figure P13.79.

(a)   If $v_s(t) = 25$ V and the 100 mF capacitor is uncharged at $t = 0$, find $v_C(t)$ for 2 sec $\leq t$.

(b)   Now suppose $v_s(t) = 25$ V and $v_C(2^-) = 10$ V. Find $v_C(t)$ for 2 sec $\leq t$.

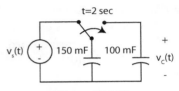

Figure P13.79

80. In the switched capacitor circuits of Figure P13.80, the switch closes at $t = 0$. Suppose $v_s(t) = 572$ mV.

(a)   For the circuit of Figure P13.80a, compute $v_{out}(t)$ for $t > 0$.

(b)   The circuit of Figure P13.80b differs from Figure P13.80a in that the $v_s(t)$-source is replaced by a 10 μF capacitor with an initial voltage $v_a(0^-) = 572$ mV while all other capacitor voltages are zero at $t = 0^-$. Compute $v_{out}(t)$ for $t > 0$.

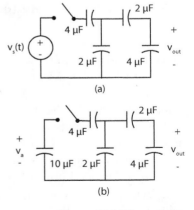

Figure P13.80

81. For the circuit in Figure P13.81 $v_1(0^-) = 0$. At time $t = 0^-$ both switches are flipped to positions A. At $t = 1$ sec, both switches move back to their original positions, and then back to positions A at $t = 2$ sec.
   (a) Determine the value of $v_1(0^+)$.
   (b) Determine the value of $v_1(2^+)$.

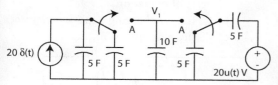

Figure P13.81

CHECK: $v_1(0^+) = 3$ V

82. Repeat Problem 81 for the circuit of Figure P13.82.

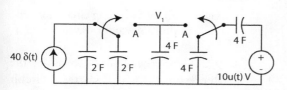

Figure P13.82

83. For the circuit shown in Figure P13.83, $v_{out}(0^-) = 0$. The switch moves from position A to position B at $t = 1$ sec, back to position A at $t = 2$ sec, and finally back to position B at $t = 3$ sec, where it remains forever.
   (a) Compute $v_{out}(2.5)$.
   (b) Compute $v_{out}(3.5)$.
   (c) Plot $v_{out}(t)$ for $0 > t > 4$ sec.

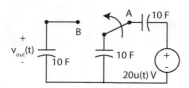

Figure P13.83

ANSWER: $v_{out}(3.5^+) = 6.25$ V

84. In the switched capacitor network of Figure P13.84, the switches S are moved to positions A at $t = 0$ and to positions B at $t = 2$. All capac-

itor voltages are zero at $t = 0^-$. Compute $v_1(t)$ and $v_2(t)$ at $t = 0^+$, $t = 2^-$, and $t = 2^+$.

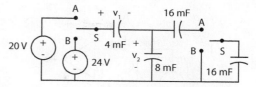

Figure P13.84

85. In the switched capacitor circuit of Figure P13.85, all capacitor voltages are zero at $t = 0^-$. Switches S are moved to positions A at $t = 0$ s and to positions B at $t = 1$ sec, and then back to A at $t = 2$ sec. Compute and plot $v_1(t)$ and $v_2(t)$ for $0 > t > 3$ sec.

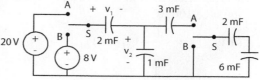

Figure P13.85

86. In Figure P13.86, $v_s(t) = -2$ V, $v_{out}(0^-) = 0$, and the switch S is in position A. At $t = 0$, S is moved to position B, after which it is moved back to A at $t = 1$ sec and then back to B at $t = 2$ sec. Compute and plot $v_{out}(t)$ for $0 > t > 3$ sec.

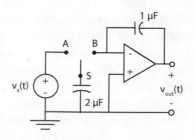

Figure P13.86

87. In Figure P13.87, suppose $v_s(t) = -2$ V, $k = 1$, and the capacitors are initially uncharged. Suppose switch S is moved to position A at $t = 0$, 2, 4, ... (even integer values) msec and to position B at $t = 1, 3, 5$ ... (odd integer values) msec.
   (a) Find $v_{out}(t)$, and plot the waveform for $0 \leq t < 20$ msec.

(b)   Repeat part (a) for $k = 0.5$.

(c)   Repeat part (a) for $k = 2$.

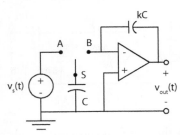

Figure P13.87

MISCELLANEOUS

88. Use the material on op amp integrator design to achieve the following input-output characteristics. In your final design, capacitors should be 100 nF.

(a) $V_{out}(s) = -\dfrac{0.5}{s}V_{a1} + \dfrac{2}{s}V_{b1}$

(b) $V_{out}(s) = -\dfrac{1}{s}V_{a1} - \dfrac{2}{s}V_{a2} + \dfrac{0.5}{s}V_{b1}$

(c) $V_{out}(s) = -\dfrac{0.25}{s}V_{a1} - \dfrac{0.5}{s}V_{a2} + \dfrac{0.75}{s}V_{b1} + \dfrac{1}{s}V_{b2}$

(d) $V_{out}(s) = -\dfrac{2}{s}V_{a1} - \dfrac{1.5}{s}V_{a2} - \dfrac{1}{s}V_{a2} + \dfrac{0.5}{s}V_{b1} + \dfrac{2}{s}V_{b2}$

89. (Sawtooth waveform generation) This problem uses some simple switching techniques in an *RC* circuit to generate an approximate sawtooth waveform. Sawtooths are common to a number of devices, such as televisions and test equipment. Consider the circuit of Figure P13.89.

(a)   Assume the circuit is initially at rest. Beginning at time $t = 0$, the switch, S, is alternately closed to the left position, A, for 1 msec and then to the right position, B, for 50 μsec. Find $v_0(t)$ for $0 \leq t \leq 1.05$ msec (one cycle of operation), and sketch the waveform.

        CHECK: $v_0(10^{-3}) \cong 19$ V

(b)   The circuit in the shaded box is a crude model of a neon lamp (costing less than a dollar) and operates as follows: (1) S is at position A when v_0 is

increasing but less than 80 V, and is moved to position B when v_0 reaches 80 V. (2) S remains at position B when v_0 is decreasing but is greater than 5 V, and is moved to position A when v_0 reaches 5 V. Find $v_0(t)$, and make a rough sketch of the waveform for one cycle of operation.

CHECK: charging time $\cong$ 4.?? msec; discharging time $\cong$ 3? μsec

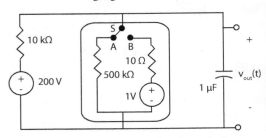

Figure P13.89 Switching circuit for generating a sawtooth waveform.

90. (Fluorescent light) Reconsider the fluorescent light starter circuit described in the text (Example 13.8). In this problem, suppose the ballast is more realistically modeled as an ideal inductor in series with a 100 Ω resistor, as shown in Figure P13.90. Using SPICE or some other circuit simulation program, compute the starting voltage, $v_C(t)$, due to the initial condition on the inductor, as depicted in Figure P13.90. Estimate the starting voltage from a plot of the response over one half-period. Compute the difference between the peak voltages with and without the 100 Ω resistor present. Is the lossless circuit of Example 13.8 a good approximation of the starter response?

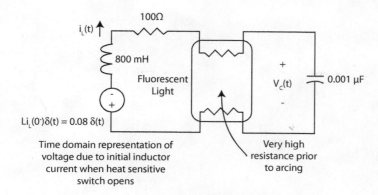

$i_L(t)$

100Ω

800 mH

Fluorescent
Light

+

$V_c(t)$

-

0.001 μF

$Li_L(0^-)\delta(t) = 0.08\ \delta(t)$

Time domain representation of
voltage due to initial inductor
current when heat sensitive
switch opens

Very high
resistance prior
to arcing

Figure P13.90 Model of fluorescent light starter circuit that includes a ballast resistance of 100 Ω. Note that in the time domain, the effect of the initial inductor current appears as an impulse in this model.

CHAPTER 14

Laplace Transform Analysis III: Transfer Function Applications

APPLICATION TO ELECTRIC MOTOR ANALYSIS

Electric motors turn fans, run air conditioners, pull trains, rotate antennas, and help us in a wide variety of ways by efficiently converting electrical energy into mechanical energy. These electromechanical devices are of two general types: ac and dc. Dc motors are ordinarily used in electric-powered transit cars, often with two motors per axle to propel the car. A typical rating of such a motor is 140 hp, 310 V, 2,500 rpm. Another type of dc motor is a high-performance dc servomotor, found in computer disk drives and microprocessor-controlled machinery. These motors are very useful in applications where starts and stops must be made quickly and accurately. As an application, we will represent a dc motor by an equivalent circuit-like model and analyze its operation using the transfer function method.

Of the several kinds of dc motors, the type most pertinent to the analysis techniques of this chapter is a permanent magnet type typically found in low-horsepower applications. They are reliable and efficient. Further, for a permanent magnet dc motor, a plot of the torque produced on the rotating shaft of the motor versus the input current to the motor is almost a linear curve. Hence, the motor has a linear circuit-like model that most nearly describes its performance over a large range of operating conditions. Since the output of the motor is a mechanical quantity, e.g., angular velocity of the motor shaft, the transfer function is the only viable modeling tool available to us at this stage of our development. Section 9 presents the circuit model of the motor and develops the transfer function analysis of its operation. Several problems at the end of the chapter extend the analysis. In more advanced courses, time domain analysis is developed and used.

CHAPTER OBJECTIVES

1. Characterize the transfer function of a circuit in terms of its poles, zeros, and gain constant.
2. Use knowledge of the pole locations of a transfer function to categorize generic kinds of responses (steps, ramps, sinusoids, exponentials, etc.) that are due to different kinds of terms in the partial fraction expansion of the transfer function.
3. Identify, categorize, define, and illustrate various classes of circuit responses, including

the zero-state (zero initial conditions) response, the zero-input response, the step response, the impulse response, the transient and steady-state responses, and the natural and forced responses.

4. Define the notion of the frequency response of a circuit, explore its meaning in terms of the transfer function, and introduce the concept of a Bode plot, which is an asymptotic graph of a circuit's frequency response.

5. Introduce the notions of frequency and magnitude scaling of circuits.

6. Illustrate the applicability of the transfer function concept to a circuit model of a dc motor.

SECTION HEADINGS

1. **Introduction**
2. **Poles, Zeros, and the *s*-Plane**
3. **Classification of Responses**
4. **Computation of the Sinusoidal Steady-State Response for Stable Networks and Systems**
5. **Frequency Response**
6. **Frequency Scaling and Magnitude Scaling**
7. **Initial- and Final-Value Theorems**
8. **Bode Plots**
9. **Transfer Function Analysis of a DC Motor**
10. **Summary**
11. **Terms and Concepts**
12. **Problems**

1. INTRODUCTION

Our experience of using Laplace transforms to calculate responses makes clear that the pole-zero structure of the transfer function sets up generic kinds of circuit behaviors: constants, ramps, exponentials, sinusoidals, exponentially modulated sinusoids, etc. Such knowledge leads to a qualitative understanding of the circuit's response. For example, if a pole is in the right half of the complex plane, then we know that the response will grow with increasing t. Identifying this kind of behavior allows us to define the notion of stability of a circuit or system. Generally, the pole-zero locations allow us to categorize and compute various special types of responses, including transient, steady-state, natural, forced, step, and impulse responses. Coupling the transfer function with the presence of initial conditions in the circuit permits us to define two further types of responses fundamental to both this text and advanced courses in circuits, systems, and control: the zero-input response (due only to the initial conditions of the circuit or system) and the zero-state response (due only to the input excitation, assuming that all initial conditions are zero).

These time domain notions are balanced by the concept of the frequency response of the circuit or system. Briefly, the frequency response is the evaluation of the transfer function, $H(s)$, for $s = j\omega$. Since $H(j\omega)$ has a magnitude and phase, the frequency response breaks down into a magnitude response and a phase response. A technique for obtaining asymptotic (straight-line) approximations (called *Bode plots*) is also outlined in this chapter.

As a final introductory remark, unless stated otherwise, all circuits in this chapter are linear and have constant parameter values. Such circuits are said to be *linear* and *time invariant*. Also, for convenience in this chapter, the symbol $\angle$ will be used to denote either of two things: (i) the angle of a complex number, $\angle(a + jb) = \arg(a + jb) = \tan^{-1}(b/a)$, with due regard to quadrant, or (ii) $\angle\phi$ = $e^{j\phi}$. The context will determine the actual usage of the symbol.

2. POLES, ZEROS, AND THE S-PLANE

In all our circuits, impedances $Z(s)$, admittances $Y(s)$, and transfer functions $H(s)$ are **rational functions** of s, i.e., they are ratios of a numerator polynomial $n(s)$, divided by a denominator polynomial, $d(s)$. Mathematically,

$$H(s) = \frac{n(s)}{d(s)} = K\frac{(s - z_1)(s - z_2)...(s - z_m)}{(s - p_1)(s - p_2)...(s - p_n)} \tag{14.1}$$

where $s = p_i$ is a **finite pole** of $H(s)$, $s = z_i$ is a **finite zero** of $H(s)$, and K is the gain constant of the transfer function. We assume that common factors of $n(s)$ and $d(s)$ have been canceled. A finite pole satisfies $H(p_i) = \infty$, which is shorthand for $H(s) \rightarrow \infty$ as $s \rightarrow p_i$; and a finite zero satisfies $H(z_i) = 0$, i.e., the transfer function takes on the value zero at each z_i. If $p_i = p_j$, $i \neq j$, the pole is a **repeated pole**. A pole repeated twice is second order; one repeated three times is third order, etc. The terminology is the same for zeros. Also, transfer functions sometimes have infinite poles or infinite zeros. If $m < n$ and $s \rightarrow \infty$, then $H(s) \rightarrow 0$, suggesting the term "zero at infinity." In such cases, $H(s)$ has a zero of order $n - m$ at infinity. If, on the other hand, $n < m$, $H(s)$ has a pole of order $m - n$ at infinity.

Out of all this terminology comes one striking fact: transfer functions, impedances, and admittances are characterized by their finite poles, their finite zeros, and their gain constant.

EXAMPLE 14.1. A transfer function $H(s)$ has poles at $s = 0$, -2, and -4, with zeros at $s = -1$ and -3. At $s = 1$, $H(s) = 4/3$. Find $H(s)$.

SOLUTION
From equation 14.1 and the given locations of its poles and zeros, the transfer function must have the form

$$H(s) = K\frac{(s - z_1)(s - z_2)}{(s - p_1)(s - p_2)(s - p_3)} = K\frac{(s + 1)(s + 3)}{s(s + 2)(s + 4)} \tag{14.2}$$

Since $H(1) = 4/3$, evaluating equation 14.2 at $s = 1$ yields

$$\frac{4}{3} = H(1) = K\frac{(1 + 1)(1 + 3)}{1(1 + 2)(1 + 4)} = K\frac{8}{15}$$

This implies that $K = 2.5$. Equation 14.2 specifies the transfer function with $K = 2.5$.

Exercises. 1. Suppose a transfer function has a zero at $s = 1$ and a pole at $s = -1$ and that as $s \rightarrow \infty$, $H(s) \rightarrow -3$. Find $H(s)$.
ANSWER: $H(s) = -3(s - 1)/(s + 1)$

2. A transfer function $H(s)$ has poles at $s = -1$ and -2 and zeros at $s = -3$ and -5. It is further known that $H(0) = 15$. Find $H(\infty)$.
ANSWER: 2

Because the essential information about transfer functions resides with the poles and zeros, a plot of these locations in the s-plane, called a *pole-zero plot*, proves informative.

EXAMPLE 14.2. Draw the pole-zero plot of the transfer function of Example 14.1.

SOLUTION
The transfer function given in equation 14.2 has the pole-zero plot shown in **Figure 14.1**.

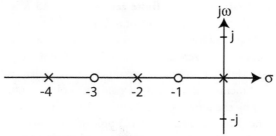

FIGURE 14.1 Pole-zero plot of $H(s)$ given by equation 14.2, where the poles are flagged by "*" and the zeros by "o." Since $s = \sigma + j\omega$, the real axis is labeled σ and the imaginary axis $j\omega$.

As a second illustration, consider the transfer function

$$H(s) = \frac{(s+1)(s+3)}{[(s+1)^2 + 1](s+2)} \tag{14.3}$$

which has the pole-zero plot shown in Figure 14.2.

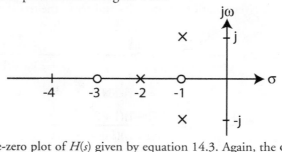

FIGURE 14.2 Pole-zero plot of $H(s)$ given by equation 14.3. Again, the σ-axis represents the real part of the pole or zero and the $j\omega$-axis represents j times the imaginary part of the pole or zero.

Plots such as those in Figures 14.1 and 14.2 communicate much about the nature of the imped-ance, admittance, or transfer function of a circuit. For example, an RC input impedance, $Z_{in}(s)$, satisfies the following properties: (i) all of its poles and zeros are on the non-positive σ-axis of the complex plane; (ii) all of its poles are simple (of multiplicity 1) with real positive residues, i.e., the coefficients in a partial fraction expansion are real and positive; (iii) $Z_{in}(s)$ does not have a pole at $s = \infty$; and (iv) poles and zeros alternate along the σ-axis. Proofs of these assertions can be found in texts on network synthesis.

Exercise. Compute the input impedance of the circuit in Figure 14.3, and show the pole-zero plot if $R_1 = R_2 = 1\ \Omega$, $C_1 = 0.25$ F, and $C_2 = 0.5$ F. Are the poles on the non-positive σ-axis? Do the poles and zeros alternate? Is there a pole at $s = \infty$?

ANSWER:
$$\frac{\frac{1}{C_1}}{s + \frac{1}{R_1 C_1}} + \frac{\frac{1}{C_2}}{s + \frac{1}{R_2 C_2}}$$

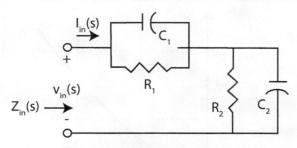

FIGURE 14.3 Series-parallel RC circuit.

More commonly, pole-zero locations provide important qualitative information about the response of the circuit. Pole locations determine the inherent, natural behavior of the circuit, and the poles are commonly called *natural frequencies*. However, the complete set of natural frequencies of the circuit may be larger than the set of poles of the transfer function. This is because there might have been a pole-zero cancellation in constructing the transfer function. The canceled pole would amount to a nat-ural frequency of the circuit that is not present in the poles of the transfer function.

The terms in a partial fraction expansion of the response establish the types of behavior present in the response. Each term has only one of several possible forms. Four very common terms are K/s, K/s^j, $K/(s - p_i)$, and $K/(s - p_i)^j$, with p_i *real* in each case. Figure 14.4 sketches each of the associat-ed responses. In Figure 14.4a, the term K/s leads to a dc response and K/s^j to a polynomial response proportional to t^{j-1}. In Figure 14.4b, the term $K/(s - p_i)$ leads to an exponential response that is increasing if $p_i > 0$ and decreasing if $p_i < 0$. Finally, in Figure 14.4c, if $p_i < 0$, the response curve has a hump.

These qualitative behaviors suggest that one important application of the transfer function is determining the "stability" of the response; i.e., under what conditions will the circuit response remain finite for all time?

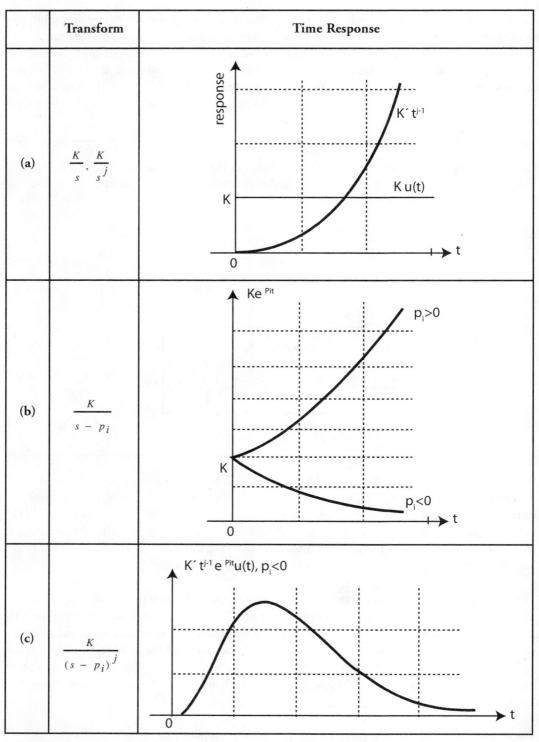

	Transform	Time Response
(a)	$\dfrac{K}{s},\ \dfrac{K}{s^j}$	
(b)	$\dfrac{K}{s - p_i}$	
(c)	$\dfrac{K}{(s - p_i)^j}$	

FIGURE 14.4 Response types common to partial fraction expansion terms. (a) The term K/s leads to a dc response and K/s^j to a polynomial response proportional to t^{j-1}. (b) The term $K/(s - p_i)$ leads to an exponential response that is increasing if $p_i > 0$ and decreasing if $p_i < 0$. (c) If $p_i < 0$, the curve has a hump.

In addition to the preceding response types, there is the sinusoidal response associated with terms of the form

$$\frac{As + B}{(s + \sigma)^2 + \omega^2}$$

Here, if $-\sigma < 0$, the response is an exponentially decaying sinusoidal, as sketched in Figure 14.5a, and if $\sigma = 0$, the purely sinusoidal response of Figure 14.5b results. If $-\sigma > 0$, the exponentially increasing sinusoidal response of Figure 14.5c occurs.

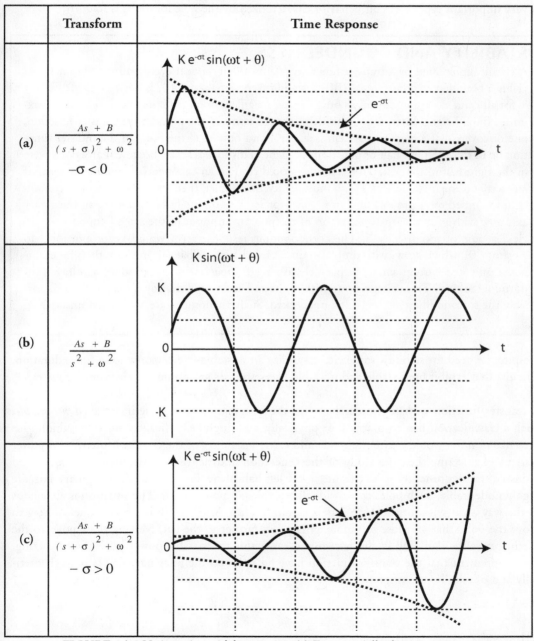

	Transform	Time Response
(a)	$\dfrac{As + B}{(s + \sigma)^2 + \omega^2}$ $-\sigma < 0$	
(b)	$\dfrac{As + B}{s^2 + \omega^2}$	
(c)	$\dfrac{As + B}{(s + \sigma)^2 + \omega^2}$ $-\sigma > 0$	

FIGURE 14.5 Various sinusoidal responses. (a) Exponentially decaying, $-\sigma < 0$. (b) Pure sinusoidal, $\sigma = 0$. (c) Exponentially increasing, $-\sigma > 0$.

Referring again to Figure 14.5, the real part of a pole, i.e., $-\sigma$, specifies the *decay rate* of the response. Often, the word *damping* is used. If $-\sigma < 0$, then the response is damped and the oscillations die out. The farther $-\sigma$ is to the left of the imaginary axis, the greater the damping. If $\sigma = 0$, there is no damping and the response is a *sustained oscillation*. If $-\sigma > 0$, the response is negatively damped, i.e., the response is *unstable* and increases without bound.

One concludes that pole locations specify the type of time domain behavior of a circuit or system. A very important type of circuit behavior characterized by the pole locations is *stability*.

STABILITY AND BOUNDEDNESS

A circuit represented by a transfer function $H(s)$ is called *stable* if every bounded input signal yields a bounded response signal (BIBO stability). A signal, say, $f(t)$, is *bounded* if $|f(t)| < K < \infty$ for all t and some constant K. In other words, a signal is bounded if its magnitude has a maximum finite height. Interpreting this definition in terms of the poles of the transfer function, one discovers that a circuit or system is stable if and only if all the poles of the transfer function lie in the open left half of the complex plane. This makes sense, because if any poles were in the right half of the plane, the response would contain an exponentially increasing term; if any were on the imaginary axis with multiplicity 2 or higher, then the response would contain an unbounded term proportional to $t^{j-1}u(t)$ for $j \geq 2$; and, finally, if there were an imaginary axis pole with multiplicity 1, excitation of the pole by an input of the same frequency would yield a pole of multiplicity 2. The corresponding response term would be proportional to $t\cos(\omega t + \theta)$, which grows with time—an unstable behavior. What this means is that, for example, a unit step current source in parallel with a 1 F capacitor would produce a voltage proportional to $tu(t)$. This voltage grows without bound and would destroy the capacitor and possibly the surrounding circuitry if left unchecked. Such phenomena are considered unstable.

Despite the need for stability, some circuits utilize an unstable-like response for a finite duration. Circuits that exhibit both stable and unstable-like responses are studied in electronics courses.

A transfer function with first-order poles on the imaginary axis is sometimes called *metastable*. Such a classification has no practical or physically meaningful significance, since the ubiquitous presence of noise would excite the mode and cause instability of the circuit. Moreover, in power systems engineering, i.e., the study of the generation and delivery of electricity to homes and industry, transfer function poles that are in the left half-plane, but close to the imaginary axis, are highly undesirable. Such poles cause wide fluctuations in power levels. The situation is analogous to the way a car without shock absorbers would bounce. Much work has been done on how to move the poles that are close to the imaginary axis farther to the left. Moving these poles to the left increases the damping in the system and maintains more stable power levels. Summarizing, the requirement that the transfer function have no poles on the imaginary axis is both theoretically and physically meaningful.

Exercises. 1. If $H(s) = V_{out}(s)/V_{in}(s) = 1/s$, find a bounded input that will make the response unbounded.

2. If $H(s) = V_{out}(s)/V_{in}(s) = 1/(s^2 + 1)$, find a bounded input that will make the response unbounded. Use MATLAB or some other program to plot the response for $0 \le t \le 10$ s.

ANSWERS: (a) $V_{in}(s) = \dfrac{1}{s^k}$ for $k \ge 1$; (b) $V_{in}(s) = \dfrac{1}{(s^2+1)^k}$ for $k \ge 1$

EXAMPLE 14.3. During a laboratory experiment, a student tried to build an inverting amplifier, as shown in Figure 14.6a. The student accidentally reversed the connection of the two input terminals and obtained the circuit of Figure 14.6b. The student was greatly surprised that the circuit did not behave as expected. Explain this phenomenon in terms of the stability theory just developed.

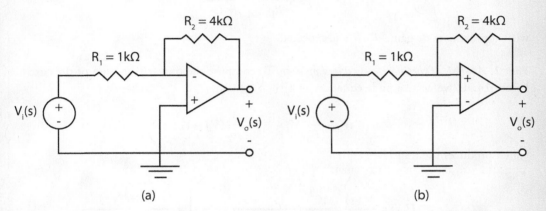

(a)   (b)

FIGURE 14.6 (a) Correct wiring of op amp circuit. (b) Accidental, improper wiring of op amp circuit.

SOLUTION
Assume that the op amp is modeled as a voltage-controlled voltage source with a finite gain of 10^4 and that there is a very small stray capacitance of 1 pF across the input terminals. Figure 14.7 illustrates the equivalent circuit model for each of the circuits in Figure 14.6.

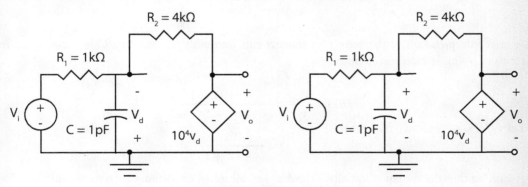

FIGURE 14.7 (a) Model of correctly wired op amp circuit.
(b) Model of incorrectly wired op amp circuit.

Part 1: *Analysis of the correctly wired op amp.* Writing a node equation at V_d to compute the transfer function of Figure 14.7a yields

$$-CsV_d + \frac{1}{R_1}(-V_d - V_i) + \frac{1}{R_2}(-V_d - V_o) = 0$$

After some algebra,

$$-\left(s + \frac{1}{CR_1} + \frac{1}{CR_2}\right)V_d - \frac{1}{CR_1}V_i - \frac{1}{CR_2}V_o = 0$$

Substituting $V_o \times 10^{-4}$ for V_d produces the stable transfer function,

$$H(s) = \frac{V_o(s)}{V_i(s)} = -\frac{10^4}{R_1C}\left(\frac{1}{s + \frac{1}{R_1C} + \frac{1 + 10^4}{R_2C}}\right) = -\frac{10^{13}}{s + 2.501 \times 10^{12}} \tag{14.4}$$

with approximate dc gain $-R_2/R_1$, as expected.

Part 2: *Analysis of the incorrectly wired op amp.* To compute the transfer function of the circuit of Figure 14.7b, we write a node equation at V_d:

$$CsV_d + \frac{1}{R_1}(V_d - V_i) + \frac{1}{R_2}(V_d - V_o) = 0$$

which produces the transfer function

$$H(s) = \frac{V_o(s)}{V_i(s)} = \frac{10^4}{R_1C}\left(\frac{1}{s + \frac{1}{R_1C} + \frac{1 - 10^4}{R_2C}}\right) = -\frac{10^{13}}{s - 2.499 \times 10^{12}} \tag{14.5}$$

The transfer function of equation 14.5 has a right half-plane pole, in contrast to that of equation 14.4. This implies that the incorrectly wired circuit is unstable, which explains the student's concern over the surprising performance of the op amp.

A brief interpretation of the zeros of a transfer function ends this section. This is best done in terms of a simple example. Suppose

$$H(s) = \frac{V_{out}}{V_{in}} = \frac{(s+1)^2 + 1}{(s+1)(s+2)(s+3)}$$

Let $v_{in}(t) = e^{-t}\sin(t)u(t)$ V, so that

$$V_{in}(s) = \frac{1}{(s+1)^2 + 1}$$

Assuming that the system is initially "relaxed," i.e., all initial conditions are zero, we obtain

$$V_{out}(s) = H(s)V_{in}(s) = \frac{1}{(s+1)(s+2)(s+3)}$$

The time response is

$$v_{out}(t) = [A_1 e^{-t} + A_2 e^{-2t} + A_3 e^{-3t}]u(t) \text{ V}$$

for appropriate constants A_i. Observe that the response dies out very quickly and does not have any term similar to the input signal. This follows because the input signal has transform poles, $s = -1 \pm j$, that coincide with the zeros of the transfer function. One can think of the pole locations in the transform of the input signal as identifying frequencies that are present in the input. Hence, the effect of these input signal frequencies (poles) is canceled out by the transfer function zeros, eliminating them from the circuit response.

3. CLASSIFICATION OF RESPONSES

In addition to the various response behaviors discussed in section 2, there are other general response classifications. Three fundamentally important general response classifications germane to all of circuit and system theory are the zero-input response, the zero-state response, and the complete response.

Zero-input response: The response of a circuit/system to a set of initial conditions with the input set to zero.

Zero-state response; The response of a circuit to a specified input signal, given that the initial conditions are all set to zero. Figure 14.8 illustrates this idea.

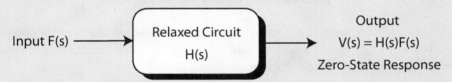

FIGURE 14.8 Relaxed circuit having transfer function $H(s)$ and zero-state response.

Complete response: The response of a circuit/system to both a given set of initial conditions and a given input signal. For linear circuits, the complete response equals the sum of the zero-input and zero-state responses.

Recall that a circuit is **linear** if, for any two inputs, $f_1(t)$ and $f_2(t)$, whose zero-state responses are $y_1(t)$ and $y_2(t)$, respectively, the response to the new input $[K_1 f_1(t) + K_2 f_2(t)]$ is $[K_1 y_1(t) + K_2 y_2(t)]$, where K_1 and K_2 are arbitrary scalars. The circuits studied in this book are linear unless otherwise stated.

The decomposition of the complete response into the sum of the zero-input and zero-state responses is important for three reasons:

1. It is defined for arbitrary input signals.

2.   The zero-state response is given by $\mathcal{L}^{-1}[H(s)F(s)]$, for the arbitrary s-domain input $F(s)$.
3.   It illustrates a proper application of the principle of superposition for linear dynamic networks having initial conditions.
     The following example illustrates point 3.

EXAMPLE 14.4. Consider two linear networks: (i) a linear resistive network, as sketched in Figure 14.9, and (ii) a linear dynamic network, as sketched in Figure 14.10.

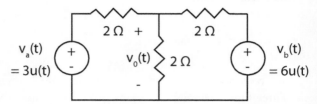

FIGURE 14.9 Linear resistive network.

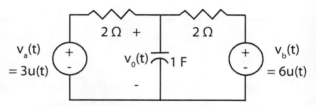

FIGURE 14.10 Linear dynamic network.

Part 1: *Response of linear resistive network of Figure 14.9.* For the resistive network of figure 14.9, the contribution to $v_o(t)$ due to $v_a(t)$ with $v_b(t) = 0$ is $v_{oa}(t) = 1u(t)$, and the contribution due to $v_b(t)$ with $v_a(t) = 0$ is $v_{ob}(t) = 2u(t)$. By superposition,

$$v_o(t) = v_{oa}(t) + v_{ob}(t) = 3u(t) \text{ V}$$

For this type of circuit there is no initial condition and the complete response consists of only the zero-state response, which decomposes into the superposition of each source acting alone.

Part 2: *Response of linear RC network of Figure 14.10.* Now consider the dynamic network of Figure 14.10. Suppose the capacitor has an initial voltage of 2 V at $t = 0$, i.e., $v_o(0) = 2$ V.
Step 1. With $v_a(t)$ applied, $v_o(0) = 2$ V, and v_b set to zero, the response is

$$v_{oa}(t) = (0.5e^{-t} + 1.5)u(t) \text{ V}$$

Step 2. With the input $v_b(t)$ applied, $v_o(0) = 2$ V, and v_a set to zero, the resulting response is

$$v_{ob}(t) = (-e^{-t} + 3)u(t) \text{ V}$$

Step 3. An *incorrect* application of superposition implies that

$$v_o(t) = v_{oa}(t) + v_{ob}(t) = (4.5 - 0.5e^{-t})u(t) \text{ V}$$

The last answer is wrong because the response due to the initial condition has been added in twice.

Step 4. A correct application of superposition would entail:
   (i)   computation of the zero-state response due to $v_a(t)$,
   (ii)   computation of the zero-state response due to $v_b(t)$, and
   (iii) computation of the zero-input response due to $v_o(0)$.
By superposition, the complete response is the sum of all three. In particular, the zero-state response due to $v_a(t)$ is

$$v_{oa}(t) = 1.5(1 - e^{-t})u(t) \text{ V}$$

and the zero-state response due to $v_b(t)$ is

$$v_{ob}(t) = 3(1 - e^{-t})u(t) \text{ V}$$

Hence, the complete zero-state response, by superposition, is the sum of $v_{oa}(t)$ and $v_{ob}(t)$. Further, the zero-input response is $2e^{-t}u(t)$. Hence, the complete response is

$$v_o(t) = v_{oa}(t) + v_{ob}(t) = (4.5 - 2.5e^{-t})u(t) \text{ V}$$

It is important to note that the transfer function is defined only for circuits whose input-output behavior is linear. In terms of the zero-state response, if a circuit has a *linear* input-output behavior characterized by a transfer function $H(s)$, then $H(s)[K_1F_1(s) + K_2F_2(s)] = K_1H(s)F_1(s) + K_2H(s)F_2(s) = K_1Y_1(s) + K_2Y_2(s)$, where $Y_i(s) = H(s)F_i(s)$ is the zero-state response of the network to $F_i(s)$. This says that the zero-state response to $[K_1f_1(t) + K_2f_2(t)]$ is $[K_1y_1(t) + K_2y_2(t)]$. Hence, the transfer function model reflects the underlying linearity of the circuit.

The complete response has a second structural decomposition in terms of the transient and steady-state responses. The notion of a periodic signal is intrinsic to these classifications. A signal $f(t)$ is **periodic** if there exists a positive constant T such that $f(t) = f(t + T)$ for all $t \geq 0$. (The restriction to $t \geq 0$ exists because our Laplace transform analysis implicitly constrains our function class to those that are zero for $t < 0$.) If a signal is periodic, there are many positive constants for which $f(t) = f(t + T)$ for all $t \geq 0$. For example, if $f(t) = f(t + T)$ for some T and for all $t \geq 0$, then it is true for $2T$, $3T$, etc. We define the **fundamental period,** often simply called the period, of $f(t)$ to be the smallest positive constant T for which $f(t) = f(t + T)$ for all $t \geq 0$. Sinusoids are periodic signals: $\sin(2\pi t) = \sin(2\pi t + 2\pi)$ with fundamental period $T = 1$. The square wave of Figure 14.11 is periodic with fundamental period $T = 2$.

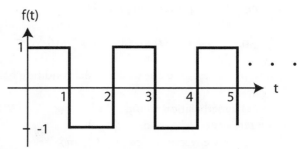

FIGURE 14.11 A periodic square wave with fundamental period $T = 2$.

This notion of periodicity and, by default, non-periodicity allows us to define the transient and steady-state responses of a circuit.

> **Steady-state response:** Those terms of the complete response that satisfy the definition of periodicity for $t \geq 0$. This includes a constant response.
>
> **Transient response:** Those terms of the complete response that are not periodic for $t \geq 0$, i.e., that do not satisfy the definition of a periodic function for $t \geq 0$. Note that a constant response satisfies the definition of a periodic function.

A circuit response may have no transient part, as illustrated by the sustained sinusoidal oscillatory response of the circuit given in Figure 14.12a. Further, the steady-state part of the response may be zero, as in the circuit of Figure 14.12b, where $v_{out}(t)$ is a damped sinusoid. If the a circuit is unstable, the transient response may blow up, overwhelming the constant or periodic part of the complete response, as in the case of $v_{out} = (e^{5t} \cos(10t) + 15)u(t)$ V, where the steady state is $15u(t)$ V. Note that "transient" here does not mean something that diminishes in importance with time.

Most circuits have both a transient and a steady-state response. When the input is constant or periodic, the circuit response approaches the steady-state response asymptotically for large t, i.e., as the transient dies out, only if the circuit is stable. For such circuits, the steady state is crucial. Further, when the input is sinusoidal, the steady-state response is easily computed via the transfer function, $H(s)$, or by the phasor method. Details of the calculation are presented in section 4.

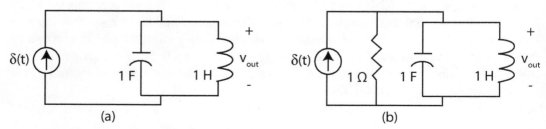

FIGURE 14.12 (a) Unstable circuit illustrating the possibility of no transient response.
(b) Stable circuit having a zero steady-state response.

EXAMPLE 14.5. Given that $i_{in}(t) = I_0 u(t)$, computing the response of the circuit of Figure 14.13 provides a simple illustration of the decomposition of the complete response into the sum of the zero-input and zero-state responses. Also, some rearrangement of the terms identifies the transient and steady-state responses.

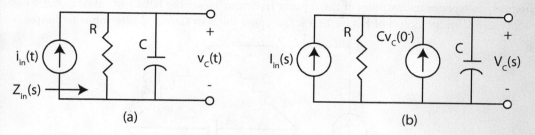

FIGURE 14.13 RC circuit for Example 14.5. (a) Time domain circuit.
(b) Frequency domain equivalent, accounting for initial condition.

SOLUTION

Step 1. *Computation of the zero-state response.* The input impedance can be viewed as a special type of transfer function. For Figure 14.13a the output $V_C(s)$ is the voltage appearing across the input current source. Hence,

$$H(s) = Z_{in}(s) = \frac{V_C(s)}{I_{in}(s)} = \frac{1}{Cs + \dfrac{1}{R}} = \frac{1}{C}\,\frac{1}{s + \dfrac{1}{RC}}.$$

Letting $i_{in}(t) = I_0 u(t)$ and $v_C(0^-) = 0$, then zero-state response is

$$\mathcal{L}^{-1}\big[H(s)I_{in}(s)\big] = \mathcal{L}^{-1}\left[\frac{RI_0}{s} - \frac{RI_0}{s + \dfrac{1}{RC}}\right] = RI_0\left(1 - e^{-\frac{t}{RC}}\right)u(t)$$

Step 2. *Computation of the zero-input and complete response.* Now, supposing that $v_C(0^-) \neq 0$, the zero-input response is the inverse transform of $H(s)[Cv_C(0^-)]$ as per Figure 14.13b. Hence, by superposition, the complete response is

$$v_C(t) = \underbrace{RI_0\left(1 - e^{-\frac{t}{RC}}\right)u(t)}_{\text{zero-state response}} + \underbrace{v_C(0^-)e^{-\frac{t}{RC}}u(t)}_{\text{zero-input response}}$$

Step 3. *Decomposition into transient and steady-state responses.* As a final point, since the input is dc, a step function, the complete response decomposes into its transient and steady-state parts as

$$v_C(t) = \underbrace{\left(v_C(0^-) - RI_0\right)e^{-\frac{t}{RC}}u(t)}_{\text{transient response}} + \underbrace{RI_0 u(t)}_{\text{steady-state response}}$$

Observe that from the above example the transient and zero-input responses are ordinarily different.

EXAMPLE 14.6. It is sometimes mistakenly said that the zero-input response contains only those frequencies represented by poles of the transfer function. To see the fallacy of this statement consider the *RC* bridge circuit of Figure 14.14. Compute the zero-state and zero-input responses.

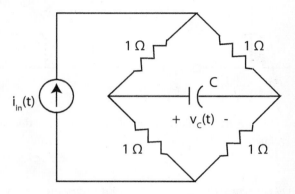

FIGURE 14.14 *RC* bridge circuit for Example 14.6.

SOLUTION
The transfer function of this circuit is $H(s) = V_{out}(s)/I_{in}(s) = 0$, which has no poles. Hence, the zero-state response is always zero. On the other hand, if $v_C(0^-) \neq 0$, then

$$v_{out}(t) = v_C(0^-) \exp\left[-\frac{t}{R_{eq}C}\right] u(t) \text{ V},$$

where $R_{eq} = 1\ \Omega$. Thus, the zero-input response is a decaying exponential whenever $v_C(0^-) \neq 0$ and $C > 0$. Notice that the transfer function has no poles. As a side remark, in this case, the zero-input response is also the transient response, with the steady-state response being zero.

The phenomenon illustrated by Example 14.6 occurs because the symmetry of the resistor values precludes excitation by the current source. Moving the current source to a different position, say, in parallel with one of the resistors, or changing the value of one of the resistors to 0.5 Ω will result in a nonzero transfer function.

EXAMPLE 14.7. This example looks at a simple initialized series *RL* circuit driven by a cosine wave, as shown in Figure 14.15. Let $i_L(t)$ denote the circuit response. Suppose $v_{in}(t) = 4\cos(t)u(t)$ V and $i_L(0^-) = 1$ A. The objective is to isolate the transient and steady-state responses from the zero-input and zero-state responses.

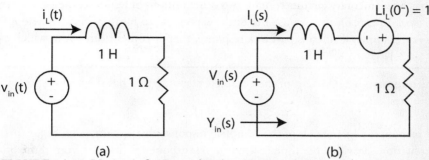

(a) **(b)**
FIGURE 14.15 *RL* circuit for Example 14.7. (a) Time domain series *RL* circuit.
(b) Frequency domain equivalent, accounting for initial condition.

SOLUTION

Using Figure 14.15b and the principle of superposition leads to the response

$$I_L(s) = Y_{in}(s)V_{in}(s) + Y_{in}(s)[Li_L(0^-)] = \frac{4s}{(s+1)(s^2+1)} + \frac{1}{s+1}.$$

The term $Y_{in}(s)V_{in}(s) = 4s/[(s+1)(s^2+1)]$ is the Laplace transform of the zero-state response, and the term $Y_{in}(s)Li_L(0^-) = 1/(s+1)$ is the transform of the zero-input response. Thus, the zero-input response is $e^{-t}u(t)$. A partial fraction expansion of the zero-state response yields

$$\frac{4s}{(s+1)(s^2+1)} = \frac{-2}{s+1} + \frac{2s+2}{s^2+1}$$

It follows that the zero-state response is $-2e^{-t}u(t) + 2[\cos(t) + \sin(t)]u(t)$. Notice that both the zero-input and the zero-state response contain a transient part, the part proportional to e^{-t}. A little rearranging shows that the complete response is

$$i_L(t) = -e^{-t}u(t) + 2[\cos(t) + \sin(t)]u(t) \text{ A}$$

which implies that the transient response of the circuit is $-e^{-t}u(t)$ and that the steady-state response of the circuit is $2[\cos(t) + \sin(t)]u(t)$ A.

Exercises. 1. An *RLC* network has transfer function $H(s) = V_{out}/I_{in} = 1/(s+1)$. If an input $i_{in}(t) = \cos(t)u(t)$ A is applied, then for very large t, $v_{out}(t)$ approaches a cosine wave of what form?
ANSWER: $0.707 \cos(t - 45°)$

2. An *RLC* network has transfer function $H(s) = V_{out}/I_{in} = 1/(s+1)$. If an input $i_{in}(t) = (1 - e^{-50t})u(t)$ A is applied, then for very large t, $v_{out}(t)$ approaches what?
ANSWER: A constant

Many books on elementary circuits contain two other notions of response: the *natural response* and *forced response*. To explain this we use the term "exponent" to mean λ in $e^{\lambda t}$, where λ is possibly complex. For example, if $\sin(t)$ is part of a response, then it comes from e^{jt}, in which case $\lambda = j$.

> **Natural response:** The portion of the complete response that has the same exponents as the zero-input response.
>
> **Forced response:** The portion of the complete response that has the same exponents as the input excitation, provided the input excitation has exponents different from those of the zero-input response.

It would seem natural to try to decompose the complete response into the sum of the natural and forced responses. Unfortunately, such a decomposition applies only when the input excitation is (i) dc, (ii) real exponential, (iii) sinusoidal, or (iv) exponentially modulated or damped sinusoidal. Further, the exponent of the input excitation, e.g., a in $f(t) = e^{at}u(t)$, must be different from the exponents appearing in the zero-input response. The natural and forced responses are properly defined only under these conditions.

The decomposition of a complete response into a natural response and a forced response is important for two reasons. First, it agrees with the classical method of solving ordinary differential equations having constant coefficients, where the natural response corresponds to the complementary function and the forced response corresponds to the particular integral. Students fresh from a course in differential equations feel quite at home with these concepts. The second reason is that the forced response is easily calculated for any of the special inputs—dc, real exponential, sinusoidal, or damped sinusoidal. For example, if the transfer function is $H(s)$ and the input is Ve^{at}, then the forced response is simply

$$H(a)Ve^{at} \tag{14.6}$$

To justify equation 14.6, note that the Laplace transform of the input is $V/(s-a)$. Since the complete response is the sum of the zero-input and zero-state responses, we have

$$\textit{Complete response} = [\textit{zero-input response}] + \mathcal{L}^{-1}\left[H(s)\frac{V}{s-a}\right]$$

The zero-input response terms all have exponents different from a, the exponent of the input. The second term, $\mathcal{L}^{-1}[H(s)\,V/(s-a)]$, has only one term with exponent equal to a. Executing a partial fraction expansion of this term yields

$$H(s)\frac{V}{s-a} = \frac{K}{s-a} + [\textit{terms corresponding to poles of } H(s)]$$

Using the residue formula to calculate K leads to $K = H(a)V$. Thus,

$$H(s)\frac{V}{s-a} = \frac{H(a)V}{s-a} + [\textit{terms corresponding to poles of } H(s)]$$

and $\mathcal{L}^{-1}[H(s)V/(s-a)]$ has a term, $H(a)Ve^{at}u(t)$, which we identify as the forced response.

By using exactly the same arguments, it is possible to show that if the input is a complex exponential function $Ve^{s_p t}$, where both V and s_p are complex numbers, then the forced response is simply

$$H(s_p)\, Ve^{s_p t}$$

A complex exponential such as Ve^{st} is a mathematical entity. It cannot be generated in the laboratory. However, the real part, $\text{Re}[Ve^{st}]$ (or the associated imaginary part), is simply an exponentially modulated sinusoidal signal, as shown in Figure 14.5, and is readily generated in a laboratory. A derivation similar to the preceding leads to the conclusion that, if the input is $\text{Re}[Ve^{st}]$, then the forced response is

$$\text{Re}[H(s)Ve^{st}]$$

This relationship of the input to the forced response prompts some textbooks to define the transfer function $H(s)$ as the ratio of the forced response to the input, under the condition that the input is a complex exponential Ve^{st}. This, however, is not natural and makes one wonder at the applicability of such a definition to the broad class of inputs for which the transfer function is most naturally defined, as covered in Chapter 13 of the text.

4. COMPUTATION OF THE SINUSOIDAL STEADY-STATE RESPONSE FOR STABLE NETWORKS AND SYSTEMS

Suppose that a transfer function $H(s)$ models a *stable* linear circuit containing a total of n capacitors and inductors. In addition, suppose there are no common factors in the numerator and denominator of $H(s)$ and that the *degree of the denominator of $H(s)$ is n*. (This means that the effect of each capacitor and inductor is included in $H(s)$.) The goal of this section is to develop the following formula: if $H(s)$ satisfies the aforementioned assumptions, and the input to the circuit has the form $A\cos(\omega t + \theta)$, then the **steady-state circuit output response** has the form

$$B\cos(\omega t + \phi)$$

$$(14.7a)$$

where the *magnitude* of the response is

$$B = A|H(j\omega)|$$

$$(14.7b)$$

and the *phase shift* is

$$\phi = \theta + \angle H(j\omega)$$

$$(14.7c)$$

Here we assume that ω is some fixed, but arbitrary, value.

From an input-output viewpoint, these formulas imply that the frequency response is the steady-state response of a circuit to sinusoids of varying frequencies. To construct this formula, suppose again that a linear circuit has a stable transfer function model $H(s)$. Suppose also that the circuit input is a sinusoid $f(t)$ whose Laplace transform is $F(s)$ with zero-state response $Y(s)$, as illustrated in Figure 14.16.

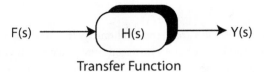

$$F(s) \longrightarrow \boxed{H(s)} \longrightarrow Y(s)$$

Transfer Function

FIGURE 14.16 Frequency domain representation of hypothetical circuit.

Since $H(s)$ is stable, all poles lie in the open left half of the complex plane. Assume that $H(s)$ has real, distinct poles labeled $p_1, \ldots, p_m$ and complex poles labeled $-\alpha_i \pm j\beta_i$, Consequently, $H(s)$ will have a partial fraction expansion containing only two types of terms: those having real poles, $p_i < 0$, and those having complex poles with negative real parts, i.e., $\alpha_i < 0$. Specifically,

$$H(s) = \underbrace{\frac{A_1}{s - p_1} + \frac{A_2}{s - p_2} + \ldots + \frac{A_m}{s - p_m}}_{\substack{real\ poles\ with \\ negative\ real\ parts}} + \underbrace{\frac{C_1 s + D_1}{(s + \alpha_1)^2 + (\beta_1)^2}}_{\substack{complex\ poles\ with \\ negative\ real\ parts}} + \cdots \qquad (14.8)$$

It is easy to account for higher-order poles.

Suppose now that the circuit is excited by a sinusoidal input of the form

$$f(t) = A\cos(\omega t + \theta) = A\frac{e^{j(\omega t + \theta)} + e^{-j(\omega t + \theta)}}{2}$$

having Laplace transform

$$F(s) = \frac{0.5 A e^{j\theta}}{s - j\omega} + \frac{0.5 A e^{-j\theta}}{s + j\omega}$$

Then a partial fraction expansion of the Laplace transform of the zero-state response, $Y(s) = H(s)F(s)$, has the form

$$Y(s) = \underbrace{\frac{\hat{A}_1}{s - p_1} + \frac{\hat{A}_2}{s - p_2} + \ldots}_{\substack{real\ poles\ with \\ negative\ real\ parts}} + \underbrace{\frac{\hat{C}_1 s + \hat{D}_1}{(s + \alpha_1)^2 + (\beta_1)^2} + \ldots}_{\substack{complex\ poles\ with \\ negative\ real\ parts}} + \underbrace{\frac{R_1}{s - j\omega} + \frac{R_2}{s + j\omega}}_{\substack{steady\text{-}state \\ contribution \\ = Y_{ss}(s)}}.$$

In the steady state, i.e., for large t, the only residues of interest are R_1 and R_2, because the part of the time response due to the other terms decays to zero with increasing t. By the usual methods of complex variables, we obtain

$$R_1 = \left[H(s)(s - j\omega) \left(\frac{0.5Ae^{j\theta}}{s - j\omega} + \frac{0.5Ae^{-j\theta}}{s + j\omega} \right) \right]_{s=j\omega}$$

$$= 0.5AH(j\omega)e^{j\theta} = 0.5A|H(j\omega)|e^{j\angle H(j\omega)}e^{j\theta}$$

$$= 0.5A|H(j\omega)|e^{j(\angle H(j\omega)+\theta)}$$

and

$$R_2 = \left[H(s)(s - j\omega) \left(\frac{0.5Ae^{j\theta}}{s - j\omega} + \frac{0.5Ae^{-j\theta}}{s + j\omega} \right) \right]_{s=j\omega}$$

$$= 0.5AH(-j\omega)e^{-j\theta} = 0.5A|H(-j\omega)|e^{j\angle H(-j\omega)}e^{-j\theta}$$

$$= 0.5A|H(-j\omega)|e^{j(\angle H(-j\omega)-\theta)}$$

But $|H(-j\omega)| = |H(j\omega)|$ and $\angle H(-j\omega) = -\angle H(j\omega)$; hence,

$$R_2 = 0.5A|H(j\omega)|e^{-j(\angle H(j\omega) + \theta)}$$

Consequently, the Laplace transform of the steady-state response when all initial conditions of the circuit are zero is

$$Y_{ss}(s) = 0.5A|H(j\omega)| \left[\frac{e^{j(\angle H(j\omega)+\theta)}}{s - j\omega} + \frac{e^{-j(\angle H(j\omega)+\theta)}}{s + j\omega} \right]$$

In fact, this is the Laplace transform of the actual steady-state response, provided that the zero-input (nonzero initial conditions) response makes no additional contribution. The zero-input response makes no contribution to the steady-state response when one or more of the following reasonable conditions on the circuit are met:

1. The network has only practical passive elements, meaning that there are always stray resistances present.
2. The circuit may have active elements in addition to passive elements, but remains stable in the sense that every capacitor voltage and every inductor current remains bounded for any bounded circuit excitation.
3. The circuit contains a total of n capacitors and inductors, and the stable transfer function, $H(s)$, has n poles.

Under conditions 1 through 3,

$$y_{ss}(t) = A|H(j\omega)| \cos(\omega t + (\angle H(j\omega) + \theta)) = B \cos(\omega t + \phi)$$

Hence, if the input to the circuit has the form $A \cos(\omega t + \theta)$, then the steady-state circuit output response has the form $B \cos(\omega t + \phi)$, where the magnitude $B = A|H(j\omega)|$ and the phase shift $\phi = \theta + \angle H(j\omega)$.

The next question concerns the numerical calculation of B and ϕ. One method is simply to evaluate $H(s)$ at $s = j\omega$. With a calculator that easily accommodates complex numbers, this is quite straightforward. An alternative method is to use the graphical technique of the next section.

At this point, it is instructive to illustrate equation 14.7 and at the same time compare it with the phasor method studied in a first course on circuit theory.

EXAMPLE 14.8. In the circuit of Figure 14.17, $\mu = 0.5$ and $v_i(t) = \cos(2t)$ V. Find $v_1(t)$ for large t.

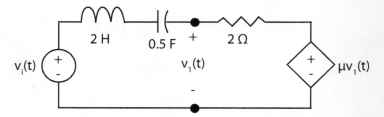

FIGURE 14.17 *RLC* circuit for steady-state computation in Example 14.8.

SOLUTION
Part 1. *Phasor method.* From the principles of phasor analysis detailed in Chapter 10, the phasor domain circuit of Figure 14.17 at $\omega = 2$ rad/sec is given by the circuit of Figure 14.18.

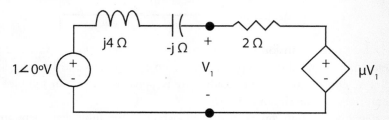

FIGURE 14.18 Phasor circuit equivalent of Figure 14.17 at $\omega = 2$.

The single node equation for $\mathbf{V}_1$ is

$$\frac{\mathbf{V}_1 - 1}{j4 - j} + \frac{\mathbf{V}_1 - 0.5\mathbf{V}_1}{2} = 0 \qquad (14.9)$$

The phasor solution to equation 14.9 is

$$\mathbf{V}_1 = \frac{1}{1 + j0.75} = 0.8\angle - 36.9°$$

Therefore, for large t

$$v_1(t) = 0.8 \cos(2t - 36.9°) \text{ V}$$

Part 2. *Laplace transform method.* The first step here is to construct the *s*-domain equivalent circuit, which is given in Figure 14.19.

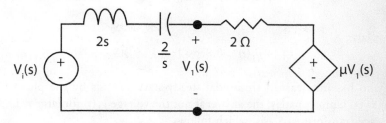

FIGURE 14.19 Frequency or *s*-domain equivalent of the circuit of Figure 14.17.

The single node equation for V_1 in the *s*-domain is

$$\frac{V_1 - V_i}{2s + \frac{2}{s}} + \frac{V_1 - 0.5V_1}{2} = 0$$

Solving for the transfer function yields

$$H(s) = \frac{V_1(s)}{V_i(s)} = \frac{2s}{(s+1)^2} \tag{14.10}$$

From equation 14.10, at $s = j\omega = j2$,

$$H(j\omega) = H(j2) = j4/(j2 + 1)^2 = 0.8 \angle{-36.9°}$$

According to equation 14.7, it follows that

$$v_1(t) = 0.8 \cos(2t - 36.9°) \text{ V}$$

In this example, the two methods give the same answers, as expected. Since complex numbers are easier to manipulate than rational functions, what is the motivation for such an analysis using the transfer function $H(s)$? Why not stay with the phasor method? The next example answers these questions.

EXAMPLE 14.9. In Example 14.8 with the circuit of Figure 14.17, let the value of μ be increased to 1.5. Find $v_1(t)$ for large *t*.

SOLUTION
Part 1. *Phasor method.* Since only the response for large *t* is desired, the problem appears to be one involving sinusoidal steady-state analysis. The phasor domain circuit of Figure 14.18 yields the single node equation

$$\frac{\mathbf{V}_1 - 1}{j4 - j} + \frac{\mathbf{V}_1 - 1.5\mathbf{V}_1}{2} = 0$$

Solving again for $\mathbf{V}_1$ yields

$$\mathbf{V}_1 = \frac{1}{1 - j0.75} = 0.8\angle 36.9°$$

Therefore, for large t,

$$v_1(t) = 0.8\cos(2t + 36.9°) \text{ V}$$

A beginner who has just learned sinusoidal steady-state analysis by the phasor method might accept this answer. Unfortunately, the answer is not the voltage $v_1(t)$ for large t! The reason is clear from the Laplace transform analysis, which follows.

Part 2: *Laplace transform method.* From the s-domain equivalent circuit of Figure 14.19,

$$\frac{V_1 - V_i}{2s + \frac{2}{s}} + \frac{V_1 - 1.5V_1}{2} = 0$$

Solving for the transfer function yields

$$H(s) = \frac{V_1(s)}{V_i(s)} = \frac{-2s}{(s-1)^2}$$

Since there are poles of $H(s)$ in the right half-plane, the circuit is unstable. As t becomes very large, the magnitude of $v_1(t)$ approaches infinity, instead of $0.8 \cos(2t + 36.9°)$ V, as calculated by the phasor method.

This analysis demonstrates that the unstable behavior of a circuit cannot be determined by the phasor method. It is desirable to know when to use a particular method in order to avoid unnecessary complicated calculations. The following guidelines help:

1. When the stability of the circuit has been assured by some means, and ω has a specific numerical value, the phasor method is the better method to use for computing the response for large t, which is also the steady-state response in this case. Circuits whose stability is guaranteed include those with only passive elements, such as resistors, capacitors, and inductors; and amplifier circuits of well-established configurations.
2. The circuit is known to be stable, but ω is variable. In this case, the $H(s)$ method is superior to the phasor method. To say the least, we need only write sL instead of $j\omega L$. There are other advantages to be gained from knowing the pole-zero plot of $H(s)$ that are not possible with the phasor method. The examples of frequency response calculations given in the next section clearly demonstrate this point.
3. If the stability of the circuit is not yet determined, then $H(s)$ should be calculated and its pole locations checked for stability. Then step 1 or step 2 should be referred to, as appropriate.

Exercises. 1. Suppose a second-order linear circuit having the transfer function

$$H(s) = \frac{V_{out}(s)}{V_{in}(s)} = \frac{s^2 - 0.5s + 5}{s^2 + 0.5s + 5.7321}$$

is driven by a sinusoidal input $v_{in}(t) = \sqrt{2}\cos(2t + 45°)u(t)$. Show that the steady-state response is given by

$$v_{out,ss}(t) = \cos(2t - 30°)$$

2. Consider the LC circuit of Figure 14.20, in which $i_L(0^-) = 0$, $v_C(0^-) = 0$, and $v_{in}(t) = 100u(t)$. Show that the largest voltage to appear across the capacitor for $t \geq 0$ is 200 V. Hint: Show that $v_C(t) = 100u(t) - 100\cos(t)u(t)$.

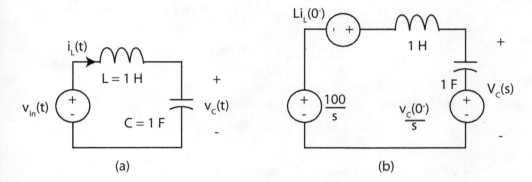

FIGURE 14.20 (a) Time domain LC circuit.
(b) Frequency domain equivalent, accounting for initial conditions.

5. FREQUENCY RESPONSE

The **frequency response** of a linear stable circuit having constant parameters characterizes the input-output behavior of the circuit to unit magnitude sinusoids, $\cos(\omega t)$, as ω varies from 0 to ∞. This extremely important concept plays a major role in the analysis and design of circuits and systems. In terms of the transfer function, the frequency response of a stable circuit is the evaluation of $H(s)$ at $s = j\omega$. In terms of phasor analysis, studied in an introductory course, the frequency response of a circuit corresponds to the ratio of the output phasor to the input phasor.

From the steady-state analysis perspective of the previous section, if an input has the form $\cos(\omega_0 t)$, then the steady-state response (i.e., the response for large t, after all transients have died out) has the form $B \cos(\omega_0 t + \phi)$, where $B = |H(j\omega_0)|$ and the phase shift $\phi = \angle H(j\omega_0)$. Here $|H(j\omega_0)|$ is the magnitude of the complex number $H(j\omega_0)$, and $\angle H(j\omega_0)$ is the angle of the complex number $H(j\omega_0)$. Thus, $H(j\omega)$, for $0 \leq \omega < \infty$, defines how a linear circuit adjusts the magnitude and phase of an input sinusoid to produce a steady-state output sinusoid of the same frequency, but possibly with a different magnitude and phase.

An example of practical importance is the specification of a stereo amplifier. Here one specifies the gain, gain-magnitude $|H(j\omega)|$, to be more or less constant from 20 Hz to 20 kHz. Why? Because musical signals are composed of sinusoids of different frequencies. Accurate amplification of the music requires that all component sinusoids be amplified with equal gain.

> **Frequency response:** The frequency response of a stable circuit or system represented by a transfer function $H(s)$ is the complex-valued function $H(j\omega)$ for $0 \leq \omega < \infty$. The magnitude (frequency) response is $|H(j\omega)|$ for $0 \leq \omega < \infty$, and the phase (frequency) response is $\angle H(j\omega)$ for $0 \leq \omega < \infty$.

A complex-valued function $H(j\omega)$ is a function such that for each value of ω, $H(j\omega)$ is a complex number. A complex number, $a_1 + jb_1$, has a polar form, $\rho_1 e^{j\phi_1}$, in which ρ_1 is the magnitude and ϕ_1 the phase angle of the number. Moreover, if $a_2 + jb_2$ is another complex number, then

$$(a_1 + jb_1)(a_2 + jb_2) = \rho_1 \rho_2 e^{j(\phi_1+\phi_2)}$$

and

$$\frac{a_1 + jb_1}{a_2 + jb_2} = \frac{\rho_1}{\rho_2} e^{j(\phi_1 - \phi_2)}$$

In polar form, the frequency response as a function of ω is

$$H(j\omega) = \rho(\omega) e^{j\,\phi(\omega)}$$

where $\rho(\omega) = |H(j\omega)|$ denotes the **magnitude response** and

$$\phi(\omega) = \angle H(j\omega) = \tan^{-1}\left(\frac{\text{Im}[H(j\omega)]}{\text{Re}[(H(j\omega)]}\right)$$

is the *angle* or *phase* of the frequency response. As in other books, *magnitude response* means the magnitude of the frequency response. Typically, frequency response computation requires a calculator or computer. We now illustrate the idea of frequency response with two so-called band-pass transfer functions in which a band of frequencies is passed with relatively little attenuation while frequencies outside the band are significantly attenuated.

EXAMPLE 14.10. Consider the two transfer functions

$$H_1(s) = \frac{0.25s}{s^2 + 0.25s + 1}$$

and

$$H_2(s) = \frac{0.0625s^2}{s^4 + 0.35355s^3 + 2.0625s^2 + 0.35355s + 1}$$

Using the MATLAB code
```
»n1 = 0.25*[1 0];
»d1 = [1 0.25 1];
»n2 = 0.0625*[1 0 0];
»d2 = [1 3.5355e-01 2.0625 3.5355e-01 1];
»w=0.2:0.005:2;
»h1 = freqs(n1,d1,w);
»h2 = freqs(n2,d2,w);
»plot(w,abs(h1),w,abs(h2))
»grid
»xlabel('Frequency r/s')
```

»ylabel('Magnitude response')
»gtext('2nd Order BP')
»gtext('4th Order BP')

we obtain the magnitude response plot given in Figure 14.21.

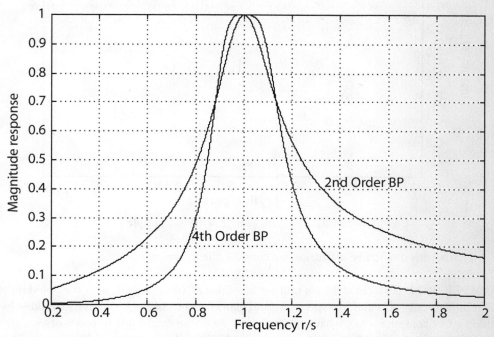

FIGURE 14.21 Magnitude responses of a second-order and a fourth-order band-pass type transfer function. The fourth-order response has steeper sides (sharper cutoff) and a flatter top.

From the transfer function and by interpolation on the plots, one observes that at $\omega = 0$ the magnitude is zero and at $\omega = \infty$ the magnitude is also zero; at $\omega = 1$ the magnitude peaks, and this frequency is called the center frequency. This is characteristic of a band-pass type of response. The fact that the response peaks at $\omega = 1$ rad/sec means that the transfer functions are "normalized." Transfer functions of practical band-pass circuits have much higher center frequencies. Such frequencies can be obtained by the technique of frequency scaling, taken up in the next section.

A very important system theoretic relationship is that of the poles and zeros of the transfer function to the magnitude and phase responses. In the above example for $H_1(s)$, the pole-zero plot is given by Figure 14.22. One immediately notices that the poles are very close to the point $j1$ on the imaginary axis with the magnitude response peaked. Further, the zeros at $\omega = 0$ and $\omega = \infty$ are where the magnitude response is zero.

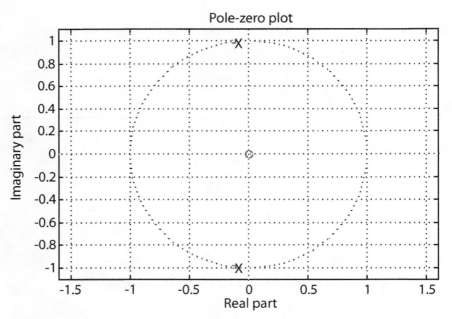

FIGURE 14.22. Pole-zero plot of $H_1(s)$ in Example 14.10.

To emphasize this qualitative discussion we consider the following example.

EXAMPLE 14.11. Two circuits have transfer functions $H_1(s)$ and $H_2(s)$, with the pole-zero plots shown in Figures 14.23a and b, respectively, and gain constants of 1. Qualitatively speaking (without doing any computations), what can we deduce about the magnitude response of each circuit? To verify our qualitative deductions we will use MATLAB to construct the exact magnitude response.

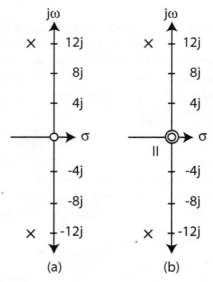

FIGURE 14.23 Pole-zero plots for Example 14.11. (a) $H_1(s)$ (b) $H_2(s)$.

SOLUTION

Part 1. *Qualitative analysis of Figure 14.23a.* This figure tells us that there are two finite poles of $H_1(s)$ near $\pm 12j$, but in the left half of the complex plane, and that there is a finite zero at the origin. So there is also a zero at $s = \infty$. Thus

    (i) $|H(j\omega)| \rightarrow 0$ as $\omega \rightarrow 0$

    (ii) $|H(j\omega)| \rightarrow 0$ as $\omega \rightarrow \infty$

    (iii) $|H(j\omega)| \approx$ maximum value as $\omega \rightarrow 12$

In the case of point (iii), we can say in general that in some neighborhood of ω near the pole, the transfer function peaks in magnitude.

Part 2. *Qualitative analysis of Figure 14.23b.* This figure tells us that there are again two finite poles of $H_2(s)$ near $\pm 12j$, but in the left half of the complex plane, and that there are two finite zeros at the origin. So there is no zero at $s = \infty$. Hence

    (i) $|H(j\omega)| \rightarrow 0$ as $\omega \rightarrow 0$

    (ii) $|H(j\omega)| \rightarrow$ constant as $\omega \rightarrow \infty$

    (iii) $|H(j\omega)| \approx$ maximum value as $\omega \rightarrow 12$

Again we cannot make a stronger general statement in point (iii) above.

Part 3. *Magnitude plots via MATLAB.* Suppose the poles are at $-0.1 \pm j12$ and we desire the actual magnitude response plot for $H_1(j\omega)$, $0 \leq \omega \leq 30$ rad/sec. To construct the plot shown in Figure 14.24, we use the following MATLAB code:

```
»w = 0:.1:20;
»n = [1 0];
»d = poly([-0.1+12*j -0.1-12*j]);
»h = freqs(n,d,w);
»plot(w, abs(h))
»grid
»xlabel('Frequency in rad/sec')
»ylabel('Magnitude H1(jw)')
```

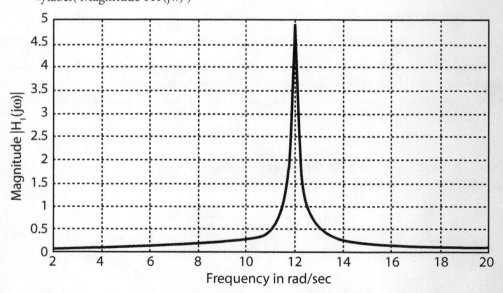

FIGURE 14.24 Magnitude frequency plot, $|H_1(j\omega)|$ vs. ω.

Exercises. 1. Suppose the poles of $H_2(s)$ are at $-0.1 \pm j12$ and that we desire the actual magnitude response plot for $0 \leq \omega \leq 30$ rad/sec. Use MATLAB to construct this plot. Verify the accuracy of the qualitative predictions.

2. Show that the transfer function

$$H(s) = \frac{s^2 + \dfrac{1}{LC}}{s^2 + \dfrac{1}{RC}s + \dfrac{1}{LC}} \tag{14.11}$$

of an unknown circuit with $R = 1\ \Omega$, $L = 0.1$ H, and $C = 1$ mF has the band-reject magnitude response depicted in Figure 14.25. Use MATLAB and the "freqs" command.

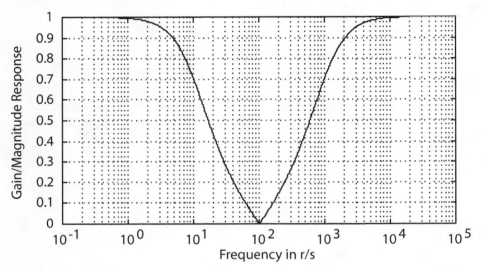

FIGURE 14.25 Plot of the magnitude response of the transfer function of equation 14.11.

The example below further illustrates this relationship with a pedagogically useful graphical technique. Mastering this technique helps concretize the meaning of magnitude and phase and reinforces the qualitative discussion above on using pole and zero locations to compute the magnitude and phase.

EXAMPLE 14.12. To better grasp the ideas of the magnitude, $|H(j\omega)|$, and the phase, $\angle H(j\omega)$, of a frequency response, suppose a transfer function has the form

$$H(s) = \frac{(s - z_1)(s - z_2)}{(s + 1)(s + 1 + j)(s + 1 - j)} \tag{14.12}$$

where $z_1 = 2j$ and $z_2 = -2j$. Figure 14.26a shows the pole-zero plot of $H(s)$.

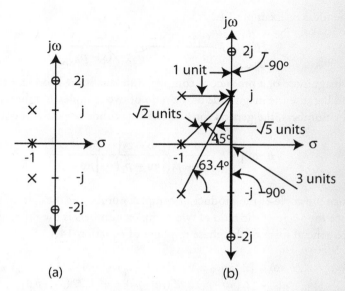

FIGURE 14.26 (a) Pole-zero plot of $H(s)$ as given by equation 14.12. (b) Measuring distances graphically from zeros and poles to the point $j1$.

SOLUTION

The plan of this example is to compute the magnitude of $H(j\omega)$ graphically for $\omega = 1$, i.e., to compute $|H(j1)|$. This computation entails the following steps.

Step 1. Observe that $j1 - z_1$ defines a complex number. Think of $j1 - z_1$ as a vector in the complex plane. Similarly, $j1 - z_2$ and $j1 - p_i$, where $p_1 = -1$ and $p_{2,3} = -1 \pm j$, define vectors. Each vector has a length that can be determined either graphically, by physically measuring the distance with a ruler, or by the Pythagorean theorem. Figure 14.26b illustrates the idea.

Step 2. Following from step 1, the magnitude of $H(j1)$ has the form

$$|H(j1)| = \frac{|j1 - j2| \times |j1 + j2|}{|j1 + 1| \times |j1 + 1 + j| \times |j1 + 1 - j|} = \frac{1 \times 3}{\sqrt{2} \times \sqrt{5} \times 1} = 0.95$$

Step 3. Suppose we wish to compute the phase or angle of $H(j1)$ graphically. In Figure 14.26b, observe that each complex number viewed as a vector $j\omega - z_i$ or $j\omega - p_i$ can be represented in the form $re^{j\psi}$, where ψ is the angle the vector makes with a horizontal line passing through its base. For example, $(j1 - j2) = -j1 = 1e^{-j(\pi/2)}$. From basic complex number theory, the angle of the product of two complex numbers is the sum of the angles, and the angle of the ratio of two complex numbers is the angle of the numerator minus the angle of the denominator. Hence, from the angles shown in Figure 14.26b,

$$\angle H(j1) = \angle(j1 - j2) + \angle(j1 + j2) - \angle(j1 + 1) - \angle(j1 + 1 + j) - \angle(j1 + 1 - j)$$
$$= -90° + 90° - 45° - 0° - 63.4° = -108.4°$$

To extend the ideas of Example 14.12, let

$$H(s) = K \frac{(s - z_1)(s - z_2)\cdots(s - z_m)}{(s - p_1)(s - p_2)\cdots(s - p_n)} \qquad (14.13)$$

Because the magnitude of a product of complex numbers is the product of the magnitudes of the numbers, and because the magnitude of the ratio of two complex numbers is the ratio of the magnitudes of the numbers, the general form of the magnitude response of equation 14.13 is

$$|H(j\omega)| = K \frac{|j\omega - z_1||j\omega - z_2|\cdots|j\omega - z_m|}{|j\omega - p_1||j\omega - p_2|\cdots|j\omega - p_n|} \qquad (14.14a)$$

Similarly, since the angle of the product of complex numbers is the sum of the angles of the numbers, and since the angle of the ratio of two complex numbers is the difference in the angles of the numbers, the general form of the phase response of equation 14.13 is

$$\angle H(j\omega) = [\angle(j\omega - z_1) + \angle(j\omega - z_2) + \cdots + \angle(j\omega - z_m) + \angle K] \qquad (14.14b)$$
$$- [\angle(j\omega - p_1) + \angle(j\omega - p_2) + \cdots + \angle(j\omega - p_n)]$$

Thus, qualitatively speaking, $H(j\omega)$ tends to have a large magnitude for $j\omega$'s near poles and a small magnitude for $j\omega$'s near zeros. As mentioned earlier, this can be used to advantage in estimating the magnitude response and phase response of a transfer function.

Exercise. Draw an estimate of the general shape of the magnitude and phase response of the Butterworth normalized low-pass transfer function

$$H(s) = \frac{1}{s^2 + \sqrt{2}s + 1}.$$

Compute the exact magnitude and phase at $\omega = 1$. What happens to the magnitude and frequency response if $H(s)$ is changed to $H_1(s) = H(s/10)$ and $H_2(s) = H(s/100)$?
ANSWERS: in random order: -45, 0.707, the general shape is the same with $H_1(j10) = H_2(j100) = 0.707$.

---

6. FREQUENCY SCALING AND MAGNITUDE SCALING

Design of filters and amplifiers often begins with a design template in which almost all parameter values are normalized. In particular, a source or load resistance is often set to 1 Ω. Also, a critical or important frequency is set to 1 rad/sec. Such circuits are called *normalized*. With the completion of a normalized design, engineers can frequency-scale to obtain realistic frequency responses and magnitude-scale to obtain reasonable impedance levels as necessary to meet power and energy restrictions.

FREQUENCY SCALING

EXAMPLE 14.13. The circuit of Figure 14.27 realizes the transfer function of equation 14.11 from the previous section for $R = 1\ \Omega$, $L = 0.1$ H, and $C = 1$ mF. We pose the following question. Suppose

$$L \rightarrow \frac{L}{K_f},\ C \rightarrow \frac{C}{K_f},$$

and R remains the same in the circuit; what happens to $H(s)$ and the frequency response plot?

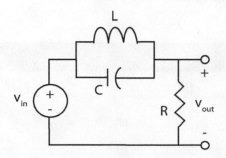

FIGURE 14.27 Band-reject type circuit that realizes the transfer function of equation 14.11.

SOLUTION

Step 1. *Calculate the circuit transfer function.* By voltage division,

$$H(s) = \frac{R}{R + \dfrac{1}{Cs + \dfrac{1}{Ls}}} = \frac{s^2 + \dfrac{1}{LC}}{s^2 + \dfrac{1}{RC}s + \dfrac{1}{LC}}$$

Step 2. *Incorporate the effect of K_f*

$$H_{new}(s) = \frac{R}{R + \dfrac{1}{\dfrac{C}{K_f}s + \dfrac{1}{\dfrac{L}{Kf}s}}} = \frac{R}{R + \dfrac{1}{C\left(\dfrac{s}{K_f}\right) + \dfrac{1}{L\left(\dfrac{s}{K_f}\right)}}}$$

$$= \frac{\left(\dfrac{s}{K_f}\right)^2 + \dfrac{1}{LC}}{\left(\dfrac{s}{K_f}\right)^2 + \dfrac{1}{RC}\left(\dfrac{s}{K_f}\right) + \dfrac{1}{LC}} = H\left(\dfrac{s}{K_f}\right)$$

We conclude that when

$$L \rightarrow \frac{L}{K_f} \text{ and } C \rightarrow \frac{C}{K_f}$$

with R unchanged, the new and old transfer functions are related in a very simple way:

$$H_{new}(s) = H_{old}\left(\frac{s}{K_f}\right)$$

Step 3. *Plot the magnitude response.* Rather than go directly to MATLAB, consider that

$$H_{new}(j\omega) = H_{old}\left(j\frac{\omega}{K_f}\right)$$

Table 14.1 depicts this relationship for specific values of ω.

TABLE 14.1

ω (rad/s)	Transfer Function Relationship
0	$H_{new}(0) = H_{old}(0)$
1	$H_{new}(j1) = H_{old}\left(j\dfrac{1}{K_f}\right)$
K_f	$H_{new}(jK_f) = H_{old}\left(j\dfrac{K_f}{K_f}\right) = H_{old}(j)$
$2K_f$	$H_{new}(j2K_f) = H_{old}\left(j\dfrac{2K_f}{K_f}\right) = H_{old}(j2)$

In general, the new transfer function is related to the original $H_{old}(s)$ through the replacement of s by s/K_f The implication of this relationship is that whatever happened at $s = j\omega_{old}$ before scaling must now happen at $s = j\omega_{new}$, where $\omega_{new} = K_f\omega_{old}$—hence the term *frequency scaling*. In terms of the plot of our specific transfer function, then, we obtain from Figure 14.25 the new plot of Figure 14.28 by inspection.

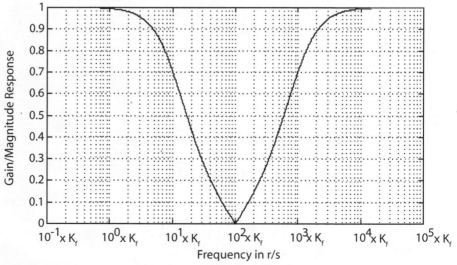

FIGURE 14.28 The new magnitude response after "frequency scaling" by K_f

For general linear circuits, we will only *state* the procedures for scaling and their effects, leaving a rigorous mathematical justification to a more advanced text on circuit theory.

Procedure for frequency scaling: To frequency-scale a linear network by a positive scale factor K_f:
1. Divide all inductances and capacitances by K_f
2. Leave all resistance values and controlled source parameters unchanged.

Effect of frequency scaling:
1. $H_{new}(s) = H(s/K_f)$

2. If $H(s) = K\dfrac{(s-z_1)\cdots(s-z_m)}{(s-p_1)\cdots(s-p_n)}$, then $H_{new}(s) = K(K_f)^{n-m}\dfrac{(s-K_f z_1)\cdots(s-K_f z_m)}{(s-K_f p_1)\cdots(s-K_f p_n)}$

    which means that $H_{new}(s)$ has zeros and poles at $K_f z_i$ and $K_f p_k$ for $i = 1, \dots , m$ and $k = 1, \dots , n$.
3. The magnitude and phase response curves of $H_{new}(j\omega)$ are those of $H(j\omega)$ with the frequency scale multiplied by K_f Conversely, the magnitude and phase of $H(j\omega)$ are the same as for $H_{new}(jK_f\omega)$.

Exercises. 1. Fill in the details of the derivation of $H_{new}(s)$ in step 2.

2. Given $H_{new}(s) = K(K_f)^{n-m}\dfrac{(s-K_f z_1)\cdots(s-K_f z_m)}{(s-K_f p_1)\cdots(s-K_f p_n)}$, verify the "converse" statement in point 3 above.

3. Given the circuit transfer function $H(s) = K\dfrac{s-z_1}{(s-p_1)(s-p_2)}$

find the new pole and zero locations if the circuit is frequency-scaled by $K_f= 1000$.
ANSWERS: $1000p_1$, $1000p_2$, and $1000z_1$

EXAMPLE 14.14. The circuit of Figure 14.29a has transfer function

$$H(s) = \frac{0.1s}{s^2 + 0.1s + 1}$$

Figure 14.29b shows the pole-zero plot and Figure 14.29c shows the magnitude response with peak value at $\omega_m = 1$ rad/sec.
(a) Frequency-scale the circuit by the factor $K_f = 10^6$ and compute the new transfer function.
(b) Compute the new pole-zero plot and the new magnitude response curve.

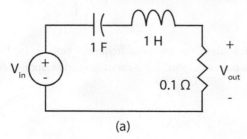

(a)

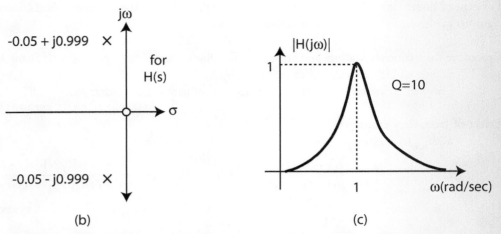

(b)   (c)

FIGURE 14.29 (a) Series resonant circuit. (b) Pole-zero plot. (c) Magnitude response.

SOLUTION

(a) Dividing the capacitance and inductance by $K_f = 10^6$ yields the circuit of Figure 14. 30a.

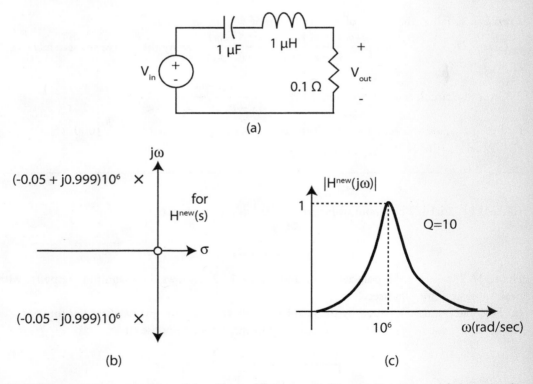

(b)   (c)

FIGURE 14.30 Demonstration of the effects of frequency scaling. (a) Frequency-scaled circuit. (b) New pole-zero plot. (c) New magnitude response curve.

Using "effects of frequency scaling" property 2, the new transfer function is

$$H_{new}(s) = H\left(\frac{s}{10^6}\right) = \frac{0.1\left(\frac{s}{10^6}\right)}{\left(\frac{s}{10^6}\right)^2 + 0.1\left(\frac{s}{10^6}\right) + 1}$$

(b) The ideas stated in part (a) are borne out by comparing the pole-zero plot and the magnitude response curve for $H_{new}(s)$, shown in Figures 14. 30b and c, with their counterparts for $H(s)$ in Figure 14.29. Notice that the pole locations have been scaled by K_f as per property 2. A direct computation in MATLAB verifies this:

```
»pnew = roots([1/1e12 0.1/1e6 1])
pnew =
 -5.0000e+04 + 9.9875e+05i
 -5.0000e+04 - 9.9875e+05i
»pold = roots([1 0.1 1])
pold =
 -5.0000e-02 + 9.9875e-01i
 -5.0000e-02 - 9.9875e-01i

»% Multiply pold by Kf and check with pnew
»1e6*pold
ans =
 -5.0000e+04 + 9.9875e+05i
 -5.0000e+04 - 9.9875e+05i
```

Also, in Figure 14.29c for $H(s)$, the peak response occurs at $\omega_{old} = 1$ rad/sec, whereas in Figure 14. 30c for $H_{new}(s)$, the peak response occurs at $\omega_{new} = K_f \omega_{old} = 10^6$ rad/sec.

At this point we round out our discussion of frequency scaling by relating it to the time domain via the time/frequency scaling property of the Laplace transform:

$$\mathcal{L}\left[h(K_f t)\right] = \frac{1}{K_f} H\left(\frac{s}{K_f}\right)$$

or, equivalently,

$$K_f \mathcal{L}\left[h(K_f t)\right] = H\left(\frac{s}{K_f}\right)$$

where $H(s)$ is the circuit transfer function and $h(t)$ is the circuit **impulse response**.

EXAMPLE 14.15. Consider the transfer function

$$H(s) = \frac{30}{(s+2)(s+5)} = \frac{10}{(s+2)} - \frac{10}{(s+5)}$$

whose impulse response is

$$h(t) = (10e^{-2t} - 10e^{-5t})u(t)$$

Find the impulse response when the transfer function is frequency-scaled by K_f

SOLUTION
The scaled transfer function is

$$H_{new}(s) = \frac{30K_f^2}{(s+2K_f)(s+5K_f)} = \frac{10K_f}{(s+2K_f)} - \frac{10K_f}{(s+5K_f)}$$

Taking the inverse Laplace transform yields

$$h_{new}(t) = \left(10K_f e^{-2K_f t} - 10K_f e^{-5K_f t}\right)u(t) = K_f\, h(K_f t)$$

Suppose for the sake of argument that $K_f = 10^3$. Let us plot $h(t)$ and $h_{new}(t)$, as has been done in Figures 14.31a and b.

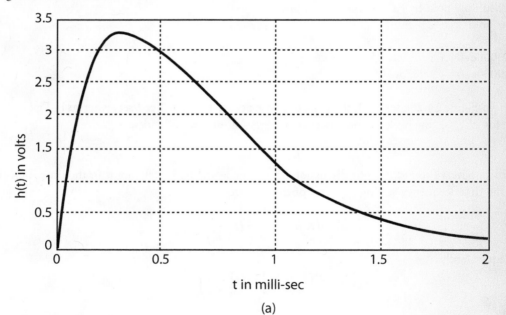

t in milli-sec

(a)

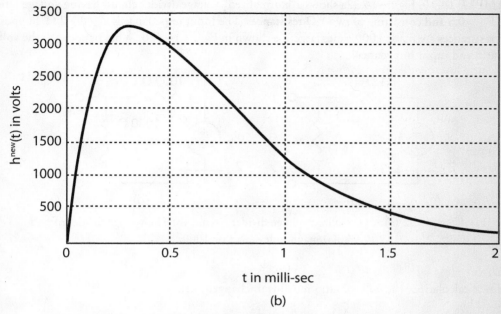

FIGURE 14.31 Illustration of the effect of frequency scaling on impulse response.
(a) Original $h(t)$. (b) $h_{new}(t)$.

Observe that the plots are structurally the same. However, the time scale in **Figure 14.30b** is now in milliseconds and the magnitude has been scaled by $K_f = 10^3$.

Exercises. 1. Use the Laplace transform time scale property to show that in general

$$h_{new}(t) = K_f h(K_f t)$$

2. In Example 14.15, suppose the poles are changed from -2 and -5 to -3 and -6, respectively. Suppose further that the transfer function is scaled by $K_f = 10$. Find the impulse response and the new impulse response after scaling.
ANSWER: $h(t) = (10e^{-3t} - 10e^{-6t})u(t)$ and $h_{new}(t) = (100e^{-30t} - 100e^{-60t})u(t)$

MAGNITUDE SCALING

Frequency scaling has allowed us to obtain realistic frequency responses from normalized circuits. Another technique in achieving a realistic design is magnitude or impedance scaling. A simple example illustrates the idea.

EXAMPLE 14.16. Figure 14.32a shows a "normalized" voltage divider circuit having voltage ratio $V_{out}/V_{in} = 0.5$ and consisting of two 1 Ω resistances. The input impedance is $Z_{in}(s) = 2$ Ω. Suppose both resistances are made 1000 times larger, as shown in Figure 14.32b. What happens to the voltage ratio and input impedance?

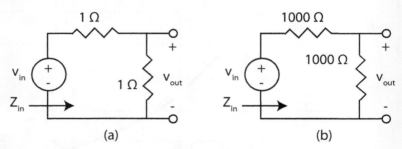

FIGURE 14.32 (a) A normalized voltage divider circuit. (b) The voltage divider of part (a) magnitude-scaled by 1000.

SOLUTION
By direct calculation, the voltage ratio remains unchanged, i.e.,

$$\left(\frac{V_{out}}{V_{in}}\right)_{new} = \frac{1000}{1000+1000} = 0.5 = \left(\frac{V_{out}}{V_{in}}\right)_{old}$$

The new input impedance is

$$Z_{in}^{new}(s) = 1000 \times 1 + 1000 \times 1 = 1000 Z_{in}^{old}(s)$$

which is 1000 times larger. The elements in Figure 14.32b are said to be *magnitude-scaled* (by 1000) from those of Figure 14.32a.

The above example motivates a more general discussion. Suppose each impedance in Figure 14.33a is scaled (multiplied) by K_m to yield the circuit of Figure 14.33b. As in Example 14.16, the voltage ratios remain the same for both circuits:

$$\left[\frac{V_{out}}{V_{in}}\right]_{(b)} = \frac{K_m Z_2}{K_m Z_1 + K_m Z_2} = \frac{Z_2}{Z_1 + Z_2} = \left[\frac{V_{out}}{V_{in}}\right]_{(a)}$$

Further, the input impedances of Figures 14.33a and b are related as

$$Z_{in}^b(s) = K_m Z_1(s) + K_m Z_2(s) = K_m Z_{in}^a(s)$$

Similar to the case in Example 14.16, the input impedance is increased by the scale factor K_m. The network of Figure 14.33b is said to be obtained from that of Figure 14.33a by magnitude scaling with scale factor K_m. If Z_1 is an inductance, then $Z_1(s) = Ls$ and $K_m Z_1(s) = (K_m L)s$, i.e., the inductance is *multiplied* by K_m. On the other hand, if Z_1 is a capacitance, then $Z_1(s) = 1/Cs$ and $K_m Z_1(s) = K_m/Cs = s/(C/K_m)$, i.e., the capacitance is *divided* by K_m.

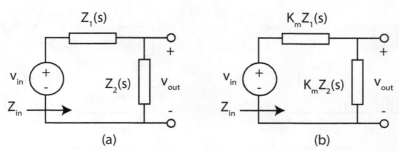

FIGURE 14.33. (a) A general voltage divider. (b) Magnitude scaling of the circuit in part (a) by K_m.

The above discussion suggests that magnitude scaling can be achieved by multiplying resistances and inductances by K_m and dividing conductances and capacitances by K_m. This and several other properties are stated next.

Procedure for magnitude scaling: To magnitude-scale a linear network by a scale factor K_m:
1. Multiply all resistances and inductances by K_m.
2. Divide all capacitances and conductances by K_m.
3. For current-controlled voltage sources (CCVS), i.e., the r_m type, multiply the parameter r_m by K_m.
4. For voltage-controlled current sources (VCCS), i.e., the g_m type, divide the parameter g_m by K_m.
5. Parameters for voltage-controlled voltage sources and current-controlled current sources remain unchanged.
6. Ideal operational amplifiers remain unchanged.

The effect of magnitude scaling on a transfer function is set forth below, with its verification left for more advanced courses on circuit theory.

Effect of magnitude scaling on transfer functions: If $H(s)$ is a voltage ratio or current ratio, magnitude scaling has no effect on $H(s)$. If $H(s)$ has units of ohms, then the magnitude-scaled network has $H_{new}(s) = K_m H(s)$. If $H(s)$ has units of siemens or mhos, then the magnitude-scaled network has $H_{new}(s) = H(s)/K_m$.

EXAMPLE 14.17. The series circuit of Figure 14.34a has input impedance

$$Z(s) = \frac{1}{s} + s + 0.1 = \frac{s^2 + 0.1s + 1}{s}$$

and transfer function

$$H(s) = \frac{V_{out}}{V_{in}} = \frac{0.1}{Z(s)} = \frac{0.1s}{s^2 + 0.1s + 1}$$

(a) Magnitude-scale the network by $K_m = 1000$.
(b) Calculate $Z_{new}(s)$ and $H_{new}(s)$ to verify the effects stated above.

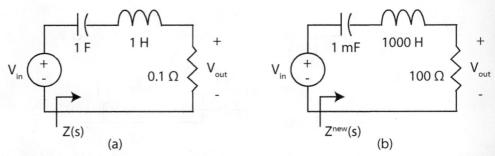

FIGURE 14.34 Magnitude scaling. (a) Original network. (b) Network magnitude scaled by 1000.

SOLUTION
(a) Following the procedure for scaling R's, L's, and C's, we obtain the scaled network shown in Figure 14.34b.
(b) For the simple circuit of Figure 14.34b, the new input impedance is

$$Z_{new}(s) = \frac{1}{0.001s} + 1000s + 100 = 1000\frac{s^2 + 0.1s + 1}{s} = 1000Z(s) = K_m Z(s)$$

By the voltage divider formula, the new transfer function is

$$H_{new}(s) = \frac{100}{1000\dfrac{s^2 + 0.1s + 1}{s}} = \frac{0.1s}{s^2 + 0.1s + 1} = H(s)$$

Because the transfer function is a voltage ratio, there is no change. These results clearly illustrate the stated effects of magnitude scaling.

Exercises. 1. Magnitude-scale the circuit of Figure 14.34a by $K_m = 5$.
ANSWER: 0.5 Ω, 5 H, 0.2 F

2. Two parallel resistors R_1 and R_2 are magnitude-scaled by K_m. Verify that $R_{eq}^{new} = K_m R_{eq}^{old}$. Hint. Use the formula for parallel resistance to compute R_{eq}^{new} and then relate that to R_{eq}^{old}.

COMBINED MAGNITUDE AND FREQUENCY SCALING

Moving from normalized circuit design to realistic circuit design ordinarily entails both magnitude and frequency scaling. This subsection provides several illustrative examples.

EXAMPLE 14.18. The amplifier circuit shown in Figure 14.35a consists of two stages. The first stage is a Sallen and Key low-pass active filter. The transfer function of the Sallen and Key circuit is

$$H_1(s) = \frac{V_a(s)}{V_{in}(s)} = \frac{1}{s^2 + s + 1}$$

The gain of the second stage is

$$H_2(s) = \frac{V_{out}}{V_a} = \frac{-2}{s+1}$$

Therefore the transfer function of the amplifier is

$$H(s) = \frac{V_{out}}{V_{in}} = \frac{V_{out}}{V_a} \frac{V_a}{V_{in}} = H_2(s)H_1(s) = \frac{-2}{(s+1)} \frac{1}{(s^2+s+1)} = \frac{-2}{s^3 + 2s^2 + 2s + 1}$$

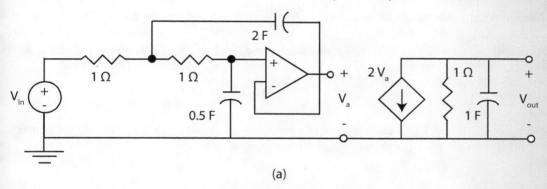

(a)

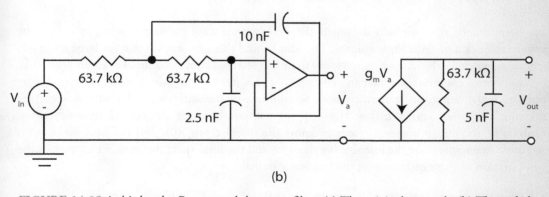

(b)

FIGURE 14.35 A third-order Butterworth low-pass filter. (a) The original network. (b) The scaled network with $K_f = 2000\pi$ and $K_m = 10,000$; $g_m = 31.4 \ \mu S$.

The overall transfer function of the two-stage amplifier is a third-order, maximally flat (Butterworth) low-pass filter, which will be studied in Chapter 19. The magnitude response of the transfer function has a 3 dB down frequency at $\omega = 1$ rad/sec, or $1/(2\pi)$ Hz $= 0.159$ Hz—an extremely non-useful audio frequency. In fact, when the 3 dB down frequency is at $\omega = 1$ rad/sec, the magnitude response is said to be *normalized*. Not only is 1 rad/sec not useful, but the element values are unsuitable for practical applications. However, using both magnitude and frequency scaling, this "reference" or normalized amplifier circuit can be made into a very practical filter.

Suppose we wish to have the 3 dB down frequency at f_{3dB} = 500 Hz and the largest capacitor at 10 nF. Such a filter could be used to direct the low-frequency content of a music signal to a woofer. Our goal requires that we frequency-scale the circuit by K_f = 1000π and magnitude-scale such that the 2 F capacitor (the largest) becomes 10 nF; i.e., K_m must satisfy

$$10 \times 10^{-9} = \frac{2}{K_f K_m} = \frac{2}{1000\pi K_m}$$

Solving for K_m yields

$$K_m = 6.366 \times 10^4$$

The scaled circuit meeting the requirements is shown in Figure 14.35b, where g_m = 31.4 μS.

Exercise. A circuit has transfer function $H(s) = \dfrac{V_{out}(s)}{I_{in}(s)} = \dfrac{cs}{s^2 + as + b}$, with a = 0.1, b = 4 and c = –40.

After both magnitude and frequency scale changes of K_m = 4 and K_f = 2, compute the new a, b, and c.

ANSWER: in random order: –320, 0.2, and 16

Examples 14.15 and 14.16 illustrate one of several reasons for scaling a linear network. Chapter 19, on elementary filter design, will detail further applications.

7. INITIAL- AND FINAL-VALUE THEOREMS

In system theory, and especially in control theory, engineers want the output signal of a circuit or system to track a given reference signal. The idea behind the term *track* is that for large t, the reference signal and the circuit output are more or less indistinguishable. To accomplish this, design/control engineers generate an error signal, $e(t) = y(t) - y_{ref}(t)$, where $y(t)$ is the circuit output and $y_{ref}(t)$ is a given reference signal. Since much of the analysis is done in the frequency domain, one often knows $E(s) = Y(s) - Y_{ref}(s)$ without knowing the related time functions. Ordinarily, the design engineer needs to know the initial error, $e(0)$, and the final error, $e(\infty)$. Available to engineers are the initial-value theorem and the final-value theorem, which permit the computation of these quantities in the frequency domain.

Initial-value theorem: Let $\mathcal{L}[f(t)] = F(s)$ be a strictly proper rational function of s, i.e., the numerator and denominator of F are both polynomials in s, with the degree of the numerator less than that of the denominator. Then

$$\lim_{s \to \infty} sF(s) = f(0^+) \tag{14.15}$$

Proof: The quantity $sF(s)$ suggests a derivative operation on $f(t)$. Specifically,

$$sF(s) - f(0^-) = \mathcal{L}\left[\frac{d}{dt}f(t)\right]$$

Applying the definition of the Laplace transform to the right-hand side and taking limits as s approaches infinity implies that

$$\lim_{s \to \infty} [sF(s) - f(0^-)] = \lim_{s \to \infty} \left[\int_{0^-}^{0^+} \dot{f}(t)e^{-st}dt + \int_{0^+}^{\infty} \dot{f}(t)e^{-st}dt \right]$$

(14.16)

$$= \lim_{s \to \infty} \left[\int_{0^-}^{0^+} \dot{f}(t)e^{-st}dt \right] + \lim_{s \to \infty} \left[\int_{0^+}^{\infty} \dot{f}(t)e^{-st}dt \right],$$

where the dot over the function $f(t)$ indicates the derivative of the function. Observe that

$$\lim_{s \to \infty} \left[\int_{0^+}^{\infty} \dot{f}(t)e^{-st}dt \right] = \int_{0^+}^{\infty} \dot{f}(t) \lim_{s \to \infty} (e^{-st})dt = 0$$

and that, because e^{-st} is continuous at $t = 0$,

$$\lim_{s \to \infty} \left[\int_{0^-}^{0^+} \dot{f}(t)e^{-st}dt \right] = \left[\lim_{s \to \infty} e^{-s0} \right] \int_{0^-}^{0^+} \dot{f}(t)dt = f(0^+) - f(0^-).$$ (14.17)

Hence, the left-hand side of equation 14.16 equals the right-hand side of equation 14.17. Equation 14.15 follows from equating these two sides and canceling the $f(0^-)$ terms in both.

EXAMPLE 14.19. The Laplace transform of a capacitor voltage is given by

$$V_C(s) = \frac{2}{s} - \frac{1}{5s+2}$$

Find the initial capacitor voltage $v_C(0^+)$.

SOLUTION
By direct application of the initial-value theorem,

$$v_C(0^+) = \lim_{s \to \infty} sV_C(s) = \lim_{s \to \infty} \left[2 - \frac{s}{5s+2} \right] = 2 - \frac{1}{5} = 1.8 \text{ V}$$

EXAMPLE 14.20. Let the Laplace transform of the velocity of a certain projectile be given by

$$V(s) = \frac{500s + 20}{s(5s+20)(10s+1)}$$

Find the initial velocity, $v(0^+)$, and the initial acceleration, $a(0^+)$.

SOLUTION
To find the initial velocity, we directly apply the initial-value theorem:

$$v(0^+) = \lim_{s \to \infty} sV(s) = \lim_{s \to \infty} \left[\frac{500s + 20}{(5s+20)(10s+1)} \right] = 0$$

Since acceleration is the derivative of velocity, and the initial velocity is zero, from the time differentiation property of Table 12.2, assuming the velocity is continuous at $t = 0$,

$$A(s) = sV(s) - v\left(0^+\right) = \frac{500s + 20}{(5s + 20)(10s + 1)}$$

From the initial-value theorem in standard units,

$$a(0^+) = \lim_{s \to \infty} \left[\frac{500s^2 + 20s}{(5s + 20)(10s + 1)} \right] = \lim_{s \to \infty} \left[\frac{500 + \dfrac{20}{s}}{\left(5s + \dfrac{20}{s}\right)\left(10s + \dfrac{1}{s}\right)} \right] = 10$$

Exercise. Suppose $F(s) = (8s + 2)/(2s^2 + 8s + 3)$. Find $f(0^+)$.

ANSWER: 4

The initial-value theorem and the examples that follow it illustrate the computation of initial values. To compute final values we use the next theorem.

Final-Value Theorem: Suppose $F(s)$ has poles only in the open left half of the complex plane, with the possible exception of a single-order pole at $s = 0$. Then

$$\lim_{s \to 0} sF(s) = \lim_{t \to \infty} f(t).$$

(14.18)

Proof: The condition of the theorem, i.e., the condition that $F(s)$ has poles only in the left half of the complex plane, with the possible exception of a first-order pole at the origin, guarantees that the limit on the right side of equation 14.18 exists. This is because a partial fraction expansion of $F(s)$ leads to a time function $f(t)$ that is a sum of exponentially decaying signals and at most one constant signal. Since the right-hand limit is well defined,

$$\lim_{s \to 0} \left[sF(s) - f(0^-) \right] = \lim_{s \to 0} \left[\int_{0^-}^{\infty} \dot{f}(t) e^{-st} dt \right] = \int_{0^-}^{\infty} \dot{f}(t) dt$$

$$= \left(\lim_{t \to \infty} f(t) \right) - f(0^-).$$

This implies equation 14.18.

EXAMPLE 14.21. As in Example 14.20, suppose the velocity, $v(t)$, of a certain projectile has Laplace transform

$$V(s) = \frac{500s + 20}{s(5s + 20)(10s + 1)}$$

with the Laplace transform of the acceleration, $a(t)$, given by

$$A(s) = sV(s) - v\left(0^+\right) = \frac{500s + 20}{(5s + 20)(10s + 1)}$$

Find the final values of $v(t)$ and $a(t)$ if possible.

Solution

$V(s)$ and $A(s)$ have poles that meet the conditions of the final-value theorem. Hence,

$$\lim_{t\to\infty} v(t) = \lim_{s\to 0} sV(s) = \lim_{s\to 0}\left[\frac{500s + 20}{(5s + 20)(10s + 1)}\right] = 1$$

and

$$\lim_{t\to\infty} a(t) = \lim_{s\to 0} sA(s) = \lim_{s\to 0}\left[\frac{500s^2 + 20s}{(5s + 20)(10s + 1)}\right] = 0$$

Observe that a constant final velocity implies a zero acceleration as these expressions indicate.

EXAMPLE 14.22. What if the conditions of the final-value theorem are not met? What would go wrong? A simple example illustrates the problem. Let

$$F(s) = \frac{1}{s^2 + 1}$$

which corresponds to $f(t) = \sin(t)u(t)$. Then

$$\lim_{s\to 0} sF(s) = \lim_{s\to 0}\frac{s}{s^2 + 1} = 0,$$

but

$$\lim_{t\to\infty} f(t) = \lim_{t\to\infty} \sin(t)u(t)$$

is undefined, i.e., it does not exist. The theorem, however, presupposes that both limits exist. Again, the condition of poles in the left half complex of the plane with at most one pole at the origin is necessary and sufficient for both limits to exist.

Exercises. 1. If $F(s) = (6s + 10)/(2s^2 + 4s)$, then, by the final value theorem, $f(t)$ approaches what value for large t?
ANSWER: 2.5

2. The Laplace transform of a signal, $5y(t - 2)u(t - 2)$, is

$$\frac{15e^{-2s}(s^2 + s - 2)}{s(s^2 + 5s + 6)}$$

Find the value of $y(t)$ for very large t.
ANSWER: −5

8. BODE PLOTS

Section 5 described the use of the poles and zeros of $H(s)$ to compute the frequency response of a circuit. In this regard, Hendrik Bode developed a technique for computing approximate or asymptotic frequency response curves. These so-called Bode plots can be quickly drawn by hand. A description of the technique requires the introduction of some terms widely used in the engineering literature.

Let $H(s)$ be a transfer function that is a dimensionless voltage ratio or a current ratio. As explained in section 4, for sinusoidal steady-state analysis, one replaces s by $j\omega$ to study the circuit's magnitude response, $|H(j\omega)|$, and phase response, $\angle H(j\omega)$. For convenience, let $|H(j\omega)|$ be a voltage gain, $|V_2/V_1|$. The *gain in dB* (decibels), denoted by H_{dB}, is defined by the equation

$$H_{dB}(j\omega) \equiv 20 \log_{10}|H(j\omega)| \tag{14.19}$$

For convenience, whenever we write $\log(x)$, we will mean $\log_{10}(x)$. Solving for $|H(j\omega)|$ in equation 14.19 yields the inverse relationship

$$|H(j\omega)| = 10^{0.05 H_{dB}(j\omega)} \tag{14.20}$$

Table 14.2 presents some pairs of $|H|$ and H_{dB}. Thus, instead of saying that $|V_2|$ is 10 times $|V_1|$, we may say that V_2 is 20 dB *above* V_1, or that V_1 is 20 dB *below* V_2. Similarly, to say that V_2 is 3 dB above V_1 means that $|V_2|$ is 1.414 times $|V_1|$.

TABLE 14.2 Transfer Function Gain in Magnitude and in Decibels

$|H|$	1	$\sqrt{2}$	2	10	100	1000
H_{dB}	0	$\cong 3$	$\cong 6$	20	40	60

Exercise. A certain amplifier has a dc gain of 80 dB. What is the actual voltage gain?
ANSWER: 10^4

One of the reasons for using the dB terminology is that it simplifies the analysis and design of multistage amplifiers. Suppose an amplifier has three stages with voltage gains equal to 20, 100, and 10, respectively. The overall voltage gain is the *product* of the gains of each individual stage, which is $20 \times 100 \times 10 = 20{,}000$. Using the dB specification, the overall gain in dB is the *sum* of the dB gains of the individual stages. This is $(26 + 40 + 20) = 86$ dB. It is easy to justify this claim. First,

$$|H| = |H_1| \times |H_2| \times |H_3|$$

Taking the logarithm of both sides and multiplying by 20 yields

$$20 \log|H| = 20 \log|H_1| + 20 \log|H_2| + 20 \log|H_3|$$

or

$$H_{dB} = H_{1,dB} + H_{2,dB} + H_{3,dB}$$

This summation has pronounced advantages for repetitive calculations at many frequency points, as when plotting a magnitude response such as equation 14.14a. We could, of course, convert this equation to an equation having all terms in dB. However, with an eye toward a further simplification, it is desirable to first rewrite $H(s)$ in a slightly different, but equally general, form, namely

$$H(s) = Ks^{\alpha} \frac{\left(\dfrac{s}{-z_1}+1\right)\left(\dfrac{s}{-z_2}+1\right)\cdots}{\left(\dfrac{s}{-p_1}+1\right)\left(\dfrac{s}{-p_2}+1\right)\cdots} \tag{14.21}$$

where $\{p_1, p_2, \ldots\}$ are those poles of $H(s)$ that are not at the origin and $\{z_1, z_2, \ldots\}$ are those zeros of $H(s)$ that are not at the origin; if α is positive (negative), then $H(s)$ has α zeros (poles) at the origin. For example, a transfer function

$$H(s) = 1.2\frac{(s+50)(s+200)}{(s+8)(s+600)}$$

has the equivalent form

$$H(s) = 2.5\frac{\left(\dfrac{s}{50}+1\right)\left(\dfrac{s}{200}+1\right)}{\left(\dfrac{s}{8}+1\right)\left(\dfrac{s}{600}+1\right)} \tag{14.22}$$

Observing that $H_{dB} = 10 \log_{10}|H(j\omega)|^2$, setting $s = j\omega$ in equation 14.21, and noting that the magnitude squared of a complex number is the imaginary part squared plus the real part squared yields

$$H_{dB}(\omega) = |K|_{dB} + \left[\omega^{\alpha}\right] dB + \left[\left(\frac{\omega}{z_1}\right)^2+1\right]^{0.5}_{dB} + \left[\left(\frac{\omega}{z_2}\right)^2+1\right]^{0.5}_{dB} + \cdots \tag{14.23}$$

$$- \left[\left(\frac{\omega}{p_1}\right)^2+1\right]^{0.5}_{dB} - \left[\left(\frac{\omega}{p_2}\right)^2+1\right]^{0.5}_{dB} - \cdots.$$

Equation 14.23 suggests that we may compute the dB vs. ω curve for each term on the right-hand side and graphically sum the curves to obtain the desired H_{dB} vs. ω curve. However, each individual curve is reasonably sketched by using $\log(\omega)$ instead of ω as the independent variable. This amounts to using semi-log paper to plot the dB vs. ω curves. Such a plot, with a linear scale for the dB values on the vertical axis and a logarithmic scale for ω on the horizontal axis, is called a **Bode magnitude plot**, in honor of its inventor. Similarly, a plot of $\angle H(j\omega)$ vs. ω, with a linear scale for $\angle H(j\omega)$ and a log scale for ω, is called a **Bode phase plot**. Note that, because of the logarithmic scale for the ω-axis, the actual distance on the paper between $\omega = 1$ and $\omega = 10$ is the same as that between $\omega = 0.1$ and $\omega = 1$. (See Figure 14.36a.) Note also that the $\omega = 0$ point will not appear on the graph, because $\log(\omega)$ approaches $-\infty$ as ω approaches 0.

With $\log(\omega)$ chosen as the independent variable, the plot of each term in equation 14.23 either is exactly a straight line or is a curve having two straight line **asymptotes**. This is illustrated in Figure 14.36.

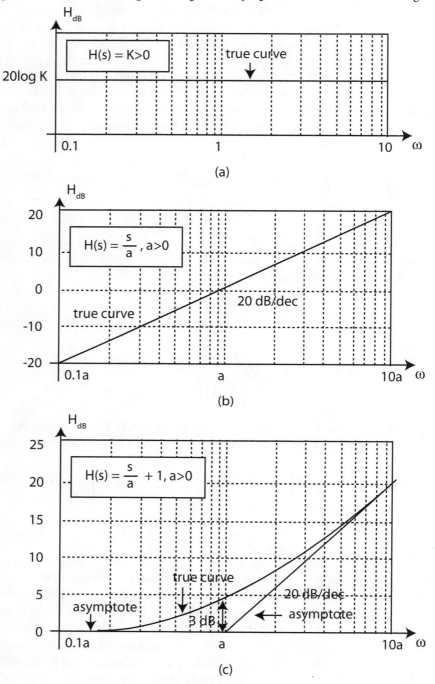

FIGURE 14.36 Bode magnitude plots for three basic transfer functions.
(a) $H(s) = K > 0$. (b) $H(s) = s/a$, $a > 0$. (c) $H(s) = s/a + 1$, $a > 0$.

In Figures 14.36b and 14.36c, the rising asymptote has a slope of 20 dB/decade, which means that along this line, an increase in frequency by a factor of 10 causes an increase in gain of 20 dB. Here, the word *decade* (abbreviated dec) simply means "10 times." Another way to express the same slope is to indicate the increase of gain in dB when the frequency is doubled, or increased by an **octave**, in music terminology. It is easy to see that 20 dB/dec is equivalent to 6 dB/octave. In general, if a frequency, ω_2, is d decades above another frequency, ω_1, then, by definition, $(\omega_2/\omega_1) = 10^d$. Conversely, if $(\omega_2/\omega_1) = r$, then we say that ω_2 is $\log(r)$ decades above ω_1.

In Figure 14.36c, the left asymptote is a horizontal line and hence has a slope of 0 dB/dec. The point where the two asymptotes intersect is called the **breakpoint**, and the corresponding frequency is called the *break frequency* or *corner frequency*.

The derivations of the true curves and asymptotes in Figures 14.36a and 14.36b are very simple and are left as exercises. For figure 14.36c,

$$H(s) = \frac{s}{a} + 1$$

in which case

$$|H(j\omega)| = \left|\frac{j\omega}{a} + 1\right| = \sqrt{\left(\frac{\omega}{a}\right)^2 + 1}$$

and

$$|H(j\omega)|_{dB} = 20\log\left[\sqrt{\left(\frac{\omega}{a}\right)^2 + 1}\right]$$

For $\omega \ll a$,

$$|H(j\omega)|_{dB} = 20\log\left[\sqrt{1}\right] = 0$$

indicating that $|H(j\omega)|_{dB}$ approaches the left asymptote in the figure. For $\omega \gg a$,

$$|H(j\omega)|_{dB} = 20\log\left[\sqrt{\left(\frac{\omega}{a}\right)^2 + 1}\right] \cong 20\log\left[\frac{\omega}{a}\right] = 20\log(\omega) - 20\log(a)$$

indicating that $|H(j\omega)|_{dB}$ approaches the right asymptote in the figure. The two asymptotes intersect at the point $(\omega = a, |H(j\omega)|_{dB} = 0)$. At this corner frequency, the largest error, 3 dB, occurs between the true value of $|H(j\omega)|_{dB}$ and the value read from the asymptotic curve. The error at twice or half the corner frequency is about 1 dB.

The following variations of the three basic Bode magnitude plots of Figure 14.36 are easily derived:
1. If $H(s) = (s/a)^n$, the Bode magnitude is similar to that shown in Figure 14.36b, except that the slope is now $20n$ dB/dec. The curve still passes through the point $(\omega = a, |H(j\omega)|_{dB} = 0)$.
2. If $H(s) = (s/a + 1)^n$, the Bode magnitude is similar to that shown in figure 14.36c, except that the right asymptote has a slope of $20n$ dB/dec. The breakpoint is still at $(\omega = a, |H(j\omega)|_{dB} = 0)$, and the error at the corner frequency is $3n$ dB. If n is negative, the right asymptote points downward.

Let us now consider the Bode plot for a general transfer function $H(s)$. After expressing $H(s)$ in the form of equation 14.21, we can draw the asymptotes for each term in equation 14.23 with the aid of Figure 14.26. The asymptotes for the H_{dB} vs. $\log(\omega)$ curve can then be constructed very easily by graphically summing the individual asymptotes. Since the asymptote for each term in equation 14.23 is a *piecewise linear curve*, the graphical sum of all the asymptotes is also a piecewise linear curve. Accordingly, it is not necessary to calculate the sum of dB values at a large number of frequencies. If there are n break frequencies, then the summation need only be carried out at these frequencies and for the slopes of the leftmost and the rightmost segments of the piecewise linear asymptote. The following example illustrates this procedure.

EXAMPLE 14.23. Obtain the asymptotes for the Bode magnitude plot of $H(s)$ of equation 14.22 rewritten below:

$$H(s) = 2.5 \frac{\left(\dfrac{s}{50}+1\right)\left(\dfrac{s}{200}+1\right)}{\left(\dfrac{s}{8}+1\right)\left(\dfrac{s}{600}+1\right)}$$

SOLUTION
There are four break frequencies. Rewriting $H(s)$ as a product of five factors with break frequencies appearing in *ascending order* yields

$$H(s) = 2.5\left[\frac{s}{8}+1\right]^{-1}\left[\frac{s}{50}+1\right]\left[\frac{s}{200}+1\right]\left[\frac{s}{600}+1\right]^{-1}$$

$$= H_1 \times H_2 \times H_3 \times H_4 \times H_5.$$

(14.24)

Figure 14.37a shows the asymptotes for the five individual terms in equation 14.22, and Figure 14.37b shows the asymptotes for H_{dB}.

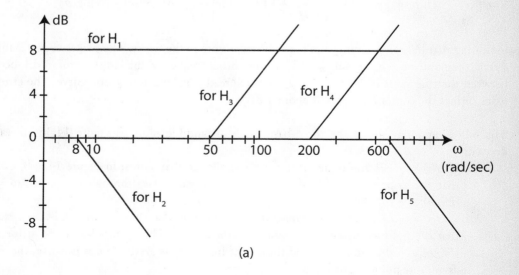

(a)

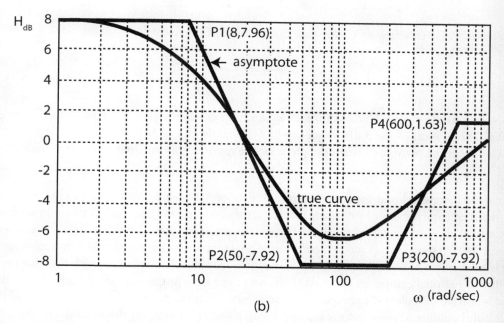

FIGURE 14.37 Asymptote and Bode magnitude for $H(s)$ of equation 14.22 or, equivalently, equation 14.24. (a) Bode plots for H_1 through H_5. (b) Bode plot for H_{dB} equal to sum of those for H_1 through H_5.

The calculation of the asymptote in Figure 14.37b proceeds as follows:

1. The slope of the leftmost segment, i.e., the segment to the left of $\omega = 8$, is obviously zero, from Figure 14.37a.
2. There is a breakpoint P_1 at $\omega = 8$. The only factor contributing to a dB value at this frequency is $H_1(s) = 2.5$ for which $H_{1,dB} = 20 \log(2.5) \cong 8$.
3. There is a breakpoint P_2 at $\omega = 50$. Since $H_2(s)$ contributes -20 dB/dec to the slope for $\omega > 8$, and since 50 rad/sec is 0.796 decade above 8 rad/sec (from $\log(50/8) = 0.796$), the dB value corresponding to P_2 is

$$8 - 20 \times 0.796 = -7.92$$

4. There is a breakpoint P_3 at $\omega = 200$. Since $H_3(s)$ contributes an *additional* 20 dB/dec to the slope for $\omega > 50$, resulting in a slope of zero, we have -7.9 dB for P_3.
5. There is a breakpoint P_4 at $\omega = 600$. Since $H_4(s)$ contributes an *additional* 20 dB/dec to the slope for $\omega > 600$, resulting in a slope of 20 dB/dec, and since 600 rad/sec is 0.477 decade above 200 rad/sec (because $\log(600/200) = 0.477$), the dB value corresponding to P_4 is

$$-7.92 + 20 \times 0.477 = 1.63$$

6. Finally, consider the slope of the rightmost segment. Since $H_5(s)$ contributes an *additional* -20 dB/dec to the slope for $\omega > 600$, the resulting slope of the rightmost segment is zero.

The complete specification of the piecewise linear asymptote is shown in Figure 14.37b.

Exercises. 1. Given a first-order low-pass characteristic of the form

$$H(s) = \frac{V_{out}}{V_{in}} = \frac{K}{1 + \dfrac{s}{a}}$$

show that (i) the dc gain is K and (ii) for $s = ja$, i.e., at $\omega = a$, the gain is 3 dB down from the dc value. This ω is called the **3 dB frequency** or 3 dB bandwidth of $H(s)$.
2. Construct the piecewise linear asymptote for the transfer function $H(s) = 40s/[(s + 2)(s + 20)]$.

Once the piecewise linear asymptote for the Bode plot has been constructed, the true curve can be sketched approximately by noticing that the error at each corner frequency is about $3n$ dB (provided that the two neighboring corner frequencies are more than five times larger or smaller). Figure 14.37b shows such a rough sketch. In the pre-computer era, the ability to draw such a curve by hand—even a crude approximation—was considered valuable. Nowadays, with the wide availability of personal computers and CAD software, one might just as well get the exact plot without bothering to look at the straight-line asymptotes. From the perspective of circuit analysis, the value of the ability to construct the asymptotes for a Bode plot is greatly diminished, but the technique is still important for its application in the design of feedback control systems. Such an application utilizes both the Bode magnitude plot and the Bode phase plot. Some background in control systems is required for one to understand the use of the Bode technique in this kind of application. We will relegate the discussion of the topic to a more advanced course in feedback control. Our objective in mentioning it in this text is twofold: (1) to introduce the definitions of some commonly used terms, such as *decibels*, *decade*, and *octave*, and (2) to demonstrate a highly systematic procedure for adding up several piecewise linear curves to obtain a desired curve, as described in Figure 14.37.

9. TRANSFER FUNCTION ANALYSIS OF A DC MOTOR

A permanent magnet dc motor, an electromechanical device, converts direct current or voltage into mechanical energy. The shaft of the motor rotates freely on bearings. Mounted on the shaft within the housing is a wire coil called the *armature winding* of the motor. Surrounding the coil are permanent magnets that interact with a magnetic field produced when a current flows through the armature winding. This interaction of magnetic fields forces the shaft to rotate in a process of energy conversion. Here current flowing through the armature coils rotating through the magnetic field produced by the permanent magnets encasing the coils produces a torque on the motor shaft, which drives a load. Power is delivered from the source to the load.

The electromechanical characteristics of the permanent magnet dc motor have a simple circuit-like model amenable to Laplace transform analysis. The model is given in Figure 14.38 and consists of an adjustable dc voltage source in series with a resistor R_a, an inductance L_a, and a device labeled "motor." Here, R_a represents the resistance present in the armature winding, and L_a represents the equivalent inductance of the wire coil. The device labeled "motor" has the current $i_a(t)$ as an input and the angular velocity $\omega(t)$ of the rotating shaft as an output. The interaction between the electrical part of the model and the mechanical part of the model occurs at the location of this symbol. The voltage $v_m(t)$ is an induced voltage proportional to the angular velocity $\omega(t)$. Because $\omega(t)$ is not a circuit variable, classical notions of impedance, admittance, voltage gain, etc., do not fit the problem, whereas the more general notion of transfer function does, forcing us to move slightly beyond the confines of circuit theory proper to analyze the system.

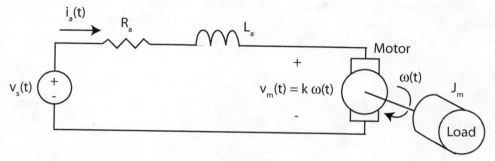

FIGURE 14.38 Equivalent circuit-like model of permanent magnet dc motor.

The torque $T(t)$ produced on the rotating shaft by the current flowing through the armature coils is proportional to the armature current $i_a(t)$, i.e.,

$$T(t) = ki_a(t) \tag{14.25}$$

The mechanical rotation of the motor affects the electrical portion of the system by inducing a voltage $v_m(t)$. This voltage is proportional to the motor's rotational speed, or angular velocity, $\omega(t)$, i.e.,

$$v_m(t) = k_1 \omega(t) \tag{14.26}$$

Since the motor converts electrical energy to mechanical energy, conservation of energy dictates that the constant of proportionality be equal to the same constant that relates torque and current for a lossless motor. Specifically, electrical power-in must equal mechanical power-out or, in equation form,

$$v_m(t)i_a(t) = T(t)\omega(t)$$

which forces $k_1 = k$.

Our first goal is to find the transfer function of the motor, $H(s) = \mathcal{L}[\omega(t)]/V_s(s)$. For convenience, let $\Omega(s)$ denote $\mathcal{L}[\omega(t)]$. As a first step, sum the voltages around the loop of elements in the circuit model of Figure 14.38. This results in the differential equation

$$v_s(t) = R_a i_a(t) + L_a \frac{di_a(t)}{dt} + k\omega(t) \tag{14.27}$$

Assuming zero initial conditions, the Laplace transform of equation 14.27 is

$$V_s(s) = (R_a + L_a s)I_a(s) + k\Omega(s) \tag{14.28}$$

From basic mechanics, the differential equation governing the mechanical portion of the system is

$$T(t) = J_m \frac{d\omega(t)}{dt} + B\omega(t) \tag{14.29}$$

where J_m is the moment of inertia of the combined armature, rotating shaft, and load; B is the coefficient of friction; and $T(t)$ is the torque produced by the motor to turn the load. Recalling equation 14.25, $T(t) = ki_a(t)$, the Laplace transform of equation 14.29, assuming zero initial conditions, is

$$kI_a(s) = (J_m s + B)\Omega(s) \tag{14.30}$$

Solving equation 14.30 for $I_a(s)$, substituting the result into equation 14.28, and then solving for $\Omega(s)$ leads to the expression

$$\Omega(s) = \frac{\dfrac{k}{L_a J_m}}{s^2 + \left(\dfrac{B}{J_m} + \dfrac{R_a}{L_a}\right)s + \dfrac{R_a B + k^2}{L_a J_m}} V_s(s) \tag{14.31}$$

Equation 14.31 characterizes the pertinent dynamics of the permanent magnet dc motor and allows one to find the angular velocity of the motor shaft as a function of time for given inputs.

To see how the motor responds to step inputs, suppose $v_s(t) = Ku(t)$. As objectives, let us find (1) the steady-state value, or final angular velocity, of the shaft and (2) the steady-state value of the armature current. The final speed of the shaft is important because, for example, one needs a fixed speed for the rotation of a compact disk or a fan. The final or steady-state current is important for determining the power needed from the source.

If $v_s(t) = Ku(t)$, then $V_s(s) = K/s$. It follows from equation 14.31 that

$$\Omega(s) = \frac{\dfrac{k}{L_a J_m}}{s\left(s^2 + \left(\dfrac{B}{J_m} + \dfrac{R_a}{L_a}\right)s + \dfrac{R_a B + k^2}{L_a J_m}\right)} K \tag{14.32}$$

Applying the final value theorem to equation 14.32 implies that

$$\omega_{ss} = \lim_{t \to \infty} \omega(t) = \frac{k}{R_a B + k^2} K$$

To isolate the armature current $I_a(s)$, again apply the final-value theorem to determine the steady-state value of $i_a(t)$. Combine equations 14.30 and 14.31 to obtain

$$I_a(s) = \frac{\dfrac{1}{L_a}s + \dfrac{B}{L_a J_m}}{s^2 + \left(\dfrac{B}{J_m} + \dfrac{R_a}{L_a}\right)s + \dfrac{R_a B + k^2}{L_a J_m}} V_s(s) \tag{14.33}$$

Equation 14.33 allows us to find the armature current as a function of time for a given input voltage. As above, if the input is a step, i.e., if $v_s(t) = Ku(t)$, it follows that

$$I_a(s) = \frac{\dfrac{1}{L_a}s + \dfrac{B}{L_a J_m}}{s\left(s^2 + \left(\dfrac{B}{J_m} + \dfrac{R_a}{L_a}\right)s + \dfrac{R_a B + k^2}{L_a J_m}\right)} K. \tag{14.34}$$

Once again, application of the final-value theorem to this expression leads to the value of the steady-state armature current:

$$i_{a,ss} = \lim_{t \to \infty} i_a(t) = \frac{B}{R_a B + k^2} K$$

The preceding analysis suggests the utility of the Laplace transform as a tool for analyzing the dynamic behavior of electromechanical systems. In fact, system transfer functions of the form of equations 14.31 and 14.33 are often starting points for further analysis. Extensions of the above analysis can be found in the homework exercises.

10. SUMMARY

This chapter has expanded the notion of transfer function from its definition in Chapter 13 into a tool for modeling not only circuits, but other practical systems, such as a dc motor. The transfer function characterizes circuit or system behavior by the location of its poles and zeros. For example, if a transfer function has a pole on the imaginary axis or in the right half-plane, the associated circuit or system is said to be *unstable*, because there is an input or, possibly, simply an initial condition (as in the case of a second-order pole on the $j\omega$-axis) that will excite the pole and cause the response to grow without bound. Further, the ubiquitous presence of noise will always excite poles on the imaginary axis and cause the response to be unstable.

This chapter categorized various types of responses: zero-state and zero-input responses, natural and forced responses, transient and steady-state responses, etc. Recall that the zero-state response is the response to an input assuming zero initial conditions, which is the inverse Laplace transform of the product of the transfer function and the Laplace transform of the input excitation. Recall that the zero-input response is due only to initial conditions on the capacitors and/or inductors of the circuit. The complete response for linear circuits having constant parameter values is simply the sum of the zero-input and zero-state responses. This decomposition generalizes to the broad class of linear systems studied in advanced courses. Under reasonable conditions, other decompositions are possible, such as a decomposition into the natural and forced responses or transient and steady-state responses. Other important responses are the impulse and step responses.

For stable circuits, the frequency response provides important information about the circuit. Recall that frequency response is a plot of the magnitude and phase of $H(j\omega)$ as ω varies from 0 to ∞. The Bode plot is a plot of gain in dB vs. frequency represented using a log scale. In this context it is relatively straightforward to construct an asymptotic approximation using straight line segments, from which the actual plot is easily sketched by hand. The information in such a plot tells us how a circuit behaves when excited by sinusoids of different frequencies.

Lastly, this chapter has introduced the initial- and final-value theorems, which provide a quick means of computing the initial and final values of a time function from knowledge of its Laplace transform. Such theorems have wide application in control system analysis, as evidenced in our analysis of the dc servo motor.

TERMS AND CONCEPTS

3 dB frequency for low-pass characteristics: the frequency at which the gain is 3 dB down from the dc gain.

Asymptote: a limiting straight-line approximation to a curve.

Bode magnitude plot: a plot of gain in dB vs. frequency represented on a log scale.

Bode phase plot: a plot of phase in degrees vs. frequency represented on a log scale.

Bounded: the condition where a signal $f(t)$ satisfies $|f(t)| < K < \infty$ for all t; i.e., it has a maximum, finite height.

Breakpoint: the point at which two asymptotes of a Bode plot intersect.

Complete response: the total response of a circuit to a given set of initial conditions and a given input signal.

Corner (break) frequency: frequency at which two asymptotes of a Bode plot intersect.

Decade: a frequency band whose endpoint is a factor of 10 larger than its beginning point.

Decibel (dB): a log-based measure of gain equal to $20 \log_{10}|H(j\omega)|$.

Final-value theorem: a theorem stating the following: suppose $F(s)$ has poles only in the open left half of the complex plane, with the possible exception of a single-order pole at $s = 0$. Then the limit of $f(t)$ as $t \rightarrow \infty$ equals the limit of $sF(s)$ as $s \rightarrow 0$.

Forced response: the portion of the complete response that has the same exponent as the input excitation, provided the input excitation has exponents different from those of the zero-input response.

Frequency response: measure of circuit behavior to unit magnitude sinusoids, $\cos(\omega t)$, as ω varies from 0 to ∞. Equal to the evaluation of the transfer function $H(s)$ at $s = j\omega$ for all ω.

Frequency scaling: for a linear passive network, dividing all inductances and capacitances by a factor, K_f, while keeping all resistance values unchanged.

Fundamental period of periodic $f(t)$: the smallest positive number T such that $f(t) = f(t + T)$.

Impulse response: assuming all initial conditions are zero, if the circuit or system input is $\delta(t)$, then the resulting $y(t)$ is called the *impulse response*. The inverse transform of the transfer function equals the impulse response.

Initial-value theorem: a theorem stating the following: let $\mathcal{L}[f(t)] = F(s)$ be a strictly proper rational function of s; i.e., the numerator and denominator of $F(s)$ are both polynomials in s, with the degree of the numerator less than that of the denominator . Then $f(0^+)$ is the limiting value of $sF(s)$ as $s \rightarrow \infty$.

Linear circuit: circuit such that for any two inputs $f_1(t)$ and $f_2(t)$, whose zero-state responses are $y_1(t)$ and $y_2(t)$, respectively, the response to the new input $[K_1 f_1(t) + K_2 f_2(t)]$ is $[K_1 y_1(t) + K_2 y_2(t)]$, where K_1 and K_2 are arbitrary scalars.

Magnitude response: the magnitude of the frequency response as a function of ω.

Magnitude scaling: for a linear network, multiplying all resistances and inductances by a factor, K_m, and dividing all conductances and capacitances by K_m; for dependent sources, this means all parameters having ohms as units would be multiplied by K_m and those having siemens as units would be divided by K_m dimensionless parameters are left alone.

Natural response: the portion of the complete response that has the same exponents as the zero-input response.

Octave: a frequency band whose endpoint is twice as large as its beginning point.

Op amp open-loop gain: the gain of the op amp when no feedback paths to the input terminals are present.

Periodic $f(t)$: function satisfying the condition that there exists a positive constant, T, such that $f(t) = f(t + T)$ for all $t \geq 0$.*

Phase response: the angle of the frequency response as a function of ω.

Piecewise linear curve: an unbroken curve composed of straight-line segments.

Pole (simple) of rational function: zero of order 1 in denominator polynomial.

Poles (finite) of rational function: zeros of denominator polynomial.

Ramp function, $r(t)$: integral of unit step function having the form $Ktu(t)$ for some constant K.

Rational function: ratio of two polynomials; a polynomial is a rational function.

Stable transfer function: a transfer function for which every bounded input signal yields a bounded response signal; i.e., all poles are in open left half of the complex plane.

Steady-state response: that part of the complete response which either is constant or satisfies the definition of periodicity for $t \geq 0$.

Step response: the response of the circuit to a step function, assuming all initial conditions are zero.

Transient response: those terms of the complete response that are neither constant nor periodic for $t \geq 0$; i.e., the transient response does not satisfy the definition of a periodic function for $t \geq 0$.

Zero-input response: the response of a circuit to a set of initial conditions with the input set to zero.

Zero-state response: the response of a circuit to a specified input signal, given that the initial conditions are all set to zero.

Zeros of rational function: values that make the numerator polynomial zero.

---

* This (nonstandard) definition has been adapted for one-sided Laplace transform problems.

PROBLEMS

POLES AND ZEROS

1. For the pole-zero diagram shown in Figure P14.1, $H(0) = 8$.
   (a) Compute the transfer function $H(s)$.
   (b) Compute the impulse and step responses. Check that the impulse response is the derivative of the step response.
   (c) If the input is $10e^{-at}u(t)$, find the positive number a such that the response does not have a term of the form $Ke^{-at}u(t)$. Find the response under this condition.

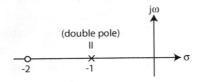

Figure P14.1

2. Consider the pole-zero plot of a transfer function $H(s)$ given in Figure P14.2.
   (a) If the dc gain is -10, find $H(s)$.
   (b) Compute the impulse response.
   (c) Compute the step response.
   CHECK: Your answer to (b) should be the derivative of your answer to (c), since the delta function is the derivative of the step function.
   (d) If the input is $10e^{-at}u(t)$, find the positive number a such that the response does not have a term of the form $Ke^{-at}u(t)$. Find the zero-state response under this condition.

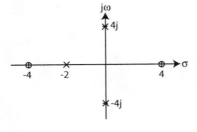

Figure P14.2

3. Consider the pole-zero plot of the transfer function $H(s)$ given in Figure P14.3.
   (a) If the dc gain is 2, find $H(s)$.
   (b) Compute the impulse response. Identify the steady-state and transient parts of the response.
   (c) Compute the step response. Identify the steady-state and transient parts of the response.
   CHECK: Your answer to (b) should be the derivative of your answer to (c), since the delta function is the derivative of the step function.
   (d) If the input is $10e^{-at}u(t)$, find the positive number a such that the response does not have a term of the form $Ke^{-at}u(t)$. Find the zero-state response under this condition. Identify the transient and steady-state responses.

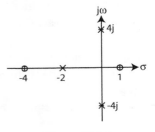

Figure P14.3

4. Consider the pole-zero plot of a transfer function $H(s)$ given in Figure P14.4.
   (a) If the gain $H(s)$ is 4.8 at $s = 1$, find $H(s)$.
   (b) Compute the impulse response. Identify the steady-state and transient parts of the response.
   (c) Compute the step response. Identify the steady-state and transient parts of the response.
   CHECK: Your answer to (b) should be the derivative of your answer to (c), since the delta function is the derivative of the step function.
   (d) If the input is
   $$\frac{10s}{s^2 + \omega_1^2},$$
   find ω_1 such that the response does not have a purely sinusoidal term.

Find the zero-state response under this condition. Identify the transient and steady-state responses.

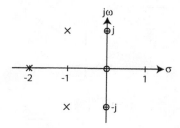

Figure P14.4

5. The pole-zero plot of the transfer function $H(s)$ is given in Figure P14.5.
   (a) If the dc gain is -1, find $H(s)$.
   (b) Compute the impulse response.
   (c) Compute the step response. Identify the transient and steady-state responses.
   (d) Compute the response to the input $\sin(4t)u(t)$.
   (e) If the input is $e^{-at}u(t)$, find the positive number a such that the response does not have a term of the form $Ke^{-at}u(t)$. Find the zero-state response under this condition.

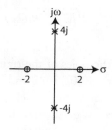

Figure P14.5

6. The transfer function $H(s)$ of a particular active circuit has poles at $-1 \pm 2j$ and -2. At $s = -1$, the transfer function gain is found to be 8, and $H(\infty)$ is known to be finite. When this circuit is excited by the input waveforms $v_{in}(t) = \sin(2t)u(t)$ V and $v_{in}(t) = u(t)$ V, it is found that the output is zero after a long time.
   (a) Find the zeros of the transfer function.
   (b) Construct $H(s)$ and then compute the step response.

(c) If the input to the circuit is $v_{in}(t) = 2tu(t)$, use MATLAB to compute the partial fraction expansion and the resulting time response. Sketch the approximate response for large t.
(d) Repeat part (c) for $v_{in}(t) = 2t^2u(t)$.

7. The input impedance of each network shown in Figure P14.7 is $Z(s) = \dfrac{s-a}{s+a}$ with $a > 0$. Which network is stable and why? Compute the step response of the stable network. Identify the transient and steady-state parts of the response.

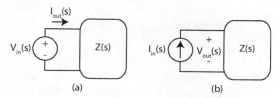

(a)   (b)

Figure P14.7

8. Reconsider the linear active networks of Figure P14.7, which have input impedance function $Z(s)$ with poles at $s = -1, -3 \pm j4$ and zeros at $s = 2, \pm j2$. It is known that $Z(0) = 8$.
   (a) Write $Z(s)$ as the ratio of two polynomials in s.
   (b) If the network of Figure P14.7b has zero initial conditions and $i_{in}(t) = 2u(t)$ A, write down the general form of $V_{out}(s)$ and then $v_{out}(t)$. In writing down the forms, simply leave the constants as literals without calculating numbers. Does the output remain finite as $t \to \infty$?
   (c) If the network of Figure P14.7a has zero initial conditions with input $v_{in}(t) = 20u(t)$ V, find the general form of $I_{out}(s)$ and then $i_{out}(t)$. Again leave the constants in these forms as literals. Does the output remain finite as $t \to \infty$?

9. Repeat Problem 8 for when the zeros of $Z(s)$ are $s = -2, \pm j2$. Also, answer the following questions:

(a) Which configuration shows a stable circuit?

(b) If one of the configurations is an unstable circuit, what input will cause the zero-state response to grow without bound as $t \rightarrow \infty$?

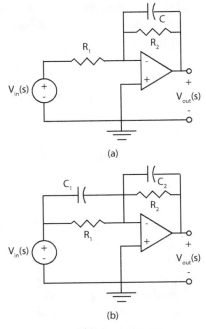

(a)

(b)

Figure P14.12

10. Reconsider Figure P14.7a with impedance $Z(s)$ having poles at $s = 2, 6 \pm j8$ and zeros at $s = -1, -2, -4$. It is known that $Z(0) = 32\ \Omega$.

(a) Write $Z(s)$ as the ratio of two polynomials of s.

(b) If the network has zero initial conditions, and $v_{in}(t) = 5u(t)$ V is applied to the two terminals, find the current $i_{out}(t)$ flowing into the network for $t \geq 0$.

11. Show the pole-zero plot of each of the following transfer functions, and determine whether the system (circuit) is stable. Hint: Use the MATLAB command "roots" to obtain the zeros and poles of each transfer function, even though there are methods that can determine stability without calculating the exact pole locations.

(a) $H_1(s) = \dfrac{2s - 12}{s^2 + 8s + 12}$

(b) $H_2(s) = \dfrac{1.5s^2 - 0.5s + 1.25}{s^3 + 2s^2 + 7.5s + 6.5}$

(c) $H_3(s) = \dfrac{1.5s^2 + 2.25s + 4.4062}{s^3 + 4.3125s + 5.3125}$

(d) $H_4(s) = \dfrac{2s^3 + 2.5s^2 - 17s + 220.5}{s^4 + 7s^3 + 55s^2 + 569s + 520}$

12. For each of the op amp circuits of Figure P14.12, find the transfer functions and the poles and zeros in terms of the indicated literals.

13. Consider the circuit of Figure P14.13.

(a) Show that the transfer function is

$$H(s) = \dfrac{\dfrac{s}{R_1 C_2}}{s^2 + \left(\dfrac{1}{R_1 C_1} + \dfrac{1}{R_2 C_2} + \dfrac{1}{R_1 C_2}\right)s + \dfrac{1}{R_1 C_1 R_2 C_2}}$$

(b) If $C_1 = C_2 = 1$ mF, find R_1 and R_2 so that the poles of the transfer function are at $s = -0.21922$ and -2.2808. The problem of finding element values to realize a given transfer function is called the *synthesis problem.*

ANSWER: $(R_1, R_2) = (4\ k\Omega, 0.5\ k\Omega)$ or $(1\ k\Omega, 2\ k\Omega)$

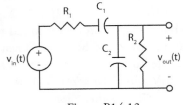

Figure P14.13

STABILITY PROBLEMS

14. For each of the circuits below, compute the indicated transfer function and determine the range of g_m or r_m for which the circuit as modeled by the transfer function is stable in the BIBO sense.

   (a) For the circuit of Figure P14.14a the transfer function is $H(s) = \dfrac{V_{out}(s)}{V_{in}(s)}$, $R = 4\ \Omega$, $C = 0.25$ F.

   (b) For the circuit of Figure P14.14b the transfer function is $H(s) = \dfrac{V_{out}(s)}{V_{in}(s)}$.

   (c) For the circuit of Figure P14.14c the transfer function is $H(s) = \dfrac{V_{out}(s)}{I_{in}(s)}$.

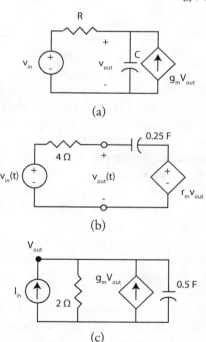

(a)

(b)

(c)

Figure P14.14

15. Given that $H(s) = \dfrac{V_{out}}{V_{in}}$ in the circuit in Figure P14.15, where $C = 2$ F and $g_m = 0.5$ S, find the *complete* range of R for which the circuit is stable.

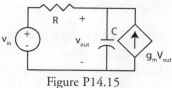

Figure P14.15

16. (a) The pole-zero plot of a transfer function is given in Figure P14.16a. Construct an input that will cause the response to be unbounded (unstable).

   (b) Repeat part (a) for the pole-zero plot of Figure P14.16b.

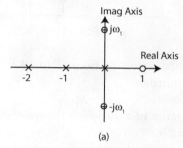

(a)

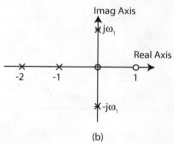

(b)

Figure P14.16

17. The circuit in Figure P14.17 is to be stabilized in the BIBO sense.

   (a) If $R = 2\ \Omega$, determine the complete range of a required for stability.

   (b) If $a = 0.5$, determine the complete range of R required for stability.

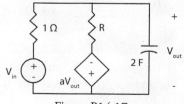

Figure P14.17

18. (a) Find $H(s) = \dfrac{V_{out}}{I_{in}}$ in the circuit of
Figure P14.18; then find the *complete*
range g_m for which the circuit is stable, assuming $R = 1 \; \Omega$.

(b) Now suppose that $g_m = -2$ S. Find the
complete range R for which the circuit
is stable.

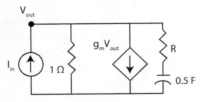

Figure P14.18

19. For the circuit of Figure P14.19 find the
transfer function $H(s) = \dfrac{V_o(s)}{V_i(s)}$
and the range of a for which the transfer function is stable.

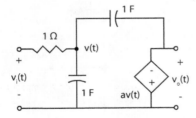

Figure P14.19

RESPONSES AND CLASSIFICATIONS

20. Consider the circuit in Figure P14.20.
(a) Show that the transfer function is

$$H(s) = \dfrac{\dfrac{1}{R_1 C}}{s + \dfrac{1}{R_{eq} C}} = \dfrac{8}{s+10}$$

when $R_1 = 50 \; \Omega$, $R_2 = 200 \; \Omega$, $R_{eq} = R_1$
$// R_2$, and $C = 2.5$ mF.
(b) Suppose $v_{in}(t) = 10 \sin(5t)u(t)$ V, and
$v_C(0^-) = 8$ V.

(1) Find the zero-state response, the
zero-input response, and the complete response.
(2) Find the steady-state response and
the transient response.
(3) Find the forced response and the
natural response.

(c) Repeat part (b) for $v_{in}(t) = 20e^{-10t}u(t)$
V and $v_C(0^-) = 0$. Is the forced
response well defined? Hint: Use
MATLAB's command "[r,p,k] =
residue(n,d)" to compute the partial
fraction expansions.

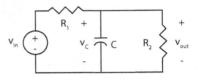

Figure P14.20

21. Find the complete response for $t \geq 0$ for the
circuit of Figure P14.20 for each of the following circuit conditions. (If you properly utilize
the results of Problem 20 and the principle of
linearity, the answers to this problem can be
written down directly.)
(a) $v_C(0^-) = 6$ V, $v_{in}(t) = 5 \sin(5t)u(t)$ V
(b) $v_C(0^-) = 16$ V, $v_{in}(t) = 8e^{-10t}u(t)$ V

22. Consider the network of Figure P14.22.
Suppose $C = 0.1$ F, $v_{in}(t) = 10u(t)$ V, and $v_C(0^-)$
$= 4$ V; a decomposition of the complete
response $v_{out}(t)$ has been found to be
zero-state response $= 10(1 - e^{-2t})u(t)$ V
zero-input response $= 4e^{-2t}u(t)$ V
(a) Determine the transfer function.
(b) If $v_{in}(t) = 10 \cos(10t)u(t)$ and $v_C(0^-)$ is
changed to 8 V, compute the complete
response $v_{out}(t)$ for $t \geq 0$. Identify the
transient and steady-state parts of the
response.

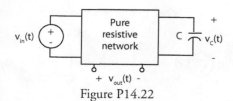

Figure P14.22

23. In the circuit shown in Figure P14.23, R_1 = 5 Ω, R_2 = 20 Ω, C = 0.05 F, $v_{s1}(t)$ = $25u(t)$ V, $v_{s2}(t)$ = $20e^{-4t}u(t)$ V, and $v_C(0^-)$ = 10 V.

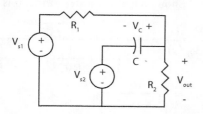

Figure P14.23

(a) Find $v_{out}(t)$ for $t \geq 0$ by the Laplace transform method.

(b) Write $v_{out}(t)$ as the sum of two components: the zero-state response and the zero-input response.

(c) Separate the zero-state response into one term due to v_{s1} and another term due to v_{s2}.

(d) Separate the complete response into the steady-state response and the transient response.

(e) Utilize the answers of previous parts to write down the complete response for $t \geq 0$ for the following two cases (all values in volts):

	$v_C(0^-)$	$v_{s1}(t)$	$v_{s2}(t)$
Case 1	20	$10u(t)$	$15e^{-4t}u(t)$
Case 2	5	$20u(t)$	$30e^{-4t}u(t)$

24. Consider the circuit of Figure P14.24a. Suppose C = 25 mF, R_1 = 25 Ω, R_2 = 100 Ω, and $R_{eq} = R_1 // R_2$.

(a) Show that the transfer function is

$$H(s) = \frac{sC + \dfrac{1}{R_1}}{sC + \dfrac{1}{R_1} + \dfrac{1}{R_2}} = \frac{s + \dfrac{1}{CR_1}}{s + \dfrac{1}{CR_{eq}}} = \frac{s+1.6}{s+2}$$

Draw its pole-zero plot.

(b) Express the $v_{in}(t)$ given in Figure P24.b as a sum of possibly shifted simple functions such as steps, ramps, etc. Then find the zero-state response.

(c) Draw the equivalent frequency domain circuit, and find the zero-input response when $v_C(0^-) = -1$ V. Hint: The response of an undriven first-order RC circuit is $v_C(0^-)e^{-t/\tau}$, where $\tau = R_{eq}C$.

(d) Compute the complete response.

(e) If $v_{in}(t)$ were changed to the input shown in Figure P14.24c, what would be the response of the circuit? (Hint: What is the relationship between Figures P14.24b and P14.24c? Note that differentiation and integration are linear operations when the initial conditions are zero and all circuit parameters are constant. For linear constant parameter circuits, then, a linear operation on the input induces the same linear operation on the response, provided that the initial conditions are zero and the circuit parameter values are constant. Combine this concept with the structure of the decomposition to obtain the answer.)

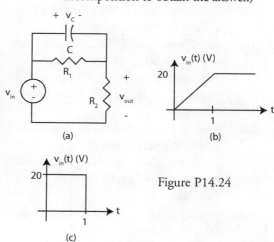

Figure P14.24

25. For the circuit of Figure P14.25, $C_1 = 0.2$ mF and $C_2 = 0.5$ mF. The initial conditions are $v_{C1}(0^-) = 80$ mV and $v_{C2}(0^-) = 0$. The input is $v_{in}(t) = 0.4u(t)$ V.
   (a) Find the transfer function
   $$H(s) = \frac{V_{out}(s)}{V_{in}(s)}.$$
   (b) Find the zero-state response.
   (c) Find the zero-input response.
   (d) Find the complete response.
   (e) Specify the general form of the natural response.
   (f) Specify the transient and steady-state responses.

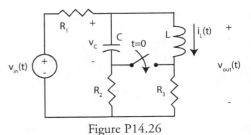

Figure P14.26

27. Consider the circuit of Figure P14.27.
   (a) Show that the transfer function is
   $$H(s) = \frac{V_{out}(s)}{V_{in}(s)} = \frac{\dfrac{1}{LC}}{s^2 + \left(\dfrac{R_1}{L} + \dfrac{1}{R_2 C}\right)s + \left(1 + \dfrac{R_1}{R_2}\right)\dfrac{1}{LC}}$$
   $$= \frac{16}{s^2 + 4s + 20}$$
   when $R_1 = 1$ Ω, $R_2 = 4$ Ω, $C = 0.125$ F, and $L = 0.5$ H.
   (b) Find the zero-state response to the input $v_{in}(t) = 10u(t)$ V.
   (c) Draw an equivalent s-plane circuit that accounts for the initial conditions. There are four possibilities, but one is superior.
   (d) Find the zero-input response for $v_C(0^-) = 0$ and $i_L(0^-) = 1$ A.
   (e) Find the zero-input response for $i_L(0^-) = 0$ and $v_C(0^-) = 4$ V.
   (f) Find the complete response for $v_{in}(t) = 5u(t)$, $v_C(0^-) = 2$ V, and $i_L(0^-) = -2$ A. Hint: Use linearity.

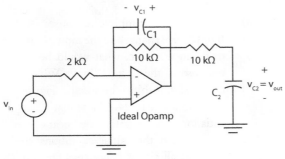

Figure P14.25

26. Consider the circuit in Figure P14.26. Suppose $R_1 = 10$ Ω, $R_2 = 30$ Ω, $R_3 = 15$ Ω, $C = 1/40$ F, and $L = 8$ H. Suppose $v_{in}(t) = v_1(t) + v_2(t) = -25u(-t) + 25u(t)$ V (plot this input function so that you know its shape).
   (a) Compute $v_C(0^-)$ and $i_L(0^-)$.
   (b) Compute $v_C(t)$ and $i_L(t)$ for $t > 0$; this is the complete response.
   (c) Identify the part of the response due only to the initial conditions at $t = 0^-$ that result from $-25u(-t)$; one could think of this as the zero-input response.
   (d) Identify the part of the response due only to the part of the input for $t > 0$, i.e., due to $25u(t)$ assuming zero initial conditions; one could think of this as the zero-state response.
   (e) Identify the transient and steady-state parts of the complete response.

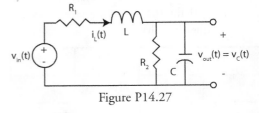

Figure P14.27

28. The purpose of this problem is to illustrate the computation of the zero-input response using the transfer function concept. This requires proper application of the equivalent initial condition circuits and the use of a transfer function for each initialized dynamic element. Consider the circuit of Figure P14.28a.

(a) Find the transfer function $H_1(s)$ if the input is a properly directed current source connected between nodes 1 and 2. The value of the current source, $Cv_C(0^-)$, accounts for the initial capacitor voltage $v_C(0^-)$.

(b) Find the transfer function $H_2(s)$ if the input is a properly directed current source connected between nodes 1 and 3. The value of this current source, $\dfrac{i_L(0^-)}{s}$, accounts for the initial inductor current.

(c) For the circuit of Figure P14.28b, utilize the transfer functions obtained in parts (a) and (b) to find

  (i) the zero-input response for $v_{out}(t)$ for $t \geq 0$, assuming initial conditions $v_C(0^-) = 4$ V and $i_L(0^-) = 5$ A;

  (ii) the zero-state response for $v_{out}(t)$ for $t \geq 0$; and

  (iii) the complete response for $v_{out}(t)$ for $t \geq 0$.

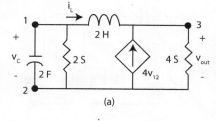

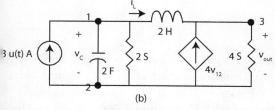

Figure P14.28

STEADY-STATE CALCULATION

29. For the pole-zero diagram of
$$H(s) = \frac{V_{out}(s)}{V_{in}(s)}$$
shown in Figure P14.29 $H(0) = 4$. Compute the sinusoidal steady-state response to $v_{in}(t) = \sin(2t)u(t)$ V.

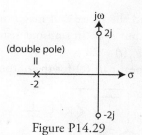

Figure P14.29

30. For the pole-zero diagram of
$$H(s) = \frac{V_{out}(s)}{V_{in}(s)}$$
shown in Figure P14.30 $H(2) = 10$. Compute the sinusoidal steady-state response to $v_{in}(t) = 2\cos(4t)u(t)$ V. Also find zero-state response.

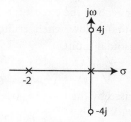

Figure P14.30

ANSWER: 0

31. Find the phase and magnitude in sinusoidal steady state for each of the given transfer functions when excited by the indicated sinusoidal input.

(a) $H(s) = \dfrac{V_{out}(s)}{V_{in}(s)} = \dfrac{2s+6}{s^2+4s+16}$

  is excited by the input $v_{in}(t) = 2\cos(2t + 45°)$ V.

(b) Suppose $H(s) = \dfrac{(s+1)(s-1)}{s(s+4)}$ has output $v_{out}(t) = 3\cos(3t + 45°)$ V. Find the sinusoidal input $v_{in}(t)$ that gives rise to this output.

(c) Let $L = 0.2$ H, $C = 0.25$ F, and $R_1 = R_2 = R = 1$ Ω in the circuit of Figure P14.31. Compute the transfer function
$$H(s) = \frac{V_C(s)}{V_{in}(s)} = \frac{??}{s^2 + \dfrac{2}{RC}s + \dfrac{1}{LC}}.$$

Find the phase and magnitude of the output $v_C(t)$ in sinusoidal steady state to $v_{in}(t) = 20 \cos(4t)$ V.

(d) Repeat part (c) for the output $i_L(t)$.

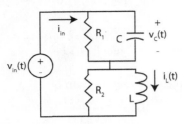

Figure P14.31

32. Find the phase and magnitude in sinusoidal steady state for each of the circuit transfer functions (or circuits) below when excited by the indicated sinusoidal input.

(a) $H(s) = \dfrac{V_{out}(s)}{V_{in}(s)} = \dfrac{16s + 96}{s^2 + 8s + 64}$

is excited by the input $v_{in}(t) = 4 \cos(4t + 45°)$ V.

(b) Suppose $H(s) = 4\sqrt{5}\,\dfrac{(s+2)(s-2)}{(s+2)^2 + 16}$

has output $v_{out}(t) = 25 \cos(3t + 100°)$ V. Find the sinusoidal input $v_{in}(t) = K \cos(\omega t + \theta)$ that gives rise to this output.

(c) Consider the circuit of Figure P14.32. Suppose $L = 1$ H, $R_1 = 15\,\Omega$, $C = 0.01$ F, and $R_2 = 5\,\Omega$. Suppose $v_{in}(t)$, the voltage + to − across the current source, is the desired output. The transfer function

$H(s) = \dfrac{V_{in}(s)}{I_{in}(s)} = Z_{in}(s) = \dfrac{??}{s^2 + \dfrac{?}{L}s + \dfrac{1}{LC}}$

Find the phase and magnitude of the output cosine, $v_{out}(t) = K \cos(\omega t + \theta)$, in sinusoidal steady state to $i_{in}(t) = 200 \cos(10t)$ mA.

(d) Repeat part (c) for the output $v_C(t)$.

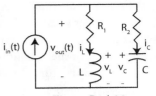

Figure P14.32

33. Consider the circuit of Figure P14.33. Given that $C_2 = 1$ F, find C_1, R_1, and R_2 so that the transfer function is $H(s) = -5\,\dfrac{s + 0.2}{s + 4}$.

(a) Find the steady-state response $v_{out}(t) = K \cos(2t + \theta)$ when $v_{in}(t) = 2 \sin(2t)u(t)$.

(b) Now find the complete response to the input $v_{in}(t) = 2 \sin(2t)u(t)$ and identify the transient and steady-state parts, assuming no initial conditions are present.

(c) Verify that your answer to part (a) and your steady-state answer to part (b) coincide.

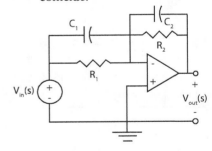

Figure P14.33

34. The pole-zero plot of a certain RLC network transfer function $H(s) = V_{out}/V_{in}$ is shown in Figure P14.34. If the input is a sinusoidal voltage of 1 V amplitude, then, in the steady state, the output voltage has the greatest amplitude at approximately what ω? At what approximate value of ω does the amplitude of the output voltage dip to its lowest value? Give qualitative, not quantitative, justification for your answers.

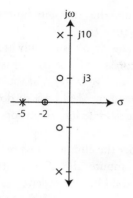

Figure P14.34

35. A linear circuit with a transfer function

$$H(s) = \frac{V_{out}}{V_{in}} = \frac{2s+4}{s^2+5s+6}$$

has an input $v_{in}(t) = 4\cos(2t + 45^o)$. Compute the magnitude and phase (in degrees) of the output of the circuit in the steady state.

36. A stable active circuit has the transfer function

$$H(s) = \frac{V_{out}}{V_{in}} = 4\frac{s^2+4}{s^2+2s+5}$$

Compute the phase and the magnitude of the steady-state response for each of the following inputs:

(a) $v_{in}(t) = 4\cos(2t)$ V
(b) $v_{in}(t) = 4\cos(4t)$ V

37. A stable circuit has the transfer function

$$H(s) = \frac{V_{out}(s)}{V_{in}(s)} = \frac{s^2-0.5s+5}{s^2+0.5s+5.7321}$$

The input to the circuit is $v_{in}(t) = 2\sqrt{2}\cos(2t + 30^o)$ V. Knowing that the response has the form $A\cos(2t + \phi)$, find A and ϕ.

38. A stable active circuit has the transfer function

$$H(s) = \frac{V_{out}(s)}{V_{in}(s)} = \frac{16s^2+44s+128}{s^3+8s^2+36s+80}$$

Compute the phase and the magnitude of the steady-state response for each of the following inputs:

(a) $v_{in}(t) = 20\cos(4t + 45^o)$ V
(b) $v_{in}(t) = 4\cos(40t)$ V
(c) $v_{in}(t) = 4\cos(2t)$ V

Suggestion: One can use the command "freqs(n,d,w)" for a rational function.

39. Following are the transfer functions of some linear networks that contain controlled sources. The networks are initially at rest, i.e., with zero initial conditions. An input $v_{in}(t) = 8\cos(2t)u(t)$ V is applied at $t = 0$. Using the residue command in MATLAB to generate the partial fraction expansion, determine the approximate form of the output, $v_{out}(t)$, for very large values of t. Hint: For stable circuits, what part of the response is meaningful, and for unstable circuits, what part of the response is dominant?

(a) $H_1(s) = \dfrac{-3.75s+5}{s^2+4s+3}$

(b) $H_2(s) = \dfrac{2.5s-3}{s^2+2s+5}$

(c) $H_3(s) = \dfrac{7s^2+s+4}{s^3+s^2+9s+9}$

(d) $H_4(s) = \dfrac{14s^2-23s+20}{s^3+s^2+7.25s+18.5}$

(e) $H_5(s) = \dfrac{7s^2+s+7.75}{s^3+3s^2+11.25s+18.5}$

(f) $H_6(s) = \dfrac{s^3-5.5s^2+14s-12}{s^3+5.5s^2+14s+12}$

ANSWER: (b) The result follows from the following MATLAB code:

```
»din = [1 0 4];
»nin = 8*[1 0];
»n = [2.5 -3];
»d = [1 2 5];
»nn = conv(n,nin);
»dd = conv(d,din);
```

```
»[r,p,k] = residue(nn,dd)
r =
   -4.0000e+00 - 7.0000e+00i
   -4.0000e+00 + 7.0000e+00i
    4.0000e+00 + 4.0000e+00i
    4.0000e+00 - 4.0000e+00i
p =
   -1.0000e+00 + 2.0000e+00i
   -1.0000e+00 - 2.0000e+00i
   -7.7716e-16 + 2.0000e+00i
   -7.7716e-16 - 2.0000e+00i
k = []
»K = 2*abs(r(4))
K = 1.1314e+01
»Phi = atan2(imag(r(4)),real(r(4)))*180/pi
Phi = -4.5000e+01
```

(Apply item 19 of Table 12.1.)

📖40. Following are the transfer functions of some linear networks that contain controlled sources. The networks are initially at rest, i.e., with zero initial conditions. An input $v_{in}(t) = 8 \sin(40t)u(t)$ V is applied at $t = 0$. Using the residue command in MATLAB to generate the partial fraction expansion, determine the approximate form of the output, $v_{out}(t)$, for very large values of t. Hint: For stable circuits, what part of the response is meaningful, and for unstable circuits, what part of the response is dominant?

(a) $\quad H_1(s) = \dfrac{80}{s^2 + 4s + 3}$

(b) $\quad H_2(s) = \dfrac{80}{s^2 + 4s}$

(c) $\quad H_1(s) = \dfrac{80(s^2 - 1)}{s^3 + 2s + 1601s}$

(d) $\quad H(s) = \dfrac{s^2 + 8s + 1616}{s^4 + 15s^3 + 83s^2 + 199s + 170}$

📖41. In Figure 14.17 of Example 14.8, let $\mu = 1.4$ and recall that $v_i(t) = \cos(2t)$ V.

(a) Find the transfer function $H(s) = V_1(s)/V_i(s)$.

(b) If the circuit is initially at rest, find an expression for $v_1(t)$ that is valid for all t (small or large). Use the residue command in MATLAB to do the needed partial fraction expansion.

42. Consider the circuit of Figure P14.42.

(a) Compute the transfer function. (It should be first order.)

(b) Find an expression for the natural response of the circuit.

(c) Suppose that for $t = 0^-$, each capacitor voltage is 1 V and the inductor current is zero. Compute the zero-input response. Observe that this response has a steady-state part. This phenomenon illustrates why equations 14.7 require more than just a stable transfer function as an underlying assumption. It is necessary that the transfer function include all the dynamics of the circuit, which is not the case in the circuit of Figure P14.42.

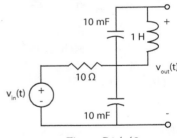

Figure P14.42

43. The circuit shown in Figure P14.43 has zero initial conditions, $v_C(0^-) = 0$, and $i_L(0^-) = 0$.

(a) Find the transfer function $H(s) = V_C(s)/V_{in}(s)$.

(b) If $v_{in}(t) = V_0 u(t)$, find $V_C(s)$ and $v_C(t)$, and determine the largest voltage that can appear across the capacitor. At what times does this occur? This phenomenon has application in the design of insulators that support transmission lines. For example, a 100 kV voltage transmission line would need insulators that can handle at least 200 kV.

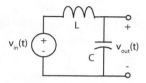

Figure P14.43

44. Suppose the voltage source in Problem 43 is changed to

$$10 \sin \left(\frac{0.9}{\sqrt{LC}} t \right) u(t) \text{ V.}$$

For $t \geq 0$, determine a tight upper bound on the largest voltage that can appear across the capacitor.
ANSWER: 100 V

45. In the circuit of Figure P14.45, all stored energy is zero at $t = 0$. The current source is $i_{in}(t) = I_0 u(t)$ A. Determine $v_{out}(t)$. What is the largest voltage that can appear across the capacitor?

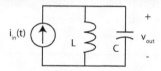

Figure P14.45

46. At $t = 0$, the energy stored in the LC elements in Figure P14.46 is nonzero. The current source is $i_{in}(t) = \cos(2t)u(t)$ A. Is the circuit stable? Determine the behavior of $v_{out}(t)$ for large t.

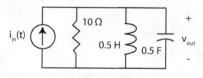

Figure P14.46

47. This problem illustrates the condition of steady state, but not sinusoidal steady state. For the circuit shown in Figure P14.47, the op amp is assumed to be ideal.
   (a) Find the transfer function $H(s)$.
   (b) If $v_C(0^-) = 0$ and $v_{in}(t) = \delta(t) + \delta(t -$

1) V, find $v_{out}(t)$ for $t \geq 0$ by the Laplace transform method.
(c) If the impulses are applied to the circuit every second, i.e., if $v_{in}(t) = [\delta(t) + \delta(t-1) + \delta(t-2) + \cdots]$ V, find the output $v_{out}(t)$ for $n < t < (n + 1)$ when the integer n becomes very large (i.e., after the circuit reaches a steady state). Sketch the waveform.

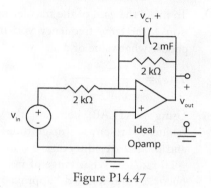

Figure P14.47

FREQUENCY RESPONSE

48. Draw to scale the pole-zero plot for $H(s) = (s + 1)/[s(s^2 + s + 10]$, and graphically compute the magnitude and phase of $H(j\omega)$ for $\omega = 0.2$, 0.5, 1, and 10. What are the limiting values of the magnitude and phase, i.e., for $\omega = \pm\infty$?

49. Draw to scale (on graph paper) the pole-zero plot of the transfer function

$$H(s) = \frac{\left((s+1)^2 + 16\right)\left(s^2 - 1\right)}{\left((s+1)^2 + 4\right)\left((s+1)^2 + 36\right)},$$

and determine graphically the magnitude and phase of $H(j\omega)$ for $\omega = 0, 2, 4, 6$, and ∞. Do your answers make sense physically? Explain.

50. (This problem focuses on a qualitative understanding of poles and zeros of a frequency response.) You have determined the pole-zero plot of a band-reject filter as shown in Figure P14.50 (or so you think, according to the qualitative boardwalk suggestions of the "professor").

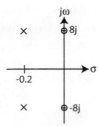

Figure P14.50

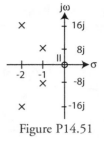

Figure P14.51

(a) To make the gain of the transfer function 1 for large frequency you must choose what value of K in

$$H(s) = K \frac{(s - z_1) \cdots}{(s - p_1) \cdots}?$$

(b) Using MATLAB, compute the frequency response (magnitude and phase) over the range $0 \leq \omega \leq 20$ rad/sec. What range of frequencies are rejected, approximately?

51. (This problem focuses on a qualitative understanding of the poles and zeros of a frequency response.) You have come up with the pole-zero plot of Figure P14.51 for a possible band-pass filter with a second-order zero at the origin.

(a) To make the gain of the transfer function 1 for $\omega = 6$ you must choose what value of K in

$$H(s) = K \frac{(s - z_1) \cdots}{(s - p_1) \cdots}?$$

(b) Using MATLAB, compute the frequency response (magnitude and phase) over the range $0 \leq \omega \leq 20$ rad/sec.

(c) Is the filter of the band-pass type? What range of frequencies are passed, approximately? What ranges of frequencies are rejected, approximately?

52. Figure P14.52 shows the magnitude frequency response of a transfer function $H(s)$ whose numerator and denominator are both fourth order. Qualitatively determine the structure of $H(s)$ approximately.

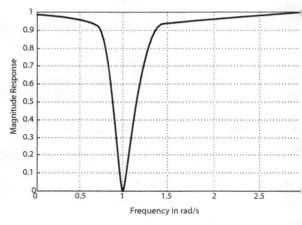

Figure P14.52

ANSWER:　$$\frac{(s^2 + 1)^2}{\left[(s + 0.1)^2 + 1 \right]^2}$$

53. Figure P14.53 shows the magnitude frequency response of a transfer function $H(s)$.

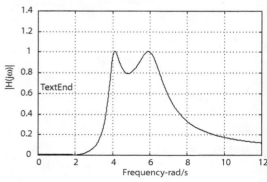

Figure P14.53

Which of the following is the best candidate for $H(s)$?

(1) $K\dfrac{s}{(s+0.5+j5)(s+0.5-j5)}$

(2) $K\dfrac{(s+j2)(s-j2)}{(s+0.5+j5)(s+0.5-j5)}$

(3) $K\dfrac{(s+j2)(s-j2)}{s(s+0.5+j5)(s+0.5-j5)}$

(4) $K\dfrac{(s+j2)(s-j2)}{(s+0.3+j4)(s+0.3-j4)(s+0.8+j6)(s+0.8-j6)}$

(5) $K\dfrac{s(s+j2)(s-j2)}{(s+0.3+j4)(s+0.3-j4)(s+0.8+j6)(s+0.8-j6)}$

(6) $K\dfrac{(s+j2)(s-j2)}{s(s+0.3+j4)(s+0.3-j4)(s+0.8+j6)(s+0.8-j6)}$

(7) $K\dfrac{(s+j2)(s-j2)}{s(s+0.8+j6)(s+0.8-j6)}$

ANSWER: (5)

54. The circuit in Figure P14.54 is to be frequency and magnitude scaled so that the value of $H(s)$ at $s = j5$ becomes the value of $H_{new}(s)$ at $s = j1000$. If the final value of the capacitor is to be 1 mF, then what is the new final value of the inductor, L_{new} (in H)?

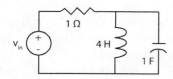

Figure P14.54

55.(a)   Plot, using MATLAB or the equivalent, the magnitude and phase responses for $0 \le \omega \le 5$ rad/sec for the second-order normalized low-pass Butterworth filter transfer function

$$H(s) = \dfrac{1}{s^2 +1.414s +1}$$

(b)   Denormalization to achieve a different 3 dB down point is done by frequency scaling. Plot, using MATLAB or the

equivalent, the magnitude and phase responses $0 \le \omega \le 10{,}000\pi$ rad/sec for the second-order low-pass Butterworth filter

$$I(s)=\dfrac{1}{\left(\dfrac{s}{2000\pi}\right)^2 +1.414\left(\dfrac{s}{2000\pi}\right)+1}.$$

which has a 3 dB down point at 1000 Hz.

(c) Referring to part (b), find the steady-state phase and magnitude of the output of the Butterworth filter when the input is (i) $\cos(500\pi t)$, (ii) $\cos(2000\pi t)$, and (iii) $\cos(4000\pi t)$.

(d) Do your answers to part (c) make sense? Why?

Remark: The *gain*, or the magnitude of $H(j\omega)$, is typically given in decibels and computed as gain (dB) = 20 $\log_{10}|H(j\omega)|$.

56. (a)   Plot, using MATLAB or the equivalent, the magnitude and phase responses for the second-order normalized high-pass Butterworth filter transfer function

$$H(s) = \dfrac{s^2}{s^2 +1.414s +1}$$

over the range $0 \le \omega \le 5$ rad/sec.

(b)   Repeat part (a) for the denormalized high-pass filter

$$H(s)=\dfrac{\left(\dfrac{s}{2000\pi}\right)^2}{\left(\dfrac{s}{2000\pi}\right)^2 +1.414\left(\dfrac{s}{2000\pi}\right)+1}$$

over the range $0 \le \omega \le 10{,}000\pi$ rad/sec.

(c)   Referring to part (b), find the steady-state phase and magnitude of the output of the Butterworth filter when the input is (i) $\cos(500\pi t)$, (ii) $\cos(2000\pi t)$, and (iii) $\cos(4000\pi t)$.

57. Frequency responses are typically plotted as $|H(j\omega)|$ vs. $\log_{10}|\omega|$ or as gain = $20\log_{10}|H(j\omega)|$ dB vs. $\log_{10}\omega$. These plots are commonly called Bode plots.

(a) Repeat parts (a) and (b) of Problem 55, except plot only the gain in dB vs. $\log_{10}|\omega|$.

(b) Repeat parts (a) and (b) of Problem 56, except plot only the gain in dB vs. $\log_{10}|\omega|$.

58.(a) Plot, using MATLAB or the equivalent, the magnitude and phase responses for $0 \le \omega \le 5$ rad/sec for the second-order normalized low-pass Chebyshev filter transfer function

$$H(s) = \frac{0.5012}{s^2 + 0.6449s + 0.7079}$$

(b) Denormalization to achieve a different 3 dB down point is done by frequency scaling. Plot, using MATLAB or the equivalent, the magnitude and phase responses $0 \le \omega \le 10{,}000\pi$ rad/sec for the second-order low-pass Chebyshev filter

$$H(s) = \frac{0.5012}{\left(\dfrac{s}{2000\pi}\right)^2 + 0.6449\left(\dfrac{s}{2000\pi}\right) + 0.7079}$$

which has a 3 dB down point at 1000 Hz.

(c) Referring to part (b), find the steady-state phase and magnitude of the output of the Chebyshev filter when the input is (i) $\cos(500\pi t)$, (ii) $\cos(2000\pi t)$, and (iii) $\cos(4000\pi t)$.

(d) Do your answers to part (c) make sense? Why?

Remark: The *gain*, or the magnitude of $H(j\omega)$, is typically given in decibels and computed as gain (dB) = $20\log_{10}|H(j\omega)|$.

59.(a) Plot, using MATLAB or the equivalent, the magnitude and phase responses for the second-order nor-

malized high-pass Chebyshev filter transfer function

$$H(s) = \frac{0.5012s^2}{0.7079s^2 + 0.6449s + 1}$$

over the range $0 \le f \le 1$ Hz.

(b) Repeat part (a) for the denormalized high-pass Chebyshev filter

$$H(s) = \frac{0.5012\left(\dfrac{s}{2000\pi}\right)^2}{0.7079\left(\dfrac{s}{2000\pi}\right)^2 + 0.6449\left(\dfrac{s}{2000\pi}\right) + 1}$$

over the range $0 \le f \le 1000$ Hz.

(c) Referring to part (b), find the steady-state phase and magnitude of the output of the Chebyshev filter when the input is (i) $\cos(500\pi t)$, (ii) $\cos(2000\pi t)$, and (iii) $\cos(4000\pi t)$.

60. The circuit in Figure P14.60, with $R = 1\ \Omega$, $C = 1$ F, and $L = 2$ H, has a so-called third-order Butterworth low-pass characteristic. The transfer function of the filter is

$$H(s) = \frac{0.5}{s^3 + 2s^2 + 2s + 1}$$

(a) Calculate the poles of $H(s)$. Verify that their magnitude is 1, meaning that they lie on a circle of radius 1 in the complex plane.

(b) Using MATLAB or the equivalent, plot the frequency response for $0 \le f \le 1$ Hz.

ANSWER:
w = [0:0.01:1]*π
n = 0.5;
d = [1 ????]
h = freqs(n,d,w);
plot(w/(2*pi), abs(h))

(c) If K_m is chosen so that $R = 10\ \Omega$, find $K_f= (??)\pi$ so that $C = 15.915\ \mu$F. What is the new value of L? The filter transfer function becomes

$$H(s) = \frac{0.5}{\left(\dfrac{s}{(??)\pi}\right)^3 + 2\left(\dfrac{s}{(??)\pi}\right)^2 + 2\left(\dfrac{s}{(??)\pi}\right) + 1}$$

Plot the new frequency response (magnitude and phase) for $0 \le f \le 1000$ Hz.

(d) From your plots in part (c), determine approximately the magnitude and phase of the steady-state output voltage when the input is (i) $10\cos(20\pi t)$, (ii) $10\cos(200\pi t)$, (iii) $10\cos(1000\pi t)$, (iv) $10\cos(2000\pi t)$, (v) $10\cos(5000\pi t)$.

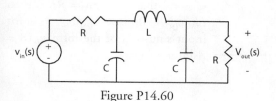

Figure P14.60

61. The circuit of Figure P14.61 has transfer function

$$H(s) = \frac{\dfrac{s}{R_1 C_2}}{s^2 + \left(\dfrac{1}{R_1 C_1} + \dfrac{1}{R_2 C_2} + \dfrac{1}{R_1 C_2}\right)s + \dfrac{1}{R_1 C_1 R_2 C_2}}$$

With $C_1 = C_2 = 1$ mF, $R_1 = 4$ kΩ, and $R_2 = 0.5$ kΩ, plot the magnitude of the frequency response; i.e., plot $|H(j\omega)|$ as a function of ω over the range $0 \le \omega \le 5$ rad/sec and $0 \le \omega \le 50$ rad/sec

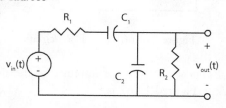

Figure P14.61

62. The circuit of Figure P14.62 is called an *all-pass* network. This means that the circuit does not alter the magnitude of a sinusoidal waveform, but it does introduce a new phase shift depending on the frequency of the sinusoid.

(a) Assuming an ideal op amp, show that the transfer function is

$$H(s) = -\frac{s - \dfrac{1}{RC}}{s + \dfrac{1}{RC}}$$

(b) Assuming that $R = 20$ kΩ and $C = 1$ μF, use MATLAB or the equivalent to plot the magnitude and phase (in degrees) responses for $0.1 \le \omega \le 1000$ rad/sec. From the magnitude response, it is apparent why this circuit is called an all pass network.

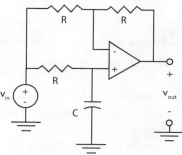

Figure P14.62

⌨63. Figure P14.63 is a circuit that realizes a band-pass characteristic. Use SPICE to determine the frequency response of the circuit and verify that it is a band-pass characteristic. Note that the computation of the transfer function for this circuit is quite involved.

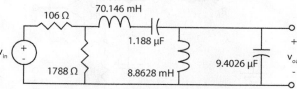

Figure P14.63

⌨64. A certain band-pass circuit has transfer function

$$H(s) = \frac{0.0846\left(\left(\frac{s}{100}\right)^5 + 139.90\left(\frac{s}{100}\right)^3 + 0.9989\left(\frac{s}{100}\right)\right)}{s^6 + 2.7996\left(\frac{s}{100}\right)^5 + 13.051\left(\frac{s}{100}\right)^4 + 17.271\left(\frac{s}{100}\right)^3 + 13.051\left(\frac{s}{100}\right)^2 + 2.7991\left(\frac{s}{100}\right) + 1}$$

Plot the magnitude frequency response in dB on a semilog scale to verify a band-pass characteristic. Find an appropriate range of ω.

65. The circuit of Figure P14.65 is the Sallen and Key low-pass filter, which can be used to eliminate unwanted high-frequency noise.
   (a) Show that the transfer function is

$$H(s) = \frac{V_{out}}{V_{in}} = \frac{\dfrac{K}{R_1 R_2 C_1 C_2}}{s^2 + \left(\dfrac{1}{R_1 C_1} + \dfrac{1}{R_2 C_1} + \dfrac{1-K}{R_2 C_2}\right)s + \dfrac{1}{R_1 R_2 C_1 C_2}}$$

where $K = 1 + R_B/R_A$. Hint: You will need to write three node equations at the indicated nodes.

   (b) Now let $C_1 = C_2 = 1$ F, $R_1 = R_2 = R_A = 1$ Ω, and $R_B = 0.8$ Ω. These values represent a normalized design. Use MATLAB or the equivalent to plot the magnitude and phase responses for $0 \le \omega \le 5$ rad/sec.

tle attenuation. This can easily be seen, since the transfer function has the form

$$H(s) = \frac{V_o(s)}{V_{in}(s)} = \frac{s^2 + \dfrac{1}{R^2 C^2}}{s^2 + \dfrac{4}{RC}s + \dfrac{1}{R^2 C^2}}$$

Observe that if $s = j\omega_0 = j/RC$, then $H(j\omega_0) = 0$, i.e., $V_o(j\omega_0) = 0$ for input sinusoids of the form $v_{in}(t) = A\sin(\omega_0 t + \varphi)$.

   (a) Use nodal analysis to derive the transfer function.
   (b) Show that the poles (zeros of the denominator) are $(-2 \pm \sqrt{3})/RC$.
   (c) Let $R = 5/\pi \cong 1.5915$ Ω and $C = 0.01$ F. Plot the magnitude and phase responses over the range 0 to 200 Hz.

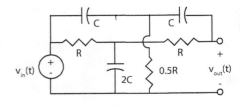

Figure P14.66 Twin-T RC circuit.

67. Suppose the impulse response of the network in Figure P14.67 is

$$v_{out}(t) = [2e^{-t} + 3\cos(2t) - 1.5\sin(2t)]u(t) \text{ V}$$

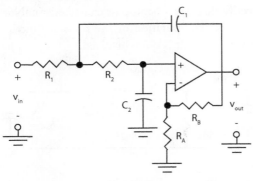

Figure P14.65 Sallen and Key low-pass filter.

66. Figure P14.66 depicts the so-called twin-T RC circuit. This network is often used as a band-reject filter, i.e., a circuit that stops or rejects signals at certain frequencies and allows signals at all other frequencies to pass with lit-

   (a) Compute the transfer function, $H(s)$.
   (b) Find all poles and zeros, and draw the pole-zero plot.
   (c) If there is no initial stored energy in the network, i.e., no initial conditions,

compute the network response to $v_{in}(t) = \sin(t)u(t)$ V. Does your answer make sense? Why or why not?

(d) Now suppose $v_{in}(t) = \cos(2t)u(t)$ V. For large t, show that $v_{out}(t) \approx Kt \cos(2t + \phi)$ V for appropriate K and ϕ. Does the response remain finite as $t \to \infty$? Is the network stable or unstable?

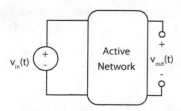

Figure P14.67

68. A time-shift differentiator circuit has the property that its zero-state output is always equal to the derivative of its input shifted in time by some $T > 0$. Hence, its impulse response is $h(t) = \delta'(t - T)$, a shifted version of the derivative of the delta function. Suppose $T = 1$ and the input to the circuit is given by $f(t)$, as sketched in Figure P14.68. Compute the zero-state response $y(t)$ of the circuit at $t = 2.5$ and $t = 6$ seconds.

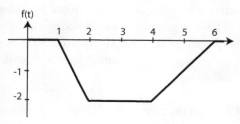

Figure P14.68

ANSWER: –2, 1

69. The Laplace transform of the response of a relaxed circuit to a ramp $v_1(t) = r(t - 1)$ V is

$$Y(s) = \frac{e^{-2s}}{s^3}$$

(a) Find the response of this same relaxed circuit to the input $v_2(t) = u(t)$ V.

(b) Find the response of this same relaxed circuit to the input $v_3(t) = \delta(t)$ V.

(c) Find the response of this same relaxed circuit to the input $v_4(t) = 5\delta(t - 2)$ V.

70. For this problem you are to use the properties of Laplace transforms as listed in Chapter 12. Suppose the Laplace transform of the impulse response, $h(t)$, of a circuit is given by

$$H(s) = \frac{s+2}{(s+1)(s+3)}$$

Find $H_1(s)$ and $H_2(s)$ when

(a) $h_1(t) = h(4t)$

(b) $h_2(t) = \exp(-4t)h(4t - 8)$

INITIAL AND FINAL VALUE PROBLEMS

71. (a) The Laplace transform of a signal $f(t)$ is

$$\frac{(5s-1)(4s-5)(6s-2)}{s(2s+1)(3s+2)(5s+4)}$$

Find the values of $f(0^+)$ and $f(\infty)$.

(b) Repeat part (a) when the Laplace transform of a signal $f(t-2)u(t-2)$ is

$$e^{-2s}\frac{(10s-1)(8s-5)(12s-2)}{s(5s+1)(3s+2)(s+4)}$$

72. The output, $v_{out}(t)$, of a particular circuit is engineered to track a reference signal, $v_{ref}(t)$. After the circuit overheated, it was found that

$$\text{Error}(s) = V_{ref}(s) - V_{out}(s) = \frac{1}{s} - \frac{1}{s} \times \frac{4s^2 - 4s - 4}{s^2 + 2s + 2}$$

Find the error, $\text{error}(t) = v_{ref}(t) - v_{out}(t)$, for large t. What was the initial error, $\text{error}(0^+)$? CHECK: for large t, error is 3

73. Given the following functions $F_i(s)$, find $f_i(0^+)$ and $f_i(\infty)$ for those cases where the initial-value and final-value theorems are applicable (a root-finding program is useful in this endeavor).

(a) $F_1(s) = \dfrac{21s + 5}{0.1s^2 + 4s + 3}$

(b) $F_2(s) = \dfrac{2s^3 + 7s^2 + s + 4}{s(s^3 + s^2 + 7s + 6)}$

(c) $F_3(s) = \dfrac{s^2 + 4s + 3}{s(s^4 + 5s^3 + 5s^2 + 4s + 4)}$

74. The capacitor voltages, $v_C(t)$, of various networks have the Laplace transforms given below. Use only the initial- and final-value theorems to determine $v_C(0^+)$ and $v_C(\infty)$ for each of the transforms. If either theorem is not applicable, explain why.

(a) $\dfrac{(14s - 1)^2}{s^2(7s + 2)}$

(b) $\dfrac{(2s - 15)(2 - e^{-6s})}{s(s + 10)(1 - e^{-3s})}$

(c) $200\dfrac{(0.1s - 1)(0.2s + 1)}{s(s^2 + 144)}$

(d) $\dfrac{(16s - 1)}{(2s + 1)^2}\left[1 + \dfrac{4s - 1}{2s + 1} + \left(\dfrac{2s - 1}{2s + 1}\right)^2 + \left(\dfrac{4s - 1}{2s + 1}\right)^3\right]$

75. The capacitor voltages, $v_C(t)$, of various networks have the Laplace transforms given below. Use only the initial- and final-value theorems to find $i_C(0^+)$ and $i_C(\infty)$ for each of the transforms given that $C = 0.2$ F. If either theorem is not applicable, explain why.

(a) $\dfrac{10}{s(s + 2)}$

(b) $20\dfrac{(0.1s - 1)}{s^2(2s + 25)}$

(c) $20\dfrac{(s - 2)}{s(2s + 25)^2}$

APPLICATIONS AND BODE TECHNIQUES

76. Obtain the asymptotes for the Bode plot of

$$H(s) = 10\dfrac{\left(\dfrac{s}{100} + 1\right)\left(\dfrac{s}{400} + 1\right)}{\left(\dfrac{s}{16} + 1\right)\left(\dfrac{s}{1200} + 1\right)}$$

Then verify your asymptotic plots by using MATLAB to generate the true Bode plot.

77. Obtain the asymptotes for the Bode plot of

$$H(s) = 10\dfrac{\left(\dfrac{s}{16} + 1\right)\left(\dfrac{s}{1200} + 1\right)}{\left(\dfrac{s}{100} + 1\right)\left(\dfrac{s}{400} + 1\right)}$$

Then verify your asymptotic plots by using MATLAB to generate the true Bode plot.

78. Obtain the asymptotes for the Bode plot of

$$H(s) = 10\dfrac{\left(\dfrac{s}{16} + 1\right)^2\left(\dfrac{s}{1200} + 1\right)^2}{\left(\dfrac{s}{100} + 1\right)^2\left(\dfrac{s}{400} + 1\right)^2}$$

Then verify your asymptotic plots by using MATLAB to generate the true Bode plot. Note that the shape of your asymptotic Bode plot resembles that of Problem 77 except that the slopes are steeper.

79. Obtain the asymptotes for the Bode plot of

$$H(s) = 10\dfrac{\left(\dfrac{s}{20} + 1\right)^2\left(\dfrac{s}{2400} + 1\right)}{\left(\dfrac{s}{200} + 1\right)^2\left(\dfrac{s}{800} + 1\right)}$$

Then verify your asymptotic plots by using MATLAB to generate the true Bode plot.

➤80. Suppose a dc motor is modeled as per Figure 14.38 of the text. The motor parameters are $R_a = 25$ Ω, $J_m = 0.005$ kg-m^2, $k = 0.02$ N-m/A, $B = e^{-5}$ N-m-s, and $L_a = 100$ mH. Calculate the steady-state angular velocity of the motor and the steady-state armature current for the input $10u(t)$ V.

➤81. Consider the dc motor modeled in Figure 14.38 of the text.

   (a) Using the same parameters as in Problem 80 with $v_s(t) = 5u(t)$ V, compute the armature current and the angular velocity as a function of time.

   (b) Change R_a to 50 Ω, and recompute the armature current as a function of time.

   (c) What effect does R_a have on the steady-state speed for a step input of a fixed amplitude?

   (d) What effect does R_a have on the rate at which the armature current approaches a steady state for a step input of fixed amplitude?

➤82.(a) For the dc motor of Figure 14.38 of the text, suppose $k = 0.05$ N-m/A and $R_a = 50$ Ω. Plot the steady-state current as a function of B as B ranges from 0 to ∞.

   (b) Using $R_a = 25$ Ω and $B = e^{-4}$ N-m-s, plot ω_{ss} as k ranges from 0 to ∞. Recall that increasing k increases the torque per ampere of armature current. Explain why increasing k reduces ω_{ss}.

CHAPTER 15

Time Domain Circuit Response Computations: The Convolution Method

AVERAGING BY A FINITE TIME INTEGRATOR CIRCUIT

Sometimes one must compute the average of some quantity, such as the average value of the light intensity on a solar cell over the last fifteen minutes, or the average value of the temperature in a room over the last hour, or the average value of a voltage over the last 50 milliseconds. When such an average is updated continuously, it is termed a running average. The idea is that the readout of the device that averages these quantities always produces an updated value valid over a specified prior time interval. If a voltage represents the value of the quantity to be averaged, then one can build a circuit whose output voltage is the required average. This is done by observing that an average value of a continuous time variable is simply the integral of the variable over the proper time interval, divided by the length of the time interval. A device that integrates a variable over the last, say, T seconds is called a finite time integrator. As an application of the ideas of this chapter, we will look at a finite time integrator circuit and how it can be used to compute the average value of a quantity. The convolution concept directly leads to the required transfer function of such a circuit.

CHAPTER OBJECTIVES

1. Introduce the notion of the convolution of two signals.
2. Using the notion of convolution, develop a technique of time domain circuit response computation that is the counterpart of the transfer function approach in the frequency domain, presented in Chapter 14. In particular, we seek to show that the convolution of an input excitation with the impulse response of a circuit or system produces the zero-state circuit or system response.
3. Develop objective 2 from two angles: first, from a strict time domain viewpoint, and second, as a formal theorem relating convolution to the transfer function approach.
4. Develop graphical and analytical methods—in particular, an algebra—for evaluating the convolution of two signals.
5. Illustrate various properties of convolution that are pertinent to block diagrams of series, parallel, and cascade interconnections of circuits or systems modeled by transfer functions.

SECTION HEADINGS

1. INTRODUCTION

At the beginning of Chapter 12, we claimed that circuit response computations could take place in either the time domain or the s-domain. Yet, except for the solution of some very elementary differential equations, circuit response computations have relied almost exclusively on the Laplace transform technique. This chapter develops and explores the time domain counterpart of the Laplace transform method by introducing the notion of the convolution of two signals to produce a third signal. We then show that the time domain convolution of an input excitation with the circuit's impulse response yields the zero-state circuit response. Pictorially, the idea is expressed in Figure 15.1, which is an update of Figure 12.3.

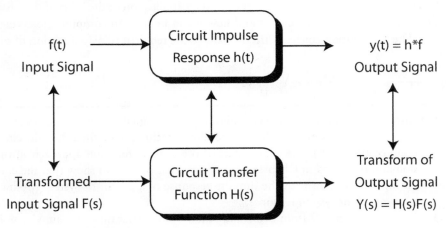

FIGURE 15.1 Diagram showing the symmetric relationship of time and frequency domain analyses. The upper part of the diagram asserts that the convolution, denoted $h * f$, of the input signal $f(t)$ with the circuit impulse response $h(t)$ produces a third signal, $y(t)$. This third signal is the zero-state response, which equals $\mathcal{L}^{-1}[H(s)F(s)]$.

Justification of the diagram of Figure 15.1 begins with the formal introduction of the notion of convolution. Mathematically, the **convolution** of two functions $h(t)$ and $f(t)$, denoted by $h * f$ or $h(t) * f(t)$, results in a third function through the integration process,

$$y(t) = \int_{-\infty}^{\infty} h(t - \tau) f(\tau) d\tau \tag{15.1}$$

The second step in verifying the diagram is proving that the convolution of the input signal with the impulse response produces the zero-state circuit response. There are two approaches. One is to work strictly in the time domain and construct the actual zero-state response from the impulse response and an arbitrary input excitation. The second approach is to prove that $\mathcal{L}^{-1}[H(s)F(s)]$ is the convolution of the signal $f(t)$ with the impulse response $h(t)$. Because of the Laplace transform development of the last three chapters, this direction seems the most painless and will be taken up in section 3, after an introduction to the basic ideas of convolution in section 2. Section 4 will look at the graphical method of convolution, which helps in visually grasping the definition. Section 5 will describe a convolution algebra, which yields a harvest of shortcuts for evaluating the convolution of certain types of signals. Section 6 looks at the computation of circuit responses using the convolution method, and here the averaging circuit is developed. Section 7 overviews various properties of convolution. The respective properties lend themselves to different structures (e.g., parallel or cascade) for designing interconnected circuits. Section 8 describes the rudiments of the construction of the zero-state response of a circuit working strictly in the time domain toward the development of a convolution integral.

Before closing this introduction, we should consider the question, *why is convolution important?* One reason is that it allows engineers to directly model the input-output behavior of circuits and general systems in the time domain, just as transfer functions model circuit behavior in the *s*-domain.

A second reason is that circuit diagrams are sometimes unavailable or even get lost. How would one generate a circuit model for analysis? One way is to display the impulse response on a CRT and approximate $h(t)$ by some interpolation function such as the staircase approximation illustrated in Figure 15.2. The process of constructing an approximate impulse response of an unknown circuit or system is called *system identification* and is a vibrant area of research. By storing the measured impulse response data as a table in a computer, one can numerically compute the zero-state response of the circuit or system to an arbitrary input signal. This numerical simulation process lets an engineer investigate a circuit's behavior off-line. For example, simulating a circuit destined for use in a hazardous environment provides a cost-effective means of evaluating its performance in a simulated environment. Such performance evaluations often identify needed design improvements prior to the constructing and testing of prototypes.

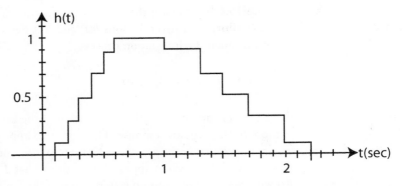

FIGURE 15.2 A rectangular approximation to a hypothetical impulse response obtained from measured data on an unknown circuit or system.

A third important reason for studying convolution in the context of circuits relates to a deficiency of the one-sided Laplace transform: function segments that are nonzero for $t < 0$ are ignored by the one-sided Laplace transform technique. Hence, time domain convolution offers a direct means of computing the circuit response when signals are nonzero over the entire time interval, $-\infty < t < \infty$.

2. DEFINITION, BASIC PROPERTIES, AND SIMPLE EXAMPLES

As mentioned, convolution is an integral operation between two functions to produce a third; i.e., the convolution of two functions $h(t)$ and $f(t)$ produces a third function $y = h * f$. One might expect that the convolution $h * f$ equals the convolution $f * h$, i.e., the operation of convolution is **commutative**. In fact, this is the case. To emphasize this property, we restate equation 15.1 in its more general form:

CONVOLUTION

The convolution of two signals $f(t)$ and $h(t)$ produces a third signal, $y(t)$, defined according to the formula

$$y(t) = \int_{-\infty}^{\infty} h(t-\tau)f(\tau)d\tau = \int_{-\infty}^{\infty} h(\tau)f(t-\tau)d\tau \qquad (15.2)$$

which is well defined, provided that the integral exists. This formula emphasizes the property that convolution is a commutative operation, i.e., $h * f = f * h$.

The equivalence expressed in equation 15.2 comes about in a straightforward manner by the change of variable $\tau' = t - \tau$. In addition to being commutative, convolution is **associative**,

$$h * (f * g) = (h * f) * g \qquad (15.3)$$

and **distributive**,

$$h * (f + g) = h * f + h * g \qquad (15.4)$$

Exercises. 1. Verify equation 15.2 using the change of variable $\tau' = t - \tau$.

2. Illustrate the associative property of equation 15.3 with a specific example.

3. Verify the distributive property of equation 15.4. Hint: Integration is distributive.

4. Verify that $a(h \ast f) = (a f) \ast h = f \ast (ah)$. Hint: Use equation 15.2.

Another very useful property of convolution is the so-called **time shift property**: if $h(t) \ast f(t) = g(t)$, then

$$h(t - T_1) \ast f(t - T_2) = g(t - T_1 - T_2) \tag{15.5}$$

Some simple examples serve to demonstrate the actual calculation process. These examples will utilize the **sifting property** of the delta function: if $h(t)$ is continuous at $t = T$, then

$$h(T) = \int_{T^-}^{T^+} h(\tau)\delta(T - \tau)d\tau = \int_{0^-}^{0^+} h(T - \tau)\delta(\tau)d\tau$$

EXAMPLE 15.1. Compute the convolution of an arbitrary continuous function $f(t)$ with $\delta(t)$.

SOLUTION

By the definition of equation 15.2 and the sifting property of the delta function,

$$y(t) = f \ast \delta = \int_{-\infty}^{\infty} f(t - \tau)\delta(\tau)d\tau = f(t - \tau)\big]_{\tau=0} = f(t) \tag{15.6}$$

Example 15.1 makes the point that $\delta(t)$ acts like an identity for convolution, i.e., it always returns the function with which it is convolved. The next example indicates that the convolution of an arbitrary continuous function $f(t)$ with $a\delta(t - T)$ produces a scaled and shifted version of $f(t)$ as given below in equation 15.7.

EXAMPLE 15.2. Compute the convolution of an arbitrary continuous function $f(t)$ with $a\delta(t - T)$.

SOLUTION

By the definition of equation 15.2 and the sifting property of the delta function,

$$y(t) = f(t) \ast \left[a\delta(t - T)\right] = a\int_{-\infty}^{\infty} f(t - \tau)\delta(\tau - T)d\tau = a f(t - \tau)\big]_{\tau=T} = af(t - T) \tag{15.7}$$

Exercise. Check the associative property of convolution by computing $f_1 \ast f_2 \ast f_3$ for

(i) $f_1(t) = \delta(t - 1)$, $f_2(t) = u(t) - u(t - 2)$, and $f_3(t) = \delta(t - 3)$

(ii) $f_1(t) = \delta(t - T_1)$, $f_2(t) = u(t) - u(t - 2)$, and $f_3(t) = \delta(t + T_2)$

Let us apply the results of the above two examples to a simple staircase function.

EXAMPLE 15.3. Compute the convolution of the function $f(t)$ shown in Figure 15.3a with $h(t)$ $= 2\delta(t) - 2\delta(t - 1)$.

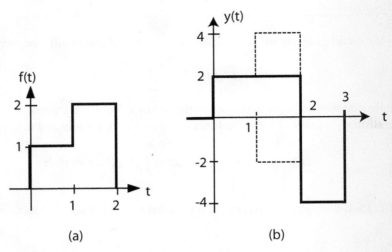

FIGURE 15.3 (a) Function $f(t)$ for Example 15.3. (b) $y = f * h$.

SOLUTION
By the definition of equation 15.1 and the sifting property of the delta function,

$$y(t) = f * h = 2\int_{-\infty}^{\infty} f(t - \tau)\delta(\tau)d\tau - 2\int_{-\infty}^{\infty} f(t - \tau)\delta(\tau - 1)d\tau$$
$$= 2 f(t - \tau)]_{\tau=0} - 2 f(t - \tau)]_{\tau=1} = 2 f(t) + (-2f(t - 1))$$

(15.8)

Graphically combining the result given in equation 15.8 yields the relation in Figure 15.3b.

Exercise. Compute the convolution of the function $f(t)$ shown in Figure 15.3a with $h(t) = 2\delta(t) -$ $2\delta(t - 2)$.
ANSWER: $2f(t) - 2f(t - 2) = g(t)$

EXAMPLE 15.4. Find the convolution of $u(t)$ with itself, i.e., $u(t) * u(t)$.

SOLUTION
By the definition of equation 15.2,

$$u(t) * u(t) = \int_{-\infty}^{\infty} u(t - \tau)u(\tau)d\tau = \int_{0}^{\infty} u(t - \tau)d\tau$$

(15.9)

since

$$u(\tau) = \begin{cases} 1 & \tau \geq 0 \\ 0 & \tau < 0 \end{cases}$$

We now break equation 15.9 into two regions: (i) $\tau > t$, in which case $u(t - \tau) = 0$, making equation 15.9 zero; and (ii) $0 \leq \tau \leq t$, in which case $u(t - \tau) = 1$. When $0 \leq \tau \leq t$, equation 15.9 reduces to

$$u(t) * u(t) = \int_0^t u(t - \tau) d\tau = \int_0^t d\tau = tu(t) = r(t) \tag{15.10}$$

In other words, $u(t) * u(t) = r(t)$, the ramp function.

We now apply the time shift property of convolution to the result of Example 15.4.

EXAMPLE 15.5. Find $y = h * f$ when $h(t) = u(t)$ and $f(t) = u(t + 1) - u(t - 1)$, as shown in Figures 15.4a and b.

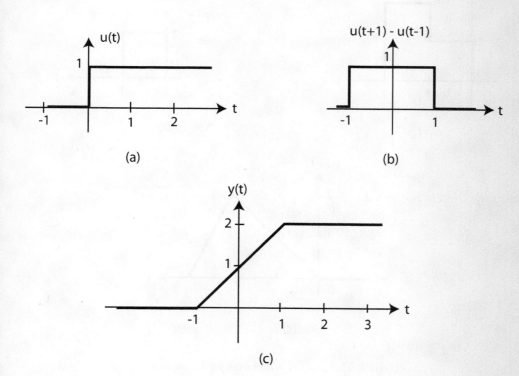

(a)

(b)

(c)

FIGURE 15.4 (a) $h(t) = u(t)$, the step function. (b) $f(t) = u(t + 1) - u(t - 1)$.
(c) Resulting $y(t)$ for the convolution $h*f$.

SOLUTION

Using the distributive law of convolution,

$$y(t) = u(t)*[u(t + 1) - u(t - 1)] = u(t)*u(t + 1) - u(t)*u(t - 1) \qquad (15.11)$$

Now using the time shift property of convolution and equation 15.10, we conclude that

$$y(t) = u(t) * u(t + 1) - u(t)*u(t - 1) = r(t + 1) - r(t - 1)$$

which is plotted in Figure 15.4c. We note that this result was achieved without any direct integration.

EXAMPLE 15.6. Compute the convolution $y = h * f$ of the two waveforms $h(t)$ and $f(t)$ in Figure 15.5.

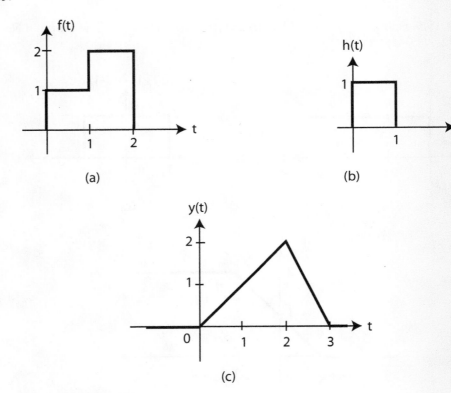

FIGURE 15.5 (a), (b) Waveforms for Example 15.6.
(c) $y = h * f$

SOLUTION

We first note that $f(t) = u(t) + u(t - 1) - 2u(t - 2)$ and $h(t) = u(t) - u(t - 1)$. Using the distributive law of convolution,

$$y(t) = [u(t) + u(t-1) - 2u(t-2)]*[u(t) - u(t-1)]$$
$$= [u(t) + u(t-1) - 2u(t-2)]*u(t) - [u(t) + u(t-1) - 2u(t-2)]*u(t-1)$$

Now using the time shift property of convolution and equation 15.10, with a further application of the distributive law, we conclude that

$$y(t) = [r(t) + r(t-1) - 2r(t-2)] - [r(t-1) + r(t-2) - 2r(t-3)] = r(t) - 3r(t-2) + 2r(t-3)$$

which is plotted in Figure 15.5c. We again note that this result was achieved without any direct integration.

EXAMPLE 15.7. Compute the convolution $y = h * f$ of the signals $h(t) = u(-t)$ and $f(t) = e^{-t}[u(t) - u(t-T)]$, as given in Figure 15.6.

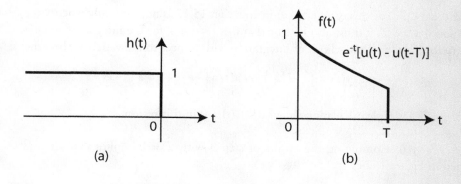

(a)   (b)

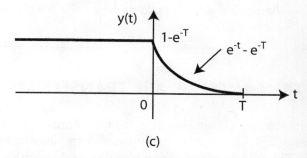

(c)

FIGURE 15.6 (a), (b) Functions $f(t)$ and $h(t)$ for the convolution of Example 15.7. (c) The resulting convolution.

SOLUTION

Step 1. *Apply definition and adjust limits of integration.* Applying the definition of equation 15.2 yields

$$y(t) = h * f = \int_{-\infty}^{\infty} h(t-\tau)f(\tau)d\tau = \int_{-\infty}^{\infty} u(\tau-t)e^{-\tau}[u(\tau) - u(\tau-T)]d\tau$$

Since $u(\tau) - u(\tau - T)$ is nonzero only for $0 \le \tau \le T$, the lower and upper limits of integration become 0 and T, respectively:

$$y(t) = h * f = \int_0^T e^{-\tau} u(\tau - t) d\tau \qquad (15.12)$$

Step 2. *Determine the regions of t over which the integral is to be evaluated.* From the limits of integration and the $u(\tau - t)$, there are three regions of interest: (i) $t < 0$, (ii) $0 \le t \le \tau < T$, and (iii) $T \le t$.

Step 3. *Evaluate the convolution integral, equation 15.12, over the given regions.*

Case 1: $t < 0$. Here, $t < 0$ implies that $\tau - t \ge 0$ over $0 \le \tau \le T$. Hence, $u(\tau - t) = 1$ over $0 \le \tau \le T$, and

$$y(t) = \int_0^T e^{-\tau} d\tau = 1 - e^{-T}$$

Case 2: $0 \le t < T$. For this case, $u(\tau - t)$ in equation 15.12 is nonzero only when $\tau \ge t$. Hence, in the region $0 \le t < T$, it must also be true that $0 \le t \le \tau < T$ for the integral of equation 15.12 to be nonzero. Thus, the lower limit of integration with respect to the variable τ becomes t:

$$y(t) = \int_t^T e^{-\tau} u(\tau - t) d\tau = \left[-e^{-\tau} \right]_t^T = e^{-t} - e^{-T}$$

Case 3: $t \ge T$. A simple calculation shows that $y(t) = 0$ in this region.

Step 4. *Plot y(t).* Combining the results of step 3 with $T = 1$ implies that $y(t)$ has the graph sketched in Figure 15.6c.

Exercise. Find the convolution, say, $y(t)$, of the signal $f(t) = e^{-at} u(t)$ with $h(t) = Ku(-t)$.
ANSWER: $Ka^{-1} u(-t) + Ka^{-1} e^{-at} u(t)$

---

3. CONVOLUTION AND LAPLACE TRANSFORMS

As claimed in the introduction, circuit analysis in the time domain by convolution and circuit analysis in the frequency domain by Laplace transformation are equivalent in terms of zero-state response calculations. The purpose of this section is to rigorize the equivalence between the time and frequency domain analysis methods by formally showing that $\mathcal{L}[h * f] = H(s)F(s)$. Section 8 presents a time domain rendition.

> # CONVOLUTION THEOREM
> Suppose $f(t) = 0$ and $h(t) = 0$ for $t < 0$. Then
>
> $$\mathcal{L}\,[h(t) * f(t)] = H(s)F(s) \qquad\qquad (15.13a)$$
>
> i.e., convolution in the time domain is equivalent to multiplication of transforms in the frequency domain; or equivalently,
>
> $$h(t) * f(t) = \mathcal{L}^{-1}[H(s)F(s)] \qquad\qquad (15.13b)$$

The justification of this theorem proceeds as follows:

Step 1. Given equation 15.13a, the first step is to write down the transform of $h(t) * f(t)$. Specifically,

$$\mathcal{L}[h(t) * f(t)] = \int_{0^-}^{\infty} \left(\int_{0^-}^{\infty} h(t - \tau)u(t - \tau)f(\tau)d\tau \right)e^{-st}\,dt \qquad (15.14)$$

A couple of points are in order: (i) the inner integral, surrounded by parentheses, represents the convolution $h(t) * f(t)$; and (ii) the presence of the step function $u(t - \tau)$ is added as an aid to emphasize the fact that $h(t) = 0$ for $t < 0$.

Step 2. The goal at this point is to manipulate the integral of equation 15.14 into a form that can be identified as the product of the Laplace transforms of two functions, i.e., as $H(s)F(s)$. The only possible strategy is to interchange the order of integration and group appropriate terms. Note that the Re[s] must be chosen sufficiently large to ensure the existence of the Laplace transforms of both $h(t)$ and $f(t)$. Under certain conditions that are typically met, it is possible to interchange the order of integration within a common domain of convergence of $H(s)$ and $F(s)$. Interchanging the order and regrouping the t-dependent terms inside a single parenthetical expression produces

$$\mathcal{L}[h(t) * f(t)] = \int_{0^-}^{\infty} f(\tau)\left(\int_{0^-}^{\infty} h(t - \tau)u(t - \tau)e^{-st}\,dt \right)d\tau \qquad (15.15)$$

Step 3. Observe that the interior integral, surrounded by parentheses, in equation 15.15 is simply the Laplace transform of a time-shifted $h(t)$, i.e., $\mathcal{L}[H(t - \tau)u(t - \tau)] = e^{-\tau s}H(s)$. Substituting $e^{-\tau s}H(s)$ for the interior integral leads to the desired equivalence:

$$\mathcal{L}[h(t) * f(t)] = \int_{0^-}^{\infty} f(\tau)H(s)e^{-s\tau}\,d\tau = H(s)\int_{0^-}^{\infty} f(\tau)e^{-s\tau}\,d\tau = H(s)F(s)$$

This theorem asserts the equivalence of convolution of one-sided signals with multiplication of their transforms in the s-domain. For our purposes, $h(t)$ assumes the role of the impulse response of our circuit and $f(t)$ the role of the input excitation. Accordingly, the convolution of the impulse response of a circuit or system with an input signal, a time domain computation, equals the inverse transform of the product of the respective Laplace transforms. In other words, the diagram of Figure 15.1 is correct, as claimed under the conditions of the theorem.

Exercises. 1. An unknown relaxed linear system has impulse response $h(t) = u(t) - u(t - 1)$. Find the response to the input signal $f(t) = 2u(t - 1)$ using convolution and check your answer using the Laplace transform method.
ANSWER: $2r(t - 1) - 2r(t - 2)$

2. An unknown relaxed linear system has impulse response $h(t) = (a - b)e^{-at}u(t), a,b > 0$ and input signal $f(t) = e^{-bt}u(t)$. Find the response $y(t) = h(t) * f(t)$.
ANSWER: $(e^{-bt} - e^{-at})u(t)$

The conditions of the theorem are somewhat restrictive in terms of computing circuit responses strictly in the time domain. Specifically, it is the one-sided Laplace transform that does not recognize function segments over the negative real axis—hence the condition on the input excitation that $f(t) = 0$ for $t < 0$. This restriction does not lend itself to the computation of initial conditions and circuit responses due to input signals extending back in time to $t = -\infty$. In general, the convolution of an input excitation with a circuit's impulse response presupposes no such restriction. However, justification of the computation of zero-state responses due to input excitations extending back in time to $t = -\infty$ cannot be based on the convolution theorem of the one-sided Laplace transform. A justification of the time domain convolution approach to computing circuit response is reserved for the last section of this chapter, due to its complexity. Nevertheless we will use the result as necessary, such as in the next exercise.

Exercise. The transfer function of a particular system is

$$H(s) = \frac{2}{s + 0.2}$$

Compute the convolution of the impulse response with the input $v(t)$, shown in Figure 15.7.
ANSWER: $10[1 - e^{-0.2(t + 1)}]u(t + 1) - 10[1 - e^{-0.2(t - 1)}]u(t - 1)$

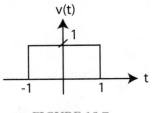

FIGURE 15.7

---

4. GRAPHICAL CONVOLUTION AND CIRCUIT RESPONSE COMPUTATION

The convolution integration formula, although explicit, has many layers of interpretation. **Graphical convolution** is a pen-and-pencil technique for determining the convolution integral of simple, squarish waveforms. The technique often leads to a more penetrating insight into the convolution integral.

There are four key ideas in the graphical procedure: flip, shift, multiply, area. The following is a detailed description of the procedure.

To compute $y(t)$ given by the convolution integral of equation 15.1 or 15.2, for a specific value of $t = T$, perform the following steps:
1.   Plot $h(\tau)$ vs. τ and $f(\tau)$ vs. τ curves.
2.   Flip the $f(\tau)$ curve about the vertical axis ($\tau = 0$) to obtain the $f(-\tau)$ vs. τ curve.
3.   Shift the $f(-\tau)$ curve to the right by the amount T to obtain the $f(T-\tau)$ vs. τ curve.
4.   *Multiply.* Plot the product $h(\tau)f(T-\tau)$ vs. τ curve.
5.   *Area.* Calculate the area beneath the $h(\tau)f(T-\tau)$ vs. τ curve for $-\infty < \tau < \infty$. The result is $y(T)$.

We make the following remarks with regard to the above steps:
(1)   The roles of $h(\tau)$ and $f(\tau)$ may be interchanged because of the commutative property of convolution, equation 15.2. In other words, we may flip and shift $h(\tau)$ instead of $f(\tau)$. Usually we flip and shift the simpler waveform.
(2)   In step 3, if T is negative, the shift is actually to the left.
(3)   As T is varied from $-\infty$ to $+\infty$, a complete plot of $y(t)$ is obtained. In shifting and finding the area, we often have to divide T into separate intervals, because each interval may require a different formula to compute the area beneath the $h(\tau)f(T-\tau)$ vs. τ curve.

With regard to step 2, for each t, $h(t-\tau)$ is a **shifted horizontal flip** of $h(\tau)$: as t moves from $-\infty$ to ∞, $h(t-\tau)$ moves along the τ-axis from $\tau = -\infty$ to $\tau = \infty$; a simple illustration is $h(t-\tau) = u(t-\tau)$, which is sketched in Figure 15.8. Part (a) of the figure shows $u(\tau)$, part (b) depicts $u(-\tau)$, and part (c) plots $u(t-\tau)$, which slides to the right along the τ-axis as t increases. For comparison, we can consider the function $h(t-\tau) = u(t-1-\tau)$, whose three forms are given in Figures 15.8d through f.

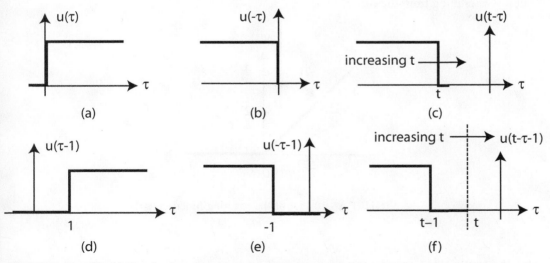

FIGURE 15.8 (a)–(c) For the function $h(t-\tau) = u(t-\tau)$, plots of (a) $u(\tau)$, (b) $u(-\tau)$, and (c) $u(t-\tau)$. (d)–(f) For the function $h(t-\tau) = u(t-\tau-1)$, plots of (d) $u(\tau-1)$, (e) $u(-\tau-1)$, and (f) $u(t-\tau-1)$.

EXAMPLE 15.8. Graphically compute the convolution $y = h * f$ of the two waveforms $h(t) = u(-t)$ and $f(t) = 2u(-t)$ shown in Figure 15.9.

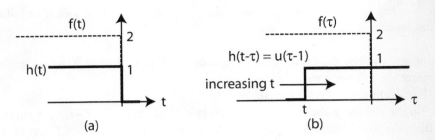

(a)   (b)

FIGURE 15.9 Waveforms for Example 15.8.

SOLUTION
The graphical solution to this convolution depends on a partitioning of the time line into special segments over which the graphical convolution is easily done. There are two regions to consider: (1) $-\infty < t \le 0$ and (2) $t > 0$.

Step 1. *Consider the region $t > 0$.* Figure 15.9b shows that $h(t - \tau)f(\tau) = 0$ for all $t > 0$ and all τ. Thus the convolution integral is zero and $y(t) = 0$, $t > 0$.

Step 2. *Consider the region $-\infty < t \le 0$.* From Figure 15.9b, the product $h(t - \tau)f(\tau) = 1 \times 2 = 2$ for $t \le \tau \le 0$ and zero elsewhere. Consequently, the area under the nonzero portion of the product functions $h(t - \tau)f(\tau)$ is

$$y(t) = \int_t^0 h(t - \tau)f(\tau)d\tau = 2\int_t^0 d\tau = -2t$$

Step 3. *Combine the foregoing calculations into a plotted waveform.* Figure 15.10 shows the function $y(t)$ resulting from the convolution.

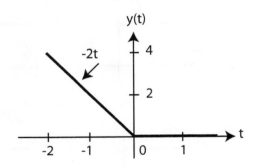

FIGURE 15.10 Plot of the resulting function, $y(t)$.

Exercise. In Example 15.8, suppose $f(t)$ is changed to $f(t) = 2u(t) - 2u(t-1)$. Find $y(t)$ at $t = 0.5$, 1.5, and 2.5 sec by the graphical convolution method.
ANSWER: 1, 1, 0

Another, more complex example will end our illustration of the graphical convolution technique.

EXAMPLE 15.9. Compute the convolution $y(t)$ of the triangular pulse $h(t)$ with the square pulse $f(t)$ as sketched in Figure 15.11.

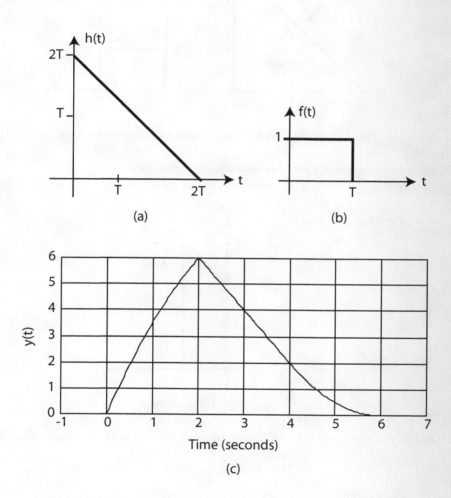

FIGURE 15.11 Convolution of triangular signal (a) with square pulse (b) to produce the signal (c) for $T = 2$ for Example 15.9.

The goal, of course, is to graphically evaluate the convolution integral of equation 15.2 using the following steps:

1. Since τ is the variable of integration, draw $h(t - \tau)$ and $f(\tau)$ on the τ-axis.
2. Evaluate the product $h(t - \tau)f(\tau)$ for various regions of t.
3. Compute the area under the product curve for each region determined in step 2.

Step 1. *Draw $h(t - \tau)$ and $f(\tau)$ on the τ-axis for $t < 0$ and compute the area of their product.* Figure 15.12 shows $h(t - \tau)$ and $f(\tau)$ on the τ-axis. From the figure, it is clear that $h(t - \tau)f(\tau) = 0$ for $t < 0$; hence, $y(t) = 0$ in the first region.

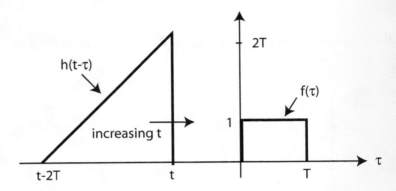

FIGURE 15.12 Graph of $h(t - \tau)$ and $f(\tau)$ on τ-axis for $t < 0$.

Step 2. *Consider the region $0 \leq t < T$, as illustrated in Figure 15.13.*

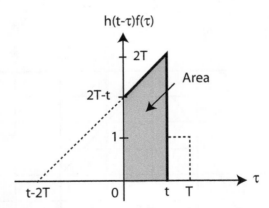

FIGURE 15.13 Graph of $h(t - \tau)f(\tau)$ on the τ-axis for $0 \leq t < T$.

The shaded area of the figure is the difference between the area of the large triangle, defined as

$$\text{Area A} = 0.5(2T)^2$$

and the area of the smaller triangle to the left of the vertical axis, defined as

$$\text{Area B} = 0.5(2T - t)^2$$

Hence,

$$y(t) = \text{Area} = \text{Area A} - \text{Area B} = 2Tt - 0.5t^2$$

for $0 \le t < T$. Alternatively, one may use the formula for the area of a trapezoid, i.e., the average height times the base, in which case one immediately obtains $0.5(2T + 2T - t)t = \text{Area} = 2Tt - 0.5t^2$.

Step 3. *Now consider the region $T \le t < 2T$, as depicted in Figure 15.14.*

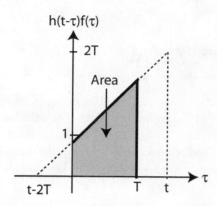

FIGURE 15.14 Graph of $h(t - \tau)f(\tau)$ on the τ-axis for $T \le t < 2T$.

In this figure, the shaded area, which determines $y(t)$, is again the difference of two triangular areas, specifically,

$$y(t) = 0.5[T - (t - 2T)]^2 - 0.5[2T - t]^2 = 2.5T^2 - Tt$$

for $T \le t < 2T$.

Step 4. Figure 15.15 shows the next region, $2T \le t < 3T$. Another straightforward calculation yields

$$y(t) = 0.5(3T - t)^2 = 4.5T^2 - 3Tt + 0.5t^2$$

for $2T \le t < 3T$.

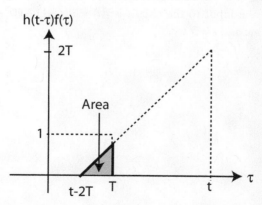

FIGURE 15.15 Graph of $h(t - \tau)f(\tau)$ on the τ-axis for $2T \le t < 3T$.

Step 5. *Consider the region $3T \le t$. Here, the product $h(t - \tau)f(\tau) = 0$, in which case $y(t) = 0$ for t
$> 3T$.*

In sum,

$$y(t) = \begin{cases} 0, & t < 0 \\ 2Tt - 0.5t^2, & 0 \le t < T \\ 2.5T^2 - Tt, & T \le t < 2T \\ 4.5T^2 - 3Tt + 0.5t^2, & 2T \le t < 3T \\ 0, & t \ge 3T \end{cases}.$$

A plot of $y(t)$ appears in Figure 15.11c for $T = 2$.

Exercises. 1. Repeat the calculations of the preceding example, except flip and shift $f(t)$ instead of
$h(t)$. Here, one looks at $f(t - \tau)$ sliding through $h(\tau)$. The calculations should be easier and the
result the same.
2. Find the output of a linear and relaxed circuit with input $f(t)$ and impulse response $h(t)$ (shown
in Figures 15.16a and b) at time $t = 2.5$ seconds. Hint: Use graphical convolution.
ANSWER: 1

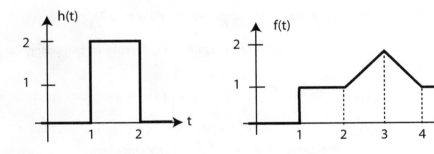

FIGURE 15.16

3. The impulse response of a particular circuit is approximately measured on a scope as illustrated in Figure 15.17. If the input to the circuit is $f(t) = u(t - 1) - u(t - 2)$, find the value of the
response, $y(t)$, at $t = 3$ sec and $t = 3$ sec.
ANSWERS: 2, 1

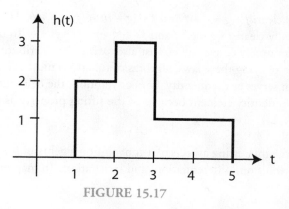

FIGURE 15.17

4. For $h(t)$ and $f(t)$ as sketched in Figure 15.18, find $y(1)$ for the convolution $y(t) = h(t) * f(t)$.
ANSWER: 4

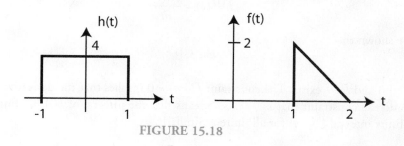

FIGURE 15.18

Sometimes the foregoing graphical techniques prove difficult to execute. Nevertheless, an under-standing of them offers fundamental insight into the meaning of the convolution integral. A use-ful set of techniques for quickly evaluating convolution integrals arises from the properties of a convolution algebra, discussed in the next section.

5. CONVOLUTION ALGEBRA

A set of functions, together with operations called addition and multiplication, is called an *alge-bra*, assuming certain conditions are satisfied. The set of all functions that can be convolved with each other also constitutes an algebra with respect to the operations of addition (+), and convolu-tion (*). This set, together with the two operations, is called a **convolution algebra**. In this con-text, operations such as differentiation and integration are inverses of each other. For example, integrating a function and then differentiating the result returns the original function. Within the convolution algebra, the convolution $f * g$ is equivalent to the convolution of the integral of f with the derivative of g. The advantage here is that, by successive integration and differentiation, it is often possible to reduce an apparently difficult convolution to a simpler one.

For a set of functions to be an **algebra** with respect to + and *, several arithmetic operations must hold. In particular, + and * must be both commutative and associative. The commutativity and associativity of + is clear: $f + g = g + f$ and $f + (g + h) = (f + g) + h$. The commutativity and associativity of $*$ is equally clear: $f * g = g * f$ and $f * (g * h) = (f * g) * h$. To be an algebra, the set of all functions that are mutually convolvable must also satisfy the distributive law, $f * (g + h) = f * g + f * h$. In addition to obeying these laws, algebras of functions must contain identity elements. For +, the zero function serves as the identity. For convolution, the delta function plays this role. The delta function is an identity element because of the sifting property as set forth in Example 15.2 and equation 15.7.

For our purposes, the interesting aspects of a convolution algebra of functions rests with the interrelationship of convolution, differentiation, and integration. To map out this kinship, we use the following notation

$$f^{(-1)}(t) = \int_{-\infty}^{t} f(\tau)d\tau \tag{15.16}$$

and

$$h^{(1)}(t) = \frac{dh(t)}{dt} \tag{15.17}$$

It can be shown that

$$f * h = f^{(1)} * h^{(-1)} \tag{15.18}$$

if $f(-\infty) = 0$ and $h^{(-1)}$ exists. The constraint $f(-\infty) = 0$ implies that the derivative of $f(t)$ is zero at $t = -\infty$ and the constraint that $h^{(-1)}$ exists means that the integral of $h(t)$ has finite area over the semi-infinite interval $(-\infty, t]$ for all finite t. Similarly,

$$f * h = f^{(-1)} * h^{(1)} \tag{15.19}$$

if $h(-\infty) = 0$ and $f^{(-1)}$ exists.

EXAMPLE 15.10. Find the convolution $y = f * h$ for $f(t) = u(t) + u(t-1)$ and $h(t) = u(t) - u(t-1)$, as sketched in Figures 15.5a and b.

SOLUTION
The goal is to use equation 15.18 to evaluate the convolution, i.e., $f * h = f^{(1)} * h^{(-1)}$, where $f^{(1)}(t) = \delta(t) + \delta(t-1) - 2\delta(t-2)$ and $h^{(-1)}(t) = r(t) - r(t-1)$, as presented in Figure 15.19.

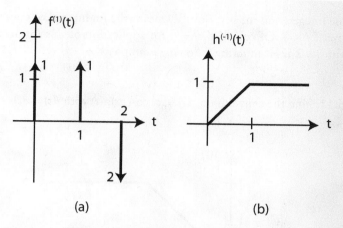

FIGURE 15.19 (a) The derivative of $f(t) = u(t) + u(t-1) - 2u(t-2)$.
(b) The integral of $h(t) = u(t) - u(t-1)$, as given in Example 15.10.

Since $f^{(1)}(t) = \delta(t) + \delta(t-1) - 2\delta(t-2)$, a sum of impulse functions, the sifting property of the impulse function implies

$$y(t) = f^{(-1)}(t) * h^{(-1)}(t) = h^{(-1)}(t) + h^{(-1)}(t-1) - 2h^{(-1)}(t-1) \tag{15.20}$$

With the picture of $h^{(-1)}(t)$ given in Figure 15.19b, the right-hand side of equation 15.20 can be interpreted as a graphical sum of (shifted) versions of $h^{(-1)}(t)$, as illustrated in Figure 15.20.

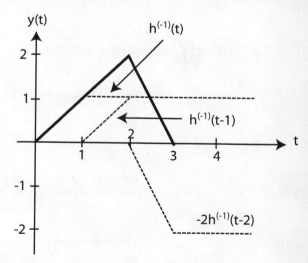

FIGURE 15.20 $y(t)$ equal to superposition of shifted replicas of $h^{(-1)}(t)$.

Equations 15.18 and 15.19, as illustrated in the preceding example, are special cases of more general formulas. Specifically, let $y = f * h$. Then

$$y^{(j+k)} = f^{(j)} * h^{(k)} \tag{15.21}$$

where j and k are integers and the notation $f^{(j)}$ means the jth integral of f over $[-\infty, t]$ if $j < 0$, and the jth derivative if $j > 0$. Of course, $f^{(0)} = f$. An application of this formula to the special case where $j = -k$ with $j = 2$ is given in the following example.

EXAMPLE 15.11. Find the convolution $g(t) = \pi^2 \cos(\pi t)u(t)$ with $f(t) = r(t) - r(t-2)$, as sketched in Figure 15.21.

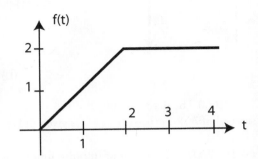

FIGURE 15.21 Waveform for Example 15.11.

SOLUTION
Some preliminary thought suggests that evaluation of the convolution integral might proceed more smoothly via equation 15.21; i.e.,

$$f * g = f^{(2)} * g^{(-2)}$$

where

$$g^{(-1)}(t) = \pi^2 \int_{0^-}^{t} \cos(\pi q)dq = \pi \sin(\pi t)u(t)$$

and

$$g^{(-2)}(t) = \pi^2 \int_{0^-}^{t} \sin(\pi q)dq = [1 - \cos(\pi t)]u(t)$$

Differentiating $f(t)$ twice leads to

$$f^{(2)}(t) = \delta(t) - \delta(t-2)$$

Hence,

$$f * g = f^{(2)} * g^{(-2)} = [1 - \cos(\pi t)]u(t) - [1 - \cos(\pi(t-2))]u(t-2)$$

But $[1 - \cos(\pi(t-2))]u(t-2)$ is just a right shift by one period of $[1 - \cos(\pi t)]u(t)$. Therefore,

$$f * g = f^{(2)} * g^{(-2)} = [1 - \cos(\pi t)]u(t)u(2-t)$$

This function is plotted in Figure 15.22.

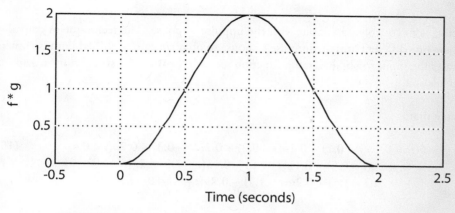

FIGURE 15.22 Graph of the convolution $f*g$ for Example 15.11.

Convolution algebra and graphical convolution lend themselves to a second application of the convolution technique: the computation of circuit responses from a staircase approximation to a circuit impulse response. If a circuit schematic is lost, such an approximation could result from a CRT readout of the circuit impulse response measured in a laboratory. The following example illustrates this application.

EXAMPLE 15.12. Suppose the schematic diagram of a very old linear circuit is lost. However, the circuit impulse response is measured in the laboratory and approximated by the staircase waveform of Figure 15.23. If the input to the circuit is $v_{in}(t) = 100u(t)$, compute the output voltage, $v_{out}(t)$, at $t = 0$ and at $t = 0.5$ sec.

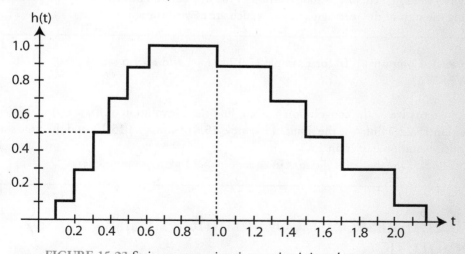

FIGURE 15.23 Staircase approximation to circuit impulse response.

SOLUTION

The objective is to convolve the input with the impulse response. The technique of convolution algebra where one differentiates the impulse response to obtain a sum of shifted impulse functions and integrates the input to obtain a ramp function seems to be the most straightforward approach for this calculation.

First, observe that

$$
\begin{aligned}
h(t) = & \ 0.1u(t - 0.1) + 0.2u(t - 0.2) + 0.2u(t - 0.3) + 0.2u(t - 0.4) \\
& + 0.2u(t - 0.5) + 0.1u(t - 0.6) - 0.1u(t - 1) - 0.2u(t - 1.3) \\
& - 0.2u(t - 1.5) - 0.2u(t - 1.7) - 0.2u(t - 2) - 0.1u(t - 2.2)
\end{aligned} \tag{15.22}
$$

Hence,

$$
\begin{aligned}
h'(t) = & \ 0.1\delta(t - 0.1) + 0.2\delta(t - 0.2) + 0.2\delta(t - 0.3) + 0.2\delta(t - 0.4) \\
& + 0.2\delta(t - 0.5) + 0.1\delta(t - 0.6) - 0.1\delta(t - 1) - 0.2\delta(t - 1.3) \\
& - 0.2\delta(t - 1.5) - 0.2\delta(t - 1.7) - 0.2\delta(t - 2) - 0.1\delta(t - 2.2)
\end{aligned}
$$

Now, since the integral of the input is $100r(t)$, we can compute the output voltage as

$$
\begin{aligned}
v_{out}(t) = & \ 10r(t - 0.1) + 20r(t - 0.2) + 20r(t - 0.3) + 20r(t - 0.4) \\
& + 20r(t - 0.5) + 10r(t - 0.6) - 10r(t - 1) - 20r(t - 1.3) \\
& - 20r(t - 1.5) - 20r(t - 1.7) - 20r(t - 2) - 10r(t - 2.2)
\end{aligned}
$$

At $t = 0$, $v_{out}(0) = 0$, and at $t = 0.5$, $v_{out}(0.5) = 4 + 6 + 4 + 2 = 16$.

Of course, it is possible to obtain the solution just as easily in this case using the graphical method. Simply flip the $v_{in}(t)$ curve and slide it through the $h(t)$ curve. The area under the product curve is simply the sum of the rectangular areas, which are easy to compute.

Exercises. 1. Compute $v_{out}(t)$ for Example 15.12 at $t = 1$ and $t = 1.5$ sec.
ANSWERS: 65, 106

2. As an alternative to the convolution algebra, find the convolution of $f(t) = u(t)$ with $h(t)$ given in equation 15.22. Hint: Use the result of Example 15.4 or equation 15.10, and the time shift theorem of convolution, equation 15.5.
ANSWER: Replace the step functions in equation 15.22 with ramp functions.

---

6. CIRCUIT RESPONSE COMPUTATIONS USING CONVOLUTION

This section contains a series of examples that illustrate the convolution approach to computing zero-state circuit responses.

EXAMPLE 15.13. Consider the *RC* circuit of Figure 15.24, whose impulse response is $h(t) = e^{-t}u(t)$. If the input is $v_{in}(t) = e^{at}u(-t)$, find $v_{out}(T)$ when $T = 0$ and $T > 0$ for $a = 0$ and $a > 0$.

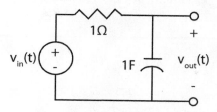

FIGURE 15.24 *RC* circuit for Example 15.13.

SOLUTION

Case 1: a = 0. Here, $v_{in}(t) = u(-t)$ V. Since the capacitor voltage, $v_{out}(t)$, is the convolution of the input with the impulse response,

$$v_{out}(T) = \int_{-\infty}^{\infty} h(T - \tau)v_{in}(\tau)d\tau = \int_{-\infty}^{\infty} h(\tau)v_{in}(T - \tau)d\tau$$

$$= \int_{-\infty}^{\infty} e^{-\tau}u(\tau)u(\tau - T)d\tau = \int_{0}^{\infty} e^{-\tau}u(\tau - T)d\tau$$

Consider the case of $T = 0$. Then

$$v_{out}(0) = \int_{0}^{\infty} e^{-\tau}u(\tau)d\tau = 1$$

This makes sense because with $a = 0$, $v_{in}(t) = u(-t)$ V is a 1 V constant, and the capacitor looks like an open circuit at $T = 0$. On the other hand if $T > 0$,

$$v_{out}(T) = \int_{0}^{\infty} e^{-\tau}u(\tau - T)d\tau = \int_{T}^{\infty} e^{-\tau}u(\tau - T)d\tau = e^{-T}$$

This also makes sense physically because at $T = 0$, the initial capacitor voltage is 1 V and the capacitor discharges with a time constant of 1 sec since there is no further nonzero input.

Case 2: a > 0. Similarly to case 1,

$$v_{out}(T) = \int_{-\infty}^{\infty} h(\tau)v_{in}(T - \tau)d\tau = \int_{-\infty}^{\infty} e^{-\tau}u(\tau)e^{a(T-\tau)}u(\tau - T)d\tau$$

$$= e^{aT}\int_{0}^{\infty} e^{-(1+a)\tau}u(\tau - T)d\tau$$

Consider the case of $T = 0$. Then

$$v_{out}(0) = \int_{0}^{\infty} e^{-(1+a)\tau}d\tau = \frac{1}{1+a}$$

This initial voltage depends on the history of the excitation and can be computed only by convolution because the circuit is not in steady state at $T = 0$.

For $T > 0$, then, since the input is zero, we know that

$$v_{out}(T) = v_C(0)e^{-T}u(T) = \frac{e^{-T}}{1+a}u(T)$$

Exercises. 1. For case 1 in Example 15.13, find $v_{out}(T)$ for $T > 0$.
ANSWER: 1

2. For case 2 in Example 15.13, find $v_{out}(T)$ for $T > 0$.

ANSWER: $\dfrac{e^{aT}}{1+a}$

3. In Figure 15.24, suppose the resistor has value $R > 0$ and the capacitor $C > 0$. Show that the impulse response of the circuit is

$$h(t) = \frac{1}{RC}e^{-\frac{t}{RC}}u(t)$$

4. For the general impulse response computed in Exercise 3, suppose $v_{in}(t) = e^{at}u(-t)$ V. Compute $v_{out}(0)$ when $a = 0$ and $a > 0$

ANSWERS: 1, $\dfrac{\dfrac{1}{RC}}{\dfrac{1}{RC}+a}$

EXAMPLE 15.14. The goal of this example is to design a circuit that computes the running average of a voltage $v_{in}(t)$ over the interval $[t - T, t]$ given a specification of the necessary impulse response, $h(t)$.

SOLUTION
A circuit that computes a running average must have the input-output relation

$$v_{out}(t) = \frac{1}{T}\int_{t-T}^{t} v_{in}(\tau)d\tau$$

From our development of convolution, such a circuit must have an impulse response $h(t)$ satisfying the relationship

$$v_{out}(t) = \int_{-\infty}^{\infty} h(t-\tau)v_{in}(\tau)d\tau = \frac{1}{T}\int_{t-T}^{t} v_{in}(\tau)d\tau$$

Now, $h(t - \tau)$ must be a window function that captures the portion of $v_{in}(t)$ over the interval $t - T \le \tau \le t$. Figure 15.25 depicts the proper forms of $h(t - \tau)$ and $h(t)$.

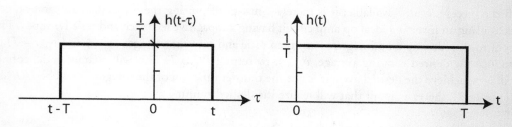

FIGURE 15.25 (a) The window function $h(t - \tau)$. (b) The impulse response $h(t)$.

The circuit design problem reduces to developing a circuit that integrates the function segment $v_{in}(\tau)$ over $t - T \le \tau \le t$. To achieve this integration, note that

$$\int_{-\infty}^{t} v_{in}(\tau)d\tau = \int_{-\infty}^{t-T} v_{in}(\tau)d\tau + \int_{t-T}^{t} v_{in}(\tau)d\tau$$

This relationship leads to

$$\int_{t-T}^{t} v_{in}(\tau)d\tau = \int_{-\infty}^{t} v_{in}(\tau)d\tau - \int_{-\infty}^{t-T} v_{in}(\tau)d\tau \qquad (15.23)$$

For the second integral on the right-hand side of equation 15.23, let $\lambda = \tau + T$. Then $d\lambda = d\tau$, $\tau = \lambda + T$, and

$$\int_{-\infty}^{t-T} v_{in}(\tau)d\tau = \int_{-\infty}^{t} v_{in}(\lambda - T)d\lambda \qquad (15.24)$$

Since both λ and τ are dummy variables of integration, we may replace λ by τ and rewrite equation 15.23 as

$$\int_{t-T}^{t} v_{in}(\tau)d\tau = \int_{-\infty}^{t} \left[v_{in}(\tau) - v_{in}(\tau - T) \right]d\tau \qquad (15.25)$$

where $v_{in}(\tau - T)$ is simply a delayed replica of $v_{in}(\tau)$ and where, for practical reasons, we can replace the lower limit of $-\infty$ by t_0, the time the actual circuit turns on.

As a convenience, we will define a device called an **ideal delay of T seconds**, whose input is $v_{in}(t)$ and whose output is $v_{in}(t - T)$. Figure 15.26 shows the ideal delay as a device having infinite input impedance combined with a dependent voltage source whose output is a delayed version of the input. Such a device can be achieved by storing the values of $v_{in}(t)$ in a digital computer or, for small T, by the use of an analog delay line.

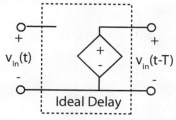

FIGURE 15.26 Depiction of a device called an ideal delay of T seconds whose output is a delayed replica of the input.

All the pieces are now available; it is merely a matter of putting them together. Integration can occur using an inverting ideal op amp circuit having a capacitive feedback and resistive input. The input to this ideal op amp integrator can then scale and sum the voltages $-v_{in}(t)$ and $v_{in}(t-T)$ to produce the desired running average, $v_{out}(t)$, by setting $RC = T$. This will guarantee the correct scaling to achieve the desired average, since the transfer function of the integrator will be $1/(RCs)$. Figure 15.27 shows a circuit that will realize the desired running average.

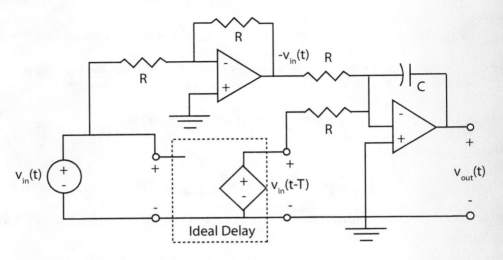

FIGURE 15.27 Op amp circuit that produces a running average of the input waveform, $v_{in}(t)$, provided that $RC = T$.

For the circuit of Figure 15.27, the input drives the first inverting op amp circuit and also feeds the ideal delay. The second op amp circuit is an ideal inverting integrator whose inputs are $-v_{in}(t)$ and $v_{in}(t-T)$. With $RC = T$ the output is the required running average.

7. CONVOLUTION PROPERTIES REVISITED

From the perspective of the impulse response theorem, the convolution properties of commutativity, associativity, and distributivity have important implications in terms of circuit and system configurations. For example, if $h_1(t)$ and $h_2(t)$ are the impulse responses of two systems, then commutativity says that $h_1 * h_2 = h_2 * h_1$. This means that the order of a cascade of circuits or systems is mathematically irrelevant, provided that there is no loading between the circuits. The idea is illustrated theoretically in Figure 15.28.

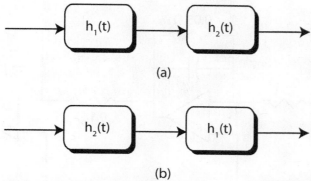

(a)

(b)

FIGURE 15.28 Interchanging the order of impulse responses, in which the equivalence of parts (a) and (b) follows from the commutativity of convolution.

Although mathematically parts (a) and (b) of Figure 15.28 produce the same result, practical considerations often dictate a more careful realization process. For example, one circuit may have a high input impedance while the other has both a low input impedance and a low output impedance. In this case we would put the first high-input-impedance circuit at the front end and the other circuit at the output end.

A cascade op amp realization of a transfer function,

$$H(s) = \frac{100}{(s+1)(s+20)}$$

(15.26)

illustrates commutativity nicely. A designer may use either Figure 15.28a or Figure 15.28b to realize $H(s)$. Magnitude scaling, say by $K_m = 10^4$, will yield more realistic resistor and capacitor values. Recall that in magnitude scaling, $R_{new} = K_m R_{old}$, $C_{new} = C_{old}/K_m$, and $L_{new} = K_m L_{old}$.

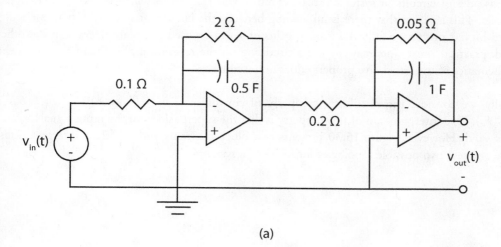

(a)

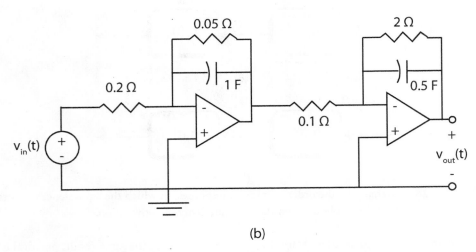

(b)

FIGURE 15.29 Realization of interchanging the order of cascaded circuits.

Exercise. Verify that each of the circuits of Figure 15.29 realizes the transfer function of equation 15.26. Magnitude-scale by $K_m = 10^5$ to obtain more realistic parameter values. If only 1 μF capacitors were available, you would need two scale factors—K_{m1} for the first op amp stage and K_{m2} for the second op amp stage. What are the two scale factors?

ANSWERS: Multiply each resistor by 10^5 and divide each capacitor by 10^5 to obtain the new values. If only 1 μF capacitors are available, then for Figure 15.29b, $K_{m1} = 10^6$ and $K_{m2} = 0.5 \times 10^6$. For Figure 15.29a, the two scale factors are interchanged.

Theoretically speaking, the associative property, $h_1 * (h_2 * h_3) = (h_1 * h_2) * h_3$, means that multiple cascades of circuits or systems can be combined or realized in whatever order the designer chooses. This assumes that there is no loading between the circuits or systems. Op amp circuits called buffers or voltage followers having gains of 1 can be used to isolate stages. On the other hand, practical constraints may impose a condition on the realization of a circuit for which the mathematics of the associative property does not account.

Finally, we consider the distributive property of convolution: $h_1 * (h_2 + h_3) = (h_1 * h_2) + (h_1 * h_3)$. One interpretation of this property is that the superposition of the input signals h_2 and h_3 is valid. However, Figure 15.30 presents two block diagrams with different interpretations. Here one sees two possible topologies for realizing a system.

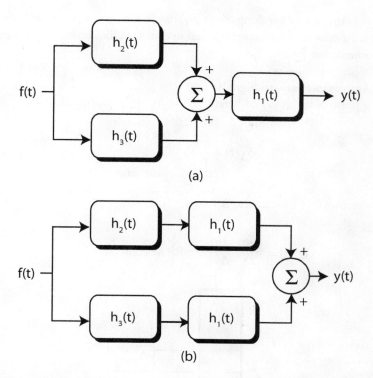

(a)

(b)

FIGURE 15.30 Two possible block diagram interpretations of the distributive law for convolution.

Exercise. In terms of reliability, i.e., possibly continued partial operation in the face of a circuit failure, suggest reasons why one realization in Figure 15.30 would be superior to the other. In terms of minimum number of components, suggest reasons why one realization would be better than the other.

---

8. TIME DOMAIN DERIVATION OF THE CONVOLUTION INTEGRAL FOR LINEAR TIME-INVARIANT CIRCUITS

As mentioned, a deficiency in the one-sided Laplace transform technique is its inability to deal with signals whose time dependence may extend back to $-\infty$. This section develops the zero-state system response as the convolution of a not necessarily one-sided input excitation with the impulse response of the circuit or system. Throughout the development, we will assume that the circuit or system under consideration is linear, i.e., composed of linear circuit elements. This implies that the zero-state response of the circuit satisfies the conditions of **linearity**; i.e., if the zero-state response to the excitation $f_i(t)$ is $y_i(t)$ for $i = 1, 2$, then the zero-state response to the input excitation $a_1 f_1(t) + a_2 f_2(t)$ is $a_1 y_1(t) + a_2 y_2(t)$. In addition, we assume that the circuit or system is **time invariant**, i.e., each circuit element is characterized by constant parameter values. Mathematically, this means that if $f(t)$ is the input to a circuit element and $y(t)$ the zero-state response of the circuit element, then $y(t - T)$ is the response to $f(t - T)$ for all T and all possible input signals $f(t)$. Intuitively speaking, time invariance means that if we shift the input in time, then the associated zero-state response is shifted in time by a like amount. These properties underlie the development that follows.

Rectangular Approximations to Signals

Let us define a pulse of width Δ and height $1/\Delta$ as $\delta_\Delta(t)$. In particular,

$$\delta_\Delta(t) = \begin{cases} \dfrac{1}{\Delta} & 0 < t < \Delta \\[2mm] 0 & \textit{otherwise} \end{cases} \qquad (15.27)$$

Figure 15.31 sketches $\delta_\Delta(t)$ for several Δ values.

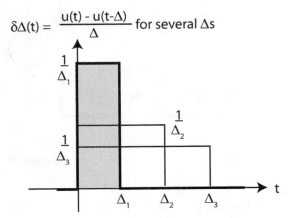

$$\delta\Delta(t) = \frac{u(t) - u(t-\Delta)}{\Delta} \text{ for several } \Delta s$$

FIGURE 15.31 $\delta_\Delta(t)$ for several Δs.

Shifting the pulse of equation 15.27 yields $\delta_\Delta(t - k\Delta)$, as represented in Figure 15.32.

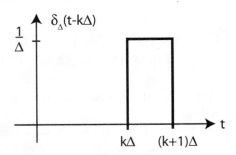

FIGURE 15.32 $\delta_\Delta(t - k\Delta) = \delta_\Delta(t - t_k)$ is a shifted version of the pulse $\delta_\Delta(t)$, where $t_k = k\Delta$.

For convenience, let $t_k = k\Delta$ so that $\delta_\Delta(t - k\Delta) = \delta_\Delta(t - t_k)$. Figure 15.33 shows a rectangular approximation, $\hat{v}(t)$, to a continuous waveform $v(t)$.

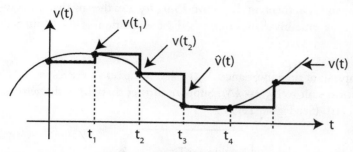

FIGURE 15.33 Rectangular approximation, $\hat{v}(t)$, to a continuous signal $v(t)$.

Expressing the rectangular approximation indicated in the figure analytically, using shifted versions of the pulse functions defined in equation 15.27, leads to the infinite summation

$$\hat{v}(t) = \sum_{k=-\infty}^{\infty} v(k\Delta)\delta_\Delta(t - k\Delta)\Delta = \sum_{k=-\infty}^{\infty} v(t_k)\delta_\Delta(t - t_k)\Delta \qquad (15.28)$$

Hence, for sufficiently small Δ, it follows from equation 15.28 that

$$v(t) \approx \hat{v}(t) = \sum_{k=-\infty}^{\infty} v(t_k)\delta_\Delta(t - t_k)\Delta \qquad (15.29)$$

One concludes that if $v(t)$ is continuous, then

$$v(t) = \lim_{\Delta \to 0} \sum_{k=-\infty}^{\infty} v(t_k)\delta_\Delta(t - t_k)\Delta = \int_{-\infty}^{\infty} v(\tau)\delta(t - \tau)d\tau \qquad (15.30)$$

where we have interpreted the delta function as a limit of short-duration pulses whose height is inverse to the width so that the area is constant:

$$\delta(t) = \lim_{\Delta \to 0} \delta_\Delta(t)$$

Observe that the right-hand integral of equation 15.30 is precisely the convolution $v(t) * \delta(t)$

Computation of Response for Linear Time-Invariant Systems

Suppose $h_\Delta(t)$ is the zero-state response of a linear time-invariant circuit to the pulse $\delta_\Delta(t)$. If the circuit's impulse response satisfies smoothness conditions, i.e., if it has sufficient continuous derivatives, then the circuit's impulse response is the limit of the $h_\Delta(t)$ values as Δ goes to zero. In particular,

$$h(t) = \lim_{\Delta \to 0} h_\Delta(t) = \lim_{\Delta \to 0} \int_{0^-}^{\infty} h(t - \tau)\delta_\Delta(\tau)d\tau = \int_{0^-}^{\infty} h(t - \tau)\delta(\tau)d\tau, \qquad (15.31)$$

where the fact that

$$h_\Delta(t) = \int_{0^-}^{\infty} h(t - \tau)\delta_\Delta(\tau)d\tau$$

follows from the convolution theorem. Equation 15.31 restates the law that the zero-state response of an input to a linear time-invariant circuit is the convolution of the input with the impulse response.

Now, by the assumption of time invariance, $h_\Delta(t - k\Delta)$ is the zero-state response of a well-behaved linear time-invariant circuit to $\delta_\Delta(t - k\Delta)$. Suppose further that $y(t)$ is the zero-state response of the same circuit to $v(t)$. (See Figure 15.34.)

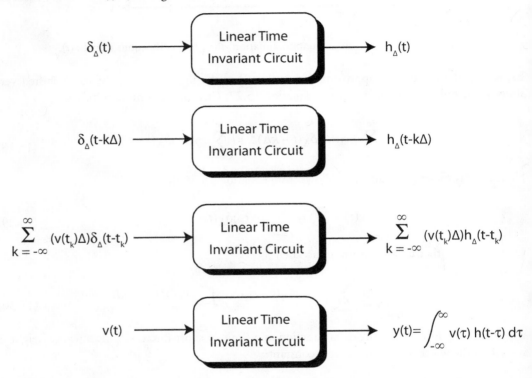

FIGURE 15.34 Zero-state responses to a particular linear time-invariant system, showing the framework of the derivation. Note that the bottom condition follows because as $\Delta \to 0$, $\Delta \to d\tau$, $t_k \to \tau$, and $\Sigma \to \int$.

It is now possible to use the approximation for $v(t)$ given in equation 15.31 to generate an approximation to $y(t)$ in terms of a summation of terms of the form $v(k\Delta)h_\Delta(t - k\Delta)\Delta$. Taking the limit as $\Delta \to 0$ will yield $y(t)$ as the convolution of $v(t)$ and $h(t)$.

To derive this, note that for each k, $v(k\Delta) = v(t_k)$ is a scalar. Hence, the zero-state response to $v(t_k)\delta_\Delta(t - t_k)\Delta$ is $v(t_k)h_\Delta(t - t_k)\Delta$. By the linearity assumption, which implies superposition, the zero-state circuit response to $\hat{v}(t)$, equation 15.29, is

$$\hat{y}(t) = \sum_{k=-\infty}^{\infty} h_\Delta(t - t_k)v(t_k)\Delta \tag{15.32}$$

In the limit, as Δ approaches zero, t_k approaches a continuous variable τ and $\Delta \to d\tau$. Hence, if the impulse response is sufficiently smooth—i.e., if it has sufficient continuous derivatives—then

$$y(t) = \lim_{\Delta \to 0} \sum_{k=-\infty}^{\infty} v(k\Delta)h_\Delta(t - k_\Delta)\Delta = \int_{-\infty}^{\infty} v(\tau)h(t - \tau)d\tau \qquad (15.33)$$

Thus, we conclude that for a linear time-invariant circuit, the zero-state response $y(t)$ to an input excitation $v(t)$ is the convolution of the input $v(t)$ with the impulse response $h(t)$. We will refer to equation 15.33 as the **impulse response theorem**,[1] which says that *the zero-state response of a linear time-invariant circuit or system to a (possibly two-sided) input signal is the convolution of the input with the impulse response of the circuit.*

9. SUMMARY

This chapter has introduced the concept of the convolution of two signals. The convolution can be evaluated analytically (by direct computation of the convolution integral) or graphically. Often, by applying the properties of convolution algebra, it is possible to implement shortcuts for calculating the convolution of two signals, resulting in the simplification of the analytical calculation or of the graphical calculation.

Using the notion of convolution, the chapter developed a technique of computing circuit responses in the time domain. This technique is the direct counterpart of computing the transfer function in the frequency domain, the approach presented in Chapter 14. Using the convolution approach, one can compute the zero-state response of a circuit excited by signals that extend back in time to $-\infty$, something not directly possible with the one-sided Laplace transform. However, for one-sided signals—which constitute the great majority of signals that are relevant to circuit analysis—the convolution and Laplace transform approaches are completely equivalent, as demonstrated by the convolution theorem. The chapter gave an example of designing a circuit to compute a running average. In addition, it presented an application of the convolution technique to the computation of circuit responses for a circuit whose impulse response is approximated on a CRT. Future courses will expand the seeds planted in this chapter. For example, convolution is pertinent to an understanding of radar techniques, commonly used to identify speeding motorists.

10. TERMS AND CONCEPTS

Algebra: a set of functions with respect to two operations, + and *, satisfying the commutative, associative, and distributive laws. In addition, there must be an identity with respect to each operation. For addition, the zero function serves as the identity. For convolution, the delta function plays this role. The delta function is an identity element because of its sifting property.

Associativity: for convolution, the property $h * (f * g) = (h * f) * g$.

Commutativity: for convolution, the property $h * f = f * h$.

Convolution: an integration process between two functions to produce a third, new function in accordance with equation 15.1 or 15.2.

Convolution algebra: the algebra of functions with respect to the operations of addition (+) and convolution (*).

Convolution theorem: for one-sided waveforms $h(t)$ and $f(t)$, $\mathcal{L}[h(t) * f(t)] = H(s)F(s)$.

Distributivity: for convolution, the property $h * (f + g) = h * f + h * g$.

Graphical convolution (flip-and-shift method): a pen-and-pencil technique for determining the convolution integral of simple, squarish waveforms.

Ideal delay of T seconds: waveform with input $f(t)$ and output $f(t - T)$, a delayed replica of $f(t)$.

Impulse response theorem: theorem stating that the zero-state response of a linear time-invariant circuit or system to a (possibly two-sided) input signal is the convolution of the input with the impulse response of the circuit.

Linearity: property whereby, if the zero-state response to the excitation $f_i(t)$ is $y_i(t)$ for $i = 1, 2$, then the zero-state response to the input excitation $a_1 f_1(t) + a_2 f_2(t)$ is $a_1 y_1(t) + a_2 y_2(t)$.

Running average: average that is updated continuously.

Sifting property: the property of a delta function for simplifying an integral, i.e.,

$$f(T) = \int_{-\infty}^{+\infty} f(t)\delta(t - T)dt,$$ provided that $f(T)$ is continuous at T.

Time invariance: property such that, if $f(t)$ is the input to a circuit element and $y(t)$ is the zero-state response to the circuit element, then $y(t - T)$ is the response to $f(t - T)$ for all T and all possible input signals $f(t)$.

Zero-state response: response of a circuit or system to an input when all initial stored energy is zero, i.e., all initial conditions are zero.

---

[1] The derivation of this result is, of course, not rigorous. A rigorous justification is given as theorem 4 of Sandberg's "Linear Maps and Impulse Responses," *IEEE Transactions. on Circuits and Systems*, vol. 35, no. 2, February 1988, pp. 201-206.

Problems

CONVOLUTION BY INTEGRAL

1. Let $f(t) = K_1 \delta(t - T_1)$, $T_1 > 0$. Compute and plot in terms of K_1 and T_1 the results of the following convolutions:

    (a) $f(t) * u(t + 2T_1)$

    (b) $f(t) * r(t + T_1)$

    (c) $f(t) * [u(t + T_1) - u(t - 4T_1)]$

    (d) $f(t) * [r(t + 2T_1) - r(t - 2T_1)]$

    (e) $f(t) * [r(t + T_1) - r(t - 2T_1)]$

2. Let $f(t) = 2\delta(t + 2) - 2\delta(t - 2)$. Compute and plot the following convolutions:

    (a) $f(t) * u(t)$

    (b) $f(t) * r(t)$

    (c) $f(t) * [u(t) - u(t - 4)]$

    (d) $f(t) * [r(t) - r(t - 2)]$

    (e) $f(t) * [r(t) - r(t - 2) - u(t - 4)]$

3. Let $f(t) = 2\delta(t + 2) - 2\delta(t - 2)$. Compute and plot the following convolutions:

    (a) $f(t) * \cos(\pi t)u(t)$

    (b) $f(t) * \sin(\pi t)u(t)$

4. Compute and plot the results of each of the following convolutions:

    (a) $u(t) * u(t - 2)$

    (b) $u(t - 1) * u(t - 2)$

    (c) $u(t) * [u(t) - u(t - 4)]$

    (d) $u(t + 2) * [u(t) - u(t - 2)]$

    (e) $[u(t + 2) - u(t)] * [u(t) - u(t - 2)]$

5. Use the convolution integral to compute the convolutions in parts (a) and (b). Then use the results for (a) and (b) to compute the remaining convolutions. (Parts (a) and (b) can be used to solve many of the subsequent problems also.)

    (a) $u(t) * r(t)$

    (b) $r(t) * r(t)$

    (c) $u(t) * [r(t) - r(t - 4)]$

    (d) $r(t) * [r(t) - r(t - 4)]$

    (e) $[u(t + 2) - u(t)] * [u(t) - u(t - 2)]$

    (f) $[r(t + 2) - r(t)] * [r(t) - r(t - 2)]$

    (g) $f_1(t) * f_2(t)$ for the functions in Figure P15.5

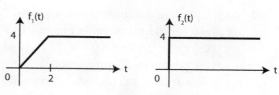

Figure P15.5

6. Let $f(t) = Ke^{-at}u(t)$, $a > 0$. Use the convolution integral to compute the convolution in part (a). Then use the results for (a) along with the various convolution properties, especially the time shift property, to compute the remaining convolutions.

    (a) $u(t) * f(t)$

    (b) $u(t + T) * f(t)$

    (c) $[u(t + T) - u(t - T)] * f(t)$

    (d) $[u(t + T_1) - u(t - T_2)] * f(t - T_3)$, $T_1 > 0$, $T_2 > 0$, $T_3 > 0$

    (e) $u(t + T) * [f(t) - e^{-aT}f(t - T)]$

7. Let $f_1(t) = K_1 e^{-at}u(t)$, $a > 0$, and $f_2(t) = K_2 e^{-bt}u(t)$, $b > 0$. Use the Laplace transform method to compute the following convolutions:

    (a) $f_1(t) * f_1(t)$

    (b) $f_1(t) * f_2(t)$

Now use the results for parts (a) and (b) and the various convolution properties to compute the following convolutions:

    (c) $f_1(t + T) * f_1(t - T)$, $T > 0$

    (d) $f_1(t + T) * f_2(t - T)$

    (e) $f_1(t + T_1) * f_2(t - T_2)$, $T_1, T_2 > 0$

8. Compute and draw the convolution $y(t) = f(t) * v(t)$ of Figure P15.8 with $v(t) = \delta(t + 4) - \delta(t + 2)$.

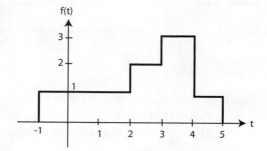

Figure P15.8

9. (a) Compute the convolution $y(t) = f(t) *$ $v(t)$ for the functions in Figure P15.9 using the fact that $r(t) = u(t) * u(t)$ and the properties of convolution.

    (b) Repeat the calculation in part (a) using graphical convolution for $T = 1$ and $K = 1$.

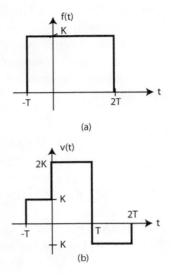

(a)

(b)

Figure P15.9

10. (a) Compute the convolution of $y(t) = f(t)$ $* v(t)$ in Figure P15.10 using the fact that $r(t) = u(t) * u(t)$ and the properties of convolution.

    (b) Repeat the calculation in part (a) using graphical convolution for $T = 1$.

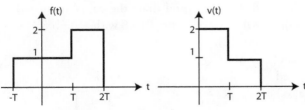

Figure P15.10

11. (a) Compute the convolution of $y(t) = f(t)$ $* v(t)$ in Figure P15.11 using the fact that $r(t) = u(t) * u(t)$ and the properties of convolution.

    (b) Repeat the calculation in part (a) using graphical convolution for $K = 2$.

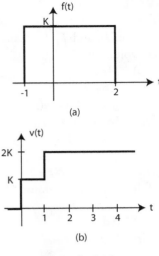

(a)

(b)

Figure P15.11

12. (a) Compute the convolution $y(t) = f(t) *$ $v(t)$ of the functions in Figure P15.12. Hint: Use the results of Problem 5 for $u(t) * r(t)$.

    (b) Repeat the calculation in part (a) using graphical convolution for $A = 2$ and $T = 1$.

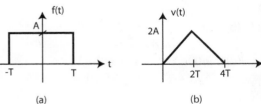

(a)

(b)

Figure P15.12

13. (a) Compute the convolution $y(t) = f(t) * v(t)$ of the functions in Figure P15.13. Hint: Use the results of Problem 5.

    (b) Repeat the calculation in part (a) using graphical convolution for $A = 2$ and $T = 1$.

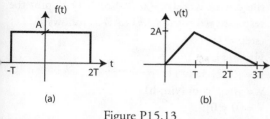

(a)   (b)

Figure P15.13

14. Find the convolution of $f(t)$ in Figure P15.14 with

(a) $v(t) = \delta(t) + \delta(t + 2)$ (Hint: Use graphical convolution.)

(b) $h(t) = \delta(t + 1) + \delta(t - 1)$ (Hint: Use graphical convolution.)

(c) $i(t) = u(t + 3) - u(t + 1)$

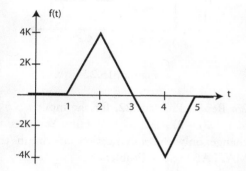

Figure P15.14

15.(a)   Let $h(t) = 4u(t)$. Find $y(t) = h(t) * f_1(t)$ with $f_1(t)$ as shown in Figure P15.15a. Sketch $y(t)$. Verify your answer using the Laplace transform method.

(b)   Repeat part (a) for $f_1(t)$ in Figure P15.15b.

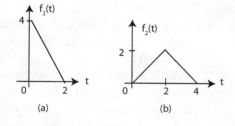

(a)   (b)

Figure P15.15

16. If $f(t) = K_1 e^{at}u(-t)$, $a > 0$, and $h(t) = K_2 u(t)$, compute $y(t) = h(t) * f(t)$.

17. If $f(t) = K_1 u(-t)$, and $h(t) = K_2 e^{-at}u(t)$, $a > 0$, $a > 0$, compute $y(t) = h(t) * f(t)$.

18. Compute the impulse response, $h(t)$, of the circuit of Figure P15.18. Express $h(t)$ in terms of $b = 1/(RC)$. Suppose $f_1(t) = K_1 e^{at}u(-t)$, $a > 0$, and $f_2(t) = K_2 e^{-at}u(t)$, $a > 0$ and $a \neq b$. Compute each of the indicated convolutions using equation 15.2.

(a) $y_1(t) = h(t) * u(-t)$, and sketch for $-\infty < t < \infty$.

(b) $y_2(t) = h(t) * f_1(t)$, and sketch for $-\infty < t < \infty$.

(c) $y_3(t) = h(t) * f_2(t)$, and sketch for $-\infty < t < \infty$.

(d) $y_4(t) = h(t) * [f_1(t) + f_2(t)]$, and sketch for $-\infty < t < \infty$.

(e) $y_5(t) = h(t) * [u(-t) + f_2(t)]$, and sketch for $-\infty < t < \infty$.

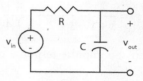

Figure P15.18

19. A particular active circuit has the transfer function

$$H(s) = \frac{K}{s + a}$$

where $a > 0$. Suppose the input to the circuit is $v(t)$, shown in Figure P15.19 with $T > 0$. Using convolution techniques, compute the response $y(t)$.

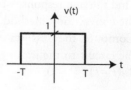

Figure P15.19

20. Consider the RLC circuit in Figure P15.20.

(a)   Compute the transfer function $H(s)$, and express $H(s) = \dfrac{K_1}{s + a} + \dfrac{K_2}{s + b}$

using the indicated element values. Then compute the impulse response $h(t)$.

(b) Compute $y(t) = h(t) * v_{in}(t)$, where $v_{in}(t) = v(t)$ is given in Figure P15.19.

(c) If the input voltage is $v_{in}(t) = 5u(-t)$ V, compute $y(t) = h(t) * v_{in}(t)$.

(d) Repeat (c) if the input voltage is $v_{in}(t) = 5e^{2t}u(-t)$ V.

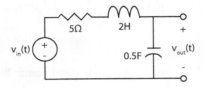

Figure P15.20

21. The transfer function of a particular time-invariant linear circuit is

$$H(s) = \frac{2}{s+1} - \frac{2}{s+2} + \frac{4}{s+4}$$

(a) Find the impulse response of the circuit.

(b) Find the transfer function of the circuit as a ratio of polynomials by using the residue command of MATLAB as follows:
   (i)   define the array p = [-1, -2,-4];
   (ii) define the array r = [2,-2,4];
   (iii) define k = 0;
   (iv) use the command "[n,d] = residue(r,p,k)" to obtain the coefficients of the numerator and denominator polynomials.

(c) Find the step response of the circuit by the convolution method.

(d) Compute the zero-state response of the circuit to the input $f(t) = 8u(-t) + 8u(t-T)$, $T > 0$.

GRAPHICAL CONVOLUTION

22. The impulse response of a particular circuit is measured approximately on an oscilloscope, as illustrated in Figure P15.22. If the input to

the circuit is $v_{in}(t) = u(t) -u(t-1)$ V, plot the response $y(t)$ using MATLAB as follows:

```
tstep = 1;
vin = [1];
h = [0, 2, 3, 1, 1];
y = tstep*conv(vin, h);
y = [0 y 0];
t = 0:tstep:tstep*(length(vin)+length(h));
plot(t,y)
grid
```

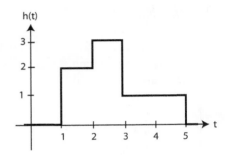

Figure P15.22

23. Repeat Problem 22, for the input $v_{in}(t) = u(t) + u(t-2) - 2u(t-4)$ V. Hint: You need to change only the specification of vin in the MATLAB code of Problem 22.

24. This problem formalizes the use of the MATLAB code above to compute the convolution of two piecewise constant time functions. The steps below develop a formula for the convolution of two staircase time functions $v(t)$ and $h(t)$ (shown in Figures P15.24a and b) using polynomial multiplication. In general, we assume $h(t)$ and $v(t)$ are zero for negative t and have a finite nonzero duration.

(a) Let $y(t) = h(t) * v(t)$ and let the time step $T = 1$. Using Example 15.4 and the convolution properties, argue that $y(t)$ has a piecewise linear structure as shown in Figure P15.24c. Find the breakpoint coordinates y_1, y_2 y_3 and y_4, first in terms of the literal levels v_0, v_1, h_0, h_1, and h_2, and then in terms of their actual numerical values.

(b) Define three polynomials in x

using the coefficients from Figures P15.24a, b, and c as follows:

$$p(x) = v0 \, x + v1$$
$$q(x) = h0 \, x^2 + h1 \, x + h2$$
$$r(x) = y1 \, x^3 + y2 \, x^2 + y3 \, x + y4$$

Show that $r(x) = p(x)q(x)$ with y1 through y4 assigned the values computed in part (a).

Remark: The result of step (b) indicates that the coefficients [y1, y2, y3, y4] in $r(x)$ or the break-point values in $y(t)$ of Figure P15.24c can be obtained from the coefficients [v0 v1] and [h0 h1 h2] by polynomial multiplication. The MATLAB command "conv" performs the desired polynomial multiplication. In particular, the code for obtaining the plot of Figure P15.24c is

```
v = [ 2 4];
h= [ 3 -2 1];
T = 1;
tstep = T;
y = [0 conv(v,h)*tstep 0];
% The additional beginning and
ending zeros are added to indicate
% that the initial and final values of
the convolution are zero, due to the
% finite duration assumption.
t = 0: tstep : tstep* (length(v) +
length(h));
plot(t, y)
grid
```

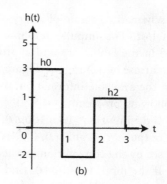

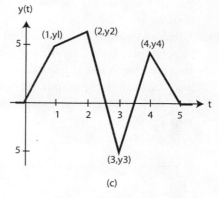

Figure P15.24

25. Repeat Problem 24 for the waveforms in Figure P15.25. When using the MATLAB code, it is necessary to account for $f_1(t)$ being nonzero for negative t. This is easily accomplished simply by shifting $f_1(t)$ to the right by one unit and then doing the indicated convolution; the proper result is obtained by a left shift of one unit.

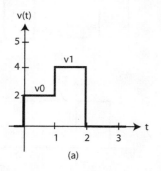

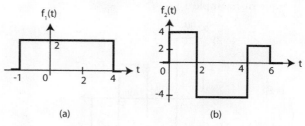

Figure P15.25

26. The schematic diagram of a very old linear circuit is lost. The impulse response, $h(t)$, is measured in the laboratory and approximated by the staircase waveform of Figure P15.26. Based on the available information, solve each of the following problems.

   (a) If the input is $v_{in}(t) = 100u(t)$, find the output $v_{out}(t)$ at $t = 0, 0.5, 1$, and 1.5 sec by the convolution method.

   (b) If the input is $100tu(t)$, find the output at $t = 1$ by the method of your choice.

   (c) Use the MATLAB code of Problem 24 to solve part (a). Hint: Although the step function is not of finite duration, after 2.2 seconds the output, $y(t) = h(t) * v_{in}(t)$, does not change; also, the time step $T = 0.1$ sec.

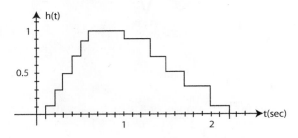

Figure P15.26

27. A crude approximation to the impulse response of an active RLC circuit is given by $h(t)$, as sketched in Figure P15.27a. If the input signal is $v(t)$, as given in Figure P15.27b, find the response $y(t)$ using MATLAB and the method developed in Problem 24. Plot the response for $0 \le t \le 6$ sec. Verify using the technique of Example 15.4.

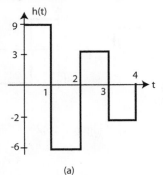

(a)

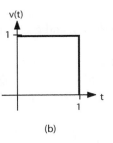

(b)

Figure P15.27

CONVOLUTION ALGEBRA

28. (a) Using the techniques of convolution algebra, find $f_3(t) = f_1(t) * f_2(t)$, and plot $f_3(t)$ vs. t, where $f_1(t)$ and $f_2(t)$ are as shown in Figures P15.28a and b. Hint: Graphically integrate $f_2(t)$.

   (b) Repeat part (a) with $f_1(t)$ changed to the waveform of Figure P15.28c for $T = 2$ and $A = 2$.

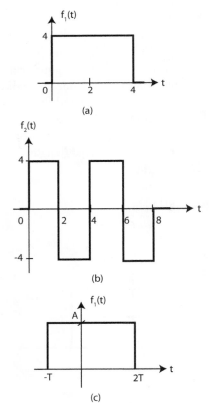

(a)

(b)

(c)

Figure P15.28

29. Repeat Problem 28(a) with $f_2(t)$ as given in Figure P15.29. Hint: Graphically integrate $f_2(t)$.

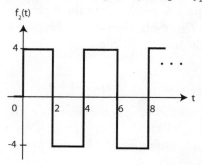

Figure P15.29

30. Suppose the impulse response of a circuit is given by $h(t) = a^2 e^{-at} u(t)$ and the input is a pulse $v_{in}(t) = u(t + T) - u(t - T)$, $T > 0$ V. Find $v_{out}(t) = h(t) * v_{in}(t)$ using convolution algebra techniques.

31. Use convolution algebra techniques to determine $y(t) = h(t) * f(t)$ for each of the functions $f(t)$ given in Figure P15.31 where $h(t) = \pi^2 \cos(\pi t) u(t)$. Plot the resulting $y(t)$ using MATLAB or its equivalent.

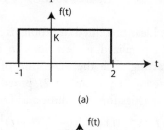

(a)

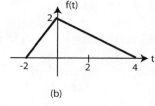

(b)

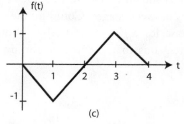

(c)

Figure P15.31

32. Consider Figure P15.32.
   (a) Find the transfer function $H(s)$ and the impulse response $h(t)$ of the circuit in Figure P15.32a.
   (b) Find the step response using convolution algebra methods.
   (c) Find $v_{out}(t)$ due to the rectangular pulse input in Figure P15.32b. Hint: Again, use convolution algebra methods, and sketch the output waveform for $T = 2\pi \sqrt{(LC)}$..

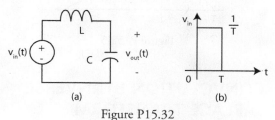

(a)   (b)

Figure P15.32

33. For the circuit of Figure P15.32a, find the convolution $y(t) = h(t) * f(t)$ for each $f(t)$ given in Figures 15.31b and c.

34. Consider the circuit presented in Figure P15.34a.
   (a) Find the transfer function $H(s)$.
   (b) Compute the impulse response $h(t)$.
   (c) Use only convolution algebra techniques to compute the response $v_{out}(t)$ for the input $i_{in}(t)$ in Figure P15.34b.
   (d) Plot your answer.

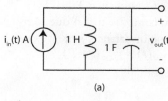

(a)

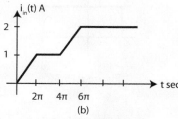

(b)

Figure P15.34

35. Compute the convolution of $y(t) = h(t) *$ $f(t)$ for each $f(t)$ below, given the $h(t)$ of Figure P15.35.

  (a) $f(t) = \omega^2 \cos(\omega t) u(t)$
  (b) $f(t) = \omega^2 \sin(\omega t) u(t)$
  (c) $f(t) = a^2 e^{-at} u(t)$

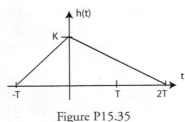

Figure P15.35

CONVOLUTION BY INVERSE LAPLACE TRANSFORM

36. This problem illustrates a trick for using the Laplace transform method for computing responses when the input is nonzero for $-T_0 \le t < 0$ when $T_0 > 0$. Suppose a circuit has the transfer function

$$H(s) = \frac{V_{out}}{V_{in}} = \frac{2}{s+2}$$

Suppose the input to the circuit is $v_{in}(t)$, shown in Figure P15.36.

  (a)   Define $\hat{v}_{in}(t) = v_{in}(t-T)$ and compute $\hat{V}_{in}(s)$.

  (b)   Compute $\hat{V}_{out}(s)$ due to the input $\hat{V}_{in}(s)$ and then $\hat{v}_{out}(t)$.

  (c)   To compute $v_{out}(t)$ due to $v_{in}(t)$, use time invariance, i.e.,
$$v_{out}(t) = \hat{v}_{out}(t+T).$$

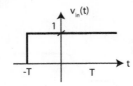

Figure P15.36

ANSWER: (c) $v_{out}(t) = \left(1 - e^{-2(t+T)}\right) u(t+T)$ V

37. This problem repeats the trick of Problem 36 of using the Laplace transform method to compute responses when the input is nonzero for $-T \le t < 0$ when $T > 0$. Suppose a circuit has the transfer function

$$H(s) = \frac{V_{out}}{V_{in}} = \frac{A}{s+a}$$

Suppose the input to the circuit is $v_{in}(t)$, shown in Figure P15.37.

  (a)   Define $\hat{v}_{in}(t) = v_{in}(t-T)$ and compute $\hat{V}_{in}(s)$.

  (b)   Compute $\hat{V}_{out}(s)$ due to the input $\hat{V}_{in}(s)$ and then $\hat{v}_{out}(t)$.

  (c)   To compute $v_{out}(t)$ due to $v_{in}(t)$ use time invariance, i.e.,
$$v_{out}(t) = \hat{v}_{out}(t+T).$$

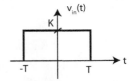

Figure P15.37

38. The ideas presented in Problems 36 and 37 can be generalized as developed in this problem. Starting with the definition of convolution integral, equation 15.2, prove the time shift properties for convolution as follows:

  (a) $f(t) * g(t) = \left[f(t - T_1) * g(t) \right]_{t=t+T_1}$

  (b) $f(t) * g(t) = \left[f(t - T_1) * g(t - T_2) \right]_{t=t+T_1}.$

39. Two active circuits with impulse responses $h_1(t) = 2e^{-2t} u(t)$ and $h_2(t) = 24te^{-10t} u(t)$ are cascaded as shown in Figure P15.39. No loading occurs between stages. Compute the zero-state response for $v_{in}(t) = u(t + 2)$ V. Would it be advantageous to use the convolution method here?

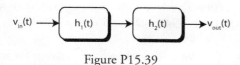

Figure P15.39

40. This problem shows the advantage of the Laplace transform method over the convolution method in the context of a very common situation. After solving this problem, try evaluating the convolution integrals, but don't spend much time on this. Consider Figure P15.40, which shows the cascade of three circuits with indicated impulse responses $h_1(t) = 2u(t)$, , $h_2(t) = 20e^{-5t}u(t)$, and $h_3(t) = 2tu(t)$.

(a) Find the impulse response of the cascade, i.e., $h(t) = h_1(t) * h_2(t) * h_3(t)$.

(b) Find the step response of the cascade.

Hint: You might want to use the residue command in MATLAB to compute the partial fraction expansions.

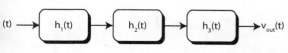

Figure P15.40

41. Consider the circuit of Figure P15.41.

(a) Determine the value of $v_{out}(0^-)$ from the input $v_{in}(t) = 10e^{10t}u(-t)$ V. Can this be done using Laplace transform techniques? If so, how?

(b) Find the transfer function

$$H(s) = \frac{V_{out}(s)}{V_{in}(s)}$$

and the impulse response $h(t)$ in terms of R and C.

(c) Suppose $R = 10\ \Omega$ and $C = 10$ mF. Given $v_{out}(0^-)$ computed in part (a), find the response due to the input $v_{in}(t) = te^{-t}u(t)$ using the Laplace transform method. Evaluate the advantages of this method over the time domain convolution method.

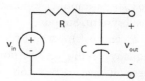

Figure P15.41

42. Consider the circuit of Figure P15.42, which has the transfer function

$$H(s) = \frac{\dfrac{s}{R_1 C_2}}{s^2 + \left(\dfrac{1}{R_1 C_1} + \dfrac{1}{R_2 C_2} + \dfrac{1}{R_1 C_2}\right)s + \dfrac{1}{R_1 C_1 R_2 C_2}}$$

(a) Let $C_1 = 0.5$ mF, $C_2 = 1$ mF, $R_1 = 2$ kΩ, and $R_2 = 1$ kΩ. Find the poles and zeros of $H(s)$, a partial fraction expansion of $H(s)$, and the resulting impulse response, $h(t)$.

(b) Find $v_{out}(0^-)$ when $v_{in}(t) = 54te^{2t}u(-t)$ V. Then find the (zero-input) response due to $v_{out}(0^-)$.

(c) Find the zero-state response when $v_{in}(t) = 54te^{-2t}u(t)$ V. Would the time domain convolution method be desirable for this calculation?

(d) Find the complete response, given your answers to parts (b) and (c).

(e) If $v_{out}(0^-)$ were doubled and $v_{in}(t) = 72te^{-2t}u(t)$ V, find the complete response without doing any further calculations.

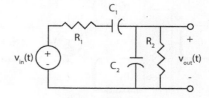

Figure P15.42

MISCELLANEOUS

43. Figure P15.43 shows a configuration for an interconnection of active circuits. The function

inside each box is the impulse response of the sub-circuit or subsystem. Suppose $h_1(t) = 4u(t)$, $h_2(t) = 4u(t-2)$, and $h_3(t) = \pi\cos(\pi t)u(t)$.

    (a)   Using any of the convolution techniques you have learned, compute and sketch the overall impulse response.

    (b)   Using any of the convolution techniques you have learned, find $y(t)$ for $f(t) = 6u(t)$.

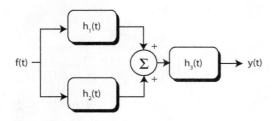

Figure P15.43

44. Repeat Problem 43 for $h_2(t) = 4\delta(t-2)$ and with the other impulse responses remaining the same.

45. Figure P15.45 shows a configuration for an interconnection of active circuits. The function inside each box is the impulse response of the sub-circuit or subsystem. Suppose $h_1(t) = 4u(t)$, $h_2(t) = 5e^{-2t}u(t)$, $h_3(t) = 10e^{-4t}u(t)$, and $h_4(t) = \delta(t)$. Using any of the convolution techniques you have learned, compute and plot (using MATLAB) the composite impulse response of the configuration.

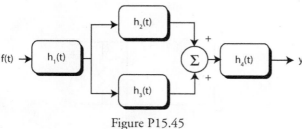

Figure P15.45

46. Repeat Problem 45 with $h_1(t)$ changed to $h_1(t) = 4[u(t) - u(t-4)]$. Having done Problem 45, can you do this problem without any further calculations?

47. Repeat Problem 45 for $h_1(t) = 4u(t)$, $h_2(t) = 5\delta(t)$, $h_3(t) = 5\delta(t-2)$, and $h_4(t) = 2e^{-2t}u(t)$.

48. Figure P15.48 shows a configuration for an interconnection of active circuits. The function inside each box is the impulse response of the circuit. Suppose $h_1(t) = 4\delta(t)$, $h_2(t) = 4\delta(t-1)$, $h_3(t) = 4\delta(t-2)$, $h_4(t) = 4\delta(t-4)$, and $h_5(t) = 2\cos(\pi t)u(t)$.

    (a)   Using any of the convolution techniques you have learned, compute and sketch the overall impulse response of each configuration.

    (b)   Using any of the convolution techniques you have learned, find $y(t)$ for $f(t) = 6u(t)$.

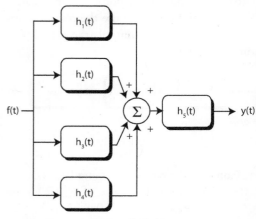

Figure P15.48

49. Let $f(t)$ be an arbitrary signal having a well-defined Laplace transform. Let

$$h(t) = \sum_{k=0}^{\infty} \delta(t - kT)$$

be a so-called impulse train. Find $L[f(t)h(t)]$. Note that if $e^{sT} = z$, a complex variable, then your answer should be of the form

$$\sum_{k=0}^{\infty} f(kT)z^{-k}$$

which is the so called Z-transform of the sequence $\{f(kT) \mid k = 0,1,2,...\}$.

50. The circuit of Figure P15.50 is initially at rest. The input

$$v_{in}(t) = \sum_{k=0}^{\infty} \delta(t - kT)$$

is a periodic impulse train with $T = 1$ sec.

  (a) Show that the impulse response is $h(t) = 0.5e^{-0.5t}u(t)$.

  (b) Find the exact solution of $v_{out}(t)$ for $0 < t < 1$.

  (c) Find the exact solution of $v_{out}(t)$ for $1 < t < 2$

  (d) Repeat part (c) for the interval $4 < t < 5$. You should wind up with the expression

$$v_{out}(t) = 0.5e^{-0.5(t-4)}(1 + e^{-0.5} + (e^{-0.5})^2 + (e^{-0.5})^3 + (e^{-0.5})^4)$$

  (e) Simplify the expression computed in part (d) by making use of the sum formula for a geometric series:

$$\sum_{k=0}^{n-1} \lambda^k = \frac{1 - \lambda^n}{1 - \lambda}$$

provided $\lambda \neq 1$.

  (f) Find the solution of $v_{out}(t)$ for $n < t < (n + 1)$. Make use of the techniques set forth in parts (d) and (e) to simplify your expression.

  (g) Sketch the waveform for large n. This is the so-called steady-state solution.

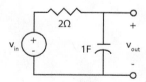

Figure P15.50

51. Consider the circuit of Figure P15.50 again. Let the input $v_{in}(t)$ be the impulse train shown in Figure P15.51, i.e., analytically

$$v_{in}(t) = \sum_{k=0}^{\infty} \delta(t + kT)$$

with $T = 1$. Find $v_{out}(t)$ for $t > 0$ and plot the resulting waveform.

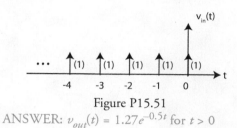

Figure P15.51

ANSWER: $v_{out}(t) = 1.27e^{-0.5t}$ for $t > 0$

16

Band-Pass Circuits and Resonance

HOW A TOUCH-TONE PHONE SIGNALS THE NUMBERS DIALED

Calling friends and others by phone occurs daily. When we dial a number, the information is sent to the central office by one of two methods: fast tone dialing or the much slower pulse dialing. For example, electronic processing of the pulse-dialed long-distance number 555-555-5555 requires about 11 seconds, while electronic processing of the tone-dialed takes only about 1 second.

The keypad of an ordinary touch-tone phone has 12 buttons arranged in four rows and three columns, as shown in the following diagram:

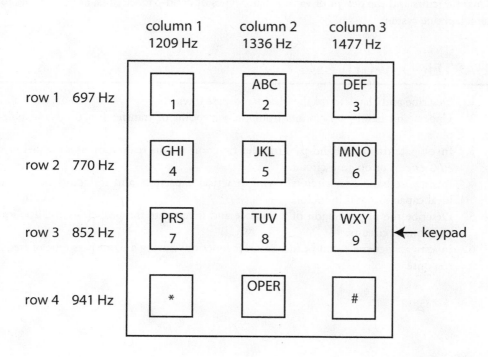

Pressing any one button generates two tones, with the frequencies selected by an electronic circuit inside the telephone set. For example, pressing the number 5 generates tones at 770 and 1336 Hz. The row and column arrangements and the dual-tone method permit the representation of 10 digits (0, ... , 9) and two symbols (*, #) using only seven tones. These seven tones are divided into two groups: the low-frequency group, from 697 to 941 Hz, and the high-frequency group, from 1209 to 1477 Hz.

Such tones are easily produced by an *LC* resonant circuit. The four tones in the low-frequency group are produced by connecting a capacitor to four different taps of a single coil (inductor). A similar connection generates the three tones in the high-frequency group. When a button is pressed to the halfway point, a dc current from the central office is sent through the coil in the tank circuit. When the button is fully pressed, the dc current is interrupted. This action initiates sinusoidal oscillations in the *LC* resonant tank circuit at a frequency inversely proportional to $\sqrt{LC}$. The presence of small resistances causes the oscillations of the tank circuit to die out. However, pressing the button fully also connects the tank circuit to a transistor circuit that replenishes the lost energy and sustains the oscillations.

At the central office, the equipment used to detect the presence of the tones and to identify their frequencies is much more sophisticated. Two filters are required, one for each of the frequency groups. Each filter must pass the frequencies within ±2% of their nominal values (697 to 941 Hz for one filter and 1209 to 1477 Hz for the other) and reject the signal if the frequencies are outside of ±3% limits. The output tone from each filter is then processed digitally to determine its frequency.

The concepts and methods developed in this chapter will allow us to understand the properties of resonant circuits and the design of various basic types of band-pass circuits, as used in the touch-tone telephone system.

CHAPTER OBJECTIVES

1. Describe and characterize the ideal band-pass filter.
2. Understand band-pass circuits from the viewpoint of transfer functions and pole-zero plots.
3. Investigate the basic band-pass transfer function and its realization as a parallel or series *RLC* circuit or op amp circuit.
4. Investigate band-pass circuits having practical capacitors and inductors in contrast to ideal capacitors and inductors
5. Describe the phenomenon of resonance and investigate the properties and applications of resonant circuits.
6. Investigate general second-order transfer functions having a band-pass type of frequency response.

SECTION HEADINGS

1. INTRODUCTION

How is it possible to listen to a favorite radio station by merely pushing a button or two or simply turning a dial? Why do some very expensive receivers have very clear reception, while with some cheaper models other stations chatter in the background? What circuitry inside the radio makes this difference? The ability to clearly select a particular broadcast station depends on the design of an internal band-pass circuit. Such a circuit will pass signals within a narrow band of frequencies while rejecting or significantly attenuating signals outside of that band. To understand why this is important, note that audio signals have *significant* frequency components up to about 3 kHz for voice and up to about 15 kHz for high-fidelity music. These frequencies are far too low for wireless transmission. In (wireless) AM radio transmission, the audio signal modulates the amplitude of a carrier signal that is suitable for wireless transmission. The carrier signal is a high-frequency sinusoidal waveform between 500 kHz and 1650 kHz. The modulated waveform contains many frequency components centered about the carrier signal frequency, but extending over a range of frequencies equal to twice the highest audio frequency. For example, the radio station WBAA, at Purdue University, has a carrier frequency of 920 kHz and occupies a band from approximately 915 to 925 kHz. To select WBAA from all the carrier signals received by a radio requires a good band-pass filter to pass the frequency band of 915 to 925 kHz while rejecting signals outside this band. This chapter introduces the idea and properties of a band-pass filter.

In its simplest form, a band-pass circuit consists of only one capacitor, one inductor, and one resistor, connected either in series or in parallel. In the first half of the text, we analyze simple *RLC* circuits where we emphasize (1) transient behavior under dc excitation and (2) sinusoidal steady-state behavior at a single frequency. This chapter investigates the behavior of circuits over bands of sinusoidal frequencies. Many useful results may be obtained with the phasor and impedance concepts studied earlier. However, rapid advances in technology have made it possible to have a band-pass circuit without any of the usual *RLC* circuit components. Therefore, a study of the band-pass property of a transfer function $H(s)$ dominates the material of this chapter. The resulting analysis is readily applicable to general linear systems, whether they be electrical, mechanical, or otherwise. Although transfer functions underlie our approach, circuit realizations with ideal and practical components illustrate all the basic concepts and properties.

2. THE IDEAL BAND-PASS FILTER

This section sets forth the ideal band-pass characteristic and shows how a basic second-order transfer function can be used to approximate the ideal. Specifically, let $H(s)$ be a voltage gain or some other type of network function. As we know, the curves for $|H(j\omega)|$ vs. ω and $\angle H(j\omega)$ vs. ω are called the *magnitude* (frequency) response and *phase* response, respectively. Figure 16.1a shows an *ideal band-pass* (rectangular) magnitude response curve. Here, "ideal" means that all frequency components of the input signal within the range $\omega_1 < \omega < \omega_2$ are amplified with equal gain (in magnitude), and all frequency components outside of the range are totally eliminated from the output. Actually, for a band-pass circuit to pass a signal with frequency components in the range $\omega_1 < \omega < \omega_2$ without distortion, there is also a requirement on the phase response that is ordinarily studied in a course on signal analysis.

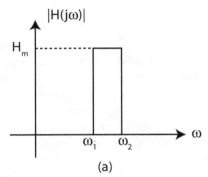

(a)

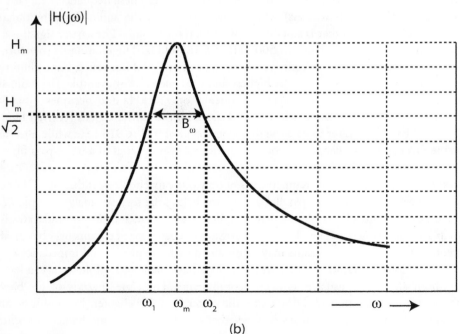

(b)

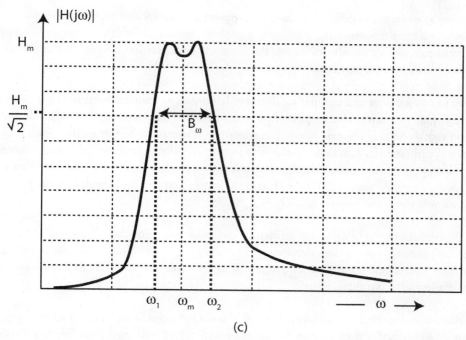

(c)

FIGURE 16.1 Definitions of peak frequency, ω_m, and bandwidth, B_ω. (a) Ideal band-pass characteristic. (b) Approximate band-pass characteristic of simple *RLC* circuit. (c) Band-pass characteristic of a more complex circuit.

Unfortunately, an ideal rectangular band-pass characteristic is not realizable by a rational transfer function or any circuit. One can, however, approximate the ideal characteristic with a simple tuned circuit whose transfer function produces a bell-shaped magnitude response curve as illustrated in Figure 16.1b. With more complex circuits and more complex transfer functions, one can improve the approximation as shown in Figure 16.1c. (How to do so is a topic to be studied in advanced courses.) The bell-shaped curve of Figure 16.1b has several important features. The frequency at which $|H(j\omega)|$ reaches its maximum value, H_m, is called the **peak frequency** and is denoted by ω_m. The two side frequencies at which $|H(j\omega)|$ is $1/\sqrt{2}$ of its maximum value are called the **half-power frequencies,** denoted ω_1 and ω_2. The term "half-power" comes from the fact that if the output is a voltage across a fixed resistance, then a drop in voltage by the factor $1/\sqrt{2}$ means a drop in power equal to $\left(1/\sqrt{2}\right)^2 = 0.5$.

Half-power frequencies are also called 3 dB (down) frequencies. Recall from Chapter 14 that dB gain is defined as

$$\text{dB } (gain) = 20\log_{10}|H(j\omega)|$$

If a gain of is reduced by the factor $1/\sqrt{2}$, then the dB reduction in gain is given by

$$-3\text{dB} = 20\log_{10}\left|\frac{1}{\sqrt{2}}\right|$$

Hence the terminology "3 dB down."

For obvious reasons, ω_1 is called the **lower half-power frequency** and ω_2 the **upper half-power frequency,** and their difference $B_\omega = \omega_2 - \omega_1$ is the half-power bandwidth (or simply **bandwidth**) of the band-pass circuit/transfer function. Band-pass circuits are designed so that all frequencies of interest fall within the pass band $[\omega_1, \omega_2]$.

One way of categorizing and comparing different band-pass circuits/transfer functions is by their **selectivity,** i.e., a circuit's relative capability to discriminate between frequencies inside the pass band and signals outside the pass band. The selectivity is measured by the **quality factor,** Q, of a band-pass circuit/transfer function. The quality factor, Q, is the ratio of the (geometric) center frequency ($\sqrt{\omega_1 \omega_2}$) to the bandwidth, B_ω. For second–order circuits and transfer functions with a bell-shaped magnitude response (Figure 16.1b), the center frequency and peak frequency ω_m coincide, as we will show. In this case $Q = Q_p = \omega_m / B_\omega$.

where Q_p denotes the pole Q, to be defined in equation 16.1. For the circuit realization of the transfer function, Q is sometimes denoted by Q_{cir} or $Q_{circuit}$. A high-Q circuit passes only a very narrow band of frequencies relative to $\sqrt{\omega_1 \omega_2}$ or ω_m, whereas a low-Q circuit has a broad (pass) band and a less selective characteristic.

It is important to note that the concepts of ω_m, B_ω, and Q of a circuit are all based on the magnitude function $|H(j\omega)|$ and, therefore, on how the transfer function $H(s)$ is defined. Even for the same circuit, these values are different when the output is associated with different branches or when the input is changed from a voltage source to a current source. Further, for the investigation of the frequency-selective characteristic of the circuit, the foregoing definition of Q is most appropriate because it directly assesses the sharpness of the magnitude response curve. As such the definition allows Q to be experimentally determined. In certain other applications, where only one fixed frequency is of interest, there is another definition of a circuit's Q based on an energy relationship that is inadequate for general band-pass circuit design.

3. THE BASIC BAND-PASS TRANSFER FUNCTION AND ITS CIRCUIT REALIZATIONS

The most basic second-order band-pass transfer function having the bell-shaped curve of Figure 16.1b has a pair of complex poles and a single zero at the origin, i.e.,

$$H(s) = K\frac{s}{(s - p_1)(s - p_2)} = K\frac{s}{(s + \sigma_p - j\omega_d)(s + \sigma_p + j\omega_d)}$$

$$(16.1)$$

$$= K\frac{s}{s^2 + 2\sigma_p s + \sigma_p^2 + \omega_d^2} = K\frac{s}{s^2 + 2\sigma_p s + \omega_p^2} = K\frac{s}{s^2 + \dfrac{\omega_p}{Q_p}s + \omega_p^2}$$

Figure 16.2 illustrates the pole-zero plot of the transfer function of equation 16.1 as well as the relationships among the various parameters. Note that there is a finite zero at the origin. Further, because the numerator has degree strictly less than the denominator, there is also a zero at $s = \infty$. In both cases the magnitude of the transfer function is zero.

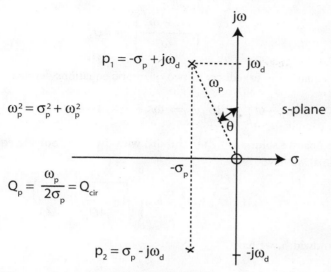

FIGURE 16.2 Pole-zero plot of a band-pass $H(s)$ with a single zero at origin as per equation 16.1.

Our next goal is to derive the peak frequency ω_m, the maximum gain $H_m = |H(j\omega)|_{max} = |H(j\omega_m)|$, the half-power frequencies ω_1 and ω_2, B_ω, the circuit Q, and the frequency at which the angle of $H(j\omega)$ is zero. The first step in this derivation is to write down $H(j\omega)$:

$$H(j\omega) = K\frac{j\omega}{(j\omega)^2 + j2\omega\sigma_p + \omega_p^2} = \frac{K}{2\sigma_p + j\left(\dfrac{\omega^2 - \omega_p^2}{\omega}\right)} \tag{16.2a}$$

whose peak magnitude response is

$$\max_\omega |H(j\omega)| = \frac{|K|}{\min_\omega\left(\sqrt{4\sigma_p^2 + \left(\dfrac{\omega_p^2 - \omega^2}{\omega}\right)^2}\right)} = \frac{|K|}{\sqrt{4\sigma_p^2 + \left(\dfrac{\omega_p^2 - \omega^2}{\omega}\right)^2}}\Bigg]_{\omega=\omega_p} = \frac{|K|}{2\sigma_p} \tag{16.2b}$$

Note that since the numerator is a constant, the maximum occurs when the denominator is a minimum, i.e., when $\omega = \omega_p$. Therefore we conclude that

$$\omega_m = \omega_p \tag{16.3}$$

Noting that $Q_p = \dfrac{\omega_p}{2\sigma_p}$, we further conclude

$$H_m \equiv |H(j\omega)|_{max} = |H(j\omega_m)| = \frac{|K|}{2\sigma_p} = \frac{|K|Q_p}{\omega_m} \tag{16.4}$$

To find the half-power frequencies ω_1 and ω_2, the maximum gain must be reduced by the factor $1/\sqrt{2}$. Considering equation 16.2b, this occurs at those ω's for which

$$\left(\frac{\omega_p^2 - \omega^2}{\omega}\right)^2 = 4\sigma_p^2 \tag{16.5}$$

This fourth-degree equation in ω reduces to two quadratic equations defined by

$$\omega_p^2 - \omega^2 = \pm 2\sigma_p \omega \;\Rightarrow\; \omega^2 \mp 2\sigma_p \omega - \omega_p^2 = 0 \tag{16.6}$$

We denote the two positive solutions by ω_1 and ω_2, with $\omega_2 > \omega_1$. Solving equation 16.6 using the quadratic formula, we have

$$\omega_{1,2} = \mp\sigma_p + \sqrt{\sigma_p^2 + \omega_p^2} = \omega_p \left(\sqrt{1 + \frac{1}{4Q_p^2}} \mp \frac{1}{2Q_p} \right). \tag{16.7}$$

To compute the bandwidth, we have

$$B_\omega = \omega_2 - \omega_1 = 2\sigma_p = \frac{\omega_p}{Q_p}. \tag{16.8}$$

Exercise. Show that for high-Q_p transfer functions/circuits (say $Q_p \geq 8$),

$$\omega_{1,2} \cong \omega_p \mp \frac{B_\omega}{2} \tag{16.9}$$

We can deduce one more important property from the above equations. When equation 16.5 holds, the real and imaginary parts of the denominator in equation 16.2a are equal. Hence the denominator has an angle of $\pm 45°$. Thus the angles of

$$H(j\omega_i) = \frac{K}{2\sigma_p + j\left(\dfrac{\omega_i^2 - \omega_p^2}{\omega_i}\right)}$$

($i = 1, 2$) are $\pm 45°$ for $K > 0$, and $\pm 45° - 180°$ for $K < 0$.

Exercises. 1. Given the transfer function $H(s) = \dfrac{s}{s^2 + 2s + 256}$, find H_m, ω_m, B_ω, and $\omega_{1,2}$.
ANSWERS: $\omega_m = 16$ rad/sec, $H_m = 0.5$, $B_\omega = 2$, $\omega_1 = 15.031$, and $\omega_2 = 17.031$

2. Suppose that $Q_p = 8$ and $\omega_p = 1000$ rad/sec. Find the exact $\omega_{1,2}$ from equation 16.7, the B_ω from equation 16.8, and the approximate $\omega_{1,2}$ from equation 16.9. Now compute the magnitude of the percent errors between the approximate and exact values of $\omega_{1,2}$.
ANSWERS: in random order: 125, 939.45, 1064.5, 1062.5, 937.5, 0.1833, 0.2077

3. Given equation 16.9, show that for a high-Q_p transfer function/circuit (say $Q_p \geq 8$),

$$\omega_p \cong 0.5(\omega_1 + \omega_2)$$

i.e., for the high-Q case the geometric and arithmetic means of ω_1 and ω_2 are approximately equal.

The next two questions are: (i) what is the geometric center frequency of the magnitude response and (ii) is it equal to ω_m? Here, using the square of the geometric center frequency,

$$\omega_1\omega_2 = \left(-\sigma_p + \sqrt{\sigma_p^2 + \omega_p^2}\right)\left(\sigma_p + \sqrt{\sigma_p^2 + \omega_p^2}\right) = -\sigma_p^2 + \sigma_p^2 + \omega_p^2 = \omega_p^2$$

we conclude that for the above transfer function and associated circuit realizations, the geometric center frequency, the pole frequency, and the peak frequency coincide, i.e.,

$$\sqrt{\omega_1\omega_2} = \omega_p = \omega_m$$

As mentioned earlier, the circuit/transfer function Q is the ratio of the center frequency to the bandwidth. Given equations 16.8 and 16.10, we have

$$Q = Q_{cir} = Q_p = \frac{\omega_m}{B_w} \tag{16.11}$$

Finally, the nonzero frequency for zero phase shift occurs when the transfer function is purely real, i.e, when the imaginary part is zero. This occurs at $\omega = \omega_p$; hence

$$\text{zero phase-shift frequency} = \omega_p \tag{16.12}$$

EXAMPLE 16.1. Compute ω_m, $|H(j\omega)|_{max} = |H(j\omega_m)|$, Q, B_w, ω_1, ω_2 (exact and approximate values) for the simple parallel RLC circuit shown in Figure 16.3a where the current source is the input and the voltage across the input nodes is the output. Then compute and label the magnitude response and verify that it has the bell shape of Figure 16.1b.

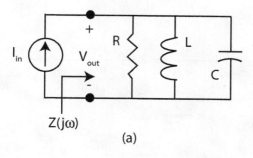

(a)

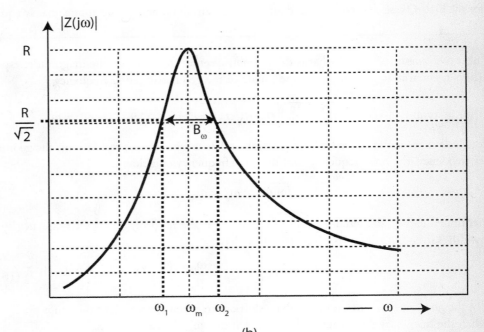

(b)

FIGURE 16.3 (a) A parallel *RLC* circuit. (b) Magnitude response curve.

SOLUTION

The key to the solution of the example is the computation of the transfer function as follows:

$$H(s) = \frac{V_{out}(s)}{I_{in}(s)} = Z(s) = \frac{1}{Y(s)} = \frac{1}{Cs + \dfrac{1}{R} + \dfrac{1}{Ls}} = \frac{\dfrac{1}{C}s}{s^2 + \dfrac{1}{RC}s + \dfrac{1}{LC}}$$

Then, according to equation 16.1,

$$H(s) = K\frac{s}{s^2 + 2\sigma_p s + \omega_p^2} = K\frac{s}{s^2 + \dfrac{\omega_p}{Q_p} + \omega_p^2} = \frac{\dfrac{1}{C}s}{s^2 + \dfrac{1}{RC}s + \dfrac{1}{LC}}$$

we conclude that $\sigma_p = \dfrac{1}{2RC}$, $\omega_P = \dfrac{1}{\sqrt{LC}}$, and

$$Q = Q_p = \frac{\omega_p}{2\sigma_p} = \frac{RC}{\sqrt{LC}} = R\sqrt{\frac{C}{L}} \qquad (16.13a)$$

Thus from equations 16.3 and 16.4,

$$\omega_m = \omega_p = \frac{1}{\sqrt{LC}} \qquad (16.13b)$$

and

$$H(j\omega_m) = \frac{|K|}{2\sigma_p} = R \qquad (16.13c)$$

Note that this implies that $H(j\omega_m) = H_m = |Z(j\omega)|_{max} = R$.

To compute the bandwidth B_ω, equation 16.8 implies that

$$B_\omega = 2\sigma_p = \frac{1}{RC} \tag{16.14}$$

Similarly from equations 16.7 and 16.9, the exact half-power frequencies are

$$\omega_{1,2} = \mp\frac{1}{2RC} + \sqrt{\left(\frac{1}{2RC}\right)^2 + \frac{1}{LC}} \tag{16.15}$$

while the approximate half-power frequencies for the high-Q_p case are

$$\omega_1 \cong \frac{1}{\sqrt{LC}} - \frac{1}{2RC}, \quad \omega_2 \cong \frac{1}{\sqrt{LC}} + \frac{1}{2RC}. \tag{16.16}$$

Finally, to obtain the frequency response we note the values of the above computations, and the fact that at $\omega = 0$ and $\omega = \infty$ the magnitude of the frequency response is zero, and the bell-shaped curve of Figure 16.3b results.

Exercises. 1. Compute the exact values of ω_m, ω_1, ω_2, B_ω, and Q for the circuit of Example 16.1 when $R = 2.5$ kΩ, $L = 0.1$ H, and $C = 0.1$ µF.
ANSWERS: $\omega_m = 10^4$ rad/sec, $\omega_1 = 12.198 \times 10^3$, $\omega_2 = 8.198 \times 10^3$, $B_\omega = 4 \times 10^3$, and $Q = 2.5$

The preceding example and exercise demonstrate that for high-Q circuits ($Q \geq 8$), there is really no need to use the exact equation 16.15 to compute ω_1 and ω_2, as the much simpler estimates given by equation 16.16 are sufficiently close to the true answers. Indeed, sometimes it is convenient to use the approximate formulas when $Q \geq 6$.

In many practical circuits, the independent source could be a voltage source in series with a resistor. Before applying any of the foregoing formulas, it is necessary to transform the circuit into the form of Figure 16.3a by the use of the Norton equivalent circuit studied in a first course on circuits. The resistance R in Figure 16.3a then is not a physical resistor, but rather, the equivalent resistance of several resistances in parallel. The following example illustrates this reformulation.

EXAMPLE 16.2. In the circuit of Figure 16.4a, an independent voltage source V_{in} in series with an internal resistance $R_s = 40$ kΩ models a real-world sinusoidal excitation. Suppose $L = 20$ mH, $C = 0.05$ µF, and $R_L = 10$ kΩ.
    (a)   Find the exact values of ω_m, B_ω, and Q.
    (b)   Estimate the values of ω_1 and ω_2.
    (c)   Find $|H(j\omega)|_{max}$.

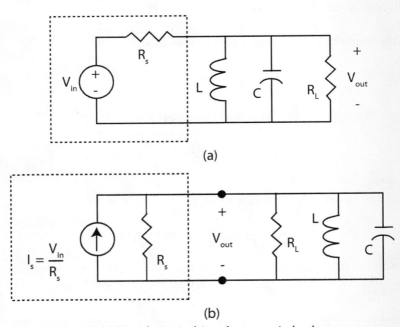

(a)

(b)

FIGURE 16.4 (a) Tuned circuit driven by a practical voltage source.
(b) Equivalent circuit showing Norton equivalent for the source.

SOLUTION

The solution proceeds by using the formulas developed in Example 16.1 after replacing the practical V_{in}-R_s source by its Norton equivalent and identifying R in Example 16.1 as the parallel combination of R_s and R_L. By the usual formula for two parallel resistors,

$$R = \frac{R_s R_L}{R_s + R_L} = \frac{40,000 \times 10,000}{40,000 + 10,000} = 8,000 \ \Omega$$

This results in the transfer function

$$H(s) = \frac{V_{out}}{V_{in}} = \frac{V_{out}}{I_s} \times \frac{I_s}{V_{in}} = \frac{\dfrac{1}{C}s}{s^2 + \dfrac{1}{RC}s + \dfrac{1}{LC}} \times \frac{1}{R_s} = \frac{\dfrac{1}{R_s C}s}{s^2 + \dfrac{1}{RC}s + \dfrac{1}{LC}}$$

Notice the new value: $K = \dfrac{1}{R_s C}$.

(a)   From equations 16.13b, 16.14, and 16.13c,

$$\omega_m = \omega_p = \frac{1}{\sqrt{LC}} = \frac{1}{\sqrt{0.02 \times 0.05 \times 10^{-6}}} = 31,622.77 \ \text{rad/s}$$

$$B_\omega = \frac{1}{RC} = \frac{1}{8,000 \times 0.05 \times 10^{-6}} = 2,500 \ \text{rad/s}$$

and

$$Q = \frac{\omega_m}{B_\omega} = \frac{31,622.77}{2,500} = 12.6$$

(b)   Using the approximate equations for high-Q circuits, we obtain the half-power frequencies as

$$\omega_1 \cong \omega_m - \frac{B_\omega}{2} = 31,622.77 - 1,250 = 30,372.77 \text{ rad/s}$$

and

$$\omega_2 \cong \omega_m + \frac{B_\omega}{2} = 31,622.77 + 1,250 = 32,872.77 \text{ rad/s}$$

(c)   Now we use equation 16.4 to obtain

$$H(j\omega_m) = \frac{|K|}{2\sigma_p} = \frac{1}{R_s C B_\omega} = \frac{R}{R_s} = 0.2$$

Notice that we did not use equation 16.13c because the input in Example 16.1 is a current source and not the voltage source of Figure 16.4a. Hence the maximum value is not equal to R.

This example demonstrates that putting an external resistance in parallel with the LC tank circuit reduces the value of R, which in turn causes a larger bandwidth and thus a lower circuit Q while keeping the peak frequency ω_m unaffected.

Exercise. Repeat Example 16.2 with the element values changed to $R_s = 36$ kΩ, $L = 40$ mH, $C = 0.25$ µF, and $R_L = 4$ kΩ.
ANSWERS: 10,000 rad/sec, 1111.11 rad/sec, 9, 9444.44 rad/sec, 10,555.55 rad/sec.

The examples so far have illustrated only the analysis of parallel RLC circuits. In the design of a parallel-tuned circuit, we must also pay attention to other factors, such as available component sizes, desired voltage gain, and cost. In practice, design specifications ordinarily impose a small number of constraints relative to the number of circuit parameters to be determined. Consequently, realistic design problems usually do not have a unique answer, as illustrated in the next example.

EXAMPLE 16.3.

Design a parallel RLC circuit, as shown in Figure 16.4a, to have a magnitude response with $f_m = 200$ kHz and a bandwidth of 20 kHz. Only inductors in the range 1 to 5 mH are available. The source has a resistance $R_s = 50$ kΩ.

SOLUTION

For the circuit of Figure 16.4a there is a restriction on the available inductors. Hence, we keep L as a variable, subject to the condition that $0.001 \le L \le 0.005$ H. Using the specified peak frequency and the fact that $\omega_m^2 = \omega_p^2 = \dfrac{1}{LC}$,

$$C = \frac{1}{\omega_m^2 L} = \frac{1}{4\pi^2 \times 4 \times 10^{10} \times L} = 6.33 \times 10^{-13} L$$

From the specified bandwidth and equation 16.14,

$$R = \frac{1}{B_\omega C} = \frac{4\pi^2 \times 4 \times 10^{10} \times L}{2\pi \times 2 \times 10^4} = 1.257 \times 10^7 L$$

As explained in Example 16.2, R is the parallel combination of R_s and R_L, i.e.,

$$\frac{1}{R} = \frac{1}{R_L} + \frac{1}{R_s},$$

or

$$\frac{1}{R_L} = \frac{1}{R} - \frac{1}{R_s}$$

Once a specific value of L is chosen, we can calculate successively the values of C, R, and R_L. Since R, which is the parallel combination of R_L and R_s, must be no greater than $R_s = 5 \times 10^4$ Ω, the upper limit for L is

$$L_{max} = \frac{R_{max}}{1.257 \times 10^7} = \frac{50,000}{1.257 \times 10^7} = 0.00398 \text{ H.}$$

Numerical values corresponding to the extreme values of L are given in Table 16.1.

TABLE 16.1

L (mH)	C (pF)	R (kΩ)	R_L (kΩ)
1	633	12.57	16.77
3.98	159	50	infinite

Table 16.1 clearly shows that there is no unique answer to the design problem. The freedom in choosing a value for L in the range 1 to 3.98 mH can be utilized to accommodate another design specification, such as a value for H_m.

Exercises. 1. In Example 16.3, find the maximum and minimum values of $|H(j\omega_m)|$.
ANSWER: 0.251, 1.0

2. In Example 16.3, if the bandwidth requirement is changed to 10 kHz, determine the minimum and maximum possible values of L.
ANSWER: 1.0 mH and 1.99 mH

Dual to the parallel *RLC* of Figure 16.3a is the series *RLC* of Figure 16.5, which has a voltage input and a current output. Although we can use duality, to infer frequency response behavior, we prefer the direct transfer function approach.

EXAMPLE 16.4. For the series *RLC* circuit of Figure 16.5, let I_s be the desired output. Find

(a) The transfer function $H(s) = \dfrac{I_s(s)}{V_s(s)}$,
(b) The peak frequency ω_m
(c) The bandwidth, B_ω
(d) The circuit Q
(e) The half-power frequencies, ω_1 and ω_2 (exact and approximate values)
(f) $H_m = |H(j\omega)|_{max} = |H(j\omega_m)|$
(g) The frequency at which the angle of $H(j\omega)$ is zero

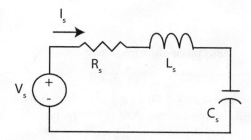

FIGURE 16.5 Series *RLC* circuit having a band-pass transfer function.

SOLUTION
(a) *Find the transfer function.* By inspection,

$$H(s) = \frac{I_s(s)}{V_s(s)} = \frac{1}{Z_{in}(s)} = \frac{1}{R_s + L_s s + \dfrac{1}{C_s s}} = \frac{\dfrac{1}{L_s}s}{s^2 + \dfrac{R_s}{L_s}s + \dfrac{1}{L_s C_s}} \tag{16.17}$$

$$= K\frac{s}{s^2 + 2\sigma_p s + \omega_p^2} = K\frac{s}{s^2 + \dfrac{\omega_p}{Q_p}s + \omega_p^2}$$

Thus, $K = 1/L_s$, $2\sigma_p = R_s/L_s$, $\omega_p = 1/\sqrt{L_s C_s}$, and $\omega_p/Q_p = R_s/L_s$.

(b)–(f) Inspecting the transfer function of equation 16.17, we immediately have

Peak frequency: $\qquad \omega_m = \omega_p = \dfrac{1}{\sqrt{L_s C_s}} \tag{16.18}$

Circuit Q: $\qquad Q_{cir} = Q_p = \dfrac{\omega_p}{B_w} = \dfrac{\omega_p L_s}{R_s} = \dfrac{1}{R_s}\sqrt{\dfrac{L_s}{C_s}} \tag{16.19}$

Half-power bandwidth: $B_\omega = \dfrac{R_s}{L_s}$ (16.20)

Half-power frequencies (see equation 16.7):

$$\omega_{1,2} = \omega_p\left(\sqrt{1 + \frac{1}{4Q_p^2}} \pm \frac{1}{2Q_p}\right) \approx \frac{1}{\sqrt{L_sC_s}} \pm \frac{R_s}{2L_s} \qquad (16.21)$$

where the approximation for the high-Q_p case is

$$\omega_{1,2} \approx \omega_m \mp \frac{B_\omega}{2} = \frac{1}{\sqrt{L_sC_s}} \pm \frac{R_s}{2L_s} \qquad (16.22)$$

From equation 16.4,

$$H_m = \frac{|K|}{2\sigma_p} = \frac{|K|Q_p}{\omega_m} = \frac{1}{R_s} \qquad (16.23)$$

Finally, from equation 16.12,

$$\text{zero phase-shift frequency} = \omega_p$$

Exercise. For Example 16.4, suppose $R_s = 5\ \Omega$, $L_s = 1$ mH, $C_s = 0.1\ \mu$F, and V_s has a fixed magnitude. Find ω_m, B_ω, Q_{cir}, and the approximate half-power frequencies.
ANSWER: $\omega_m = 10^5$ rad/sec, $B_\omega = 5000$ rad/sec, $Q_{cir} = 20$, 97,500 rad/sec, and 102,500 rad/sec.

The above derivations and calculations yield formulas and numbers. To add some meaning to the concept of circuit Q and its relationship to bandwidth, we provide a plot of the normalized $|H(j\omega)|$ vs. ω curve for different values of $Q_{cir} = Q_p$ in Figure 16.6. The ordinate is the ratio $|H(j\omega)|/|H(j\omega)|_{max}$, while the abscissa shows the ratio ω/ω_m. These curves are called **universal resonance curves** because they are applicable to parallel *RLC* circuits, to series *RLC* circuits, or to any system having a transfer function of the form of equation 16.1. Observe that as Q increases, the bandwidth decreases, indicating a better selectivity.

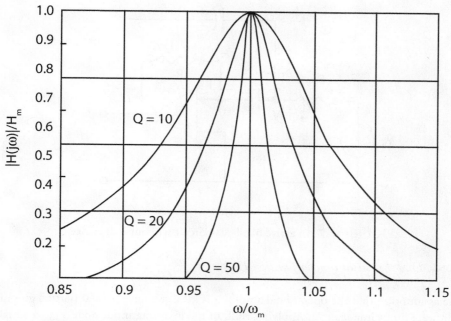

FIGURE 16.6 Normalized magnitude response for equation 16.1.

To conclude this section, we present an example of an active band-pass circuit that avoids the use of inductors. The band-pass circuit illustrated here is only one of more than a dozen configurations in use. This example illustrates the possibility of eliminating inductances while producing the same kind of frequency response as the parallel or series *RLC* circuit. You can learn a lot more about these active filters in a more advanced course.

EXAMPLE 16.5. The operational amplifier in Figure 16.7 is assumed to be ideal.

1. Find the transfer function $H_{cir}(s) = V_{out}(s)/V_{in}(s)$.
2. A band-pass circuit is to have peak frequency $\omega_m = 1000$ rad/sec and a bandwidth $B_\omega = 100$ rad/sec. Find Q for the desired transfer function. Then realize (find values for R_1, R_2, C_1, and C_2) for the normalized transfer function

$$H_{norm}(s) = \frac{Ks}{s^2 + \dfrac{1}{Q}s + 1}$$

   under the condition that $C_1 = C_2$.
3. Magnitude- and frequency-scale the circuit to achieve the correct center frequency with $C_1 = C_2 = 1\ \mu F$.
4. Verify the frequency response with a SPICE simulation of the real circuit.

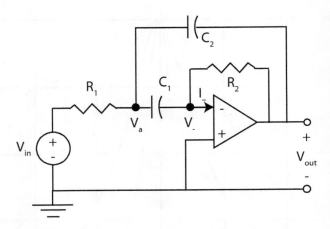

FIGURE 16.7 An active band-pass circuit without inductance.

SOLUTION 1. *Find the circuit transfer function* $H_{cir}(s) = \dfrac{V_{out}(s)}{V_{in}(s)}$.

From the assumption that the operational amplifier is ideal, we have $V_- = 0$ (virtual ground) and $I_- = 0$ (infinite input impedance). Applying KCL to the inverting input node V_-, we obtain

$$sC_1 V_a + \frac{V_{out}}{R_2} = 0,$$

which yields

$$V_a = -\frac{1}{sC_1 R_2} V_{out}$$

Next, we apply KCL to node V_a to obtain

$$\frac{V_a - V_{in}}{R_1} + sC_1 V_a + sC_2(V_a - V_{out}) = 0$$

Substituting the previous expression for V_a into this equation, regrouping terms, and solving for V_{out} results in the desired transfer function:

$$H_{cir}(s) = \frac{V_{out}(s)}{V_{in}(s)} = \frac{\dfrac{-s}{R_1 C_2}}{s^2 + \left(\dfrac{1}{R_2 C_1} + \dfrac{1}{R_2 C_2}\right)s + \dfrac{1}{R_1 R_2 C_1 C_2}} \tag{16.24}$$

This is precisely the form of equation 16.1.

2. *Computation of normalized transfer function to be realized.* The desired transfer function is

$$H(s) = \frac{V_{out}(s)}{V_{in}(s)} = \frac{Ks}{s^2 + B_\omega s + \omega_m^2} = \frac{Ks}{s^2 + \left(\dfrac{\omega_m}{Q}\right)s + \omega_m^2} = \frac{Ks}{s^2 + 100 s + (1000)^2}$$

We now frequency-scale $s \rightarrow \omega_m$ with $\omega_m = 1000$ rad/sec. Thus the normalized transfer function is

$$H_{nor}(s) = \frac{K}{\omega_m} \times \frac{s}{s^2 + \left(\dfrac{1}{Q}\right)s + 1} = \frac{K}{1000} \times \frac{s}{s^2 + 0.1s + 1}$$

3. *Computation of element values for normalized transfer function.* Equating the circuit transfer function with the normalized transfer function, we have

$$\frac{1}{R_1 R_2 C_1 C_2} = 1 \text{ rad / sec} \qquad (16.25)$$

and

$$\frac{1}{Q} = \frac{1}{R_2 C_1} + \frac{1}{R_2 C_2} = 0.1 \text{ rad / sec.} \qquad (16.26)$$

Under the condition that

$$C_1 = C_2 = 1 \text{ F}$$

we have, from equation 16.26,

$$\frac{1}{Q} = \frac{2}{R_2} = 0.1 \Rightarrow R_2 = 20 \ \Omega$$

From equation 16.25 we obtain

$$\frac{1}{R_1 R_2 C_1 C_2} = \frac{1}{R_1 \, 20} = 1 \Rightarrow R_1 = 0.05 \ \Omega$$

Note also that

$$H_m = |H(j\omega_m)| = \frac{\dfrac{1}{R_1 C_2}}{\dfrac{1}{R_2 C_1} + \dfrac{1}{R_2 C_2}} = \frac{R_2}{2R_1} = \frac{20}{0.1} = 200$$

4. *Frequency and magnitude scaling.* As per the problem statement, we desire $\omega_m = 1000$ rad/sec. Hence we frequency-scale with $K_f = 1000$. It is further required that $C_{1new} = C_{2new} = 1 \ \mu\text{F}$. Hence, we solve the following for K_m:

$$C_{new} = \frac{C_{old}}{K_m K_f} \Rightarrow K_m = \frac{C_{old}}{C_{new} K_f} = \frac{1}{10^{-6} \times 10^3} = 1000$$

It follows that

$$C_{1new} = C_{2new} = 1 \ \mu\text{F}, \ R_{1new} = K_m R_{1old} = 50 \Omega, \ R_{2new} = 1000 R_{2old} = 20 \Omega$$

The ratio of R_2 to R_1 is very large and unrealistic. It turns out that this circuit is best suited to low-Q transfer functions.

5. *Verify frequency response.* To verify the frequency response for a realistic implementation of the above circuit, we consider a SPICE simulation using the standard 740 operational amplifier as shown in Figure 16.8. Observe that the maximum value of the magnitude response is 200, as expected, and that the bandwidth is about 16 Hz, which translates to about 100 rad/sec.

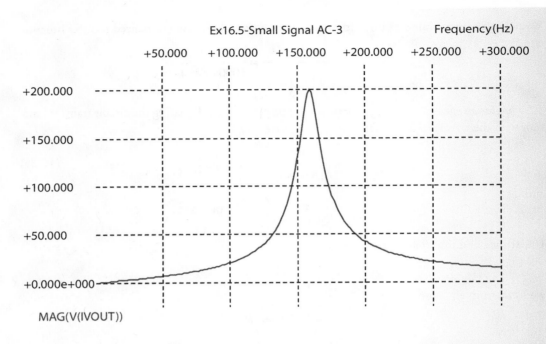

FIGURE 16.8 Magnitude response plot resulting from a
SPICE simulation of the 740 operational amplifier.

Exercises. 1. In Example 16.5, find the circuit Q and the approximate half-power frequencies $\omega_{1,2}$.
ANSWER. 10,950 rad/sec and 1050 rad/sec

2. In Example 16.5, if $C_1 = 0.5$ μF and $C_2 = 1$ μF, find the values of R_1 and R_2 required to meet the specifications on ω_m and B_ω.
ANSWER: $R_1 = 66.67$ Ω, $R_2 = 30$ kΩ

Throughout this section we have assumed the use of ideal inductors and capacitors. Practical inductors have complex circuit models to account for real-world behavior. The next section takes up an approximate analysis of circuits containing simple models of practical inductors and capacitors.

4. BAND-PASS CIRCUITS WITH PRACTICAL COMPONENTS

Quality Factor of L and C Components

How can we analyze band-pass circuits in the presence of practical (non-ideal) inductors and capacitors? Practical inductors and capacitors have models consisting of their ideal cousins and other ideal ("parasitic") elements to account for losses and coupling effects. Figure 16. 9a illustrates a simple model of a practical inductor for low to medium frequencies while Figure 16.9b is a reasonable model for high frequencies.

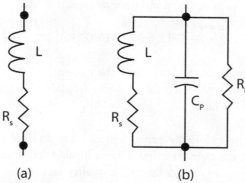

FIGURE 16.9 Two models of an inductor. (a) For low to medium frequencies. (b) For high frequencies.

The primary parameter is, of course, the inductance L. The remaining elements, R_s, R_p, and C_p, account for undesirable yet unavoidable practical effects and are called parasitic. Since an inductor usually consists of a coil of wire, R_s represents the wire's resistance. Also, a capacitance is present between adjacent turns of wire. Hence C_p models this parasitic capacitance. The resistance R_p accounts for the energy loss in the magnetic core material (if present) inside the coil. Complex models such as Figure 16.9b, although important, if used for every inductor would unduly complicate the analysis of a band-pass circuit. Fortunately, for low to medium frequencies (up to a few megahertz), the simpler model of Figure 16.9a suffices and hence underlies the material that follows.

Figure 16.10 shows two models of a practical capacitor. Again, the primary parameter here is the capacitance C; R_p, R_s, and L_s are "parasitics." The *leakage resistance*, R_p, accounts for the energy loss in the dielectric; the inductance L_s and resistance R_s are due mainly to the connecting wires of the capacitor. At frequencies above $1/\sqrt{L_s C}$, the capacitor actually behaves as an inductor! For frequencies of up to a few megahertz, the simpler model of Figure 16.10a suffices, and it is used for the analyses of this text.

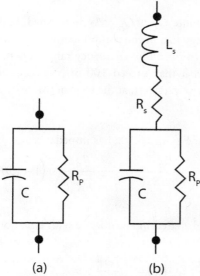

FIGURE 16.10 Two models of a practical capacitor. (a) For low to medium frequencies. (b) For high frequencies.

How close is a practical inductor (Figure 16.9a) or a practical capacitor (Figure 16.10a) to the ideal? The so-called element quality factor provides a quantitative measure. To develop the quality factor, consider that each practical inductor or capacitor has an impedance

$$Z(j\omega) = \text{Re}\{Z(j\omega)\} + j\,\text{Im}\{Z(j\omega)\} = R(\omega) + jX(\omega)$$

For both the practical inductor and capacitor, the frequency-dependent reactance X is the primary parameter of concern, whereas R represents the unavoidable "parasitic effect." In the ideal case, R is not present. Hence, the ratio $\dfrac{|X|}{|R|}$ provides a measure of how close the network model is to an ideal inductor or capacitor. The larger the ratio, the better the element behaves as an ideal inductor or capacitor. (Ideally, $R = 0$ and the ratio is infinite.) This suggests defining the **quality factor** associated with a practical inductor or capacitor having impedance $Z(j\omega)$ as

$$Q_Z(\omega) = \frac{|X(\omega)|}{|R(\omega)|} \tag{16.27}$$

The inclusion of ω in equation 16.27 is to emphasize the fact that Q_Z depends on the frequency of operation. The subscript Z indicates a generic impedance and may be replaced by more specific descriptors such as "coil" or "capacitor." Unlike the Q of a circuit, which depends on the values of the elements of the circuit and on the circuit's configuration, the quality factor Q_Z of a practical inductor or capacitor varies with the operating frequency ω and remains unchanged irrespective of its connection in the circuit. Any element with a finite ratio given in equation 16.27 is termed a *lossy component,* as are all real-world components.

For the practical inductor model of Figure 16.9a, equation 16.27 reduces to

$$Q_L(\omega) = \frac{\omega L}{R_s} \tag{16.28}$$

where we may synonymously denote Q_L as Q_{coil}. As mentioned, higher Q_L implies a better-quality coil in the sense that the energy loss in the component is smaller. Infinite Q_L represents a pure inductance L, which is lossless. In the audio frequency range, Q_L may vary from 5 to 20, whereas in the radio frequency range, it may exceed 100 in practical applications. Although R_s here varies with frequency (due to the skin effect), at the level of this text, we treat R_s as a constant independent of ω.

Similarly, the capacitor model of Figure 16.10a has impedance

$$Z(j\omega) = \frac{1}{\dfrac{1}{R_p} + j\omega C} = \frac{R_p}{1 + j\omega R_p C} = \frac{R_p}{1 + \left(\omega R_p C\right)^2}\left(1 - j\omega R_p C\right) \equiv R(\omega) + jX(\omega)$$

Hence, the factor Q_C or Q_{cap} calculated in accordance with equation 16.27 is

$$Q_C(\omega) = \frac{|X(\omega)|}{R(\omega)} = \omega R_p C \tag{16.29}$$

A higher Q_C implies a better-quality capacitor, in the sense that the energy loss in the device is smaller and closer to ideal. Infinite Q_C represents an idealized, lossless capacitor modeled by a pure capacitance. In practice, Q_C is usually much greater than Q_L, meaning that Q_L is more critical for circuit performance, i.e., Q_C is often assumed to be infinite. The reciprocal of Q_C is called the *dissipation factor* of the capacitor and is denoted by d_C. A lower dissipation factor means a better-quality capacitor.

The determination of Q_L and Q_C requires specification of the operating frequency ω. If the value of ω is unspecified, the analysis of a band-pass (or tuned) circuit proceeds under the assumption that $Q_L = Q_L(\omega_0)$ and $Q_C = Q_C(\omega_0)$, where $\omega_0 \equiv 1/\sqrt{LC}$. We will discuss the meaning of ω_0 in section 5, on resonance.

Reduction of Band-Pass Circuits to Approximate Series or Parallel RLC Circuits

When one uses the practical inductor and capacitor models of Figures 16.9a and 16.10a, the series *RLC* and parallel *RLC* band-pass circuits with practical sources have the more complex configurations of Figures 16.11a and b, respectively. Since they are no longer series or parallel *RLC*, their transfer function is not the ideal band-pass type of equation 16.1. Hence the associated formulas for peak frequency, bandwidth, etc. are not directly applicable.

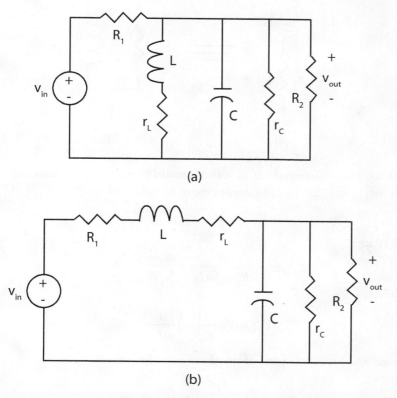

(a)

(b)

FIGURE 16.11 (a) Model of a parallel tuned circuit using practical inductor and capacitor models. (b) Model of a series tuned circuit using practical inductor and capacitor models.

Topologically speaking, the circuits of Figure 16.11 are series-parallel because the input impedance "seen" by the source consists of a sequence of series connections and parallel connections of
simple networks. Exact analysis of such series-parallel band-pass circuits to obtain ω_m and B_ω is
cumbersome, especially in light of a simpler, more efficient method widely used by engineers to
compute approximate solutions. The approximate analysis relies on the conversion between a
series circuit and an equivalent parallel circuit at a particular frequency. This conversion process
depends on the component quality factors developed in the previous subsection. See Problem 85
for the development of this equivalence. The next example illustrates the conversion process for
an inductor.

EXAMPLE 16.6. When the ideal components of a parallel RLC circuit are modeled with a practical inductor and a practical capacitor, the circuit is no longer parallel. However, by converting
the series inductor model to an "equivalent" parallel model, we can proceed with our standard
analysis. To illustrate this conversion, consider the practical inductor model of Figure 16.12a and
the "equivalent" parallel configuration in Figure 16.12b. The goal of this example is to find R_p and
L_p in terms of L_s and R_s at a particular ω.

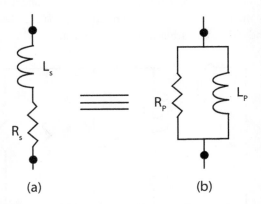

FIGURE 16.12 Conversion of an inductor model from (a) a series connection to
(b) a parallel connection at a fixed frequency.

SOLUTION
For Figure 16.12a,

$$Y_s(j\omega) = \frac{1}{R_s + j\omega L_s} = \frac{R_s - j\omega L_s}{R_s^2 + \omega^2 L_s^2}$$

For Figure 16.12b,

$$Y_p(j\omega) = \frac{1}{R_p} - j\frac{1}{\omega L_p}$$

Equating the real and imaginary parts of the above two admittances, we obtain

$$R_p = R_s + \frac{\omega^2 L_s^2}{R_s} = R_s\left(1 + Q_L^2(\omega)\right) \qquad (16.30a)$$

and

$$L_p = \frac{R_s^2 + \omega^2 L_s^2}{\omega^2 L_s} = L_s \left(1 + \frac{1}{Q_L^2(\omega)} \right) \tag{16.30b}$$

where

$$Q_L(\omega) = \frac{\omega L_s}{R_s}$$

At any particular frequency ω, if $Q_L(\omega)$ is sufficiently large (say $Q_L(\omega) \geq 8$), then equations 16.30a and b suggest that

$$R_p \cong Q_L^2(\omega) R_s = Q_L(\omega) \times \omega L \tag{16.31a}$$

and

$$L_p \cong L_s \tag{16.31b}$$

Conclusion: In a "parallel" RLC with a practical inductor, we can replace the practical inductor by its parallel counterpart valid in a neighborhood of a single frequency and analyze the circuit to obtain approximate values of peak frequency, bandwidth, etc.

Exercise. A 2 mH coil purchased from an electronic parts store has a Q_{coil} of 50 at 100 kHz. Find the element values in the series representation and the parallel representation at 100 kHz as shown in Figure 16.35.
ANSWERS: $L_s = 2$ mH, $R_s = 25.13$ Ω, $L_p \cong 2$ mH, $R_p = 62.83$ kΩ

A similar derivation can be done for the practical capacitor model of Figure 16.13a. The details of the derivation are left as a homework problem. The exact conversion equations for a specific ω are

$$C_s = C_p \left(1 + \frac{1}{Q_C^2(\omega)} \right) \tag{16.32a}$$

and

$$R_s = R_p \left(\frac{1}{1 + Q_C^2(\omega)} \right) \tag{16.32b}$$

As before, for high Q_C (say ≥ 8), these equations reduce to the simpler forms of

$$C_s \cong C_p \tag{16.33a}$$

and

$$R_s \cong \frac{R_p}{Q_C^2(\omega)} = \frac{1}{Q_C(\omega) \times \omega C} \tag{16.33b}$$

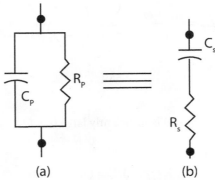

FIGURE 16.13 Conversion of a capacitor model from (a) a parallel connection to (b) a series connection exact at a fixed frequency.

Table 16.2 summarizes the various conversion formulas for both parallel to series and series to parallel inductor and capacitor models.

TABLE 16.2 Conversion of Models for Practical Inductors and Capacitors

Original Circuit	Exact Equivalent Circuit at ω_0	Approximate Equivalent Circuit, for High Q ($Q_L > 8$ and $Q_C > 8$) and ω within $(1 \pm 0.05)\,\omega_0$
L, R_s $Q_L(\omega_0) = \dfrac{\omega_0 L}{R_s}$	$R_s(1 + Q_L^2)$ $\quad L\left(1 + \dfrac{1}{Q_L^2}\right)$	$Q_L^2 R_s = Q_L \omega_0 L$ $\quad L$
R_p, C $Q_C(\omega_0) = \omega_0 R_p C$	$C\left(1 + \dfrac{1}{1 + Q_C^2}\right)$ $\quad R_p\left(\dfrac{1}{1 + Q_C^2}\right)$	C $\quad \dfrac{R_p}{Q_C^2} = \dfrac{1}{Q_C \omega_0 C}$
R_p, L $Q_L(\omega_0) = \dfrac{R_p}{\omega_0 L}$	$\dfrac{L}{1 + 1/Q_L^2}$ $\quad \dfrac{R_p}{1 + Q_L^2}$	L $\quad \dfrac{R_p}{Q_L^2} = \dfrac{\omega_0 L}{Q_L}$
C, R_s $Q_C(\omega_0) = \dfrac{1}{\omega_0 R_s C}$	$(1 + Q_C^2)R_s$ $\quad \dfrac{C}{1 + 1/Q_C^2}$	$Q_C^2 R_s = \dfrac{Q_C}{\omega_0 C}$ $\quad C$

The next example illustrates the application of the conversion formulas to a "practical" circuit to obtain approximate band-pass characteristics.

EXAMPLE 16.7. The circuit of Figure 16.14a contains practical components that make the circuit not amenable to the formulas for ideal series and parallel *RLCs* developed earlier. For example, the sinusoidal source is represented by an independent voltage source V_{in} in series with an internal resistance $R_s = 40$ kΩ. The practical capacitor has a value of $C = 0.05$ µF and a dissipation factor of $d_C = 0.01$ at ω_0. Here ω_0 represents the undamped (no resistance) natural frequency of the circuit, i.e.,

$$\omega_0 = \frac{1}{\sqrt{LC}}.$$

The practical coil has an inductance of 20 mH and $Q_L(\omega_0) = 40$. The external load resistance is $R_L = 10$ kΩ. Our goal is to find approximate values of ω_m, B_ω, Q_{cir}, ω_1, ω_2, and H_{max}.

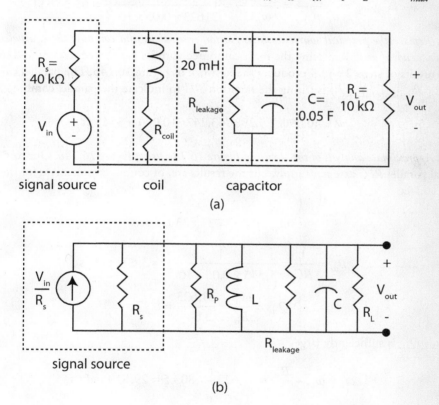

FIGURE 16.14 Approximate analysis of a high-*Q* circuit. (a) Original circuit. (b) Approximate parallel *RLC* circuit.

Solution

Step 1. *Compute ω_0.*

$$\omega_0 = \frac{1}{\sqrt{LC}} = \frac{1}{\sqrt{0.02 \times 0.05 \times 10^{-6}}} = 31,623 \text{ rad / s}$$

Step 2. *Find the parallel equivalent circuit values for the practical inductor.* Since $Q_L(\omega_0) = 40$, from equations 16.31 and column 3 of Table 16.1,

$$R_p = Q_L(\omega_0)\omega_0 L = 40 \times 31{,}623 \times 0.02 = 25{,}298 \ \Omega$$

and

$$L_p \cong L$$

Step 3. *Represent the capacitor model by a parallel RC.* First, the dissipation factor tells us that

$$Q_C(\omega_0) = \frac{1}{d_C} = 100$$

From row 3 and column 4 of Table 16.1, we note that the parallel resistance associated with the capacitor is the so-called leakage resistance

$$R_{\text{leakage}} = \frac{1}{d_C \omega_o C} = \frac{Q_C(\omega_o)}{\omega_o C} = \frac{100}{31623 \times 0.05 \times 10^{-6}} = 63{,}247 \ \Omega$$

Step 4. *Replace the practical source with its Norton equivalent and compute the equivalent parallel resistance, denoted as R.* Replacing the practical source with its Norton equivalent and incorporating the results of steps 2 and 3 produces the network of Figure 16.14b. The parallel combination of R_s, R_p, R_{leakage}, and R_L is (using the notation "//" to indicate the parallel combination)

$$R = 40{,}000//25{,}298//63{,}247//10{,}000 = 5545 \ \Omega$$

Step 5. *Approximate analysis of the circuit of Figure 16.14a using Figure 16.14b.* Our formulas from the ideal parallel *RLC* case now apply, but the results are, of course, approximate for the circuit of Figure 16.14a:

$$\omega_m \cong \omega_o = 31{,}623 \ \text{rad/sec}$$

$$B_\omega = \frac{1}{RC} = \frac{1}{5{,}545 \times 0.05 \times 10^{-6}} = 3{,}607 \ \text{rad/s},$$

$$Q_{cir} = \frac{\omega_m}{B_\omega} = \frac{31{,}623}{3{,}607} = 8.77,$$

and since Q_{cir} is sufficiently large,

$$\omega_1 \cong \omega_m - \frac{B_\omega}{2} = 31{,}623 - 1{,}803.5 = 29{,}819 \ \text{rad/s},$$

$$\omega_2 \cong \omega_m + \frac{B_\omega}{2} = 31{,}623 + 1{,}803.5 = 33{,}426 \ \text{rad/s}$$

Finally,

$$H_{max} = H(j\omega_m) = \frac{1}{R_s C B_w} = \frac{1}{40 \times 10^3 \times 0.05 \times 10^{-6} \times 3607} = 0.13862$$

5. THE RESONANCE PHENOMENON AND RESONANT CIRCUITS

The term "resonance" has different meanings in different disciplines. From *Webster's Collegiate Dictionary*, in the field of engineering, resonance refers to the phenomenon of "a vibration of large amplitude in a mechanical or electrical system caused by a relatively small periodic stimulus of the same or nearly the same period as the natural vibration period of the system." In this section, we shall investigate this notion of resonance and its manifestation in *RLC* circuits, in both cases utilizing the theories studied in previous chapters. The main applications of resonant circuits are for the filtering and tuning purposes. Additionally, resonant circuits can be used to transform a resistance from one value to another value (at a single frequency) to achieve maximum power transfer. This "matching" application is discussed in the last subsection.

The Resonance Phenomenon

A child on a swing knows by instinct when to flex his or her knees in synchrony with the swing's pendulum motion to make the swing go higher and higher. This activity illustrates the phenomenon of resonance, in which small and quick leg movements at just the right moment produce the large pendulum-like motion of the swing. If the child stops his or her leg movements, the swinging motion gradually dies down.

Let us approximate the pendulum motion of the swing using a linear system with transfer function

$$H(s) = \frac{K\omega_d}{(s + \sigma_p)^2 + \omega_d^2} \tag{16.34a}$$

and having the associated impulse response

$$h(t) = Ke^{-\sigma_p t} \sin(\omega_d t) u(t) \tag{16.34b}$$

Mathematically, the periodic leg movements are modeled by the periodic impulse function

$$x(t) = A[\delta(t) + \delta(t - T) + \delta(t - 2T) + ...] \tag{16.35}$$

where $T = 2\pi/\omega_d$. This choice coincides with the period of the natural response of the system transfer function. The build-up in magnitude can be seen quickly through superposition. The contribution to the output, say $y(t)$, due to the first impulse at $t = 0$ in equation 16.35 is

$$y_0(t) = AKe^{-\sigma_p t} \sin(\omega_d t) u(t) \tag{16.36}$$

Figure 16.15a shows the form of this equation. Note that successive positive peaks $\{V_0, aV_0, a^2V_0, a^3V_0, ...\}$ decrease geometrically by $a = e^{-\sigma_p t}$. In fact, the waveform of each period replicates the waveform of the prior period scaled by the factor a.

From time invariance, the contribution to the output $y(t)$ due to the second impulse at $t = T$ in equation 16.35 is simply $y_0(t - T)$, as shown in Figure 16.15b. Similarly, for the third input impulse at $t = 2T$, the response is $y_0(t - 2T)$, as shown in Figure 16.15c.

From superposition, the response over $[0, 3T)$ is simply the sum

$$y_0(t) + y_0(t - T) + y_0(t - 2T)$$

i.e., for $0 \le t < 3T$. This sum is illustrated in Figure 16.15d.

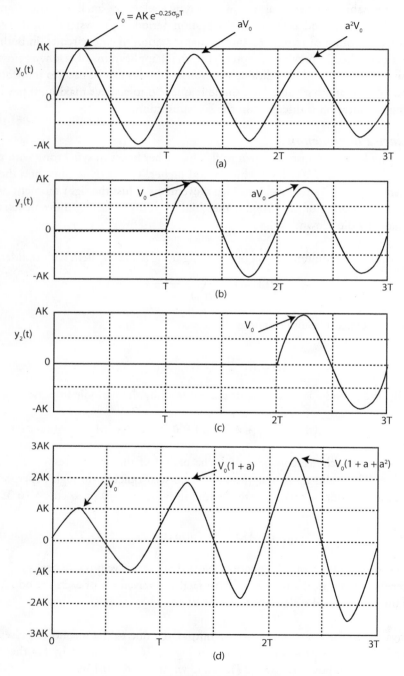

FIGURE 16.15

In Figure 16.15d we observe that successive positive peaks are given by $\{V_0, (1 + a)V_0, (1 + a + a^2)V_0\}$. In general one can show that the nth positive peak is given by

$$\left(1 + a + a^2 + \cdots + a^{n-1}\right)V_0 = V_0 \sum_{k=0}^{n-1} a^k = \begin{cases} n & a = 1 \\[2mm] \dfrac{1 - a^n}{1 - a} & a \neq 1 \end{cases} \tag{16.37}$$

For our purposes $a = e^{-\sigma_p t} < 1$ since $\sigma_p > 0$. Hence as n gets large,

$$\lim_{n \to \infty} \frac{1 - a^n}{1 - a} = \frac{1}{1 - a}$$

This means that the waveform reaches a steady-state periodic response in which the positive peaks have value

$$\frac{V_0}{1 - a}$$

Exercises. 1. Consider the slightly damped second-order system with transfer function

$$H(s) = \frac{4}{(s + 0.1)^2 + 4}$$

having a periodic input

$$x(t) = 2[\delta(t) + \delta(t - \pi) + \delta(t - 2\pi) + \text{---}]$$

(a) Find the first five positive peaks.
(b) Find the positive peak in steady state.
ANSWERS: (a) 3.698, 7.116, 10.28, 13.2, 15.9; (b) 48.96

2. Suppose in the previous exercise σ_p is changed from 0.1 to zero, i.e.,

$$H(s) = \frac{4}{s^2 + 4}$$

(a) Find the first five peaks for the input of the previous exercise.
(b) Is the system stable? Why or why not?
ANSWERS: (a) 4, 8, 12, 16, 20. (b) The system is unstable because there is a pole on the imaginary axis of the complex plane.

Another interesting application of resonance is in product security in stores. The security tag is an *RLC* circuit with very small *R*. The circuit is excited by a "radio" wave at its resonant frequency by the security panels in front of the exit doors. If the circuit has not been destroyed at the checkout counter, it begins to resonate as one approaches the exit and transmits a signal back to a detecting device at an amplitude much higher than the original transmitted signal. This sets off an alarm.

The easiest way to understand this phenomenon is by way of frequency response. Let us consider the second-order transfer function

$$H(s) = \frac{4}{(s + 0.1)^2 + 4}$$

The magnitude frequency response is given in Figure 16.16. Notice the sharp peak at about 2 rad/sec, which is 10 times the dc gain of about 1. Thus sinusoidal inputs at frequencies close to 2 rad/sec produce a steady-state response with magnitude almost 10 times larger. This is precisely the type of resonance phenomenon that occurs with the security tags at stores.

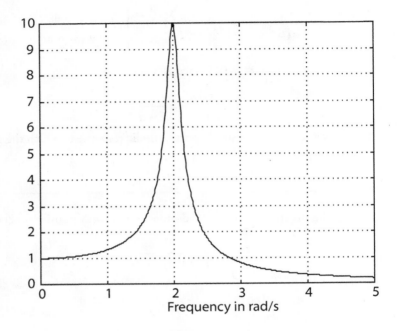

FIGURE 16.16

Series and Parallel Resonant Circuits

The simplest stable linear circuits capable of producing the resonance phenomenon are the series and parallel resonant circuits shown in Figures 16.17a and b, respectively.

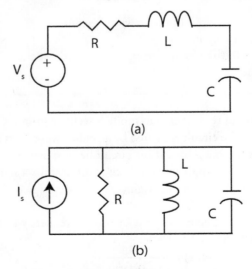

(a)

(b)

FIGURE 16.17 Series (a) and parallel (b) resonant circuits.

Using the frequency response approach of Figure 16.16, we see that small inputs in these circuits can produce large outputs when the conditions are right: (i) a high-Q circuit and (ii) input signal frequency and the peak frequency coincide. To develop a circuit-theoretic perspective on resonance we take a more basic approach using the sinusoidal steady-state analysis ideas of Chapter 10.

For any circuit containing one inductance L and one capacitance C, denote by ω_0 the frequency at which the two **reactances** $X_C = -1/\omega_0 C$ and $X_L = \omega_0 L$ have equal magnitudes, i.e.,

$$\frac{1}{\omega_o C} = \omega_o L, \tag{16.38}$$

in which case

$$\omega_o = \frac{1}{\sqrt{LC}} \tag{16.39}$$

In Chapter 9, on second-order RLC circuits, $\omega_0 = 1/\sqrt{LC}$ is called the **undamped natural frequency**. The name stems from the fact that if all resistive elements are absent (i.e., the circuit is undamped), then a parallel or series connection of L and C produces a natural response of the form $K \cos(\omega_0 t + \theta)$. In the jargon, the parallel LC circuit of Figure 16.17b is called a **tank circuit**, and ω_0 is called the **tank frequency**.

Since $jX_C = -j/\omega_0 C$ and $jX_L = j\omega_0 L$ have opposite signs and equal magnitudes, if the elements L and C are connected in series, then at ω_0 the resulting impedance, $jX_C + jX_L = -j/\omega_0 C + j\omega_0 L = 0$ and hence is equivalent to a short circuit. Similarly, if the elements L and C are connected in parallel, then at ω_0 the resulting admittance is zero and hence equivalent to an open circuit. These properties are illustrated in Figure 16.18.

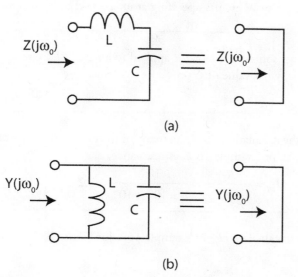

(a)

(b)

FIGURE 16.18 (a) Series- and (b) parallel-connected LC elements with $\omega_0 = \frac{1}{\sqrt{LC}}$.

With the short circuit and open circuit equivalents shown in Figure 16.18, we can easily deduce the following properties for the series and parallel resonant circuits of Figure 16.17 operating at : $\omega_0 = \frac{1}{\sqrt{LC}}$.

(1)   The magnitude of the impedance seen by the voltage source in the series RLC of Figure 16.17a is minimum, and the impedance is a pure resistance equal to R.

(2)   The magnitude of the impedance seen by the current source in the parallel RLC of Figure 16.17b is maximum, and the impedance is a pure resistance equal to R.

(3)   The magnitude of the voltage across C or L in the series RLC of Figure 16.17a is $Q_{cir}(\omega_0)$ = $\omega_0 L/R$ times the magnitude of the source voltage.

(4)   The magnitude of the current through L or C in the parallel RLC of Figure 16.17b is $Q_{cir}(\omega_0) = \omega_0 RC$ times the magnitude of the source current.

To establish property (3), we use voltage division and property (1):

$$|V_C(j\omega_0)| = \frac{\left|\frac{1}{j\omega_0 C}\right|}{R}|V_s| = \frac{|j\omega_0 L|}{R}|V_s| = |V_L(j\omega_0)| = \frac{\omega_0 L}{R}|V_s| = Q_{cir}(\omega_0)|V_s|$$

A similar derivation yields property (4).

Exercise. Derive property (4) using current division and property (2).

Properties (1) and (2) lead to a general definition of resonant frequency, denoted ω_r. Specifically ω_r is the frequency at which the source sees an impedance or an admittance that is purely real, i.e., purely resistive, despite the presence of capacitors and inductors. For the series and parallel RLC circuits

$$\omega_r = \omega_0 = \frac{1}{\sqrt{LC}}.$$

For more general circuits, $\omega_r \neq \omega_0$. Circuits operating at ω_r are said to be at resonance.

Exercise. A sinusoidal voltage source at 1 MHz is applied to a series RLC circuit. If L = 300 μH, R = 5 Ω, and C is adjustable, what value of C produces resonance?
ANSWER: 84.4 pF

Using property (2) above, the calculation of H_m in many of the earlier examples can be done with very little effort. For example, in Example 16.2, at resonance,

$$H_m = \frac{R_L}{R_s + R_L} = \frac{10^4}{4 \times 10^4 + 10^4} = 0.2$$

Exercise. Use property (2) to calculate H_m in Example 16.4 for Figure 16.5.
ANSWER: $1/R_s$

The resonance condition $\omega_r = \omega_0 = 1/\sqrt{LC}$ can be achieved by varying one of the three parameters ω_0, C, or L. When C or L or both are adjusted to achieve resonance, the circuit is often called a **tuned circuit**. The next example illustrates how this tuning can be used in a practical application.

EXAMPLE 16.8. Figure 16.19 displays an amplifier model containing a VCCS with $g_m = 2$ mS (milli-siemens) and $R_L = 20$ kΩ. The applied sinusoidal voltage, $V_{in}(j\omega)$, has a magnitude of 0.1 V at 10 MHz. The load is modeled by the parallel combination of R_L and the 40 pF capacitor; the capacitance accounts for such real-world phenomena as wiring capacitance, the device input capacitance, and other embedded capacitances. This capacitance cannot be removed from the circuit and often has deleterious effects on the amplifier performance.

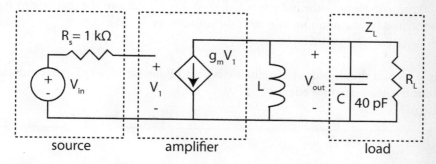

FIGURE 16.19 Amplifier circuit illustrating the application of the tuned circuit concept to eliminate undesirable capacitive effects.

The example objectives are

(a) With the load connected directly as shown (without L), find the magnitude of the output voltage.

(b) If an inductance L is connected across the load to tune out the effect of the capacitance, find the value of L and the resulting $|V_{out}|$ that will show that the amplifier gain at 10 MHz is greatly increased.

SOLUTION

(a) At 10 MHz, the load impedance is

$$Z_L(j2\pi10^7) = \frac{1}{0.00005 + j2\pi \times 10^7 \times 40 \times 10^{-12}} = 397.8\angle -88.9° \ \Omega.$$

Therefore, since $|V_1| = |V_{in}| = 0.1$ V, the magnitude of the output voltage is

$$|V_{out}| = 0.1 \times 0.002 \times 397.8 = 0.0796$$

Here the voltage gain is $0.0796/0.1 = 0.796$ due to the low impedance of C at the high operating frequency.

(b) By tuning out the effect of the capacitance, this poor gain response can be eliminated. The inductance needed to tune out the capacitance is calculated from equation 16.39:

$$L = \frac{1}{\omega_0^2 C} == \frac{1}{4\pi^2 \times 10^{14} \times 40 \times 10^{-12}} = 6.33 \times 10^{-6} \ \text{H}$$

With a 6.33 µH inductor connected across the load, the parallel LC behaves like an open circuit at 10 MHz and the load looks like a pure resistance of 20 kΩ to the amplifier. The new output voltage magnitude is

$$|V_{out}\,(j\omega_0)| = 0.1 \times 0.002 \times 20{,}000 = 4 \text{ V}$$

with a resultant voltage gain of $4/0.1 = 40$.

Exercises. 1. Suppose the frequency of the input is changed from 10 MHz to 5 MHz in Example 16.8. Find (a) the gain without the inductor connected, (b) the value of the inductor to tune out the capacitance, and (c) the resulting gain with the capacitance tuned out.
ANSWERS: (a) 3.18, (b) 25.33 µH, (c) 40

2. In Example 16.8, if the embedded capacitance is 63.3 pF instead of 40 pF, what inductance is needed to tune out the capacitive effects?
ANSWER: 4 µH

Series-Parallel Resonant Circuits

For parallel and series *RLC* circuits, $\omega_r = \omega_0$. For series-parallel circuits containing only one *L* and one *C*, when it exists $\omega_r \neq \omega_0$ in general, but it may not exist at all. The next example illustrates the point.

EXAMPLE 16.9. Find the resonant frequency, ω_r, and the input impedance at ω_r for the circuit shown in Figure 16.20.

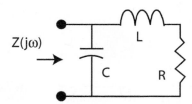

FIGURE 16.20 A variation of a parallel resonant circuit.

SOLUTION
Step 1. *Calculate the admittance "looking into" the input node pair of the circuit of Figure 16.20.* By the usual techniques,

$$Y(j\omega_r) = \frac{1}{Z(j\omega_r)} = j\omega_r C + \frac{1}{R + j\omega_r L} = j\omega_r C + \frac{R - j\omega_r L}{R^2 + (\omega_r L)^2}. \qquad (16.40)$$

Step 2. *Set the imaginary part to zero.* Resonance occurs when *Y* is real, i.e., when

$$\text{Im}\{Y\} = \omega_r C - \frac{\omega_r L}{R^2 + (\omega_r L)^2} = 0.$$

Solving for ω_r and then expressing it as a function of $\omega_0 = 1/\sqrt{LC}$ yields

$$\omega_r = \sqrt{\frac{1}{LC} - \frac{R^2}{L^2}} = \omega_0 \sqrt{1 - \frac{CR^2}{L}} \qquad (16.41)$$

The rightmost term shows how the resonant frequency is scaled away from the parallel or series ideal cases where $\omega_r = \omega_0$.

Step 3. To obtain the values of the admittance and impedance at resonance, substitute this value of ω_r into equation 16.40 to obtain

$$Y(j\omega_r) = \frac{RC}{L} \quad \text{and} \quad Z(j\omega_r) = \frac{L}{RC}. \tag{16.42}$$

Three conclusions can be drawn from equations 16.40 through 16.42:

1. If $(CR^2)/L > 1$, then there is no real solution for ω_r; this means that the source voltage and source current cannot be in phase at any frequency.
2. If $(CR^2)/L < 1$, then there is a unique nonzero resonant frequency ω_r that is strictly smaller than ω_0.
3. When ω_r exists, i.e., there is a real solution to equation 16.41, then at $\omega = \omega_r$ the source "sees" a pure resistance, the value of which equals (L/RC) and is greater than R.

Exercises. 1. Fill in the details of step 3 above.
2. In Figure 16.20, let $L = 1$ H and $C = 1$ F. Find the resonant frequency ω_r for (a) $R = 0.8\ \Omega$, and (b) $R = 2\ \Omega$. Compute the resistance seen by the "source" in each case.
ANSWERS: (a) 0.6 rad/sec, 1.25 Ω; (b) ω_r does not exist.

Results similar to equations 16.40 through 16.42 can be derived for the circuit of Figure 16.21.

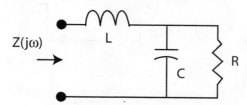

FIGURE 16.21 A variation of a series resonant circuit.

Exercises. 1. For the circuit of Figure 16.21, let $G = 1/R$. Show that the input impedance is

$$Z(j\omega) = \frac{G}{G^2 + (\omega C)^2} + j\left(\omega L - \frac{\omega C}{G^2 + (\omega C)^2}\right)$$

2. Again for Figure 16.21, show that the resonant frequency is

$$\omega_r = \omega_0\sqrt{1 - \frac{L}{CR^2}} \tag{16.43}$$

3. Finally, show that the input impedance at the resonant frequency is

$$Z(j\omega_r) = \frac{L}{RC} \tag{16.44}$$

In the above example and exercise, we concluded that

> When ω_r exists, i.e., there is a real solution to equation 16.41, then at $\omega = \omega_r$ the source "sees" a pure resistance, the value of which equals (L/RC).

This property finds application in maximum power transfer from source to load in fixed-frequency situations, as illustrated by the next example.

EXAMPLE 16.10. The output stage of a certain radio transmitter is represented by a 1 MHz sinusoidal voltage source having a fixed magnitude of 50 V_{rms} and an internal resistance of 100 Ω as shown in Figure 16.22a. A load resistance R_L models an antenna connected to the transmitter also shown in the figure. The purpose of this example is to show how a **matching network** based on the principle of resonance can be designed to maximize the power delivered to the antenna.

(a) If R_L is adjustable, find the value of R_L yielding the maximum average power P_L absorbed by the load. What is the value of (P_L)max?

(b) If $R_L = 20\ \Omega$ in Figure 16.22a, find the value of P_L.

(c) Suppose that R_L is fixed at 20 Ω, but a coupling network consisting of LC elements is inserted between the source and load to increase the power P_L as shown in Figure 16.22b. Choose values for L and C in the circuit so that $(P_L)_{max}$ of part (a) is again obtained.

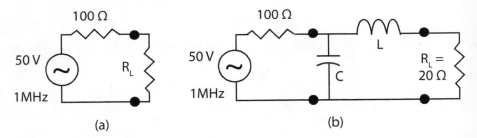

FIGURE 16.22 Matching load to source using a resonant circuit. (a) Load connected directly to source. (b) A coupling network designed to maximize the load power.

SOLUTION

(a) From the maximum power transfer theorem,

$$R_L = R_s = 100\ \Omega$$

and

$$(P_L)_{max} = \frac{V_s^2}{4R_s} = \frac{2,500}{400} = 6.25\ \text{W.}$$

(b) With $R_L = 20\ \Omega$, by voltage division

$$V_L = \frac{20}{100 + 20}\ 50 = 8.33\ \text{V,}$$

in which case

$$P_L = \frac{(8.33)^2}{20} = 3.472\ \text{W.}$$

(c) If we can make the impedance at the input terminals of the LC coupling network equal to $(100 + j0)$ Ω, then maximum power will be drawn from the source. Since LC elements consume zero average power, the same maximum power will be delivered to the load resistance. The resonant circuit shown in Figure 16.22b provides a possible design. The "LCR" load circuit is the one analyzed in Example 16.9. Hence, to calculate the element values, we use equations 16.41 and 16.42 as follows:

and

$$\omega_r^2 = \left(2\pi \times 10^6\right)^2 = \frac{1}{LC} - \frac{20^2}{L^2} \text{ (from equation 16.41, squared)}$$

$$Z(j\omega_r) = 100 = \frac{L}{20C} \text{ (from equation 16.42)}$$

Solving these equations simultaneously results in $L = 6.37$ μH and $C = 3.18$ nF.

Some remarks about this design are in order:

1. $f_o = 1/(2\pi\sqrt{LC}) = 1.12$ MHz, and $f_r = 1$ MHz $\neq f_o$. The 20 Ω resistance is transformed into a 100 Ω resistance at $f = f_r$ (not at $f = f_o$).
2. Should the source resistance be smaller than the fixed load resistance, C is moved to be in parallel with R_L. In that case, equations 16.43 and 16.44 are used in place of equations 16.41 and 16.42.

Exercise. Redesign the coupling network in Example 16.10 if the resistors are $R_s = 300$ Ω and $R_L = 50$ Ω.

ANSWERS: 17.79 μH and 1.18 nF

A variation on the computation of the resonant frequency is the design of a circuit to achieve a desired resonant frequency using a variable capacitor; this is the design that underlies the tuning of many AM radios.

EXAMPLE 16.11. Consider the series RLC of Figure 16.23. Here, the voltage source has a fixed magnitude $|V_s|$ and a *fixed frequency* ω. With R and L fixed, we seek the value of the variable capacitance C that maximizes the magnitude of the voltage across the capacitor.

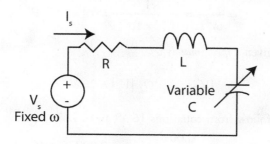

FIGURE 16.23 Adjusting C for maximum output voltage.

The first step is to compute the magnitude of the voltage across the capacitor using voltage division:

$$|V_C| = \frac{|V_s|}{\sqrt{R^2 + \left(\omega L - \dfrac{1}{\omega C}\right)^2}} \frac{1}{\omega C}$$

Maximizing $|V_C|$ is equivalent to maximizing $|V_C|^2$ or minimizing $\dfrac{|V_s|^2}{|V_C|^2}$. To obtain this last expression, we square the above expression for $|V_C|$ and rewrite it as

$$f(C) = \frac{|V_s|^2}{|V_C|^2} = (\omega C R)^2 + (\omega^2 L C - 1)^2$$

To minimize $f(C)$, set $f'(C)$ to zero, i.e.,

$$f'(C) = 2C(\omega R)^2 + 2(\omega^2 LC - 1)\omega^2 L = 0$$

Solving for C produces

$$C = \frac{L}{R^2 + (\omega L)^2} = \frac{1}{L\omega^2}\left[\frac{1}{\dfrac{R^2}{L^2\omega^2} + 1}\right]$$

Although this value of C produces a maximum capacitor voltage, the circuit is *not in resonance*, as the value of $1/\sqrt{LC}$ is not equal to the signal source frequency ω. However, given this equation for C, for a high-Q circuit ($\omega L/R > 8$), the condition is practically the same as $1/\sqrt{LC} = \omega$.

EXAMPLE 16.12. In Figure 16.23, let $|V_s| = 1$ V, $\omega = 10^5$ rad/sec, $R = 5\ \Omega$. Let C be variable. Find $|V_C|_{max}$ and the corresponding value of C for each of the following cases: (a) $L = 1$ mH; (b) $L = 100\ \mu$H.

SOLUTION

(a) For $L = 1$ mH, $Q_L = \omega L/R = 10^5 \times 10^{-3}/5 = 20$. This is a high-$Q$ circuit. Therefore, C is given *approximately* by

$$C = \frac{1}{\omega^2 L} = \frac{1}{10^{10} \times 10^{-3}} = 0.1 \times 10^{-6}\ \text{F},$$

and $|V_C|_{max}$ is given *approximately* by

$$|V_C|_{max} = Q_L |V_s| = 20\ \text{V}$$

Exact solutions follow from equations 16.48 and 16.47:

$$C = \frac{0.001}{5^2 + (10^5 \times 0.001)^2} = 0.09975 \times 10^{-6}\ \text{F},$$

and

$$|V_C|_{max} = \frac{1}{\sqrt{5^2 + \left(10^5 \times 0.001 - \frac{1}{10^5 \times 0.09975 \times 10^{-6}}\right)^2}} \times \frac{1}{10^5 \times 0.09975 \times 10^{-6}} = 20.025 \text{ V}$$

Plainly, the approximate solutions are very close to the exact solutions.

(b) For $L = 100$ μH, $Q_L = \omega L/R = 10^5 \times 10^{-4}/5 = 2$. This is a low-$Q$ circuit, requiring the use of equations 16.48 and 16.47. Here,

$$C = \frac{0.0001}{5^2 + (10^5 \times 0.0001)^2} = 0.8 \times 10^{-6} \text{ F},$$

and

$$|V_C|_{max} = \frac{1}{\sqrt{5^2 + \left(10^5 \times 0.0001 - \frac{1}{10^5 \times 0.8 \times 10^{-6}}\right)^2}} \times \frac{1}{10^5 \times 0.8 \times 10^{-6}} = 2.236 \text{ V}.$$

Exercise. For part (b) of Example 16.12, compute $\omega_0 = 1/\sqrt{LC}$. Is this value equal to the signal frequency? Why or why not?
ANSWER. $\omega_0 = 1.118 \times 10^5$ rad/sec

---

6. GENERAL STRUCTURE OF THE BAND-PASS TRANSFER FUNCTION WITH ONE PAIR OF COMPLEX POLES

With the experience gained from the analysis of the first five sections, we now present a general transfer function approach to the analysis of band-pass type circuits. The circuits considered in this section will contain only one inductance and one capacitance. Hence any associated network function will have at most a *second-degree* polynomial in s as the denominator and numerator. Thus, the general transfer function $H(s)$ (which includes the impedance function $Z(s)$ and the admittance function $Y(s)$ as special cases) has the *biquadratic* form

$$H(s) = \frac{n(s)}{d(s)} = \frac{a_2 s^2 + a_1 s + a_o}{s^2 + 2\sigma_p s + \omega_p^2}. \qquad (16.45)$$

A reasonably sharp band-pass characteristic requires that $H(s)$ have complex poles, i.e., $\sigma_p < \omega_p$. The finite zeros of $H(s)$, which are roots of $n(s) = 0$, may be real or complex. The case of complex zeros, corresponding to more advanced filter characteristics such as the inverse Chebyshev or elliptic types, is beyond the scope of this text. For practical reasons we focus on the case where $H(s)$ has one real zero or no zero.

When $H(s)$ has one real zero and complex poles, then equation 16.45 reduces to

$$H(s) = \frac{a_1 s + a_o}{s^2 + 2\sigma_p s + \omega_p^2} = \frac{a_1 s + a_o}{s^2 + \frac{\omega_p}{Q_p} s + \omega_p^2}, \qquad (16.46)$$

with the pole-zero plot shown in Figure 16.24.

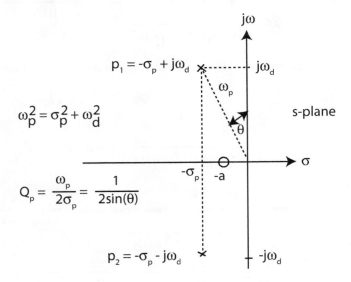

FIGURE 16.24 Pole-zero plot of the transfer function $H(s)$ of equation 16.46.

Equation 16.46 contains the usual quality factor, Q_p, mathematically called the **pole Q**, defined as

$$Q_p \equiv \frac{\omega_p}{2\sigma_p} = \frac{1}{2\sin(\theta)}, \qquad (16.47a)$$

$$\theta = \arcsin\left(\frac{\sigma_p}{\omega_p}\right), \qquad (16.47b)$$

where the angle θ is as shown in Figure 16.24. Relative to the pole-zero plot, Q_p measures how close the pole is to the $j\omega$-axis: a higher Q_p means a smaller θ, implying a pole closer to the $j\omega$-axis. As we already know, Q_p is related to the circuit Q, i.e., Q_{cir}, and serves as a quick estimate of the sharpness of the response curve. For some special cases, $Q_p = Q_{cir}$.

In Figure 16.24, another new quantity, ω_d, appears. To ascertain the meaning of ω_d, recall from Chapter 15 that the impulse response of a system characterized by equation 16.46 has the form

$$h(t) = \mathcal{L}^{-1}\{H(s)\} = Ke^{-\sigma_p t}\cos(\omega_d + \theta)$$

The waveform $h(t)$ is a damped sinusoid, and the quantity ω_d (not ω_p) specifies the frequency of oscillation. For this reason, ω_d is referred to as the **damped oscillation frequency**.

For a transfer function of the form of equation 16.46, our goal is to determine several key quantities that are indicative of the circuit's behavior: ω_m, the bandwidth B_ω, H_m, and the half-power frequencies.

Case 1. *No finite zeros.* When the transfer function of equation 16.46 has no finite zero, then it reduces to

$$H(s) = \frac{K}{s^2 + \dfrac{\omega_p}{Q_p}s + \omega_p^2} = \frac{K}{s^2 + 2\sigma_p s + \omega_p^2} \tag{16.48}$$

The pole-zero plot of this $H(s)$ differs from that of Figure 16.24 only in that the single zero is now absent. The series *RLC* circuit of Figure 16.25a with the capacitor voltage as the output and the parallel *RLC* of Figure 16.25b with the inductor current as the output both fall into this category.

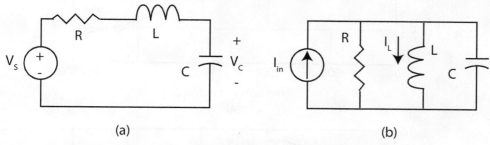

(a)   (b)

FIGURE 16.25 (a) Series *RLC* with capacitor voltage as output.
(b) Parallel *RLC* with inductor current as output.

The transfer function of equation 16.48 can represent either a low-pass or a band-pass characteristic depending on the value of Q_p and the value of $H(0)$. In fact, the transfer function can display both characteristics as well as intermediate behaviors. For example, if

$$H(s) = \frac{(1000)^2}{s^2 + \dfrac{1000}{\sqrt{2}}s + (1000)^2}$$

then the maximally flat low-pass characteristic of Figure 16.26a results.

On the other hand, if

$$H(s) = \frac{(1000)^2}{s^2 + \dfrac{1000}{1.31}s + (1000)^2}$$

then something between a low-pass and a band-pass characteristic results, as shown in Figure 16.26b. Here low frequencies are still passed, yet the characteristic has a selectivity property resulting from the pole Q of 1.31. The ratio of the maximum gain to dc gain is $\sqrt{2}$. This means that the peak is 3 dB above the dc gain. Now, with increasing pole Q the selectivity goes up, as does the maximum gain, and the characteristic looks more and more like a pure band-pass. Finally, if

$$H(s) = \frac{(1000)^2}{s^2 + \dfrac{1000}{10}s + (1000)^2}$$

then the approximate band-pass characteristic of Figure 16.26c results. Here the pole Q is 10 and the ratio of the maximum gain to dc gain is also 10. Although low frequencies are not attenuated, the characteristic is highly selective and frequencies near ω_p are highly amplified, so that for all practical purposes the characteristic is identified as band-pass.

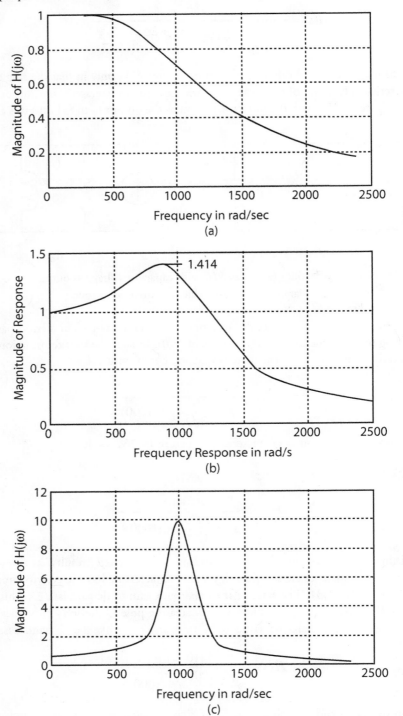

FIGURE 16.26 (a) Low-pass characteristic. (b) Moderate Q_p, resulting in characteristic exhibiting both low-pass and band-pass behavior. (c) High-Q_p case, showing a marked band-pass characteristic.

The important aspects of the curves in Figure 16.26 are ω_m, H_m, and B_ω. We first compute ω_m and then H_m. At the peak value, the derivative of $|H(j\omega)|$ is zero. Setting the derivative of $|H(j\omega)|$ to zero and solving for ω_m produces the exact formula[1],

$$\omega_m = \omega_p \sqrt{1 - \frac{1}{2Q_p^2}} = \omega_p \sqrt{\cos(2\theta)} \tag{16.49}$$

To compute H_m, consider

$$H_m^2 = \frac{K^2}{\left(\omega_p^2 - \omega_m^2\right)^2 + \dfrac{\omega_p^2 \omega_m^2}{Q_p^2}} \Bigg|_{\omega_m = \omega_p \sqrt{\cos(2\theta)}} = \frac{K^2}{\left(\omega_p^2 - \omega_p^2 \cos(2\theta)\right)^2 + \omega_p^4 4 \sin^2(\theta)\cos(2\theta)}$$

$$= \frac{K^2}{\omega_p^4 4 \sin^2(\theta)\cos^2(\theta)} = \frac{K^2}{\omega_p^4 \sin^2(2\theta)}$$

Hence

$$H_m = \frac{K}{\omega_p^2 \sin(2\theta)} = \frac{H(0)}{\sin(2\theta)} \tag{16.50}$$

Similar derivations with more complex algebra yield the half-power frequencies,

$$\omega_{1,2} = \omega_p \sqrt{\cos(2\theta) \mp \sin(2\theta)} \tag{16.51}$$

and the bandwidth,

$$B_\omega = \omega_p \sqrt{2\left(\cos(2\theta) - \sqrt{\cos(4\theta)}\right)} \tag{16.52}$$

provided there exist real solutions.

Referring back to equation 16.49, if $Q_p < 1/\sqrt{2} = 0.707$ or, equivalently, if $\theta > 45°$, then there is no real solution for ω_m. In this case, although the poles are complex, the magnitude response displays no peak at any frequency. Rather, $H_m = |H(j0)|$ and $|H(j\omega)|$ decreases *monotonically* to zero as ω increases, as demonstrated in Figure 16.26a.

When Q_p is greater than $1/\sqrt{2}$ or, equivalently, if $\theta < 45°$, the magnitude response of equation 16.48 starts from a nonzero value at $\omega = 0$, rises to the peak value at $\omega = \omega_m$, and finally decreases to zero as $\omega \to \infty$. This behavior is illustrated in Figures 16.26b and c. If Q_p is only slightly greater than 0.707, then the magnitude response is essentially that of the low-pass type, with a small hump in the pass band.

For the high-Q_p case ($Q_p \geq 8$), the magnitude response near ω_m approximates that of a pure band-pass circuit. The preceding exact expression for ω_m reduces to $\omega_m \cong \omega_p$ for the high-Q_p case.

Similarly, for high Q_p, $B_\omega \cong 2\sigma_p$, $H_m = \dfrac{|K|Q_p}{\omega_p^2}$, and $\omega_{1,2} \cong \omega_p \mp \sigma_p$. In fact, the error of these approximations is less than 0.5%.

Exercise. For $Q_p = 8$, $\omega_p = 1000$ rad/sec, and $K =$ in equation 16.48, compute the exact and approximate values of ω_m, B_ω, and H_m. Then compute the percentage error in the approximation:

$$\frac{|exact - approximate|}{|exact|} \times 100$$

ANSWERS: Exact values are $\omega_m = 996.09$ rad/sec, $B_\omega = 125.5$ rad/sec, and $H_m = 8.0157$. Approximate values are ω_m 1000 rad/sec, $B_\omega = 125$ rad/sec, and $H_m = 8$. Percentage errors are, respectively, 0.39%, 0.39%, and 0.196%.

EXAMPLE 16.13. Consider the *Sallen and Key* active network of Figure 16.27, which can be used to realize the transfer function of equation 16.48. As per a homework exercise in Chapter 14, the transfer function is

$$H(s) = \frac{V_{out}}{V_{in}} = \frac{\dfrac{1}{R_1 R_2 C_1 C_2}}{s^2 + \left(\dfrac{1}{R_1 C_1} + \dfrac{1}{R_2 C_1}\right)s + \dfrac{1}{R_1 R_2 C_1 C_2}}$$

Suppose $C_1 = 2 \ \mu F$, $C_2 = 5$ nF, $R_1 = R_2 = 10 \ k\Omega$.
    (a)   Compute ω_p, Q_p, and K of equation 16.48.
    (b)   Compute exact values of ω_m and H_m.
    (c)   Compute approximate values of B_ω, ω_1, and ω_2.
    (d)   Plot the magnitude response curve of the transfer function.
    (e)   Simulate the circuit using SPICE.

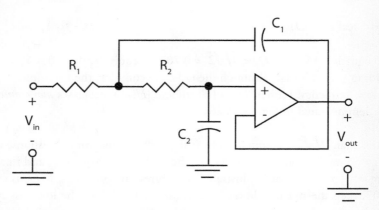

FIGURE 16.27 Sallen and Key active network for realizing the transfer
function of equation 16.48.

SOLUTION

For the given parameter values,

$$H(s) = \frac{V_{out}}{V_{in}} = \frac{(1000)^2}{s^2 + \dfrac{1000}{10}s + (1000)^2}$$

(a) By inspection $\omega_p = 1000$ rad/sec, $Q_p = 10$, and $K = (\omega_p)^2 = 10^6$.

(b). From equation 16.48,

$$\omega_m = \omega_p\sqrt{1 - \frac{1}{2Q_p^2}} = 997.5 \text{ rad/sec.}$$

Numerically evaluating $H(s)$ at $s = j\omega_m$, we obtain $H_m = 10.013$ via the following MATLAB code:

```
»n = wp^2;
»d = [1 wp/Qp wp^2];
»Hm = abs(polyval(n,j*wm))/abs(polyval(d,j*wm))
Hm = 1.0013e+01
```

(c) The approximate values are given by

$$B_\omega \cong 2\sigma_p = 100, \; \omega_1 \cong \omega_p - \sigma_p = 900, \; \omega_2 \cong \omega_p + \sigma_p = 1100,$$

all in rad/sec.

(d) To obtain the magnitude response for the given transfer function, as shown in Figure 16.28, we use the following MATLAB code:

```
»n = wp^2;
»d = [1 wp/Qp wp^2];
»f = logspace(1,3,600);
»w = 2*pi*f;
»h = freqs(n,d,w);
»semilogx(f,abs(h))
»grid
»xlabel('Frequency in Hz')
»ylabel('Magnitude H(jw)')
```

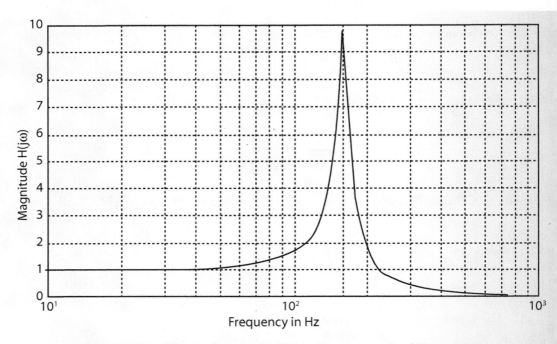

FIGURE 16.28 Magnitude response of high-Q_p active circuit of Figure 16.27.

(e) A SPICE simulation yields the corresponding plot in Figure 16.29.

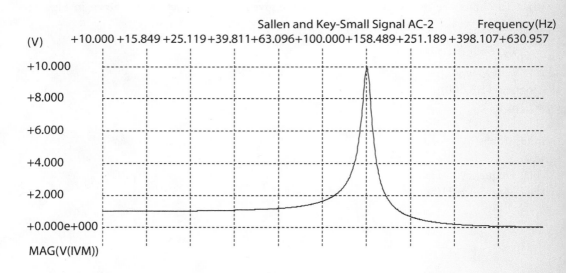

FIGURE 16.29 SPICE simulation of active circuit of Figure 16.27.

Exercise. Suppose $C_1 = 0.2$ µF, $C_2 = 0.5$ nF, and $R_1 = R_2 = 10$ kΩ. Find ω_p, Q_p, and K of equation 16.48, along with exact values of ω_m and H_m. Find approximate values of B_ω, ω_1, and ω_2.
ANSWERS: $\omega_p = 10^4$ rad/sec, $Q_p = 10$, $K = 10^8$, $\omega_m = 9975$ rad/sec, $H_m = 10.013$, $B_\omega \cong 10^3$ rad/sec, $\omega_1 \cong 9000$ rad/sec, $\omega_2 \cong 11{,}000$ rad/sec

Case 2. *A single zero off the origin.* In equation 16.46 with $a_0 \neq 0$, $a_1 \neq 0$, and a_0 and a_1 of the same sign, a zero is present in the left half-plane. A second form of the band-pass transfer function of equation 16.16 is

$$H(s) = K\frac{s+a}{s^2 + \dfrac{\omega_p}{Q_p}s + \omega_p^2} = K\frac{s+a}{s^2 + 2\sigma_p s + \omega_p^2} \tag{16.53}$$

Figure 16.32 sketches the pole-zero plot for this transfer function. Intuitively speaking, the closer the zero is to the origin, the more the magnitude response resembles the response of the case with a single zero at the origin. For the transfer function of equation 16.53, derivations of *exact* values of ω_m and the zero-phase-shift frequency are possible. (See the homework problems.) The results are

$$\omega_m = \sqrt{-a^2 + \sqrt{\left(\omega_p^2 + a^2\right)^2 - \left(2\sigma_p a\right)^2}}\,, \tag{16.54}$$

$$\omega \text{ (for zero phase shift)} = \sqrt{\omega_p^2 - 2\sigma_p a} \tag{16.55}$$

No exact expressions are available for the half-power frequencies and the bandwidth B_ω. For the case of high Q_p and $a \ll \omega_p$, *approximate* answers are

$$\omega_m \cong \omega_p, \; Q_{cir} \cong Q_p, \; H_m \cong \frac{|K|}{2\sigma_p}, \text{ and } B_\omega \cong 2\sigma_p \quad \text{for } Q_p > 8 \tag{16.56}$$

To see why, observe that

$$|H(j\omega)| = \left| K\frac{j\omega + a}{(j\omega)^2 + j2\omega\sigma_p + \omega_p^2} \right| = \left| K\frac{j\omega}{(j\omega)^2 + j2\omega\sigma_p + \omega_p^2} \right| \times \left| \frac{j\omega + a}{j\omega} \right| \tag{16.57a}$$

$$= \left| \frac{K}{2\sigma_p + j\left(\dfrac{\omega^2 - \omega_p^2}{\omega}\right)} \right| \times \left| 1 + \frac{a}{j\omega} \right| \tag{16.57b}$$

The second factor is approximately 1 for frequencies near ω_p by virtue of our assumption $a \ll \omega_p$. Hence the properties of the magnitude response reduce approximately to those of the first factor in equation 16.57b. These properties are those of a single zero at the origin. Hence for high-Q_p circuits, the relations of equation 16.56 approximate the single zero at the origin case.

Exercise. Use equation 16.57b to show that when $Q_p > 8$ and $a << \omega_p$, $H_m \cong \dfrac{|K|}{2\sigma_p}$.

When exact values are desired, one must resort to SPICE or MATLAB to obtain the frequency response from which values can be graphically determined. SPICE has the advantage of not having to compute the transfer function of the circuit; MATLAB requires this computation.

7. SUMMARY

This chapter began with a study of simple series and parallel *RLC* band-pass circuits. Because of its generality, we set forth a transfer function approach to the analysis and design of these band-pass circuits. Specifically, we first investigated a second-order transfer function with a single zero at the origin for which

$$H(s) = K\,\frac{s}{s^2 + \dfrac{\omega_p}{Q_p} + \omega_p^2} = K\,\frac{s}{s^2 + 2\sigma_p s + \omega_p^2}$$

For this transfer function we derived various formulas for determining band-pass parameters, such as the peak frequency,

$$\omega_m = \omega_p$$

the peak value,

$$|H(j\omega)|_{\max} = |H(j\omega_m)| = \frac{|K|}{2\sigma_p} = \frac{|K|Q_p}{\omega_m},$$

and the bandwidth,

$$B_\omega = 2\sigma_p = \frac{\omega_p}{Q_p}.$$

These formulas have a special form for the parallel *RLC*:

$$\omega_m = \omega_o = \frac{1}{\sqrt{LC}}$$

which is also the resonant frequency of the associated impedance; the bandwidth is found to be

$$B_\omega = \omega_2 - \omega_1 = \frac{1}{RC}$$

while the circuit Q is

$$Q = \frac{\omega_m}{B_\omega} = \omega_o RC = R\sqrt{\frac{C}{L}}$$

The resonance phenomenon of a second-order *RLC* circuit was then set forth from a frequency domain perspective. One application was the design of a matching network that produces maximum power transfer, at a single frequency, from a source with fixed internal resistance to a fixed resistance load.

After this we took up the consideration of more general second-order band-pass transfer functions—for example, those with a single zero off the origin,

$$H(s) = K \frac{s+a}{s^2 + \dfrac{\omega_p}{Q_p} s + \omega_p^2} = K \frac{s+a}{s^2 + 2\sigma_p s + \omega_p^2},$$

or with no finite zero:

$$H(s) = \frac{K}{s^2 + \dfrac{\omega_p}{Q_p} s + \omega_p^2} = \frac{K}{s^2 + 2\sigma_p s + \omega_p^2}$$

Other cases are left to more advanced courses on filter design.

8. TERMS AND CONCEPTS

Active band-pass circuit: a circuit containing operational amplifiers and no inductors that achieves a band-pass characteristic.

Band-pass circuit: a circuit that passes signals within a band of frequencies while rejecting other frequency components outside of the band.

Bandwidth (3 dB bandwidth), B_ω: $\omega_2 - \omega_1$, the difference between the two half-power frequencies.

Damped oscillation frequency, ω_d: frequency given by the condition that if the transfer function of a second-order linear circuit has complex poles at $s = -\sigma_p \pm j\omega_d$, then the impulse response has the form $K\cos(\omega_d t + \theta)$. The constant ω_d is called the *damped natural frequency*.

Half-power frequencies: see lower and upper half-power frequencies.

***LC* resonance frequency:** frequency at which the reactances of L and C have the same magnitude; equals $1/\sqrt{LC}$ rad/sec. (See also tank frequency.)

Lower half-power frequency, ω_1: the radian frequency below the center frequency at which the magnitude response is 0.707 times the maximum value.

Matching network: an *LC* network that transforms a resistance R_L into a resistance of a different, specified value at one frequency or a band of frequencies.

Peak frequency (center frequency), ω_m: the radian frequency at which the magnitude response curve reaches its peak.

Q_p (pole Q): for a pair of complex poles $s_{1,2} = \omega_p e^{\pm j(\theta + \pi/2)}$, $Q_p \equiv 1/(2\sin\theta)$.

Quality factor Q (Q_{cir}) of a band-pass circuit: the ratio of the center frequency to the bandwidth, i.e., $Q = \omega_m / B_\omega$.

Quality factor Q_L (Q_{coil}) of a coil: for a coil modeled by an inductance L in series with a resistance R_s, $Q_L = \omega L / R_s$ and is frequency dependent.

Quality factor Q_C (Q_{cap}) of a capacitor: for a capacitor modeled by a capacitance C in parallel with a resistance R_p, $Q_C = \omega C R_p$ and is frequency dependent.

Quality factor Q_z of a reactive component: for an impedance expressed as $Z = R + jX$, $Q_z = |X|/R$.

Reactance: in sinusoidal steady-state analysis, the imaginary part of an impedance. For L, the reactance is ωL; for C the reactance is $-1/(\omega C)$.

Resonance frequency, ω_r: the unique radian frequency at which the input impedance of a two-terminal linear circuit becomes purely resistive.

Selectivity of a band-pass circuit: The circuit Q, defined as the ratio of the center frequency to the bandwidth. A higher Q corresponds to better selectivity.

Susceptance: in sinusoidal steady-state analysis, the imaginary part of an admittance. For C, the susceptance is ωC. For L, the susceptance is $-1/(\omega L)$.

Tank circuit: the parallel connection of an inductor and a capacitor. In the idealized case (no resistance), the total energy stored in a tank circuit remains constant, although there is a continuous interchange of the energy stored in the various components.

Tank frequency, ω_0: defined as $1/\sqrt{LC}$ in this text, regardless of the connection of the single L and single C with other components in the circuit.

Tuned circuit: a second-order circuit containing one inductance and one capacitance, at least one of which is adjustable to reach a condition of near resonance.

Undamped natural frequency: the natural frequency of a circuit consisting of lossless inductors and capacitors. For the case of one inductor and one capacitor connected together, this frequency is the same as the LC resonance frequency or the tank frequency and is equal to $1/\sqrt{LC}$.

Universal resonance curve: a normalized magnitude response curve of a band-pass transfer function having one pair of complex poles and a single zero at the origin. The magnitude is normalized with respect to the maximum gain, and the frequency is normalized with respect to the center frequency.

Upper half-power frequency, ω_2: the radian frequency above the center frequency at which the magnitude response is 0.707 times the maximum value.

---

[1] We would like to thank Les Axelrod of the Illinois Institute of Technology for providing derivations that led to this formula and those of the bandwidth and half-power frequencies.

Problems

BASIC BAND-PASS TRANSFER FUNCTION PROBLEMS

1. Fill in the details of the derivation of

$$\omega_{1,2} = \pm\sigma_p + \sqrt{\sigma_p^2 + \omega_p^2} = \omega_p\left(\sqrt{1 + \frac{1}{4Q_p^2}} \pm \frac{1}{2Q_p}\right)$$

from

$$H(j\omega) = K\frac{j\omega}{(j\omega)^2 + j2\omega\sigma_p + \omega_p^2} = \frac{K}{2\sigma_p + j\left(\frac{\omega^2 - \omega_p^2}{\omega}\right)}$$

2. Suppose the basic second-order band-pass transfer function of equation 16.1 has poles at $-1 \pm j3$ and the gain at ω_m is 2. Find K, $H(s)$, ω_m, H_m, Q, B_ω, ω_1, and ω_2. Now plot the frequency response using MATLAB and verify your answers.
CHECK: $\omega_1 = 2.317$ rad/sec

3. Suppose the basic second-order band-pass transfer function of equation 16.1 has poles at $-80 \pm j1599$ and the gain at ω_m is 6.25. Find K, $H(s)$, ω_m, H_m, Q, B_ω, exact and approximate values for ω_1 and ω_2, and the relative percent error in the approximate computation of ω_2. Using MATLAB, plot the magnitude response and verify your calculations.
CHECK: $\omega_2 = 1681$ rad/sec while the approximate value of ω_2 is 1679 rad/sec

4. Suppose a basic band-pass transfer function has $\omega_m = \sqrt{82}$ rad/sec, $H_m = 10$, and $B_\omega = 2$ rad/sec.
   (a) Find $H(s)$ and its poles.
   (b) Find ω_1 and ω_2. Using MATLAB, plot the magnitude response and verify your calculations.
   (c) If $H(s) = \dfrac{V_{out}(s)}{I_{in}(s)}$, find a parallel circuit realization of $H(s)$.
   (d) Use SPICE to obtain the magnitude frequency response and compare it with your answer in part (b).

CHECK: poles are $p_{1,2} = -1 \pm j9$

5. Suppose a basic band-pass transfer function has $\omega_m = \sqrt{8200}$ rad/sec, $H_m = 10$, and $B_\omega = 8$ rad/sec.
   (a) Find $H(s)$, its poles, and Q_p.
   (b) Find approximate values for ω_1 and ω_2. Using MATLAB, plot the magnitude response and verify your calculations.
   (c) If $H(s) = \dfrac{V_{out}(s)}{I_{in}(s)}$, find a parallel circuit realization of $H(s)$.
   CHECK: $Q_p = 11.32$

6. Suppose a series RLC circuit has $\omega_m = 80$ rad/sec, $Q_p = 8$, $C = 1/80$ F, and output $I_L(s)$ with voltage source input $V_{in}(s)$.
   (a) Write down the transfer function for the series RLC circuit in terms of R, L, and C.
   (b) Find the values of R and L.
   (c) Find H_m and approximate values for ω_1 and ω_2.
   (d) Use SPICE to obtain the magnitude (frequency) response and verify your answers to part (c).
   CHECK: $R = 0.125$ □

7. Suppose a series RLC circuit has $\omega_m = 100$ rad/sec, $Q_p = 4$, $C = 0.01$ F, and output $I_{out}(s)$ with voltage source input $V_{in}(s)$.
   (a) Write down the transfer function for the series RLC circuit in terms of R, L, and C.
   (b) Find the values of R and L.
   (c) Find H_m and approximate values for ω_1 and ω_2.
   (d) Use SPICE to obtain the magnitude (frequency) response and verify your answers to part (c).
   CHECK: $L = 0.01$ H and $H_m = 4$ S

8. Consider the circuit of Figure P16.8 for which $R_s = 40$ kΩ, $R_L = 10$ kΩ, $L = 10$ mH, and $C = 1$ μF.

(a) Find the *exact* value of the maximum voltage gain and the corresponding frequency (in Hz).

(b) Find the *exact* 3 dB bandwidth B_ω (in Hz).

(c) Find the circuit Q.

(d) Find *approximate* values of the upper (ω_2) and lower (ω_1) half-power frequencies (in Hz).

(e) Use MATLAB or its equivalent to plot $|H(j\omega)|$ vs. ω.

(f) What is the new bandwidth if the input is changed from V_{in} to an independent current source I_{in}?

(g) If the circuit is frequency scaled by $K_f = 10$, find the new values of B_ω, ω_2, and ω_1.

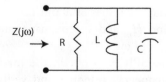

Figure P16.8

9. The parallel *RLC* circuit of Figure P16.9 has $\omega_m = 1$ Mrad/sec, $Z(\omega_m) = 20$ kΩ, and $|Z(0.9\omega_m)| = 10$ kΩ. Find the transfer function, $H(s) = Z(s)$, R, L, C, B_ω, Q, and approximate values for ω_1 and ω_2.

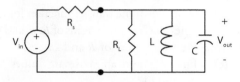

Figure P16.9

10. The equivalent circuit of a radio frequency amplifier in an AM receiver is shown in Figure P16.10, where $g_m = 2$ mS, $R_s = 100$ Ω, and $R_1 = 10$ kΩ.

(a) Find $H(s)$ in literal form.

(b) If $L = 100$ μH, $f_m = 1040$ kHz, and $B_f = 10$ kHz, find C, R, and H_m.

(c) If $f_m = 920$ kHz, $B_f = 10$ kHz, and $C = 250$ pF, find L, R, and H_m.

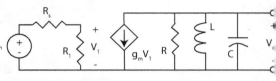

Figure P16.10

11. For this problem consider the circuit of Figure P16.11, in which $L = 300$ μH. The following additional ideal components are available:

Variable capacitors (in pF): 20–200, 30–200, 30–300, 40–400

Resistors: all values

(a) Using these components, design a parallel resonant circuit such that:

(i) The circuit can be tuned from 550 to 1650 kHz (standard AM broadcast band).

(ii) When the circuit is tuned to 920 kHz (WBAA, at Purdue), the bandwidth is 20 kHz.

(b) With the components as selected in part (a), find the bandwidth when the circuit is tuned to the low end and then the high end of the AM broadcast band.

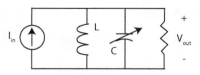

Figure P16.11

12. For this problem again consider the circuit of Figure P16.11. The following ideal components are available:

Variable capacitor (in pF): 30–300
Resistors: all values

(a) Using these components, find the range of allowable inductance so that the circuit can be tuned from 550 to 1650 kHz (standard AM broadcast band) as follows:
   (i) Find the largest value of L (L_{large}) allowable. Hint: For the smallest value of C, the inductor must be chosen so that the circuit can be tuned to 1650 kHz.
   (ii) Find the smallest value of L (L_{small}) allowable. Hint: For the largest value of C, the inductor must be chosen so that the circuit can be tuned to 550 kHz.

(b) For L_{small}, find the range of capacitance utilized in the tuning.

(c) For L_{large}, find the range of capacitance utilized in the tuning.

CHECK: 30 pF ≤ C ≤ 270 pF

(d) For L = 295 µH, find R so that when the circuit is tuned to 920 kHz (WBAA, Purdue), the bandwidth is 20 kHz.

(e) With the components as selected as in part (d), what is the bandwidth when the circuit is tuned to the low end and then to the high end of the AM broadcast band?

13. For the two-terminal parallel *RLC* circuit in Figure P16.13, $H(s) = Z(s)$, the peak frequency f_m = 10 kHz, and the bandwidth B_f = 3 kHz. If C = 0.1 µF, find the corresponding values of L, R, and the circuit Q.

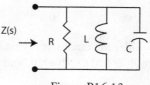

Figure P16.13

ANSWER: (a) 2.5 mH and 530 Ω

14. For the two-terminal circuit in Figure P16.14, $H(s) = Z(s)$, the peak frequency f_m = 10 kHz, and the bandwidth B_f = 3 kHz. If C = 0.1 µF, find the corresponding values of L, R, and the circuit Q.

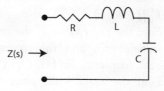

Figure P16.14

CHECK: Q = 3.333

15. Consider the circuit of Figure P16.15, for which R_s = 40 kΩ, R_L = 10 kΩ, L = 10 mH, and C = 1 µF.

(a) Find the transfer function
$$H(s) = \frac{V_{out}(s)}{V_{in}(s)}$$ and its poles.

(b) Find ω_m, H_m, Q, and B_ω.

(c) Find f_m and B_f (the corresponding quantities, in Hz, to ω_m and B_ω).

(d) Find approximate values of the upper and lower half-power frequencies in rad/sec and Hz.

(e) Use MATLAB or the equivalent to plot $|H(j\omega)|$ vs. ω.

(f) What is the new bandwidth if the input is changed from V_{in} to an independent current source I_{in}?

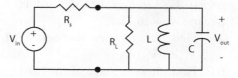

Figure P16.15

ANSWERS: $\omega_m = 10^4$ and $Q = 80$

16. Again consider the circuit of Figure P16.15. Now suppose $R_s = 100\ \Omega$, $Q = 10$, and $f_m = 10$ kHz.

    (a) If $R_L = \infty$, find C and L.

    (b) Now suppose $R_L = 1000\ \Omega$. Again find C and L.

CHECK: (a) $C = 1.59\ \mu\text{F}$

17. For the series resonant circuit shown in Figure P16.17, $R_1 = 40\ \Omega$, $L = 0.8$ H, $C = 1.25$ μF, and $R_2 = 160\ \Omega$.

    (a) Find the transfer function $H(s) = V_{out}(s)/V_{in}(s)$.

    (b) Find the *exact* value of the maximum voltage gain and the corresponding peak frequency (in Hz).;

    (c) Find the *exact* 3 dB bandwidth (in Hz).

    (d) Find the *exact* values of the upper and lower half-power frequencies.

    (e) Find Q of the circuit.

    (f) Plot the magnitude response using MATLAB or the equivalent.

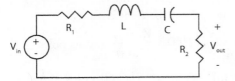

Figure P16.17

ANSWERS: (b) 0.8 at 159.15 Hz; (c) 39.79 Hz; (d) 180.29, 140.51 Hz; (e) 4

18. Repeat Problem 17 for $R_1 = 80\ \Omega$, $L = 0.25$ H, $C = 0.5\ \mu$F, and $R_2 = 320\ \Omega$.

CHECK: $B_\omega = 3200$ rad/sec.

19. Reconsider the series circuit shown in Figure P16.17. Let $R_1 = 4\ \Omega$, $L = 0.1$ H, $C = 0.1\ \mu$F, and $R_2 = 16\ \Omega$.

    (a) Find the transfer function $H(s) = V_{out}(s)/V_{in}(s)$.

    (b) Find the *exact* value of the maximum voltage gain and the corresponding

peak frequency (in Hz).

    (c) Find the *exact* 3 dB bandwidth (in Hz).

    (d) Find Q of the circuit.

    (e) Find approximate values of the upper and lower half-power frequencies.

    (f) Plot the magnitude response using MATLAB or the equivalent.

CHECK: (e) 9.9 and 10.1 krad/sec.

20. For the circuit in Figure P16.20 $R_f = 1$ kΩ, $L = 10$ mH, $C = 0.1$ mF, $R_{in} = 100\ \Omega$, and the op amp is ideal.

    (a) Construct the transfer function $H(s) = V_{out}(s)/V_{in}(s) = -Z_f(s)/Z_{in}(s)$ in terms of the circuit elements R_{in}, R_f, L, and C, and put it in the general form

$$H(s) = K\frac{s}{s^2 + 2\sigma_p s + \omega_p^2}.$$

    (b) Find the values of K, ω_m, the circuit Q, and $H_m = |H(j\omega_m)|$.

    (c) Compute the value of the half-power bandwidth B_ω and the half-power frequencies ω_1 and ω_2. (Approximate values are acceptable.)

    (d) Sketch the pole-zero diagram that represents the circuit, and note the *exact* locations of all the poles and zeros.

    (e) If $v_{in}(t) = 100\sin(10^3 t)$ mV, determine the magnitude of $v_{out}(t)$ in steady state.

    (f) Use SPICE or the equivalent to generate the magnitude response plot for $0 \leq f \leq 500$ Hz.

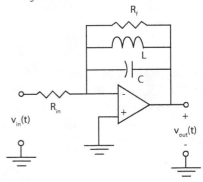

Figure P16.20

CHECK: $Q = 100$

21. For the circuit shown in Figure P16.21:
    (a) Find the transfer function $H(s) = \dfrac{I_L(s)}{V_s(s)}$.
    (b) Find ω_m, H_m, B_ω, and Q.
    (c) Now suppose $f_m = 200$ Hz and $B_f = 20$ Hz are desired. One has available a 1 H inductor, a 10 μF capacitor, and arbitrary resistors.
        (i) Determine the necessary value of β.
        (ii) Determine the value of R_s.

ANSWER: (b) $\omega_m = \sqrt{\dfrac{\beta+1}{LC}}$, $H_m = \dfrac{1}{R_s}$, $B_\omega = \dfrac{R_s}{L}$

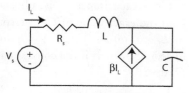

Figure P16.21

22. For the circuit shown in Figure P16.22:
    (a) Find the transfer function $H(s) = V_R/I_{in}$.
    (b) Find ω_m, H_m, B_ω, and Q.
    (c) Now suppose $f_m = 200$ Hz and $B_f = 20$ Hz are desired. One has available a 1 H inductor, a 10 μF capacitor, and arbitrary resistors.
        (i) Determine the necessary value of μ.
        (ii) Determine the value of R.

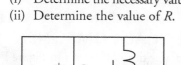

Figure P16.22

CHECK: $R = 795.77\ \Omega$

23. Design the active band-pass circuit in Figure P16.23 to have $f_m = 1000$ Hz and $B_f = 12.5$ Hz. The final value of C_1 for the actual circuit should be 0.1 μF. Hint: Follow the procedure described in Example 16.5. Use SPICE or its equivalent to generate the magnitude response plot for $900 \le f \le 1200$ Hz.

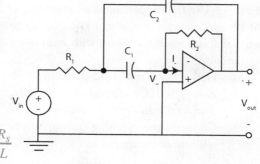

Figure P16.23

24. For the circuit transfer function of Problem 23,

$$H_{cir}(s) = \frac{V_{out}(s)}{V_{in}(s)} = \frac{-\dfrac{s}{R_1 C_2}}{s^2 + \left(\dfrac{1}{R_2 C_1} + \dfrac{1}{R_2 C_2}\right)s + \dfrac{1}{R_1 R_2 C_1 C_2}}$$

show that

(i) $H_m = \left(\dfrac{C_1}{C_2} + 1\right)Q^2$ and (ii) $\dfrac{R_1}{R_2} = \dfrac{Q^2}{H_m^2} \times \dfrac{C_1}{C_2}$.

Hence, for high-Q transfer functions, this circuit is undesirable.

25. Reconsider the active band-pass circuit of Figure P16.23. The filter is to pick out the midrange of a typical audio speaker, in which case $f_1 = 500$ Hz, $f_2 = 3200$ Hz, and $f_m = 1265$ Hz. The gain H_m is to be 10. Find B_f and the circuit parameters. Hint: Follow the procedure described in Example 16.5. Use SPICE or its equivalent to generate the magnitude response plot for $1 \le f \le 5000$ Hz.
CHECK: $Q < 1$ and $R_1 > 2$, slightly

26. Consider the band-pass circuit of Figure P16.26, which contains two op amps rather than the one in Problem 25.
    (a) Show that the transfer function of the circuit is

$$H(s) = \frac{\dfrac{2G_3}{C}s}{s^2 + \dfrac{G_3}{C}s + \left(\dfrac{G_2}{C}\right)^2}$$

Hints: (i) Use the properties of ideal op amps. (ii) Use voltage division across G_1. (iii) Write two nodal equations and solve with unknowns V_{out} and V_1.

(b) Design an active band-pass circuit to have $f_m = 1000$ Hz and $B_f = 12.5$ Hz. The final value of C should be 0.1 μF.

(c) Compare the resulting resistor values with those computed in Problem 25.

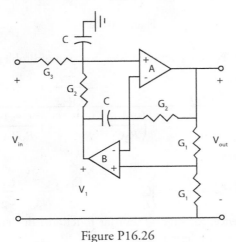

Figure P16.26

27. Consider again the band-pass circuit of Figure P16.26. In Problem 16.26 it was shown that

$$H(s) = \frac{\dfrac{2G_3}{C}s}{s^2 + \dfrac{G_3}{C}s + \left(\dfrac{G_2}{C}\right)^2}$$

(a) Design an active band-pass circuit to have peak frequency $\omega_m = 1000$ rad/sec and bandwidth $B_\omega = 100$ rad/sec. The final value of C should be 1 μF.

(b) Compare the resulting resistor values with those computed in Example 16.5.

28. Consider the circuit in Figure P16.28.

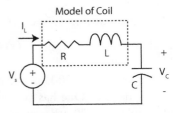

Figure P16.28

(a) Find $H_1(s) = I_L/V_s$ and $H_2(s) = V_C/V_s$.

(b) For $H_1(s)$, find the exact values of ω_m, $H_1(\omega_m)$, B_ω, and Q.

(c) For $H_2(s)$, consider the case where the ac voltage source has fixed V_s and ω, but the capacitance C is adjustable. Find the exact value of C (in terms of R, L, and ω) such that $|V_C|$ is maximized. Show that if the coil has a high Q, then

$$|V_C|_{max} \cong Q_{coil}|V_s|$$

This result provides a practical way of measuring Q_{coil}. It also provides a practical means for generating a very high short-duration voltage from a relatively small voltage; this has application in ignition circuits.

BAND-PASS CIRCUITS WITH PRACTICAL COMPONENTS

29. Consider the RLC circuit in Figure P16.29, in which $R_p = 50$ Ω, $L = 1/9$ H, $R_s = 0.08$ Ω, and $C = 1/9$ F.

(a) Find the coil Q at $\omega_0 = \dfrac{1}{\sqrt{LC}}$.

(b) Find an approximate parallel representation of the coil near ω_0.

(c) Find the circuit Q.

(d) If the circuit is high Q, compute approximate values for H_m, B_ω, ω_m, ω_1, and ω_2.

(e) Suppose $i_{in}(t) = 2\cos(\omega_m t)$ A. Find, approximately, $v_{out}(t)$ (in volts) in steady state.

ANSWERS: (a) 12.5; (c) 10; (d) $v_{out}(t) = 20$ $\cos(\omega_m t)$ V

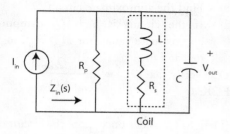

Coil

Figure P16.29

30. Repeat Problem 29 for R_p = 3750 Ω, L = 0.1 mH, R_s = 4 Ω, C = 10 nF, and $i_{in}(t)$ = 2 $\cos(\omega_m t)$ mA.
CHECK: Q = 15 and $v_{out}(t)$ = 3 $\cos(\omega_m t)$ V

31. Consider the RLC circuit in Figure P16.31, in which R_s = 20 kΩ, L = 0.5 mH, R_{series} = 20 Ω, and C = 0.5 nF.
   (a) Find the coil Q at $\omega_0 = \frac{1}{\sqrt{LC}}$.
   (b) Find an approximate parallel representation of the coil near ω_0.
   (c) Find the approximate circuit Q.
   (d) If the circuit is high Q, compute approximate values for H_m, B_ω, ω_m, ω_1, and ω_2.

ANSWERS: (a) 12.5; (c) 10; (d) $v_{out}(t)$ = 20 $\cos(\omega_m t)$ V

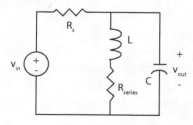

Figure P16.31

32. Repeat Problem 31 for R_s = 40 kΩ, L = 4 mH, R_{series} = 100 Ω, and C = 0.25 nF.
CHECK: 1 > H_m > 0.5

33. Consider the circuit of Figure P16.33. Suppose R_c = 2 kΩ, L = 1 mH, and C = 10 μF.

(a) Find the capacitor Q at $\omega_0 = \frac{1}{\sqrt{LC}}$.
(b) Find an approximate series representation of the capacitor near ω_0.
(c) Find the approximate value of Q.
(d) If the circuit is high Q, compute approximate values for H_m, B_ω, ω_m, ω_1, and ω_2.

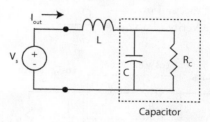

Capacitor

Figure P16.33

CHECK: ω_1 = 9975 rad/sec

34. Consider the circuit of Figure P16.34. Suppose R_s = 0.6 Ω, L = 1 mH, R_C = 2.5 kΩ, and C = 0.1 μF.
   (a) Find the capacitor Q at $\omega_0 = \frac{1}{\sqrt{LC}}$.
   (b) Find an approximate series representation of the capacitor near ω_0.
   (c) Find the approximate circuit Q.
   (d) If the circuit is high Q, compute approximate values for H_m, B_ω, ω_m, ω_1, and ω_2.

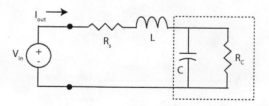

Figure P16.34

CHECK: Q_{cir} = 100

35. Repeat Problem 34 for R_s = 0.32 Ω, L = 16 μH, R_C = 800 Ω, and C = 0.25 μF.
CHECK: Q_{cir} = 20

36. Consider the circuit of Figure P16.36. Suppose L = 0.5 mH, C = 1.25 μF, Q_L = 20, and Q_C = 80.

(a) Find R_s and R_p.

(b) Convert the capacitor to a series model at ω_0. Denote the resulting series resistance by R_{series}.

(c) Find the approximate circuit Q.

(d) If the circuit is high Q, compute approximate values for H_m, B_ω, ω_m, ω_1, and ω_2.

(e) Verify that $Q_{cir} \cong \dfrac{Q_{coil}Q_{cap}}{Q_{coil} + Q_{cap}}$.

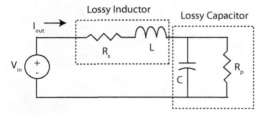

Figure P16.36

CHECK: $Q_{cir} = 16$

37. Again consider the circuit of Figure P16.36. Verify that for high Q_{coil} and high Q_{cap} at ω_0,

$$Q_{cir} \cong \frac{Q_{coil}Q_{cap}}{Q_{coil} + Q_{cap}}$$

38. Now consider the circuit of Figure P16.38. Verify that for high Q_{coil} and high Q_{cap} at ω_0,

$$Q_{cir} \cong \frac{Q_{coil}Q_{cap}}{Q_{coil} + Q_{cap}}$$

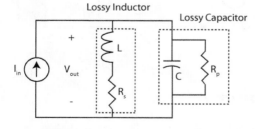

Figure P16.38

39. Consider again the circuit of Figure P16.38. Suppose $L = 0.5$ mH, $C = 1.25$ μF, $Q_L = 30$, and $Q_C = 60$ at ω_0.

(a) Find R_s and R_p.

(b) Convert the inductor to a parallel model at ω_0. Denote the resulting series resistance by R_{para}.

(c) Find the approximate circuit Q.

(d) If the circuit is high Q, compute approximate values for H_m, B_ω, ω_m, ω_1, and ω_2.

(e) Verify that $Q_{cir} \cong \dfrac{Q_{coil}Q_{cap}}{Q_{coil} + Q_{cap}}$.

CHECK: $Q_{cir} = 20$

40. (Design) Consider once more the circuit of Figure P16.38. Recall that for high-Q components,

$$Q_{cir} \cong \frac{Q_{coil}Q_{cap}}{Q_{coil} + Q_{cap}}$$

Suppose you have been asked by your supervisor to use this circuit in a band-pass design with $f_m = 1$ MHz and a bandwidth of 20 kHz and a lossy capacitor with $C = 200$ pF and having $Q_{cap} = 100$.

(a) Determine the circuit Q, i.e., Q_{cir}.

(b) Find the necessary Q_{coil} to achieve the desired Q_{cir}.

(c) Find the inductance L of the lossy coil and then find R_s as shown in Figure P16.38.

*(d) Suppose we now desire to double the bandwidth by adding a resistor, R_{source}, in parallel with the lossy coil (i.e., in parallel with the current source). What is the proper value of R_{source}?

CHECK: (c) $R_s < 8\ \Omega$

41. Repeat Problem40 under the conditions $f_m = 1.6$ MHz, a bandwidth of 30 kHz, a lossy capacitor with $C = 100$ pF, and $Q_{cap} = 200$.

42. (Design) Consider the circuit of Figure P16.42. Recall that for high-Q components,

$$Q_{cir} \cong \frac{Q_{coil}Q_{cap}}{Q_{coil} + Q_{cap}}$$

Suppose you have been asked by your supervisor to use this circuit in a band-pass design with $f_m = 1$ MHz, a bandwidth of 20 kHz, and a

lossy inductor $L = 100$ μH having $Q_{coil} = 100$.
  (a) Determine the circuit Q, i.e., Q_{cir}.
  (b) Find the necessary Q_{cap} to achieve the desired Q_{cir}.
  (c) Find the capacitance C of the lossy capacitor and then find R_p as given in Figure P16.42.
  (d) Suppose we now desire to double the bandwidth by adding a resistor, R_{source}, in series with the lossy coil (i.e., in series with the voltage source). What is the proper value of R_{source}?

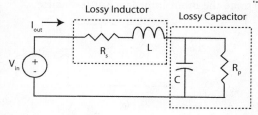

Figure P16.42

CHECK: (c) 130 kΩ > R_p > 115 kΩ

43. (a) Repeat Problem 42 (except for part (d)) under the conditions $f_m = 1.6$ MHz, a bandwidth of 25 kHz, and a lossy inductor $L = 100$ μH having $Q_{coil} = 100$.
   (b) Now suppose you were only able to purchase a capacitor having $Q_{cap} = 200$; you must achieve the required bandwidth by adding a resistor in parallel with the lossy capacitor. What is the value of this new parallel resistance?

CHECK: $Q_{cir} = 64$

44 Consider the amplifier circuit shown in Figure P16.44.
   (a) Find Q of the coil at $\omega = 10^5$ rad/sec.
   (b) Represent the coil by a parallel RL circuit that is valid for frequencies near 10^5 rad/sec using column 3 of Table 16.2.
   (c) Find *approximate* values of the 3 dB bandwidth,

and the corresponding frequency (in Hz).
  (d) Check your results by doing a SPICE (or equivalent) simulation of your circuit.

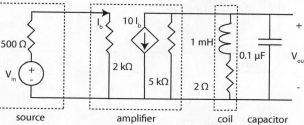

Figure P16.44

ANSWERS: (a) 50; (b) 1 mH, 50 kΩ; (c) 636.6 Hz, 10, 15.91 kHz

45. Repeat Problem 44 for the circuit shown in Figure P16.45, where $R_1 = 500$ Ω, $R_2 = 2$ kΩ, $R_3 = 5$ kΩ, $R_4 = 2$ Ω, $L = 1$ mH, $C = 1$ μH, and $a = 25$.

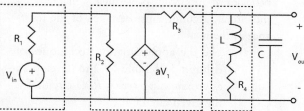

Figure P16.45

CHECK: 650 Hz > B_f > 630 Hz

46. Consider the circuit in Figure P16.46.
   (a) Find the input admittance of the circuit, $Y_{in}(s)$, and compute its poles and zeros.
   (b) Determine ω_m and the approximate 3 dB bandwidth and half-power frequencies of the circuit.
   (c) Determine Q_{coil} and Q_{cap} at ω_0. Then use column 3 of Table 16.2 to find the approximate bandwidth and half-power frequencies. Compare the result with that obtained in part (b).

(d)   Check your results by doing a SPICE (or equivalent) simulation of your circuit.

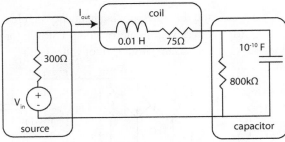

Figure P16.46

47. Consider the circuit in Figure P16.47, which contains a non-ideal capacitor, a non-ideal inductor, and a meter to measure the current response, $i_{out}(t)$. The 1 Ω resistor representing the meter is a precision resistor. The voltage across the resistor, $v_{out}(t)$, equals the current through the resistor. Thus, a practical way of measuring current is by measuring the voltage across a small resistance in the circuit. If the resistance of the meter is sufficiently small, it should have little effect on the behavior of the circuit. Nevertheless, in analyzing the circuit, account must be taken of the resistance of the meter.

(a)   Use the approximation techniques of column 3 of Table 16.2 to develop an approximating series RLC circuit for the given circuit.

(b)   Compute approximate values for B_ω, Q, ω_r, ω_m, ω_1, and ω_2.

(c)   At resonance, find the approximate steady-state current response, $i_{out}(t)$, when the input voltage is 10 $\sin(\omega_r t)u(t)$.

(d)   Check your results by doing a SPICE (or equivalent) simulation of your circuit.

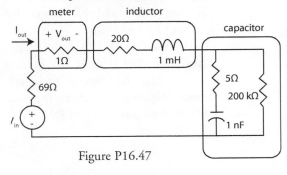

Figure P16.47

ANSWERS: (b) $B_\omega \cong 10{,}000$ rad/sec, $Q \cong 10$, $\omega_r \cong \omega_m \cong 10^6$ rad/sec, $\omega_1 \cong 950{,}000$ rad/sec, $\omega_2 \cong 1{,}050{,}000$ rad/sec; (c) $v_{out}(t) \cong 0.1 \sin(\omega_r t)$ V

RESONANT CIRCUITS WITH APPLICATIONS

48. For each two-terminal circuit in Figure P16.48, the resonant frequency $f_r = 10$ kHz. If $C = 0.1$ μF and $R = 1$ kΩ, find the corresponding values of L. For each circuit, determine the input impedance at resonance.

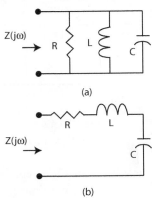

Figure P16.48

ANSWERS: (a) 2.5 mH; (b) 2.5 mH

49. For each circuit in Figure P16.49,

(a)   Find the resonant frequency, ω_r, in terms of R_1, R_2, L, and C. Verify that $\omega_r = \omega_0$. (This always follows when L and C are either in parallel or in series.)

(b)   Now find the input impedance at ω_r in terms of R_1, R_2, L, and C. Verify that the input impedance is independent of L and C at ω_r. (Again, this always follows when L and C are either in parallel or in series.)

(c)   In the case of Figure P16.49a, find $v_{out}(t)$ in steady state in terms of R_1, R_2, and V_m if $v_{in}(t) = V_m \sin(\omega_r t)$. In the case of Figure P16.49b, find $i_{out}(t)$ in steady state if $i_{in}(t) = I_m \sin(\omega_r t)$.

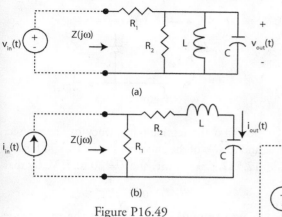

(a)

(b)

Figure P16.49

50. Figure P16.50 displays an amplifier model containing a VCCS with g_m = 4 mS (milli-siemens) and R_L = 40 kΩ. Suppose the applied sinusoidal voltage, $V_{in}(j\omega)$, has a magnitude of 0.2 V at 20 MHz.

(a) With the load connected directly as shown (without L), find the magnitude of the output voltage.

(b) If an inductance L is connected across the load to tune out the effect of the capacitance, find the value of L and the resulting $|V_{out}|$ that will show that the amplifier gain at 10 MHz is greatly increased.

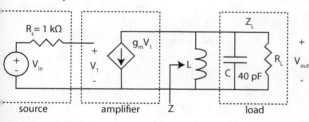

Figure P16.50

51. Figure P16.51 displays an amplifier model containing a VCCS with g_m = 4 mS (milli-siemens) and R_L = 20 kΩ. Observe that the amplifier has an input capacitance of 0.2 nF. The inductor L is used to tune out this effect.

(a) Without L find the magnitude of the voltage gain at 10 MHz.

(b) If an inductance L is connected as

shown, find the value of L to maximize the voltage gain $|V_{out}(j\omega)$ / $V_{in}(j\omega)|$ at f = 10 MHz. Verify that the amplifier gain at 10 MHz is greatly increased.

(c) Plot the frequency responses of the circuit over 1 MHz ≤ f ≤ 10 MHz with L and without L.

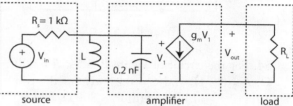

Figure P16.51

52. Find the resonant frequency ω_r in rad/sec, and $Z(j\omega_r)$ of the circuit in Figure P16.52 for R = 2.8 Ω, C = 0.2 mF, and L = 20 mH. Verify that

$$\omega_r \neq \omega_0 = \frac{1}{\sqrt{LC}}.$$

Now compute $Z_{in}(s)$ and plot the magnitude and phase responses from $0.75\omega_r$ to $1.25\omega_r$. Verify that at ω_r, the phase angle of $Z_{in}(j\omega_r)$ is zero.

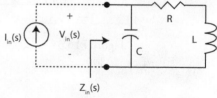

Figure P16.52

ANSWERS: 480 rad/sec, 35.71 Ω

53. Consider the circuit of Figure P16.53. The CL part of the circuit can be thought of as a matching network when the values are properly chosen. In case 1 we will see that maximum power transfer is achieved, whereas in case 2 it is not.

(a) Find the resonant frequency, ω_r and $Z(j\omega_r)$ for the cases
(i) R_L = 80 Ω
(ii) R_L = 60 Ω

(b) Now suppose $V_m = 250\ V_{rms}$ with $\omega = \omega_r$. Then compute the average power delivered to R_L for each of the cases in part (a).

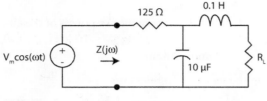

Figure P16.53

54. For the circuit of Figure P16.54, $R = 800$ Ω, $L = 0.2$ H, and $C = 0.25$ μF. For the two cases (i) $R_L = 1$ kΩ and (ii) $R_L = 800$ Ω, find ω_r and at resonance, find $v_C(t)$, the voltage across the capacitor in the steady state, due to the input $i_{in}(t) = 10\ \cos(\omega_r t)$ mA. Finally, find the average power delivered to R_L in each case.

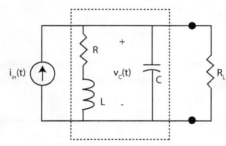

Figure P16.54

ANSWER: *Case 1:* 2000 rad/sec, 5 cos(2000t) V, 12.5 mW

55. This problem develops formulas similar to equations 16.41 and 16.42 for the *RLC* circuit of Figure P16.55.
(a) Prove that, for the circuit shown in Figure P16.55,

$$\omega_r = \sqrt{\frac{1}{LC} - \frac{1}{R^2 C^2}} = \omega_0 \sqrt{1 - \frac{L}{CR^2}}$$

and

$$Z(j\omega_r) = \frac{L}{CR}.$$

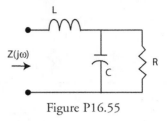

Figure P16.55

56. Find the resonant frequency ω_r, in rad/sec, and $Z(j\omega_r)$ of the circuit in Figure P16.56 for $R = 12.5$ Ω, $C = 0.2$ mF, and $L = 20$ mH. Verify that

$$\omega_r \neq \omega_0 = \frac{1}{\sqrt{LC}}.$$

Now compute $Z_{in}(s)$ and plot the magnitude and phase responses from $0.5\omega_r$ to $2\omega_r$. Verify that at ω_r, the phase angle of $Z_{in}(j\omega_r)$ is zero.

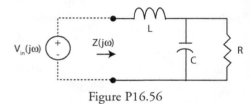

Figure P16.56

57. Consider the circuit of Figure P16.57 for $R_s = 80$ Ω, $R_L = 125$ Ω, $C = 2$ μF, and $L = 20$ mH. The *LC* part of the circuit can be thought of as a matching network when the values are properly chosen.
(a) Find the resonant frequency ω_r, $Z_1(j\omega_r)$, and $Z(j\omega_r)$.
(b) Now suppose $V_m = 250\ V_{rms}$ with $\omega = \omega_r$. Compute the average power delivered to R_L. Is it possible to deliver more power to R_L by varying the values of L and C?

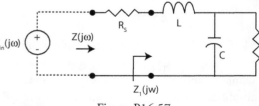

Figure P16.57

58. A two-terminal network has input impedance

$$Z_{in}(s) = \frac{s+a}{s^2 + 4s + 8}$$

(a) If $a = 1$, find ω_r and $Z_{in}(j\omega_r)$.
(b) Find the range of $a \geq 0$ for which the impedance has a real resonant frequency.

ANSWERS: (a) 2 rad/sec and 0.25 Ω; (b) $0 \leq a < 2$

59. For the circuit shown in Figure P16.59, where

$$Y(s) = \frac{20s}{s^2 + 100},$$

find the value of C that makes the circuit resonant at $\omega = 30$ rad/sec and then find the value of $Y_{in}(j30)$.

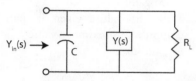

Figure P16.59

60. This problem uses an LC coupling network to maximize power to a load. The LC coupling network always has a series connection for L; the "parallel" capacitor is always closest to the larger resistance—in this case the source resistance. Problem 62 will consider the general case. Suppose the voltage source in Figure P16.60 has value $v_{in}(t) = 100\sqrt{2}\cos(2\pi \times 10^6 t)$ V. Compute the values of L and C such that the average power delivered to the load resistance R_L is maximized. What is P_{max}?

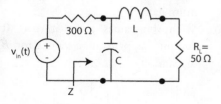

Figure P16.60
ANSWERS: 17.79 µH, 1,186 pF, 8.3333 W

61. This problem uses an LC coupling network to maximize power to a load. The LC coupling network always has a series connection for L; the "parallel" capacitor is always closest to the larger resistance—in this case the load resistance. Problem 62 will consider the general case. Suppose the voltage source in Figure P16.61 has value $v_{in}(t) = 100\sqrt{2}\cos(2\pi \times 10^6 t)$ V. Compute the values of L and C such that the average power delivered to the load resistance R_L is maximized. What is P_{max}?

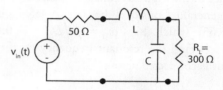

Figure P16.61
ANSWERS: 17.9 µH, 1,186 pF, 50 W

62. Equations 16.41 through 16.44, or those derived in Problem 55, can be combined to produce a set of design formulas for a lossless network that matches two unequal resistances at a single frequency. The matching network consists of only one capacitance C and one inductance L, as shown in Figure P16.62. At a specified frequency ω, it is desired to have matching at both ends, i.e.,

$$Z_1(j\omega) = R_1 + j0 \text{ and } Z_2(j\omega) = R_2 + j0$$

Let R_{small} denote the smaller of (R_1, R_2) and R_{large} the larger. Prove that
(a) C should be connected in parallel with R_{large}, and L should be connected between the two top terminals and thus in series with R_{small}.
(b) The element values are given by

$$L = \frac{1}{\omega}\sqrt{R_{small}\left(R_{large} - R_{small}\right)}$$

and

$$C = \frac{1}{\omega R_{large}}\sqrt{\frac{R_{large} - R_{small}}{R_{small}}}.$$

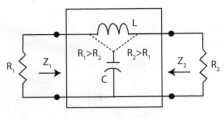

Figure P16.62

63. The purpose of this problem is to show that the resonant frequency ω_r depends on the choice of the input terminals and that $\omega_r \neq \omega_0$ in general. Consider the circuit of Figure P16.63, which has $\omega_0 = 1/\sqrt{LC} = 0.25$ rad/sec. Find the resonant frequency ω_r if the input is connected across
  (a) A and B
  (b) B and C
  (c) A and C

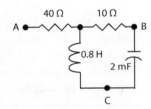

Figure P16.63

ANSWERS: 25, $12.5\sqrt{3} = 21.65$, 28.86, all in rad/sec

64. The circuit of Figure P16.64 has $\omega_0 = 1/\sqrt{LC} = 25$ rad/sec. The purpose of this problem is to show that the resonant frequency ω_r depends on the choice of the input terminals and that $\omega_r \neq \omega_0$ in general. Find the resonant frequency ω_r and the equivalent impedance seen at the terminals if the source is connected across
  (a) E and F
  (b) D and E

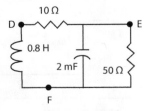

Figure P16.64

ANSWERS: $12.5\sqrt{3}/8 = 21.65$ rad/s & 22.2 Ω, $5\sqrt{21} = 22.91$ rad/s, 4.44 Ω.

BAND-PASS TRANSFER FUNCTIONS WITH NO ZEROS OR A SINGLE ZERO OFF THE ORIGIN

65. Consider the circuit of Figure P16.65.
  (a) With $V_s(s)$ as the input and $V_C(s)$ the output, compute the transfer function $H(s)$.
  (b) If $R = 5$ Ω, $L = 0.1$ H, and $C = 10$ μF, verify that the circuit is high Q_p.
  (c) Compute ω_m exactly and verify that $\omega_m \approx \omega_p$. Then compute the maximum gain.
  (d) With the values given in part (b), compute approximate values for B_ω, ω_1, and ω_2.

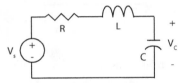

Figure P16.65

66. Consider the circuit of Figure P16.66.
  (a) With $I_{in}(s)$ as the input and $I_L(s)$ the output, compute the transfer function $H(s)$.
  (b) If $R = 4$ kΩ, $L = 0.1$ H, and $C = 10$ μF, verify that the circuit is high Q_p.
  (c) Compute ω_m exactly and verify that $\omega_m \approx \omega_p$. Then compute the maximum gain, H_m.
  (d) With the values given in part (b), compute approximate values for B_ω, ω_1, and ω_2.

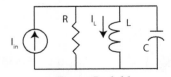

Figure P16.66

67. Consider the Sallen and Key active network of Figure P16.67, which realizes the transfer function

$$H(s) = \frac{V_{out}}{V_{in}} = \frac{\dfrac{1}{R_1 R_2 C_1 C_2}}{s^2 + \left(\dfrac{1}{R_1 C_1} + \dfrac{1}{R_2 C_1}\right)s + \dfrac{1}{R_1 R_2 C_1 C_2}}$$

Suppose $C_1 = 0.2\ \mu F$, $C_2 = 0.5\ nF$, $R_1 = R_2 = 50\ k\Omega$.

(a) Compute ω_p, Q_p, and K of equation 16.48.

(b) Compute ω_m and H_m.

(c) Find approximate values of B_ω, ω_1, and ω_2.

(d) Plot the magnitude response curve of the transfer function.

(e) Simulate the circuit using SPICE.

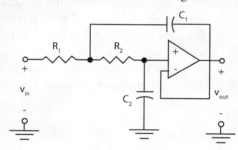

Figure P16.67

68. Consider the Sallen and Key circuit of Problem 67. Suppose $C_1 = 50\ nF$, $C_2 = 3.125$ nF, $R_1 = R_2 = 8\ k\Omega$.

(a) Compute ω_p, Q_p, and K of equation 16.48. Verify that the circuit is not high Q.

(b) Compute the exact values of ω_m and H_m.

(c) Find the exact values of B_ω, ω_1, and ω_2.

(d) Plot the magnitude response curve of the transfer function.

69. For the circuit shown in Figure P16.69:

(a) Show that the transfer function is

$$H(s) = \frac{\dfrac{1}{LC}}{s^2 + \left(\dfrac{1}{R_L C} + \dfrac{R_s}{L}\right)s + \dfrac{1}{LC}\left(1 + \dfrac{R_s}{R_L}\right)}$$

(b) Recall that $\omega_0 = 1/\sqrt{LC}$. By the use of equation 16.49, show that

$$\omega_m = \omega_0 \sqrt{1 - \frac{1}{2LC}\left(R_s^2 C^2 + \frac{L^2}{R_L^2}\right)} =$$

$$\frac{1}{\sqrt{LC}}\sqrt{1 - \frac{1}{2LC}\left(R_s^2 C^2 + \frac{L^2}{R_L^2}\right)}$$

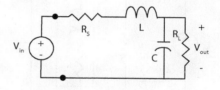

Figure P16.69

70. Consider a transfer function with a single zero off the origin, i.e.,

$$H(s) = K\frac{s + a}{s^2 + 2\sigma_p s + \omega_p^2}$$

(a) Derive equation 16.54, i.e.,

$$\omega_m = \sqrt{-a^2 + \sqrt{\left(\omega_p^2 + a^2\right)^2 - \left(2\sigma_p a\right)^2}},$$

Hint: Instead of maximizing $|H(j\omega)|$, try to maximize $|H(j\omega)|^2$, and consider ω^2 as the independent variable.

(b) Derive equation 16.55, i.e.,

$$\omega \text{ (for zero phase shift)} = \sqrt{\omega_p^2 - 2\sigma_p a}$$

Hint: Write $1/H(s)$ as $[(1/K + F(s)]$, set $\text{Im}\{F(j\omega)\} = 0$, and solve for ω.

71. Consider the circuit shown in Figure P16.71.

(a) Show that the transfer function is

$$H(s) = Z_{in}(s) = \frac{\frac{1}{C}\left(s + \frac{R_s}{L}\right)}{s^2 + \left(\frac{1}{R_pC} + \frac{R_s}{L}\right)s + \left(1 + \frac{R_s}{R_p}\right)\frac{1}{LC}}$$

(b) By the use of equation 16.54 or part (a) of Problem 65, show that

$$\omega_m = \sqrt{-\frac{R_s^2}{L^2} + \frac{1}{LC}\sqrt{\left(1 + \frac{2R_s}{R_p} + \frac{2R_s^2 C}{L}\right)}}$$

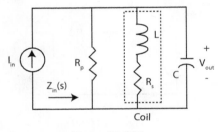

Coil

Figure P16.71

72. The *RLC* circuit of Figure P16.71 has transfer function

$$H(s) = Z_{in}(s) = \frac{\frac{1}{C}\left(s + \frac{R_s}{L}\right)}{s^2 + \left(\frac{1}{R_pC} + \frac{R_s}{L}\right)s + \left(1 + \frac{R_s}{R_p}\right)\frac{1}{LC}}$$

(a) If $R_p = 50\ \Omega$, $L = 1/9$ H, $R_s = 0.08\ \Omega$, and $C = 1/9$ F, verify that the circuit is high Q_p and that $a << \omega_p$.
(b) Compute ω_m exactly and verify that $\omega_m \cong \omega_p$. Then, compute the maximum gain.
(c) Compute approximate values for B_ω, ω_1, and ω_2.
(d) Suppose that at ω_m, $|I_{in}(j\omega_m)| = 1$ A. Determine, approximately, the value of $|V_{out}(j\omega_m)|$ (in volts).

ANSWERS: (b) $\omega_m = 9.0071$ rad / sec; (d) 10 V

73. The *RLC* circuit of Figure P16.71 has transfer function

$$H(s) = Z_{in}(s) = \frac{\frac{1}{C}\left(s + \frac{R_s}{L}\right)}{s^2 + \left(\frac{1}{R_pC} + \frac{R_s}{L}\right)s + \left(1 + \frac{R_s}{R_p}\right)\frac{1}{L}}$$

Let $R_p = 100$ kΩ, $L = 0.225$ H, $R_s = 1000\ \Omega$, and $C = 0.5$ μF.

(a) Compute ω_p and verify that the circuit is low Q_p.
(b) Compute ω_m exactly and verify that ω_m is quite different from ω_p.
(c) Plot the frequency response for $0 \le \omega \le 4000$ rad/sec. Compute the maximum gain and verify the result from the plot. Also compute the zero of the transfer function as $-z_1$.
(d) Define the coefficients of the numerator and denominator of your transfer function in MATLAB as n = [1 z1]/C and d = [1 ? ?]. Now use the commands below to compute the impulse and step responses of your transfer function.

```
sys=TF(n,d)
impulse(sys)
pause
step(sys)
```
ANSWERS: wp = 2.9963e+03, Q = 6.7115e-01, z1 = 4.4444e+03, wm = 1.0128e+03

74. Again consider the circuit of Figure P16.71, in which $L = 1$ mH. This inductor has $Q_{coil} = 40$ at 100 kHz. The magnitude response is to have a peak at $f_m \cong 100$ kHz.

(a) Specify the value of the capacitor C.
(b) Specify the value of R_p so that the bandwidth is approximately 10 kHz.
(c) What is Q_{cir} ($\cong Q_p$)?

ANSWERS: 2533 pF, 8378 Ω, 10

75. Consider the circuit in Figure P16.75. Let $L = 1$ H, $C = 1$ F, $R_L = 0.08$ Ω, $R_C = 0.02$ Ω, and $R_s = 40$ Ω.

  (a) Find the transfer function $H(s) = V_{out}(s)/V_{in}(s)$.

  (b) Find Q_p and approximate answers for ω_m, B_ω, H_{max}, ω_1, and ω_2.

  (c) Obtain a magnitude frequency response plot to graphically verify your answers.

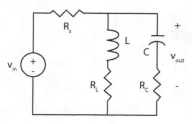

Figure P16.75

76. The analysis of the non-series-parallel circuit shown in Figure P16.76 requires writing node or loop equations. Because there are no series-parallel connections, one cannot apply the equivalents of Table 16.2.

  (a) Let $Y_1 = 1/R_1$, $Y_2 = 1/R_2$, $Y_3 = 1/R_3$, $Y_4 = 1/L_4s$, $Y_5 = C_5s$. Using nodal analysis, show that the transfer function is

$$H(s) = \frac{V_{out}}{I_{in}} =$$

$$\frac{Y_1(Y_3 + Y_4) + Y_2(Y_1 + Y_4)}{Y_2(Y_1 + Y_4)(Y_3 + Y_5) + Y_3Y_4(Y_1 + Y_5) + Y_1Y_5(Y_3 + Y_4)}$$

$$= \frac{s + 0.05}{s^2 + 0.12038s + 1.009}$$

  (b) Obtain the pole-zero plot of $H(s)$.

  (c) Find ω_p and verify that the circuit is high Q_p.

  (d) Find approximate values of ω_m, B_ω, H_m, ω_1, and ω_2.

  (e) Check your results by doing a SPICE (or equivalent) simulation of your circuit.

  (f) Frequency-scale the circuit by $K_f = 1000$ and $K_m = 100$. Show the new

element values and repeat parts (c) and (d) for the scaled circuit.

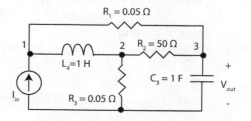

Figure P16.76

MISCELLANEOUS

77. Consider the idealized (tank) circuit of Figure P16.77. The moment the inductor current passes through zero with positive slope is taken as the reference point, $t = 0$. At this time instant the capacitance voltage is E volts.

  (a) Find $v_C(t)$ and $i_L(t)$ for $t \geq 0$ by the Laplace transform method.

  (b) Find the energy stored in C as a function of t.

  (c) Find the energy stored in L as a function of t.

  (d) Show that the total energy stored in the LC tank is constant and is equal to $0.5CE^2$.

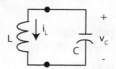

Figure P16.77

78. Consider the circuit in Figure P16.78, which contains a non-ideal inductor and a variable capacitor. Suppose $L = 100$ μH and $R_C = 5.4$ Ω.

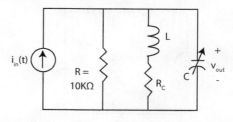

Figure P16.78

(a) Determine the range of the capacitance C, i.e., $C_0 \leq C \leq C_1$, such that the circuit can be tuned to resonance over the AM radio band (from 550 kHz to 1650 kHz).

(b) When the circuit is tuned to 550 kHz and 1650 kHz, determine the circuit Q's, the two bandwidths, and the lower and upper half-power frequencies for each bandwidth.

79. This problem illustrates the conceptual differences among the various frequencies encountered in this chapter. For practical high-Q circuits, the numerical values of these frequencies are all very close. To see the differences, we choose a low-Q circuit. Consider the circuit shown in Figure P16.79.

(a) Show that

$$H(s) = \frac{\dfrac{1}{R_sC}\left(s+\dfrac{R}{L}\right)}{s^2+\left(\dfrac{1}{R_sC}+\dfrac{R}{L}\right)s+\left(1+\dfrac{R}{R_s}\right)\dfrac{1}{LC}}.$$

(b) If $R_s = 5/3\ \Omega$, $L = 1$ H, $R = 0.8\ \Omega$, and $C = 1$ F, show that:
The LC tank frequency $\omega_0 = 1$ rad/sec.
The resonant frequency $\omega_r = 0.6$ rad/sec.
The peak frequency
$\omega_m = \sqrt{29/25} = 1.077$ rad/sec.
The pole frequency
$\omega_p = \sqrt{1.48} = 1.2165$ rad/sec.
The natural damped frequency $\omega_d = 0.995$ rad/sec.

Hints:
1. To find ω_r, use equation 16.41
2. To find ω_m, use equation 16.54.

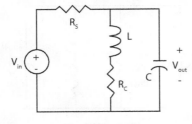

Figure P16.79

80. Consider the circuit of Figure P16.80, in which $R_p = 10\ \Omega$, $L = 1$ H, $R_s = 0.8\ \Omega$, and $C = 1$ F.

(a) Show that the transfer function $H(s)$ is that given in Problem 71.

(b) Find the *exact* values of ω_0, ω_p, ω_d, ω_r, and ω_m. Since this is a low-Q circuit, do not use the high-Q approximations.

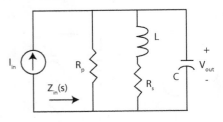

Figure P16.80

ANSWERS: (b) $\omega_0 = 1$, $\omega_p = 1.0392$, $\omega_d = 0.9367$, $\omega_r = 0.6$, and $\omega_m = 0.9602$, all in rad/sec.

81. Suppose the current source in the circuit of Problem 80 is $i_{in}(t) = 2\cos(t)$ A and that C is variable. Find the value of C such that $|\mathbf{V}_{out}|$ is maximum. Verify that ω_0 is not equal to the source frequency in this case.

82. (Experimental measurement of Q) Consider the circuit in Figure P16.82, in which $C = 0.1\ \mu\text{F}$, $L = 1$ mH, and $R_s = 5/\pi\ \Omega$.

(a) Find the transfer function $H(s) = V_{out}(s)/I_{in}(s)$.

(b) Compute the poles and zeros of $H(s)$. From these, approximately compute ω_m, B_ω, and Q_p. Is this a high-Q circuit?

(c) The impulse response for the circuit is of the form

$$h(t) = Ae^{-at}\cos(\omega t) + Be^{-at}\sin(\omega t).$$

Show that $|A| \gg |B|$ for the high-Q case and, therefore, $h(t) \cong Ae^{-at}\cos(\omega t)$.

(d) Plot $h(t) \cong Ae^{-at}\cos(\omega t)$ using MATLAB or its equivalent. Show that the peak will decrease to $1/e$ of the first peak in approximately Q/π cycles of oscillations. (This is how one experimentally determines Q.)

(e) Find approximate values of ω_m, H_m, B_ω, and Q by changing the series L-R to an approximate parallel L-R connection. Do the results agree with those obtained in part (a) by the transfer function approach?

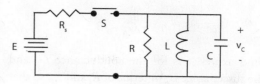

L
I_{in}
+
V_{out}
-
C
R_s
Coil

Figure P16.82

83. The switch S in Figure P16.83 has been closed for a long time and is opened at $t = 0$.

(a) Use the Laplace transform method of Chapter 14 to show that if $Q = \omega_o RC > 0.5$, then for $t > 0$, the capacitance voltage is

$$v_C(t) = V_m e^{-at} \sin(\omega_d t + \theta)$$

where

$$a = \frac{\omega_o}{2Q}$$

and

$$\omega_d = \omega_o \sqrt{1 - \frac{1}{4Q^2}}$$

(b) Show that the peak amplitude of the damped sinusoidal waveform of part (a) decreases to $1/e = 0.368$ of the highest peak approximately after Q/π cycles if Q is large (and, hence, $\omega_o \cong \omega_d$).

R_s
S
E
R
L
C
+
V_C
-

Figure P16.83

84. The opening section of this chapter discussed the generation of dial tones by resonant circuits. The circuit for generating the three tones in the high-frequency group is shown in Figure P16.84. Using the results of Problem 71, explain how the tone generation circuit works. Your explanation will include an oscillation of an undriven LC circuit that is coupled to the transistor circuit inside the box by transformers studied in the next chapter.

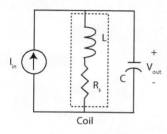

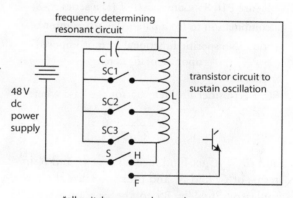

frequency determining
resonant circuit
C
SC1
SC2
SC3
S
H
F
L
48 V
dc
power
supply
transistor circuit to
sustain oscillation
*all switches open when no button
is pressed
*pressing a button in column k
closes the switch SCK
*when any button is pressed halfway,
S is closed to contact H; when the
button is fully pressed, S moves away
from H and makes contact with F

Figure P16.84

85. For the circuit of Figure P16.85a, we have the impedance

$$Z(j\omega) = R_s + jX_s$$

and for the circuit of Figure 16.85b, we have the admittance

$$Y(j\omega) = G_p + jB_p$$

Find expressions for the conductance G_p and the susceptance B_p in terms of R_s and X_s, so that the two circuits have the same impedance at a single frequency ω. Then find expressions for R_s and X_s in terms of G_p and B_p under the same condition.

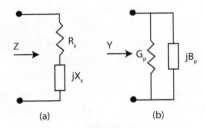

(a)   (b)

Figure P16.85 Conversion of (a) a series R_s-X_s combination to (b) a parallel G_p-B_p combination. No specific functional form is imposed upon X_s or B_p here.

86. Reconsider Problem 85 for the case where

$$jX_s = \frac{1}{jC_s\omega}$$

and $jB_p = jC_p\omega$. Find formulas for C_s and R_s in terms of C_p and R_p and thus specialize the formulas of Problem 85 to the case of a transforming a parallel RC to a series RC at a single frequency ω. Now do the converse.

C H A P T E R *17*

Magnetically Coupled Circuits and Transformers

WHAT IS INSIDE THE AC ADAPTOR?

Most electronic equipment operates with dc power sources. For portable equipment, such as a cordless phone and a cordless electric drill, batteries supply the dc power. Using non-rechargeable batteries becomes expensive. Furthermore, replacing batteries in special equipment is a task not easily handled by ordinary consumers. These two factors have prompted manufacturers to install rechargeable lithium-ion batteries in portable equipment. By connecting several batteries in series, the available dc voltage may range from 1.5 to 12 V. Whenever the battery runs low, it must be recharged.

Recharging a battery requires a low dc voltage source (1.5–12V). An adaptor houses a device called a *transformer* that changes the 110 V ac voltage at the household outlet to a much lower ac voltage. The lower ac voltage is then rectified to become a dc voltage that charges the battery. Some adaptors contain the transformer only, while others may also contain the rectifier circuit.

Typical specifications appearing on the casing of an adaptor may be as follows:

model: AC9131
input:
ac   120 V, 60Hz, 6 W
output:
ac   3.3 V. 500 mA

model: KX-A10
input:
ac   120 V, 60Hz, 5 W
output:
dc   12 V, 100 mA

The concepts and methods developed in this chapter will allow us to understand how a transformer works to change the ac voltage level and also to perform some other important functions in electronic equipment.

CHAPTER OBJECTIVES

1. Understand how the mutual inductance M, between two inductances L_1 and L_2, accounts for an induced voltage in each inductor due to the change of current in the other inductor.
2. Develop a systematic method for writing time domain and frequency domain equations for circuits containing mutual inductances.
3. Understand why the mutual inductance is less than or equal to the geometric mean of the individual self-inductances using an energy perspective.
4. Expand the repertoire of basic circuit elements to include ideal transformers, and learn how to analyze circuits containing ideal transformers.
5. Learn how to model a pair of coupled inductors by an ideal transformer and at most two self-inductances.
6. Learn how a practical transformer can be modeled by an ideal transformer and some additional RL elements.
7. Investigate some important applications of transformers and coupled inductors in power engineering and communication engineering.

SECTION HEADINGS

1. Introduction
2. Mutual Inductance and the Dot Convention
3. Differential Equation, Laplace Transform, and Phasor Models of Coupled Inductors
4. Analysis of Coupled Circuits with Open-Circuited Secondary
5. Analysis of Coupled Circuits with Terminated Secondary
6. Coefficient of Coupling and Energy Calculations
7. Ideal Transformers
8. Models for Practical Transformers
9. Coupled Inductors Modeled with an Ideal Transformer
10. Summary
11. Terms and Concepts
12. Problems

1. INTRODUCTION

You may recall from your high school or grade school science class that if iron filings are sprinkled on a piece of paper and a magnet is moved around beneath the paper, the iron filings move in concert with the magnet because the magnetic field induces a force on the iron filings. Similar to the magnet and the iron filings, a changing current in one coil that is very close to another coil induces a voltage across the terminals of the other coil.

Figure 17.1a shows two unconnected coils of wire in close proximity. Figure 17.1b shows two unconnected wire coils wound around a single ferromagnetic core. In both cases, a voltage source excites coil 1 while coil 2 is left open-circuited. Experimental evidence shows that a change in the current i_1 generates a voltage v_2, called the *induced voltage*, across the open circuit; the induced voltage is proportional to the rate of change of i_1. Each pair of coils in Figures 17.1a and 17.1b is said to be *magnetically coupled*.

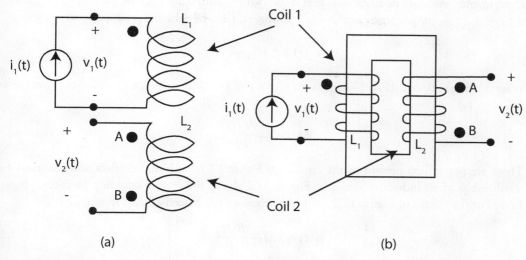

(a)   (b)

FIGURE 17.1 Induced voltage in coupled coils. (a) Two coils in close proximity. (b) Two coils wound on the same ferromagnetic core.

How does one quantitatively account for magnetic coupling? The strategy is to introduce a new circuit quantity called **mutual inductance** for coupled coils; specifically, similar to the *v-i* relationship of a single coil, the induced voltage satisfies the equation

$$v_2 = \pm M_{21} \frac{di_1}{dt} \tag{17.1}$$

where $M_{21} > 0$ is the proportionality constant called the **mutual inductance** from coil 1 to coil 2, and the sign, here ±, depends on the relative winding directions of the coils. Dot markings indicate the relative winding directions. With reference to Figure 17.1, a dot is placed on coil 1 for reference; if the dot on coil 2 is in position A, the sign on equation 17.1 is +, and if the dot is in position B, the sign is −. A description of the general dot convention is presented in the next section.

The situation illustrated in Figure 17.1b is motivated by an extremely important magnetically coupled device called a **transformer**, which is used to transform voltages and currents from one level to another. In electric power systems, transformers are used to step up ac voltages from 10 kV at a generating station to over 240 kV for the purpose of transmitting electric power efficiently over long distances. At a customer's site, such as a home, transformers step these high voltage levels down to 220 V or 110 V for safe, everyday uses. In addition, transformers have numerous uses in electronic systems, including (1) stepping ac voltages up or down, (2) isolating parts of a circuit from dc voltages, and (3) providing impedance level changes to achieve maximum power transfer between devices, and tuning circuits to achieve a resonant behavior at a particular frequency. After the basic analysis methods are set forth, some examples will illustrate these uses.

2. MUTUAL INDUCTANCE AND THE DOT CONVENTION

Experimental evidence demonstrates that if the two coils in Figure 17.1 are *stationary,* the induced voltage, $v_2(t)$, is proportional to the *rate of change* of $i_1(t)$, i.e., the induced voltage

$$v_2(t) = \pm M_{21} \frac{di_1(t)}{dt},$$

as set forth in equation 17.1. Note, however, that coil 1 with inductance L_1 continues to act as an inductor for which

$$v_1(t) = L_1 \frac{di_1(t)}{dt}.$$

There are two effects present in the circuits of Figure 17.1: an induced effect and the usual *v-i* relationship of an inductor. Similarly, Figure 17.2 shows the reverse coupling to that of Figure 17.1. For the circuit of Figure 17.2, with the reference dot placed at the top of coil 2,

$$v_1(t) = \pm M_{12} \frac{di_2(t)}{dt} \tag{17.2}$$

where + would be used if the dot on coil 1 were in position A and − if in position B. As in figure 17.1, the dots indicate the relative directions of the windings of the two coils. Also as before, the second coil continues to act as an inductor, for which

$$v_2(t) = L_2 \frac{di_2(t)}{dt}.$$

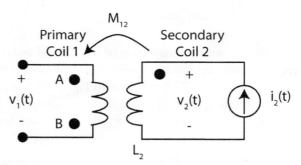

FIGURE 17.2. Coupling from coil 2 to coil 1 (winding directions not explicitly shown).

As will be verified in a later section, $M_{21} = M_{12} > 0$. Hence we designate the positive constant

$$M \triangleq M_{21} = M_{12} \tag{17.3}$$

as the mutual inductance (in henries) of the **coupled inductors.**

Figure 17.3 shows a composite of Figures 17.1a and 17.2 where currents are present in both coil 1 and coil 2. Four effects are now present in the circuit: two induced effects and the two usual self-inductance effects. Linearly superimposing the effects (superposition), we obtain the equations of the mutually coupled inductors:

$$v_1(t) = L_1 \frac{di_1}{dt} \pm M \frac{di_2}{dt} \tag{17.4a}$$

$$v_2(t) = \pm M \frac{di_1}{dt} + L_2 \frac{di_2}{dt} \tag{17.4b}$$

Two questions remain: (i) When is the sign positive and when is the sign negative? (ii) How is the value of M determined (experimentally)?

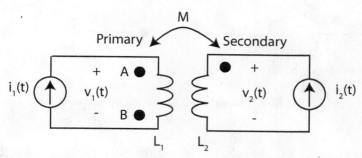

FIGURE 17.3. Coupled coils with current excitations present on primary (coil 1) and secondary (coil 2).

The following rule, identified with equation numbers, governs the choice of sign for the induced voltage.

RULE FOR THE INDUCED VOLTAGE DROP DUE TO MUTUAL INDUCTANCE

> The voltage drop across one coil, from the dotted terminal to the undotted terminal, equals M times the derivative of the current through the other coil, from the dotted terminal to the undotted terminal. (17.5a)

Or, equivalently,

> The voltage drop across one coil, from the undotted terminal to the dotted terminal, equals M times the derivative of the current through the other coil, from the undotted terminal to the dotted terminal. (17.5b)

With reference to Figure 17.3, if the dot is in position A, all signs are positive, whereas if the dot is in position B, the sign on M is negative. This rule gives the voltage drop due to the mutual inductance. To obtain the total voltage drop across an inductor that is coupled to another, one must add in the voltage drop induced by the self-inductance of the individual coil, which depends on whether the labeling is consistent with the passive sign convention.

EXAMPLE 17.1. For the configurations of Figure 17.4, determine the pair of equations that specify the relationship between the voltages and currents.

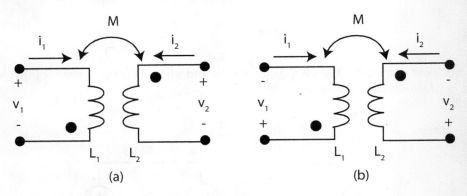

FIGURE 17.4. Two scenarios for setting up coil equations.

SOLUTION

First we consider Figure 17.4a. The voltage $v_1(t)$ and the current $i_1(t)$ as well as the voltage $v_2(t)$ and the current $i_2(t)$ satisfy the passive convention. For each coil acting alone,

$$v_k(t) = L_k \frac{di_k}{dt}$$

for $k = 1, 2$. However, the voltage induced in coil 1 by the current in coil 2 is negative relative to the indicated polarity on $v_1(t)$ as per rule 17.5b, i.e., $i_2(t)$ enters the dotted terminal so that $-i_2(t)$ can be viewed as entering the undotted terminal. Hence

$$v_1(t) = L_1 \frac{di_1}{dt} + M \frac{d(-i_2)}{dt} = L_1 \frac{di_1}{dt} - M \frac{di_2}{dt}$$

Using rule 17.5a, the same arguments apply to coil 2, in which case

$$v_2(t) = M \frac{d(-i_1)}{dt} + L_2 \frac{di_2}{dt} = -M \frac{di_1}{dt} + L_2 \frac{di_2}{dt}$$

Now we consider Figure 17.4b. Here observe that neither pair (v_1, i_1) nor (v_2, i_2) satisfy the passive sign convention; hence

$$v_k(t) = -L_k \frac{di_k}{dt}$$

for $k = 1, 2$. However, the voltage induced in coil 1 by the current $i_2(t)$ satisfies rule 17.5a. Hence

$$v_1(t) = -L_1 \frac{di_1}{dt} + M \frac{di_2}{dt}$$

On the other hand, the voltage induced in coil 2 by $i_1(t)$ satisfies rule 17.5b. Hence

$$v_2(t) = M \frac{di_1}{dt} - L_2 \frac{di_2}{dt}$$

Exercise. For the configurations of Figure 17.5, determine the pair of equations that specify the relationship between the voltages and currents.

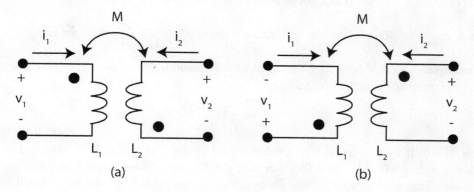

M   M

(a)   (b)

FIGURE 17.5. Two more scenarios for setting up coil equations.

ANSWERS: (a) $v_1(t) = L_1 \dfrac{di_1}{dt} - M \dfrac{di_2}{dt}$ and $v_2(t) = -M \dfrac{di_1}{dt} + L_2 \dfrac{di_2}{dt}$; (b)

$$v_1(t) = -L_1 \frac{di_1}{dt} + M \frac{di_2}{dt} \text{ and } v_2(t) = -M \frac{di_1}{dt} + L_2 \frac{di_2}{dt}$$

EXAMPLE 17.2. This example presents the procedures for marking the dots on an unmarked pair of coupled inductors and for determining the value of *M*. Consider the configuration of Figure 17.6, in which a current source is exciting terminal A of coil 1 of the coupled inductors with unknown *M*.

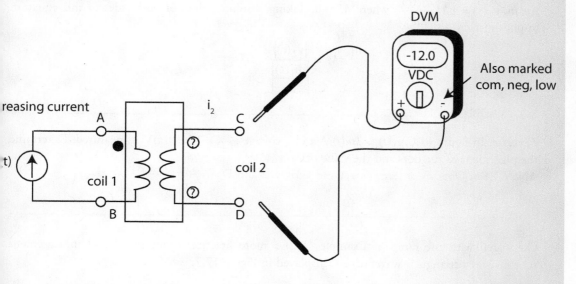

DVM

-12.0
VDC

Also marked
com, neg, low

reasing current

A i_2 C

t)

coil 1   coil 2

B   D

FIGURE 17.6. Diagram for determining dot placement and mutual inductance.

Part 1: Dot Placement

Step 1. *Place a dot on the terminal at which $i_1(t)$ enters.*
Observe that we have inserted a dot at terminal A.

Step 2. *Apply an increasing current $i_1(t)$, i.e., a current for which $\dfrac{di_1(t)}{dt} > 0$ for all t.*

For example, one could set $i_1(t) = 10tu(t)$ mA for 10 seconds. We know that $i_1(t)$ induces a voltage at the terminals of coil 2 according to

$$v_2(t) = \pm M \frac{di_1(t)}{dt},$$

where $M > 0$. If we put the leads of a voltmeter across the terminals C-D as suggested in Figure 17.6, the reading will either be $v_2(t) > 0$ or $v_2(t) < 0$. Suppose the reading is $v_2(t) < 0$. If we reverse the leads of the voltmeter by putting them across the terminals D-C, the reading will have the opposite sign, i.e., $v_2(t) > 0$.

Step 3. *Reconnect the voltmeter leads until $v_2(t) > 0$, i.e., the reading is positive. Place a dot on the terminal of coil 2 for which the voltmeter lead is marked + (or "plus" or "high" or "pos"), i.e., at the terminal of higher potential.*

For the situation described above, the dot would be placed at terminal D. However, the dots on coils 1 and 2 could be simultaneously moved to the opposite terminals of each coil without changing the relative information they convey. (Problem 1 confirms this statement.)

Part 2: Determining M

Again, $v_2(t) = \pm M \dfrac{di_1(t)}{dt}$, where $M > 0$. Taking absolute values of both sides of this equation implies that

$$M = \frac{|v_2(t)|}{\left| \dfrac{di_1(t)}{dt} \right|}$$

Exercise. In figure 17.6, $i_1(t) = 2tu(t)$ A and a voltage $v_{CD} = 10u(t)$ mV is measured. Determine the placement of the dots and the value of M.
ANSWERS: Dots are at terminals A and C or at B and D; $M = \dfrac{0.01}{2} = 0.005$ H.

Compared with the ramp of Example 17.2, a more practical input signal and measurement scheme uses a triangular waveform, as displayed in Figure 17.7.

EXAMPLE 17.3. Figure 17.7 shows a circuit and two waveforms, $i_1(t)$ and $v_2(t)$, as might be displayed on an oscilloscope. Determine the placement of a dot either at position C or at position D, and the value of the mutual inductance.

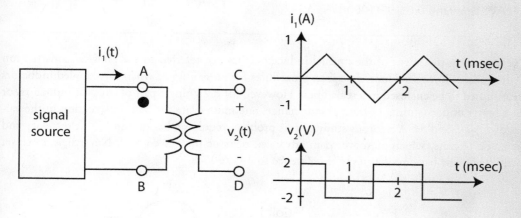

FIGURE 17.7 Circuit and waveforms for Example 17.3.

SOLUTION

For the time interval $0 < t < 0.5$ msec, $i_1(t)$ is a ramp function and $v_2(t)$ is constant. The information is similar to that given in Example 17.2 and the solution method is the same. First we place the dot at terminal A as the current enters A. The current i_1 is increasing, and v_2 is positive. We must now determine if the dot goes at terminal C or D. Since the current $i_1(t)$ enters the dotted terminal and is increasing over $0 < t < 0.5$ msec, its derivative is positive over $0 < t < 0.5$ msec. Also, the voltage $v_2(t)$ is positive for $0 < t < 0.5$ msec with the indicated polarities. Hence according to rule 17.5a, the dot goes at terminal C and

$$v_2(t) = M \frac{di_1}{dt}$$

To determine M, consider that the measured values during $0 < t < 0.5$ msec give $v_2 = 2$ V and $di_1/dt = 1/0.0005 = 2000$ A/sec. Thus, $M = 2/2000 = 0.001$ H.

EXAMPLE 17.4. In the circuit of Figure 17.7, suppose the dot positions are at A and C. If $i_1(t) = 2(1 - e^{-100t})u(t)$ A, find $v_2(t)$.

SOLUTION

In this case, $\dfrac{di_1}{dt} = 200e^{-100t}u(t)$. From Example 17.3, $M = 0.001$ H. Hence

$$v_2(t) = M \frac{di_1}{dt} = 0.2e^{-100t}u(t) \text{ V}$$

Exercise. In the circuit of Figure 17.7 with the dots in positions A and C, if $i_1(t) = 0.01$ $\sin(1000t)u(t)$ A, find $v_2(t)$ for $t > 0$.

ANSWER: $v_2(t) = 0.01 \, \cos(1000t)u(t)$ A

The preceding treatment of the mutual inductance has not referred to the physical construction of the coils, although we have described procedures for measuring M when the coupled inductors are assumed to be enclosed in a sealed box. However, for designing a pair of coupled inductors, or for a better understanding of mutual inductance, one must relate the coil construction to the values of L_1, L_2, and M. A rigorous study of this problem requires a background in field theory and magnetic circuits, which are covered in advanced texts or physics courses. Nevertheless, we set forth here a few basic properties with reference to Figure 17.8.

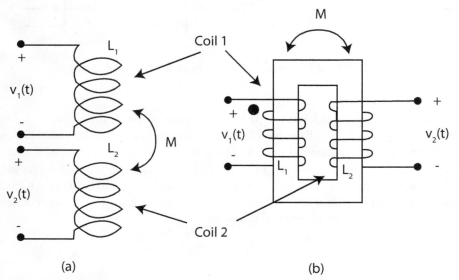

FIGURE 17.8. (a) Coupled coils in close proximity. (b) Coils coupled through magnetic core.

1. In Figures 17.8a and b, the number of turns for each of coils 1 and 2, respectively, is N_1 and N_2. Then the self- and mutual inductances have *approximately* the ratio

$$L_1 : L_2 : M = N_1^2 : N_2^2 : N_1 N_2$$

2. If two coils/inductors are placed in a nonmagnetic medium (e.g., air), bringing the inductors closer together increases the value of M.
3. If one inductor of a pair is rotated, then a larger value of M results when the axes of the inductors are parallel to each other. The smallest value of M occurs when the axes are perpendicular to each other.
4. Changing the core on which the two inductors are wound from a nonmagnetic material (e.g., air, plastic) to a ferromagnetic material may increase the values of L_1, L_2, and M by a factor of several thousand.

As a final note, the development above presupposes linearity. If the two inductors are placed in a nonmagnetic medium, this holds true. If the inductors are coupled through a ferromagnetic medium (e.g., an iron core), then the linear relationships of equation 17.4 hold only if both currents are sufficiently small that the magnetic medium avoids *saturation*, a phenomenon discussed in other courses or more advanced texts. Our investigations consider only the linear case.

3. DIFFERENTIAL EQUATION, LAPLACE TRANSFORM, AND PHASOR MODELS OF COUPLED INDUCTORS

Figure 17.9 shows a pair of coupled inductors. As developed in section 2, since each coil continues to act as an inductor, the terminal voltage depends on the derivative of the current through the coil plus an additional induced voltage due to the changing current in the other coil. This led to the set of differential equations 17.4, repeated below as 17.6, where the plus sign is used when the secondary dot is in position A and the minus sign is used when the secondary dot is in position B.

$$v_1(t) = L_1 \frac{di_1}{dt} \pm M \frac{di_2}{dt} \tag{17.6a}$$

$$v_2(t) = \pm M \frac{di_1}{dt} + L_2 \frac{di_2}{dt} \tag{17.6b}$$

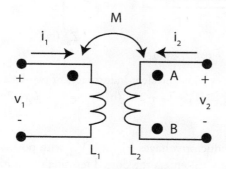

FIGURE 17.9 A pair of coupled inductors.

Assuming that there is no internal stored energy in the coupled inductors at time $t = 0$, then the Laplace transform of equations 17.6 yields

$$V_1(s) = L_1 s I_1 \pm M s I_2(s) \tag{17.7a}$$

$$V_2(s) = \pm M s I_1(s) + L_2 s I_2(s) \tag{17.7b}$$

Equations 17.7 represent the *s*-domain model of the coupled inductors. Further, if one is concerned with the sinusoidal steady state, then replacing *s* by *jω* yields the following equations for analysis using the phasor method:

$$\mathbf{V}_1 = j\omega L_1 \mathbf{I}_1 \pm j\omega M \mathbf{I}_2 \tag{17.8a}$$

$$\mathbf{V}_2 = \pm j\omega M \mathbf{I}_1 + j\omega L_2 \mathbf{I}_2, \tag{17.8b}$$

For equations 17.7 and 17.8, the plus sign is used when the secondary dot is in position A and the negative sign otherwise. These three sets of equations constitute the core of the examples and analyses presented in the remainder of the chapter. Throughout, the s-domain method is the preferred method, but each pair of equations has its specific uses.

EXAMPLE 17.5. A pair of coupled inductors are connected in two different ways, as shown in Figure 17.10. Find the input impedances, $Z_{in1}(s)$ and $Z_{in2}(s)$, and the corresponding equivalent inductances, L_{eq1} and L_{eq2}.

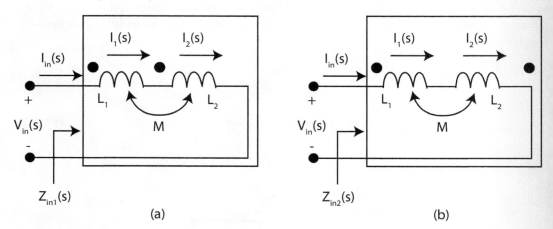

FIGURE 17.10 Impedances and equivalent inductances of two series-connected inductors.
(a) $L_{eq1} = L_1 + L_2 + 2M$. (b) $L_{eq2} = L_1 + L_2 - 2M$.

SOLUTION

For Figure 17.10a, label the inductor voltages V_{L1} and V_{L2} with positive reference on the left side of each inductor. Observe that $I_{in}(s)$ enters the dotted terminals of both coils. Hence $I_{in}(s) = I_1(s) = I_2(s)$. Directly applying equations 17.7, we obtain

$$V_{in}(s) = V_{L1}(s) + V_{L2}(s) = (L_1 s + Ms)I_{in}(s) + (Ms + L_2 s)I_{in}(s)$$
$$= (L_1 + L_2 + 2M)sI_{in}(s)$$

It follows that

$$Z_{in1}(s) = \frac{V_{in}(s)}{I_{in}(s)} = (L_1 + L_2 + 2M)s = L_{eq1}s$$

implying that

$$L_{eq1} = (L_1 + L_2 + 2M)$$

Similarly, for Figure 17.10b, with the same voltage definitions, we observe that again $I_{in}(s) = I_1(s) = I_2(s)$. Also note that $I_2(s)$ enters the undotted terminal. Applying equations 17.7, we obtain

$$V_{in}(s) = V_{L1}(s) + V_{L2}(s) = (L_1 s - Ms)I_{in}(s) + (-Ms + L_2 s)I_{in}(s)$$
$$= (L_1 + L_2 - 2M)sI_{in}(s)$$

It follows that

$$Z_{in2}(s) = \frac{V_{in}(s)}{I_{in}(s)} = (L_1 + L_2 - 2M)s = L_{eq2}s$$

implying that

$$L_{eq2} = (L_1 + L_2 - 2M)$$

Exercise. Suppose the circuits of Figures 17.10a and b are connected in parallel. If $L_1 = 40$ mH, $L_2 = 60$ mH, and $M = 25$ mH, find the equivalent inductance of the parallel connection.
ANSWER: 37.5 mH

The preceding example implies that

$$L_{eq1} - L_{eq2} = 4M \qquad (17.9)$$

This relationship suggests another way of determining M and dot markings from measurements. If an instrument for measuring self-inductance is available, we can use the instrument to measure L_{eq1} and L_{eq2}; from equation 17.9, the difference is $4M$.

4. ANALYSIS OF COUPLED CIRCUITS WITH OPEN-CIRCUITED SECONDARY

In the next example we compute the voltage across the secondary of a coupled inductive circuit using a high-resistance voltmeter.

EXAMPLE 17.6. In the circuit of Figure 17.11, assume that the meter resistance R_m is very large and looks like an open circuit to the secondary. The switch S is closed at $t = 0$. Find $i_1(t)$ and $v_2(t)$.

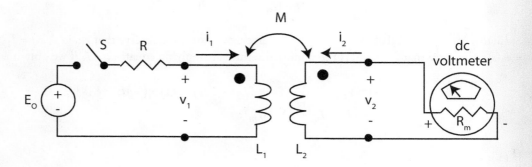

FIGURE 17.11 A coupled inductor circuit that might be found in a lab to determine dot position.

SOLUTION

Since the secondary looks like an open circuit, $I_2 = 0$ and $V_2 = MsI_1$. Hence if we find I_1 we can easily find V_2.

Applying KVL to the primary loop, using equation 17.7a and the fact that $I_2 = 0$ produces

$$V_1 = \frac{E_0}{s} - RI_1 = L_1 s I_1 \Rightarrow I_1 = \frac{E_0}{s(L_1 s + R)} = \frac{E_0/R}{s} - \frac{E_0/R}{s + \dfrac{R}{L_1}}$$

Therefore

$$i_1(t) = \frac{E_0}{R}\left(1 - e^{-\frac{R}{L_1}t}\right) u(t)$$

Since

$$V_2 = MsI_1 = Ms\frac{E_0}{s(L_1 s + R)} = \frac{ME_0}{(L_1 s + R)} = \frac{ME_0/L_1}{s + \dfrac{R}{L_1}}$$

It follows that

$$v_2(t) = \frac{ME_0}{L_1} e^{-\frac{R}{L_1}t} u(t)$$

Note that $v_2(t) \geq 0$ by virtue of the way the meter leads were connected to the secondary. Hence we conclude that the dot is indeed on the upper terminal of the secondary.

Exercise. Suppose in the circuit of Example 17.6 $E_0 = 10$ V, $R = 2$ Ω, the time constant of the circuit is 1 msec, and the voltage $v_2(t_1)$ at $t_1 = 1$ msec is 3679 V. Find L_1, M, and the ratio $M : L_1$ to see the factor by which the small, safe voltage E_0 is amplified when the switch is closed. The point here is that when switching with inductors occurs, a small voltage may produce a dangerously high voltage. ANSWERS: $L_1 = 2$ mH, $M = 2$ H, and $M : L_1 = 1000$

An interesting application to older car ignition systems also uses an open-circuited secondary to produce a very high voltage from a small one to fire a spark plug.

EXAMPLE 17.7. Figure 17.12a shows an automobile ignition system found on older cars while figure 17.12b shows a simplified equivalent circuit model. Today's ignition systems use electronic switching. Specifically, the block with the condensor (capacitor) and ignition point is replaced by something referred to as an "ignition module." The module contains a power transistor circuit to perform the switching action electronically, with the trigger timing typically actuated by a sensor that measures the position of the cam shaft either optically or magnetically.

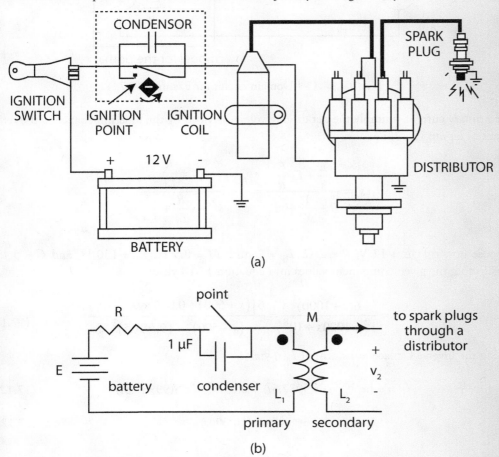

FIGURE 17.12 (a) An automobile ignition system. (b) Simplified equivalent circuit model.

In Figure 17.12a, the ignition coil is a pair of inductors wound on the same iron core, which creates a strong coupling between the coils. The **primary** coil is connected to the battery, while the **secondary** coil is connected to the spark plugs or load. The primary has a few hundred turns of heavy wire, the secondary about 20,000 turns of very fine wire. When the ignition point (or contact) opens by cam action, a voltage exceeding 20,000 V is induced across the secondary, causing the spark plug to fire. The generation of a high voltage to cause the spark plug to fire is accomplished by a basic *RLC* circuit containing a switch that represents the point of the ignition system.

Since the secondary is open-circuited, it has no effect on the solution for the primary current. Let us do the analysis using the equivalent circuit model of Figure 17.12b. Suppose that the switch has been closed for a long time. Accordingly, at $t = 0^-$, we have $i_1 = E/R = 12$ A. Using the model for an initialized inductor given in Figure 14.18 results in the *s*-domain equivalent circuit of figure 17.13.

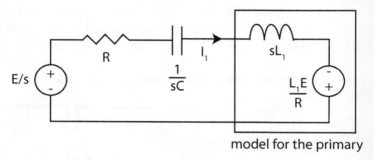

model for the primary

FIGURE 17.13 *s*-Domain circuit for Example 17.7.

The primary current is simply the net driving voltage divided by the total impedance in the series circuit of Figure 17.13, i.e.,

$$I_1(s) = \frac{\dfrac{E}{s} + L_1 \dfrac{E}{R}}{R + \dfrac{1}{sC} + sL_1} = \frac{E}{R} \times \frac{s + \dfrac{R}{L_1}}{s^2 + s\dfrac{R}{L_1} + \dfrac{1}{L_1 C}} \cdot \quad (17.10)$$

Suppose now that $E = 12$ V, $R = 2\ \Omega$, $L_1 = 2$ mH, $M = 0.5$ H, $L_2 = 130$ H, and $C = 5\ \mu F$. Substituting the given component values into equation 17.10 yields

$$I_1(s) = \frac{6(s + 1000)}{s^2 + 1000s + 10^8} = \frac{6[(s + 500) + 0.005 \times 9{,}987.5]}{(s + 500)^2 + (9{,}987.5)^2} \quad (17.11)$$

Taking the inverse Laplace transform of $I_1(s)$ yields

$$i_1(t) = 6e^{-500t} \cos(9987.5t) + 0.03e^{-500t} \sin(99987.5t) \quad (17.12a)$$

$$\cong 6e^{-500t} \cos(10{,}000t) \quad (17.12b)$$

Having obtained $i_1(t)$, we calculate $v_2(t)$ from the basic relationship of **equation 17.6**, using the plus sign for the dot positions:

$$v_2(t) = M \frac{di_1}{dt}$$

$$= 0.5 \times 6 \left[-500 e^{-500t} \cos(10{,}000t) - 10^4 e^{-500t} \sin(10{,}000t) \right]$$

$$\cong -30{,}000 e^{-500t} \sin(10{,}000t) \text{ V, for } t > 0.$$

From this expression, the voltage $v_2(t)$ reaches a magnitude of 30,000 V in about 157 μsec (one-fourth of a cycle of the oscillations). This voltage is high enough to cause the spark plug to fire. After the spark plug fires, the secondary is no longer an open circuit; the above methodology ceases to hold during the firing of the spark plug.

Exercise. Compute the energy stored in L_1 just prior to the switch across the capacitor opening.
ANSWER. 0.036 J

In going from equation 17.12a to 17.12b, we neglected the second term in 17.12a, retaining only the first. With the practical component values used in ignition circuits, this approximation is usually valid. In terms of equation 17.11, the approximation is as follows:

$$I_1(s) = \frac{E}{R} \times \frac{\left(s + \dfrac{R}{2L_1}\right) + \dfrac{R}{2L_1}}{s^2 + \dfrac{R}{L_1}s + \dfrac{1}{L_1 C}} \cong \frac{E}{R} \times \frac{\left(s + \dfrac{R}{2L_1}\right)}{s^2 + \dfrac{R}{L_1}s + \dfrac{1}{L_1 C}} \qquad (17.13)$$

Applying the inverse Laplace transform to equation 17.13 yields

$$I_1(t) \cong \frac{E}{R} e^{-\sigma_p t} \cos(\omega_d t) u(t)$$

where

$$\sigma_p = \frac{R}{2L_1} \quad \text{and} \quad \omega_d = \sqrt{\frac{1}{L_1 C} - \left(\frac{R}{2L_1}\right)^2} \cong \frac{1}{\sqrt{L_1 C}}$$

For the first few cycles of oscillations, the value of $-\sigma_p t$ is nearly zero and the value of the exponential factor is nearly 1. Therefore, $i_1(t)$ may be further simplified to

$$i_1(t) \cong \frac{E}{R} \cos\left(\frac{t}{\sqrt{L_1 C}}\right) \quad \text{for small } t > 0$$

Finally, we compute $v_2(t)$ from equation 17.6:

$$v_2(t) = M \frac{di_1}{dt} = M \frac{E}{R} \left[\frac{-1}{\sqrt{L_1 C}} \sin\left(\frac{t}{\sqrt{L_1 C}}\right) \right] u(t)$$

for small $t > 0$. Thus,

$$\left| v_2(t) \right|_{\text{max}} \cong M \frac{E}{R} \frac{1}{\sqrt{L_1 C}} = \frac{M}{L_1} QE \tag{17.14}$$

approximates the maximum value of v_2, where

$$Q = \sqrt{\frac{L_1}{C}} \frac{1}{R}$$

is the quality factor of the series $RL_1 C$ circuit. (See Chapter 16.) Later we will show that M/L_1 is approximately equal to the ratio of the number of turns of the secondary to that of the primary.

Equation 17.14 is a simple formula for estimating the maximum voltage that occurs at the secondary. It shows that although the battery voltage is only 12 V, what appears at the secondary for a brief moment is very much higher due to the switching action. The voltage is stepped up due to two factors: (1) the Q of the series $RL_1 C$ circuit and (2) the turns ratio of the ignition coil. From equation 17.14, a smaller R produces a higher voltage across the secondary. But a small R causes a larger current to flow in the primary circuit and therefore shortens the life of the breaker point. In practice, when the engine is running, a resistance wire or an actual resistor is placed in the primary circuit to limit the amount of current flow through the breaker point. The capacitor (condenser) serves a similar purpose—that of protecting the breaker point by suppressing the arc that results when the point opens.

Exercise. The ignition circuit of Figure 17.12b has $E = 12$ V, $L_1 = 1$ mH, and $C = 0.01$ μF. If the total resistance in the primary circuit is $R = 8\ \Omega$ and $M/L_1 = 50$, estimate the maximum voltage appearing at the open-circuited secondary when the switch opens.
ANSWER: About 23.7 kV

---

5. ANALYSIS OF COUPLED CIRCUITS WITH TERMINATED SECONDARY

To begin this section, we would like to develop some general formulas that make the analysis of doubly terminated coupled inductors straightforward. By "doubly terminated" we mean the configuration of Figure 17.14 where there is a source and source impedance connected to a load impedance through a pair of coupled inductors.

EXAMPLE 17.8.

Consider the circuit of Figure 17.14. Show that

(i) $Z_{in}(s) = L_1 s - \dfrac{M^2 s^2}{L_2 s + Z_2}$

(ii) $Z_{tot}(s) = Z_1(s) + Z_{in}(s) = Z_1(s) + L_1 s - \dfrac{M^2 s^2}{L_2 s + Z_2}$

(iii) $G_{v2} = \dfrac{V_2}{V_1} = \dfrac{Z_2 M}{\left(L_1 L_2 - M^2\right)s + L_1 Z_2}$

(iv) the voltage gain $G_{v1} = \dfrac{V_1}{V_s} = \dfrac{Z_{in}}{Z_{in} + Z_1}$

(v) the overall voltage gain $G_v = \dfrac{V_2}{V_s} = \dfrac{Z_{in}}{Z_{in} + Z_1} \times \dfrac{Z_2 M}{\left(L_1 L_2 - M^2\right)s + L_1 Z_2}$

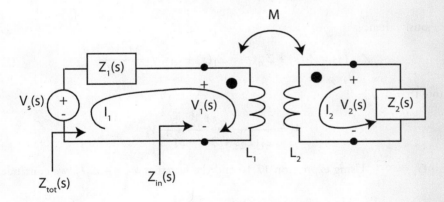

FIGURE 17.14. Coupled inductor circuit.

SOLUTION

Part 1: Find $Z_{in}(s) = \dfrac{V_1(s)}{I_1(s)}$ by two loop equations in the s-domain, converting to **matrix form**, and solving.

Step 1. For the primary loop,

$$V_1 = L_1 s I_1 + M s I_2$$

For the secondary loop

$$0 = V_2 + Z_2 I_2 = M s I_1 + L_2 s I_2 + Z_2 I_2 = M s I_1 + (L_2 s + Z_2) I_2$$

In matrix notation,

$$\begin{bmatrix} L_1s & Ms \\ Ms & L_2s + Z_2 \end{bmatrix}\begin{bmatrix} I_1 \\ I_2 \end{bmatrix} = \begin{bmatrix} V_1 \\ 0 \end{bmatrix}$$

Step 2. *Solve for I_1 and I_2.* Using the well-known formula for the inverse of a 2 * 2 matrix, we have

$$\begin{bmatrix} I_1 \\ I_2 \end{bmatrix} = \begin{bmatrix} L_1s & Ms \\ Ms & L_2s + Z_2 \end{bmatrix}^{-1}\begin{bmatrix} V_1 \\ 0 \end{bmatrix}$$

$$= \frac{1}{L_1s(L_2s + Z_2) - M^2s^2}\begin{bmatrix} L_2s + Z_2 & -Ms \\ -Ms & L_1s \end{bmatrix}\begin{bmatrix} V_1 \\ 0 \end{bmatrix}$$

Hence

$$I_1 = \frac{L_2s + Z_2}{L_1s(L_2s + Z_2) - M^2s^2}V_1$$

which implies that

$$Z_{in}(s) = \frac{V_1}{I_1} = L_1s - \frac{M^2s^2}{L_2s + Z_2} \tag{17.15a}$$

Part 2: Obviously, then,

$$Z_{tot}(s) = Z_1(s) + Z_{in}(s) = Z_1(s) + L_1s - \frac{M^2s^2}{L_2s + Z_2} \tag{17.15b}$$

Also, we note that

$$I_2 = \frac{-M}{L_1(L_2s + Z_2) - M^2s}V_1 \tag{17.16}$$

Part 3: Find $G_{v2} = \dfrac{V_2}{V_1}$. Using expression 17.16 and the fact that $V_2 = -Z_2I_2$, we conclude that

$$G_{v2} = \frac{V_2}{V_1} = \frac{-Z_2I_2}{V_1} = \frac{Z_2M}{L_1(L_2s + Z_2) - M^2s} = \frac{Z_2M}{\left(L_1L_2 - M^2\right)s + L_1Z_2} \tag{17.17}$$

Part 4: Find $G_{v1} = \dfrac{V_1}{V_s}$. By voltage division

$$V_1 = \frac{Z_{in}}{Z_{in} + Z_1}V_s \quad \Rightarrow \quad G_{v1} = \frac{Z_{in}}{Z_{in} + Z_1} \tag{17.18}$$

Part 5: Find $G_v = G_{v1}G_{v2}$.

$$G_v = G_{v1}G_{v2} = \frac{Z_{in}}{Z_{in} + Z_1} \times \frac{Z_2M}{\left(L_1L_2 - M^2\right)s + L_1Z_2} \tag{17.19}$$

Exercise. In Example 17.8 find the current gain ratio, $\dfrac{I_2}{I_1}$.

ANSWER: $\dfrac{-Ms}{L_2s + Z_2}$

Note that the answer can be obtained directly be writing a loop equation for the secondary.

The next example applies the foregoing development to a specific case and at the same time expands the developed formulas to new cases.

EXAMPLE 17.9

Consider the circuit of Figure 17.15, in which $v_{in}(t) = 60\sqrt{2}\cos(10t)$ V, $R_L = 10\ \Omega$, $R_s = 5\ \Omega$, $C = 0.01$ F, $L_1 = 2$ H, $L_2 = 0.5$ H, and $M = 0.5$ H. Find (i) $Z_{in}(j10)$, (ii) $V_1(j10)$, (iii) $V_2(j10)$, (iv) $I_1(j10)$, (v) $I_2(j10)$, and (vi) the average power $P_L(j10)$ delivered to the C-R_L load.

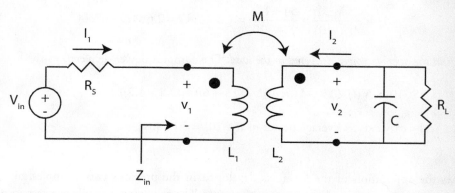

FIGURE 17.15 Coupled inductor circuit.

SOLUTION

First we note that $\omega = 10$ rad/sec. Also, using the effective value for the phasor voltage of the source yields $V_{in} = 60\ \angle\ 0^o$ V. Thus all calculated voltage and current phasors are effective values.

Step 1. *Find* $Z_{in}(j10)$. Observe that

$$Z_2(j10) = \frac{1}{Cj10 + \dfrac{1}{R_L}} = \frac{1}{0.1j + 0.1} = \frac{10}{1+j} = 5 - 5j$$

From equation 17.15 and the fact that $M = 0.5$ H, we have

$$Z_{in}(j10) = L_1 j10 - \frac{-M^2 100}{0.5 \times j10 + 5 - 5j} = L_1 j10 + 20M^2 = 5 + j20$$

Step 2. *Find* $\mathbf{V}_1(j10)$. Using voltage division,

$$\mathbf{V}_1(j10) = \frac{Z_{in}(j10)}{Z_{in}(j10) + R_s} \mathbf{V}_{in}(j10) = \frac{5 + j20}{5 + j20 + 5} 60 = 6\frac{5 + j20}{1 + j2} = 54 + j12 = 21.87\angle 12.53^o$$

Step 3. *Find* $\mathbf{V}_2(j10)$. From equation 17.17,

$$\mathbf{V}_2 = G_{v2}\mathbf{V}_1 = \frac{Z_2 M}{\left(L_1 L_2 - M^2\right)j10 + L_1 Z_2}\mathbf{V}_1 = \frac{0.5(5 - 5j)}{7.5j + 10 - 10j}(54 + j12) = 18 - 6j = 18.97\angle -18.44^o$$

Step 4. *Find* $\mathbf{I}_2(j10)$. From equation 17.16,

$$\mathbf{I}_2 = \frac{-M}{(L_1 L_2 - M^2)j10 + L_1 Z_2}\mathbf{V}_1 = \frac{-0.5}{10 - 2.5j}(54 + j12) = -2.4 - 1.2j = 2.68\angle -153.4^o$$

Step 5. *Find* $\mathbf{I}_1(j10)$. From Ohm's law,

$$\mathbf{I}_1 = \frac{\mathbf{V}_1}{Z_{in}(j10)} = \frac{54 + 12j}{5 + j20} = 1.2 - 2.4j = 2.683\angle -63.44^o$$

Step 6. *Find the average power delivered to the load.* The complex power delivered to the load is

$$\mathbf{S}(j10) = \mathbf{V}_2(j10) \times [-\mathbf{I}_2(j10)]^* = (18 - 6j) \times (2.4 - 1.2j) = 36 - j36$$

Taking the real part yields an average power of $P_L(j10) = 36$ watts.

Sometimes the application of the formulas developed in the previous two examples, although straightforward, is not the simplest route to the answer. The following example is a case in point.

EXAMPLE 17.10. Find the steady-state components of $v_1(t)$ and $v_2(t)$ at the frequency 1 rad/sec for the circuit of Figure 17.16, in which $v_{in}(t) = \cos(t)u(t)$ V. Note that because a resistance is present, the circuit responses will contain both a transient and a steady-state component.

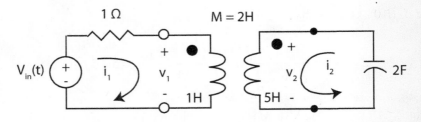

FIGURE 17.16 Circuit for loop analysis of coupled inductor.

SOLUTION

Since only the steady-state responses are required, we use the phasor method and write two loop equations given that $\omega = 1$ rad/sec and $\mathbf{V}_{in} = 1 \angle 0^o$ V.

The two loop equations are

$$\mathbf{V}_{in} = 1 = \mathbf{I}_1 + j\mathbf{I}_1 + j2\mathbf{I}_2 = (1 + j)\,\mathbf{I}_1 + j2\mathbf{I}_2$$

and

$$0 = \frac{1}{j2}\mathbf{I}_2 + j5\mathbf{I}_2 + j2\mathbf{I}_1 = j2\mathbf{I}_1 + j4.5\mathbf{I}_2$$

Putting this in matrix form, we have

$$\begin{bmatrix}1\\0\end{bmatrix} = \begin{bmatrix}1+j & j2\\j2 & j4.5\end{bmatrix}\begin{bmatrix}\mathbf{I}_1\\\mathbf{I}_2\end{bmatrix} \Rightarrow \begin{bmatrix}\mathbf{I}_1\\\mathbf{I}_2\end{bmatrix} = \frac{1}{-0.5+j4.5}\begin{bmatrix}j4.5 & -j2\\-j2 & 1+j\end{bmatrix}\begin{bmatrix}1\\0\end{bmatrix} = \frac{1}{-0.5+j4.5}\begin{bmatrix}j4.5\\-j2\end{bmatrix}$$

It follows that

$$\mathbf{V}_1 = \mathbf{V}_{in} - \mathbf{I}_1 = 1 - \frac{j4.5}{-0.5+j4.5} = \frac{0.5}{0.5-j4.5} = \frac{1}{1-j9} = 0.11043\angle 83.66^o$$

$$\mathbf{V}_2 = -\frac{1}{j2}\mathbf{I}_2 = -\frac{1}{j2}\times\frac{-j2}{-0.5+j4.5} = \frac{1}{-0.5+j4.5} = \frac{-2}{1-j9} = 0.22086\angle -96.34^o$$

Observe that

$$\mathbf{V}_2 = -2\mathbf{V}_1$$

in which case

$$v_{2,ss} = 0.22086\cos(t-96.34^o)\text{ V and } v_{1,ss} = 0.11043\cos(t+83.66^o)\text{ V}$$

Exercise. In Example 17.7, write the simultaneous equations in matrix form and then solve by Cramer's rule. Which method is easier?

Example 17.10 is a drastic case contrived to bring up an underlying property: the dot markings for coupled inductors (L_1, L_2, M) determine the $\pm$ sign in the equation relating v_2 to di_1/dt and the equation relating v_1 to di_2/dt. No intrinsic relations between the polarities of v_1 and v_2 are conveyed by the **dot convention**. For most practical circuits, however, it is true that the voltage drops of coupled inductors from the dotted terminal to the undotted terminal are in phase or nearly in phase.

As a practical example to end this section, we analyze the circuitry typical of the front end of an AM radio receiver.

EXAMPLE 17.11. Figure 17.17a shows circuitry typical of the front end of an AM radio. Figure 17.17b shows a simplified *RLCM* model of this circuitry. Here, V_s and the resistance $R_1 = 300\ \Omega$ together represent the antenna. Typical parameter values might be $R_2 = R_{in} = 14.7\ \text{k}\Omega$, $L_1 = 50$ nH, $L_2 = 2450$ nH, $M = 350$ nH, and $C = 104.5$ pF. Find the transfer function

$$H(s) = \frac{V_C(s)}{V_s(s)}.$$

Then using the methods developed in Chapter 16, compute ω_m, $|H(j\omega)|_{max} = |H(j\omega_m)|$, Q, B_ω, ω_1, ω_2, and the maximum value and the bandwidth of the magnitude response (i.e., the curve of $|V_C/V_s|$ vs. ω).

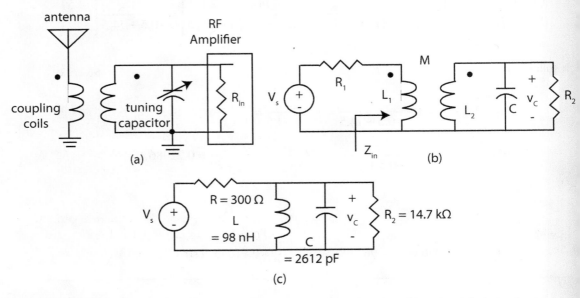

FIGURE 17.17 Circuits for Example 17.11. (a) The original circuit. (b) A simplified circuit model in which $L_1 L_2 = M^2$. (c) A design without coupled inductors.

SOLUTION
The first step in the solution is to observe that $L_1 L_2 = M^2$. Then from equation 17.19, the transfer function is

$$H(s) = \frac{V_C(s)}{V_s(s)} = \frac{Z_{in}}{Z_{in} + Z_1} \times \frac{Z_2 M}{\left(L_1 L_2 - M^2\right)s + L_1 Z_2} = \frac{Z_{in}}{Z_{in} + R_1} \times \frac{M}{L_1} \qquad (17.20)$$

where $Z_1 = R_1$ and $\quad Z_2 = R_2 // C = \dfrac{\dfrac{1}{C}}{s + \dfrac{1}{R_2 C}}$. Also, from equation 17.15,

$$Z_{in}(s) = L_1 s - \frac{M^2 s^2}{L_2 s + Z_2} = \frac{\left(L_1 L_2 - M^2\right) s^2 + L_1 Z_2 s}{L_2 s + Z_2} = \frac{L_1 Z_2 s}{L_2 s + Z_2} \tag{17.21}$$

Substituting the expression for $Z_{in}(s)$ in equation 17.21 into equation 17.20, then dividing through by Z_2 and simplifying, we obtain

$$H(s) = \frac{\dfrac{M}{CR_1 L_2} s}{s^2 + \left(\dfrac{L_1}{CR_1 L_2} + \dfrac{1}{R_2 C}\right) s + \dfrac{1}{L_2 C}} = \frac{4.5568 \times 10^6 s}{s^2 + 1.3020 \times 10^6 s + \left(6.2497 \times 10^7\right)^2} \tag{17.22}$$

$$= \frac{Ks}{s^2 + B_w s + \omega_m^2} = \frac{Ks}{s^2 + \dfrac{\omega_m}{Q} s + \omega_m^2}$$

By inspection, then,

$$\omega_m = 6.2497 \times 10^7 \text{ rad/s}, \; B_\omega = 13.02 \times 10^6 \text{ rad/sec}, \; H_m = \frac{K}{B_\omega} = 3.5, \; Q = 48,$$

and

$$\omega_{1,2} \cong \omega_m \mp 0.5 B_\omega = 61.85 \times 10^6, \; 63.15 \times 10^6 \text{ rad/s}.$$

Instead of the coupled inductors of Figures 17.17a and b, the same ω_m and B_ω can be obtained with the parallel resonant circuit of Figure 17.17c. Following the design method described in Example 16.3, we find the required element values to be $L = 98$ nH and $C = 2612$ pF. However, the maximum voltage gain would have a much lower value, $|H(j\omega_m)| = 14{,}700/(14{,}700 + 300) = 0.98$, compared with 3.5 for the coupled circuit of Figure 17.17a.

The higher voltage gain achieved in Figure 17.17b can be explained by the concept of maximum power transfer. A routine analysis would reveal that at $\omega = \omega_m$, the input impedance "seen" by the source is a pure resistance equal to the source resistance R_1, i.e.,

$$Z_{in}(j\omega_m) = \frac{L_1 Z_2 j\omega_m}{L_2 j\omega_m + Z_2} = 300 \; \Omega$$

Hence the source's impedance of R_1 sees a load Z_{in} equal to itself. Thus, maximum power has been extracted from the source. Since the inductors and the capacitor together form a lossless coupling network (see the homework problems), the same maximum power is transferred to the load resistor R_2. In Figure 17.17c, the reflected load impedance at $j\omega_m$ is not 300 Ω, and hence maximum power is not transferred.

Figure 17.17b contains two inductors and one capacitor. Accordingly, one would normally consider it a third-order circuit. Yet the denominator of equation 17.22 is only a second-degree polynomial in s. This simplification is due to the condition $M^2 = L_1 L_2$. The significance of this condition will be discussed in the next section.

The general analysis given above is quite useful for complex circuits. However, for many applications the two coils have a common terminal, as shown in Figure 17.18a. Figures 17.18b and c show the **T-** and **π-equivalent circuits** that often allow the application of series-parallel techniques to simplify the analysis. With the common terminal the coupled inductors have only three accessible terminals instead of four. Such an arrangement is called a *three-terminal device*. The reader might observe that if $M > L_1$ or $M > L_2$, one of the three inductances in the T- or π-equivalent circuit may have a negative value. This negative inductance appears in a mathematical model and is not the inductance of a physical component. Also, the equivalent circuits shown in Figures 17.18b and c are for the specific dot locations indicated in Figure 17.18a. A change of one dot location in Figure 17.18a will result in a change in the sign in front of M in Figures 17.18b and c.

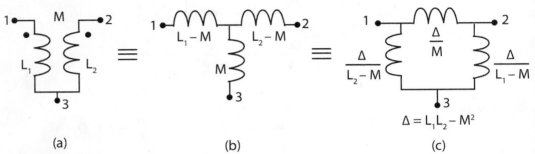

(a)   (b)   (c)

FIGURE 17.18 (a) Coupled coils with a common terminal. (b) T-equivalent circuit. (c) π-equivalent circuit.

Verification of the parameter values in the equivalent circuits in terms of L_1, L_2, and M is left to the homework problems. We now illustrate their use with a simple equivalent inductance example.

EXAMPLE 17.12. In the circuit of Figure 17.19, all initial conditions are zero. If $i_{in}(t) = u(t - 1)$ A, find the response, $v_{out}(t)$.

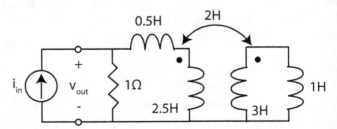

FIGURE 17.19 *RLM* circuit to be simplified using the T-equivalent.

SOLUTION

The key is to find the input impedance seen by the current source. Apply the T-equivalent circuit to the coupled inductors of Figure 17.18b. Using series-parallel techniques, the equivalent inductance in parallel with the 1 Ω resistance is $L_{eq} = 2$ H. Thus

$$V_{out}(s) = \frac{2s}{2s+1}I_{in}(s) = \frac{2e^{-s}}{2s+1} = \frac{e^{-s}}{s+0.5}$$

Therefore $v_{out}(t) = e^{-0.5(t-1)}u(t-1)$ V.

Exercises. 1. Repeat Example 17.12 with the dot on the 3 H inductor moved to the bottom. ANSWER: Same as in example.

2. Find the input impedance $Z_{in}(s)$ for the circuit of Figure 17.20. In the circuit, $k = \dfrac{M}{\sqrt{L_1 L_2}}$ is called the coupling coefficient, to be studied in the next section.

ANSWER: $\dfrac{s}{2s + 0.5}$

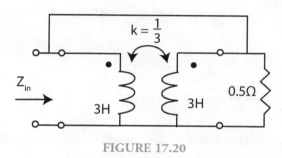

FIGURE 17.20

6. COEFFICIENT OF COUPLING AND ENERGY CALCULATIONS

The first part of this section justifies our assumption that $M_{12} = M_{21} = M$. One can justify that $M_{12} = M_{21} = M$ by the principles of magnetic circuits, but this is beyond the scope of this text. Our justification stems from the physical property that a pair of stationary coupled coils cannot generate average power. We will also show that the mutual inductance M has upper bound $\sqrt{L_1 L_2}$; i.e., the mutual inductance can never exceed the geometric mean of the self-inductances.

Justification of $M_{12} = M_{21} = M$

This property is a consequence of the principles of electromagnetic field theory, which are beyond the scope of this text. To make our approach accessible to the beginning student, we build our justification on the passivity principle for inductors.

THE PASSIVITY PRINCIPLE FOR INDUCTORS

A pair of stationary coupled inductors is a passive system; i.e., they cannot generate energy and, hence, cannot deliver average power to any external network.

Our technique is a so-called proof by contradiction; i.e., to show that $M_{12} = M_{21} = M$ is equivalent to showing that if $M_{12} \neq M_{21}$, then the passivity principle is violated. As a first step, suppose that $M_{12} \neq M_{21}$. Then, instead of having equations 17.6, the differential equations for the coupled inductors with the standard dot locations must take the form

$$v_1 = L_1 \frac{di_1}{dt} + M_{12} \frac{di_2}{dt} \qquad (17.23a)$$

$$v_2 = M_{21} \frac{di_1}{dt} + L_2 \frac{di_2}{dt} \qquad (17.23b)$$

Let us apply $i_1 = \sin(t)$ and $i_2 = \cos(t)$ to the inductors. From equations 17.23, the terminal voltages are

$$v_1 = L_1 \frac{d}{dt} \sin(t) + M_{12} \frac{d}{dt} \cos(t) = L_1 \cos(t) - M_{12} \sin(t)$$

and

$$v_2 = M_{21} \frac{d}{dt} \sin(t) + L_2 \frac{d}{dt} \cos(t) = M_{21} \cos(t) - L_2 \sin(t)$$

The total instantaneous power delivered to the coupled inductors is the sum of the powers delivered to the inputs, i.e., $p(t) = v_1(t)i_1(t) + v_2(t)i_2(t)$. Therefore

$$p(t) = v_1 i_1 + v_2 i_2 = L_1 \cos(t)\sin(t) - M_{12} \sin^2(t) + M_{21} \cos^2(t) - L_2 \sin(t)\cos(t) \quad (17.24)$$

To calculate P_{ave}, the average power delivered to the coupled inductors, we use the identities $\sin(t)\cos(t) = 0.5 \sin(2t)$, $\sin^2(t) = 0.5[1 - \cos(2t)]$, and $\cos^2(t) = 0.5[1 + \cos(2t)]$. It follows immediately that the first and the last terms in equation 17.24 make no contribution to P_{ave}, whereas the terms involving M_{12} and M_{21} lead to

$$P_{ave} = 0.5(M_{21} - M_{12}) \qquad (17.25)$$

This result shows very clearly that if $M_{12} > M_{21}$, then the excitations $i_1 = \sin(t)$ and $i_2 = \cos(t)$ will lead to a negative P_{ave}, violating the passivity principle. Similarly, if $M_{12} < M_{21}$, then the new (transposed) excitations $i_1 = \cos(t)$ and $i_2 = \sin(t)$ will again lead to a negative P_{ave}, violating the passivity principle. Therefore, we conclude that $M_{12} = M_{21}$.

With $M_{12} = M_{21} = M$, the average power, P_{ave}, is always zero for arbitrary sinusoidal excitations.

CALCULATION OF STORED ENERGY

Having proved that $M_{12} = M_{21} = M$, we shall now show that there is a limit to the value of M that is attainable once L_1 and L_2 are specified. Again we use the passivity principle and the fact that stored energy is the integral of the instantaneous power.

Consider the coupled inductors shown in Figure 17.9. The voltage-current relationships at the terminals are given by

$$v_1(t) = L_1 \frac{di_1}{dt} \pm M \frac{di_2}{dt} \qquad (17.26a)$$

and

$$v_2(t) = \pm M \frac{di_1}{dt} + L_2 \frac{di_2}{dt} \qquad (17.26b)$$

with the upper sign (+) for the dot in position A and the lower sign (−) for the dot in position B. Let us assume that the inductor currents are initially zero (at $t = 0$). In this state, there is no energy stored in the system.

Recall that the **energy stored** by any device over a time period $[0, T]$ is the integral of the instantaneous power over the interval. For a pair of coupled inductors with no initial stored energy, we have

$$W(T) = \int_0^T (v_1 i_1 + v_2 i_2) dt = \int_0^T \left[\left(L_1 \frac{di_1}{dt} \pm M \frac{di_2}{dt} \right) i_1 + \left(\pm M \frac{di_1}{dt} + L_2 \frac{di_1}{dt} \right) i_2 \right] dt$$

$$= \frac{1}{2} L_1 i_1^2(T) + \frac{1}{2} L_2 i_2^2(T) \pm \int_0^T M \left(i_1 \frac{di_2}{dt} + i_2 \frac{di_1}{dt} \right) dt$$

$$= \frac{1}{2} L_1 i_1^2(T) + \frac{1}{2} L_2 i_2^2(T) \pm M \int_{i_1(0)i_2(0)}^{i_1(T)i_2(T)} d(i_1 i_2)$$

$$= \frac{1}{2} L_1 i_1^2(T) + \frac{1}{2} L_2 i_2^2(T) \pm M i_1(T) i_2(T) \qquad (17.27)$$

where $d(i_1 i_2)$ is the total derivative of the product $i_1 i_2$ and is equal to $i_1 di_2 + i_2 di_1$.

For the specific scenario above, we apply driving sources to the inductors to bring the currents up to $i_1(T) = I_1$ and $i_2(T) = I_1$. Then, the energy *delivered* to the inductors during the time interval $(0, T)$ is, by equation 17.27,

$$W(T) = \frac{1}{2} L_1 I_1^2 + \frac{1}{2} L_2 I_2^2 \pm M I_1 I_2 \qquad (17.28a)$$

A couple of things about this result are worth noting:
1. The final integral in equation 17.27 depends only on the final values, which in the case of equation 17.28a are I_1 and I_2. The exact waveforms of $i_1(t)$ and $i_2(t)$ during $0 \leq t \leq T$ are immaterial.
2. The energy $W(T)$ delivered by the sources during $0 \leq t \leq T$ is not lost, but merely *stored* in the system.

To grasp property 2, we may adjust the sources so that the currents are brought back from I_1 and I_2 at $t = T$ to zero at some $t = T' > T$. Then the energy *delivered* by the sources to the inductors during $T < t < T'$ may be calculated in a similar manner, to obtain

$$W(T') - W(T) = -\frac{1}{2}L_1 I_1^2 - \frac{1}{2}L_2 I_2^2 \mp M I_1 I_2 \tag{17.28b}$$

with the upper (−) sign for the dot in position A and the lower (+) sign for the dot in position B in Figure 17.9. Equation 17.28b is precisely the negative of equation 17.28a. Thus, all of the energy delivered by the sources during $0 \leq t \leq T$ has been returned to the sources during $T \leq t \leq T'$. For this reason, the energy given by equation 17.27 is called the *stored energy*. Another way of recovering the stored energy is described in Problem 48. The physics of the situation shows that the energy is stored in the magnetic field produced by the currents in the inductors.

Upper Bound for M and the Coefficient of Coupling

The energy $W(T)$ must be nonnegative for arbitrary values of I_1 and I_2. Otherwise, the inductors will be *generating* energy during the time interval $0 \leq t \leq T$, which would violate the passivity principle. To ensure a nonnegative $W(T)$ for all I_1 and I_2, the values of L_1, L_2, and M must satisfy the inequality

$$M^2 \leq L_1 L_2 \tag{17.29a}$$

or

$$M \leq \sqrt{L_1 L_2} \tag{17.29b}$$

To show this, we rewrite equation 17.27 in the form

$$W(T) = \frac{1}{2}I_2^2\left[L_1\left(\frac{I_1}{I_2}\right)^2 \pm 2M\left(\frac{I_1}{I_2}\right) + L_2\right] = \frac{1}{2}I_2^2\left[L_1 x^2 \pm 2Mx + L_2\right] \equiv \frac{1}{2}I_2^2 f(x) \tag{17.30}$$

where $x \equiv I_1/I_2$ is a current ratio. Equation 17.30 shows that $W(T)$ is negative whenever $f(x)$ is negative. Now, $f(x)$ is a second-degree polynomial in x with a positive coefficient for the x^2 term. Consequently, the curve of $f(x)$ vs. x will be a parabola opening upward. From analytic geometry, depending on the sign of the discriminant $D \equiv (M^2 - L_1 L_2)$, the curve may or may not intersect the $f(x) = 0$ axis, as illustrated in Figure 17.21. From the figure, it is obvious that if $D > 0$, there will be some current ratio that yields a negative $f(x)$ and hence a negative $W(T)$, again violating the passivity principle. Therefore, $D = (M^2 - L_1 L_2) \leq 0$, which yields equation 17.29.

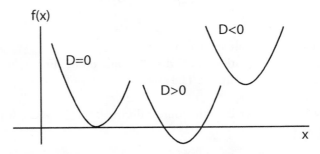

FIGURE 17.21 A plot of the stored energy vs. the current ratio x.

The degree to which M approaches its upper bound $\sqrt{L_1 L_2}$ is expressed by a positive number called the **coefficient of coupling**, defined as

$$k \equiv \frac{M}{\sqrt{L_1 L_2}}.$$

(17.31)

From equations 17.30 and 17.31,

$$0 \le k \le 1$$

(17.32)

When $k = 0$, M is also zero, and the inductors are uncoupled. When $k = 1$, the inductors have *unity coupling*, an idealized situation impossible to realize in practice.

EXAMPLE 17.13. For both situations of Figure 17.9, suppose that $L_1 = 5$ H, $L_2 = 20$ H, $M = 8$ H, $i_1(T) = I_1 = 2$ A, and $i_2(T) = I_2 = 4$ A. Find:
1. The coupling coefficient k.
2. The stored energy at $t = T$.

SOLUTION
1. For both dot positions, $k = \dfrac{M}{\sqrt{L_1 L_2}} = \dfrac{8}{\sqrt{5 \times 20}} = 0.8$.

2. For the dot in position A in Figure 17.9,

Stored energy $= 0.5 L_1 I_1^2 + 0.5 L_2 i_2^2 + M I_1 I_2$

$= 0.5 \times 5 \times 2^2 + 0.5 \times 20 \times 4^2 + 8 \times 2 \times 4 = 234$ joules

For the dot in position B,

Stored energy $= 0.5 L_1 I_1^2 + 0.5 L_2 I_2^2 - M I_1 I_2$

$= 0.5 \times 5 \times 2^2 + 0.5 \times 20 \times 4^2 - 8 \times 2 \times 4 = 106$ joules

---

Exercise. Repeat Example 17.13 with $L_1 = 4$ H, $L_2 = 16$ H, $M = 6$ H, $I_1 = 4$ A, and $I_2 = -2$ A.
ANSWERS: in random order: 112, 0.75, 16

EXAMPLE 17.14. In the circuit of Figure 17.22, $i_1(T) = I_1 = 6$ A. Find the minimum value of the stored energy and the corresponding value of $i_2(T) = I_2$ A.

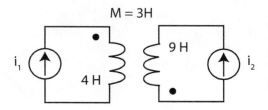

FIGURE 17.22 Coupled inductors for calculating the stored energy in Example 17.14.

SOLUTION

From equation 17.27,

$$W = 0.5L_1I_1^2 + 0.5L_2I_2^2 - MI_1I_2$$

$$= 0.5 \times 4 \times 36 + 0.5 \times 9 \times I_2^2 - 3 \times 6 \times I_2$$

$$= 4.5I_2^2 - 18I_2 + 72 \triangleq f(I_2)$$

Following the standard method in calculus for finding the maximum and minimum, we set dW/dI_2 to zero and solve for I_2:

$$\frac{dW}{dI_2} = \frac{df}{dI_2} = 9I_2 - 18 = 0$$

This yields $I_2 = 2$ A, and the corresponding minimum stored energy $W_{min} = 18 - 36 + 72 = 54$ J.

Exercise. Repeat Example 17.14 for $I_1 = 12$ A.
ANSWERS: 4 A, 216 J

7. IDEAL TRANSFORMERS

Developing the Equations for an Ideal Transformer

Two coupled *differential equations* containing three parameters L_1, L_2, and M characterize the coupled coils of Figure 17.9. By imposing two idealized conditions on these parameters, the pair of differential equations can be approximated by a pair of algebraic equations.

> **Idealization 1:** The coupled inductors have unity coupling; i.e., $M^2 = L_1L_2$, and the coupling coefficient $k = 1$.

> *Effect of Idealization 1:* With $k = 1$, the coupled coils have the **voltage transformation property**

$$v_1(t) = \frac{L_1}{M} v_2(t) = \frac{M}{L_2} v_2(t) = a v_2(t) \tag{17.33}$$

where a is a constant and both v_1 and v_2 are the *voltage drops from dotted (position A) to undotted terminals* of the coils, as per figure 17.9.

To derive the condition of equation 17.33, note that the constraint $M^2 = L_1 L_2$ implies that $L_1/M = M/L_2$. Let the ratio

$$\frac{L_1}{M} = a$$

Then $L_1 = aM$ and $M = aL_2$. Substitute these relationships into equations 17.7. This leads to the following sequence of equalities:

$$v_1(t) = L_1 \frac{di_1}{dt} + M \frac{di_2}{dt} = aM \frac{di_1}{dt} + aL_2 \frac{di_2}{dt} = a\left(M \frac{di_1}{dt} + L_2 \frac{di_2}{dt} \right) = a v_2(t)$$

Although the idealization of a unity coupling coefficient is not achievable in practice, coupling coefficients near unity are achievable by winding the turns of two inductors very closely together so that nearly all the flux that links one coil also links the other coil. The constant $a = L_1/M = M/L_2 = \sqrt{L_1 / L_2}$ in equation 17.33 reduces to the ratio of the numbers of turns, denoted by N_1 and N_2, of the two coils, i.e., $a = N_1/N_2$. Thus a is called the *turns ratio*. When $k = 1$, it is possible to show that

$$L_1 : L_2 : M = N_1^2 : N_2^2 : N_1 N_2$$

With unity coupled coils, equation 17.33 indicates that the voltages $v_1(t)$ and $v_2(t)$, both *from dotted (position A of Figure 17.9) to undotted terminals,* always have the same polarity. With coupling less than unity, it is possible for v_1 and v_2 to have opposite polarities at some time instants, as shown in Example 17.10.

Idealization 2: In addition to unity coupling, the coupled coils have infinite mutual and self-inductances.

Effect of Idealizations 1 and 2: With $k = 1$ and idealization 2, the pair of coils has the **current transformation property**

$$i_2(t) = -a i_1(t) \tag{17.34}$$

where both i_1 and i_2 are the *currents entering the dotted terminals* of the coils, as per Figure 17.9, dot in position A.

To derive equation 17.34, we again make use of the unity coupling condition $L_1/M = M/L_2 = a$, and rewrite equations 17.7 as

$$v_1 = L_1 \frac{di_1}{dt} + M \frac{di_2}{dt} = M\left[\frac{d}{dt}\left(\frac{L_1}{M} i_1 + i_2 \right) \right] \tag{17.35a}$$

Reasonably assuming that the voltage $v_1(t)$ is bounded, i.e., there is some finite constant K_v such that $|v_1(t)| \le K_v$ for all t, then

$$\lim_{M \to \infty} \left[\frac{d}{dt}\left(\frac{L_1}{M} i_1 + i_2 \right) \right] \to 0$$

A similar relationship holds for

$$v_2 = M \frac{di_1}{dt} + L_2 \frac{di_2}{dt} = L_2 \frac{d}{dt}\left(\frac{M}{L_2} i_1 + i_2 \right) = L_2 \left[\frac{d}{dt}\left(a i_1 + i_2 \right) \right] \tag{17.35b}$$

where if $|v_2(t)| \le K_v$ for all t,

$$\lim_{L_2 \to \infty} \left[\frac{d}{dt}\left(a i_1 + i_2 \right) \right] \to 0$$

In other words, as $L_1 \to \infty$, $L_2 \to \infty$, and $M \to \infty$, equations 17.35a and 17.35b each lead to

$$\frac{d}{dt}\left(a i_1 + i_2 \right) = 0 \tag{17.36a}$$

whose solution is

$$a i_1(t) + i_2(t) = C \tag{17.36b}$$

for some constant C. Now assume that the coils are unenergized prior to the application of excitations; i.e., at some time $t = t_0$ in the past, $i_1(t_0) = i_2(t_0) = 0$, which must be true for any real circuit. It follows that the constant C in equation 17.36b is zero, and consequently $i_2(t) = -a i_1(t)$. We remark that this derivation is valid for non-dc voltages and currents. The negative sign in equation 17.33 implies that at any time, if a current *enters* one coil at the dotted terminal, then the current in the other coil must *leave* the dotted terminal.

The condition of infinite inductances (L_1, L_2, M) is another idealization that is not realizable in practice. However, this condition can be *approximated* by using a magnetic material with very high permeability as the common core for the two coils.

THE IDEAL TRANSFORMER

Two coupled coils satisfying the relationships

$$\frac{v_1(t)}{v_2(t)} = \frac{N_1}{N_2} = a \text{ and } \frac{i_1(t)}{i_2(t)} = -\frac{N_2}{N_1} = -\frac{1}{a}$$

are said to be an **ideal transformer**, as shown in Figure 17.23a.

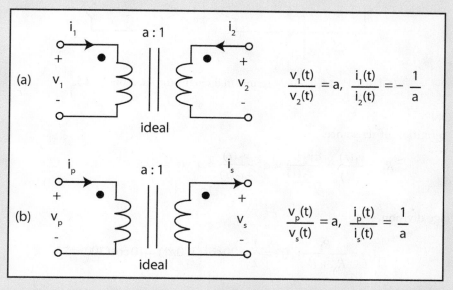

$$\frac{v_1(t)}{v_2(t)} = a, \quad \frac{i_1(t)}{i_2(t)} = -\frac{1}{a}$$

$$\frac{v_p(t)}{v_s(t)} = a, \quad \frac{i_p(t)}{i_s(t)} = \frac{1}{a}$$

FIGURE 17.23 Symbol and defining equations for an ideal transformer. (a) Both currents enter the dotted terminals. (b) i_p enters the dotted terminal and i_s leaves the dotted terminal.

In Figure 17.23 two vertical bars serve as a reminder of the presence of a ferromagnetic core in the physical device. The word "ideal" may or may not appear in the schematic diagram. Again, the mathematical model of an ideal transformer depends only on the turns ratio $a : 1$ and the relative dot positions. To avoid the negative sign in the current relationship, an alternative labeling of voltages and currents as shown in Figure 17.23b may be used. The subscript p stands for the **primary** coil, which is connected to a power source, and s for the **secondary** coil, which is connected to a load. Note that i_p is entering at the dotted terminal and i_s is leaving the dotted terminal. The notation of Figure 17.23b is more commonly used in the study of electric power flow.

One important simplification resulting from the idealizations is that an ideal transformer is characterized by two *algebraic* equations in terms of its terminal voltages and currents through a single parameter a, the turns ratio. This is to be contrasted with a pair of coupled coils, characterized by two differential equations containing three parameters, L_1, L_2, and M.

EXAMPLE 17.15. For the circuit of Figure 17.24, suppose $R_s = 10\ \Omega, a = 0.1$, $R_L = 1000\ \Omega$, and $v_s(t) = 20\cos(300\pi t)$ V. Find R_1, $v_1(t)$, $i_1(t)$, $v_2(t)$, and $i_2(t)$.

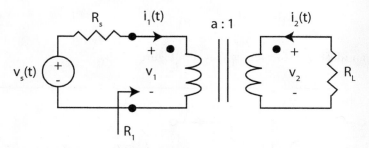

FIGURE 17.24. Doubly terminated circuit for Example 17.15.

SOLUTION

From the definition of resistance,

$$R_1 = \frac{v_1(t)}{i_1(t)} = \frac{av_2(t)}{-\dfrac{i_2(t)}{a}} = a^2\frac{v_2(t)}{-i_2(t)} = a^2 R_L = 0.01 \times 1000 = 10\ \Omega$$

Using voltage division,

$$v_1(t) = \frac{R_1}{R_s + R_1}v_s(t) = \frac{1}{2} \times 20\cos(300\pi t) = 10\cos(300\pi t)\ \text{V}$$

It follows that

$$i_1(t) = \frac{v_1(t)}{R_1} = \frac{10\cos(300\pi t)}{10} = \cos(300\pi t)\ \text{A}$$

Using the ideal transformer equations,

$$v_2(t) = \frac{v_1(t)}{a} = 100\cos(300\pi t)\ \text{V and } i_2(t) = -ai_1(t) = -0.1\cos(300\pi t)\ \text{A}$$

Exercises. 1. If in Example 17.15 $R_s = 0$, find $v_1(t)$, $i_1(t)$, $v_2(t)$, , and $i_2(t)$.
ANSWERS. 10 Ω, 20 cos(300πt) V, 2 cos(300πt) A, 200 cos(300πt) V, –0.2 cos(300πt) A

2. If in Example 17.15 $R_s = 10\ \Omega$ and $R_L = 4000\ \Omega$, find R_1, $v_1(t)$, $i_1(t)$, $v_2(t)$, and $i_2(t)$.
ANSWERS. 40 Ω, 16 cos(300πt) V, 0.4 cos(300πt) A, 160 cos(300πt) V,–0.04cos(300πt) A

3. In Example 17.15, find the average power absorbed by R_L.
ANSWER. 5 watts

The next example generalizes Example 17.15.

EXAMPLE 17.16. For the circuit of Figure 17.25, find input impedance $Z_{in}(s)$ and then the transfer functions

$$H_1(s) = \frac{V_1(s)}{V_s(s)}, \ H_2(s) = \frac{V_2(s)}{V_1(s)}, \ H_3(s) = \frac{V_2(s)}{V_s(s)}, \text{ and } H_4(s) = \frac{I_2(s)}{V_s(s)}$$

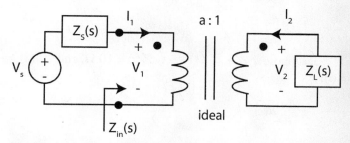

FIGURE 17.25 Doubly terminated circuit for Example 17.16.

SOLUTION
Using the definition of an impedance and the properties of the ideal transformer,

$$Z_{in}(s) = \frac{V_1(s)}{I_1(s)} = \frac{aV_2(s)}{-\dfrac{I_2(s)}{a}} = a^2 \frac{V_2(s)}{-I_2(s)} = a^2 Z_L(s)$$

Therefore, by voltage division,

$$H_1(s) = \frac{V_1(s)}{V_{in}(s)} = \frac{Z_{in}(s)}{Z_s(s) + Z_{in}(s)} = \frac{a^2 Z_L(s)}{Z_s(s) + a^2 Z_L(s)}$$

Further, by the properties of the ideal transformer,

$$H_2(s) = \frac{V_2(s)}{V_1(s)} = \frac{1}{a}$$

in which case

$$H_3(s) = \frac{V_2(s)}{V_s(s)} = H_1(s)H_2(s) = \frac{a Z_L(s)}{Z_s(s) + a^2 Z_L(s)}$$

Finally,

$$H_4(s) = \frac{I_2(s)}{V_s(s)} = \frac{-V_2(s)}{Z_L(s)V_s(s)} = \frac{-1}{Z_L(s)} H_3(s) = \frac{-a}{Z_s(s) + a^2 Z_L(s)}$$

Exercises. 1. Repeat the derivation of $Z_{in}(s)$ in Example 17.16 for the case where the dot on the secondary is at the bottom.

ANSWER: $Z_{in}(s) = a^2 Z_L(s)$

2. In Example 17.16, suppose $Z_L(s) = \dfrac{1}{Cs}$ and $Z_s(s) = R + L_s$. Find the frequency ω_0 at which the primary is resonant, i.e., the source sees a pure resistance.

ANSWER. $\dfrac{a}{\sqrt{LC}}$

3. In Example 17.16, suppose $a = 2$, $Z_L(s) = \dfrac{1}{s}$, $Z_s(s) = 10 \ \Omega$, and $V_s(s) = \dfrac{10}{s}$. Find $I_1(s)$ and $i_1(t)$.

ANSWER: $\dfrac{1}{s+0.4}$, $e^{-0.4t} u(t)$

The calculation of $Z_{in}(s)$ in Example 17.16 is so important (especially in maximum power transfer calculations) that we call it the **impedance transformation property** of an ideal transformer.

IMPEDANCE TRANSFORMATION PROPERTY

If the secondary of an ideal transformer is terminated in a load impedance $Z_L(s)$, then the impedance looking into the primary is

$$Z_{in}(s) = a^2 Z_L(s) \tag{17.37}$$

where a is the turns ratio taken in the direction from source to load. (See Figure 17.26; the dots are not marked on the figure because their positions are immaterial for this application.)

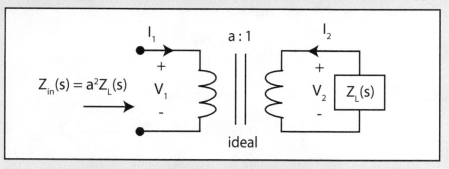

FIGURE 17.26 Impedance transformation by an ideal transformer.

Exercise. In Figure 17.26, suppose $a = 0.1$ and $Z_L(s) = \dfrac{10s}{s^2 + 10}$. Find $Z_{in}(s)$.

ANSWER. $\dfrac{0.1s}{s^2 + 10}$

EXAMPLE 17.17. For the circuit of Figure 17.25, find the instantaneous power delivered by the ideal transformer and the energy stored by the ideal transformer.

SOLUTION

With conversion of the voltages and currents to the time domain, the power delivered to ideal transformer is by definition

$$p(t) = v_1 i_1 + v_2 i_2 = \left(a v_2\right)\left(-\frac{i_2}{a}\right) + v_2 i_2 = 0$$

Hence, considered as a single unit, an ideal transformer neither generates nor consumes instantaneous power: whatever instantaneous power is received at one side must transfer to the other side. Furthermore, since $p(t)$ is identically zero, its integral with respect to t is also zero. Thus, an ideal transformer cannot store any energy.

In summary, the instantaneous power delivered to one side of an ideal transformer is transferred to whatever loads the other side, and as a result no energy is stored in the ideal transformer.

Exercises. 1. In Figure 17.26 suppose the primary of the transformer has $a = 0.1$ and is connected to a source having $Z_{th}(s) = 10\ \Omega$. Find $Z_L(s)$ for maximum power transfer to the load.
ANSWER. $Z_L(s) = 1\ k\Omega$

2. In Figure 17.26 suppose the primary of the transformer again has $a = 0.1$ and is connected to a source having $Z_{th}(j100) = 10 + j50\ \Omega$. Find $Z_L(j100)$ for maximum power transfer to the load.
ANSWER. $Z_L(j100) = 1000 - j5000\ \Omega$

Limitations of the Idealization

Although mathematically direct, the derivation of equation 17.34 does not provide insight into the limits of practical transformer design and operation. Let us reconsider the second idealization using the s-domain coupled transformer equations,

$$V_1(s) = L_1 s I_1(s) + M s I_2(s)$$
$$V_2(s) = M s I_1(s) + L_2 s I_2(s)$$

Divide the first equation by Ms and the second equation by $L_2 s$. To observe the frequency-dependent behavior, set $s = j\omega$ to obtain

$$\frac{V_1(j\omega)}{Mj\omega} = \frac{L_1}{M}I_1(j\omega) + I_2(j\omega) = aI_1(j\omega) + I_2(j\omega)$$

$$\frac{V_2(j\omega)}{L_2 j\omega} = \frac{M}{L_2}I_1(j\omega) + I_2(j\omega) = aI_1(j\omega) + I_2(j\omega)$$

Practical transformers have two customary properties: (i) the frequency content of the voltages and currents are band limited, meaning that the transformer is guaranteed to operate only over a restricted frequency range, $0 < \omega_{min} \leq \omega \leq \omega_{max}$, that does not include dc; and (ii) maximum voltages are specified, i.e., $|V_1(j\omega)| \leq V^{max}$ and $|V_2(j\omega)| \leq V^{max}$. Under these practical conditions we see that it is necessary to make M and L_2 (and since $a = L_1 / M$, L_1 also) sufficiently large over the frequency range $\omega_{min} \leq \omega \leq \omega_{max}$ so that

$$\left|\frac{V_1(j\omega)}{M\omega}\right| \cong 0 \text{ and } \left|\frac{V_2(j\omega)}{L_2\omega}\right| \cong 0 .$$

Thus we will achieve the current transformation property $I_1(j\omega) = -I_2(j\omega) / \alpha$ over the frequency range of interest provided the voltage magnitudes are less than V^{max}.

Some Practical Applications of Ideal Transformers

We now indicate some practical applications of ideal transformer properties.

EXAMPLE 17.18. In Figure 17.27 an ideal transformer steps down the voltage of a 2400 V$_{rms}$ source to power 10 incandescent lamps in parallel, each drawing 0.5 A.
   (a)   Find the voltage (magnitude) across each lamp, and
   (b)   find the current (magnitude) delivered by the source.

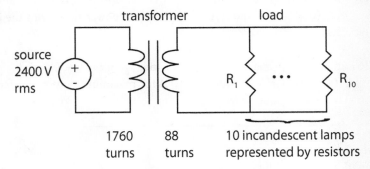

FIGURE 17.27 A transformer is used to step down a source voltage to meet the lamp specs.

SOLUTION

Since only magnitudes are involved in this problem, dot positions on the transformer and reference directions for voltages and currents are immaterial. The turns ratio is $a = 1760/88$. Since 2400 $= a|v_{load}|$, the voltage across each lamp is

$$|v_{load}| = \frac{88}{1,760} \, 2,400 = 120 \text{ V}_{rms}$$

Since 10 lamps use 0.5 A each, the total current used is 5 A. Using the current transformation property of the ideal transformer,

$$|i_{source}| = \frac{88}{1,760} \times 50 = 0.25 \text{ A}$$

Exercise. In the circuit of Figure 17.26, represent each lamp by a 200 Ω resistance. Suppose the source is limited to delivering a maximum of 3 kVA. How many lamps in parallel can be connected to the secondary of the transformer?
ANSWER: 41

EXAMPLE 17.19. Figure 17.28 shows a simplified model of an audio amplifier containing an ideal transformer. The input voltage is 1 V$_{rms}$ at 2 kHz. The load is a loudspeaker, represented by a 4 Ω resistance.
   (a)   Find the average power delivered to the 4 Ω load if it is connected directly to the amplifier (i.e., with the transformer removed and the resistor connected across A-B).
   (b)   With the transformer and load connected as per Figure 17.28, with turns ratio $a = 5$, find the average power delivered to the load.
   (c)   If the turns ratio a is adjustable, what value allows maximum power transfer to the load? What is the value of the maximum power?

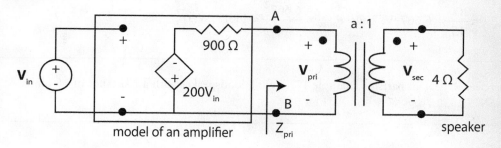

FIGURE 17.28 An ideal transformer used for maximum power transfer.

SOLUTION
(a) *No transformer and resistor connected across A-B.* First observe that the magnitude of the current through the 4 Ω resistor is

$$|\mathbf{I}_{4\Omega}| = \frac{200|\mathbf{V}_{in}|}{900 + 4} = 0.2212 \text{ A}$$

Therefore, the average power absorbed is

$$P_{4\Omega} = (0.2212)^2 \times 4 = 0.1958 \text{ W}$$

(b) *Transformer reconnected, a* = 5. At the primary of the transformer,

$$Z_{pri} = a^2 \times 4 = 100\ \Omega$$

From the voltage divider formula, the voltage across the primary is

$$\left|\mathbf{V}_{pri}\right| = \frac{100}{900 + 100}\left(200\left|\mathbf{V}_{in}\right|\right) = 20\ \text{V}$$

The voltage across the secondary is

$$\left|\mathbf{V}_{sec}\right| = \frac{20}{5} = 4\ \text{V}$$

Therefore,

$$P_{4\Omega} = \frac{4^2}{4} = 4\ \text{W}$$

(c) For the maximum power transfer, the turns ratio *a* should match the secondary impedance to that of the primary; i.e., the 4 Ω resistance reflected back to the primary should be 900 Ω to match the internal resistance of the amplifier. Hence

$$900 = a^2 \times 4 \Rightarrow a = 15$$

With *a* = 15, the reflected impedance is 900 Ω and $\left|\mathbf{V}_{pri}\right|$ = 100 V, meaning that

$\left|\mathbf{V}_{sec}\right| = \dfrac{100}{15} = 6.667$ V. Thus $P_{max} = \dfrac{6.667^2}{4} = 11.11$ W.

Exercise. Repeat all parts of Example 17.19 for a loudspeaker with a resistance of 16 ohms.
ANSWERS: 0.7628 W, 9.467 W, 11.11 W

---

8. MODELS FOR PRACTICAL TRANSFORMERS

The ideal dc voltage source was one of the first basic circuit elements studied in this text. "Ideal" means that it maintains a constant voltage for arbitrary loads connected across its terminals. Such a voltage source does not exist in the real world. However, because an ideal dc source in series with a resistance will approximately represent practical sources such as a battery, its use is very important. Similarly, although an ideal transformer does not exist in the real world, it can approximately represent a practical transformer with the addition of some other ideal circuit elements.

Because of the ferromagnetic core used in its construction, a practical transformer is inherently a nonlinear device. Nevertheless, a first level of approximation represents a practical transformer, by the circuit of either Figure 17.29b or Figure 17.29c. Figure 17.29a is ideal, whereas Figure 17.29b contains extra inductances to account for leakage fluxes and other magnetic phenomena. These inductances give rise to a finite usable bandwidth for a practical transformer. Figure 17.29c shows additional resistances to account for internal power losses. Thus the ideal transform in conjunction with other circuit elements can be used to approximate a real transformer. Specifically, transformers do not work for dc or for very high-frequency signals. The circuit model of Figure 17.29b makes this behavior clear: the inductance L_m causes a short circuit at dc, and the inductances L_p and L_s produce open circuits at high frequencies. A second level of approximation would begin to account for nonlinearities, a topic left for more advanced courses.

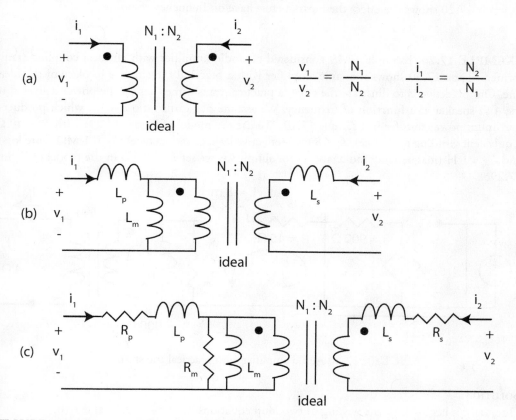

FIGURE 17.29 Linear transformer models. (a) Ideal case. (b) Extra inductances to account for leakage fluxes and other magnetic phenomena. (c) Extra resistances to account for internal power losses.

In Figures 17.29b and c, the parameters L_p and L_s are called primary and secondary winding leakage inductances, respectively. They are due to that part of the magnetic flux that links with one winding but not the other. L_m is called the magnetizing inductance and is due to that part of the magnetic flux that links both windings. Practically speaking, $L_m \gg L_p$ and $L_m \gg L_s$.

The model of Figure 17.29c incorporates the resistances R_p and R_s, which are the resistances of the primary and secondary windings, respectively. R_m accounts for the power loss in the iron core due to hysteresis and eddy currents. These resistances as well as the inductances L_p, L_s, and L_m can be calculated from knowledge of the physical layout and materials properties used in the construction of the transformer. This task usually requires a large set of design formulas and (empirical) charts. After a transformer is constructed, these parameters can be experimentally determined. Books on ac machinery set forth such experimental techniques.

From a circuit's perspective, once the model parameters in Figure 17.30 are known, the steady-state analysis of a circuit containing practical transformers requires no more than the phasor method or the Laplace transform method for its analysis. Two examples will now be given. Example 17.20 shows the effect these parameters have on frequency response.

EXAMPLE 17.20. Example 17.19 considered an audio amplifier with an ideal coupling transformer. Figure 17.30 shows the same amplifier with a practical transformer in place of the ideal one. Our objective is to illustrate the effect a practical transformer has on the power delivered to the 4 Ω speaker as a function of frequency. We assume our turns ratio $a = 15$, which produced maximum power transfer in Example 17.19. We use the model of Figure 17.29c with $R_p = 40$ Ω (equivalent winding resistance), $L_p = 8$ mH (effective leakage inductance), $R_m = 1$ MΩ (core loss), and $L_m = 1$ H (magnetizing inductance). For simplicity, we set $L_s = R_s = 0$ in the model of Figure 17.29c.

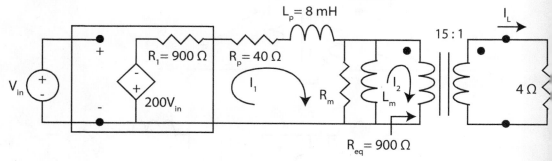

FIGURE 17.30 Audio amplifier with practical transformer.

SOLUTION
Our analysis begins with the writing of two loop equations in terms of $I_1(s)$ and $I_2(s)$:

$$-200V_{in} = \left(R_1 + R_p + L_p s + \frac{R_m L_m s}{R_m + L_m s} \right) I_1 - \frac{R_m L_m s}{R_m + L_m s} I_2$$

Note that since $R_m = 1$ MΩ and $L_m = 1$ H,

$$\frac{R_m L_m s}{R_m + L_m s} = \frac{L_m s}{1 + \frac{L_m}{R_m} s} \cong L_m s$$

for $s = j\omega$ and ω in the audio frequency range 0 to $2\pi \times 20$ krad/sec. Hence

$$-200 V_{in} = (R_1 + R_p + (L_p + L_m)s)I_1 - L_m s I_2$$

For the second loop, again neglecting R_m, we have

$$0 = -L_m s I_1 + (R_{eq} + L_m s)I_2$$

In matrix form

$$\begin{bmatrix} -200 V_{in} \\ 0 \end{bmatrix} = \begin{bmatrix} \left(R_1 + R_p + (L_p + L_m)s\right) & -L_m s \\ -L_m s & (R_{eq} + L_m s) \end{bmatrix} \begin{bmatrix} I_1 \\ I_2 \end{bmatrix}$$

From Cramer's rule

$$I_2 = \frac{-200 L_m s}{L_p L_m s^2 + \left((R_1 + R_p)L_m + R_{eq}(L_p + L_m)\right)s + (R_1 + R_p)R_{eq}} V_{in}$$

Our purpose is to show how average power to the load varies as a function of frequency in a practical transformer. To this end, for $v_{in}(t) = \sqrt{2}\cos(\omega t)$ V, the power delivered to the load is the power delivered to the primary of the ideal transformer, which is

$$P_L(\omega) = R_{eq}|I_2(j\omega)|^2$$

Recall that the effective value of the input voltage is 1 V. To obtain the desired plot of $P_L(f)$ (displayed in Figure 17.31), consider the following MATLAB code.

```
»R1 = 900; Rp = 40; Lm = 1; Lp = 0.016;
»Req = 900;
»num = [-200 0];
»den = [Lp*Lm ((R1+Rp)*Lm+Req*(Lp+Lm)) (R1+Rp)*Req];
»w = linspace(0,2*pi*1e4,500);
»I2 = freqs(num,den,w);
»I2mag = abs(I2);
»PL = Req*I2mag.^2;
»plot(w/(2*pi),PL)
»grid
»xlabel('Frequency in Hz')
»ylabel('Average Power to Load')
```

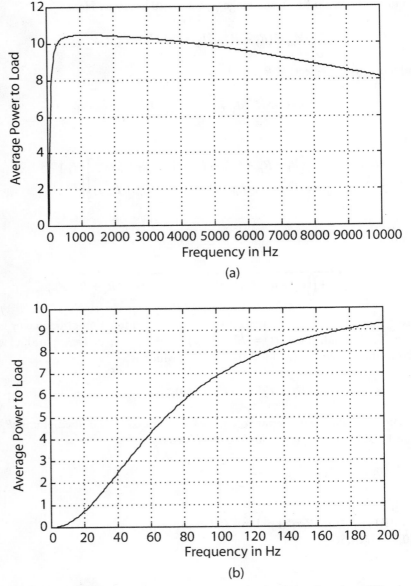

FIGURE 17.31 Plot of the power delivered to the speaker load as a function of frequency in Hz. (a) Overall response, showing sharp decrease with increasing frequency. (b) Low-frequency response, showing significant reduction in power transfer.

The plot shows that the power is down from the peak by about 0.5 W at 4500 Hz, and, at 10 kHz, the power delivered to the load is down about 2.5 W. Also note that a practical transformer does not operate at dc and at very low frequencies, as demonstrated in Figure 17.31b.

In the next example we consider voltage drops in commercial power lines using practical transformers.

EXAMPLE 17.21. The practical transformer in Figure 17.32a is designed for operation at 60 Hz and 1100/220 V, i.e., ideally a 5:1 step-down transformer. Using the model of Figure 17.29c, we represent the circuit in Figure 17.32b with the following parameters: R_p = 0.050 Ω (the primary winding resistance), R_s = 0.002 Ω (secondary winding resistance), $X_p = \omega L_p$ = 0.4 Ω (primary leakage reactance), $X_s = \omega L_s$ = 0.016 Ω (secondary leakage reactance), and $X_m = \omega L_m$ = 250 Ω (magnetization reactance). R_m is very large and its effect is neglected. If the load draws 100 A at a power factor of 0.6 lagging, i.e., I_2 = 100 ∠ −53.13° A, and the load voltage is V_2 = 220 ∠ 0° V, what is the magnitude of the needed source voltage to achieve the desired load voltage-current values in the presence of non-idealities of the transformer?

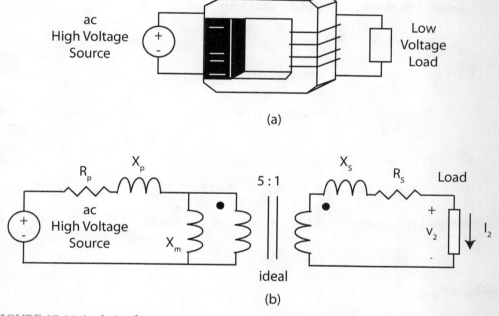

FIGURE 17.32 Analysis of a circuit containing a practical transformer. (a) A practical transformer for stepping down ac voltage. (b) A circuit model utilizing an ideal transformer.

SOLUTION
Using phasors for all voltages and currents, we have

Load voltage = V_2 = 220 ∠0° V

Load current = I_2 = 100 ∠−53.13° = 60 − j80 A

(from cos(53.13°) = 0.6, the given power factor).

Voltage drop across the impedance $R_s + jX_s$ = 0.002 + j0.016 Ω

$$= (60 − j80)(0.002 + j0.016) = 1.4 + j0.8 = 1.612∠29.7° \text{ V}$$

Voltage across the secondary of the ideal transformer

$$= (1.4 + j0.8) + 220 = 221.4 + j0.8 = 221.4\angle 0.207^\circ \text{ V}$$

Since $\dfrac{N_1}{N_2} = \dfrac{1,100}{220} = 5$ (given turns ratio), we have

Voltage across the primary of the ideal transformer

$$= 5 \times 221.4\angle 0.207^\circ = 1107\angle 0.207^\circ \text{ V}$$

Current through the primary of the ideal transformer

$$= 0.2 \times (60 - j80) = 12 - j16 \text{ A}$$

Current through the magnetizing inductance:

$$= \frac{1,107\angle 0.207^\circ}{j250} = 4.428\angle -89.8^\circ = 0.015 - j4.428 \text{ A}$$

Current through the primary winding:

$$= (12 - j16) + (0.015 - j4.428) = 23.7\angle -59.54^\circ \text{ A}$$

Voltage drop across the impedance $R_p + jX_p = 0.05 + j0.4 \ \Omega$

$$= 23.7\angle -59.54^\circ \times (0.05 + j0.4) = 9.55\angle 23.34^\circ \text{ V}$$

Therefore, the source voltage is

$$9.55\angle 23.34^\circ + 1107\angle 0.207^\circ = 1115.8\angle 0.4^\circ \text{ V}$$

Thus, the magnitude of the source voltage is 1115.8 V, which is 15.8 V higher (to overcome the non-ideal effects) than what would be needed for an ideal transformer.

9. COUPLED INDUCTORS MODELED WITH AN IDEAL TRANSFORMER

When a circuit contains coupled inductors, loop analysis is a natural way to analyze the circuit. One then solves the resulting *simultaneous equations* by any of the techniques studied earlier. Although very general and systematic, such methods have extensive mathematical operations that obscure the essential physical properties of the circuit. In this section, we shall present some **models for a pair of coupled inductors** that utilize an ideal transformer. Since the three basic properties (voltage, current, and impedance transformations) of an ideal transformer are relatively easy to comprehend, using such models in place of a pair of coupled inductors helps us to more easily understand the circuit behavior, without complicated mathematics.

As a first case, consider a pair of inductors with **unity coupling** (i.e., $k = 1$, or $M^2 = L_1L_2$), as shown in Figure 17.33a. Figures 17.33b and 17.33c show two equivalent circuits, each consisting of one inductor and one ideal transformer whose turns ratio depends on the values of L_1 and L_2.

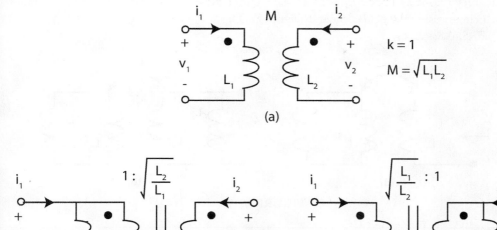

FIGURE 17.33 Ideal transformer models for unity coupled inductors. (a) A pair of unity coupled inductors. (b) A model consisting of one inductor and one ideal transformer. (c) An alternative model.

The proof of the equivalence is easy. The v-i relationships for the circuit of Figure 17.33a are given by differential equations 17.6. We need only show that the circuits of Figures 17.33b and 17.33c have the same v-i relationships. Consider Figure 17.33b first. In this case, $M = \sqrt{L_1L_2}$. Using the current transformation property of the ideal transformer, we obtain

$$v_1 = L_1 \frac{d}{dt}\left[i_1 + \sqrt{\frac{L_2}{L_1}}\, i_2\right] = L_1 \frac{di_1}{dt} + \sqrt{L_1L_2}\, \frac{di_2}{dt} = L_1 \frac{di_1}{dt} + M\frac{di_2}{dt}$$

Next, using the voltage transformation property,

$$v_2 = \sqrt{\frac{L_2}{L_1}}\, v_1 = \sqrt{L_1L_2}\, \frac{di_1}{dt} + L_2 \frac{di_2}{dt} = M\frac{di_1}{dt} + L_2 \frac{di_2}{dt}$$

These two equations are exactly the same as equations 17.6. Hence, the circuits of Figure 17.33a and Figure 17.33b are equivalent. A similar derivation proves the equivalence of the circuits of Figures 17.33a and 17.33c.

Exercises. 1. Verify the equivalence of the circuits of Figures 17.33a and 17.33c.

2. In Figures 17.34a and b, let $L_1 = 2.4$ H, $L_2 = 6$ mH, $L_2 = 3$ mH, and $L_4 = 0.2$ H. Find $Z_{in}(s)$ for each circuit using the equivalences in Figure 17.33.
ANSWERS. 0.8s, 0.16s.

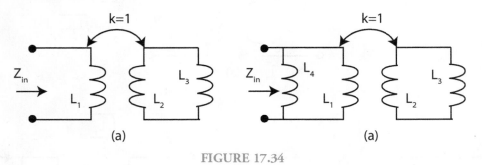

FIGURE 17.34

EXAMPLE 17.22. The circuit of Figure 17.35a has a unity coupling coefficient. Find the bandwidth, the center frequency, and the maximum voltage gain.

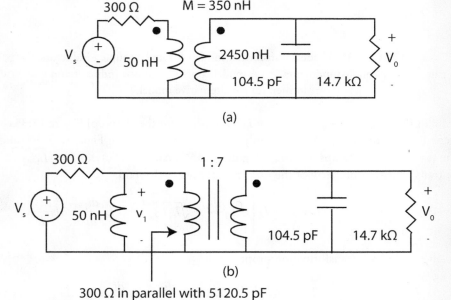

300 Ω in parallel with 5120.5 pF

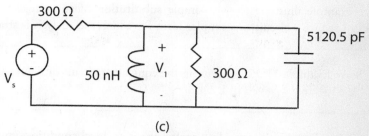

(c)

FIGURE 17.35 Analysis of a unity coupled circuit. (a) A circuit containing unity coupled inductors. (b) An equivalent circuit utilizing ideal transformers. (c) Equivalent circuit after reflecting load impedances to primary.

SOLUTION

The coefficient of coupling is $k = M/\sqrt{L_1 L_2} = 350/\sqrt{50 \times 2{,}450} = 1$. Replacing the coupled coils with the equivalent circuit of Figure 17.33b yields the circuit of Figure 17.35b. From the impedance transformation property of an ideal transformer, looking into the primary we see an impedance $1/49$ times the load impedance. Therefore, looking into the primary, we see a resistance of $14{,}700/49 = 300 \ \Omega$ in parallel with a capacitance of 49×104.5 pF $= 5120.5$ pF.

Figure 17.35c captures the new equivalent circuit. The band-pass characteristics of $H(s) = \dfrac{V_1(s)}{V_s(s)}$

follow the analysis done in Example 16.2. Consistent with the notation of Example 16.2, we have

$$R = 300//300 = 150 \ \Omega, \ L = 50 \text{ nH}, \ C = 5{,}120.5 \text{ pF}$$

$$\omega_m = \omega_p = \frac{1}{\sqrt{LC}} = \frac{1}{\sqrt{50 \times 10^{-9} \times 5{,}120.5 \times 10^{-12}}} = 62.5 \times 10^6 \text{ rad/sec}$$

$$B_\omega = \frac{1}{RC} = \frac{1}{150 \times 5{,}120.5 \times 10^{-12}} = 1{,}302 \times 10^6 \text{ rad/sec}$$

$$|H(j\omega)|_{max} = \frac{R}{R_s} = 0.5$$

From the voltage transformation property of an ideal transformer, V_o is simply 7 times V_1. So the only quantities affected for the transfer function

$$H_o(s) = \frac{V_o(s)}{V_s(s)} = 7H(s),$$

are the maximum values:

$$|H_o(j\omega)|_{max} = 7|H(j\omega)|_{max} = 3.5$$

The above example illustrates how a simple substitution of an equivalent model for coupled inductors can reduce a circuit to a simpler form for which the analysis is straightforward.

Exercise. Solve Example 17.22 again using the equivalent circuit of Figure 17.33c.
ANSWERS: The same as given in Example 17.22, of course.

As mentioned earlier, a unity coupling coefficient is an ideal condition impossible to achieve in practice. It is desirable, therefore, to modify the models of Figure 17.33 to account for a coupling coefficient $k < 1$. The resulting equivalent circuits are shown in Figure 17.36.

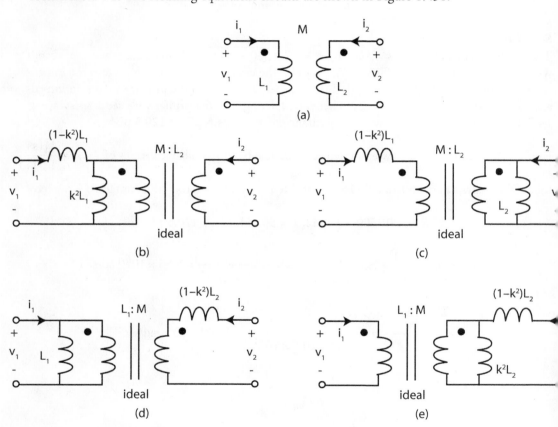

FIGURE 17.36 Four different models for coupled coils with $k \leq 1$, each consisting of one ideal transformer and two self-inductances. (a) Coupled inductor with $k < 1$. (b) One model using an ideal transformer and two inductances. (c), (d), (e) Three alternative models.

Exercise. In Figures 17.37a and b, $L_1 = 2.4$ H, $L_2 = 6$ mH, $L_3 = 3$ mH, $L_4 = 1.6$ H, and $k = \dfrac{1}{\sqrt{2}}$. Find $Z_{in}(s)$ for the circuits.
ANSWERS. $1.6s$, $0.8s$

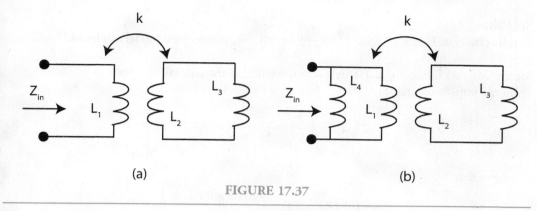

FIGURE 17.37

The equivalent circuits of Figure 17.36 are derived as follows. Since $L_1 L_2 > M^2$, it is possible to subtract a small inductance from L_1 such that the remaining inductance, L'_1, satisfies the condition $L'_1 L_2 = M^2$. In other words, the new inductance, L'_1 together with L_2 and M, forms a unity coupling system. Since $L'_1 = M^2/L_2 = k^2 L_1$, the inductance to be subtracted is equal to $L_1 - L'_1 = (1 - k^2)L_1$. This inductance must be added back in series with L'_1 to obtain a model for the original coupled inductors. The models of Figures 17.36b and c result. Repeating the process on L_2 yields the models of Figures 17.36d and e. A total of four equivalent circuits is possible. Each consists of two *uncoupled* inductors and one ideal transformer. Clearly, when $k = 1$, the models in Figure 17.36 reduce to those in Figure 17.33.

As an application of the models of Figure 17.36, let us reconsider Example 17.10.

EXAMPLE 17.23. Find the steady-state components of $v_1(t)$ and $v_2(t)$ at the frequency 1 rad/sec for the circuit of Figure 17.38, in which $v_{in}(t) = \cos(t)u(t)$ V. Note that because a resistance is present, the circuit responses will contain both a transient and a steady-state component.

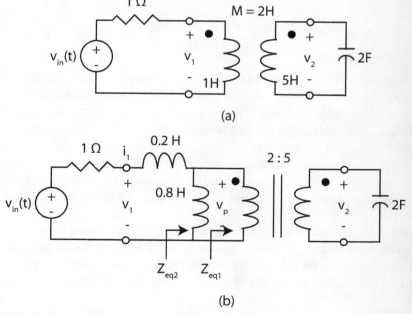

FIGURE 17.38 Steady-state analysis of coupled inductors using the ideal transformer model.

SOLUTION

For the circuit of Figure 17.38a, $k^2 = \dfrac{M^2}{L_1 L_2} = 0.8$. Substituting the model of Figure 17.36b

for the coupled inductors in Figure 17.38a, we obtain the equivalent circuit of Figure 17.38b. From the impedance transformation property of the ideal transformer,

$$Z_{eq1}(j) = \left(\frac{2}{5}\right)^2 \frac{1}{j2} = \frac{4}{50\,j} = -j0.08 \ \Omega$$

It follows that

$$Z_{eq2}(j) = 0.8\,j \parallel Z_{eq1}(j) = \frac{-j0.08 \times 0.8\,j}{-j0.08 + 0.8\,j} = -0.0889\,j \ \Omega$$

Using the voltage divider property,

$$\mathbf{V}_p = \frac{-j0.0889}{1 + 0.2\,j - j0.0889} \mathbf{V}_{in} = 0.088345\angle -96.34^o \ \mathrm{V}$$

Further,

$$\mathbf{V}_1 = \frac{0.2\,j - j0.0889}{1 + 0.2\,j - j0.0889} \mathbf{V}_{in} = 0.11042\angle 83.66^o \ \mathrm{V}$$

Using the ideal transformer property,

$$\mathbf{V}_2 = 2.5\mathbf{V}_p = 0.22086 \angle -96.34^o \ \mathrm{V}$$

This result agrees with the solution given in Example 17.10.

SUMMARY

This chapter has examined the phenomenon of induced voltage in one inductor caused by a change of current in another inductor. A new parameter called the *mutual inductance* (M) between the coils was introduced. M was defined as a constant and is present in the coupled differential and s-domain equation models of the coupled inductors. As illustrated in the chapter, M can be determined experimentally. This treatment has avoided digressing into the study of magnetic circuits, which in fact underlies a rigorous development of the concepts in this chapter. From the circuit analysis perspective, this mathematical treatment is adequate. However, for a deeper understanding of the physical phenomena, one must study the principles of magnetic circuits.

Of foremost importance in analyzing a circuit containing mutual inductance is the formulation of correct time domain or frequency domain equations—in particular, the correct signs for the mutual terms. For this reason, considerable time was spent on the dot convention. Once the equations are correctly written, we may use any of the techniques (loop or node analysis) studied earlier to analyze the circuit.

A proof of $M_{12} = M_{21} = M$ was given that made use of the passivity principle and some trigonometric identities. With the establishment of this equality, computing the energy stored in the coupled inductors follows from simple integration. The expression for the stored energy and the passivity principle then led to an upper bound for the value of the mutual inductance, namely $M \leq \sqrt{L_1 L_2}$. The coefficient of coupling was then defined as $k = M/\sqrt{L_1 L_2}$.

An ideal transformer was defined as a device satisfying both the voltage transformation and current transformation properties. For practical transformers, these two properties hold only approximately—for example, over specified voltage and current ranges as well as over specified frequency ranges.

Transformers have broad applicability. For example, in power engineering, transformers are used to step up or step down the voltages. In communication engineering, transformers are used to change a load impedance for the purpose of maximum power transfer.

Although ideal transformers can only be approximated in the real world, they nevertheless remain an important basic circuit element because a practical transformer or a pair of coupled inductors can be modeled by an ideal transformer and some additional R, L, and/or C elements. The use of such models simplifies many analysis problems and gives physical insight into the operation of a circuit.

TERMS AND CONCEPTS

Coefficient of coupling: usually denoted by k, equal to $M/\sqrt{L_1 L_2}$.

Coupled inductors (coils): two inductors having a mutual inductance $M \neq 0$.

Current transformation property: for unity coupled inductors having infinite inductances, the ratio $|i_1(t) : i_2(t)|$ is a constant equal to the turns ratio N_2/N_1.

Dot convention: a commonly used marking scheme for determining the polarity of induced voltages.

Energy stored in a pair of coupled inductors: $0.5 L_1 I_1^2 + 0.5 L_2 I_2^2 \pm M I_1 I_2$ joules, with the + sign for the case where both currents enter (or leave) dotted terminals and the − sign for the case where one current enters a dotted terminal and the other leaves the dotted terminal.

Ideal transformer: two network branches satisfying both the voltage transformation and current transformation properties exactly.

Impedance transformation property: when the secondary of an ideal transformer is terminated in an impedance $Z(s)$, the input impedance across the primary is equal to $(N_1/N_2)^2 Z(s) = a^2 Z(s)$.

Models for a pair of coupled inductors: a pair of coupled inductors can be represented by one ideal transformer and one inductance for the case $k = 1$. For coupling coefficients k less than 1, the representation requires one ideal transformer and two self-inductances.

Mutual inductance: a real number, usually denoted by M, that determines the induced voltage in one coil due to the change of current in another coil.

π-equivalent for coupled inductors: if two inductors have one terminal in common, then the three-terminal coupled inductors are equivalent to three uncoupled inductors (one of which may have a negative inductance) connected in the π-form.

Primary: The winding (coil) in a transformer that is connected to a power source.

Secondary: The winding (coil) in a transformer that is connected to a load.

T-equivalent for coupled inductors: if two inductors have one terminal in common, then the three-terminal coupled inductors are equivalent to three uncoupled inductors (one of which may have a negative inductance) connected in the T-form.

Transformer: a practical device that satisfies approximately the voltage transformation and current transformation properties.

Unity coupling: coefficient of coupling $k = 1$.

Voltage transformation property: for unity coupled inductors, the ratio $|v_1(t) : v_2(t)|$ is a constant equal to the turns ratio $N_1/N_2 = a$.

Problems

DOT PLACEMENT, M, AND BASIC EQUATIONS

1. For each of the circuits in Figure P17.1, determine the proper signs on the equations for coupled inductors:

$$v_1(t) = \pm L_1 \frac{di_1}{dt} \pm M \frac{di_2}{dt}$$

$$v_2(t) = \pm M \frac{di_1}{dt} \pm L_2 \frac{di_2}{dt}$$

Observe that some of the labelings do not conform to the passive sign convention.

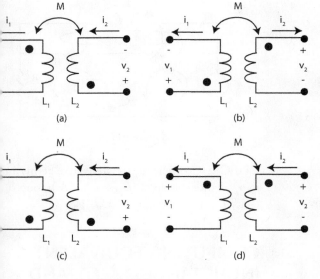

Figure P17.1

2. For the circuit of Figure P17.2, $M = 2$ H; compute $v_1(t)$ when $i_2(t) = 2t \cos(20t)u(t)$ mA for the dot in position A and then in position B.

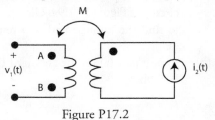

Figure P17.2

3. For each circuit shown in Figure P17.3:
   (a) Express $v_x(t)$ and $v_y(t)$ in terms of $i_a(t)$ and $i_b(t)$ for the dot in position C.
   (b) Obtain the s-domain equations that contain the initial currents $i_a(0^-)$ and $i_b(0^-)$ by taking the Laplace transform of the equations of part (a).
   (c) Repeat parts (a) and (b) with the dot moved to position D.

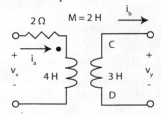

Figure P17.3

4. For the circuit shown in Figure P17.4:
(a) Express $v_x(t)$ and $v_y(t)$ in terms of $i_a(t)$ and $i_b(t)$ for the dot in position C.
(b) Obtain the s-domain equations containing the initial $i_a(0^-)$ and $i_b(0^-)$ by taking the Laplace transform of the equations of part (a).
(c) Repeat parts (a) and (b) with the dot moved to position D.

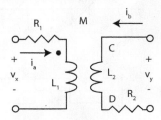

Figure P17.4

5. In the circuit of Figure P17.5, if $i_1(t) = 4tu(t)$ A, a voltage $v_{CD} = 40u(t)$ mV is observed. Determine the placement of the dots and the value of M. Repeat if $v_{CD} = -80u(t)$ mV is observed.

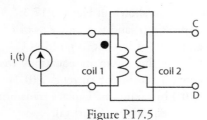

Figure P17.5

6. Consider the laboratory setup of Figure P17.6a.

   (a) The two waveforms, $i_1(t)$ and $v_2(t)$, in Figures P17.6b and c are shown as displayed on an oscilloscope attached to the circuit of Figure P17.6a. Determine the placement of the dots and the value of the mutual inductance.

   (b) Now suppose the signal source produces $i_1(t) = 2(1 - e^{-100t})u(t)$ A. Find $v_2(t)$.

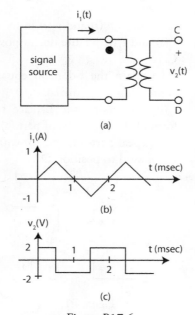

Figure P17.6

7. Consider the circuit shown in Figure P17.7. The 2 H inductor is short-circuited. The currents in the coupled inductors for $t \geq 0$ are $i_1(t) = 2e^{-t}$ A and $i_2(t) = e^{-t}$ A.

   (a) Find the mutual inductance M (in henries).

(b) Compute $v_2(t)$ for $t > 0$.

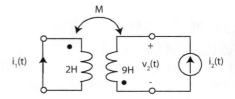

Figure P17.7

ANSWERS: (a) 4 H; (b) $-e^{-t}$ V

8. Write three mesh equations in the time domain for the circuit shown in Figure P17.8. Be particularly careful about the signs of induced voltages due to the mutual inductance. Apply the rule given by equation 17.6 if you have any doubt about the signs.

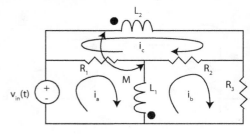

Figure P17.8

CHECK: Coefficients for derivative terms in i_a, i_b, i_c order: loop a, $(L_1, -L_1, -M)$; loop b, $(-L_1, L_1, M)$, loop c, $(-M, M, L_2)$

COMPUTING EQUIVALENT INDUCTANCES, $Z_{in}(s)$, AND RESPONSES FOR SIMPLE CIRCUITS

9. (a) Find the equivalent impedance of the circuit of Figure P17.9a. Use the result to find the impulse and step responses.

   (b) Repeat part (a) for the circuit of Figure P17.9b.

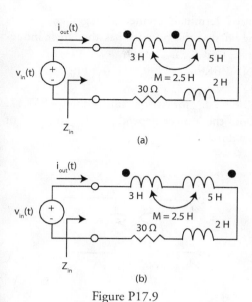

(a)

(b)

Figure P17.9

10. (a) Find the equivalent impedance of the circuit of Figure P17.10 when
    (i)   the dot is in position A
    (ii) the dot is in position B
    (b) Use the result of part (a) to find the transfer function with the dot in position A.
    (c) Use the result of part (b) to find the impulse and step responses with the dot in position A.
    (d) If $v_s(t) = 2 \cos(2t)$ V, find the steady-state value of $v_{out}(t)$.

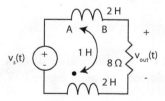

Figure P17.10

11. For the circuit shown in Figure P17.11, find (a) the equivalent inductance, L_{eq}, of the coupled coils; (b) the transfer function of the circuit; (c) the step response; (d) the resonant frequency of the circuit; and (e) the bandwidth and Q of the circuit.

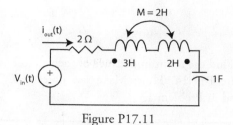

Figure P17.11

ANSWERS: (a) $L_{eq} = 1$ H, (c) $i_{out}(t) = te^{-t}u(t)$ A, (d) 1 rad/sec, (e) $B_\omega = 2$ rad/sec, $Q = 0.5$

12. Consider the circuit shown in Figure P17.12.
    (a) Determine the input impedance and input admittance.
         CHECK: admittance poles at -0.5 and -3
    (b) Determine the transfer function

$$H(s) = \frac{V_{out}(s)}{V_{in}(s)}$$

    (c) If $v_{in}(t) = 30(1 - e^{-0.5t})u(t)$ V, compute $v_{out}(t)$ using the "ilaplace" command in MATLAB.

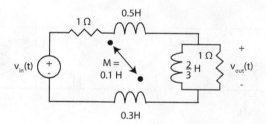

Figure P17.12

13. Figure P17.13 shows the two ways to connect a pair of coupled inductors in parallel. For each case, find L_{eq}.

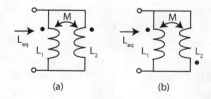

(a)   (b)

Figure P17.13

ANSWERS: (a) $(L_1L_2 - M^2)/(L_1 + L_2 - 2M)$;
(b) $(L_1L_2 - M^2)/(L_1 + L_2 + 2M)$

14. Find the input impedance of each circuit in Figure P17.14. Hint: Make use of each L_{eq} found in Problem 13.

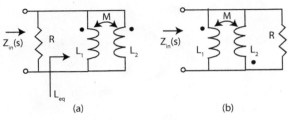

(a)   (b)

Figure P17.14

ANSWERS: (a) $Z_{in}(s) = \dfrac{RL_{eq}s}{R + L_{eq}s}$;

(b) Same as (a) with different L_{eq}

15. For each of the circuits in Figure P17.15, $R = 1\ \Omega$, $L_1 = 2$ H, $L_2 = 10$ H, $M = 2$ H. For figure P17.15a, $C = 0.8$ mF; for figure P17.15b, $C = 2$ F. Find the values of ω at which resonance occurs. For each case find $Z_{in}(j\omega_r)$

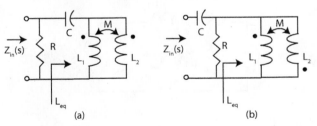

(a)   (b)

Figure P17.15

ANSWERS: (a) 25 rad/sec, 0 Ω; (b) 1 rad/sec, $Z_{in}(j\omega_r) = 0.5\ \Omega$

16. A handy dandy henry counter is used to measure the inductance of a pair of coupled inductors in various configurations. After three experiments the results are as follows: equivalent inductances are (i) 3.7 H after series-aiding connection; (ii) 2.5 H after series-opposing connection (dotted to dotted connection); and (iii) 121/250 H after parallel connection with

dotted terminals connected together. If it is known that the primary has the larger inductance, find L_1, L_2, and M.
ANSWERS: 2.6 H, 0.5 H, 0.3 H

17. Find L_{eq} for each circuit of Figure P17.17. Does the answer depend on the positions of the dots?

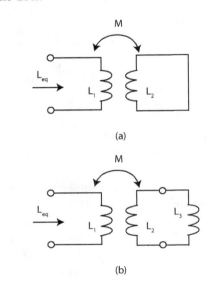

(a)

(b)

Figure P17.17

ANSWERS: (a) $L_1 - \dfrac{M^2}{L_2}$,

independent of dot positions;

(b) $L_1 - \dfrac{M^2}{L_2 + L_3}$,

independent of dot positions

18. Find the input admittance of each of the coupled inductor circuits in Figure P17.18.

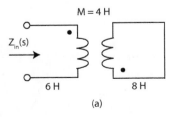

(a)

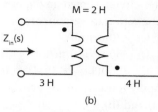

Figure P17.18

ANSWERS: (a) $\dfrac{1}{4s}$, (b) $\dfrac{1}{2s}$

19. Find the input impedance $Z_{in}(s)$ of the circuit shown in Figure P17.19. Does the answer depend on the positions of the dots?

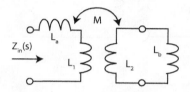

Figure P17.19

ANSWER: $s\left[L_a + L_1 - \dfrac{M^2}{L_2 + L_b}\right]$,

independent of the dot positions

20. Find the zero-state response $i_{out}(t)$ to the input $v_{in}(t) = 6u(t)$ V for the circuit of Figure P17.20.

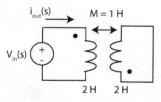

Figure P17.20

21. For the circuit shown in Figure P17.21, find Z_{in} (s) and the value of C (in F) that causes resonance to occur at $\omega = 2$ rad/sec.

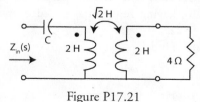

Figure P17.21

ANSWER:
$C = 1/6$ F; $Z_{in}(s) = \dfrac{Cs^2(s+4) + s + 2}{Cs(s+2)}$

22. For the circuit of Figure P17.8, find

$$Z_{in}(s) = \dfrac{V_{in}(s)}{I_a(s)}$$

when $R_1 = R_2 = R_3 = 1\ \Omega$, $L_1 = L_2 = 2$ H, and $M = 1$ H. Hint: After writing loop equations with numbers, consider using the symbolic toolbox in MATLAB (or the equivalent) to evaluate determinants in a Cramer's rule solution.

ANSWER: $Z_{in}(s) = \dfrac{9s^2 + 6s + 1}{3s^2 + 10s + 3}$

GENERAL ANALYSIS OF CIRCUITS WITH COUPLED INDUCTORS

23. For the circuit shown in Figure P17.23, $v_{in}(t) = 10e^{-2t}u(t)$ V.

   (a)   If a dot is placed at position A, compute the zero-state response.
   (b)   If a dot is placed at position B, compute the zero-state response.

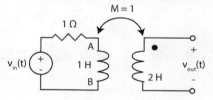

Figure P17.23

24. The switch S in the circuit of Figure P17.24 is closed at $t = 0$. If $v_{in}(t) = 10$ V, find $v_2(t)$.

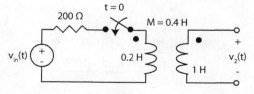

Figure P17.24

25. If a 900 Ω resistor is connected across the secondary of the circuit of Figure P17.24 and

$v_{in}(t) = 10$ V, find the voltage gain

$$G_v(s) = \frac{V_2(s)}{V_{in}(s)}$$

and then compute $v_2(t)$, $t \geq 0$. Use the formulas developed in Example 17.8 or, alternatively, write two mesh equations and solve by Cramer's rule.

CHECK: $G_v = \dfrac{9000s}{(s+500)(s+9000)}$

26. Consider the circuit shown in Figure P17.26.

   (a) Suppose the circuit is relaxed at $t = 0^-$. If $v_{in}(t) = 12u(t)$ V, find the time constant of the circuit and $v_2(t)$ for $t > 0$.

   (b) Suppose $v_{in}(t) = -12u(-t)$ V; find $v_2(t)$ for $t > 0$.

   (c) Suppose $v_{in}(t) = -6u(-t) + 6u(t)$ V; find $v_2(t)$ for $t > 0$. Hint: Use linearity.

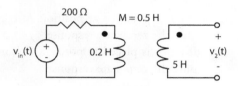

Figure P17.26

27. For the circuit shown in Figure P17.27, $i_{in}(t) = 100u(t)$ mA.

   (a) If a dot is placed at position A, compute the zero-state response.

   (b) If a dot is placed at position B, compute the zero-state response.

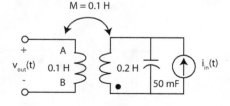

Figure P17.27

ANSWER (b): $v_{out,zs}(t) = 0.1 \sin(10t)u(t)$ V

28. For the circuit shown in Figure P17.28, compute the zero-state response for

   (a) $i_{in}(t) = 30u(t)$ mA

   (b) $i_{in}(t) = 30e^{-4t}u(t)$ mA

   (c) $i_{in}(t) = 60(1 - e^{-4t})u(t)$ mA. Hint: Apply linearity to the answers to parts (a) and (b).

   (d) Suppose $v_C(0^-) = 30$ V and all other initial conditions are zero. Find the zero-input response.

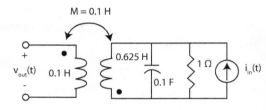

Figure P17.28

ANSWER (d): $v_{out}(t) = 1.6e^{-2t}u(t) - 6.4e^{-8t}u(t)$ V

29. In the circuit shown in Figure P17.29, all initial conditions are zero at $t = 0^-$.

   (a) If $i_{in}(t) = 2u(t - 1)$ A, compute the response, $v_{out}(t)$. Hint: Use the result of Problem 19.

   (b) Find $i_1(t)$.

   (c) Find $i_2(t)$

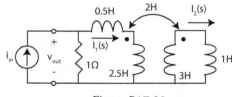

Figure P17.29

ANSWER: (b) $i_1(t) = 2(1 - e^{-0.5(t-1)})u(t-1)$ A

30. Consider the circuit shown in Figure P17.30, having zero initial stored energy. Let $R_1 = 9$ Ω.

   (a) Find $Z_{in}(s)$.

   (b) Find $Z_{tot}(s) = R + Z_{in}(s)$.

   (c) Find $G_{v1} = \dfrac{V_1}{V_{in}}$.

(d) Find $G_{v2} = \dfrac{V_2}{V_1}$.

(e) Find $G_v = \dfrac{V_2}{V_{in}}$.

(f) Find the response, $v_2(t)$, to the input
$v_{in}(t) = 16.81e^{-40t}u(t)$ V.

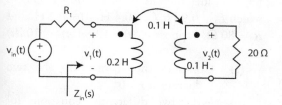

Figure P17.30

31. Suppose that $i_{in}(t) = 3\sqrt{20}\cos(t)u(t)$ A
and the initial stored energy is zero in the circuit shown in Figure P17.31.

(a) Compute the transfer function,
$$H(s) = \frac{V_2(s)}{I_{in}(s)},$$
using the symbolic toolbox in MAT-
LAB as follows:
(i) Write down the mutual inductance equations in s, a set of two
equations in V_1, V_2, I_1, I_2.
Rewrite each equation in the
form $0 = ? \times V_1 + ? \times I_1 + ? \times V_2 +$
$? \times I_2$.
(ii) Use KCL and Ohm's law to write
the relationship in the form $? \times I_{in}$
$= ? \times V_1 + ? \times I_1$.
(iii) Similarly construct the terminal
constraint equation of the form 0
$= ? \times V_2 + ? \times I_2$.
(iv) Write the preceding four equations in matrix form, $Ax = b$.
(v) Use MATLAB as follows:

```
syms s Iin A b Y H
A = "Coefficient Matrix in part 4"
% Y is the coefficient matrix A with the column
corresponding to V_2 replaced by the % vector b
with I_in (s) set to 1.
```

```
Y = .......
H = det(Y)/det(A)
simple(H)
```

CHECK: $H(s) = \dfrac{-3s}{(s+1)(s+3)}$

(b) Find the impulse response, $h(t)$, again
using the symbolic toolbox in MAT-
LAB:

```
syms t h
h = ilaplace(H)
```

(c) Using the residue command or
"ilaplace", show that the zero-state
response, $v_2(t)$, is

$v_2(t) = 18.112e^{-3t}u(t) - 10.06e^{-t}u(t) +$
$9\cos(t - 153.4°)u(t)$

(d) Identify the steady-state and transient
responses.

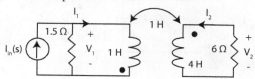

Figure P17.31

32. Consider the circuit of Figure P17.32.
Before the closing of the switch at $t = 0$, the
output voltage is zero. The input voltage is
$v_{in}(t) = 12\sqrt{2}$ V.

(a) Write two loop equations and find the
matrix $Z(s)$ such that

$$Z(s)\begin{bmatrix} I_1(s) \\ I_2(s) \end{bmatrix} = \begin{bmatrix} ? & ? \\ ? & ? \end{bmatrix}\begin{bmatrix} I_1(s) \\ I_2(s) \end{bmatrix} = \begin{bmatrix} V_{in}(s) \\ 0 \end{bmatrix}$$

(b) Using Cramer's rule find $I_2(s)$. Then
determine $V_2(s)$.

(c) Find $v_2(t)$.
CHECK: $v_2(t) = 8(e^{-0.4t} - e^{-t})u(t)$

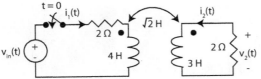

Figure P17.32

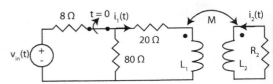

Figure P17.34

33. Consider the *RLC* circuit shown in Figure P17.33.
   (a) Find L_{eq}, $Y_{in}(s)$, and $Z_{in}(s)$.
   (b) Find the poles and zeros of $Z_{in}(s)$. Find the impulse response.
   (c) Find the zero-input response if $v_C(0^-) = 10$ V with all other initial conditions zero.
   (d) Find the zero-state response if $i_{in}(t) = \cos(100t)u(t)$ A.
   (e) Find the complete response of the circuit.

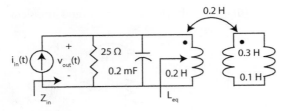

Figure P17.33

34. Consider the circuit of Figure P17.34, in which $L_1 = 0.9$ H, $L_2 = 0.6$ H, $M = 0.2$ H, and $R_2 = 100$ Ω. Let $v_{in}(t) = 120$ V. The switch S has been closed for a long time (i.e, the circuit has reached equilibrium) and is then opened at $t = 0$.
   (a) Compute the currents in the two inductors at $t = 0^-$.
   (b) After the switch is opened, write two differential equations for the coupled circuits in terms of $i_1(t)$ and $i_2(t)$.
   (c) Take the Laplace transform of the equations computed in part (b) accounting for the initial condition.
   (d) Find $v_2(t)$ for $t \geq 0$.

35. Consider the circuit of Figure P17.35, in which $L_1 = 0.9$ H, $L_2 = 0.4$ H, $M = 0.2$ H, and $R_2 = 80$ Ω. Suppose $v_{in}(t) = 120u(-t) - 120u(t)$ V.
   (a) Compute the currents in the two inductors at $t = 0^-$.
   (b) Write two differential equations for the coupled circuits in terms of $i_1(t)$ and $i_2(t)$.
   (c) Take the Laplace transform of the equations computed in part (b), accounting for the initial condition.
   (d) Find $v_2(t)$ for $t \geq 0$.
   (e) Now suppose $v_{in}(t) = 12u(-t) - 12e^{-t}u(t)$ V. Find $v_2(t)$ for $t \geq 0$.

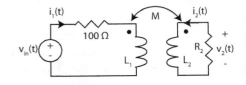

Figure P17.35

36. The two circuits shown in Figure P17.36 have the same transfer function

$$H(s) = \frac{V_{out}(s)}{V_{in}(s)}.$$

Find the values of the three uncoupled inductances assuming $M = 2$ H. Hint: Use the T-equivalent circuit of Figure 17.18 with the indicated labeling in Figure P17.36a; node 3 is the common terminal of the coupled inductors.

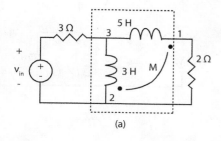

(a)

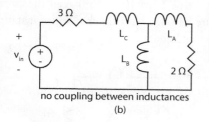

no coupling between inductances

(b)

Figure P17.36

ANSWER: $L_C = 2$ H, $L_A = 3$ H, and $L_B = 1$ H

COUPLING COEFFICIENT PROBLEMS AND ENERGY CALCULATIONS

37. Consider the circuit in Figure P17.37. Find the value of C such that the voltage gain is zero at $\omega = 333.33$ rad/sec. Hint: Use the T-equivalent circuit.

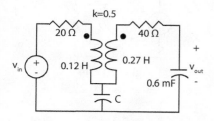

Figure P17.37

ANSWER: 0.1 mF

38. For the circuit shown in Figure P17.38, determine the coupling coefficient k such that at $\omega = 10^4$ rad/sec, the voltage gain V_{out}/V_{in} is zero. Hint: Use the T-equivalent circuit for coupled coils and recall that a series LC behaves as a short circuit at $\omega = 1/\sqrt{LC}$.

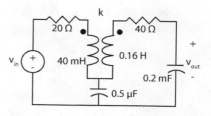

Figure P17.38

ANSWER: $k = 0.25$

39. Consider the circuit of Figure P17.39. Find the value of C such that the voltage gain is zero at $\omega = 333.33$ rad/sec. Use the π-equivalent circuit of Figure 17.18c.

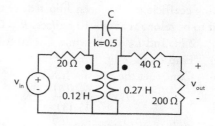

Figure P17.39

ANSWER: 33.33 μF

40. For the circuit shown in Figure P17.40, find the coupling coefficient k such that at $\omega = 10^4$ rad/sec, the voltage gain V_{out}/V_{in} is zero. Use the π-equivalent circuit of Figure 17.18c.

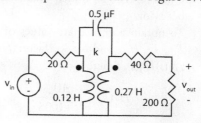

Figure P17.40

ANSWER: $k = 0.946$

41. Consider the circuit of Figure P17.41.
   (a) Find the mutual inductance M so that the coupling coefficient $k = 1$.
   (b) Obtain the transfer function using the formulas of Example 17.8. Make use of your answer to part (a).

(c)   Obtain values for ω_m, B_ω, Q, and H_m.

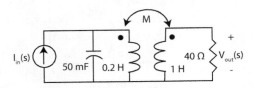

Figure P17.41

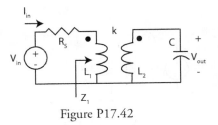

Figure P17.42

ANSWER: (c) B_ω = 2.5 rad/sec, $Q = \dfrac{\omega_m}{B_\omega} = 4$ ω_m = 10 rad/sec, H_m = 17.888

42. In the circuit shown in Figure P17.42, the coupling coefficient k has been adjusted for the circuit to be resonant at ω = 10 rad/sec. R_s = 0.1 Ω, L_1 = 2 H, and L_2 = 1 H while C = 20 mF.
(a)   Find k.
(b)   Construct the transfer function
$$H(s) = \frac{V_{out}(s)}{V_{in}(s)}$$
and then determine its poles and zeros. Hint: Use the T-equivalent for coupled coils to simplify calculations.
(c)   For a sense of the circuit behavior, obtain a magnitude frequency response (using MATLAB) of the circuit for 1 rad/sec $\leq \omega \leq$ 20 rad/sec. Observe the band-pass characteristic behavior.
(d)   To obtain approximate values of ω_m, B_ω, Q, H_m, and $\omega_{1,\,2}$, factor $H(s) = H_1(s)H_2(s)$, where
$$H_1(s) = \frac{10}{s+0.05}.$$
Note that $|H_1\,(j10)| \cong 1$ so that the 2nd order characteristic of $H_2(s)$ approximately characterizes the band-pass behavior.

ANSWERS: (a) $\sqrt{2}$,, (d) $B_\omega \cong$ 0.05 rad/sec, $Q \cong$ 200, H_m = 100

43. Let $i(t)$ = 2 cos(10t + 35º) A in the circuit in Figure P17.43. Find the maximum instantaneous stored energy.

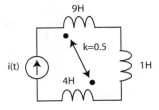

Figure P17.43

44. In the circuit of Figure P17.44, L_1 = 0.8 H, L_2 = 0.45 H, M = 0.175 H, R = 12 Ω. If $v_s(t)$ = 30cos(10t) V, find the maximum instantaneous steady-state power delivered to R when
(a)   A dot is in position A.
(b)   A dot is in position B.

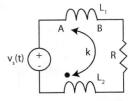

Figure P17.44

ANSWER: (a) 48 watts, (b) 27 watts

45. Consider the coupled inductors shown in Figure P17.45.
(a)   If I_1 = 2 A and I_2 = –3 A, find the stored energy.
(b)   If I_1 = 2 A, find the value of I_2 that will minimize the stored energy W. What is the value of W_{min}?
(c)   Plot W as a function of I_2 for $-3 \leq I_2 \leq 3$ A using MATLAB or the equivalent.

(d) Determine the coefficient of coupling k.

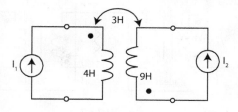

Figure P17.45

ANSWERS: (a) 66.5 J, (b) 2/3 A and 6 J, (d) 0.5

46. For the coupled inductors of Figure P17.46, $i_1(t) = A \cos(\omega t + \theta)$ and $i_2(t) = B \cos(\omega t + \varphi)$. What is the period T in terms of ω? Regardless of whether the dots are at A-B or A-C, show that the average power delivered to the coupled inductors is zero; i.e., show that

$$P_{ave} = \frac{1}{T}\int_0^T \left(v_1 i_1 + v_2 i_2\right) dt = \frac{1}{T}\left[\frac{1}{2}L_1 i_1^2(t) + \frac{1}{2}L_2 i_2^2(t) \pm M i_1(t) i_2(t)\right]_0^T = 0$$

Hints:

1. Write down the coupled inductor equations in the time domain with ± for dots in B and C.

2. Recall that $P_{ave} = \frac{1}{T}\int_0^T \left(v_1 i_1 + v_2 i_2\right) dt$;

    substitute and simplify.

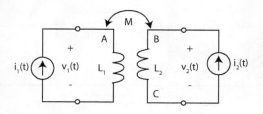

Figure P17.46

47. Reconsider the circuit of Figure P17.46 with dots at A-B. At $t = 0$, suppose the initial currents are $i_1(0) = I_1$ and $i_2(0) = I_2$. Also suppose $i_1(T) = 0 = i_2(T)$. Between 0 and T, $i_1(t)$ and $i_2(t)$ have arbitrary waveforms. Show that the energy delivered by the inductors to the current sources during the interval $0 \le t \le T$ equals $(0.5L_1 I_1^2 + 0.5L_2 I_2^2 + M I_1 I_2)$. If the dots are now moved to A-C, show that the final result is $(0.5L_1 I_1^2 + 0.5L_2 I_2^2 - M I_1 I_2)$.

48. In contrast to Problem 47, where the coupled inductors delivered power to the sources, we can connect two resistors R_1 and R_2 to the inductors and let the current decrease exponentially to zero. In particular, for the circuit of Figure P17.48, it is possible to show that the energy delivered by the inductors to the resistors during $0 \le t \le \infty$ equals $(0.5L_1 I_1^2 + 0.5L_2 I_2^2 \pm M I_1 I_2)$ depending on the position of the dots. Demonstrate the validity of this assertion for the specific parameter values and initial currents $L_1 = 10$ H, $L_2 = 2$ H, $M = 3$ H, $R_1 = R_2 = 1\ \Omega$, $i_1(0) = 1$ A, and $i_2(0) = -3$ A, with dots in position A-B, as follows:

(a) Calculate the stored energy at $t = 0$.
(b) Calculate $i_1(t)$ and $i_2(t)$ for $t > 0$.
(c) Evaluate the integral

$$W(0, \infty) = \int_0^\infty \left(v_1 i_1 + v_2 i_2\right) dt$$

and compare with the result of part (a).

CHECK: (a) 5 J, (b) $i_1(t) = e^{-t}u(t)$ A and $i_2(t) = -3e^{-t}u(t)$ A

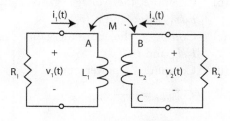

Figure P17.48

49. Mr. A claimed he had constructed a pair of coupled inductors with $L_1 = 10$ H, $L_2 = 8$ H, and $M = 9$ H. Rebut his claim by showing a specific set of (I_1, I_2) for which the stored energy is negative.

ANALYSIS OF CIRCUITS CONTAINING IDEAL TRANSFORMERS

50. In the circuit of Figure P17.50, $R_s = 150\ \Omega$, $R_1 = 600\ \Omega$, $R_L = 12\ k\Omega$, and $v_{in}(t) = 5\sqrt{2}\cos(\omega t)$ V, where $\omega = 2000\pi$ rad/sec.

   (a) Find the turns ratio, $n_1 : n_2$ for maximum power transfer to R_L.
   (b) Given your answer to part (a), find $v_1(t)$ and $v_{out}(t)$.
   (c) Given your answer to part (a), find $i_1(t)$ and $i_2(t)$.
   (d) Find the power delivered to the load.

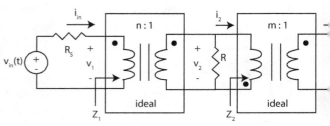

Figure P17.52

ANSWER: (a) $Z_2 = 10\ k\Omega$, $Z_1 = 500\ k\Omega$; (b) 0.625, 0.1, –0.1, –0.00625; (c) 5/1, –50/1, (d) 0.5

53. Repeat Problem 52 for $m = 20$ and $n = 5$. CHECK: $G_{power} = 0.2$

54. Repeat Problem 52 for $R_L = 300\ k\Omega$, $R = 10\ k\Omega$, $R_s = 100\ \Omega$, $n = 0.2$, and $m = 0.1$. CHECK: $Z_1 = 80\ \Omega$, $i_3/i_{in} = -0.013333$

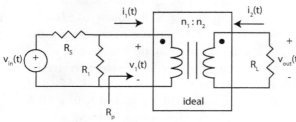

Figure P17.50

51. Repeat Problem 50 for $R_s = 400\ \Omega$, $R_1 = 1200\ \Omega$, $R_L = 3\ \Omega$, and $v_{in}(t) = 10\sqrt{2}\cos(\omega t)$ V, where $\omega = 2000\pi$ rad/sec.

52. In the circuit of Figure P17.52, $R_L = 100\ \Omega$, $R = 10\ k\Omega$; $R_s = 300\ k\Omega$, and $m = n = 10$.
   (a) Compute Z_2 and Z_1.
   (b) Compute the voltage gains v_1/v_{in}, v_2/v_1, v_3/v_2, and v_3/v_{in}.
   (c) Compute the current gains i_2/i_{in} and i_3/i_{in}.
   (d) Compute the power gain,

$$G_{power} = \frac{v_3 i_3}{v_1 i_{in}}.$$

55 In the audio amplifier circuit shown in Figure P17.55, both transformers are ideal, $R_s = 50\ \Omega$, and $v_s(t) = 0.1\sqrt{2}\cos(\omega t)$ V. The 4 Ω resistor represents a loudspeaker load.
   (a) Find the turns ratios a and b such that the average power delivered to the 4 Ω loudspeaker is as large as possible. What is the maximum power?
   (b) The turns ratios are those determined in part (a) for maximum power transfer. Now suppose that a loudspeaker with 16 Ω resistance is used in place of the 4 Ω speaker. What is the power consumed by the 16 Ω speaker?

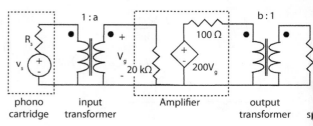

Figure P17.55

ANSWER: (a) 20, 5, 100 W; (b) 64 W

56. The circuit shown in Figure P17.56 crudely models an audio amplifier circuit. Set

$$K = \frac{20}{3}.$$

Each 8 Ω resistor models a tweeter, and each 16 Ω resistor models a woofer. Suppose the left and right speakers each consume 80 watts of average power. Determine:

(a) The turns ratios a and b such that the average power delivered to the speakers is as large as possible.

(b) The voltage across and current through (rms values) each woofer and tweeter.

(c) i_g and then i_s (rms values).

(d) The power delivered by the dependent source.

(e) The power delivered by the independent source.

(f) The power gain of the circuit, i.e., the ratio of the power delivered to the speaker load to the power delivered by the independent source.

such, it can be represented approximately by an ideal transformer in conjunction with some inductances and resistances. This problem illustrates the use of such a model for analysis purposes.

A certain calculator that normally uses four 1.5 V batteries comes with an adapter whose approximate model is shown in Figure P17.57a, where $L_a = 0.9$ H, $L_b = 9.6$ H, $n = 20$, and $R = 200$ Ω.

(a) If the calculator is not connected to the adapter output, but the adapter is plugged into a household ac outlet of 110 V_{rms}, what average power is consumed by the adapter?

(b) If the output of the adapter is accidentally short-circuited, what are the magnitudes (rms) of the ac current drawn from the wall outlet and through the short circuit?

(c) If a typical load, represented by a 50 Ω resistor, is connected to the output of the adapter, what are the approximate magnitudes (rms values) of the voltage across and the current through the load?

(d) If the adapter is treated as a pair of coupled inductors as shown in Figure P17.57b, determine the parameters L_1, L_2, and M.

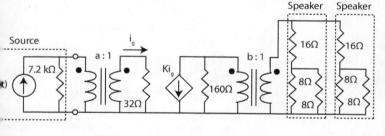

Figure P17.56

CHECK: $b = 4$, $v_{tweeter} = 8 \, V_{rms}$, $P_{gain} = 27.778$

57. Today almost every electronic gadget comes with its own ac adapter. An ac adapter's main function is to step down the household ac voltage of 110 V_{rms} to, say, 3 V_{rms} or 6 V_{rms} before conversion to a low dc voltage by the use of a rectifier. An adapter is basically a (practical) transformer. As

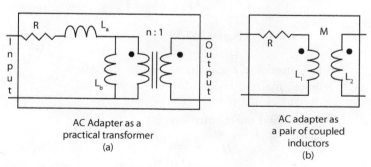

AC Adapter as a practical transformer
(a)

AC adapter as a pair of coupled inductors
(b)

Figure P17.57

ANSWER: (a) 0.154 watts; (b) 0.279 A, 5.59
A, 15.6 watts; (c) 4.98 V, 0.0996 A; (d) 10.5 H,
24 mH, 0.48 H

58. Consider the circuit shown in Figure
P17.58 with dots in positions A-B.
  (a) Compute the impulse response.
  (b) Compute the step response.
  (c) If the capacitor is initialized at $v_{out}(0^-)$
      = 16 V, find the zero-input response.
  (d) Compute the zero-state response to
      $v_{in}(t) = 20 \cos(2t)$ V. Identify the tran-
      sient and steady-state responses.
  (e) Repeat steps (a)-(d) with the dots in
      position A-C.

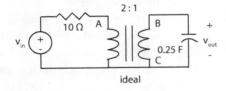

ideal

Figure P17.58

59. Consider the circuit shown in Figure
P17.59.
  (a) Compute the impulse response.
  (b) Compute the step response.
  (c) Compute the zero-state response to
      $v_{in}(t) = 26 \cos(2t)$ V. Identify the tran-
      sient and steady-state responses.

Figure P17.59

60. Consider the circuit of Figure P17.60.
Suppose the transformer dots are in positions
A-B.
  (a) Find the transfer function
      $$H(s) = \frac{V_{out}(s)}{V_{in}(s)}.$$
  (b) Compute the zero-state response,

$v_{out}(t)$, when $v_{in}(t) = 10u(t)$ V.
  (c) Find the steady-state response to $v_{in}(t)$
      = 20 cos(20t) V using the phasor
      method.
  (d) Repeat steps (a) through (c) with the
      dots in positions A-C.

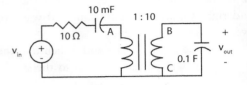

Figure P17.60

61. Figure P17.61 contains a linear circuit with
a pair of coupled inductors. Suppose $k = 1$
(unity coupling) and

$$\frac{L_1}{M} = \frac{M}{L_2} = a = 0.1$$

The circuit is in the sinusoidal steady state and
has known phasor values $\mathbf{V}_1 = 1$ V and $\mathbf{I}_1 = -j$
A.
  (a) If M is infinite (i.e., an ideal trans-
      former) and $\omega = 10$ rad/sec, find the
      magnitude of the current ratio $\mathbf{I}_2/\mathbf{I}_1$.
      Compare your answer with the value a
      = 0.1.
      ANSWER: 0.1
  (b) Repeat part (a) for $M = 10$ H and $\omega =$
      10 rad/sec. Hint: Use the phasor equa-
      tions in section 7 to calculate the cur-
      rent ratio $\mathbf{I}_2/\mathbf{I}_1$.
      ANSWER: 0.09
  (c) Repeat part (a) for $M = 10$ H and $\omega =$
      120π rad/sec.
      ANSWER: 0.099735, practically the same as
      for $a = 0.1$

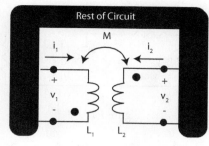

Figure P17.61

62. In the circuit of Figure P17.62, $R_s = 0.2\ \Omega$ and $b = 10$.

(a) Compute the transfer function. Note there are no finite zeros. Find the pole locations of the transfer function. From the pole locations determine the type (impulse/step) of response (critically damped, overdamped, or underdamped).

(b) Compute the impulse and step responses.

(c) Find ω_p and Q_p. Is Q_p high? If high Q_p, find approximate values for ω_m, B_ω, and the half-power frequencies $\omega_{1,2}$. Recall equations 16.47,

$$Q_p \equiv \frac{\omega_p}{2\sigma_p}.$$

Plot the magnitude response of the transfer function to verify your answers.

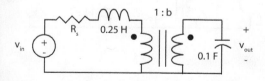

Figure P17.62

63. Repeat Problem 62 for $R_s = 2\ \Omega$. This will be a low-Q_p case. Recall equation 16.47:

$$Q_p \equiv \frac{\omega_p}{2\sigma_p} = \frac{1}{2\sin(\theta)} \quad \text{and} \quad \theta = \arcsin\left(\frac{\sigma_p}{\omega_p}\right),$$

and for exact answers to part (c) recall equations 16.49 through 16.52.

64. Consider the circuit shown in Figure P17.64.

(a) Find the output impedance, $Z_{out}(s)$, as a function of the turns ratio a and the resistance R.

(b) Find the Thevenin equivalent circuit in terms of a, R, and $V_{in}(s)$.

(c) Let $R = 13\ \Omega$ and $a = 2$. Suppose the output is terminated in a series LC with $L = 0.1$ H and $C = 0.1$ mF. Compute the transfer function $H(s) = V_{out}(s)/V_{in}(s)$.

(d) Compute $v_{out}(t)$ for $v_{in}(t) = 15u(t)$. Hint: Use "ilaplace" in the symbolic toolbox of MATLAB.

(e) Given the parameter values of part (c) with $v_{in}(t) = 3\sqrt{10}\ \sin(\sqrt{10}\ t)$ V, find the zero-state response, $v_{out}(t)$. Identify the transient and steady-state responses.

CHECK:

$$v_{out}(t) = \left(-3.2298e^{-400t} + 5.1181e^{-250t} -\right.$$
$$\left.1.8883\cos(10\sqrt{10}t) + 9.0948\sin(10\sqrt{10}t)\right)u(t)$$

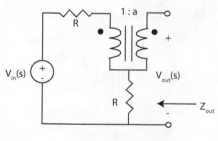

Figure P17.64

65. For the circuit shown in Figure P17.65, find $Z_{th}(s)$, $V_{oc}(s)$, and the Thevenin equivalent circuit.

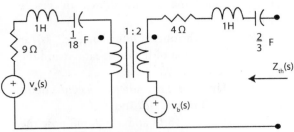

Figure P17.65

ANSWER: $V_{oc}(s) = 2V_a(s) + V_b(s)$ and

$$Z_{th}(s) = 40 + 5s + \frac{73.5}{s}$$

66. In the circuit shown in Figure P17.66, $v_{in}(t) = 5\cos(2t)$ V. Find $i_a(t)$, $i_b(t)$, and $i_c(t)$.

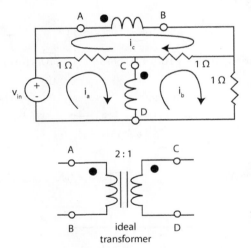

Figure P17.66

67. In the circuit of Figure P17.67, suppose R_s = 10 Ω, R_1 = 40 Ω, and R_L = 2 Ω.
    (a) If the load is identified as the terminals A-B, find a for maximum power transfer.
    (b) If the load is identified as terminals C-D (i.e., R_L is the load), find a for maximum power transfer.

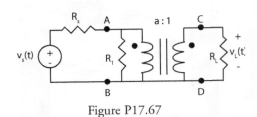

Figure P17.67

68. In the circuit of Figure P17.68, R_s = 25 Ω, R_1 = 144 Ω, R_L = 256 Ω, and $v_s(t) = 10\sqrt{2}\cos(10^5 t)$ V.
    (a) Suppose $a_1 = 1$ (equivalent to a straight-through connection with dc isolation). Find a_2 for maximum power transfer to R_L. For this value of a_2, find $P_{L,ave}$ and $|\mathbf{V}_L|$.
    (b) Suppose $a_2 = 1$ (equivalent to a straight-through connection with dc isolation). Find a_1 for maximum power transfer to R_L. For this value of a_1, find $P_{L,ave}$ and $|\mathbf{V}_L|$.
    (c) Suppose a_1 and a_2 are both adjustable. Find a set of values of a_1 and a_2 so that $P_{L,ave}$ is greater than the values computed in parts (a) and (b). Find this value of $P_{L,ave}$.
    (d) What is the maximum value of $P_{L,ave}$ if a_1 and a_2 are both adjustable?

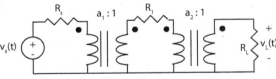

Figure P17.68

ANSWER: (a) 0.1479 watt, (b) 0.64 watt, (c) for a_2 = 2, a_1 = 0.1463, 0.87671 watt, (d) 1 watt

69. Consider the network configuration in Figure P17.69. The source voltage $v_s(t) = \sqrt{2}\cos(\omega t)$ V, where $\omega = 10^5$ rad/sec.
    (a) Design a lossless network N such that maximum average power is transferred to the 10 Ω load at $\omega = 10^5$ rad/sec for

each of the cases cited below.

Case 1: N consists of one inductor L and one ideal transformer. You will need to determine the configuration of the inductor and ideal transformer as well as the values of L and the turns ratio a.

Case 2: N consists of a pair of coupled inductors, in which case you must determine the values of L_1, L_2, and M.

Case 3: N is simply an LC network.

(b) For each case compute the transfer function

$$H(s) = \frac{V_{10\Omega}(s)}{V_s(s)}.$$

Plot the magnitude response for $10^3 \leq \omega \leq 10^7$ rad/sec using a semilog scale. Which circuit has the better performance over the range of frequencies and why?

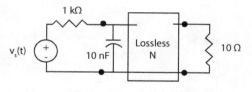

Figure P17.69

70. Consider the resonant circuit shown in Figure P17.70 containing a pair of coupled inductors. Suppose $R_s = 20\ \Omega$, $L_1 = 1.5$ mH, $M = 3$ mH, $L_2 = 6$ mH, and $R_L = 320\ \Omega$.

(a) Find the coupling coefficient k.

(b) Model the coupled inductors using one of the equivalent circuits in Figure 17.33. Find the impedance $Y_1(s)$ and then $Y_1(j\omega)$ as a function of C.

(c) Determine the value of C that leads to a resonant frequency of $\omega_r = 10$ krad/sec.

(d) Find the voltage gain at resonance; i.e., find $V_{out}(j\omega_r)/V_{in}(j\omega_r)$.

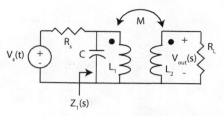

Figure P17.70

71. The front end of a radio receiver uses a band-pass circuit as shown in Figure P17.71a consisting of a capacitor and coupled coils whose mutual inductance is determined by the tap position. For analysis purposes, this circuit is modeled using the equivalent circuit of Figure P17.71b. Since the circuit is connected to an RF amplifier, we will assume that the amplifier input impedance is infinite to avoid loading. Our goal is to obtain the frequency response of this circuit and verify that it is a band-pass circuit. Assume $C = 100$ pF and $R_s = 500\ \Omega$.

(a) Compute the coupling coefficient k.

(b) Find the transfer function $H(s) = V_{out}(s)/V_{in}(s)$. Suggestion: Use one of the models of Figure 17.33 for the coupled coils.

(c) Compute ω_m, $|H(j\omega_m)|$, B_ω, and Q of the circuit.

Remark: We could have designed a parallel resonant circuit to achieve the frequency-selective property of the circuit in Figure P17.71a. However, because of the mutual inductance, we can achieve a much larger voltage gain than that achievable by a simple parallel RLC. This becomes clear when one uses a model for the coupled inductors of Figure P17.71b; this model contains an ideal (step-up) transformer having a turns ratio of 1:10.

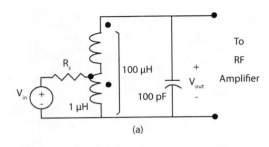

(a)

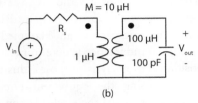

(b)

Figure P17.71

ANSWERS: (a) 1; (c) $\omega_m = 10^7$ rad/sec, $B_\omega =$
2×10^5 rad/sec, $Q = 50$, and $H_m = 10$

APPLICATIONS

72. Recently, coupled coils have been manufactured on printed circuit boards by printing two
spiral coils on opposite sides of a board. Such
coils are called coreless PCB transformers and
can be used to effect electrical isolation as well as
energy and signal transfer. This problem illustrates the use of the models developed in this
chapter for an approximate analysis of the frequency response of coreless PCB transformers.

Figure P17.72 shows an equivalent circuit of a
PCB transformer for which $L_1 = L_2 = 600$ nH,
$M = 300$ nH, and $R_1 = R_2 = 1.2\ \Omega$, which represent the winding resistances. Suppose the second coil is terminated in a capacitance of $C =$
600 pF in parallel with a resistance of $R_L =$
2000 Ω.

(a) Find the coupling coefficient k.
(b) Construct a model for the coupled
     inductors using Figure 17.35 of the
     text.
(c) For a first analysis, assume that the
     effect of R_L and the shunt inductance

(CHECK: 600 nH) in the model constructed in part (b) can be neglected.
Find approximate values of ω_m,
$|H(j\omega_m)|$, B_ω, and Q for the voltage
gain $H(s) = V_{out}/V_{in}$. Plot the magnitude frequency response for 2 MHz <
f < 16 MHz.

Hint: Reflect the voltage source and 1.2 Ω
resistance to the right side of the ideal transformer in the equivalent circuit model of part
(b). That is to say, replace the circuit to the left
of the ideal transformer by its Thevenin equivalent.

(d) Including the effects of R_L and the
     600 nH shunt inductance, use a cir-
     cuit simulation program to find more
     accurate values for ω_m, $|H(j\omega_m)|$, B_ω,
     and Q of part (c) and compare.

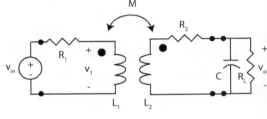

Figure P17.72

ANSWERS: $k = 0.5$, (b) Lshunt= 600 nH,
Lseries= 450 nH, turns ratio a = 2, (c) $\omega_m =$
6.086 Mrad/sec, $B_\omega = 3.333$ Mrad/sec, $Q =$
18.26, $H_m = 9.13$

73. Figure P17.73 shows two series RLC circuits physically isolated but magnetically coupled. The input is a sinusoidal voltage source
$v_{in}(t) = V_s \cos(\omega t)$. We will investigate some
steady-state voltage gains. This investigation
has application to biomedical implants, where
the coupled inductors represent a magnetic
coupling between an implant and an external
circuit.

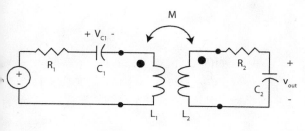

Figure P17.73

If the coupling coefficient $k = 0$ (or $M = 0$), then the steady-state output magnitude is $\mathbf{V}_{out} = 0$. Further, as shown in Chapter 16, $|\mathbf{V}_{C1}| = Q_1 V_s$. As k increases from zero to its maximum value of 1, $|\mathbf{V}_{C1}|$ decreases monotonically, whereas $|\mathbf{V}_{out}|$ increases to a maximum value and then decreases. Exact formulas for several critical values will now be derived.

(a) Show that if the two series RLC circuits in the primary and secondary are identical, i.e., $L_1 = L_2 = L$, $C_1 = C_2 = C$, $R_1 = R_2 = R$, and

$$\omega = \omega_0 = \frac{1}{\sqrt{LC}},$$

then

$$\left|\frac{\mathbf{V}_{out}}{\mathbf{V}_{in}}\right| = \frac{kQ^2}{1+(kQ)^2}$$

and

$$\left|\frac{\mathbf{V}_{C1}}{\mathbf{V}_{in}}\right| = \frac{Q}{1+(kQ)^2}$$

(b) Show that if k is the only adjustable parameter and $Q \geq 1$, then

$$\max_k \left|\frac{\mathbf{V}_{out}}{\mathbf{V}_{in}}\right| = \frac{Q}{2}$$

when

$$k = \frac{1}{Q},$$

and at this degree of coupling

$$\left|\frac{\mathbf{V}_{C1}}{\mathbf{V}_{in}}\right| = \frac{Q}{2}$$

(c) Suppose $L = 1$ mH, $R = 5\ \Omega$, $C = 0.1\mu$ F, $\omega = 100$ krad/sec, and $Vs = 0.1$ V. Find the value of the mutual inductance M needed to maximize $|\mathbf{V}_{out}|$. What is the maximum value of $|\mathbf{V}_{out}|$?

ANSWERS: (c) $Q = 20$, $M = 0.05$ mH, $V_{out,max} = 1$ V

74. Example 17.7 provided an approximate analysis of a car ignition system. This problem asks for a SPICE simulation of the example. Reconsider the circuit of Figure 17.12b without the switch. Show that the initial current through the primary of the coupled inductors is $i_{L1}(0^-) = 12$ A. Since the secondary is open-circuited, there is no initial current there. Generate a SPICE simulation for $0 \leq t \leq 1$ msec. Verify that $|v_2(t)|_{max} \cong 36 \times 10^3$ V.

75. In the car ignition circuit of Figure 17.12b, suppose the resistance and capacitance values are changed to 2 Ω and 0.25 μF. Find approximately the maximum voltage appearing cross the spark plug.
ANSWER: 36,000 V

76. Again consider the car ignition circuit of Figure 17.12b. Suppose the resistance is 2 Ω, $C = 1$ μF, and $L_1 = 10$ mH; find the approximate value of the mutual inductance M to achieve $|v_2(t)|_{max} = 38 \times 10^3$ V. What is the approximate minimum corresponding value of L_2, the inductance of the secondary of the coil? Verify your circuit operation using SPICE.

C H A P T E R *18*

Two-Ports

THE AMPLIFIER: A PRACTICAL TWO-PORT

An actress speaks into a microphone. Speakers instantly replicate her voice, which resounds throughout the auditorium. What happens between the speaking and the hearing? A microphone produces a voltage signal that changes in proportion to the tenor and loudness of the voice of the actress. Amplifiers magnify this changing voltage signal perhaps a hundred or a thousand times to drive speakers whose cones reverberate in proportion to the changing voltage signal. The cones then cause the air to vibrate intensely, also in proportion to the tenor and loudness of the actress's voice. Her words become heard by thousands because of the amplifier.

Amplifier circuits are found in instrumentation and a huge number of appliances. In radios, radio frequency amplifiers first magnify signals from an antenna. Special circuitry then extracts the audio portion from these antenna signals. Other circuitry amplifies the audio to drive speakers. Video signals from a video cassette recorder are amplified by special circuits for connection to a monitor. Amplifier circuitry enhances signals coming from sensors in various manufacturing processes. Repeater circuits, among other things, amplify phone signals whose magnitudes have attenuated during microwave transmission. There are a large number of other applications of amplifiers.

From the preceding discussion, one can surmise that an amplifier circuit has an input signal and an output signal. This configuration is represented by a device called a *two-port* that has an input port for the input signal and an output port for the output signal. The following figure represents the idea (one of the homework problems asks for an analysis of this amplifier):

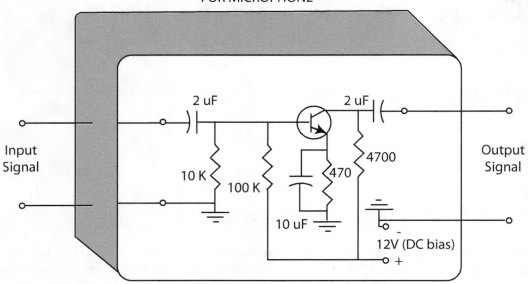

TRANSISTOR AMPLIFIER CIRCUIT
FOR MICROPHONE

Often the circuit between the ports is highly complex. This chapter looks at shorthand methods for analyzing the input-output properties of two-ports without having to deal directly with a possibly highly complex circuit internal to the two-port. The chapter will provide a variety of methods for analyzing amplifier circuits.

CHAPTER OBJECTIVES

1. Provide a general setting for one-port analysis.
2. Extend one-port analysis to the analysis of two-ports.
3. Describe the input-output properties of two-ports in terms of four sets of characteristic parameters: impedance, or *z*-parameters; admittance, or *y*-parameters; hybrid, or *h*-parameters; and transmission, or *t*-parameters.
4. Develop specific formulas for the input impedance, output impedance, and voltage gain of two-ports driven by a practical source voltage and terminated by a load impedance.
5. Introduce and interpret the notion of reciprocity in terms of the various two-port parameters.

SECTION HEADINGS

1. **Introduction**
2. **One-Port Networks**
3. **Two-Port Admittance Parameters**
4. ***y*-Parameter Analysis of Terminated Two-Ports**
5. **Two-Port Impedance Parameters**
6. **Impedance and Gain Calculations of Terminated Two-Ports Modeled by *z*-Parameters**

1. INTRODUCTION

Figure 18.1 shows a general *one-port,* whose two terminals satisfy the property that for any voltage V_1 across them, the current entering one terminal, say, I_1, equals the current leaving the second terminal. A resistor is a one-port: the current entering one terminal equals the current leaving the other terminal. A capacitor and an inductor are also one-ports. A *general* one-port contains any number of interconnected resistors, capacitors, inductors, and other devices. In a one-port, only the relationship between the port voltage and current is of interest. For example, the port voltage and current in a resistor, capacitor, and inductor satisfy the relationships $v_R = Ri_R$, Cdv_c/dt $= i_C$ and $Ldi_L/dt = v_L$, respectively. Practically speaking, one-ports are macroscopic device models emphasizing input-output properties rather than detailed internal models. Thévenin and Norton equivalent circuits determine one-port models when only a pair of terminals of a network is of interest.

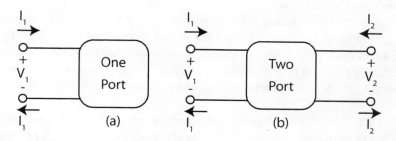

FIGURE 18.1 (a) General one-port. (b) General two-port.

A two-port is a linear network having two pairs of terminals, as illustrated in Figure 18.1b. Each terminal pair behaves as a port; i.e., the current entering one terminal of a port equals the current leaving the second terminal of the same port for all voltages across the port. Coupling networks such as transformers have two pairs of terminals, each of which behaves as a one-port. Hence, transformers are two-port devices. In modeling a two-port, one must define a relationship among four variables. Different groupings of current and voltage variables lead to different types of characterizing parameters. For example, **admittance parameters** (termed *y*-parameters) relate the two (input) voltages, V_1 and V_2, to the (output) port currents, I_1 and I_2. **Impedance parameters** (termed *z*-parameters) relate the two-port (input) currents to the two-port (output) voltages. Other types of parameters investigated in this chapter are **hybrid or *h*-parameters**, and **transmission or *t*-parameters**.

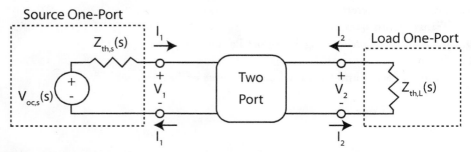

Source One-Port

FIGURE 18.2 A general two-port connected to a source one-port (represented by a Thevenin equivalent) and terminated in a load one-port (represented by a Thevenin equivalent impedance).

In practice, two-ports often represent coupling devices in which a source delivers energy to a load through the two-port network as suggested in Figure 18.2. For example, stereo amplifiers take a small low-power audio signal and increase its power so that it will drive a speaker system. Determining and knowing ratios such as the voltage gain, current gain, and power gain of a two-port is very important in dealing with a source that delivers power through a two-port to a load. This chapter develops various formulas for computing these gains for each type of two-port parameter.

2. ONE-PORT NETWORKS

Basic Impedance Calculations

As mentioned in the introduction, the current entering one terminal of a one-port, illustrated in Figure 18.1, must equal the current leaving the second terminal for any voltage V_1 across the terminals. We begin study of one-ports by exploring two impedance calculations of a transistor amplifier circuit that are pertinent to basic electronic analysis. The impedance or admittance seen at a port is fundamental to the behavior of a network to which a one-port or two-port is connected.

EXAMPLE 18.1. The circuit of Figure 18.3 is pertinent to a simplified model of a common-collector stage of a transistor amplifier circuit. Specifically, the one-port of Figure 18.3 models the **input impedance**. In a common-collector amplifier stage, the impedance Z_1 is very large and is often neglected, i.e., we assume $|Z_1| \approx \infty$ over the useful bandwidth of the circuit. The goal of this example is to compute Z_{in} and interpret Z_{in} in terms of a transistor current gain parameter denoted by β.

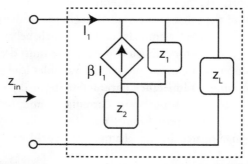

FIGURE 18.3 One-port representing the input characteristic of a common-collector transistor amplifier circuit.

SOLUTION

We attach a hypothetical voltage source, say, V_{in}, across the port terminals, top to bottom being plus to minus, to induce a hypothetical current I_1. Neglecting Z_1, from KCL,

$$I_1 + \beta I_1 - \frac{V_{in}}{Z_L} = 0$$

Since the input impedance is the ratio of the input voltage to the input current,

$$Z_{in} = \frac{V_{in}}{I_1} = (\beta + 1) Z_L$$

Thus, for a large β, say, 150, the input impedance can reach very high levels for reasonably sized impedances Z_L. When amplifying voltage signals, we desire to have the ratio of the internal source impedance to the amplifier input impedance be small. Conversely, the amplifier input impedance should be large relative to the internal source impedance.

Exercises. 1. In the circuit of Figure 18.3, suppose $\beta = 100$, $Z_1 = \infty$, and the load resistance $Z_L = 16 \ \Omega$. If a sinusoidal voltage $v_{in}(t) = 15\sqrt{2} \ \sin(5000t)$ V is connected to the input terminals, find the average powers P_{in} and P_L and the power gain.
ANSWER: 139 mW, 14.06 W, 101

2. Consider the circuit of Figure 18.4. The one-port shown here models the output impedance of a common-collector stage of a transistor amplifier circuit. Show that

$$Z_{out} = \frac{V_{out}}{I_{out}} = \frac{Z_1}{(\beta + 1)}.$$

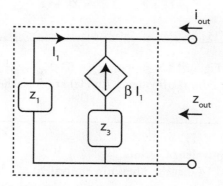

FIGURE 18.4 One-port representing the output characteristic of a common-collector transistor amplifier circuit.

3. In the circuit of Figure 18.4, suppose $\beta = 99$ and $Z_1 = 1$ kΩ. If a current source $I_s = 200$ mA is connected to the one-port terminals, find $|V_{out}|$ and $|I_1|$.
ANSWER: 2 V, 2 mA

A second example that is common to basic electronic analysis depicts a circuit used in the analysis of field-effect transistors (FETs).

EXAMPLE 18.2. Figure 18.5a represents a simplified model of the input impedance of a field-effect transistor circuit. As in Example 18.1, we neglect the large impedance in parallel with $g_m V_1$. Compute the input impedance Z_{in} in terms of Z_1, Z_2, Z_3, and g_m.

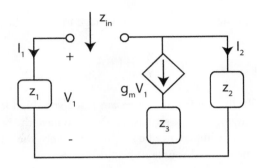

FIGURE 18.5. Simplified input impedance model of a field-effect transistor (FET) circuit.

SOLUTION
As illustrated in Figure 18.5, assume a hypothetical voltage source V_{in} (plus to minus is right to left) has been attached across the port terminals. At the bottom node, $I_1 + g_m V_1 + I_2 = 0$. Since $V_1 = Z_1 I_1$ and $I_2 = (V_{in} + V_1)/Z_2$, it follows that

$$I_1 + g_m Z_1 I_1 + \frac{V_{in}}{Z_2} + \frac{Z_1 I_1}{Z_2} = 0 \ .$$

Now, since $Z_{in} = \dfrac{V_{in}}{I_{in}}$,

$$Z_{in} = Z_1 + (1 + g_m Z_1)\,Z_2 = Z_2 + (1 + g_m Z_2)\,Z_1.$$

Exercise. In the circuit of Figure 18.5, suppose $g_m = 0.002$ S, $Z_2 = 500\ \Omega$, and Z_1 is the impedance of a 0.1 μF capacitor. Find the frequency (in Hz) at which the magnitude of Z_{in} is 707 Ω.
ANSWER: 1003.2 Hz

Thevenin and Norton Equivalent Circuits
Recall from Chapters 6 and 14 that the **Thevenin equivalent** of a network, as seen from a pair of terminals (i.e., from a one-port), is a voltage source in series with the Thevenin equivalent impedance Z_{th}. Z_{th} is simply the equivalent impedance of the one-port when all internal independent sources are set equal to zero. The value of the voltage source, V_{oc} equals that voltage appearing at the open-circuited port terminals. A source transformation on the Thevenin equivalent produces the **Norton equivalent**, which is simply a current source in parallel with the Thevenin impedance. The value of the current source is $I_{sc} = V_{oc}/Z_{th}$. This is the current that would pass through a short circuit on the port terminals.

EXAMPLE 18.3. Consider the circuit ofFfigure 18.6. Find the Thevenin and Norton equivalent circuits.

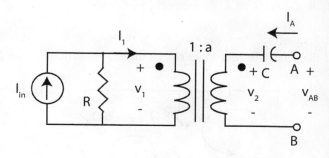

FIGURE 18.6 Transformer circuit for Example 18.3.

SOLUTION

The voltage across A-B satisfies

$$V_{AB} = \frac{1}{Cs} I_A + V_2 = \frac{1}{Cs} I_A + aV_1$$

To determine the Thevenin equivalent, we express V_1 in terms of I_{in} and I_A. From Ohm's law and the current relationship $I_1 = -aI_A$,

$$V_1 = R(I_{in} - I_1) = RI_{in} + aRI_A$$

Therefore,

$$V_{AB} = \left(\frac{1}{Cs} + Ra^2 \right) I_A + aRI_{in} \triangleq Z_{th}(s)I_A + V_{oc}(s)$$

Thus $Z_{th}(s) = \frac{1}{Cs} + Ra^2$, $V_{oc}(s) = aRI_{in}$, and $I_{sc}(s) = \frac{V_{oc}(s)}{Z_{th}(s)} = \frac{aRI_{in}}{\frac{1}{Cs} + Ra^2}$, as set forth in Figures 18.7a and b.

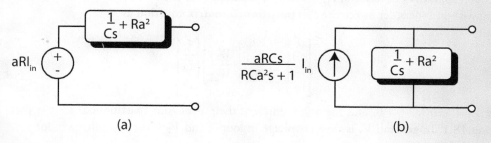

FIGURE 18.7 (a) Thevenin and (b) Norton equivalent circuits of Figure 18.6.

Exercise. Suppose $R = 100\ \Omega$, $a = 5$, $C = 10$ mF, and $i_{in}(t) = 4u(t)$ mA. Find V_{oc}, and the poles and zeros of Z_{th}.
ANSWER: $V_{oc} = 10/s$; pole at $s = 0$ and zero at $s = -0.04$.

A technique known as **matrix partitioning** often simplifies the calculation of Thevenin and Norton equivalent circuits. Example 18.4 illustrates the technique.

EXAMPLE 18.4. Find the Thevenin equivalent circuit for the network of Figure 18.8.

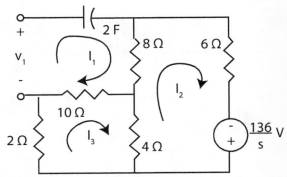

FIGURE 18.8 A three-loop *RC* circuit for Example 18.4.

SOLUTION

Step 1. *Construct the loop equations.* By inspection, the loop equations for the circuit satisfy

$$
\begin{bmatrix} V_1 \\ \hline \dfrac{136}{s} \\ 0 \end{bmatrix} = \begin{bmatrix} \dfrac{1}{2s}+18 & -8 & -10 \\ \hline -8 & 18 & -4 \\ -10 & -4 & 16 \end{bmatrix} \begin{bmatrix} I_1 \\ \hline I_2 \\ I_3 \end{bmatrix} \tag{18.1}
$$

The goal is to find V_1 in terms of I_1; the matrix equation 18.1 is partitioned accordingly. To avoid dealing with numbers, let us rewrite this **partitioned matrix** equation as

$$
\begin{bmatrix} V_1 \\ V_2 \\ V_3 \end{bmatrix} = \begin{bmatrix} W_{11} & W_{12} \\ \hline W_{21} & W_{22} \end{bmatrix} \begin{bmatrix} I_1 \\ I_2 \\ I_3 \end{bmatrix} \tag{18.2}
$$

where $V_2 = 136/s$, $V_3 = 0$, and the W_{ij}'s represent their analogous (partitioned) counterparts in equation 18.1. In general, V_2 is the net voltage in loop 2 and V_3 is the net voltage in loop 3.

Step 2. *Solve the partitioned matrix equations for V_1 in terms of I_1 and the vector $[V_2, V_3]^T$.* The matrix equation 18.2 may be rewritten as a set of two equations, namely

$$
V_1 = W_{11}I_1 + W_{12}\begin{bmatrix} I_2 \\ I_3 \end{bmatrix} \tag{18.3}
$$

and

$$\begin{bmatrix} V_2 \\ V_3 \end{bmatrix} = W_{21}I_1 + W_{22}\begin{bmatrix} I_2 \\ I_3 \end{bmatrix}. \tag{18.4}$$

Since W_{22} is invertible (for all passive networks), solving equation 18.4 for the vector $[I_2 \ I_3]^T$ yields

$$\begin{bmatrix} I_2 \\ I_3 \end{bmatrix} = W_{22}^{-1}\begin{bmatrix} V_2 \\ V_3 \end{bmatrix} - W_{22}^{-1}W_{21}I_1 \tag{18.5}$$

Substituting equation 18.5 into equation 18.3 produces

$$V_1 = \left[W_{11} - W_{12}W_{22}^{-1}W_{21} \right] I_1 + W_{12}W_{22}^{-1}\begin{bmatrix} V_2 \\ V_3 \end{bmatrix} \tag{18.6}$$

MATLAB or its equivalent, or possibly a symbolic manipulation software package, allows the matrices in this equation to be conveniently computed. Comparing equation 18.6 with the structure of a Thevenin equivalent circuit,

$$V_1 = Z_{th}I_1 + V_{oc}$$

allows us to conclude that the Thevenin impedance is

$$Z_{th} = \left[W_{11} - W_{12}W_{22}^{-1}W_{21} \right] \tag{18.7}$$

and the open-circuit voltage is

$$V_{oc} = W_{12}W_{22}^{-1}\begin{bmatrix} V_2 \\ V_3 \end{bmatrix} \tag{18.8}$$

Step 3. *Compute Z_{th} and V_{oc}.* From equation 18.7,

$$Z_{th}(s) = \frac{36s+1}{2s} - \begin{bmatrix} -8 & -10 \end{bmatrix}\begin{bmatrix} 18 & -4 \\ -4 & 18 \end{bmatrix}^{-1}\begin{bmatrix} -8 \\ -10 \end{bmatrix} = 5.265\frac{s+0.095}{s} \ \Omega$$

and from equation 18.8,

$$V_{oc} = \begin{bmatrix} -8 & -10 \end{bmatrix}\begin{bmatrix} 18 & -4 \\ -4 & 16 \end{bmatrix}^{-1}\begin{bmatrix} \frac{136}{s} \\ 0 \end{bmatrix} = -\frac{84}{s}.$$

Figure 18.9 shows the resulting Thevenin equivalent circuit.

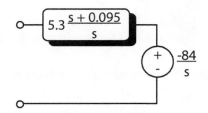

FIGURE 18.9 Thevenin equivalent circuit of Figure 18.8.

Exercise. Suppose the capacitor in Example 18.4 is changed to a 2 Ω resistor. Compute the new $Z_{th}(s)$.

ANSWER: $Z_{th}(s) = 7.2647$ Ω via the MATLAB code

»W11 = 20; W12 = [-8 -10];
»W21 = [-8; -10]; W22 = [18 -4;-4 16];
»Zth = W11 - W12*inv(W22)*W21

3. TWO-PORT ADMITTANCE PARAMETERS

Basic Definitions and Examples
As mentioned earlier, two-port representations allow us to deal only with the terminal voltages and currents of the two-port, depicted in Figure 18.1b and repeated in Figure 18.10.

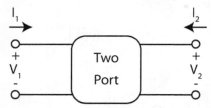

FIGURE 18.10 Standard two-port configuration having four external variables: I_1, I_2, V_1, and V_2.

ADMITTANCE PARAMETERS

Throughout this and later discussions, assume that the two-port of Figure 18.10 has no internal independent sources and that all dynamic elements are initially relaxed, i.e., have no initial conditions. Under these assumptions, the admittance parameters or y-parameters of a two-port are expressions for the terminal currents, I_1 and I_2, in terms of the port voltages, V_1 and V_2, i.e.,

$$\begin{bmatrix} I_1 \\ I_2 \end{bmatrix} = \begin{bmatrix} y_{11} & y_{12} \\ y_{21} & y_{22} \end{bmatrix} \begin{bmatrix} V_1 \\ V_2 \end{bmatrix} \qquad (18.9a)$$

or, in scalar notation,

$$I_1 = y_{11}V_1 + y_{12}V_2$$
$$I_2 = y_{21}V_1 + y_{22}V_2 \qquad (18.9b)$$

Using either of these sets of equations, one can define each admittance parameter, y_{ij}, as follows:

$$y_{11} = \left.\frac{I_1}{V_1}\right|_{V_2=0} \qquad y_{12} = \left.\frac{I_1}{V_2}\right|_{V_1=0} \qquad (18.10)$$

$$y_{21} = \left.\frac{I_2}{V_1}\right|_{V_2=0} \qquad y_{22} = \left.\frac{I_2}{V_2}\right|_{V_1=0}$$

Since each admittance is defined with regard to a shorted terminal voltage $V_i = 0$, the y_{ij} are often called **short-circuit admittance parameters**. The unit of an admittance parameter is S.

Some examples will illustrate convenient methods for computing the y-parameters.

EXAMPLE 18.5. Compute the short-circuit admittance parameters of the circuit in Figure 18.11.

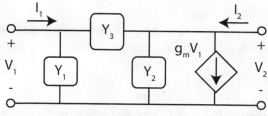

FIGURE 18.11 Circuit for Example 18.5.

SOLUTION

The overall strategy is to write equations for I_1 and I_2 in terms of V_1 and V_2 using nodal analysis. Accordingly, by inspection,

$$I_1 = Y_1\,V_1 + Y_3\,(V_1 - V_2) = (Y_1 + Y_3)\,V_1 - Y_3\,V_2$$

and

$$I_2 = g_m V_1 + Y_2 V_2 + Y_3 (V_2 - V_1) = (g_m - Y_3) + (Y_2 + Y_3) V_2$$

In matrix form, these equations are

$$\begin{bmatrix} I_1 \\ I_2 \end{bmatrix} = \begin{bmatrix} Y_1 + Y_3 & -Y_3 \\ g_m - Y_3 & Y_2 + Y_3 \end{bmatrix} \begin{bmatrix} V_1 \\ V_2 \end{bmatrix} \triangleq \begin{bmatrix} y_{11} & y_{12} \\ y_{21} & y_{22} \end{bmatrix} \begin{bmatrix} V_1 \\ V_2 \end{bmatrix}$$

in which case

$$y_{11} = Y_1 + Y_3, y_{12} = -Y_3, y_{21} = g_m - Y_3, \text{ and } y_{22} = Y_2 + Y_3.$$

Exercise. Suppose the controlled current source in Figure 18.11 is reconnected across Y_3 with the arrow pointing to the left. Find the new y-parameters.

ANSWER: $y_{11} = Y_1 + Y_3 - g_m, y_{12} = -Y_3, y_{21} = g_m - Y_3,$ and $y_{22} = Y_2 + Y_3$

In the next example, we combine the method of matrix partitioning with the use of nodal equations to compute the y-parameters.

EXAMPLE 18.6. Compute the y-parameters for the circuit of Figure 18.12.

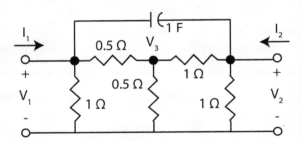

FIGURE 18.12 Three-node circuit for Example 18.6.

SOLUTION
Computation of the y-parameters will again proceed by the method of nodal analysis. Consider, for example, node 1, in which the current I_1 must be determined. I_1 has the form

$$I_1 = Y_{11} V_1 + Y_{12} V_2 + Y_{13} V_3$$

where each variable is understood to be a function of s. The coefficient Y_{11} is simply the sum of the admittances incident at node 1. The coefficient Y_{12} is simply the negative of the sum of the admittances between nodes 1 and 2, and Y_{13} is the negative of the sum of the admittances between nodes 1 and 3. Similarly, $I_2 = Y_{21} V_1 + Y_{22} V_2 + Y_{23} V_3$, where Y_{22} is the sum of the admittances incident at node 2, etc. Hence, in matrix form, the node equations of the circuit of Figure 18.12 are

$$\begin{bmatrix} I_1 \\ I_2 \\ 0 \end{bmatrix} = \begin{bmatrix} s+3 & -s & \vdots & -2 \\ -s & s+2 & \vdots & -1 \\ \hline -2 & -1 & \vdots & 5 \end{bmatrix} \begin{bmatrix} V_1 \\ V_2 \\ V_3 \end{bmatrix} \equiv \begin{bmatrix} W_{11} & \vdots & W_{12} \\ \hline W_{21} & \vdots & W_{22} \end{bmatrix} \begin{bmatrix} V_1 \\ V_2 \\ V_3 \end{bmatrix} \tag{18.11}$$

This nodal equation matrix is symmetric, because the *RLC* network contains no dependent sources. Using the method of matrix partitioning introduced by equation 18.6 yields

$$\begin{bmatrix} I_1 \\ I_2 \end{bmatrix} = \begin{bmatrix} W_{11} - W_{12}W_{22}^{-1}W_{21} \end{bmatrix} \begin{bmatrix} V_1 \\ V_2 \end{bmatrix} = \left(\begin{bmatrix} s+3 & -s \\ -s & s+2 \end{bmatrix} - \frac{1}{5} \begin{bmatrix} 2 \\ 1 \end{bmatrix} \begin{bmatrix} 2 & 1 \end{bmatrix} \right) \begin{bmatrix} V_1 \\ V_2 \end{bmatrix}$$

$$= \begin{bmatrix} s+2.2 & -(s+0.4) \\ -(s+0.4) & s+1.8 \end{bmatrix} \begin{bmatrix} V_1 \\ V_2 \end{bmatrix}.$$

Exercise. The circuit of Figure 18.12 is modified by adding a 2 Ω resistor in parallel with the capacitor. Find the new *y*-parameters.

ANSWER: $\begin{bmatrix} s+2.7 & -(s+0.9) \\ -(s+0.9) & s+2.3 \end{bmatrix}$

A last example couples a transformer with a resistive π-network.

EXAMPLE 18.7. Compute the *y*-parameters of the circuit in Figure 18.13.

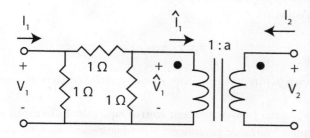

FIGURE 18.13 A resistive π-network coupled to an ideal
transformer circuit for Example 18.7.

SOLUTION

Find the y-parameters by using nodal analysis in conjunction with the ideal transformer equations.
First, at port 1,

$$I_1 = V_1 + \left(V_1 - \hat{V}_1 \right) = 2V_1 - \hat{V}_1 = 2V_1 - \frac{1}{a}V_2 \tag{18.12}$$

Now considering that $aI_2 = -\hat{I}$, a node equation at the primary of the transformer yields

$$I_2 = -\frac{1}{a}\hat{I}_1 = -\frac{1}{a}\left[-\hat{V}_1 + \left(V_1 - \hat{V}_1\right)\right] = -\frac{1}{a}V_1 + \frac{2}{a^2}V_2,\qquad(18.13)$$

where the last equality uses the relationship $a\hat{V}_1 = V_2$. Putting equations 18.12 and 18.13 in **matrix** form yields the *y*-parameter relationship

$$\begin{bmatrix} I_1 \\ I_2 \end{bmatrix} = \begin{bmatrix} 2 & -\dfrac{1}{a} \\ -\dfrac{1}{a} & \dfrac{2}{a^2} \end{bmatrix}\begin{bmatrix} V_1 \\ V_2 \end{bmatrix}.$$

Exercise. In the circuit of Figure 18.13, the top resistor is changed from 1 Ω to 0.25 Ω. With the turns ratio $a = 2$, find the new *y*-parameters.

ANSWER: $\begin{bmatrix} 5 & -2 \\ -2 & 1.25 \end{bmatrix}$ S.

Two–Dependent Source Equivalent Circuit

The key to engineering analysis rests with the interpretation of appropriate mathematical equations. The key to two-port analysis is the interpretation of the two-port equations. Take, for example, the first admittance equation,

$$I_1 = y_{11}V_1 + y_{12}V_2$$

One circuit-theoretic interpretation of this equation has the port current I_1 equal to the port voltage V_1 times an admittance y_{11} in parallel with a voltage-controlled current source $y_{12}V_2$. A similar interpretation of the equation

$$I_2 = y_{21}V_1 + y_{22}V_2$$

yields an admittance branch y_{22} in parallel with a voltage-controlled current source $y_{21}V_1$.

These interpretations lead to the **two–dependent source equivalent circuit** of a two-port represented by the short-circuit admittance parameters. (See Figure 18.14b.) This equivalent circuit aids the computation of input and output impedances and voltage gain formulas. Note: In this chapter and later chapters the standard resistance symbol often is used to designate a general impedance or admittance.

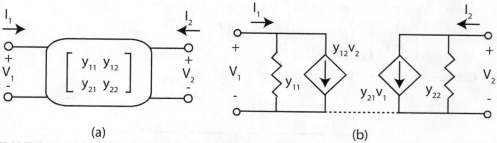

(a)   (b)

FIGURE 18.14 (a) Short-circuit admittance parameters. (b) Their two–dependent source equivalent circuit. The dotted line at the bottom of (b) indicates that the two halves are not necessarily connected.

4. Y-PARAMETER ANALYSIS OF TERMINATED TWO-PORTS

This section takes up the task of analyzing **terminated two-ports** modeled by y-parameters. A two-port is terminated when it has source/load admittances. Any circuit or system in which a source provides an excitation signal to an interconnection network that modifies the signal and drives a load impedance can be represented by a terminated two-port. Such a scenario is common to numerous real-world systems. For example, the utility industry delivers power to a home from a generating facility through a transmission network, and a telephone system delivers voice information from the phone through a transmission network to a receiver.

Input and Output Admittance Calculations
The input and output admittances of a terminated two-port are important for determining power transfer and for various gain computations. In what follows, we will show two different methods for computing the input admittance of the terminated two-port illustrated in Figure 18.15. Computation of the output admittance is left as an exercise.

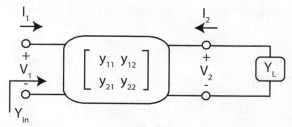

FIGURE 18.15 Two-port terminated by a load admittance Y_L.

The first method for computing $Y_{in} = I_1/V_1$ is a matrix method. Recall equation 18.9a,

$$\begin{bmatrix} I_1 \\ I_2 \end{bmatrix} = \begin{bmatrix} y_{11} & y_{12} \\ y_{21} & y_{22} \end{bmatrix} \begin{bmatrix} V_1 \\ V_2 \end{bmatrix}$$

Using the terminal conditions imposed by the load Y_L in Figure 18.15, we obtain

$$I_2 = -Y_L V_2$$

Incorporating this terminal condition into the two-port y-parameter equation yields

$$\begin{bmatrix} I_1 \\ -Y_L V_2 \end{bmatrix} = \begin{bmatrix} y_{11} & y_{12} \\ y_{21} & y_{22} \end{bmatrix} \begin{bmatrix} V_1 \\ V_2 \end{bmatrix} \Rightarrow \begin{bmatrix} I_1 \\ 0 \end{bmatrix} = \begin{bmatrix} y_{11} & y_{12} \\ y_{21} & y_{22} + Y_L \end{bmatrix} \begin{bmatrix} V_1 \\ V_2 \end{bmatrix}.$$

Using Cramer's rule to solve for V_1 in terms of I_1 results in

$$V_1 = \frac{\det\begin{bmatrix} I_1 & y_{12} \\ 0 & y_{22} + y_L \end{bmatrix}}{\det\begin{bmatrix} y_{11} & y_{12} \\ y_{21} & y_{22} + Y_L \end{bmatrix}} = \frac{(y_{22} + Y_L) I_1}{y_{11}(y_{22} + Y_L) - y_{12} y_{21}}.$$

Because Y_{in} is the ratio of I_1 to V_1, the **input admittance** of the two-port of Figure 18.15 is

$$Y_{in} = \frac{I_1}{V_1} = y_{11} - \frac{y_{12} y_{21}}{y_{22} + Y_L}. \qquad (18.14)$$

EXAMPLE 18.8. Derive the input admittance of equation 18.14 using the two–dependent source equivalent of Figure 18.16. Here we avoid the solution of simultaneous equations while increasing insight into the operation of the two-port.

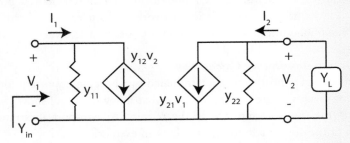

FIGURE 18.16 Input admittance calculation using two–dependent source equivalent circuit.

SOLUTION
With regard to the right side of Figure 18.16, the current $y_{21} V_1$ must equal the current through the parallel admittances y_{22} and Y_L, i.e.,

$$y_{21} V_1 = - (y_{22} + Y_L) V_2$$

Hence,

$$V_2 = - \frac{y_{21}}{y_{22} + Y_L} V_1 \qquad (18.15)$$

Equation 18.15 says that the voltage V_2 equals the current, $-y_{21} V_1$, times the impedance, $\frac{1}{(y_{22} + Y_L)}$. Now consider the left side of Figure 18.16. Here

$$I_1 = y_{11} V_1 + y_{12} V_2 = \left(y_{11} - \frac{y_{12} y_{21}}{y_{22} + Y_L} \right) V_1$$

Again, $Y_{in} = \frac{I_1}{V_1}$ leads to the same formula as equation 18.14.

Exercise. Let Y_s denote a source admittance. Show that the **output admittance** of Figure 18.17 is

$$Y_{out} = \frac{I_2}{V_2} = y_{22} - \frac{y_{12}y_{21}}{y_{11} + Y_s}. \qquad (18.16)$$

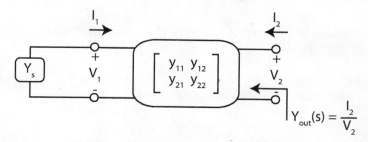

FIGURE 18.17 Input-terminated two-port for output admittance calculation.

Gain Calculations

Our objective now is to derive a formula for the voltage gain of a doubly terminated two-port, as illustrated in Figure 18.18. Again, the y_{ij} symbols denote general admittances rather than the traditional conductances.

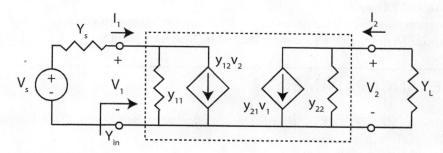

FIGURE 18.18 Doubly terminated two-port driven by voltage source, V_s; y_{ij} symbols denote general admittances.

The specific goal is to derive the voltage gain formula

$$G_V = \frac{V_2}{V_s} = \left(\frac{Y_s}{Y_s + Y_{in}} \right) \left(\frac{-y_{21}}{y_{22} + Y_L} \right). \qquad (18.17)$$

The overall gain calculation breaks down into two cascaded gain calculations as follows:

$$G_V = \frac{V_2}{V_s} = \frac{V_2}{V_1} \times \frac{V_1}{V_s} \qquad (18.18)$$

Computation of the gain, $G_{v1} = \dfrac{V_1}{V_s}$, follows directly from voltage division:

$$G_{v1} = \frac{V_1}{V_2} = \frac{Z_{in}}{Z_{in} + Z_s} = \frac{Y_s}{Y_s + Y_{in}} \qquad (18.19a)$$

To compute the gain $G_{v2} = \dfrac{V_2}{V_1}$ from directly from equation 18.15,

$$V_2 = -\frac{y_{21}}{y_{22} + Y_L} V_1$$

which implies

$$G_{v2} = \frac{V_2}{V_1} = -\frac{y_{21}}{y_{22} + Y_L} \tag{18.19b}$$

Equation 18.17 follows by substituting equations 18.19a and 18.19b into equation 18.18.

Exercises. 1. In the circuit of Figure 18.18, $y_{11} = 5$, $y_{12} = -0.2$, y $_{21} = 50$, $y_{22} = 1$ (all in mS), $R_s = 1$ kΩ and $R_L = 2$ kΩ. Find Z_{in} and $G_V = V_2/V_s$.
ANSWER: $Z_{in} = 85.7$ Ω, $G_V = -2.632$

2. Compute the current gain, $(-I_2)/I_1$, of the circuit of Figure 18.18.

ANSWER: $\left(\dfrac{-y_{21}}{Y_{in}} \right) \left(\dfrac{Y_L}{Y_L + y_{22}} \right)$.

---

5. TWO-PORT IMPEDANCE PARAMETERS

Definition and Examples

The z-parameters, or impedance parameters, relate s-domain currents to s-domain voltages, as one would expect. The z-parameters are the inverse of the y-parameters in most cases.

IMPEDANCE PARAMETERS

For the two-port of Figure 18.10, the z-parameters z_{ij} relate the port currents to the port voltages according to the matrix equation

$$\begin{bmatrix} V_1 \\ V_2 \end{bmatrix} = \begin{bmatrix} z_{11} & z_{12} \\ z_{21} & z_{22} \end{bmatrix} \begin{bmatrix} I_1 \\ I_2 \end{bmatrix}, \tag{18.20}$$

under the assumption of zero initial conditions and no internal independent sources. Therefore, from equation 18.20, each individual z-parameter is defined according to the formulas

$$z_{11} = \left. \frac{V_1}{I_1} \right|_{I_2 = 0} \qquad z_{12} = \left. \frac{V_1}{I_2} \right|_{I_1 = 0}$$

$$\tag{18.21}$$

$$z_{21} = \left. \frac{V_2}{I_1} \right|_{I_2 = 0} \qquad z_{22} = \left. \frac{V_2}{I_2} \right|_{I_1 = 0}$$

Since each z_{ij} is defined with one of the ports open-circuited, i.e., $I_1 = 0$ or $I_2 = 0$, the z_{ij}'s are called **open-circuit impedance parameters***. Their unit is the ohm.*

EXAMPLE 18.9

Compute the z-parameters for the circuit of Figure 18.19.

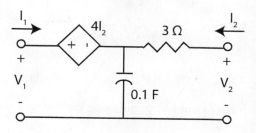

FIGURE 18.19 A simple *T*-circuit for computation of z-parameters in Example 18.9.

SOLUTION

Rather than apply the z-parameter definitions of equation 18.21, we will use mesh analysis to obtain equation 18.20 directly.

Step 1. A loop equation at port 1 yields

$$V_1 = 4I_2 + \frac{10}{s}(I_1 + I_2) = \frac{10}{s}I_1 + \left(4 + \frac{10}{s}\right)I_2 \qquad (18.22)$$

Step 2. Similarly, a loop equation at port 2 produces

$$V_2 = 3I_2 + \frac{10}{s}(I_1 + I_2) = \frac{10}{s}I_1 + \left(3 + \frac{10}{s}\right)I_2 \qquad (18.23)$$

Step 3. By inspection of the right-hand sides of equations 18.22 and 18.23,

$$Z = \begin{bmatrix} z_{11} & z_{12} \\ z_{21} & z_{22} \end{bmatrix} = \begin{bmatrix} \dfrac{10}{s} & \dfrac{4s+10}{s} \\ \dfrac{10}{s} & \dfrac{3s+10}{s} \end{bmatrix}$$

In the next example, we utilize the technique of matrix partitioning to compute the z-parameters of a π-network.

EXAMPLE 18.10. Compute the z-parameters of the π-network of Figure 18.20.

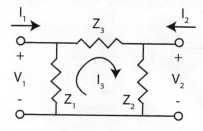

FIGURE 18.20. π-network for Example 18.10.

SOLUTION

It is straightforward to write the following three loop equations:

$$V_1 = Z_1 I_1 - Z_1 I_3$$
$$V_2 = Z_2 I_2 + Z_2 I_3$$

and

$$0 = -Z_1 I_1 + Z_2 I_2 + (Z_1 + Z_2 + Z_3) I_3$$

Putting these equations in matrix form and partitioning the matrix appropriately yields

$$\begin{bmatrix} V_1 \\ V_2 \\ \hline 0 \end{bmatrix} = \begin{bmatrix} Z_1 & 0 & -Z_1 \\ 0 & Z_2 & Z_2 \\ \hline -Z_1 & Z_2 & Z_1 + Z_2 + Z_3 \end{bmatrix} \begin{bmatrix} I_1 \\ I_2 \\ \hline I_3 \end{bmatrix}$$

Hence, using the matrix partitioning formula derived in Example 18.4, we obtain

$$\begin{bmatrix} z_{11} & z_{12} \\ z_{21} & z_{22} \end{bmatrix} = \begin{bmatrix} W_{11} - W_{12} W_{22}^{-1} W_{21} \end{bmatrix}$$

$$= \begin{bmatrix} Z_1 & 0 \\ 0 & R_2 \end{bmatrix} - \frac{1}{Z_1 + Z_2 + Z_3} \begin{bmatrix} -Z_1 \\ Z_2 \end{bmatrix} \begin{bmatrix} -Z_1 & Z_2 \end{bmatrix}$$

$$= \begin{bmatrix} Z_1 & 0 \\ 0 & Z_2 \end{bmatrix} - \frac{1}{Z_1 + Z_2 + Z_3} \begin{bmatrix} Z_1^2 & -Z_1 Z_2 \\ -Z_1 Z_2 & Z_2^2 \end{bmatrix}$$

$$= \frac{1}{Z_1 + Z_2 + Z_3} \begin{bmatrix} Z_1(Z_2 + Z_3) & Z_1 Z_2 \\ Z_1 Z_2 & Z_2(Z_1 + Z_3) \end{bmatrix}$$

Exercises. 1. In Example 18.10, suppose $Z_1 = Z_2 = 1\ \Omega$ and Z_3 is a 1 H inductor. Find the z-parameters.

ANSWER: $\begin{bmatrix} z_{11} & z_{12} \\ z_{21} & z_{22} \end{bmatrix} = \frac{1}{s+2} \begin{bmatrix} s+1 & 1 \\ 1 & s+1 \end{bmatrix}$

2. Now suppose that Z_1 and Z_2 are changed to 1 F capacitors and Z_3 is a 1 H inductor. Find the new z-parameters.

ANSWER: $\begin{bmatrix} z_{11} & z_{12} \\ z_{21} & z_{22} \end{bmatrix} = \frac{1}{s(s^2 + 2)} \begin{bmatrix} s^2 + 1 & 1 \\ 1 & s^2 + 1 \end{bmatrix}$

Relationship to y-Parameters

Since the z-parameters relate port currents to port voltages and the y-parameters relate port voltages to port currents, one might expect that the z-parameter matrix and the y-parameter matrix are inverses of each other. Specifically, if

$$\begin{bmatrix} V_1 \\ V_2 \end{bmatrix} = \begin{bmatrix} z_{11} & z_{12} \\ z_{21} & z_{22} \end{bmatrix} \begin{bmatrix} I_1 \\ I_2 \end{bmatrix},$$

then perhaps it follows that

$$\begin{bmatrix} I_1 \\ I_2 \end{bmatrix} = \begin{bmatrix} z_{11} & z_{12} \\ z_{21} & z_{22} \end{bmatrix}^{-1} \begin{bmatrix} V_1 \\ V_2 \end{bmatrix} = \begin{bmatrix} y_{11} & y_{12} \\ y_{21} & y_{22} \end{bmatrix} \begin{bmatrix} V_1 \\ V_2 \end{bmatrix}$$

This relationship is valid provided $z_{11}z_{22} - z_{21}z_{12} \neq 0$ and $y_{11}y_{22} - y_{21}y_{12} \neq 0$, in which case

and

$$\begin{bmatrix} z_{11} & z_{12} \\ z_{21} & z_{22} \end{bmatrix}^{-1} = \begin{bmatrix} y_{11} & y_{12} \\ y_{21} & y_{22} \end{bmatrix} \tag{18.24a}$$

$$\begin{bmatrix} z_{11} & z_{12} \\ z_{21} & z_{22} \end{bmatrix} = \begin{bmatrix} y_{11} & y_{12} \\ y_{21} & y_{22} \end{bmatrix}^{-1} \tag{18.24b}$$

Exercise. A certain two-port has z-parameters $\begin{bmatrix} z_{11} & z_{12} \\ z_{21} & z_{22} \end{bmatrix} = \dfrac{1}{s+2} \begin{bmatrix} s+1 & 1 \\ 1 & s+1 \end{bmatrix}$.

Find the y-parameters. Can you construct a three-element passive circuit that has these y-parameters?

ANSWER: ; $\begin{bmatrix} y_{11} & y_{12} \\ y_{21} & y_{22} \end{bmatrix} = \dfrac{1}{s} \begin{bmatrix} s+1 & -1 \\ -1 & s+1 \end{bmatrix}$; see Figure 18.20 with $Z_1 = Z_2 = 1\ \Omega$ and Z_3 changed to a 1 H inductor.

Despite this inverse relationship, some circuits have z-parameters but not y-parameters, and vice versa, as illustrated by the following example.

EXAMPLE 18.11. Compute the z-parameters of the circuit of Figure 18.21. Do the y-parameters exist?

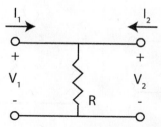

FIGURE 18.21 Resistive two-port having z-parameters but not y-parameters.

SOLUTION

By inspection, the z-parameters of the circuit of Figure 18.21 are

$$\begin{bmatrix} V_1 \\ V_2 \end{bmatrix} = \begin{bmatrix} R & R \\ R & R \end{bmatrix} \begin{bmatrix} I_1 \\ I_2 \end{bmatrix} \triangleq \begin{bmatrix} z_{ij} \end{bmatrix} \begin{bmatrix} I_1 \\ I_2 \end{bmatrix}$$

The z-parameter matrix, $[z_{ij}]$, is singular, since $\det[z_{ij}] = R^2 - R^2 = 0$. Because the $[z_{ij}]$ matrix does not have an inverse, the circuit has no y-parameters. One can check y_{11} directly to verify this claim. Consider Figure 18.22. Because $V_2 = 0$, there is also a short circuit across V_1, making the ratio

$$y_{11} = \frac{I_1}{V_1}\bigg|_{V_2=0}$$

undefined.

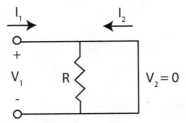

FIGURE 18.22. Equivalent circuit for computing y_{11} in which port 2 is shorted, so that $V_2 = 0$.

The Two–Dependent Source Equivalent Circuit

As with the y-parameters, the z-parameters have a two–dependent source equivalent circuit interpretation, illustrated in Figure 18.23b. Consider first the equation

$$V_1 = z_{11}I_1 + z_{12}I_2$$

Here, V_1 equals the sum of two voltages: $z_{11}I_1$ plus the voltage due to a current-controlled voltage source given by $z_{12}I_2$. This is precisely the left-hand portion of Figure 18.23b. A similar interpretation follows for $V_2 = z_{21}I_1 + z_{22}I_2$, yielding the right-hand side of Figure 18.23b.

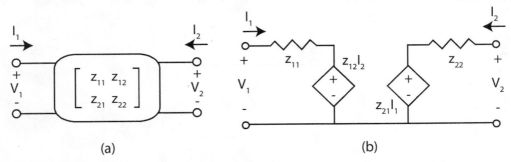

(a)   (b)

FIGURE 18.23 (a) z-parameters for modeled network. (b) Two–dependent source equivalent circuit.

This equivalent circuit proves useful for computing voltage gains and input and output impedances of terminated two-ports. It should also be noted that there are other, equivalent circuits that interpret the z-parameters. A similar remark can be made for y-parameters.

6. IMPEDANCE AND GAIN CALCULATIONS OF TERMINATED TWO-PORTS MODELED BY Z-PARAMETERS

Input and Output Impedance Computations

Earlier we derived the formula for the input admittance of a terminated two-port in terms of the y-parameters, leaving the output admittance calculation as an exercise. Here, using z-parameters, we derive the output impedance of the terminated two-port of Figure 18.24 and leave the input impedance calculation as an exercise. Specifically, our first step is to derive the **output impedance** formula

$$Z_{out} = z_{22} - \frac{z_{12}z_{21}}{z_{11} + Z_s}$$

(18.25)

as a function of the network z-parameters and the source impedance.

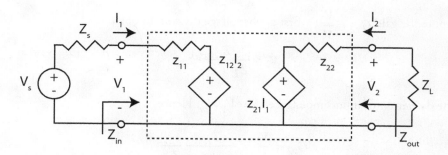

FIGURE 18.24 Terminated two-port modeled by z-parameters.

Beginning on the right-hand side of Figure 18.24,

$$V_2 = z_{22}I_2 + z_{21}I_1$$

(18.26)

In calculating Z_{out}, which is the Thevenin impedance seen by the load, the independent voltage source, V_s, is set to zero. Hence,

$$-(Z_s + z_{11})\, I_1 = z_{12}\, I_2$$

Solving for I_1 and substituting into equation 18.26 yields the output impedance formula of equation 18.25.

Exercises. 1. Repeat the preceding derivation using a matrix method and Cramer's rule.
2. Derive the following formula for the input impedance of a terminated two-port:

$$Z_{in} = z_{11} - \frac{z_{12}z_{21}}{z_{22} + Z_L} \tag{18.27}$$

Gain Calculations

The next phase of our two-port analysis is to repeat the y-parameter derivation of the voltage gain of the two-port in the context of z-parameters. Specifically, our aim is to compute

$$G_V = \frac{V_2}{V_s} = \frac{V_2}{V_1}\frac{V_1}{V_s} \equiv G_{v2}G_{v1} \tag{18.28}$$

The ratio $G_{v1} = V_1/V_s$ follows from voltage division:

$$G_{v1} = \frac{V_1}{V_s} = \frac{Z_{in}}{Z_{in} + Z_s} \tag{18.29}$$

To compute the gain $G_{v2} = \dfrac{V_2}{V_1}$ first apply voltage division to obtain

$$V_2 = \frac{Z_L}{Z_L + z_{22}} z_{21}I_1$$

From the definition of input impedance, $I_1 = V_1/Z_{in}$. Hence,

$$G_{v2} = \frac{V_2}{V_1} = \frac{Z_L}{Z_L + z_{22}} \frac{z_{21}}{Z_{in}} \tag{18.30}$$

Substituting equations 18.29 and 18.30 into equation 18.28 yields the voltage gain:

$$G_V = G_{v2}G_{v1} = \left(\frac{Z_L}{z_{22} + Z_L}\right)\left(\frac{z_{21}}{Z_{in} + Z_S}\right) \tag{18.31}$$

An application of this formula and its derivation to a cascaded network of two-ports (two transistor amplifier circuits) appears in the next example.

EXAMPLE 18.12. Consider the network of Figure 18.25, which represents a two-stage (transistor) amplifier configuration. Each stage utilizes the same transistor in a different circuit configuration. The first stage is an amplification stage that will amplify a small source voltage to a much larger one. The second stage is an impedance-matching stage used to match the load to the output impedance of the amplifier circuitry to achieve maximum power transfer, at least approximately. The z-parameters, in ohms, for each stage in Figure 18.25 are given by

$$Z_1 = \begin{bmatrix} 350 & 2.667 \\ -10^6 & 6667 \end{bmatrix}; \ Z_2 = \begin{bmatrix} 1.0262 \times 10^6 & 6,791 \\ 1.0258 \times 10^6 & 6,794 \end{bmatrix} \qquad (18.32)$$

(a)   Compute the input impedances, Z_{in2} and Z_{in}.

(b)   Compute the voltage gain, V_{out}/V_s.

(c)   Check the matching of the load and output impedance of the amplifier circuit.

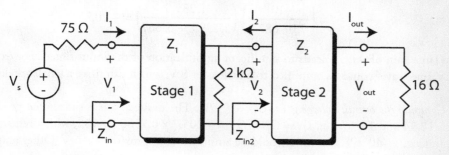

FIGURE 18.25 Two-stage amplifier network.

SOLUTION

Step 1. *Determine Z_{in2}.* A straightforward application of equation 18.27 using the z-parameters of stage 2 produces the impedance Z_{in2}:

$$Z_{in2} = z_{11} - \frac{z_{12}z_{21}}{z_{22} + Z_L} = 1.0262 \times 10^6 - \frac{6791 \times 1.0258 \times 10^6}{6794 + 16} \qquad (18.33)$$

$$= 3262 \ \Omega$$

Step 2. *Compute the voltage gain, $G_{v2} = V_{out}/V_2$, for stage 2.* Equation 18.30 applies here. In particular,

$$G_{v2} = \frac{V_{out}}{V_2} = \frac{Z_L}{Z_L + z_{22}} \frac{z_{21}}{Z_{in2}} = \frac{16\left(1.0258 \times 10^6\right)}{(16 + 6,794) \, 3,262} = 0.7388 \qquad (18.34)$$

The gain here is small, but remember that this stage's real purpose is impedance matching, not amplification. By proper choice of the z-parameters, the output impedance will approximately match that of the load. This allows us to dispense with an expensive and bulky impedance-matching transformer.

Step 3. *Compute Z_{in} for stage 1.* Here, observe that Z_{in2} (equation 18.33) in parallel with the 2 kΩ resistor between the two stages acts as a load to stage 1. Z_{in2} in parallel with 2 kΩ becomes the load to stage 1, denoted $Z_{L1} = 1239.8 \ \Omega$.

The input impedance seen at the front end of stage 1 follows from equation 18.27:

$$Z_{in} = z_{11} - \frac{z_{12}z_{21}}{z_{22} + Z_{L1}} = 350 + \frac{2.667 \times 10^6}{6,667 + 1,239.8} = 687.3 \ \Omega \qquad (18.35)$$

Step 4. *Compute the voltage gain, $G_{v1} = V_2/V_s$, for stage 1.* Using the result of equation 18.35 and applying equation 18.29 yields

$$G_{v1} = \frac{V_2}{V_s} = \left(\frac{z_{21}}{Z_s + Z_{in}}\right)\left(\frac{Z_{L1}}{Z_{L1} + z_{22}}\right) \tag{18.36}$$

$$= \left(\frac{-10^6}{75 + 687.3}\right)\left(\frac{1,239.8}{1,239.8 + 6,667}\right) = -205.7.$$

Here the large gain of stage 1 leads to significant amplification of the input signal. For example, a −40 mV sine wave would be amplified to a little over 8 V, which can drive a small speaker.

Step 5. *Compute the overall voltage gain, $G_V = V_{out}/V_s$.* The desired gain is simply the product of equations 18.34 and 18.36, i.e., $G_V = G_{v1} \times G_{v2} = -205.7 \times 0.7388 = -152$, which remains fairly large. Indeed, a −40 mV sine wave would be amplified to approximately 6 V. Other amplification stages could be added to further increase the overall gain.

Step 6. *Verify that the load matches the amplifier circuitry to a reasonable degree.* In this task, one first computes the output impedance, Z_{out1}, of stage 1 using equation 18.25. The parallel combination of Z_{out} with 2 kΩ, denoted $Z_{s2} = 1.732$ kΩ, becomes the source impedance to stage 2. It is then easy to compute the output impedance of stage 2, again using equation 18.25. The answer is $Z_{out} = 17$ Ω. The details are left as an exercise.

Exercises. 1. For the circuit of Figure 18.25, verify that the output impedance equals 17 Ω.
2. In Figure 18.25, call the 2 kΩ resistor R_2. Find a new value of R_2 so that Z_{out} exactly equals 16 Ω. ANSWER: 1.783 kΩ
3. Suppose the 75 Ω source resistance in Example 18.12 is changed to 300 Ω, which would represent a twin line connection between an ideal voltage source and the first amplifier stage. Redo the example. ANSWER: Numerical values are obtained using the following MATLAB code:

```
Rs = 300;
zz11 = 1.0262e6; zz12 = 6791;
zz21 = 1.0258e6; zz22 = 6794;
z11 = 350; z12 = 2.667; z21 = -1e6;
z22 = 6667; R2 = 2e3; ZL = 16;
Zin2 = zz11 - zz12*zz21/(zz22 + ZL)
G2 = (zz21/Zin2)*ZL/(ZL+zz22)
ZL1 = R2*Zin2/(R2+Zin2)
Zin = z11 - z12*z21/(z22 + ZL1)
G1 = (z21/(Rs+Zin))*ZL1/(ZL1+z22)
Gv = G1*G2
Zout1 = z22-z12*z21/(z11+Rs)
Zs2 = R2*Zout1/(R2+Zout1)
Zout = zz22-zz12*zz21/(zz11+Zs2)
```

---

7. HYBRID PARAMETERS

Basic Definitions and Equivalences

As we have seen, some circuits have y-parameters but not z-parameters, and vice versa. A circuit element that has neither is the ideal transformer.

EXAMPLE 18.13. This example shows that the ideal transformer of Figure 18.26 has neither z- nor y-parameters. From the definition of an ideal transformer (Chapter 17), $V_2 = aV_1$ and $I_1 = -aI_2$. Clearly, V_1 and V_2 cannot be expressed as functions of I_1 and I_2, nor can I_1 and I_2 be expressed as functions of V_1 and V_2. Hence, an ideal transformer has neither y- nor z-parameters.

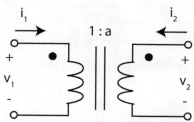

FIGURE 18.26 Ideal transformer, having neither y- nor z-parameters.

Two-port circuits having neither z- nor y- parameters require an alternative modeling technique. The hybrid parameters offer one of several alternatives.

HYBRID PARAMETERS

Hybrid parameters, h_{ij}, are a cross between y- and z-parameters: a voltage V_1 and a current I_2 are outputs, with I_1 and V_2 as inputs. Specifically, if the two-port of Figure 18.10 contains no internal independent sources and has no initial stored energy, then the hybrid parameters are defined by the matrix equation

$$\begin{bmatrix} V_1 \\ I_2 \end{bmatrix} = \begin{bmatrix} h_{11} & h_{12} \\ h_{21} & h_{22} \end{bmatrix} \begin{bmatrix} I_1 \\ V_2 \end{bmatrix} \qquad (18.37)$$

This mixture of variables actually arises as a simplification of a model of a common emitter configuration of a bipolar transistor. From the context of equation 18.37, h_{11} has units of ohms, h_{12} and h_{21} are dimensionless, and h_{22} has units of S.

As with both y- and z-parameters, we interpret equation 18.37 as a two–dependent source equivalent circuit, as illustrated in Figure 18.27.

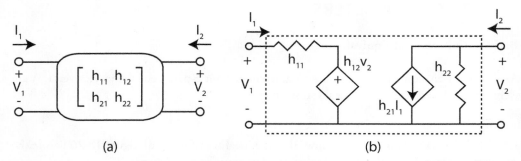

FIGURE 18.27 (a) Hybrid parameters. (b) Two–dependent source equivalent circuit.

Exercise. Justify the two–dependent source equivalent circuit interpretation of Figure 18.27; i.e., apply KVL and KCL to the circuit of Figure 18.27b to derive equation 18.37.

The definition of each h-parameter follows directly from either the preceding equivalent circuit or from equation 18.37. For example,

$$h_{11} = \frac{V_1}{I_1}\bigg|_{V_2=0} \tag{18.38a}$$

Because h_{11} is the ratio of an input voltage to an input current, it is an input impedance. Since $V_2 = 0$, h_{11} is termed the **short-circuit input impedance**. Notice, however, that h_{11} is simply related to both the y- and the z-parameters as follows:

$$h_{11} = \frac{V_1}{I_1}\bigg|_{V_2=0} = \frac{1}{y_{11}} = z_{11} - \frac{z_{12}z_{21}}{z_{22}} \tag{18.38b}$$

The second h-parameter, h_{21}, is called the **short-circuit forward current gain**, since it is the ratio of I_2 to I_1 under the condition $V_2 = 0$, i.e.,

$$h_{21} = \frac{I_2}{I_1}\bigg|_{V_2=0} \tag{18.39a}$$

From the z-parameter equation $V_2 = z_{21}I_1 + z_{22}I_2$, with $V_2 = 0$, h_{21} has a simple z-parameter interpretation,

$$h_{21} = \frac{I_2}{I_1}\bigg|_{V_2=0} = -\frac{z_{21}}{z_{22}} \tag{18.39b}$$

The third h-parameter is

$$h_{12} = \frac{V_1}{V_2}\bigg|_{I_1=0} \tag{18.40a}$$

Since it is the ratio of V_1 to V_2 under the condition that port 1 is open-circuited, i.e., $I_1 = 0$, it is called the **reverse open-circuit voltage gain**. Interpreting h_{12} in terms of the y-parameters, we obtain

$$h_{12} = \frac{V_1}{V_2}\bigg|_{I_1=0} = -\frac{y_{12}}{y_{11}} \tag{18.40b}$$

Finally, we note that the **open-circuit output admittance** is

$$h_{22} = \frac{I_2}{V_2}\bigg|_{I_1=0} \tag{18.41a}$$

and has units of S. The word "open-circuit" suggests a z-parameter interpretation. Considering that $V_2 = z_{21}I_1 + z_{22}I_2$, if $I_1 = 0$, then

$$h_{22} = \frac{I_2}{V_2}\bigg|_{I_1=0} = \frac{1}{z_{22}} = y_{22} - \frac{y_{12}y_{21}}{y_{11}} \tag{18.41b}$$

This relationship is similar (notice the subscripts) to equation 18.38b, which determines the short-circuit input impedance.

We will return to these equivalences later, after we gain some computational experience.

Computation of h-Parameters

The first example of this subsection demonstrates a circuit that has neither z- nor y-parameters.

EXAMPLE 18.14. Consider the two-port of Figure 18.28, whose front end is a short circuit and whose secondary is an open circuit. Thus it has neither z- nor y-parameters. Our objective is to compute the h-parameters.

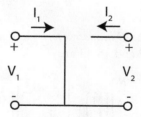

FIGURE 18.28 Simple two-port with h-parameters but neither z- nor y-parameters.

SOLUTION.

By inspection, the h-parameters are

$$\begin{bmatrix} V_1 \\ I_2 \end{bmatrix} = \begin{bmatrix} 0 & 0 \\ 0 & 0 \end{bmatrix} \begin{bmatrix} I_1 \\ V_2 \end{bmatrix}$$

The second example illustrates the computation of h-parameters for an ideal transformer circuit.

EXAMPLE 18.15. Find the *h*-parameters of the two-port in Figure 18.29.

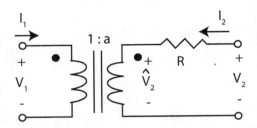

FIGURE 18.29 Ideal transformer circuit for Example 18.15.

SOLUTION
Step 1. *Construct an equation for V_1.* From the primary and secondary voltage relationship of an ideal transformer,

$$V_1 = \frac{1}{a}\hat{V}_2$$

By KVL at port 2, $= V_2 - RI_2$, in which case

$$V_1 + \frac{R}{a}I_2 = \frac{1}{a}V_2 \tag{18.42}$$

Step 2. *Construct an equation for I_2 in terms of the other variables.* From the primary and secondary current relationship of an ideal transformer,

$$I_2 = \frac{-1}{a}I_1 \tag{18.43}$$

Step 3. *Write equations 18.42 and 18.43 in matrix form, and solve for V_1 and I_2 in terms of I_1 and V_2.* In matrix form, equations 18.42 and 18.43 are

$$\begin{bmatrix} 1 & \dfrac{R}{a} \\ 0 & 1 \end{bmatrix}\begin{bmatrix} V_1 \\ I_2 \end{bmatrix} = \begin{bmatrix} 0 & \dfrac{1}{a} \\ -\dfrac{1}{a} & 0 \end{bmatrix}\begin{bmatrix} I_1 \\ V_2 \end{bmatrix} \tag{18.44}$$

Solving for the vector $[V_1 \; I_2]^T$ produces the *h*-parameter equation

$$\begin{bmatrix} V_1 \\ I_2 \end{bmatrix} = \begin{bmatrix} 1 & \dfrac{R}{a} \\ 0 & 1 \end{bmatrix}^{-1}\begin{bmatrix} 0 & \dfrac{1}{a} \\ -\dfrac{1}{a} & 0 \end{bmatrix}\begin{bmatrix} I_1 \\ V_2 \end{bmatrix} = \begin{bmatrix} \dfrac{R}{a^2} & \dfrac{1}{a} \\ -\dfrac{1}{a} & 0 \end{bmatrix}\begin{bmatrix} I_1 \\ V_2 \end{bmatrix}$$

Exercise. In the circuit of Figure 18.29, suppose the resistor R is connected in parallel (instead of in series) with the secondary winding of the ideal transformer. Find the new h-parameters.

ANSWER: $h_{11} = 0$, $h_{12} = 1/a$, $h_{21} = -1/a$, $h_{22} = 1/R$

EXAMPLE 18.16. Find the h-parameters of the circuit of Figure 18.30.

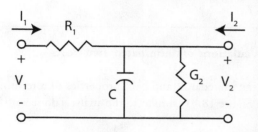

FIGURE 18.30 Simple circuit for illustrating h-parameter definitions.

SOLUTION

Step 1. *Find h_{11} using equation 18.38a.* With $V_2 = 0$, i.e., port 2 shorted,

$$h_{11} = \frac{V_1}{I_1}\bigg|_{V_2=0} = R_1$$

Step 2. *Find h_{21} using equation 18.39a.* Again with $V_2 = 0$, i.e., port 2 shorted, the current $I_2 = -I_1$ since all current flows through the short circuit. Hence

$$h_{21} = \frac{I_2}{I_1}\bigg|_{V_2=0} = -1$$

Step 3. *Find h_{12} using equation 18.40a.* With $I_1 = 0$, i.e., port 1 open-circuited, $V_1 = V_2$ since there is no current through R_1. Hence,

$$h_{12} = \frac{V_1}{V_2}\bigg|_{I_1=0} = 1$$

Step 4. *Find h_{22} using equation 18.41a.* Again with $I_1 = 0$, $I_2 = (C_S + G_2)V_2$, in which case

$$h_{22} = \frac{I_2}{V_2}\bigg|_{I_1=0} = Cs + G_2$$

In summary,

$$[h_{ij}] = \begin{bmatrix} R_1 & 1 \\ -1 & Cs + G_2 \end{bmatrix}$$

Exercises. 1. If all parameters in Example 18.16 are 1 with proper units, find the *h*-parameters.

ANSWER: $\begin{bmatrix} 1 & 1 \\ -1 & s+1 \end{bmatrix}$

2. If the capacitor in Example 18.16 becomes an *L* H inductor, find the new *h*-parameters.

ANSWER: $\left[h_{ij} \right] = \begin{bmatrix} R_1 & 1 \\ -1 & \dfrac{1}{Ls} + G_2 \end{bmatrix}$

Impedance and Gain Calculations of Terminated Two-Ports

This subsection analyzes the impedance and gain properties of a terminated two-port characterized by *h*-parameters as in Figure 18.31, similar to the analysis done with both the *z*- and *y*-parameters.

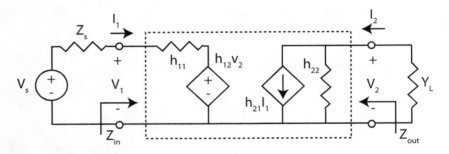

FIGURE 18.31 Doubly terminated hybrid equivalent circuit of a two-port.

Recall that $Z_{in}(s) = \dfrac{V_1}{I_1}$. From the right half of Figure 18.31, $h_{21}I_1 = I_2 - h_{22}V_2 = -(h_{22} + Y_L)V_2$.

It follows immediately that

$$V_2 = -\frac{h_{21}}{h_{22} + Y_L} I_1 \tag{18.45}$$

From the left-hand side of Figure 18.31, $V_1 = h_{11}I_1 + h_{12}V_2$. Substituting for V_2 using equation 18.45 implies that the **input impedance** is

$$Z_{in} = \frac{V_1}{I_1} = h_{11} - \frac{h_{12}h_{21}}{h_{22} + Y_L} \tag{18.46}$$

Exercise. Show that the output admittance is given by

$$Y_{out} = \frac{I_2}{V_2} = h_{22} - \frac{h_{12}h_{21}}{h_{11} + Z_s} \tag{18.47}$$

Knowledge of the input and output impedances/admittances permits us to derive various gain formulas in terms of the h-parameters. For example, consider again the left half of Figure 18.31. Since the input impedance is known from equation 18.46,

$$I_1 = \frac{1}{Z_{in}} V_1 \tag{18.48a}$$

Thus, from equations 18.45 and 18.48a,

$$V_2 = -\frac{h_{21}}{h_{22} + Y_L} I_1 = -\frac{h_{21}}{h_{22} + Y_L} \times \frac{1}{Z_{in}} V_1 \tag{18.48b}$$

implying the voltage gain formula,

$$G_{v2} = \frac{V_2}{V_1} = \frac{-h_{21}}{Z_{in}(h_{22} + Y_L)} . \tag{18.49}$$

Voltage division at the front end of Figure 18.31 yields the other voltage gain formula,

$$G_{v1} = \frac{V_1}{V_s} = \frac{Z_{in}}{Z_{in} + Z_s} . \tag{18.50}$$

The overall voltage gain is the product of equations 18.49 and 18.50, i.e.,

$$G_V = \frac{V_2}{V_s} = G_{v1}G_{v2} = -\frac{1}{(Z_{in} + Z_s)} \frac{h_{21}}{(h_{22} + Y_L)} \tag{18.51}$$

Exercise. Compute the current gain I_{out}/I_1.

ANSWER: $\dfrac{I_2}{I_1} = \dfrac{Y_L h_{21}}{h_{22} + Y_L}$

---

8. TRANSMISSION PARAMETERS

Transmission or t-parameters were first used by power system engineers for transmission line analysis and are still so used today. They are sometimes called ABCD parameters.

TRANSMISSION PARAMETERS

The t-parameter representation has the matrix relationship

$$\begin{bmatrix} V_1 \\ I_1 \end{bmatrix} = \begin{bmatrix} t_{11} & t_{12} \\ t_{21} & t_{22} \end{bmatrix} \begin{bmatrix} V_2 \\ -I_2 \end{bmatrix} \tag{18.52}$$

with the matrix, $T = [t_{ij}]$ called the t-parameter matrix. As with the y-, z-, and h-parameters, the entries t_{ij} are defined as follows:

$$t_{11} = \left.\frac{V_1}{V_2}\right|_{I_2=0} ; \quad t_{12} = \left.\frac{V_1}{-I_2}\right|_{V_2=0}$$

$$t_{21} = \left.\frac{I_1}{V_2}\right|_{I_2=0} ; \quad t_{22} = \left.\frac{I_1}{-I_2}\right|_{V_2=0} \tag{18.53}$$

The matrix equation 18.52 leads directly to the relationships of equations 18.53 by setting the appropriate quantity, I_2 or V_2, to zero.

In computing a single t_{ij} with equations 18.53, some care must be exercised in exciting the circuit. By definition,

$$t_{11} = \left.\frac{V_1}{V_2}\right|_{I_2=0}$$

The ordinary interpretation of this equation is: apply an input V_2 and find an output V_1 under the condition that $I_2 = 0$. Then t_{11} is the ratio of the Laplace transform of the response V_1 to that of the input V_2, i.e., a **reverse voltage gain** when port 2 is open-circuited. This situation causes a predicament: an independent voltage source for V_2 causes a current I_2 to flow. To circumvent this predicament, we use the slightly modified formula

$$t_{11} = \left.\frac{1}{\dfrac{V_2}{V_1}}\right|_{I_2=0}$$

The quantity $V_2/V_1]_{I_2} = 0$ (the forward voltage gain when port 2 is open-circuited) suggests that we excite port 1 by V_1 with port 2 open-circuited, which forces $I_2 = 0$. It is then straightforward to calculate t_{11} as the inverse of $V_2/V_1]_{I_2} = 0$. Similar interpretations must be made with regard to the other defining formulas in 18.53.

A simple example illustrates t-parameter calculations.

EXAMPLE 18.17. Consider again an ideal transformer circuit, shown in Figure 18.32. Here $V_2 = aV_1$ and $I_1 = a(-I_2)$. This leads to the t-parameter matrix

$$\begin{bmatrix} V_1 \\ I_1 \end{bmatrix} = \begin{bmatrix} \dfrac{1}{a} & 0 \\ 0 & a \end{bmatrix} \begin{bmatrix} V_2 \\ -I_2 \end{bmatrix}$$

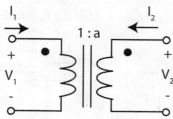

FIGURE 18.32 Simple transformer circuit for Example 18.17.

Input and output impedance calculations for terminated two-ports modeled by t-parameters do not follow the usual pattern. Nevertheless, the two-port of Figure 18.33 has **input impedance**

$$Z_{in} = \frac{t_{11}Z_L + t_{12}}{t_{21}Z_L + t_{22}}$$

(18.54)

and **output impedance**

$$Z_{out} = \frac{t_{22}Z_s + t_{12}}{t_{21}Z_s + t_{11}}$$

(18.55)

The derivation of these results is left as a homework problem.

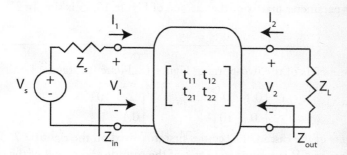

FIGURE 18.33 Terminated two-port modeled by t-parameters.

Exercise. The two-port of Figure 18.33 has t-parameters $t_{11} = 0.0025$, $t_{12} = 500$ Ω, $t_{21} = 3.125 \times 10^{-8}$ S, and $t_{22} = 0.00625$. If $Z_L = 200$ kΩ and $Z_s = 20$ kΩ, find Z_{in} and Z_{out}.
ANSWER: 80 kΩ, 200 kΩ

One of the most important characteristics of t-parameters is the ease with which one can use them to determine the overall t-parameters of cascaded two-ports, as illustrated in the next example.

EXAMPLE 18.18. Compute the t-parameters of the cascaded two-port of Figure 18.34 in terms of T_1 and T_2, the two-port parameter matrices of the first and second sections, respectively.

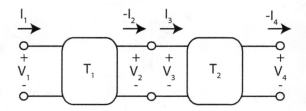

FIGURE 18.34 Cascade of two two-ports modeled by t-parameters.

SOLUTION

From the definition of the t-parameters for each two-port,

$$\begin{bmatrix} V_1 \\ I_1 \end{bmatrix} = T_1 \begin{bmatrix} V_2 \\ -I_2 \end{bmatrix} \quad \text{and} \quad \begin{bmatrix} V_3 \\ I_3 \end{bmatrix} = T_2 \begin{bmatrix} V_4 \\ -I_4 \end{bmatrix}.$$

But since $V_2 = V_3$ and $-I_2 = I_3$, it follows that

$$\begin{bmatrix} V_1 \\ I_1 \end{bmatrix} = T_1 \begin{bmatrix} V_2 \\ -I_2 \end{bmatrix} = T_1 \begin{bmatrix} V_3 \\ I_3 \end{bmatrix} = T_1 T_2 \begin{bmatrix} V_4 \\ -I_4 \end{bmatrix}$$

implying that the t-parameter matrix of the cascade of Figure 18.35 is simply $T_{new} = T_1 T_2$.

Exercise. The t-parameters of the two-ports in Figure 18.34 are (in standard units)

$$T_1 = \begin{bmatrix} 0.1 & 0 \\ 0 & 10 \end{bmatrix} \quad \text{and} \quad T_2 = \begin{bmatrix} 2 & 10 \\ 0.1 & 1 \end{bmatrix}$$

Find the t-parameters of the cascade (T_1 comes first, followed on the right by T_2). Then find V_2 when a voltage source of 4 V is applied to port 1 of the cascade and port 2 of the cascade is open-circuited.

ANSWER: $T = \begin{bmatrix} 0.2 & 1 \\ 1 & 10 \end{bmatrix}$, $V_2 = 20$ V

---

9. GENERAL RELATIONS AMONG TWO-PORT PARAMETERS

The h-, y-, t-, and z-parameters are interrelated. This subsection derives several relationships, with others left to the homework problems. The complete set of results is summarized in Table 18.1 for reference purposes.

To express the h-parameters in terms of the z-parameters, first note that

$$\begin{bmatrix} V_1 \\ I_2 \end{bmatrix} = \begin{bmatrix} h_{11} & h_{12} \\ h_{21} & h_{22} \end{bmatrix} \begin{bmatrix} I_1 \\ V_2 \end{bmatrix} \quad \text{and} \quad \begin{bmatrix} V_1 \\ V_2 \end{bmatrix} = \begin{bmatrix} z_{11} & z_{12} \\ z_{21} & z_{22} \end{bmatrix} \begin{bmatrix} I_1 \\ I_2 \end{bmatrix}$$

The trick is to rewrite the z-parameter equations so that V_1 and I_2 are on the left with I_1 and V_2 on the right:

$$V_1 - z_{12}I_2 = z_{11}I_1$$
$$z_{22}I_2 = z_{21}I_1 - V_2$$

Writing these two equations in matrix form yields

$$\begin{bmatrix} 1 & -z_{12} \\ 0 & z_{22} \end{bmatrix} \begin{bmatrix} V_1 \\ I_2 \end{bmatrix} = \begin{bmatrix} z_{11} & 0 \\ -z_{21} & 1 \end{bmatrix} \begin{bmatrix} I_1 \\ V_2 \end{bmatrix} \tag{18.56}$$

Solving equation 18.56 for the vector $[V_1 \ I_2]^T$ under the proviso that $z_{22} \neq 0$ yields

$$\begin{bmatrix} V_1 \\ I_2 \end{bmatrix} = \begin{bmatrix} 1 & -z_{12} \\ 0 & z_{22} \end{bmatrix}^{-1} \begin{bmatrix} z_{11} & 0 \\ -z_{21} & 1 \end{bmatrix} \begin{bmatrix} I_1 \\ V_2 \end{bmatrix} \tag{18.57}$$

$$= \frac{1}{z_{22}} \begin{bmatrix} z_{11}z_{22} - z_{12}z_{21} & z_{12} \\ -z_{21} & 1 \end{bmatrix} \begin{bmatrix} I_1 \\ V_2 \end{bmatrix}$$

Thus, we have used matrix methods to directly compute the h-parameters in terms of the z-parameters under the condition that $z_{22} \neq 0$:

$$\begin{bmatrix} h_{11} & h_{12} \\ h_{21} & h_{22} \end{bmatrix} = \frac{1}{z_{22}} \begin{bmatrix} z_{11}z_{22} - z_{12}z_{21} & z_{12} \\ -z_{21} & 1 \end{bmatrix} \tag{18.58}$$

All other relationships are derived in a similar manner. For example, to express y-parameters in terms of h-parameters, one must rewrite the h-parameter equations so that I_1 and I_2 appear on the left-hand side with V_1 and V_2 on the right. Then using matrix form and inverting the appropriate matrix under the condition of a nonzero determinant produces the desired result.

Table 18.1 specifies the interrelationships among all the parameters studied thus far.

TABLE 18.1. Interrelations among Two-Port Parameters, Where γ Can Stand for z, y, h, or t in $\Delta\gamma = \gamma_{11}\,\gamma_{22} - \gamma_{12}\,\gamma_{21}$

	z-Parameters	y-Parameters	h-Parameters	t-Parameters
z-Parameters	$\begin{bmatrix} z_{11} & z_{12} \\ z_{21} & z_{22} \end{bmatrix}$	$\begin{bmatrix} \dfrac{y_{22}}{\Delta y} & \dfrac{-y_{12}}{\Delta y} \\[2mm] \dfrac{-y_{21}}{\Delta y} & \dfrac{y_{11}}{\Delta y} \end{bmatrix}$	$\begin{bmatrix} \dfrac{\Delta h}{h_{22}} & \dfrac{h_{12}}{h_{22}} \\[2mm] -\dfrac{h_{21}}{h_{22}} & \dfrac{1}{h_{22}} \end{bmatrix}$	$\begin{bmatrix} \dfrac{t_{11}}{t_{21}} & \dfrac{\Delta t}{t_{21}} \\[2mm] \dfrac{1}{t_{21}} & \dfrac{t_{22}}{t_{21}} \end{bmatrix}$
y-Parameter	$\begin{bmatrix} \dfrac{z_{22}}{\Delta z} & \dfrac{-z_{12}}{\Delta z} \\[2mm] \dfrac{-z_{21}}{\Delta z} & \dfrac{z_{11}}{\Delta z} \end{bmatrix}$	$\begin{bmatrix} y_{11} & y_{12} \\ y_{21} & y_{22} \end{bmatrix}$	$\begin{bmatrix} \dfrac{1}{h_{11}} & \dfrac{-h_{12}}{h_{11}} \\[2mm] \dfrac{h_{21}}{h_{11}} & \dfrac{\Delta h}{h_{11}} \end{bmatrix}$	$\begin{bmatrix} \dfrac{t_{22}}{t_{12}} & \dfrac{-\Delta t}{t_{12}} \\[2mm] \dfrac{-1}{t_{12}} & \dfrac{t_{11}}{t_{12}} \end{bmatrix}$
h-Parameter	$\begin{bmatrix} \dfrac{\Delta z}{z_{22}} & \dfrac{z_{12}}{z_{22}} \\[2mm] -\dfrac{z_{21}}{z_{22}} & \dfrac{1}{z_{22}} \end{bmatrix}$	$\begin{bmatrix} \dfrac{1}{y_{11}} & \dfrac{-y_{12}}{y_{11}} \\[2mm] \dfrac{y_{21}}{y_{11}} & \dfrac{\Delta y}{y_{11}} \end{bmatrix}$	$\begin{bmatrix} h_{11} & h_{12} \\ h_{21} & h_{22} \end{bmatrix}$	$\begin{bmatrix} \dfrac{t_{12}}{t_{22}} & \dfrac{\Delta t}{t_{22}} \\[2mm] \dfrac{-1}{t_{22}} & \dfrac{t_{21}}{t_{22}} \end{bmatrix}$
t-Parameter	$\begin{bmatrix} \dfrac{z_{11}}{z_{21}} & \dfrac{\Delta z}{z_{21}} \\[2mm] \dfrac{1}{z_{21}} & \dfrac{z_{22}}{z_{21}} \end{bmatrix}$	$\begin{bmatrix} \dfrac{-y_{22}}{y_{21}} & \dfrac{-1}{y_{21}} \\[2mm] \dfrac{-\Delta y}{y_{21}} & \dfrac{-y_{11}}{y_{21}} \end{bmatrix}$	$\begin{bmatrix} \dfrac{-\Delta h}{h_{21}} & \dfrac{-h_{11}}{h_{21}} \\[2mm] \dfrac{-h_{22}}{h_{21}} & \dfrac{-1}{h_{21}} \end{bmatrix}$	$\begin{bmatrix} t_{11} & t_{12} \\ t_{21} & t_{22} \end{bmatrix}$

Exercises. 1. Use the code below to create an m-file in MATLAB for conversion of z-parameters to h-parameters. Verify that Z = [1 2;3 4] produces H = [–0.5 0.5;–0.75 0.25].

```
% convert z parameters to h parameters
function [h, h11,h12,h21,h22] = ztoh(z)
z11 = z(1,1); z12=z(1,2);z21=z(2,1);z22=z(2,2);
deltaz = z11*z22-z12*z21;
h11 = deltaz/z22;
h12 =z12/z22;
h21 = -z21/z22;
h22 = 1/z22;
h = [ h11 h12; h21 h22];
```

2. Use the code below to create an m-file in MATLAB for conversion of z-parameters to t-parameters. Verify that Z = [1 2;3 4] produces T = [1/3 –2/3;1/3 4/3].

```
%converting z to t paramters
function [t,t11,t12,t21,t22] = ztot(z)
z11 =z(1,1); z12=z(1,2); z21=z(2,1); z22=z(2,2);
deltaz = z11*z22 - z12*z21;
t11 = z11/z21;
t12 = deltaz/z21;
t21 = 1/z21;
t22 = z22/z21;
t = [ t11 t12; t21 t22];
```

3. Write m-files for the remaining items in the conversion table for your own future use.

10. RECIPROCITY

Writing node equations for an ordinary linear circuit leads to a matrix equation having the form

$$\begin{bmatrix} I_1 \\ \vdots \\ I_n \end{bmatrix} = \begin{bmatrix} y_{11} & \cdots & y_{1n} \\ \vdots & \cdots & \vdots \\ y_{n1} & \cdots & y_{nn} \end{bmatrix} \begin{bmatrix} V_1 \\ \vdots \\ V_n \end{bmatrix} = \mathbf{Y} \begin{bmatrix} V_1 \\ \vdots \\ V_n \end{bmatrix}$$

Often the node admittance matrix $\mathbf{Y} = [y_{ij}]$ is symmetric, i.e., $y_{ij} = y_{ji}$ for $i \neq j$. Such networks are termed "reciprocal."

RECIPROCAL NETWORKS

Any circuit that has a symmetric coefficient matrix either in a nodal equation or loop equation representation is said to be reciprocal. Further, a two-port represented by either z-parameters or y-parameters is said to be reciprocal if $z_{12} = z_{21}$ or $y_{12} = y_{21}$.

From Chapter 3 we know that circuits without dependent sources have symmetric coefficient matrices in both the nodal and loop equation representations. On the other hand, the symmetry of the z-parameters and y-parameters is typically lost when dependent sources are present. In general, we can prove that *a circuit containing R's, L's, C's, and transformers, but no dependent sources or independent sources, is a reciprocal network*. From the definition of a reciprocal two-port, we can conclude further that if the hybrid and/or t-parameters exist, then from Table 18.1

$$h_{12} = -h_{21}$$

and/or

$$t_{11}t_{22} - t_{12}t_{21} = 1$$

Conversely, any two-port that has parameters satisfying these conditions is said to be reciprocal.

Exercise. Recall that the relationship between the t-parameters and z-parameters is

$$\begin{bmatrix} z_{11} & z_{12} \\ z_{21} & z_{22} \end{bmatrix} = \begin{bmatrix} \dfrac{t_{11}}{t_{21}} & \dfrac{\Delta t}{t_{21}} \\ \dfrac{1}{t_{21}} & \dfrac{t_{22}}{t_{21}} \end{bmatrix}.$$

Show that if $z_{12} = z_{21}$, then $\Delta t = 1$, where Δ denotes "determinant." Then show the converse, i.e., if $\Delta t = 1$, then $z_{12} = z_{21}$.

Proving that any two-port created from a reciprocal network has symmetric z-parameters is straightforward. We write the loop equations in matrix form with V_1, I_1, V_2, and I_2 being the voltages and currents of ports 1 and 2. Since the underlying network is reciprocal by assumption, its loop equation has a symmetric coefficient matrix partitioned as shown:

$$\begin{bmatrix} V_1 \\ V_2 \\ V_3 \\ \vdots \\ V_n \end{bmatrix} = \begin{bmatrix} z_{11} & z_{12} & z_{13} & \cdots & z_{1n} \\ z_{12} & z_{22} & z_{23} & \cdots & z_{2n} \\ z_{13} & z_{23} & z_{33} & \cdots & z_{3n} \\ \vdots & \vdots & \vdots & \ddots & \vdots \\ z_{1n} & z_{2n} & z_{3n} & \cdots & z_{nn} \end{bmatrix} \begin{bmatrix} I_1 \\ I_2 \\ I_3 \\ \vdots \\ I_n \end{bmatrix}$$

Solving for $[V_1 \ V_2]^T$ by the method of matrix partitioning yields

$$\begin{bmatrix} V_1 \\ V_2 \end{bmatrix} = \begin{bmatrix} W_{11} - W_{12}W_{22}^{-1}W_{21} \end{bmatrix} \begin{bmatrix} I_1 \\ I_2 \end{bmatrix} = \begin{bmatrix} z_{11} & z_{12} \\ z_{21} & z_{22} \end{bmatrix} \begin{bmatrix} I_1 \\ I_2 \end{bmatrix}$$

where the W_{ij} matrices are defined in the obvious way. W_{11} is symmetric. Since the inverse of the symmetric matrix W_{22} is symmetric, and since the sum of two symmetric matrices is symmetric, the resulting z-parameters are also symmetric. The symmetry of the y-parameters follows by the symmetry of the inverse of a symmetric matrix.

We now set forth physical interpretations of reciprocity.

Reciprocity Interpretation 1: Consider a reciprocal two-port, N. If the voltage inputs, V_{in}, are the same as in Figures 18.35a and b, then by reciprocity the zero-state short-circuit responses, I_{2a} and I_{1b}, coincide.

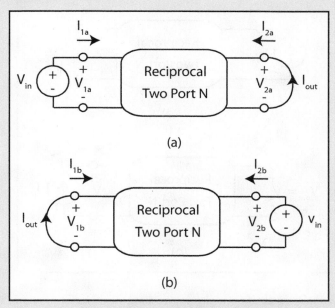

(a)

(b)

FIGURE 18.35 Equivalence of short-circuit zero-state responses induced by voltage sources for reciprocal networks.

What reciprocity interpretation 1 says is that if one applies a voltage at port 1 and measures the short-circuit current at port 2 with an ideal ammeter (zero meter resistance), then applying the same voltage at port 2 would result in measurement of the same short-circuit current at port 1.

Conversely, we can show that if reciprocity interpretation 1 is true, then the y-parameters are symmetric. To see this, observe that the configuration of Figure 18.35a implies

$$y_{21} = \frac{I_{out}}{V_{in}}\bigg|_{V_{2a}=0} = \frac{I_{2a}}{V_{1a}}\bigg|_{V_{2a}=0}$$

and the configuration of Figure 18.35b implies

$$y_{12} = \frac{I_{out}}{V_{in}}\bigg|_{V_{1b}=0} = \frac{I_{1b}}{V_{2b}}\bigg|_{V_{1b}=0}$$

Thus one must have $y_{12} = y_{21}$, i.e., symmetric y-parameters.

Reciprocity Interpretation 2: Consider a reciprocal two-port, N. As illustrated in Figure 18.36, if I_{in} is the same in Figures 18.36a and 18.36b, then the open-circuit zero-state responses V_{2a} and V_{1b} coincide.

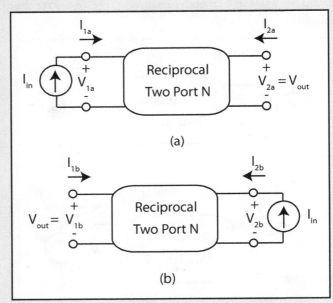

(a)

(b)

FIGURE 18.36 Equivalence of open-circuit zero-state responses for a reciprocal network.

Reciprocity interpretation 2 says that if one injects a current at port 1 and measures the voltage at port 2 with an ideal voltmeter (infinite input resistance), injecting the same current at port 2 would result in measurement of the same voltage at port 1.

Exercise. Assuming reciprocity interpretation 2 is true, show that it follows that $z_{12} = z_{21}$, i.e., the z-parameters are symmetric.

Reciprocity Interpretation 3: Consider a reciprocal two-port, N. As illustrated in Figure 18.37, I_{in} leads to a zero-state response denoted $-I_{2a}$, as shown in Figure 18.37a, and V_{in} leads to a zero-state response denoted V_{out}, as in Figure 18.37b. For a reciprocal two-port, the short-circuit current ratio in Figure 18.37a is equal to the open-circuit voltage ratio in Figure 18.37b.

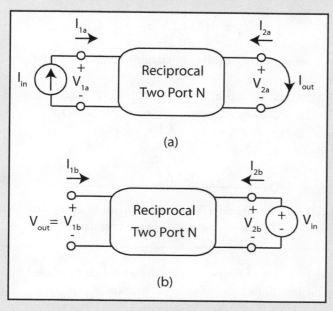

(a)

(b)

FIGURE 18.37 Reciprocal two-ports illustrating *h*-parameter anti-symmetry.

Exercise. Show that if reciprocity interpretation 3 is assumed true, it follows that $h_{12} = -h_{21}$; i.e., the *forward short-circuit current gain,* $\dfrac{I_{out}}{I_{in}}$, equals the *reverse open-circuit voltage gain,* $\dfrac{V_{out}}{V_{in}}$.

The three reciprocity interpretations have rigorous proofs. They however, are beyond the scope of the text.

Finally, a two-port that is reciprocal has an equivalent circuit representation with no dependent sources. For example, suppose a reciprocal two-port has the *z*-parameter representation

$$V_1 = z_{11} I_1 + z_{12} I_2$$

and

$$V_2 = z_{12} I_1 + z_{22} I_2$$

in which case

$$V_1 = (z_{11}-z_{12})\, I_1 + z_{12}\, (I_1+I_2) \tag{18.59a}$$

and

$$V_2 = z_{12}\, (I_1+I_2) + (z_{22}-z_{12})\, I_2 \tag{18.59b}$$

Equations 18.59 have the so-called T-equivalent circuit interpretation given by Figure 18.38.

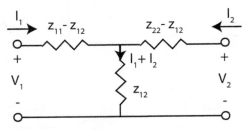

FIGURE 18.38 z-parameter T-equivalent circuit of a reciprocal 2-port.
The resistors represent general impedances.

Exercise. Suppose the two-port of Figure 18.39 is reciprocal and modeled by y-parameter equations. Compute Y_1, Y_2, and Y_3 in terms of the y-parameters, y_{11}, y_{12}, y_{22}.
ANSWER: $-y_{12}$, $y_{11} + y_{12}$, $y_{22} + y_{12}$

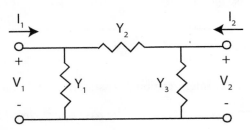

FIGURE 18.39 y-parameter π-equivalent circuit of a reciprocal two-port.
The resistors represent general impedances.

---

11. PARALLEL, SERIES, AND CASCADED CONNECTIONS OF TWO-PORTS

A general linear two-port has four external terminals for connection to other networks as illustrated in Figure 18.40a. When the input and output ports have a common terminal, only three external terminals are available, as indicated in Figure 18.40b. Such a two-port is typically called a *common-ground two-port*, although the common terminal is not necessarily grounded in the sense of being connected to earth.

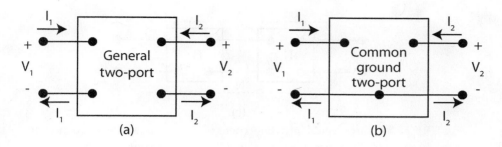

FIGURE 18.40 (a) A general two-port. (b) A common-ground two-port.

Interconnecting two common-ground two-ports N_a and N_b, forms a new two-port N. Figure 18.41 shows three typical interconnection structures: parallel, series, and cascade. To avoid over-crowding of symbols, Figure 18.41 omits all voltage and current reference labels. These labels are understood to conform to those in Figure 18.40a. In particular, note that at each port the current entering one terminal must equal the current leaving the other terminal for the two-port parameters to be valid or meaningful.

An interconnection of two-ports has a new set of z-, y-, h-, or t-parameters obtained very simply from the individual two-port parameters. The interconnected two-ports of Figure 18.41 have new parameters computed from those of N_a and N_b as follows:

1. For the parallel connection of Figure 18.41a,

$$\mathbf{Y} = \mathbf{Y}_a + \mathbf{Y}_b \tag{18.60}$$

2. For the series connection of Figure 18.41b,

$$\mathbf{Z} = \mathbf{Z}_a + \mathbf{Z}_b \tag{18.61}$$

3. For the cascade connection of Figure 18.41c,

$$\mathbf{T} = \mathbf{T}_a \mathbf{T}_b \tag{18.62}$$

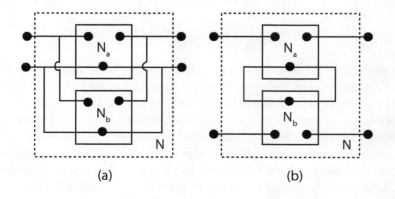

(a)   (b)

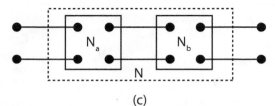

(c)

FIGURE 18.41 Three typical interconnections of two-ports. (a) Parallel connection.
(b) Series connection. (c) Cascade connection.

where **Y**, **Z**, and **T** denote the admittance, impedance, and transmission parameter matrices, respectively, and the subscripts a and b refer to the networks N_a and N_b, respectively.

A derivation of these formulas is straightforward. For example, from the definitions of N_a and N_b,

$$I_a = Y_a V_a \text{ and } I_b = Y_b V_b$$

where

$$V_a = \begin{bmatrix} V_{1a} \\ V_{2a} \end{bmatrix}, \qquad I_a = \begin{bmatrix} I_{1a} \\ I_{2a} \end{bmatrix},$$

and similarly for the voltage and current vectors of N_b. From Figure 18.41a, $V_a = V_b = V$ and $I = I_a + I_b$. A substitution yields $I = (Y_a + Y_b)V$, which verifies equation 18.60.

To verify equation 18.61, consider Figure 18.41b. Here $V_a = Z_a I_a$ and $V_b = Z_b I_b$. But $V = V_a + V_b$ and $I_a = I_b = I$. By direct substitution $V = (Z_a + Z_b)I$, verifying equation 18.61.

Equation 18.62 was derived earlier in this chapter.

The derivation of equations 18.60 and 18.62 is easily extended to the case of more than two two-ports:

1. If two or more common-ground two-ports are connected in parallel, then

$$Y = Y_a + Y_b + Y_c + \cdots \tag{18.63}$$

2. If two or more two-ports, common-ground or not, are connected in cascade, then

$$T = T_a T_b T_c \cdots \tag{18.64}$$

Equations 18.62 and 18.64 for the cascade connection hold whether or not the component two-ports are of the common-ground type. However, equations 18.60 and 18.63 for parallel connections *in general* hold only for common-ground two-port connections, as shown in Figure 18.41a. Similarly, the series connection equation 18.61 holds only for the case illustrated in Figure 18.41b. Series connection of two general two-ports (Figure 18.40a) or series connection of more than two common-ground two-ports (Figure 18.40b) requires an ideal transformer for coupling, as demonstrated in the homework problems. Examples 18.19 and 18.20 explain why equations 18.60 and 18.61 fail when two non-common-ground two-ports are connected together.

EXAMPLE 18.19. This example illustrates the difficulty with a non-common-ground series connection. Consider the two-port shown in Figure 18.42, which is a series connection of two component two-ports. The z-parameters of the individual two-ports are given by

$$Z_a = \begin{bmatrix} 2 & 1 \\ 1 & 2 \end{bmatrix} \text{ and } Z_b = \begin{bmatrix} R_1 + 2 & 1 \\ 1 & R_3 + 2 \end{bmatrix}.$$

(a) Show that $Z \neq Z_a + Z_b$ when $R_1 = 6\ \Omega$ and $R_2 = 3\ \Omega$.
(b) Show that $Z = Z_a + Z_b$ if $R_1 = R_2 = 0$.
(c) Justify the statements of parts (a) and (b).

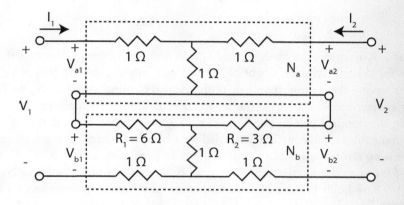

FIGURE 18.42 A series connection that causes difficulty.

SOLUTION
(a) Observe that the parallel connection of the $6\ \Omega$ and $3\ \Omega$ resistors is $2\ \Omega$. Thus, by direct calculation, the z-parameter matrix of the interconnected two-port is

$$Z = \begin{bmatrix} 6 & 4 \\ 4 & 6 \end{bmatrix} \neq Z_a + Z_b = \begin{bmatrix} 2 & 1 \\ 1 & 2 \end{bmatrix} + \begin{bmatrix} 8 & 1 \\ 1 & 5 \end{bmatrix} = \begin{bmatrix} 10 & 2 \\ 2 & 7 \end{bmatrix}$$

(b) With $R_1 = R_2 = 0$,

$$Z_b = \begin{bmatrix} 2 & 1 \\ 1 & 2 \end{bmatrix}$$

On the other hand, by direct calculation, the z-parameter matrix of the interconnected two-port is

$$Z = \begin{bmatrix} 4 & 2 \\ 2 & 4 \end{bmatrix} = Z_a + Z_b = \begin{bmatrix} 4 & 2 \\ 2 & 4 \end{bmatrix}$$

(c) In part (a), $Z \neq Z_a + Z_b$, because, after the interconnection, neither N_a nor N_b acts as a two-port, as defined in Figure 18.40. This can be understood by inspecting Figure 18.43, with the indicated the loop currents.

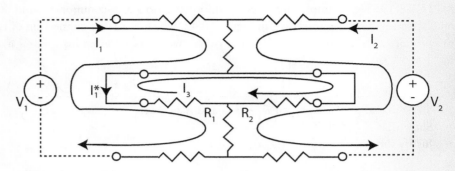

FIGURE 18.43 A nonzero I_3 leads to $I_1 \neq I_1^*$.

With R_1 and R_2 nonzero, the mesh current I_3 is in general nonzero. Observe that $I_1^* = I_1 - I_3 \neq I_1$. Hence, for the left terminal pair of N_a, the current entering the top terminal does not equal the current leaving the bottom terminal. With unequal terminal currents, N_a no longer has a z-parameter characterization, because the z-parameter definition requires equal currents entering and leaving the terminal pair, as per Figure 18.40. On the other hand, if $R_1 = R_2 = 0$ in Figure 18.43, then the third mesh equation is satisfied for arbitrary values of the mesh currents I_1, I_2, and I_3, as there is no resistance at all in the third mesh. In particular, let $I_3 = 0$, in which case $I_1^* = I_1$, and the z-parameter characterization of N_a is valid. Similar arguments hold for N_b. If $R_1 = R_2 = 0$, equation 18.61 holds because Figure 18.43 now has the same interconnection as depicted in Figure 18.41b.

EXAMPLE 18.20. This example illustrates the problem of a non-common-ground parallel connection. Figure 18.44 shows two two-ports connected in parallel. Before the connection, each two-port has y-parameters

$$Y_a = Y_b = \begin{bmatrix} 0.7 & -0.2 \\ -0.2 & 0.7 \end{bmatrix} \text{ S.}$$

After the connection, by direct calculation, the new two-port has y-parameter matrix

$$Y = \begin{bmatrix} 1.625 & -0.625 \\ -0.625 & 1.625 \end{bmatrix} \text{ S.}$$

Clearly, $Y \neq Y_a + Y_b = 2Y_a$ in this case.

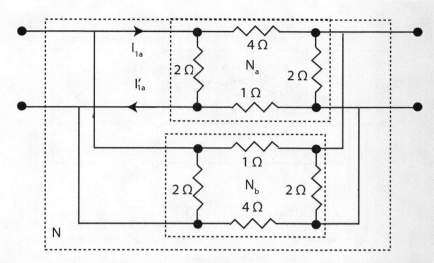

FIGURE 18.44 New two-port N subject to invalid application of equation 18.60 for non-common-ground two-ports in parallel.

The reason for the failure of equation 18.60 under these circumstances is the same as for the circuit of Example 18.19. If a voltage source is applied to port 1 of N, we find that the currents I_{1a} and I'_{1a} are not equal. Thus, N_a cannot continue to be characterized by a set of two-port y-parameters in forming N.

There are, however, some special cases of non-common-ground two-ports for which equation 18.60 holds for a parallel connection. The following is one example.

EXAMPLE 18.21. This example illustrates how to achieve a parallel interconnection of two general two-ports of Figure 18.40a so that equation 18.63 remains valid. Reconsider the two-ports N_a and N_b of Example 18.20, which have y-parameter matrices $\mathbf{Y}_a$ and $\mathbf{Y}_b$, respectively. Suppose a 1:1 ideal transformer is placed at the front end of N_a in Figure 18.45; call this new two-port N_a^*.

(a)   Show that the y-parameter matrix of N_a^* is $\mathbf{Y}_a$.

(b)   Show that the y-parameter matrix of the interconnection of N_a^* and N_b is $\mathbf{Y} = \mathbf{Y}_a + \mathbf{Y}_b$, even though N_a^* and N_b are not common-ground two-ports.

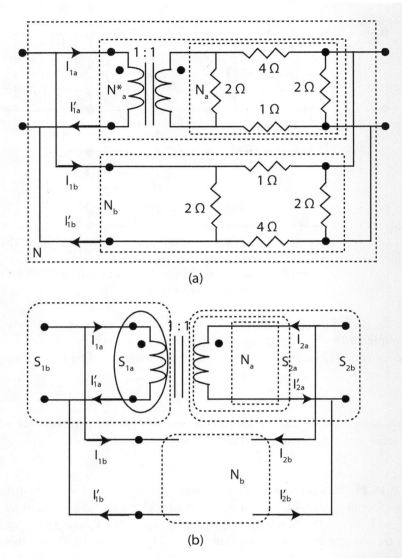

(a)

(b)

FIGURE 18.45 Equation 18.63 holds for these non-common-ground two-ports. (a) Parallel connection of two general two-ports. (b) Justification of equal currents at two terminals of each port.

SOLUTION

(a) The y-parameters of N_a^* are the same as for N_a because the ideal 1:1 transformer with indicated dot positions forces N_a^* and N_a to have the same port currents and voltages at port 1. In fact, if N_a^* and N_a were enclosed in a box with only the leads observable, the two-port properties would be identical.

(b) To show that the y-parameters of N (the interconnection of N_a^* and N_b) are $\mathbf{Y} = \mathbf{Y}_a + \mathbf{Y}_b$, we must first show that

(i) $I'_{1a} = I_{1a}$ and $I'_{1b} = I_{1b}$
(ii) $I'_{2a} = I_{2a}$ and $I'_{2b} = I_{2b}$

To confirm these equalities, consider Figure 18.45b, which represents Figure 18.45a with all the nonessential components removed to avoid overcrowding figure. Four Gaussian surfaces are drawn, S_{1a}, S_{1b}, S_{2a}, and S_{2a}. Note that all Gaussian surfaces go through the core of the transformer. Recall that KCL holds for a Gaussian surface: the algebraic sum of the currents entering (leaving) the surface is zero. This immediately asserts the validity of statements (i) and (ii) above. Hence, within the interconnection, the two-ports N_a^* and N_b continue to act as individual two-ports. Therefore, equation 18.60 remains valid, i.e.,

$$\mathbf{Y} = \mathbf{Y_a} + \mathbf{Y_b} = \begin{bmatrix} 1.4 & -0.4 \\ -0.4 & 1.4 \end{bmatrix}$$

This example extends directly to multiple parallel interconnections. Although the precise conditions for the applicability of equations 18.60 and 18.61 to the parallel and series connections of non-common-ground two-ports are known, they are not practical enough to be included here. Our emphasis is on interconnections of common-ground two-ports, which occur most often in practice.

12. SUMMARY

This chapter presented a unified setting for one-port analysis while providing a comprehensive extension to two-ports. Two-ports are common to numerous real-world systems such as the utility power grid that delivers power to a home from a generating facility through a transmission network. Another representative two-port is a telephone system that delivers a speaker's voice to a listener by sending a converted electrical signal through a transmission network. The characterization of a two-port for such systems is done through their input-output properties. Four sets of characterizing parameters were developed: **impedance** or z-parameters, **admittance** or y-parameters, **hybrid** or h-parameters, and **transmission** or t-parameters. In order to analyze various aspects of a system characterized by a two-port, formulas for computing the input impedance/admittance, the output impedance/admittance, the voltage gain, etc. were derived. Quantities such as voltage and power gain are very important aspects of amplifier analysis and design, as illustrated in Example 18.12, which depicts a two-stage transistor amplifier configuration. Although Example 18.12 utilized the medium of z-parameters, the more customary medium for transistor amplifier design is h-parameters.

Conditions and formulas for parallel connection of two-ports were presented in terms of y-parameters while series connections were studied using z-parameters. Formulas for determining the transmission parameters of cascades of two-ports were also developed. In addition, the chapter introduced and interpreted the notion of reciprocity in terms of the different two-port parameters. Reciprocal circuits generally contain only R's, L's, C's and transformers. Under certain restricted conditions a reciprocal network may contain a dependent source, as the homework problems will investigate.

13. TERMS AND CONCEPTS

Admittance or *y*-parameters: descriptive two-port parameters in which the port currents are functions of the port voltages.

Hybrid parameters or *h*-parameters: descriptive two-port parameters in which V_1 and I_2 are expressed as functions of I_1 and V_2.

Impedance or *z*-parameters: descriptive two-port parameters in which the port voltages are functions of the port currents.

Input admittance: the admittance seen at port 1 of a possibly terminated two-port.

Input impedance: the impedance seen at port 1 of a possibly terminated two-port.

Matrix partitioning: the partitioning of a matrix set of equations into groups to obtain a simplified solution in terms of the partitioned submatrices.

Norton equivalent of one-port: a current source in parallel with the Thevenin impedance.

Open-circuit impedance parameters: the impedance or *z*-parameters.

Open-circuit output admittance: the hybrid parameter h_{22}.

Output admittance: the admittance seen at port 2 of a two-port possibly terminated by a source impedance.

Output impedance: the impedance seen at port 2 of a two-port possibly terminated by a source impedance.

Partitioned matrix: a matrix that is partitioned into submatrices for easier solution of sets of equations.

π-equivalent circuit: equivalent circuit of a reciprocal two-port containing three general impedances in the form of π, as in Figure 18.39.

Reciprocal network: a network whose node equations or loop equations have a symmetric coefficient matrix.

Reciprocal two-port: $z_{12} = z_{21}$ or $y_{12} = y_{21}$.

Reverse open-circuit voltage gain: the hybrid parameter h_{12}.

Short-circuit admittance parameters: the admittance or *y*-parameters.

Short-circuit forward current gain: the hybrid parameter h_{21}.

Short-circuit input impedance: the hybrid parameter h_{11}.

T-equivalent circuit: equivalent circuit of a reciprocal two-port having three general impedances in a T shape as in Figure 18.38.

Terminated two-port: a two-port attached to a load impedance and a source with, in general, a nonzero impedance.

Thevenin equivalent of a one-port: Voltage source in series with the Thevenin impedance.

Transmission or *t*-parameters: parameters where V_1 and I_1 are expressed as functions of V_2 and $-I_2$.

Two–dependent source equivalent circuit: equivalent circuit for a two-port containing impedances/admittances and two dependent sources.

Problems

ONE-PORTS

1. In Figure P18.1, suppose $V_s = 100$ V, $Z_L = 20$ Ω, $R_s = 1$ kΩ, and $\beta = 49$. Find Z_{in} of the one-port. Then find V_1 and the power to the one-port. Does Z_2 have any effect on the answers?

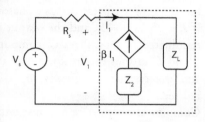

Figure P18.1

CHECK: 50 V

2. In Figure P18.2, suppose $V_s = 100$ V, $Z_1 = 2500$ Ω, $R_s = 50$ Ω, and $\beta = 49$. Find Z_{in} of the one-port. Then find V_1 and the power delivered to the one-port. Does Z_3 have any effect on the answers?

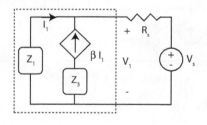

Figure P18.2

CHECK: 50 W

3. Consider the circuit of Figure P18.3. Suppose $C = 0.1$ mF, $v_C(0^-) = 10$ V, $Z_1 = 1$ kΩ, $Z_2 = 2$ kΩ, $Z_3 = 1$ kΩ, and $g_m = 4$ mS.
   (a) Find the input impedance Z_{in}.
   (b) Find $v_C(t)$ for $t \geq 0$.

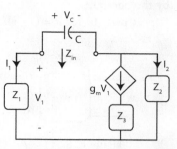

Figure P18.3

CHECK: (a) 11 kΩ

4. Consider the circuit of Figure P18.4. Suppose $i_{in}(t) = 10\sqrt{2} \cos(1000t)$ A, $a = 10$, $R = 10$ Ω, and $C = 1$ μF.
   (a) Find the Thevenin equivalent seen by the load.
   (b) Find the load impedance for maximum average power transfer and compute the resulting average power.
   (c) Find a series RL load that achieves maximum power transfer.

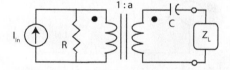

Figure P18.4

CHECK: (c) 1 kΩ resistor in series with 1 H inductor

5. The loop equations in standard units describing the one-port of Figure P18.5 are given by

$$\begin{bmatrix} V_1 \\ 0 \\ 0 \end{bmatrix} = \begin{bmatrix} 10a & 1 & -a \\ a & 0.5 & 0 \\ -1 & 0 & 0.5 \end{bmatrix} \begin{bmatrix} I_1 \\ I_2 \\ I_3 \end{bmatrix}$$

   (a) Find the input impedance of the one-port as a function of a. Hint: Consider Cramer's rule.
   (b) If $a = 2$ and a voltage source is connected so that $v_1(t) = 12\sqrt{2} \cos(2t)$ V, find the average power consumed

by the one-port.

(c)   Repeat part (b) for $a = (1+ j4/3)$ at ω $=2$ rad/sec.

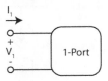

Figure P18.5

ANSWER: (a) $6a\,\Omega$, (b) $12\,W$, (c) $8.64\,W$

6. Consider the circuit in Figure P18.6.

(a)   Find the Thevenin equivalent circuit using the method of matrix partitioning.

(b)   If the circuit is terminated in a 1.2 Ω resistor, find $v_{out}(t)$ for $t \geq 0$ assuming the initial inductor current is zero.

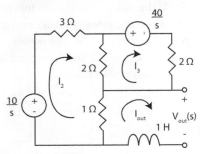

Figure P18.6

ANSWER: (a) $Z_{th} = s + 0.8$, $V_{oc} = -2/s$.

7. Consider the one-port circuit of Figure P18.7.

(a)   Find the Norton equivalent using the methods of matrix partitioning and node analysis.

(b)   Making use of the information obtained in part (a), compute the impulse and step responses of the circuit.

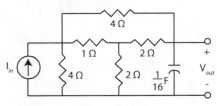

Figure P18.7

CHECK: $I_{sc} = 0.5\,I_{in}$

8. Consider the circuit of Figure P18.8. In solving this problem, fully utilize the properties of an ideal transformer.

(a)   Find the Thevenin equivalent circuit as a function of the turns ratio, b, and the resistance, R.

(b)   If $i_{in}(t) = \cos(10t + 45°)$ A, find the Norton equivalent circuit with the Norton source represented as a phasor.

(c)   If $R = 25\,\Omega$, $b = 2$, and the output is terminated in a parallel LC, compute L and C so that the bandwidth is 10 and the resonant frequency $\omega_r = 5$ rad/sec.

(d)   Compute the zero-state response with the above input and element values. Identify the steady-state part. If the input cosine frequency were changed to 100 rad/sec, what would happen to the steady-state magnitude?

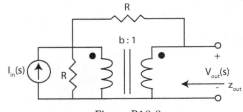

Figure P18.8

ANSWERS: $V_{oc} = \dfrac{bR}{2b^2 - 2b + 1}I_{in}$ and

$Z_{th} = Z_{out} = \dfrac{R}{2b^2 - 2b + 1}$; $C = 0.02$ F and $L = 2$ H

y-PARAMETERS

9. Find the y-parameters of the two-port in Figure P18.9.

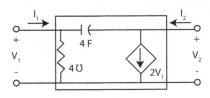

Figure P18.9

10. Consider the circuit of Figure P18.10, where G_1 and G_2 denote conductances.
   (a) Compute the short-circuit admittance parameters.
   (b) Suppose port 1 is short-circuited and a voltage $v_2(t)$ is applied to port 2. Find $I_1(s)$ and $I_2(s)$, in terms of the literals and $V_2(s)$.

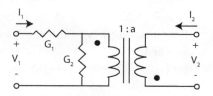

Figure P18.10

11. Consider the circuit of Figure P18.11, in which $a = 2$, $R_1 = 2\ \Omega$, $R_2 = 16\ \Omega$, $R_3 = 320\ \Omega$, and $R_4 = 80\ \Omega$.
   (a) Compute the short-circuit admittance parameters.
   (b) If port 2 is short-circuited and
$$V_1(s) = \frac{64}{s^2 + 16}, \text{ find } I_1(s) \text{ and } I_2(s).$$
   (c) If port 2 is terminated in a 240 Ω resistor, $R_1 = 6\ \Omega$, $R_2 = 60\ \Omega$, and
$$V_1(s) = \frac{64}{s^2 + 16}, \text{ find } I_1(s) \text{ and } I_2(s).$$

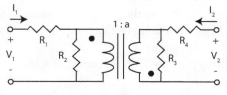

Figure P18.11

12. The two-port of Figure P18.12 has y-parameter matrix
$$[y_{ij}] = \begin{bmatrix} y_{11} & y_{12} \\ y_{21} & y_{22} \end{bmatrix}.$$

Compute Y_1, Y_2, Y_3, and g_m in terms of the y-parameters.

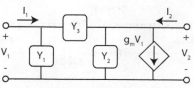

Figure P18.12.

ANSWERS: in random order: $(y_{21} - y_{12})$, $(y_{11} + y_{12})$, $-y_{12}$, $(y_{22} + y_{12})$

13. Consider the circuit of Figure P18.13.
   (a) Compute the short-circuit admittance parameters,
$$[y_{ij}] = \begin{bmatrix} y_{11} & y_{12} \\ y_{21} & y_{22} \end{bmatrix}.$$
   (b) Reverse the process and compute Y_1, Y_2, Y_3, and g_m in terms of the y-parameters.
   (c) Suppose $Y_1 = Y_2 = Y_3 = (s + 1)$,
$$V_1(s) = \frac{2}{(s+1)^2}$$
   and port 2 is short-circuited. Find $I_1(s)$, $I_2(s)$, $i_1(t)$, and $i_2(t)$.

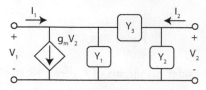

Figure P18.13

14. In the circuit of Figure P18.14, $R_1 = 2\ \Omega$, $R_2 = 2\ \Omega$, $R_3 = 2\ \Omega$, $C_1 = 0.5$ F, $C_2 = 0.25$ F, and $C_3 = 0.5$ F.
   (a) Compute the y-parameters using the method of matrix partitioning
   (b) If port 2 is short-circuited, find $I_1(s)$, $I_2(s)$, $i_1(t)$, and $i_2(t)$ assuming $v_1(t) = 10u(t)$ V.

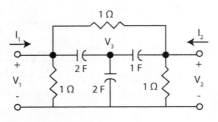

Figure P18.14

15. The terminated two-port configuration in Figure P18.15 is characterized by *y*-parameters.

Suppose $\dfrac{V_1}{V_s} = \dfrac{1}{s+2}$.

(a) Find y_{11} if R_s = 10 Ω, y_{21} = −1 S, y_{12} = 0.03 S, y_{22} = 0.2 S, and R_L = 10 Ω.

(b) If $v_s(t)$ = 30$u(t)$ V, find the power, $p_L(t)$, absorbed by the 10 Ω load at port 2.

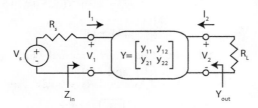

Figure P18.15

16. Consider the circuit of Figure P18.16.
   (a) Compute the *y*-parameters of the boxed two-port.
   (b) If port 2 is terminated in a 2 S resistor, compute Y_{in}, Z_{in}, and the voltage gain

$$G_v = \dfrac{V_2(s)}{V_1(s)} .$$

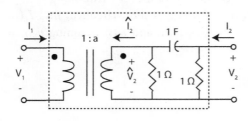

Figure P18.16

17. Reconsider the circuit of Figure P18.15, in which $\begin{bmatrix} y_{11} & y_{12} \\ y_{21} & y_{22} \end{bmatrix} = \begin{bmatrix} 4 & -0.1 \\ 60 & 1 \end{bmatrix}$ mS ,

R_s = 1 kΩ, and R_L = 2 kΩ.
   (a) Compute Y_{in}, Z_{in}, Y_{out}, and Z_{out}.
   (b) Find the voltage gain

$$G_v = \dfrac{V_2}{V_s} .$$

   (c) If $v_s(t)$ = 9$u(t)$ V, find the power absorbed by R_L.

ANSWERS: (a) Y_{in} = 8 mS, Z_{in} = 125 Ω, Y_{out} = 2.2 mS; (b) −4.44; (c) 0.8 watt

18. Reconsider the two-port of Figure P18.15, in which R_s = 10 Ω, y_{21} = 2 S, y_{12} = 0.02 S, y_{22} = 0.2 S, R_L = 10 Ω, and the voltage gain is

$$\dfrac{V_1}{V_s} = 0.6 .$$

   (a) Find y_{11}.
   (b) If $v_s(t)$ = 10$u(t)$ V, find the power absorbed by R_L .

CHECK: (b) 160 watts

19. Consider the circuit of Figure P18.19, in which R_1 = 2 Ω, R_2 = 8 Ω, R_3 = 32 Ω, β = 3, and μ = 4.
   (a) Compute the *y*-parameters of the boxed two-port.
   (b) Compute the input admittance, $Y_{in}(s)$, of the complete circuit.
   (c) Compute $Y_{out}(s)$.
   (d) If $i_s(t)$ = 5$u(t)$ A, compute $v_1(t)$.

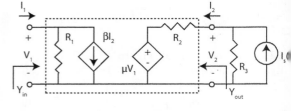

Figure P18.19

20. In a laboratory, you are asked to determine the admittance parameters of a circuit. You

decide to short-circuit port 2, place a unit step current source at port 1, and measure the port voltage, $v_1 = (1 - e^{-4t})u(t)$ V, and the port 2 current, $i_2(t) = -e^{-3t}u(t)$ A. Knowing that this is sufficient to determine at most two of the parameters, you then break the short circuit and terminate port 2 with a 1 Ω resistor and measure the new step responses as $v_1 = (1 - e^{-4t} + te^{-4t})u(t)$ V and $i_2(t) = -e^{-7t}u(t)$ A.

(a) Compute the y-parameters of the two-port.

(b) If port 2 is terminated in a 1 Ω resistor, find the input impedance seen at port 1.

(c) If port 2 is terminated in a 1 Ω resistor and driven at port 1 by a current source $i_1(t) = \cos(t)u(t)$ A, compute the steady-state magnitude of the gain, $\left|\dfrac{V_2}{I_1}\right|$.

21. This problem shows that one can simulate an inductor using an active two-port terminated by a capacitor. For the circuit of Figure P18.21, $C_1 = 125\ \mu F$, $C_2 = 0.8$ F, $y_{11} = y_{22} = 0$, and $y_{21} = -y_{12} = 4$ S.

(a) Compute the input impedance $Z_{in}(s)$. What is the equivalent L_{eq} seen at port 1?

(b) Determine the resonant frequency ω_r.

(c) If $R = 25$ Ω is placed in parallel with C_1, find the new resonant frequency ω_r.

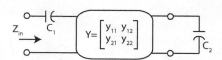

Figure P18.21

CHECK: (a) $L_{eq} = 50$ mH, (b) 400 rad/sec, (c) 240 rad/sec

22. For the circuit of Figure P18.22, $R_s = 1$ Ω, $R_L = 1$ Ω, and $C_L = 1$ F.

(a) Find the y-parameters.

(b) Find the voltage gain, $G_{V2} = \dfrac{V_2(s)}{V_s(s)}$.

(c) Compute the impulse and step responses of the terminated

network assuming $v_2(t)$ is the output.

(d) If $v_s(t) = 10\sin(2t)u(t)$ V, compute the steady-state and transient responses assuming $v_2(t)$ is the output. Use the residue command in MATLAB to compute the partial fraction expansion.

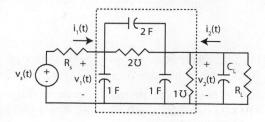

Figure P18.22

CHECK: (b) $0.25\ /\ (s + 1)$, (d) transient response $e^{-t}u(t)$ V, steady-state response $(-\cos(2t) + 0.5\sin(2t))u(t)$ V

23. Figure P18.23 represents a two-stage amplifier. Suppose in mS

$$Y_1 = \begin{bmatrix} 2 & -0.64 \\ 25 & 1 \end{bmatrix} \quad Y_2 = \begin{bmatrix} 0.4 & -0.007 \\ 7.5 & 0.025 \end{bmatrix}$$

$R_s = 150$ Ω, $R_L = 2$ kΩ, and $R_1 = 2$ kΩ. Find the voltage gain

$$G_v = \frac{V_{out}}{V_{in}}.$$

Although the solution may be obtained by solving a set of six simultaneous equations, a much better method that also gives more insight into the performance of the amplifier, and works for any number of stages, is to proceed as follows:

(a) Find the input admittances Y_{in2} and Y_{in1}.

(b) Find the voltage gains of the stages successively, starting from the source end. Use this information to find the overall voltage gain,

$$G_v = \frac{V_{out}}{V_{in}}.$$

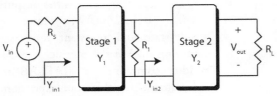

Figure P18.23

ANSWERS: (a) Y_{in1} = 10 mS, Y_{in2} = 500 μS;
(b) G_V = 500

24. Consider the switched circuit of Figure P18.24, in which C = 0.25 F and R_L = 1 Ω. For t < 0.25 sec the switch is in position A, and for $t \geq 0.25$ sec the switch is in position B. Suppose the y-parameter matrix is

$$Y(s) = \begin{bmatrix} -1 & 1 \\ 1 & -1 \end{bmatrix} \text{ S.}$$

(a) With the switch in position A, find

$$\frac{V_2(s)}{V_1(s)}.$$

(b) If $v_1(t)$ = $-4u(t)$ V, find $v_2(t)$, $0 \leq t <$ 0.25 sec.
(c) Is the circuit stable? Explain your reasoning.
(d) Find $v_2(0.25^-)$.
(e) For t $\geq$ 0.25 sec, after the switch has moved to position B, draw and label the frequency domain equivalent circuit that will allow one to compute $V_3(s)$.
(f) Compute $Z_1(s)$ if b = 2.
(g) Compute $v_3(t)$ for $t \geq 0.25$ sec.

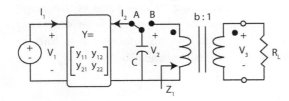

Figure P18.24
CHECK: (d) $v_2(0.25)$ = 3.463 V, (g) $v_3(t)$ = $3.436e^{-(t-0.25)}u(t-0.25)$ V

z-PARAMETERS

25. (a) Compute the z-parameters of the mutually coupled inductors in Figure P18.25 when (i) the dots are in positions A-B and (ii) the dots are in positions A-C.
    (b) Compute the y-parameters of the two-port by inverting the z-parameters matrix. Do the y-parameters exist if the coupling coefficient k = 1?

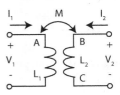

Figure P18.25

26. Consider the circuit of Figure P18.26.
    (a) Compute the open-circuit impedance parameters.
    (b) If port 2 is terminated in $Z_2(s)$, find the input impedance, $Z_{in}(s)$.
    (c) If port 1 is open-circuited, $Z_1(s)$ is a 1 H inductor, $Z_2(s)$ is a parallel combination of a 1 Ω resistor and a 0.5 F capacitor, and

$$I_2(s) = \frac{2K}{s^2+4},$$

find $v_1(t)$ and $v_2(t)$ in steady state.

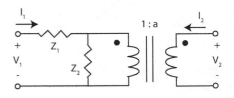

Figure P18.26
CHECK: (c) $v_{2sss}(t)$ = 0.707Ka^2sin(2t – 45°) V

27. Consider the circuit of Figure P18.27, in which R_1 = 35 Ω, R_2 = 50 Ω, R_3 = 800 Ω, R_4 = 100 Ω, and a = 4.

(a) Compute the z-parameters.
(b) Find the y-parameters by matrix inversion using, for example, MATLAB.
(c) If port 1 is open-circuited and

$$I_2 = \frac{8}{s^2 + 16},$$

find $V_1(s)$ and $V_2(s)$.

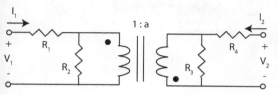

Figure P18.27

CHECK: (a) $Z = \begin{bmatrix} 60 & ?? \\ -100 & ?? \end{bmatrix} \Omega$

28. Consider the two-port of Figure P18.28.
(a) Compute the open-circuit impedance parameters in terms of the Z_i and r_m.
(b) If the two-part has z-parameter matrix

$$\begin{bmatrix} z_{11} & z_{12} \\ z_{21} & z_{22} \end{bmatrix},$$

find Z_1, Z_2, Z_3, and r_m in terms of the z-parameters.
(c) If $Z_1(s) = Z_2(s) = s$, $Z_3(s) = 1$, $r_m = -5$, port 2 is short-circuited, and $v_1(t) = 10u(t)$ V, find $i_1(t)$ and $i_2(t)$.

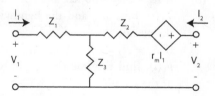

Figure P18.28

CHECK: $\begin{bmatrix} Z_1 + Z_3 & Z_3 \\ ?? & ?? \end{bmatrix}$.

29. Use the method of matrix partitioning to find the z-parameter matrix of the circuit in Figure P18.29.

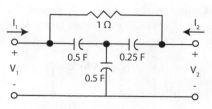

Figure P18.29

30. Consider the circuit of Figure P18.30, in which $Z_1(s) = 5\ \Omega$, $Z_2(s) = 10\ \Omega$, $Z_3(s) = \dfrac{10}{s}$, $Z_L = 10\ \Omega$, and $r_m = 10$.
(a) Compute the z-parameters.
(b) Compute $Z_{in}(s)$.
(c) If $v_1(t) = 10u(t)$ V, find $i_1(t)$ and $i_2(t)$.

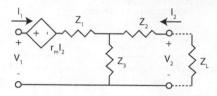

Figure P18.30

31. The circuit of Figure P18.31 has $Z_L = 10\ \Omega$, and z-parameter matrix

$$[z_{ij}] = \frac{20s}{s^2 + 100} \begin{bmatrix} 1 & 1 \\ 1 & 1 \end{bmatrix} \Omega$$

(a) Use the result of Problem 28 part (b) to draw an equivalent circuit having these z-parameters.
(b) The circuit is resonant at what value of ω_r?
(c) Compute the Q of the circuit.
(d) Plot the magnitude and phase response of $Z_{in}(j\omega)$, using a program such as MATLAB.

Figure P18.31

32. Consider the circuit of Figure P18.32. Suppose $v_s(t) = 20\sqrt{2}\cos(\omega t)u(t)$ V, $Z_s = 1$ Ω, $Z_L = 1$ Ω, and the z-parameter matrix is

$$\mathbf{Z} = \begin{bmatrix} s + \dfrac{0.5}{s} & \dfrac{0.5}{s} \\[2mm] \dfrac{0.5}{s} & s + \dfrac{0.5}{s} \end{bmatrix} \Omega$$

(a) Find $Z_{in}(s)$ and $Z_{out}(s)$.
(b) Find $V_2(s)$ and the average power absorbed by Z_L.
(c) For $\omega = 0$ and $\omega = 1$ rad/sec, find the average powers absorbed by Z_L.
(d) Using the equivalent circuit of Figure 18.38, construct a circuit having the given z-parameters.

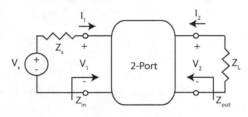

Figure P18.32

ANSWERS:

(a) $Z_{in}(s) = Z_{out}(s) = \dfrac{2s^3 + 2s^2 + 2s + 1}{(?)s^2 + (?)s + (?)}$;

(b) $H(s) = \dfrac{0.5}{s^3 + 2s^2 + 2s + 1}$ and

$|H(j\omega)|^2 = \dfrac{0.25}{1 + \omega^{(?)}}$; (c) 100 W and 50W

(c) 100 W and 50W

33. Consider the two-port configuration of Figure P18.33 having z-parameter matrix

$$\mathbf{Z} = \begin{bmatrix} \alpha R_0 & \alpha^2 R_0 \\ 2\alpha R_0 & 26\alpha^2 R_0 \end{bmatrix}$$

(a) Determine the value of b that yields maximum power transfer from the

output of the two-port to the load.
(b) What is the maximum average power absorbed by $R_L = 2$ Ω when $i_s(t) = 2\sqrt{2}\cos(2t)$ A, $\alpha = 2$, and $R_0 = 50$ Ω?

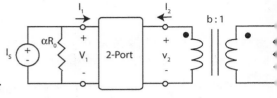

Figure P18.33

CHECK: (a) $(??)\sqrt{\dfrac{R_0}{??}} = 5\alpha\sqrt{\dfrac{R_0}{R_L}}$; (b) 2 watts

34. The two-port shown in Figure P18.34 is called a "gyrator." A gyrator terminated in a capacitor behaves like an inductor at its input terminals. This is one of the many ways to simulate inductors in the design of an "active filter." Suppose the active two-port in Figure P18.34, built with resistors and ideal operational amplifiers, has z-parameters

$$\begin{bmatrix} 0 & 1000 \\ -1000 & 0 \end{bmatrix} \Omega \text{ and } C_1 = 100 \text{ pF.}$$

(a) If $C_2 = 10$ nF, find $Z(s)$, the impedance looking into port 1. Is the result a surprise to you?
(b) If $R_1 = R_2 = 100$ kΩ and the output is $V_1(s)$, find ω_m and the bandwidth of the circuit.
(c) If $R_1 = R_2 = 10$ kΩ and the output is $v_1(t)$, find the impulse response, $h(t)$, of the circuit.
(d) Find the y-parameters of the two-port, using the property that the Z-matrix is the inverse of the Y-matrix (and vice versa).
(e) If $R_1 = R_2 = 10$ kΩ, and the output is $v_2(t)$, what is the impulse response? Hint: Make use of the results of parts (c) and (d), and the two–dependent source equivalent circuit in terms of the y-parameters.

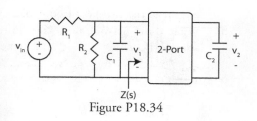

Figure P18.34

35. The stages in the circuit of Figure P18.35 have z-parameter matrices

$$= \begin{bmatrix} 2 & 0 \\ -10^3 & 20 \end{bmatrix} k\Omega \text{ and } Z_2 = \begin{bmatrix} 62.582 & 1.2075 \\ 63.75 & 1.25 \end{bmatrix} k\Omega$$

respectively.

(a) Compute the input impedances, Z_{in2} and Z_{in1}.
(b) Compute the voltage gain, V_{out}/V_s.
(c) Compute the power gain,

$$P_{gain} = \frac{v_{out} i_{out}}{v_1 i_1}.$$

(d) Check the matching of the load and output impedance of the amplifier circuit.

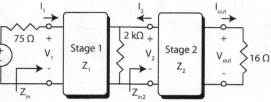

Figure P18.35

CHECK: $Z_{in} = 2 \text{ k}\Omega$ and $\dfrac{V_{out}}{V_s} = -9.8147$

36. Consider the cascaded two-port in Figure P18.36, in which $R_s = 10 \text{ }\Omega$, $R_1 = 20 \text{ }\Omega$, and $R_L = 4 \text{ }\Omega$.

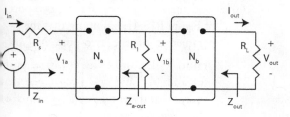

Figure P18.36

The two-port N_a has y-parameters and the two-port N_b has z-parameters as given below (with standard port labeling and units):

$$[y_a] = \begin{bmatrix} 0.1 & 0.1 \\ -0.2 & 0.1 \end{bmatrix}; \quad [z_b] = \begin{bmatrix} 36 & 2 \\ 40 & 4 \end{bmatrix}$$

(a) Find Z_{out}.
(b) Find Z_{in}.
(c) Find the gains

$$G_1 = \frac{V_{1a}}{V_{in}}, \ G_2 = \frac{V_{1b}}{V_{1a}}, G_3 = \frac{V_{out}}{V_{1b}}, \text{ and}$$

$$G_4 = \frac{V_{out}}{V_{in}}.$$

(d) Find the power gain of the circuit, i.e., the ratio

$$P_{gain} = \frac{v_{out} i_{out}}{v_{in} i_{in}}.$$

37. This problem shows how to establish several equivalent circuits for a pair of coupled inductors using z-parameters. Each equivalent circuit consists of one ideal transformer and two inductances. The analysis of a coupled circuit with the use of such equivalent circuits is very often more illuminating than writing and solving simultaneous equations.

(a) Find the z-parameters of the two-port N_1 of Figure P18.37a.
(b) Show that two-port N_2 of Figure P18.37b has the same z-parameters as N_1.
(c) Show that two-port N_3 of Figure P18.37c has the same z-parameters as N_1.
(d) Use the equivalent circuit N_2 of part (b) and the properties of an ideal transformer to find ω_m, the bandwidth, and

$$H_m = \left| \frac{V_o}{V_{in}} \right|_{max}$$

approximate values for the half-power frequencies for the coupled tuned circuit of Figure P18.37d.

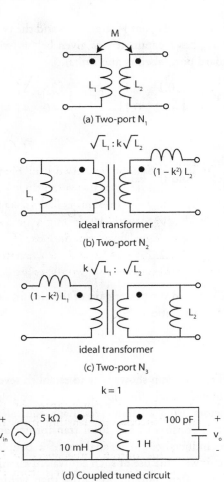

(a) Two-port N_1

(b) Two-port N_2

(c) Two-port N_3

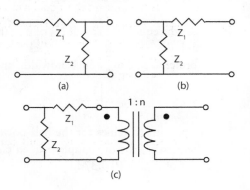

(d) Coupled tuned circuit

Figure P18.37

h-PARAMETERS

38. Find the *h*-parameters for each two-port shown in Figure P18.38 assuming the standard labeling and units.

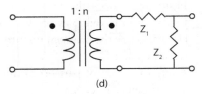

(d)

Figure P18.38

39. Find the *h*-parameters for each two-port shown in Figure P18.39, assuming standard port voltage and port current labeling, with $R = 2$ kΩ, $C = 0.1$ mF, and $n = 0.1$.

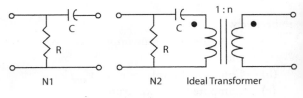

Figure P18.39

40. Find the *h*-parameters for each two-port shown in Figure P18.40, assuming standard port voltage and port current labeling, with $R = 2$ kΩ, $C = 0.1$ mF, and $n = 10$.

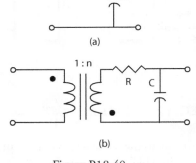

Figure P18.40

41.(a) Find the *h*-parameters for the two-port of Figure P18.41a assuming that the *h*-parameters of the two-port N_1 are

$$\begin{bmatrix} \hat{h}_{11} & \hat{h}_{12} \\ \hat{h}_{21} & \hat{h}_{22} \end{bmatrix}.$$

Hint: What does the transformer do to the output variables of the boxed circuit? Break the problem up into two separate parts.

(b)   Apply the result of part (a) to the circuit of Figure P18.41b.

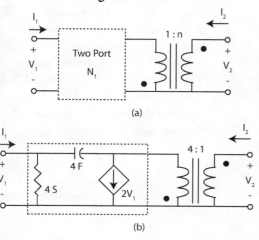

(a)

(b)

Figure P18.41

42. The h-parameters of the two-port of Figure P18.42 are $h_{11} = 250 \ \Omega$, $h_{12} = 2.5 \times 10^{-3}$, $h_{21} = 125$, $h_{22} = 2.25 \ \text{mS}$, $Z_s = 1 \ \text{k}\Omega$, and $Z_L = 500 \ \Omega$.

(a)   Find Z_{in} and Z_{out}.

(b)   Find the gain $G_v = \dfrac{V_2}{V_s}$.

(c)   Find the power gain of the circuit, i.e., the ratio

$$P_{gain} = \frac{-v_2 i_2}{v_1 i_1}.$$

(d)   Suppose a capacitance of 5 μF is connected across port 2. Suppose $v_s(t) = 10\sqrt{2} \cos(400t)$ V. Find $v_2(t)$ in steady state and the average power absorbed by $Z_L = 500 \ \Omega$. Hint: Obtain the Thevenin equivalent circuit seen by the Z_L and C combination.

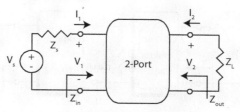

Figure P18.42

43. Consider the amplifier network of Figure P18.43. Stage 1 is a common-emitter stage that drives stage 2, the common-collector stage. Such an amplifier combination might be used to drive a low-impedance load. Suppose $Z_s = 2$ kΩ, $Z_m = 3$ kΩ, $Z_L = 64 \ \Omega$, and the h-parameters in standard units of the two stages are

$$\mathbf{H}_1 = \begin{bmatrix} 2000 & 0 \\ 50 & 0.05 \times 10^{-3} \end{bmatrix}$$

$$\mathbf{H}_2 = \begin{bmatrix} 1000 & 0.966 \\ -51 & 0.8 \times 10^{-3} \end{bmatrix}$$

(a)   Find the input impedances of the stages successively, starting from the load end.

(b)   Find the output impedances of the stages successively, starting from the source end.

(c)   Find the overall voltage gain,

$$G_v = \frac{V_{out}}{V_s}.$$

(d)   A 1 μF capacitor is inserted in series with Z_s to prevent dc voltage in the power supply (not shown in the diagram) from entering the signal source. Because of this capacitance, low-frequency signals will be amplified less. Determine the frequency (in Hz) at which the magnitude

$$\left| \frac{V_{out}}{V_s} \right| = 0.707 \times \text{Max Value}$$

Hint: Analyze the simple circuit consisting of V_s, Z_s, C, and Z_{in}.

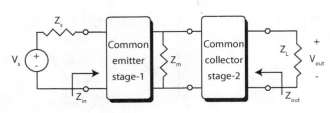

Figure P18.43

44. Reconsider Problem 43. If Z_m is adjustable, find the value of Z_m so that $Z_{out} = Z_L = 64\ \Omega$. Then find Z_{in} and

$$G_v = \frac{V_{out}}{V_s}.$$

45. For Figure P18.45, the y- and h-parameter matrices (in standard units) of two-ports N_1 and N_2, respectively, are

$$\mathbf{Y}_{N1} = \begin{bmatrix} 0.02 & 0.001 \\ 2.5 & 0.2 \end{bmatrix} \quad \mathbf{H}_{N2} = \begin{bmatrix} 10 & 0.05 \\ -125 & 0.5 \end{bmatrix}$$

If $Z_L = 8\ \Omega$ and $Z_s = 25\ \Omega$, find the input impedance and the voltage gains,

$$G_{v1} = \frac{V_1}{V_s} \quad \text{and} \quad G_{v2} = \frac{V_L}{V_1}$$

for the cascaded two-port.

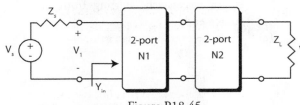

Figure P18.45

CHECK: $G_{v1} = 0.8$ and $G_{v2} = -100$

46. Consider the terminated two-port configuration in Figure P18.46, with h_{ij} indicating the two-port h-parameters. Suppose that (i) the current through the admittance, Y_L, equals the current through h_{22}; (ii) the current gain

$$G_I = \frac{I_2}{I_1} = 100;$$

and (iii) the source resistance is $Z_s = 8\ \text{k}\Omega$.

(a) Compute h_{22} in terms of Y_L and possibly other h-parameters. Explain your reasoning.

(b) Derive a formula for the current gain, G_p in terms of the h-parameters and Y_L.

(c) Find h_{21}.

(d) Suppose now that the source, $I_s - Z_s$, at the front end of the two-port is briefly disconnected and a voltage, $v_2(t) = 10u(t)$ V is applied to port 2. The measurement $v_1(t) = -5u(t)$ V is made. Compute h_{12}.

(e) Suppose the source is reconnected to the two-port. If the current gain

$$\frac{I_1}{I_s} = 0.8$$

and if $Y_L = 0.25$ S, determine Z_{in} and h_{11}.

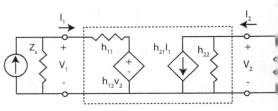

Figure P18.46

CHECK: $h_{21} = 200$, $h_{12} = -0.5$, $h_{11} = 1800$ Ω, $Z_{in} = 2\ \text{k}\Omega$

47. In this problem you are to design an amplifier circuit represented by the doubly terminated equivalent circuit shown in Figure P18.47. This means you will be given certain amplifier specifications that will allow you to determine the parameters of the amplifier circuit.

Amplifier specifications: $R_L = 2\ \Omega$ and $R_s = 40\ \Omega$.

(i)   When I_1 is zero, the ratio

$$\frac{V_1}{V_2} = 0,$$

when a source is applied to port 2.

(ii)   There must be maximum power transfer from the amplifier output to the load under the condition that $Z_{out}(s) = 800\ \Omega$.

(iii) $\dfrac{V_1}{V_s} = \dfrac{25}{26}$

(iv)   The voltage gain $\dfrac{V_2}{V_1} = -100$.

Given these specifications:

(a)   Compute h_{12}.

(b)   Compute Y_{out}, h_{22}, and the turns ratio a.

(c)   Compute h_{11}.
        Compute the input impedance Z_{in}

(d)   Compue h_{21}

(e)   Compute the ratio of the power delivered to the load R_L to the power *delivered* to the amplifier, $V_1 I_1$.

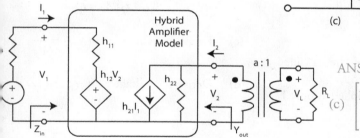

Figure P18.47

48. Repeat Problem 47 with the amplifier specification

$$\frac{V_1}{V_2} = 0.01$$

when $I_1 = 0$.

ANSWER: $[h_{ij}] = \begin{bmatrix} 2 \times 10^3 & 0.01 \\ 490.38 & 3.6538 \times 10^{-3} \end{bmatrix}$

(in standard units)

t-PARAMETERS

49.(a)   Verify that the *t*-parameters of the circuit of Figure P18.49a are

$$T = \begin{bmatrix} 1 & Z_1 \\ 0 & 1 \end{bmatrix}.$$

(b)   Verify that the *t*-parameters of the circuit of Figure P18.49b are

$$T = \begin{bmatrix} 1 & 0 \\ Y_2 & 1 \end{bmatrix}.$$

(c)   Compute the *t*-parameters of Figures P18.49c and d.

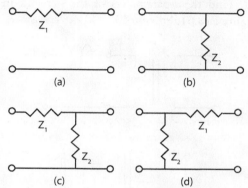

Figure P18.49

ANSWERS:

(c) $\begin{bmatrix} 1+\dfrac{Z_1}{Z_2} & Z_1 \\ Y_2 & 1 \end{bmatrix}$; (d) $\begin{bmatrix} 1 & Z_1 \\ Y_2 & 1+\dfrac{Z_1}{Z_2} \end{bmatrix}$.

50.(a)   Show that the *t*-parameters of the ideal transformer of Figure P18.50a are

$$T = \begin{bmatrix} n & 0 \\ 0 & 1/n \end{bmatrix}.$$

(b)   Given the answer to part (a), compute the *t*-parameters of the circuits of Figure P18.50b and c.

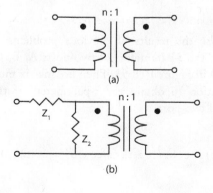

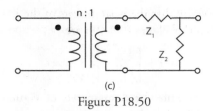

(c)

Figure P18.50

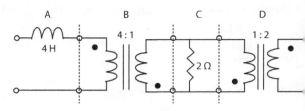

Figure P18.53

51. Find the t-parameters of the network in Figure P18.51.

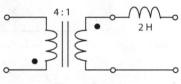

Figure P18.51

ANSWER: $\begin{bmatrix} -4 & 0 \\ 0 & -\dfrac{1}{4} \end{bmatrix} \begin{bmatrix} 1 & 2s \\ 0 & 1 \end{bmatrix} = \begin{bmatrix} -4 & -8s \\ 0 & -0.25 \end{bmatrix}$

52.(a) Find the t-parameters of the cascaded network in Figure P18.52.
   (b) If a 14 volt source is applied at port 1, compute $v_2(t)$.

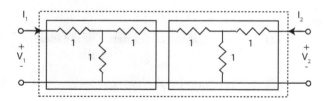

Figure P18.52

ANSWER: (a) $\begin{bmatrix} 2 & 3 \\ 1 & 2 \end{bmatrix}^2 = \begin{bmatrix} 7 & 12 \\ 4 & 7 \end{bmatrix}$.

53. Use the results of previous problems to obtain the t-parameters of two-ports A, B, C, and D in Figure P18.53. Then use matrix multiplication to obtain the t-parameters of the overall two-port.

54. The two-port of Figure P18.54 is described by t-parameters.
   (a) Derive the input impedance relationship
   $$Z_{in} = \frac{t_{11}Z_L + t_{12}}{t_{21}Z_L + t_{22}}.$$
   (b) Derive the output impedance relationship
   $$Z_{out} = \frac{t_{22}Z_s + t_{12}}{t_{21}Z_s + t_{11}}.$$
   (c) Derive the voltage gain $G_{v1} = \dfrac{V_1}{V_s}$.
   (d) Derive the voltage gain $G_{v2} = \dfrac{V_2}{V_1}$.

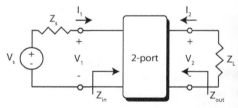

Figure P18.54

55. Suppose a two-port has both z-parameters and t-parameters. Compute the t-parameters in terms of the z-parameters. Hint: Rewrite the z-parameter equations in the form
$$M_1 \begin{bmatrix} V_1 \\ I_1 \end{bmatrix} = M_2 \begin{bmatrix} -I_2 \\ V_2 \end{bmatrix}$$
and invert the appropriate matrix to obtain
$$[t_{ij}] = \begin{bmatrix} \dfrac{z_{11}}{z_{21}} & \dfrac{\Delta z}{z_{21}} \\ \dfrac{1}{z_{21}} & \dfrac{z_{22}}{z_{21}} \end{bmatrix}$$

where $\Delta z = z_{11}z_{22} - z_{12}z_{21}$.

56. For the circuits of Figure P18.56, suppose the *t*-parameters in standard units of the two-port N_1 are

$$T_{N1} = \begin{bmatrix} -0.1 & -0.1 \\ -0.0015 & -0.005 \end{bmatrix},$$

$R = 200\ \Omega$, $L = 0.5$ H, and $n = 5$.

   (a)   Find the *t*-parameters of the cascaded two-port for each configuration.

   (b)   Find the steady-state $i_1(t)$ under the conditions that $v_1(t) = 10\cos(100t)$ V and each two-port is terminated in a parallel RC circuit with $R = 10\ \Omega$ and $C = 1$ mF.

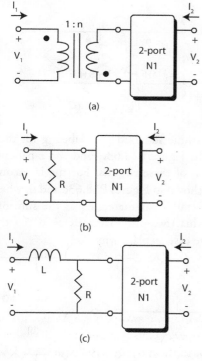

Figure P18.56

57. A certain high-voltage transmission line operating at 60 Hz is represented by a two-port with the following *t*-parameters: $t_{11} = 0.895 + j0.022$, $t_{12} = 40 + j180\ \Omega$, $t_{21} = -2.617 \times 10^{-5} + j1.102 \times 10^{-3}$ S, and $t_{22} = 0.895 + j0.022$.

   (a)   If the receiving end (port 2) draws a current $-I_2 = 361\angle 0$ A at $V_2 = 115,200\angle 0$ V, find the sending end (port 1) voltage V_1, the current I_1, the power delivered by the source, and the power loss in the transmission line.

   (b)   If $V_1 = 134,000\angle 0$ V, and a resistive load represented by a 500 Ω resistor is connected to the receiving end (port 2), find the power delivered by the source and the magnitude of the load voltage.

ANSWER: (a) $V_1 = 135,550\angle 29.87°$ V, $I_1 = 347.36\angle 22.86°$ A, 46.733 megawatts, 5.146 megawatts; (b) 35.331 megawatts, 127.96 kV.

PARAMETER CONVERSION AND INTERCONNECTION OF TWO-PORTS

58. A two-port N has known *z*-, *y*-, and *h*-parameters. A new two-port N_{new} is formed by adding one single impedance Z or admittance Y to N in various ways. Prove the following relationships between the old and new -two-port parameters.

   (a)   If Z is connected in series with port 1 of N, then $z_{11,new} = z_{11} + Z$ and $h_{11,new} = h_{11} + Z$. Other *z*- and *h*-parameters remain the same.

   (b)   If Z is connected in series with port 2 of N, then $z_{22,new} = z_{22} + Z$. Other *z*-parameters remain the same.

ANSWERS: (a) $[t_{ij}] = \begin{bmatrix} 0.02 & 0.02 \\ 0.0075 & 0.025 \end{bmatrix}$; $[t_{ij}] = \begin{bmatrix} -0.1 & -0.1 \\ -0.002 & -0.0055 \end{bmatrix}$;

$[t_{ij}] = \begin{bmatrix} -0.001s - 0.1 & 0.00275s - 0.1 \\ -0.002 & -0.0055 \end{bmatrix}$; (b) $i_{1a}(t) = 4.666\cos(100t + 8.84°)$ A, $i_{1b}(t) = 0.23618\cos(100t + 6.977°)$ A, $i_{1c}(t) = 0.1627\cos(100t - 46.86°)$ A.

(c) If Y is connected in parallel with port 2 of N, then $y_{22,new} = y_{22} + Y$ and $h_{22,new} = h_{22} + Y$. Other y- and h- parameters remain the same.

(d) If Y is connected in parallel with port 1 of N, then $y_{11,new} = y_{11} + Y$. Other y-parameters remain the same.

59. A common-ground two-port N has known z- and y-parameters. A new two-port N_{new} is formed by adding one single impedance Z or admittance Y to N in two different ways. Prove the following relationships between the old and new -two-port parameters.

(a) If Y is connected across the top terminals of N, then $y_{11,new} = y_{11} + Y$, $y_{22,new} = y_{22} + Y$, $y_{12,new} = y_{12} - Y$, and $y_{21,new} = y_{21} - Y$.

(b) If Z is connected in series with the common terminal of N, then $z_{11,new} = z_{11} + Z$, $z_{22,new} = z_{22} + Z$, $z_{12,new} = z_{12} + Z$, and $z_{21,new} = z_{21} + Z$.

60. Assume standard labeling in Figure P18.60, in which the inner two-port labeled N has z-parameter matrix

$$\begin{bmatrix} 4 & 5 \\ 3 & 4 \end{bmatrix} \Omega.$$

(a) Find the y-parameters of the overall two-port N* and then find the z-parameter matrix.

(b) Now suppose the ports are connected to current sources, in which case $i_1(t) = i_2(t) = 15u(t)$ A. Assuming zero initial conditions, find $v_1(t)$ and $v_2(t)$. What is the response if $i_1(t) = i_2(t) = 5u(t)$ A? Hint: Use linearity.

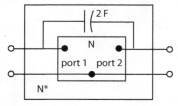

Figure P18.60

61. Compute the y-parameters of the two-port of Figure P18.61.

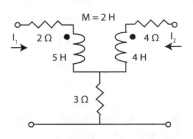

Figure P18.61

62. Compute the y-parameters of the two-port of Figure P18.62.

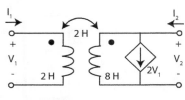

Figure P18.62

63. Assume standard port labeling for the circuits in Figure P18.63. Find the z-parameters of each of these circuits. For the interconnections shown in Figures P18.63c and d, when do the overall z-parameters equal the sum of the individual two-port parameters? When this is *not* the case, explain why not.

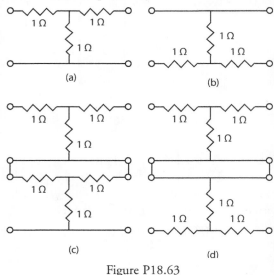

Figure P18.63

64. Compute any one set of the z-, y-, and h-parameters for each of the circuits of Figure P18.64. Then obtain the remaining two sets by the use of the conversion table (Table 18.1).

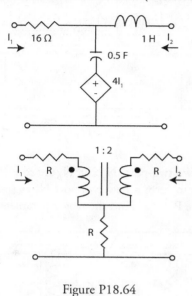

Figure P18.64

65. For the circuit of Figure P18.65, find the z-parameters of each two-port, N_a and N_b, and then the overall z-parameters of the interconnected two-port.

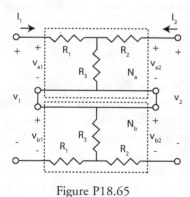

Figure P18.65

66. For the circuit of Figure P18.66, find the y-parameters of each two-port, N_a and N_b, and then the overall z-parameters of the interconnected two-port.

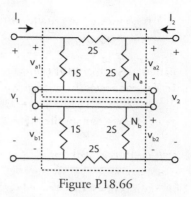

Figure P18.66

67. Find the z-parameters of the interconnected circuit shown in Figure P18.67, in which $L = 1$ H, assuming standard units for the y-parameters.

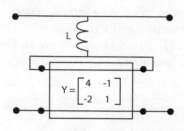

Figure P18.67

ANSWER: $[z_{ij}] = \begin{bmatrix} s+0.5 & s+0.5 \\ s+1.0 & s+2.0 \end{bmatrix} \Omega.$

68. Repeat Problem 67 with the inductor changed to a 0.5 F capacitor.

69. Find the z- and y-parameters of the interconnected network shown in Figure P18.69, in which $R_1 = 2\ \Omega$, $R_2 = 2\ \Omega$, $R_3 = 1\ \Omega$, and

$$Z = \begin{bmatrix} 2 & 1 \\ 0 & -2 \end{bmatrix} \Omega.$$

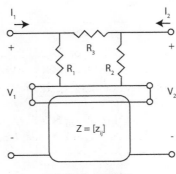

Figure P18.69

ANSWER: $[z_{ij}]_{new} = \begin{bmatrix} 3.2 & 1.8 \\ 0.8 & -0.8 \end{bmatrix} \Omega.$

70. The two-port N_a is connected to a 1:1 ideal transformer to form N_a^*. Then N_a^* and N_b are connected in series to form a new two-port N as shown in Figure P18.70, in which $R_1 = 2\ \Omega$, $R_2 = 2\ \Omega$, $R_3 = 3\ \Omega$, and $R_4 = 2\ \Omega$. Find the z-parameters of N.

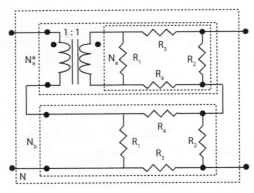

Figure P18.70

71. Two two-ports N_a and N_b are connected in series to form a new two-port N, as shown in Figure P18.71, in which $n = 2$, $R_1 = 4\ \Omega$, $R_2 = 4\ \Omega$, $R_3 = 2\ \Omega$, $R_4 = 8\ \Omega$, $R_5 = 2\ \Omega$, $R_6 = 2\ \Omega$, $R_7 = 1\ \Omega$, and $R_8 = 4\ \Omega$, Find the new z-parameters.

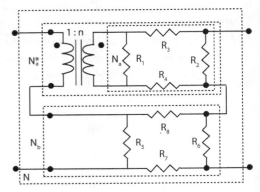

Figure P18.71

72. Two two-ports N_a and N_b are connected in series to form a new two-port N, as shown in Figure P18.72. Find the new z-parameters.

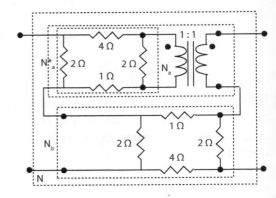

Figure P18.72

73. Consider the two-port in Figure P18.73, which depicts a transistor amplifier stage for a microphone. Suppose that the transistor has h-parameters given as $h_{11} = 4.2$ kΩ, $h_{12} = 0$, $h_{21} = 150$, and $h_{22} = 0.1$ mS.

(a)   Assume the input and output capacitors are short circuits at the frequency of interest and that the capacitor across the 470 Ω resistor is also a short circuit. Determine the h-parameters of the overall two-port.

(b)   Given your answer to part (a), determine the value of a for maximum power transfer.

(c)   Given your answer to part (a), determine the voltage and power gain of the overall two-port.

(d)   Plot the frequency response of the amplifier as a function of $f = 2\pi\omega$ for $0 \leq f \leq 20{,}000$ kHz using Spice.

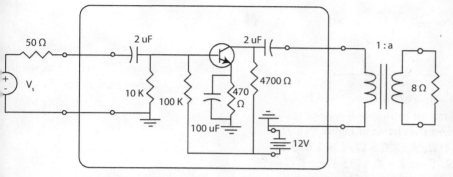

Figure P18.73

RECIPROCITY

74. The two-port in Figure P18.74a has port 1 attached to an ideal current source. It is known that the two-port consists only of R's, L's, C's, and transformers. If $i_1(t) = \delta(t)$, your oscilloscope shows a waveform having the analytic expression $v_2(t) = 3e^{-t} + 5e^{-t} \cos(t - 30°)$ V for $t \geq 0$. Figure P18.74b shows the same two-port with different terminations.

(a)   If $i_2(t) = u(t)$ A, find the zero-state response, $v_1(t)$.

(b)   If the input is changed so that $i_2(t) = 10 \cos(500t)u(t)$ A, find the steady-state response, $v_1(t)$.

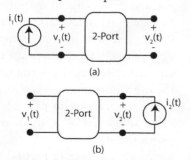

(a)

(b)

Figure P18.74

ANSWER: (b) $v_{1,ss}(t) = 0.1466 \cos(500t - 90°)$ V.

75. Consider the circuit of Figure P18.75, in which $\mu = 0.2$, $R_1 = 1\ \Omega$, and $R_2 = 4\ \Omega$.

(a)   For what value of r_m is the two-port reciprocal?

(b)   For the value of r_m found in part (a), compute the circuit's z- and h-parameters.

(c)   If port 2 is terminated in a 2 Ω–1 mF parallel RC combination, and $v_1(t) = 20\sqrt{2} \cos(2000t)$ V, find $v_2(t)$ in steady state. Use the value of r_m found in part (b).

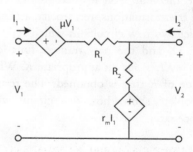

Figure P18.75

ANSWER: (a) $r_m = 1\ \Omega$;

(b) $[z_{ij}] = \begin{bmatrix} 7.5 & 5.0 \\ 5.0 & 4.0 \end{bmatrix}$, $[h_{ij}] = \begin{bmatrix} 1.25 & 1.25 \\ -1.25 & 0.25 \end{bmatrix}$;

(c) $v_2(t) = 10 \cos(2000t - 45°)$ V.

76. Consider the circuit of Figure P18.76, in which $R_s = R_L = 1 \, \Omega$, $L = \sqrt{2}$ H and $C = \sqrt{2}$ F. The transfer function

$$H(s) = \frac{n(s)}{d(s)} = \frac{?}{s^2 + \sqrt{2}s + 1}.$$

Use all the theorems you have learned in this course (including reciprocity) to construct an alternative circuit that has the same $d(s)$.

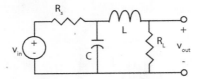

Figure P18.76

77. Again consider the circuit of Figure P18.76, in which $R_s = 1 \, \Omega$, $R_L = 0.5 \, \Omega$, $L = 1.673$ H and $C = 0.8966$ F, or $L = 0.4483$ H and $C = 3.3466$ F. The transfer function is

$$H(s) = \frac{V_{out}(s)}{V_{in}(s)} = \frac{\dfrac{1}{3}}{s^2 + \sqrt{2}s + 1}$$

Use the theorems you learned in this course (source transformations, reciprocity, scaling) to construct two new circuits in which $R_s = 1 \, \Omega$, $R_L = 2 \, \Omega$, and the new transfer function is $H_{new}(s) = KH(s)$ for an appropriate K. What is the value of K that is obtained? The result of this problem shows how one circuit with a resistance ratio

$$\frac{R_L}{R_s} = a$$

can be derived from another with

$$\frac{R_L}{R_s} = \frac{1}{a}.$$

CHECK: $L = 3.3466$ H or $L = 0.8966$ H in series R_s with $K = 2$

C H A P T E R *19*

Principles of Basic Filtering

LOUDSPEAKERS AND CROSSOVER NETWORK

In a stereo system, a power amplifier is connected to a pair of speakers. The most common type of speaker system consists of one or more drivers enclosed in a wooden box. The amplifier feeds a signal to each driver's voice coil. The voice coil sits within a field produced by a permanent magnet and is attached to a heavy paper or plastic cone. When excited by a current from the power amplifier, the coil interacts with the magnetic field of the permanent magnet and vibrates. The coil pushes and pulls the speaker's cone, which, in turn, proportionately moves the air, making sound waves.

The high and low frequencies that make up music place opposite requirements on a loudspeaker. A good low-frequency speaker should be large, in order to push a lot of air. A good high-frequency speaker should be light, in order to move back and forth rapidly.

A two-way speaker system consists of a small, light tweeter to handle the treble signals and a large woofer to handle the bass. A better system is a three-way system, with a third, midrange speaker to handle the frequencies in the middle. The magnitude response of a typical two-way system is shown in the following figure.

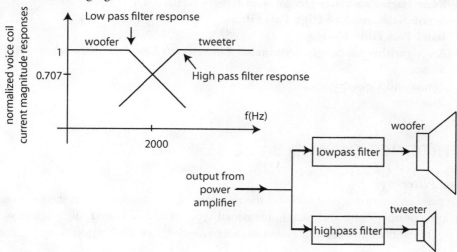

A *crossover network* (low-pass or high-pass filter) separates the frequencies so that the woofer receives the low-frequency content of the music and the tweeter the higher-frequency content. In the magnitude response plot illustrated in the figure, both curves have the same 3 dB frequency at 2000 Hz. This frequency is called the *crossover frequency*. This chapter explores the basic design principles and realizations of low-pass and high-pass filters. Some band-pass filtering is also discussed. Some simple crossover circuits are considered in the problems section.

CHAPTER OBJECTIVES

1. Introduce the meaning and (brickwall) specification of low-pass filters in terms of dB loss.
2. Set forth the maximally flat Butterworth magnitude response and associated Butterworth transfer function.
3. Present a step-by-step design algorithm for finding the filter order and associated Butterworth transfer function.
4. Present basic passive and active circuits that realize a Butterworth transfer function.
5. Set forth the properties of the Butterworth transfer function.
6. Introduce high-pass filter design through frequency transformation.
7. Present basic passive and active circuits that realize a Butterworth high-pass filter.
8. Introduce an algorithm for the design of a band-pass filter using frequency transformations.

SECTION HEADINGS

1. **Introduction and Basic Terminology**
2. **Low-Pass Filter Basics**
3. **Butterworth Solution to the Approximation Problem**
4. **Butterworth Passive Realization**
5. **Active Realization of Low-Pass Butterworth Filters**
6. **Input Attenuation for Active Circuit Design**
7. **Properties of the Butterworth Loss Function**
8. **Basic High-Pass Filter Design with Passive Realization**
9. **Active Realization of High-Pass Filters**
10. **Band-Pass Filter Design**
11. **An Algorithm for Singly Terminated Butterworth Low-Pass Networks**
12. **Summary**
13. **Terms and Concepts**
14. **Problems**

1. INTRODUCTION AND BASIC TERMINOLOGY

Types of Filtering

Filtering plays an important role in circuit theory, as well as in communication theory, image processing, and control. There are three fundamental types of filters: analog, digital, and switched capacitor. Digital filter analysis and design requires knowledge of the Nyquist sampling theorem.

Switched capacitor networks/filters, an idea introduced in Chapter 13, are something of a hybrid between analog and digital filters. Both are beyond the scope of this chapter, which takes a circuit's viewpoint on some basic analog filtering concepts and techniques.

Analog filters process the actual input waveform with circuits composed of discrete components such as resistors, capacitors, inductors, and op amps. Analog filters are of two types, passive and active. Passive analog filters are composed only of resistors, capacitors, and inductors. Active analog filters consist of resistors, capacitors, and op amps or other types of active elements.

Basic Terminology

A **filter** is a device (often an electrical circuit) that shapes or modifies the frequency content (spectrum) of a signal or waveform. We represent a filter by a transfer function $H(s)$ whose frequency response is $H(j\omega)$. The gain, **gain magnitude**, or **frequency response magnitude** is $|H(j\omega)|$. If $H(j\omega)$ is normalized so that its maximum gain is 1, then the **gain in dB** (decibels) is $G_{dB}(\omega) = 20 \log_{10}|H(j\omega)|$. In this chapter we will assume that the maximum value of $|H(j\omega)|$ has been normalized to 1.

An important frequency called the **cutoff frequency** (or frequencies), also called the **half-power point** (or points), is that frequency, denoted ω_c, for which

$$|H(j\omega_c)|^2 = \frac{1}{2}, \text{ in which case the gain is } |H(j\omega_c)| = \frac{1}{\sqrt{2}} = 0.70714.$$

Since power is proportional to voltage squared or current squared (gain squared) for a fixed load resistance, we arrive at the terminology of half-power. This is also called the **3dB down point** because $G_{dB}(\omega_c) = 10 \log_{10}|H(j\omega_c)|^2 = 10 \log_{10}(0.5) = -3$ dB.

Often, design specifications for a filter are expressed in terms of attenuation or loss rather than gain. This results in several definitions that are dual to the gain-related definitions above. The loss function, denoted $\hat{H}(j\omega)$, is the reciprocal of the transfer function. Thus, the **attenuation**, or **loss magnitude**, is

$$|\hat{H}(j\omega)| = \frac{1}{|H(j\omega)|}$$

It follows that the filter loss or attenuation in dB is

$$A(\omega) = dB \text{ loss} = -20 \log_{10}|H(j\omega)| = 20 \log_{10}|\hat{H}(j\omega)| \qquad (19.1)$$

With these definitions established, the chapter will examine low-pass (LP), high-pass (HP), and-band pass (BP) filters, leaving the study of band-reject filters to higher-level texts. A **low-pass filter** is a device (typically a circuit) offering very little attenuation to the low-frequency content (low-frequency spectrum) of signals while significantly attenuating (blocking) the high-frequency content of those signals. **High-pass filters** do the opposite: they block the low-frequency content and allow the high-frequency content of a signal to pass through. Finally, band-pass filters, as described in Chapter 16, allow a band of frequencies to pass while significantly attenuating those outside the band. Interestingly, the general practical design of all such filters, especially for the high-order case, is generally based on a low-pass prototype design; the low-pass prototype is transformed into a high-pass or band-pass type using a frequency transformation.

As a first step in exploring the design of such filters, we describe Butterworth (low-pass) transfer functions. Butterworth was the name of the English engineer who first developed this special class of transfer functions in his paper "On the Theory of Filter Amplifiers."[1] The next step is presentation of an algorithm that adapts these basic Butterworth transfer functions into ones that meet a given set of LP filter design specifications. Once the transfer function is known, an engineer must implement this transfer function as a passive or active circuit. This chapter will also outline some techniques for generating passive and active realizations.

2. LOW-PASS FILTER BASICS

As mentioned, a low-pass filter allows the low-frequency content of a signal to pass with little attenuation while significantly blocking the high-frequency content. The immediate question is, why use a low-pass filter? One possible answer is that the noise in a noisy signal often has most of its energy in the high-frequency range. For example, so-called white noise has a constant frequency spectrum. Hence, low-pass filtering a sinusoidal signal corrupted by white noise will generally result in a "cleaned-up" information signal, as illustrated in Figure 19.1. In Figure 19.1, the thickness of the curves represents the infiltration of noise. By using the low-pass filter, the curve is "sharpened" by reducing the high-frequency noise content.

The suspension system of a car is a low-pass filter: slow, rhythmic (low-frequency) road variations are permitted, while the effects of chuckholes and bumps (high frequencies) are "filtered out."

FIGURE 19.1 The effect of low-pass filtering on a noisy signal.

Whenever LP filters are needed, an engineer must provide design specifications. Historically this was done using a **low-pass filter (brickwall) specification,** as illustrated in Figure 19.2. Two pairs of numbers (ω_p, A_{max}) and (ω_s, A_{min}) characterize this brickwall specification, where

1. ω_p equals the pass-band edge frequency,
2. $0 \leq \omega \leq \omega_p$ is called the pass-band,
3. A_{max} is maximum dB attenuation permitted in the pass-band,
4. ω_s is the stop band edge frequency.
5. $\omega_s \leq \omega$ defines the stop band,
6. A_{min} is the minimum allowable dB attenuation in the stop band, and
7. $\omega_p \leq \omega \leq \omega_s$ defines the transition band.

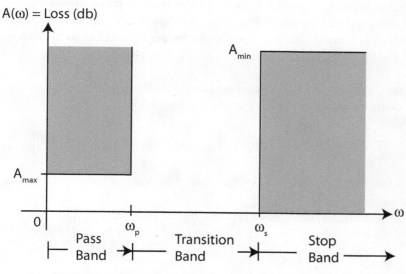

FIGURE 19.2 Brickwall specification of low-pass filter.

The shaded region in Figure 19.2 represents a brickwall. The attenuation of the filter in dB, i.e., $A(\omega) = -20 \log_{10}|H(j\omega)| = 20 \log_{10}|\hat{H}(j\omega)|$, must reside outside the shaded region. Finding a (normalized) filter transfer function that meets the brickwall specs is called the **approximation problem.** Once the approximation problem is solved by identification of the proper normalized transfer function, the next step is to construct a circuit realization of the normalized transfer function and then to frequency-scale to obtain the proper pass-band edge frequency, and finally to magnitude-scale to obtain the proper impedance levels.

The simplest technique for solving the approximation problem is with 3dB normalized Butterworth transfer functions whose squared magnitude responses are given by

$$|H(j\omega)|^2 = \frac{1}{1 + \varepsilon^2 \left(\dfrac{\omega}{\omega_p}\right)^{2n}} = \frac{1}{1 + \left(\dfrac{\omega}{\omega_c}\right)^{2n}} \qquad (19.2)$$

where $\omega_c = \dfrac{\omega_p}{\varepsilon^{(1/n)}}$ is the 3 dB down point, or cutoff frequency, of the filter, and ε is a to-be specified constant. As mentioned earlier, the term "3 dB down point" arises here because, for all n,

$$10 \log_{10}\left[1 + \left(\frac{\omega}{\omega_c}\right)^{2n}\right]_{\omega=\omega_c} = 10 \log_{10}[2] = 3 \text{ dB}$$

i.e., there is 3 dB of loss at $\omega = \omega_c$. When $\varepsilon = 1$ in equation 19.2, the pass-band edge frequency ω_p and cutoff frequency ω_c coincide. If we now define the *normalized frequency* as $\Omega = \dfrac{\omega}{\omega_c}$, the magnitude response of equation 19.2 becomes

$$|H_{3dBNLP}(j\Omega)|^2 = \frac{1}{1 + \Omega^{2n}} \qquad (19.3)$$

Equation 19.3 denotes the *n*th-order 3 dB *normalized Butterworth magnitude response*. The words "3 dB normalized" refer to the fact that at $\Omega = 1$, the loss is 3 dB (the gain is –3 dB). Remember that the actual filter transfer function depends on a proper choice of ω_c or ε, which we will clarify shortly.

Before proceeding further, we ask a critical question: *Does this kind of representation make sense?* To answer this question, note that the dB loss of equation 19.3 is

$$A(\Omega) = loss(dB) = 20\log_{10}\left|\widehat{H}_{3dBNLP}(j\Omega)\right| = 10\log_{10}\left[1 + \Omega^{2n}\right] \tag{19.4}$$

Plotting this function for various *n*'s as a function of normalized frequency, Ω, for a normalized stop band edge frequency, $\Omega_s = 3.5$, $A_{max} = 3$ dB, and $A_{min} = 20$ dB yields the polynomial curves in Figure 19.3. Clearly, for higher *n*, we can have higher values of A_{min}. These curves can be made to lie outside the brickwalls of Figure 19.2 and hence can validly be used to meet a LP filter specification.

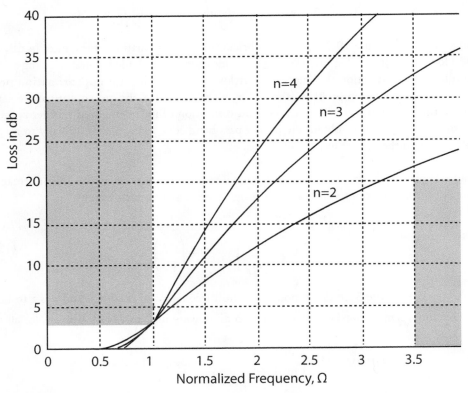

FIGURE 19.3 Plot of the normalized Butterworth magnitude responses for *n* = 2, 3, 4.

3. BUTTERWORTH SOLUTION TO THE APPROXIMATION PROBLEM

From Figure 19.3, we see that as n increases, the pass-band magnitude response becomes flatter and the transition to stop band becomes steeper. This property suggests that there is an appropriate value of n to meet a given set of brickwall specifications (ω_p, A_{max}) and (ω_s, A_{min}). The idea is to use equation 19.2 and these specs to determine a proper value for the filter order n, a proper cutoff frequency ω_c, and the normalized transfer function.

Computation of Filter Order n and ω_c

To compute the filter order, observe that n must satisfy the pass-band and stop band edge frequency constraints. From equations 19.1 and 19.4, at the stop band edge frequency,

$$A_{min} \leq A(\omega_s) = 10\log_{10}\left|\hat{H}\left(j\omega_s\right)\right|^2 = 10\log_{10}\left[1+\left(\frac{\omega_s}{\omega_c}\right)^{2n}\right]$$

After some algebra, we obtain

$$\left(\frac{\omega_s}{\omega_c}\right)^{2n} \geq 10^{0.1A_{min}} - 1 \tag{19.5a}$$

Similarly, at the pass-band edge frequency,

$$A_{max} \geq A(\omega_p) = 10\log_{10}\left|\hat{H}\left(j\omega_p\right)\right|^2 = 10\log_{10}\left[1+\left(\frac{\omega_p}{\omega_c}\right)^{2n}\right]$$

Again, after some algebra we obtain

$$\left(\frac{\omega_p}{\omega_c}\right)^{2n} \leq 10^{0.1A_{max}} - 1 \tag{19.5b}$$

Dividing the left and right sides of equation 19.5a by the left and right sides of equation 19.5b (this maintains the inequality of equation 19.5a) yields

$$\left(\frac{\omega_s}{\omega_p}\right)^{2n} \geq \frac{10^{0.1A_{min}} - 1}{10^{0.1A_{max}} - 1} \tag{19.6}$$

Solving for n produces the *order formula*

$$n \geq \frac{\log_{10}\left(\sqrt{\dfrac{10^{0.1A_{min}} - 1}{10^{0.1A_{max}} - 1}}\right)}{\log_{10}\left(\dfrac{\omega_s}{\omega_p}\right)} \tag{19.7}$$

Thus n can be any integer satisfying inequality 19.7. Usually one takes the smallest such n.

Once n is picked, there is a permissible range of ω_c given by

$$\omega_{c,min} = \frac{\omega_p}{\sqrt[2n]{10^{0.1A_{max}} - 1}} \leq \omega_c \leq \frac{\omega_s}{\sqrt[2n]{10^{0.1A_{min}} - 1}} = \omega_{c,max} \qquad (19.8)$$

To derive the range given in equation 19.8, we reconsider equations 19.5. Taking the $2n$th root of both sides of equation 19.5a and solving for ω_c yields the right side of equation 19.8. On the other hand, taking the $2n$th root of equation 19.5b and solving for ω_c yields the left side of equation 19.8.

Exercise. Verify the mathematical details in the above paragraph on the derivation of equation 19.8.

EXAMPLE 19.1. Suppose we are given the brickwall specs (f_p, A_{max}) = (500 Hz, 2 dB) and (f_s, A_{min}) = (2000 Hz, 30 dB). Find (a) the minimum filter order n, (b) $\omega_{c,min}$ and $\omega_{c,max}$, (c) and the normalized LP squared magnitude function, and then (d) plot the magnitude responses for the cases where $\omega_c = \omega_{c,min}$ and $\omega_c = \omega_{c,max}$ over the frequency range $0 \leq 2\pi f \leq 2\pi \times 2500$.

Step 1. Find the minimum filter order n from equation 19.7. Here

$$n \geq \frac{\log_{10}\left(\sqrt{\frac{10^{0.1A_{min}} - 1}{10^{0.1A_{max}} - 1}}\right)}{\log_{10}\left(\frac{\omega_s}{\omega_p}\right)} = \frac{\log_{10}\left(\sqrt{\frac{10^3 - 1}{10^{0.2} - 1}}\right)}{\log_{10}(4)} = 2.68$$

implying that the minimum filter order is 3. This can also be accomplished in MATLAB as follows:

»n=buttord(wp,ws,Amax,Amin,'s')
n = 3

Step 2. Using equation 19.8 with $n = 3$,

$$\omega_{c,min} = \frac{\omega_p}{\sqrt[2n]{10^{0.1A_{max}} - 1}} = \frac{2\pi \times 500}{\sqrt[6]{10^{0.2} - 1}} = 2\pi \times 547 = 3435 \text{ rad/s}$$

and

$$\omega_{c,max} = \frac{\omega_s}{\sqrt[2n]{10^{0.1A_{min}} - 1}} = \frac{2\pi \times 2000}{\sqrt[6]{10^3 - 1}} = 2\pi \times 633 = 3975 \text{ rad/s}$$

Step 3. Since the order is 3, we have from equation 19.2

$$|H(j\omega)|^2 = \frac{1}{1 + \left(\dfrac{\omega}{\omega_c}\right)^6}$$

Step 4. Plotting this function over the frequency range $0 \leq 2\pi f \leq 2\pi \times 2500$ can be achieved with the following MATLAB code. The resulting plot is displayed in Figure 19.4.

```
»f = 0:4:2500;
»h1 = sqrt(1. ./(1 + (2*pi*f ./wcmin).^6));
»h2 = sqrt(1. ./(1 + (2*pi*f ./wcmax).^6));
»plot(f,-20*log10(h1),f,-20*log10(h2))
»grid
```

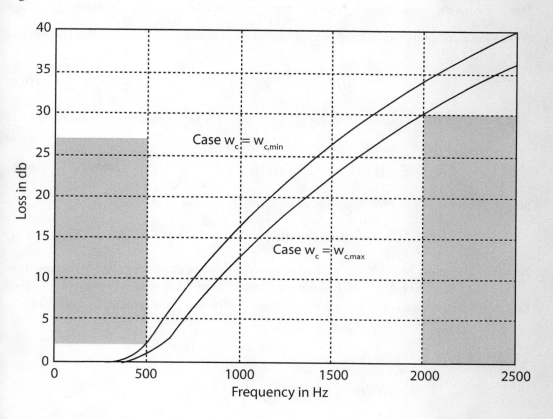

FIGURE 19.4. Plot of third-order LP filter response using $\omega_{c,min}$ and $\omega_{c,max}$.

Observe that with the choice of $\omega_c = \omega_{c,min}$, the magnitude response curve passes through the pass-band edge frequency with loss exactly equal to A_{max}, leaving more than adequate loss at ω_s. On the other hand, with the choice of $\omega_c = \omega_{c,max}$, the magnitude response curve passes through the stop band edge frequency with loss exactly equal to A_{min} with less than the maximal allowable loss at ω_p. In practice, to allow for element tolerances, one would choose ω_c somewhere in between.

Exercises. 1. Verify that if $\omega_c = \omega_{c,min}$, the loss $A(\omega_p) = A_{max}$ and $A(\omega_p) \geq A_{max}$.
2. Verify that if $\omega_c = \omega_{c,max}$, the loss $A(\omega_s) = A_{min}$ and $A(\omega_p) \leq A_{max}$.

Computation of the Normalized Butterworth H(s)

The objective here is to derive the normalized Butterworth transfer function $H_{3dBNLP}(s)$ whose magnitude satisfies the 3 dB normalized Butterworth magnitude response. Mathematically the question is, how do we go from the squared magnitude response $\left|H_{3dBNLP}(j\Omega)\right|^2 = \dfrac{1}{1+\Omega^{2n}}$ to the 3 dB normalized transfer function denoted $H_{3dBNLP}(s)$?

Step 1. Recall the loss function $\left|\hat{H}(j\Omega)\right| = \dfrac{1}{\left|H(j\Omega)\right|}$. From the preceding discussion, $\hat{H}(j\Omega)$ is a 3 dB normalized Butterworth loss function (polynomial) if and only if it has a Butterworth loss magnitude, i.e.,

$$\left|\hat{H}_{3dBNLP}(j\Omega)\right|^2 = 1 + \Omega^{2n} \tag{19.9}$$

Step 2. *Find* $\hat{H}(s)$ *such that at* $s = j\Omega$, $\left|\hat{H}_{3dBNLP}(j\Omega)\right|^2 = 1 + \Omega^{2n}$. *Observe that if* $s = j\Omega$, *then* $\Omega = -js$ *and*

$$\left|\hat{H}(j\Omega)\right|^2 = \hat{H}(j\Omega)\,\hat{H}(-j\Omega) = \hat{H}(s)\hat{H}(-s) = 1 + (js)^{2n} = 1 + (-s^2)^n \tag{19.10}$$

Step 3. *Find the zeros of* $\hat{H}(s)$. From equation 19.10, $\hat{H}(s)\,\hat{H}(-s)$ must satisfy

$$\hat{H}(s)\,\hat{H}(-s) = 1 + (-s^2)^n = 0 \tag{19.11}$$

Because a filter must be stable, the left-half complex plane zeros of equation 19.11 will determine $\hat{H}(s)$. Since the zeros of $\hat{H}(s)$ and $\hat{H}(-s)$ are symmetric about the vertical axis, the zeros of $\hat{H}(-s)$ lie in the right half plane; i.e., the zeros of $\hat{H}(s)$ are the left half plane zeros of $1 + [-s^2]^n = 0$.

EXAMPLE 19.2. In the first-order case,

$$\hat{H}(s)\,\hat{H}(-s) = 1 + (-s^2)^1 = 1 - s^2 = (1 + s)(1 - s)$$

Here the zeros are $s = -1$ (left half plane) and $s = 1$ (right half plane). Thus $\hat{H}(s) = s + 1$ and $\hat{H}(-s) = 1 - s$. It follows that

$$H_{3dBNLP}(s) = \frac{1}{s+1}.$$

EXAMPLE 19.3. In the second-order case,

$$\hat{H}(s)\,\hat{H}(-s) = 1 + (-s^2)^2 = 1 + s^4$$

To find $\hat{H}(s)$ we need to identify the left half plane zeros of $1 + s^4$ and form the associated polynomial. Using MATLAB, we obtain

```
»zeros = roots([1 0 0 0 1])
zeros =
```

-7.0711e-01 + 7.0711e-01i
-7.0711e-01 - 7.0711e-01i
7.0711e-01 + 7.0711e-01i
7.0711e-01 - 7.0711e-01i

Note that two zeros are in the right half complex plane and two in the left. The two in the left half plane determine the Butterworth loss function. For illustration we provide the zero plot of Figure 19.5. Again, note the symmetry of the zeros with respect to the imaginary axis.

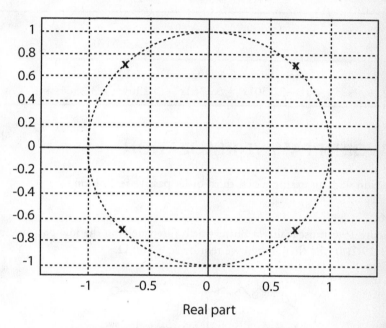

Real part

FIGURE 19.5 Plot of the zeros of $\hat{H}(s)\,\hat{H}(-s) = 1 + s^4$.

As per our earlier statement, from stability considerations we pick the left half plane zeros to form $\hat{H}_{3dBNLP}(s)$, i.e.,

»Hhat = poly(zeros(1:2))
Hhat =
1.0000e+00 1.4142e+00 1.0000e+00

We conclude that

$$H_{3dBNLP}(s) = \frac{1}{s^2 + \sqrt{2}s + 1}$$

For higher-order cases, the procedure of Example 9.3 can again be used. The results are given in Table 19.1, which presents the 3 dB normalized Butterworth loss functions. In practice, one never computes the actual transfer function of a filter that meets a set of non-normalized specs. Rather, one realizes the normalized loss or transfer function and then magnitude- and frequency-scales to obtain the proper circuit. This process is illustrated in the next section.

TABLE 19.1 Normalized Butterworth
Loss Functions, $n = 1, ..., 5$.

n	$\widehat{H}_{3dBNLP}(s)$
1	$s + 1$
2	$s^2 + \sqrt{2}s + 1$
3	$(s + 1)(s^2 + s + 1) = s^3 + 2s^2 + 2s + 1$
4	$(s^2 + 0.76537s + 1)(s^2 + 1.84776s + 1)$ $= s^4 + 2.6131s^3 + 3.4142s^2 + 2.6131s + 1$
5	$(s + 1)(s^2 + 0.61803s + 1)(s^2 + 1.61803s + 1)$ $= s^5 + 3.2361s^4 + 5.2361s^3 + 5.2361s^2 + 3.2361s + 1$

4. BUTTERWORTH PASSIVE REALIZATION

We begin this section with an example of a third-order passive realization.

EXAMPLE 19.4

The circuit of Figure 19.6 must realize a Butterworth filter meeting the low-pass brickwall speci
fication of Figure 19.7. In the final design we require $R_s = 100\ \Omega$.

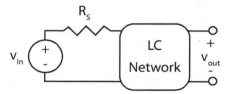

FIGURE 19.6 Structure of a filter driven by a practical source.

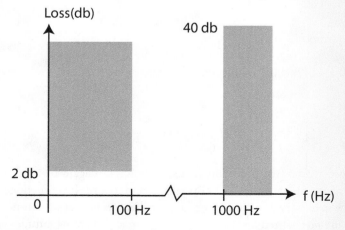

FIGURE 19.7 Low-pass brickwall specification of Example 19.4.

SOLUTION

From the brickwall specification of Figure 19.7, we have ($\omega_p = 200\pi$, $A_{max} = 2$ dB) and ($\omega_s = 2000\pi$, $A_{min} = 40$ dB).

Step 1. *Compute the filter order.* Here

$$n \geq \frac{\log_{10}\left(\sqrt{\dfrac{10^{0.1A_{min}} - 1}{10^{0.1A_{max}} - 1}}\right)}{\log_{10}\left(\dfrac{2000\pi}{200\pi}\right)} = 2.1164$$

This implies that the minimum filter order is $n = 3$.

Step 2. *Compute ω_c.* Recall equation 19.8:

$$\omega_{c,min} = \frac{\omega_p}{\sqrt[2n]{10^{0.1A_{max}} - 1}} \leq \omega_c \leq \frac{\omega_s}{\sqrt[2n]{10^{0.1A_{min}} - 1}} = \omega_{c,max}$$

By historical convention, we will use $\omega_{c,min}$, in which case

$$\omega_c = \omega_{c,min} = \frac{\omega_p}{\sqrt[2n]{10^{0.1A_{max}} - 1}} = \frac{200\pi}{0.91449} = 687 = 218.7\pi \text{ rad/s ec}$$

Step 3. Looking up the third-order normalized Butterworth loss function and inverting to obtain the transfer function yields

$$H_{3dBNLP}(s) = \frac{1}{(s+1)(s^2 + s + 1)} = \frac{1}{s^3 + 2s^2 + 2s + 1} \tag{19.12}$$

Step 4. Choose a candidate passive circuit and obtain its transfer function, $H_{cir}(s) = \dfrac{V_{out}(s)}{V_{in}(s)}$. We wind up with the circuit of Figure 19.8. Note that we will adjust the source resistance later with magnitude scaling.

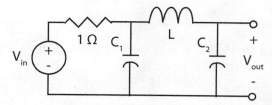

FIGURE 19.8 Third-order filter circuit for Example 19.4.

We obtain the transfer function using nodal analysis. At the top node of C_1,

$$\left(C_1 s + \frac{1}{Ls} + 1\right) V_{C1} - \frac{1}{Ls} V_{out} = V_{in} \tag{19.13a}$$

At the output node,

$$-\frac{1}{Ls}V_{C1} + \left(C_2 s + \frac{1}{Ls} \right) V_{out} = 0 \tag{19.13b}$$

Solving equation 19.13b for V_{C1} and eliminating V_{C1} from equation 19.13a yields

$$H_{cir}(s) = \frac{V_{out}(s)}{V_{in}(s)} = \frac{\dfrac{1}{LC_1 C_2}}{s^3 + \dfrac{1}{C_1}s^2 + \dfrac{C_1 + C_2}{LC_1 C_2}s + \dfrac{1}{LC_1 C_2}} \tag{19.14}$$

Step 4. *Equate the coefficients of $H_{cir}(s)$ and $H_{3dBNLP}(s)$ to solve for C_1, C_2, and L.* Equating the coefficients in equations 19.14 and 19.12 yields

$$\frac{1}{C_1} = 2 \, , \, \frac{1}{LC_1 C_2} = 1, \text{ and } \frac{C_1 + C_2}{LC_1 C_2} = 2$$

Clearly, $C_1 = 0.5$ F and $\dfrac{2}{LC_2} = 1$ implies $LC_2 = 2$. Then

$$\frac{C_1 + C_2}{LC_1 C_2} = \frac{0.5 + C_2}{1} = 2 \Rightarrow C_2 = 1.5 \text{ F and } L = \frac{4}{3} \text{ H}$$

In summary, $C_1 = 0.5$ F, $C_2 = 1.5$ F, and $L = 4/3$ H.

Step 5. *Frequency-scale the circuit to obtain the desired, and magnitude-scale to obtain the specified source resistance of 100 Ω.* We frequency-scale by $K_f = \omega_c = \omega_{c,min} = 687 = 218.7\pi$ and magnitude-scale by $K_m = 100$. Thus, $R_{s,new} = 100 \, R_s = 100 \, \Omega$

$$C_{1,new} = \frac{C_1}{K_m K_f} = 7.2773 \text{ }\mu\text{F}, \, C_{2,new} = \frac{C_2}{K_m K_f} = 21.832 \text{ }\mu\text{F, and } L_{new} = \frac{K_m L}{K_f} = 0.19406 \text{ H}.$$

This completes the design.

Exercise. Reconsider Example 19.4. If R_s is to be 50 Ω and $\omega_c = \omega_{c,max}$, recompute the final element values.
ANSWERS: 7.3872 µF, 22.162 µF, and 0.0492 H

In the above example, there is no load attached to the filter. A situation in which there is no source resistance but a load resistance is given in the problems at the end of the chapter. In addition, there is a problem containing both a source and a load resistance.

Here we have used a coefficient matching technique that is manageable for orders 1, 2, and 3. For higher orders the method is unwieldy and we must use the methods of network synthesis studied in other courses or resort to normalized filter tabulations given in filter handbooks.

5. ACTIVE REALIZATION OF LOW-PASS BUTTERWORTH FILTERS

For active realization of Butterworth filters we use the factored form of the transfer functions given in Table 19.1. For example, we would represent the third-order normalized Butterworth transfer function as the product

$$H_{3dBNLP}(s) = \frac{K_a}{s^2 + s + 1} \times \frac{K_b}{s + 1} \triangleq H_a(s)H_b(s) \tag{19.15}$$

where $K_a K_b = 1$. The gains K_a and K_b are present in equation 19.15 to allow the individual stages to have dc gains different from 1, as is sometimes necessary. The overall transfer function is realized as a cascade of

$$H_a(s) = \frac{K_a}{s^2 + s + 1} \quad \text{and} \quad H_b(s) = \frac{K_b}{s + 1}$$

with the constraint $K_a K_b = 1$. All transfer functions beyond first order require one or more second-order stages in cascade for their realization. The use of a basic second-order active circuit is key to the realization of any of these second-order sections.

The Sallen and Key Second-Order Section

One of the common single–op amp second-order circuits is the Sallen and Key (S&K) configuration illustrated in Figure 19.9. As part of the realization process we must first compute its second-order transfer function. Denote the circuit transfer function by $H_{SK}(s)$.

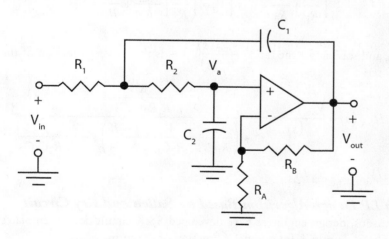

FIGURE 19.9 Sallen and Key low-pass second-order section.

To derive $H_{SK}(s)$, note that the properties of an ideal op amp force the voltage across R_A to be V_a. Using voltage division,

$$V_a = \frac{R_A}{R_A + R_B} V_{out}$$

or equivalently,

$$V_{out} = \frac{R_A + R_B}{R_A} V_a = \left(1 + \frac{R_B}{R_A} \right) V_a \equiv K V_a$$

This observation permits us to simplify the diagram of Figure 19.9 to the equivalent circuit of Figure 19.10.

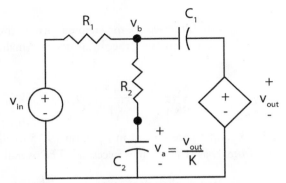

FIGURE 19.10 Equivalent circuit of the Sallen and Key low-pass filter of Figure 19.9; V_b identifies node b and V_a identifies node a.

Construction of $H_{SK}(s)$ now proceeds by nodal analysis. At node a—the node between C_2 and R_2,

$$-\frac{1}{R_2}V_b + \left(\frac{1}{R_2} + C_2 s\right)\frac{V_{out}}{K} = 0 \tag{19.16}$$

Similarly at node b,

$$\left(\frac{1}{R_1} + \frac{1}{R_2} + C_1 s\right)V_b = \left(\frac{1}{R_2}\right)\frac{V_{out}}{K} + \frac{V_{in}}{R_1} + C_1 s V_{out} \tag{19.17}$$

Eliminating V_b and solving for the ratio $\dfrac{V_{out}(s)}{V_{in}(s)}$ yields the transfer function of the Sallen and Key low-pass filter:

$$H_{SK}(s) = \frac{V_{out}(s)}{V_{in}(s)} = \frac{\dfrac{K}{R_1 R_2 C_1 C_2}}{s^2 + \left(\dfrac{1}{R_1 C_1} + \dfrac{1}{R_2 C_1} + \dfrac{1-K}{R_2 C_2}\right)s + \dfrac{1}{R_1 R_2 C_1 C_2}} \tag{19.18}$$

where $K = V_{out}/V_a = 1 + R_B/R_A$

Butterworth LP Design Algorithm Based on Sallen and Key Circuit

For practical reasons, design engineers have developed S&K circuit design templates based only on the circuit Q and dc gain K for a transfer function of the form

$$H(s) = \frac{K}{s^2 + \dfrac{1}{Q}s + 1} \tag{19.19}$$

The design template derives from equating the denominator coefficients of equations 19.18 and 19.19. The dc gain can be adjusted using other methods such as input attenuation (to be described later). In particular, equating the denominator coefficients of equations 19.18 and 19.19 requires that

$$\frac{1}{Q} = \frac{1}{R_1C_1} + \frac{1}{R_2C_1} + \frac{1-K}{R_2C_2}$$

(19.20a)

and

$$1 = \frac{1}{R_1R_2C_1C_2}$$

(19.20b)

The solutions to equations 19.20a and 19.20b are not unique because there are two equations in five unknowns. This means that we can impose up to three additional constraints to produce different solutions. Different solutions produce the proper filtering action, but have different behaviors in terms of the sensitivity of the frequency response to variations nominal resistance and capacitor values. Also, different designs have different ratios of element values that may have practical significance.

One popular and robust design is the *Saraga design*. For the (normalized) Saraga design, the three additional constraints imposed on the solution of equations 19.20 are $C_2 = 1$, $C_1 = \sqrt{3}Q$, and $\dfrac{R_2}{R_1} = \dfrac{Q}{\sqrt{3}}$, which were chosen to minimize certain sensitivities in the circuit performance. Solving equations 19.20a and 19.20b for R_1, R_2, and K using these additional constraints yields $C_2 = 1$, $C_1 = \sqrt{3}Q$, $R_1 = Q^{-1}$, $R_2 = 1/\sqrt{3}$, $K = 4/3$, $R_B = R_A/3$. The final circuit realization of equation 19.19 is given in Figure 19.11. Again we note that the dc gain $K = 4/3$ will need to be adjusted by other means, to be described. Frequency and magnitude scaling are necessary to achieve proper cutoff frequencies and impedance levels.

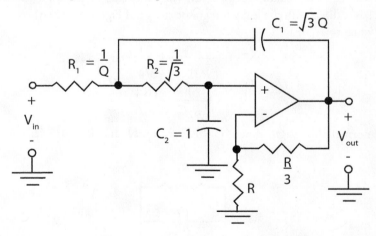

FIGURE 19.11 Normalized Saraga design of Sallen and Key circuit for realizing the transfer function of equation 19.19.

With this design in hand, we now redo Example 19.4 using an active realization.

EXAMPLE 19.5. This example presents an active realization of a third-order Butterworth filter with $\omega_c = 687 = 218.7\pi$ rad/s ec. Recall from equation 19.15 that we factor the third-order Butterworth transfer function as a product:

$$H_{3dBNLP}(s) = \frac{K_a}{s^2 + s + 1} \times \frac{K_b}{s+1} \triangleq H_a(s)H_b(s)$$

where $K_a K_b = 1$. Later, we will frequency- and magnitude-scale to obtain proper cutoffs and impedance levels. For

$$H_a(s) = \frac{K_a}{s^2 + s + 1} = \frac{K_a}{s^2 + \dfrac{1}{Q}s + 1}$$

we have $Q = 1$. The Saraga design element values are $C_2 = 1$ F, $C_1 = \sqrt{3}$ F, $R_1 = 1$ Ω, $R_2 = 1/\sqrt{3}$ Ω, $K = K_a = 4/3$, and $R_B = R_A/3$. These element values realize the transfer function

$$H_{a,cir}(s) = \frac{4/3}{s^2 + s + 1}$$

To obtain the correct $H_{3dBNLP}(s)$, we realize the transfer function

$$H_b(s) = \frac{K_b}{s+1} = \frac{3/4}{s+1} \tag{19.21}$$

Observe that the choice of K_b makes $K_a K_b = 1$, the correct overall dc gain. The transfer function of equation 19.21 is realized with the leaky integrator circuit of Figure 19.12, whose transfer function $H_{b,cir}(s)$ set equal to $H_b(s)$ is

$$H_{b,cir}(s) = \frac{\dfrac{1}{R_1 C}}{s + \dfrac{1}{R_2 C}} = \frac{3/4}{s+1} = H_b(s)$$

To complete the normalized design, choose $C = 1$ F, $R_2 = 1$ Ω, and $R_1 = 4/3$ Ω.

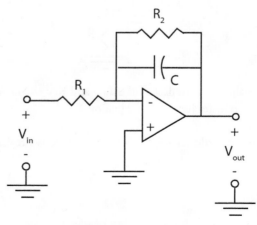

FIGURE 19.12. Leaky integrator circuit.

Combining the second-order and first-order sections of the normalized design produces the circuit of Figure 19.13.

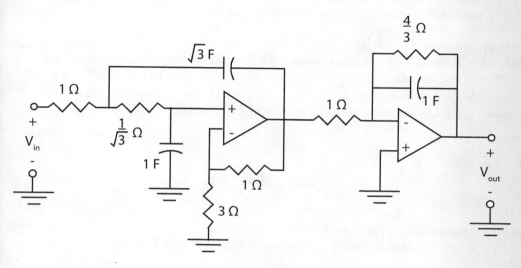

FIGURE 19.13. Realization of third-order 3 dB NLP Butterworth transfer function.

In the last step of our design we frequency-scale by $K_f = 218.7\pi$. If the smallest capacitor is to be 100 nF, then

$$K_m = \frac{C_{old}}{K_f C_{new}} = \frac{1}{218.7\pi 10^{-7}} = 14.555 \times 10^3$$

Hence the final parameter values for the second-order section are

$C_2 = 100 \ nF, \ C_1 = 173.2 \ nF, \ R_1 = 14.56 \ k\Omega, \ R_2 = 8.4 \ k\Omega, \ R_A = 30 \ k\Omega, \ and \ R_B = 10 \ k\Omega$

Similarly, for the first-order section $C = 100$ nF, $R_2 = 14.55$ kΩ, and $R_1 = 19.4$ kΩ. This leads to the final circuit design given in Figure 19.14.

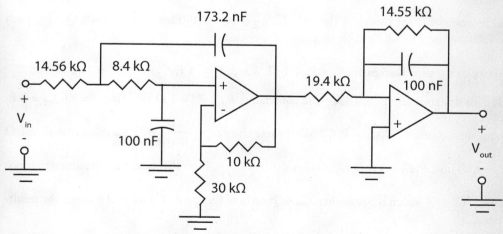

FIGURE 19.14. Final third-order Butterworh LP design, for which $\omega_c = 687 = 218.7\pi$ rad/sec.

This design assumes the filter is driven by a voltage source with a very small resistance; otherwise the source impedance must be considered part of the front end series resistance. Also, we could have magnitude-scaled each section separately (but chose not to).

Other Sallen and Key Designs

In addition to the Saraga design, there are two other S&K normalized designs that we will set forth here. The first is termed **Design A:** $R_A = \infty$, $R_B = 0$, $K = 1$, $R_1 = R_2 = 1\ \Omega$, $C_1 = 2Q$, and $C_2 = \dfrac{1}{2Q}$. To obtain design A we normalize $K = 1$ and $R_1 = R_2 = 1\ \Omega$. Using these three constraints and equating the coefficients of equations 19.19 and 19.18 leads to

$$H_{SKA}(s) = \frac{K}{s^2 + \dfrac{1}{Q}s + 1} = \frac{\dfrac{1}{C_1 C_2}}{s^2 + \dfrac{2}{C_1}s + \dfrac{1}{C_1 C_2}} \tag{19.22}$$

Matching the denominator coefficients produces $C_1 = 2Q$ and $C_2 = 1/2Q$. Since the dc gain with $K = 1$ is 1, there is no need to modify the overall circuit gain. Figure 19.15 illustrates design A.

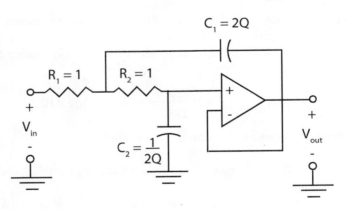

FIGURE 19.15 Design A for realizing the transfer function of equation 19.19.

Notice, however, that for Design A the ratio $C_1/C_2 = 4Q^2$ is 400 for a circuit with $Q = 10$. Such a large variation may be undesirable in a practical circuit.

A third design we term **Design B:** $R_A = R_B = 1\ \Omega$, $K = 2$, $C_1 = 1$ F, $C_2 = 1/Q$ F, $R_1 = 1\ \Omega$, and $R_2 = Q\ \Omega$. This design is computed using the constraints $R_A = R_B = 1\ \Omega$ (yielding $K = 2$), $C_1 = 1$ F, and equal time constant, $R_1 C_1 = R_2 C_2$. With these choices, $\dfrac{1}{R_1 R_2 C_1 C_2} = 1$ implies that $R_1 = 1\ \Omega$ and $R_2 = Q$. Thus, $R_1 C_1 = 1 = R_2 C_2$ implies that $C_2 = \dfrac{1}{Q}$. Here the maximum parameter ratio is $R_2/R_1 = C_1/C_2 = Q$, which is better than the $4Q^2$ ratio of Design A. Figure 19.16 shows the resulting circuit.

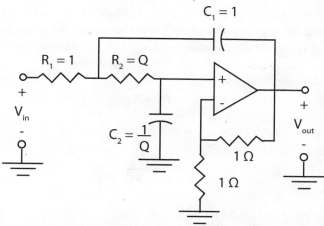

FIGURE 19.16 **Design B** for realizing the normalized transfer function of equation 19.19, assuming standard units.

6. INPUT ATTENUATION FOR ACTIVE CIRCUIT DESIGN

In design A of the previous section, the gain of the filter is $K = 1$. In design B, the filter gain is $K = 2$. In Example 19.5, the gain of the Saraga design ($K = 4/3$) was corrected by the subsequent first-order section. If the filter were second order with a dc gain of 1, then the circuit gain would need to be modified for both Design B and the Saraga design. In order to reduce a high gain, maintain the filtering properties, and keep as many parameter values as possible at their original design values, we use a technique known as *input attenuation*. In this technique, the front-end resistor is replaced with a voltage divider circuit, as illustrated in Figure 19.17.

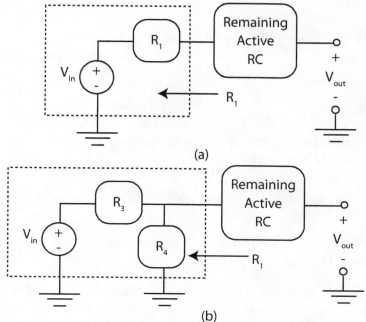

FIGURE 19.17 Illustration of input attenuation concept. (a) Original circuit. (b) Circuit with input attenuation.

Figure 19.17a represents the original active network, while Figure 19.17b represents the modified network. For the filtering characteristics to remain invariant, the impedance looking into the dashed boxes must remain at R_1 for both circuits. Thus, the parallel combination of R_3 and R_4 must equal R_1, i.e.,

$$R_1 = \frac{R_3 R_4}{R_3 + R_4} \qquad (19.23)$$

In addition, if the new gain is to be αK, $\alpha < 1$, instead of K, then

$$\alpha = \frac{R_4}{R_3 + R_4} \qquad (19.24)$$

Solving equations 19.23 and 19.24 for R_3 and R_4 yields

$$R_3 = \frac{R_1}{\alpha} \quad \text{and} \quad R_4 = \frac{R_1}{1 - \alpha} \qquad (19.25)$$

Thus, one can reduce the gain of the Sallen and Key low-pass circuit via the simple technique of input attenuation.

7. PROPERTIES OF THE BUTTERWORTH LOSS FUNCTION

As mentioned earlier, the Butterworth loss function has a maximally flat pass-band response in the sense that as many derivatives as possible of the loss magnitude response are zero at $\Omega = 0$. This is often termed **maximally flat**. In the case of the nth-order Butterworth loss function, it is possible to show that

$$\frac{d^k}{d\Omega^k} \left| \hat{H}(j\Omega) \right| = 0, \qquad \text{for } k = 1, 2, ..., 2n - 1 \qquad (19.26)$$

at $\Omega = 0$ and

$$\frac{d^{2n}}{d\Omega^{2n}} \left| \hat{H}(j\Omega) \right| = 0.5 (2n)! \neq 0 \qquad (19.27)$$

at $\Omega = 0$. This is consistent with the notion of being maximally flat. To verify equations 19.26 and 19.27, observe that if $\Omega^{2n} << 1$, then

$$\left| \hat{H}(j\Omega) \right| = \sqrt{1 + \Omega^{2n}} = 1 + \frac{1}{2} \Omega^{2n} - \frac{1}{8} \Omega^{4n} + \cdots \cong 1 + \frac{1}{2} \Omega^{2n} \qquad (19.28)$$

Equations 19.26 and 19.27 follow after differentiation of equation 19.28 $2n$ times and evaluation of each derivative at $\Omega = 0$.

Several closing remarks are now in order. First, the cutoff frequency is the half-power point, or the 3 dB down point. The terminology follows because $|H(j0)| = 1$ (the gain at dc is unity) and

$$|H(j1)| = \frac{1}{\sqrt{2}} = 0.707 \, ,$$

yielding half power for normalized Butterworth gain functions $H(s)$. Second, for large Ω,

$$|H(j\Omega)| = \frac{1}{\Omega^n}$$

Hence as Ω becomes large, the gain $G_{dB}(\Omega)$ decreases according to

$$-n \, [20 \, \log_{10}(\Omega)]$$

indicating that the gain rolls off with a slope proportional to the number of poles, n, of the gain function. Specifically, the slope equals $-20n$ dB per decade. These statements are also valid if ω replaces Ω.

8. BASIC HIGH-PASS FILTER DESIGN WITH PASSIVE REALIZATION

High-pass filters invert the frequency characteristics of low-pass filters. A high-pass filter is a device—usually a circuit—that significantly attenuates the low-frequency content of a signal while passing the high-frequency content with minimal attenuation. Figure 19.18 illustrates typical brickwall specifications for a high-pass filter. Here frequencies above the pass-band edge frequency ω_p have little attenuation, while frequencies below ω_s are significantly attenuated—precisely the inverse function of a low-pass filter. In fact, low-pass and high-pass specifications are related by a simple inversion of ω. In particular, we define a **HP (high-pass) to LP (low-pass) frequency transformation** as

$$\Omega = \frac{\omega_p}{\omega} \tag{19.29}$$

This frequency transformation applied to the brickwall specifications of Figure 19.18 yields the set of normalized low-pass specifications given by Figure 19.19.

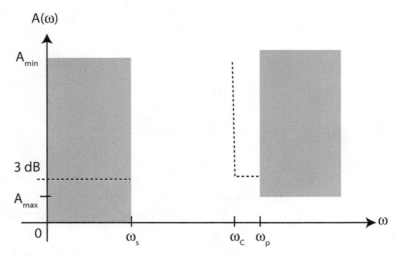

FIGURE 19.18 Typical brickwall characteristic of a high-pass filter.

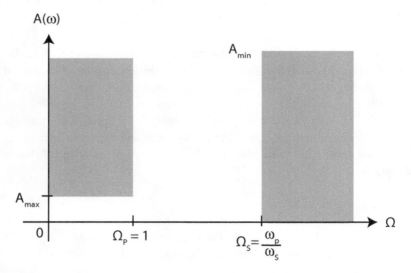

FIGURE 19.19 Normalized low-pass equivalent specifications derived from high-pass specifications

Given the normalized LP (NLP) equivalent specs, one finds the Butterworth filter order accord
ing to the usual formula, equation 19. 7, reproduced as equation 19.30:

$$
n \geq \frac{\log_{10}\left(\sqrt{\dfrac{10^{0.1A_{min}} - 1}{10^{0.1A_{max}} - 1}}\right)}{\log_{10}\left(\dfrac{\Omega_s}{\Omega_p}\right)} = \frac{\log_{10}\left(\sqrt{\dfrac{10^{0.1A_{min}} - 1}{10^{0.1A_{max}} - 1}}\right)}{\log_{10}\left(\Omega_s\right)}
\qquad (19.30)
$$

By convention we take the filter order to be the smallest integer n satisfying equation 19.30. Fo
a passive realization, once n is known, one would realize the resulting NLP Butterworth transfe

function. For example, suppose $n = 2$; the second-order 3 dB NLP Butterworth transfer function could be realized by (among others) the circuit of Figure 19.20, whose transfer function is

$$H_{cir}(s) = \frac{\dfrac{1}{Cs}}{Ls + R_s + \dfrac{1}{Cs}} = \frac{\dfrac{1}{LC}}{s^2 + \dfrac{R_s}{L}s + \dfrac{1}{LC}}$$

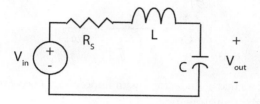

FIGURE 19.20 Circuit to realize second-order NLP Butterworth filter.

As with the development of equation 19.8, once n is picked, there is a permissible range of normalized Ω_c given by

$$\Omega_{c,min} = \frac{1}{\sqrt[2n]{10^{0.1A_{max}} - 1}} \le \Omega_c \le \frac{\Omega_s}{\sqrt[2n]{10^{0.1A_{min}} - 1}} = \Omega_{c,max} \qquad (19.31)$$

where $\Omega_s = \omega_p/\omega_s$. Once Ω_c is chosen, the Butterworth transfer function that meets the NLP equivalent specs is found as

$$H_{NLP}(s) = H_{3dBNLP}\left(\frac{s}{\Omega_c}\right)$$

This is, of course, equivalent to frequency scaling a realization of the Butterworth filter by $K_f = \Omega_c$, as illustrated in Figure 19.21.

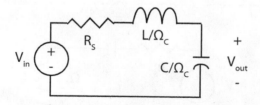

FIGURE 19.21 Circuit that realizes $H_{NLP}(s)$.

The actual HP transfer function is related to the other "normalized" transfer functions as follows:

$$H_{HP}(s) = H_{NLP}\left(\frac{\omega_p}{s}\right) = H_{3dBNLP}\left(\frac{\omega_p}{\Omega_c s}\right) = H_{3dBNLP}\left(\frac{\omega_c}{s}\right)$$

This is equivalent to doing a special frequency transformation on the circuit elements of Figure 19.21. Specifically

$$s \to \frac{\omega_p}{s} \qquad\qquad (19.32)$$

is called the **LP to HP frequency transformation** and has a natural interpretation in terms of the inductors and capacitors of a passive network. In particular, a capacitor C with impedance $\frac{1}{Cs}$ becomes an inductor of value $\frac{1}{C\omega_p}$, i.e.,

$$\frac{1}{Cs} \to \frac{1}{C\left(\frac{\omega_p}{s}\right)} = \frac{1}{C\omega_p}s$$

A similar substitution shows that an inductor L with impedance Ls becomes a capacitor of value $\frac{1}{L\omega_p}$. Figure 19.22 illustrates the transformation.

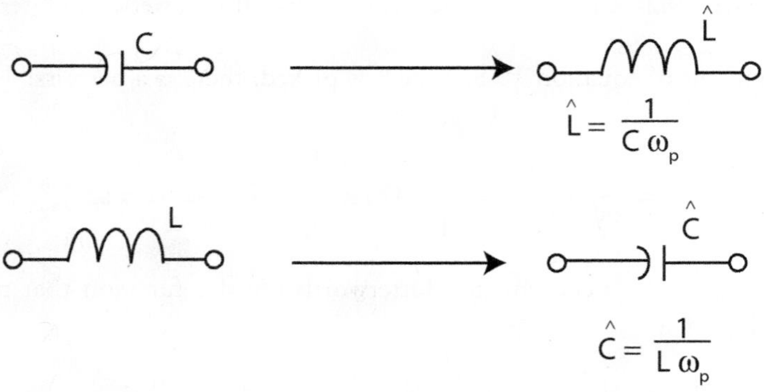

$$\hat{L} = \frac{1}{C\omega_p}$$

$$\hat{C} = \frac{1}{L\omega_p}$$

FIGURE 19.22 Circuit element change under the LP to HP transformation.

Hence the LP circuit of Figure 19.21 becomes the HP circuit of Figure 19.23.

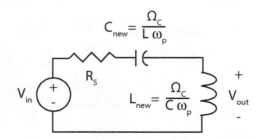

$$C_{new} = \frac{\Omega_c}{L\omega_p}$$

$$L_{new} = \frac{\Omega_c}{C\omega_p}$$

FIGURE 19.23. HP circuit derived from the NLP realization of Figure 19.21. Observe that

$$C_{new}L = L_{new}C = \frac{\Omega_c}{\omega_p}.$$

It remains only to magnitude-scale this circuit to obtain an acceptable passive HP filter. We illustrate the above ideas in the following example.

EXAMPLE 19.6. Design a minimum-order Butterworth filter meeting the high-pass brickwall specifications of Figure 19.24, in which the loss magnitude curve is to pass through the pass-band corner.

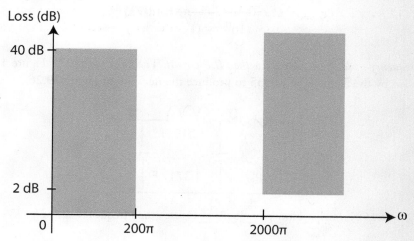

FIGURE 19.24 High-pass filter specifications for Example 19.6.

SOLUTION

Step 1. *Compute the equivalent NLP specifications.* Using $\Omega = \dfrac{2000\pi}{\omega}$, the HP to NLP transformation, the NLP specifications are easily computed as

$$(\Omega_p = 1, A_{max} = 2 \text{ dB}) \text{ and } (\Omega_s = 10, A_{min} = 40 \text{ dB})$$

Step 2. *Determine the Butterworth transfer function.* As these specifications correspond to those of Example 19.3, one computes the filter order $n = 3$. From Table 19.1, the third-order Butterworth transfer function is

$$H_{3dBNLP}(s) = \frac{1}{\left(s^2 + s + 1\right)(s + 1)} = \frac{1}{s^3 + 2s^2 + 2s + 1}$$

Step 3. *Realize the third-order 3 dB NLP transfer function.* According to Example 19.4, the 3 dB NLP Butterworth transfer function can be realized by the passive *RLC* circuit of Figure 19.8, with values assigned as in Figure 19.25.

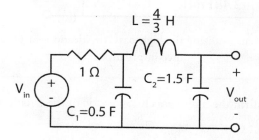

FIGURE 19.25 Third-order 3 dB NLP filter circuit of Example 19.4 having 3 dB down point at 1 rad/sec.

Step 4. *Compute the normalized 3 dB down frequency.* The 3 dB frequency Ω_c of the actual NLP equivalent filter is calculated from equation 19.31, using the left inequality:

$$\Omega_c = \frac{1}{\left(10^{0.2} - 1\right)^{1/6}} = 1.0935$$

Step 5. *Frequency-scale by Ω_c to obtain the NLP circuit.* The NLP circuit of Figure 19.25 is frequency-scaled by the factor $K_f = 1.0935$ to produce the network of Figure 19.26.

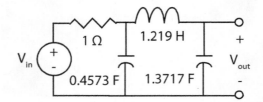

FIGURE 19.26 Normalized low-pass circuit of Figure 19.25 frequency-scaled by $K_f = 1.0935$.

Step 6. *Apply the LP to HP circuit element transformation detailed in Figure 19.22.* Converting the circuit of Figure 19.26 to the required HP circuit yields the network of Figure 19.27.

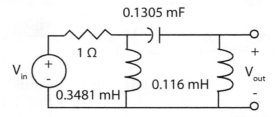

FIGURE 19.27 High-pass circuit meeting the specifications of Figure 19.24.

Step 7. *Impedance-scale.* To obtain a 10 Ω source resistance, we impedance-scale by the factor K_m = 10 to obtain the final realization shown in Figure 19.28.

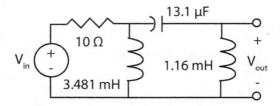

FIGURE 19.28 Passive circuit realizing the high-pass specifications of Figure 19.24 with 10 Ω source impedance.

Exercise. Use SPICE to confirm the high-pass characteristic of the circuit of Figure 19.28.

The preceding example completes our basic discussion of passive high-pass filtering.

9. ACTIVE REALIZATION OF HIGH-PASS FILTERS

Active realization of high-pass filters ordinarily proceeds with the cascading of first- and second-order sections, as done earlier. There is a Sallen and Key second-order high-pass section obtained with an interchange of resistors and capacitors in the S&K low-pass configuration. As with all active circuits, inductors are not used. We illustrate this interchange of resistors and capacitors as an approach to high-pass design with the following example.

EXAMPLE 19.7. Let us reconsider Example 19.6, in which $(\omega_p, A_{max}) = (2000\pi, 2\ dB)$ and $(\omega_s, A_{min}) = (200\pi, 40\ dB)$. The filter design proceeded by converting to equivalent NLP specs, which led to a 3 dB NLP third-order transfer function,

$$H_{3dBNLP}(s) = \frac{1}{(s+1)(s^2 + s + 1)} = \frac{1}{s^3 + 2s^2 + 2s + 1}$$

One approach to HP design first realizes this 3 dB NLP transfer function as a cascade of a second-order active circuit with a first-order circuitIn Example 19.5, this resulted in the circuit of Figure 19.13, redrawn as Figure 19.29.

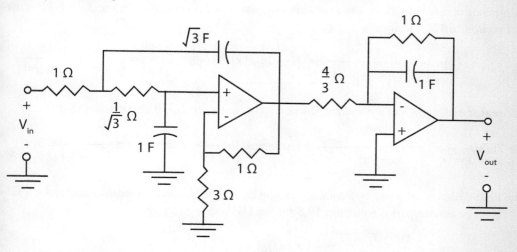

FIGURE 19.29. Realization of third-order 3 dB NLP Butterworth transfer function.

The second step in the completion of a HP design is to replace each resistor with a capacitor whose value is the reciprocal of the resistance; likewise, each capacitor is replaced by a resistor whose value is the reciprocal of the resistance value:

$$R \rightarrow C_{new} = \frac{1}{R}$$

$$C \rightarrow R_{new} = \frac{1}{C} \tag{19.33}$$

Applying the transformation of equation 19.33 to the circuit of Figure 19.29 produces the circuit of Figure 19.30, with one exception to the rule: the feedback resistors on the second-order section that connect to the inverting input node are not changed to capacitors; these resistors only set up a voltage division, so it is unnecessary to replace them.

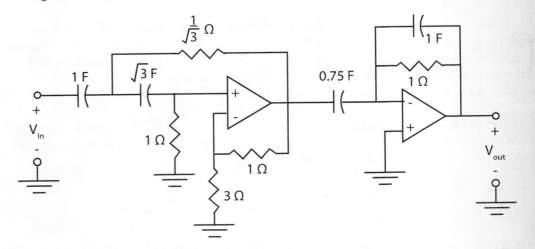

FIGURE 19.30. 3 dB normalized third-order high-pass filter in which $\Omega_{c,HP} = 1$.

In terms of transfer functions, the operation of equation 19.33 changes s in the 3 dB NLP transfer function to $1/s$. Specifically,

$$H_{3dBNLP}(s) = \frac{1}{(s+1)(s^2 + s + 1)} \rightarrow H_{3dBNHP}(s) = H_{3dBNLP}\left(\frac{1}{s}\right) = \frac{s}{(s+1)} \times \frac{s^2}{s^2 + s + 1}$$

Notice that each section of the transfer function has a (normalized) HP characteristic. For example,

$$\lim_{s \to \infty} \frac{s}{s+1} = 1 \text{ and } \lim_{s \to \infty} \frac{s^2}{s^2 + s + 1} = 1 \text{ whereas } \lim_{s \to 0} \frac{s}{s+1} = 0 \text{ and } \lim_{s \to 0} \frac{s^2}{s^2 + s + 1} = 0.$$

The final design is achieved by frequency-scaling by $K_f = \omega_{c,HP}$ and impedance-scaling by a factor K_m. The counterpart of equation 19.8 for the HP case is given by

$$\omega_s \sqrt[2n]{10^{0.1A_{min}} - 1} \le \omega_{c,HP} \le \omega_p \sqrt[2n]{10^{0.1A_{max}} - 1} \tag{19.34}$$

(The derivation of equation 19.34 is assigned as a homework problem.) By convention, frequency scaling places the magnitude response through the pass-band edge, in which case

$$\omega_{c,HP} = \omega_p \sqrt[2n]{10^{0.1A_{max}} - 1} = 5745.9 \text{ rad/sec}$$

and $K_f = 5745.9$.

Let us suppose that the smallest capacitor is to be 10 nF. Then

$$K_m = \frac{C_{old}}{K_f C_{new}} = 13.05$$

Hence the final circuit design is given by Figure 19.31.

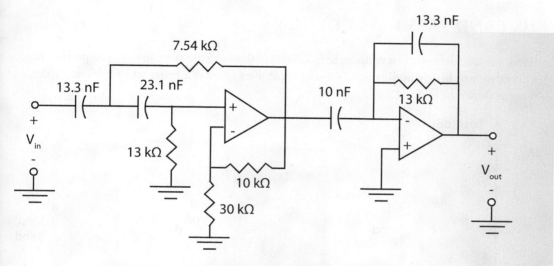

FIGURE 19.31. Final design of third-order high-pass filter.

The circuit of Figure 19.31 presumes that an ideal voltage source is input to the filter. If the source has an internal impedance, this impedance will affect the pole locations and the overall performance of the high-pass circuit. To circumvent this problem, one may insert a voltage follower circuit between the practical source and the input to the filtering circuit of Figure 19.31, as illustrated in Figure 19.32.

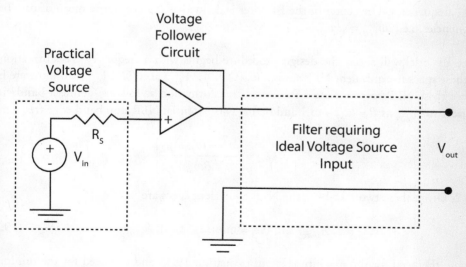

FIGURE 19.32. Practical source in series with voltage follower circuit.

It is possible to justify the **RC to CR transformation** through the use of well-known networks theorems. However, such justification, which is found in more advanced texts on filter design, is beyond the scope of our current endeavors. There are other HP design algorithms that do not utilize the transformation of 19.33, but these are left to texts on filter design.

10. BAND-PASS FILTER DESIGN

Band-pass circuits were first considered in Chapter 16, mostly from an analysis perspective. In this section we will briefly outline the design of a BP filter to meet a given set of brickwall specifications as shown in Figure 19.33.

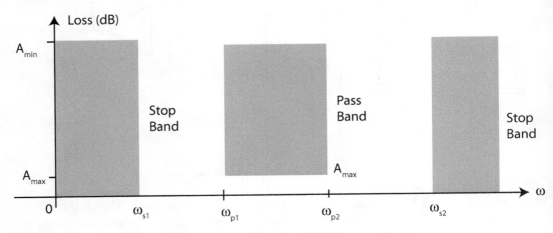

FIGURE 19.33. Typical brickwall BP filter specifications.

In Figure 19.33 ω_{p1} and ω_{p2} are the pass-band edge frequencies while ω_{s1} and ω_{s2} are the stop band edge frequencies. The center of the BP filter is defined as the geometric mean of pass-band edge frequencies, i.e., $\omega_{cen} = \sqrt{\omega_{p1}\omega_{p2}}$.

Given the BP brickwall specs, the design procedure begins with a frequency transformation to convert these specs to equivalent NLP specs, i.e., $(\Omega_p = 1, A_{max})$ and (Ω_s, A_{min}). To illustrate how to generate these NLP specs we use a frequency transformation as follows: define a **bandwidth with respect to** A_{max} as $B = \omega_{p2} - \omega_{p1}$ and define two potential NLP stop band edge frequencies as

$$\Omega_{si} = \left| \frac{\omega_{si}^2 - \omega_{p1}\omega_{p2}}{B\omega_{si}} \right| \tag{19.35}$$

for $i = 1,2$. Given these two numbers, the NLP equivalent specs are

$$(\Omega_p = 1, A_{max}) \text{ and } (\Omega_s = \min[\Omega_{s1}, \Omega_{s2}], A_{min}) \tag{19.36}$$

To illustrate the need for the magnitude sign in equation 19.35 and the need for the minimum function in equation 19.36, let us do a simple example by breaking apart the formula of equation 19.35 and putting it back together. Suppose $\omega_{p1} = 2$ rad/sec and $\omega_{p2} = 8$ rad/sec with $\omega_{s1} = 1$

rad/sec and ω_{s2} = 20 rad/sec. Then $\omega_{cen} = \sqrt{\omega_{p1}\omega_{p2}} = \sqrt{16} = 4$ rad/sec. Let us evaluate a part of equation 19.35 in which B is not present and the absolute value signs are gone. Thus

$$\Omega_{pi} = \frac{\omega_{pi}^2 - \omega_{p1}\omega_{p2}}{\omega_{pi}} = \begin{cases} -6 & i=1 \\ 6 & i=2 \end{cases}$$

and

$$\Omega_{si} = \frac{\omega_{si}^2 - \omega_{p1}\omega_{p2}}{\omega_{si}} = \begin{cases} -15 & i=1 \\ 19.2 & i=2 \end{cases}$$

One can view this partial transformation as the generation of two distinct LP filters: one defined on negative frequency ($\Omega_{p,neg} = -6$, $\Omega_{s,neg} = -15$) and one defined on positive frequency ($\Omega_{p,pos} = 6$, $\Omega_{s,pos} = 19.2$). Of course, we cannot really have two distinct LP filters, one for negative frequency and another for positive frequency. Our mathematics requires that the magnitude responses of each filter be symmetric because magnitude is an even function of ω. Thus the only way to properly interpret the above transformation is that it implicitly generates two distinct low-pass filters, one of which is more stringent than the other. Specifically, the "negative" filter is really a LP filter with edge frequencies ($\Omega_{p1} = 6$, $\Omega_{s1} = 15$), and the positive filter is really a LP filter with edge frequencies ($\Omega_{p2} = 6$, $\Omega_{s2} = 19.2$). Observe that the ratio

$$\frac{\Omega_{si}}{\Omega_{pi}} = \begin{cases} 2.5 & i=1 \\ 3.2 & i=2 \end{cases}$$

differs for the two filters. The first filter is more stringent than the second, prompting the need for a minimum function in equation 19.36. Since our edge frequencies are always specified on the positive axis and since the magnitude response is symmetric, we insert the absolute value signs in equation 19.35. Finally, in order to have $\Omega_{pi} = 1$ we divide by $B = \omega_{p2} - \omega_{p1}$ so that the edge frequencies correspond to a NLP filter.

EXAMPLE 19.8. A BP filter has the specs ω_{s1} = 26 krad/sec, ω_{s2} = 37 krad/sec, ω_{p1} = 30.6 krad/sec, ω_{p2} = 32.6 krad/sec, A_{max} = 3 dB, and A_{min} = 14 dB. Design a minimum-order Butterworth passive BP filter meeting these specs, assuming the filter is driven by a voltage source in series with a 10 kΩ source resistance.

SOLUTION

Step 1. *Find the equivalent NLP specs for this filter.* Using MATLAB, we have

```
»ws1 = 26e3; ws2 = 37e3; wp1 = 30.6e3;
»wp2 = 32.6e3; Amax = 3; Amin = 14;
»K = wp1*wp2
K =
997560000
»B = wp2 - wp1
B =
2000
»ws = [ws1 ws2];
»Wsi = abs((ws .^2 - K) ./(B*ws))
```

Wsi =
  6.1838e+00 5.0195e+00
»Ws = min(Wsi)
Ws =
  5.0195e+00

Step 2. *Find the filter order.*

»n = buttord(1,Ws,Amax,Amin,'s')
n = 1

Step 3. *Find a realization of the NLP filter.* Since A_{max} = 3 dB, the first-order Butterworth NLP
has transfer function $H_{NLP}(s) = \dfrac{1}{s+1}$, whose passive circuit realization is given in Figure 19.34.

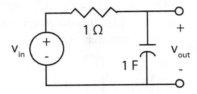

FIGURE 19.34. Realization of the NLP transfer function.

Step 4. *Realize the given BP filter.* The algorithm to generate the BP filter from the NLP filter is
based on the frequency transformation

$$s \;\rightarrow\; \frac{s^2 + \omega_{p1}\omega_{p2}}{Bs} \qquad (19.37)$$

This frequency transformation causes a change in the capacitive and inductive circuit elements of
the NLP circuit. From equation 19.37, the impedance of an inductor changes as follows:

$$Ls \rightarrow L\left(\frac{s^2 + K}{Bs}\right) = \frac{L}{B}s + \frac{1}{\left(\dfrac{B}{LK}\right)s} \qquad (19.38)$$

The expression on the right of equation 19.38 is a sum of impedances. This sum, then, is a circuit composed of an inductor in series with a capacitor as shown in Figure 19.35.

FIGURE 19.35. LP to BP transformation on an inductor.

Similarly, from equation 19.37, the admittance of a capacitor changes as follows:

$$Cs \rightarrow C\left(\frac{s^2 + K}{Bs}\right) = \frac{C}{B}s + \frac{1}{\left(\frac{B}{CK}\right)s} \qquad (19.39)$$

The expression on the right is a sum of admittances. Hence under the frequency transformation, a capacitor becomes a parallel LC combination as illustrated in Figure 19.36.

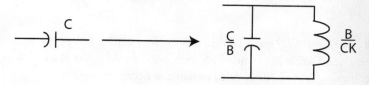

FIGURE 19.36. LP to BP transformation on a capacitor.

Applying the transformation of Figure 19.36 to the NLP circuit of Figure 19.34 yields the circuit of Figure 19.37, where

$$\frac{1}{B} = 10^{-4} \text{ and } \frac{B}{K} = 2.005 \times 10^{-6}.$$

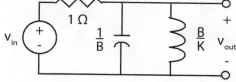

FIGURE 19.37. BP circuit without magnitude scaling.

Step 5. *Magnitude-scale to obtain the correct source resistance.* With $K_m = 10^4$, we conclude that $C_{bp} = 50$ nF and $L_{bp} = 20$ mH. The final circuit realization is illustrated in Figure 19.38.

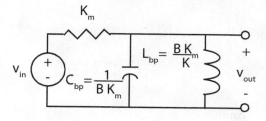

FIGURE 19.38. Final BP design.

The final transfer function of the circuit is

$$H_{BP}(s) = H_{NLP}(s)\Big]_{s=\frac{s^2+K}{Bs}} = \frac{Bs}{s^2 + Bs + K}$$

This transfer function and circuit show a clear similarity to the material developed in Chapter 16. In contrast to the above procedure, the development of Chapter 16 works only for $n = 1$ in the equivalent NLP circuit, which results in a BP circuit of order 2. If the order of the NLP equiva-

lent were 2 or higher, the BP order would be at least 4 and the background from Chapter 1
would prove inadequate. The next example illustrates how design is done for higher-order circuit

EXAMPLE 19.9. A BP filter has the specs ω_{s1} = 1500 rad/sec, ω_{s2} = 6000 rad/sec, ω_{p1} = 300
rad/sec, ω_{p2} = 4000 rad/sec, A_{max} = 2 dB, and A_{min} = 20 dB. Design a passive BP filter meetin
these specs that is driven by a voltage source having an internal resistance of 100 Ω.

SOLUTION
Part 1. *Find equivalent NLP specs.* Define $K = \omega_{p1}\omega_{p2} = 12 \times 10^6$ as the square of the center fre
quency of the filter. Define the bandwidth with respect to A_{max} as $B = \omega_{p2} - \omega_{p1} = 1000$ rad/se
We use the following MATLAB code to obtain the necessary numbers.

```
ws1 = 1500; ws2 = 6000;wp1 = 3000; wp2 = 4000;
B = wp2 - wp1
K = wp1*wp2
ws = [ws1 ws2];
Wsi = abs((ws .^2 - K) ./(B*ws))
Ws = min(Wsi)
```

The MATLAB output is

```
B = 1000
K = 12000000
Wsi = 6.5000e+00 4.0000e+00
Ws = 4
```

Therefore, the equivalent NLP specs are

$(\Omega_p = 1, A_{max} = 2)$ and $(\Omega_s = 4, A_{min} = 20)$

These NLP specs are then used to design a NLP transfer function.

Part 2. *Find an equivalent NLP Butterworth transfer function and its realization.* In MATLAB, w
have

```
»Wp = 1; Ws = 4; Amax = 2; Amin = 20;
»n = buttord(Wp,Ws,Amax,Amin,'s')
n = 2
```

From Table 19.1, the second-order Butterworth filter is given by

$$H_{3dBNLP}(s) = \frac{1}{s^2 + \sqrt{2}s + 1}$$

(19.40

This does not have the correct A_{max} at $\Omega = 1$. We will adjust this by frequency-scaling the circuit that realizes $H_{3dBNLP}(s)$. A circuit that realizes the second-order Butterworth transfer function is given in Figure 19.39.

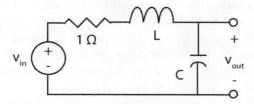

FIGURE 19.39. Second-order circuit.

The circuit of Figure 19.39 has transfer function

$$H_{cir}(s) = \frac{\dfrac{1}{LC}}{s^2 + \dfrac{1}{L}s + \dfrac{1}{LC}} \qquad (19.41)$$

By equating the denominators of equations 19.40 and 19.41, we obtain $L = \dfrac{1}{\sqrt{2}}$ H and $C = \sqrt{2}$ F.

Next we frequency-scale so that we obtain the correct A_{max} spec at $\Omega = 1$. For this we use

$$\Omega_c = \Omega_{c,min} = \frac{1}{\sqrt[2n]{10^{0.1A_{max}} - 1}} = 1.1435 \text{ rad/sec}$$

in which case $L_{new1} = \dfrac{L}{\Omega_c} = 0.61838$ H and $C_{new1} = \dfrac{C}{\Omega_c} = 1.2368$ F.

Thus the circuit that realizes the NLP characteristic with a loss of A_{max} at $\Omega = 1$ or, equivalently, 3 dB of loss at $\Omega_c = 1.1435$ is given in Figure 19.40.

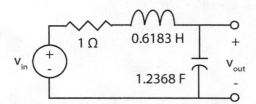

FIGURE 19.40. Second-order circuit.

Note that the circuit of Figure 19.40 has transfer function

$$H_{NLP}(s) = H_{3dBNLP}\left(\frac{s}{\Omega_c}\right) = \frac{1}{\left(\dfrac{s}{1.1435}\right)^2 + \sqrt{2}\left(\dfrac{s}{1.1435}\right) + 1} = \frac{1.3076}{s^2 + 1.6171s + 1.3076} \qquad (19.42)$$

Part 3. *Realize the passive band-pass filter.* To compute the desired BP transfer function, one would replace s in $H_{NLP}(s)$ as follows:

$$s \rightarrow \frac{s^2 + \omega_{p1}\omega_{p2}}{Bs}$$

One could then attempt to realize the BP transfer function directly. This is numerically unwise and, fortunately, unnecessary. A simple, numerically sound procedure is to replace each inductor and capacitor in the NLP realization by an equivalent circuit representing the above frequency transformation. These equivalent circuits were developed in equations 19.38 and 19.39, and illustrated in Figures 19.35 and 19.36, respectively. The substitution scheme is repeated in Figure 19.41.

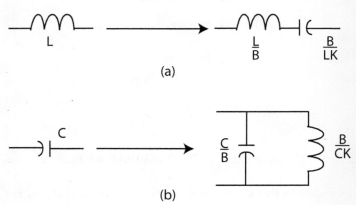

(a)

(b)

FIGURE 19.41. Illustration of the LP to BP transformation in terms of inductors and capacitors.

Given the transformations of Figure 19.41, the following MATLAB code produces the desired BP circuit parameter values, assuming the final source resistance is 100 Ω.

```
» % Insert L, C, B, K, and Km
»L = 0.61838; C = 1.2368;
»B = 1000; K = 12000000; Km = 100;

» % Compute Kf
»Kf = sqrt(K)
Kf = 3.4641e+03

» % Compute BP circuit parameters associated with L
»L1bp = Km*L/B
L1bp = 6.1838e-02
»C1bp = B/(L*K)/Km
C1bp = 1.3476e-06

» % Compute BP circuit parameters associated with C
»C2bp = C/(B*Km)
C2bp = 1.2368e-05
»L2bp = Km*B/(C*K)
L2bp = 6.7378e-03
```

The final circuit is given in Figure 19.42, where $R_s = 100\ \Omega$, $L_{1bp} = 61.8$ mH, $C_{1bp} = 1.35$ µF, $L_{2bp} = 6.74$ mH, and $C_{2bp} = 12.4$ µF.

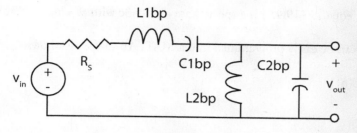

FIGURE 19.42. Final BP filter realization.

This concludes our illustration of the passive BP filter design procedure.

Active BP design is beyond the scope of this text, but coverage can be found in more advanced texts.

11. AN ALGORITHM FOR SINGLY TERMINATED BUTTERWORTH LOW-PASS NETWORKS

In 1937 E. L. Norton published a paper in the *Bell Systems Technical Journal*.[2] The paper contains explicit formulas for the LC element values in a singly terminated 3 dB normalized Butterworth LP filter where the load or source resistance is 1 ohm. Our goal in this section is to present these formulas without derivation. Similar formulas have been developed by other researchers in the context of network synthesis, a subject for which many texts are available for further reference.

For an nth order 3 dB normalized Butterworth filter with a single 1 Ω termination, the sequential formulas for the LC elements are

$$a_1 = \sin\left(\frac{\pi}{2n}\right) \tag{19.43a}$$

$$a_2 = \frac{\sin\left(\dfrac{3\pi}{2n}\right)\sin\left(\dfrac{\pi}{2n}\right)}{a_1 \cos^2\left(\dfrac{\pi}{2n}\right)} \tag{19.43b}$$

$$\vdots$$

$$a_m = \frac{\sin\left(\dfrac{(2m-1)\pi}{2n}\right)\sin\left(\dfrac{(2m-3)\pi}{2n}\right)}{a_{m-1}\cos^2\left(\dfrac{(m-1)\pi}{2n}\right)} \tag{19.43c}$$

$$\vdots$$

$$a_n = n \sin\left(\frac{\pi}{2n}\right) \qquad (19.43d)$$

It turns out that equation 19.43d is a special case of 19.43c with $m = n$.

These formulas can be easily programmed into a MATLAB m-file as follows:

```
function buttLC(n)
nn = 2*n;
a = zeros(1,n);
a(1)=sin(pi/nn);
for m = 2:1:n;
        x = a(m-1)*(cos((m-1)*pi/nn))^2;
        a(m) = sin((2*m-1)*pi/nn)*sin((2*m-3)*pi/nn)/x;
end
Elvalues = a'
```

To illustrate the use, suppose the Butterworth filter order is 4. Then

```
»buttLC(4)
Elvalues =
 3.8268e-01
 1.0824e+00
 1.5772e+00
 1.5307e+00
```

Given these values, it remains to interpret them as inductances or capacitances. Two rules govern the realization:

   (i)   There is never a shunt element in parallel with a voltage source.
   (ii) There is never a dangling element at the load end.

Hence for the fourth-order filter above there are two possible circuit realizations, given in Figure 19.43.

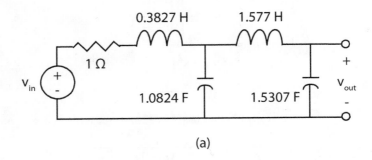

(a)

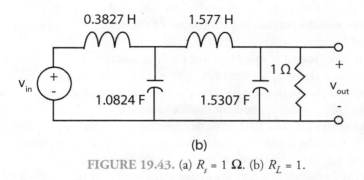

(b)

FIGURE 19.43. (a) $R_s = 1\ \Omega$. (b) $R_L = 1$.

All that remains in the design procedure is frequency and magnitude scaling, which are covered in earlier sections of the chapter.

12. SUMMARY

This chapter has covered the basics of Butterworth LP, HP, and BP filter design. Such design techniques build on a set of filter specifications requiring that the desired filter magnitude response lie outside certain brickwall regions. Finding transfer functions that meet a set of brickwall constraints is called the *approximation problem*. This chapter developed algorithms for finding Butterworth transfer functions for the LP, HP, and BP cases. In addition, basic passive realizations were presented as well as active circuit realizations, except in the BP case. In the active case, the focus was the Sallen and Key low-pass and high-pass topologies. Both the passive and active types of realization are built around the coefficient-matching technique, which associates the coefficients in the circuit transfer function with the coefficients of the desired transfer function; one then solves the resulting equations for appropriate circuit parameter values.

There are, of course, many other types of filter transfer functions; Chebyshev, inverse Chebyshev, and elliptic are other well-known types. Also, in addition to low- and high-pass filters, there are band-pass, band-reject, and magnitude and phase equalizers. To add to the richness of the area of filtering, there are analog passive, analog active, recursive digital, non-recursive digital, and switched capacitor implementations of all of these filter types. The preceding exposition is merely a drop in a very large and fascinating bucket of filter design challenges.

13. TERMS AND CONCEPTS

dB bandwidth of BP filter: the difference between the two frequencies of a BP filter at which the gain is 3 dB down from the maximum gain; if the maximum gain is 1, these frequencies correspond to a gain of $1/\sqrt{2}$ this is the most common reference for the meaning of bandwidth.

dB bandwidth of LP filter: the 3 dB down frequency, ω_c.

active realization: a realization consisting of op amps, R's, and C's.

approximation problem: the problem of finding a transfer function having a magnitude response that meets a given set of brickwall filter specifications.

Attenuation (dB): the loss magnitude expressed in dB, i.e., $A(\omega) = -20 \log_{10}|H(j\omega)|$.

BP to LP frequency transformation: the generation of equivalent normalized LP specs from a given set of BP specs.

Bandwidth with respect to A_{max}: the difference between the two frequencies of a BP filter at which the gain is down by a value of A_{max} from the maximum gain.

Band-pass filter: a filter that passes any frequency within the band $\omega_{p1} \leq \omega \leq \omega_{p2}$ while significantly attenuating frequency content outside this band.

Coefficient-matching technique: a method of determining circuit parameter values by matching the coefficients of the transfer function of the circuit to those of the desired transfer function and solving the resulting equations for the circuit parameters.

Cutoff frequency: the frequency at which the magnitude response of the filter is 3 dB down from its maximum value.

Filter: a circuit or device that significantly attenuates the frequency content of signals in certain frequency bands and passes the frequency content within certain other, user-specified frequency bands in the sinusoidal steady state.

Frequency response magnitude: magnitude of the transfer function as a function of $j\omega$, i.e., $H(j\omega)$.

Gain in dB: $20 \log_{10}|H(j\omega)|$ or $10 \log_{10}|H(j\omega)|^2$.

Gain magnitude: frequency response magnitude.

Half-power point: the point at which the magnitude response curve is 3 dB down.

High-pass filter: a filter that significantly attenuates the low-frequency content of signals and passes the high-frequency content.

HP to LP frequency transformation: a transformation that converts a given set of high-pass brickwall specifications to an equivalent set of low-pass specifications.

Loss magnitude: the inverse of the gain magnitude, i.e., $\dfrac{1}{|H(j\omega)|}$.

Low-pass (brickwall) filter specification: a filter specification requiring that the desired filter magnitude response lie outside certain "brickwall" regions.

Low-pass filter: a filter that passes the low-frequency content of signals and significantly attenuates the high-frequency content.

LP to BP frequency transformation: a technique for converting a passive LP filter to a BP filter by converting L's to series LC circuits and C's to parallel LC circuits.

LP to HP frequency transformation: a technique for converting a passive LP filter to a HP filter by converting L's to C's and C's to L's.

Maximally flat: property of a filter at a point ω wherein the magnitude response has a maximum number of zero derivatives.

Normalized Butterworth loss functions: a set of Butterworth transfer functions, ordered by degree, having 3 dB loss at the normalized frequency $s = j\Omega = j1$.

Passive analog filter: a filter composed only of resistors, capacitors, inductors, and transformers.

RC to CR transformation: a technique for translating a low-pass active filter to a high-pass active filter in which resistors become capacitors and capacitors become resistors.

---

[1] *Wireless Engineer*, vol. 7, 1930, pp. 536–541.
[2] E. L. Norton, "Constant Resistance Networks with Applications to Filter Groups," *Bell Systems Technical Journal* vol. 16, 1937, pp. 178–196.

Problems

LOW-PASS BASICS

1. Filters can have magnitude responses quite different from the maximally flat Butterworth magnitude response. For each of the plots in Figure P19.1, identify whether it is HP, LP, or BP.

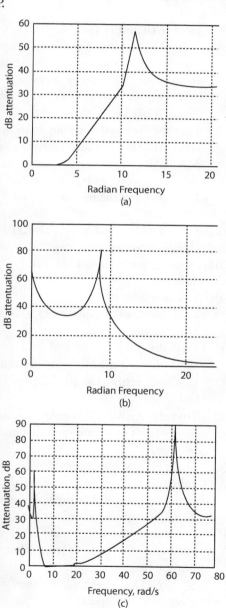

Figure P19.1

2. Figures P19.2 (a) and (c) show an approximate gain magnitude response associated with different transfer functions $H(s)$. Figures P19.2 (b) and (d) show new gain magnitude responses whose transfer functions are related to the $H(s)$ of (a) and (c) by scaling. Properly label the horizontal axis of each graph in (b) and (d).

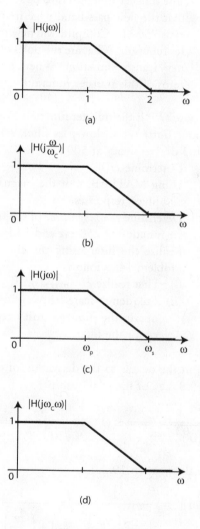

Figure P19.2

3. A NLP filter transfer function

$$H_{NLP}(s) = \frac{0.65378}{s^2 + 0.80381643s + 0.82306043}$$

has a pass-band edge frequency (A_{max} = 2 dB) at 1 rad/sec.

(a) Use MATLAB to plot the magnitude response of the transfer function and verify that the magnitude response is 2 dB down from its maximum of 1.

(b) Use the "roots" command in MATLAB to compute the poles of $H_{NLP}(s)$.

(c) The transfer function is to be scaled so that the new pass-band edge frequency is 750 Hz. Compute the new transfer function. What are the poles of this new transfer function? Where are the zeros of this transfer function?

4. Suppose $H(s)$ is the transfer function of a second-order Butterworth low-pass filter, $H(0)$ = 1, with 3 dB frequency at 500 Hz.

(a) Determine $H(s)$.

(b) Using MATLAB, plot the magnitude and phase responses.

(c) Determine the *magnitude* of the transfer function at 250 Hz and 1 kHz.

(d) Realize the filter using the circuit of Problem 14 as follows:
(i) First realize $H_{3dBNLP}(s)$.
(ii) Frequency-scale the circuit to obtain the proper cutoff frequency for the filter.

5. Fill in the details to the derivation of equation 19.8, i.e., of the relationship

$$\frac{\omega_p}{\sqrt[2n]{10^{0.1A_{max}} - 1}} \leq \omega_c \leq \frac{\omega_s}{\sqrt[2n]{10^{0.1A_{min}} - 1}}$$

6. Recall equation 19.2, i.e.,

$$|H(j\omega)|^2 = \frac{1}{1 + \varepsilon^2 \left(\dfrac{\omega}{\omega_p}\right)^{2n}} = \frac{1}{1 + \left(\dfrac{\omega}{\omega_c}\right)^{2n}}$$

This equation specifies a relationship between the coefficient ε, ω_p, and ω_c. Show that for any LP brickwall specification the range of ε is given by

$$\frac{\sqrt{10^{0.1A_{min}} - 1}}{\left(\dfrac{\omega_s}{\omega_p}\right)^n} = \varepsilon_{min} \leq \varepsilon \leq \varepsilon_{max} = \sqrt{10^{0.1A_{max}}}$$

7. Suppose $H_1(s)$ and $H_2(s)$ are second- and third-order NLP Butterworth transfer functions, respectively, with $\omega_p = \omega_c$ = 1 rad/sec and $H(0)$ = 1.

(a) Find $H_1(s)$ and $H_2(s)$.

(b) Using MATLAB and the "freqs" command, plot the magnitude frequency response of $H_1(s)$ and $H_2(s)$ for $0 \leq \omega \leq 5$ rad/sec on the same graph. Properly label your plots.

(c) Find the step responses for both systems using the "residue" command in MATLAB to compute the partial fraction expansion, and then use MATLAB to sketch the response curves.

LOW-PASS APPROXIMATION

8. Find a minimum-order Butterworth transfer function meeting the brickwall specs in Figure P19.8 as follows:

(a) Find the filter order.

(b) Find the range of allowable ω_c and the range of allowable f_c.

(c) Find the poles, zeros, and gain constant of the 3 dB NLP transfer function and verify that they lie on the unit circle.

(d) Find the 3 dB NLP transfer function as a product of second- and first-order sections.

(e) Using $\omega_{c,min}$ find the zeros, poles, and gain constant of the non-normalized transfer function. Do the poles lie on a circle about the origin?

(f) Accurately plot (using MATLAB) the magnitude response of your filter in terms of gain and gain in dB for $0 \leq f \leq 1.2f_s$. Does your filter meets the given brickwall specs?

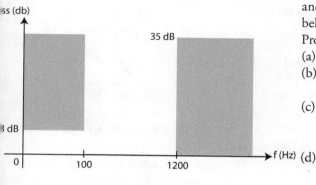

Figure P19.8

ANSWER: Enter the following MATLAB code to obtain the solution.

```
fp = 100; fs = 1200;Amax = 0.3; Amin = 35;
n = buttord(fp,fs,Amax,Amin,'s')
fcmin = fp/((10^(0.1*Amax)-1)^(1/(2*n)))
fcmax = fs/((10^(0.1*Amin)-1)^(1/(2*n)))
wcmin = 2*pi*fcmin
wcmax = 2*pi*fcmax
wc = wcmin;
fc = fcmin;
[z,p,k] = buttap(n)
zplane(p)
grid
pause
znew = z*wc
pnew = p*wc
knew = k*wc^n
f = 0:fc/50:1.2*fs;
h =
freqs(knew*poly(znew),poly(pnew),2*pi*f);
plot(f,abs(h))
grid
pause
plot(f,20*log10(abs(h)))
grid
```

9. Repeat parts (e) and (f) of Problem 8 using $\omega_{c,max}$ instead of $\omega_{c,min}$. Modify the MAT-LAB code of Problem 8 as necessary.

10. Find a minimum-order Butterworth filter meeting the specs $(A_{max}, f_p) = (1 \text{ dB}, 75 \text{ Hz})$

and $(A_{min}, f_s) = (45 \text{ dB}, 450 \text{ Hz})$ as detailed below. Hint: Modify the MATLAB code of Problem 8.

(a) Find the filter order.

(b) Find the range of allowable ω_c and the range of allowable f_c.

(c) Find the poles, zeros, and gain constant of the NLP transfer function and verify that they lie on the unit circle.

(d) Find the NLP transfer function as a product of the second- and first-order sections.

(e) Using $\omega_{c,min}$, find the zeros, poles, and gain constant of the non-normalized transfer function. Do they lie on the unit circle;

(f) Accurately plot (using MATLAB) the magnitude response of your filter in terms of gain and gain in dB for $0 \leq f \leq 1.2f_s$. Does your filter meets the given brickwall specs?

11. Repeat parts (e) and (f) of Problem 10 using $\omega_{c,max}$ instead of $\omega_{c,min}$. Modify the MATLAB code of Problem 10 as necessary.

12. The specs $(A_{max}, f_p) = (2 \text{ dB}, 100 \text{ Hz})$, $(A_{min1}, f_{s1}) = (20 \text{ dB}, 500 \text{ Hz})$, and $(A_{min2}, f_{s2}) = (40 \text{ dB}, 1000 \text{ Hz})$ are associated with the brickwall specifications of Figure P19.12.

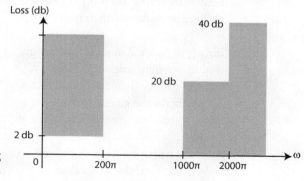

Figure P19.12

(a) Find the minimum filter order meeting these specs.

(b)   Find the range of allowable ω_c and the range of allowable f_c.

(c)   Find the poles, zeros, and gain constant of the NLP transfer function.

(d)   Find the NLP transfer function as a product of second- and first-order sections.

(e)   Using $\omega_{c,min}$, find the zeros, poles, and gain constant of the non-normalized transfer function.

(f)   Now find the transfer function as a product of first- and second-order sections.

(g)   Accurately plot (using MATLAB) the magnitude response of your filter in terms of gain and gain in dB for $0 \le f \le 1.2f_s$. Does your filter meet the given brickwall specs?

PASSIVE LOW-PASS REALIZATION

13. The circuit of Figure P19.13 is to realize a maximally flat second-order low-pass filter with 3 dB frequency f_c = 1000 Hz.

(a)   Show that the circuit transfer function is

$$H_{cir}(s) = \frac{\dfrac{1}{LC}}{s^2 + \dfrac{1}{R_L C}s + \dfrac{1}{LC}}$$

(b)   With R_L = 1 Ω, compute L in H and C in F to realize the normalized second-order Butterworth transfer function.

(c)   If R_L is to be 1 kΩ in your final design, choose a magnitude scale factor, K_m, and the proper frequency scale factor, K_f, to obtain the given filter specifications. Then compute the new values of L and C.

(d)   As an alternative design, suppose C is to be 1 μF. Find the new values of K_m and K_f to achieve the original filter specifications. What are the new values of L and R_L?

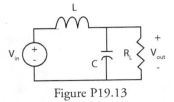

Figure P19.13

14. The circuit of Figure P19.14 is to realize a maximally flat second-order low-pass filter with 3 dB frequency f_c = 500 Hz.

(a)   Show that the transfer function is

$$H_{cir}(s) = \frac{\dfrac{1}{LC}}{s^2 + \dfrac{R_s}{L}s + \dfrac{1}{LC}}$$

(b)   With R_s = 1 Ω, compute L in H and C in F to realize the normalized second-order Butterworth transfer function.

(c)   If R_s is to be 10 Ω in your final design, choose a magnitude scale factor, K_m, and the proper frequency scale factor, K_f, to obtain the given filter specifications. Then compute the new values of L and C.

(d)   As an alternative design, suppose C is to be 1 μF. Find the new values of K_m and K_f to achieve the original filter specifications. What are the new values of L and R_s?

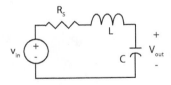

Figure P19.14

15. Consider the circuit in Figure P19.15.

(a)   Show that the transfer function of the circuit is

$$H_{cir}(s) = \frac{V_{out}}{V_{in}} = \frac{\dfrac{1}{LC}}{s^2 + \left(\dfrac{1}{CR_L} + \dfrac{R_s}{L}\right)s + \dfrac{1+R_s/}{LC}}$$

(b) If $R_s = 2\ \Omega$ and $R_L = 8\ \Omega$, find two solutions (L, C) so that $H(s)$ has a second-order Butterworth (maximally flat) response with 3 dB down (from its maximum gain) frequency equal to 1 rad/sec.

(c) Magnitude- and frequency-scale both networks resulting from part (b) so that in the new networks the smaller resistance is 2 kΩ, and the 3 dB frequency is 5 kHz.

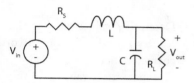

Figure P19.15

16. With $R_s = R_L = 1\ \Omega$ in the circuit of Figure P19.15, the transfer function becomes

$$H_{cir}(s) = \frac{\dfrac{1}{LC}}{s^2 + \left(\dfrac{1}{C} + \dfrac{1}{L}\right)s + \dfrac{2}{LC}}$$

which can be used to realize the normalized second-order Butterworth characteristic

$$H(s) = \frac{K}{s^2 + \sqrt{2}s + 1}$$

for $K = 0.5$. Suppose such a filter is to have cutoff frequency $f_c = 3500$ Hz. (In this case, the filter is to have magnitude 3 dB down from its maximum gain at f_c.)

(a) Find values of L and C to realize the normalized transfer function.

(b) If the final value of the resistors is to be 1 kΩ, compute the new values of L and C that realize the desired filter.

(c) Alternatively, suppose the final value of the capacitor is to be 10 nF. Compute the new values of L and C that realize the desired filter.

17. Consider the circuit of Figure P19.17.

(a) Show that the transfer function is

$$H(s) = \frac{V_{out}}{V_{in}} = \frac{\dfrac{1}{R_s LC}}{s^2 + \left(\dfrac{1}{R_s C} + \dfrac{1}{L}\right)s + \dfrac{1 + 1/R_s}{LC}}$$

(b) If $R_s = 2\ \Omega$, find two sets of parameter values that realize the second-order normalized Butterworth gain function.

(c) Magnitude- and frequency-scale both networks resulting from part (b) so that in the new networks the smaller resistance is 2 kΩ, and the 3 dB frequency is 5 kHz.

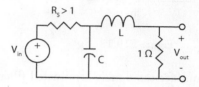

Figure P19.17

ANSWERS: (b) $(C, L) = (0.4483, 3.3460)$ or $(1.6730, 0.8966)$ in F and H; (c) $(C, L) = (26.6$ nF, 57.1 mH) or (7.1 nF, 0.213 H), R_s becomes 4 kΩ

18. Consider the second-order low-pass transfer function

$$H(s) = \frac{V_{out}}{V_{in}} = \frac{K}{\left(\dfrac{s}{\omega_p} + 1\right)^2}$$

where $\omega_p = 10^5$ rad/sec.

(a) Find $|H(j\omega)|$ and ω_c (the 3 dB down frequency). Is it true that $\omega_c = \omega_p$ in this case?

(b) Find the impulse response and step response.

(c) Using the RLC topology shown in Figure P19.18a, realize $H(s)$ given that $R_L = 10$ kΩ in your final design. What is the value of K in your solution?

Hint: First set $\omega_p = 1$ and then frequency- and magnitude-scale to obtain the final answer.

(d) Using the *RLC* topology shown in Figure P19.18b, realize $H(s)$ given that $R_s = 100\ \Omega$ in your final design. What is the value of K in your solution?

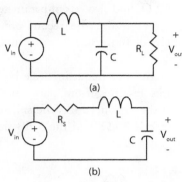

(a)

(b)

Figure P19.18

19. Consider the doubly terminated filter circuit of Figure P19.19

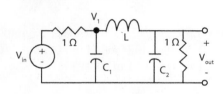

Figure P19.19

(a) Show that the transfer function of the circuit is

$$H_{cir}(s) = \frac{V_{out}(s)}{V_{in}(s)} = \frac{\dfrac{1}{LC_1C_2}}{s^3 + \dfrac{C_1+C_2}{C_1C_2}s^2 + \dfrac{C_1+C_2+L}{LC_1C_2}s + \dfrac{2}{LC_1C_2}}$$

(b) By equating the coefficients of the denominator of $H_{cir}(s)$ with the denominator of the third-order 3 dB NLP Butterworth loss function, show that the only possible realization is $C_1 = C_2 = 1$ F and $L = 2$ H.

(c) If the source and load resistance are to be 1 kΩ and $f_c = 1500$ Hz (meaning the

gain is down 3 dB from its maximu value), find the new element values.

ANSWER: $C_{1new} = C_{2new} = 0.106\ \mu F$ and L_{ne} $= 0.2122$ H.

20. Consider Figure P19.20.

(a) Show that the transfer function is

$$\frac{\dfrac{1}{L_1L_2C}}{s^3 + \left(\dfrac{1}{L_1}+\dfrac{1}{L_2}\right)s^2 + \left(\dfrac{L_1+L_2+C}{L_1L_2C}\right)s}$$

(b) Determine values so that the circu realizes a third-order Butterworth ga characteristic.

(c) Find parameter values of a third-ord low-pass Butterworth filter havir cutoff 20 kHz and resistor values of kΩ.

(d) Use SPICE to verify the frequenc response.

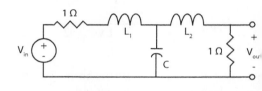

Figure P19.20

ANSWERS: (b) $L_1 = L_2 = 1$ H, $C = 2$ F; (15.9 nF and 15.9 mH.

21. Consider the two-port configuratior shown in Figure P19.21.

(a) Prove that for Figure P19.21a

$$\frac{V_{out}}{V_{in}} = \frac{z_{21}}{z_{11}+R_s}$$

(b) Prove that for Figure P19.21b

$$\frac{V_{out}}{V_{in}} = \frac{-y_{21}}{y_{22} + G_L}$$

(c)   Making use of the results of part (a), design a second-order Butterworth low-pass filter having $\omega_p = \omega_c = 1$ rad/sec, $R_s = 1\ \Omega$, and $R_L^t = \infty$. Hint: Divide the numerator and denominator of $H(s)$ by s, and then equate the result with the gain expression to obtain two z-parameters. Next, review the z-parameters of a two-port consisting of one series element and one shunt element (a special case of a T- or π-network). Finally, put all results together to design the two-port.

(d)   Making use of the results of part (b), associated with Figure 19.21b, design a second-order Butterworth low-pass filter having $\omega_c = 1$ rad/sec, $R_s = 0\ \Omega$ and $R_L = 1\ \Omega$.

(e)   Frequency- and magnitude-scale the circuits of parts (c) and (d) so that the new filters have $\omega_c = 5000$ rad/sec, and the single resistance in the circuit is 1 kΩ.

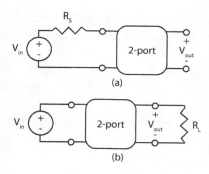

(a)

(b)

Figure P19.21

22. (a)   Find the 3 dB NLP Butterworth transfer function meeting the specs $(\omega_p, A_{max}) = (2\pi500, 2\ dB)$ and $(\omega_s, A_{min}) = (2\pi2000, 20\ dB)$. Then find $\omega_{c,min}$ and $\omega_{c,max}$.

(b)   Using one of the circuits with a non-zero source resistance from the earlier homework problems, realize the 3 dB

NLP transfer function obtained in part (a).

(c)   Magnitude- and frequency-scale the circuit so that the gain is 3 dB down at $\omega_{c,min}$ and the source resistance is 100 Ω.

(d)   Plot the magnitude response in terms of dB gain vs. f (in Hz).

ACTIVE LOW-PASS DESIGN

23. Consider the circuit of Figure P19.23, which has a practical source driving a leaky integrator circuit.

(a)   Compute the circuit transfer function using $R_{eq} = R_s + R_1$.

(b)   Realize the first-order normalized Butterworth transfer function

$$H_{3dBNLP}(s) = \frac{V_{out}(s)}{V_{in}(s)} = \frac{1}{s+1}$$

i.e., compute values for R_{eq}, R_2, and C. Hint: Let $C = 1$ F.

(c)   If the 3 dB down frequency is to be $f_p = 3500$ Hz, find K_m so that the capacitor is 1 nF in your design. Compute the new resistor values and then determine R_1 if the source resistance is $R_s = 500\ \Omega$.

CHECK: $R_1 = 45$ kΩ

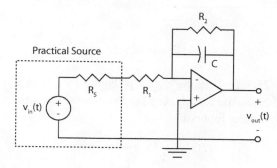

Figure P19.23

24. Consider the circuit of Figure P19.24 as a candidate for realizing the second-order low-pass transfer function

$$H(s) = \frac{V_{out}}{V_{in}} = \frac{K}{\left(\dfrac{s}{\omega_p} + 1\right)^2}$$

(a) Compute the transfer function $H_{cir}(s)$ of the circuit of Figure P19.24.

(b) For a normalized design, $\omega_p = 1$ rad/sec and $K = -10$, choose element values to realize the given transfer function; i.e., find all circuit parameter values.

(c) If $\omega_p = 10^5$ rad/sec and the capacitors are to be 1 nF, choose K_f and K_m and compute the final circuit parameter values. Simulate using SPICE to verify your design.
CHECK: $R_1 = 1$ kΩ

(d) Find the 3 dB down frequency, ω_c, in terms of ω_p.
CHECK:
$\omega_c = 0.6436\,\omega_p$

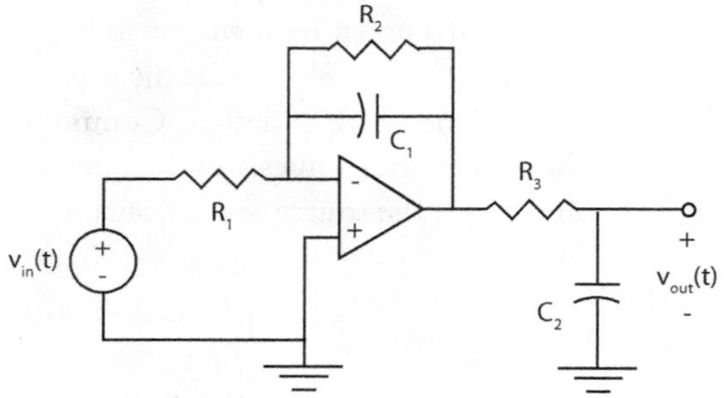

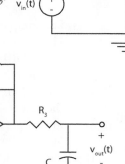

Figure P19.24

25. Consider the circuit of Figure P19.25 as a candidate for realizing the third-order low-pass transfer function

$$H(s) = \frac{V_{out}}{V_{in}} = \frac{K}{\left(\dfrac{s}{\omega_p} + 1\right)^3}$$

(a) Compute the transfer function $H_{cir}(s)$ = $H_{stage1}(s)\,H_{stage2}(s)\,H_{stage3}(s)$ of the circuit of Figure P19.25.

(b) For a normalized design, $\omega_p = 1$ rad/sec and $K = 10$, choose element values to realize the given transfer function; i.e., find all circuit parameter values.

(c) If $\omega_p = 10^5$ rad/sec and the capacitors are to be 1 nF, choose K_f and K_m and compute the final circuit parameter values. Simulate using SPICE to verify your design. (The solution is not unique, which is often the case for practical design problems.)

(d) Find the 3 dB down frequency, ω_c, in terms of ω_p.
CHECK: $\omega_c = 0.5098\omega_p$

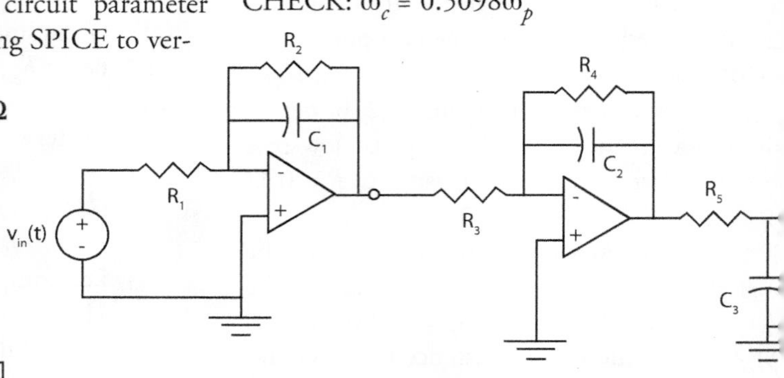

Figure P19.25

26. A second-order normalized Butterworth filter can be realized by the Sallen and Key circuit described in section 5. In the final design, the filter is to have a dc gain of 1, a 3 dB down frequency of $f_c = 1000$ Hz, and largest capacitance of 10 nF.

(a) Using design A and input attenuation if necessary, determine the appropriate Sallen and Key circuit.

(b) Using design B and input attenuation if necessary, determine the appropriate Sallen and Key circuit.

(c) Using the Saraga design and input attenuation if necessary, determine the appropriate Sallen and Key circuit.

Hint: It is often useful to generate an Excel spreadsheet to do the relevant calculations, especially for multiple designs such as these; this allows one to set forth the normalized and final designs along with K_m and K_f

27. Realize the filter of Problem 8 as a cascade of a Sallen and Key circuit (Saraga design) and a first-order active circuit such as the leaky integrator. The 3 dB down point should be at $\omega_{c,min}$. The largest capacitor should be 50 nF. (Hint: It is often useful to generate an Excel spreadsheet to do the relevant calculations, especially for multiple designs.) Verify the frequency response of your design using SPICE.

28. Repeat Problem 27, except use design A instead of the Saraga design for the Sallen and Key portion.

29. Realize the filter of Problem 10 as a cascade of two Sallen and Key circuits. The 3 dB down point should be at ω_{cmin}. The largest capacitor should be 0.1 µF.

30. Repeat Problem 29, except use design B instead of the Saraga design for the Sallen and Key portion.

31. In addition to Butterworth filter transfer functions, there are Chebyshev filters, which have a faster transition from pass-band to stop band. This problem investigates the implementation of a Chebyshev LP filter transfer function. Recall that a second-order transfer function

$$H_{cir}(s) = \frac{K}{s^2 + \dfrac{1}{Q}s + 1}$$

can be realized by the Saraga design of a Sallen and Key circuit.

   (a)   Determine the new values of R_1, R_2, C_1, and C_2 that will realize the poles of a normalized second-order Chebyshev LP transfer function

$$H_{NLP}(s) = \frac{0.65378}{s^2 + 0.80382s + 0.82306}$$

Hint: Frequency-scale, $s \rightarrow \omega_0 s$ where $\omega_0^2 = 0.82306$. What is the resulting transfer function?

   (b)   What is the dc gain, K, of the circuit? Modify the circuit to achieve the desired gain of the transfer function.

   (c)   If the actual pass-band edge frequency is to be 7 kHz and C_2 is to be 0.05 µF, determine the new element values.

   (d)   Plot the loss magnitude response of $H(s)$ in part (a) in dB. Determine A_{max}.
CHECK: $A_{max} = 2$ dB

32. Repeat Problem 31 using design A of the Sallen and Key circuit.

33. Repeat Problem 31 using design B of the Sallen and Key circuit.

HIGH-PASS PASSIVE DESIGN

34. A second-order Butterworth HP filter has 3 dB down point at $f_c = 2000$ Hz. The second-order Butterworth NLP prototype is given in Figure P19.34 and has transfer function

$$H_{cir}(s) = \frac{\dfrac{1}{LC}}{s^2 + \dfrac{1}{R_L C}s + \dfrac{1}{LC}}$$

   (a)   With $R_L = 1\ \Omega$, compute L in H and C in F to realize the normalized second-order prototype.
CHECK: $C = 1/\sqrt{2}$ F

   (b)   Using the results of part (a), construct a NHP circuit with $R_L = 1\ \Omega$ and $\Omega_{c,HP} = 1$ rad/sec. This is the so-called 3 dB normalized HP filter (3 dB NHP).

   (c)   Now construct the final HP design with $R_L = 100\ \Omega$.

   (d)   Do a SPICE simulation to verify your design.

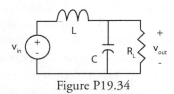

Figure P19.34

35. A third-order Butterworth HP filter has 3 dB down point at f_c = 5000 Hz. The third-order Butterworth NLP prototype is given in Figure P19.35 and has transfer function

$$H_{cir}(s) = \frac{V_{out}(s)}{V_{in}(s)} = \frac{\dfrac{1}{LC_1C_2}}{s^3 + \dfrac{1}{C_1}s^2 + \dfrac{C_1+C_2}{LC_1C_2}s + \dfrac{1}{LC_1C_2}}$$

(a) Compute L in H and C_1 and C_2 in F to realize the 3 dB normalized third-order prototype.
CHECK: C_1 = 1.5 F and $L = \dfrac{4}{3}$ H

(b) Using the results of part (a), construct a NHP circuit with $\Omega_{c,HP}$ = 1 rad/sec. This is the so-called 3 dB normalized HP filter (3 dB NHP).

(c) Now construct the final HP design if the source resistance is to be 100 Ω.

(d) Do a SPICE simulation to verify your design.

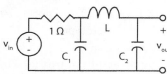

Figure P19.35

36. A second-order high-pass Butterworth loss function has stop band edge frequency f_s = 1 kHz, pass-band edge frequency f_p = 7 kHz, and cutoff frequency f_c = 5500 Hz.

(a) Determine the attenuation in dB at f_p and f_s.

(b) Realize the second-order 3 dB NLP Butterworth filter using the circuit of Figure P19.36; i.e., compute values for L and C.

(c) Using the results of part (b), construct a NHP circuit with the source and load resistances equal to 1 Ω an $\Omega_{c,HP}$ = 1 rad/sec. This is the so-called 3 dB normalized HP filter (3 dB NHP).

(d) Given your answer to part (c), construct an appropriate high-pass filter circuit whose resistor values are 1 kΩ

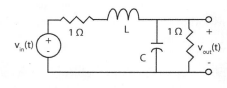

Figure P19.36

37. The circuit of Figure P19.37 realizes third-order NLP Butterworth filter. Determine a third-order high-pass filter circuit, with A_{max} = 2 dB at f_p = 5 kHz so that the largest capacitor equals 100 nF, as follows:

(a) Determine ε_{max} and $\Omega_{c,min}$ for the NLP filter.

(b) Compute ω_c for the desired HP filter. Does your answer make sense? Think about this carefully.

(c) Realize the 3 dB NLP transfer function using the circuit of Figure P19.37. Set L = 2 H and make C_1 = C_2. Recall from Problem 19 that the transfer function of the circuit is

$$H_{cir}(s) = \frac{V_{out}(s)}{V_{in}(s)} = \frac{\dfrac{1}{LC_1C_2}}{s^3 + \dfrac{C_1+C_2}{C_1C_2}s^2 + \dfrac{C_1+C_2+L}{LC_1C_2}s + \dfrac{}{L}}$$

(d) Construct the appropriate high-pass circuit with the largest capacitor equal to 100 nF.

(e) Verify the frequency response of your design using SPICE.

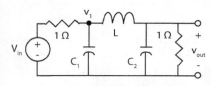

Figure P19.37

38. The brickwall specs of a HP filter are shown in Figure P19.38, in which f_s = 500 Hz, f_p = 600 Hz, A_{max} = 2 dB, and A_{min} = 25 dB. The filter is to be driven by a voltage source having a source resistance of 100 Ω. There is no load resistance. At the pass-band edge frequency the attenuation should be exactly 2 dB. Develop a passive HP filter circuit to solve this problem.

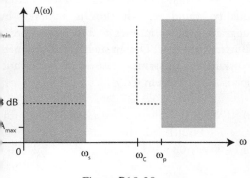

Figure P19.38

39. Repeat Problem 38 assuming that the filter is driven by a voltage source with negligible source resistance and that there is a load resistance that in the final design is to be 75 Ω. Consider using the NLP prototype of Problem 4.

40. The brickwall specs of the HP filter in Figure P19.38 now apply to f_s = 500 Hz, f_p = 600 Hz, A_{max} = 2 dB, and A_{min} = 30 dB. The filter is to be driven by a voltage source having a source resistance of 250 Ω. There is no load resistance. At the pass-band edge frequency the attenuation should be exactly 2 dB. Develop a passive HP filter circuit to meet these specifications. Hint: Use the circuit of Figure 19.8, having transfer function of equation 19.14.

41. Reuse the specs of Problem 40, except that there is to be negligible source resistance and a load resistance of 75 Ω.

(a) Compute the transfer function, say $H_{cir}(s)$, of the circuit in Figure P19.41 in terms of L_1, L_2, and C.

(b) Determine values so that the circuit realizes a third-order Butterworth 3 dB NLP Butterworth gain function.

(c) Determine the HP filter.

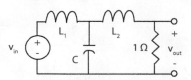

Figure P19.41

HIGH-PASS ACTIVE REALIZATION

42. Consider the set of HP specs for which (f_c, A_{max}) = (2400 Hz, 3dB) and (f_c, A_{min}) = (600 Hz, 30dB). Following the procedure of Example 19.7, design an active HP filter meeting these specs so that the smallest capacitor in the final design is 10 nF.

43. Consider the set of HP specs for which (f_p, A_{max}) = (2400 Hz, 2dB) and (f_s, A_{min}) = (600 Hz, 30dB). Following the procedure of Example 19.7, design an active HP filter meeting these specs so that the smallest capacitor in the final design is 10 nF. The magnitude response should go through the pass-band edge. Hint: See equation 19.34.

44. The Sallen and Key Saraga design HP filter of Figure P19.44 realizes the normalized HP filter transfer function

$$H(s) = \frac{Ks^2}{s^2 + \frac{1}{Q}s + 1}$$

where K = 1 + R_B/R_A. A second-order HP Butterworth filter is to have f_c = 3 kHz.

Compute the parameter values of the HP filter of Figure P19.44. The smallest resistor should be 10 kΩ. Verify the frequency response of your design using SPICE.

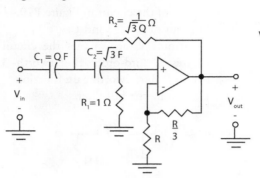

Figure P19.44

45. Again consider the circuit of Figure P19.44.
   (a) Cascading two sections together, realize a fourth-order normalized HP Butterworth transfer function.
   (b) If the smallest resistor is to be 10 kΩ and f_c = 3 kHz, find the new parameter values for your design.

46. You have a set of tweeters that feature great sound reproduction for frequencies above 5 kHz. As a lark, you decide to build an active high-pass Butterworth filter that will isolate highs from a particular audio signal. The specs you decide on are (f_p = 5 kHz, A_{max} = 3 dB) and (f_s = 1.5 kHz, A_{min} = 40 dB). Determine the filter as a product of second-order active Sallen and Key circuits. Minimum resistor values should be 20 kΩ in a Saraga design.

MISCELLANEOUS

47. A certain audio amplifier has a very low internal resistance. It is therefore approximately represented by an ideal voltage source. The 8 Ω woofer and tweeter each may be approximately represented by a resistance of 8 ohms. Design the simple crossover network shown in Figure P19.47 such that the crossover frequency is 2000 Hz.

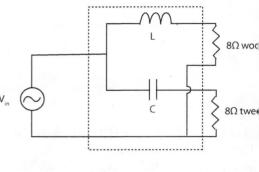

Figure P19.47

ANSWER: L = 636 µH, C = 9.95 µF

48. The crossover network of the previou problem provides first-order Butterworth low pass and high-pass response curves. Better quality sound reproduction is achieved by upgrading the responses to the second-orde Butterworth type. Design such a crossover net work having the same crossover frequency and loads as in Problem 47.

ANSWER: For the woofer, L = 0.90032 mH C = 7.0337 µF. For the tweeter, L = 0.9003 mH, C = 7.0337 µF.

CHAPTER 20

Brief Introduction to Fourier Series

CHAPTER OBJECTIVES

1. Introduce the concept of and calculation procedure for the Fourier series of a periodic signal.
2. Describe the relationship between the complex and the real Fourier series representations.
3. Set forth and discuss basic properties of the Fourier series.
4. Show how the basic Fourier series properties can be used to compute the Fourier series of a wide range of signals from a few basic ones.

SECTION HEADINGS

1. **Introduction**
2. **The Fourier Series: Trigonometric and Exponential Forms**
3. **Additional Properties and Computational Shortcuts for the Fourier Series Representation**
4. **Summary**
5. **Terms and Concepts**
6. **Problems**

INTRODUCTION

Non-sinusoidal periodic waveforms are an important class of signals in electrical systems. Some prominent examples are the square waveform used to clock a digital computer and the sawtooth waveform used to control the horizontal motion of the electron beam of a cathode ray TV picture tube. Non-sinusoidal periodic functions also have importance for non-electrical systems. In fact, the study of heat flow in a metal rod led the French mathematician J. B. J. Fourier to invent the trigonometric series representation of a periodic function. Today the series bears his name. The Fourier series of a periodic signal exciting a linear circuit or system leads to a simplified understanding of the effect of the system on the original periodic signal. This idea is briefly explained in the next few paragraphs and illustrated in Example 20.1.

When a periodic input excites a linear circuit, there are many ways to determine the steady-state output. Using a Fourier series method of analysis, the input is first resolved into the sum of a dc component and infinitely many ac components at harmonically related frequencies. For example, a 1 kHz square wave voltage with zero mean and a 0.5π V peak-to-peak value, with $f(t) = (u(t) -2u(t-0.5T) + u(t-T))$, $T = 1$ msec,

$$v_s(t) = 0.25\pi \sum_{n=0}^{\infty} f(t - nT)$$

(20.1a)

has the Fourier series representation

$$v_s(t) = \sum_{n=0}^{\infty} \frac{(-1)^n \sin\big((2n+1)\omega_o t\big)}{2n+1} u(t)$$

$$= \left[\sin(\omega_o t) - \frac{1}{3}\sin(3\omega_o t) + \frac{1}{5}\sin(5\omega_o t) - \dots \right] u(t)$$

(20.1b)

where $\omega_0 = 2\pi \times 1000$ rad/sec. In the Fourier series representation (which we will later develop), we observe that the signal has a zero average dc value and harmonically related frequency components

$$\frac{(-1)^n \sin\big((2n+1)\omega_o t\big)}{2n+1}, \; n = 0, 1, 2, \dots .$$

By linearity and superposition, the effect of the linear system on $v_s(t)$ can be determined by summing the time domain effects of the linear system on each component in the Fourier series.

For steady-state calculations, let $\mathbf{V}_{s,n}$, be a phasor representing the $(2n + 1)$th harmonic of the Fourier series representation of $v_s(t)$. Then, in steady state, the effect of the system on this term is given by $\mathbf{V}_{o,n} = H(j(2n + 1)\omega_0)\mathbf{V}_{s,}$, assuming that the transfer function $H(s)$ is stable. Again by linearity, the actual time domain output is then computed for each $\mathbf{V}_{o,n}$. The resulting time functions are then summed to obtain the *steady-state* part of the complete response. Example 20.1 illustrates the particulars.

EXAMPLE 20.1. Figure 20.1b shows a simple RC circuit ($R = 1$ Ω, $C = 1$ F) and a square wave input voltage (Figure 20.1a) with $E = 30\pi$ V and $T = 4$ sec. Find the first four components of the output voltage $v_o(t)$ in the steady state.

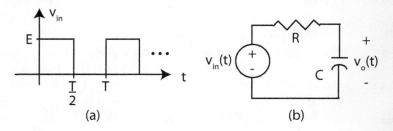

(a)   (b)

FIGURE 20.1 Series RC circuit excited by square wave, used to demonstrate calculation of the steady-state response.

SOLUTION

Step 1. *Determine the Fourier series representation of* $v_{in}(t)$. The fundamental frequency of $v_{in}(t)$ is $f_0 = 1/T = 0.25$ Hz or $\omega_0 = 2\pi f_0 = 0.5\pi$ rad/sec. As shown in the next section, $v_{in}(t)$ has the infinite (Fourier) series representation

$$v_{in}(t) = 30\pi \sum_{n=0}^{\infty} \left[u(t - nT) - u(t - (n+0.5)T) \right]$$

$$= \left[15\pi + 60 \sin(\omega_0 t) + 20 \sin(3\omega_0 t) + 12 \sin(5\omega_0 t) + ... \right] u(t)$$

$$= \left[15\pi + 60 \sum_{n=0}^{\infty} \frac{\sin\left((2n+1)\omega_o t\right)}{2n+1} \right] u(t)$$

For this signal the average dc value is 15π. The remaining terms in the Fourier series are harmonically related sinusoids of decreasing magnitude.

Step 2. *Find the circuit transfer function.* The stable transfer function of the circuit is

$$H(s) = \frac{V_O}{V_{in}} = \frac{1}{s+1}$$

Step 3. *Determine the magnitude and phase of* $H(s)$ *at* $s = j\omega$. For sinusoidal steady-state analysis, set $s = j\omega$ to obtain

$$H(j\omega) = \frac{1}{j\omega + 1} = H_m e^{i\theta}$$

in which case

$$H_m = \frac{1}{\sqrt{\omega^2 + 1}}, \quad \theta = -\tan^{-1}(\omega)$$

Step 4. *Find steady-state responses to all components of* $v_{in}(t)$. Table 20.1 lists the steady-state responses to several components of $v_{in}(t)$.

TABLE 20.1. Several Fourier Series Components of $v_{in}(t)$ and the Corresponding Magnitude and Phase of the Response

ω	0	ω_0	$3\omega_0$	$5\omega_0$
Input magnitude	15π	60	20	12
Input angle*	0	0	0	0
H_m	1	0.5370	0.2075	0.1263
θ (degrees)	0	−57.52	−78.02	−82.74
Output magnitude	15π	32.22	4.150	1.516
Output angle	0	−57.52	−78.02	−82.74

*Angles are in reference to sine functions in this example.

Step 5. *Apply superposition to obtain the steady-state portion of the complete response.* Neglecting har monics of seventh order and higher, the approximate steady-state solution is

$$v_o(t) = 15\pi + 32.22\sin(\omega_0 t - 57.52°) + 4.15\sin(3\omega_0 t - 78.02°) + 1.516\sin(5\omega_0 - 82.74°) + .$$

This response shows the effect of the system on each component of the input signal and how it i turn affects the overall steady-state output response. Since the time constant of the circuit is 1 se this steady-state response more or less constitutes the actual response for $t > 5$ sec.

Exercise. Use MATLAB to plot the approximate waveform for $v_o(t)$ for $0 \leq t \leq 8$ sec, based or Table 20.1. From the plot identify $[v_o(t)]_{max}$ and $[v_o(t)]_{min}$. Alternatively, you can use the ma and min commands in MATLAB.
ANSWER: 82.013 V and 12.235 V

In Example 20.1, the first step of the solution was to represent a periodic waveform as a sum o sinusoidal components, called the *Fourier series*. Section 2 covers the definition and basic proper ties of Fourier series. Section 3 describes several shortcuts for computing the Fourier coefficient and identifies other important properties. Since, in practice, only a finite number of terms can be considered, the Fourier series method yields only an *approximate* solution.

Because many mathematical and engineering handbooks have extensive tables of Fourier series o different waveforms, it is convenient to use these tables in much the same way as one uses a tabl of integrals or a table of Laplace transforms. The Fourier series of some basic signals are provid ed later, in Table 20.3. The use of this table, together with some properties and shortcuts dis cussed in section 3, make the study of the Fourier series method much more palatable to begin ning students of circuit analysis.

2. THE FOURIER SERIES: TRIGONOMETRIC AND EXPONENTIAL FORMS

BASICS

A signal $f(t)$ is **periodic** if, for some $T > 0$ and all t,

$$f(t + T) = f(t) \tag{20.2}$$

T is the period of the signal. The **fundamental period** is the smallest positive real number T_0 for which equation 20.2 holds; $f_0 = 1/T_0$ is called the **fundamental frequency** (in hertz) of the sig nal; $\omega_0 = 2\pi f_0 = 2\pi/T_0$ is the fundamental *angular* frequency (in rad/sec). The sinusoidal wave form of an ac power source and the square wave form used to clock a digital computer are com mon periodic signals. Figure 20.2 shows a portion of a hypothetical periodic signal.

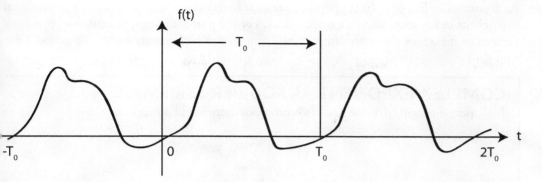

FIGURE 20.2 A hypothetical periodic signal.

THE REAL TRIGONOMETRIC FOURIER SERIES

Under conditions that are ordinarily satisfied by signals encountered in engineering practice, a periodic function $f(t)$ has a decomposition as a sum of sinusoidal functions

$$f(t) = \frac{a_o}{2} + \sum_{n=1}^{\infty} \left(a_n \cos(n\omega_o t) + b_n \sin(n\omega_o t) \right) \tag{20.3}$$

Equivalently,

$$f(t) = d_o + \sum_{n=1}^{\infty} d_n \cos(n\omega_o t + \theta_n) \tag{20.4}$$

where

$$d_o = \frac{a_o}{2} \tag{20.5a}$$

and

$$d_n = \sqrt{a_n^2 + b_n^2}, \quad \theta_n = \tan^{-1}\left(-\frac{b_n}{a_n} \right) \tag{20.5b}$$

Both infinite series, equations 20.3 and 20.4, are called the **trigonometric Fourier series** representations of $f(t)$. We note that equations 20.5a and 20.5b follow from the trigonometric identity

$A\cos(x) + B\sin(x) = \sqrt{A^2 + B^2}\ \cos\left(x + \tan^{-1}\left(-\frac{B}{A} \right) \right)$ with due regard to quadrant.

From equation 20.4, observe that d_0 is the **average value** of $f(t)$. In electric circuit analysis, d_0 refers to the dc component of $f(t)$. The first term under the summation sign, $d_1\cos(\omega_0 t + \theta_1)$, is called the **fundamental component** (or *first harmonic*) of $f(t)$, with amplitude d_1 and phase angle θ_1. The term $d_2 \cos(2\omega_0 t + \theta_2)$ is called the *second harmonic* of $f(t)$, with amplitude d_2 and phase angle θ_2, and similarly for the other terms $d_n \cos(n\omega_0 t + \theta_n)$, which are the nth harmonics as indicated by the term $n\omega_0$.

As illustrated in Example 20.1, given any periodic function $f(t)$, it is important to determine th coefficients in equation 20.3 or equation 20.4. For the purpose of easier calculation, it is advar tageous to introduce the equivalent complex Fourier series representation of the periodic func tion $f(t)$.

COMPLEX EXPONENTIAL FOURIER SERIES

For a periodic signal $f(t)$, the so-called complex exponential Fourier series is

$$f(t) = \sum_{n=-\infty}^{\infty} c_n e^{jn\omega_o t} \qquad (20.6a)$$

where it can be shown that

$$c_n = \frac{1}{T_0} \int_{t_0}^{t_0+T_o} f(t) e^{-jn\omega_0 t} dt \qquad (20.6b)$$

Since we have two (allegedly) equivalent forms, let us now develop the relationship between th real and complex forms of the Fourier series. Recall the Euler identities:

$$\cos(x) = \frac{e^{jx} + e^{-jx}}{2} \quad \text{and} \quad \sin(x) = \frac{e^{jx} - e^{-jx}}{2j}$$

Then equating the real and complex forms, we have

$$f(t) = \sum_{n=-\infty}^{\infty} c_n e^{jn\omega_o t} = \frac{a_o}{2} + \sum_{n=1}^{\infty} \left(a_n \cos(n\omega_0 t) + b_n \sin(n\omega_0 t) \right)$$

$$= \frac{a_o}{2} + \sum_{n=1}^{\infty} \left(a_n \frac{e^{jn\omega_o t} + e^{-jn\omega_o t}}{2} + b_n \frac{e^{jn\omega_o t} - e^{-jn\omega_o t}}{2j} \right)$$

$$= \frac{a_o}{2} + \sum_{n=1}^{\infty} \left(0.5(a_n - jb_n) e^{jn\omega_o t} + 0.5(a_n + jb_n) e^{-jn\omega_o t} \right)$$

Equating coefficients yields

$$c_n = 0.5(a_n - jb_n) \qquad (20.7$$

and

$$c_{-n} = 0.5(a_n + jb_n) = c_n^* \qquad (20.7$$

where c_n^* is the complex conjugate of c_n. Equivalently,

$$a_n = 2\text{Re}(c_n) \qquad (20.8$$

$$b_n = -2\text{Im}(c_n) \qquad (20.8$$

$$d_0 = c_0 = \frac{a_0}{2} \tag{20.8c}$$

$$d_n = 2|c_n|, \; n = 1, 2, 3,\ldots \tag{20.8d}$$

$$\theta_n = \angle \, c_n, \; n = 1, 2, 3,\ldots \tag{20.8e}$$

In the real trigonometric Fourier series equations 20.3 and 20.4, the summation is over positive integer values of n, whereas in the complex exponential Fourier series the summation extends over integers n such that $-\infty < n < \infty$. While each term in equation 20.3 or equation 20.4 has a waveform displayable on an oscilloscope, each individual term in the complex exponential Fourier series lacks such a clear physical picture. However, two conjugate terms in the complex exponential Fourier series always combine to yield a real-time signal $d_n\cos(n\omega_0 t + \theta_n)$.

To develop equation 20.6b for the coefficient c_n, we multiply both sides of equation 20.6a by $e^{-jn\omega_0 t}$ to obtain

$$f(t)e^{-jn\omega_0 t} = \sum_{k=-\infty}^{\infty} \left(c_k e^{jk\omega_0 t} \right) e^{-jn\omega_0 t}$$

Integrating over one entire period, $[t_0, t_0 + T_0]$, produces

$$\int_{t_0}^{t_0+T_o} f(t)e^{-jn\omega_0 t}\,dt = \int_{t_0}^{t_0+T_o} \sum_{k=-\infty}^{\infty} \left(c_k e^{jk\omega_0 t} \right) e^{-jn\omega_0 t}\,dt$$

$$= \sum_{k=-\infty}^{\infty} \left\{ \int_{t_0}^{t_0+T_o} \left(c_k e^{jk\omega_0 t} e^{-jn\omega_0 t} \right) dt \right\} = \sum_{k=-\infty}^{\infty} \left\{ \int_{t_0}^{t_0+T_o} c_k e^{j(k-n)\omega_0 t}\,dt \right\} \tag{20.9}$$

Because $\omega_0 T_0 = 2\pi$ and $e^{j2\pi} = 1$, the integral in equation 20.9 becomes

$$\int_{t_0}^{t_0+T_o} c_k e^{j(k-n)\omega_0 t}\,dt = \begin{cases} 0 & \text{for } k \neq n \\ c_n T_0 & \text{for } k = n \end{cases} \tag{20.10}$$

Substituting equation 20.10 into equation 20.9 and dividing by T_0 yields

$$c_n = \frac{1}{T_0}\int_{t_0}^{t_0+T_o} f(t)e^{-jn\omega_0 t}\,dt \tag{20.11}$$

The lower limit of integration, t_0, can be any real number, but is usually chosen to be 0 or $-T_0/2$, whichever is more convenient. In addition, T_0 will sometimes be written simply as T.

A hand computation of coefficients would proceed by first computing c_n for $n = 0, 1, \ldots$, using equation 20.11. One would then obtain a_n and b_n using equations 20.8a and b, and d_n and θ_n by equations 20.8c, d, and e. Other formulas are available for obtaining the Fourier coefficients by integrals involving sine and cosine functions. However, equation 20.11 is preferred because an integration involving exponential functions is often simpler than an integration involving sinusoidal functions.

EXAMPLE 20.2. Find the trigonometric Fourier series for the square wave signal of Figure 20.3.

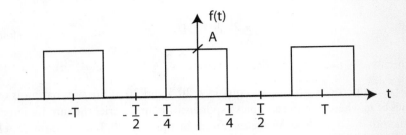

FIGURE 20.3 A square wave signal denoted $f(t)$.

SOLUTION
The fundamental period of $f(t)$ is $T_0 = T$. By inspection, the average (dc) value of $f(t)$ is

$$c_0 = d_0 = \frac{a_0}{2} = \text{average value of } f(t) = \frac{A}{2}$$

To calculate other Fourier coefficients, choose $t_0 = -T/2$. Equation 20.11 yields, for $n \neq 0$,

$$c_n = \frac{1}{T}\int_{-\frac{T}{2}}^{\frac{T}{2}} f(t)e^{-jn\omega_0 t}\,dt = \frac{1}{T}\int_{-\frac{T}{4}}^{\frac{T}{4}} Ae^{-jn\omega_0 t}\,dt = \frac{A}{T}\left[e^{-jn\omega_0 t}\right]_{-\frac{T}{4}}^{\frac{T}{4}}$$

(20.12)

$$= \frac{A}{\pi n}\sin\left(n\frac{\pi}{2}\right), n = 1, 2, \ldots$$

Thus our signal has the complex Fourier series of equation 20.6 with c_n given by equation 20.12 and

$$c_0 = d_0 = \frac{a_0}{2} = \frac{A}{2}$$

To obtain the real Fourier series, from equation 20.8,

$$a_n = 2\,\text{Re}(c_n) = \frac{2A}{\pi n}\sin\left(n\frac{\pi}{2}\right), b_n = 0$$

Substituting these coefficients into equation 20.3 yields the following trigonometric Fourier series for the square wave of Figure 20.3:

$$f(t) = \frac{A}{2} + \frac{2A}{\pi}\left(\cos(\omega_0 t) - \frac{1}{3}\cos(3\omega_0 t) + \frac{1}{5}\cos(5\omega_0 t) - \ldots\right)$$

(20.13)

TWO PROPERTIES OF THE FOURIER SERIES

After computing the Fourier series of a periodic signal $f(t)$, it is straightforward to find the Fourier series of a related periodic signal $g(t)$ whose plot is a translation of the plot of $f(t)$. A **translation of a plot** is horizontal and/or vertical movement of the plot without any rotation. A translation of the waveform in the vertical direction causes a change in the dc level and affects only the coefficients a_0, d_0, and c_0. A translation of the waveform in the horizontal direction causes a time shift that changes only the angles θ_n and has no effect on the amplitude d_n. We state the relationship formally as follows.

Translation property of Fourier series: Let c_n be the coefficients of the exponential Fourier series of a periodic function $f(t)$, and $\hat{c}_n$ be those for another periodic function $g(t)$. If $g(t)$ is a translation of $f(t)$ consisting of a dc-level increase K and a time shift (delay) to the right by t_d, then

$$g(t) = f(t - t_d) + K \tag{20.14a}$$

$$\hat{c}_0 = \hat{d}_0 = c_0 + K = d_0 + K = \frac{a_0}{2} + K \tag{20.14b}$$

$$\hat{c}_n = c_n e^{-jn\omega_0 t_d}, \quad n = \pm 1, \pm 2, \ldots \tag{20.14c}$$

$$\hat{d}_n = d_n = 2\left|c_n e^{-jn\omega_0 t_d}\right| = 2\left|c_n\right|, \quad n = \pm 1, \pm 2, \ldots \tag{20.14d}$$

$$\hat{\theta}_n = \angle\hat{c}_n = \left(\theta_n - n\omega_0 t_d\right), \quad n = 1, 2, 3, \ldots \tag{20.14e}$$

The proof of this property is straightforward and is left as an exercise. Note that equation 20.14c indicates that a time shift of the signal affects only the phase angles of the harmonics; the amplitudes of all harmonics remain unchanged.

Exercise. Suppose $f(t)$ in Example 20.2 has $A = 30$ and suppose $g(t) = f(t) - 10$. Find the coefficients a_n, b_n, and d_n of the Fourier series of $g(t)$.
ANSWER: $d_0 = 0.5a_0 = 5$; all other coefficients are unchanged, i.e., $d_n = a_n = \dfrac{2A}{\pi n}\sin\left(n\dfrac{\pi}{2}\right)$ because $b_n = 0$.

We will now use the translation property to obtain the Fourier series of a square wave that is a translation of Figure 20.3.

EXAMPLE 20.3

The square wave $g(t)$ shown in Figure 20.4 is anti-symmetrical with respect to the origin. Find the complex and real Fourier series for $g(t)$ given the Fourier series (equation 20.12) of $f(t)$ depicted in Figure 20.3.

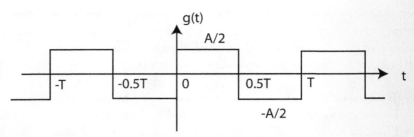

FIGURE 20.4 A square wave anti-symmetrical with respect to the origin.

SOLUTION

The curve $g(t)$ is a translation of the waveform $f(t)$ of Figure 20.3. Specifically,

$$g(t) = f(t - t_d) + K = f\left(t - \frac{T}{4}\right) - \frac{A}{2}$$

By equations 20.14 and 20.13,

$$\cos(n\omega_0 t) \rightarrow \cos(n\omega_0(t - t_d)) = \cos(n\omega_0 t - n\omega_0 t_d) = \cos\left(n\omega_0 t - \frac{n\omega_0 T}{4}\right) = \cos\left(n\omega_0 t - n\frac{\pi}{2}\right)$$

resulting in the desired Fourier series,

$$g(t) = \frac{2A}{\pi}\left\{\cos\left(\omega_0 t - \frac{\pi}{2}\right) - \frac{1}{3}\cos\left(3\omega_0 t - \frac{3\pi}{2}\right) + \frac{1}{5}\cos\left(5\omega_0 t - \frac{5\pi}{2}\right) - \dots\right\}$$

$$= \frac{2A}{\pi}\left(\sin(\omega_0 t) + \frac{1}{3}\sin(3\omega_0 t) + \frac{1}{5}\sin(5\omega_0 t) + \dots\right) \qquad (20.15$$

For the case of a square wave signal, by choosing the time origin and dc level properly, the result ant plot displays a special kind of symmetry that results in the disappearance of all sine terms c all cosine terms. The square waves of Figures 20.3 and 20.4 are special cases of the periodic func tions amenable to such simplifications. The general case is given by the following statement.

Symmetry properties of the Fourier series
(1) If a periodic function $f(t)$ is an **even function**, i.e., if $f(t) = f(-t)$, then its Fourier serie has only cosine terms and possibly a constant term.
(2) If a periodic function $f(t)$ is an **odd function**, i.e., $f(t) = -f(-t)$, then its Fourier serie has only sine terms.

The plot of an even function is symmetrical about the vertical axis. Examples of even functions include $\cos(\omega t)$ and the square wave of Figure 20.3. The plot of an odd function is anti-sym metrical about the vertical axis. Examples of odd functions are $\sin(\omega t)$ and the square wave of Figure 20.4. The proofs of the symmetry properties are left as homework problems.

ercise. Suppose $g(t)$ in Figure 20.4 from Example 20.3 is shifted into $q(t) = g(t - 0.25T)$. Find coefficients a_n, b_n, and d_n of the Fourier series of $q(t)$.

NSWER: $d_0 = a_0 = 0$, and for $n \neq 0$, $d_n = a_n = -\dfrac{2A}{\pi n}\sin\left(n\dfrac{\pi}{2}\right)$, $b_n = 0$.

simplify the calculation of the Fourier coefficients, we should *attempt* to relocate the time ori- or change the dc level so that the new function $g(t)$ displays even or odd function symmetry. is may not be possible for an arbitrary periodic signal. When it is possible, we will calculate Fourier coefficients of the new function $g(t)$, which has only cosine terms or sine terms, and en use the translation property to obtain the Fourier coefficients for the original function $f(t)$.

waveform of particular importance in signal analysis is the periodic rectangular signal shown in gure 20.5. The fundamental period is T, and the pulse width is βT. The constant β is called **duty cycle**, usually expressed as a percentage of T. The square wave of Figure 20.3 is a spe- l case of the rectangular wave of Figure 20.5 with a 50% duty cycle.

AMPLE 20.4

d the trigonometric Fourier series for the rectangular waveform of Figure 20.5.

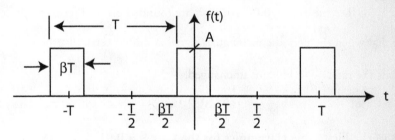

FIGURE 20.5 A periodic pulse train.

LUTION

e procedure is almost identical to that used in Example 20.2 for a square wave. By inspection, average value of $f(t)$, F_{av}, is

$$d_0 = \frac{a_0}{2} = F_{av} = \beta A \tag{20.16a}$$

calculate the other Fourier coefficients, choose $t_0 = -T/2$. Equation 20.10 then yields

$$c_n = \frac{1}{T}\int_{-\frac{T}{2}}^{\frac{T}{2}} f(t)e^{-jn\omega_0 t}dt = \frac{1}{T}\int_{-\frac{\beta T}{2}}^{\frac{\beta T}{2}} Ae^{-jn\omega_0 t}dt = \frac{A}{-jn\omega_0 T}\left[e^{-jn\omega_0 t}\right]_{-\frac{\beta T}{2}}^{\frac{\beta T}{2}}$$

$$= \frac{A}{\pi n}\sin(n\beta\pi), \quad n = 1, 2, \dots \tag{20.16b}$$

From equation 20.7, the coefficients of the trigonometric series are

$$a_n = 2\,\text{Re}(c_n) = \frac{2A}{\pi n}\sin(n\beta\pi) \tag{20.16c}$$

$$b_n = 0 \tag{20.16d}$$

Specifying the coefficients completes the determination of the Fourier series, i.e.,

$$f(t) = \frac{a_o}{2} + \sum_{n=1}^{\infty} a_n \cos(n\omega_o t) = \beta A + \frac{2A}{\pi}\sum_{n=1}^{\infty}\left(\frac{\sin(n\beta\pi)}{n}\right)\cos(n\omega_o t)$$

A very important conclusion about the rectangular wave can be drawn by examining equation 20.16: as the ratio of the pulse width to the period becomes very small, the magnitudes of the fundamental and all harmonic components converge to twice the average (dc) values. To see this recall that $\sin(x)/x$ approaches 1 as x approaches 0. From equation 20.16c, we may rewrite a_n as

$$a_n = \frac{2A}{\pi n}\sin(n\beta\pi) = 2\beta A\frac{\sin(n\beta\pi)}{n\beta\pi}.$$

It follows that $a_n \rightarrow 2\beta A$ as $\beta \rightarrow 0$.

To give some concrete feel to this property, Table 20.2 gives the ratios of $\dfrac{d_n}{F_{av}} = \dfrac{|a_n|}{F_{av}}$ and $\dfrac{d_n}{d_1}$

for $n = 1, ..., 9$ for the case of $\beta = 0.01$. Answers are rounded off to three digits after the decimal point. Note that when the periodic rectangular signal is shifted vertically, the ratio $\dfrac{d_n}{F_{av}} = \dfrac{|a_n|}{F_{av}}$

is affected, but the ratio $\dfrac{d_n}{d_1}$ remains unchanged.

TABLE 20.2

Amplitudes of the First Nine Harmonics for the Case β = 0.01

n	1	2	3	4	5	6	7	8	9
$\dfrac{d_n}{F_{av}} = \dfrac{|a_n|}{F_{av}}$	2.000	1.999	1.998	1.995	1.992	1.989	1.984	1.979	1.974
$\dfrac{d_n}{d_1}$	1.000	1.000	0.999	0.998	0.996	0.994	0.992	0.990	0.987

The constant d_n property holds approximately true when the pulse width is a very small fraction of the period T_0. *Even if a waveform is not rectangular*, if the pulse width is very small compared to its period, then the nearly constant property of $\dfrac{d_n}{F_{av}}$ and $\dfrac{d_n}{d_1}$ continues to hold, as long as the pulse is of a single polarity. For example, consider the periodic short pulse shown in Figure 20.6. In calculating the Fourier coefficients c_n of this pulse using equation 20.11, the limits of the integral, originally $(t_0,\ t_0 + T)$, are changed to $(-\beta T/2,\ \beta T/2)$. As β approaches zero, the factor $e^{-jn\omega_0 t}$ in the integrand has a value very close to 1 in the new time interval, as long as n, the harmonic order being considered, is not very high. Therefore, for *pulses of narrow width* we have

$$c_n = \frac{1}{T}\int_{-\frac{\alpha T}{2}}^{\frac{\alpha T}{2}} f(t)e^{-jn\omega_0 t}\,dt \cong \frac{1}{T}\int_{-\frac{\alpha T}{2}}^{\frac{\alpha T}{2}} f(t)\,dt = F_{av} \qquad (20.17a)$$

Thus in terms of the d_n coefficients, from equation 20.8, again for pulses of narrow width we have

$$d_n = 2|c_n| \cong 2F_{av}, \quad n = 1,\ 2,\dots \qquad (20.17b)$$

This result is pertinent to the approximate analysis of a rectifier circuit covered in other texts or in the second edition of this text.

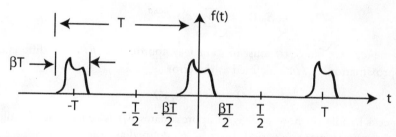

FIGURE 20.6 Periodic positive short pulses of an arbitrary periodic waveform.

To this point, we have calculated the Fourier coefficients only for some very simple periodic signals. The evaluation of the integral in equation 20.11 becomes much more involved when the signal $f(t)$ is not rectangular. Fortunately, many mathematical and engineering handbooks now include comprehensive tables of Fourier series. From a utility point of view, one may use these tables much the same as one uses a table of integrals or a table of Laplace transforms. In effect, the need to carry out the integration in equation 20.11 is not compelling in practice.

In many applications, it is important to know the average power of a (periodic) signal and the magnitude of its various harmonics. From equation 20.5, the **effective value** of the dc component is $|d_0|$, and those for the fundamental and various harmonics are $d_n/\sqrt{2}$, $n = 1, 2, \dots$. It is easy to show that the effective value, or the rms value, of $f(t)$ is

$$F_{eff} = \sqrt{d_0^2 + \frac{1}{2}d_1^2 + \frac{1}{2}d_2^2 + \dots} = \sqrt{\sum_{n=-\infty}^{\infty} |c_n|^2} \qquad (20.18a)$$

where the d_i coefficients are from the Fourier series of equation 20.4. If $f(t)$ (current or voltage) drives a 1 ohm resistor, then the average power absorbed by the resistor is $P_{ave} = F_{eff}^2 \times 1 = F_{eff}^2$. Hence we say that the average power of a periodic signal $f(t)$ represented by a Fourier series is given by

$$P_{ave, f(t)} = F_{eff}^2 = d_0^2 + \frac{1}{2}d_1^2 + \frac{1}{2}d_2^2 + \ldots = \sum_{n=-\infty}^{\infty} |c_n|^2 \qquad (20.18b)$$

The relationships indicated in equations 20.18 are often termed *Parseval's theorem*.

The information on the phase angle is important when one wishes to construct the time domain response in steady state. For the time domain problem, the Fourier series method yields only an approximate solution, because one can only sum a finite number of terms in the series.

Convergence of the Fourier Series

Convergence of the Fourier series is an intricate mathematical problem, the details of which are beyond the scope of this text. On the other hand, it is important to be aware of the ways in which the Fourier series may or may not converge to a given $f(t)$. Our discussion is not comprehensive, but is adequate for our present purposes.

To begin, we define a partial sum of terms of the complex Fourier series of a function $f(t)$ as

$$S_N(t) = \sum_{k=-N}^{N} c_k e^{jk\omega_0 t}$$

From our experience thus far, $S_N(t)$ must in some way approximate $f(t)$. The difference between $f(t)$ and its approximation, $S_N(t)$, is defined as the error

$$e_N(t) = f(t) - S_N(t)$$

One way to get a handle on how well $S_N(t)$ approximates $f(t)$ is to use the so-called integral squared-error magnitude over one period, $[t_0, t_0 + T]$, defined as

$$E_N = \int_{t_0}^{t_0 + T} |e_N(t)|^2 \, dt$$

This is often called the energy in the error signal, as energy is proportional to the integral of the squared magnitude of a function. It turns out that for functions having a Fourier series, the choice of the Fourier coefficients minimizes E_N for each N. Further, for such functions, $E_N \to 0$ as $N \to \infty$, i.e., the energy in the error goes to zero as N becomes large. This does not mean that at each t, $f(t)$ and its Fourier series are equal; it merely means that the energy in the error goes to zero.

Continuous and piecewise continuous periodic functions have Fourier series representations. A piecewise continuous function, such as a square wave, is a function that (1) has a finite number of discontinuities over each period but is otherwise continuous, and (2) has well-defined right- and left-hand limits as the function approaches a point of discontinuity. For piecewise continuous functions, it turns out that the Fourier series converges to a value halfway between the values of the left- and right-hand limits of the function around the point of discontinuity. Even so, $E_N \to 0$ as $N \to \infty$ for piecewise continuous functions.

There are many functions that are not piecewise continuous and yet have a Fourier series. A set of conditions that is sufficient, but not necessary, for a function to have a Fourier series representation is called the *Dirichlet conditions*.

Dirichlet conditions

Condition 1. Over any period, $[t_0, t_0+T]$, $f(t)$ must have the property that

$$\int_{t_0}^{t_0 + T} |f(t)| dt < \infty$$

In the language of mathematics, this means that $f(t)$ is *absolutely integrable*. The consequence of this property is that each of the Fourier coefficients c_n is finite, i.e., the c_n exist.

Condition 2. Over any period of the signal, there must be only a finite number of minima and maxima. In other words, functions like $\sin(1/t)$ are excluded. In the language of mathematics, a function that has only a finite number of maxima and minima over any finite interval is said to be of *bounded variation*.

Condition 3. Over any period, $f(t)$ can have only a finite number of discontinuities.

As mentioned, at points of discontinuity, the Fourier series will converge to a value midway between the left- and right-hand values of the function next to the discontinuity. There may be other differences as well. Despite these differences, the energy between the function $f(t)$ and its Fourier series representation is zero; i.e., E_N, with N approaching ∞, goes to zero. Thus, for all practical purposes, the functions are identical. This practical equivalence allows us to analyze how a circuit responds to a signal $f(t)$ by analyzing how the circuit responds to each of its Fourier series components.

3. ADDITIONAL PROPERTIES AND COMPUTATIONAL SHORTCUTS FOR THE FOURIER SERIES REPRESENTATION

If a periodic function $f(t)$ is known only at some sampled points, e.g., by measurements, then its Fourier coefficients must be calculated by the use of numerical methods. On the other hand, if $f(t)$ has an analytic expression, then its Fourier coefficients can often be calculated from equation 20.11. The properties discussed below are of great value in simplifying the calculation of Fourier coefficients. Their proofs are fairly straightforward and are left as homework problems.

The linearity property: Let $f_1(t)$ and $f_2(t)$ be periodic with fundamental period T. If $f(t)$ = $K_1 f_1(t) + K_2 f_2(t)$, then the Fourier coefficients of $f(t)$ may be expressed in terms of those of $f_1(t)$ and $f_2(t)$ according to the following formulas:

$$c_n = K_1 c_{1n} + K_2 c_{2n} \tag{20.19a}$$

$$a_n = K_1 a_{1n} + K_2 a_{2n}, \tag{20.19b}$$

and

$$b_n = K_1 b_{1n} + K_2 b_{2n} \tag{20.19c}$$

In general,

$$d_n \neq K_1 d_{1n} + K_2 d_{2n}$$

unless the angle θ_n is the same for all n.

This property allows us to easily obtain the Fourier series of a sum of periodic signals when the Fourier series of the individual signals is already known.

DEFINITION
A periodic function $f(t)$ is said to be **half-wave symmetric** if

$$f(t - 0.5T) = -f(t) \text{ for all } t$$

In words, $f(t)$ is half-wave symmetric if a half-period shift of the plot combined with a flip about the horizontal axis results in the identical function $f(t)$. Some simple examples of half-wave symmetric functions are $\sin(\omega t)$, $\cos(\omega t)$, and the square wave of Figure 20.4.

> **The half-wave symmetry property:** A half-wave symmetric periodic function $f(t)$ contains only odd harmonics.

The waveform of Figure 20.4 illustrates this property.

> **The derivative/integration property.** Let $f^{(k)}(t)$ denote the kth derivative of a periodic function $f(t)$. Then the Fourier coefficients, $c_n^{(k)}$, of $f^{(k)}(t)$ of satisfy

$$c_n^{(k)} = \left(jn\omega_0 \right)^k c_n \text{ for all } n \tag{20.20a}$$

$$c_n = \frac{c_n^{(k)}}{\left(jn\omega_0 \right)^k} \text{ for all } n \text{ except } n = 0 \tag{20.20b}$$

Except for the constant term, all terms of $f^{(k)}(t)$ derive from those of $f(t)$ by differentiating k times conversely, all terms of $f(t)$ derive from those of $f^{(k)}(t)$ by k-fold (indefinite) integration. The exclusion of the constant term in the relationships poses no difficulty, since the constant is simply th average value of the periodic function.

Again, these properties help to simplify the calculation of Fourier coefficients. In fact, these prop erties make it possible to compute the Fourier series for the waveforms given later in Table 20. without carrying out the integration of equation 20.11. Achieving this, however, depends on firs finding the Fourier series of a waveform for which computing c_n by equation 20.11 is extremel easy. We will illustrate several such calculations in Examples 20.5 through 20.8. During this devel opment, the following trigonometric identities will prove useful:

$$\cos(x) - \cos(y) = -2\sin\left(\frac{x + y}{2}\right) \sin\left(\frac{x - y}{2}\right) \tag{20.21a}$$

$$\cos(x)\cos(y) = 0.5[\cos(x + y) + \cos(x - y)] \tag{20.21b}$$

EXAMPLE 20.5. Find the Fourier series for the periodic impulse train $f_\delta(t)$ shown in Figure 20.7.

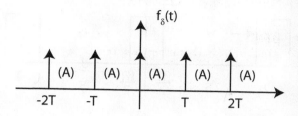

FIGURE 20.7 A periodic impulse train.

SOLUTION

Using the sifting property of an impulse function together with equation 20.10 yields

$$c_n = \frac{1}{T} \int_{-\frac{T}{2}}^{\frac{T}{2}} A\delta(t)e^{-jn\omega_0 t}\,dt = \frac{A}{T} \tag{20.22a}$$

for all n. Hence

$$f_\delta(t) = \frac{A}{T} \sum_{n=-\infty}^{\infty} e^{jn\omega_0 t} = \frac{A}{T} + \frac{2A}{T} \sum_{n=1}^{\infty} \cos(n\omega_0 t) \tag{20.22b}$$

Equation 20.22b states that, for a periodic impulse train, all harmonic components have magnitude equal to twice the average value. This is the limiting case of the short pulse property stated in section 2. The next example uses the derivative and integral properties of the Fourier series to develop an alternative derivation of the Fourier series for a periodic rectangular pulse train, derived earlier by the use of equation 20.11.

EXAMPLE 20.6. Find the Fourier series for the periodic rectangular pulses $f_p(t)$ shown in Figure 20.8a. (This corresponds to item 2 in Table 20.3.)

SOLUTION

Figures 20.8a and b show $f_p(t)$ and its derivative, $f_p'(t)$. The latter may be written as the sum of two shifted impulse trains:

$$f_p'(t) = f_\delta\left(t + \frac{\beta T}{2}\right) - f_\delta\left(t - \left(\frac{\beta T}{2}\right)\right) \tag{20.23}$$

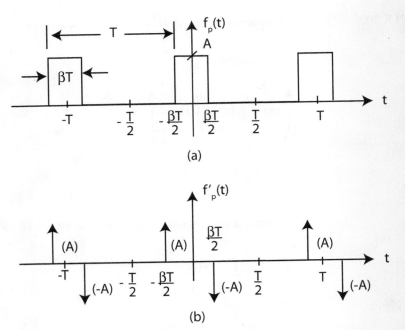

FIGURE 20.8 (a) Periodic rectangular pulse train and (b) its derivative. The parenthetical values, (A) and (−A), next to each impulse in part (b) denote the weight of that impulse, i.e., its area.

Using the time shift property (translation in the horizontal direction), together with equation 20.22b and 20.21a,

$$f_p'(t) = \sum_{n=1}^{\infty} \frac{2A}{T}\left(\cos\left(n\omega_0 t + n\beta\pi\right) - \cos\left(n\omega_0 t - n\beta\pi\right)\right)$$

(20.24)

$$= \sum_{n=1}^{\infty} \frac{-4A}{T}\left(\sin\left(n\omega_0 t\right)\sin\left(n\beta\pi\right)\right)$$

Applying the derivative/integration property to equation 20.24 yields

$$f_p(t) = \left[f_p(t)\right]_{ave} + \sum_{n=1}^{\infty} \frac{4A}{n\omega_0 T}\left(\cos\left(n\omega_0 t\right)\sin\left(n\beta\pi\right)\right)$$

(20.2)

$$= \alpha A + \sum_{n=1}^{\infty} \frac{2A\sin\left(n\beta\pi\right)}{n\pi}\cos\left(n\omega_0 t\right)$$

The result agrees, of course, with equation 20.16.

XAMPLE 20.7. Find the Fourier series for the half-wave rectified sine wave $f_{hs}(t)$ shown in gure 20.9a (item 10 in Table 20.3).

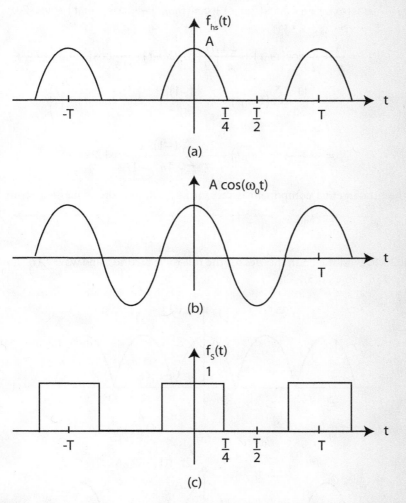

(a)

(b)

(c)

FIGURE 20.9 (a) Half-wave rectified sine wave as the product of two functions: (b) a cosine wave and (c) a square wave.

OLUTION

he periodic function $f(t)$ of Figure 20.9a may be viewed as the product of the sinusoidal wave $\cos(\omega_0 t)$ and the square wave $f_s(t)$, shown in Figures 20.9b and c, respectively. Using the Fourier ries for $f_s(t)$ given by equation 20.13, we have

$$f_{hs}(t) = \left[A\cos(\omega_0 t)\right] f_s(t)$$

$$= A\cos(\omega_0 t)\left\{\frac{1}{2} + \frac{2}{\pi}\left(\cos(\omega_0 t) - \frac{1}{3}\cos(3\omega_0 t) + \frac{1}{5}\cos(5\omega_0 t) - \ldots\right)\right\} \tag{20.26}$$

Applying equation 20.21b to each product in equation 20.26 yields

$$f_{hs}(t) = \frac{A}{2}\cos(\omega_0 t) + \frac{A}{\pi}\left\{\cos(2\omega_0 t) + \cos(0\omega_0 t) - \frac{1}{3}\cos(4\omega_0 t) + \cos(2\omega_0 t) - \ldots\right\}$$

$$= \frac{A}{\pi} + \frac{A}{2}\cos(\omega_0 t) + \frac{2A}{\pi}\left(\frac{1}{3}\cos(2\omega_0 t) - \frac{1}{15}\cos(4\omega_0 t)\right.$$

$$\left. + \frac{1}{35}\cos(6\omega_0 t) + \ldots + \frac{(-1)^{n+1}}{4n^2-1}\cos(2n\omega_0 t) + \ldots\right)$$

$$= \frac{A}{\pi} + \frac{A}{2}\cos(\omega_0 t) + \frac{2A}{\pi}\sum_{n=1}^{\infty}\frac{(-1)^{n+1}}{4n^2-1}\cos(2n\omega_0 t) \qquad (20.27$$

Note that the fundamental component is present in $f_{hs}(t)$, and the remaining terms are all eve
harmonics.

EXAMPLE 20.8. Find the Fourier series for the full-wave rectified sine wave $f_{fs}(t)$ shown in Figur
20.10a (item 9 in Table 20.3).

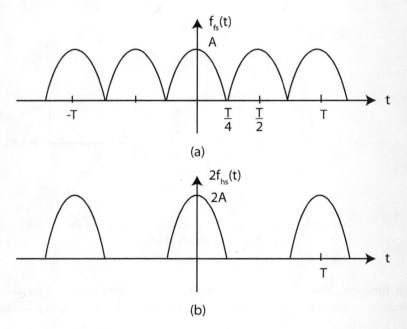

(a)

(b)

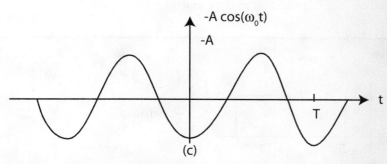

FIGURE 20.10 (a) Full-wave rectified sine wave as the sum of the two signals in (b) and (c).

SOLUTION

One approach is to apply the same technique as in Example 20.7. Specifically,

$$f_{fs}(t) = A \cos(\omega_0 t)[2f_s(t) - 1]$$

where $f_s(t)$ is the square wave of Figure 20.9c. Alternatively, to avoid repeating all the arithmetic, note that $f_{fs}(t)$ is the sum of the two waveforms shown in Figures 20.10b and c; i.e.,

$$f_{fs}(t) = 2f_{hs}(t) - A \cos(\omega_0 t) \tag{20.28}$$

Substituting equation 20.27, the Fourier series $f_{hs}(t)$, into equation 20.28 yields the desired Fourier series:

$$f_{fs}(t) = \frac{2A}{\pi} + \frac{4A}{\pi}\left(\frac{1}{3}\cos(2\omega_0 t) - \frac{1}{15}\cos(4\omega_0 t) + \frac{1}{35}\cos(6\omega_0 t) \right. \tag{20.29}$$

$$\left. \cdots - \frac{(-1)^{n+1}}{4n^2 - 1}\cos(2n\omega_0 t) + \ldots \right)$$

The derivative property is particularly useful for tackling periodic piecewise linear waveforms. Piecewise linear waveforms consist of straight-line segments. From differentiating once, or at most twice, impulses appear. The integration given by equation 20.11 is trivial if the integrand contains a shifted impulse function. This fact, together with the derivative and integral properties, reduce the task of calculating the Fourier coefficients for piecewise linear waveforms to some complex number arithmetic. These examples and several other commonly encountered periodic waveforms have Fourier series as given in Table 20.3. Engineering and mathematical handbooks contain much more comprehensive tables. Of course, when a waveform does not appear in a table, the Fourier coefficients must be computed manually or numerically.

TABLE 20.3

1. Square wave 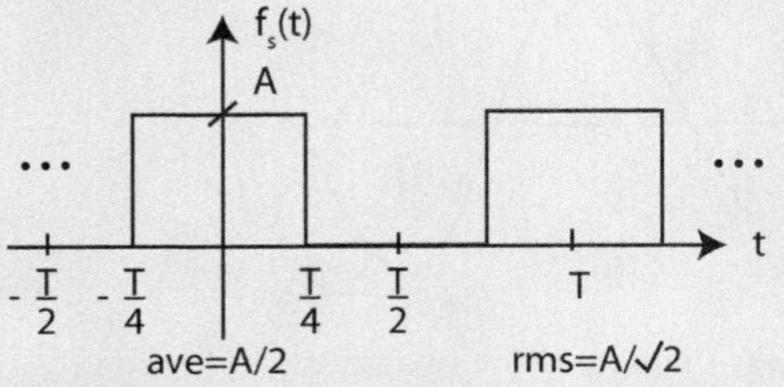	$f(t)=\frac{A}{2}+\frac{2A}{\pi}\left(\cos(\omega_0 t)-\frac{1}{3}\cos(3\omega_0 t)+\frac{1}{5}\cos(5\omega_0 t)-...\right)$ Note: $\lvert a_n \rvert \approx 1/n$
2. Rectangular wave 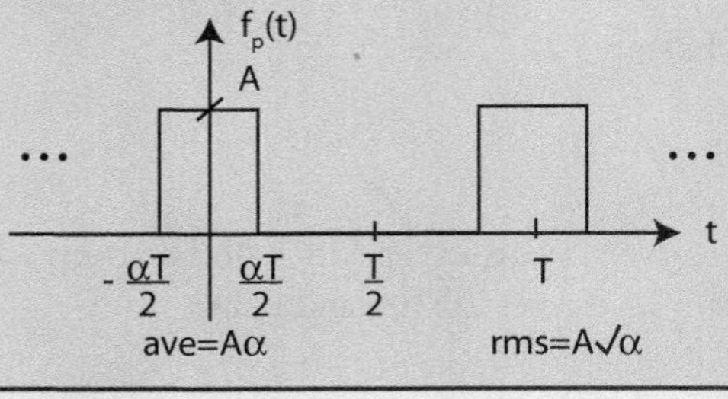	$f(t)=\beta A+\frac{2A}{\pi}\sum_{n=1}^{\infty}\left(\frac{\sin(n\beta\pi)}{n}\right)\cos(n\omega_o t)$
3. Triangular wave 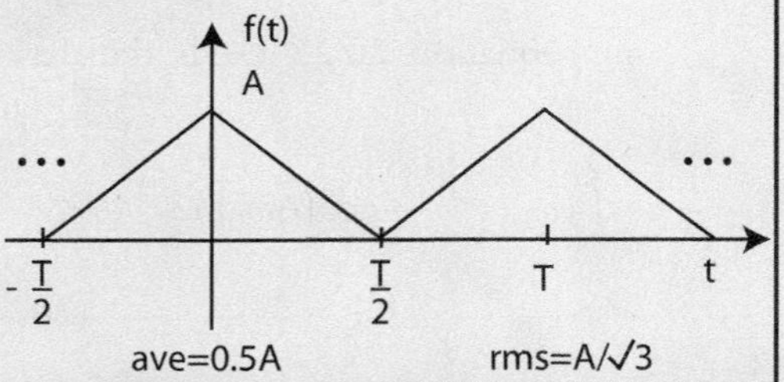	$f(t)=\frac{A}{2}+\frac{4A}{\pi^2}\left(\cos(\omega_0 t)+\frac{\cos(3\omega_0 t)}{9}+\frac{\cos(25\omega_0 t)}{25}+...\right)$
4. Isoceles triangular wave 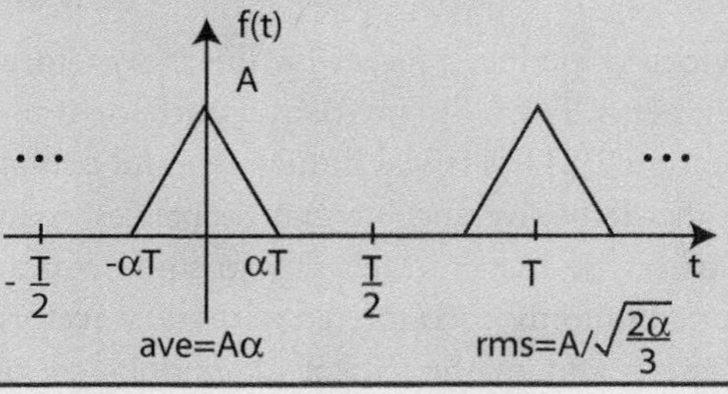	$0.5a_0=\alpha A,\ a_n=\frac{2A}{\alpha\pi^2}\left[\frac{\sin(\alpha n\pi)}{n}\right]^2$ $b_n=0$
5. Sawtooth wave 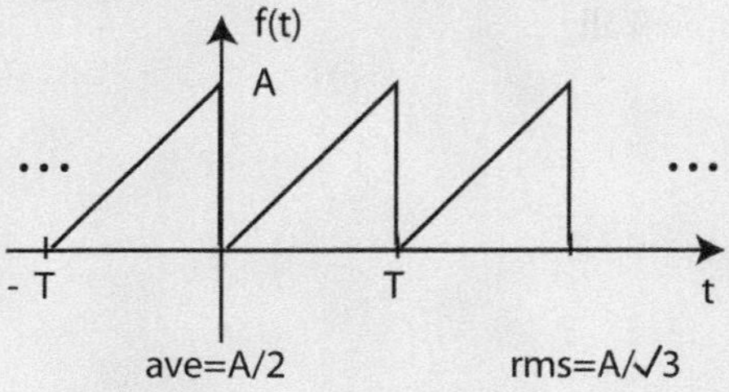	$f(t)=\frac{A}{2}-\frac{A}{\pi}\left(\sin(\omega_0 t)+\frac{\sin(2\omega_0 t)}{2}+\frac{\sin(3\omega_0 t)}{3}+...\right)$ Note : $\lvert b_n \rvert \approx 1/n$

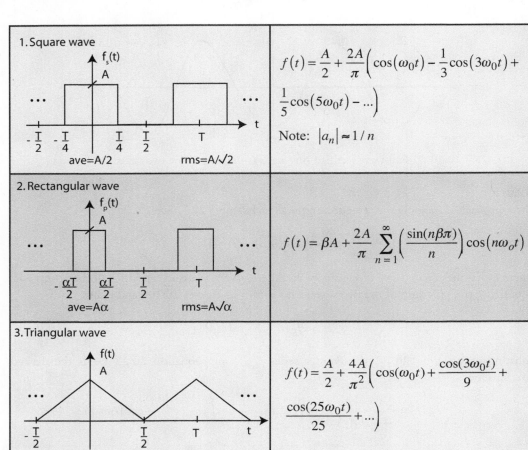

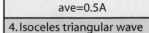

6. Clipped sawtooth wave 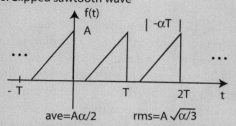 ave=$A\alpha/2$ rms=$A\sqrt{\alpha/3}$	$d_0 = F_{av} = 0.5\alpha A$ $d_n = \dfrac{A}{\alpha\pi^2 n^2}\left[\sin^2(n\alpha\pi) + n\alpha\pi\left(n\alpha\pi - \sin(2n\alpha\pi)\right)\right]^{0.5}$ θ_n : too complicated for inclusion
7. Asymmetrical triangular wave 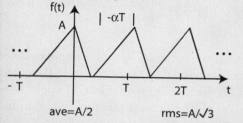 ave=$A/2$ rms=$A/\sqrt{3}$	$d_0 = F_{av} = 0.5A$ $d_n = \dfrac{A}{\alpha(1-\alpha)\pi^2 n^2}\sin(n\alpha\pi)$ θ_n : too complicated for inclusion
8. Symmetrical trapezoidal wave 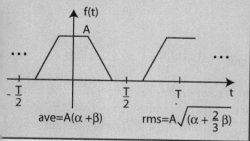 ave=$A(\alpha+\beta)$ rms=$A\sqrt{(\alpha+\frac{2}{3}\beta)}$	$0.5a_0 = F_{av} = (\alpha+\beta)A$ $a_n = (\alpha+\beta)A\dfrac{\sin(n\beta\pi)}{n\beta\pi} \times \dfrac{\sin\left(n(\alpha+\beta)\pi\right)}{n(\alpha+\beta)\pi}$ $b_n = 0$
9. Full-wave rectified sine wave $f(t) = \lvert A\cos\omega t\rvert$ 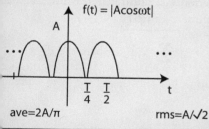 ave=$2A/\pi$ rms=$A/\sqrt{2}$	$T_0 = 0.5T, \quad \omega_0 = 2\omega$ $f(t) = \dfrac{4A}{\pi}\left(0.5 + \dfrac{\cos(2\omega t)}{3} - \dfrac{\cos(4\omega t)}{15} + \dfrac{\cos(6\omega t)}{35}\right.$ $\left. \dots + \dfrac{(-1)^{n+1}\cos(2n\omega t)}{4n^2 - 1} + \dots \right)$
10. Half-wave rectified sine wave $f(t) = A\cos\omega t\, f_s(t)$ 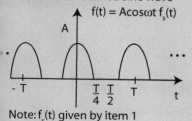 Note: $f_s(t)$ given by item 1 ave=A/π rms=$A/2$	$f(t) = \dfrac{A}{\pi} + \dfrac{A}{2}\cos(\omega_0 t) +$ $\dfrac{2A}{\pi}\sum_{n=1}^{\infty}\dfrac{(-1)^{n+1}}{4n^2 - 1}\cos(2n\omega_0 t)$

4. SUMMARY

Given that many mathematical and engineering handbooks have extensive lists of the Four
series of common signals, this chapter has taken a practical approach to the calculation of t
Fourier series and its application to circuit analysis. The idea is to use tables such as Table 20.3
the same way engineers have come to use integral tables. The keys to using such tables for the cor
putation of the Fourier series of a waveform are the various properties that allow one to conver
known series into one that fits a new waveform. The idea here is to express the new waveform
a translation of, a linear combination of, a k-fold derivative of, or a k-fold integral of signals wi
known Fourier series as in Table 20.3, or any mixture of these operations. The Fourier coefficier
of the new signal can then be expressed in terms of the Fourier coefficients of signals with knov
Fourier series.

Knowledge of the Fourier coefficients of a signal such as the output of an audio amplifier allo
one to investigate phenomena including the distortion introduced by the amplifier. In the case
a dc power supply, such knowledge allows us to characterize the degree of unwanted ripple in t
output of a rectifier circuit. In addition, the application of Fourier series plays an important r
in the computation of steady-state circuit responses to periodic input signals.

5. TERMS AND CONCEPTS

Average value of periodic $f(t)$: the term d_0 in equation 20.4; also referred to as the dc compo
nent of $f(t)$.

Derivative/integration property: Let $f^{(k)}(t)$ denote the kth derivative of a periodic function $f($
Then the Fourier coefficients of the kth derivative, $c_n^{(k)}$, satisfy $c_n^{(k)} = (jn\omega_0)^k c_n$ for all
and conversely, if $f(t)$ is the kth integral of $f^{(k)}(t)$, then $c_n = c_n^{(k)}/(jn\omega_0)^k$ for all n exce
0.

Duty cycle: for a rectangular signal having fundamental period T as illustrated in Figure 20.5, t
duty cycle is the constant β that determines the pulse width βT.

Effective value (rms value): $F_{eff} = \sqrt{d_0^2 + \frac{1}{2}d_1^2 + \frac{1}{2}d_2^2 + ...} = \sqrt{\sum_{n=-\infty}^{\infty} |c_n|^2}$, where the d_n

coefficients are from the Fourier series of equation 20.4 and the c_n are from equati
20.6b.

Even function: $f(t) = f(-t)$. The plot of an even function is symmetrical about the vertical axis.

Exponential (complex) Fourier series: decomposition of $f(t)$ into a sum of complex exponenti
as given in equations 20.6.

Fundamental component (first harmonic): the first term under the summation sign in equa
tion 20.4, $d_1 \cos(\omega_0 t + \theta_1)$, having amplitude d_1 and phase angle θ_1.

Fundamental frequency (in hertz): $f_0 = \dfrac{1}{T_0}$, where T_0 is the fundamental period. Note:
$\omega_0 = 2\pi f_0 = \dfrac{2\pi}{T_0}$ is the fundamental angular frequency in rad/sec.

Fundamental period: the smallest positive real number T_0 for which $f(t + T_0) = f(t)$.

half-wave symmetric: refers to a periodic function $f(t)$ that satisfies $f(t - 0.5T) = -f(t)$.

half-wave symmetry property: A half-wave symmetric periodic function $f(t)$ contains only odd harmonics.

linearity property: let $f_1(t)$ and $f_2(t)$ be periodic with fundamental period T. If $f(t) = K_1 f_1(t) + K_2 f_2(t)$, then the Fourier coefficients of $f(t)$ may be expressed in terms of those of $f_1(t)$ and $f_2(t)$ according to the formulas $c_n = K_1 c_{1n} + K_2 c_{2n}$, $a_n = K_1 a_{1n} + K_2 a_{2n}$, and $b_n = K_1 b_{1n} + K_2 b_{2n}$. In general, $d_n \neq K_1 d_{1n} + K_2 d_{2n}$ unless the angle θ_n is the same for all n.

odd function: $f(t) = -f(-t)$.

periodic signal, $f(t)$: A signal whose waveform repeats every T seconds. Mathematically, for some $T > 0$, and all t, , $f(t + T) = f(t)$ where T is the period of the signal.

symmetry properties of Fourier series: (1) If a periodic function $f(t)$ is an even function, then its Fourier series has only cosine terms and possibly a constant term. (2) If a periodic function $f(t)$ is an odd function, then its Fourier series has only sine terms.

translation of a plot: horizontal and/or vertical movement of the plot of a function without any rotation.

translation property of Fourier series: if $f(t)$ is a translation of $f(t)$ consisting of a dc level increase K and a time shift (delay) to the right by t_d, then $g(t) = f(t - t_0) + K$. See equation 20.14.

trigonometric Fourier series: representation of a periodic signal $f(t)$ in terms of sines and cosines, as given in equations 20.3 or 20.4.

Problems

ANSWER: (a) $c_n = 0.5/(\log_e 2 + j2n\pi)$; (b) $f($
$= -0.7213 + 0.158 \cos(2\pi t - 83.7^o) +$
$0.0795 \cos(4\pi t - 86.84^o) + \ldots$

FOURIER SERIES COEFFICIENTS BY INTEGRATION

1. Find the Fourier series for the periodic impulse trains shown in Figure P20.1. Carry out the integration given by equation 20.10 and write the Fourier series in the form of ··· equations 20.3.

3. Repeat Problem 2 for the signal of Figur P20.3.

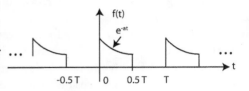

Figure P20.3

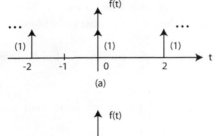

(a)

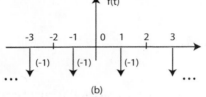

(b)

Figure P20.1

ANSWER: (a) $0.5 + \cos(\pi t) + \cos(2\pi t) + \cos(3\pi t) + \cos(4\pi t) + \ldots$; (b) $-0.5 + \cos(\pi t) - \cos(2\pi t) + \cos(3\pi t) - \cos(4\pi t) + \cos(5\pi t) \ldots$

4. Using trig identities, compute the Fourie series coefficients d_n and c_n for the func tions

   (a) $f(t) = \cos(4t) \sin(2t)$
   (b) $f(t) = \sin^2(4t) \cos^2(8t)$
   (c) $f(t) = [2 + 1.5 \sin(500t)$
            $- 2 \cos(2000t)]\cos(10^6 t)$

FOURIER SERIES USING PROPERTIES WITHOUT INTEGRATION

5. Construct the Fourier series of item 5 i Table 20.3;, i.e., find the Fourier series of th sawtooth waveform in Figure P20.5. Hin Note that $f'(t) = A - Af_\delta(t)$. Use equatio 20.22b for $f_\delta(t)$.

2. For the periodic function $f(t)$ shown in Figure P20.2, $T = 1$ second and $a = \log_e 2 = 0.693$.

   (a)   Find the coefficient c_n by carrying out the integration given by equation 20.6b.
   (b)   Write the first three terms of the Fourier series (see equation 20.4).

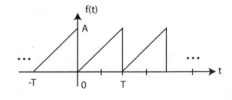

Figure P20.5

6. Find the Fourier coefficients d_n and θ_n of th periodic function shown in Figure P20.6. Hin Make use of the result of Problem 5.

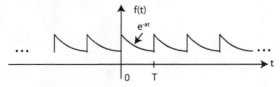

Figure P20.2

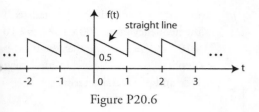

Figure P20.6

ANSWER: $f(t) = 0.75 + 0.159 \sin(2\pi t) + 0.0795 \sin(4\pi t) + \cdots$

. Construct item 3 of Table 20.3; i.e., compute the Fourier series for the triangular wave shown in Figure P20.7. Hint: $f'(t)$ is a square wave whose Fourier series was calculated before.

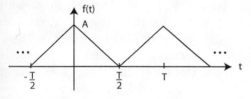

Figure P20.7

. Consider the isosceles triangular wave shown in Figure P20.8, which is item 4 of Table 20.3. Compute the Fourier series of this waveform utilizing the results for items 2 and 3 of Table 20.3. Hint: $f(t)$ is the product of (1) the waveform of item 2 with proper height and (2) the waveform of item 3 shifted down by a suitable amount.

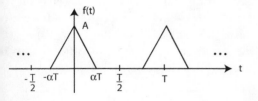

Figure P20.8

. Consider the clipped sawtooth wave shown in Figure P20.9, which is item 6 of Table 20.3. Compute the Fourier series of this waveform utilizing the results of items 2 and 5 of Table 20.3. Hint: $f(t)$ is the product of (1) the wave-

form of item 2 with proper height and (2) the waveform of item 5 shifted down by a suitable amount.

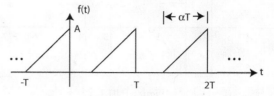

Figure P20.9

10. Find the Fourier series for the waveform shown in Figure P20.10. Hint: Make use of the result of Problem 9.

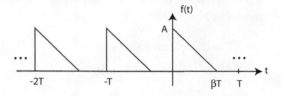

Figure P20.10

11. Consider the asymmetrical triangular wave shown in Figure P20.11, which is item 7 of Table 20.3. Let $T = 1$ and $\alpha = 0.25$. Find the Fourier coefficients d_0, d_1, and d_2. Hint: Differentiating $f(t)$ twice results in periodic impulse trains.

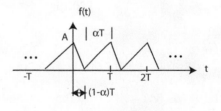

Figure P20.11

12. Find the Fourier series coefficients c_0, c_1, and c_2 for the periodic waveform shown in Figure P20.12. Hint: The period $T = 4$, and you should express $f(t)$ as the sum of several shifted square waves.

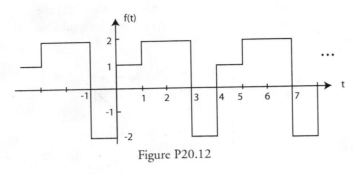

Figure P20.12

$\omega_c = 10^6$ rad/sec and $\omega_m = 10^3$ rad/sec. For the frequency range $990 \times 10^3 < \omega < 1010 \times 10^3$ rad/sec, assume that the band-pass amplifier has the following ideal magnitude and phase characteristics:

$$|H(j\omega)| = 10$$
$$\text{and}$$
$$\angle H(j\omega) = -\frac{\omega - 10^6}{5000} \times 45^o$$

(a) Express $v_{in}(t)$ as the sum of all of its frequency components.

(b) Express $v_{out}(t)$ as the sum of all of its frequency components.

(c) If $v_{out}(t)$ is expressed in the form $v_{out}(t) = g(t) \times \cos(\omega_0 t)$ V, show that $g(t) = 10f(t - t_d)$, where $t_d = 78.54$ sec. The interpretation of this result is that the envelope of $v_{out}(t)$ is an ampli-fied and delayed version of the enve-lope of $v_{in}(t)$; the amplification is 10 and time delay is 78.54 μsec.

SINUSOIDAL STEADY-STATE ANALYSIS WITH MULIT-FREQUENCY INPUTS

13. The input to the low-pass filter circuit shown in Figure P20.13 is

$$v_{in}(t) = 200 + 200\sqrt{2} \cos(377t) + 60\sqrt{2} \cos(3 \times 377t + 30^o)$$
$$+ 80\sqrt{2} \cos(5 \times 377t + 50^o) \text{ V}$$

(a) Find $v_{out}(t)$ at steady state.

(b) Find the rms value of $v_{out}(t)$ and the average power absorbed by the 10 kΩ resistor.

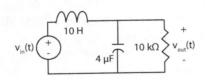

Figure P20.13

ANSWER: (a)

$$v_{out}(t) = 200 + \sqrt{2}\left[42.6\cos(\omega t - 175.4^o) + \right.$$
$$\left. 1.192\cos(3\omega t - 148.7^o) + 0.56\cos(5\omega t - 129.2^o)\right] \text{ V}$$

with $\omega = 377$ rad/sec; (b) 204.5 V, 4.18 W

14. Consider the circuit of Figure P20.14. The input to this ideal band-pass amplifier is an amplitude-modulated waveform given by $v_{in}(t) = f(t) \times \cos(\omega_c t)$ V, where

$$f(t) = 2[1 - 0.2 \cos(\omega_m t) + 0.1 \cos(2\omega_m t)]$$

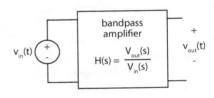

Fig P20.14

ANSWER: (a) $v_{in}(t) = 0.2 \cos((\omega_c - \omega_m)t)$ 0.1 $\cos((\omega_c - 2\omega_m)t) + 2 \cos(\omega_c t) + 0.2 \cos((\omega_c + \omega_m)t) + 0.1 \cos((\omega_c + 2\omega_m)t)$ V; (b) $v_{out}(t)$ 2 $\cos((\omega_c - \omega_m)t + 4.5^o) + \cos((\omega_c - 2\omega_m)t$ 9°) + 20 $\cos(\omega_c t) + 2 \cos((\omega_c + \omega_m)t$ 4.5°) + $\cos((\omega_c + 2\omega_m)t - 9^o)$ V

15. Consider the circuit of Figure P20.15, which $C = 1$ F (initially relaxed), $R = 1.443$ $1/ln(2)$ Ω, and $i_{in}(t)$ is a sequence of impulse currents $Q\delta(t - nT)$, $T = 1$ sec, $Q = 1$ coulomb and $n = 0, 1, 2, ...$

(a) Show that the response *due to the first impulse current alone is* $v_{out}(t)$

$e^{-0.693t} = (0.5)^t$ V for $0 < t < \infty$; $v_{out}(t)$ starts with 1 V at $t = 0$ and then is halved for every second of elapsed time.

(b) Making use of linearity and the result of part (a), show that for $1 < t < 2$, $v_{out}(t) = 1.5(0.5)^{(t-1)}$ V. Note that only the first and second impulses have an effect on this time interval.

(c) Show that for $n < t < n + 1$, $v_{out}(t) = [2 - (0.5)^n](0.5)^{(t-n)}$ V.

(d) For very large $n < t < n + 1$, show that $v_{out}(t) = 2(0.5)^{(t-n)}$ V and that the waveforms for the time intervals $(n, n + 1)$ and $(n + 1, n + 2)$ are identical except for a time delay of n. This means $v_{out}(t)$ has reached the steady state.

(e) Show that for arbitrary values of R, C, Q, and T the output in steady state is

$$v_{out}(t) = \frac{Q}{C} \frac{1}{1 - e^{-\frac{T}{RC}}} e^{-\frac{t-nT}{RC}}$$

for $nT < t < (n + 1)T$.

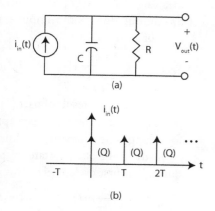

(a)

(b)

Figure P20.15

15. Consider again the circuit of Figure 20.15. If the impulse current train has been applied for a long time (theoretically since $t = \infty$), that is, $n = -\infty, \ldots, -2, -1, 0, 1, 2, \ldots$

(a) Show that for the time interval $(0, T)$ $v_{out}(t)$ is given by

$$v_{out}(t) = \frac{Q}{C} \frac{1}{1 - e^{-\frac{T}{RC}}} e^{-\frac{t}{RC}}$$

(b) For this and the remaining parts, use the special set of values

$$R = \frac{1}{ln(2)} = 1.442 \ \Omega$$

$C = 1$ F, $Q = 1$ coulomb, and $T = 1$ sec. Show that the steady-state response for the first cycle is $v_{out}(t) = 2e^{-0.693t} = 2(0.5)^t$ V, for $0 < t < 1$. Sketch the waveform for the first cycle.

(c) Using the transfer function

$$H(s) = \frac{V_{out}(s)}{I_{in}(s)} = \frac{1}{Cs + \frac{1}{R}}$$

and the Fourier series for $i_{in}(t)$, equation 20.22, find the Fourier series for $v_{out}(t)$. Write the answer to include up to the third harmonic.

(d) Find the average power (approximate value) dissipated the resistor. Since only a finite number of the terms in the infinite series are included in the calculation, the answer is only approximate. What is the error when only harmonics up to the third are considered? To determine this error, compute the exact average power using the result of part (b).

CHECK: (c) $v_{out}(t) = 1.4427 + 0.3164 \cos(2\pi t - 87.7047°) + 0.1589 \cos(4\pi t - 86.8428°) + 0.1060 \cos(6\pi t - 87.88°)$ V; (d) 2.109 watts, exact answer is 2.164 watts, error = 2.53%

17. The *LC* resonant circuit in Figure P20.17a is initially relaxed and the impulse train

$$i_{in}(t) = Q \sum_{n=0}^{\infty} \delta(t - nT)$$

shown in Figure P20.17b is applied. The natu-
ral frequency of the tank circuit is $f_0 = \dfrac{1}{2\pi\sqrt{LC}}$
and the period is $T_0 = \dfrac{1}{f_0} = 2\pi\sqrt{LC}$.

(a)   Show that if the frequency of the
      impulse train, $f = \dfrac{1}{T}$, is the same as
      the resonant frequency $f_0 = \dfrac{1}{2\pi\sqrt{LC}}$,
      then $i_L(t)$ grows without bound.
      Specifically, show that the inductor
      current in the time interval $(n-1)T < t < nT$ is given by $i_L(t) = n \times \omega_0 Q \sin(\omega_0 t)$.

(b)   Show that if $f = f_0/m$ for any integer m,
      the inductor current again grows
      unbounded.

(c)   Show that if $f = 2f_0$, then $i_L(t)$ is peri-
      odic and remains bounded. Sketch
      two cycles of $i_L(t)$.

(d)   Show that in general if $T = (2m + 1)(0.5T_0)$ or, equivalently,
      $$f = \frac{2f_0}{2m+1},$$
      then $i_L(t)$ is periodic. Sketch two
      cycles of $i_L(t)$.

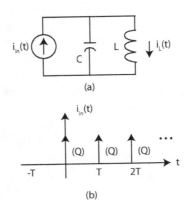

(a)

(b)

Figure P20.17

18. The LC resonant circuit in Figure P20.18a
is initially relaxed and the impulse train

$$i_{in}(t) = Q \sum_{n=0}^{\infty} \delta(t - nT)$$

shown in P20.18b is applied. This problem
(and circuit) differs from the previous one only
in the presence of the resistance R. When the
value of R is large enough, the impulse respons
is a damped sinusoid. The purpose of this problem is to show that if the damping effect of R i
slight and the input impulse train "synchro-
nizes" with the damped sinusoids of the
impulse response, then the output voltage mag
nitude, although finite, can reach a very larg
value.

(a)   Find $H(s) = \dfrac{V_{out}(s)}{I_{in}(s)}$. If $H(s)$ has a pa
      of poles at $s_{1,2} = -\sigma \pm j\omega_d$, then th
      impulse response has the form $h(t)$
      $e^{-\sigma t}[A\cos(\omega_d t) + B\sin(\omega_d t)]u(t)$. Fin
      σ and ω_d in terms of R, L, and C.
      Note that (i) $\omega_d \neq \omega_0 = \dfrac{1}{\sqrt{LC}}$, and
      (ii) the waveform of every cycle (tim
      interval of length $T_d = \dfrac{2\pi}{\omega_d}$) differs
      from its adjacent cycle by only a mu
      tiplicative constant $\alpha = e^{-\sigma T_d}$.

(b)   Show that if the input impulse train sy
      chronizes with the damped oscillation
      meaning that $T = T_d = 2\pi/\omega_d$, the
      the output due to the impulse tra
      $i_{in}(t)$ of Figure P20.18b is
      $$v_{out}(t) = Q\frac{1-\alpha^n}{1-\alpha}h\big(t - (n-1)T_d\big)$$
      for $(n-1)T_d < t < nT_d$

(c)   Deduce from the result of part (b) th
      if the impulse current has been appli
      for a long time (theoretically from
      $-\infty$), then the steady-state respor
      during the first cycle is given by
      $$v_{out}(t) = \frac{Q}{1-\alpha}h(t) \text{ for } 0 < t < T_d$$
      which shows that as α comes ve
      close to 1 (meaning very little dam
      ing), the magnitude of the output c
      reach a very large value.

(d)   If $R = 0.7213\ \Omega$, $C = 1$ F, and L
      0.025 H, in which case $\alpha = 0.5$, fi
      $H(s)$ and the impulse response h

Further, if $Q = 1$ coulomb and $T = 1$ sec, show that $v_{out}(t)$ in steady state is given by (consider item 18 of Table 12.1)

$$v_{out}(t) = 2e^{-0.6931t}[\cos(2\pi t) - 0.1103\sin(2\pi t)]\text{ V for }0 < t < 1$$

Plot this waveform using MATLAB or its equivalent. Justify that this is indeed the steady-state value by showing

$$v_{out}(0^+) - v_{out}(1^-) = \frac{Q}{C} = 1\text{ V}.$$

(e) If $R = 9.747\ \Omega$, $C = 1$ F, $L = 0.02533$ H, $Q = 1$ coulomb and $T = 1$ sec, in which case $\alpha = 0.95$, find the steady-state response for $0 < t < 1$ sec.

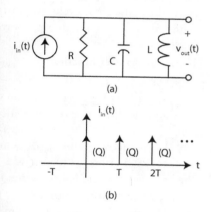

(a)

(b)

Figure P20.18

CHECK: (d) $H(s) = \dfrac{s}{s^2 + 1.3863s + 39.96}$;

e) $H(s) = \dfrac{s}{s^2 + 0.102586s + 34.481049}$

nd $v_{out}(t) = 20e^{-0.0513t}[\cos(2\pi t) - 0082\sin(2\pi t)]$ V for $0 < t < 1$

). Consider the first-order RC circuit of Figure 20.19a, where the input $v_{in}(t)$ is a square wave ith period T, as shown in Figure P20.19b. om the analysis methods discussed in Chapter the output $v_{out}(t)$ consists of exponential rises d decays, as illustrated in Figure P20.19c.

(a) If the input has been applied for a long time, then the circuit reaches steady state and $a_1 = a_0$, as shown in Figure P20.19c. Using this fact and equation 8.19a, show that in steady state,

$$v_{out,min} = V_{min} + \frac{V_{max} - V_{min}}{1 + e^{\frac{0.5T}{RC}}}$$

$$v_{out,max} = V_{max} - \frac{V_{max} - V_{min}}{1 + e^{\frac{0.5T}{RC}}}$$

(b) In Example 20.1, find the exact values of $v_{out,min}$ and $v_{out,max}$ by using the results of part (a). Then plot $v_{out}(t)$ for one complete cycle using MATLAB or its equivalent.

(c) In the exercise following Example 20.1, four frequency components of the Fourier series are used to approximate the input, and it is found that $v_{out,min} = 12.235$ V and $v_{out,max} = 82.013$ V. What are the percentage errors of these approximate answers?

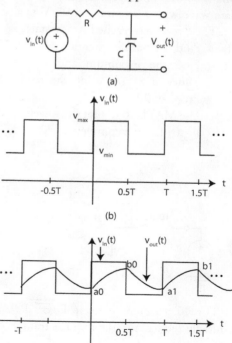

(a)

(b)

(c)

Figure P20.19

ANSWERS:(b) 11.235V, 83.013 V, (c) 8.9 %, −1.2%

20. The method of Problem 19 can be extended to computation of the steady-state response of any first-order linear network to a square wave input. Let the transfer function of the first-order linear network be

$$H(s) = \frac{\mathcal{L}[\text{Output}(t)]}{\mathcal{L}[\text{Input}(t)]} = \frac{H(0)}{1+\tau s}$$

and the input be the square wave of Figure P20.19b. Show that in steady state,

$$\text{Output}(t)_{min} = H(0)\left[V_{min} + \frac{V_{max}-V_{min}}{1+e^{\frac{0.5T}{\tau}}}\right]$$

$$\text{Output}(t)_{max} = H(0)\left[V_{max} + \frac{V_{max}-V_{min}}{1+e^{\frac{0.5T}{\tau}}}\right]$$

21. Consider the leaky integrator circuit of Figure P20.21. The input $v_{in}(t)$ is a 1 kHz square wave with zero mean and 2 V peak to peak. The circuit has reached steady state.
   (a) Find the transfer function $H(s)$.
   (b) Find the maximum and minimum values of $v_{out}(t)$, using the results of Problem 20.
   (c) Use MATLAB to plot the waveform of $v_{out}(t)$ for one complete cycle.

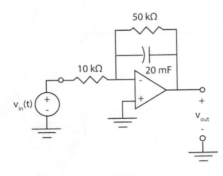

Figure P20.21

ANSWERS: (a) $H(s) = \dfrac{5}{1000s+1}$; (b) ±0.245 V

22. Consider the circuit of Figure P20.22, in which $C = 1$ F, $R = 1//ln(2) = 1.443\ \Omega$, $T = 4$ sec, $V_1 = 15$ V, $\beta = 0.25$, $v_C(0^-) = v_{out}(0^-) = 1$ V, and $v_{in}(t)$ is a sequence of rectangular pulses as shown in Figure P20.22b.
   (a) Show that for the first cycle, $0 < t < 4$ sec, the complete response is $v_C(t) = [15 − 15(0.5)^t] + (0.5)^t = 15 − 14(0.5)^t$ V for $0 < t < 1$ sec and $v_C(t) = 8(0.5)^{t-1}$ V, for $1 < t < 4$ sec. Sketch the $v_C(t)$ waveform for the first cycle $0 < t < 4$ sec.
   (b) Show that $v_{out}(4^+) = v_{out}(4^-) = v_{out}(0^+) = 1$ V.
   (c) Use the result of part (b) to verify that the $v_C(t)$ waveform for any subsequent cycle, $nT < t < (n + 1)T$ is identical to that of the first cycle except for a time shift, i.e., the first-cycle waveform found above is the steady-state waveform
   (d) Using the transfer function

$$H(s) = \frac{V_{out}(s)}{V_{in}(s)} = \frac{\frac{1}{RC}}{s+\frac{1}{RC}}$$

and the Fourier series for $v_{in}(t)$, item 2 of Table 20.3 , find the magnitudes of the harmonics of the Fourier series for $v_{out}(t)$, up to the third harmonic.
   (e) Find the average power (approximate value) dissipated in the resistor. Since only a finite number of the infinite series is included in the calculation, the answer is only approximate. What is the error when only harmonics up to the third are considered? To determine this error, compute the exact average power using the result of part (a).

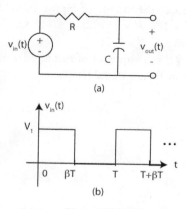

(a)

(b)

Figure P20.22

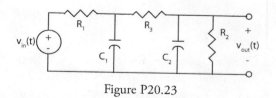

Figure P20.23

23. Consider the second-order circuit of Figure P20.23, in which $R_1 = 1\,\Omega$, $R_2 = 8\,\Omega$, $R_3 = 6\,\Omega$, $C_1 = 0.25$ F, and $C_2 = 0.125$ F.

(a) Derive the transfer function

$$H(s) = \frac{\dfrac{16}{3}}{s^2 + 7s + 10}.$$

(b) If the input voltage is a square wave shown in Figure 20.4 with peak-to-peak voltage 18 V and $T = 1$ sec, find the Fourier series for $v_{out}(t)$ and list the first four components.

(c) Find the approximate average power delivered to R_2 considering only the first four components.

(d) Assuming the input has been applied for a long time so that the circuit is in steady state,

find the expression for $v_{out}(t)$ during the time interval $(0, T)$. Use MATLAB to plot the waveform.   Hint: Expand $H(s)$ into partial fractions. Represent the system as the parallel connection of two first-order subsystems. Then use the result of Problem 17 to solve each subsystem. Finally, combine the solutions in time domain to obtain the expression for $v_{out}(t)$.

I N D E X